23rd International Federation of Heat Treatment and Surface Engineering Congress 2016

Savannah, Georgia, USA
18 - 21 April 2016

ISBN: 978-1-5108-2370-9

Printed from e-media with permission by:

Curran Associates, Inc.
57 Morehouse Lane
Red Hook, NY 12571

Some format issues inherent in the e-media version may also appear in this print version.

Printed by Curran Associates, Inc. (2016)

For permission requests, please contact ASM International
at the address below.

ASM International
9639 Kinsman Road
Materials Park, Ohio 44073

Phone: (440) 338-5151

memberservicecenter@asminternational.org

Additional copies of this publication are available from:

Curran Associates, Inc.
57 Morehouse Lane
Red Hook, NY 12571 USA
Phone: 845-758-0400
Fax: 845-758-2633
Email: curran@proceedings.com
Web: www.proceedings.com

International Federation of Heat Treatment and Surface Engineering Congress 2016

Proceedings of the
23rd IFHTSE Congress

April 18–21, 2016
Hyatt Regency Savannah
Savannah, Georgia, USA

Organized by

INTERNATIONAL FEDERATION FOR HEAT TREATMENT AND SURFACE ENGINEERING
IFHTSE

HTS
Heat Treating Society
ASM INTERNATIONAL

ASM
INTERNATIONAL

Published by
ASM International
Materials Park, Ohio 44073-0002
www.asminternational.org

IFHTSE 2016 Organizing Committee

Conference Chair

Dr. D. Scott MacKenzie
Houghton International Inc., FASM

Domestic Committee Members

Mr. Gord Montgomery Montgomery & Associates	Dr. Lynn Ferguson Dante-Solutions, Inc.
Mr. Rob Goldstein Fluxtrol, Inc.	Dr. Rozalia Papp Air Liquide
Dr. Christian Moreau Concordia University	Mr. Andy Banka AirFlow Sciences
Dr. George Totten Totten & Associates	Dr. David Furrer Pratt & Whitney
Dr. Rick Sisson WPI	Dr. Robert Gaster John Deere
Dr. Erhan Ulvan Acuren, Inc.	Mr. John Duggan Park-Heatbath
Mr. David Guisbert Guisbert & Associates	Prof. Arvind Agarwal Florida University
Dr. Stefan Hock General Secretary, IFHTSE	

International Committee Members

Mr. Gerard Bell	Prof. Lauralice Canale
Prof. Rafael Colas	Dr. Sabine Denis Henri Poincare University, LSG2M
Dr. Hanshan Dong	Dr. Imre Felde
Mr. Udo Fiorini S+F Editora	Dr. Kiyoshi Funatani IMST
Prof. Weimin Gao	Prof. Jianfeng Gu Shanghai Jiao Tong University
Prof. Haken Gur	Dr. Patrick Jacquot Bodycote Belguim
Dr. Maciej Korecki Seco-Warwick Poland	Prof. Young-kook Lee
Prof. Bozidar Liscic	Prof. Thomas Luebben
Prof. Bernardo Hernandes Morales	Prof. Rosa Otero Universidade de São Paulo
Prof. Narayan Prabhu	Prof. Schneider Reinhold TU Upper Austria Wels
Prof. Barbara Rivolta Politecnico di Milano	Prof. Y.M. Rong
Dr. Satyam Sahay John Deere Pune	Prof. Bozo Smoljan
Dr. Paul Stratton Matscribe	Ms. Eva Troell, Swerea
Dr. Thomas Waldenmaier Bosch	Mr. Ole West
Dr. Emilia Wolowiec-Korecki Lodz University of Technology	Prof. Hans-Werner Zoch IWT

IFHTSE 2016 Reviewers

Extended Abstract and Manuscript Reviewers

Mr. Aquil Ahmad Eaton, Retired	Mr. Andy Banka AirFlow Sciences
Dr. Lynn Ferguson Dante-Solutions, Inc.	Dr. David Furrer Pratt & Whitney
Dr. Robert Gaster John Deere	Mr. Rob Goldstein Fluxtrol, Inc.
Mr. David Guisbert Guisbert & Associates	Dr. D. Scott MacKenzie Houghton International Inc.
Dr. Rozalia Papp Air Liquide	

Contents

Keynote

Very Short and Very Long Heat Treatments 1
H.K.D.H. Bhadeshia, University of Cambridge

Alloying Element Nitride Development in Ferritic Fe-Based Materials upon Nitriding 11
T. Steiner and E.J. Mittemeijer
Max Planck Institute for Intelligent Systems

Nitriding

The Effects of Process Parameters on Nitrogen Infiltration and the Controlling of Nitrogen Concentration in Vacuum Nitriding 19
M. Okumiya[1], Y. Tsunekawa[1], J.-H. Kong[1], and S.-G. Kim[2]
(1) Toyota Technological Institute
(2) Korea Institute of Industrial Technology

Gas Phase Deep Nitriding of 33CrMoV12-9 Steel—Microstructural Characterization of the Precipitates 25
O. Skiba[1], J. Dulcy[1], G. Marcos[1], T. Czerwiec[1], A. Courleux[2], and S. Becquerelle[2]
(1) Université de Lorraine
(2) Hispano-Suiza, groupe Safran

Effect of Plasma Nitriding on Corrosion Behavior of AISI 4330 Low Alloy Steel 31
V. Patil, J. Sherkar, A. Supare, and R. Singh
Bharat Forge Ltd.

Low Temperature Nitriding of Strain Induced Martensite and Laser Quenched Austenite 38
V. Leskovšek[1], M. Godec[1], and P. Kogej[2]
(1) Institute of Metals and Technology
(2) RLS Merilna Tehnika d.o.o.

Influence of Strain Hardening on Gaseous Nitriding of Steels 46
B. Guillot, S. Jégou, and L. Barrallier
Arts et Métiers ParisTech

Low Temperature Surface Hardening of Stainless Steel—The Role of Plastic Deformation 51
F. Bottoli, F.N. Jespersen, J.H. Hattel, G. Winther, T.L. Christiansen, and M.A.J. Somers
Technical University of Denmark

Effects of Different Microstructural and Hardness Gradients Generated by Single and Combined Surface Treatments with a Nitriding Top Layer 55

A. Buchwalder[1], H.-J. Spies[1], N. Klose[1], A. Jung[1], and R. Zenker[2]
(1) TU Bergakademie Freiberg
(2) Zenker Consult

Optimization of Gaseous Nitriding of Steels by Multiphysics Modelling 63

S. Jégou[1], L. Barrallier[1], and G. Fallot[2]
(1) Arts et Métiers ParisTech
(2) Airbus Helicopters

Vacuum Processes and Technology

In-Line, High-Volume, Low-Distortion, Precision Case Hardening for Automotive, Transmission, and Bearing Industry 71

M. Korecki[1] and E. Wolowiec-Korecka[2]
(1) Seco/Warwick S.A.
(2) Lodz University of Technology

Quenching and Quenchants

Press Quench Process Design for a Bevel Gear Using Computer Modeling 78

Z. Li[1], A.M. Freborg[1], B.L. Ferguson[1], P. Ding[2], and M. Hebbes[2]
(1) Dante Solutions, Inc.
(2) ANSYS, Inc.

Intensive Quenching of AISI 5160 Spring Steel Using Small Scale Laboratory Equipment—A Progress Report 88

L.M. Albano[1], J.G. Lúcio[2], F.S. Misina[1], V. Leskovisek[3], G.E. Totten[4], and L.C.F. Canale[1]
(1) University of São Paulo
(2) Rassini NHK
(3) University of Ljubljana
(4) Portland State University

Ionic Liquids for Immersion Quenching of Aluminium Alloys 94

M. Beck, M. Reich, O. Keßler, C. Neise, U. Kragl, M. Ahrenberg, and C. Schick
University of Rostock

Beer, Blood and Urine—Mythological Quenchants of Ancient Blacksmiths 101

D.S. MacKenzie and G. Graham
Houghton International, Inc.

Selection of Oil Quenchants for Heat Treating Processes 110

D.S. MacKenzie, Houghton International, Inc.

Why Fine Filtering for Quench Oil Systems is a Must! 118

J. Pfaffenberger, C.C. Jensen Inc.

Innovative Quenching Process for Pipe and Bar 123
P. della Putta[1], M. Fabro[2], K. Sawamiphakdi[3], D. Bader[3], and S.J. Nalawadi[3]
(1) SMS Meer SpA
(2) SMS U.S.A. LLC
(3) Timken Steel

A New Probe to Measure and Evaluate the Real Quenching Intensity Based on Temperature Gradients 132
B. Liscic[1], B. Matijevic[1], G.E. Totten[2], and L.C.F. Canale[3]
(1) University of Zagreb
(2) Portland State University
(3) Universidade de Sao Paulo

Reliability and Process Control

Pre-Oxidation Treatment to Increase the Service Life of Heat Treatment Furnace Conveyor Belts Made of AISI 330Cb Industrial Application 139
C. Combe[1], P.F. Cardey[1], H. Buscail[2], C. Issartel[2], F. Riffard[2], and R. Rolland[2]
(1) Centre Technique des Industries de la Mécanique
(2) Laboratoire Vellave sur l'Elaboration et l'Etude des Matériaux

Effect of Microstructure on the Fatigue Crack Growth Behavior in Low Carbon Steel 146
D. Watanuki[1], H. Li[2], and P. Bowen[2]
(1) NSK Ltd.
(2) The University of Birmingham

Life Extension of Furnace and Fixture Alloys by Surface Engineering in Carburizing Atmospheres 153
A. Wang and R.D. Sisson, Jr.
Worcester Polytechnic Institute

Advanced Thermal Processing

Design of Advanced Tempering Diagrams of Hot Work Tool Steel Using Multifunctional CNPTB Specimen 161
B. Podgornik , G. Puš, V. Leskovšek, and M. Godec
Institute of Metals and Technology

Microstructural Evolution in Microalloyed Steels with High-Speed Thermomechanical Bar and Rod Rolling 168
R. Cryderman, B. Whitely, and J. Speer
Colorado School of Mines

Hot Hydroforging of Lightweight Bimaterial Gears and Hollow Products **175**
B. Chavdar[1], R. Goldstein[2], and L. Ferguson[3]
(1) Eaton
(2) Fluxtrol Inc.
(3) Dante Solutions Inc.

Effect of Processing Route on Microstructure and Impact Toughness of Duplex Stainless Steel **184**
A. Powar, A. Gujar, N. Manthani, V. Pawar, and R. Singh
Bharat Forge Ltd.

Effect of Electron Beam Welding on Microstructure and Mechanical Properties of Ti-6Al-4V Alloy **190**
S. Thakare[1], N. Prabhu[2], P. Kattire[1], V. Bhavar[1], and R.K.P. Singh[1]
(1) Bharat Forge Limited
(2) IIT Bombay

Induction Hardening Shafts of Shafts and Shaft-Like Components **196**
V. Rudnev, W. West, A. Goodwin, and S. Fillip
Inductoheat Inc.

Effects of the Inductive Hardening Process on the Martensitic Structure of a 50CrMo4 Steel **203**
A. Vieweg, G. Ressel, P. Prevedel, S. Marsoner, and R. Ebner
Materials Center Loeben Forschung GmbH

Minimizing Case/Core Initiated Failures in Induction Hardened Components **210**
M.A. Pershing, G. Raab, and W. Li
Caterpillar Inc.

Induction Camshaft Hardening Technology with Practically Undetectable Distortion **217**
G. Doyon, V. Rudnev, J. Maher, R. Minnick, and G. Desmier
Inductoheat Inc.

Local Hardening of Ductile Cast Iron—The Importance of Temperature **223**
D.E. Guisbert, M.H. Guisbert, and D.A. Guisbert
QA Metallurgical Services, LLC

Effect of the Surface Morphology and Surface Finish on the Corrosion Resistance of Ti6Al4V via Direct Metal Laser Sintering **229**
Y. Xu, Y. Lu, J. Liang, and R.D. Sisson, Jr.
Worcester Polytechnic Institute

Future Trends of Low Temperature Surface Hardening of Stainless Steel **235**
A. Karl[1], M. Wägner[1], and P. Jacquot[2]
(1) Bodycote, Landsberg
(2) Bodycote, Pusignan

Computer Simulation of Induction Thermal Processing of Ferrous and Nonferrous Alloys 242
C. Russell and V. Rudnev
Inductoheat Inc.

Modeling of the Heating Sequences of Lightweight Steel/Aluminum Bimaterial Billets for Hot Forging and Hot Hydroforging 248
R. Goldstein[1], B. Chavdar[2], and L. Ferguson[3]
(1) Fluxtrol Inc.
(2) Eaton
(3) Dante Solutions, Inc.

Influence of Synthesis on the Mechanical Properties of Sintered and Nanoporous Copper 256
A. Moore, D. Carroll, R. Behera, C. Deo, and A. Antoniou
Georgia Institute of Technology

Grain Refinement Strategies in High-Strength Cast Steel 262
D.C. Van Aken, T.O. Webb, and J.D. Green
Missouri University of Science and Technology

Modeling the Mechanical Properties of Quenched and Tempered Martensitic Steels 273
X. Cai, Y. Lu, H. Yu, and R.D. Sisson, Jr.
Worcester Polytechnic Institute

Thermodynamic and Kinetic Simulation and Experimental Results Homogenizing Advanced Alloys 279
P.D. Jablonski, J.J. Licavoli, and J.A. Hawk
United States Department of Energy

Improvement in Creep Rupture Ductility of Creep Strength Enhanced Ferritic Steel by New Heat Treatment Process with Intermediate Tempering 289
K. Kimura and K. Sawada
National Institute for Materials Science

Evaluation of Mechanical and Fatigue Properties of Weld Deposit and Its Comparison with that of Die Steel 295
S.S. Patil, S.S. Kirwai, S. Kumar, A.R. Patil, R.P. Singh, and J.B. Mane
Bharat Forge Ltd.

Cryogenic Processing

Sub-Zero Celsius Treatment—A Promising Option for Future Martensitic Stainless Steels 304
M. Villa, T.L. Christiansen, and M.A.J. Somers
Technical University of Denmark

Residual Stress Symposium

Modeling and Measurement of Residual Stresses and Machining Distortions in Aircraft Engine Disks 313

M.G. Glavicic[1], R. Ress, III[1], S. Srivatsa (retired)[2], J. Castle[3], B.K. Chun[4], J.Y. Oh[4], W.-T. Wu[4], T.M. Holden[5], B. Clausen[6], D.W. Brown[6], T. Sisneros[6], M. Knezevic[7], and J. Simmons[8]
(1) Rolls-Royce Corporation
(2) GE Aviation, Cincinnati
(3) Boeing Research and Technology
(4) Scientific Forming Technologies Corporation
(5) Northern Stress Technologies
(6) Los Alamos National Laboratory
(7) University of New Hampshire
(8) Air Force Research Laboratories

Distortion Potential of Cold Forming Prior to Induction Heat Treatment Processes 325

D. Nadolski, A. Schulz, and H.-W. Zoch
University of Bremen

Prediction of Quench-Induced Residual Stresses in Al-Cu-Mg Impeller Forgings 332

D. Carron[1], N. Chobaut[2], J.-M. Drezet[2], and P. Saelzle[3]
(1) Université Bretagne Sud
(2) Ecole Polytechnique Fédérale de Lausanne
(3) ABB Turbo Systems Ltd.

Study of Residual Stresses after Steel Quenching with Vegetable Oils 341

C. Civera[1], B. Rivolta[2], R.L.S. Otero[3], J.G. Lúcio[3], L.C.F. Canale[3], and G.E. Totten[4]
(1) Siderval Spa
(2) Politecnico di Milano
(3) Universidade de Sao Paulo,
(4) Portland State University

Parallel PSO Method for Estimation Heat Transfer Coefficients 348

I. Felde[1], S. Szénási[1], G. Pintér[1], W. Shi[2], R. Colas[3], and O. Zapata-Hernández[3]
(1) Óbuda University
(2) Tsinghua University
(3) Universidad AutÓnoma de Nuevo LeÓn

Residual Stress and Bending Fatigue Strength in Carburized and Quench Hardened Pyrowear 53 Steel Gears 354

B.L. Ferguson, A.M. Freborg, and Z. Li
Dante Solutions, Inc.

The Development of a Process-Structure-Properties-Performance Map for Aluminum-Zinc-Magnesium-Copper Alloys Used In Aircraft Applications 362

A.N. Goulding, R.W. Neu, and T.H. Sanders, Jr.
Georgia Institute of Technology

Metallurgical Aspects of Distortion and Residual Stresses in Heat Treated Parts 372

D.S. MacKenzie, Houghton International, Inc.

Surface Structuring as a Method to Influence on Rewetting during Liquid Quenching 383
N. Kozlov and O. Kessler
University of Rostock

Mathematical Modeling and Computer Simulation of Nonmonotonic Quenching 391
B. Smoljan, D. Iljkić, and L. Štic
University of Rijeka

Using Heat Treat Simulation to Characterize Sensitivity of Quench Hardening Response in Hot Mill Steel Work Rolls 398
A.M. Freborg, Z. Li, and B.L. Ferguson
Dante Solutions, Inc.

Materials Characterization

Martensite Lattice Parameter Measured by Modern X-Ray Diffraction in Fe-C Alloy 404
Y. Lu, H. Yu, X. Cai, Y. Rong, and R.D. Sisson
Worcester Polytechnic Institute

In Situ X-Ray Diffraction Investigations of Martensitic Transformation and Austenite Modifications during Quenching of Low Alloy Steels 410
J. Epp, University of Bremen

Hardness-Toughness Relationship of a Boron Alloyed Quench and Tempering Steel after Different Heat Treatment Cycles 418
R.S.E. Schneider[1], K. Steineder[1], Y. Toshima[2], and M. Okumiya[2]
(1) University of Applied Sciences Upper Austria
(2) Toyota Technological Institute

Close Relations between Metallic Materials, Heat Treatment and Thermochemical Treatment on the Microstructure—Contribution of Metallographic Examinations Performed on Industrial Parts through Several Case Studies 425
P. Jacquot, B. Stauder, N. Jacquot, and G. Fürst
Bodycote

Microstructural Analysis of Carbides in Steel by Electrochemical Extraction Technique 434
H. Yu, Y. Lu, X. Cai, R.D. Sisson, Jr.
Worcester Polytechnic Institute

Study of Phase Transformations during Tempering of AISI L6 Tool Steel by Means of Dilatometry and High Temperature X-Ray Diffraction 440
D.I. Quiñones-Salinas[1], R.D. Mercado-Solís[1], L.A. Leduc-Lezama[1], F. Fernandez-Guzman[2], and R. Cerda-Rojas[2]
(1) Universidad Autónoma de Nuevo León, México
(2) Frisa Forjados S.A. de C.V.

Hardenability Characterization and Microstructural Evolution in Heavy Gauge Plate Steels 445
I. Vieira and E. De Moor
Colorado School of Mines

Influence of Cooling and Heating Rates on the Thermal Properties of Ni-Ti Shape Memory Alloy 451
B.B. Fraj and Z. Tourki
Université de Sousse

Thermal Diffusivity in Cast Aluminum Alloys 457
H. Arenas-García[1], A.G. Esmeralda[1], R. Colás[1], A.F. Rodríguez[2], and J. Talamantes-Silva[2]
(1) Universidad Autónoma de Nuevo León
(2) Nemak México, S.A. de C.V.

Calorimetry for the Study of Phase Transformations during Tempering of AISI L6 Low Alloy Tool Steel 464
D.I. Quiñones-Salinas[1], R.D. Mercado-Solís[1], L.A. Leduc-Lezama[1], F. Fernandez-Guzman[2], and R. Cerda-Rojas[2]
(1) Universidad Autonoma de Nuevo León
(2) Frisa Forjados S.A. de C.V.

Influence of UNSM Treatment Temperature on Microstructure and Mechanical Properties of Inconel 600 471
J.-H. Kim, A. Amanov, Y.-S. Pyun, and H.-D. Kim
Sun Moon University

Influence of Heat Treatment on Fatigue and Fracture Toughness Behavior of AISI 4140 Steel 473
P. Jadhav[1], R. Mane[1], M. Ukhande[1], S. Arangi[1], and R.C. Prasad[2]
(1) Bharat Forge Ltd.
(2) Indian Institute of Technology Bombay

Coating Technology

Manufacture, Examination and Application of Modern Composite Materials with a Nanocarbon Dispersion Phase 481
T. Babul and A. Olbrycht
Institute of Precision Mechanics

Carburizing and Nitrocarburizing

Simulation of The Effect of Temperature and Time on the Tempering Behavior of Carburized Steels 487
Y. Kim, X. Cai, L. Zhang, H. Yu, and R.D. Sisson, Jr.
Worcester Polytechnic Institute

Fluidized-Bed Carburizing in Chemically Active Powders of C22 Steel 493
A. Ciski[1], T. Babul[1], K. Lankiewicz[1], J. Nižňanská[2], and P. Šuchmann[2]
(1) Institute of Precision Mechanics
(2) Comtes FHT a.s.

An Enhancement to the Low Pressure Carburizing Simulation 499
L. Zhang and R.D. Sisson, Jr.
Worcester Polytechnic Institute

Pre-Nit Technology for Accelerated Vacuum Carburizing 504
M. Lister, Seco-Warwick

Tribology and Wear of Engineered Surfaces

Influence of Ultrasonic Nanocrystal Surface Modification Treatment Temperature on Microstructure and Tribological Behavior of Bearing Steel 511
A. Amanov, J.-H. Kim, and Y.-S. Pyun
Sun Moon University

The Erosion Behavior of Die-Steel Salt-Bath-Nitirocarburized with Lithium-Iron Compound Oxide Layer in Molten Al-Si-Cu Alloy 513
H. Ishizuka, Y. Hirai, and Y. Watanabe
Parker Netsushori Kogyo Co., Ltd.

Poster Session

Development of a Laboratory Intensive Quenching System 523
L.M. Albano[1], F.S. Misina[1], G.E. Totten[2], and L.C.F. Canale[1]
(1) University of São Paulo
(2) Portland State University

Effects of Quench and Partitioning Heat Treatment in Mechanical Properties and the Microstructural Characterization of AISI 300M 529
R.G. Gregui[1], L.C.F. Canale[2], and D.A. Ludewigs[3]
(1) Instituto Federal de São Paulo
(2) Universidade de São Paulo
(3) Durferrit Company – HEF Group

Effect of Quenching and Tempering in Microalloyed Steels Containing Boron and Titanium 536
A.I. Filho[1], R.V. da Silva[1], W.M. Souza[2], J.B.R. Martins[2], L.C. Casteletti[3], and G.E. Totten[4]
(1) Instituto Federal de Educação
(2) ArcelorMittal Tubarão
(3) Universidade de São Paulo
(4) Portland State University

Characterization of BN/TiBN/TiN Multilayer Coatings Prepared by Pulsed DC Plasma-Enhanced Chemical Vapor Deposition 541
K. Kawata and T. Kidachi
Oriental Engineering Co., Ltd.

Application of Nanostructured Copper-Polymer Catalyst for Activation of Gas Nitriding of Steels 547
L.G. Petrova, I.M. Papisov, I.S. Belashova, V.A. Aleksandrov, G.Y. Ostaeva, A.I. Papisova, and A.E. Perekrestov
Moscow Automobile and Road Construction State Technical University

Effect of Deep Cryogenic Treatment on Fracture Toughness of 300M Steel Using Circumferentially Notched KIC Test Specimens 554
D. Schuller[1] and V. Leskovšek[2]
(1) Universidade de São Paulo
(2) Institute of Metals and Technology

Uphill Quenching of Aluminum Alloys—A Process Review 560
W.S. Mattos[1], G.E. Totten[2], D.S. Mackenzie[3], L.C.F. Canale[1], and T. Croucher[4]
(1) University of São Paulo
(2) Portland State University
(3) Houghton International Inc.
(4) Metallurgical Service

Wear Behavior of Layers Produced by Boriding and PVD plus Boriding Treatments on AISI H13 Tool Steel 566
R.G. Pereira[1], G.C. Rêgo[1], L.C. Casteletti[1], A.N. Lombardi[2], and G.E. Totten[3]
(1) Universidade de São Paulo
(2) Federal Technological University of Parana
(3) Portland State University

Author Index 569

International Federation of Heat Treatment and Surface Engineering 2016
Proceedings of the 23rd IFHTSE Congress
April 18–21, 2016, Savannah, Georgia, USA

Very Short and Very Long Heat Treatments

H. K. D. H. Bhadeshia
University of Cambridge, Materials Science and Metallurgy
Pembroke Street, Cambridge CB3 0FS, U. K.
hkdb@cam.ac.uk

Abstract

There are significant processes emerging where the heat treatment times can be as long as 10 days or shorter than 100 ms. In both of these examples, extraordinary properties are achieved which have made the treatments commercially viable. Heat treatment involves the thermally activated movement of atoms into lower energy configurations. Therefore, the time-scale of the heat treatment depends on the choreography of the atomic motions, whether the motions are coordinated or not. It is demonstrated that remarkable structures can be generated by controlling these mechanisms, including structures that have never before been observed or generated on a scale that is large enough to exploit.

Introduction

Steels are blessed with the ability to undergo a variety of solid-state phase transformations; pure iron, without any alloying elements, exists in at least three different crystal structures, body-centred cubic, face-centred cubic and hexagonal close-packed [1, 2]. These can be induced to transform into one another either by breaking all the bonds and rearranging the atoms into the new pattern, or by homogeneously deforming the structures into new forms. The addition of carbon opens up a huge scenario of transformation products, with the palette dramatically expanded when substitutional solutes are added. Steels can also be thermo-mechanically processed to alter properties over a very wide range without changing the chemical composition.

Heat treatment is a subset of thermomechanical processing and relies entirely on the thermally activated movement of atoms. In steels, it has two roles, the first being to generate microstructures by the transformation of austenite and the second to help relax the resulting microstructure through tempering reactions. The focus of this paper is on the former, so it is appropriate to begin with a brief discussion of the atomic mechanisms which determine the time scales of phase transformations.

Transformation mechanisms

The transformation products of austenite can be classified into two categories, those which involve long-range diffusion and others in which the change in crystal structure is achieved by a macroscopically homogeneous deformation [3]. The former mechanism is known as reconstructive and the latter displacive (Fig. 1).

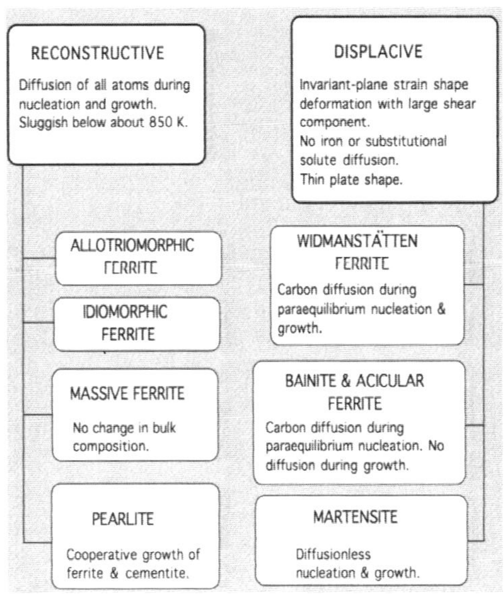

Figure 1: Classification of transformation products as a function of the atomic mechanism of transformation.

The formation of allotriomorphic, idiomorphic and

massive ferrite, and of pearlite, requires the diffusion of all elements including iron [4]. The movement of iron or substitutional solutes occurs by a vacancy mechanism and hence can be slow at low temperatures. As illustrated in Fig. 2, the scale of the microstructure must diminish as the transformation temperature is reduced [5], which is an advantage since both strength and toughness are improved when the structure is refined.

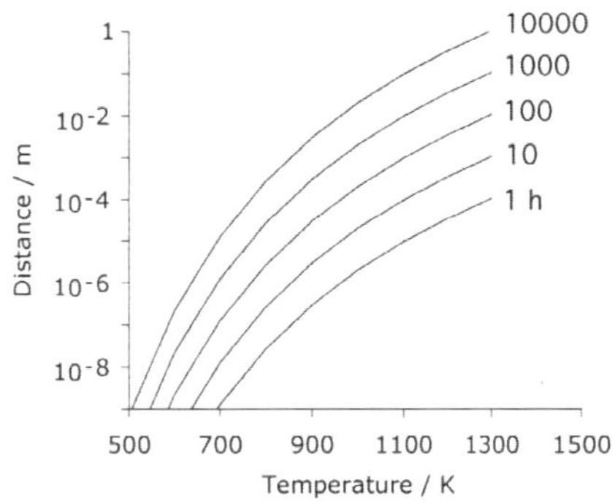

Figure 2: Diffusion distance of an iron atom in austenite as a function of temperature and time. The estimate is made using $2\sqrt{Dt}$ where D is the diffusion coefficient of iron in austenite [6] and t is the time.

However, it is evident that the time required to achieve a given scale increases dramatically as the temperature is reduced, making it impractical to design structures which are fine by reducing the transformation temperature. To summarise, rapid heat treatments are impossible with reconstructive transformations if the ultimate goal is to achieve fine microstructures.

The second class of transformations does not require the diffusion of atoms (Fig. 1) in substitutional sites and hence can be much more rapid. The change in crystal structure is achieved by a physical deformation which can be detected experimentally as a permanent strain [7]. Interstitial atoms may in certain cases be mobile, but their order or disorder does not influence the nature of the deformation [8].

The motion of the interface during displacive transformation is in principle limited only by the speed of sound in the metal, some thousands of m s^{-1}, which is in contrast to the highest solidification velocities in pure nickel at some 80 m s^{-1}. In practice the growth velocity of, for example martensite, may not get as high as the speed of sound in the metal, being limited by damping due to the emission of dislocations or other dissipating phenomena.

Diffusion of solute, whether this is substitutional or interstitial must always slow the rate of transformation. Diffusional transformations will therefore tend to be faster in pure materials. The rapid transformation of austenite, involving diffusion, can therefore only occur at elevated temperatures. There is one exception to this, as described below.

Many commercial processes cause the steel to revert into the austenitic condition. The transformation of low-temperature ferrite into high-temperature austenite differs from the case where the latter transforms during cooling. Transformation during cooling follows C-curve kinetics in which the overall transformation rate goes through a maximum as a function of the undercooling below the equilibrium temperature. This is because diffusion coefficients decrease but the magnitude of the driving force ΔG increases as the temperature is reduced. In contrast, both the diffusivity and driving force for austenite formation increase as a ferritic microstructure is superheated. The rate of *diffusion-controlled* transformation increases indefinitely as the temperature is raised, Fig. 3 [9].

Since the transformation rate during heating accelerates indefinitely at the temperature is raised, the transformation times tend towards zero. One consequence of this is that in many steel diffusion is unavoidable during heating, so incredibly high heating rates are needed in order to induce displacive transformation during heating. Austenite formation can be induced in incredibly small time periods using laser heating [10–13].

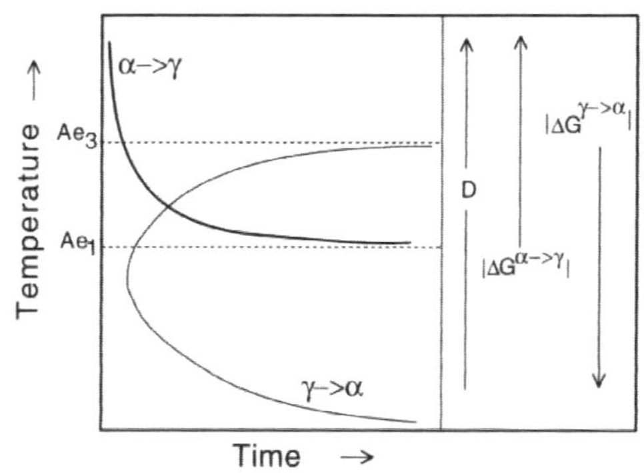

Figure 3: The TTT (time temperature transformation) curves for the $\gamma \rightarrow \alpha$ reaction, and for the $\alpha \rightarrow \gamma$ reverse transformation. The Ae_3 and Ae_1 terms represent the temperatures at which austenite formation begins and ends respectively, under equilibrium heating conditions..

Long Transformation Times

Slow reaction rates can be an advantage when dealing with strong steels which are destined for critical applications such as shafts. This is because the austenite does not need to be cooled rapidly, thus avoiding the development of quench stresses [14? –17].

Temperature gradients naturally occur in large samples when they are cooling from the austenite phase field to an isothermal transformation temperature. The transformation then does not occur at the same instant in all locations within the component being heat-treated. A slow reaction rate would ensure that the steel reaches a uniform temperature before transformation begins, thus making the material resistant to the development of internal stresses due to a heterogeneous distribution of temperature.

Long transformation times also permit the manufacture of components which are large in all three dimensions; this is why processes such as the manufacture of metallic glasses by rapid quenching are limited to thin ribbons.

A slow rate can be achieved either by increasing the hardenability of the steel or by transforming at a temperature where solid-state transformations become sluggish. We present here an example where both hardenability and suppressed-transformation temperatures are exploited in order to reduce the rate of transformation.

It would be nice to have a strong material which can be used for making components which are large in all their dimensions, and which does not require mechanical processing or rapid cooling to reach the desired properties. The following conditions are required to achieve this:

- The material must not rely on perfection to achieve its properties. Strength can be generated by incorporating a large number density of defects such as grain boundaries and dislocations, but the defects must not be introduced by deformation if the shape of the material is not to be limited.
- Defects can be introduced by phase transformation, but to disperse them on a sufficiently fine scale requires the phase change to occur at large undercoolings (large free energy changes). Transformation at low temperatures also has the advantage that the microstructure becomes refined.
- A strong material must be able to fail in a safe manner. It should be tough.
- Recalescence limits the undercooling that can be achieved. Therefore, the product phase must be such that it has a small latent heat of formation and grows at a rate which allows the ready dissipation of heat.

Certain discoveries have shown that carbide-free bainite can satisfy these criteria [18]. Bainite and martensite are generated from austenite without diffusion by a displacive mechanism. Not only does this lead to solute-trapping but also a huge strain energy term [9, 19, 20], both of which reduce the heat of transformation. The growth of individual plates in both of these transformations is fast, but unlike martensite, the overall rate of reaction is much smaller for bainite. This is because the transformation propagates by a sub-unit mechanism in which the rate is controlled by nucleation rather than growth [21, 22]. This mitigates recalescence [23].

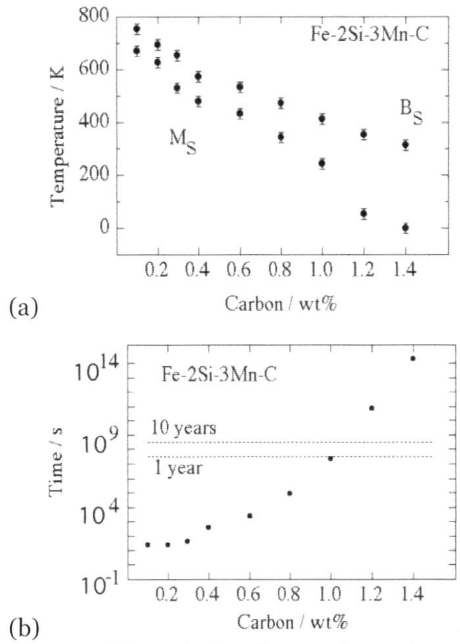

Figure 4: (a) Calculated transformation start temperatures in Fe-2Si-3Mn wt% steel as a function of the carbon concentration. (b) The calculated time required to initiate bainite [24].

The theory of the bainite transformation allows the estimation of the lowest temperature at which bainite can be induced to grow. Such calculations are illustrated in Fig. 4a, which shows how the bainite-start (B_S) and martensite-start (M_S) temperatures vary as a function of the carbon concentration, in a particular alloy system. There is in principle no lower limit to the temperature at which bainite can be generated. On the other hand, the rate at which bainite forms slows down dramatically as the transformation temperature is reduced (Fig. 4b). It may take hundreds or thousands of years to generate bainite at room temperature. For practical purposes, the carbon concentration has to be limited to about 1 wt% for the case illustrated.

An alloy has been designed in this way, with the approximate composition Fe-1C-1.5Si-1.9Mn-0.25Mo-1.3Cr-0.1V wt%, which on transformation at 200C, leads to bainite plates which are only 20-40 nm thick. The slender plates of bainite are dispersed in stable carbon-enriched austenite which, with its face-centred cubic lattice, buffers the propagation of cracks (Fig. 5).

(a)
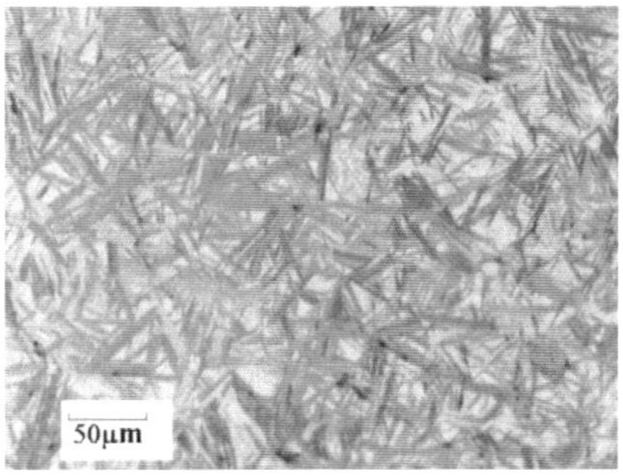

(b)
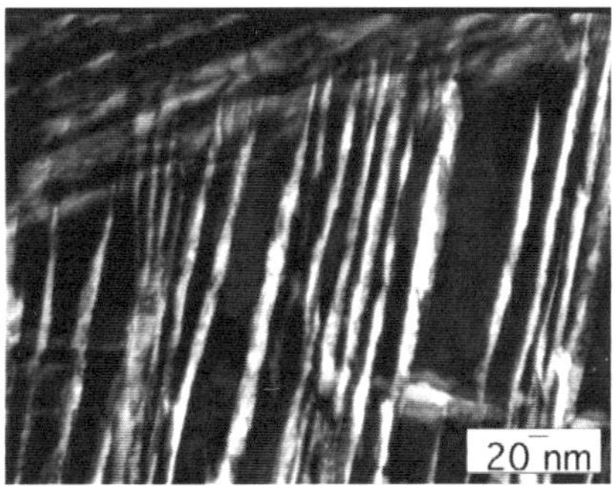

Figure 5: Bainite obtained by transformation at 200C. (a) Optical micrograph. (b) Transmission electron micrograph (Mateo and Bhadeshia).

Compromises

There is often a balance needed between the need to obtain fine structures by phase transformation at low temperatures, and the time taken to obtain the structure. Khare et al. [25] have examined this dilemma in the context of carbide-free bainite.

Excluding the TRIP-assisted steels where the dominant phase is allotriomorphic ferrite [26–30], there are two classes of strong steels that have been developed based on a microstructure consisting of bainitic ferrite, carbon-enriched retained austenite and perhaps some martensite. The first consists of steels whose properties are achieved by continuous cooling transformation [31–38]. The second (discussed above) is where the hardenability is controlled so that the transformation temperature is dramatically suppressed, giving slender plates of bainitic ferrite, under isothermal conditions [18, 39–47]. The transformation time in the latter case can, however, be greater than ten days at temperatures as low as 125°C making the process suitable for large components; strength values in excess of 2000 MPa can routinely be achieved.

There is an industrial requirement for the mass production of strong steel-sheet a few mm thick using a continuous process which is rapid and in which the transformation must occur over a narrow temperature range around 350°C, making the process essentially isothermal. The transformation temperature is fixed by the production process and the transformation time must be limited to ten minutes. The ultimate tensile strength should be approximately 1400 MPa with a ductility of about 20%. Bainitic steels that do not include carbides are at the forefront of such technologies.

Whether or not such a steel can be successful depends on the ability to produce it on a large scale using continuous processing, which may or may not include isothermal steps but the duration must be in minutes rather than hours or days.

The short isothermal transformation time is therefore a vital design criterion and to assess whether this can be achieved, data were collected from reported isothermal transformation experiments on carbide-free bainite [18, 48–51]. The data are plotted in Fig. 6. They correspond to a variety of silicon-rich steels with the carbide-free bainite; much of the variation in the transformation temperature is due to differences in the carbon concentrations of the alloys. The transformation time increase dramatically as the carbon suppresses the transformation temperature. The curves in Fig. 6 are generated by subjecting the data to neural network analysis [52–54] and the upper and lower curves in each plot represent $\pm 1\sigma$ modelling uncertainties. Note that the limited data does not deter from using a neural network based on a Bayesian framework in order to identify and plot the uncertainty bounds illustrated. The Bayesian approach used [52, 53] avoids the overfitting which may lead to a modelling of noise in the data.

The shortest time plotted on Fig. 6 is 30 min, but the trend indicates that by controlling the chemical composition, it should be feasible to design a steel to comply with the required 10 minutes in the temperature range where bainite, and to achieve the required strength. The elongation data show a great deal of scatter so a different, percolation based procedure [55] is used to estimate the expected ductility; the details are in [25].

A systematic alloy design procedure was implemented, including neural network analysis, thermodynamic and kinetic predictions and elementary mechanical property estimates to achieve the 1400 MPa with a ductility of about 20%, while avoiding empirical experiments [25]. The algorithms needed to do this are freely available on

www.msm.cam.ac.uk/map/mapmain.html

This led to a prediction that a steel with the chemical composition Fe-0.35C-1.5Si-0.5Al-0.25Mo-1.2Mn-1Cr-1Ni-0.1V-1Co-1Wwt% would have sufficient hardenability to avoid elevated temperature transformations when cooled at a rate of 5°C s^{-1} in the critical temperature range 700-500 °C; these cooling parameters are consistent with large scale production facilities. The calculated martensite-start temperature is about 330°C so that bainite can be obtained during the continuous cooling through the temperature range B_S-M_S.

Judging from Fig. 6, the alloy should achieve the design strength and elongation in the hot-rolled condition. The material was made, reheated to 1200°C for one hour followed by hot-rolling into 30 mm plate. Fig. 7 shows that the alloy conceived entirely by theoretical design, has almost achieved the properties required at the first attempt.

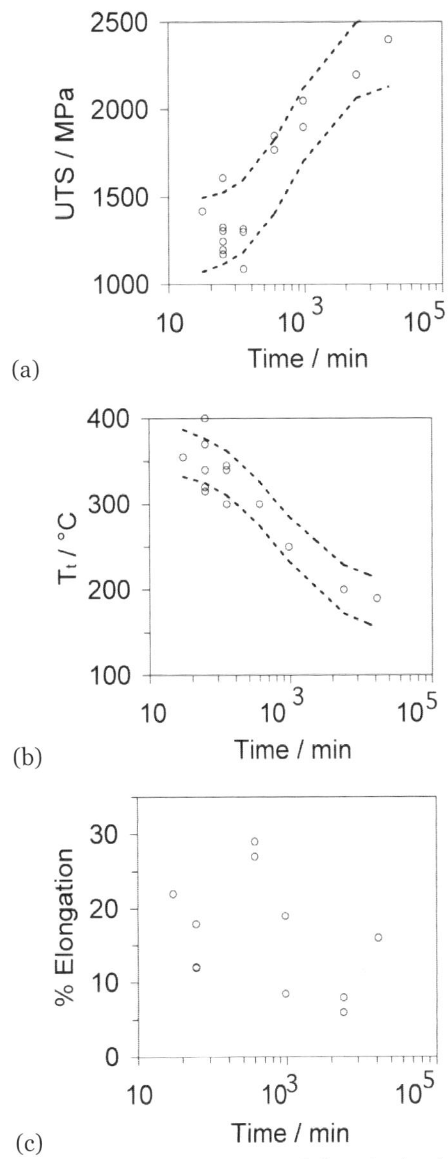

Figure 6: Published experimental data (points) [18, 48–51] as a function of transformation time. The curves represent $\pm 1\sigma$ confidence limits obtained by subjecting the data to a neural network analysis. (a) Ultimate tensile strength (UTS) as a function of isothermal transformation time. (b) Transformation temperature (T_t) as a function of isothermal transformation time. (c) Elongation as a function of isothermal transformation time.

Flash Processing

It has been argued above that compromises are necessary when attempting to comply with the capabilities of industrial production facilities. However, there is a story where the trends described in Fig. 6 have been bypassed in a most elegant and innovative manner.

It appears that some steels can be induced into a bainitic microstructure in tens of milliseconds, using *flash processing* [56]. In the original work, a strip shaped sample of steel of typical composition 0.2C-0.3Si-0.7Mn-0.5Cr-0.5Ni-0.2Mo-0.2Cuwt% was passed through an oxygen-propane fired system that applies

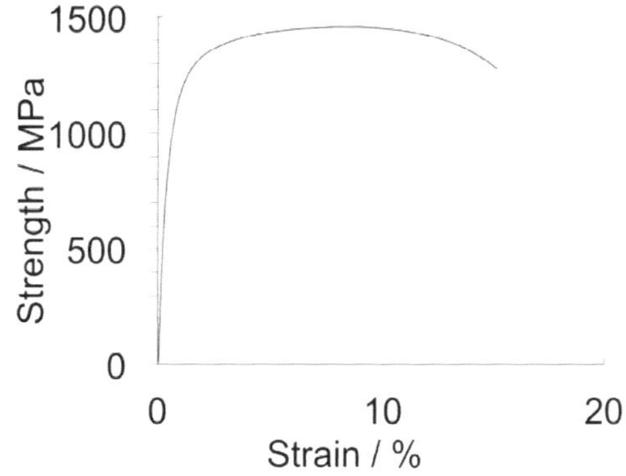

Figure 7: Engineering stress-strain curve of carbide-free bainitic steel

heat directly to the steel strip as it passes through the equipment. The rapidly heated 1.5 mm thick strip was then directly water quenched. The process naturally is associated with a high productivity, resulting in steel with a yield strength in the range 786-1269 MPa and elongation in the range 37%, depending largely on the chemical composition.

The original concept has now been generalised to a variety of steels, and the process has been engineered into a more sophisticated form [57]; a schematic diagram is shown in Fig. 8a. Processing times are now of the order of just 10 s and product dimensions up to 6.35 × 609 × 3657 mm. As the steel passes through an induction coil it is heated at a rate of about 400 °C s^{-1} to between 1000-1100 °C, and shortly afterwards quenched into water; short duration tempering is also available to improve ductility. The ultimate tensile strength and ductility combinations achieved are shown in Fig. 8b, bearing in mind that some of the alloy microstructures obtained are mixtures of bainite and martensite, the actual proportions depending on the chemical composition and process control.

The microstructure evolves rapidly during the quench, inconsistent at first sight from the kinetics of transformation expected from the alloy composition. There are two reasons for this [62]. The austenite that forms during rapid heating consumes the ferrite first, and cementite particles only partly dissolve, and even when they dissolve, the resulting carbon remains localised in the vicinity of the original cementite. This is a consequence of the short duration of the heat treatment. It leaves much of the austenite depleted in carbon, and hence transforms rapidly on quenching.

An alternative scenario is one that leaves some of the carbides undissolved during very rapid austenitisation; the austenite grain size is then refined and the

(a)

(b)

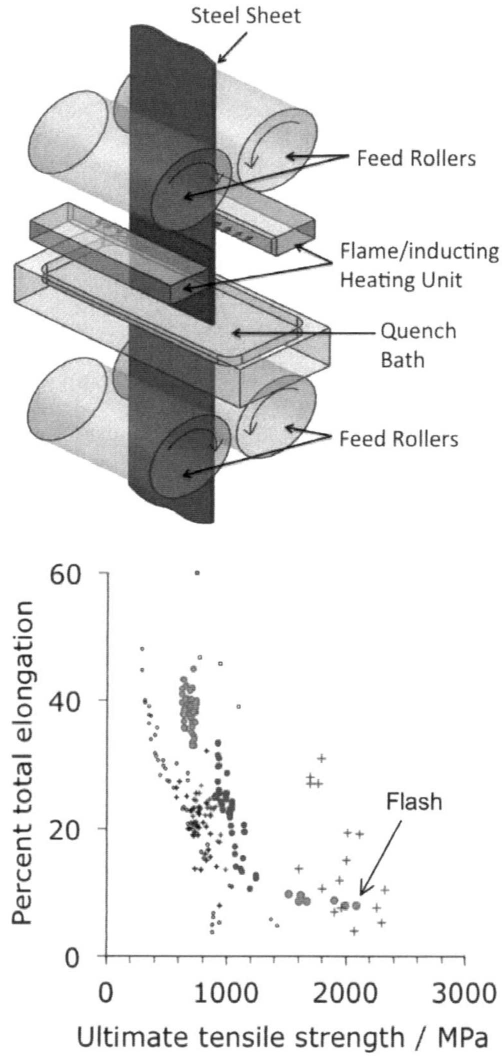

Figure 8: (a) Schematic illustration of the modern variant of flash processing equipment, courtesy of Gary Cola. (b) A comparison of the properties achieved in flash processed steel, against standard automotive steels and nanostructured bainite. Data from [57–61].

resulting finer bainite and martensite microstructure on quenching apparently gives a good combination of properties, i.e. a stronger steel without compromising ductility [63].

Flash processed steel has huge potential, for example in the automotive industry and in the manufacture of armour. In the standard tests used to assess vehicle door side-impact, where a 320 kg mass is thrown against the test piece at 5 m s^{-1}, flash processed tubing demonstrated a 20% increase in the bending energy absorbed relative to an industry standard boron-steel [57]. In resisting armour piercing bullets, it outperformed more costly aluminium, magnesium and titanium alloys. Like all high-strength steels, welding can be problematic because of the inevitable formation of a soft region in the heat-affected zone [64].

Some prototype automotive components produced using the flash process are illustrated in Fig. 9.

(a) (b)

Figure 9: Steel components produced using the flash process. (a) Designed for energy absorption during a crash. (b) A pillar used in the car frame.

Summary

We have illustrated how steels continues to inspire entirely novel phenomena which are so genuine that they enter commercial production at a breathtaking pace. But more than that, the new steels that are entering the market have been developed on the basis of theory and critical experiments rather than by trial and error. The theoretical basis for steel design is now so strong that the subject leads all others in materials science when it comes to dealing with and exploiting the complexity of solid-state phase transformations and heat treatments.

References

1. H. K. D. H. Bhadeshia, and R. W. K. Honeycombe: Steels: Microstructure and Properties, 3rd edition: Butterworth-Heinemann, London, 2006.

2. H. K. D. H. Bhadeshia: Physical Metallurgy, chap. 21: North-Holland: Elsevier, 5th ed., 2014:2157–2214.

3. J. W. Christian: Theory of Transformations in Metals and Alloys, Part I: 3 ed., Oxford, U. K.: Pergamon Press, 2003.

4. H. K. D. H. Bhadeshia: 'Diffusional formation of ferrite in iron and its alloys', *Progress in Materials Science*, 1985, 29, 321–386.

5. H. K. D. H. Bhadeshia: 'Very short and very long heat treatments in the processing of steel', *Materials and Manufacturing Processes*, 2010, 25, 1–6.

6. J. Fridberg, L.-E. Tornda¨hl, and M. Hillert: 'Diffusion in iron', *Jernkontorets Annaler*, 1969, 153, 263–276.

7. J. W. Christian: 'Deformation by moving interfaces', *Metallurgical Transactions A*, 1982, 13, 509–538.

8. J. W. Christian: 'The origin of surface relief effects in phase transformations', In: V. F. Zackay, and H. I. Aaronson, eds. *Decomposition of austenite by diffusional processes.* New York, USA: Interscience, 1962:371–386.

9. H. K. D. H. Bhadeshia: Bainite in steels: theory and practice: 3rd ed., Leeds, U.K.: Maney Publishing, 2015.

10. T. Reti, G. Bagyinszki, I. Felde, B. Vero, and T. Bell: 'Prediction of as-quenched hardness after rapid austenitisation and cooling of surface hardened steels', *Computational Materials Science*, 1999, 15, 101–112.

11. R. C. Reed, Z. Shen, T. Akbay, and J. M. Robinson: 'Laser pulse heat treatment: application to reaustenitisation from ferrite/cementite mixtures', *Materials Science & Engineering A*, 1997, 232, 140–149.

12. M. Y. Wei, and C. Chen: 'Predicting case depth in tempered steels hardened via laser processing', *Materials Science and Technology*, 1994, 10, 69–73.

13. J. R. Bradley, and S. Kim: 'Laser transformation hardening of a high purity Fe-C-Cr alloy', *Scripta Materialia*, 1989, 23, 131–136.

14. P. J. Withers, and H. K. D. H. Bhadeshia: 'Residual stress part 1 – measurement techniques', *Materials Science and Technology*, 2001, 17, 355–365.

15. P. J. Withers, and H. K. D. H. Bhadeshia: 'Residual stress part 2 – nature and origins', *Materials Science and Technology*, 2001, 17, 366–375.

16. M. Burtchen, M. Hunkel, T. Lu¨bben, F. Hoffmann, and H.-W. Zoch: 'Simulation of quenching treatments on bearing components', *Strojniski Vestnik-Journal of Mechanical Engineering*, 2009, 55, 155–159.

17. H. Surm, F. Frerichs, T. Lu¨bben, F. Hoffmann, and H.-W. Zoch: 'Distortion of rings due to inhomogeneous temperature distribution', *Materialwissenschaft und Werkstofftechnik*, 2012, 43, 29–36.

18. F. G. Caballero, H. K. D. H. Bhadeshia, K. J. A. Mawella, D. G. Jones, and P. Brown: 'Very strong, low-temperature bainite', *Materials Science and Technology*, 2002, 18, 279–284.

19. J. W. Christian: 'Accommodation strains in martensite formation, the use of the dilatation parameter', *Acta Metallurgica*, 1958, 6, 377–379.

20. J. W. Christian: 'Thermodynamics and kinetics of martensite', In: G. B. Olson, and M. Cohen, eds. *International Conference on Martensitic Transformations ICOMAT '79*. Massachusetts, USA: Alpine Press, 1979:220–234.

21. R. F. Hehemann: 'The bainite transformation', In: H. I. Aaronson, and V. F. Zackay, eds. *Phase Transformations*. Materials Park, Ohio, USA: American Society of Materials, 1970:397–432.

22. H. K. D. H. Bhadeshia: 'Solute-drag, kinetics and the mechanism of the bainite transformation', In: A. R. Marder, and J. I. Goldstein, eds. *Phase Transformations in Ferrous Alloys*. Ohio, USA: TMS-AIME, 1984:335–340.

23. T. Yokota, C. Garcia-Mateo, and H. K. D. H. Bhadeshia: 'Formation of nanostructured steel by phase transformation', *Scripta Materialia*, 2004, 51, 767–770.

24. H. K. D. H. Bhadeshia: 'A thermodynamic analysis of isothermal transformation diagrams', *Metal Science*, 1982, 16, 159–165.

25. S. Khare, K. Y. Lee, and H. K. D. H. Bhadeshia: 'Carbide-free bainite: Compromise between rate of transformation and properties', *Metallurgical & Materials Transactions A*, 2010, 41A, 922–928.

26. O. Matsumura, Y. Sakuma, and H. Takechi: 'Enhancement of elongation by retained austenite in intercritical annealed 0.4C-1.5Si-0.8Mn steel', *Transactions of the Iron and Steel Institute of Japan*, 1987, 27, 570–579.

27. O. Matsumura, Y. Sakuma, and H. Takechi: 'TRIP and its kinetic aspects in austempered 0.4C-1.5Si-0.8Mn steel', *Scripta Metallurgica*, 1987, 27, 1301–1306.

28. Y. Sakuma, O. Matsumura, and H. Takechi: 'Mechanical-properties and retained austenite in intercritically heat-treated bainite-transformed steel and their variation with Si and Mn additions', *Metallurgical & Materials Transactions A*, 1991, 22, 489–498.

29. P. J. Jacques: 'Transformation-induced plasticity for high strength formable steels', *Current Opinion in Solid State and Materials Science*, 2004, 8, 259–265.

30. B. DeCooman: 'Structure-properties relationship in TRIP steels containing carbide-free bainite', *Current Opinion in Solid State and Materials Science*, 2004, 8, 285–303.

31. N. Jin, and P. Clayton: 'Effect of microstructure on rolling/sliding wear of low carbon bainitic steels', *Wear*, 1997, 202, 202–207.

32. F. G. Caballero, H. K. D. H. Bhadeshia, K. J. A. Mawella, D. G. Jones, and P. Brown: 'Design of novel high-strength bainitic steels: Part I', *Materials Science and Technology*, 2001, 17, 512–516.

33. F. G. Caballero, H. K. D. H. Bhadeshia, K. J. A. Mawella, D. G. Jones, and P. Brown: 'Design of novel high-strength bainitic steels: Part II', *Materials Science and Technology*, 2001, 17, 517–522.

34. Z. G. Yang, and H. S. Fang: 'An overview on bainite formation in steels', *Current Opinion in Solid State and Materials Science*, 2005, 9, 277–286.

35. H. S. Fang, Z. L. Tan, and B. Z. Bai: 'Characteristics of Mn-series bainitic steels and its recent development', *Journal of Iron and Steel Research International*, 2005, 12, 1–10.

36. F. G. Caballero, M. J. Santofimia, C. Capdevila, C. Garcia-Mateo, and C. G. de Andr´es: 'Design of advanced bainitic steels by optimisation of TTT diagrams and T_0 curves', *ISIJ International*, 2006, 46, 1479–1488.

37. M. R. Zhang, and H. C. Gu: 'Microstructure and properties of carbide free bainite railway wheels produced by programmed quenching', *Materials Science and Technology*, 2007, 23, 970–974.

38. F. G. Caballero, C. G. Mateo, J. Chao, M. J. Santofimia, C. Capdevila, and C. G. de Andr´es: 'Effects of morphology and stability of retained austenite on the ductility of TRIP-aided bainitic steels', *ISIJ International*, 2008, 48, 1256–1262.

39. C. Garcia-Mateo, F. G. Caballero, and H. K. D. H. Bhadeshia: 'Development of hard bainite', *ISIJ International*, 2003, 43, 1238–1243.

40. P. M. Brown, and D. P. Baxter: 'Hyper-strength bainitic steels', In: *Materials Science and Technology 2004*. Warrendale, Pennsylvania, USA: TMS, 2004:433–438.

41. C. Garcia-Mateo, F. G. Caballero, and H. K. D. H. Bhadeshia: 'Acceleration of low-temperature bainite', *ISIJ International*, 2003, 43, 1821–1825.

42. M. Peet, S. S. Babu, M. K. Miller, and H. K. D. H. Bhadeshia: 'Three-dimensional atom probe analysis of carbon distribution in low-temperature bainite', *Scripta Materialia*, 2004, 50, 1277–1281.

43. C. Garcia-Mateo, M. Peet, F. G. Caballero, and H. K. D. H. Bhadeshia: 'Tempering of a hard mixture of bainitic ferrite and austenite', *Materials Science and Technology*, 2004, 20, 814–818.

44. K. Goa, L. D. Wang, M. Zhu, J. D. Chen, Y. J. Shi, and M. K. Kang: 'Refinement of grain and enhancement of impact toughness for low-alloying ultra-high strength bainite steels', *Acta Metallurgica Sinica*, 2007, 43, 315–320.

45. H. K. D. H. Bhadeshia: 'Large chunks of very strong steel', *Materials Science and Technology*, 2005, 21, 1293–1302.

46. H.-S. Yang, and H. K. D. H. Bhadeshia: 'Designing low-carbon, low-temperature bainite', *Materials Science and Technology*, 2008, 24, 335–342.

47. F. C. Zhang, T. S. Wang, P. Zhang, C. L. Zhang, B. Lv, M. Zhang, and Y. Z. Zhang: 'A novel method for the development of a low-temperature bainitic microstructure in the surface layer of low-carbon steel', *Scripta Materialia*, 2008, 59, 294–296.

48. H. K. D. H. Bhadeshia, and D. V. Edmonds: 'Bainite in silicon steels: a new composition-property approach II', *Metal Science*, 1983, 17, 420–425.

49. L. C. Chang: 'The rolling/sliding wear performance of high silicon carbide-free bainitic steels', *Wear*, 2005, 258, 730–743.

50. C. G. Mateo, and F. G. Caballero: 'Ultrahigh-strength bainitic steels', *ISIJ International*, 2005, 45, 1736–1740.

51. V. T. T. Miihkinen, and D. V. Edmonds: 'Tensile deformation of two experimental high-strength bainitic low alloys steels containing silicon', *Materials Science and Technology*, 1987, 3, 432–440.

52. D. J. C. MacKay: 'Practical Bayesian framework of backpropagation networks', *Neural Computation*, 1992, 4, 448–472.

53. D. J. C. MacKay: 'Bayesian non-linear modelling for the energy prediction competition', *ASHRAE Transactions*, 1994, 100, 1053–1062.

54. H. K. D. H. Bhadeshia: 'Neural networks in materials science', *ISIJ International*, 1999, 39, 966–979.

55. H. K. D. H. Bhadeshia: 'The first bulk nanostructured metal', *Science and Technology of Advanced Materials*, 2013, 14, 014202.

56. G. M. Cola Jr.: 'Properties of bainite nucleated by water quenching in 80 ms', In: T. Furuhara, and K. Tsuzaki, eds. *1st International Symposium on Steel Science*. Tokyo, Japan: Iron and Steel Institute of Japan, 2007:187–190.

57. D. J. Tung, T. Lolla, S. S. Babu, and G. M. Cola Jr.: 'Flash processing produces unique combination of properties in steels', *Advanced Materials and Processes*, 2013, 171, 48–50.

58. S. Sadagopan, C. Wong, M. Huang, B. Yan, and D. Urban: 'Formability characterisation of new generation of high strength steels': Tech. Rep. TRP 0012, American Iron and Steel Institute, Pittsburgh, USA, 2003.

59. H. K. D. H. Bhadeshia: 'Problems in the calculation of transformation texture in steels', *ISIJ International*, 2010, 50, 1517–1522.

60. O. Grässel, L. Kruger, G. Frommeyer, and L. W. Meyer: 'High strength Fe-Mn-(Al, Si) TRIP/TWIP steels development-properties-application', *International Journal of Plasticity*, 2000, 16, 1391–1409.

61. H. L. Yi, K. Y. Lee, and H. K. D. H. Bhadeshia: 'Extraordinary ductility in Al-bearing δ-TRIP steel', *Proceedings of the Royal Society A*, 2011, 467, 234–243.

62. T. Lolla, G. Cola, B. Narayanan, B. Alexandrov, and S. S. Babu: 'Development of rapid heating and cooling (flash processing) process to produce advanced high strength steel microstructures', *Materials Science and Technology*, 2011, 27, 863–875.

63. M. J. Holzweissig, J. Lackmann, S. Konrad, M. Schaper, and T. Niendorf: 'Influence of short austenitization treatments on the mechanical properties of low-alloy steels for hot forming applications', *Metallurgical & Materials Transactions A*, 2015, ?, DOI: 10.1007/s11661-015-2907-3.

64. B. Hanold, S. S. Babu, G. Cola, S. Ream, B. Nagy, and B. Victor: 'Investigation of HAZ softening in laser welding of AHS/high hardness steels', In: S. Babu, H. K. D. H. Bhadeshia, C. E. Cross, S. A. David, T. DebRoy, J. N. DuPont, T. Koseki, and S. Liu, eds. *Trends in Welding Research*. Ohio, U. S. A.: ASM International, 2013:15–20.

International Federation of Heat Treatment and Surface Engineering 2016
Proceedings of the 23rd IFHTSE Congress
April 18–21, 2016, Savannah, Georgia, USA

Alloying element nitride development in ferritic Fe-based materials upon nitriding

T. Steiner[1,2]* *and E.J. Mittemeijer*[1,3]
1 Max Planck Institute for Intelligent Systems (formerly Metals Research), Heisenbergstrasse 3, 70569 Stuttgart, Germany
2 Robert Bosch GmbH, Robert-Bosch-Strasse 2, 71701 Schwieberdingen, Germany
3 Institute for Materials Science, University of Stuttgart, Heisenbergstrasse 3, 70569 Stuttgart, Germany
*tobias.steiner2@de.bosch.com

1. Introduction

Nitriding is a thermochemical surface treatment, by which nitrogen is introduced from an outer gas atmosphere, plasma, or salt bath into the surface adjacent regions of a component [1]–[3]. The nitrogen absorbed at the surface will diffuse inwardly leading to the development of the so-called *diffusion zone*. At sufficiently high chemical potential of nitrogen in the nitrogen-donating medium (the "nitriding atmosphere"), iron may react directly with nitrogen leading to the development of a layer of iron nitride(s) on the specimen surface, the so-called *compound layer*, on top of the diffusion zone. This layer is generally composed of grains of γ'-Fe_4N_{1-x} and ε-Fe_3N_{1+x} [2].

In pure iron and in plain carbon steels, upon slow cooling (from the nitriding temperature) or prolonged aging at lower temperatures, precipitation of γ' nitride from the nitrogen-supersaturated ferrite matrix can occur in the diffusion zone with a characteristic needle shape [4]. Also, at temperatures smaller than say 150 °C the formation of an intermediate precipitate α"-$Fe_{16}N_2$ can occur [5], [6].

Phase transformations can also occur in the compound layer upon cooling after nitriding. Thus, γ' may precipitate in ε (giving rise to the typical "staircase microstructure") [7] and α-Fe may precipitate (as needles) in γ' [8].

If the nitrided iron-based material contains alloying elements (*Me*) with an affinity for nitrogen stronger than that of iron for nitrogen, the nitrogen in the diffusion zone will interact with the alloying element. In the case of *ferritic* materials (nitriding temperatures in the range 510-580 °C), this usually means the precipitation of more or less fine alloying element nitrides. Examples of such papers (with *Me* as indicated): Ti [9]–[11], V [12]–[16], Cr [17]–[24], Mo [9], [25]–[27], Al [28]–[33], Si [34]–[37] and Mn [38]–[42]. In the case of *austenitic* materials a strongly nitrogen-supersaturated (up to 25 at.%) solid solution may develop at sufficiently low temperatures (say 450 °C) to suppress the mobility of the alloying element *Me* and thus the precipitation of alloying element nitride particles (pp. 169-170 in Ref. [43] and Ch. 14 in Ref. [3]).

In the present paper the focus lies on the development of alloying element nitrides in the ferritic diffusion zone *during the nitriding process*.

2. Strength of the alloying element (*Me*)-nitrogen (N) interaction

In order to determine the nitriding response that can be expected for *Me*, with nitrogen, N, in iron-based materials containing the nitride forming alloying element, knowledge of the strength of interaction of the alloying elements present in the system is helpful. Alloying elements having a strong nitrogen affinity are likely to form very stable compounds with nitrogen, i.e. the nitrides of these alloying elements will have a strongly negative Gibbs energy of formation from the elements. Alloying elements commonly present in steel and which can form nitrides in the matrix are Ti, V, Al, Si, Cr and Mo (see Table 7.1 in Ref. [3]). Alloying elements as Ni, Co, or Cu can only form nitrides at chemical potentials of nitrogen even higher than those required for the development of iron nitrides (at the surface) and are therefore not considered here.

If the MeN_x particle forms within a grain of the ferrite matrix, it is constrained by the surrounding matrix and strains arise due to the volumetric mismatch of the nitride precipitate and the ferrite matrix (see Table 1) and the crystallographic constraints as exposed at the precipitate/matrix interface. The total volumetric mismatch can be considered by comparing the molar volumes of the alloying element nitride, v_{MeNx}, and the molar volume of Fe in the ferritic matrix, v_{Fe}, according to:

$$\frac{\Delta v}{v} = \frac{v_{MeN_x} - v_{\text{Fe}}}{v_{\text{Fe}}} \cdot 100\% \qquad (1)$$

recognizing that upon precipitation one mole MeN_x develops from one mole *Me* atoms orginially dissolved in the nitrogen-less, parent ferrite lattice.

The strain energy, which arises during precipitation, may slow the nitride-precipitation process considerably. A striking case occurs for *Me* = Si: although the precipitation of Si_3N_4 is accompanied by the release of a large amount of chemical Gibbs energy (210 $kJmol^{-1}$ per mole N_2 [3]), its development occurs at the expense of an even larger amount of strain energy, if full elastic accommodation of the misfit would occur (523 $kJmol^{-1}$ per mole N_2 [3]). This leads to silicon reacting only extremely slowly with the inwardly diffusing nitrogen [36].

In order to characterize this interplay of release of chemical Gibbs energy, ΔG_{chem}, and expense of strain energy, ΔG_{strain}, an interaction parameter I was proposed, defined as [3], [44]:

$$I = -\frac{\Delta G_{chem}}{\Delta G_{strain}} \qquad (2)$$

It allows to arrange the reactivities of various *Me* with N in binary ferritic alloys in roughly this order: Ti > V > Al > Cr > Si > Mo. Note that in binary Fe-Mn alloys, complex phase transformation reactions take place [38]–[42], which have led to its exclusion from this series.

3. Crystal structures of MeN_x

The most common composition and crystal structure of the alloying element nitride (as it (can) occur(s) for *Me* = Ti, V, Al, and Cr) is described by the stoichiometry *Me*N and the rock salt (NaCl) crystal structure. Yet, a large variety of crystal structures occurs for phases in the binary *Me*-N systems; an overview of nitride phases that can form in steels is presented in Ref. [45]. That listing is not exhaustive: for example, the δ-MoN phase that develops by discontinuous precipitation (DP) in binary Fe-Mo alloys [27], has a complex hexagonal crystal structure, amorphous Si_3N_4 is found to precipitate in Fe-Si alloys [35], [36], and a ternary, layered hexagonal $CrMoN_2$ nitride forms by DP upon prolonged nitriding of ternary Fe-Cr-Mo alloys [46], etc.

Table 1: Lattice parameters (a, c), number of Me atoms per unit cell (z), and volume misfit ($\frac{\Delta v}{v}$) between the nitride and the pure ferrite matrix, for various alloying element nitride phases.

nitride	crystal lattice	lattice parameters [Å]	z	$\frac{\Delta v}{v}$	Ref.
TiN	cubic	a=4.2417	4	62%	[47]
VN	cubic	a=4.139	4	51%	[47]
AlN	hexagonal	a=3.11 c=4.979	4	77%	[48]
Mo_2N	cubic	a=4.163	4	53%	[47]
CrN	cubic	a=4.14	4	51%	[47]
Cr_2N	hexagonal	a=4.795 c=4.469	6	26%	[49]
Si_3N_4	hexagonal	a=7.71 c=2.908	6	112%	[50]
δ_3-MoN	hexagonal	a=5.63 c=5.736	8	67%	[51]
$CrMoN_2$	hexagonal	a=2.85 c=15.63	6	56%	[46]

4. Precipitation morphology and microstructure

The crystal structures of the nitride precipitates determine their shape and crystallographic orientation in the ferrite matrix. The most widely encountered crystal structure is the NaCl-type crystal structure. TiN, VN, cubic AlN, CrN, and Mo_2N (with 50 % vacancies on the N sublattice) all precipitate in this modification. The good matching of the NaCl-type crystal structure to the bcc ferrite-matrix crystal structure in the Baker-Nutting orientation relationship (OR) [52] determines the shape of the precipitates: they develop as finely distributed, usually nano-sized platelets on the $\{100\}_{\alpha\text{-Fe}}$ planes of the ferrite matrix, (see Figure 1). As they form largely homogeneously throughout the ferrite matrix, this microstructure is commonly referred to as continuously precipitated (CP) [3], [43]. The fine distribution of nitride precipitates, their large volume mismatch and coherent nature of the platelet faces are the main origins of the high hardness and large residual microstresses that develop in the diffusion zone upon nitriding.

Figure 1: TEM bright field (left), dark field (right) micrographs, and corresponding selected area diffraction pattern (SADP), of an Fe-2.0 at.%Cr alloy specimen nitrided at 580 °C for 385 h with a nitriding potential of 0.1 atm$^{-1/2}$: The aperture for dark field imaging, employing the intersection of the streaked diffraction spots of CrN of two orientation variants, is indicated by the white circle; white arrows indicate the positions of intensity maxima along the streaks pertaining to 200_{CrN}-type spots. From Ref. [53].

In binary Fe-Cr [18]–[20], [22], Fe-V [15] and Fe-Mo [27] alloys, as well as in ternary Fe-Cr-Mo [46], [54] and quaternary Fe-Cr-Mo-V [55] alloys, prolonged aging may lead to the development of a much coarser lamellar microstructure consisting of alternating ferrite and nitride lamellae, (see Figure 2). This microstructure forms by a discontinuous, moving grain-boundary mediated mechanism [56], [57].

In Fe-Cr and Fe-V alloys a discontinuous *coarsening* (DC) reaction occurs, which is driven by the reduction of precipitate -matrix interfacial area, loss of nitrogen supersaturation of the matrix and (misfit) stress relaxation [18]. The DC reaction can be expressed in the following way:

$$\alpha' + \beta' \rightarrow \alpha + \beta \qquad (3)$$

where α' denotes the supersaturated, strained ferrite matrix, β' denotes the fine, coherent CP nitride particles, α the relaxed ferrite matrix and β the coarsened nitride phase.

In binary Fe-Mo and ternary Fe-Cr-Mo alloys a discontinuous *precipitation* (DP) reaction occurs, in which not only coarsening (as for DC) occurs, but which also involves a phase transformation of the initially precipitated, intermediate CP nitride phase to the thermodynamically stable nitride phase of different composition and crystal structure: δ-MoN forms from γ-Mo_2N in binary Fe-Mo alloys and $CrMoN_2$ forms from $(Cr_zMo_{1-z})N_{(1+z)/2}$ in low Cr/Mo-ratio Fe-Cr-Mo alloys [46], [54]. The additional contribution of the chemical energy released by formation of the stable nitride phase accelerates the DP reaction as compared to the DC reaction [54]. The reaction DP can be expressed in the following way:

$$\alpha' + \beta \rightarrow \alpha + \delta \quad (3)$$

where α' denotes the supersaturated, strained ferrite matrix, β denotes the intermediate CP nitride particles, α the relaxed ferrite matrix and δ the thermodynamically stable DP nitride phase.

The consequence of the formation of the coarsened microstructure by both types of discontinuous reactions is a considerable loss of hardness and residual stresses in these regions, see e.g. Refs. [24], [27], [54].

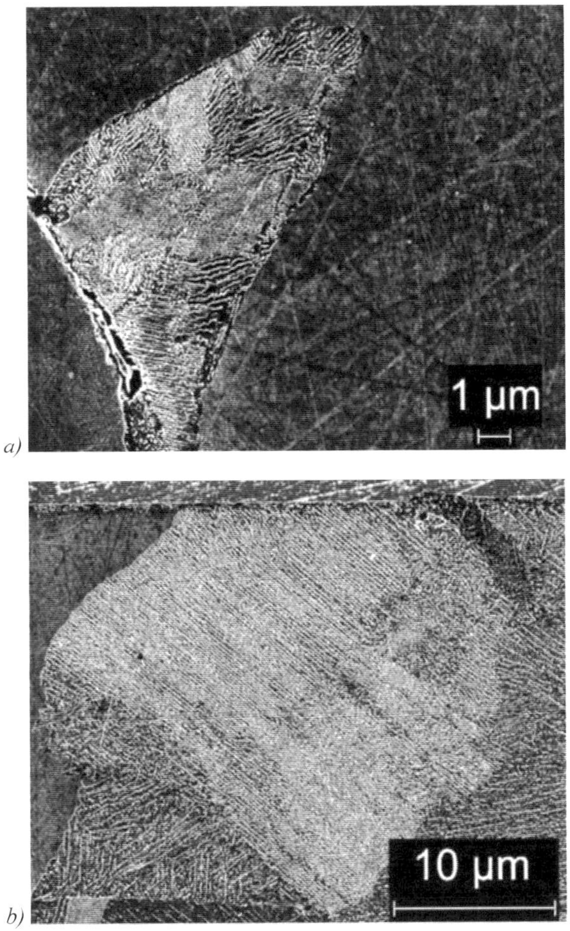

Figure 2: SEM micrographs recorded from the cross-sections of a) a DC region in Fe-1.81s at%Cr-0.25 at.%Mo alloy after nitriding at 580 °C for 72 h with a nitriding potential of 0.1 atm$^{-1/2}$. b) DP regions in Fe-1.04 at%Cr-1.01 at.%Mo alloy after nitriding at 580 °C for 72 h with a nitriding potential of 0.1 atm$^{-1/2}$. From Ref. [54].

5. Development of nitrogen-content depth profiles

The strength of the *Me*-N interaction also has predictive power for the shape of the nitrogen-content depth profiles: the stronger the interaction, the more immediate the dissolved alloying element will react with the inwardly diffusing nitrogen. Two extreme cases can be identified [29] (see Figure 3): i) The nitrogen-content depth profiles of strong nitride forming elements have a rectangular shape of high (saturation-level) nitrogen content near the surface and (after an abrupt transition) nil at depths larger than the nitrided region, as the alloying element reacts practically instantaneously with nitrogen upon its arrival at a certain depth in the specimen; only if all alloying element at the depth considered has precipitated, the inwardly diffusing nitrogen can reach larger depths. This behavior is observed, for example, for the precipitation of TiN [9]–[11], VN [12]–[16], and CrN [17]–[24] at not too high nitriding temperatures. ii) In case of weak nitride formers, the nitrogen-content is of practically constant level throughout the depth of the specimen (of finite thickness), and this nitrogen level rises only slowly with proceeding precipitation (as a bucket that becomes filled with liquid). This behavior is for example observed for the precipitation of Mo_2N [27], hexagonal AlN [58], and especially Si_3N_4 [36]). In practice, for cases of non-extreme *Me*-N interaction, nitrogen-content depth profiles with more or less shallow nitrided case/unnitrided core transitions occur.

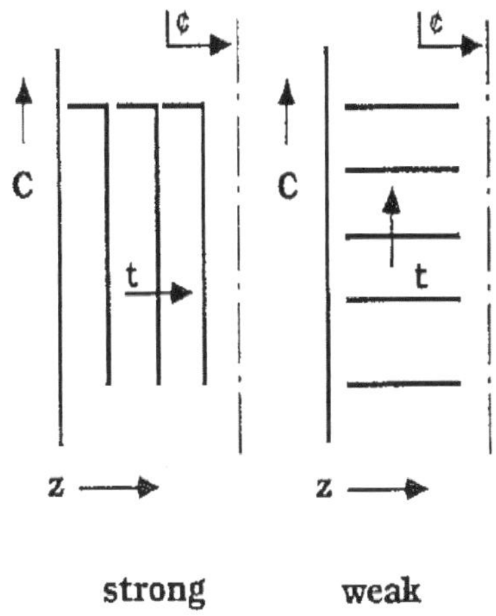

Figure 3: Schematic temporal evolution of the nitrogen-content-depth profiles of weak (left) and strong (right) nitride forming elements. From Ref. [29].

Upon comparing the nitriding behaviors of Fe-Mo and Fe-Si alloys, an interesting observation can be made: although the Si-N interaction parameter (~0.4, [3]) is higher than that of Mo-N (~0.2) (Si_3N_4 is a chemically very stable nitride but with a large volume misfit, whereas Mo_2N is a chemically relatively less stable nitride but with a moderate volume misfit), the nitrogen-content depth profiles of nitride Fe-Si alloys are more shallow than those of Fe-Mo alloys, indicating that in fact Si is the weaker nitride forming element. The origin of this effect likely lies in the crystallography of the respective nitrides: on one hand Mo_2N has the NaCl-type crystal structure (with, as a special feature, only 50 % of the octahedral interstices of the Mo lattice occupied by N) and can thus adopt the favorable Baker-Nutting OR with the ferrite matrix allowing its comparatively easy nucleation. Si_3N_4, with its complex hexagonal crystal structure, on the other hand, can precipitate in the bcc matrix only with great difficulty, and surprisingly its precipitation occurs by initially forming an amorphous (this is highly unusual for precipitation of a solid second phase in a solid, supersaturated matrix) metastable [35] precipitate which can later, or at more elevated temperature, crystallize [36], [37] (see Figure 4).

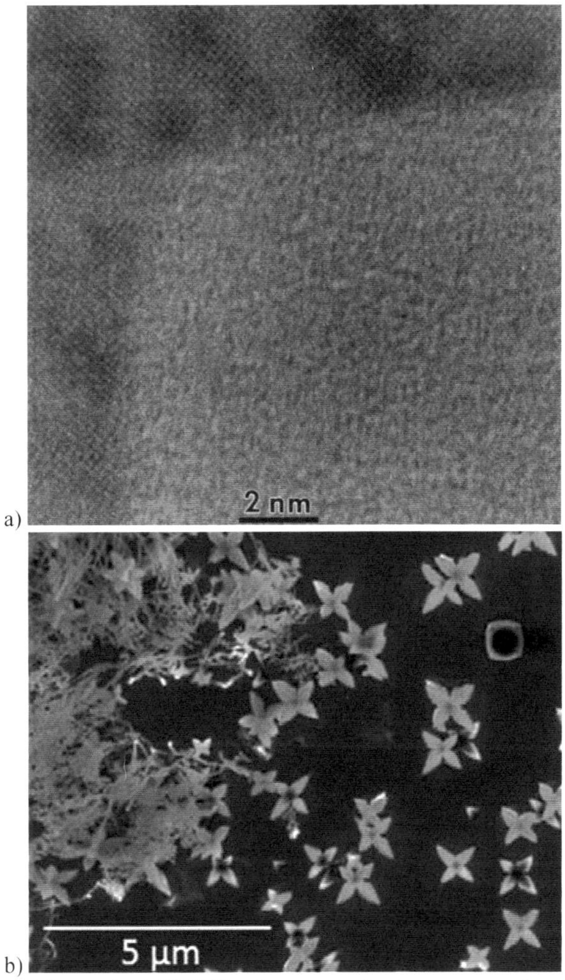

Figure 4: a) HRTEM micrograph showing an amorphous Si-nitride particle (bottom right) in the ferrite matrix (near 100 zone axis) of an Fe-1.6 at.%Si alloy specimen nitrided at 550 °C for 112 h with a nitriding potential of 0.078 atm$^{-1/2}$. From Ref. [35]; b) SEM micrograph recorded from the cross-section of an Fe-4.5 at.%Si alloy specimen nitrided at 650 °C for 48 h with a nitriding potential of 0.02 atm$^{-1/2}$ showing amorphous, octapod-shaped Si_3N_4 precipitates (right) and finer, crystalline α-Si_3N_4 precipitates. From Ref. [37].

6. Ternary alloys

In steels often more than one (dissolved) alloying element is present that can interact with nitrogen such that nitride precipitation can occur. To date, the nitriding behavior of a number of ternary model alloys have been investigated systematically: Fe-Cr-Al [59]–[62], Fe-Cr-Ti [63], Fe-Cr-V [64], Fe-Cr-Mo [46], [54], Fe-Cr-Si [65], Fe-V-Si [65], [66], Fe-Ti-Si [65], Fe-V-Al [64], Fe-Ti-Al [64], and Fe-Ti-Mo [67], and of one quaternary alloy [55]. A focus of interest here is the question: how do the alloying elements Me_1 and Me_2, if present together, interact with N: do mixed $(Me_1,Me_2)N_x$ nitrides occur, and if not, how are the separate precipitation processes of Me_1N_x and Me_2N_x affected?

In many cases, the precipitation of mixed $(Me_1,Me_2)N_x$ nitrides was found to occur: thus (Cr,Al)N [59]–[62], (Cr,Ti)N [63], (Cr,V)N [64], and $(Cr,Mo)N_x$ [46], [54] were detected.

However, the mixed nitrides are metastable with respect to decomposition into Me_1N and Me_2N: as a consequence of kinetic constraints the system accepts a less than maximal gain (i.e. release) of (total) Gibbs energy by allowing the more slowly precipitating element to be taken up in the initially precipitating nitride particles. The change of the precipitate/matrix volume misfit for the mixed nitride as compared to the binary nitride can but need not play a dominant role here: in the case of TiN, the uptake of Cr, of smaller atomic size, leads to a favorable, i.e. decreased, volume misfit; in the case of CrN the uptake of Mo, of larger atomic size, leads to an unfavorable, i.e. increased, volume misfit; for the development of (Cr,Al)N, see Refs. [59]–[62]. In all these cases mixed nitrides are formed.

In those systems where a stable ternary nitride phase can exists that does not occur in the binary boundary systems, such as $CrMoN_2$ [46], $MnSiN_2$ [68], and CrTaN [69], these phases may also be found in ferrite/steels.

If the precipitation kinetics of the two nitride forming elements are, separately, (very) largely different, as e.g. V and Al [64], Ti and Al [64], and Ti and Mo [67], the initially forming nitride precipitates so fast that then separate precipitation of binary nitrides is observed. The slower nitride forming element may not be incorporated (in significant amounts) in time or the Me_1-Me_2 interaction is unfavorable. Such behavior may especially occur if the crystal structures of the binary, boundary nitrides of the nitride forming elements are very different. This for example holds for Fe-Ti/V/Cr-Si alloys [65], [66]. In this case the *sequential* precipitation of the binary nitrides is observed. Then the presence of the nitride particles of the faster forming element may have an accelerating (TiN), or decelerating (VN) effect on the subsequent precipitation of the more slowly reacting alloying element, due to a faster ot slower, respectively, coherent → incoherent transition for the Me_1(Ti,V,Cr)N/α-Fe interface. The subsequent precipitation of amorphous Si_3N_4 particles occurs preferentially at the incoherent Me_1N/α-Fe interface [65] (see Figure 5).

7. The role of carbon

Despite the obvious practical significance and the vast number of studies on the nitriding behavior of technical steels, only very few fundamental studies have been devoted to understanding the role of carbon in the diffusion zone [70], [71]. The major consequences for the microstructure of the nitrogen diffusion zone of the presence of carbon are known, but the operating mechanisms are only crudely understood.

In general, as follows from their Gibbs energies of formation, carbides are less stable than nitrides. For example, at 773 K (500 °C), the Gibbs energy of formation of

ΔG^f = -268 kJmol^{-1} per mole Ti for TiN and it is -174 kJmol^{-1} per mole Ti for TiC [72]. Thus carbides present in the to be nitrided matrix (e.g. as the result of a preceding tempering treatment) can become replaced by nitrides upon nitriding. This is observed indeed [70], [71], [73], see also [74].

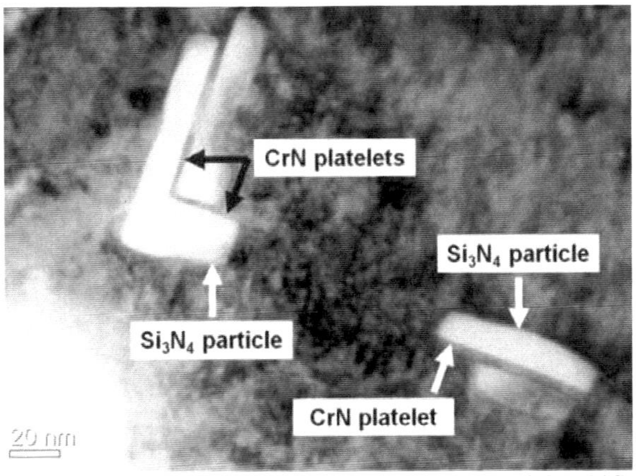

Figure 5: TEM bright field of an Fe-2.0 at.%Cr-2.0 at.%Cr alloy specimen nitrided at 580 °C for 648 h with a nitriding potential of 0.1 atm$^{-1/2}$. Amorphous Si_3N_4 particles grow along the faces of CrN platelets. From Ref. [65].

The thereby released carbon now is free to diffuse in the specimen. If a compound layer can form on the specimen surface, i.e. at sufficiently high chemical potentials of nitrogen (and carbon) in the surrounding atmosphere, it may become incorporated into the compound layer. If no compound layer can form on the specimen surface, carbon may escape into the atmosphere. In a constantly renewed purely nitriding atmosphere (involving a chemical potential of carbon equal to nil), thermodynamics dictate that all carbon should leave the specimen.

Especially at larger depths in the specimen, the carbon released by the nitride formation may lead to local supersaturation of carbon such that carbon precipitates as cementite, Fe_3C, at easy nucleation sites, such as prior austenite grain boundaries. The compressive macrostress developing upon nitriding in the diffusion zone parallel to the surface imposes a preferred development of the cementite on grain boundaries more or less parallel to the surface, as this allows to (partially) accommodate this macrostress state (see Figure 6).

8. Conclusions

Control of the properties of nitrided, iron-based components requires understanding of, at least, the parameters governing the alloying element (*Me*) nitride (*Me*N_x) particle constitution, crystal structure, and morphology, the nitrogen-content depth profile and the nitriding kinetics.

The strength of the interaction of *Me* and N, at the background of the observed phenomena, can be crudely characterized by (the ratio of) the chemical Gibbs energy driving the precipitation of *Me*N_x and the counteracting misfit-strain energy. The (strived for) crystal structure of the nitride may be largely incompatible with that of the ferrite matrix, leading to precipitation of metastable nitrides of non-equilibrium crystal structure or even of amorphous nature.

The kinetics of nitride precipitation can be drastically influenced by the developing misfit-strain fields, and the rate of their relaxation by the coherent → incoherent transition of the *Me*N/α-Fe interfaces.

In the past, fundamental research focused on the nitriding behavior of binary iron-based alloys. Only in recent years, attention has been devoted to the nitriding of ternary (in one case even quaternary) iron-based alloys. Evidently, the nitriding response of the latter alloys is not a simple extrapolation of that of the binary alloys. The road to full control of the nitriding process has not at all been trod till the end.

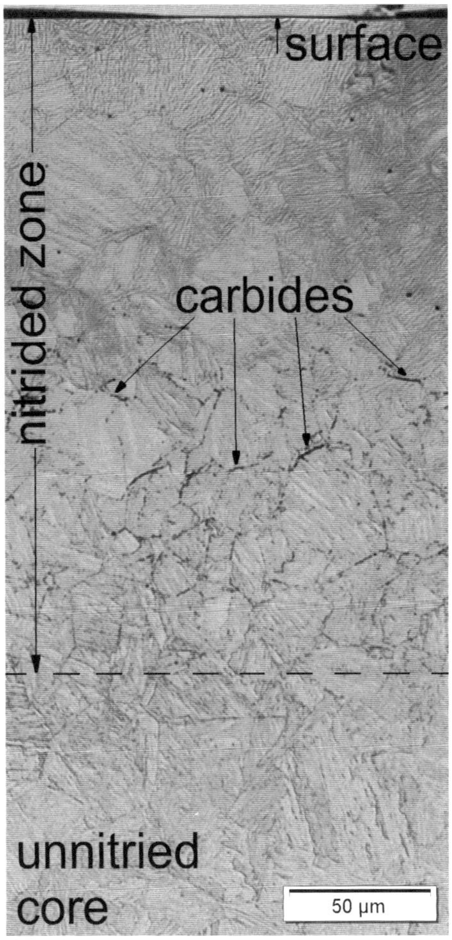

Figure 6: Light optical micrograph of the surface adjacent region of a previously martensitically hardened Fe-1.75 at.%Cr-0.25 at.%Mo-0.61 at.%C alloy specimen nitrided at 580 °C for 3 h with a nitriding potential of 0.1 atm$^{-1/2}$. Close to the nitrided surface a decarburized region is found, below which a region with cementite development along prior austenite grain boundaries parallel to the specimen surface occurs. From Ref. [75].

9. References

[1] C. M. Cotell, J. A. Sprague, and F. A. Smidt, Eds., ASM Handbook Volume 5: Surface Engineering, ASM International (Materials Park, 1994).

[2] E. J. Mittemeijer, "Fundamentals of Nitriding and Nitrocarburizing," in ASM Handbook, Volume 4A, Steel Heat Treating Fundamentals and Processes, vol. 4A, J. Dosset and G. E. Totten, Eds. ASM International (Materials Park, 2013), pp. 619–646.

[3] E. J. Mittemeijer and M. A. J. Somers, Eds., Thermochemical Surface Engineering of Steels, 1st Ed. Woodhead Elsevier (20159.

[4] E. J. Mittemeijer, A. B. P. Vogels, and P. J. Schaaf, "Morphology and lattice distortions of nitrided iron and iron-chromium alloys and steels," *J. Mater. Sci.*, vol. 15 (1980), pp. 3129–3140.

[5] E. J. Mittemeijer, A. B. P. Vogels, and P. Van Der Schaaf, "Aging at room temperature of nitrided α-iron," *Scr. Metall.*, vol. 14 (1980), pp. 411–416.

[6] W. T. M. Straver, H. C. F. Rozendaal, and E. J. Mittemeijer, "Consequences of the Heterogeneous Nitriding of α-Iron: Dislocation Production and Oriented Precipitation," *Metall. Trans. A*, vol. 15 (1984), pp. 627–637.

[7] H. Göhring, S. Kante, A. Leineweber, and E. J. Mittemeijer, "Microstructural development and crystallographic properties of decomposing Fe–N–C compound layers," *Int. J. Mater. Res. (formerly Z. Metallkd.)*, (2016), doi:10.3139/146.111330.

[8] M. A. J. Somers and E. J. Mittemeijer, "Development and relaxation of stress in surface layers; Composition and residual stress profiles in γ′-Fe4N1− x layers on α-Fe substrates," *Metall. Trans. A*, vol. 21 (1990), pp. 189–204.

[9] G. P. Huffman and H. H. Podgurski, "Mössbauer study of nitrided Fe-Mo and Fe-Ti alloys," *Acta Metall.*, vol. 23 (1975), pp. 1367–1379.

[10] D. S. Rickerby, S. Henderson, A. Hendry, and K. H. Jack, "Overview no. 51 Structure and thermochemistry of nitrided iron-titanium alloys," *Acta Metall.*, vol. 34 (1986), pp. 1687–1699.

[11] D. S. Rickerby, A. Hendry, and K. H. Jack, "Low-temperature aging of nitrided Fe-Ti alloys," *Acta Metall.*, vol. 34 (1986), pp. 1925–1932.

[12] M. Pope, P. Grieveson, and K. H. Jack, "Nitride precipitation in ferritic iron-vanadium alloys," *Scand. J. Metall.*, vol. 2 (1973), pp. 29–34.

[13] S. Bor and Ö. E. Atasoy, "The nitriding of Fe-V alloys," *Metall. Trans. A*, vol. 8 (1977), pp. 975–979.

[14] J.-D. Kamminga and G. C. A. M. Janssen, "Calculation of nitrogen depth profiles in nitrided Fe-Mn and Fe-V," *Surf. Coatings Technol.*, vol. 200 (2005), pp. 909–912.

[15] S. S. Hosmani, R. E. Schacherl, and E. J. Mittemeijer, "Nitriding behavior of Fe–4wt%V and Fe–2wt%V alloys," *Acta Mater.*, vol. 53 (2005), pp. 2069–2079.

[16] N. E. Vives Díaz, S. S. Hosmani, R. E. Schacherl, and E. J. Mittemeijer, "Nitride precipitation and coarsening in Fe–2.23at.% V alloys: XRD and (HR)TEM study of coherent and incoherent diffraction effects caused by misfitting nitride precipitates in a ferrite matrix," *Acta Mater.*, vol. 56 (2008), pp. 4137–4149.

[17] B. Mortimer, P. Grieveson, and K. H. Jack, "Precipitation of nitrides in ferritic iron alloys containing Chromium," *Scand. J. Metall.*, vol. 1 (1972), pp. 203–209.

[18] P. M. Hekker, H. C. F. Rozendaal, and E. J. Mittemeijer, "Excess nitrogen and discontinuous precipitation in nitrided iron-chromium alloys," *J. Mater. Sci.*, vol. 20 (1985), pp. 718–729.

[19] R. E. Schacherl, P. C. J. Graat, and E. J. Mittemeijer, "Gaseous nitriding of iron-chromium alloys," *Zeitschrift für Met.*, vol. 93 (2002), pp. 468–477.

[20] M. Sennour, P. H. Jouneau, and C. Esnouf, "TEM and EBSD investigation of continuous and discontinuous precipitation of CrN in nitrided pure Fe-Cr alloys," *J. Mater. Sci.*, vol. 9 (2004), pp. 4521–4531.

[21] M. Sennour, C. Jacq, and C. Esnouf, "Mechanical and microstructural investigations of nitrided Fe-Cr layers," *J. Mater. Sci.*, vol. 39 (2004), pp. 4533–4541.

[22] G. Miyamoto, A. Yonemoto, Y. Tanaka, T. Furuhara, and T. Maki, "Microstructure in a plasma-nitrided Fe–18 mass% Cr alloy," *Acta Mater.*, vol. 54 (2006), pp. 4771–4779.

[23] S. S. Hosmani, R. E. Schacherl, and E. J. Mittemeijer, "Nitrogen absorption by Fe-1.04at.%Cr alloy: uptake of excess nitrogen," *J. Mater. Sci.*, vol. 43 (2008), pp. 2618–2624.

[24] N. E. Vives Díaz, R. E. Schacherl, L. F. Zagonel, and E. J. Mittemeijer, "Influence of the microstructure on the residual stresses of nitrided iron–chromium alloys," *Acta Mater.*, vol. 56 (2008), pp. 1196–1208.

[25] R. Wagner and S. S. Brenner, "Morphology and chemistry of internally nitrided Fe-3at.% Mo," *Acta Metall.*, vol. 26 (1978), pp. 197–206.

[26] M. M. Yang and A. D. Krawitz, "Resistometric study of Fe-V and Fe-Mo nitrided by constant activity aging," *Metall. Trans. A*, vol. 15 (1984), pp. 1545–1554.

[27] H. Selg, E. Bischoff, S. R. Meka, R. E. Schacherl, T. Waldenmaier, and E. J. Mittemeijer, "Molybdenum-Nitride Precipitation in Recrystallized and Cold-Rolled Fe-1 at. pct Mo Alloy," *Metall. Mater. Trans. A*, vol. 44 (2013), pp. 4059–4070.

[28] H. H. Podgurski, R. Oriani, F. Davis, J. Li, and Y. Chou, "Nitrogenation of Fe-Al alloys. Pt. 2. The Adsorption and solution of nitrogen in nitrogenated Fe-Al alloys," *Trans. Metall. Soc. AIME*, vol. 245 (1969), pp. 1603–1608.

[29] M. H. Biglari, C. M. Brakman, M. A. J. Somers, W. G. Sloof, and E. J. Mittemeijer, "On the internal nitriding of deformed and recrystallized foils of Fe-2 at% Al," *Zeitschrift für Met.*, vol. 84 (1993), pp. 124–131.

[30] M. H. Biglari, C. M. Brakman, and E. J. Mittemeijer, "Crystal structure and morphology of AlN precipitating on nitriding of an Fe-2at.% Al alloy," *Philos. Mag. A*, vol. 72 (1995), pp. 1281–1299.

[31] M. H. Biglari, C. M. Brakman, E. J. Mittemeijer, and S. Van Der Zwaag, "Analysis of the nitrogen absorption isotherms of cold-rolled Fe-2 at.% Al specimens with different AlN precipitate dimensions," *Philos. Mag. A*, vol. 72 (1995), pp. 931–947.

[32] M. Sennour and C. Esnouf, "Contribution of advanced microscopy techniques to nano-precipitates characterization: case of AlN precipitation in low-carbon steel," *Acta Mater.*, vol. 51 (2003), pp. 943–957.

[33] S. R. Meka, E. Bischoff, S. S. Hosmani, and E. J. Mittemeijer, "Interrelationships of defects, nitride modification and excess nitrogen in nitrided Fe-4.75 at.% Al alloy," *Int. J. Mater. Res.*, vol. 105 (2014), pp. 1057–1066.

[34] G. R. Booker and J. Norbury, "A New Nitride Precipitate in Iron-Silicon Alloys," *Nature*, vol. 184 (1959), pp. 1311–1312.

[35] E. . Mittemeijer, M. . Biglari, A. . Böttger, N. . van der Pers, W. . Sloof, and F. . Tichelaar, "Amorphous precipitates in a crystalline matrix; precipitation of amorphous Si3N4 in α-Fe," *Scr. Mater.*, vol. 41 (1999), pp. 625–630.

[36] S. R. Meka, K. S. Jung, E. Bischoff, and E. J. Mittemeijer, "Unusual precipitation of amorphous silicon nitride upon nitriding Fe–2at.%Si alloy," *Philos. Mag.*, vol. 92 (2012), pp. 1435–1455.

[37] S. R. Meka, E. Bischoff, B. Rheingans, and E. J. Mittemeijer, "Octapod-shaped, nanosized, amorphous precipitates in a crystalline ferrite matrix," *Philos. Mag. Lett.*, vol. 93 (2013), pp. 238–245.

[38] J. Baird, "Precipitation of nitrides in iron-manganese-nitrogen alloys," *J. Iron Steel Inst.*, (1966), pp. 1122–1130.

[39] J. F. Enrietto, "The Solubility and Precipitation of Nitrides in Alpha-Iron Containing Mangenese," *Trans. Metall. Soc. AIME*, vol. 224 (1961), pp. 43–48.

[40] M. Gouné, A. Redjaïmia, T. Belmonte, and H. Michel, "Identification and characterization of a novel Mn–N nitride formed in Fe–Mn–N alloy," *J. Appl. Crystallogr.*, vol. 36 (2003), pp. 103–108.

[41] M. Gouné, T. Belmonte, A. Redjaïmia, P. Weisbecker, J. . Fiorani, and H. Michel, "Thermodynamic and structural studies on nitrided Fe–1.62%Mn and Fe–0.56%V alloys," *Mater. Sci. Eng. A*, vol. 351 (2003), pp. 23–30.

[42] S. R. Meka, A. Chauhan, T. Steiner, E. Bischoff, P. K. Ghosh, and E. J. Mittemeijer, "Generating duplex microstructures by nitriding; nitriding of iron based Fe–Mn alloy," *Mater. Sci. Technol.*, (2015), doi: 10.1179/1743284715Y.0000000098.

[43] E. J. Mittemeijer, Fundamentals of Materials Science, Springer (Berlin Heidelberg, 2011).

[44] M. A. J. Somers, R. M. Lankreijer, and E. J. Mittemeijer, "Excess nitrogen in the ferrite matrix of nitrided binary iron-based alloys," *Philos. Mag. A*, vol. 59 (1989), pp. 353–378.

[45] D. H. Jack and K. H. Jack, "Invited review: carbides and nitrides in steel," *Mater. Sci. Eng.*, vol. 11 (1973), pp. 1–27.

[46] T. Steiner, S. R. Meka, B. Rheingans, E. Bischoff, T. Waldenmaier, G. Yeli, T. L. Martin, P. A. J. Bagot, M. P. Moody, and E. J. Mittemeijer, "Continuous and discontinuous precipitation in Fe-1 at.%Cr-1 at.%Mo alloy upon nitriding; crystal structure and composition of ternary nitrides," *submitted for publication* (2016).

[47] ICDD, "JCPDS-international Centre for Diffraction Data, PCPDFWIN," *JCPDS-international Cent. Diffr. Data, PCPDFWIN*, 2002.

[48] W. Paszkowicz, S. Podsiadło, and R. Minikayev, "Rietveld-refinement study of aluminium and gallium nitrides," *J. Alloys Compd.*, vol. 382 (2004), pp. 100–106.

[49] S. Eriksson, "Röntgenundersökningar över systemet järn-krom-kväve," *Jernkontorets Ann.*, vol. 118 (1934), pp. 530–543.

[50] P. Yang, H.-K. Fun, I. A. Rahman, and M. I. Saleh, "Two phase refinements of the structures of α-Si3N4 and β-Si3N4 made from rice husk by Rietveld analysis," *Ceram. Int.*, vol. 21 (1995), pp. 137–142.

[51] A. Y. Ganin, L. Kienle, and G. V. Vajenine, "Synthesis and characterisation of hexagonal molybdenum nitrides," *J. Solid State Chem.*, vol. 179 (2006), pp. 2339–2348.

[52] R. G. Baker and J. Nutting, "The tempering of a Cr-Mo-V-W and a Mo-V steel," *Iron Steel Inst. Spec. Rep.*, vol. 64 "Precipitation processes in steels" (1959), pp. 1–22.

[53] T. Steiner, M. Akhlaghi, S. R. Meka, and E. J. Mittemeijer, "Diffraction-line shifts and broadenings in continuously and discontinuously coarsening precipitate-matrix systems: coarsening of initially coherent nitride precipitates in a ferrite matrix," *J. Mater. Sci.*, vol. 50 (2015), pp. 7075–7086.

[54] T. Steiner, S. R. Meka, E. Bischoff, T. Waldenmaier, and E. J. Mittemeijer, "Nitriding of ternary Fe-Cr-Mo alloys; role of the Cr/Mo-ratio," *submited for publication* (2016).

[55] C. W. Kang, S. R. Meka, R. E. Schacherl, and E. J. Mittemeijer, "Microstructure and Kinetics of Nitride Precipitation in a Quaternary Iron-Based Model Fe-2.82 at. pct Cr-0.13 at. pct Mo-0.18 at. pct V Alloy," *Metall. Mater. Trans. A*, vol. 46 (2014), pp. 328–336.

[56] D. B. Williams and E. P. Butler, "Grain boundary discontinuous precipitation reactions," *Int. Mater. Rev.*, vol. 26 (1981), pp. 153–183.

[57] I. Manna, S. K. Pabi, and W. Gust, "Discontinuous reactions in solids," *Int. Mater. Rev.*, vol. 46 (2001), pp. 53–91.

[58] M. H. Biglari, C. M. Brakman, E. J. Mittemeijer, and S. van der Zwaag, "The kinetics of the internal nitriding of Fe-2 at. pct Al alloy," *Metall. Mater. Trans. A*, vol. 26 (1995), pp. 765–776.

[59] K. S. Jung, R. E. Schacherl, E. Bischoff, and E. J. Mittemeijer, "Nitriding of ferritic Fe–Cr–Al alloys," *Surf. Coatings Technol.*, vol. 204 (2010), pp. 1942–1946.

[60] A. R. Clauss, E. Bischoff, R. E. Schacherl, and E. J. Mittemeijer, "Phase transformation of mixed Cr 1− x Al x N nitride precipitates in ferrite," *Philos. Mag.*, vol. 89 (2009), pp. 565–582.

[61] A. R. Clauss, E. Bischoff, S. S. Hosmani, R. E. Schacherl, and E. J. Mittemeijer, "Crystal Structure and Morphology of Mixed Cr1−x Al x N Nitride Precipitates: Gaseous Nitriding of a Fe-1.5 Wt Pct Cr-1.5 Wt Pct Al Alloy," *Metall. Mater. Trans. A*, vol. 40 (2009), pp. 1923–1934.

[62] M. Akhlaghi, S. R. Meka, E. A. Jägle, S. J. B. Kurz, E. Bischoff, and E. J. Mittemeijer, "Low temperature nitriding of recrystallized and cold-rolled ferritic Fe-Cr-Al alloy," *submitted for publication* (2016)..

[63] K. S. Jung, S. R. Meka, R. E. Schacherl, E. Bischoff, and E. J. Mittemeijer, "Nitride Formation and Excess Nitrogen Uptake After Nitriding Ferritic Fe-Ti-Cr Alloys," *Metall. Mater. Trans. A*, vol. 43 (2011), pp. 934–944.

[64] G. Miyamoto, S. Suetsugu, K. Shinbo, and T. Furuhara, "Surface Hardening and Nitride Precipitation in the Nitriding of Fe-M1-M2 Ternary Alloys Containing Al, V, or Cr," *Metall. Mater. Trans. A*, vol. 46 (2015), pp. 5011–5020.

[65] B. Schwarz, P. J. P. J. J. P. Rossi, L. Straßberger, F. Jörg, S. R. Meka, E. Bischoff, R. E. Schacherl, and E. J. Mittemeijer, "Coherency strain and precipitation kinetics: crystalline and amorphous nitride formation in ternary Fe–Ti/Cr/V–Si alloys," *Philos. Mag.*, vol. 94 (2014), pp. 1–22.

[66] B. Schwarz, S. R. Meka, R. E. Schacherl, E. Bischoff, and E. J. Mittemeijer, "Nitriding of iron-based ternary Fe–V–Si alloy: The precipitation process of separate nitrides," *Acta Mater.*, vol. 76 (2014), pp. 394–403.

[67] S. Henderson, "Solute interactions in nitrided iron alloys," Ph. D. Thesis, University of Newcastle upon Tyne, 1976.

[68] W. Roberts, P. Grieveson, and K. H. Jack, "Precipitation of silicon nitrides and manganese-silicon nitrides in steel," *J. Iron Steel Inst.* (1972), pp. 931–937.

[69] H. K. Danielsen, J. Hald, and M. A. J. Somers, "Atomic resolution imaging of precipitate transformation from cubic TaN to tetragonal CrTaN," *Scr. Mater.*, vol. 66 (2012), pp. 261–264.

[70] P. C. Van Wiggen, H. C. F. Rozendaal, and E. J. Mittemeijer, "The nitriding behaviour of iron-chromium-carbon alloys," *J. Mater. Sci.*, vol. 20 (1985), pp. 4561–4582.

[71] Y. Tomio, S. Kitsuya, K. Oh-ishi, K. Hono, G. Miyamoto, and T. Furuhara, "Plasma Nitriding Behavior of Fe-C-M (M = Al, Cr, Mn, Si) Ternary Martensitic Steels," *Metall. Mater. Trans. A*, vol. 45 (2013), pp. 239–249.

[72] M. W. J. Chase, NIST-JANAF Thermochemical Tables, 4th ed., American Chemical Society (1998).

[73] P. F. Colijn, E. J. Mittemeijer, and H. C. F. Rozendaal, "Light-Microscopical Analysis of Nitrided or Nitrocarburized Iron and Steels," *Z. Mettalkd.*, vol. 74 (1983), pp. 620–627.

[74] C. Ginter, L. Torchane, J. Dulcy, M. Gantois, A. Malchère, C. Esnouf, and T. Turpin, "A new approach to hardening mechanisms in the diffusion layer of gas nitrided α-alloyed steels. Effects of chromium and aluminium: experimental and simulation studies," *La Metall. Ital.*, vol. 7–8 (2006), pp. 29–35.

[75] T. Pott, T. Steiner, and E. J. Mittemeijer, "Nitriding of Fe-Cr-Mo-C alloys," *in preparation* (2016).

International Federation of Heat Treatment and Surface Engineering 2016
Proceedings of the 23rd IFHTSE Congress
April 18–21, 2016, Savannah, Georgia, USA

The Effects of Process Parameters on Nitrogen Infiltration and the Controlling of Nitrogen Concentration in Vacuum Nitriding.

Masahiro Okumiya, Yoshiki Tsunekawa, Jung-Hyun Kong
Toyota Technological Institute, Nagoya, Japan
okumiya@toyota-ti.ac.jp

Sang-gweon Kim
Korea Institute of Industrial Technology, Incheon, South Korea

Abstract

In vacuum nitriding, the effect of process parameters such as treating temperature and NH_3 concentration of infiltration gas to furnace on nitrogen concentration of specimen and the controlling of nitrogen concentration of the specimen are investigated. The decomposition of NH_3 in furnace is increased as the treating temperature is increased, therefore the nitrogen concentration of the specimen is decreased. The residual NH_3 concentration is increased as the treating temperature is decreased, therefore the infiltration nitrogen content to specimen is increased. If the residual NH_3 concentration is too high, the voids are generated and the nitrogen concentration decreases. The retention time in the furnace is decreased as the total gas flow rate is increased, therefore the residual NH_3 concentration increase and the nitrogen concentration increases. The number of NH_3 molecules per unit time, which contact on the steel surface can affect the infiltration amount of nitrogen. In other words, the infiltration nitrogen amount is according to the product of the residual NH_3 concentration by the exhaust gas flow rate.

Introduction

Steel is used for variety fields and is an essential material, because it is low-cost, abundant, and covers long range of strength. Heat treatment was applied to steel for mechanical structure to improve mechanical character. Heat treatment, such as gas carburizing, gas carbonitriding, vacuum carburizing, gas nitriding, soft gas nitriding, and high-frequency induction hardening are the processes of choice for surface hardening[1], [2], [3]. Recently vacuum heat treatment attracts significant attention.

Vacuum carburizing has been studied and it is now practical use. Vacuum carburizing has several advanced characteristics for conventional gas carburizing[4], [5]. In this process, it makes possible to restrain intergranular oxidation, shorten the processing time and reduce amount of exhaust gas. Moreover, to improve quenching character and, wear resistance and corrosion resistance, the vacuum nitriding also is applied to the steel in addition to vacuum carburizing. Nitrogen infiltration in to the steel increases the hardnability and tempering softening resistance. Therefore low alloy steel such as SPCC or SPHC can be used for surface heat treatment specimen. On the other hand if the nitrogen concentration of heat treated specimen increases retained austenite will remain after quench or void which is the hole of nitrogen is de-nitriding. To the author's knowledge, report of research on vacuum nitriding is very few and there are a lot of research tasks.

In this study, the effect of process parameters with treating temperature, premixed ammonia on the nitrogen infiltration amount and the gases composition inside of furnace was investigated.

Experimental

Material

The material investigated in this study was the carbon steel for machine structural use JIS-S15C. The chemical composition is shown in Table 1. Specimen shape is 70 mm in length, 35 mm in width and 0.25 mm in thickness. They were ultrasonically cleaned in an acetone before vacuum nitriding.

Table 1: Chemical composition of the carbon steel for machine structural use JIS-S15C in mass%

C	Si	Mn	P	S	Cr	Ni	Al
0.160	0.172	0.41	0.015	0.016	0.033	0.014	0.028

Experimental equipment and treating conditions

Figure 1 shows the schematic diagram of experimental apparatus. The experimental conditions are shown in Figure 2. And process parameters are show in Table 2.

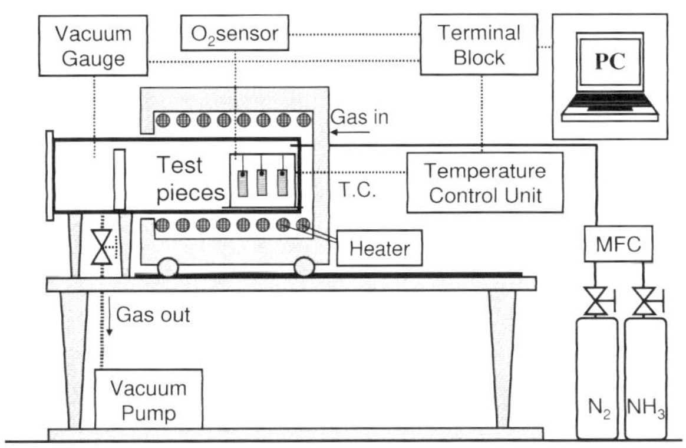

Figure1: Schematic diagram of experimental apparatus.

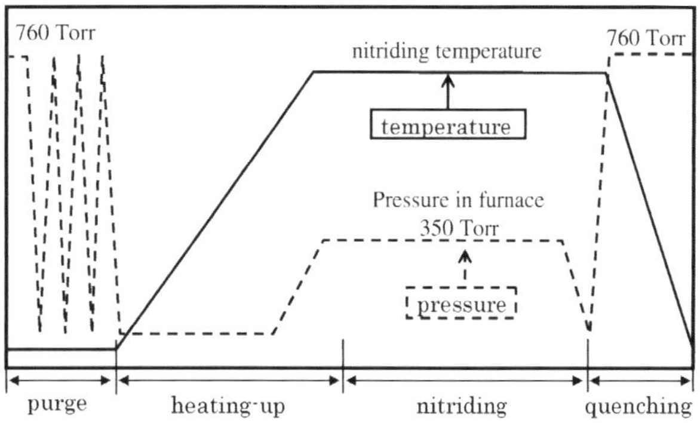

Figure 2: Experimental conditions

Table 2: Peocess parameter

Treating temperature [°C]	680, 700, 730, 750, 780, 800, 820
Pressure in the furnace [hPa]	467
NH_3 flow rate [L/min]	1.4
N_2 flow rate [L/min]	3.6
Treating time [h]	1.0

After furnace purge with nitrogen was operated 3 times, the pressure was decreased, and then heating started. After the treatment temperature was reached, it is preserved, when upon the ammonia and nitrogen gases were supplied and nitriding was performed for 1.0 h. After nitriding was over, the pressure was decreased, and nitrogen was introduced to increased furnace pressure, the specimen was took out from furnace and immersed in the water and quenching was performed. In addition, the concentration of ammonia in the evacuated gases from oil free scroll vacuum pump was analyzed by neutralization titration. After the vacuum nitriding, the specimen was washed out by acetone, and then measurements of nitrogen content by combustion method, nitrided layer phase identification by XRD, as well as structural observations by microscope were performed.

Result and Discussion

Effects of treating temperature on retained ammonia and nitrogen content of specimen

Relation between temperature in furnace and retained ammonia is shown in Figure 3. Retained ammonia decreases with the increase of temperature. That is because thermo chemical decomposition of ammonia was accelerated by increase of temperature.

The nitrogen content in relation to temperature in furnace with premixed ammonia in 28 vol.% is shown in Figure 4. INitrogen content decreases with the increase of temperature in furnace. The retained ammonia decreases with the increase of temperature in furnace in Fig. 3, therefore the nitrogen content decreases with the increase of temperature in furnace.

Morphology and phase of modified layer

Figure 5 shows the nital etched cross-sectional microstructures of substrate with nitrided temperature at 680, 700, 730, 750, 780 and 820 °C, and premixed ammonia in 28 vol.%. Figure 5

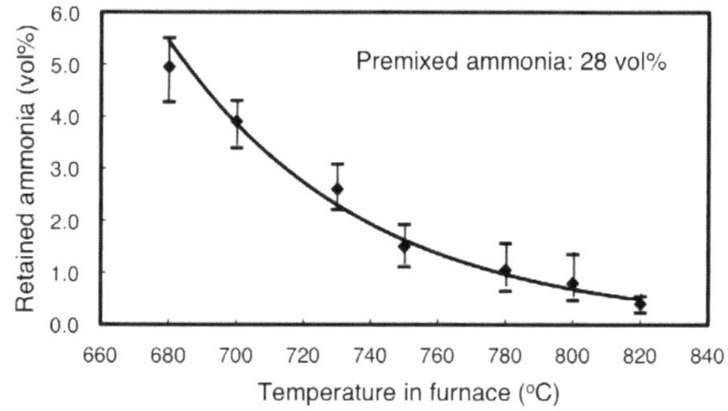

Figure 3: Relation between temperature in furnace and retained ammonia

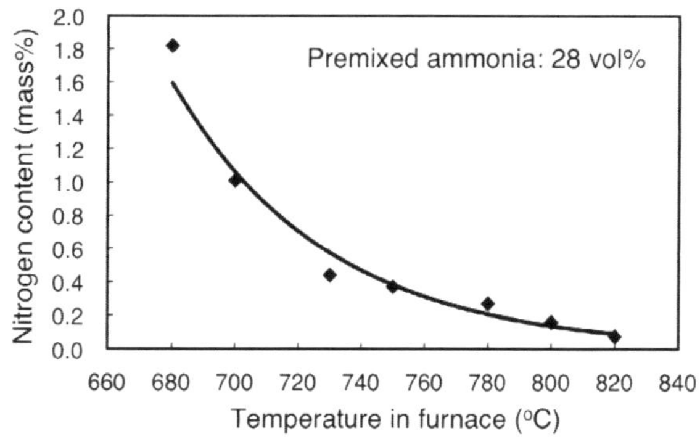

Figure 4: Relation between temperature in furnace and nitrogen content.

shows the X-ray diffraction patterns of substrate with nitrided temperature at 680, 700, 750 and 820 °C, and premixed ammonia in 28 vol.%.

In Figure 5, the surface layers which nitrided at 680 and 700 °C are consisted of white layers, and Figure 6 shows that these layers are Fe_2N and Fe_3N. At 680 and 700 °C, voids and white compound layer was formed on surface of specimen. In Figure 5, at elevated temperature more than 730 °C, there is no formation of voids and compound layer, the surface layers, which nitrided at 730 and 750 °C are consisted one layer, and Figure 6 shows that these layers are martensite. In Fig. 4, the grain growths occur in surface at 780 and 820 °C by heating into high temperature and nitrogen infiltration. In Figure 6, the α phase is observed at 800 °C by X-ray diffraction, this specimen nitrogen content is low, therefore the $Fe_{2\sim3}N$ or martensite phase is not observed.

Figure 7 shows the optical micrograph without nital etching showing the cross-section of heat-treated layer with nitrided temperature at 680 °C, and premixed ammonia in 28 vol.%. Voids formation was confirmed. Under the high ammonia atmosphere, great amount of nitrogen infiltrate to steel and then void formation in austenite grain boundaries by precipitation of molecular nitrogen gas occurs.

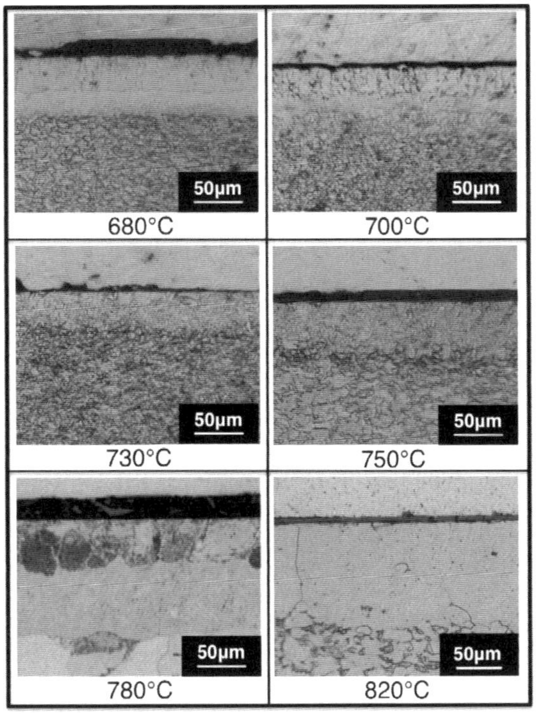

Figure 5: Optical micrographs showing the cross-section of heat treated layer with nitrided temperature at 680, 700, 730, 750, 780 and 820 °C, and premixed ammonia in 28 vol.% (Nital etched).

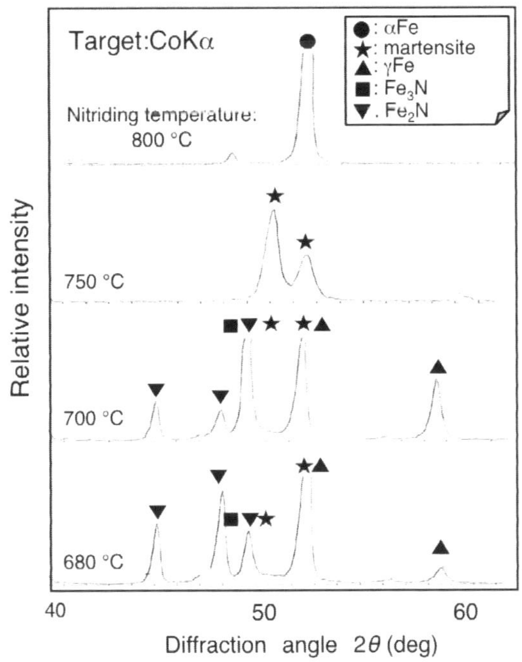

Figure 6: X-ray diffraction patterns of heat-treated layer with nitrided temperature at 680, 700, 750 and 820 °C, and premixed ammonia in 28 vol.%.

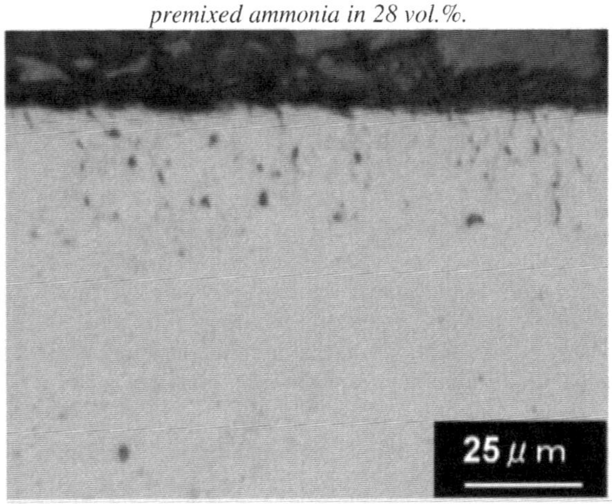

Figure 7: Optical micrograph without nital etching showing the cross-section of heat-treated layer with nitrided temperature at 680 °C, and premixed ammonia in 28 vol.%.

Effect of premixed ammonia content on retained ammonia

Effect of premixed ammonia content on retained ammonia was investigated with premixed ammonia in 100 and 28 vol.%, nitrided temperature at 680, 730, 750, 780 and 800 °C, furnace pressure at 467 Pa and nitriding time on 1.0 h. Relation between nitriding temperature and retained ammonia in each premixed ammonia was shown in Figure 8. The retained ammonia, which the premixed ammonia is 100 vol.%, also decreases with the increase of temperature in furnace. Below 750 °C, the retained ammonia, which premixed ammonia is 100 vol.% is more than that in 28 vol.%, above 780 °C, each retained ammonia in 100 and 28 vol.% is same.

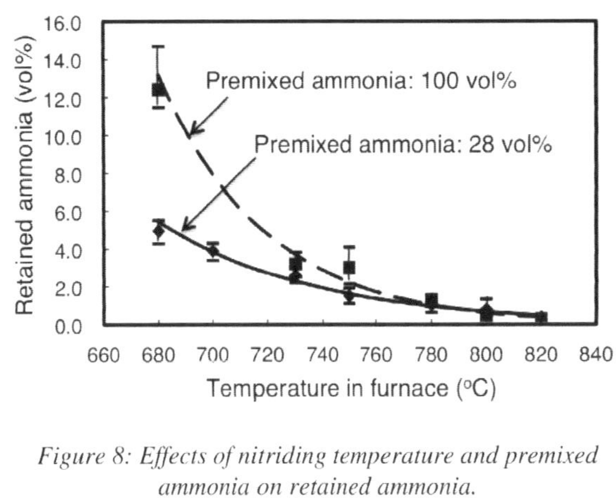

Figure 8: Effects of nitriding temperature and premixed ammonia on retained ammonia.

Effect of premixed ammonia content on nitrogen content of specimen

Effect of premixed ammonia content on nitrogen content of specimen was investigated with premixed ammonia in 100 and

28 vol.%, nitrided temperature at 680, 730, 750, 780 and 800 °C, furnace pressure at 467 Pa and nitriding time on 1.0 h. Relation between nitriding temperature and nitrogen content of specimen in each premixed ammonia was shown in Figure 9. The nitrogen content increases with the decrease of temperature in furnace except at 680 °C and 100 vol.%, because the retained ammonia increases with the decrease of temperature in furnace in Figure 8. The optical micrograph without nital etching showing the cross-section of heat-treated layer with nitrided temperature at 680 °C, and premixed ammonia in 28 vol.% is shown in Figure 10. Voids formation was confirmed. In this condition, the retained ammonia is extremely high and then the void formation is occurred. If the nitriding temperature is low, the premixed ammonia must be diluted by nitrogen, for prevent the void formation.

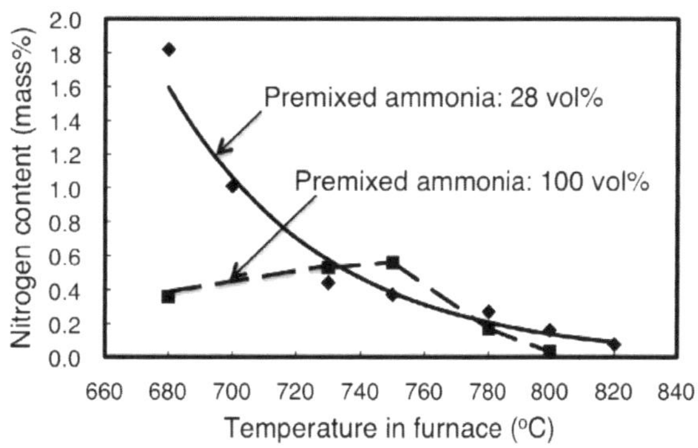

Figure 9: Effects of nitriding temperature and premixed ammonia on nitrogen content.

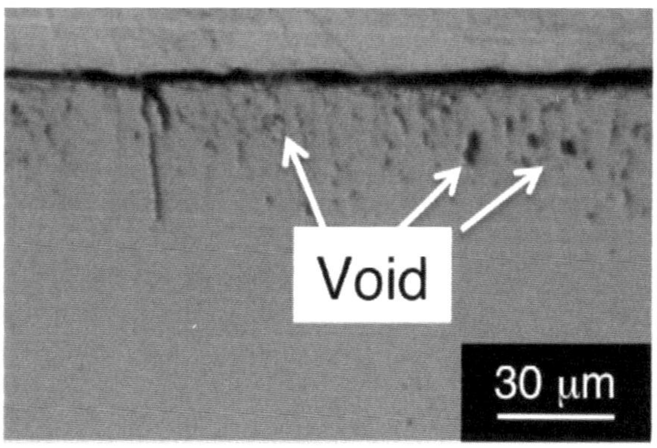

Figure 10: Optical micrograph without nital etching showing the cross-section of heat-treated layer with nitrided temperature at 680 °C, and premixed ammonia in 100 vol.%.

Effect of total gas flow rate on retained ammonia

Effect of total gas flow rate on retained ammonia was investigated with premixed ammonia in 30 vol.%, nitrided temperature at 750 °C, furnace pressure at 467 Pa, nitriding time on 1.0 h and total gas flow rate in 1.5, 3.0, 4.5 and 6.0 L/min. Relation between total gas flow rate and retained ammonia is shown in Figure 11. In this figure, retained ammonia increases with the increase of total gas flow rate. Although ammonia decomposes at high temperatures, to reach the equilibrium value takes time. In furnace pressure is constant at 467 Pa, the residence time of ammonia decreases with the increase of total gas flow rate, therefore retained ammonia increases.

Effect of retained ammonia on nitrogen content

Relation between retained ammonia and nitrogen content is shown in Figure 12. Process parameters are same at Figure 11, the premixed ammonia is 30 vol.%, the nitrided temperature is 750 °C, the furnace pressure is 467 Pa, the nitriding time is 1.0 h and total gas flow rate are 1.5, 3.0, 4.5 and 6.0 L/min. Nitrogen contents with retained ammonia at about 0.8 and about 1.6 vol.% are nearly equal. The nitrogen content increases dramatically with the increase of retained ammonia from about 1.6 vol.% to 1.9 vol.%. Consequently, the control of nitrogen content using retained ammonia value is difficult.

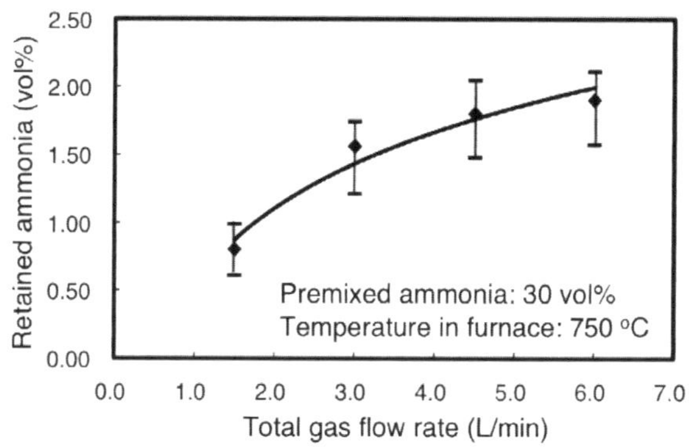

Figure 11: Relation between total gas flow rate and retained ammonia.

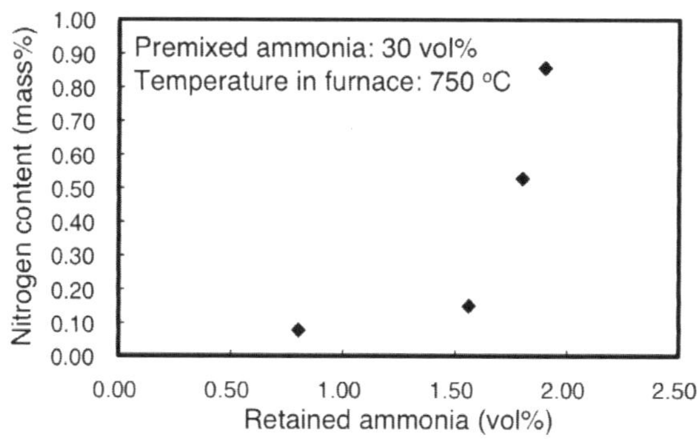

Figure 12: Relation between retained ammonia and nitrogen content.

Control of the nitrogen content and the promotion of nitrogen infiltration

The diffusion rate increases with the increase of temperature, so high temperature nitriding is effective in this point. However, the retained ammonia decreases with the increase of temperature. In order to investigate the process parameter, which suppresses the decomposition of ammonia and infiltrate the nitrogen effectively at high temperature, experiments were performed with nitrided temperature at 820 °C, furnace pressure at 467 Pa, nitriding time on 1.0 h, ammonia gas flow rate in 1.4 and 2.0 l/min and nitrogen gas flow rate in 0, 3.6 and 5.0 L/min. The aim of this study is controlling the nitrogen content in the range of less than 0.4 mass% at 820 °C, because the void does not occur in this condition. Relation between retained ammonia and nitrogen content various ammonia gas flow rate and nitrogen gas flow rate is shown in Figure 13. The ammonia gas flow rate and nitrogen gas flow rate are indicated beside each plot. Nitrogen content increases with the increase of the ammonia gas flow rate and nitrogen gas flow rate. Unfortunately, there is no relation between nitrogen content and retained ammonia. As in the case of Fig.11, the control of the nitrogen content by retained ammonia is difficult.

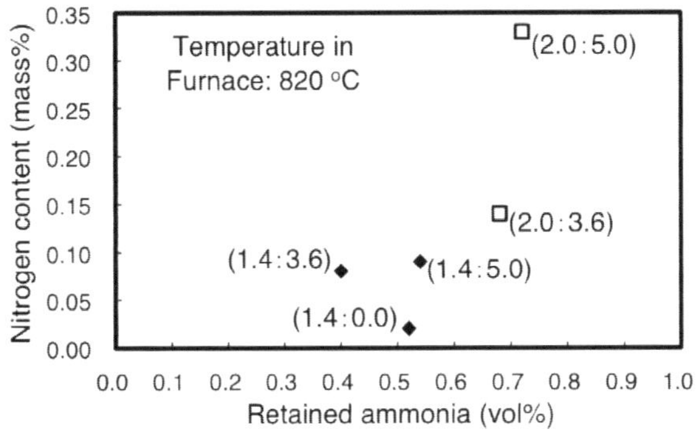

Figure 13: Relation between retained ammonia and nitrogen content various ammonia gas flow rate and nitrogen gas flow rate. The ammonia gas flow rate and nitrogen gas flow rate indicate beside the each plot.

Infiltration of nitrogen into the steel is carried out using active nitrogen, which is generated by ammonia decomposition when the molecule ammonia is adsorbed on the surface of steel. Thus, the number of ammonia molecules, which adsorb on the heated steel surface per unit time affects the nitrogen content. In this study, the stirring is not use, so the number of ammonia molecules, which adsorb on the heated steel surface per unit time is affected by retained ammonia and gas flow in the furnace. The gas flow in the furnace has a relation with exhaust gas flow. The exhaust gas flow rate is calculated by premixed nitrogen gas, retained ammonia and decomposed ammonia, which are nitrogen and hydrogen. Because, 1 mol of ammonia generates the 1/2 mol nitrogen and 3/2 mol hydrogen, the total gas amount is 2 mol. Therefore the exhaust gas flow is calculated by premixed nitrogen gas flow rate, premixed ammonia gas flow rate and retained ammonia ratio. Calculate equation of the exhaust gas flow is as follows:

$$F_E = F_{N2} + \frac{F_{NH3} \times R_{NH3}}{100} + F_{NH3} \times \left(1 - \frac{R_{NH3}}{100}\right) \times 2$$

F_E: Exhaust gas flow (L/min)

F_{N2}: Premixed nitrogen gas flow rate (L/min)

F_{NH3}: Premixed ammonia gas flow rate (L/min)

R_{NH3}: Retained ammonia volume fraction (vol.%)

The number of ammonia molecules, which adsorb on the heated steel surface per unit time (1 min) is affected by the product of retained ammonia volume fraction and exhaust gas flow rate. Relation between the product of retained ammonia volume fraction and exhaust gas flow rate and nitrogen content various ammonia gas flow rate and nitrogen gas flow rate is shown in Figure 14. As shown in Figure 14, by using the product of the exhaust gas flow rate and residual ammonia volume fraction, the nitrogen content of steel is controlled more correctly.

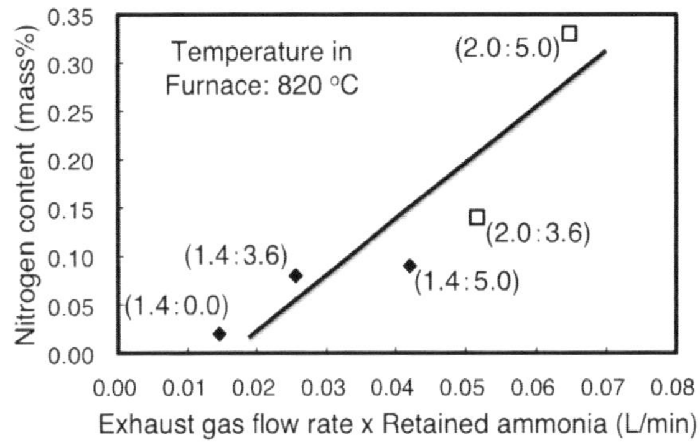

Figure 14: Relation between the product of retained ammonia volume fraction and exhaust gas flow rate and nitrogen content various ammonia gas flow rate and nitrogen gas flow rate. The ammonia gas flow rate and nitrogen gas flow rate indicate beside the each plot.

Conclusions

The effect of process parameters with treating temperature, premixed ammonia on the nitrogen infiltration amount and the gases composition inside of furnace was investigated. Following results are obtained.

- Retained ammonia decreases with the increase of treating temperature by thermo chemical decomposition.
- Nitrogen content in the steel decreases with the increase of treating temperature.
- In high ammonia atmosphere, great amount of nitrogen infiltrate to steel and then void formation in austenite grain boundaries by precipitation of molecular nitrogen gas occurs.
- White layer, which identified Fe_2N and Fe_3N by XRD was formed on surface of the specimen at 680 °C. At elevated temperature more than 730 °C, there is no formation of voids and compound layer.
- The retained ammonia, which the premixed ammonia is 100 vol.%, also decreases with the increase of temperature in furnace. Below 750 °C, the retained ammonia, which premixed ammonia is 100 vol.% is more than that in 28

vol.%, above 780 °C, each retained ammonia in 100 and 28 vol.% is same.

- With the nitrided temperature at 680 °C, and premixed ammonia in 28 vol.%, the retained ammonia is extremely high and then the void formation is occurred. If the nitriding temperature is low, the premixed ammonia must be diluted by nitrogen, for prevent the void formation.
- In constant furnace pressure, the residence time of ammonia decreases with the increase of total gas flow rate, therefore retained ammonia increases.
- Nitrogen contents with retained ammonia at about 0.8 and about 1.6 vol.% are nearly equal. The nitrogen content increases dramatically with the increase of retained ammonia from about 1.6 vol.% to 1.9 vol.%. Consequently, the control of nitrogen content using retained ammonia value is difficult.
- By using the product of the exhaust gas flow rate and residual ammonia volume fraction, the nitrogen content of steel is controlled more correctly.

References

[1] T. Holm: Heat Treatment '73(1975), p.125.

[2] R. Davis, C. G. Smith: Metal Progress, 114(1978), p.40.

[3] J. Slycke, T. Ericsson: Journal of Heat Treating, 2(1981), p.3.

[4] T. Ushigome, M. Kobayashi, H. Hashimoto, and et.al: Journal of the Japan Society for Heat Treatment, 47(2007), p.12.

[5] K. Isokawa, K. Takada and T. Katoh: Journal of the Japan Society for Heat Treatment, 22(1982), p.163.

International Federation of Heat Treatment and Surface Engineering 2016
Proceedings of the 23rd IFHTSE Congress
April 18–21, 2016, Savannah, Georgia, USA

Gas phase deep nitriding of 33CrMoV12-9 steel: microstructural characterization of the precipitates

O. Skiba, J. Dulcy, G. Marcos, T. Czerwiec
Institut Jean Lamour (UMR 7198) CNRS - Université de Lorraine – Parc de Saurupt CS 50840 F- 54011 Nancy Cedex, France

A. Courleux, S. Becquerelle
Hispano-Suiza, groupe Safran, division transmission de puissance, 18, boulevard Louis Seguin, 92707 Colombes, France

Abstract

Better knowledge of the precipitation mechanisms occurring in complex systems is needed for increasing efficiency of models intended to predict hardness and residual stress field. Aim of this study is to bring information about the precipitations formed during nitriding treatment, thus 33CrMoV12-9 low alloyed steel has been deep nitrided in a gaseous atmosphere at a temperature of 520°C during more than 100 h with a nitriding potential > 3 $atm^{-1/2}$.The nitrided layer has been characterized through micro-hardness measurements, scanning electron microscopy observations. Concentration profiles of C and N has been determined with electron probe microanalysis. Transmission electronic microscopy, elements mapping with an EDX system and EELS analyses of the near surface area has been conducted revealing existence of 2 different complex nitrides: Cr-rich nitrides - substituted mostly by Mn and in lesser proportion Mo and V and – and V-rich nitrides - containing Mo.

Introduction

Gaseous nitriding is a superficial thermochemical process based on the diffusion of atomic nitrogen through the surface of the piece. It results in strong hardening and creates a compressive residual stresses field increasing the mechanical properties (hardness, wear and corrosion resistance) of the treated pieces. The treatment leads to small dimensional deformations of the treated part, thus it is commonly used in automotive and aeronautic industry for highly loaded workpieces such as gears and shafts. The treatment is usually carried out between 480°C and 580°C on iron-based alloy in the ferritic (Fe-α) domain in an atmosphere composed of ammonia and cracked ammonia (N_2 + H_2). Ammonia is in equilibrium with the metal surface and a heterogeneous reaction occurs, providing the atomic nitrogen. In presence of H_2 in the atmosphere it is possible to define the nitriding potential K_N ($P(NH_3)/P(H_2)^{3/2}$, with P(X) the partial pressure of species X). Knowledge of the latter and the treatment temperature permit the prediction of the general structure of the nitrided layer thanks to the Lehrer's diagram [1] valid for pure iron. In the general case of iron alloys, the layer is composed by a compound layer of iron nitrides $Fe_{2-x}N$ (ε) and/or $Fe_{4-x}N$ (γ') at the extreme surface. For highly loaded applications, this layer could lead to failures due to its brittleness and is therefore mechanically or chemically removed. Right under this white layer is the diffusion layer. Numeral studies have been conducted on the behavior and precipitation mechanism occurring within this layer in binary systems Fe-Cr [2-4], Fe-V [5-8], Fe-Mn [8] and ternary systems Fe-Cr-Al [9], Fe-C-Cr [10-11] and fewer on complex systems such as industrial steel grades [12-14]. In those systems the diffusion layer consists in a solid solution of N in the Fe-α matrix and nitrides precipitates. Atomic nitrogen reacting with alloying elements in solution leads to a finely dispersed alloying elements nitrides. The latter precipitate in platelet-like shape in semi-coherency with the matrix according to the Baker-Nutting orientation relationship [15]:

$$\{100\}_{\alpha} // \{100\}_{CrN} ; \langle 100\rangle_{\alpha} // \langle 110\rangle Cr_N$$

This precipitation results in the observed hardening of the layer by opposing the dislocation movements and the volume difference between the nitrides and the matrix leads to the creation of the compressive residual stresses [10]. Several studies have put in evidence a peculiar phenomenon concerning the nitrogen uptake known as "excess nitrogen" [2-6]. The sample takes an amount of nitrogen higher than awaited by taking into account the nitrogen solubility limit in the ferritic matrix and the necessary amount of nitrogen needed to precipitate all the alloying elements. In industrial steels, carbides resulting from the quenching and tempering treatments also react with the nitrogen and undergo a transformation in globular nitrides. Those nitrides do not act as a hardening phase. Released carbon can either diffuses towards the un-nitrided core or precipitates at the grains boundaries in the form of cementite arrays [13]. As explained, the nitriding process of steels is based on a coupled diffusion-precipitation phenomenon. Mechanisms have been thoroughly investigated in binary and ternary alloys but for industrial steel grade, better knowledge of the precipitation mechanisms is still needed for increasing efficiency of models intended to predict hardness and residual stress field. Thus the aim of this study focuses on the microstructural state of such complex system. The retained grade is the 33CrMoV12-9 steel, which is commonly used in aeronautic industry for power transmission applications.

Sample preparation & characterization

Nitriding procedure

Before nitriding, a cylinder of 33CrMoV12-9 (chemical composition given in Tab 1) 25 mm in diameter and 20 mm in length was quenched and tempered at 620°C during 3 h, and then cooled under air until ambient temperature to obtain a martensitic microstructure. Two slices, 4 mm thick, were cut The samples are directly quenched in water at the end of the treatment.

Table 1: Standard composition of the 33CrMoV12-9 grade steel

	C	Si	Mn	P	S	Cr	Ni	V	Fe
wt. % min	0.29	0.1	0.4	-	-	2.8	-	0.15	Bal.
wt. % max	0.36	0.4	0.7	0.015	0.01	3.3	0.3	0.35	

Micro-hardness measurements

Cross sections of the samples were mounted in an epoxy resin and polished until ¼ µm with silica suspension before micro-hardness measurements were carried out. In order to check the homogeneity of the nitrided layer, 4 micro-hardness profiles were randomly performed over a depth of 1000 µm at a rate of one indentation every 50 µm. Micro-hardness profiles were carried out under a load of 200 g for a dwelling time of 10 sec. Conventional nitriding depth is defined as the depth where the hardness value equals the core hardness + 100 Hv. The samples were then etched with Nital 4% solution in order to reveal the microstructure by optical micrographs.

X-ray diffraction (XRD) analysis

XRD analysis was carried out on a D8 advance from Bruker Corporation device in the Bragg-Brentano configuration using Cu-K_α (λ = 0.154 nm) radiation. Analyses were performed on the surface of the as-nitrided samples.

Electron microscopy and probe

Samples for Scanning Electron Microscopy (SEM) and Electron Probe Micro Analysis (EPMA) were mounted in a conductive metallic resin and polished down to 0.25 µm using a silica suspension. Special attention has been given to obtain planar surfaces on the samples using a specimen holder for controlled material removal. EPMA analysis was carried out on a JEOL JXA 8530-F to determine the concentration depth profiles of C and N. Points of analysis (1µm^2) were made every 10 µm for a total depth of 1000 µm. SEM observations were carried out on a QUANTA FEG 650 (FEI) at a tension of 15 kV both in secondary and back-scattered electrons. Samples were etched with Nital 4% solution in order to reveal the microstructure.

The average value of the standard composition of the steel given in table 1 and the measured depth concentration profiles of C and N were then reintroduced in thermodynamics simulation software in order to estimate the volume fractions and composition of the different phases with respect to the depth of analysis.

and mechanically polished with SiC paper down to 1200. The slices were then cleaned with ethanol and simultaneously nitrided in a tubular furnace during more than 100 h at 520°C in a gaseous atmosphere of ammonia and cracked ammonia ($N_2 + H_2$). The mass flows were chosen in order to obtain a K_N > 3 atm$^{-1/2}$. According to the Lehrer's diagram for pure iron, the conditions of the treatment correspond to a ε / γ' / α layer configuration.

Transmission Electron Microscopy (TEM) observations were carried on a JEOL ARM 200F. Thin foils were extracted at a depth of 50 µm from a cross section of the sample using a Focused Ion Beam (FIB) allowing the control of the extraction depth. Electron Energy Loss Spectroscopy analyses were carried out with a GIF Quantum ER 965. EDX analysis were performed using a JEOL DRY SD 30 GV with a resolution of 129eV for 20 000 counts on the Mn-Kα1 ray.

Results & Discussion

Initial microstructure

After tempering, the samples show a typical martensitic steel microstructure in laths with the presence of precipitates of alloying elements within the laths as well as at the boundaries [14]. Those precipitates have been identified as $M_{23}C_6$ carbides with Face Centered Cubic (fcc) structure. They are derived from the $Cr_{23}C_6$ structure where the Cr is substituted by Fe and Mo atoms. Other carbides, containing high amount of V with a fcc structure, have also been put in evidence. They are of M_4C_3 carbides type, derived from the V_4C_3 structure. For this study, the volume fractions of the different phases have been calculated for a tempering temperature of 620°C and with the average between minimum and maximum mass fraction of the elements of the standard composition (Tab 1) with thermodynamic simulation software. According to Tab 2, a very slight amount of M_7C_3 carbides are also present. These results must be taken with caution since the simulation software provides only the existent species at the equilibrium state. The kinetics is not taken into account. The value of the initial hardness of the alloy has been averaged on 30 randomly performed indentations on a slice of the cylinder. A hardness of 394 Hv has been obtained.

Table 2: Table 2: Calculated volume fractions of phases. The values for each element is the average of the standard composition given in table 1

Phases	Fe-α	M23C6	M4C3	M7C3
Volume fraction (%)	94.5	4.92	0.52	0.06

After nitriding

Fig 1 shows the concentration depth profiles of C and N together with the hardness profile. N-concentration profile shows a value around 6 wt. % near the surface which corresponds to the iron nitride $Fe_{4-x}N$ γ' in the compound layer. N-concentration drops at around 1.4 wt. % in the diffusion layer and decreases continuously until around 0.8 wt. % at 500 µm before dropping abruptly to 0 at 650 µm. C-concentration profile shows strong variations attributed to the punctual character of the analyses. Cementite precipitates are probably embedded in the analyzed volume (≈ 1 μm^3). Nevertheless, this profile shows a decarburization phenomenon within the diffusion layer. In contrary, right after the diffusion front of N, carbon concentration is higher than the initial content in the untreated alloy. This can be explained by the diffusion of C that occurs during the treatment. There seem to have a double phenomenon of carbon diffusion depending on the depth considered. Close to the surface the carbon either precipitates into cementite arrays or escape in the atmosphere leading to a superficial decarburizing. In depth, the carbon diffuses towards the un-nitrided core leading to an increase in the carbon content right at the diffusion front. Both mechanisms will have an effect on the phases present in the diffusion layer as can be seen on Fig 2. Hardness profile follows fairly well the N-concentration profile. The profile shows a conventional nitriding depth of 600 μm with a surface hardness of around 850 Hv. Core hardness remains the same as the alloy hardness prior treatment. Hardening is deduced from the core hardness subtracted from the hardness after treatment. At the beginning of the diffusion layer, hardening is about 450 Hv.

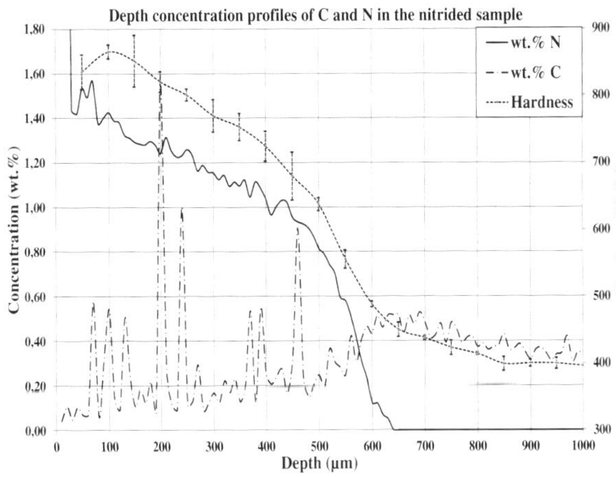

Figure 1: Hardness and concentration depth profiles of C and N, determined by EPMA

Fig. 2 represent the volume fractions of the different phases with respect to depth calculated with the measured concentration depth profile of C and N and the mean value of the standard chemical composition of Table 1. The simulation begins in the diffusion layer. It appears that the volume fraction of nitrides follows also fairly well the N-concentration profile. According to the calculation, those nitrides are mostly chromium nitrides of general formula CrN. While the N-content of the layer is higher than 0.8 wt.%, no other carbide than cementite is present. Thermodynamically, nitrides are more stable than carbides [16] so the initial carbides transform into smaller nitrides in a grape-shape disposition and the carbon is released in the ferritic matrix. Due to the treatment temperature it continues diffusing in the direction of the core or segregate at the grain boundaries in form of cementite arrays. Such cementite arrays are highlighted in Fig 3 and seem to be located at the grains boundaries. These precipitates seem to have a preferential orientation, which may be related to the stress field created following the change of volume between the matrix and nitrides.

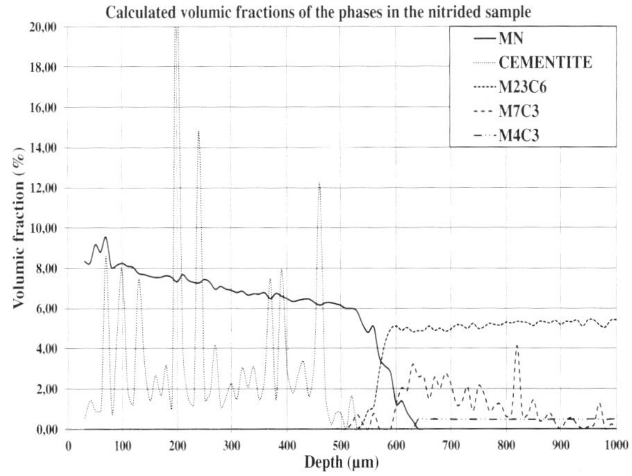

Figure 2: Calculated volume fraction of the different phases with respect to the depth

Reappearance of carbides alloying elements solely takes place when the nitrogen content is less than 0.8 wt. %. The nitrogen content is not enough anymore to precipitate all the alloying element in the form of nitride. $M_{23}C_6$ and vanadium carbides reach rapidly their equilibrium volume fraction in the un-nitrided part. Volume fraction of M_7C_3 shows strong variations and reaches a higher fraction than calculated prior treatment in the quenched sample. This is the consequence of the carbon retro-diffusion towards the core.

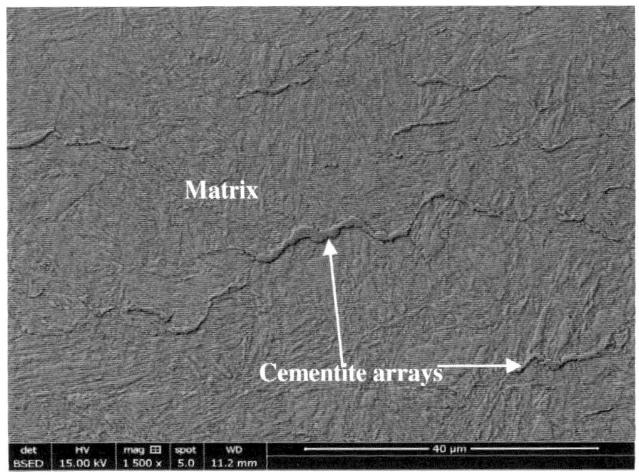

Figure 3: SEM micrograph of etched (with Nital 4%) nitrided layer (at 100 μm below the surface) revealing the martensitic microstructure as well as cementite arrays (in relief)

The XRD diffractogram of the as-nitrided samples shows the presence of iron nitrides at surface, which is expected according to Lehrer's diagram. One can also notice the slight presence of oxides of general formula AB_2O_4 (with A and B metal(s) element) of spinel structure (Fig 4). This may be due to the quenching operation right at the end of the treatment.

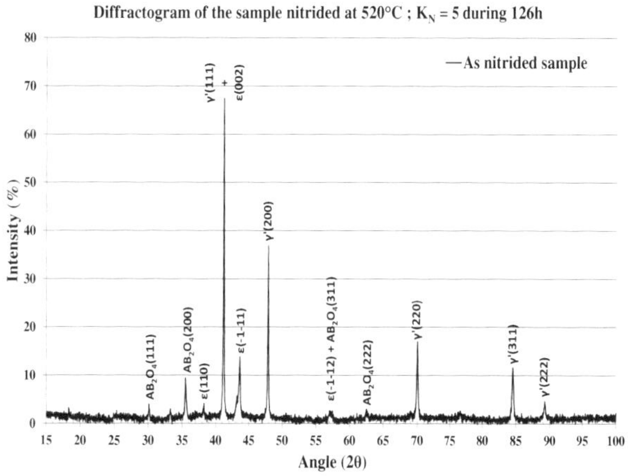

Figure 4: XRD Diffractogram of the as-nitrided sample showing the presence of a compound layer consisting in iron nitrides ε and γ' as well as oxides

TEM observations on a thin foil reveal the laths structure and put in evidence globular precipitates within the matrix located at the grains boundaries and within the laths (Fig 5). Diffraction pattern given in Fig 6 corresponds to a ferritic grain diffracting according to the axis zone $[011]_\alpha$. One notices slight streaks between the spots of the $<200>_\alpha$ direction. Those streaks are indications of very thin precipitates present in the matrix. To obtain the dark field micrograph, the aperture has been placed on one of those streaks, thus revealing finely dispersed precipitates.

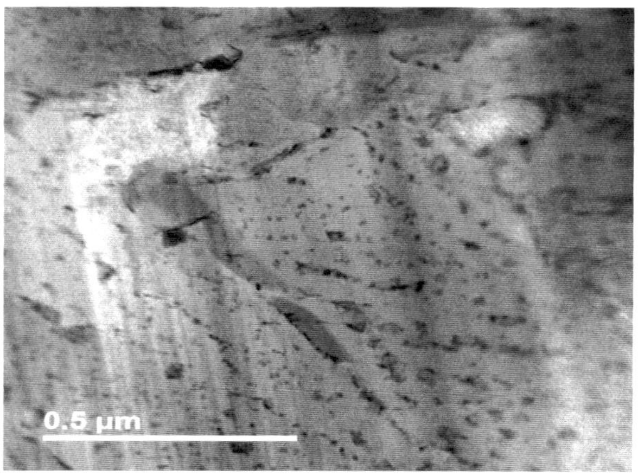

Figure 5: Bright field of a thin foil extracted at 50 µm below the surface revealing intra- and intergranular precipitates in the martensitic structure

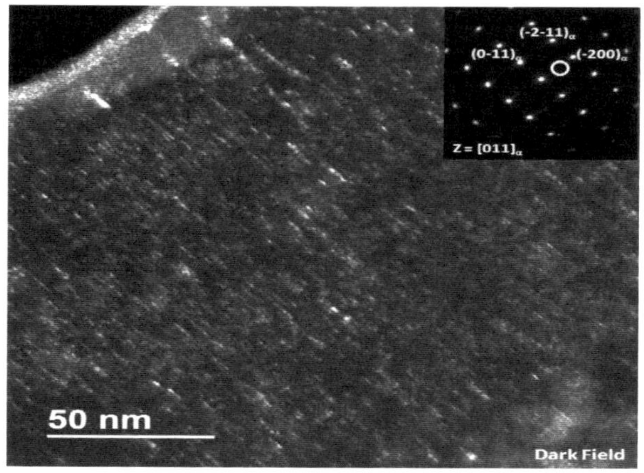

Figure 6: Diffraction pattern of ferrite in a grain in axis zone [011]. The white circle highlights the marks used to take the corresponding dark field image, which puts in evidence a finely dispersed precipitation within the ferritic grain

The corresponding EDX analyses of the Fig 5 are given in Fig 8 One can notice two types of globular precipitates: Cr-rich nitrides - very present - and V-rich nitrides - only present in the bottom left corner of the V element map. According to the mapping element of Mo, Mn and V, the Cr-rich nitrides seem to be substituted mostly by Mn and in lesser proportion by Mo and V, whereas the V-rich ntirides seem to contain more Mo and almost no Mn. At the location of those nitrides, ferritic matrix shows depletion in iron. Several studies have shown the possibility to substitute the metal element of nitrides by other nitrides forming element such as Mo, V, Fe and Al. In [12,14], emphasis has been put on Fe, Mo and V substitution in the CrN precipitates but in this study it appears that the substitution is mostly due to Mn. Considering that the Mn K-level emissions Kα1 (5,90 keV) and Kβ1 (6,49 keV) overlap respectively with the Kβ1 emission of Cr (5,94 keV) and the Kα1 emission of Fe (6,40 keV), substitution of Cr by Mn in chromium nitrides has not been put in evidence in these studies. Here, according to the EDX analysis (Fig) and the EELS spectra (Fig 7) obtained on globular nitrides embedded in the ferritic, it is confirmed that slight presence of Mn is present within the nitride.

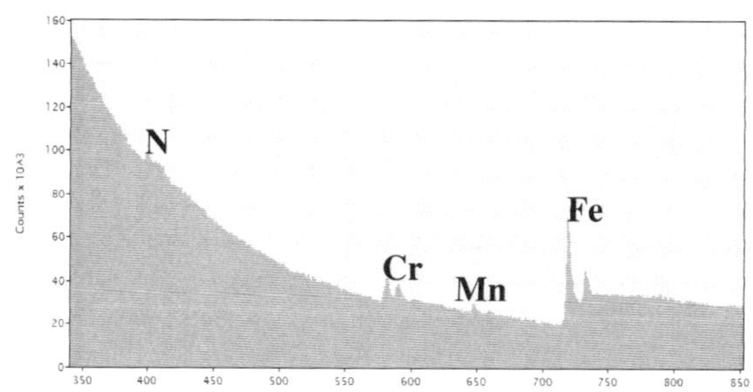

Figure 7: EELS spectra of a precipitate embedded in the ferritic matric containing N and slight amount of Mn

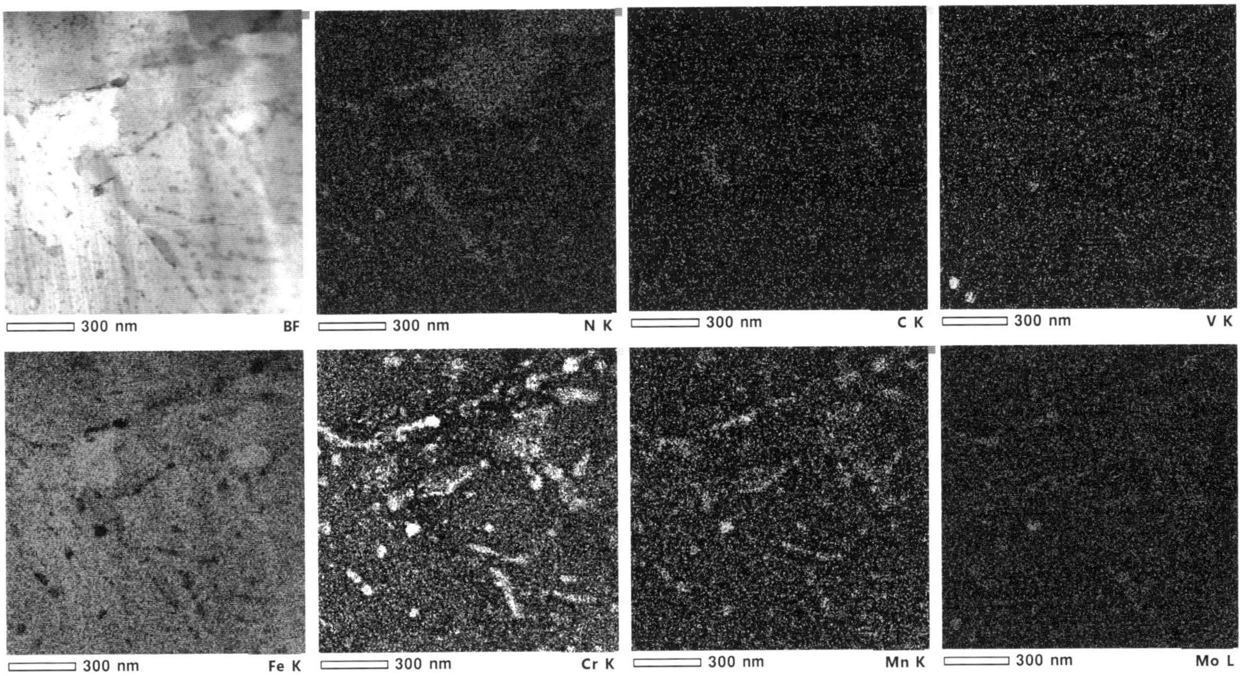

Figure 8: EDX analyses corresponding to Fig 5. Two different kinds of nitrides are present on these element maps: Cr-rich nitrides - very present - are substituted by Mn, Mo and V and V-rich nitrides - only present in the bottom left corner of the V element map - are only substituted by Mo

Conclusion

For 33CrMoV12-9 steel it has been shown that gaseous deep nitriding leads to a strong hardening of the treated layer. Diffraction pattern of a matrix grain shows presence of slight streaks between some spots. A dark field micrograph centered on one of those streaks put in evidence presence of finely dispersed precipitates within the grain, which can be held responsible for the strong hardening observed throughout the layer.

The initial microstructure contains carbides of alloying elements resulting from the tempering operation. Those carbides are mostly derived from the $Cr_{23}C_6$ and in a lesser proportion V_4C_3. After nitriding slight decarburization in the near surface area is observed. This phenomenon is accompanied by the transformation of the tempering carbides in more stable nitrides and precipitation of cementite arrays at the grain boundaries. EDX analysis clearly shows presence of Cr-rich and V-rich nitrides. Both have different chemical composition. The Cr-rich nitrides are mostly substituted by Mn and in a lesser proportion by Mo and V, whereas the V-rich nitrides seem to be substituted by Mo only. Existence of non-stoechiometric nitrides is already known but presence of Mn-element in those was not yet emphasized.

Quantification of this substitution is still to be done and confronted with thermodynamics simulation and the obtained results should lead to a better understanding of the precipitation mechanisms occurring during nitriding treatment of complex system such as 33CrMoV12-9 steel grade.

References

[1] Lehrer, E, "Über das Eisen-Wasserstoff-Ammoniak-Gleichgewicht-Elektrochem.", *Zeitschrift für Elektrochemie und angewandte Physikalische Chemie,* Vol. 36 (1930), pp 383-392

[2] Hekker, P. M. *et al*, "Excess nitrogen and discontinuous precipitation in nitride iron-chromium alloys", *Journal of Materials Science*, Vol. 20, No. 3 (1985), pp. 718-729

[3] Sennour, M. *et al.*, "TEM and EBSD investigations of continuous and discontinuous precipitation of CrN in nitrided pure Fe-Cr alloys", *Journal of Materials Science,* Vol. 39 (2004), pp.

[4] Jessner, P *et al.*, "Investigations of the nitrided subsurface layers of an Fe-Cr alloy", Ultramicroscopy 109 (2009), pp. 530-534

[5] Hosmani, S. S. *et al.*, "Nitriding behavior of Fe-4wt%V and Fe-2wt%V alloys", *Acta Materialia,* Vol. 53 (2005), pp. 2069-2079

[6] Hosmani, S. S. *et al.*, "Nitrogen uptake of a Fe-V alloy: Quantitative analysis of excess nitrogen", *Acta Materialia,* Vol. 54 (2006), pp. 2783-2792

[7] Hosmani, S. S. *et al.*, "Nitride precipitation and coarsening of in Fe-2.23 at% V alloys: XRD and (HR)TEM study of coherent and incoherent diffraction effects caused by misfitting nitride precipitates in a ferrite matrix ", *Acta Materialia,* Vol. 56 (2008), pp. 2783-2792

[8] Gouné, M. *et al.*, “Modelling of diffusion-precipitation in nitrided alloyed iron”, *Thin Solid Films*, Vol 377-378 (2000), pp. 543-549
[9] Jung, K. S. *et al.*, “Nitriding of ferritic Fe-Cr-Al alloy”, *Surface and Coatings technology*, Vol. 204 (2010), pp. 1942-1946
[10] Jegou, S. *et al.*, “Phase transformations and induced volume changes in a nitrided ternary Fe-3%Cr-0.345%C alloy”, *Acta Materialia (2010)*, pp. 2666-2676
[11] Van Wiggen, P. C. *et al.*, “The nitriding behavior of iron-chromium-carbon alloys”, *Journal of Materials Science 20* (1985), pp. 4561-4582
[12] Ginter, C. *et al.*, “A new approach of hardening mechanisms in the diffusion layer of gas nitrided α-alloyed steels”, *Proc 2nd International Conference on Heat Treatment and Surface Engineering in Automotive Industry* (2005), Riva Del Garda, June 2005, pp. 29-35
[13] Barrallier, L. *et al.*, “Morphology of intergranular cementite arrays in nitride chromium-alloyed steels”, *Materials Science and Engineering A*, Vol. 393 (2005), pp. 247-253
[14] Locquet, J. N. *et al*, "Complete TEM investigation of a nitrided layer for a Cr alloy steel", *Microscopy, Microanalysis and Microstructure*, Vol. 8 (1997), pp. 335-352
[15] Baker, R. G. *et al.*, “Precipitation process in steels”, *The Iron and Steels Institute Journal special report*, Vol. 64 (1959), p.1
[16] Goldschmidt, H. J. *et al.*, Interstitial alloys, Butterwoth and Co. Ldt (1967)

International Federation of Heat Treatment and Surface Engineering 2016
Proceedings of the 23rd IFHTSE Congress
April 18–21, 2016, Savannah, Georgia, USA

Effect of Plasma Nitriding on Corrosion Behavior of AISI4330 Low Alloy Steel

Vinaykumar Patil[1], Jagdish Sherkar[1], Ashish Supare[1], Dr. Rajkumar Singh[1].
[1]Bharat Forge Ltd (Kalyani Center for Technology and Innovation), Pune, Maharashtra state, India.
vinaykumar.patil@bharatforge.com, jagdish.sherkar@bharatforge.com, Ashish.supare@bharatforge.com, Rajkumarsingh@bharatforge.com,

Abstract

Plasma nitriding is a surface treatment process, which is widely used to improve wear, fatigue and corrosion resistance of the industrial parts. In this study, the corrosion behavior of plasma nitrided and untreated AISI4330 steel samples was compared. Plasma nitriding of conventionally heat treated AISI4330 steel samples were performed at 480 °C for 5, 10 and 14h and the gathered properties were assessed by evaluating hardness profile, compound layer thickness, case depth and surface roughness by using microhardness tester, scanning electron microscopy, optical microscopy, surface profilometer and X-ray diffraction. It has been found that the thickness of the compound layer increases with increasing treatment time. The surface roughness, the surface hardness, the hardness profile, the case depth, the compound layer thickness, and phase content of the compound layer depend on nitriding parameters. The corrosion tests were carried out with Potentiodynamic polarization technique in 3.5 % NaCl solution. The result shows that, the corrosion resistance and surface hardness of the AISI4330 steel increased with an increase in plasma nitriding time. The plasma nitriding raises the surface hardness up to 710 Hv. The reasons for increase in corrosion resistance have been discussed in this paper.

Keywords: - AISI4330 Steel, Plasma Nitriding, Corrosion.

1. Introduction

Plasma nitriding is a process of surface hardening using glow discharge technology to introduce nascent (elemental) nitrogen to the surface of a metal part for subsequent diffusion into the material. In vacuum, high-voltage electrical energy is used to form plasma, through which nitrogen ions are accelerated to impinge on the workpiece. Nitrogen ions are combined with alloying elements such as chromium to form a fine dispersion of alloy nitrides. This ion bombardment heats the workpiece, cleans the surface, and provides active nitrogen [1]. Nitrogen diffusion modifies surface and near surface microstructure producing hard layers with altered mechanical properties. The layer on the plasma nitrided steel consists of two parts, compound layer and diffusion layer. The outer and thinner layer consists of mainly γ'-Fe_4N and/or ε-$Fe_{2,3}N$ inter-metallic phases. Therefore, it is commonly called as 'compound layer' [2]. On the other hand, it is also referred to as 'white layer' because of its white color after Nital etch on the microscopy. Thickness and phase content of compound layer depend on the treatment parameters such as specific gas composition, time and temperature [1]. The carbon content of the steel has also a significant influence on the formation of the compound layer [4]. It is important for some applications that the compound layer may be comprised of γ' -Fe_4N or ε-$Fe_{2,3}N$ phases or a mixture of both.

The layer, beneath the compound layer, is known as 'diffusion layer' which consists of mainly interstitial atoms in solid solution and fine coherent nitride precipitates when the solubility limit is reached. Diffusion layer thickness is also referred to as 'case depth'. The hardness and the depth of the diffusion layer depend on the amount of nitride former elements in the steel, the initial microstructure of the material, the time and the temperature of the plasma nitriding treatment and the specific gas composition used in plasma nitriding system [7].

The plasma nitriding process is more economical than all the other conventional nitriding processes such as gas nitriding and salt bath nitriding because of faster nitrogen diffusion, lower nitriding temperature, shorter treatment time and lower gas consumption. Easier control of compound and diffusion layer formation and also suppressed compound layer formation are reported to be the other advantages of plasma nitriding [6]. The plasma nitriding process, which is extensively used to improve the surface properties, wear, fatigue, corrosion, friction properties, and also load bearing capacity of dynamically loaded components, is widely used in the manufacturing and machining industries primarily in engine components, tools for hot/cold work, machine tools, gears and forging dies.

The properties of a nitrided steel component are determined by both the core strength and the structural characteristics of the compound layer and the diffusion layer [8]. While the compound layer determines the corrosion resistance [9], the diffusion layer determines the strength of the component, as well as its fatigue resistance [8].

The compound layer can have either a positive effect of high hardness or a negative effect because of brittleness. The presence of compound layers on nitrided components can improve the resistance to scuffing, seizure and galling under lubricated conditions. Since the outer porous part of the compound layer is able to absorb a quantity of oil, the compound layer leads a reduction in the friction coefficient [10].

After plasma nitriding, the corrosion resistance increases due to the formation of a compound layer consisting of highly dense γ′ -Fe_4N and ε-$Fe_{2,3}N$ nitrides. The ε-$Fe_{2,3}N$ nitride exhibits better corrosion resistance when compared with γ′ -Fe_4N nitride due to its crystalline structure and higher nitrogen content [11]. Although the compound layer provides the materials with good physical properties against wear and corrosion, in the case of a very hard and brittle layer.

The surfaces become rougher due to the nitrogen ion bombardment and the formation of a nitride layer [12]. The degree of the increase in the surface roughness varies with the plasma nitriding parameters such as gas compositions, time, temperature and applied voltage [12].

There have been many investigations aiming the prediction of the resulting structures of plasma nitrided components and their service behavior. However, the studies on the effects of plasma nitriding on corrosion behavior of AISI4330 steel are very scarce. In this study, AISI4330 steel was plasma nitrided under different process conditions. The process variables investigated include time (5, 10 and 14h) and temperature (480°C). The effect of plasma nitriding process parameters on corrosion resistance were investigated by using Potentiodynamic polarization technique and other properties such as surface roughness, hardness profile, diffusion layer thickness, compound layer thickness and phase content of compound layer were investigated by using a microhardness tester, scanning electron microscopy, optical microcopy, surface profilometer and X-ray diffraction (XRD).

2. Material and Experimental Studies

The material used in this study was low alloy AISI4330 steel and the chemical composition of the steel was given in Table 1. AISI4330 steel was received in hot-rolled condition and austenitized at 850°C, oil quenched and double tempered at 630°C and 580°C. The hardness of AISI4330 steel was obtained up to 390 HV. The square specimen's (10 mm x 10 mm and 5 mm thick) made from AISI4330 steel were used for testing. The surface of the samples exposed to plasma nitriding were mechanically polished up to 1 μm diamond paste and 0.5 micron alumina paste in order to obtain a final mirror finish of a mean roughness Ra value better than 0.01μ and cleaned in acetone. After polished, the specimens placed in an industrial DC glow discharge plasma nitriding furnace.

Plasma nitriding was carried out in a large working volume with necessary arrangements of vacuum system and auxiliary heating. The specimens were employed as the cathode. Plasma nitriding of samples were performed at 480°C for 5, 10 and 14h. Temperatures of specimens were measured by thermocouples, which were embedded in the specimens. The treatment time was accounted after treatment temperature was reached. After plasma nitriding, the specimens were cooled to the room temperature in the vacuum chamber. The composition, surface roughness, microstructure and hardening effect of the nitrided specimens were characterized by using metallographic examination, XRD, surface profilometer and microhardness tests.

Table 1: Chemical Composition of AISI4330 Steel

Elements	Weight (%)
C	0.30
Ni	2.60
Cr	1.20
Si	0.15
Mn	0.45
S	0.010 max
P	0.015 max
Mo	0.50
V	0.12
Al	0.035 max
Fe	balance

Surface profilometer was used to determine of the surface roughness before and after plasma nitriding.

Scanning Electron microscopy and optical microscopy was used to characterize the microstructures. Cross sections from nitrided specimens were ground, polished and etched using Nital 2% immersion reagent. The compound layer thickness was measured, after etching.

X-ray diffraction (XRD) was used to determine the phases present in the modified layer.

Microhardness measurements were used to determine diffusion layer thickness. The surface hardness and the microhardness profile of plasma-nitrided specimens were evaluated using a microhardness tester.

Potentiodynamic polarization technique was used to investigate the corrosion characteristic of the treated samples. The test was performed in 3.5 wt. % NaCl aqueous solutions at room temperature in a flat Pyrex glass cell using the specimens as the working electrode, saturated calomel as reference electrode and platinum as counter electrode. The corroded area of the specimens was about 0.20 cm^2. The surface morphologies of the corroded specimens were examined by SEM. The results of the corrosion tests were evaluated using corrosion rate, Ecorr and Icorr values obtained from the polarization curves.

3. Results and Discussion

The specimens are substantially rougher, as would be expected from the formation of a nitride layer due to the nitrogen ion bombardment [12]. The degree of the increase in the surface roughness varies with the plasma nitriding conditions [13]. The results of the surface roughness measurements are given in Table 2. After the plasma nitriding process, the surface roughness increased up to 0.16 μm while it was 0.01 μm initially. Also, the test results show that higher surface roughness values can be obtained with increasing plasma nitriding time. For instance, surface roughness value is increased from 0.10 μm to 0.16 μm as a result of increasing plasma nitriding time, from 5h to 14h at 480°C. Surface roughness at higher treatment time is mainly due to more nitrogen bombardment on to the surface of specimen.

The plasma nitriding treatment time, temperature and gas composition as well as the amount of C in the steel influence the kinetics of the formation of the compound layer [1–6]. The presence of a continuous and uniform compound layer could be observed in Fig. 1. The compound layer thickness increases with increasing time, see in Fig. 2. When the treatment time was increased from 5h to 14h, the compound layer thickness increased from 2.4μm to 5.9μm at 480°C.

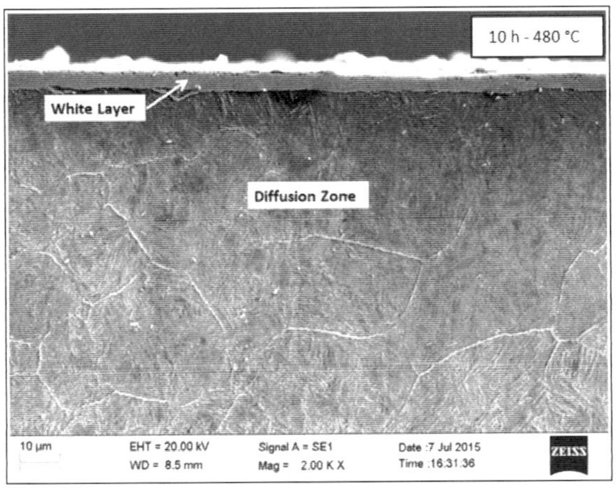

Figure 1: Scanning Electron micrographs of plasma nitride AISI4330 steel (2000x) at 10h-480°C.

The unnitrided specimen shows only α-Fe phase, whereas the plasma nitrided specimen exhibits additional phases, ε-$Fe_{2-3}N$ and/or γ'-Fe_4N, due to different nitrides [9]. According to these patterns, it was obvious that the intensity of α -Fe phase progressively decreases and the intensity of the nitride phases progressively increase with increasing plasma nitriding time [9]. Fig. 3 shows typical X-ray diffraction spectra for untreated and plasma nitrided specimens. The Fig. 3a shows result of untreated specimen and Fig. 3b, 3c, 3d for 5h, 10h and 14h at 480°C. The unnitrided specimen shows only α-Fe phase, see in Fig. 3a whereas Fig. 3b, 3c and 3d shows, the intensity of the ε-$Fe_{2-3}N$ phase gradually decreases with increasing nitriding time, while the intensity of the γ'-Fe_4N phase increases with increasing nitriding time. Thus, the proportion of γ'-Fe_4N to ε-$Fe_{2-3}N$ -nitride phases increased with increasing nitriding time at 480 °C. The higher amount of ε-$Fe_{2-3}N$ in the beginning of the process and gradual decrease in the amount of ε-$Fe_{2-3}N$ in the compound layer could be ascribed to the decrease in carbon content of the steel due to sputtering [14]. The analysis shows that two nitrides, ε-$Fe_{2-3}N$ and γ'-Fe_4N, are present at 480 °C.

Beneath the compound layer, in the diffusion zone, nitrogen is dissolved in ferrite at nitriding temperature and forms nitride precipitates during cooling as a result of decreasing solubility of nitrogen [8]. The hardness decreases from surface to core, since the concentration of metal nitrides decreases towards core [3].

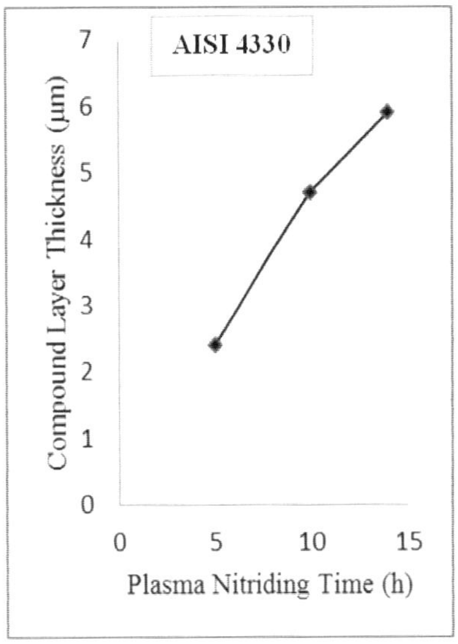

Figure 2: The relationship between compound layer thickness and treatment time at different conditions.

Table 2: Surface roughness of unnitrided and plasma nitrided AISI4330 at different conditions.

Treatment Conditions	Untreated	5h-480°C	10h-480°C	14h-480°C
Average Surface Roughness, Ra (μm)	0.01	0.10	0.12	0.16

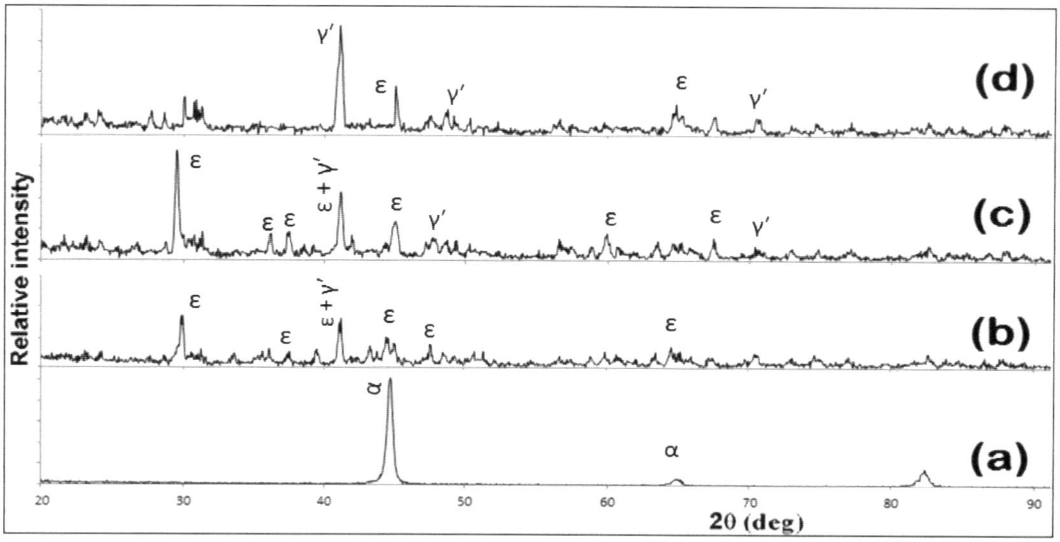

Figure 3: X-ray diffraction patterns of plasma nitrided AISI4330 steel for (a) untreated (b) 5h-480°C (c) 10h-480°C (d) 14h-480°C

The hardness profiles of the plasma nitrided specimens are given in Fig. 4.

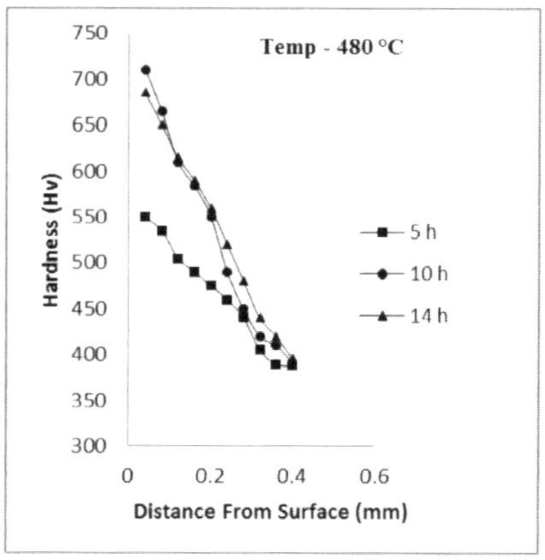

Figure 4: Microhardness profiles of AISI4330 steel, plasma nitrided at different conditions

The surface hardness measurements of the samples treated at different times are given Table 3. The surface hardness of AISI4330 steel is increased upto two times by the plasma nitriding process. It is due to the formation of thicker nitride layer at higher time and temperature [4]. The case depth is more at higher time because of faster diffusion of nitrogen. The hardness of compound or white layer is much higher than the diffusion zone. The surface hardness of plasma nitrided specimens is in the range of 550–710 HV, while the core hardness remains unchanged, 390 HV. The surface hardness increases to reach a maximum value at 10h for 480°C, and then lower values are obtained in longer treatment period, 14h. The reduced surface hardness is obtained at longer nitriding time because of the formation of large nitride precipitates [7]. At longer times, the precipitates particles are larger in size and more prone to coarsening, leading to a lower precipitate density and hence lower hardness [13]. The Fig. 5 represents the relation between hardness in Hv and the plasma-nitriding time.

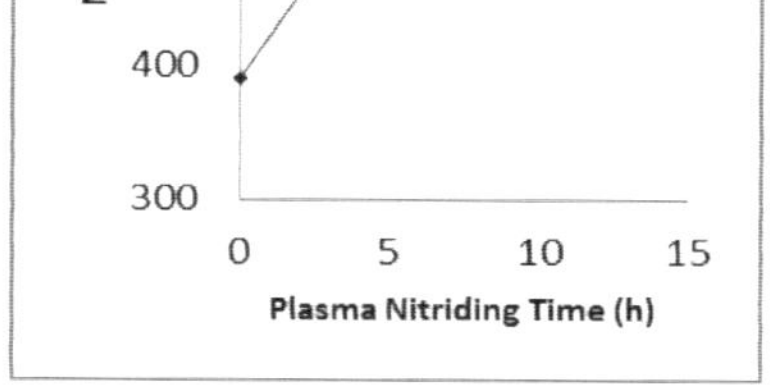

Figure 5: Effect of plasma nitriding time and surface hardness.

Corrosion rate, corrosion potential and corrosion current data of the untreated and plasma nitrided samples are displayed in Table 3. The SEM micrographs in Fig. 6 and 7a-c show the corroded surface of untreated and nitrided specimens for different plasma nitriding time at 480°C. It is seen that corrosion has happened on both untreated and nitrided samples. In comparison with the nitrided specimens, the more corrosion is observed in untreated specimen.

The result of the EDAX analysis carried out for untreated specimen as shown in fig.6. We observed the presence of the trace amount of oxygen in the spectra which has causes pitting corrosion, it was confirmed that the compound layer prevents formation of the oxide layer [1].

Figure 6: SEM micrographs with EDAX of corroded surfaces of untreated specimens

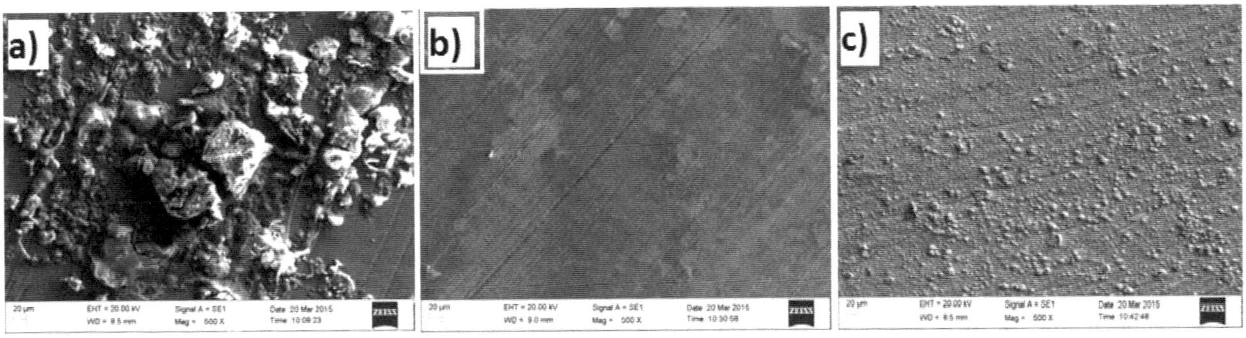

Figure 7: SEM micrographs of the corroded surfaces of the treated specimens a, b, c) nitride at 5, 10 and 14h respectively at 480°C

Table 3: The surface hardness and corrosion test results of AISI4330 Samples

Sample	Compound (white) Layer (μm)	Case Depth (μm)	Surface Hardness (Hv)	Ecorr (V)	Icorr (i/A)	Corrosion Rate (mm/Year)
Untreated	-	-	390	-0.706	1.330e-5	0.160
5h – 480 °C	2.4	280	550	-0.707	5.086e-6	0.114
10h – 480 °C	4.7	340	710	-0.401	3.631e-6	0.043
14h – 480 °C	5.9	430	685	-0.505	7.032e-6	0.083

Potentiodynamic polarization curves of the untreated and nitrided specimens are studied using 3.5 wt. % NaCl environment are shown in Fig. 8. The untreated sample shows an active behavior confirming a low corrosion potential and high anodic current. There is a general and uniform corrosion observed on the corroded samples. Nitriding treatment improved the corrosion resistance of AISI4330 steel in the test solution. The data extracted from the polarization curves are listed in Table 3. The corrosion current density of plasma nitrided samples at 10h - 480°C was more positive than other samples including the untreated sample, hence it can be said that nitriding treatment improved the corrosion resistance of AISI4330 in the test solution. However after 14h the corrosion resistance decreases this is due to formation of γ' in the compound layer which is not effective for corrosion resistance [11].

The corrosion results in Table 3 indicate that nitriding at 5h could not effectively improve the corrosion resistance of AISI4330 steel. This can be attributed to the thin compound layer produced at this time. In conclusion, the maximum corrosion potential, the minimum corrosion rate and the highest passive corrosion behavior refers to the sample treatment of nitriding at 10h and 480 °C.

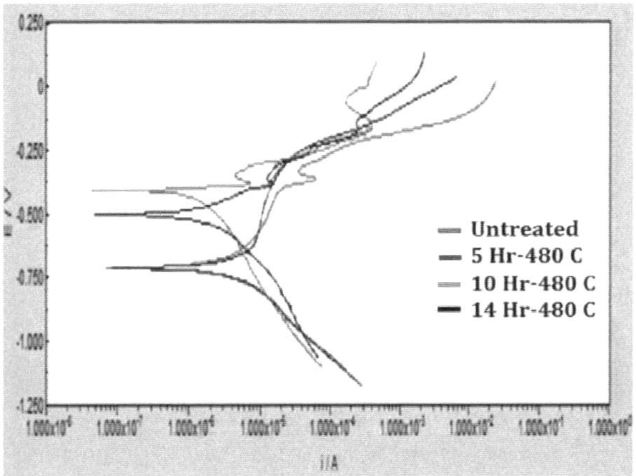

Figure 8: Potentiodynamic polarization curves of the untreated sample and nitrided samples

Conclusions

The microstructure and the corrosion behaviour of low alloy AISI4330 steel nitrided by plasma were studied as a function of concentration of plasma nitriding time.

1. The surface roughness is increased by plasma nitriding and the degree of the increase depends on the plasma nitriding conditions.
2. The compound layer thickness and the diffusion layer thickness increase with increasing time.
3. The surface hardness of plasma nitrided specimen increases with increasing time, but the reduced surface hardness is observed at longer nitriding time because of the formation of larger nitride precipitates which are less effective in increasing hardness than smaller ones. It can be said that the surface hardness depends on the nitriding time when the temperature is constant. The surface hardness increases in the first 10h when it reaches a maximum value and then decreases at 14h because of the formation of coarser nitride precipitates. The maximum surface hardness is obtained at 480°C at 10h.
4. Specimens with the maximum hardness showed good corrosion resistance.
5. The highest corrosion resistance belonged to the sample that had been plasma nitriding at 10h and 480 °C. However after 14h the corrosion resistance decreases due to the formation of γ' in compound layer which is not effective for corrosion resistance.

Acknowledgments

The author gratefully acknowledge the extended support provided to this work by KCTI (Kalyani Centre for Technology and Innovation) for providing financial funding, laboratory and library facilities. The author also acknowledges the support provided by Bharat Forge Limited, Pune and DSIR (Department of Scientific and Industrial Research), Government of India. Finally, the author would like to express special thanks and gratitude to review committee and top management of Bharat Forge Limited for grating the permission to publish/present the research work. Special thanks to BFL colleagues for encouragement and helped in writing this paper.

References

[1] M. Ebrahimi, M. Heydarzadeh Sohi, A. Honarbakhsh Raouf, F. Mahboubi, "Effect of plasma nitriding temperature on the corrosion behaviour of AISI 4140 steel before and after oxidation" Surface and Coatings Technology 205 (2010), Elsevier publication, pp. 261–266.

[2] Celik A, Karadeniz S. "Investigation of compound layer formed during ion nitriding of AISI4140 steel". Surf Coat Technol 1996;80:283–6.

[3] Jagdish M. Sherkar, Dr. Rajkumar P. Singh, Vinay Patil, Dr. Raju Kadam, "Influence of Plasma (Ion) Nitriding Time and Temperature on the Mechanical Properties of AISI4330 Low Alloy Steel", AES-ATEMA'2014 Eighteen International Conference on Advances and Trends in Engineering Materials and their Applications.

[4] N. Yasavol, F. Mahboubi, "The effect of duplex plasma nitriding-oxidizing treatment on the corrosion resistance of AISI4130 steel", Materials and Design 38 (2012), Elsevier publication, pp. 59–6.

[5] S. Syed, "Atmospheric Corrosion of Materials", Emirates Journal for Engineering Research, 11 (1), 1-24 (2006).

[6] O.Belahssen, A.Chala, S. Benramache, D. Bensahal, F. Chabane, "Microstructure and Corrosion Behavior of Steel Nitrided By Plasma", Journal of Science and Engineering Vol. 2 (1), 2013, 7-12.

[7] Spalvins T. Tribological and microstructural characteristics of ion-nitrided steels. Thin Solid Films 1983;108:157–63.

[8] Ashrafizadeh F. Influence of plasma and gas nitriding on fatigue resistance of plain carbon (CK 45) steel. Surf Coat Technol 2003;173–174:1196–200.

[9] Wen DC. Microstructure and corrosion resistance of the layers formed on the surface of precipitation hardenable plastic mold steel by plasma-nitriding. Appl Surf Sci 2009;256:797–804.

[10] Karamıs MB. Tribological behavior of plasma nitrided 722M24 material under dry sliding conditions. Wear 1991;147:385–99.

[11] Oliveira SD, Tschiptschin AP, Pinedo CE. Simultaneous plasma nitriding and ageing treatments of precipitation

hardenable plastic mould steel. Mater Des 2007;28:1714–8.
[12] Blawert C, Mordike BL, Huchel U, Stramke S, Collins GA, Short KT, et al. Surface treatment of nitriding steel 34CrAlNi7: a comparison between pulsed plasma nitriding and plasma immersion ion implantation. Surf Coat Technol 1998;98:1181–6.
[13] Alsaran A. Determination of tribological properties of ion nitrided AISI4140 steel. Mater Charact 2003;49:171–6.
[14] Kurny ASW, Mallya RM, Rao MM. A study on the nature of the compound layer formed during the ion nitriding of En40B steel. Mater Sci Eng 1986;78:95–100.
[15] Subroto Mukherjee , Alphonsa Joseph, Ghanshyam Jhala1, Satyapal M, A. S. Khanna, Pratipal Rayjada, Narendra Chauhan, Raja V. S, "Effect of Plasma Nitriding and Nitrocarburising Process on the Corrosion Resistance of Grade 2205 Duplex Stainless Steel", The International Conference on Plasma Surface Engineering, September 10-14, 2012, in Garmisch-Partenkirchen, Germany.
[16] Satyapal.M, A.S. Khanna, Alphonsa Joseph, "Effect of Temperature on the Plasma Nitriding of Duplex Stainless Steels", International Journal of Engineering and Innovative Technology (IJEIT) Volume 2, Issue 11, May 2013.
[17] James, M. O., O'Brien & Associates, Dan, G. (1991), ASM Handbook, Heat Treating, vol.4. American Society for Metals, Materials Park, Ohio, P.944-953.

International Federation of Heat Treatment and Surface Engineering 2016
Proceedings of the 23rd IFHTSE Congress
April 18–21, 2016, Savannah, Georgia, USA

Low temperature nitriding of strain induced martensite and laser quenched austenite

Vojteh Leskovšek
Institute of metals and technology, Lepi pot 11, 1000 Ljubljana, Slovenia
*e-mail address:*vojteh.leskovsek@imt.si, Tel.:+386 1 470 18 70; fax.: +386 1 470 19 39

Matjaž Godec
Institute of metals and technology, Lepi pot 11, 1000 Ljubljana, Slovenia
e-mail address:matjaz.godec@imt.si, Tel.:+386 1 470 19 01; fax.: +386 1 470 19 39

Peter Kogej
RLS Merilna tehnika d.o.o. Poslovna cona Žeje pri Komendi, Pod vrbami 2, SI-1218 Komenda, Slovenia
e-mail address:peter.kogej@rls.si, Tel.:+386 1 527 21 00; fax.: +386 1 527 21 29

Abstract

The deformation behaviour and microstructure characteristics of 304L steel during bar extrusion at different strain after deep cryogenic treatment (DCT) and after low temperature plasma nitriding were investigated. According to the enhance of magnetic and mechanical properties after low temperature nitriding in combination with isothermal treatment at 400°C, it is assumed that both, ε-martensite and α'-martensite are present in deformation microstructure, indicating the simultaneous stress-induced-transformation and strain-induced-transformation of austenite. The effects of the laser surface treatment and local appearance of non-magnetic phase due to α'→ γ transformation after laser surface treatment are investigated. Analyses with Vickers hardness, light microscopy, SEM/EDS/EBSD of cross-sections of bars with laser marked, after low temperature plasma nitriding in combination with isothermal treatment were performed. EDS maps show high concentration of nitrogen in both alternating hard surface layers γ_N and α'_N (expanded austenite and expanded martensite), but no significant higher concentration of chromium or iron. The results allow concluding that the interaction of deep-cryogenic treatment, plastic deformation, phase transitions, and low temperature pulse plasma nitriding in combination with isothermal treatment as well as laser surface quenching are promising area as they enable to explore new structures and find new applications.

Introduction

It is well known that austenitic stainless steels cannot be hardened by heat treatments. On the other hand, cold working in combination with a DCT can make such steels harden [1]. The hardness of 304L austenitic stainless steel that results from cold working in combination with the DCT formation of strain-induced martensite (α') in austenitic stainless steel is above 460 HV_{10}, with the martensite acting as an elastic reinforcing phase with higher stress than the austenite (γ) tensile strength, which deforms plastically [1]. Such a material has many potential applications in architecture, chemical processing plants, and as a magnetic/non-magnetic substrate for encoder scale [2], etc. Strain-induced martensite formation is a unique feature of austenitic stainless steels, and two types of martensites can form spontaneously in such steels.

The studies [3, 4], which are confirmed in ref. [1], show that the transformation sequence with plastic straining is fcc → hcp → bcc, with the phases being denoted as γ → ε → α', respectively. The intersections of the stacking faults or the hcp bands act as nuclei for the α', and if their formation is accelerated, the rate of α' nucleation is also accelerated. The hcp phase can not act as an intermediate phase if the composition is changed and the stacking-fault energy is increased; however, this sequence was found to be applicable in this study. It has also been shown [5] that the stacking-fault energy in austenitic stainless steel decreases with decreasing temperature. The formation of strain-induced martensite also occurs at room temperature, but it requires large strains to increase the thermodynamic driving force. The α' phase is thermodynamically much more stable than the paramagnetic martensite phase (ε). The ε phase forms before the α' phase during the cold working of 304L stainless steel. At high deformations the amount of previously formed ε phase decreases with the increasing deformation because the α' martensite grows at its expense, i.e., for high levels of deformation the α' phase predominates [6].

The objective of this study was to explore the combination of a DCT, plasticity, and low-temperature plasma nitriding together with an isothermal treatment and phase transitions, in this order, which increases the amount of formed magnetic α' phase, thereby improving the magnetic contrast when applying a non-magnetic γ marking on the substrate using laser surface quenching [1]. The effects of a DCT combined with the severe deformation of 304L steel, laser surface quenching and low-temperature plasma nitriding together with an isothermal treatment are shown on the formation and characteristics of the nitrided layer. This was evaluated by measuring the surface microhardness and the depth of the nitrided layer formed by nitrogen diffusing into the substrate.

Experimental methods

Material, cold working and DCT
The samples used in the investigation were made from a bar (ϕ18×3000mm) of AISI 304L stainless steel.

The bar samples, labelled as $C_{-196°C}$, were machined with the axis in the direction of the bar's extrusion, then annealed at 1050°C for 30 min in a vacuum furnace, and quenched to room temperature in a flow of high-purity nitrogen to obtain a fully austenitic initial microstructure. The bar extrusion of the $C_{-196°C}$ samples was performed using an experimental drawing bench with eleven exchangeable cold-drawing matrices with different diameters (from 16.72 to 12.19 mm) at liquid-nitrogen temperature [1].

After each reduction, the samples were immersed in liquid nitrogen to equalize the temperature of the drawn bars with the temperature of the liquid nitrogen.

Laser surface treatment
With laser hardening or quenching the high-power laser beam quickly irradiates the specimen's surface in order to rapidly increase its temperature to higher than the phase-transformation point and lower than the melting point. After the laser beam is switched off, the heated region is cooled quickly by the cooling base and hardened, or it is quenched by itself and the specimen surface is hardened or softened and its performance is changed.

In the case of strain-induced martensite being formed in plastically deformed, DCT, austenitic stainless steel the quenching effect includes the local appearance of the non-magnetic phase, a decrease of the surface hardness due to the $\alpha' \rightarrow \gamma$ transformation and a change of the depth and the width of the quenching layer [1].

The experiments were performed on rods (ϕ18 × 80mm) taken from bar samples of the $C_{-196°C}$ group with the highest content of martensitic phase (α'), using a YLR-400-AC ytterbium-fibre laser machine with a nominal output power of 400 W. The power of the laser in the experiment was set to ~100 W, with the parameters required to obtain an approximate depth of 350 μm and the width of the quenched layer in the shape of a semicircle in the form of a raster, perpendicular to the direction of the bar's extrusion [7].

Low-temperature nitriding
Surface-engineering technologies can produce hard or modified layers on the surface of 304L stainless steel that have good anti-corrosive and wear-resistance properties. It has been shown that the surface treatment of these materials based on pulse plasma nitriding provides an effective way to improve the wear resistance without losing its corrosion resistance, but only if the maximum treatment temperature is below 450°C. In this case, a micrometre-or-two-thick surface layer composed of nitrogen-saturated fcc phase with outstanding properties is formed. This phase, which is commonly referred to as expanded austenite (γ_N), is hard and has good wear properties as well as maintaining or even enhancing the corrosion resistance of the austenitic stainless steel [8-11]. Unfortunately, this layer is so thin (due to the slow diffusion of nitrogen in austenite at low temperature) that the load-bearing capability of the surface is limited to mild wear phenomena. The diffusion of nitrogen in steels depends on many factors, including lattice structure, grain size, chemical composition, defect density, etc.

Ferkel et al. [12, 13] reported that the severe deformation of materials can enhance the nitrogen diffusion in pure iron and highly alloyed steel, owing to an increased grain-boundary area and dislocation density. Most of the austenitic stainless steels can undergo martensitic transformations, caused by strain-induced transformations, because of cold working at room and liquid-nitrogen temperatures.

The thermochemical treatment of the plastically deformed, DCT, austenitic, stainless-steel specimens after laser quenching was performed in a Metaplas Ionon HZIW 600/1000 reactor equipped with a convection heating system and an internal gas/water heat-exchanger for rapid cooling. The convection and plasma heating of the specimens to the processing temperature took approximately 3.5 hours. The soaking time at the nitriding temperature, which is at the same time also the isothermal treatment, at 400 °C was for 10 hours. Before introducing the process gas, the chamber was evacuated to a pressure of 2 Pa. The 4-kHz pulse plasma was produced in a total pressure of 3.3 hPa for nitriding, with a negative bias potential of 490 V being applied to the specimens, which acts as the cathode. The specimens were nitrided in a 75 vol-% H_2: 25 vol-% N_2 gas mixture at a nitriding temperature of 400 °C.

Optical and SEM microscopy and Vickers hardness
Optical microscopy, SEM and SEM-based electron-backscatter-diffraction (EBSD) analyses were performed on the metallographic samples to confirm the presence of the magnetic phase (α') as well as the γ, γ_N or α'_N. EBSD is a widely used technique for obtaining crystallographic information on the micro- and nanoscales. Currently, EBSD can be used to map crystallographic parameters, such as the phase type and fraction, the grain size, the orientation distribution, and even the degree of deformation [14].

The microstructure was examined in an optical microscope (Nikon Microphot FXA) and the EBSD data were collected in a FE-SEM JEOL JSM 6500F field-emission scanning electron microscope using an HKL Nordlys II EBSD camera and Channel 5 software. The EBSD mapping analyses were performed with samples tilted by 70° using a 15-kV accelerating voltage and a 1.5-nA probe current. The EBSD map was generated in steps of 0.4 μm using 613 × 4459 grids with 4 × 4 binning and employing 25 reflectors and minimum 5 and maximum 6 band detection for the best discrimination between the bcc and fcc phases. The analysed area was sufficient to be statistically representative and the selected scanning step was sufficiently small to obtain a clear morphology of the microstructure.

The Vickers hardness HV_{10} and microhardness $HV_{0.01}$ were measured on the same metallographic samples as used for the determination of the relative magnetic phase content (α') by employing a Vickers Tukon 2100 B hardness tester. The relative volume fractions of the transformed martensite at different stages of the deformation process were assessed using a Ferritescope MP-30 [1].

Results and discussion

Vickers hardness HV_{10} and Relative magnetic phase (α') content

The HV_{10} hardness after vacuum annealing at 1050°C for 30 min. and then quenching in a flow of nitrogen to room temperature was 160 HV_{10} on the $C_{-196°C}$ bar samples. The initial microstructure in the bars was fully austenitic (Fig. 1).

Figure 1: Initial microstructure in the $C_{-196°C}$ bars after vacuum annealing at 1050°C for 30 min. and quenching to room temperature.

The diagram in (Fig. 2) shows the correlation between the relative magnetic phase (α') content of the $C_{-196°C}$ samples immersed after each reduction in a bath of liquid nitrogen, together with the Vickers hardness [1].

Due to the relatively high correlation coefficient between the relative content of magnetic phase (α') and the Vickers hardness, the measurement of hardness after the treatment described above could also provide a quick indication of the presence of the α' phase in the microstructure of the deformed DCT 304L steel.

A further increase in the amount of martensite α' in the deformed 304L steel, as reported in ref. [15], could be obtained after annealing at 450 °C, i.e., a few degrees above the austenite start (A_s).

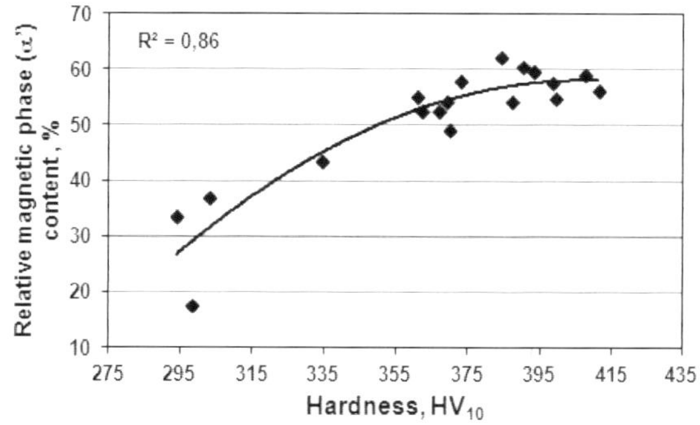

Figure 2: Correlation between the relative magnetic phase (α') content of the $C_{-196°C}$ samples immersed after each reduction in a bath of liquid nitrogen and the Vickers hardness [1].

However, there is no generally agreed explanation for this effect. Mangonon and Thomas [16] found that the increase of α' at 400 °C was due to the nucleation of new particles, rather than the growth of existing α' particles, but their explanation is refuted by Guy et al. [17]. They attributed the martensite increase at 400 °C to the growth of existing α' laths, possibly due to the relaxation of the α'→ γ interfaces. It was also suggested [18] that the increase of α' in the range 300–400°C is due to the precipitation of carbides that locally increase M_s and so promote additional α' formation. This suggestion was also refuted by Guy et al. [17], since they did not find any carbide particles. Tavares et al. [15] observed, using isothermal treatments, an increase in the amount of martensite α' content at 330°C and 400 °C, and found at 550 °C a small decrease in the martensite content, which suggests an isothermal reversion. At 560 °C a decrease in the magnetization was also observed [15], indicating that the reaction α'→ γ can progress at a fixed temperature and was not totally athermal. This is in agreement with the observation of Guy et al. [17], that the transformation at 550 °C in a 18 Cr–8 Ni steel occurred after an incubation period of about 10 min and then progressed by an isothermal diffusion mechanism. The first stage occurred by an athermal shear mechanism.

These findings are very interesting for maximizing the increase of the magnetic phase and the mechanical properties (hardness, etc.) of the deformed DCT 304L steel and avoiding any decrease of the magnetization as well as the hardness due to the reaction α'→ γ, which may progress at a temperature of approximately 550 °C.

Metallography and microhardness $HV_{0.01}$

A metallographic analysis was performed on the $C_{-196°C}$ specimens after cumulative true strains of 0.34 and 0.40 [1] and the laser surface treatment (Fig. 3).

Figure 3: Example of raster performed after the laser surface treatment on rod $C_{-196°C}$

Fig. 4a shows optical micrographs of a cross-section of the pulse-plasma-nitrided, 304L stainless-steel sample with the periodic and uniform semi-circular laser markings of a non-magnetic (γ) – magnetic (α') substrate at a temperature of 400°C for 10 hours, in which a clear separation can be seen between the martensite and austenite regions in the substrate with a continuous and uniform nitride layer. Characteristic profile of the laser marking at a depth of approximately 200 µm and for a width of 350 µm in the quenched raster in the shape of an approximate semicircle on the substrate for an encoder scale, clearly shows, that the temperature on the specimen's surface was higher than the phase-transformation point and lower than the melting point, (Fig. 4a).

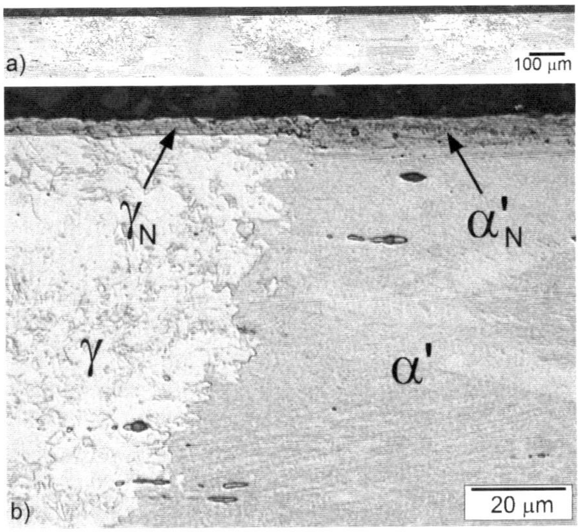

Figure 4a: Micrographs of the nitrided layer formed by the pulse plasma nitriding of the AISI 304L strain-induced martensitic (bulk) and the local non-magnetic phase austenitic stainless steel at 400°C for 10 hours, detail from Fig. 4a.

At a higher magnification (Fig. 4b) it is possible to see the difference in the thickness and the microstructure of the nitride layer above the strain-induced martensite (bulk) and the local non-magnetic phase due to the α'→ γ transformation after the laser surface treatment. The thickness of the compound layer for the two different fields is different. For the field with martensite (α'), the thickness of the compound layer (α'_N) is 7–8 µm, while the compound layer (γ_N) above the field with austenite (γ) is 3.5–4.1 µm. From (Fig. 4b) it can be assumed that the difference in the thickness of the compound layers is neither a function of the nitriding temperature or time, but depends only on the kinds of phases (α' and γ) treated under the same conditions.

The compound layer thickness above the martensite field is two times thicker than that above the austenite field. It is well known that the diffusion coefficient of nitrogen in martensite is much higher than that in austenite [19], so a martensite layer provides a path for rapid nitrogen transportation in the material. On the other hand, defect densities (dislocations, grain boundaries) are also important factors that enhance the diffusion of nitrogen in materials. Severe deformation can create much higher defect densities in the strain-induced martensite (bulk). Therefore, the nitrogen diffusion depth in 304L of strain-induced martensite is greater than that of the surface-laser-quenched, strain-induced martensite that transforms in austenite.

The transition of the compound layer from an austenitic microstructure to a martensitic microstructure is smooth and without cracks. It can also be assumed that a highly deformed martensite with high residual stresses accelerates the nitrogen diffusion, with the consequence of the two-times-thicker compound layer over the martensitic microstructure in comparison with the thickness of the compound layer over the austenitic microstructure (Fig 4b).

The laser-surface-treated rods that were plasma nitrided at 400°C due to the isothermal treatment also showed an increase of the hardness by 120 HV_{10} and ~7.5 % of relative magnetic phase (α') content in the bulk microstructure. As mentioned, the transformation sequence as a result of the plastic strain is γ → ε → α'. Therefore, we assume that an increase in the hardness and the percentage of relative magnetic phase (α') content in the bulk microstructure can be attributed to the isothermal transformation during plasma nitriding of the paramagnetic martensite phase (ε) in the α' ferromagnetic martensite.

Fig. 5 shows the direction of the microhardness depth-profile measurements in the martensitic and austenitic fields of the $C_{-196°C}$ specimens after low-temperature pulse plasma nitriding in combination with an isothermal treatment for 10 hours on 304L stainless steel.

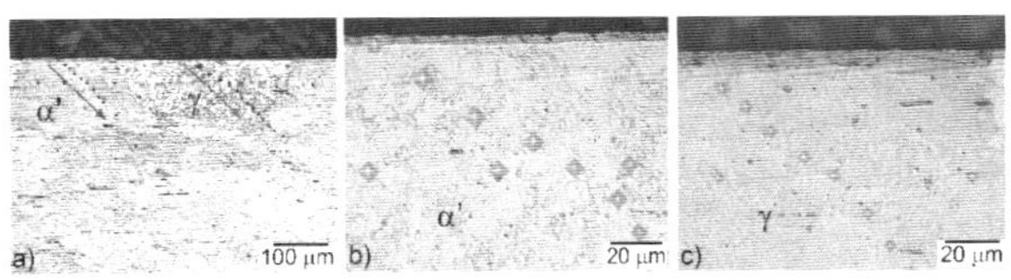

Fig.5: Direction of the microhardness depth-profile measurement in the martensitic and austenitic fields of the $C_{-196°C}$ specimen (a), (b and d) details from micrograph 5a.

Fig. 6 shows the variation of the depth-profile microhardness in the martensitic and austenitic fields of the $C_{-196°C}$ specimen.

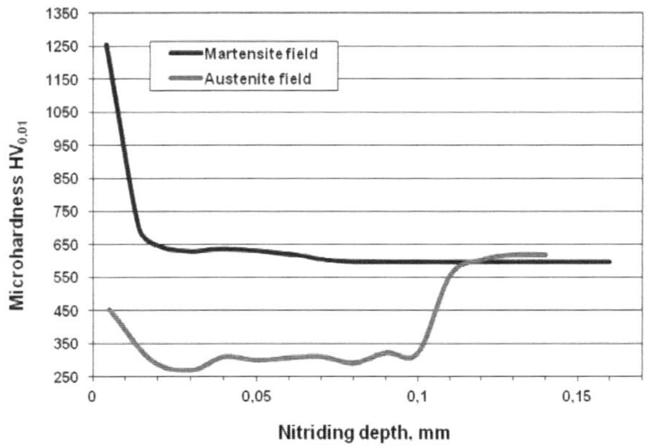

Figure 6: Microhardness depth profile in the martensitic and austenitic fields of the $C_{-196°C}$ specimen

The microhardness depth profile of the low-temperature pulse plasma nitriding in combination with the isothermal treatment sample shows a phase-dependent behaviour. This is because the nitrogen at the nitriding temperature penetrates deeper into the martensitic phase in comparison to the austenitic phase and forms a thicker modified layer above the martensitic phase. The surface hardness is greatly increased due to the formation of nitrided layers and the increase of the nitrogen concentration on both kinds of microstructure. It is clear that the value of the microhardness increased from about 620 $HV_{0.01}$ for the initial martensitic phase up to about 1250 $HV_{0.01}$, and from ~295 $HV_{0.01}$ up to ~ 525 $HV_{0.01}$ for the initial austenite phase at a depth of 4μm below the surface after the plasma nitriding at 400°C.

The surface microhardness is ~1320 $HV_{0.01}$. The hardness measurements indicated that the strain-induced martensite results in an increase of the microhardness of the nitrided layers for the low-temperature nitriding, as shown in (Fig. 6). The effective nitriding depth (Nht) in the martensite field is~40 μm, while in the austenite field it is between 5 and 18 μm. This confirms that the nitriding layer consists of a compound layer and a diffusion zone. The diffusion zone is at least two times thicker in the strain-induced martensite than in the austenite. At a depth of between 0.1 mm and 0.12 mm the microhardness profile obtained in the austenite field shows a gradual increase in the hardness in the transition from an austenite to a martensite microstructure. Such a transition hardness from one microstructure to another is, from the point of view of stress conditions, favourable, since in the bars under load it reduces the potential for crack initiation and rapid crack propagation in a harder and less ductile microstructure such as strain-induced martensite.

The hardness of the nitrided layer is determined by two hardening mechanisms: solid-solution hardening and precipitation hardening. Solid-solution hardening is the most important for austenitic stainless steels and precipitation hardening is the predominant mechanism for the same steels with a strain-induced martensite [19]. According to the results of S. J. Ji et al. [20], the diffraction pattern of 304L steel without a severely deformed layer nitrided within the temperature range 380°C to 420°C exhibits a series of broad peaks, which shift to lower 2θ values than those for normal austenite, which corresponds to a fcc structure that is similar to the nitrogen-expanded austenite phase previously reported by other researchers [21-26].

The broad peaks labelled γ_N (expanded austenite or S phase) at a smaller 2θ angle compared with those of the austenite peaks from the original substrate beneath the modified layer are associated with the nitrogen-rich expanded austenite phase. The peak positions of this nitrogen-rich expanded austenite phase change with the nitrogen concentration doped in the nitrided layer. This indicates that the major constituent in the nitrided layers has a fcc structure similar to the substrate austenite, but with an expanded lattice. The peak-position shift of the expanded austenite increased with nitriding temperature up to 420°C. At this temperature, the expanded austenite may have saturated in terms of nitrogen. It is also clear from the spectra that the intensity of the original austenite decreased considerably with the increasing temperature of the nitriding. It is expected that, as the temperature of the processing increases, a thicker nitrided layer is obtained, which in turn will result in an increasing contribution from the γ_N phase. Furthermore, the diffraction peaks from the nitrided layers were significantly broader than those obtained from the untreated substrate. Therefore, according to ref [20], it can be concluded that γ_N is the only phase present after plasma nitriding at low temperature.

Ref [20] also shows that for the samples with a pre-deformed surface layer, the diffraction patterns are different from their counterparts. After severe deformation, the surface zone consists mainly of martensite with a small part of retained austenite due to a strain-induced phase transformation. This martensite phase has a strong preferred (110) orientation. In the X-ray diffraction patterns of the samples with different nitriding temperatures, no clear evidence indicates that the expanded austenite with nitrogen (γ_N) was found in the nitride layer. It was also difficult to divide the diffraction peak in the diffraction-angle range from 45° to 55°. These nitride layers may consist predominantly of $(Fe, Cr, Ni)_{2-3}N$ or nitrogen-containing martensite (α'_N) or their mixtures. From the diffraction peaks obtained on these samples, the severely deformed layer has major effects on the nitrogen diffusion and phase formation in the nitrided layer.

SEM and EBSDE of low-temperature plasma nitriding

SEM/EDS/EBSD cross-sections of the low-temperature plasma nitriding bar with laser-marked areas were performed. EDS maps of nitrogen, chromium and iron in both surface layers γ_N and α'_N (expanded austenite and expanded martensite) are shown in (Fig. 7).

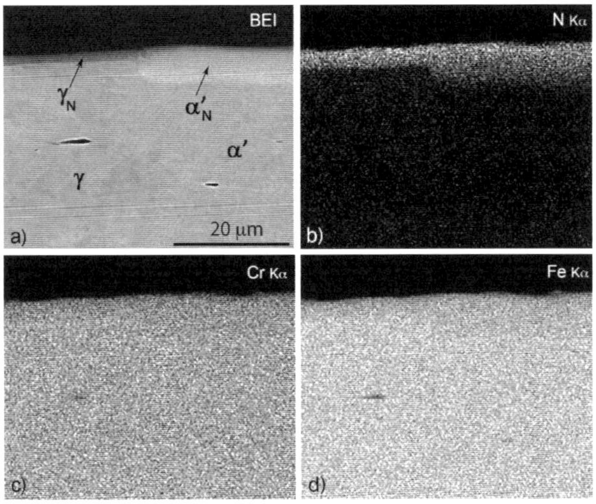

Fig. 7: Backscattered-electron image and corresponding X-ray images of N, Cr and Fe of low-temperature plasma nitriding bar with laser-marked areas.

EDS maps (Fig. 7) show the high concentration of nitrogen in both surface layers γ_N and α'_N (expanded austenite and expanded martensite) and no significantly higher concentration of chromium or iron. There is a slight depletion of chromium in both surface layers. The backscattered-electron image shows the different thicknesses of the surface layers formed on the austenite and martensite. There are no carbo-nitride precipitates observed at that magnification in both layers as well as in the martensite and austenite. EDS line spot analyses were carried out from the surface to the interior at a distance of 15.5 µm, indicating a greater enrichment of nitrogen in both layers in comparison to the bulk material.

The nitrogen distribution follows the different thicknesses of the surface layers (Fig. 8).

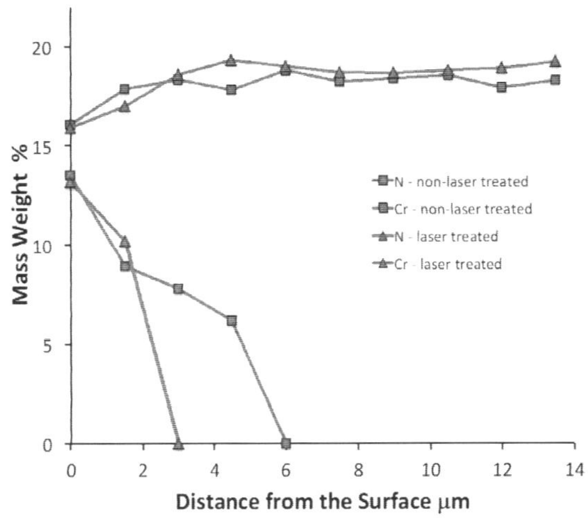

Fig. 8: Spot line EDS analyses of Cr and N concentrations from the surface towards the inner region as marked in Fig.9.

Fig. 8 shows that the amount of nitrogen is similar in both layers, but drops much more rapidly in the thinner layer. There is no proof of chromium enrichment in either surface layer and, as observed from the mappings, there is a slight depletion of chromium, most probably due to the enrichment of nitrogen. There is slightly larger amount of chromium in the martensite phase in comparison with the austenite phase.

Fig. 9 shows a backscattered-electron image on the border between the austenite and martensite phases and the EDS line spot analyses are marked with crosses over both regions of austenite and martensite.

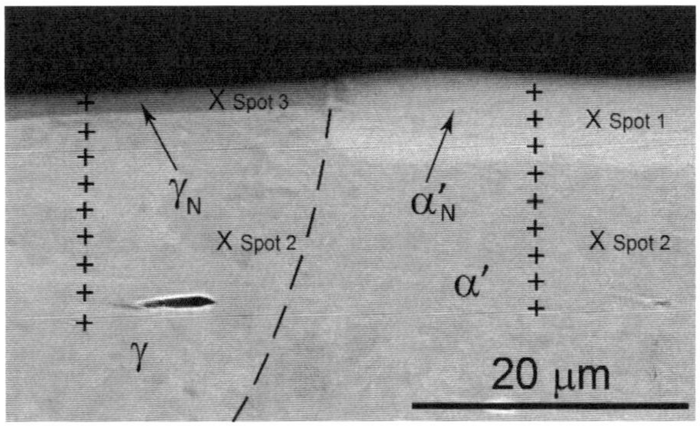

Fig. 9: SEM image of the cross-section of the deformed and laser marked DCT 304L steel after low-temperature plasma nitriding, from which a clear separation can be seen between the modified surface region and the substrate (see also Fig. 4b). The figure shows the marked spots of the EDS line analyses and the four spots for the EBSD analysis.

In (Fig. 9) the EBSD spot analyses are marked for the four different microstructures. The γ_N and α'_N layers are clearly visible, but the thickness of the γ_N fields is obviously different to that of the α'_N. The thickness of the α'_N is 7–8 µm, while the γ_N layer above the field with austenite (γ) is 3.5–4.1µm. The thinner surface layer of expanded austenite over the austenite phase is darker in the backscatter image compared to the thicker surface layer of the expanded martensite above the high-stress-induced transformed martensite. The electron-channelling contrast describes the effect of the interaction of energetic electrons with crystalline materials and can be used to detect the crystallographic properties of the sample [27]. The different contrast of both thin layers can be due to a different chemistry, which was not proven by the EDS analysis, but can also be due to the channelling effect in relation to the orientation contrast. There is no difference in the chemistry and most probably no significant difference in the preferred crystallographic orientation. The explanation for such a huge difference in the contrast between both layers most probably lies in the different internal stresses caused by the incorporation of nitrogen atoms in the crystal structures. The hardness measurements in (Fig.6) prove this, i.e., the hardness of the expanded martensite is over 1250 HV, while the expanded austenite layer has a hardness of 450 HV. It is known that the fcc structure can incorporate a larger quantity of nitrogen atoms in the interstitial sites, while, on the other hand, the bcc structure has fewer interstitial sites. Therefore, the stress in the bcc is much higher and this leads to a lighter colour in the backscattered image and more diffuse Kikuchi patterns.

The Kikuchi patterns on the surface-treated area are sharper with less stress, and the thinner surface layer was found to be austenite fcc with an expanded crystal structure (Fig. 10).

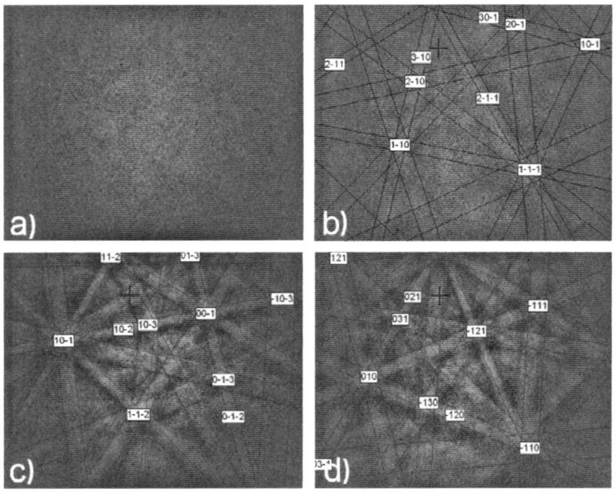

Fig. 10: EBSD Kikuchi patterns reveals that (a) the layer formed over the martensite has a very high stress, which is reflected in the highly diffuse pattern – spot 1 in Fig.9, (b) strain-induced martensite with a moderately high stress – spot 2 in Fig. 9, (c) the layer formed over austenite is fcc – identical to the austenite with nitrogen incorporated into the crystal structure – spot 3 in Fig. 9, (d) laser-treated martensite phase transformed into austenite fcc – spot 4 in Fig.9.

The matrix phase is strain-induced martensite, reflected in the moderately highly diffused Kikuchi patterns. The Kikuchi patterns of the thin surface layer over the martensite occurred during low-temperature nitriding are so blurred that no phase could be resolved.

From research to innovation

The combination of a deep-cryogenic treatment, plasticity, low-temperature pulse plasma nitriding with an isothermal treatment and phase transitions ($\varepsilon \rightarrow \alpha'$), in this order, increases the amount of formed magnetic α' phase, thereby improving the magnetic contrast when applying a non-magnetic γ marking on the substrate by laser surface quenching [1].

The measurements of the Hall probe on the $C_{-196°C}$ rod (Fig. 11a) showed that the amplitude of the signal after the low-temperature pulse plasma nitriding in combination with the isothermal treatment at 400°C was the same in comparison with the amplitude of the signal of the conventional encoder bar, with the grooves made by turning and filled by electrolytically deposited, environmentally questionable, hard chromium (Fig. 11b).

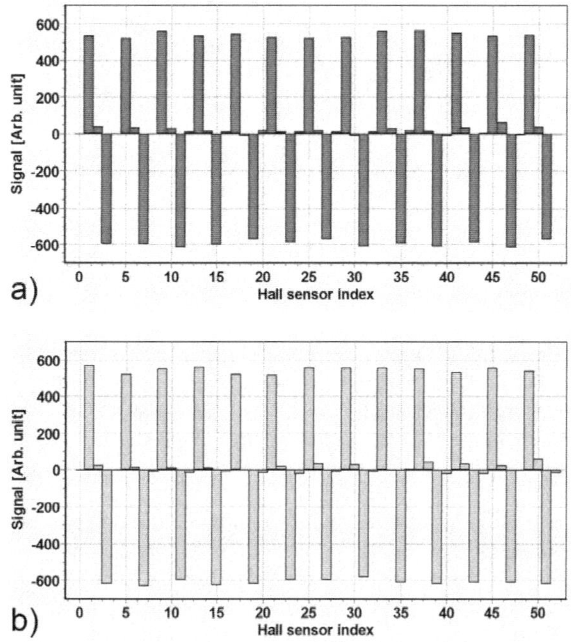

Fig.11: Measurements of the Hall probe on rod $C_{-196°C}$, (a) the heat-affected-zones laser surface treatment, (b) the grooves made by turning.

From the amplitude of the signal of the Hall probe it can be concluded that the alternating hard surface layers γ_N and α'_N do not adversely affect the magnetic properties on the substrate for the encoder scale.

Furthermore, due to the higher hardness of the austenite as well as the martensite bulk microstructure obtained after the deep-cryogenic treatment and plastic deformation, as well as the relatively thicker layer obtained after low-temperature nitriding in combination with the isothermal treatment, in comparison to the low-temperature nitriding of the same grade of stainless steel with an initial austenite microstructure, it could be expected that the load-bearing capacity of the surface, which is in the latter case limited to mild wear phenomena, will significantly increase.

Conclusions

The presented results allow us to conclude that the interaction of a deep-cryogenic treatment, plastic deformation, phase transitions, and low-temperature pulse plasma nitriding, in combination with an isothermal treatment as well as laser surface quenching, is a promising area for future work as they allow us to explore new structures and find new applications.

In addition, it provides a route to develop materials with an exceptional combination of magnetic properties, strength, ductility, as well as wear and corrosion resistance.

Acknowledgement

The authors gratefully acknowledge the financial support of the Slovenian Ministry of Economic Development and Technology, within the framework of RIP 09, for the project "Absolute position measurement systems AMS-P". The authors would like to thank Dr. Matjaž Torkar, Dr. Borivoj Šuštaršič and Prof. Dr. Franc Vodopivec, from the Institute of Metals and Technology of Ljubljana, for discussions and helpful remarks during the investigation. We would like to express our thanks to Janez Novak from the RLS Merilna tehnika d.o.o., Poslovna cona Žeje pri Komendi, Pod vrbami 2, SI 1218 Komenda, Slovenia for their help during the experimental part of this work.

References

[1] V. Leskovšek, M. Godec, P. Kogej, Strengthening via the formation of strain-induced martensite and the effects of laser marking on the microstructure of austenitic stainless steel, Metall. Mater. Trans. A, 45A (2014) 2819-2826.

[2] P. Kogej, V. Leskovsek, Manufacturing method for producing magnetic substrate for encoder involves mechanically working substrate comprising austenite grade of stainless steel, Patent Number:WO2012117230-A; Patent Assignee: RLS Merilna tehnika d.o.o.; Renishaw PLC

[3] P. L. Magnonom, G. Thomas, Structure and properties of thermal-mechanically treated 304 stainless steel, Metall. Trans. 1, (1970)1587-1594.

[4] J. A. Venables, The martensite transformation in stainless steel, Philos. Mag. 7 (1962) 35-44.

[5] F. Lecroisey, A. Pineau, Martensitic transformation induced by plastic deformation in the Fe-Ni-Cr-C system, Metall. Trans.32 (1972) 391-400.

[6] M. Milada, N. Zreibaa, F. Elhalouanib, C. Baradaib, The effect of cold work on structure and properties of AISI 304 stainless steel, J. Mater. Process. Tech. 203 (2008) 80-85.

[7] M. Jezeršek, A. Babnik, J. Možina, Laser micro-record information in the metal surface, University of Ljubljana, Faculty of Mechanical Engineering, (2010) 1-67.

[8] W. Liang, Surface modification of AISI 304 austenitic stainless steel by plasma nitriding, Applied Surface Science 211 (2003) 308–314

[9] X.Yun-tao, L. Dao-xin, H. Dong, Improvement of erosion and erosion–corrosion resistance of AISI420 stainless steel by low temperature plasma nitriding, Applied Surface Science 254 (2008) 5953–5958.

[10] B.C.E.S. Kurelo, G.B. de Souza, S.L.R. da Silva, F.C. Serbena, C.E. Foerster, C. Alves, Plasma nitriding of HP13Cr supermartensitic stainless steel, Applied Surface Science 349 (2015) 403–414.

[11] J.P. Riviere, P. Meheust, J.P. Villain, Wear resistance after low-energy high-flux nitrogen implantation of AISI 304L stainless steel, Surf. Coat. Technol. 158–159 (2002) 647-652.

[12] H. Ferkel, M. Glatzer, Y. Estrin, R.Z. Valiev, RF plasma nitriding of a severely deformed high alloyed steel, Scr. Mater. 46 (2002) 623-628.

[13] H. Ferkel, M. Glatzer, Y. Estrin, R.Z. Valiev, C. Blawert, RF plasma nitriding of severely deformed iron-based alloys, Mater. Sci Eng. A, 348 (2003) 100-110.

[14] M. Godec, B. Šetina Batič, Dj. Mandrino, A. Nagode, V. Leskovšek, D. S. Škapin, M. Jenko, Characterization of the carbides and the martensite phase in powder-metallurgy high-speed Steel, Mater. Charact. 61 (2010) 452-458.

[15] S. S. M. Tavares, D. Fruchart, S. Miraglia, A magnetic study of the reversion of martensite α'in a 304 stainless steel, Journal of Alloys and Compounds, 307 (2000) 311-317.

[16] P. L. Magnonom, G. Thomas, Structure and properties of thermal-mechanically treated 304 stainless steel, Met. Trans. 16 (1970) 1587-1594.

[17] K. B. Guy, E.P. Butler; D. R. F. West, Reversion of bcc α' martensite in Fe-Cr-Ni austenitic stainless steels. Met. Sci. 174 (1983)167-176.

[18] A. F. Padilha, L. C. Guedes, Aços inoxidáveis austeniticos, São Paulo, Hemus Editora Limitada, 1994:170.

[19] J. Franclin, P. Hill, and C. Allen, The Effect of Heat Treatment on Retained Austenite in a 1 % Carbon-Chromium Bearing Steel, Heat Treat. Met. 2 (1979) 46-50.

[20] S.J. Ji, L. Wang, J.C. Sun, Z.K. Hei, The effects of severe surface deformation on plasma nitriding of austenitic stainless steel, Surf. Coat. Technol. 195 (2005) 81-84.

[21] G. A. Collins, R. Hutchings, K. T. Short, J. Tendys, X. Li, M. Samandi, Nitriding of austenitic stainless steel by plasma immersion ion implantation, Surf. Coat. Technol. 74-75 (1995) 417-424.

[22] E. Menthe, K. T. Rie, J. W. Schultze, S. Simson, Structure and properties of plasma-nitrided stainless steel, Surf. Coat. Technol. 74 (1995) 412-416.

[23] Y. Sun, X.Y. Li, T. Bell, X-ray diffraction characterisation of low temperature plasma nitrided austenitic stainless steels, J. Mater. Sci. 34 (1999) 4793-4802.

[24] C. Blawert, A. Weisheit, B. L. Mordike, F. M. Knoop, Plasma immersion ion implantation of stainless steel: austenitic stainless steel in comparison to austenitic-ferritic stainless, Surf. Coat. Technol. 85 (1996) 15.

[25] S. Leigh, M. Samandi, G. A. Collins, K. T. Short, P. Martin, L.Wielunski, The influence of ion energy on the nitriding behaviour of austenitic stainless steel, Surf. Coat. Technol. 85 (1996) 37.

[26] M. J. Baldwin, M. P. Fewell, S. C. Haydon, S. Kumar, G. A. Collins, K. T. Short, J. Tendys, Rf-plasma nitriding of stainless steel, Surf. Coat. Technol. 98 (1998) 1187-1191.

[27] D. E. Newbury et al., Advanced Scanning Electron Miscopy and X-Ray Microanalysis, Springer Science +Business Media New York 1986.

International Federation of Heat Treatment and Surface Engineering 2016
Proceedings of the 23rd IFHTSE Congress
April 18–21, 2016, Savannah, Georgia, USA

Influence of strain hardening on gaseous nitriding of steels

Benjamin GUILLOT, Dr. Sébastien JÉGOU, Prof. Laurent BARRALLIER
MSMP Laboratory, Arts&Métiers ParisTech, Aix-en-Provence, France.
benjamin.guillot@ensam.eu, sebastien.jegou@ensam.eu, laurent.barrallier@ensam.eu

Abstract

Machining is known to introduce plastic straining and residual stress gradients in surface of mechanical parts. If not controlled, the mechanical state of a surface can influence adsorption and diffusion of atoms during subsequent chemical surface treatments, such as nitriding. In the present work, the influence of strain hardening on gaseous nitriding of steels is studied. The nitriding process was carried out using a laboratory set-up enabling the control of nitriding parameters (time, temperature, nitriding potential) on 33CrMoV12-9 steel samples. Homogeneously strain hardened samples are obtained by controlled tensile tests, enabling to nitride samples at different levels of strengthening. Results are compared to a reference state, obtained from tensile specimen not submitted to strain-hardening, as well as an industrially representative sample nitrided using the same conditions. Nitrided samples are characterized by SEM observations, hardness measurements and carbon and nitrogen in-depth concentration profiles.

Introduction

Gaseous nitriding of steels plays a leading role over surface treatments in automotive and aeronautic industries.
Iron-based alloys involve catalytic decomposition of NH_3-rich atmosphere between 450 and 590 °C [1]. Diffusion of adsorbed nitrogen atoms can occur up to 1 mm in depth below the treated surface, depending on nitriding conditions (temperature T, time t, and nitriding potential K_N). The nitriding potential is defined as:

$$K_N = \frac{P_{NH_3}}{P_{H_2}{}^{3/2}}$$

where P_{NH3} and P_{H2} represents the partial pressures of ammonia and hydrogen respectively.

Nitrided surfaces are defined by a compound layer composed of ε-$Fe_{2\text{-}3}N$ and/or γ'-Fe_4N over a diffusion zone composed of finely dispersed alloying elements nitride MN (M= Cr, V, …) into a ferritic matrix with interstitially dissolved nitrogen and cementite at grain boundaries of prior austenite.
Influence of nitriding potential and temperature was described by Lehrer in the case of pure iron [2]. It results in surface strength hardening and the generation of compressive residual stresses ensuring a better resistance to fatigue and corrosion [3].

Gaseous nitriding is a very sensitive process and it is possible to fail nitriding while mastering nitriding parameters [4]. The phenomenon responsible for this prevention can be split in two categories: surface and bulk phenomenon.
The first category is known as catalytic deactivation and concerns the first few atomic layers, such as geometrical parameters (e.g. surface roughness) and surface chemical interaction (e.g. sulfur) [5]. The second group concerns mechanical interactions (residual stress and strain hardening) that can be induced by machining for example, and modifies microstructure up to hundredths of micrometers [6]. The following work only concerns the influence of strain hardening on nitriding. It is of great interest, as Tong showed that a strong defect density can significantly modify the thermodynamic stability of iron nitrides in the case of binary iron-based alloys, leading to the possible formation of a compound layer at low temperatures [7]. Kinetics of nitrides precipitation is generally enhanced by strain-hardening in the case of binary iron based-alloys [8-9]. From a more industrial point of view, it was shown that the presence of strong strain-hardening and surface recrystallization induced by cold burnishing of an AISI D2 tool steel before gaseous nitriding leads to a deeper diffusion of nitrogen and to higher microhardness values at considered depths [10]. However, the process of burnishing remains very restraining and could have a different influence on nitriding than a classical machining, like turning. The present work focuses on the influence of strain hardening on gaseous nitriding on an industrial steel grade 33CrMoV12-9 dedicated to nitriding.

Composition (wt.%)					
C	Cr	Mo	V	Mn	Fe
0.30	2.97	0.91	0.28	0.51	balance

Table 1: Chemical composition of the studied 33CrMoV12-9 steel grade.

Experimental details

Material parameter

The material used for the study is a 33CrMoV12-9 steel. It was austenitized at 920 °C for 90 min, quenched and tempered at 640 °C for 90 min under nitrogen atmosphere. The measured chemical composition of the alloy is detailed in Table 1. Chemical composition was obtained on several samples by spark emission spectroscopy.

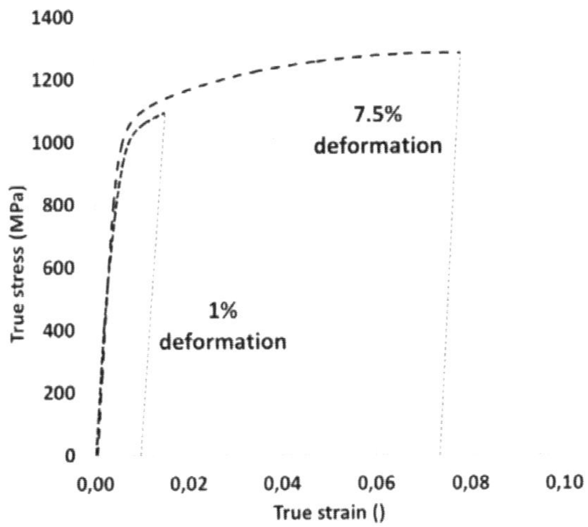

Figure 1: Tensile tests of 33CrMoV12-9 steel specimens stopped at different levels of strain hardening.

Strain hardening

Tensile specimens were carried out in order to study influence of strain hardening on the nitriding of steel. Specimens were wire and waterjet cut to an effective length of 60 mm. Strain hardening was obtained by tensile tests at two different levels of plastic deformation, 1 % and 7.5 %. No necking was observable after tensile testing of the samples.
Reference samples were taken from undeformed tensile specimen as well as samples that were submitted to industrial surface preparation, mainly based on milling, followed by grinding. These last samples were chosen as their microstructure is more representative of a structure industrially generated.

Specimens were cut in 18 × 12 × 5 mm^3 parallelepipeds before nitriding.

Nitriding

Prior to nitriding, samples were grinded using a 1200-grit abrasive paper, followed by manual degreasing.
Gaseous nitriding was carried out in a Setsys Evolution thermogravimetric analyser from Setaram Instrumentation.
Before nitriding, a preliminary procedure based on vacuum and N_2 purges prevents samples from detrimental oxidation during the heating to the nitriding temperatures.
Nitriding was performed using a combination of NH_3, N_2 and H_2 at 520 °C with a K_N = 3.7 $atm^{-1/2}$ during 300 min (5 hours). The total gas flow rate of 200 $ml.min^{-1}$ leads to a gas velocity of 10 $mm.s^{-1}$, ensuring a constant nitriding potential at the surface of the samples.
Heating and cooling of the samples were done at a rate of 10 $°C.min^{-1}$, under N_2 atmosphere.

During the treatment of samples, the dynamics of fluids might disturb the signal of the mass change. In order to remove any influence of the gas flow on the mass gain, a sample covered with copper, insensitive to ammonia decomposition was submitted to the same nitriding conditions. In data presented in Figure 2, the corresponding flux associated mass variation was subtracted to all experimental data, to only keep the nitriding associated mass variation.

Material analyses

Microstructure observations were carried out using a FEG-SEM JEOL JSM 7001F and backscattered electrons, with a 15 kV acceleration voltage. To this purpose, samples were grinded and polished down to 0.5 μm diamond solution, cleaned by ultrasounds during 5 min and etched with a 2%-nital solution.
Hardness measurements were carried out on a Leica VMHT device with a load of 2 N during 15 s.
Residual stress analysis was carried out by X-ray diffraction on a Siemens D500 diffractometer equipped with a linear detector and employing Cr-Kα radiation on the {211} diffracting plane of α-Fe. The $\sin^2(\Psi)$ method was used to determine the residual stresses (σ_{xx}-σ_{zz}) in α-Fe.

Results

The stress-strain curves obtained by tensile testing are provided in Figure 1. The heat-treated material proved to have a yield stress of 1000 MPa, an ultimate stress of 1220 MPa, a maximum strain of about 15%, and a Young modulus of 212 GPa, which is coherent with standards. Both strain-hardening tests are consistent with the behavior law.
Several analyses of residual stress by X-ray diffraction at the surface of both tensile specimens before cutting were carried out, leading to a repetitive compressive residual stress of 150 MPa.

Thermogravimetric analysis

Mass gain profiles of nitrided samples are given in Figure 2.

The kinetics of mass gain vary during the first hours depending on the level of strain-hardening. The higher the strain-hardening, the faster the nitriding is. However, a steady state appears after nearly 150 min for all samples and the mass gain obtained after 5 h of nitriding are similar and close to 1.3 $mg.cm^{-2}$.

The unstrained reference sample shows a similar behavior to the low strain hardened sample, in terms on mass gain kinetics and final mass gain.

The sample obtained by industrial surface preparations shows a different behavior. The mass gain kinetics is higher during the first two hours. The final mass gain is close to 1.5 $mg.cm^{-2}$, which is more than all other samples. However, it also reaches a steady state after 200 min, leading to a constant mass gain of 0.16 $mg.cm^{-2}.h^{-1}$ until the end of the treatment.

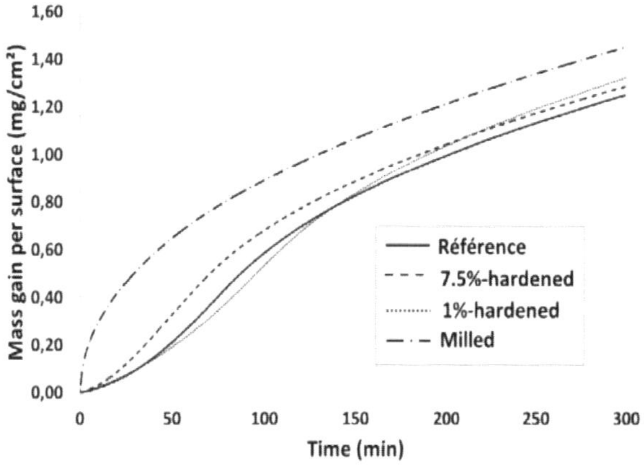

Figure 2: Mass gain of 33CrMoV12-9 steel specimen nitrided 5 h at 520 °C and 3.70 atm$^{-1/2}$

Microstructure observation

The comparison of the microstructure of strain-hardened samples shows that the nitrided depth is similar between all samples (Fig. 3). The compound layers are characterized by a thickness of about 10 µm.

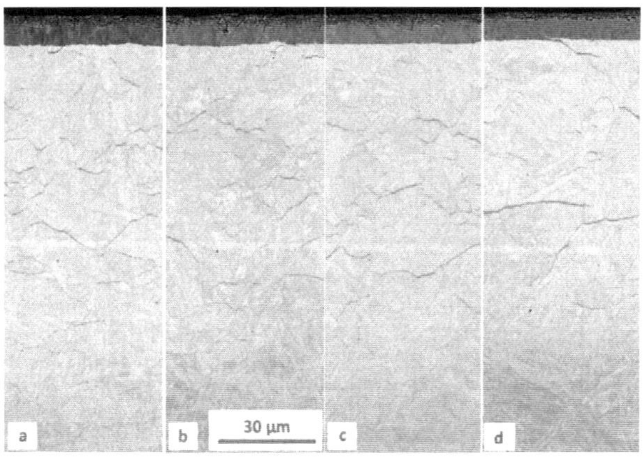

Figure 3: Scanning electron micrographs of nitrided 33CrMoV12-9 specimens. Comparison between the a) reference, b) 1% hardened, c) 7.5% hardened, and d) industrially prepared samples nitrided at 520 °C and 3.70 atm$^{-1/2}$ during 5 h.

Hardness profiles

Hardness profiles are given on Figure 4. They are all similar, including the milled sample. The bulk hardness is equal for all samples and reaches a value of 375 HV0.2. The effective depth, defined as the core hardness majored by 100 HV0.2, is equal to 140 µm for most samples, except the industrially prepared one, reaching an effective depth of 130 µm.

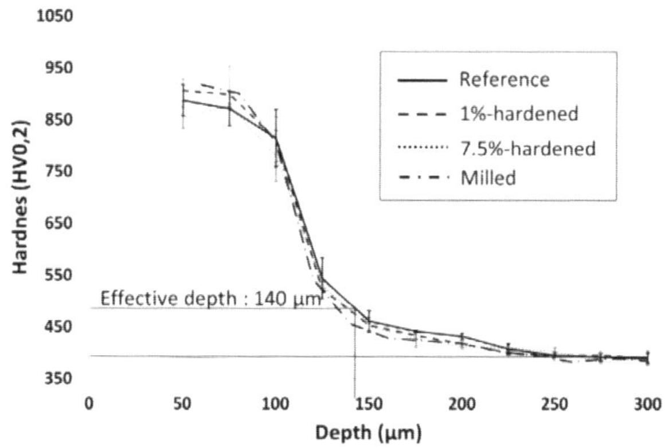

Figure 4: Hardness profiles of 33CrMoV12-9 specimens nitrided at 550 °C and 3.70 atm$^{-1/2}$ during 5 h.

Nitrogen and Carbon concentration

Nitrogen and carbon content in-depth profiles are given in Figure 5.
Surface measurements of nitrogen content cannot be provided due to a spectrometer measure limit of 1.3 wt.% exceeded by far in the compound layer.

Chemical content profiles are all similar and in agreement with hardness profiles and microstructure observation.

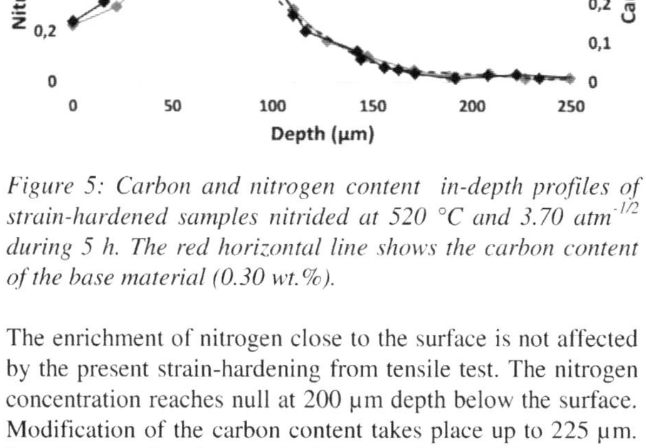

Figure 5: Carbon and nitrogen content in-depth profiles of strain-hardened samples nitrided at 520 °C and 3.70 atm$^{-1/2}$ during 5 h. The red horizontal line shows the carbon content of the base material (0.30 wt.%).

The enrichment of nitrogen close to the surface is not affected by the present strain-hardening from tensile test. The nitrogen concentration reaches null at 200 µm depth below the surface. Modification of the carbon content takes place up to 225 µm. Samples are also characterized by a carbon enrichment up to 0.34 wt.% at the nitrogen diffusion front, at 160 µm.

Discussion

Mass gain and loss

In the present work, the industrially prepared samples showed a faster nitriding kinetics and higher final mass gain. However, nitrogen and carbon content analyses of all samples indicate similar concentration gradients that results in no difference in hardness measurements.

In order to determine the origin of the final mass gain difference, one can calculate the mass gain and loss attributed to the diffusion layer by considering experimental nitrogen and carbon content profiles. A density of 7.8 $mg.mm^{-3}$ was used in order to convert dimensions into mass.
For the carbon content, considering a bulk value of 0.30 wt.% and calculating the difference between measured and bulk values at each depth, a mass loss of approximately 0.39 $mg.cm^{-2}$ for the whole sample is estimated.

For the mass gain of nitrogen in the diffusion layer, a null bulk concentration value is used for the calculation. This leads to a nitrogen mass gain of 0.85 $mg.cm^{-2}$ in the diffusion layer from experimental nitrogen in-depth profiles.

The final mass gain associated with the diffusion layer is thus equal to 0.46 $mg.cm^{-2}$. According to the final mass gain obtained from experimental data, this leads to a calculated compound layer mass gain of 0.84 $mg.cm^{-2}$ for the tensile specimens and 1.04 $mg.cm^{-2}$ for the industrially prepared sample. It is then possible to calculate the theoretical thickness of the compound layer for different samples, estimating the mass gain of a compound layer, depending on its thickness.

It is known that the mass percentage of nitrogen in ε-$Fe_{2-3}N$ and/or γ'-Fe_4N varies from 11 to 6 wt.%. Considering a linear decrease of the nitrogen content in the compound layer, which supposes an equal thickness of ε-$Fe_{2-3}N$ and γ'-Fe_4N, the thickness of the compound layer is estimated as 12.7 µm for the tensile specimen and 15.7 µm for the industrially prepared sample.

It was not possible to validate these theoretical calculations by observations of the compound layer as the difference of thickness calculated is too small compared to the variation of the compound layer along the sample and the damages due to sample preparation.

However, considering the similar nitrogen and carbon content gradients, it can be concluded that the faster mass gain observed for the industrially prepared sample can then be attributed to a faster growth of the compound layer.

Theoretical influence of strain-hardening on nitriding

Strain-hardening of steels is expected to have two potentials effects on gaseous nitriding. The first effect concerns the impact of strain-hardening on nitrides formation.

Considering alloy composition for this study, nitrides formed are mainly Cr, Mo and V nitrides. Influence of defect density was studied by Selg on binary Fe-Mo alloy [11], for cold-rolled and recrystallized specimen. It was observed that nitriding kinetics is higher in the case of cold-rolled specimens. This was explained by the modification of nitrides precipitation by the high density of defects. In the case of recrystallized specimen, no Mo nitride precipitation was observed. As the solubility of Mo in the γ' layer is very low, its content in the ferritic matrix slows the formation of the compound layer, limiting the nitriding kinetics. However, in the case of cold-rolled specimen, the high-density of defects leads to the precipitation of hexagonal MoN and cubic Mo_2N. This precipitation leads to an easier formation of γ'-layer that overruns Mo nitrides instead of trying to form in a ferritic matrix, saturated with dissolved Mo.

In order to understand the influence of the density of defects on nitride precipitation, Biglari studied the nitriding of Fe-2wt.%Al samples obtained under different conditions, including cold-rolling and recrystallization [12]. The increase of the nitrogen content due to nitriding of recrystallized specimen leads to a precipitation of thermodynamically stable hexagonal, wurtzite type, AlN nitride. However, in the case of nitrided cold-rolled specimen, the presence of NaCl type AlN precipitates was observed.

The presence of such a thermodynamically unstable nitride was supposed to be possible due to the misfit conditions associated with their formation. The stable hexagonal nitride formation is associated with a large uptake of misfit-strain energy while precipitation of cubic AlN nitride is associated with a low increase of misfit-strain energy. Hexagonal and cubic nitride chemical Gibbs free energy of formation was estimated as -288 $kJ.mol^{-1}$ and -69 $kJ.mol^{-1}$, respectively [12]. In the case of strain-hardened specimen, the presence of cubic AlN is possible if the increase of chemical Gibbs free energy associated with their formation is compensate with the decrease of misfit-strain energy of the matrix. It was also shown in different study that the content of aluminum strongly modifies the influence of strain-hardening [13].

In the present work, the mass gain kinetics increases with strain-hardening of tensile samples, which can be attributed to a faster precipitation of MoN and Mo_2N nitrides at the surface, due to a higher defect-density in the strain-hardened samples, leading to an easier formation of Fe_4N and a faster growth of compound layer. However, the levels of strain-hardening of the tensile specimen remain too low to observe a significant improve of nitrogen or carbon gradient, microstructure or mechanical properties of the diffusion layer. In most of the work studying influence of strain-hardening on nitriding, samples are cold-rolled to obtain a thickness reduction of 70 to 80% [12]. However, the strain-hardening of tensile samples is high enough in our work to have a non-meaningless improve of the mass gain kinetics.

In the case of the industrially prepared sample, the second potential expected effect of strain-hardening on nitriding, concerning surface recrystallization of the sample, should be considered. It is assumed that no recrystallization occurs for the tensile specimen, considering their low level of strain-hardening. However, for the industrially prepared sample, a recrystallization linked to surface milling can be expected, and could have a strong influence on nitriding kinetics. In the case of cathode sputtering before gaseous nitriding for example, which imposes strong lattice deformation at the surface of samples and strain-hardening, the process has proved to strongly increase the kinetics of white layer's nitride formation [14].

It should also be considered that tensile test is a monotonic test, while milling may impose cyclic plastic loadings to the sample and so, cumulative strains. Moreover, the mechanical loading in the case of milling induces more shearing. The higher mass gain kinetics of the milled sample tends to confirm such mechanical differences.

It should finally be considered that a single nitriding potential was used in this work. It may be assumed that the influence of a low strain-hardening would be more effective using a lower nitriding potential case. In order to verify this hypothesis, more experiments are planned, with the same levels of hardening, but lower nitriding potentials.

Conclusions

Influence of strain-hardening on gaseous nitriding of 33CrMoV12-9 steels has been studied. Two levels of strain-hardening were studied, 1 % and 7.5 % deformations, and compared to industrially representative samples. Such a level of hardening slightly modifies early mass gain kinetics but not the final mass gain, after 5 h of nitriding at 520°C. The faster kinetics is attributed to a faster growth of the compound layer due to a modification of nitride precipitation by strain hardening. However, it does not induce any differences in terms of neither microstructure, nor grain boundary precipitation, nor in terms of hardening. It appears that static loading tests are not well adapted for strain hardening study on gaseous nitriding of steels.

References

[1] Grabke, H.J., "Reaktionen von Ammoniak, Stickstoff und Wasserstoff and der Öberfläche von Eisen", Berichte der Bunsengesellchaft, Bd. 72, No. 4 (1968), pp. 533-541.

[2] Lehrer, E., "Uber das Eisen-Wasserstoff-Ammoniak-Gleichgewicht", vol. 36 (1930), pp. 383-392.

[3] Jack, D.H. *et al.*, "Effect of process variables on the fatigue resistance of nitrided steels", Heat treatment (1979), The Metal Society, pp. 169-177.

[4] Irretier, O. *et al.*, "Influence of contaminants on gas nitriding", Surface Engineering, vol. 12, No. 4 (1996), pp. 331-334.

[5] Arabczyk, W. *et al.*, "Poisoning of iron catalyst by sulfur", Catalysis Today, vol. 124 (2007), pp. 43-48.

[6] Haase, B. *et al.*, "Formation of reaction layers in steel machining: impact on surface treatment processes", Surface Engineering, vol. 15, No. 3 (1999), pp. 242-248.

[7] Tong, W.P., *et al.*, "Nitriding iron at lower temperatures", Science, vol. 299 (2003), pp. 686-686.

[8] Tong, W.P., *et al.*, "The formation of ε-$Fe_{2\text{-}3}N$ phase in a nanocrystalline Fe", Scripta Materialia, vol. 50 (2004), pp. 647-650.

[9] Brahmi, A. *et al.*, "Study of aluminum nitride precipitation in pure Fe-Al-N alloy by thermoelectric power measurement", Acta Materialia, vol. 45, No. 5 (1997), pp. 1889-1897.

[10] Tobola, D. *et al.*, "Structure and properties of burnished nitrided AISI D2 tool steel", Materials Science, vol. 21, No. 4 (2015), pp. 511-516.

[11] Selg, H. *et al.*, "Defect-dependant nitride surface layer development upon nitriding of Fe-1 at.% Mo alloy", Philosophical Magazine, vol. 93 (2013), pp. 2133-2160.

[12] Biglari, M.H. *et al.*, "Analysis of the nitrogen absorption isotherms of cold-rolled Fe-2 at.% Al specimens with different AlN precipitate dimensions", Philosophical magazine A, vol. 72, No. 4 (1995), pp. 931-947.

[13] Meka, S.R. *et al.*, "Interrelationships of defects, nitride modification and excess nitrogen", International Journal of Materials research, vol. 105 (2014), pp. 1057-1066.

[14] Baranowska, J., "Influence of surface pretreatment on case formation during gaseous nitriding", Surface and Coatings Technology, vol. 125 (2000), pp. 30-34.

International Federation of Heat Treatment and Surface Engineering 2016
Proceedings of the 23rd IFHTSE Congress
April 18–21, 2016, Savannah, Georgia, USA

Low temperature surface hardening of stainless steel; the role of plastic deformation

Federico Bottoli, Freja N. Jespersen, Jesper H. Hattel, Grethe Winther, Thomas L. Christiansen, Marcel A.J. Somers
Technical University of Denmark, Department of Mechanical Engineering
Produktionstorvet b. 425, 2800 Kgs. Lyngby, Denmark

somers@mek.dtu.dk

Abstract

Thermochemical surface engineering by nitriding of austenitic stainless steel transforms the surface zone into expanded austenite, which improves the wear resistance of the stainless steel while preserving the stainless behavior. As a consequence of the thermochemical surface engineering, huge residual stresses are introduced in the developing case, arising from the volume expansion that accompanies the dissolution of high interstitial contents in expanded austenite. This work addresses two aspects of the role of plastic deformation on the case developing during low temperature nitriding:

- plastic deformation of metastable austenitic stainless steels leads to the development of strain-induced martensite, which compromises the uniformity and the homogeneity of the expanded austenite zone.
- during low temperature surface engineering composition *and* stress profiles develop. On numerical modelling the evolution of composition and stress profiles from the processing parameters temperature, time and gas composition it is necessary to include elastic-plastic accommodation of the composition-induced strains associated with lattice expansion.

Introduction

Thermochemical surface engineering of stainless steels by low temperature nitriding (LTN) has developed over the last 30 years from a curiosum obtained by plasma nitriding or ion implantation [1] into a commercially applied gaseous treatment that can be accurately controlled by control of the temperature, gas composition and treatment duration. Since the temperature range of LTN is below 450 °C, the nitriding performance is very sensitive for the as-delivered condition of the steel. In this respect plastic deformation of austenitic stainless steel resulting from the shaping technology applied to obtain the component geometry is the rule rather than the exception. Nevertheless the influence of (severe) plastic deformation on the nitriding behavior and the resulting microstructure appears largely uninvestigated in the scientific literature. Recent work by our group [2,3] has shown that the development of strain-induced martensite in metastable austenitic stainless steels, as the widely applied AISI 304 and 316 types, leads to the development of chromium nitrides, CrN, in the expanded austenite zone, which negatively affects the corrosion properties. Also the presence of strain-induced martensite in the microstructure (together with austenite) leads to a significantly thicker case depth upon nitriding. On the other hand, for deformed stable austenitic stainless steel, wherein no strain-induced martensite forms, no CrN precipitates in the expanded austenite zone during LTN, while the thickness of the case is independent of the degree of deformation [2]. In the present extended abstract the effect of high temperature solution nitriding (HTSN) is verified as a remedy against strain-induced martensite formation during plastic deformation and to prevent CrN formation and irregular case depths. For full details of this research work the reader is referred to a forthcoming publication [4].

It has been documented earlier [5] that the case of expanded austenite developing during LTN is plastically deformed as a consequence of the enormous strains building up by the lattice expansion associated with nitrogen dissolution, i.e. composition-induced strains. No plastic deformation has so far been reported to low temperature carburized (LTC) stainless steel, which is in accordance with the less pronounced lattice expansion due to lower interstitial contents of carbon as compared to nitrogen. The elastic-plastic accommodation of lattice strains appears to be important to incorporate in a diffusion model that predicts composition- and stress-depth profiles in the expanded austenite zone. Only an elastic-plastic approach coupled with stress-induced diffusion and solid-solution strengthening by interstitials leads to realistic composition- and stress-depth profiles [6,7]. The extent of plastic deformation in the expanded austenite zone will be discussed in this extended abstract, as well as the difference between LTN and LTC with respect to plastic deformation.

LTN of deformed HTSN austenitic stainless steel [4]

AISI 316 with nominal composition (in wt.%) 0.07 %C, 0.05 %N, 17 %Cr, 10.55 %Ni, 2 % Mo, 1.6 % Mn, 0.4 %Si and a thickness of 0.45 mm was solution nitrided at 1150 °C for 4 h in 0.3, 0.6 or 0.9 bar N_2, such that a homogeneous nitrogen distribution was obtained over the thickness of the sheet. An isopleth, calculated with ThermoCalc showing the austenite phase field at high temperature and the isobaric nitrogen solubility lines is given in Fig. 1. Combustion analysis of the nitrogen content confirmed an excellent correspondence between the predicted and the measured nitrogen contents in the steel [4]. Tensile test specimens in as-received and HTSN conditions were plastically deformed to equivalent strain levels

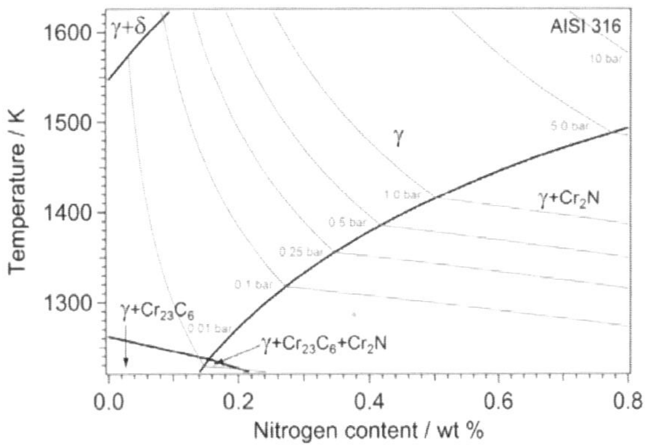

Figure 1: Isopleth showing the phase stability of austenite in the temperature range 1223-1623K (950-1350 °C) vs. nitrogen content in the solid state. Superimposed on the isopleths, N_2 isobars are given that would provide the equilibrium nitrogen contents [4].

of 0.15, 0.30 and 0.42 through tensile straining in the direction parallel to the rolling direction of the plate, at a strain rate of $6.67*10^{-3}s^{-1}$. Tensile testing curves and hardness achieved by plastic deformation are shown in Fig.2.

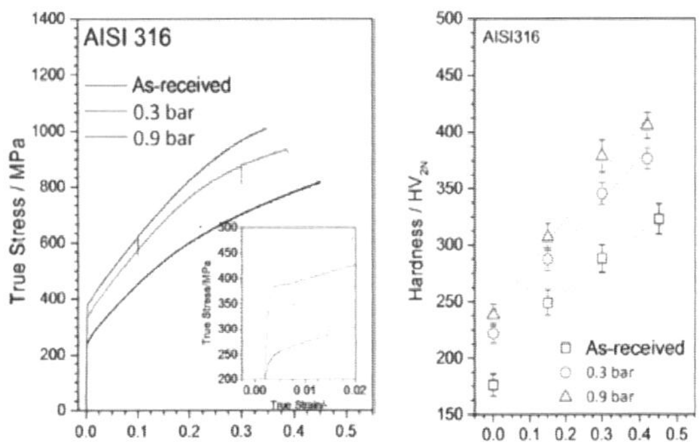

Figure 2: True stress (left) and hardness (right) vs. true strain for AISI 316 in as-received condition and after high-temperature solution nitriding treatment at 0.3 bar and 0.9 bar N_2 pressure. Results for 0.6 bar are in-between those for 0.3 and 0.9 bars. Inserts show an enlargement of the transition from elastic to plastic region [4].

Clearly, the dissolution of nitrogen in austenite leads to an increase in both yield strength, hardness and work hardening rate. X-ray diffraction showed that, as opposed to the as-received state, neither of the HTSN treated samples developed strain-induced martensite during tensile straining. This confirms that nitrogen is a strong austenite stabilizer.

HTSN treated samples were subjected to LTN at 703K (430 °C) for 20h in a partly dissociated ammonia atmosphere. Prior to LTN the sample surfaces were ground and polished (unto 3 μm diamond) and finally electropolished to remove additional deformation induced by mechanical preparation. Surface activation prior to LTN was performed in-situ with a proprietary treatment. The microstructure of the LTN treated samples in as HTSN treated and HTSN and deformed condition are shown in Fig. 3. Clearly, the expanded austenite zones in as HTSN treated and in additionally deformed conditions are featureless and appear identical.

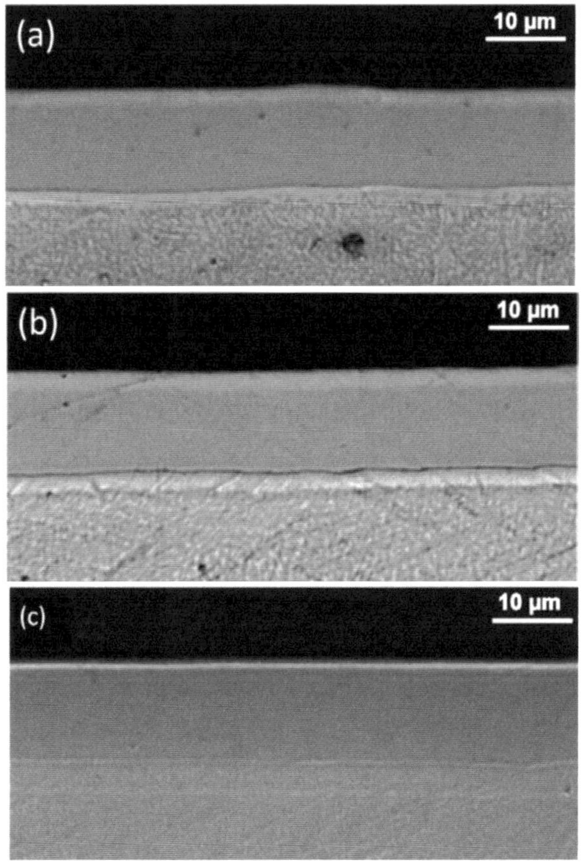

Figure 3: Cross sectional micrographs of AISI 316 after LTN treatment in undeformed HTSN 0.3 bar (a), deformed HTSN 0.3 bar ε=0.42 (b) and deformed HTSN 0.9 bar ε=0.42 (c) conditions [4].

X-ray diffraction (XRD) analysis showed that the only phases present in the cases developing during LTN are expanded austenite and ε-$(Fe,Cr)_2N_{1-z}$. The latter phase is present at the surface and appears light in the micrographs (cf. previous work [8]). The distribution of nitrogen after LTN was investigated with Glow Discharge Optical Emission Spectroscopy (GD-OES) for the series of deformed HTSN samples and is shown for the 0.3 bar and 0.9 bar series in Fig. 4. Evidently, the nitrogen profiles are identical for the investigated LTN conditions, irrespective of the N_2 pressure applied during HTSN treatment, and independent of the degree of deformation. The results in Figs. 3 and 4 clearly demonstrate that the stabilization of austenite in the steel is key for obtaining reproducible thickness of the expanded austenite zone and preventing the development of CrN. Moreover, an increase in the hardness of the austenitic stainless steel by a combination of solution hardening (dissolution of nitrogen during HTSN) and work hardening during subsequent deformation, contributes to the load-bearing capacity necessary to subtend the hard expanded austenite zone. It is mentioned that the combination of HTSN and subsequent LTN was previously patented [9].

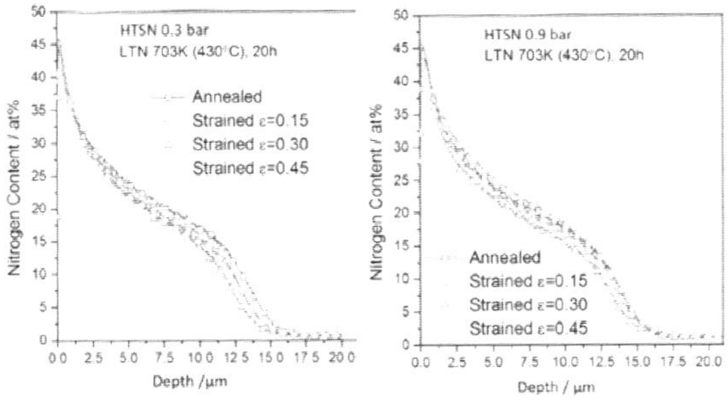

Figure 4: GD-OES nitrogen depth profiles for LTN treated samples strained to various degrees of equivalent strained after HTSN treatment at 0.3 and 0.9 bar. Within experimental accuracy the nitrogen-depth profiles are identical [4].

The role of plasticity in numerical modelling of LTN [7]

Numerical modelling of composition- and stress-depth profiles over the expanded austenite case has been attempted by various research groups. A survey of these attempts can be found in [7]. At first sight the modelling of the evolution of the nitrogen concentration profile over the expanded austenite case during LTN appears a diffusion problem that can be approached by straightforwardly solving Fick's second law for 1-dimensional diffusion of nitrogen into the steel. However, the diffusion problem becomes quite complicated if the developing residual stresses of several GPa's and associated steep stress gradients are taken into account. High stresses affect in particular the equilibrium solubility of nitrogen, while steep stress gradients affect the diffusion kinetics. A realistic numerical model that incorporates the hitherto observed features of LTN, takes into account:

- the concentration dependence of the diffusion coefficient of nitrogen in (stress-free) expanded austenite;
- nitrogen trapping by chromium atoms, which enhances the nitrogen content in the expanded austenite zone;
- kinetics of the transfer of nitrogen from the gas atmosphere to the solid state;
- the effect of residual stresses and their gradients on the solubility of nitrogen and the diffusive flux;
- elastic-plastic accommodation of the strain evoked by lattice expansion due to nitrogen dissolution;
- solid-solution strengthening of (expanded) austenite by the dissolution of nitrogen.

A numerical model incorporating all these features was published recently [6,7]. In this model it was demonstrated that in order to predict the level of residual stress of about 5 GPa in compression along with a realistic composition profile, both elastic-plastic accommodation of composition-induced strains and solid-solution strengthening are of utmost importance. The relation between yield strength and nitrogen concentration as given in Fig.5 was determined from *i.* the dependence of yield strength on hardness (cf. Fig. 2) and *ii.* typical hardness-depth profiles along with concentration-depth profiles.

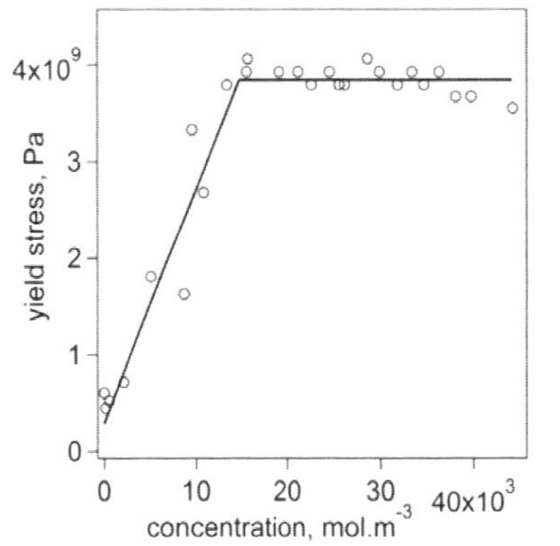

Figure 5: Dependence of the yield stress in austenite on the nitrogen concentration. The maximum yield stress reflects the constant hardness in expanded austenite zone [7].

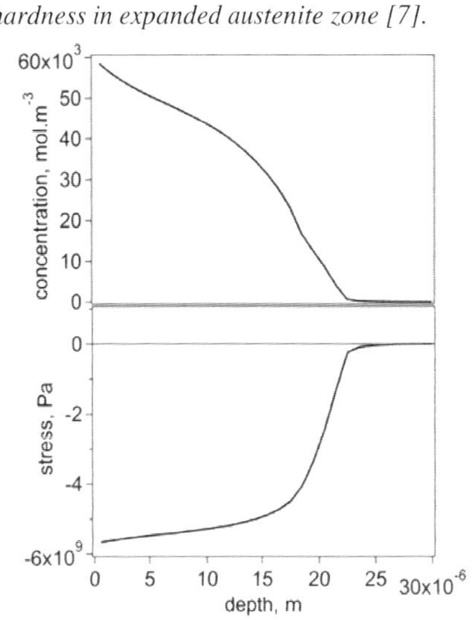

Figure 6: Calculated nitrogen concentration and residual stress profiles for typical LTN conditions (445 °C, 22 h in almost pure NH_3; for calculation details see [7]).

The concentration and residual stress profiles obtained for typical LTN conditions, in this case 445 °C, 22h and almost pure NH_3, yielded the results displayed in Fig. 6. The evolution of the case depth of the expanded austenite zone and the thickness of the plastic zone in expanded austenite is shown in Fig. 7 for two values of the nitriding potential, $K_N=pNH_3/(pH_2)^{3/2}$ (pJ is the partial pressure of gas J in the gas), which is proportional to the thermodynamic activity of nitrogen in the gas mixture. Evidently a large portion of the lattice expansion is accommodated plastically. A higher nitriding potential in the gas mixture, corresponding to a higher nitrogen concentration in expanded austenite is associated with a larger depth of the plastic zone. A reduction of the plastic zone depth can be achieved by lowering the nitriding potential.

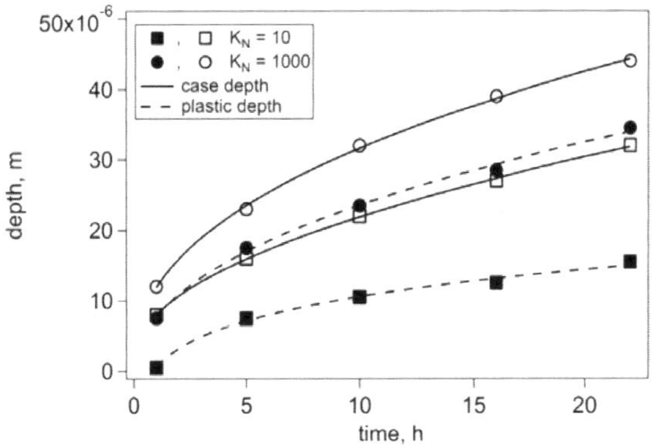

Figure 7: Evolution of expanded austenite case depth (drawn lines) and plastic depth (dashed lines) for LTN at 445 °C at two nitriding potentials K_N (in atm$^{-1/2}$).

For low temperature carburizing (LTC), no plastic zone in the expanded austenite case has so far been reported. It appears that the lattice expansion on LTC is accommodated fully elastically. This can be understood from the lower interstitial content dissolved in expanded austenite. In Fig. 5 plastic deformation occurs for combinations of interstitial content and (residual) stress beyond the drawn lines. A typical concentration of carbon in expanded austenite is 20.10^3 mol.m^{-3}, while the level of compressive residual stress is generally below 3 GPa. Indeed, this combination of concentration and stress is located in the region of Fig. 5 where lattice strain is accommodated elastically, i.e. below the yield strength.

Conclusions

In this extended abstract the role of plastic deformation during low temperature nitriding (LTN) of stainless steel is addressed. It is demonstrated that a stabilization of austenite against strain-induced martensite formation during plastic deformation results in a homogeneous expanded austenite zone free of CrN, irrespective of the degree of deformation applied. Such stabilization of commercially available metastable austenitic stainless steel (typically AISI 300 series), can be achieved by high temperature solution nitriding (HTSN) prior to plastic deformation. The performance during subsequent LTN is independent of the HTSN treatment, provided that austenite is stable.

On modelling the evolution of composition and stress in the expanded austenite zone it is essential to include solid solution strengthening by dissolution of interstitials and elastic-plastic accommodation of the resulting lattice expansion. Numerical modelling demonstrates that the major part of the expanded austenite zone developing during LTN accommodates the lattice strain plastically, as opposed to low temperature carburizing.

Acknowledgement

Part of this work was financially supported by the Research Fund for Coal and Steel through the PressPerfect project.

References

[1] Zhang, Z.L., Bell, T., "Structure and Corrosion Resistance of Plasma Nitrided Stainless Steel", *Surf. Eng.*, vol. 1 (1985), pp. 131–136.

[2] Bottoli, F., Winther, G., Christiansen, T.L., Somers, M.A.J., "Influence of plastic deformation on low-temperature surface hardening of austenitic stainless steel by gaseous nitriding", *Metall. Mater. Trans. A*, vol. 46A (2015), pp. 2579-2590.

[3] Bottoli, F., Winther, G., Christiansen, T.L., Somers, M.A.J., "Influence of plastic deformation on low-temperature surface hardening of stainless steel by gaseous nitriding", Proc. Heat Treat 2015, Detroit, ASM International, pp. 612-619.

[4] Bottoli, F., Winther, G., Christiansen, T.L., Dahl, K.V., Somers, M.A.J., "Low-temperature nitriding of deformed austenitic stainless steels with various nitrogen contents obtained by prior high temperature solution nitriding", submitted for publication in *Metall. Mater. Trans. A.*

[5] Stinville, J.C., Villechaise, P., Templier, C., Rivière, J. P., Drouet, M., "Lattice rotation induced by plasma nitriding in a 316L polycrystalline stainless steel", *Acta Mater.*, 58 (2010), pp. 2814–2821.

[6] Jespersen, F.N., Hattel, J.H., Somers, M.A.J., "Modelling of composition and stress profiles in low temperature surface engineered stainless steel", Proc. Heat Treat 2015, Detroit, ASM International, pp. 606-611.

[7] Jespersen, F.N., Hattel, J.H., Somers, M.A.J., "Modelling the evolution of composition- and stress-depth profiles in austenitic stainless steels during low-temperature nitriding", *Mod. Sim. Mater. Sci. Eng.*, vol. 25 (2016), 025003 (31 pages).

[8] Fernandes, F.A.P., Christiansen, T.L., Somers, M.A.J., "Microstructural Characterization of Low Temperature Gas Nitrided Martensitic Stainless Steel Name of Publication", Proc. 28th Int. Conf. Surface Modification Technologies, Eds. T.S. Sudarshan, P. Vuoristo, H. Koivuluoto, 2015, ISBN:978-81-926196-1-3, pp. 223-233.

[9] Christiansen, T.L. Hummelshøj, T.S., Somers, M.A.J., "Forming expanded austenite and/or expanded martensite by solution hardening of cold deformed workpiece of passive alloy", WO2012 146254-A1.

International Federation of Heat Treatment and Surface Engineering 2016
Proceedings of the 23rd IFHTSE Congress
April 18–21, 2016, Savannah, Georgia, USA

Effects of Different Microstructural and Hardness Gradients Generated by Single and Combined Surface Treatments with a Nitriding Top Layer

Anja Buchwalder[1]; Heinz-Joachim Spies[1]; Normann Klose[1]; Anne Jung[1]
[1] TU Bergakademie Freiberg, Institute of Materials Engineering, 09599 Freiberg, Germany
anja.buchwalder@ww.tu-freiberg.de, +49 3731 39 3543

Rolf Zenker[1, 2]
[2] Zenker Consult, 09648 Mittweida, Germany
contact@zenker-consult.de, +49 3727 613581

Abstract

Cast iron exhibits high potential for light-weight constructions. For applications in highly loaded automotive components, however, the material's resistance to corrosion and wear has to be improved. Comparative investigations of single and combined surface treatment technologies incorporating electron beam (EB) remelting and nitriding are presented in associated previous works.

The focus of this contribution is on the nitriding mechanisms for untreated and remelted cast iron. Investigations showed great differences when compared to the well-known nitriding mechanism of the unalloyed steel C15. In this context, the paper examines the often controversial interpretation of the role of the typically high silicon content of cast iron during nitriding – including the effect of the formation of silicon nitride on the hardness-depth profile. The core interests comprise the composition and distribution of γ'-, ε-, and ε(N, C)-nitrides, as well as the thickness of the compound and diffusion layers formed after nitriding.

Finally, the potentials and limitations of the single and combined surface treatments with respect to the improvement of the tribological and corrosion load behavior of cast iron were derived based on a cross-evaluation of the results of the treatments investigated.

Introduction

Nitriding and surface remelting of gray cast irons have featured in industrial applications for many decades [1]. The progression to light-weight applications – in the automotive industry in particular – has reinvigorated the 'old' material that was cast iron. The lower density of cast irons in comparison to steels combined with their excellent, near-net shape castability with very thin wall thicknesses make them an attractive option. However, additional protection measures (surface treatments) for improving their corrosion and tribological load behavior are required. The high specific demands on the load behavior as well as on the process stability make new process technologies (i.e. duplex treatments) necessary, along with profound process knowledge concerning nitriding conditions and the nitriding mechanisms.

It is often assumed that the nitridability of cast iron is comparable to that of unalloyed steels [2]. This means that, depending on the nitriding conditions, a sandwich-type compound layer structure consisting of γ'-nitride (Fe_4N; 5.7-6.1 %N; fcc) and ε-nitride (Fe_3N; 8-11.2%N; with considerable C solubility; hex) is formed by external nitration. The diffusion layer follows below the compound layer. This so-called internal nitration is characterized by dissolved nitrogen, but mainly through nitride precipitation of iron and its alloying elements [3].

Both the formation of nitrides and the diffusion of nitrogen are influenced by the material-specific graphite as well as by the high C and Si contents of the cast irons [4]. Incomplete surface coverage with nitrides in the area of the compound layer is reflected in significantly increased surface roughness [4], and influences the level of achievable improvement of corrosion and wear behavior. The results of such comparative investigations of single and combined surface treatment technologies incorporating electron beam (EB) remelting and nitriding presented in associated previous works [5][6] clearly show that the nitriding mechanisms of both the cast iron base materials and the remelted ledeburitic surfaces differ from those of unalloyed steels.

The shape of the hardness-depth profile generated in the diffusion layer is not directly determined by the nitrogen diffusion, but strongly depends on the degree of interaction between the nitrogen and the alloying elements. Besides the stoichiometrically bonded nitrogen and the interstitially dissolved nitrogen, the so-called excess nitrogen also exists as a result of the expansion of the matrix, which in turn is a result of the precipitations [8]. Based on the typical alloying elements of steels (i.e. Cr, Ti, Mo), three different microstructural effects and, therefore, typical shapes of hardness-depth profiles can be distinguished, namely "weak", "middle", and "strong" interactions [3][9]. Some of the statements in the literature on the influence of Si on hardness are conflicting, but the majority report hardness increases that are significantly higher than for unalloyed steels [10][11].

Investigations made on the ternary system Fe-N-Si show a significantly reduction in nitrogen solubility in ferrite (0.0015 % at 600°C (875 K)) in the presence of silicon (2 %Si) [12]. In extreme cases, only external nitration is possible. Furthermore, for high Si contents and a ferritic microstructure, amorphous Si_3N_4 precipitations are detected [13][15]. The hexagonal lattice of the α- or β-Si_3N_4 differ greatly from that of the bcc-faced ferrite. Due to the high misfit-strain energy, nucleation is inhibited, resulting in a nitriding kinetic comparable with the "weak" interaction [14][15].

Based on this knowledge, the core interests of this paper are on the differences in nitriding mechanisms between unalloyed steel and cast iron as well as between the untreated and remelted cast iron. In this context, the often controversial role of the typically high silicon content of cast iron during nitriding was analyzed, including the effect of the formation of silicon nitride on the hardness-depth profile. Furthermore, the relevant wear and corrosion behavior produced by single and combined treatment will be discussed by examination of the depth-dependent microstructure within the treated surfaces and the resulting hardness-depth profiles.

Experimental

Base Materials

Conventional (GJS-400) and modified cast irons (GJL-250) were used as the base materials (BM) for these investigations, and differed in chemical composition, graphite morphology, and matrix microstructure (cf. Table 1).

For investigations regarding the specific characteristics of the nitriding mechanisms of cast irons that result from the high contents of C and Si, the unalloyed steel C15 was used as a reference material (Table 1).

Table 1: Chemical composition (GDOES), microstructure, and initial hardness of the base materials.

[wt.%]	C15 (Ref)	GJS-400	GJL-250
Fe	bal.	bal.	bal.
C	0.14	3.5	2.8
Si	0.23	2.1	1.5
Mn	0.55	0.22	0.64
P	0.006	0.026	0.030
S	0.034	0.009	0.005
Cu	-	-	0.65
Cr	-	-	0.03
matrix	ferritic	ferritic (pearlite 5%)	pearlitic (ferrite <1...5%)
graphite[1]	-	spheroidal (0 % III 4/5 + 40 % VI 5/6)	fine lamellar (10% IA + 90% ID/E)
graphite vol.-%	-	11	7.6
micro-structure (etchant: Nital)		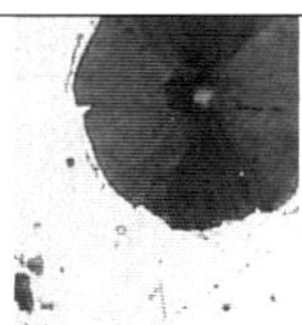	
hardness HV0.30	126 ± 4	156 ± 26	238 ± 22
λ[2] [W/mK]	50	36	48 - 44

[1] classified by DIN EN ISO 945-1
[2] firm prospect (GJL: 100-500°C (375-775 K); GJS: 330°C (605 K))

Surface treatment technologies

The present paper deals with comparative investigations of two different strategies for applying the surface treatment technologies of electron beam remelting (EBR) and gas oxinitriding (GON):

- single treatments: EBR, GON
- duplex treatment: EBR+GON

Electron beam remelting was carried out in a multi-purpose electron beam facility with an acceleration voltage of 80 kV and a beam power of 15 kW. Preheating up to a temperature of 450°C (725 K) before the actual EB remelting is essential to produce crack-free layers. This was realized with an EB field technique combined with an in-situ pyrometric temperature measurement. The meander beam deflection technique with a continuous feed rate of the sample (0.5 mm/s) and beam current of 15 mA were used to produce EBR layers with track widths of 20 mm and lengths of 80 mm. The EB-treated sample surfaces were ground and leveled before nitriding.

The gas oxinitriding process was carried out at a temperature of 540°C (815 K) for 16 h.

Characterization

The element concentration-depth profiles after single and combined surface treatments were analyzed using glow discharge optical emission spectroscopy (GDOES). Additionally, the phase composition and quantity distribution (vol.-%) was measured by X-ray diffraction (XRD) with CoK_α radiation. For the angular range of 2θ: 40-100°, the penetration depth was approx. 10 µm.

The microstructure was characterized by means of metallographic cross-sections using light optical (LOM) and scanning electron microscopy (SEM). Nital was used for standard etching of the metallographic sections. For detailed microstructural evaluation of the compound layer, Oberhoffer etching was carried out.

LOM was also used for measuring the compound and diffusion layer thicknesses (CLT and DLT, respectively) after nitriding as well as the EB remelted layer depth as an average value of a minimum of 5 single values.

Furthermore, hardness-depth profiles (HV0.1) were measured perpendicular to the surface. The hardness of the thin compound layers was measured by means of cross sections under increasing loading conditions up to a maximum load of 50 mN. The indentation hardness H_{IT} was determined by fitting the recorded unloading curve (DIN ISO 14577 1-3), which provided information on the plastic part of the hardness. This allowes conclusions regarding the Vickers hardness.

Unlubricated pin-on-disc tests (counter body [ball]: WC-Co, Ø 6 mm) were carried out with normal forces of 50, 80 and 120 N over a distance of 140 m. The subsequent calculation of the wear coefficient was done by means of measuring the abraded area across the wear track.

Comparative studies of corrosion resistance were carried out by potentiodynamic measurements in a 5% NaCl solution at room temperature. An saturated Ag/AgCl electrode was used as the reference electrode, a platinum plate as the counter electrode, and the sample as the working electrode. After determination of the open circuit potential E_{OCP} (with a measuring period of 60 min), polarization was carried out with a scan rate of 10 mV/min. Characteristic corrosion values were determined by means of potentiodynamic polarization curves: corrosion potential (E_{corr}), passive current density (i_{pass}) at E = 0.2 V, corrosion current density (i_{corr}), and the polarization resistance (R_p).

Results and Discussion

Characterization of the nitriding mechanisms for ferritic matrix microstructures (Comparison of unalloyed steel and cast iron)

The nitriding mechanisms of unalloyed and alloyed steels are well known [2]. For that reason, an unalloyed steel C15 was taken as a reference in understanding the differences in nitriding mechanisms to the cast iron GJS-400 (Table 1). Nevertheless, both materials have a mainly ferritic matrix microstructure with a maximum solubility of carbon of 0.02 wt.% and a small fraction of pearlite. In addition, the significantly higher C content of the GJS-400 (Table 1) in association with its high Si content led to graphite formation, which meant that the carbon was then unavailable for the matrix microstructure (cf. microstructure in Table 1). Furthermore, the matrix microstructure of the cast iron has a Si content 7 times higher (Table 1).

Both materials exhibited the well-known general nitriding layer structure, consisting of a small compound layer (Fig. 1, I and II) and a thicker diffusion layer.

In the case of the cast iron, the uncovered areas of the spheroidal graphite were clearly visible (Fig. 1b, III). Detailed microstructural and phase analysis indicated further distinct differences in the nitriding mechanism.

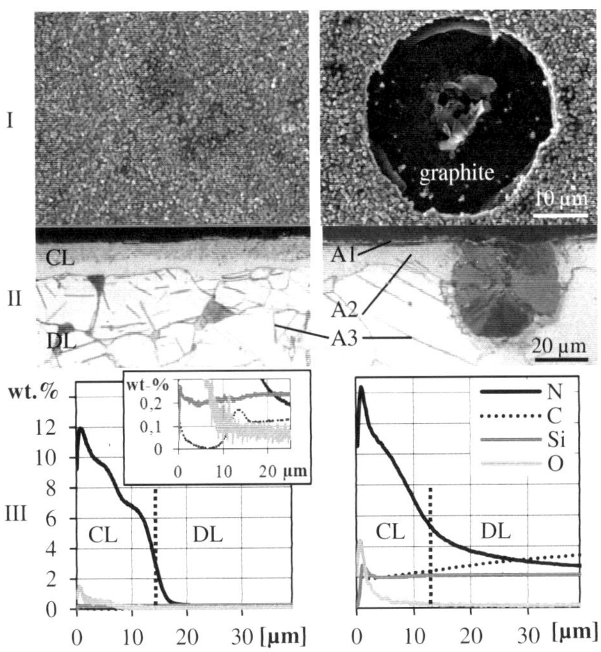

a) C15 (Ref) *b) GJS-400*

Figure 1: SEM image of the nitrided surfaces as a top view (I), cross-sections of the nitriding layers (etchant: Nital) divided into compound (CL) and diffusion layers (DL) (II), and element concentration-depth profiles measured by GDOES (III).

a) Compound layer (CL)

The phase and chemical compositions as well as the growth rate of the compound layer are determined by the chemical potentials of nitrogen and carbon [8], which are influenced by the nitriding process itself and the base material (chemical composition, microstructure).

Differing from the conventional gas nitriding process, a thin top layer consisting of iron oxide and nitride was formed after the gas oxinitriding process (Fig. 1b, II, A1). The remaining, subjacent part of the compound layer exhibited the typical iron nitrides ε and γ' (Fig. 1b, II, A2).

Despite the compound layer, the thicknesses (CLT) and the measured phase compositions (ε / γ') of both nitrided materials were in a comparable order of magnitude (Table 2). Nevertheless, the measured indentation hardness (H_{IT}) of the compound layer formed on the cast iron was 30% higher than on the unalloyed steel (Table 2). The additional formation of Si nitrides (Si_3N_4) was a plausible explanation. However, Si nitride could not be detected by means of XRD measurements to date. Further information is given under the next section.

Table 2: Characterization of nitrided layers (single treatment) on Fe-C and Fe-C-Si materials with ferritic matrix microstructures.

	Features	C15 (Ref)+ GONII	GJS-400+ GONII
CL	**CLT [µm]**	12.7 ± 1.7	11.6 ± 2.4
	wt.%N_{max} (t<5 µm)	12	14
	ε / γ' [vol.-%]	58 / 42	56 / 44 (Si_3N_4)
	H_{IT} [GPa]	7.2 ± 0.4	10.5 ± 0.8
DL	**DLT [mm]**	0.31	0.22
	wt.%N / at.-%N (d=40 µm)	0.08 / 0.54	2.6 / 8.7
	at.%Si (d=40 µm)	0.48	3.4
	SH_{max} HV0.1	200	450

In principle, the nitriding mechanism of pearlitic cast irons (GJL-250) is comparable to the description for the ferritic cast iron (GJS-400) given above. Differences in phase composition (cf. Table 3, GJL-250+GON) mainly relate to the lower diffusion coefficient of nitrogen in Fe_3C ($4.8 \cdot 10^{-15}$ m²/s at 550°C (825 K) [16]) compared to the ferrite ($1.3 \cdot 10^{-11}$ m²/s at 550°C (825 K) [16]).

b) Diffusion layer (DL)

The diffusion layer is characterized by a diffusion-controlled and continuously decreasing nitrogen gradient (Fig. 1, III). As already mentioned for the compound layer, the diffusion layer thickness (DLT) was also strongly dependent on the microstructural diffusion conditions and the limitations imposed by other alloying elements. Due to the very low matrix C content in the ferritic microstructure, the thicknesses of the diffusion layers (DLT) were, in general, higher than for the pearlitic or ledeburitic microstructures (ref. in [8]).

For the unalloyed steel, the nitrogen diffusion depth could be estimated by determining the end-point of the so-called "needle zone" (Fig. 2a). The needle-like precipitations (see A3 in Fig. 1, III) were formed during slow cooling after nitriding. With decreasing temperature, the nitrogen solubility was reduced, and iron nitride precipitations occurred.

The etched micrograph of the GJS-400 shows some differences (Fig. 2b). Normally, the nitrogen diffusion in unalloyed steels cannot be made visible by etching only the nitride precipitations (Fig. 2a).

In the case of GJS-400, the diffusion layer exhibited a darker tone after etching (Fig. 2b), and the upper part exhibited additional coarse needle-like iron-nitrides as described above for C15. Furthermore, for the nitrided cast iron, the higher nitrogen content (higher by more than one order of magnitude) in a depth beneath the surface of 40 µm is particularly apparent when compared to the unalloyed steel (cf. Table 2, Fig. 1b, III), as is the significant increase in hardness (cf. Fig. 2c).

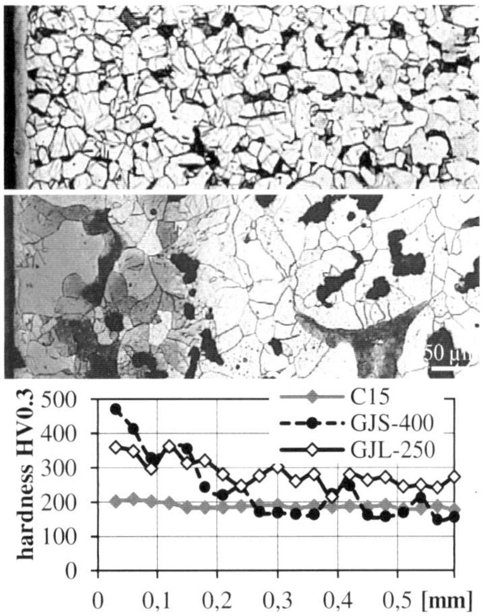

a) C15 (Ref)

b) GJS-400

c) hardness-depth profiles

Figure 2: Micrographs (etchant: Nital) and hardness-depth profiles across the diffusion layers after single treatment (GON) of different base materials.

Considering the high Si content of cast iron, all three differences mentioned resulted from the formation of Si_3N_4. This corresponds (at least in part) to investigations carried out by Meka et al. [14] using the binary alloy Fe-Si. However, the present investigations regarding cast iron (multi-component material) exhibit some differences in the nitriding mechanism.

Despite the high affinity of nitrogen to silicon, as illustrated by the very low pressure (4.8 10^{-15} bar at 550°C (825 K) [17]) compared to the iron nitrides (18 orders of magnitude higher), the interaction could be termed "weak" because of the large volume misfit with the iron matrix and the resulting inhibited nucleation of Si_3N_4 [14], as mentioned above. Consequently, this led to the strong supersaturation of the iron matrix with nitrogen over a considerable depth.

Due to the higher hardness of the CL of the nitrided GJS-400 (Table 2), it was expected that silicon nitrides would form initially. After the nucleation of iron nitrides at the surface, the resulting compound layer grew into the substrate and incorporated the Si_3N_4 precipitations. This mechanism is in contrast to that described by Meka et al. [14]. Although the formation of Si_3N_4 was not concluded, the strong nitrogen supersaturation mentioned enabled the precipitation of iron nitrides ("needles") in the upper zone of the DL during the cooling phase (A3 in Fig. 1b, II).

Tables 2 and 3 (DL) show the N and Si concentrations [at.%] at a random depth of 40 µm. At the given Si concentration of 2.1 wt.% / 3.4 at.% (Table 1, GJS-400), a N concentration of 4.5 at.% is required for the stoichiometric composition of Si_3N_4. It can be concluded, the measured high nitrogen concentration for GJS-400 of 8.7 at.% is an indication of the complete bonding of the silicon in the Si_3N_4 precipitations.

These facts are responsible for the higher hardness in both the CL and the DL.

Characterization of the nitriding mechanism in pearlitic and/or ledeburitic matrix microstructures (comparison of single and duplex treatment)

Through prior EB remelting (EBR) of the cast iron surface, graphite was nearly complete eliminated. Due to the resulting metastable solidification, a microstructure formed that consisted of pearlite (in the former primary austenite dendrites) and ledeburitic cementite (Fig. 3).

Given equal energy input, the remelted layer depth of the GJS-400 (0.88 mm) was higher than for the GJL-250 (0.75 mm), for which the higher thermal conductivity λ of GJL-250 (cf. Table 1) was deemed responsible. It resulted in a more rapid solidification and a finer morphology for the ledeburitic carbides (Fig. 3b). Apart from the carbides, the microstructural features of the intermediate areas were also important for nitrogen diffusion during nitriding.

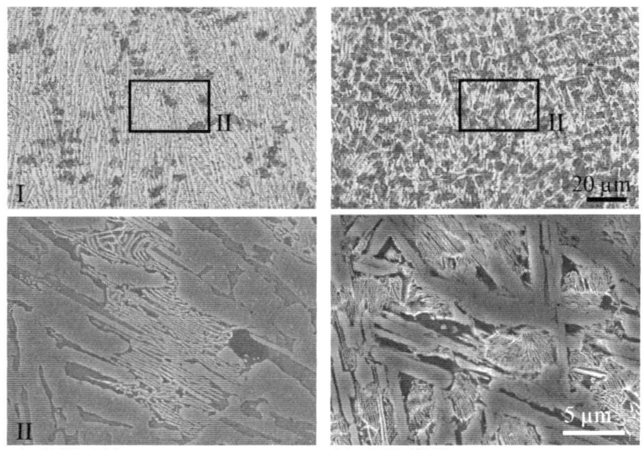

a) GJS-400 *b) GJL-250*

Figure 3: Light optical image (etchant: Nital) (I), and SEM image of ledeburite morphology after EB remelting (II).

The remelted layers exhibited a high and homogeneous hardness level of 650±30 HV0.3 over the whole remelting layer depth of 0.6 mm (after mechanical machining), as shown in Fig. 6. The heat-affected zone (HAZ) could be found underneath this layer. The pearlitic matrix microstructure (GJL-250) transformed into martensite, with an increase in hardness in this area (Fig. 6c).

The hardness of the EBR layer was not affected by the thermal cycle of the subsequent oxinitriding process (540 °C (815 K) / 16 h), which was the basis for the desired supporting effect after the duplex treatment.

a) Compound layer (CL)

After nitriding of the remelted surface (duplex treatment), the entire surface was covered with nitrides in a homogenous manner (Fig. 5b, I). At higher magnification, large cementite precipitations do not exhibit any nitrides at the surface, which would be typically formed during the first nitriding step (Fig. 5b, detail in I). Element concentration profile measured using EDX line scan across the cementite detected homogenous N profiles (Fig. 4). That suggests that only the nitride formation at the surface was inhibited but the transformation of the Fe_3C into iron nitrides took place.

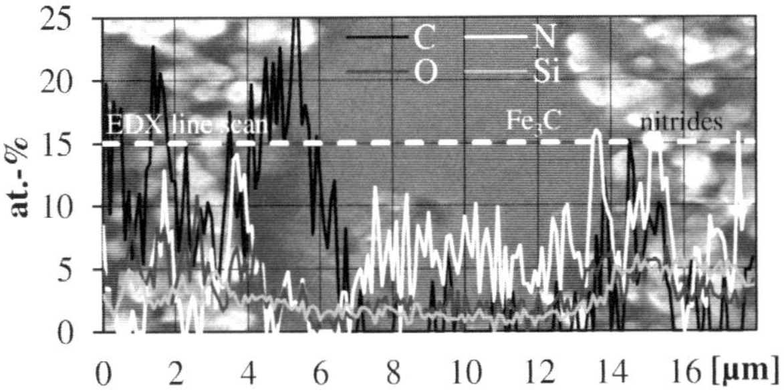

Figure 4: EDX line scan across cementite of a duplex treated surface area (GJS-400 EBR+GON).

Starting from comparable maximum nitrogen concentration levels at the surface, the nitrogen concentration within the CL decreased considerably more rapidly after duplex treatment (Fig. 5b, III) than after single treatment (Fig. 5a, III). This indicates a lower layer thickness (cf. Table 3).

Table 3: Characterization of nitrided layers after single and duplex treatment of the pearlitic GJL-250 and ferritic GJS-400.

	Features	GJL-250+ GON	GJL-250+ EBR+GON	GJS-400+ EBR+GON
CL	**CLT [µm]**	17.4 ± 3.9	10.5 ± 1.3	10.8 ± 1.1
	wt.% N_{max} (t<5 µm)	9.3	11	11
	ε/γ' [vol.-%]	76 / 24 (ε(C,N), Si_3N_4)	80 / 20 (ε(C,N), Si_3N_4)	67 / 53 (ε(C,N), Si_3N_4)
	H_{IT} [GPa]	10.4 ± 0.4	9.8 ± 0.7	12 ± 0.6
DL	**DLT [mm]**	0.16	0.07	0.10
	wt.% N / at.%N (d=40 µm)	0.9 / 3.1	1.0 / 3.5	1.6 / 5.2
	at.%Si (d=40 µm)	2.7	2.8	3.9
	SH_{max} HV0.1	360	760	760

Furthermore, a decarburization could be detected in the CL after the single as well as after the duplex treatment. However, there are both qualitative and quantitative differences (Fig. 5, III).

For the single treatment (BM+GON), starting from the diffusion front (i.e. the transition from CL to DL) the carbon content decreased continuously in the direction of the surface (min 2 wt.% C, except for the outer oxide-containing nitride layer). It corresponds to a difference of ΔC ~ 1.0 wt.% (Fig. 5a, III) and correlates with the C content of the pearlite (Fe_3C) in the matrix microstructure. This facilitates the conclusion that all carbides were transformed into iron nitrides and/or carbonitrides.

Conversely, however, a carbon accumulation occurred at the interface between CL and DL after duplex treatment (Fig. 5b, III), as known from the nitriding mechanism of alloyed steels. The carbon concentration decreased very steeply in the direction of the surface (min 0.6 wt.% C) (Fig. 5b, III).

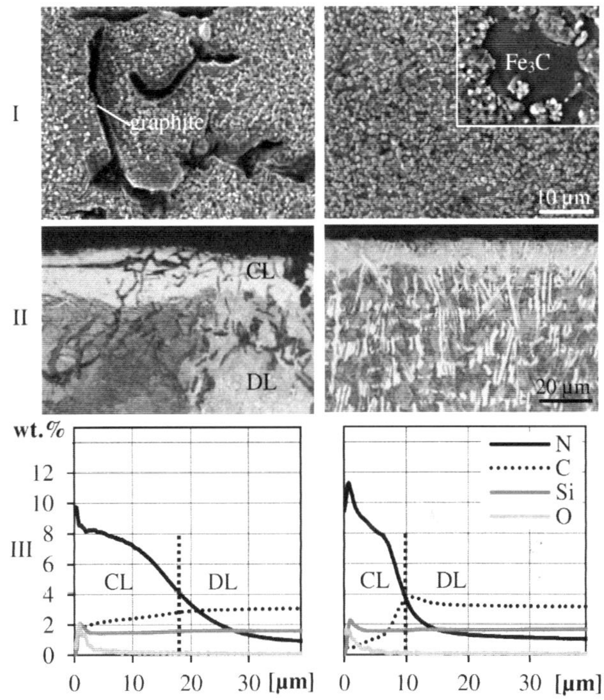

a) BM+GON b) EBR+GON

Figure 5: SEM image of the nitrided surfaces (I), cross-sections of the nitriding layers with compound (CL) and diffusion layer (DL) (etchant: Nital) (II), and element concentration-depth profiles measured by GDOES (III) of the single (a) and duplex treated GJL-250.

The nitriding mechanism in the compound layer was dependent on the local microstructure. In the pearlitic areas between the ledeburitic carbides, the same processes took place as were described for the single nitriding process of the base material. That also applied to the processes associated with the formation of Si nitride (cf. Fig. 4). Furthermore, the nitrogen diffusion inside the ledeburitic cementite was clearly limited. As a result, the CLT after duplex treatment was slightly lower than after single treatment (Table 3). Inside the cementite, a diffusion-controlled (partial) phase transformation into ε-(carbo) nitrides could be detected.

b) Diffusion layer (DL)

The investigations revealed the dependence of the initial microstructure on the remelted microstructure, on the one hand, and on the diffusion layer on the other. Whereas the GJS-400 exhibited coarse and large ledeburitic carbides (Fig. 3a), carbides of the GJL-250 were finer and shorter (Fig. 3b). EDX measurements proved the presence of Si, particularly in the intermediate areas between the carbides (cf. Fig. 4).

Fig. 6 shows the etched diffusion layers in the duplex treated surfaces. The dotted lines represent the end of the hardness increase as a criterion. Otherwise the end of the diffusion layer could not be indicated by means of etching.

Because of the graded Si_3N_4 precipitations described above, a hardness increase of approx. 150HV0.3 was obtained on top of the hardness level of the EB remelted layer (650HV0.3) (Fig. 6).

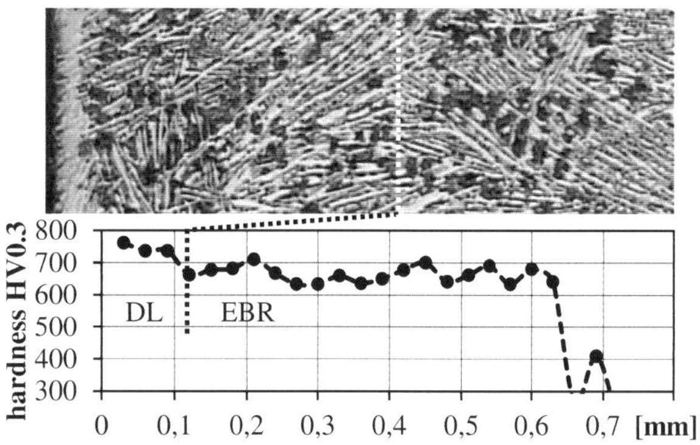

a) GJS-400

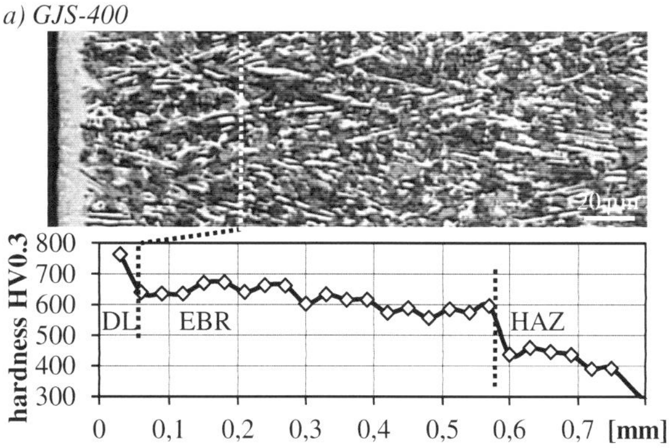

b) GJL-250

Figure 6: Micrographs (etchant: Oberhoffer) and hardness-depth profiles across the diffusion layers after duplex treatment (EBR+GON) of different cast irons.

Due to the comparable nitrogen diffusion conditions, the hardness benefit achieved for the ledeburitic microstructures (duplex treated) was comparable with that of the nitrided pearlitic cast iron GJL-250 (Fig. 2c). However, it was significantly lower than for the nitrided ferritic cast iron GJS-400 (Fig. 2c). This relation corresponded to the lower nitrogen concentration at a depth of 40 µm for the pearlitic as well as for the ledeburitic microstructures (Table 3). This means that the nitrogen content was not sufficient to achieve complete stoichiometric bonding of the silicon present to Si_3N_4. In terms of the Si concentrations (see Table 1) of the GJL-250 (1.6 wt.% / 2.8 at.%) and GJS-400 (2.1 wt.% / 3.6 at.%), approximate respective concentrations of 3.7 at.% N and 4.8 at.% N would be needed for its complete precipitation as Si_3N_4. When compared, these calculated values were lower than the actual total contents measured (cf. DL in Table 3). In addition, it should be taken into account that less than the total nitrogen content is available for silicon nitride formation. In addition, a portion of the nitrogen is dissolved interstitially, and exists as excess nitrogen. Finally, significantly higher nitrogen content is required for the complete bonding of the entire Si content, as shown for the nitrided ferritic cast iron (Table 2).

In summary and as indicated already by the poor etching behavior of the DL, the lower volume fraction of Si_3N_4 reducing the increase in hardness at greater depths (Fig. 6).

Comparative investigations regarding corrosion behavior

Fig. 7 shows exemplarily the time-dependent adjustment of the open circuit potential for single and duplex treated samples. Due to the previous cathodic cleaning procedure, an initial short (60 s) positivity of the E_{OCP} values for both materials was apparent. As the process evolved, the E_{OCP} of the nitrided cast iron (GJS-400) declined to a slightly lower level and remained constant (Fig. 7). For the pearlitic cast iron the constant level was achieved instantaneously.

In contrast, the E_{OCP} curve for the nitrided unalloyed steel C15 showed a continuous reduction of the value (Fig. 7). This could be attributed to the lack of silicon nitrides and/or carbonitrides (GJL-250) in the compound layer of the nitrided cast irons, as reported by Ebersbach et al. [19].

For the remelted and nitrided state (the duplex treatment), a displacement of the E_{OCP} curve to more positive values (cf. dotted and full lines in Fig. 7) could be observed as a result of the elimination of the graphite (Fig. 5b). Due to the similarities of the pearlitic and ledeburitic microstructures the difference is less pronounced than for the GJS-400 (Table 4).

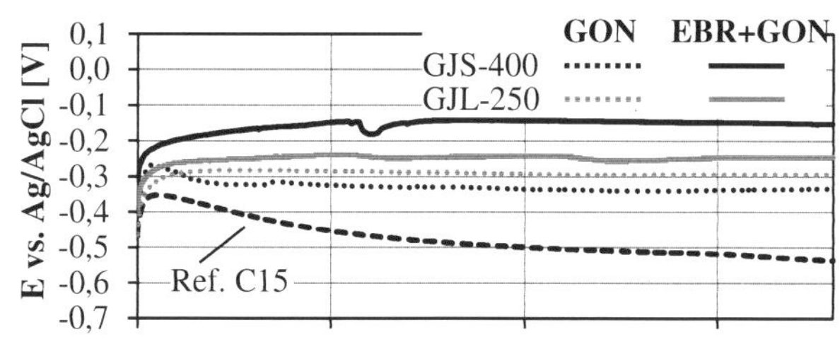

Figure 7: Time-dependent evolution of the open circuit potential E_{OCP} of single and duplex treated materials (5% NaCl).

Table 4: Selected specific corrosion values determined by means of the potentiodynamic polarization curves.

BM	State	E_{corr} [V]	i_{corr} [µA/cm²]	i_{pass} [µA/cm²]	R_p [Ω/cm²]
C15	GON	-0.59	5.1 ± 1.9	53.0 ± 4.6	7.2
GJS-400		-0.32	1.9 ± 0.8	40.9 ± 8.3	18
GJL-250		-0.30	0.3 ± 0.0	136 ± 44	107
GJS-400	EBR+ GON	-0.20	0.2 ± 0.03	13.7 ± 2.9	158
GJL-250		-0.26	0.1 ± 0.01	8.9 ± 1.4	208

Analysis of the plotted potentiodynamic polarization curves (Fig. 8) revealed an increase of all specific corrosion values for the duplex treatments when compared to the single treatment (Table 4). This means that the potentials (E_{OCP}, E_{corr}) shifted to more positive values, the current densities (i_{corr}, i_{pass}) decreased (Fig. 8b), and the polarization resistance (R_p) increased (Table 4).
The potentiodynamic polarization curves exhibited potential-dependent passivation behavior (Fig. 8). The passive area of the nitrided C15 had a broader range of existence with a lower passive current density than the nitrided GJS-400 as a result of the presence of graphite. However, the breakdown potential was independent of the material. This fact correlates with the detected corrosion phenomena.

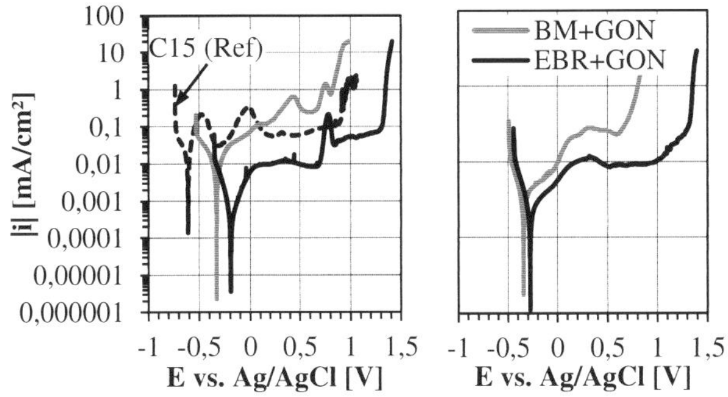

a) GJS-400 *b) GJL-250*

Figure 8: Influence of the material-specific nitriding behavior on the corrosion behavior (potentiodynamic measurements in 5% NaCl).

Comparative investigations regarding wear behavior

The wear behavior was investigated by variation of the applied load (50, 80, 120 N). In particular, the comparison between the load behavior of the nitrided (single) and the remelted and nitrided (duplex) cast irons was of interest. The material-specific load behavior was determined by the nitride layer structure and the resulting hardness gradients (Tables 2 and 3). For the merely nitrided cast irons and due to the lubricating effect of the graphite, the graphite morphology exerted an additional strong influence.
For GJL-250, the fine lamellar graphite acted as a lubrication pocket, which proved quite beneficial for the wear behavior. On the other hand, because of the limited local graphite surface area, the low hardness of the graphite did not affect the wear behavior substantially. The wear coefficient of the single treated surface (BM+GON) was very low, and exhibited only marginal dependence on the applied load (Fig. 9b).
Even at the highest normal load of 120 N, the wear depth had a range of 4.5 μm (at a maximum). Essentially, this corresponded to half of the initial CLT value, and resulted from the improved load support by the hard silicon nitride precipitations in the DL and the corresponding hardness-depth curve (Fig. 2c).
In contrast, the wear rate of the nitrided cast iron GJS-400 (BM+GON) increased progressively with increasing load (Fig. 9a).

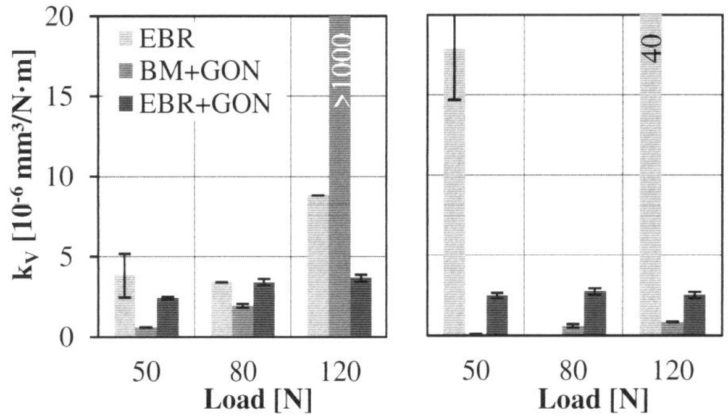

a) GJS-400 *b) GJL-250*

Figure 9: Influence of the material-specific nitriding behavior on the load-dependent wear coefficient (unlubricated pin-on-disc test).

Despite the highest hardness increase in the DL (cf. Table 2), the load support (especially for the high normal force of 120 N) was clearly limited by the low hardness of the more extensive graphite surface areas (cf. Table 1). Nevertheless, comparative investigations involving C15 reveal the superior load support of the nitrided GJS-400 because of the approx. 3-fold higher hardness in the DL. For the normal load of 80 N, the wear coefficient of the C15 ($7.8 \cdot 10^{-6}$ mm³/N·m) was approx. 4 times higher than for the GJS-400 ($1.9 \cdot 10^{-6}$ mm³/N·m). In general, the specific wear coefficient k_V of the duplex treated surfaces (EBR+GON) was almost independent of both load and material (Fig. 9). For both base materials, a very low average wear coefficient of $3 \pm 0.5 \cdot 10^{-6}$ mm³/N·m was achieved. It corresponded to penetration depths of between 3 μm (50 N) and 9 μm (120 N).
For the GJS-400 the full impact of the duplex treatment with respect to the improved load support is of benefit.

Conclusions

The understanding of the nitriding mechanism in the cast iron base material (single treatment) as well as in the EB remelted layer (duplex treatment) was the basis for optimization of the surface composite for demanding load conditions.
Compared to unalloyed steels, nitriding of untreated and previously EB remelted cast irons with high C and Si contents led to a significant increase in both the retained nitrogen content in the diffusion layer and in the hardness (CL, DL) as a result of Si_3N_4 precipitations. The hardness increase was strongly dependent on the microstructure. The greatest hardness benefit was achieved due to the complete bonding of the total Si content into Si_3N_4 precipitations for the ferritic microstructure (GJS-400). In the case of pearlitic (GJL-250) and ledeburitic (duplex treated) microstructures, only incomplete bonding of the total Si content could be confirmed, which resulted in a smaller increase in hardness.
In general, the hardness increase generated improved load support, which influenced the wear behavior. This was investigated using unlubricated pin-on-disc testing (counter body: WC-Co) and varying load conditions 50, 80, 120 N. In general the wear behavior of the duplex treated surfaces (EBR+GON)

was nearly independent of both the base material and the load, with an average wear coefficient of $3 \pm 0.5 \cdot 10^{-6}$ mm³/N·m.
By contrast, the wear behavior of the single treatments was strongly dependent on the graphite morphologies of the base materials. Even the hardness-depth profile produced for both CL and DL after nitriding of the GJS-400 is more beneficial for the load resistance and support, while the wear behavior of the nitrided GJL-250 is better. The reason for this is the fine lamellar graphite, which acts as lubrication pockets. On the other hand the low hardness of graphite was less effective.
The corrosion behavior was investigated by means of comparative potentiodynamic measurements in 5% NaCl.
All specific corrosion values of the duplex treatments that were determined were improved compared to the single treatments. It means that the potentials (E_{OCP}, E_{corr}) shifted to more positive values, the current densities (i_{corr}, i_{pass}) decreased, while the polarization resistance (R_p) increased.
In summary, the potential of the duplex treatment depends on the microstructure of the base material and the achievable properties depending on the nitriding mechanism.
In summary, it can be stated that a simultaneous improvement in both corrosion behavior and wear resistance is only possible for unalloyed cast irons by implementing a duplex surface treatment.

References

[1] Bell, T., Source Book on Nitriding, American Society for Metals (Metals Park, OH, 1977)
[2] Berns, H. and Theisen, W., Eisenwerkstoffe – Stahl und Gusseisen. Springer (Berlin, Heidelberg, New York 2008), pp. 233.
[3] Spies, H.-J. and Bergner, D.,*"Innere Nitrierung von Eisenwerkstoffen,"* HTM, Vol. 47, No. 6 (1992), pp. 346-356.
[4] Baranowska, J., "Surface quality of grey cast irons in the context of nitriding and oxygen-sulphurnitriding," *Surface and Coatings Technology,* Vol. 100-101 (1998), pp. 271-275.
[5] Buchwalder, A., Zenker, R., Rüthrich, K., Griesbach, W., Nagel, K., Hartwig, S. and Siedler, J., "Eine neue kombinierte Randschichttechnologie für hochbeanspruchte Gusseisenwerkstoffe," *HTM,* Vol. 69, No. 3 (2014), pp. 138-147.
[6] Buchwalder, A., Zenker, R., Fritzsch, K.,Rüthrich, K.; Nagel, K. and Griesbach, W., "Potenziale von ausgewählten Randschichtbehandlungsverfahren sowie deren Kombinationen zur Verbesserung des Verschleiß- und Korrosionsverhaltens von Gusseisenwerkstoffen," *Materials Science and Engineering*, Vol. 46, No. 6 (2015), pp. 550-562.
[7] Nie, X., Wang, L., Yao, Z.C., Zhang, L., and Cheng, F., "Sliding wear behaviour of electrolytic plasma nitrided cast iron and steel," *Surface & Coatings Technology*, Vol. 200 (2005), pp. 1745-1750.
[8] Spies, H. J. and Dalke, A., "Case Structure and Properties of Nitrided Steels," Krauss, G. (Ed.): Comprehensive Materials Processing; Elsevier Ltd., (2014); Vol. 12, pp. 439–488.
[9] Lightfood, B. J. and Jack, D.H., "Kinetics of nitriding with and without white-layer formation,"*Proc Heat Treatment '73*, London, 1975, pp. 59-66.
[10] Tosic, M.M. and Gligorijevi, R., "Plasma nitriding improvements of fatigue properties of nodular cast iron crankshafts,"*Materials Science and Engineering*, Vol. A140 (1991), pp. 469-473.
[11] Keller, K., "Ionitrierung von Gusseisenwerkstoffen," *HTM*, Vol. 25, No. 2 (1970), pp. 106-107.
[12] Raghavan, V., "The Fe-N-Si system," Phase diagrams of ternary iron alloys: Part I, ASM International, (Metals Park, 1987), pp. 203-205.
[13] Mittemeijer, E. J., Biglari, M. H., Böttger, A. J., van der Pers, N. M., Sloof, W. G. and Tichelaar, F. D., "Amorphous precipitates in a crystalline matrix, precipitation of amorphous Si_3N_4 in α-Fe," *ScriptaMaterialia*, Vol. 41 (1999), pp.625-630.
[14] Meka, A. R. and Mittemejer, E. J., "Abnormal nitride morphologies upon nitriding iron-based substrate," *JOM*, Vol. 65, No. 6 (2013), pp. 769-775.
[15] Meka, S.R., Bischoff, E. and Mittemeijer, E.J., Unusual precipitation of amorphous silicon nitride upon nitriding Fe–2at.%Si alloy, *Philosophical Magazine,* Vol. 92 (2012), pp. 1435-1455.
[16] Weller, M., "Point Defect Relaxations," *Materials Science Forum*, Vol. 366-368 (2001), pp. 95-137.
[17] Hütte, Taschenbuch für Eisenhüttenleute. Verlag von Wilhelm Ernst & Sohn, (Berlin 1961), pp. 83, 87.
[18] Meka, S. R., Bischoff, E., Schacherl, R. E. and Mittemeijer, E.J., "Unusual nucleation and growth of γ' iron nitride upon nitriding Fe–4.75 at.% Al alloy," *Philosophical Magazine*, Vol. 92, No. 9 (2012), pp. 1083-1105.
[19] Ebersbach, U., "Elektrochemische Korrosionsuntersuchungen an gasoxinitriertem und salzbadnitrocarburiertem Stahl in Abhängigkeit vom Aufbau der Nitrierschicht," *HTM*, Vol. 46, No. 6 (1991), pp. 339-349.

International Federation of Heat Treatment and Surface Engineering 2016
Proceedings of the 23rd IFHTSE Congress
April 18–21, 2016, Savannah, Georgia, USA

Optimization of gaseous nitriding of steels by multi-physics modelling

Dr. Sébastien JÉGOU, Prof. Laurent BARRALLIER
MSMP Laboratory, Arts et Métiers ParisTech, Aix-en-Provence, France
sebastien.jegou@ensam.eu, laurent.barrallier@ensam.eu

Dr. Guillaume FALLOT
Airbus Helicopters, Marignane, France
guillaume.fallot@airbus.com

Abstract

Gaseous nitriding is controlled by temperature, time and surface nitrogen potential. Tuning these parameters lead to the expected mechanical properties (hardness, residual stresses) and durability of treated mechanical parts. In this paper, a methodology is proposed to optimize nitriding parameters by using a multi-physics modelling of the nitriding treatment (microstructure, diffusion/precipitation mechanisms, hardness and residual stress generation). This model is particularly suited to the diffusion layer and does not apply to the compound layer. The scope concerns low-alloyed carbon steels and gaseous nitriding. The role of carbon is established and taken into account in the proposed methodology.

Introduction

High service loads, weight and cost reductions have been requesting more and more efficiency from power transmission. Process design and modelling have naturally raised significant importance in aeronautic and automotive industries in order to optimize the conception of new and complex mechanical parts.

In case of surface treatments, modelling may help optimizing process parameters in order to achieve the case depth as well as microstructure that avoid case cracking, and predicting residual stress gradients for fatigue resistance and/or macroscopic volume change (or deformation of mechanical parts) calculation. Optimization of subsequent machining may be performed in order to full fill both the final quotes of the mechanical part and desired surface mechanical properties.

Among thermochemical surface treatments of steels, gaseous nitriding is widely used because it involves low distortions of treated parts due to a lower treatment temperature (450 to 590 °C associated to slow cooling to room temperature) compared to carburizing (around 900 °C and then quenched) [1]. In return, the diffusion process at such low temperature is significantly drop off and, because of associated with phase transformations, may induce detrimental microstructure for either residual stress generation or crack propagation if not well controlled [2-4].

In case of binary iron-based alloys, gaseous nitriding is well described [5-11]. The main feature is the diffusion of nitrogen in a ferritic matrix (α-Fe). Absorption of nitrogen atoms from the outer NH_3-rich atmosphere leads to the formation of a compound layer, composed of ε-$Fe_{2\text{-}3}N$ and/or γ'-Fe_4N, followed by a diffusion layer, composed of alloying elements nitrides MN (M = Cr, V, Mo…) finely dispersed into a ferritic matrix. For carbon iron-based alloys, the diffusion layer is also characterized by the precipitation of cementite Fe_3C from the transformation of initial carbides $M_{23}C_6$/M_7C_3 [12-13]. A co-diffusion of carbon also occurs that makes more complex phase transformations kinetics at the nitrogen diffusion front where an enrichment of carbon is commonly observed [2,14]. The kinetic of the compound layer formation is more or less affected depending on the NH_3-rich atmosphere leading to mainly decarburizing of the material at low nitriding potential K_N or carbon enrichment of the compound layer at high nitriding potential [15]. Modelling of the process becomes thus complex and only few authors introduced the diffusion of carbon into modelling. Torchane *et al.* studied the influence of carbon on the kinetics of formation of the compound layer in case of ternary Fe-C-N material [16]. Depouhon *et al.* proposed a 3D modelling that allows prediction of macroscopic volume change of treated part but based on inverse fitting of diffusion and residual stress gradients in order to determine the diffusion (nitrogen and carbon) and precipitation kinetics parameters of a given treated material [17].

Based on previous developments by the present authors, this work aims to compare experimental results with a model that is able to deal with the optimization of gaseous nitriding treatment of carbon iron-based alloys [18-23]. Comparisons are performed according to several nitriding conditions (time, temperature, nitriding potential) on an industrial 33CrMoV12-9 steel grade [3].

Experimental Methods

The material of the present study is an industrial nitriding steel grade 33CrMoV12-9. Its composition is given in Table 1. Samples of 17×13×5 mm^3 were austenitized at 950 °C during 30 min and oil quenched. Annealing was performed at 590 °C during 2.5 h prior to nitriding.

Table 1: Chemical composition of the studied 33CrMoV12-9 steel grade.

Composition (wt.%)					
C	**Cr**	**Mo**	**V**	**Mn**	**Fe**
0.318	2.97	0.84	0.28	0.55	balance

Gaseous nitriding was performed in a SETSYS Evolution thermogravimetric analyser from Setaram Instrumentation. The temperature of nitriding was set to 550 °C whereas the duration and potential of nitriding were varied from 2.5 to 30 h and 0.33 to 13.77 $atm^{-1/2}$ respectively. The nitriding atmosphere was composed of a gas mixture of NH_3-N_2-H_2 using a flow rate in the furnace of 200 $mL.min^{-1}$.
The determination of nitrogen and carbon contents were achieved by optical emission spectrometry using a Spectro SPECTROMAXx MX5M BT. Four analyses are performed on four different areas (of nearly 0.5 mm diameter each by 30 µm in depth) on each treated surface. Depth profiling is obtained by successive mechanical grinding until elimination of previous sparks. Layer removal was controlled by dial indicator.
Image analysis was carried out in order to determine the surface fraction of cementite Fe_3C at grain boundaries of prior austenite. To this purpose, observations were performed on a JEOL 7001F scanning electron microscope (SEM) operating at 15 kV and equipped with an energy-dispersive X-ray detection system (EDX). Image analyses were performed on backscattered electron micrographies due to a very good Z contrast between the matrix and cementite at grain boundaries. Each image was cut in slices of 20 µm for in-depth profiling.
Hardness-depth profiles were obtained using a Leica VMHT device with a load of 2 N during 15 s. The effective depth of nitriding was industrially defined as the core hardness plus 100 HV 0.2.

Modelling of nitriding

The present modelling of nitriding is defined in one direction perpendicular to the nitrided surface and based on the coupling of diffusion in solid solution in the ferritic matrix using Fick's laws and precipitation under thermodynamic equilibrium assumption using Thermo-Calc software (version 4.0 [24] and TC-API version 7.0 [25]). This approach allows estimation of the nitrogen and carbon contents as well as the chemical composition and volume fraction of phases through the case. Calculations are performed at the nitriding temperature using Thermo-Calc Software TCFE7 Steels/Fe-alloys database version 7 [26]. From the volume fraction of each phase at each depth as well as the calculation of the volume change accompanying phase transformations, it follows that the gradient of residual stresses can be determined based on a self-consistent micro-mechanical modelling. The next sessions will only focus on diffusion and precipitation calculations. Details and parameters of the model are given elsewhere in the literature [18-23].
It has to be noticed that the model does not apply to the formation of the compound layer. However it will be shown in the next session that the growth of the diffusion layer is governed by its interface with the compound layer due to fast growth kinetics. Surface boundary conditions of the model were consequently set up from experiments to values at this interface by assuming fast establishment of a steady state of nitriding. The chemical contents of nitrogen and carbon are then set constant and equal either to 2 wt.%N and 0.2 wt.%C for high nitriding potential ($K_N > 0.33$ $atm^{-1/2}$) or 1.2 wt.%N and 0.05 wt.%C for low nitriding potential ($K_N < 0.33$ $atm^{-1/2}$).

Results

Nitrogen and carbon content in-depth profiles indicate a common trend. The longer the nitriding the deeper the case thickness is for given temperature and nitriding potential (Fig. 1). Carbon exhibits a typical co-diffusion that is characterized by, in one hand, an enrichment toward the nitrogen diffusion front and, in an other hand, a decarburization at the gas/solid interface followed by an enrichment of the compound layer for long nitriding (Fig. 2).

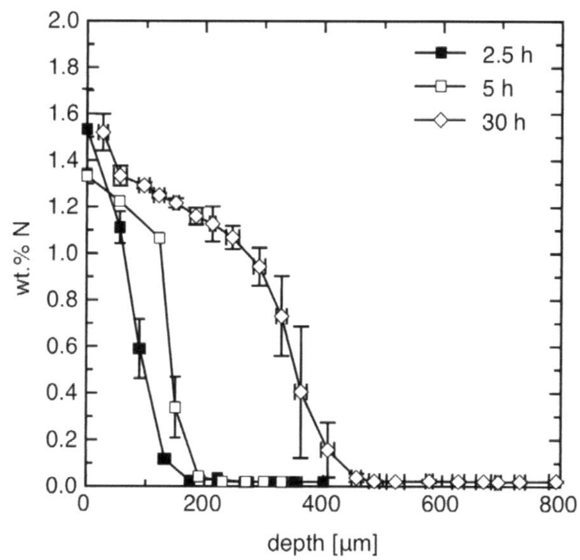

Figure 1: Nitrogen content in-depth profiles from OES analyses of a 33CrMoV12-9 steel nitrided at 550 °C and 3.65 $atm^{-1/2}$.

The evolutions of the nitrogen and carbon content in-depth profiles with the nitriding potential show that the nitrogen enrichment and decarburization of the case are strongly limited by nitriding conditions (Fig. 3 and 4). At $K_N = 0.33$ $atm^{-1/2}$ and 550 °C, the nitrogen diffusion and enrichment are non-negligibly decreased (Fig. 3). Low nitriding potential also involves deeper surface decarburization up to 100 µm below the surface against only 30 µm for higher potentials (Fig. 4).
SEM observations shows that nitriding at a low potential of $K_N = 0.33$ $atm^{-1/2}$ at 550 °C during 30 h induces no growth of a compound layer (Fig. 5.a). In the opposite, a compound layer of 11 µm thick developed at 550 °C at $K_N = 3.65$ $atm^{-1/2}$ after only 2.5 h (Fig. 5.b).
The evolution of the hardness in-depth profiles with nitriding conditions correlate observations from nitrogen content profiles (Fig. 6 and 7). The hardness levels and extent of in-depth profiles are directly related to the nitrogen diffusion and amount reacting with alloying elements to form alloying elements nitrides MN.

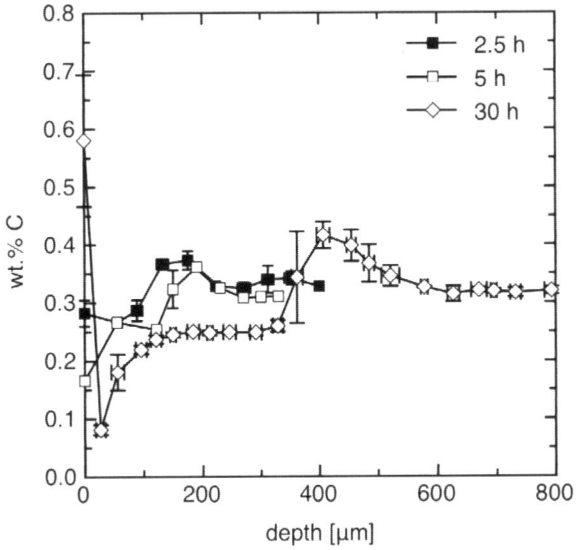

Figure 2: Carbon content in-depth profiles from OES analyses of a 33CrMoV12-9 steel nitrided at 550 °C and 3.65 $atm^{-1/2}$.

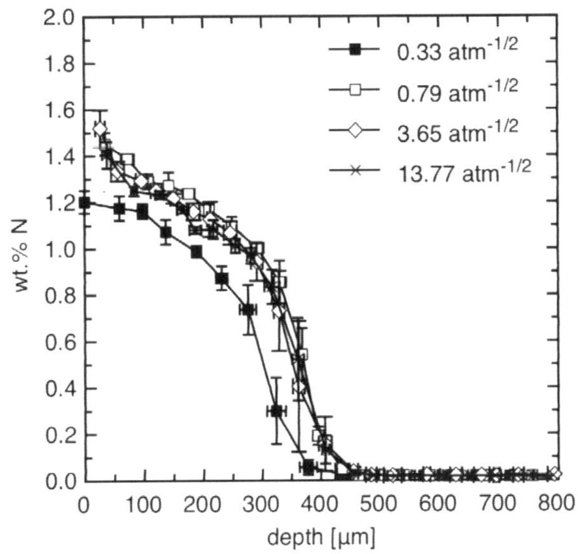

Figure 3: Nitrogen content in-depth profiles from OES analyses of a 33CrMoV12-9 steel nitrided at 550 °C during 30 h.

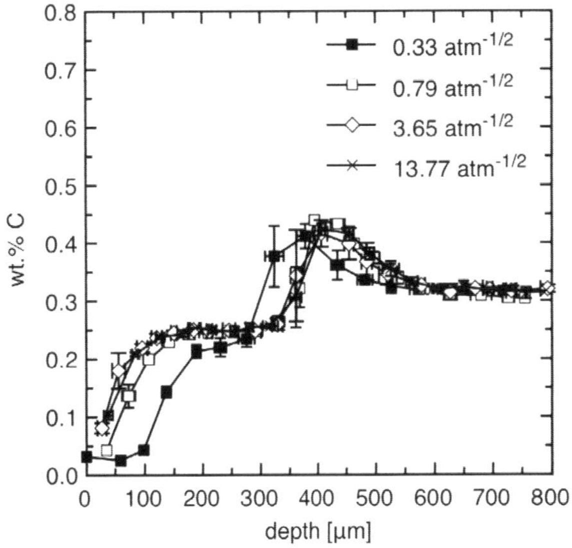

Figure 4: Carbon content in-depth profiles from OES analyses of a 33CrMoV12-9 steel nitrided at 550 °C during 30 h.

a. b.

Figure 5: SEM micrographies of a 33CrMoV12-9 steel nitrided at 550 °C and (a.) Kn = 0.33 $atm^{-1/2}$ and (b.) Kn = 0.33 $atm^{-1/2}$ during 30 h.

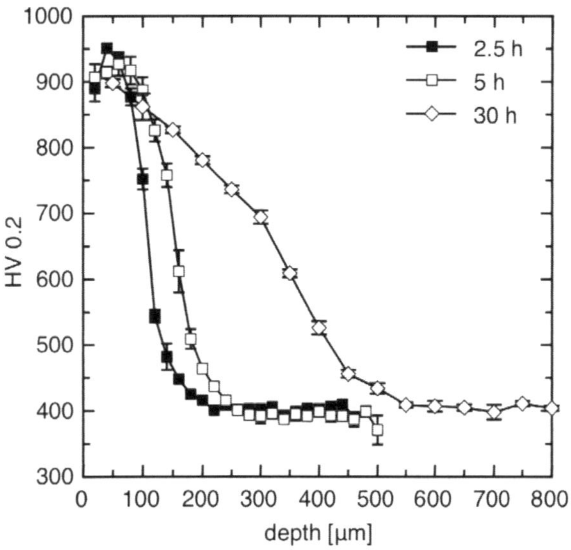

Figure 6: Hardness in-depth profiles of a 33CrMoV12-9 steel nitrided at 550 °C and 3.65 atm$^{-1/2}$.

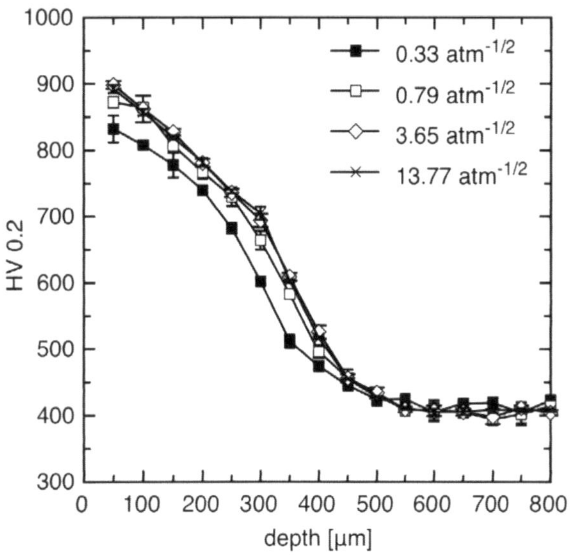

Figure 7: Hardness in-depth profiles of a 33CrMoV12-9 steel nitrided at 550 °C during 30 h.

Experimental observations show that the decarburization and growth of the diffusion layer are governed by the formation of a compound layer. The establishment of a steady state occurs in the first few hours at 550 °C and K_N = 3.65 atm$^{-1/2}$.
Plotting the effective depth of nitriding (defined as core hardness plus 100 HV 0.2) and depth where the maximum of carbon is encountered at the nitrogen diffusion front against the nitriding conditions shows a similar trend that is linear with time (Fig. 8) and nearly constant with the nitriding potential (K_N > 0.33 atm$^{-1/2}$) enabling growth of a compound layer (Fig. 9).

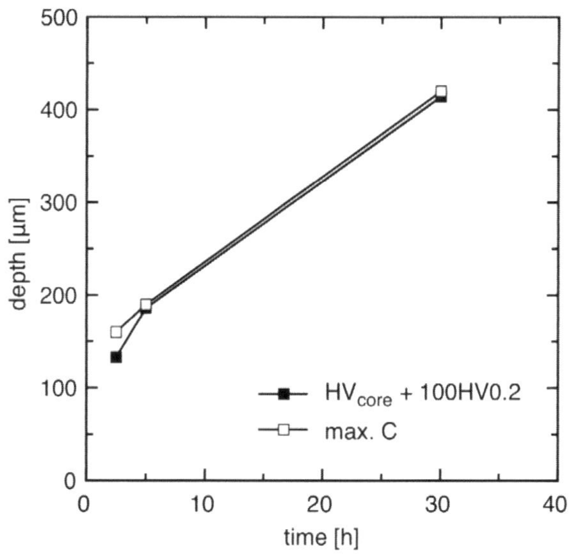

Figure 8: Relation between the effective depth and the location of the maximum carbon content of a 33CrMoV12-9 steel nitrided at 550 °C and 3.65 atm$^{-1/2}$. The effective depth of nitriding is defined as the core hardness plus 100 HV 0.2.

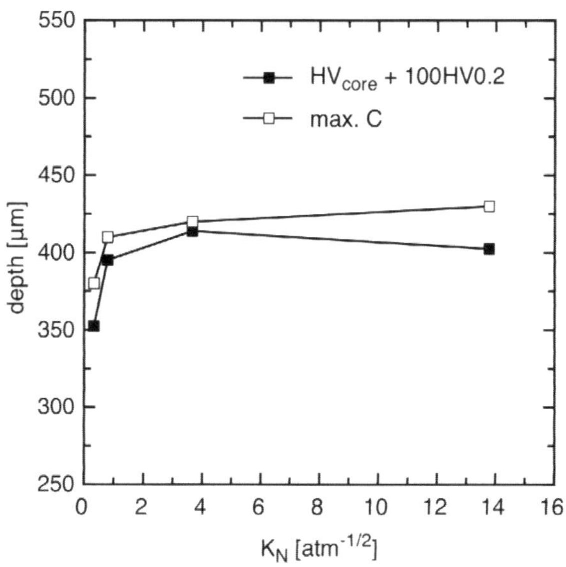

Figure 9: Relation between the effective depth and the location of the maximum carbon content of a 33CrMoV12-9 steel nitrided at 550 °C during 30 h. The effective depth of nitriding is defined as the core hardness plus 100 HV 0.2.

Comparisons with the modelling of nitriding at 550 °C and several times of treatment are given in Figures 10 and 11 respectively. Good correlations are obtained for both nitrogen and carbon. Sharp interface between the nitrogen diffusion front and the core are given by the model (Fig. 11). The carbon enrichment is also largely overestimated at the nitrogen diffusion front. They are related to the thermodynamics equilibrium assumption of the present model that is, by definition, not able to accurately track complex precipitation kinetics associated to strong diffusion gradients of both nitrogen and carbon. Results were obtained after setting up surface boundary conditions of nitrogen and carbon to fix values close to experimental contents at the outer surface for low nitriding potentials ($K_N < 0.33$ atm$^{-1/2}$ in the present case) or at the interface between the compound and diffusion layer (see § "Modelling of nitriding"). It rests on the observed fast establishment of a compound layer and steady state.

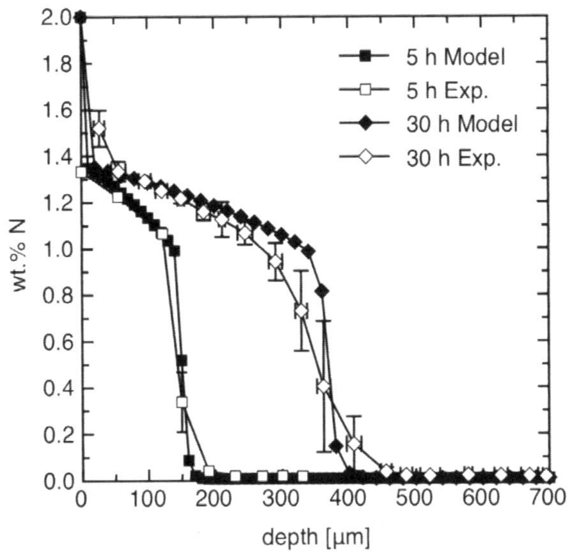

Figure 10: Comparisons between experimental and simulated nitrogen in-depths profiles of a 33CrMoV12-9 steel nitrided at 550 °C and 3.65 atm$^{-1/2}$.

Figure 12 gives a comparison between experimental and modelling content profiles in the case of a low nitriding potential that does not involve the growth of compound layer ($K_N = 0.33$ atm$^{-1/2}$). As previously, good correlations are obtained with the same limitations due to the thermodynamics equilibrium assumption. Comparisons between the depth where the maximum carbon content is found against time of nitriding from experiments and modelling are in good agreements (Fig. 13).

An interest in the present modelling also lies in the calculation of the volume fraction of each phase according to the diffusion of nitrogen and carbon. Figures 14 and 15 give the volume fraction of cementite Fe_3C as a function of the depth given by the model for two kinds of nitriding conditions (low and high nitrogen potentials). The calculated volume fraction of cementite is in good agreements with the surface fraction of cementite at grain boundaries of prior austenite grains measured by SEM image analysis.

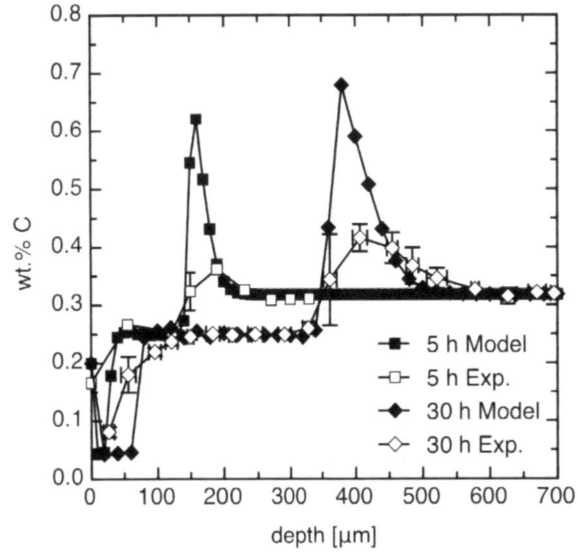

Figure 11: Comparisons between experimental and simulated carbon in-depths profiles of a 33CrMoV12-9 steel nitrided at 550 °C and 3.65 atm$^{-1/2}$.

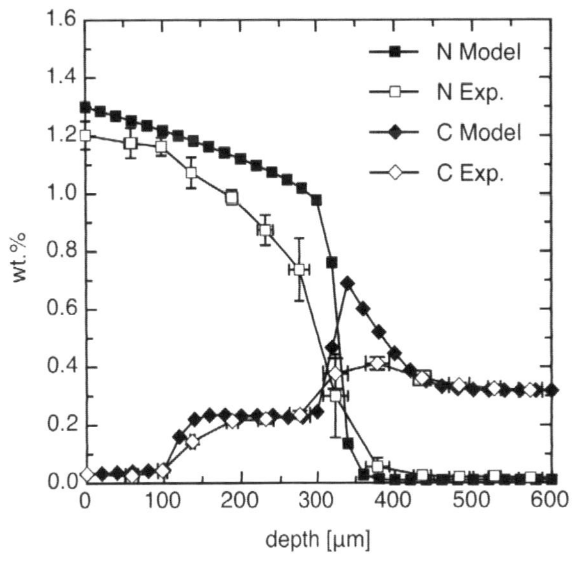

Figure 12: Comparisons between experimental and simulated nitrogen and carbon in-depths profiles of a 33CrMoV12-9 steel nitrided at 550 °C and 0.33 atm$^{-1/2}$ during 30 h.

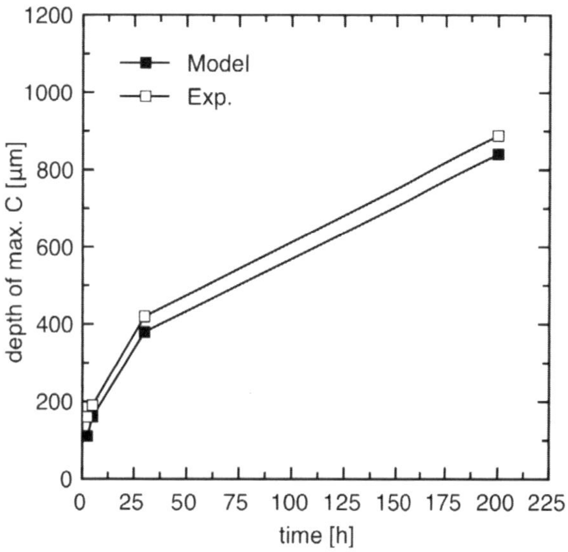

Figure 13: Comparisons between experimental and simulated depths corresponding to the maximum carbon content of a 33CrMoV12-9 steel nitrided at 550 °C and 3.65 atm$^{-1/2}$.

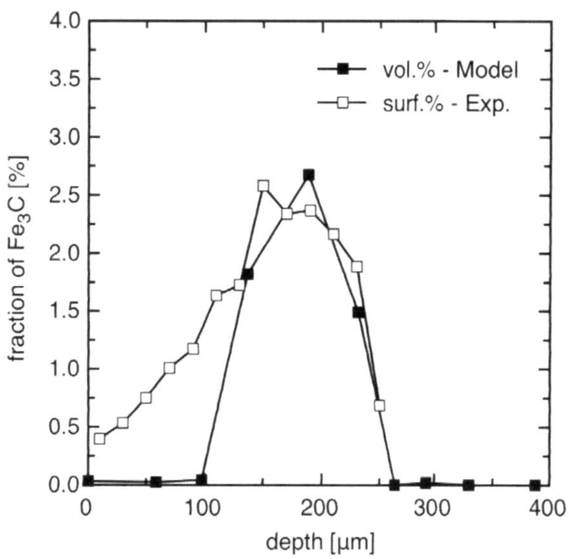

Figure 14: Comparisons between experimental and simulated fraction of cementite Fe_3C of a 33CrMoV12-9 steel nitrided at 550 °C and 0.33 atm$^{-1/2}$ during 30 h.

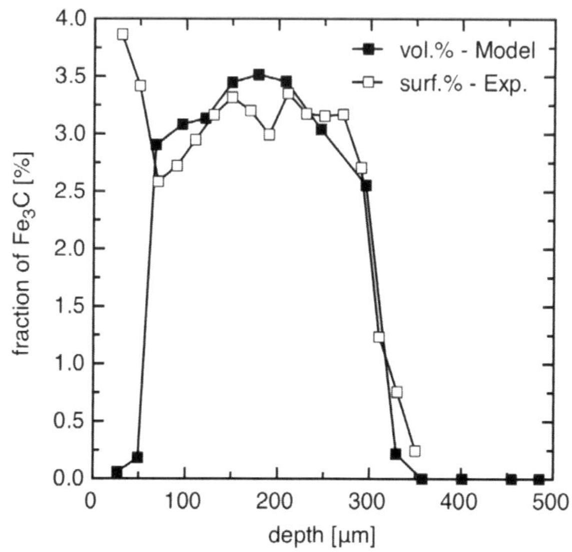

Figure 15: Comparisons between experimental and simulated fraction of cementite Fe_3C of a 33CrMoV12-9 steel nitrided at 550 °C and 3.65 atm$^{-1/2}$ during 30 h.

Discussion

Based on diffusion (Fick's laws) of both nitrogen and carbon in solid solution of a ferritic matrix and precipitation from thermodynamics calculation (Thermo-Calc package), the present model shows that a good prediction of the effective depth of nitriding as well as the fraction of phases (cementite) is obtained in case of an industrial grade dedicated to nitriding. It has to be noticed that only nitrogen and carbon surface boundary conditions were set up from experimental chemical analysis since gas/solid interactions and growth of the compound layer are not modelled. Discrepancies, mainly at the nitrogen diffusion front, are explained by the thermodynamics equilibrium assumption used for taking into account phase transformations. By definition, such assumption is not able to describe complex phase transformation kinetics due to nitrogen and carbon diffusion.

From Figures 8, 9 and 13, it follows that the nitriding conditions for a given effective depth (defined from hardness profiles) may be estimated from the present modelling based on the diffusion of carbon and its enrichment at the nitrogen diffusion front. For instance, assuming nitriding of a 33CrMoV12-9 steel at 520 °C, and using the same parameters (boundary conditions, nitrogen and carbon diffusion coefficients) set up fro nitriding at 550 °C, Figure 16 shows that good agreements are obtained between the experimental effective depth and depth where the maximum carbon content is located.

The method proposed of determining the effective depth of nitriding based on estimating the depth of the maximum content of carbon from modelling allows optimization of the treatment parameters in a process where the effective depth is the easiest measurable parameter. Coupling with residual stress calculation from precipitation modelling offers an efficient tool for time saving process development for a given application.

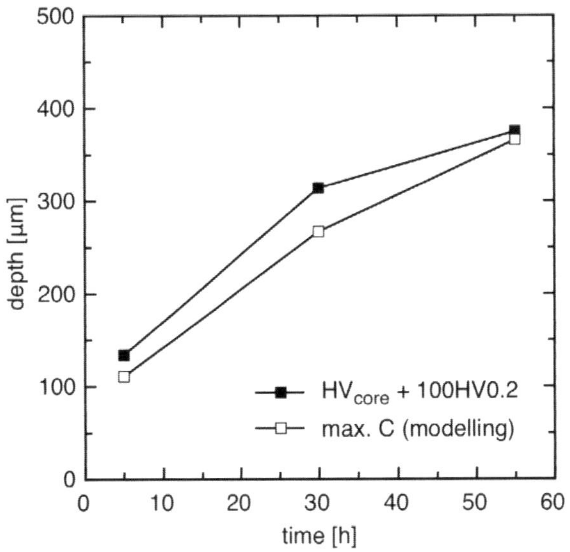

Figure 16: Comparisons between the experimental effective depth and the location of the maximum carbon content from modelling of a 33CrMoV12-9 steel nitrided at 520 °C and 3.65 atm$^{-1/2}$. The effective depth of nitriding is defined as the core hardness plus 100 HV 0.2.

Conclusions

Results from a model of nitriding of carbon low-alloyed steel was presented. Good agreements are obtained with experimental nitriding of a 33CrMoV12-9 steel grade under several treatment conditions (time, temperature and nitriding potential). The model is based on Fick's laws and thermodynamics calculations. It allows the prediction of the chemical in-depth gradients of both nitrogen and carbon as well as the fraction of phases. A method for determining the effective depth of nitriding is proposed based on consistency with the diffusion of carbon at the nitrogen diffusion front. Tuning nitriding conditions is then worth considering for a given effective depth.

References

[1] C.H. Knerr, T.C. Rose, and J.H. Filkowski, Gas Nitriding, *Heat Treating*, Vol. 4, ASM Handbook, ASM International, 1991, pp.387-409.

[2] S. Jegou, L. Barrallier, R. Kubler, M.A.J. Somers, Evolution of residual stress in the diffusion zone of a model Fe-Cr-C alloy during nitriding, *HTM J. Heat Treatm. Mat.*, Vol. 66, No. 3 (2011), pp.135-142.

[3] G. Fallot, S. Jegou, L. Barrallier, Evolution of residual stresses during short time nitriding of 33CrMoV12-9 steel grade, Advanced Materials Research, Vol. 996 (2014), pp.544-549.

[4] M. Le, F. Ville, X. Kleber, J. Cavoret, M.C. Sainte-Catherine, L. Briancon, Influence of grain boundary cementite induced by gas nitriding on the rolling contact fatigue of alloyed steels for gears, *Proc IMechE Part I: J Engineering Tribology*, Vol. 229, No. 8 (2015), pp.917-928.

[5] V.A. Phillips, A.U. Seybolt, A transmission electron microscopic study of some ion-nitrided binary iron alloys and steels, *Transaction of the Metallurgical Society of AIME*, Vol. 242 (1968), pp.2415–2422.

[6] B. Mortimer, P. Grieveson, K.H. Jack, Precipitation of nitrides in ferritic iron alloys containing chromium, *Scandinavian Journal of Metallurgy*, Vol. 1 (1972), pp203–209.

[7] K.H. Jack, Nitriding, *Heat Treatment*, No. 4 (1973), pp.39–50.

[8] M.A.J. Somers and E.J. Mittemeijer, Layer-growth kinetics on gaseous nitriding of pure iron : Evaluation of diffusion coefficients for nitrogen in iron nitrides, *Metallurgical and Materials Transactions A*, Vol. 26A (1995), pp.57–74.

[9] H. Du, J. Agren, Theoretical treatment of nitriding and nitrocarburizing of iron, *Metallurgical and Materials Transactions A*, Vol. 27A, No. 4 (1996) , pp.1073-1080.

[10] Y. Sun, T. Bell, A numerical model of plasma nitriding of low alloy steels, *Materials Science and Engineering*, Vol. A224 (1997), pp.33–47.

[11] M.A.J. Somers, Modelling nitriding of iron : from thermodynamics to residual stress, *Journal de Physique IV*, No. 120 (2004), pp.21–33.

[12] C. Ginter, L. Torchane, J. Dulcy, M. Gantois, A. Malchere, C.Eesnouf, T. Turpin, A new approach to hardening mechanisms in the diffusion layer of gas nitrided-alloyed steels. Effects of chromium and aluminium : experimental and simulation studies, *La metallurgia italiana*, Vol. 7-8 (2006), pp.29–35.

[13] J-N. Locquet, R. Soto, L. Barrallier, A. Charaï , Complete investigation of a nitrided layer for Cr alloy steel, *Microscopy Microanalysis Microstructures*, Vol. 8 (1997), pp.335-352.

[14] S. Jegou, L. Barrallier, R. Kubler, Phase transformation and induced volume changes in a nitrided ternary Fe-3%Cr-0.345%C alloy, Acta Materialia, Vol. 58 (2010), pp.2666-2676.

[15] D. Gunther, F.T. Hoffmann, T. Hirsch, P. Mayr, In-Situ measurements of residual stresses of chromium alloyed steels during a nitriding process, *ASM International 20th Heat Treating*, 9-12 October 2000.

[16] L. Torchane, P. Bilger, J. Dulcy, M. Gantois, Cinétique de croissance des couches en système polyphasé : Application aux systèmes Fe-N et Fe- N-C, *Entropie*, No. 202/203 (1997), pp.45–49.

[17] P. Depouhon, J.M. Sprauel, E. Mermoz, Prediction fo residual stresses and distortions induced by nitriding of complex 3D industrial parts, *CIRP Annals – Manufacturing Technology*, Vol. 64 (2015), pp.553-556.

[18] L. Barrallier, J.-C. Chaize, Modélisation des contraintes résiduelles dans les couches nitrurées de nuance 32CDV13, In *MAT-TEC 90, Grenoble, France*, October 1990, pp.83–88.
[19] L. Barrallier, Genèse des contraintes résiduelles de nitruration, Etude expérimentales et modélisation, PhD Thesis, Arts et Métiers ParisTech, France (1992).
[20] L. Barrallier, J. Barralis, On kinetics of gas nitriding : Case of steel, In *SMT8, Nice, France*, September 1994, pp. 621–625.
[21] S. Jegou, Influences des éléments d'alliage sur la genèse des contraintes résiduelles d'aciers nitrurées, PhD Thesis, Arts et Métiers ParisTech, France (2009).
[22] S. Jegou, R. Kubler, L. Barrallier, F. Roch, On residual stresses development during nitriding of steel: Thermochemical and time dependence. *Advanced Materials Research*, Vols. 89-91 (2010), pp.256–261.
[23] L. Barrallier, *Classical nitriding of heat treatable steel, Thermochemical Surface Engineering of Steels, Chapter 10,* Metals and Surface Engineering, Woodhead Publishing, 2014, pp.392–411.
[24] J.O. Andersson, T. Helander, L. Höglund, P.F. Shi, B. Sundman, Thermo-Calc and DICTRA, Computational tools for materials science, *Calphad*, Vol. 26 (2002), pp.273-312.
[25] Thermo-Calc Software TC-API User's Guide version 7.0, (Accessed 19 June 2014).
[26] Thermo-Calc Software TCFE7 Steels/Fe-alloys database version 7.0, (Accessed 23 Aug 2013).

International Federation of Heat Treatment and Surface Engineering 2016
Proceedings of the 23rd IFHTSE Congress
April 18–21, 2016, Savannah, Georgia, USA

In-line, High-volume, Low-distortion, Precision Case Hardening for Automotive, Transmission and Bearing Industry

Maciej Korecki
Seco/Warwick S.A. Swiebodzin, Poland
Maciej.Korecki@secowarwick.com

Emilia Wolowiec-Korecka
Lodz University of Technology, Poland
emilia.wolowiec@p.lodz.pl

Abstract

Gears and bearings are part of mechanical power transmission devices used in machinery and vehicles. They are produced on a massive scale, and for the automotive, transmission and bearing industry and global annual output reaches billions pieces.

Gears are commonly made of steel and have teeth with a hard surface and a flexible core which provides suitable strength and service life. These properties are usually obtained by heat treatment often on the basis of carburizing and quenching. In addition to the positive effect on the mechanical properties, heat treatment is related to deformation hardening, which negatively impacts the quality and significantly increases the costs of production while corrective hard machining is applied. For this reason, the aim is both to minimize distortion and to increase the precision and repeatability of the process results particularly in mass production.

A single-piece flow case hardening system based on low pressure carburizing (LPC) and high pressure gas quenching (HPGQ) will be presented. The system which allows individual adjustment to the size and shape of the particular part in order to minimize hardening distortion and ensures ideal repeatability of results throughout the parts series. It is a compact system designed for high-volume heat treatment of parts in a lean manufacturing configuration ready for implementation within machining centers.

Introduction

Global output of the automotive industry reached about 90 million vehicles in 2015. Each vehicle is fitted with a transmission gearbox. 90% of these gearboxes are traditional automatic or manual with gears being the main component. Modern transmission have multiple gears for each step as well as other parts such as drive shafts, synchronizing rings and bearings (Fig. 1). Including secondary market, global output of gears in the automotive industry is estimated to be approximately 1 billion gears per year. In addition there are millions of bearing rings which are case hardened especially for medium and large size machinery.

Fig. 1. Mechanical transmission elements.

While the automotive industry is the largest manufacturer of gears, it is not the only. There are a wide variety of gears produced for other types of machines and vehicles as well. Beyond gears, there are other steel parts that will be produced, such as several hundred million drive shafts, all requiring case hardening [1].

Case hardening by carburising is one of the most common methods of thermal, or rather thermo-chemical, treatment. It involves diffusive infiltration of carbon from the surface inwards and subsequent quenching, then tempering (Fig. 2) [2,3]. This produces a very hard and surface abrasion-resistant element, while at the same time maintaining a flexible core, capable of carrying large, impact loads. The most common steels for carburizing are the EN 16MnCr5 or 20MnCr5 (AISI/SAE 5115, 5120, 8620) grades, their equivalents, or slight modifications. Heat treating produces a typical effective case depths of 0.4-2.5 mm (0.017-0.10 in) and surface hardness of 58-62 HRC and a core hardness exceeding 30 HRC [4].

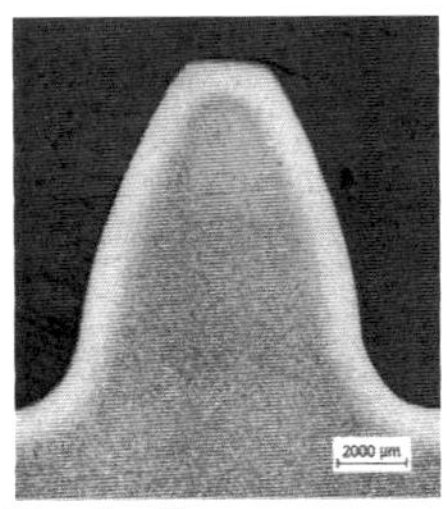

Fig. 2. Cross section of case hardened tooth.

Typically, case hardening results in the dimensional deformation of the gear caused mainly by a change of the steel structure volume resulting from austenite transformation and lack of uniformity of process parameters during the heating process, especially quenching. Distortion can also be exasperated by non-homogeneous incoming material or by stresses created during initial machining. Distortion also depends on the geometry of the part. Since the dimensional integrity of the gear, and especially the teeth, is essential for its life, efficiency, and vibration and noise reduction, any post-heat treat distortions are corrected by relevant methods, e.g., grinding. These deformation corrections are one of the most costly processes in gear manufacturing because it requires very high dimensional precision and requires costly tools for hard machining. [5-9].

An analysis of the cost for removing quenching deformations performed in 1995 by IWT (Institut fur Werkstofftechnik) Bremen for the German industry determined the cost to be approximately 850 million Euro per year in the automotive and gearbox industry and 1 billion Euro per year in the bearings industry [10]. Extrapolating this data to the global industry and taking into account the 20-year production increase, the current estimated cost to correct post-heat treatment deformation is roughly 20 billion Euro per year. It is a great burden for the industry and a great opportunity for savings.

Traditional mass production case hardening

Atmospheric Case Hardening Furnaces

The case hardening process is usually conducted in an atmospheric furnace using an endothermic atmosphere followed by an oil quench. These furnaces can vary in size from as small as a batch integral quench furnace capable of processing approximately 100 kg (220 lbs) gross per hour, up to semi- and continuous furnaces such as: pusher, rotary hearth, roller hearth furnaces capable of processing over 2,000 kg/h (4,400 lbs/h) (Fig. 3). These larger capacity continuous furnaces are typically used in the mass production automotive industry [11].

Fig. 3. The roller heart furnace.

Carburizing in endothermic atmospheres has been known for dozens of years and the process is properly controlled with oxygen probes and gas analyzers. However, it carries with it an inherent defect: intergranular oxidation (IGO), which is caused by the presence of oxidizing gases (CO_2, H_2O, O_2) in the atmosphere. Surface grinding is required to remove the dangerous IGO imperfections to the depth of as much as several dozen μm, and this removal comes at a high cost [4].

Moreover, carbon transfer in endothermic atmospheres is not especially effective (several dozen g/m^3). In order to compensate for this poor transfer, large amounts of the carbon-carrying atmosphere must be supplied to the furnace – from one to over a hundred m^3/h depending on the furnace size and surface to be carburized.

The overall cost of heat treating in an endothermic atmosphere is relatively high. Fuel is consumed not only to create the atmosphere but additional fuel is consumed to bring and keep the newly introduced atmosphere up to the appropriate temperature within the furnace. Continually introducing and heating the atmosphere consumes additional energy. This and the fact that much of the atmosphere is exhausted contributes to the systems overall inefficiency.

Endothermic atmosphere carburizing also carries with it the inherent risk of fire and explosion and numerous costly safety procedures must be followed to mitigate the risk, not to mention the environmental issues associated with emissions [12-14].

Quenching, which in most cases is done in oil, is an important stage of the case hardening process. Oil is a commonly known quenchent with known disadvantages arising from the 3-phase nature and speed of quenching and the uncontrollability of the process. If one takes into account that each phase appears at a different time in different places on the part being quenched, large and unique deformations associated with oil quenching are obvious. Additionally, parts need to be washed after quenching in special washers where chemicals are used which are increasingly problematic with respect to environmental regulations.

Continuous furnaces can occupy hundreds of square meters of valuable floor space. Because of the large size of these furnaces, changing the process parameters and stabilizing the working environment takes hours. Switching these furnaces on/off takes weeks. Any interruption of the production process results in huge energy and production losses.

Vacuum Case Hardening Furnaces

In response to the above listed weaknesses of traditional case hardening technologies, low-pressure carburizing (LPC) and high-pressure gas quenching (HPGQ) technologies were developed in the 1970's/1980's. By the 1990's, the technologies had been perfected enough to find industrial applications, including mass production industries like the automotive industry. Currently, LPC and HPGQ carried out in vacuum furnaces, have effectively replaced traditional atmosphere furnaces. These new LPC/HPGQ furnaces include batch, semi-continuous, and modular systems for mass production (Fig. 4, 5) [15].

Low-pressure carburizing overcame the weaknesses of endothermic carburizing and added a range of new possibilities. IGO was eliminated because of the total absence of oxygen in carburizing gases. Carbon transfer effectiveness increased about 20 times, considerably decreasing atmosphere consumption. Moreover, the atmosphere in the furnace is neither flammable nor explosive due to its low density. LPC is also characterized by the extraordinary capability of the atmosphere to penetrate and uniformly carburize parts with

Fig. 4. Semi-continuous, vacuum, three-chamber, LPC with oil quench furnaces.

Fig. 5. Multi chamber, rotary system, LPC with HPGQ.

shapes which make access difficult (blind holes) and densely packed loads. In addition, the processing temperature can be raised considerably resulting in shortened carburizing time – as much as 4-5 times when the temperature is raised from 925°C (1697°F) to 1040°C (1904°F) (Tab. 1.) [16].

Case depth [mm] [inch]	Carburising time					
	925°C	950°C	980°C	1000°C	1020°C	1040°C
0.50 0.02	1h23m	0h57m	0h39m	0h30m	0h24m	0h19m
1.00 0.04	5h30m	3h50m	2h35m	2h00m	1h35m	1h15m
2.00 0.08	22h00m	15h10m	10h20m	8h00m	6h10m	4h50m
Time relations	100 %	69 %	47 %	36 %	28 %	22 %

Tab.1. Carburising time of 16MnCr5 for the given case depth with criterion of 0.35%C, depending on the temp.

Furthermore, HPGQ significantly improved the outcome of the quenching process by decreasing the deformation rate. Gas quenching is a single-phase process and a more uniform process with respect to a single part. Moreover, quenching rates can be regulated freely by a change of gas pressure (density) and velocity (fan rotation), thereby making the process totally controllable. Modern HPGQ systems with nitrogen or helium under 25 bar pressure are equivalent to oil quenching. As an added benefit, gas quenching eliminates the process of washing making it a much more environmentally friendly process [17].

LPC/HPGQ vacuum furnace systems are compact, energy efficient and environmentally friendly devices. They are flexible, can be switched on and off at any time, and take about 1 hour to make production ready. Moreover, they do not require atmosphere stabilization and process parameters can be changed virtually instantaneously. The case hardening technology implemented in modern vacuum furnaces is a precise, efficient, clean and environmentally friendly process [18-23].

Characterization of Batch Case Hardening

Although there are many benefits to these new technologies (LPC & HPGQ), there is one feature that remains unchanged. Even in new LPC/HPGQ vacuum furnace systems, parts are configured and processed in batches on special fixtures (Fig. 6) and undergo the whole case hardening process in such a configuration. This means that each part in a batch is affected by the process conditions in a unique manner based on its position within the batch. Each part is affected differently regarding the heating rate, composition of the process atmosphere, and intensity and direction of cooling medium. There is no doubt that the parts in outer layers of a batch are heated more quickly and to a different temperature (according to the temperature distribution within the batch), the atmosphere around them is "richer" and they are quenched more intensely compared to the parts toward the center of the load. The result is that parts inside the batch have different physical and metallurgical properties than those on the outside of the batch, e.g., surface and core hardness, microstructure, and especially the effective case depth [24-26].

Fig. 6. Example of gears coniguration in a batch.

In an analysis of the effect of the carburizing temperature on the effective case depth, a typical parameter of the temperature uniformity was adopted: +/- 6°C (+/- 10°F) for a class II furnace in accordance with AMS 2750E. Taking as an example the temperature of 950°C (1742°F) for a 0.75 mm (0.03 in) case (for the 0.35%C criteria), the effective case depth obtained at 944°C (1731°F) was 0.72 mm (0.03 in), whereas at 956°C (1753°F), 0.78 mm (0.033 in). The difference between the effective case depths is 0.06 mm (0.003 in) (8%) for the temp. dispersion of +/- 6°C (+/- 10°F).
Similarly, the composition of the carburizing atmosphere is not uniform and it changes as the atmosphere moves toward the center of the batch with the amount of carbon decreasing gradually and, in consequence, poorer carburization of the parts at the center of the batch. The differences in the atmosphere composition (carbon potential) can be as great as 10% and the case depth can change accordingly.
The cooling rate during the quenching process also affects the case depth with a faster quench increasing the case depth and a slower quench decreasing the case depth. Non-uniformities of the cooling medium flow rate through the batch can be as high as 50% and can also significantly impacting the case depth by several more percent [17].
Due to the compounding effect of these varying parameters inherent in batch processing, it is not surprising that the industry's quality expectations are very liberal. Tolerances can be as high as 50% (0.6-0.9 mm or 0.025-0.038 in), a direct consequence of batch processing and the currently accepted tolerances of the batch processing case hardening systems.
The intensity and uniformity of parts quenched in batches is a separate issue. Intensity of quenching determines the hardness of the core and the effective case depth, and the uniformity determines the dispersion of those parameters across parts within the batch and, what is most important, the size of quenching deformations on individual parts. Batch quenching has one important common feature. Regardless of the device or the cooling agent (liquid, gas), batch quenching is a single-direction cooling (what we will call "2D"). This means that an individual part is affected by the cooling agent which flows in one specific direction (from top to bottom or from left to right, etc., with the flow not necessarily linear). In consequence, the part cooling rate is different in different places on the part. Non-uniform quenching results in temperature gradients within each part resulting in thermal stresses and a non-uniform transformation of the microstructure. This ultimately results in large deformations of the part being quenched. Quenching results are made even worse by the fact that the quenching stream within the batch is dispersed and each part is cooled differently based on its position within the batch. A critical summary of batch 2D quenching (especially oil quenching) shows that it is an uncontrolled and non-uniform process producing great deformations within each part and little consistency within the batch. Using gas as the quench media can reduce deformations within each part compared to oil due to the single-phase nature of convection cooling with gas, but all the variations within the batch still remain [27].
Batch processing also has other quality, material handling, and cost pitfalls. For example, monitoring and reporting on the case hardening process is for the entire batch and not for individual pieces within the batch. That makes it difficult, or even impossible, to introduce and implement tighter quality standards.
Material handling of batch loads is typically complicated and costly. Gears are produced individually. After being shaped, they are collected, packed, protected, and transported to the case hardening department (captive) or to an external plant (commercial) which can range from hundreds of meters to hundreds of kilometers away. The gears are then unpacked, washed and racked to form batches on fixtures designed specifically for the case hardening process. Following an oil quench, the parts are washed again, dismantled, packed, protected, and transported back to the mechanical processing department. The whole undertaking may be divided into more than ten operations and take days. These material handling costs consume considerable resources including time, materials, and money.
It is critically important to have special fixtures for case hardening and these fixtures are typically made of heat-resisting Ni-Cr alloys. The price of a medium-sized fixture starts at thousands of Euros and the lifespan ranges from 2 to 4 years. Considering the fact that the largest installations require over 100 sets to ensure continuity of the production process, the cost of fixturing alone can reach thousands of Euros a year. The fixturing also affects the amount of direct energy consumption costs because it accounts for 40-50% of the overall batch weight, thereby generating higher consumption of energy needed to heat up and cool down parts.
The current technology must be appreciated for its huge productivity and minimal per-piece cost. The technology has been used and improved for decades and sufficiently mastered. Because of the nature of the process, the ability to further develop and improve the process has hit a wall. Specifically, the increasing needs and expectations of the industry in regard to per-piece quality improvements and reporting, repeatability, flexibility, speed, process continuity, reduction of the total costs of production and neutrality to the environment cannot be satisfied with traditional methods.
A new approach to case hardening technology is needed. This new technology will not only fill the existing gap, but will provide a solid foundation for future development. Specifically, a new approach is needed that will:

- Improve accuracy, precision and repeatability
- Reduce quenching deformations
- Integrate case hardening equipment into continuous production lines.

New approach to case hardening

Single-Piece Flow Case Hardening
The vast majority of weaknesses and limitations of the current case hardening processes are associated with its batch-related nature. Therefore, to eliminate these weaknesses and limitations, it would be ideal if batch processing was eliminated and replaced with a continuous, single-piece flow model. The single-piece flow concept has been around for some time in theory, industry articles, lectures and

presentations [28,29]. Various systems, more or less in line with the idea, have been developed, but no device for mass thermal treatment has been constructed that would fully embody the idea to-date. Single-piece flow processing should mean that every single piece goes through the exact same positions and process conditions as every other piece. A system where parts are placed on trays, even in single layers, or when parts are processed individually in different process chambers, does not meet the criteria of a single-piece flow system.

UniCase Master®

Fig. 7 shows a vacuum furnace for case hardening of gears or rings using LPC and HPGQ. This system fully meets the criteria of a single-piece flow method and has all the accompanying advantages. The furnace consists of three horizontal chambers: the first one for heating up, the second for low pressure carburizing, and the third for diffusion and pre-cooling before quenching. Additionally, a separate loading chamber and a quenching chamber that doubles as an unloading chamber are connected. Parts are transported between chambers by two vertical transport elevators attached to each sides of the system.

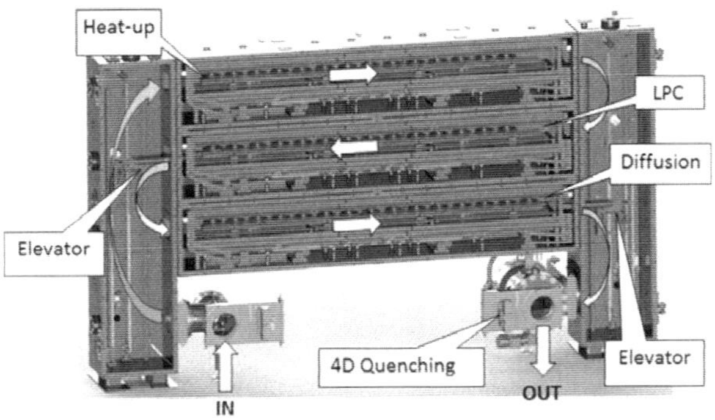

Fig.7. The Single piece flow, case hardening system by LPC and HPGQ

The single-piece flow process runs in the following manner:

- a single gear is placed inside a loading chamber,
- it is then transported and loaded into the heating chamber,
- a walking beam mechanism indexes the gear through all the positions until the gear reaches the target carburizing temperature,
- the gear is then transported and indexed through the LPC where the surface is saturated with the right amount of carbon,
- the part is then transferred and indexed through the diffusion chamber where desired carbon profile is achieved and the temperature is decreased before quenching,
- the gear is then transferred to the gas cooling chamber where it is quenched,
- the gear is then removed from the quenching chamber and is ready for temper.

Each gear follows in sequence and is processed the exact same way, with the exact same process parameters, guaranteeing the highest level of precision and repeatability.

Precision and repeatability improve because each part runs individually and is exposed to the exact same temperature and atmosphere. Single parts heat up more quickly and more uniformly due to direct radiation and the beneficial effects this direct radiation has on the more consistent conduction of heat within the gear. Moreover, any temperature gradients within the chamber are neutralized because each part goes through all the same positions. It is possible to achieve temperature uniformity as tight as +/-1°C. The same is true of the process atmosphere which directly reaches the part surface consistently. Even if the atmosphere composition varies in different places within the chamber, the average value after each gear has gone through all positions is the same. Considering the above, it is conceivable to achieve a uniform and repeatable carburized case depth in a single part and consistently throughout the batch as good as of 5% spread.

4D Quenching®

UniCase Master® also allows for significant improvements in the quenching process, specifically the reduction of distortion. This is done primarily using a high-pressure gas quenching system installed in the quenching/unloading chamber (Fig. 8).

Fig.8. 4D Quenching® chamber.

The system utilizes a proprietary arrangement of cooling nozzles that surround the part and ensures a uniform flow of cooling gas from all sides; top, bottom, and side. We refer to this as "3D" cooling. In addition, a table spins the part further enhancing quench uniformity. We refer to the spinning motion as the fourth dimension allowing us to "4D" quench gears for the best possible uniformity. The cooling nozzles allow us to achieve 15 bar quenching results – comparable to oil quenching – without the use of helium (He). Because the cooling nozzles can be adjusted to exactly fit the gears precise size, quenching is optimized and distortion significantly decreased. It is anticipated that, compared to oil quench systems, quenching deformations rates for each piece and across the entire batch will easily be cut in half.

Technical and Technological Assumptions

The UniCase Master® system was designed using the following criteria for the automotive industry:

- Part diameter/high 200/50 mm (8.0/2.0 inch)
- Weight 3.0 kg (6.6 lbs)
- Tact 60 sec

Number of parts 15 (in each process chamber)

A gear with a diameter of 200 mm (8 in) and weighing 3 kg (6.6 lbs) covers all the gears in passenger car and most truck gearboxes currently manufactured. The optimum duration of the loading-unloading cycle was 60 seconds which corresponds well to the production cycle of CNC machine tools. An analysis of the part's heating rate and the process of carburizing for 0.4-1.0 mm (0.02-0.04 in) case, required process chambers with 15 index positions. As a consequence, each part has a residence time of 15 minutes in each process chamber. It has been proven that parts can be heated-up in 15 min then the processes of carburizing and diffusion at 1040°C (1904°F), where a carburized case of 0.6 mm (0.025 in) can be obtained during 30 minute cycles. The example above is only one example. Other cycle durations and process temperatures can be used to obtain other required case depths.

Quenching of a single part in 4D Quenching® system is very effective. Having a heat transfer coefficient with nitrogen or helium at the range of 1000-2000 W/m^2K, most of the parts can be quench in a less than 60 sec.

Lean Manufacturing

UniCase Master® is intended to be installed and operated directly on the manufacturing floor next to a CNC machine and was designed so that its footprint was similar to a CNC machine (Fig. 9). It can be installed on new production floors or at sites previously occupied by other machines, including CNC machines. A newly machined gear can be introduced into and released from the case hardening system every 30-60 seconds. The system can be completely integrated into the continuous, lean production manufacturing line, thus eliminating many, if not all, batch material handling steps.

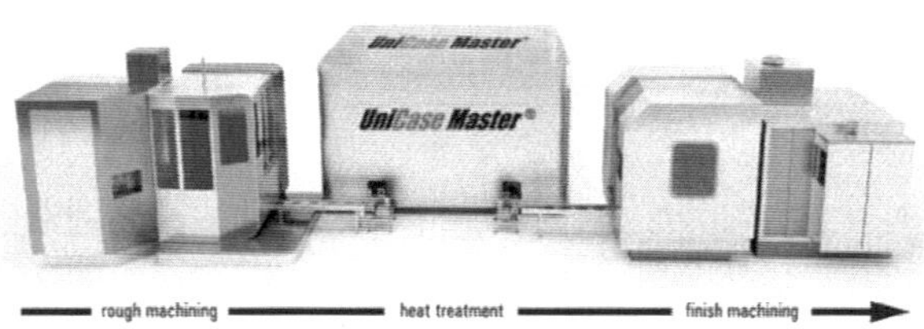

Fig.9. The single piece flow, case hardening system integrated into in-line manufacturing.

It's worth to notice that the system does not use fixtures for load racking. As noted previously, this helps reduce operating costs including the cost to purchase and replace fixtures as well as the consumption of energy.

Productivity of Carburizing

For a 200 mm (8.0 in) diameter gear indexing 15 steps within each process chamber, a gear with an effective case depth of 0.6 mm (0.025 in) can be produced every 60 seconds. The gear will spend 15 minutes in each chamber. This results in approximate production flows of:

- 1 gear/minute
- 60 gears/hour
- 1400 gears/24 hours
- 10,000 gears/week
- 43,000 gears/month, and
- approximately 500,000 gears/year with the continuous operation mode.

For gears with a smaller diameter of 100 mm (4 in) and the same process requirements, system can be configured for 30 positions and the cycle reduced to 30 seconds in each position resulting in annual output of approximately 1 million parts.

Depending on process requirements and gear production, output of the UniCase Master® system can change within a wide range, achieving several hundred thousand to as many as a million gears per year.

Conclusions

Current best available case hardening technologies, both batch and continuous, fail to meet current and future expectations of high-volume automotive gear manufacturers in terms of quality, repeatability, flexibility, and integration of production as well as environmental-friendliness, and costs, mainly because parts are arranged as batches on special fixtures for thermal treatment.

The UniCase Master® system, developed to be a real single-piece flow case hardening system, eliminates the shortcomings of the current systems and is now the new "best available technology". It meets the current and future requirements of high-volume automotive gear manufacturers and has the following features:

- Extreme precision and repeatability,
- Reduction of quenching deformations due to 4D quenching,
- Integration into continuous production lines – lean manufacturing,
- High productivity for high-volume gear manufacturers,
- Flexibility and operational speed,
- Individual part monitoring and reporting (as opposed to batch monitoring/reporting),
- Elimination of fixtures (cost, energy) and batch material handling logistics,
- Elimination of quench oils, washers and washing fluids,
- Elimination of fire and explosion hazards,
- Cleanness of the process, no effect on the environment,
- Applying cutting edge LPC and HPGQ technologies,

These properties and advantages of the UniCase Master® system revolutionize case hardening technology and pave the way to meet the increasingly stringent quality and production requirements of high-volume gear manufacturers.

Acknowledgments

The research and publication were financed by the National Centre for Research and Development as part of project no. POIR.04.01.04-00-0087/15 entitled: "Equipment for high performance and precise heat treatment with a quenching deformation reduction system for direct application in downstream production chains of mechanical gearing and bearings."

References

[1] Vasilash G. "The Auto Industry & Gear Making." *Modern Machine Shop, Vol. 9*, 2013.
[2] Marinov V. "Manufacturing Technology." 2004.
[3] Kohara Gear Industry Co. "Introduction to Gears." *Nakacho*, 2006.
[4] Parrish G. "Carburizing: Microstructures and Properties." *ASM International, Materials Park*, 1999.
[5] Krauss G. "Steels: Processing, Structure and Performance." ASM International, Materials Park, 2005.
[6] Herring D. "Grain Size and its Influence on Materials Properties." *Industrial Heating, Vol. 8*, 2005, 20-22.
[7] Herring D. "Understanding Component Failures. Part 1: Mechanisms." *Industrial Heating, Vol. 7*, 2013, 16-18.
[8] Herring D. "Understanding Component Failures. Part 2: Analysis Methods. " *Industrial Heating, Vol. 8*, 2013, 18-20.
[9] Wulpi D. "Understanding How Components Fail." *ASM International, Materials Park*, 2013.
[10] IWT (Institut fur Werkstofftechnic) "Economic consequences of distortion." *Bremen*, 1995.
[11] Herring D. "Atmosphere Heat Treatment, Volume I." *BNP Media Group*, 2014.
[12] Chandler H. "Heat Treater's Guide: Practices and Procedures for Irons and Steels. " *ASM International, Materials Park*, 1995.
[13] Herring D. "What Happens to Steel during Heat Treatment? Part One: Phase Transformations. " *Industrial Heating, Vol. 4*, 2007, 12-14.
[14] Herring D. "What Happens to Steel during Heat Treatment? Part Two: Cooling Transformations." *Industrial Heating, Vol. 6*, 2007, 12-14.
[15] Herring D. "Vacuum Heat Treating. " *BNP Medial Group*, 2014.
[16] Korecki M., Kula P., Olejnik J. "New Capabilities in HPGQ Vacuum Furnaces." *Industrial Heating, Vol. 3*, 2011.
[17] Korecki M., Olejnik J., Szczerba Z., Bazel M. "Single-chamber HPGQ Vacuum Furnace with Quenching Efficiency Comparable to Oil." *Industrial Heating, Vol. 9*, 2009, 73-77.
[18] Herring D. "Technology Trends in Vacuum Heat Treating. Part One: Markets, Processes and Applications." *Industrial Heating, Vol. 10*, 2008, 83-88.
[19] Herring D. "Technology Trends in Vacuum Heat Treating. Part Two: Processes and Applications." *Industrial Heating, Vol. 11*, 2008, 57-62.
[20] Herring D. "Technology Trends in Vacuum Heat Treating. Part Three: New Technologies and Future Developments." *Industrial Heating, Vol. 1*, 2009, 44-46.
[21] Kula P., Pietrasik R., Dybowski K., Korecki M., Olejnik J. "PreNit LPC - the Modern Technology for Automotive. " New Challanges In Heat Treatment and Surface Engineering, 2009, Dubrownik-Cavtat, Croatia, pp. 165-170.
[22] Korecki M., Kula P., Wołowiec E., Bazel M., Sut M. "Low Pressure Carburizing (LPC) and Low Pressure Nitriding (LPN) of Fuel Injection Nozzles made of Tool Steel." IFHTSE 2013, Jun. 11th-14th, 2013, Dubrovnik-Cavtat, Croatia, 69-76.
[23] Korecki M., Bazel M., Sut M., Kula P., Wolowiec-Korecka E. "LPC and LPN of Tool Steel Fuel Injection Nozzles," *Industrial Heating, Vol. 6*, 2014.
[24] Kula P., Atraszkiewicz R., Wołowiec E.: "Modern Gas Quenching Chambers supported by SimVaC Plus Hardness Application." *Industrial Processing, Vol. 3*, 2008, 55-58.
[25] MacKenzie S.D., Li Z., Ferguson B.L. "Effect of Quenchant Flow on the Distortion of Carburized Automotive Pinion Gears." International Federation for Heat Treatment and Surface Engineering—IFHTSE, 5th International Conference on Quenching and Control of Distortion, European Conference on Heat Treatment 2007, Berlin, 25–27 April 2007.
[26] MacKenzie D.S., Kumar A., Metwally H., Paingankar S., Li Z., Ferguson B.L. "Prediction of Distortion of Automotive Pinion Gears during Quenching Using CFD and FEA", *Journal of ASTM International, Vol. 6, No. 1*, 2009, 1-10.
[27] MacKenzie D.S., Totten G.E., Gopinath N., "Quenching Fundamentals—Effect of Agitation." Proceedings 10th Congress of the IFHT, T. Bell, and Mittemeijer, E. J., Eds., IOM Communications Ltd., London, England, 1999, 655–669.
[28] Heuer V., Löser K., Schmitt G., Ritter K. "One Piece Flow- Integration of Case Hardening into the Manufacturing Line." Conf. on Gears, Oct. 4th-6th, 2010, TUM Garching, Germany.
[29] IHI "In-Line Heat Treatment – Next Generation Heat Treatment Equipment," *IHI Engineering Review, Vol.44, No.2*, 2011.

International Federation of Heat Treatment and Surface Engineering 2016
Proceedings of the 23rd IFHTSE Congress
April 18–21, 2016, Savannah, Georgia, USA

Press Quench Process Design for a Bevel Gear using Computer Modeling

Zhichao Li, Andrew M. Freborg, and B. Lynn Ferguson
DANTE Solutions, Inc.
7261 Engle Road, Suite 105, Cleveland, OH 44130
Charlie.Li@Dante-Solutions.com

Peiran Ding, and Mike Hebbes
ANSYS, Inc.
900 Victors Way, Ann Arbor, MI 48018

Abstract

Press quenching is used to harden steel gears for tight dimensional control. The process is especially used when quenching thin-wall gears, face gears, and bevel gears with large dimensions. The dimensional control aims at maintaining flatness, circularity, and radial dimensional changes. The press quench process tooling design has been mainly based on experience, with trial and error used for implementation of new processes, new gear materials and configurations. Computer simulation provides a means for eliminating or at least reducing the number of try-outs that are needed to achieve the desired process results. In comparison to traditional oil quenching, the oil flow pattern and cooling rates along different gear surfaces in press quenching are more dependent on the tooling design, oil channels, and oil pump rate, and these should be considered in design of the process. Other process parameters include the die and expander loads and the locations of load application. In this paper, the gear responses during a press quench of a spiral bevel gear made of carburized AISI 9310 are assessed for reduced distortion using the commercial heat treatment software DANTE, linking with ANSYS Mechanical solver. Radial shrinkage and taper of the internal spline are observed from the trials, and the causes of the distortion are simulated and analyzed by the models. The modeling results have shown that the expanders do not effectively control the radial size change in this case. Based on the simulation results, an effective method is proposed to control the radial shrinkage of the spline by using a plug with a slightly oversized dimension, which is demonstrated by computer models.

Introduction

Heat treatment is used to improve the strength and service performance of steel components, such as gears, shafts, and bearings, etc. A combination of carburizing and quench hardening can generate beneficial compressive surface residual stresses in the hardened case, which improves the fatigue performance. [1-3] During quenching, stresses caused by the thermal gradient and phase transformations will generate plastic deformation and distortion in hardened parts. Gear components with large distortion will increase gear noise and reduce the fatigue life in service. Final machining of case hardened gears often leads to nonuniform case depth distribution, so a maximum amount of distortion caused by hardening is often specified, and parts with distortion exceeding the specification will be scrapped. Press quenching is one effective way to reduce the distortion amount from a hardening process. [4]

Heat treatment is a transient process, and the thermal gradient and phase transformations act together to cause distortion. It is both difficult and expensive to experimentally investigate the part responses during a quench hardening process, and most of these results can only be measured after heat treatment has been completed. Heat treatment results that are of common interest include the volume fractions of phases, hardness, residual stress profile, and part dimensional change. The thermal gradients during both heating and cooling work together with the phase transformations to continually change the internal stress and deformation of the part being heat treated. The material response of a specific part during the heat treatment process is difficult to document just using the final measurements. The development of heat treatment simulation software makes it possible to understand the material response during the heat treatment process, including the evolution of internal stresses and deformation, the phase transformation sequences, and the probability of cracking. HEARTS is one of the earliest finite element based quench hardening process modleing package. [5] DANTE was developed to reduce the distortion casued by heat treatmnet processes, and the processes inmclude austenitization, carburizing, quench hardening, and tempering , etc. [6-7] Recently, DANTE was developed as a set of user subroutines linking to ANSYS Mechanical, with the capabilities of modeling phase transformations, dimensional change, residual stress and hardness. [8] Computer simulation has increased the level of understanding of heat treatment processes because the events that occur during heating and cooling can be accurately modeled. [9-10] In turn, advances in computer hardware, in combination with accurate simulation, have made the design and optimization of the heat treatment processes more cost effective than traditional experimental trial-and-error methods. In this paper, DANTE and ANSYS are coupled to model the carburization and press quenching of a spiral bevel gear component. The modeling results are used to understand the causes of distortion, and further studies of the press quench models are used to improve the effectiveness of the press quench tooling design.

DANTE ANSYS Coupling and Workflow

DANTE user subroutines have been developed to link with ANSYS Workbench for modeling carburizing and quench hardening processes. ANSYS customized USERMATTH subroutine is used for carburizing and thermal analyses with phase transformations, and USERMAT with customized mechanical models is used for stress analysis. As shown in Figure 1(a), the hardening process is coupled sequentially for improved computational efficiency, and the rationale is that heat treatment processes have small deformation in general. The carburization model is executed first, and the predicted carbon distribution profile is imported to the thermal model. Once the thermal model is completed, the temperature history of the entire component will be imported to the stress model together with the carbon distribution profile. The predicted results from the stress model include microstructure phases, hardness, residual stresses, and distortion.

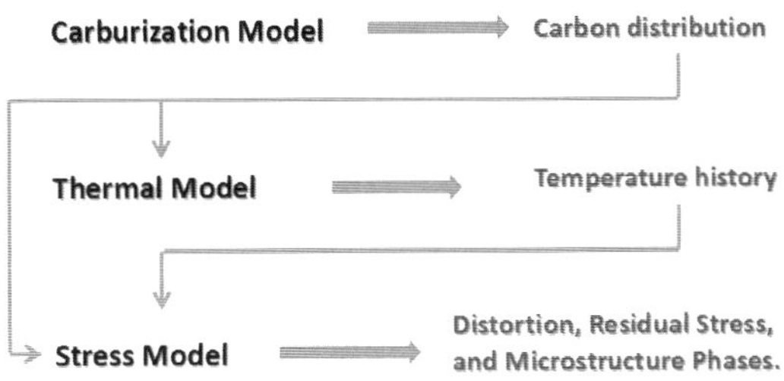

(a)

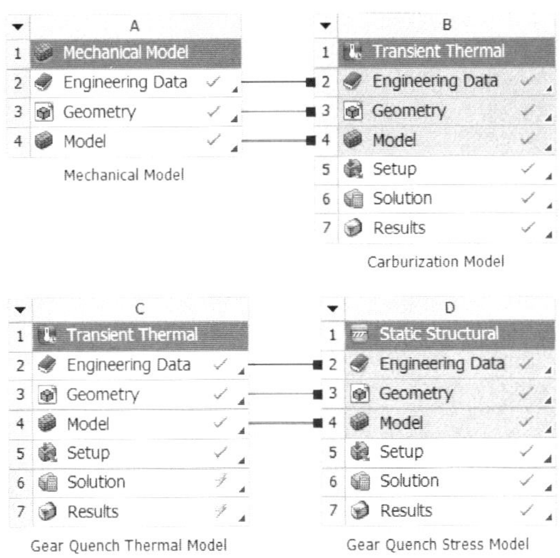

(b)

Figure 1: (a) Sequentially coupled procedure of carburizing, thermal and stress models in DANTE, and (b) implementation of the coupling procedure in ANSYS Workbench.

ANSYS Mechanical is used as the solver by linking to DANTE user subroutines with phase transformations and mechanical models. Both pre-processing and post-processing are done under ANSYS Workbench as shown in Figure 1(b). ANSYS Application Customization Toolkit (ACT) has been developed for setting up heat treatment models and post-processing using Workbench.

Phase Transformation Kinetics

Quench hardening is a highly nonlinear process due to the solid state phase transformations, and the accuracy of the phase transformation models is critical to the modeling results. The diffusive and martensitic transformation models in DANTE are described in equations (1) and (2) below.

$$\frac{d\Phi_d}{dt} = v_d(T)\Phi_d^{\alpha 1}(1-\Phi_d)^{\beta 1}\Phi_a \tag{1}$$

$$\frac{d\Phi_m}{dT} = v_m(1-\Phi_m)^{\alpha 2}(\Phi_m + \varphi\Phi_d)^{\beta 2}\Phi_a \tag{2}$$

where Φ_d and Φ_m are the volume fractions of individual diffusive phase and martensite transformed from austenite; Φ_a is the volume fraction of austenite; v_d and v_m are the mobilities of tranformation products, where v_d is a function of temperature, and v_m is a constant; $\alpha 1$ and $\beta 1$ are material related constants of diffusive transformation; $\alpha 2$, $\beta 2$ and φ are constants of martensitic transformation. For each individual phase formation, one set of transformation kinetics parameters is required.

Dilatometry data are often used to describe the phase transformation behavior of steels. Figure 2(a) is a continuous cooling dilatometry strain curve generated from the DANTE database representing the martensitic formation of AISI 9310. The horizontal axis in Figure 2(a) is temperature, and the vertical axis is strain caused by temperature change and phase transformation. The strain change due to martensitic transformation is clearly quantified.

When the dilatometry test sample cools to the martensite start (Ms) temperature, its volume expands with the crystal strucure change from austenite's face centered cubic (FCC) lattice to martensite's body centered tetragonal (BCT) lattice. Martensite's BCT structure has a lower density than austenite's FCC structrue. The strain change during transformation is a combination of thermal strain, phase transformation volume change, and strain induced by stresses generated during the transformation. The latter strain is referred to as TRansformation Induced Plasticity or TRIP. The data obtained from this specific type of dilatometry test include coefficient of thermal expansion (CTE) for austenite and martensite, martensitic transformation starting (Ms) and martensitic transformation finishing (Mf) temperatures, transformation strain, and phase transformation kinetics (transformation rate) from austenite to martensite. These data are critical to the accuracy of modeling the internal stress and deformation caused by quenching.

Diffusive transformations are also characterized by dilatometry tests. A series of dilatometry tests with different cooling rates are required to fit a full set of diffusive and martensitic phase transformation kinetics parameters. Once the full set of phase transformation kinetics paratmeters are fit from dilatometry tests, isothermal transformation (TTT) and continuous cooling transformation (CCT) diagrams can be generated for users to review. TTT/CCT diagrams are not directly used by DANTE phase transformation kinetics models, but they are useful because users can see the hardenability of the material graphically. Figure 2(b) is an isothermal transformation diagram (TTT) for 9310 steel created from the DANTE database.

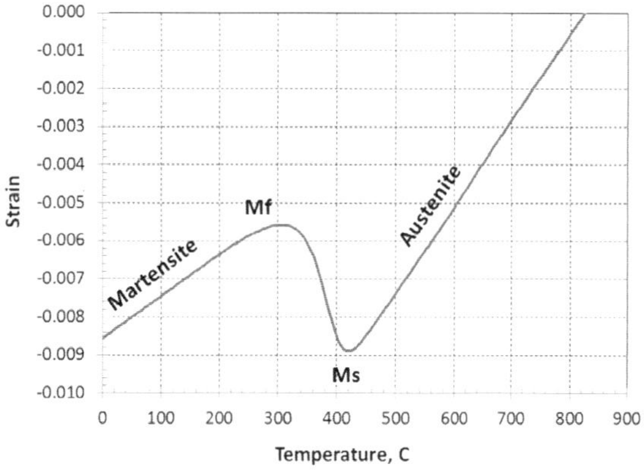

(a)

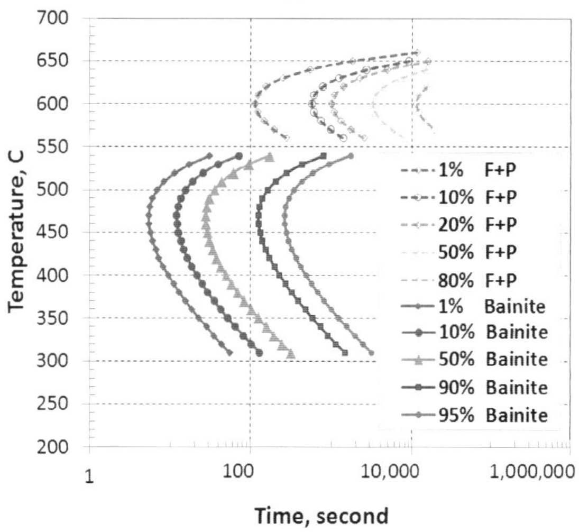

(b)

Figure 2: (a) Dilatometry strain curve during continuous cooling, and (b) TTT diagrams of AISI 9310 generated from DANTE material database.

Finite Element Model and Heat Treatment Process Description

The CAD model of the spiral bevel gear used in this study is shown in Figure 3(a). The outer diameter of the gear is 421 mm, and its height is 100 mm. Carburization and press quenching are used to harden the gear, and the major issue observed from the first hardening process try-out is radial shrinkage of the internal spline teeth. To simplify the model, the internal teeth of the gear are modeled, but the tapered spiral bevel teeth are not modeled. The hardened gear does not have oval or out-of-round distortion exceeding the specification. Therefore, a single tooth model was developed to represent the entire gear to further reduce the model size. The single tooth CAD model and the generated finite element mesh model are shown in Figure 3(b). Ten layers of fine elements were applied to a 3 mm depth in the gear surface for the purpose of catching the temperature and carbon gradients effectively during the heat treatment process. Six-node wedge elements were used in these surface layers of elements, and tetragonal elements were used inside the part where the thermal gradient was mild. The mesh was generated using ANSYS Mechanical. The finite element mesh model contained 59,862 nodes, 72,600 wedge elements, and 105,625 tetragonal elements.

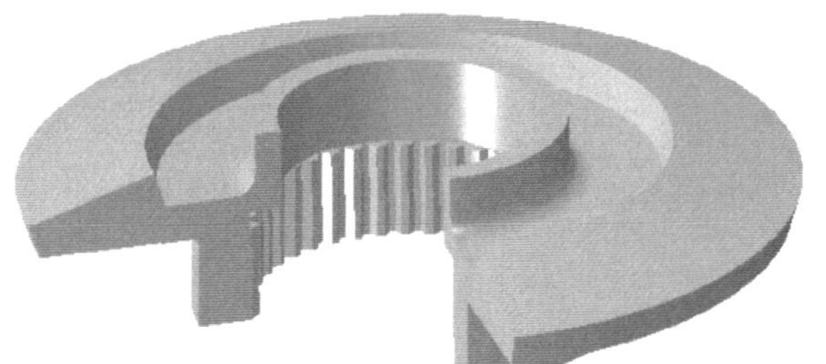

(a)

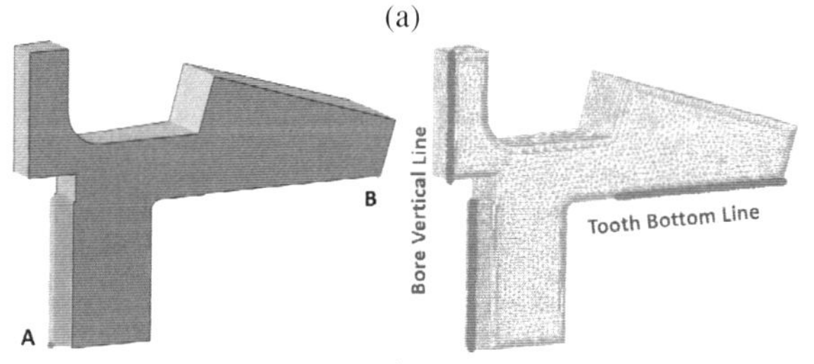

(b)

Figure 3: (a) Gear CAD model, and (b) single tooth CAD and finite element mesh models.

GLEASON press quench is often used in industry to quench harden steel parts requiring a tight dimensional tolerance. [11] In this paper, the finite element model of the press quench tooling is created by ANSYS Workbench, as shown in Figure 4 [11]. The press quench tooling include an upper outer die, an upper inner die, a lower outer die, a lower inner die, an upper expander, and a lower expander. All the tooling are modeled as rigid surfaces, and the Augmented Lagrange contact in ANSYS Mechanical is used to define the mechanical constraints between the gear and the tooling contact surfaces. During press quenching, the positions of the lower inner die and the lower outer die are fixed. Constant axial loads are applied to the upper inner die and the upper outer die to reduce the axial distortion of the spiral bevel tooth section. Radial loads are applied to the expanders to control the radial size and reduce the out-of-round distortion.

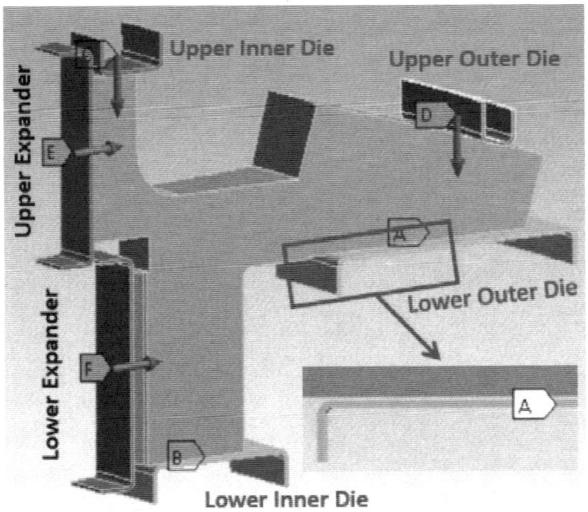

Figure 4: Press quench tooling set up used in the finite element model.

Baseline Press Quench Process Modeling

The gear is gas carburized and cooled prior to reheating and press quenching. The carburizing process is briefly described below.

- Carburization temperature: 954.5° C.
- Carburization furnace time: 54,000 s.
- Furnace carbon potential: 0.8 wt.%.

The predicted carbon distribution contour is shown in Figure 5(a), with the entire surface being carburized. The carbon distribution is important to the residual stresses and distortion obtained due to its effect on the phase transformations. The predicted carbon distribution in terms of depth from the surface is plotted in Figure 5(b). With the assumption of 0.4 wt.% carbon as the effective case depth (ECD) criterion, the predicted ECD is 1.3 mm, which meets the specification.

After carburizing, the gear is cooled to room temperature, followed by reheating and press quenching. The hardening process is briefly described below.

- Austenitization temperature: 850° C; Total furnace time: 10800 s.
- Transfer time from furnace to press quench prior to pumping oil: 15 s.
- Oil temperature: 54.5° C.
- Quenching time duration: 600 s.
- Remove and cool to room temperature (20° C).

During reheating, before the material reaches the austenitization temperature, phase transformation to austenite occurs. Snapshots of temperature and austenite distribution contours during reheating are shown in Figure 6. The thin section of the gear has a higher temperature during heating as shown in Figure 6(a), which leads to earlier phase transformation to austenite in this region. The gear surface is carburized, and the phase transformation to austenite is earlier in the carburized case due to its high carbon content as shown in Figure 6(b). Along with the phase transformation to austenite, the volume of the material decreases, which contributes to internal stresses and distortion.

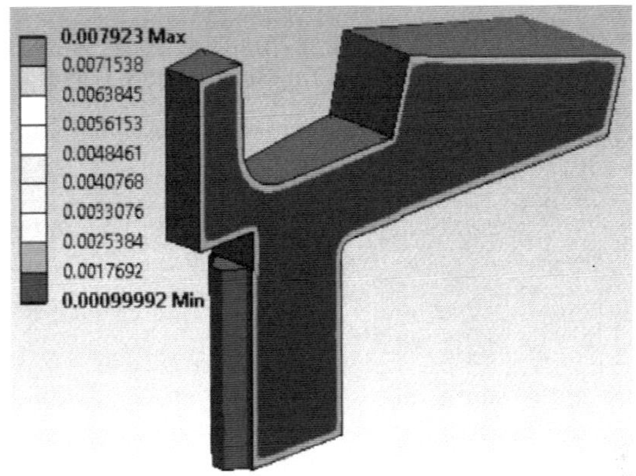

(a)

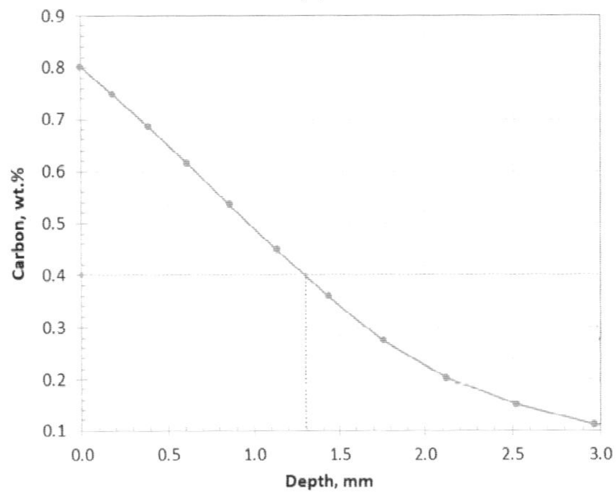

(b)

Figure 5: (a) Carbon distribution contour plot, and (b) carbon distribution in terms of depth.

After achieving full austenite and soaking in the furnace at 850° C, the gear is removed from the furnace for press quenching. The transfer time of 15 s can affect the final distortion of hardened parts, and it is important to keep the transfer time consistent. Once the gear is placed on top of the lower dies and shuttled into position, the top dies move downward to closed die positions with specific loads applied, and then oil flow begins. The quenching process continues for 600 s, then loads are released, dies are opened, and the gear is taken out and cooled to room temperature. Final steps are deep freezing to reduce retained austenite levels and low temperature tempering in a furnace. Because the deep freeze and low temperature tempering process steps have an insignificant effect on the shape change, the deep freeze and tempering processes are not modeled in this paper. If gear performance is the purpose of the model, these steps must be included in the model.

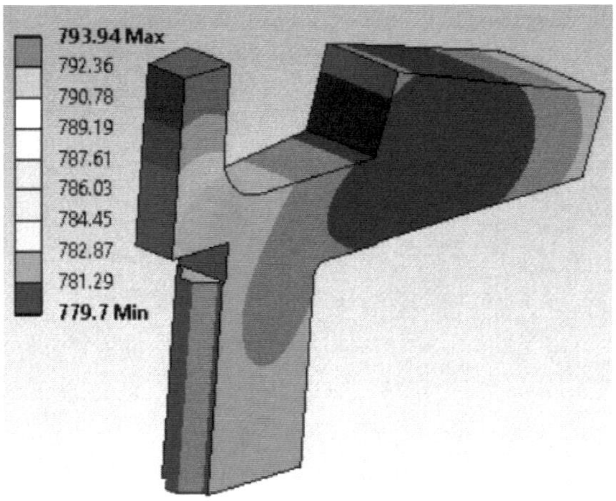

(a)

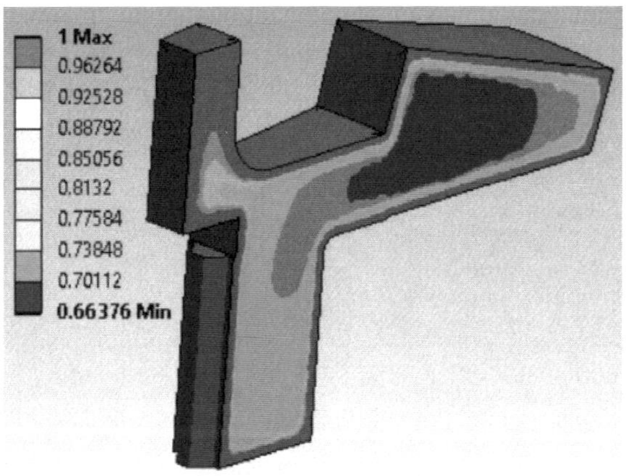

(b)

Figure 6: Phase transformation to austenite at 4085.1 s during heating: (a) temperature distribution, and (b) austenite distribution.

After the press quenched gear has cooled to room temperature, the predicted martensite and lower bainite distribution contours are shown in Figure 7(a) and 7(b), respectively. The core has about 70% martensite and 30% lower bainite, and other phases are ignorable in this location. In a location slightly deeper than the case, about 97% martensite is predicted with a small amount of lower bainite. In the carburized case, the maximum amount of retained austenite is about 15%. The existence of retained austenite in the carburized case is due to the low martensite formation starting and finishing temperatures for AISI 9310 with higher carbon content. As mentioned, the amount of retained austenite can be further reduced by deep freezing and tempering processes. In DANTE, the hardness of as-quenched parts is calculated from the obtained volume fractions of various phases and the carbon content. The predicted hardness distribution of the as-quenched gear is shown in Figure 7(c) with Rockwell C hardness unit (HRC). The hardness of the carburized case is 64 HRC, and the core is about 41 HRC, which agrees with the measured data.

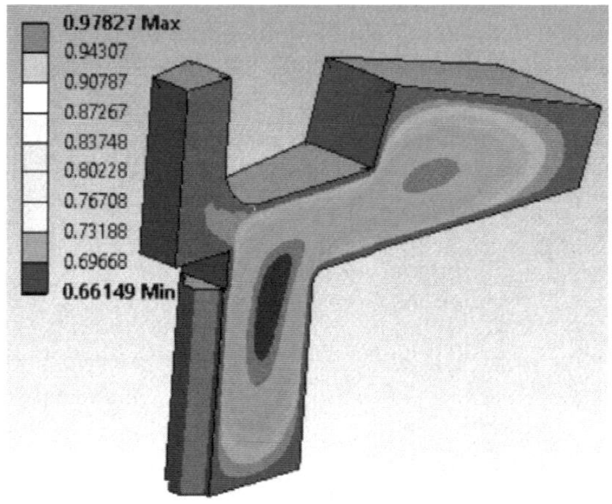

(a)

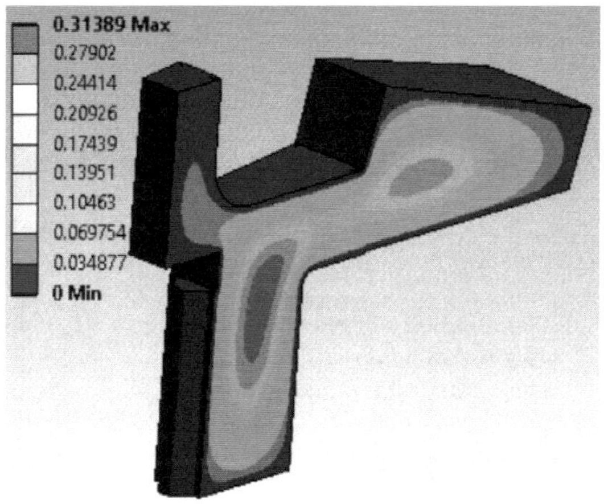

(b)

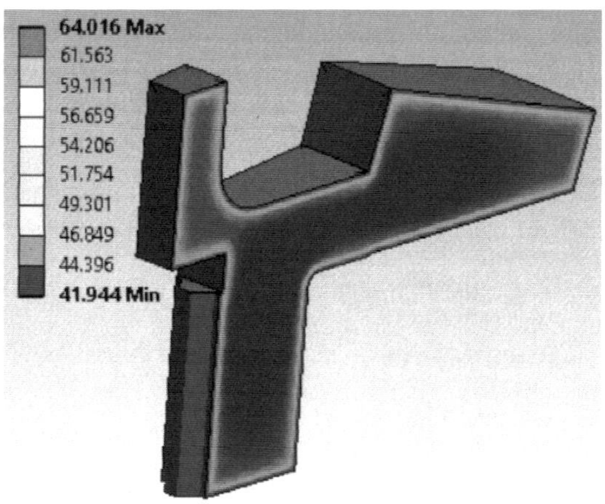

(c)

Figure 7: Phase and hardness distributions at the end of quenching: (a) martensite, (b) lower bainite, and (c) hardness.

A virtual quench hardening model without the press quench tooling constraints (free oil quench) was run with the same thermal boundary conditions applied. The purpose of this free oil quench model was to understand the effect of the tooling constraints on distortion by comparing the free quench and baseline press quench models. The free oil quench model is also helpful for setting up the position of the lower outer die to compensate for the axial distortion of the spiral bevel tooth section. The predicted radial displacement after cooling the gear to room temperature for the free quench model is shown in Figure 8(a). The internal spline tooth is predicted to shrink radially about 0.19 mm. The axial displacement is shown in Figure 8(b), and the spiral bevel teeth axial warp upward about 0.635 mm.

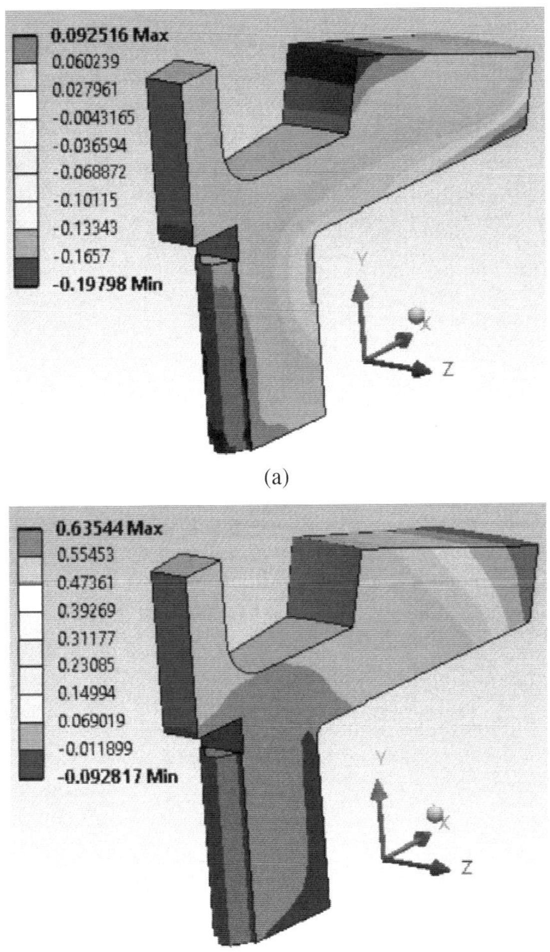

Figure 8: (a) Radial distortion, and (b) axial distortion at the end of oil quench (without the mechanical constraints and loads from the tooling).

The predicted axial distortion from the free quench model is an important value for setting up the lower die positions for the press quench. It is worth mentioning that the lower inner and the lower outer die positions are fixed during press quench, and the temperature of the tooling is negligible (remember the oil cools the tooling as well as the part) in comparison to the gear size changes during quenching due to part temperature change (heat lost to the oil) and the phase transformations. As a result of the part temperature change, the axial dimension of the gear changes between the bottom face of the spiral bevel tooth and the lower end face of the bore, and it is not expected that the lower outer die contacts the gear during the entire quench process unless extremely high die load from the upper outer die is applied. In general, the press load applied in a press quench process is much lower than the required load of causing plastic deformation of high temperature austenite. The low load applied is also a guarantee of not damaging the gear teeth locally. The axial loads are applied on the expanders to control the oval or out-of-round distortion. The axial loads applied on the upper dies are to control the axial warping of the spiral bevel teeth, together with the lower die position setting. To compensate for the predicted axial distortion of the spiral bevel teeth, as shown in Figure 8(b), the lower outer die is positioned 0.635 mm lower than the bottom face of the spiral bevel tooth at room temperature, as shown in Figure 4. In other words, if the room temperature gear is positioned on the lower dies of the press tooling, the lower end face of the bore contacts with the lower inner die, and the bottom face of the spiral bevel tooth section has a 0.635 mm gap with the lower outer die surface. This is designed to reduce the axial upward warpage predicted from the free quench hardening model.

Using the press quench tooling setting shown in Figure 4, the predicted radial and axial distortions after quenching and cooling to room temperature are shown in Figure 9. The predicted radial distortion of the internal spline tooth is approximately 0.19 mm shrinkage, which is slightly less than the value of 0.198 mm predicted by the free oil quench model. In general, the axial load applied on the expander is low, and its effect on the radial size should be insignificant. However, the expander load applied should be sufficient to control the oval or out-of-round distortion due to the concentrated radial force on the sector with lowest radial dimension of out-of-round parts. Minimal plastic strain due to the expander is desired. An expander load that is too high will cause inconsistent radial size and possible shape distortion. The predicted axial distortion of the baseline press quench model is shown in Figure 9(b), and the contour plot shows the press quench tooling is effective in reducing the axial distortion of the spiral bevel tooth section. By studying the contact history between the lower outer die and the bottom face of the spiral bevel tooth during the entire press quenching process, a conclusion can be made that the load applied on the upper outer die is much lower than the necessary load required to plastically bend the spiral bevel tooth section, even when the section is austenite at the high temperature. However, the spiral bevel tooth section can be bent downward sufficiently under a low die load when phase transformation occurs. The transformation induction plasticity (TRIP) is an important phenomenon for making the press quench process effective.

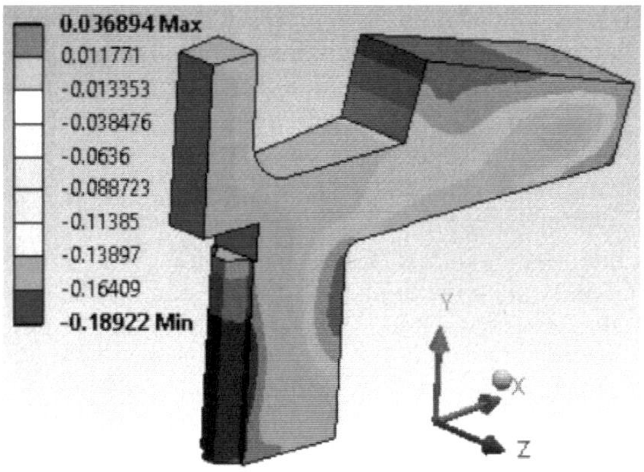

(a)

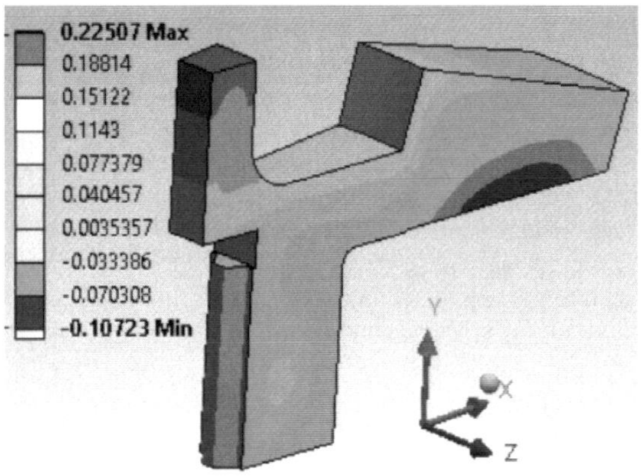

(b)

Figure 9: (a) Radial distortion, and (b) axial distortion predicted at the end of press quench.

Figure 10 shows the predicted minimum principal and the circumferential residual stresses of the hardened gear in the as-quenched condition. The carburized case is under residual compression due to the delayed martensite transformation in the case. It is also worth mentioning that the press quench tooling has an insignificant effect on the residual stress state if performed properly. The stress magnitude generated by external loads from the tooling is insignificant compared to the thermal stress and phase transformation stress generated from the quenching process.

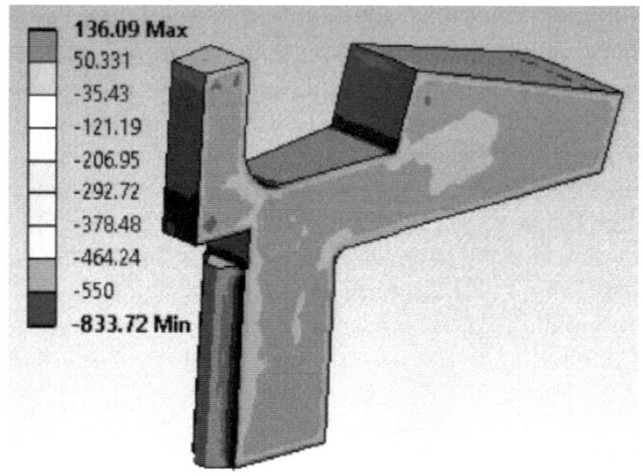

(a)

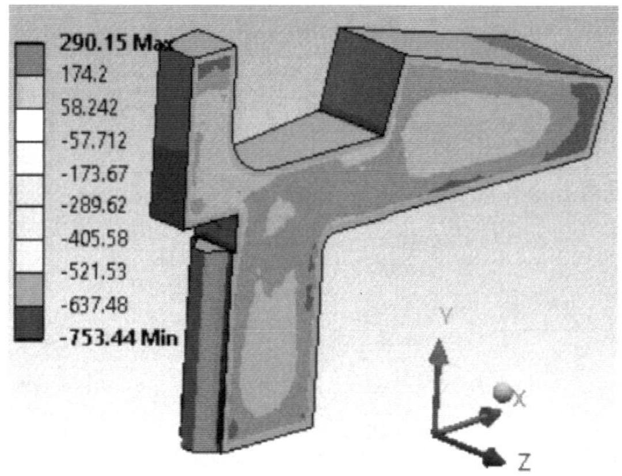

(b)

Figure 10: (a) Minimum principal residual stress, and (b) circumferential residual stress distributions predicted at the end of press quench.

Distortion Reduction by Modifying the Press Quench Set Up

The axial shrinkage of the internal spline tooth predicted from the baseline press quench model is about 0.19 mm, which is significant relative to the designed case depth and allowed final machining stock. Higher loads can be applied to the expanders to counter the radial shrinkage. However, the required load magnitude will be high, which can cause inconsistent radial size change after hardening. Too high of a load on spline tooth surface prior to transformation to martensite can damage the tooth shape. One effective method is to replace the expanders by plugs, as shown in Figure 11. In this modified press quench setting, the lower outer die and the lower inner die are kept at the same positions as those of the baseline model, and the loads applied by the upper dies are not

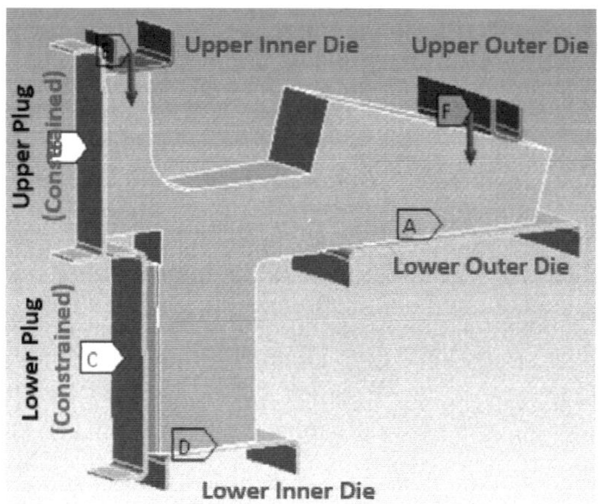

Figure 11: Modified press quench tooling set up.

changed either. As shown in Figure 11, the upper expander is replaced by an upper plug; the size of the upper plug is the same as desired bore dimension at room temperature. The lower expander is replaced by a lower plug, and the plug's radial size is 0.060 mm larger than the internal tooth tip size at the room temperature (an overclosure of 0.060 mm). After the gear is heated, enough gap will be generated between the plug and the gear due to the thermal expansion, so the plugs can be inserted into the gear for quenching.

The effectiveness of the baseline and the modified press quench settings on the radial displacements of the internal spline tooth and the bore after hardening are studied by selecting two lines of points from the bore and the internal spline tooth tip of the FEA model, as shown in Figure 3(b). The predicted radial displacements from the three models are compared in Figure 12.

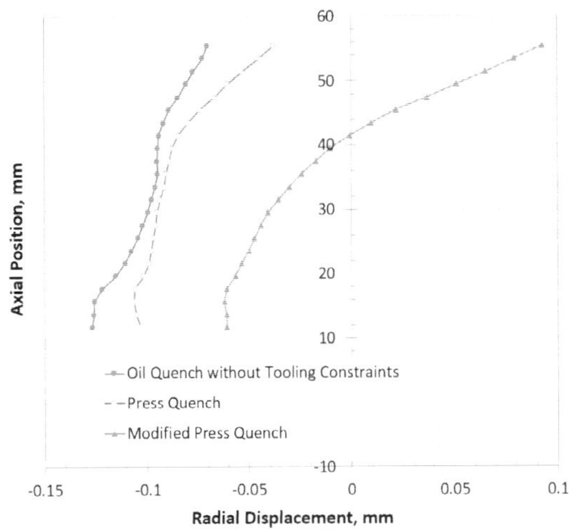

(a)

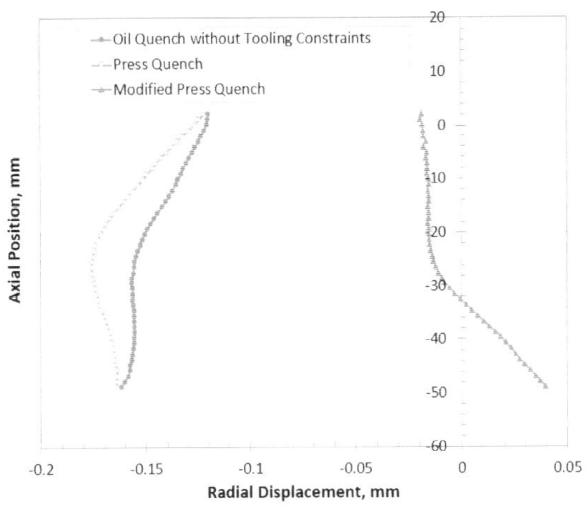

(b)

Figure 12: Comparison of the tooling effect on the bore radial distortion along (a) the upper bore line, and (b) the lower spline tooth tip line as shown in Figure 3(b).

The X-axis in Figure 12 is the radial displacement after cooling the gear to room temperature, with X = 0.0 mm representing the original position of the upper bore and the internal spline tooth tip. The Y-axis of Figure 12(a) is the axial position of the upper bore, and the Y-axis of Figure 12(b) is the axial position of the internal spline tooth tip. The free quench model and the baseline press quench model have similar radial distortions along the internal spline tooth tip and the upper bore. With the modified press quench, the radial shrinkage of the internal spline tooth is significantly reduced, as shown in Figure 12(b). The radial distortion of the upper bore is also reduced. However, the tapered distortion along the upper bore is more severe with the plug design. This can be partially caused by the part and plug interaction along the spline tooth tip.

One line of points located along the bottom face of the spiral bevel tooth section as shown in Figure 3(b) is selected to compare the axial distortion obtained from the three models, and the results are plotted in Figure 13. The X-axis is the radial position of the selected points, and Y-axis is the axial displacement, with a positive value representing warping upward. With free oil quench, the predicted axial displacement is about 0.6 mm at the end face of the spiral bevel tooth section. Both the baseline and the modified press quench models can effectively reduce the axial distortion of the spiral bevel tooth section.

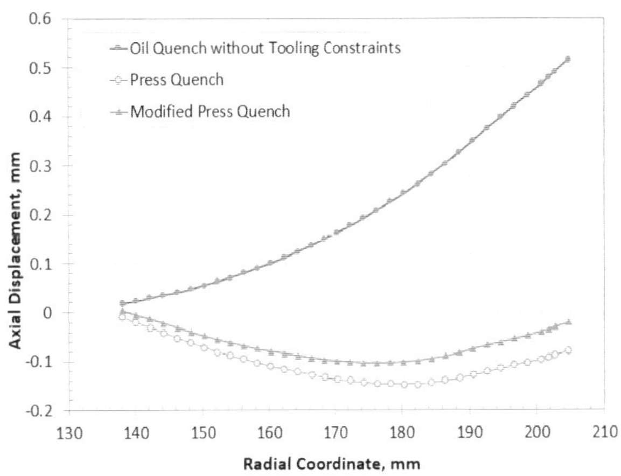

Figure 13: Comparison of the tooling effect on the axial distortion of the tooth bottom face.

Understanding the gear response during a quenching process is important to interpret the causes of distortions, and give ideas of process improvement for reduced distortion. As shown in Figure 3(b), two points are selected to plot the radial and axial displacements during the entire quenching processes. Point A is located at the lower end of the internal spline tooth tip surface, and the point B is located at the far end of the bottom face of the spiral bevel tooth.

Figure 14 plots the radial displacements of the point A for the three models. The X-axis is the time, and the Y-axis is the radial displacement. With the same thermal boundary conditions applied for the three models, the effects of the expander and the plug are clearly compared. The expander has an insignificant effect on the radial displacement. However, if the gear has an out-of-round shape during quenching before the martensitic transformation is completed, the radial force will concentrate on the least radial size section to make it round. By using the plug, the upper bore and the internal spline tooth tip surface of the gear will contact with the plug due to the thermal shrinkage. The reaction force between the plug and the gear is much higher than the force applied on the expander of the baseline press quench process, which leads to elongated circumferential plastic strain in the gear, and the gear ends up with a larger radial dimension. With further cooling, the gear expands when bulky martensitic phase transformation occurs in the gear, and the gear detaches from the plug.

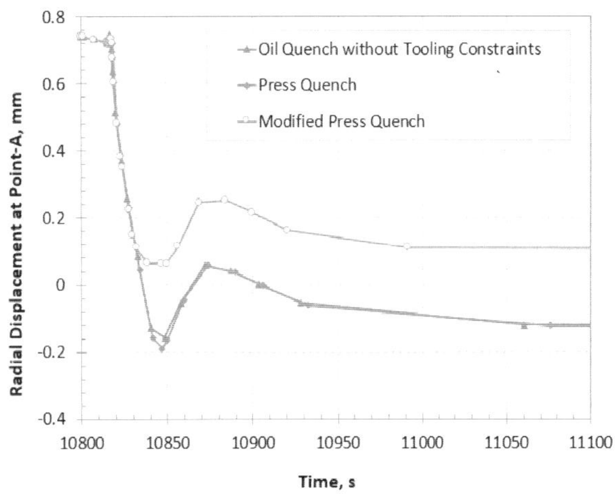

Figure 14: Radial displacement of the point-A as shown in Figure 3(b) vs. time during quenching for the three models.

Figure 15 shows the axial displacements of the point B during quenching for the three models. After furnace heating, air transfer, and before quenching, the radial displacements of point B are the same for the three models. During quenching, the same axial load is applied to the upper outer die to control the axial distortion, and no load is applied for the free quench model. Because the gear has an unbalanced geometry, the spiral bevel tooth section warps upward at the early stage of quenching. However, the amount of the upward warpage is reduced with die load applied on the upper outer die. With further cooling, the die load is sufficient to push the spiral bevel tooth section downward contacting with the lower outer die with the help of internal thermal and phase transformation stresses. It is worth mentioning again that the transformation induced plasticity (TRIP) phenomenon plays an important role in the press quenching process. When phase transformation is going on, a lower magnitude external load can be applied to introduce plastic deformation to the part.

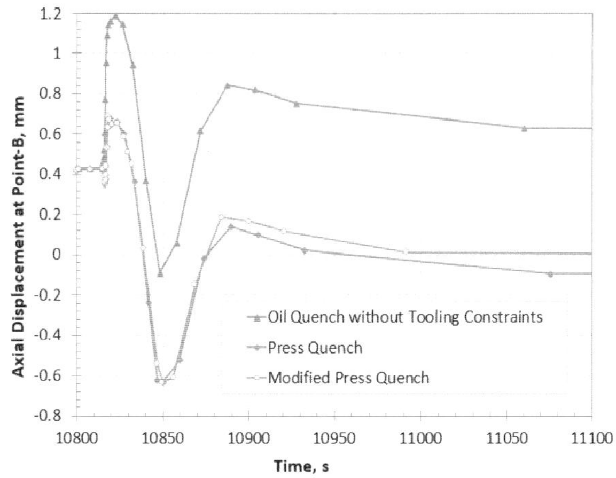

Figure 15: Axial displacement of the point-B as shown in Figure 3(b) vs. time during quenching for the three model scenarios.

Conclusions

The dimensional responses of the spiral bevel gear to press quench processes were investigated. The external loads applied to the gear by either dies or expanders are low in general to keep the gear dimensional response consistent, as well as not causing damage to the part, e.g. plastic deformation of spline or gear teeth. Due to the TRIP effect, a low magnitude of tooling loads can be sufficient to constrain the gear during quenching for improved dimensions. However, the expanders in this specific press quench design do not effectively avoid radial distortion. With the help of computer modeling, the press quench tooling was modified by replacing the expanders with plugs, and the predicted results have shown that the radial distortion of the internal tooth can be consistently and effectively controlled. In this study, straight (stepped) plugs with different radial sizes were used for the internal spline tooth section and the upper bore section. It is expected that the reaction force for a plug is much higher than that from an expander of a traditional press quench, and its effect on the neighbors (regions without plug contact) is also more severe. Plugs with tapered or profiled configurations will be designed to further reduce the tapered distortion of the upper bore location in future works.

References

[1] B.L. Ferguson, A.M. Freborg, and Z. Li, "Residual Stress and Heat Treatment – Process Design for Bending Fatigue Strength Improvement of Carburized Aerospace Gears", Journal of HTM. Härterei-technische Mitteilun, 62(6), 2007, p 279-284.

[2] Zhichao Li, Andrew M. Freborg, Bruce D. Hansen, T. S. Srivatsan, "Modeling the Effect of Carburization and Quenching on the Development of Residual Stresses

and Bending Fatigue Resistance of Steel Gears", Journal of Materials Engineering and Performance, March 2013, Volume 22, Issue 3, pp 664-672.

[3] A. Freborg, B. Ferguson, and Z. Li, "Modeling Heat Treatment for Characterizing Distortion, Residual Stress, Loading Response and Fracture in a Vacuum Carburized and Gas Quenched Steel Coupon", Proceedings from the 5th International Conference on Thermal Process Modeling and Computer Simulation, June 16–18, 2014, Gaylord Palms Resort & Convention Center, Orlando, FL, USA.

[4] Reardon, A.C. "Press Quenching," ASM Handbook, Volume 4a, Steel Heat Treating Fundamentals and Processes, J. Dossett and G. Totten, Editors, ASM International, Nov. 2013.

[5] T. Inoue, S. Nagaki, T. Kishino, and M. Monkawa, "Description of transformation kinetics, heat conduction and elastic-plastic stress in the course of quenching and tempering of some steels", Ingenieur-Archiv, Vol.50, 1981, Springer, pp 15-327.

[6] Ferguson, B.L., Freborg, A., Petrus, G., Software Simulates Quenching, Advanced Materials and Processes, H31-H36, August (2000).

[7] D. Bammann, et al., "Development of a Carburizing and Quenching Simulation Tool: A Material Model for Carburizing Steels Undergoing Phase Transformations," 2nd Int. Conference on Quenching and Control of Distortion, ASM Int., 1996.

[8] ANSYS Mechanical User's Guide, R16.

[9] Z. Li, A. Freborg, B. Ferguson, "Application of modeling to heat treat processes". Heat Treating Processes, May/June 2008.

[10] MacKenzie, Scott, Li, Zhichao and Ferguson, B. Lynn, "Effect of Quenchant Flow on the Distortion of Carburized Automotive Pinion Gears", Proceedings of the 5th International Conference and European Conference on Heat Treatment, eds. Johann Grosch, Jorg Kleff and Thomas Lubben, Berlin, Germany, pp 119-129, 2007.

[11] Reardon, A.C. "Press Quenching and Effects of Prior Thermal History on Distortion during Heat Treatment," Gear Technology, March/April, 2014, pp. 50-53.

International Federation of Heat Treatment and Surface Engineering 2016
Proceedings of the 23rd IFHTSE Congress
April 18–21, 2016, Savannah, Georgia, USA

Intensive Quenching of AISI 5160 Spring Steel Using Small Scale Laboratory Equipment: A Progress Report

Luigi M. Albano[1], João Gilberto Lúcio[2], Fernando S. Misina[1], Vojteh Leskovisek[3], George E. Totten[4], Lauralice C.F. Canale[1]

1. University of São Paulo, São Carlos, SP, Brazil
lfcanale@sc.usp.br
2. Rassini NHK, São Bernardo do Campo, SP, Brazil
jgilberto@rassini-nhk.com.br
3. University of Ljubljana, Ljubljana, Slovenia
vojteh.leskovsek@imt.si
4. Portland State University, Portland, OR, USA
GETotten@gmail.com

Abstract

Intensive quenching has been defined as "those quenching conditions that lead to uniform maximum surface compressive stresses" which provide maximum surface compressive stresses with correspondingly optimized distortion control. An earlier paper, described the design and construction of a laboratory system that provides sufficiently uniform interfacial heat transfer rates to be classified as an intensive quenching (IQ) system. This paper provides a progress report on the use of this system to intensively quench AISI 5160 spring steel and to summarize the initial properties obtained.

Keywords: Intensive quenching, spring steel, residual stress, fracture toughness

Introduction

Every metallurgist has been trained that increasing cooling rates, especially in the martensitic transformation region, leads to increasing potential for cracking as shown in Figure 1 [1]. However, since the 1920's, there have been various, often little known, industrial heat treating processes that have been designated as intense, intensive, rapid, drastic, severe, or extreme quenching or shell-hardening methods [2-8]. The essence of these methods is to harden less hardenable steels using very fast cooling rates in order to impart high compressive stresses and improve fatigue properties of the quenched component [8].

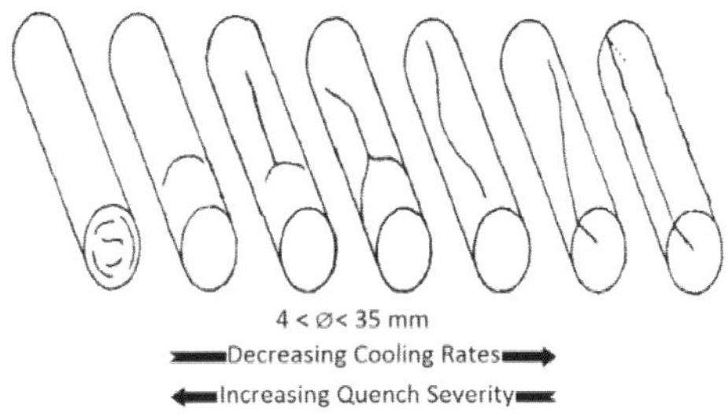

Figure 1: Quench cracking of 60SC7 steel quenched in water. Decreasing cooling rates were achieved by increasing water temperature.

In USP 1,828,325, it was reported that the depth of hardening could be controlled by spray pressure of cold water impinging on rail steel and by the speed of the steel moving through the spray zone [3]. No spray pressures and quantitative measures quench severity were provided. However, this seems to be one of the earliest published references suggesting the development of an intensive quenching process.

Kern reported that the first known example of "intense quenching" was in the production of the Ford Model T AISI 1035 rear axle [2] using a hot solution of 5% aqueous caustic soda. However, Kern also reported that the Ladish Company used an intense quenching process using vigorously agitated quench oil to produce Rolls Royce P-51 engine crankshafts. The problem with this application, when compared to the Model T application, is at least two-fold. The first is that oil is recognized to exhibit substantially less quench severity than cold water in a high-pressure spray. The second is that there is no quantitative criterion either of quench severity or to the precise meaning of "vigorous" agitation. Finally, the actual definition of the term "intense quenching" referred to by Kern is not clear from the examples provided [2].

In a series of patents published from 1967 to 1971 [5-7], somewhat greater quantification was provided for the production of various machinery parts using a process designated as a "drastic" or "extreme" quench. In these processes, pressurized tanks (accumulators or "bladders") were used to deliver very high volumes, approximately 3000 gal/min, of quenchant to selected surface areas of the part being spray or press quenched. The purpose that was stated for performing such high-volume quenchant delivery processes to the hot austenitized steel surface was to achieve maximum hardness uniformity. One patent, USP 3,515,601, reported that there was a "critical cooling rate" which was not reported.

Morio discussed the necessity of hardening of carbon or boron-containing carbon steels by a cold-water, rapid-quenching process to achieve the optimal as-quenched properties (high tensile strength and good weldability) which are superior to oil hardening [9]. However, it was recognized rapid quenching may lead to cracking and increased distortion, although some control is afforded through the heat treating process and component shape design.

Another method of minimizing cracking and distortion reported by Morio was to quench the part using sufficiently high agitation rates to eliminate film boiling on the surface which would provide a more uniform quench. The "critical cooling rate" for this process (drastic quenching) was the cooling rate, which is necessary to eliminate film boiling. The agitation rate/quench severity correlation used to calculate the critical cooling rate was taken from the traditional Grossmann Quench Severity data shown in Table 1 [9]. According to Morio, the practical limit for water quenching was 1.5-2.0, unless high-pressure sprays were employed. This approach is complicated by the absence of a definition of mild, moderate, good, etc. for agitation rate. Furthermore, it is impossible to tell by observation of "surface roll" etc. since agitation in quench tanks is notoriously non-uniform!

Table 1: Grossman Quench Severity (in^{-1}) for Various Quenching Media and Agitation

Agitation	Oil	Water	Brine
None	0.25-0.30	0.9-1.1	2.0
Mild	0.30-0.35	1.0-1.1	2.0-2.2
Moderate	0.35-0.40	1.2-1.3	
Good	0.40-0.50	1.4-1.5	
Strong	0.50-0.80	1.6-2.0	
Violent	0.8-1.1	4.0	5.0

Mei has taken the approach reported by Morio even further by stating that an agitation rate sufficient to provide a Grossman Quench Severity value of > 6.0 is required to provide an intensive quenching process [10]. However, in addition to the limitations of the use of Grossmann H-values, the approach reported by Mei is based on trial and error experimentation.

In 1964, Kobasko published the first of an extensive series of papers in which he used the term "Intensive Quenching" and showed experimental data which provided numerical evidence that showed that although it is true that increasing cooling rates result in increasing propensity for cracking as historically recognized, there does exist a critical cooling rate above which cracking propensity decreases [11]. Computer simulations were later used to validate and to develop design criteria for optimal conditions for conducting intensive quenching processes [12,13]. Subsequently, various industrial intensive quenching processes were developed and patented by Kobasko [14-17].

These historical references indicate that intensive quenching is conducted with very high ("extreme" or "violent") agitation rates with a Grossmann H-value of >6.0. Agitation is a focus of these papers and unfortunately alloys and cross-section sizes are not addressed quantitatively – with the exception of Mei's and Morio's papers [9,10]. While this information is helpful to envision an intensive quenching process, it is insufficient to properly design a quenching process for optimal results for a particular application. An invaluable reference for use in designing an intensive quenching system has been published by Kobasko, et. al. [18].

Perhaps the simplest and correct working definition of "intensive quenching" is "those conditions that lead to maximum surface compressive stresses [19, 20]. However, to properly design a system, it is important to consider together those conditions that affect the formation of maximum surface compressive stresses including: alloy, part shape, cross-section size, quenching cooling rate and final machining/grinding. By Kobasko's definition of intensive quenching, it is quite likely that, depending on the material, cross-section size and actual cooling rates achieved at the hot metal interface during quenching, many of those processes described above may either NOT be intensively quenched or they may not have possessed maximum surface compressive stresses.

An overview of the design and use of a novel laboratory system that was recently developed at the University of Sao Paulo will be provided in this discussion. This will be followed by a discussion of the use of this equipment to successfully quench AISI 5160 spring steel under intensive quenching conditions. The work discussed here was part of an MS thesis project [21].

Experimental

The composition of the AISI 5160 steel used for this work is shown in Table 2. The carbon equivalent (C_{eq}) for this steel is 0.86. Test pieces were annealed and machined to produce the test specimen illustrated in Figure 2.

Table 2: Composition of AISI 5160 Steel Used for This Work

Elemental Composition SAE 5160 (% by Wt.)					
C	*Mn*	*P*	*S*	*Si*	*Cr*
0.578	0.822	0.011	0.0080	0.269	0.705
Ni	*Mo*	*Cu*	*Al*	*N*	
0.0151	0.0028	0.259	0.028	0.0021	

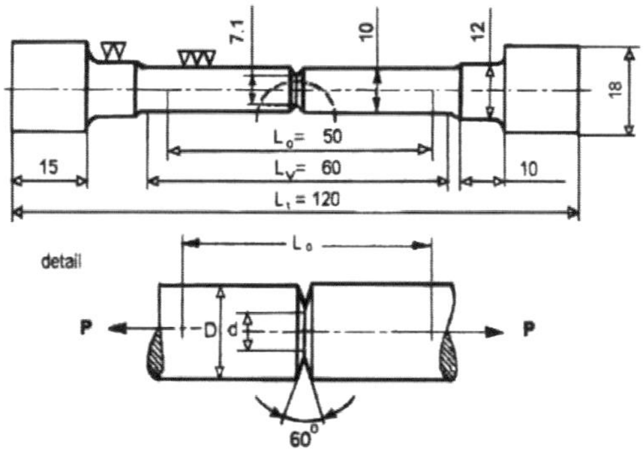

Figure 2: Drawing of steel test pieces used for intensive quenching

A fatigue pre-cracking procedure was performed where the fatigue pre-crack of 1 mm was created in the notch root of the test specimen at 1180 rpm during 5 minutes with a bending load of 450 N after which the austenization at 900°C for two (2) hours was performed [22-24]. The test coupons were heated in a steel box as shown in Figure 3. Potential oxidation was minimized by placing cast iron chips in the box before heating the covered box assembly in the furnace [25,26]. In addition, argon was injected into the heating chamber at 2 L/min during austenitization. No evidence of decarburization was observed upon removal of the test specimens from the furnace.

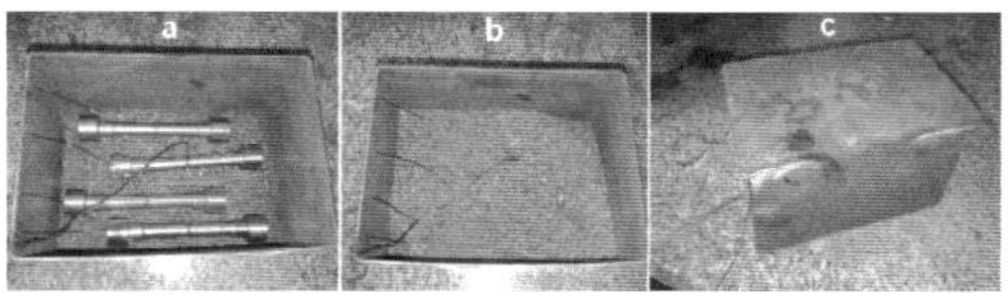

Figure 3: This figure shows: a.) typical placement in a steel box in which they will be heated in a box furnace; b.) the test specimens are packed with cast iron chips, c; the box is covered and placed into the box furnace at 900°C for austenitization.

After the austenitization of the steel test specimen was completed, they were immediately quenched by immersion into the laboratory intensive quenching (IQ) system illustrated in Figure 4. The fluid volume in the IQ quenching system shown in Figure 4 was 50 L (total fluid capacity is 60 L) and each test pieces (120 g each) was quenched individually and vertically in the center of the tank. The aqueous quenching fluid contained 9% sodium Nitrite ($NaNO_2$). For intensive quenching the total volume flow rate through the horizontal spray nozzles was 135 g/s for the left nozzle and 145 g/s for the right nozzle. The quenchant in the bath was at ambient temperature (approximately 27°C). Also, the total quenching time in the tank after immersion was 2 s, not longer, not less. After removal from the quench, the cooling process was completed in air. (Longer immersion times provided lower residual stresses.)

The quenchant used for the "conventional oil quench" was TP444 Houghton KB, a well-known "rapid" quench oil in Brazil. The austenitized test specimen was quenched into 30 L of the quench oil at 60°C. Agitation ("mild") was provided during quenching by manual movement in the oil throughout the cooling process.

Figure 4: Laboratory IQ system designed and used at the University of Sao Paulo. Agitation is provided by a flow jet at each size as shown. The jets are placed at opposite sides of the tank in a staggered arrangement so that the two jets do not impinge each other directly. The temperature of the quenchant in the bath was approximately 27°C (ambient temperature).

Optical microscopy was performed using a Zeiss Discovery Model V-8 stereoscopic system equipped with an AxioCam ERc 5s.

Rockwell hardness was determined using a Leco Model RT-240 hardness tester with a digital display.

Fracture toughness was determined using a Brazilian EMIC universal tensile testing machine with a loading of 10kN.

Figure 5 is showing regions where hardness and residual stress were measured.

Figure 5 – Sample used for tests.

Residual stress was determined by x-ray diffraction using a Rigaku Model SF-3 diffractometer.

Results and discussion

The results described here were the initial results obtained for quenching AISI 5160 steel using the new laboratory scale intensive quenching system. A summary of these results are provided in the following discussion.

1.Summary of the IQ Quenching Process

Initial results showed that if the test specimen was immediately immersed into the IQ system, it was through-quenched and the necessary hardened case characteristic of successfully intensively quenched steel was not formed. This was verified by residual stress measurements. Therefore, it was clear that a time-quench was necessary.

In this study, the time quench was performed manually using a stopwatch to control the total time the test specimen was in the IQ system before removal and subsequent air cooling. To determine total immersion time to obtain the optimal results, total immersion times of 1,2,3,4 and 5 seconds were evaluated. It was determined by optical microscopic examination that at an immersion time of 2 seconds was necessary to obtain maximum martensite transformation. Longer immersion times produced lower residual stresses so on the basis of this study, it was determined that a two second total immersion time produced the desired martensitic case and optimal residual stresses for this particular steel alloy and dimensional configuration. Similar assessments would need to be performed for other steel alloys and dimensions.

The microstructure obtained for the AISI 5160 test specimen using the 2 second total immersion time is shown in Figure 6. For comparison the microstructure of the conventionally oil quenched test specimen is also shown. This microstructural comparison shows that while both quench processes produce fully martensitic microstructure, the IQ process yielded a finer grain size, self-tempered martensitic microstructure.

Figure 6 – Comparison of the microstructure obtained by a.) IQ processing and b.) conventional oil quenching. Magnification 600x; Etchant: Nital 2%

2. Hardness and Tensile Strength

Table 3 summarizes the hardness and tensile strength obtained by both the IQ process and by conventional oil quenching. Not surprisingly, both quench processes did successfully harden the AISI 6160 steel test specimen with the IQ process producing slightly greater hardness than conventionally quenching with the accelerated (fast) oil used. Similarly, and as expected, both quench processes produced much greater tensile strength than the unhardened, as-machined test specimen. As observed with the hardness values, slightly greater tensile strength was obtained with the IQ processed AISI 5160 steel test specimen.

3. Fracture Toughness

The fracture toughness of a material containing a crack is a measure of its ability to resist brittle fracture. For proper optimization of heat treatment of a steel alloy, the knowledge of the fracture toughness value is necessary. The linear-elastic fracture toughness is determined by the stress intensity factor (K_{IC}) which is a measure of the point at which the thin crack begins to grow.

Because hardened steels typically present a high notch sensitivity it is often very difficult to create a fatigue pre-crack. This problem has been successfully addressed by the creation of the fatigue pre-crack prior to heat treatment. The experimental procedure used in this work was described in detail previously by Podgornik, *et. al.* [22]. The fracture toughness results obtained using the Podgornik procedure are shown in Table 3 where it is observed that the K_{IC} is slightly lower for the IQ process relative to the results obtained for the conventionally oil quenched test specimen. An ANOVA study suggested that this difference in K_{IC} values is probably not statistically significant. These are initial results and further work is in progress to more fully explain this observation.

Table 3: Summary of Residual Stresses, Fracture Toughness and K_{IC} Values Obtained for AISI 5160 As-Machined, Conventionally Oil Quenched and Intensively Quenched

Condition	As-Machined and Annealed	Conv. Oil Quench	Intensively Quenched
Hardness (HRC)	21.5 ± 2,1	59.2 ± 1.1	61.5 ± 1.6
Tensile Strength ($kgf.mm^{-2}$)	21.5 ± 1.5	59.2 ± 1.1	61.5 ± 1.2
K_{IC} Fracture Toughness ($MPa \cdot m^{1/2}$)	---	5.1 ± 1.1	3.8 ± 1.3
Residual Stress (MPa)	---	-50 ± 12	-916 ± 223

4. Residual Stress

Perhaps the most common observation indicating intensive quenching is the enormous compressive stresses that are obtained. A selective summary of previously reported compressive stress values reported for other IQ processed alloys is shown in Table 4. Other similar results, both experimental and simulated, can be shown as well. These data demonstrate that relative high surface compressive stresses are obtained by intensive quenching and in one example the surfaces compressive stresses are much greater than those obtained by conventional oil quenching a reasonably broad range of alloy hardenabilities. Interestingly, the compressive stresses reported in Reference [27] for 5160 steel are very close to those obtained in this work (compare to Table 3). Thus, the residual stresses obtained reported in Table 3 are a proof that the laboratory IQ system and manual quenching process described here can be used to study the effect of intensive quenching on the properties attainable for other steel compositions.

Table 4 – Examples of Published Residual Stresses for Different IQ Processed Steels

Steel	Component	Residual Stress (MPa)	Reference
5160 (Oil Quench)	Steel Rod (27 mm dia x 127 mm)	-102	[27]
5160 (IQ)	Steel Rod (27 mm dia x 127 mm)	-939	[27]
52100	Roller (75mm)	-840	[28]
52100	Roller (46 mm)	-900	[28]
4140	Kingpin (46 mm)	-563	[28]
S5	Punch (38 mm)	-750	[2]

Conclusions

The results reported here show that the small laboratory agitation system described may be used to successfully intensively quench AISI 5160 spring steel. The IQ results showed that very similar hardness and tensile strength were obtained for both the conventional oil quenched and IQ quenched test specimen, the surface microstructures were not equivalent. Although essentially complete martensitic transformation had occurred, the microstructure of the IQ sample was a tempered martensite with a significantly finer grain size. The fracture toughness was somewhat lower for the IQ processed test specimen but the results are not statistically significant. Most importantly, the IQ processed test specimen produced much higher surface compressive stresses than that obtained by oil quenching. These data show that intensive quenching was indeed obtained in the laboratory IQ test system.

Acknowledgements

The authors acknowledge CNPq (Conselho Nacional de Desenvolvimento Científico e Tecnológico) and Universidade de São Paulo (USP) and for CNPq for their financial support.

References

1. G. Beck, "Internal Stresses Associated with Heat Treatment of Alloys", *Mem. Etud. Sci. Rev. Metall.*, 1985, June, p. 269-282.
2. R.F. Kern, "Intense Quenching", *Heat Treating*, 1986, September, p. 19-23.
3. H. Kurz, "Process for Manufacture of Rails with Hardened Heads", USP 1,828,325, October 20,1931.
4. R.H. Hays, J.E. Sansom and K.D. Gladden, "Die-Quenched Crankshaft", USP 3,506,501, April 14, 1970.
5. B. Paddock, "Quench Apparatus for Providing Pulsating and Sweeping Flow of Quenching Fluid", USP 3,517,676, June 30, 1970.
6. R.H. Hays, J.E. Sansom and K.D. Gladden, "Die-Quenched Crankshaft", USP 3,589,697, June 29, 1971.
7. J.E. Sansom, "Spray Closet Quench", USP 3,515,601, June 2, 1970.
8. R.B. Liss, C.G. Massieon and A.S. McKloskey, "The Development of Heat Treat Stresses and Their Effect on Fatique Strength of Hardened Steels", SAE Technical Paper Series, Paper Number 650517, 1965.
9. A. Morio, "Water Hardening", Kinzoku Zairyou, 1977, Vol. 17., No. 3, p. 45-53.
10. M. Daming, "Intensive Quenching method for Preventing Quench Cracking", *Heat Treatment and Surface Coatings: New Processes and Application Experience Proceedings of the 7th International Congress of Heat Treatment and Technology of Surface Coating*, Moscow, Russia, December 11-14, 1990, Vol. 2, p. 62-71.
11. N.I. Kobasko, and N.I. Prokhorenko, "Cooling Rate Effect of Quenching on Crack Formation in 45 Steel", Metal Science and Heat Treatment, 1964, Vol. 6, No. 2, February, p. 104-105.
12. N.I. Kobasko, "Chapter 10.4 - Intensive Steel Quenching Methods", in *Theory and Technology of Quenching*, Eds. B. Liscic, H.M. Tensi and W. Luty, 1992, Springer-Verlag, Berlin, Germany, p. 367-389. ISBN: 0-387-52040-6
13. B.L. Ferguson, N.I. Kobasko, M.A. Aronov, and J. Powell, "Intensive Quenching of Steel Rings" *in 19th Heat Treating Society Conference Proceedings,* Eds. S. Midea and G. Pfaffmann, ASM International, Materials Park, OH, 1999 p. 355-362. ISBN: 0-87170-681-4
14. N.I. Kobasko, *Steel Quenching in Liquid Media Under Pressure*, 1980, Naukova Dumka, Kiev, Ukraine.
15. V.A. Lisovoy, N.I. Kobasko, M.V. Kindrachuk, and A.A. Khalatov, "Stamp Quenching Method", USSR Patent 4,176,788/37-02, 01.06, 1987.
16. N.I. Kobasko, Quenching Apparatus and Method for Hardening Steel Parts", USP 6,364,974, April 2, 2002.
17. N.I. Kobasko, M.A. Aronov, J.A. Powell, L.C.F. Canale and G.E. Totten, "Intensive Quenching

Process Classification and Applications", Heat Treatment of Metals, 2004, Vol. 31, No. 3, p. 51-58.
18. N.I. Kobasko, M.A. Aronov, J. Powell and G.E. Totten, *Intensive Quenching Systems: Engineering and Design*, 2010, ASTM International, West Conshohocken, PA. ISBN: 978-0-8031-7019-3
19. L.C.F. Canale, N.I. Kobasko and G.E. Totten, "Intensive Quenching: Part 1 – What Is It?", International Heat Treatment and Surface Engineering, 2007, Vol. 1, No. 1, p. 30-33.
20. L.C.F. Canale, N.I. Kobasko and G.E. Totten, "Intensive Quenching: Part 2 – Formation of Optimal Surface Compressive Stresses", International Heat Treatment and Surface Engineering, 2007, Vol. 1, No. 2, p. 60-63
21. L.L.M. Albano, "Estudo Comparativo das Propriedades Mecânicas de aço AISI 5160 Submetidos à Têmpera Convencional e Têmpera Intensiva", MS Thesis, Universidade de São Paulo, Escola de Engenharia de São Carlos, SP, Brazil, 2013. (In Portuguese). Internet: http://www.teses.usp.br/teses/disponiveis/88/88131/tde-17042013-105945/pt-br.php
22. B. Podgornik, B. Žužek, V. Leskovšek, "Experimental Evaluation of Tool Steel Fracture Toughness using Curcumferentially Notched and Precracked Tension Bar Specimen", Materials Performance and Characterization, 2014, Vol. 3, No. 3, Paper ID No. MPC20130045, p. 87-103.
23. N. Senčič and V. Leskovšek, "Fracture Toughness of the Vacuum Heat Treated Spring Steel 51CrV4", Materiali in Technologije (Materials and Technology), 2011, Vol. 45, No. 1, p. 67-73.
24. B. Ule, V. Leskovšek, B. Tuma, "Estimation of Plane Strain Fracture Toughness of AISI M2 Steel from Precracked Round Bar Specimens", Engineering Fracture Mechanics, 2000, Vol. 65, p. 559-572.
25. J.R. Davis, *ASM Specialty Handbook: Tool Materials*, ASM International, Materials Park, OH, USA p. 316. ISBN: 978-0-87170-545-7.
26. G.A. Roberts, R. Kennedy, G. Krauss, *Tool Steels, 5th Edition*, ASM International, Materials Park, OH, USA p. 232. ISBN: 978-0-87170-599-0. (inability to control decarburization)
27. C.R. Hubbard, F. Tang, M.A. Arovov, N.I. Kobasko, J.A. Powell, B.L. Ferguson, Z. Li, "Effect of Intensive Quenching on Residual Stress", in *Proceedings of 23rd ASM Heat Treating Conference*, Eds. D. Herring, R. Hill, 2005, Pittsburgh, PA, USA, ASM International, Materials Park, OH, p. 338-342. ISBN: 0-87170-832-9.
28. N.I. Kobasko, M.A. Aronov, K. Ichitani, M. Hasegawa, K. Noguchi, "High Compressive Residual Stresses in Through Hardened Steel Parts as a Function of Biot Number", in *Recent Advances in Fluid Mechanics, Heat & Mass Transfer and Biology*, Rds. Manoj K. Jha, Myriam Lazard, Azami Zaharim, Kamaruzzaman Sopian, Havard University, Cambridge, MA, USA, January 25-27, 2012, WSEAS Press, p. 35-40. ISBN: 978-1-61804-065-7.

International Federation of Heat Treatment and Surface Engineering 2016
Proceedings of the 23rd IFHTSE Congress
April 18–21, 2016, Savannah, Georgia, USA

Ionic Liquids for immersion quenching of aluminium alloys

Martin Beck, Michael Reich, Olaf Keßler
Chair of Materials Science, University of Rostock, Rostock, Germany

Christin Neise, Udo Kragl
Chair of Technical Chemistry, University of Rostock, Rostock, Germany

Mathias Ahrenberg, Christoph Schick
Polymer Physics Group, University of Rostock, Rostock, Germany

Abstract

Quenching is a very important step in the manufacturing chain of metallic components. It is commonly performed as immersion quenching in vaporising liquids. Unfortunately, this process is affected by the Leidenfrost-Effect, forming a vapour film at the component surface which can generate residual stresses and distortion. Baths of molten salts and molten metals provide a quenching without any Leidenfrost-Effect. Nevertheless, they need to be operated at elevated bath temperatures combined with a considerable effort for cleaning the quenched components. In this work Ionic Liquids (ILs) ,a new class of liquid quenching media, are investigated, showing nearly no Leidenfrost-Effect and can be used in baths at room temperature. Aluminium components have been quenched in Ionic liquid. The time-temperature curves have been recorded and the heat transfer coefficients (htc) were determined and compared to quenching in water. Additional investigations on distortion of complex aluminium profiles quenched in ionic liquid and water have been carried out. The distortion measurements were compared to quenching simulations. The results indicate a huge potential to homogenise immersion quenching.

Introduction

The manufacturing chain of metallic components consists of many process steps, whereas heat treatment is one of the most complex and important ones. Heat treatment often includes quenching which can be realised in different ways. Gas quenching is very smooth with a low amount of distortion and internal stresses, but also relatively slow. The commonly performed quenching in vaporising liquids like water or oil is very cheap and fast. Unfortunately, this process is forming a vapour film at the component surface. The thermal insulation of the component caused by this vapour film lowers the cooling rate at the beginning of the quenching. Subsequently, the vapour film collapses locally at different times, leading to different cooling rates, which can generate residual stresses and distortion. It's common knowledge, that baths of molten salts and molten metals provide a quenching without any Leidenfrost-Effect. Nevertheless, they need to be operated at elevated bath temperatures combined with a considerable effort for cleaning the quenched components.

In this work Ionic Liquids (ILs), a new class of liquid quenching media, are investigated, showing nearly no Leidenfrost-Effect and can be used in baths at room temperature. Per definition ILs have melting temperatures of less than 100 °C and some of them even possess melting points below room temperature. These liquids are industrially used since several years as solvents [1], lubricants or sealing liquids [2].

Therefore, these Ionic Liquids should be able to provide a fast and homogeneous cooling. In a previous investigation the authors demonstrated the fundamental applicability of ILs for quenching media of aluminium alloys for the very first time [3]. It was shown, that the cooling power can be increased significantly by water addition. Subsequently, also steels were successfully quenched in ionic liquids [4] and the behaviour of ILs at such a short term high temperature exposure has also been investigated by the authors [5,6]. First results indicated that quenching is afflicted only with low distortion [4]. With this work the experimental results concerning distortion by quenching shall be intensified and correlated with thermo-mechanical simulations in order to get a tool to predict the distortion and residual stresses of complex components quenched in ILs without effortful experiments.

Methods

Experimental

The detailed experimental setup of the quenching experiments is described in a previous publication of the authors [4]. Quenching experiments with aluminium cylinders were performed in order to calculate heat transfer coefficients. The cylinders (Ø 29.5 mm x 120 mm) are made of AlSi1MgMn and equipped with spark eroded holes (Ø 0.7 mm) with type K thermocouples (Ø 0.5 mm) in the sample center and 1 mm underneath the sample surface in middle axial height. The samples are heated up to annealing temperature of 470 °C and quenched in baths of 0.9 l or 15 l. Only new (not quenched before) aluminium samples have been used in order to avoid a surface conditioning effect of the sample [7]. The cooling curves of the samples and the temperature changes of the baths were recorded. Afterwards the corresponding heat transfer coefficients were calculated by an inverse thermal simulation, considering also the warming of the quenchant [8]. The samples for distortion measurements are made of AlZn8MgCu with a "T" alike profile (Figure 1). They are

equipped with one thermocouple type K (Ø 1.0 mm) in a spark eroded hole (Ø 1.2 mm) close to the sample center and an internal screw thread for adapting a sample holder. Before and after quenching samples have been measured in a 3D-coordinate measuring device (Mitutoyo Crysta-Apex C574) with an accuracy of (1.7 + 3L/ 1000) µm. The base plane has been measured with a density of 12 x 24 points along a width of 55 mm and a length of 115 mm. Afterwards for each measurement and simulation the points were referred to a best-fit-plane with the smallest sum of all square distances (for contour plots). The lateral surface of the samples has been measured after the quenching experiments with a density of 17 x 59 points on an area of 32 mm x 118 mm. For comparison of lateral planes with the initial state, four fresh samples from the same production have been measured. For the analysis of distortion three samples quenched in IL and four samples quenched in water were used.
The quenching experiments were performed with tap water and IL. A volume of 15 l of the IL [EMIm][EtSO$_4$] (1-ethyl-3-methylimidazolium ethylsulfate) with an initial bath temperature of 25 °C and a content of about 3 ma.-% deionized water has been used. For comparison, additional samples have been quenched in 15 l tap water, also with 25 °C initial bath temperature. All quenching experiments have been video recorded.

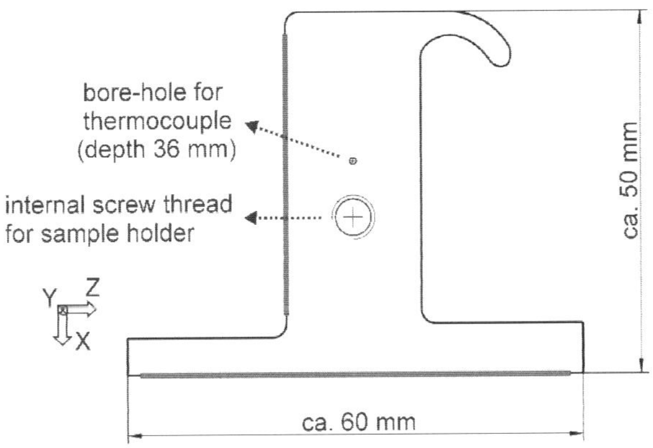

Figure 1: "T" shaped profile of samples (length 120 mm) made of AlZn8MgCu for distortion measurements at marked surfaces (bottom: base; left: lateral surface) [4].

Simulation
For the thermal simulation the density of AlZn8MgCu was set at 2.85 g/cm³ at room temperature [9]. The specific heat capacity of this batch of AlZn8MgCu was measured with a differential scanning calorimeter *Perkin Elmer Pyris 1*. The thermal conductivity was fixed at 154 W/(mK) [10]. For the mechanical simulation the thermal expansion coefficient and yield strength of the same batch of AlZn8MgCu have been determined in a quench and deformation dilatometer (*Bähr DIL 805A/D*). The yield strength of aluminium alloys is highly dependent on the cooling rate after solution annealing [11]. This behavior is caused by the strong influence of solid solution strengthening on yield strength in the cooling process directly after solution annealing. To determine the yield strength of the material as close to the real heat treatment as possible, the quenching process needs to be imitated before mechanical testing. Therefore a Newtonian cooling after solution annealing was assumed. To describe the cooling curve, the duration from 470 °C to 150 °C was used [12]. The dependency of the stress/strain curve on the cooling after solution annealing is shown in Figure 2.

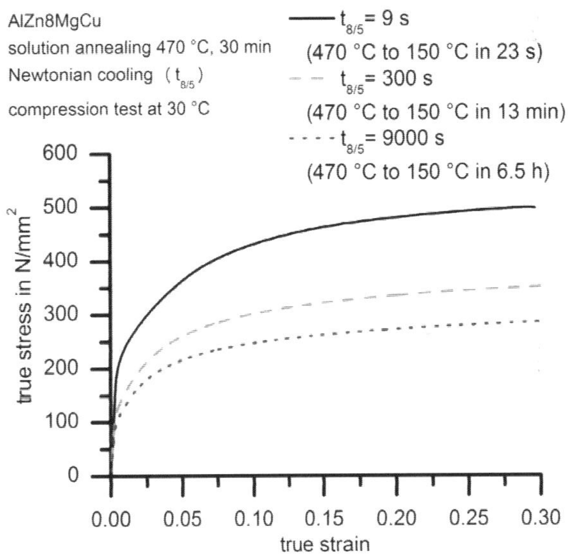

Figure 2: Flow curves of AlZn8MgCu at 30 °C after varying Newtonian cooling from solution annealing.

The experimental cooling conditions in this work are close to the relevant time range to suppress precipitation reactions between solution annealing and 150 °C [12]. Therefore the cooling was fixed (23 s to cool from 470 °C to 150 °C) and the compression test temperature was varied between 30 °C and 470 °C (Figure 3). These flow curves were used for distortion simulation. The Young's modulus was determined from the literature [13] and the Poisson's ratio is assumed to be constant at 0.33. The simulation is determined by an isotropic hardening assumption.
The simulation was performed with a 3D model of the whole sample referring to Reich and Keßler [11]. The software used for the simulation was *marc mentat* from the *MSC Software GmbH*.

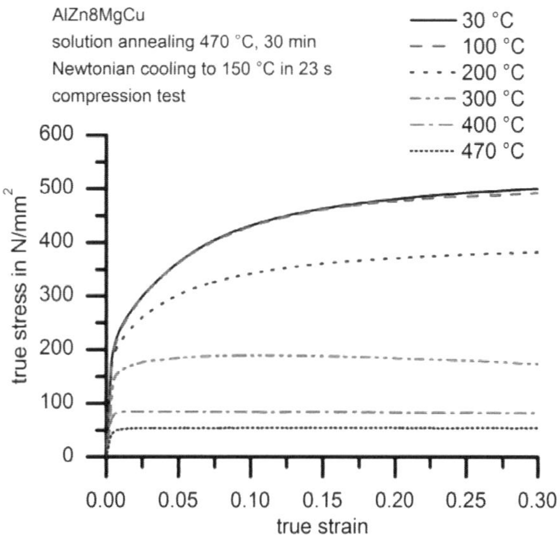

Figure 3: Flow curves of AlZn8MgCu at varying temperatures after Newtonian cooling from solution annealing.

To estimate the accuracy of the heat transfer coefficient adaption from a cylinder to a more complex component, additional quenching experiments with distortion samples equipped with six type K thermocouples (Ø 0.5 mm) were performed (Figure 4).

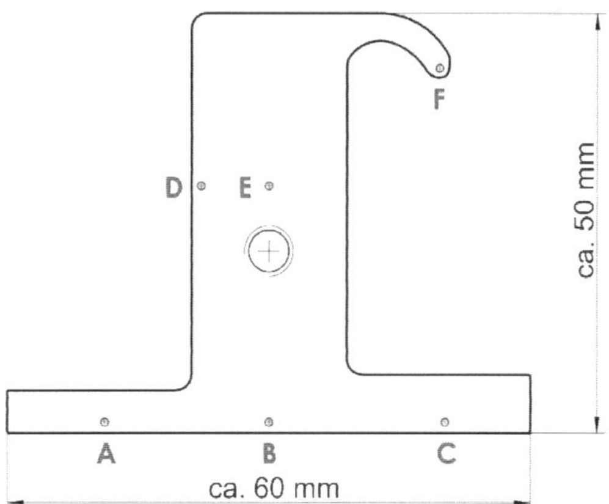

Figure 4: "T" shaped profile of samples (length 120 mm) made of AlZn8MgCu with six thermocouples (type K Ø 0.5 mm) in middle sample height (60 mm) (A, B, C, D, F 1 mm underneath sample surface).

Results and Discussion

The cooling curves of the quenched cylinders are shown in Figure 5 (top). In the case of water quenching a short film boiling occurs, visible by a low temperature drop during the first two seconds. Quenching in IL does not cause any film boiling. The calculated heat transfer coefficients for the simulation are shown in Figure 5 (bottom). The short film boiling-sequence at the very beginning of the quenching process in tap water causes an uneven curve of the heat transfer coefficient between 470 °C and 350 °C. In previous experiments, it has been shown that the lower heat transfer coefficient and the lower cooling rate in the ionic liquid is sufficient to age harden this high-strength and quench sensitive aluminium alloy 7349. Cylinders of diameter 29.5 mm and length 120 mm have been solution annealed at 470° C, quenched in water respectively ionic liquid and aged at 120° C for 24 hours. Water quenching resulted in a hardness of 207 ± 1 HV1, whereas ionic liquid quenching resulted in 197 ± 2 HV1 [4]. This slight loss in hardness will be overcompensated by the significantly lower distortion as shown below.

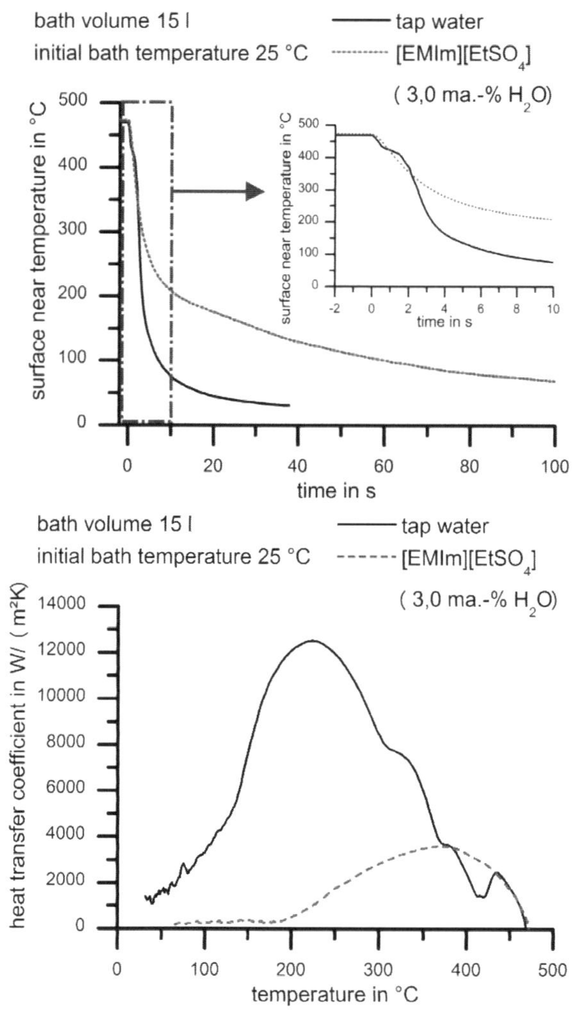

Figure 5: Cooling curves (top) and calculated heat transfer coefficients (bottom) of aluminium cylinders quenched in tap water and IL.

The 3D-measurements of the base of distortion samples in the initial state before heat treatment show a relative homogeneous bending in x-direction along the width of the sample (Figure 6 left) which could be fitted with a polynomial function. The approximated bending of the base was taken into account for the initial state of the simulation (Figure 6 right).

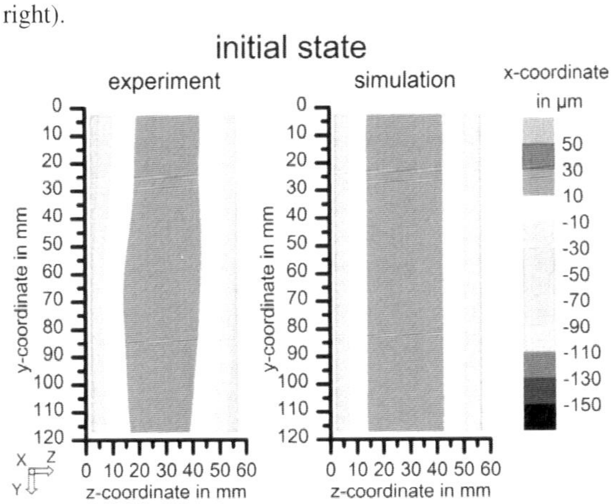

Figure 6: Contour of the base. Left: 3D measurement of an exemplary distortion sample before heat treatment. Right: Initial state of profile bending for simulation (polynomial approximation of bending at y = 60 mm of exemplary sample).

The results of the experimental and simulative contours of the base of the distortion samples after quenching in IL and water are shown in Figure 7. The experimental contour after quenching in IL shows basically the same distortion like in the initial state. Quenching in water results in a stronger bending of the upper left and lower right corner of the base. This behavior is probably caused by a local film boiling (Figure 8) which is most stable in the center of the base. The simulation of quenching in IL is similar to the experiment. The simulation after quenching in water indicates a bending of the upper left and lower right corner of the base like in the experiment, but also of the other two corners. Nevertheless, the agreement between experiments and simulations is good.

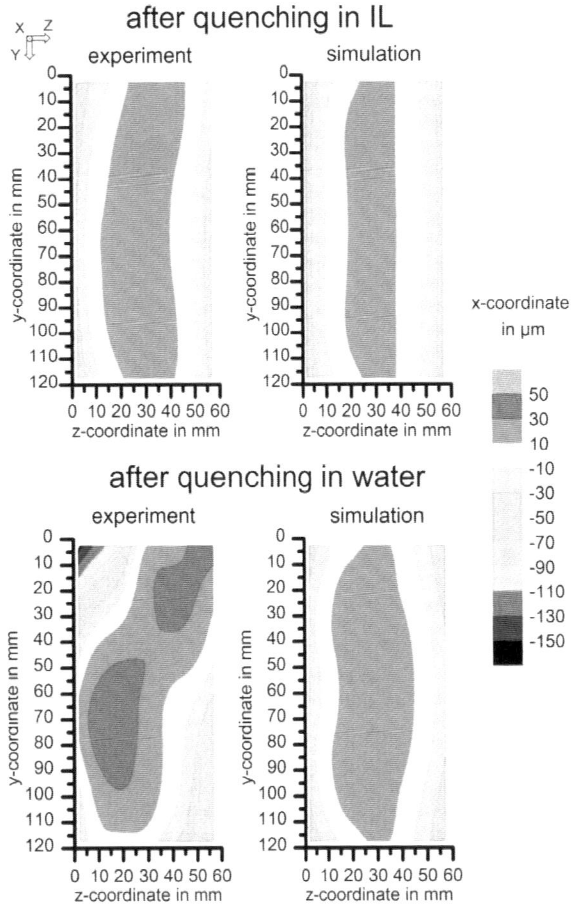

Figure 7: Contours of the bases of distortion samples - exemplary experimental results of 3D measurements versus simulation. Top: after quenching in IL. Bottom: after quenching in water.

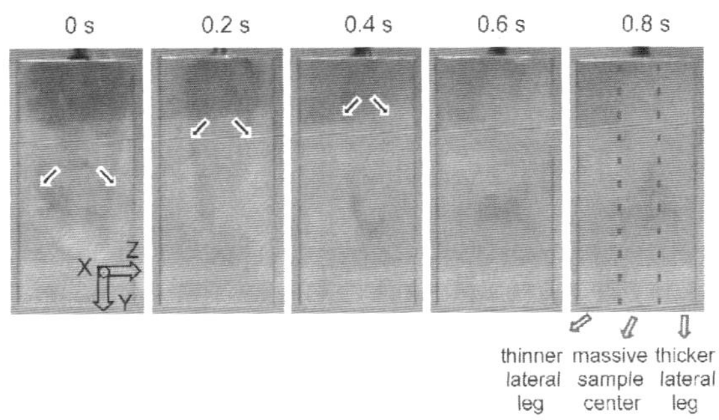

Figure 8: Photo series of quenching a distortion sample made of AlZn8MgCu in 15 l tap water with an initial bath temperature of 25 °C at different times (the white arrows are marking rewetting fronts) [4].

In order to compare experiment and simulation quantitatively, the flatness was calculated. The average experimental flatness of the base is rising slightly from about 80 µm in the initial state to about 94 µm after quenching in IL (Figure 9). Quenching in water results in a markedly greater average flatness of more than 160 µm. The simulation also shows that quenching in IL results in a lower flatness than quenching in water.

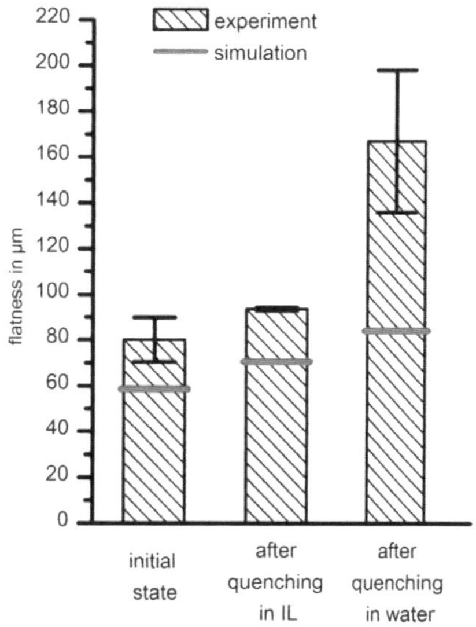

Figure 9: Flatness of the bases of distortion samples before and after heat treatment for experiments (with average and standard deviation) and simulation.

In order to further verify the simulation, a second area of the component has been measured and compared to a thermo-mechanical simulation. In Figure 10 the simulative results for the distortion of the lateral surface in z-direction of the samples after quenching in IL and water are shown. The simulation indicates a bending of the sample in negative z-direction which is more pronounced after quenching in water. The experimental results confirm a distortion in the same direction.

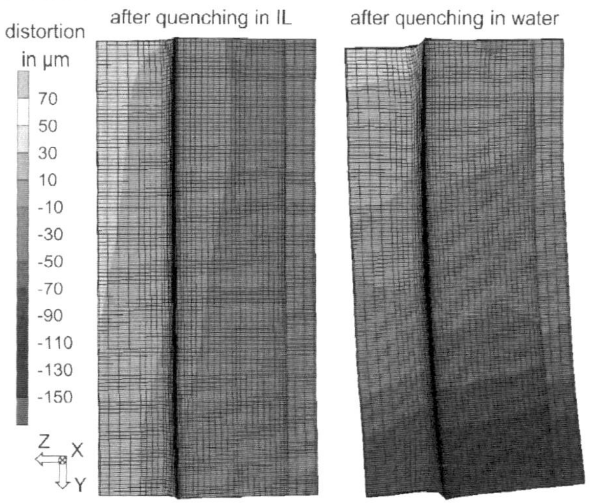

Figure 10: 3D simulation of distortion in z-direction (superelevated 50 times).

The flatness for simulation and experiment are shown in Figure 11. The average flatness of the initial state is about 30 µm, but the simulation is performed with a perfectly flat lateral surface. After quenching in IL the simulation results in a flatness of 15 µm, whereas the average experimental flatness is about 62 µm. After quenching in water the simulative flatness is about 53 µm and the experiments show an average flatness of 121 µm with a big deviation between the single measured components. Simulation and experiments show a significantly higher average flatness after quenching in water compared to quenching in IL.

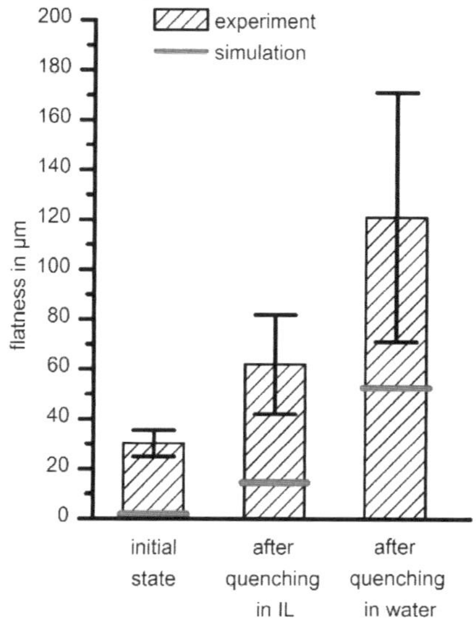

Figure 11: Flatness of the lateral surfaces of distortion samples before and after heat treatment for experiments (with average and standard deviation) and simulation (no flatness of initial state for experiment determined).

One reason, despite the initial flatness, for differences between experiment and simulation which is also valid for the simulation of the base is a heat transfer coefficient which is assumed to depend only on time and not on location (Figure 5). To verify this assumption quenching experiments with distortion samples equipped with six thermocouples (Figure 4) in different positions have been performed and the cooling curves have been compared with the simulations of a location independent heat transfer coefficient. In Figure 12 the experimental and simulated cooling curves are illustrated. The qualitative comparison between the experiment and the simulation is good. Both results indicate that the protruded part of the sample (Pos. F) is cooling fastest and the core of the sample (Pos. E) is cooling slowest. But quantitatively there are differences between experiment and simulation (Figure 13). The simulation results in a faster cooling. Especially below 350 °C the differences between experiment and simulation are obvious. In Figure 14 the cooling curves are shown for quenching the distortion sample in tap water. Both results again show that the protruded part (Pos. F) cools fastest and the core (Pos. E) cools slowest. The direct comparison (Figure 15) shows that the simulation for the protruded part is faster than the experiment. The simulation of Pos. A is in a good agreement to the experiment with a little longer lasting film boiling at the very beginning. The simulated cooling of the sample center (Pos. E) is slower than in the experiment.

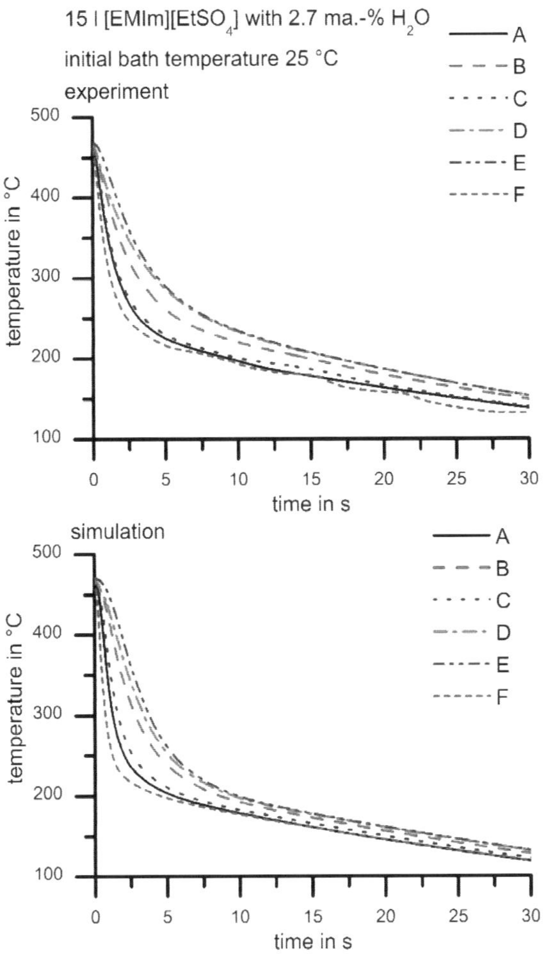

Figure 12: Experiment (top) and simulation (bottom) of quenching a distortion sample with six thermocouples in IL.

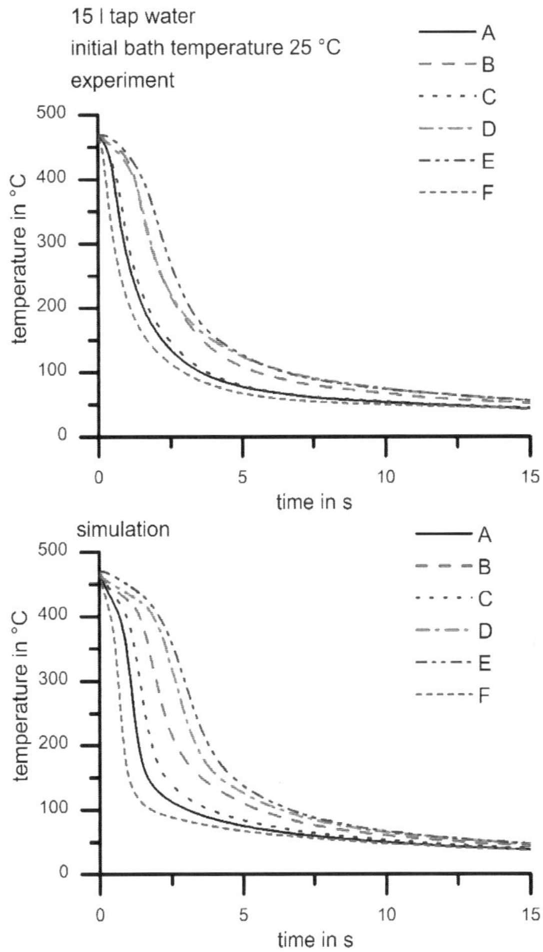

Figure 14: Experiment (top) and simulation (bottom) of quenching a distortion sample with six thermocouples in tap water.

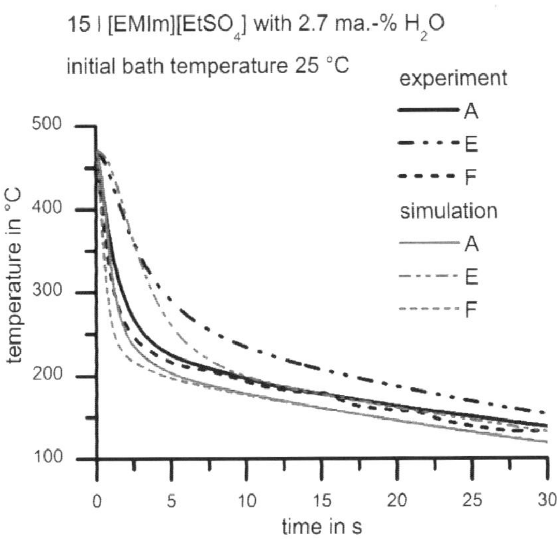

Figure 13: Comparison of experiment and result of selected positions of quenching a distortion sample with six thermocouples in IL.

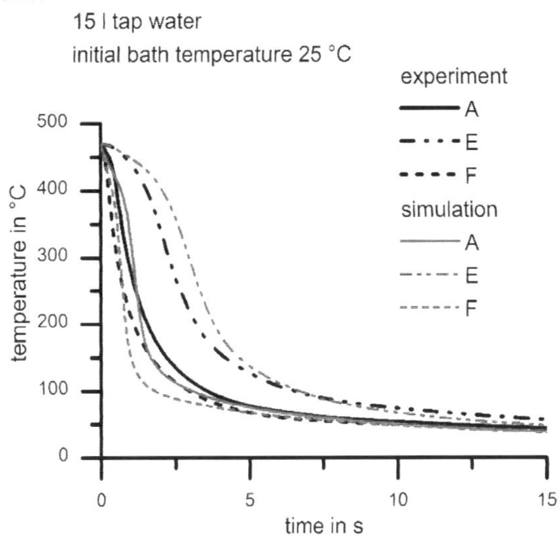

Figure 15: Comparison of experiment and result of selected positions of quenching a distortion sample with six thermocouples in tap water.

All in all the simulations are qualitatively in a good agreement with the experiments with some slight quantitatively differences. These differences could have various reasons. For the simulation only heat transfer coefficient calculations from cylinder-quenching experiments were used. So different flow conditions in the quenchant due to the markedly different shape of the profiles were not taken into account. For the simulation the heat transfer coefficient is assumed not to depend on the location of the sample surface, but in case of water quenching not every area of the distortion sample is covered by a vapour film (Figure 8). This could constitute the difference between the simulation and the experiment at Pos. A in Figure 15. Furthermore the surface structure of the cylindrical samples and the distortion samples might be different which can have a substantial influence on the cooling characteristic [14]. The assumption that the heat transfer coefficient does not depend on the location on the sample surface is a simplification that could explain the differences of distortion between experiment and simulation. Furthermore, for a better agreement between experiments and simulations residual stresses and plastic strains of the initial state must be taken into consideration in simulations.

Conclusions

Ionic Liquids (IL) are a new class of liquid quenching media with melting points below room temperature and without any macroscopic Leidenfrost-effect at low water contents. Quenching experiments of aluminium components have been performed successfully in IL proving a high thermal stability. Additionally, the distortions of a complex aluminium component quenched in water and IL have been measured and simulated. The simulated deformations have the same direction and magnitude of scale than the experiments. The results indicate a huge potential to homogenise immersion quenching.

Acknowledgments

The authors acknowledge the Fraunhofer-Application Center for Large Structures in Production Engineering AGP, Rostock, Germany, for the distortion measurements. This work is funded by the Deutsche Forschungsgemeinschaft: DFG – KE 616/18-1,-2; KR 2491/3-1,-2; SCHI 331/24-1,-2.

References

[1] P. Wasserscheid, T. Welton, Ionic Liquids in Synthesis, 2.th ed., Wiley-VCH Verlag GmbH & Co. KGaA, Weinheim, Germany, 2007.

[2] E. Schlücker, P. Wasserscheid, Ionic liquids in mechanical engineering, Chemie Ingenieur Technik 83 (2011) 1476–1484.

[3] M. Beck, C. Schmidt, M. Ahrenberg, C. Schick, U. Kragl, O. Keßler, The ideal quenching medium?: - Characterisation of ionic liquids for heat treatment of metallic components, HTM 68 (2013) 214–223.

[4] M. Beck, C. Schmidt, M. Ahrenberg, C. Schick, U. Kragl, O. Keßler, Ionic Liquids as New Quenching Media for Aluminium Alloys and Steels, HTM 70 (2015) 73–80.

[5] C. Schmidt, M. Beck, M. Ahrenberg, C. Schick, O. Keßler, U. Kragl, Room temperature ionic liquids in a heat treatment process for metals, RSC Adv 4 (2014) 55077–55081.

[6] M. Ahrenberg, M. Brinckmann, Schmelzer, J. W. P., M. Beck, C. Schmidt, O. Keßler, U. Kragl, S.P. Verevkin, C. Schick, Determination of volatility of ionic liquids at the nanoscale by means of ultra-fast scanning calorimetry, Phys Chem Chem Phys 16 (2014) 2971–2980.

[7] International Organization for Standardization, Industrial quenching oils: Determination of cooling characteristics nickel-alloy probe test method, 1st ed. ed., International Organization for Standardization, Genève, Switzerland ISO 9950.

[8] T. Lübben, H.P. Bomas, H.P. Hougardy, P. Mayr, Beschreibung der Abschreckwirkung flüssiger Abschreckmittel am Beispiel zweier Härteöle, HTM 46 (1991) 24–34.

[9] The Aluminum Association, International Alloy Designations and Chemical Composition Limits for Wrought Aluminum and Wrought Aluminum Alloys, The Aluminum Association, Inc., Arlington, 2015.

[10] ASM International Handbook Committee, ASM Handbook, Volume 02 - Properties and Selection: Nonferrous Alloys and Special-Purpose Materials, ASM International, [S.l.], 1990.

[11] M. Reich, O. Kessler (Eds.), Numerical and experimental analysis of residual stresses and distortion in different quenching processes of aluminum alloy profiles, Chicago, IL, 2012.

[12] B. Milkereit, M. Beck, M. Reich, O. Kessler, C. Schick, Precipitation kinetics of an aluminium alloy during Newtonian cooling simulated in a differential scanning calorimeter, Thermochim. Acta 522 (2011) 86–95.

[13] R.B. McLellan, T. Ishikawa, The elastic properties of aluminum at high temperatures, Journal of Physics and Chemistry of Solids 48 (1987) 603–606.

[14] N. Kozlov, O. Keßler, Influencing on liquid quenching by surface structuring, International Journal of Thermal Sciences 101 (2016) 133–142.

International Federation of Heat Treatment and Surface Engineering 2016
Proceedings of the 23rd IFHTSE Congress
April 18–21, 2016, Savannah, Georgia, USA

Beer, Blood and Urine - Mythological Quenchants of Ancient Blacksmiths

D. Scott MacKenzie, PhD, FASM
Houghton International, Inc., Valley Forge, PA USA 19482
smackenzie@houghtonintl.com

Gloria Graham
Houghton International, Inc., Valley Forge, PA USA 19482
ggraham@houghtonintl.com

Abstract

Archaeological digs have found many types of knives, with varying quality of steel and microstructure. Typically these steels are carbon steels with carbon contents on the order of 0.60%. Historically, there have been many myths concerning the quenchants used by ancient blacksmiths in the heat treatment of swords and knifes. Various liquids have been cited in the archaeometallurgical literature as quenchants. Each of these quenchants is supposed to extend to the knife special and even mythical properties. However, none have been examined for cooling curve behavior. In this paper, various quenchants are examined for typical heat transfer, and microstructure is predicted for simple steels commonly used in ancient knife making.

Introduction

In a previous paper (1), a brief review of the history of quenching was discussed. In this review the overall technology of various regions of the world were shown.

Focusing on Europe, and specifically early medieval Anglo-Saxon and Viking knife manufacture, several quenchants were identified. Some Icelandic sagas spoke of searching through many kingdoms to find the proper water to harden the sword *Ekkisax* and weapons that are hardened in blood (2). Probably the first significant work in Medieval Europe was written by Theophilus (3) (1125), a 12th century German Monk. The "Diver Arts" describe several good quenchants. His recommendations for quenchants were very specific:

"Tools are also given a harder tempering in the urine of a small, red-headed boy than in ordinary water." (3)

Other recommendations for quenchants included the urine of goats fed ferns for three days.

Giambattista della Porta (ca. 1535-1615) in his books "Natural Magic," showed an excellent understanding of the reason why many quenchants were effective, and some of the underlying principles:

"If you quench red hot iron in distilled vinegar, it will grow hard. The same will happen, if you do it into distilled urine, by reason of the salt it contains in it. If you temper it with dew, that in the month of May is found on vetches leaves, it will grow most hard. For what is collected above them, is salt, as I taught elsewhere out of Theophrastus. Vinegar, in which Salt Ammoniac is dissolved, will make a most strong temper. But if you temper Iron with Salt of Urine and saltpeter dissolved in water, it will be very hard. Or if you powder Saltpeter and Salt Ammoniac, and shut them up in a glass vessel with a long neck, in dung, or moist places, till they resolve into water, and quench the red hot Iron in the water, you shall do better. Also iron dipped into a Liquor of Quicklime, and Salt of Soda purified with a Sponge, will become extreme hard. All these are excellent things, and will do the work."

There was also an understanding of the cause of quench cracking, and the results of quenching in other than water for Porta's "*The Temper for Instruments to let blood*":

"It is quenched in oil, and grows hard, because it is tender and subtle. For should it be quenched in water, it would be wrested and broken."

Various authors, describe other quenchants including: pigeon droppings, flour, honey, olive oil and milk (4) (5) (6). Other quenchants, including urine, water and solubilized animal fats and whale oil are described by Smith (7), Biringuccio (8), Agricola (9) and others:

"Take clarified honey, fresh urine of a he-goat, alum, borax, olive oil, and salt; mix everything well together and quench therein."

Von Stahel and Eysen (1532) (10) also discussed quenchants to be used for heat treated parts:

"Take varnish, dragon's blood, horn scrapings, half as much salt, juice made from earthworms, radish juice, tallow, and vervain and quench therein. It is also very advantageous in hardening if a piece that is to be hardened is first thoroughly cleaned and well-polished."

Other modern researchers (11) (12) have indicated that oil, milk, urine and blood have been used as quenchants.

A comprehensive investigation of early Anglo-Saxon and Viking knives was conducted by Blakelock (13). She identified five types of knife manufacturing methods based on blade cross-sections (Figure 1). The alloys used can be further classified (Table 1).

Four different methods of quenching were identified (13) that would achieve different microstructures. The first method is complete immersion into a fast quenchant. This would achieve a predominately martensitic microstructure.

Using complete immersion, into a slower quenchant (Method 2) would result in a mixed microstructure of martensite, bainite, and pearlite and ferrite, depending on the speed of the quenchant and the hardenability of the steel.

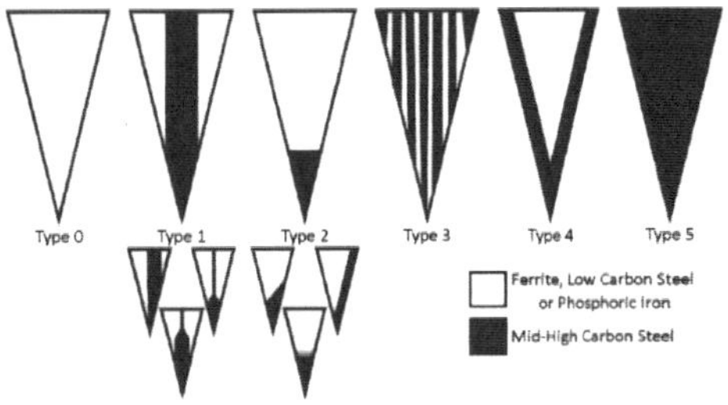

Figure 1. Knife alloy and manufacturing typology identified by Blakelock (13).

Table 1. Knife alloy based on topology (13).

Knife Topology Type	Alloy
0	Ferritic or Phosphoric Iron
1	Medium Carbon steel sandwiched between ferric, phosphoric or low carbon steel
2	Medium carbon steel welded to ferritic, phosphoric or low carbon steel
3	Piled Iron
4	Core of ferritic, phosphoric or low carbon steel, with medium carbon steel wrapped around blade
5	Medium to high carbon steel

In the third method, similar to the Japanese methods of controlled quenching, a clay mixture is applied to the surface of the knife. The thickness of the clay determines the quench rate. The clay is thinnest at the edge of the blade, and thickest at the ridge of the blade – opposite the edge. The edge is quenched with the highest heat transfer rate and produces martensite, while the ridge experiences a much milder quench and transforms to a mixture of pearlite and ferrite.

Method	Quenchant	Microstructure
Immersion	Fast	Martensite
Slow Quench	Slow	Mixture of martensite. Bainite, ferrite and pearlite
Gradient	Fast	Mixture of martensite. Bainite, ferrite and pearlite
Partial	Fast	Auto-tempered martensite

The final method is partial immersion in the quenchant, with only the knife edge immersed in the quenchant. This would produce a predominately martensitic structure, depending on the hardenability of the steel.

An extensive study of obtained microstructures from knives at a variety of locations was conducted by Blakelock (13). Over 240 different knives across England, Ireland and Europe were examined. The range of investigation covered knife manufacture over the early medieval period (AD 410-1100). The resultant summary of microstructures found are shown in Figure 2.

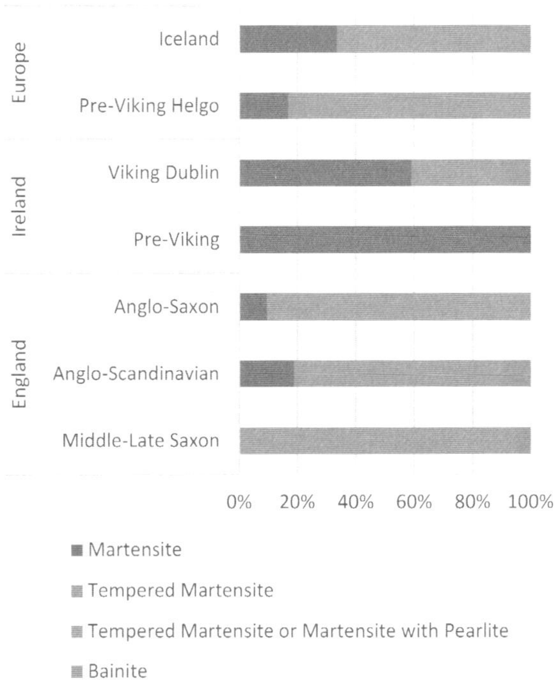

Figure 2. Classification of microstructures from knives examined by Blakelock (13).

An examination of the literature revealed investigations of microstructures and technologies which were available to ancient blacksmiths. However, no investigation of the quenchants they used has been conducted. The purpose of this investigation was to examine possible cited quenchants, and evaluate the resultant microstructures when quenching a medium to high carbon steel.

Experimental

Preparation of Quenchants:

The quenchants used in this investigation were purchased or prepared. The reference accelerated quench oil, Houghto-Quench K, was new and pulled from inventory. Water was tap water available in the laboratory.

The beer used for this evaluation was Guinness Extra Stout. There was some discussion regarding the type of beer that would have been available in the 11th-16th century. While ales are brewed from malt and yeast, beers have hops. Hops, which originated in Egypt in the first century, were cultivated in Belgium and Holland and were introduced to England in the early 15th century (14). The first reference use of hops in beers is from the 9th century (15). Since the use of hops was well established from the 9th century, it was decided to use a heavy hopped beer. Once selected, after a thorough tasting process (quality control purposes), the beer was allowed to age for 24 hours in open air. This was done to allow excess carbonation to escape and produce a "flatter" beer. It was thought that modern beers would more likely have greater carbonation than older beers.

Whole organic milk was purchased at the local market and refrigerated until use. This was done to prevent spoilage and for convenience.

While urine is readily available, because of health and safety issues, a synthetic urine was created. From a NASA study (16), the composition of human urine is shown in Table 2.

Table 2. Chemical composition of Human Urine, after Putnam (16)

Component	Formula	Amount (g/L)
Urea	CH_4N_2O	9.30
Sodium Chloride	NaCl	1.17
Potassium Chloride	KCl	0.75
Creatine	$C_4H_7N_3O$	0.67
Water	H_2O	Balance

Synthetic urine was created using the constituents listed in Table 2, with the exception of Creatine. It was not thought that the small amounts of Creatine would significantly influence the cooling curve, as the shape of the curve would be dominated by the sodium and potassium salts. Technical grade chemical were used throughout.

Blood was more of a technical and political challenge. While both human and animal blood are available, it is generally strictly controlled. There are safeguards against bio-hazards, as well as general squeamishness regarding blood in the general population. However, an understanding butcher was contacted and one liter of bovine blood was obtained. Ovine (sheep) blood was preferred as that was more common in Middle Age Europe. However ovine blood was unavailable. The blood was obtained fresh and refrigerated prior to use.

Steel used for simulation of Properties:

The steel selected for microstructure prediction was AISI 1060. This is a medium carbon steel similar to those found in many knives, and is a starting point for investigation. This steel is more difficult to fabricate than lower carbon steel grades. It is used in a variety of industrial applications such as shafting, agricultural hand tools, and springs. It is a common used in modern blacksmithing because of availability. The composition of AISI 1060 steel is found in Table 3.

AISI 1060 is a shallow hardening grade of carbon steel. The Time-Temperature-Transformation curve for this steel is shown in Figure 3. The typical Jominy data is shown in Figure 4. The iso-hardness curve for AISI 1060 is shown in Figure 5. The carbon-equivalent is 0.72%C, so it is very crack-sensitive.

Table 3. Typical composition of AISI 1060 Steel, and the composition used in this investigation.

C	Mn	Si	P	S
0.55/0.65	0.50/0.80	0.15/0.35	0.04 Max.	0.04 Max.
0.60	0.70	0.21	0.04	0.04

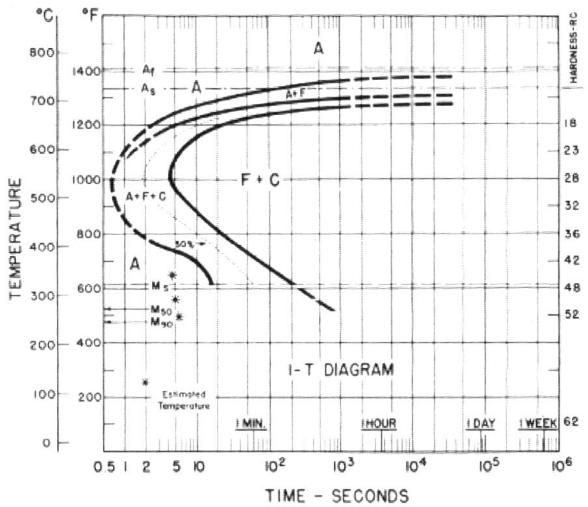

Figure 3. Typical Time-Temperature-Transformation diagram for AISI 1060 Steel (17).

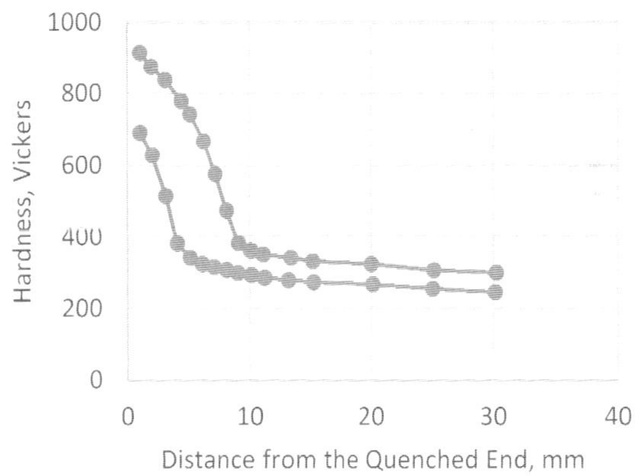

Figure 4. Typical Jominy End Quench for AISI 1060 steel.

The phase transformation properties are shown in Table 4.

While many medieval knives contained up to 1% Phosphorus, the lack of suitable Time-Temperature-Transformation diagrams prevented simulation of these alloys.

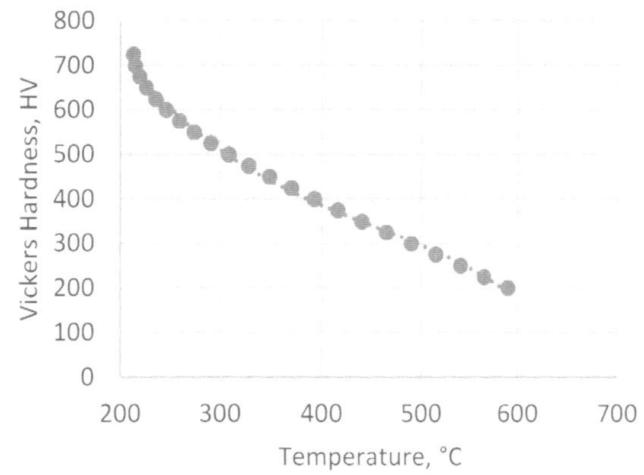

Figure 5. Iso-hardness curve based on TTT diagram.

Table 4. Critical phase transformation temperatures for AISI 1060.

Property	Symbol	°F	°C
Martensite Start	M_s	448	231
Martensite Finish	M_f	61	16
Bainite Start	B_s	851	455
Bainite Finish	B_f	635	335
γ Lower Critical (Equil.)	Ae_1	1325	718
γ Upper Critical (Equil.)	Ae_3	1378	748
γ Lower Critical (Heating)	Ac_1	1332	722
γ Upper Critical (Heating)	Ac_3	1406	763

Cooling Curves:

Cooling curves were obtained from the various quenchants according to ASTM D6200 (18). The quenchants were poured into 1-liter beakers, and were warmed to ambient temperature (22°C) using a hot plate. An Inconel 600 probe per ASTM D6200 was used, and heated to 850°C. After insuring that the probe reached the furnace temperature, the probe was withdrawn from the furnace and immersed into the quenchant. Insertion depth was kept constant per ASTM D6200. No agitation was used for the quenchants during quenching. The control (Houghto-Quench K) was quenched at 60°C.

Heat transfer coefficients and prediction of properties:

After acquisition of the cooling curve using an IVF SmartQuench (19), the temperature-dependent heat transfer coefficient was calculated using an iterative, inverse calculation as part of the IVF SQintegra package. From the heat transfer coefficient, the hardness and microstructure were predicted using an iso-thermal model (20). This prediction is part of the IVF SQintegra software package.

Results:

The cooling curves of the tested quenchants are shown in Figure 6. As can be seen by the cooling curves, the synthetic urine is a very fast quenchant and greatly exceeds the cooling rate of water. This is thought to be caused by the salts precipitating on the surface of the probe, resulting in nucleation sites for nucleate boiling to occur. This cooling curve is similar to commercial quenchants containing inorganic salts.

The water quench is similar to the synthetic urine, but much slower. Some stable vapor phase is present.

Beer showed a quench rate similar to the fast accelerated quench oil, with a similar maximum cooling rate. However, a long extended vapor phase occurred due to the residual carbonation present in the beer. This acts to stabilize the vapor phase.

Milk showed a moderate vapor phase, with a slow overall maximum cooling rate. This is thought to be the result of the butter fat present in the quenchant.

Blood showed a very unusual cooling curve. There was an initial vapor phase similar to the other quenchants, however, this was followed by a very long and prolonged stable cooling rate from approximately 750°C to 100°C. Virtually no nucleate boiling occurred. In a typical cooling curve, virtually all the quenchants reached nearly ambient temperature within 60 seconds. Blood on the other hand, at the end of the 60 seconds, the probe was at a temperature of 341°C. It took 3 minutes for the cooling curve probe to reach 80°C. This is an extremely slow cooling curve.

Upon extraction of the probe from the bovine blood, it was immediately observed that a large build-up of what appeared to be cooked fat and blood. This build-up was extensive and coated the probe to a thickness of approximately 8 mm or more. It is thought that this cooked fat and blood insulated the probe, slowing heat transfer. This insulating later drastically slowed the heat extraction rate.

A summary of the cooling curve data is shown in Table 5.

The resultant heat transfer coefficients from the inverse calculation for the various quenchants are shown in Figure 7. The results mirror the cooling curve results. This data is used in the calculation of microstructure for the various quenchants. In all cases, with the exception of blood, the maximum heat transfer occurs around 300°C. Heat transfer coefficients for blood were flat with an approximate heat transfer rate of 200 W/m^2-K. There was little change as a function of temperature. It is thought that the presence of cooked fat and blood from the blood created an insulating layer around the probe, slowing heat transfer.

In making the property and microstructure predictions, several assumptions were made. The first assumption was that the knives were austenitized at 850°C. The second assumption is that the knives were rectangular in shape, with the approximate thickness of 6.33 mm thick. Using the concept of "equivalent round" (21) for a rectangle, the equivalent round of a 6.33 mm thick plate is 9.5 mm. This reduces to an equivalent cylinder radius of 4.75 mm. This dimension of 4.75 mm was used in the calculation of properties.

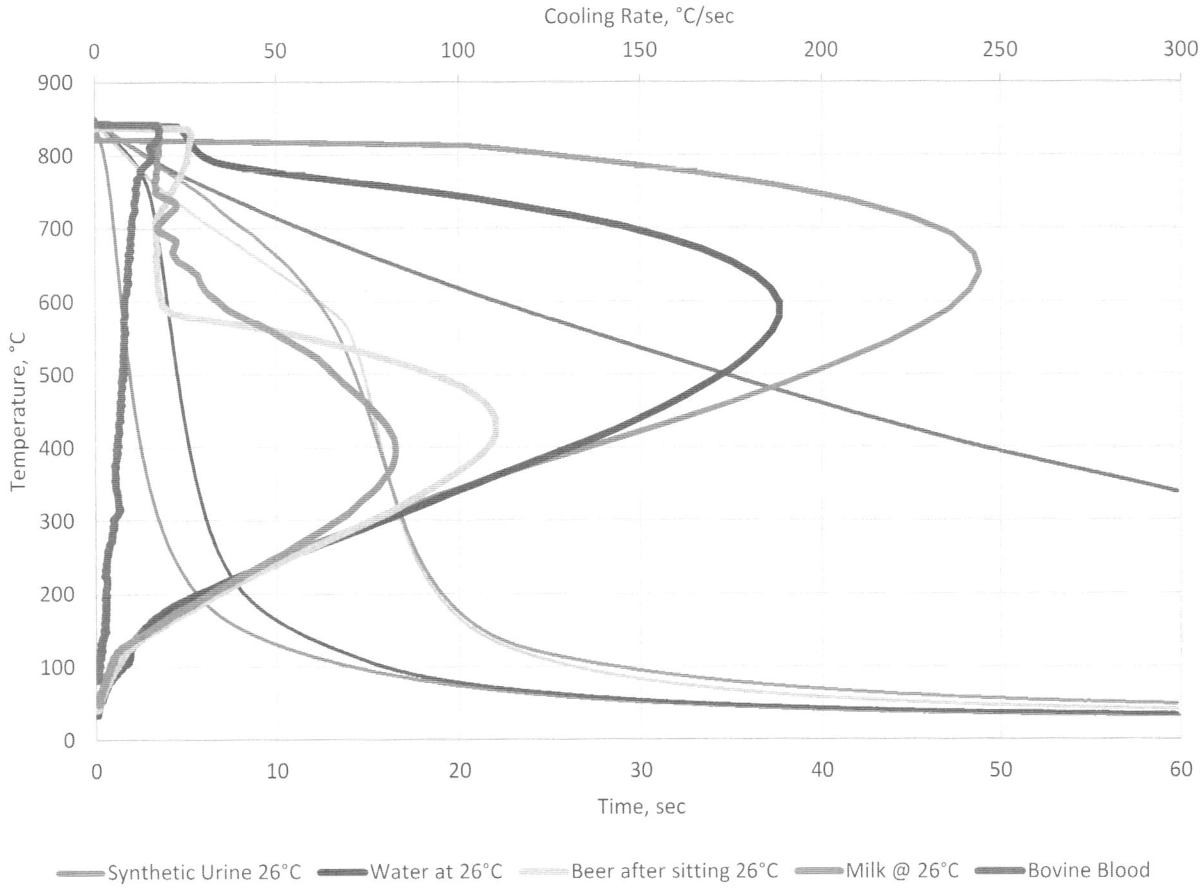

Figure 6. Cooling curves of the various quenchants at 22C. Quenchants tested per ASTM D6200.

Table 5. Summary data from cooling curve testing per ASTM D6200.

Property	Blood	Urine	Water	Beer	Milk
Max. Cooling Rate, °C/s	17.02	245.62	191.4	111.36	83.19
T @ Max. Cooling, °C	810.1	626.02	577.96	422.08	397.55
Start of Boiling, °C	889.84	842.58	798	596.96	663.02
Start of Convection, °C	753.05	165.7	138.04	79.23	98.23
Rate at 300 C, °C/s	5.16	78.91	80.09	75.86	66.93
Time to 600 °C, s	22.00	1.47	3.98	12.29	12.23
Time to 400 °C, s	49.04	2.49	5.16	15.72	15.61
Time to 200 °C, s	94.36	5.56	8.17	18.72	19
Theta 1 θ_1, °C	847.59	841.82	797.97	592.99	659.6
Theta 2 θ_2, °C	778.19	299.83	232.83	167.91	173.52

The hardness was predicted as a function of distance from the center. However, because of the high hardenability compared to the equivalent round radius, there was little difference from surface to center for a given quenchant. Each quenchant, because of the shape of the cooling curve gave different hardness results. The results of the property prediction are shown in listed in

Quenchant	Surface		Center	
	HV	HRC	HV	HRC
Houghto-Quench K	607	56	558	53
Urine	688	60	666	59
Beer	266	25	261	24
Milk	414	42	402	41
Water	688	60	645	58
Blood				

.

Microstructures were predicted using an iso-thermal model based on the Time-Temperature-Transformation (TTT) diagram for AISI 1060 steel. While this method is not completely 100% accurate, it provides a good estimate of the resultant microstructures. The resultant microstructure predictions surface to center are shown in *Figure 8* through Figure 13.

Discussion

The predicted microstructures resulting from quenching the 9.5 mm diameter round of AISI 1060 steel varied widely based on the quenchant (Figure 14). Water and synthetic urine produced nearly identical microstructure results due to the very fast overall quenching rates of each of these quenchants. Each of these quenchants produced predominately martensite in excess of 80%, with 5% bainite and 5% pearlite + ferrite at the surface.

The synthetic urine produced slightly more martensite at the surface and at the interior of the AISI 1060 round. This was accompanied by an overall reduction in the amount of bainite and ferrite + pearlite.

Synthetic urine is very similar in composition to brine quenchants. There are several brine type quenchants with similar compositions (without urea) commercially available. In these quenchants, the salts present precipitate on the surface of the probe or part, and destabilize the vapor phase. Nucleate boiling immediately initiates.

Milk, with a much slower maximum cooling rate than beer, had a much faster cooling rate in the critical temperature range of 750-550°C. This resulted in a much greater martensite content. The quench rate compared to water shows the effect of the butterfat inherent in whole milk.

Interestingly, while having nearly the same maximum cooling rate as the fast control oil, beer produced a predominately ferrite + pearlite structure (90%), with nearly equal portions of martensite (5%) and bainite (5%). This is very different from the results of the Houghto-Quench K which produced approximately 70% martensite at the surface. It is likely that the very slow initial quench and extended vapor phase due to carbonation resulted in the relatively long time in the ferrite and pearlite phase field. Significant transformation occurred. This can be understood with examination of the Continuous transformation diagram for AISI 1060 steel (Figure 15).

Blood, because of its very slow heat extraction rate and very poor cooling curve, had the lowest heat transfer coefficient of any of the quenchants. The heat transfer coefficient was nearly flat with a value of approximately 200 W/m^{2}- K. Because of the very low cooling rate and heat transfer coefficient, the predicted properties were 100% ferrite and pearlite. The hardness was also very low at the center and surface of the bar at 21 HRC (244 HV).

The two fastest quenchants, water and synthetic urine yielded a predominately martensitic structure. However, because of the speed of the quenchant, it is likely to result in cracking of the knives, especially if urine is used. This is particularly true if medium to high carbon contents are used.

To solve this issue of cracking, the use of time quenching, where the part is only immersed for a short period of time, or other methods would have to be used. Interrupted quenching, by immersing into water or urine, then quickly removing and immersing in an alternative quenchant such as oil would reduce the inherent thermal gradients and propensity for cracking.

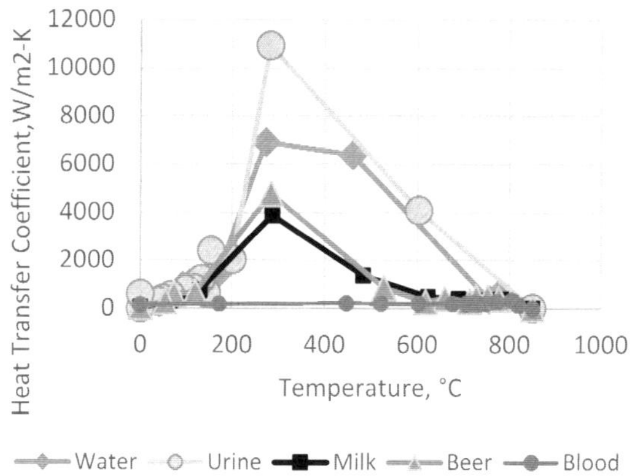

Figure 7. Results of the inverse calculation of heat transfer coefficients of the various quenchants.

Quenchant	Surface		Center	
	HV	HRC	HV	HRC
Houghto-Quench K	607	56	558	53
Urine	688	60	666	59
Beer	266	25	261	24
Milk	414	42	402	41
Water	688	60	645	58
Blood				

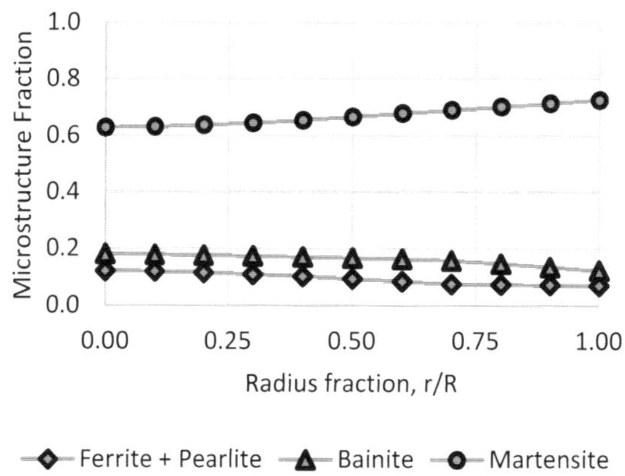

Figure 8. Microstructure prediction for Houghto-Quench K when quenching a 9.5 mm diameter round of AISI 1060 steel.

Another issue that remains to be resolved, is the quench sensitivity of the various alloys used. There is a lack of phase transformation data on the three main types of steels available to ancient blacksmiths : ferritic iron with few alloying elements (<0.1%), phosphoric iron containing more than 0.1% and upwards of 1% phosphorus, and medium to low carbon steel (0.60%+ C), with few other alloying elements (no silicon or manganese). Additional research into the composition, phase transformations and kinetics of these alloys is necessary.

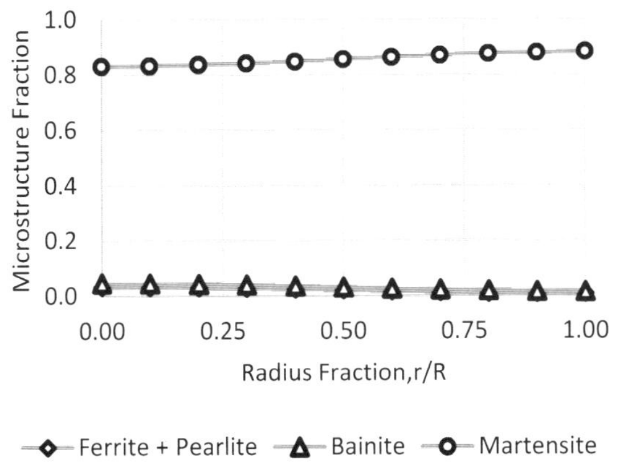

Figure 9. Microstructure prediction of synthetic urine quenching a 9.5 mm diameter round of AISI 1060 steel.

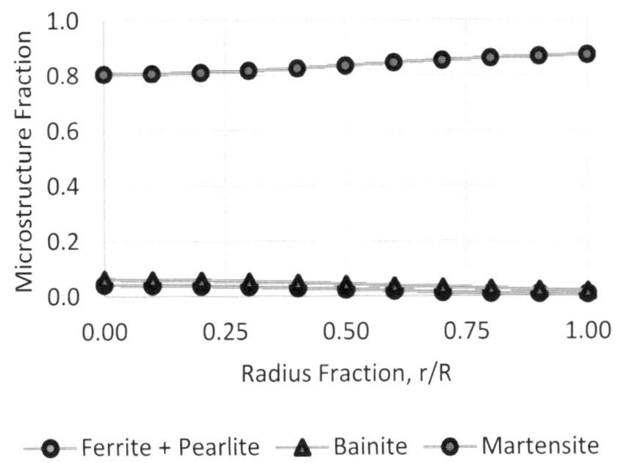

Figure 10. Microstructure prediction of water quenching a 9.5 mm diameter round of AISI 1060 steel.

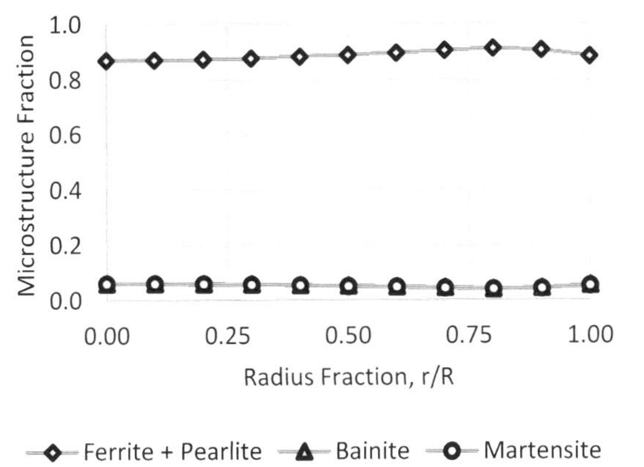

Figure 11. Microstructure prediction of beer quenching a 9.5 mm diameter round of AISI 1060 steel.

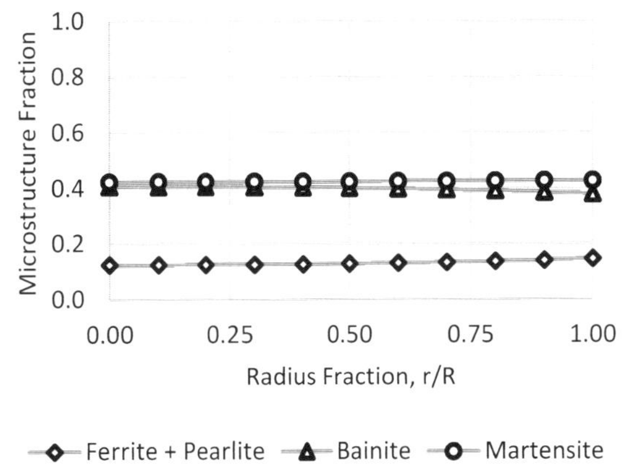

Figure 12. Microstructure prediction of milk quenching a 9.5 mm diameter round of AISI 1060 steel.

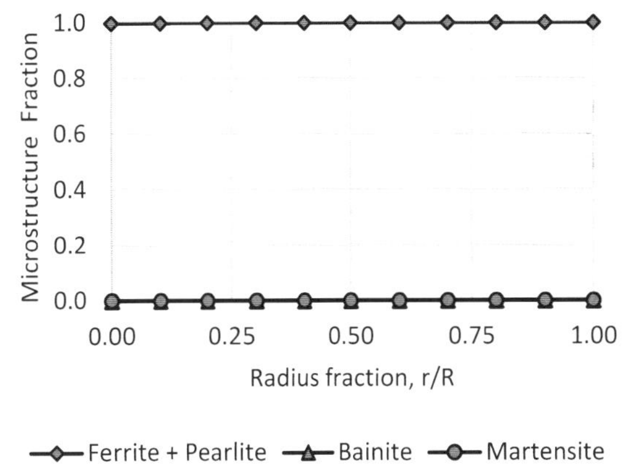

Figure 13. Microstructure prediction of bovine blood quenching a 9.5 mm diameter round of AISI 1060 steel.

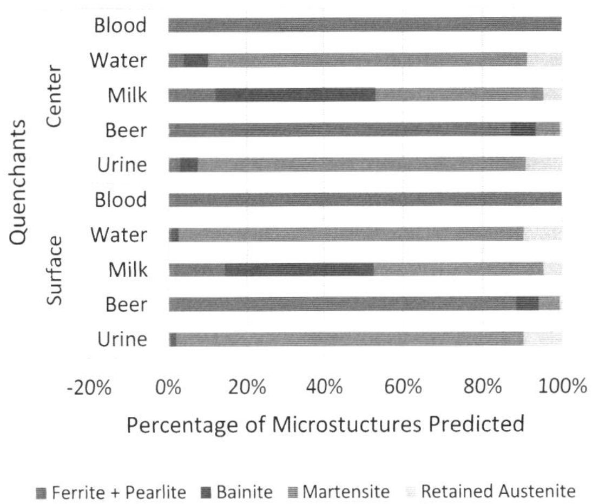

Figure 14. Summary of predicted microstructures at the surface and center of a 9.5 mm diameter round of AISI 1060 steel.

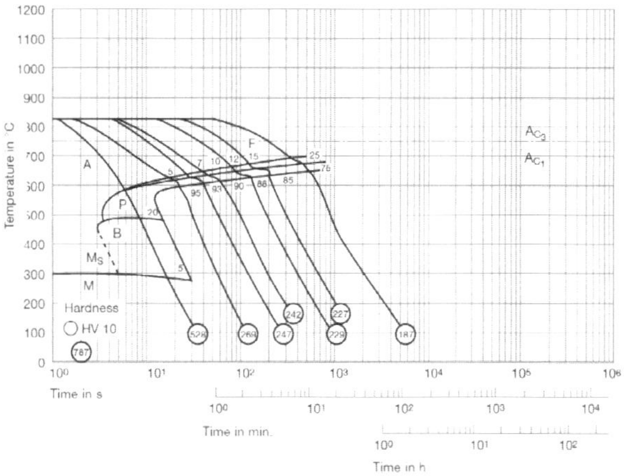

Figure 15. Continuous-Time-Transformation diagram for AISI 1060 (C60) steel (22).

Conclusions

In this work, cooling curves of various quenchants available to early blacksmiths were examined and evaluated. This was compared to modern accelerated quench oil. Properties and microstructure were predicted for these quenchants, when quenching a 9.5 mm diameter round of AISI 1060 steel.

The following conclusions can be reached regarding this investigation:

- Synthetic urine used in this investigation is a very fast quenchant and achieves the hardest microstructure of predominately martensite. Because of its speed, it is likely to result in quench cracking of anything greater than a carbon equivalent greater than 0.4%C.
- Water achieved nearly an identical microstructure and hardness as the synthetic urine. Little improvement in microstructure or hardness is achieved by increasing the quench speed above that of water, with an increased propensity for quench cracking. It is still likely that quench cracking would occur. Care would have to be taken by the blacksmith to insure that his knives or parts were quenched crack free.
- Whole milk achieved a mixed microstructure with nearly equal parts of martensite and bainite The balance of the microstructure consisted of approximately 15% of ferrite and pearlite. The cooling exhibited a faster cooling rate in the critical range of 750-550°C, minimizing the formation of non-martensitic transformation products.
- Beer showed a maximum cooling rate similar to a commercial accelerated oil quenchant. However, the temperature of maximum cooling was very low at 422°C. Beer also exhibited a very long and stable vapor phase ostensibly due to the carbonation in the beer. The CO2 from carbonation would create a very stable vapor phase, which largely contributed to the large fraction of ferrite and pearlite present at the center and surface of the AISI 1060 round.
- Blood showed an unusual cooling curve with a very prolonged slow cooling rate from 750°-100°C. This resulted in very slow cooling at very poor properties. The heat transfer coefficient was constant in this temperature range at a value of approximately 200 W/m^2-K. Properties predicted were the poorest of the quenchants investigated, with a microstructure consisting of 100% ferrite and pearlite. As a result of the very poor cooling curve of bovine blood, it is highly unlikely that this quenchant could ever be effectively used to harden steel. This is in direct contradiction of the legends concerning the use of blood as a quenchant.
- The basic methodology of using an iso-thermal microstructure is useful when comparing different quenchants , and products of different diameters.
- Additional work is needed evaluating the Time-Temperature-Transformation (TTT) and Continuous-Time-Transformation behavior of steels observed in archaeological digs. This is particularly true of high phosphorus (>0.1% P) alloys.

It is hoped that this work on possible quenchants used by ancient blacksmiths can further the understanding of the technological and cultural contribution of the first metallurgists.

Acknowledgements

I would like to thank the management of Houghton International, Inc., Valley Forge, PA for allowing me to present this work at the ASM IFHTSE 2016 Advanced Thermal Processing Congress. I would also like to thank my co-author, Gloria Graham for conducting the cooling curve testing used in this work.

References

[1] *History of Quenching.* MacKenzie, D. S. 2, 2008, Journal of Heat Treatment and Surface Engineering, Vol. 2, pp. 68-73.

[2] *De navngivne svaerd i saga, saga og folkevise.* Drachmann, A. G. Copenhagen : s.n., 1967. Filogisk-Historiske Samfund. p. 264.

[3] Theophilus. *De Diversis Artibus.* [ed.] C. R. Dodwell. London : Thomas Nelson & Sons, 1941.

[4] *A study in Pattern Welding.* Anstee, J.W. 1961, Medieval Archaeology, Vol. 5, pp. 71-93.

[5] Smith, C. S. *Sources for the History of the Science of Steel 1532-1786.* Boston : MIT Press, 1986.

[6] Haedeke, H-U. *Metalwork.* [trans.] V. Menkes. s.l. : Universe Books, 1970. p. 227.

[7] Smith, C. S. *A History of Metallography: The development of Ideas of the Structure of Metals before 1890.* Cambridge, MA : MIT Press, 1988.

[8] Biringuccio, V. *Pirotechnia.* [trans.] C. S. Smith and M. T. Gnudi. Toronto, Canada : General Publishing Company, 1990.

[9] Agricola, G. *De Re Metallica.* [trans.] Herbert Hoover. New York : Courier Cover Publications, 1950.

[10] Steiner-Welz, s., [trans.]. *Von Stahel und Eysen.* 2002.

[11] Tylecote, R. F. *Scientific Examination and Analysis of Iron Objects.* Oxford : Oxford University Press, 1990. pp. 140-154.
[12] *The Early Blacksmith.* Maddin, R. Belfast : Ulster Museum, 1987. Symposium of the UISPP Comite' pour la Sidrurgie Ancienne. pp. 7-17.
[13] Blakelock, E. PhD Dissertation: The Early Medieval Cutting Edge of Technology: An Archaeomettallurgical, technological and social study of the manufacture and use of Anglo-Saxon and Viking iron knives, and their contribution to the early medieval economy. Bradford, West Yorkshire, U.K. : University of Bradford, 2012.
[14] Unger, R. *Beer in the Middle Ages and the Renaissance.* Philadelphia : University of Pennsylvania Press, 2004.
[15] Hornsey, I. S. *A History of Beer and Brewing.* London : Royal Society of Chemistry, 2003. p. 305.
[16] Putnam, D. *Composition and Concentrative Properties of Human Urine.* McDonnell Douglas. Long Beach : NASA, 1971. CR-1802.
[17] *Metals Handbook.* Metals Park, OH : ASM, 1981.
[18] ASTM International. Standard Test Method for Determination of Cooli*ng Characteristics of Quench Oils by Cooling Curve Analysis.* Conshocken : ASTM, 2012. ASTM D6200-1 (2012).
[19] Swerea IVF AB. *SQIntegra Product Brochure.* Monldal : s.n., 2012.
[20] *New Tool for Calculation of Heat Transfer Coefficients, Microstructure and Hardness from Cooling Curve.* Troell, E., et al. [ed.] H-W Zoch. Berlin, Germany : s.n., 2007. IFHTSE 5th International Conference on Quenching and Control of Distortion.
[21] *Selection of Quenchants.* MacKenzie, D. S. 2014, Journal of International Federation of Heat Treatment and Surface Engineering, Vol. 8, pp. 8-14.
[22] Edelstahl Witten-Krefeld GMBH. *Heat Treatable Steels.* 2014.
[23] *Iron and its Alloys in the Fifth to Eleventh Centuries AD in England.* McDonnell, G. 3, s.l. : Taylor & Francis, 1989, World Archaelogy, Vol. 20, pp. 373-382.
[24] Williams, H., [trans.]. *A sixteenth-century German treatise: Von Stahel und Eysen.* 1532. Reprinted from Technical studies, vol. IV. no. 2, October, 1935.

International Federation of Heat Treatment and Surface Engineering 2016
Proceedings of the 23rd IFHTSE Congress
April 18–21, 2016, Savannah, Georgia, USA

Selection of Oil Quenchants for Heat Treating Processes

D. Scott MacKenzie, PhD, FASM

1Houghton International. Inc. Valley Forge, PA, USA 19426

smackenzie@houghtonintl.com

Abstract

Modern quenching oils offer a wide range of capability and performance. The selection of the proper quench oil is critical for proper heat treating operations. The quenchant selected must provide consistent quenching the first time, as well as over the life of the quenchant. Improper selection of quench oil can result in short oil life, or soft parts. Improper maintenance of the quench oil can result in stained parts. This paper will illustrate the selection factors in choosing a quench oil to insure proper quality parts.

Introduction

It is not known how long oils have been used in the hardening of ferrous alloys. Many types of oils have been used, including vegetable, fish and animal oils, and particular sperm whale oil have been used for quenching operations. The first petroleum-based quenching oils were developed around 1880 by E.F. Houghton in Philadelphia. Since that time, many advancements have been made in the development of quenching oils to provide highly specialized products for specific applications.

A wide range of quenching characteristics can be obtained through careful formulation and blending. High quality quenching oils are formulated from refined base stocks of high thermal stability. Selected wetting agents and accelerators are added to achieve specific quenching characteristics. The addition of complex anti-oxidant packages is included to maintain performance for long periods of continued use – particularly at elevated temperatures. Emulsifiers may be added to enable easy cleaning after quenching.

Petroleum-based quench oils can be divided into several categories, depending on the operational requirements. These requirements include quenching speed, operating temperatures and ease of removal.

The quenching speed is important because it influences the hardness and the depth of hardening. This is probably the most common method of classifying quench oils. Quench oils can be classified as normal, medium or high speed quench oils.

Normal speed quench oil*s* have relatively low rates of heat extraction, and are used in applications where the material being quenched has a high hardenability. Highly alloyed steels such as AISI 4340 or tool steels are typical examples of steels quenched in normal speed oils.

Medium speed quench oil*s* provide intermediate quenching characteristics and are widely used for medium to high hardenability applications where dependable, consistent metallurgical properties are required.

High speed quench oils are used for applications such as low hardenability alloys, carburized and carbo-nitrided components, or large cross-sections of medium hardenability steels where high rates of cooling are required to ensure maximum mechanical properties. A comparison of the quenching speeds of the different types of quench oil is shown in Figure 1.

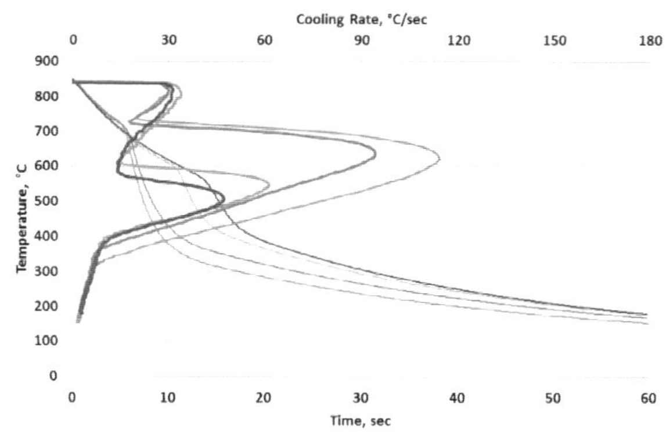

Figure 1 Comparison of Typical Cooling Curves of Several Commercial Oils.

Mar-quenching oils are a special case where the part is quenched into a quenchant at elevated temperature, typically 100-200°C. The work piece is held in the quenchant until temperature equilibrium is established throughout the section, and then air-cooled to ambient temperature.

During mar-quenching, components are quenched to an intermediate temperature close to the M_s temperature, and held at this temperature. This eliminates the temperature gradients across the surface, and consequently, during subsequent slow cooling after removal from the hot oil, transformation to Martensite occurs uniformly throughout the section. This minimizes the generation of internal stresses and reduces distortion.

Since mar-quenching oils are used at relatively high temperatures, their formulation and physical properties are different from cold quenching oils. They are formulated from very carefully selected base stocks with high oxidation resistance and thermal stability. They have high flash points and viscosities, and contain complex anti-oxidant packages to provide long life. Selection of the mar-quenching oil is based on the operating temperature and quenching characteristics. A minimum of 50°C should be maintained between the operating temperature of the oil and its flash point.

The most widely used quenchant for ferrous alloys are petroleum-based quenchants due to its favorable heat extraction characteristics. The dependence upon imported oil, price vulnerability, and contamination potential has caused suppliers to investigate alternative sources of quench oil base stock.

There are many reasons why vegetable-based oils have seen increased interest. The primary reasons are:

- Increased environmental regulation on ground and water contamination;
- Need to reduce the reliance on imported petroleum crude oils;
- Increased environmental awareness.

Canola oil, with its high oleic fat content has been found to offer the best oxidation resistance of any of the common vegetable oils (Soybean, Sunflower seed, Peanut and Palm). Canola is an oil seed crop that is grown primarily in Western Canada. Each plant produces a yellow flower, which in turn produces a seedpod. These seedpods are similar to a peapod, but are approximately one-fifth the size of a peapod. Within the pods, tiny round seeds are contained. It is these seeds that are crushed to form canola oil. These seeds are 40% oil. The seed solids are filtered from the oil, and are processed to form a high protein livestock feed.

While the environmental performance of canola-based quench oils is important, it still must perform as a quenchant. Cooling curves of canola-based quenchants show that the cooling behavior is equivalent and in some cases superior to mineral oil-based quenchants. In cooling curve tests, the canola-based quench oil showed the following characteristics when compared to a mineral oil based quenchant:

- Nearly non-existent vapor phase;
- Faster quench rates in the 1300-1100°F temperature range, where high quench rates are necessary to achieve properties;
- Slower quench rates at low (900-250°F) where low rates are desirable to minimize distortion.

The high boiling temperature of the canola oil, which is almost 166°C (300°F) above the boiling temperature of most petroleum quench oils and the higher flash point of 332°C (630°F) for the canola oil versus the petroleum-based quench oil flash temperature of 177-232°C (350-450°F) are both beneficial physical characteristics for a quench oil. The higher flash point gives greater safety during quenching, while the higher boiling temperature provides for an increased transition temperature between nucleate boiling and convection. This temperature is very important in reducing the temperature gradients, which result in residual stresses during transformation. The virtual absence of a vapor phase contributes to a uniform heat extraction and uniform heat transfer across the surface of the part.

Obtaining properties and low distortion is usually a balancing act. Often, optimal properties are obtained at the expense of high residual stresses or high distortion. Low distortion or residual stresses are usually obtained at a sacrifice in properties. Therefore, the optimum quench rate is one where properties are just met. This usually provides the minimum distortion.

Hardenability

To achieve proper strength and toughness, it is necessary to convert Austenite to Martensite, which is then tempered to form the proper tempered Martensite microstructure. To achieve this conversion of Austenite to Martensite, a rapid quench rate is required. This quench rate must be fast enough to avoid the formation of upper transformation products like Bainite and Pearlite, and convert all Austenite to Martensite. This critical quench rate just misses the "knee" of the Time-Temperature-Transformation (TTT) curve. The rate of the critical quench rate is dependent on the steel chemistry.

In practice, when a steel component is quenched, the surface cools much more rapidly than the center. This means that the surface could cool at the critical cooling rate and be fully hardened, but the center cools more slowly and forms a soft Pearlitic or Bainitic microstructure.

Hardenability is the ability of steel to through-harden. It is not the ability of the steel to get hard. In a sense, it is a measure of the critical cooling rate on the TTT curve. Increasing the hardenability decreases the critical cooling rate necessary to fully transform Austenite to Martensite.

Increasing the hardenability of steel is accomplished by increasing the alloying content of the steel. Carbon, Manganese, Chromium and Molybdenum are all effective alloying elements that increase the hardenability of the steel. These alloying elements cause a delay in transformation, by shifting the transformation curve to the right. This reduces the critical cooling rate for Martensitic transformation. However, alloying elements may be expensive and not always beneficial to other processes, such as machining, forging, etc.

Simply increasing the alloying content is not a simple panacea. The carbon content and alloying content can also have deleterious effects, by lowering the Martensite-Start Transformation Temperature (Ms). Increasing the carbon content, while shifting the TTT curve to the right, significantly lowers the Ms temperature (Table 1). Commonly used alloying elements will also shift the Ms temperature to lower temperatures. The M_s temperature of a steel can be predicted by commonly used formulas[1] such as:

$$M_s = 930 - 600\%C - 60\%Mn - 20Si\% - 50Cr - 30Ni\% - 20Mo\% - 20W\%$$

Alloying elements also increase the "effective carbon" content according to the formula :

$$C_{eq} = C + \frac{Mn}{14} + \frac{Mo}{10} + \frac{Cr}{10} + \frac{Ni}{30} + \frac{V}{6} + 3N + 20B$$

Cracking and distortion increases as the "effective carbon" content increases. Alloys become prone to distortion and cracking as the "effective carbon" exceeds 0.52%. This tendency is decreased by the proper application of quenchants. A fast enough quenchant is used that will achieve the desired properties, but slow enough that cracking or excessive distortion will not occur.

Table 1 Martensite Start Transformation Temperature (Ms) as a function of Carbon Content.

Carbon Content	M_s Temperature
0.2%	430°C
0.4%	360°C
1.0%	250°C

The shape or geometry of the part can also influence the hardness obtained (as well as the distortion of a part). This is illustrated in Table 2:

Table 2 Effect of Part Shape on the Equivalent Round for Quenching.

Shape		Equivalent Round (ER)
Solids, of Length L		
	Round	ER = 1T
	Hexagon	ER = 1.1T
	Square	ER = 1.25T
	Rectangular or Plate	ER = 1.5T
Tube (Any Section) Open Both Ends)		ER = 2T When L is less than D, consider as a plate of T thickness. When L is less than T, consider section a plate of L thickness
Restricted or Closed at One or Both Ends		ER = 2.5T (when D < 62mm) ER = 3.5T (when D > 62mm)

To summarize, the steel composition, component section thickness, and the type of quenchant all have a major influence on the properties obtained in the heat-treated condition.

Methods of Quench Oil Selection

As was discussed previously, the selection of quench oil is dependent on the part, type of furnace, and cost.

The part requirements are critical. The alloy must be considered to achieve the proper hardness and required mechanical properties. Geometry and section size is influenced by the hardenability of the alloy. This limits the section size for through hardening. The geometry of the part also influences the residual stress state and distortion. Non-uniform section sizes, sharp radii, and complex shapes can contribute to distortion and potentially part cracking. The oil must be chosen, as well as the method of fixturing, to minimize the potential of distortion and quench cracking.

The furnace also influences the choice of quenchant. If the agitation in the quench tank is poor, then fast oil must be chosen. If the part is prone to distortion, then the furnace and quench tank must be capable of heating the oil to the proper temperature for mar-tempering.

If the part is press-quenched, then the oil must be chosen that is tolerant of the presence of the inevitable infiltration of hydraulic fluid. It also should provide some lubricity to prevent excessive wear to the quench die fixtures. It also should have an excellent anti-oxidant package to satisfy provide repeatable quenching of high surface volumes with limited quenchant volumes.

The cost of the quenchant is also important. Unfortunately, this is often the only variable considered. The initial cost, as well as in-use costs, as well as the final end-of-life costs of disposal and environmental cost must be considered.

The initial costs are influenced by the quality of the base oil stock chosen and the quality and robustness of the additive package. The additive package consists of speed improvers and anti-oxidants. The quality of the anti-oxidant package also influences the in-use cost.

The in-use cost is influenced by the maintenance costs associated with the quench oil. The cost of testing to insure repeatable testing must be considered. Other maintenance costs such as filtering should be included in the calculation. The cost of additive packages (usually not required by quality quench oils) added to the quench tank as the quench oil ages is not usually added to the cost calculations because this usually falls under the maintenance budget. However, it should be added to understand the total cost of the quench oil selected.

One additional factor not usually included is the cost of cleaning the parts. Finally, the oxidative stability of the oil should be considered. If the oil tends to oxidize rapidly, this will increase the cost of cleaning, and require replacing the oil more often. This is especially true of oil used in mar-tempering applications.

Table 3. Comparison of Cooling Curve Test Methods (after Totten et al).

Variable	Method				
	ISO 9950	**AFNOR NFT-60778**	**JIS K2242**	**Z8 E 45003**	**ASTM D6200**
Country	International	France	Japan	China	USA
Probe Alloy	Inconel 600	Silver 99.999% Pure	Silver 99.999% Pure	Silver 99.999% Pure	Inconel 600
Probe Dimensions	12.5 x 60	16 x 48	10 x 30	10 x 30	12.5 x 60
Vessel Dimensions, mm	115±5 dia.	138 dia. X 99 high	300 ml beaker	300 ml beaker	115±5 dia.
Oil Volume, ml	2000	800	250	250	2000
Oil Temperature °C	40 ± 2	50 ± 2	80, 120, 160	80 ± 2	40 ± 2
Probe Temperature, °C	850 ± 5	800 ± 5	810 ± 5	810 ± 5	850 ± 5

Consumption of the oil by drag-out should also be considered in the quench oil calculations. Higher viscosity oil will tend to drag-out more, as will high surface area parts. The method of racking parts can cause the retention of higher than normal quantities of oil. Attention to racking and proper drain times can reduce consumption.

In general, there are three different situations for the selection of quench oil:

- Existing operation wanting a different quench oil because the life of the existing oil has been depleted; dissatisfaction with current supplier; or cost of existing quench oil.
- New operation similar to existing process, such as the addition of a new line for increased capacity, or a new facility at a different location processing similar parts; or
- New operation or new process with new or different configuration parts, or parts of a different alloy.

This is accomplished by one of three methods: Comparative Cooling Curves; Grossman H-Value or by the Hardening Power of the quenchant. In each case, not only must the quenching characteristics be considered, but the thermal stability of the oil should be considered.

Comparative Cooling Curves

When an existing process requires new oil, the new quenchant must be metallurgically equivalent. This means that the cooling curves or the quenching path should be similar. The cooling curve is measured by several different methods described in Table 3.

When testing oil for comparison purposes, only new oils should be chosen. This is because old oil will show oxidation and appear faster than the new oil. Further, the oil must be tested by the same method, and preferably at the same time, by the same testing laboratory to eliminate inter-laboratory testing variation.

As a general rule, oils are considered to be metallurgically equivalent if the Maximum Cooling Rate is ± 14°C/sec; the temperature at the Maximum Cooling Rate is ± 14°C, with similar viscosities and flash point. This is illustrated in Figure 2. As a general rule, if the oils meet these criteria, similar metallurgical results will be obtained.

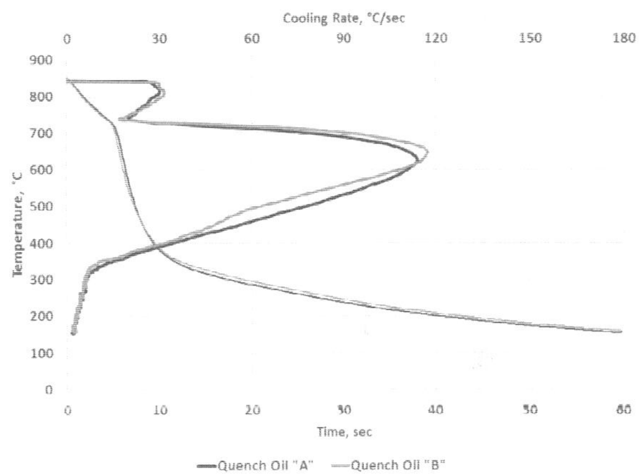

Figure 2. Comparison of Cooling Curves of different Oils.

Additional comparison can be made using critical temperatures (Figure 3) and the time to cool to specific temperatures (Figure 4).

Hardening Power

Alternatively, quenchants can be compared using cooling curve analysis, by using the Hardening Power of the quenchant[2]. In this method, three characteristic points of the cooling curve, measured by the method described in ISO 9950, are used in a formula derived by regression analysis to describe the hardening power of oil quenchants. This hardening power, for unalloyed steels, HP, is expressed as a single value as:

$$HP = 91.5 + 1.34T_{VP} + 10.88CR - 3.85T_{CP}$$

where TVP is the transition temperature between the vapor phase and the boiling phase (in °C); CR is the cooling rate at 550°C; and TCP is the transition temperature between the boiling phase and the convection phase.

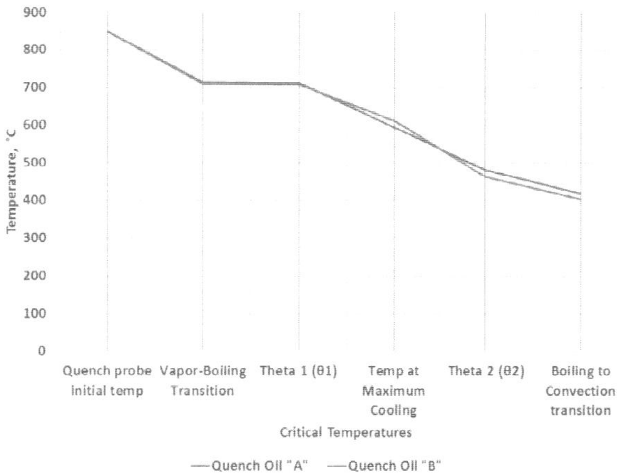

Figure 3. Comparison of Cooling Curve critical temperatures.

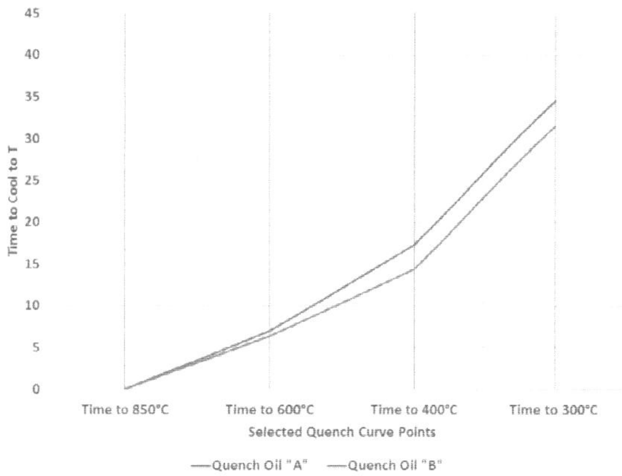

Figure 4. Comparison of times to cool to specific temperatures for two different oils.

This method allows oils to be selected by comparing the relative hardening power of each of the oils. However, for new installations, it is necessary to know the necessary cooling rate required to achieve part properties.

This method was further developed for the hardening of carburizing steels in hot oils[3]. Using the methodology described in ISO 9950, it was determined that only two critical points from the time-temperature and cooling rate curves had any significance for the measured hardness. The addition of the hardness of the Jominy End Quench at 6.5 mm increased the accuracy of the hardness prediction. In this experiment, the core hardness when martempering yielded the following equation:

$$HV30 = 135 + 0.56CR_{500} - T_{20} + 5HRC_{J=6.5}$$

Where CR_{500} is the cooling rate at 500°C; T_{20} is the temperature (°C) after 20 seconds; and $HRC_{J=6.5}$ is the hardness at a Jominy distance 6.5 mm from the quenched end. It was shown that this method accurately predicted the core hardness.

It is important to realize that different steels have different cooling requirements, and that the Harding Power equation was based on unalloyed carbon steels. Different alloys will require different regression constants[4]. However, this method has been shown to be effective in determining the relative ranking of oils for quenching. However, the Hardening Power equation should be used with caution when comparing different steels, or for general hardness predictions. Alternative equations for making hardness predictions and ranking of oil quenchants have been made by Chen and Zhou[5]

Grossman H-Value

The classic method related to the ability of a quenchant to harden steel is to determine the Grossman H-Value[6,7] or Severity of Quench. The H-Value is defined as[8]:

$$H = \frac{\alpha}{2\lambda}$$

where α is the average heat transfer coefficient at the surface of the part and λ is the thermal conductivity of the steel. For most steels, the thermal conductivity does not change appreciably over temperature or from alloy grade to grade, so it is directly approximate to the average heat transfer of the quenchant. A summary of the Severity of Quench for different media is shown in Table 4.

Table 4. Severity of Quench (H) for various quenching media.

Agitation	Air	Oil	Water	Brine
No circulation	0.02	0.25-0.30	0.90-1.0	2
Mild Circulation	-	0.30-0.35	1.0-1.1	2-2.2
Moderate Circulation	-	0.35-0.40	1.2-1.3	-
Good Agitation	-	0.40-0.50	1.4-1.5	-
Strong Agitation	0.05	0.50-0.80	1.6-2.0	-
Violent Agitation	-	0.8-1.10	4	5

The Grossman H-Value is determined experimentally by quenching a series of round bars. After quenching, the bars are sectioned, polished and etched. The 50% martensite region is determined. This is readily achieved because the transition in etching between dark and light etching corresponds to 50% martensite.

Although, this method has been used in the industry for many years, it is not without problems. The biggest issue with the application of the Grossmann H-Value, is the difficulty in quantifying agitation rates. There is really no understanding of what is meant by "mild" or "violent" agitation. Further, the different methods of quenching such as spray quenching have no equivalent to the Grossman H-Value. This method is only focused on the ability of the quenchant to harden steel, and gives no indication regarding distortion.

However, the terms describing the agitation are not quantifiable, and can result in errors. Since most oils are tested without agitation this results in a very narrow range of possible oil values. Further, when the original testing was done, the oils used in the original paper were straight oils, devoid of speed improvers. Modern oils achieve much higher quench rates (nearly double) those tested in the original paper. Previously there has been no way to correlate cooling curve data to Severity of Quench (H).

Defining *m* as the cooling rate (°C/s) at Position I, which for our purposes is the center of the ASTM D6200 cooling curve probe:

$$m = \frac{w}{T - T_m}$$

Where *w* is the cooling rate at 705°C; *T* is the temperature of interest (705°C), and T_m is the temperature of the quench media. For most cold oils, this temperature is 60°C.

Once the value of m has been determined , the Kondratjev[9] number (*Kn*) is calculated:

$$Kn = \frac{m}{m_\infty}$$

$$m_\infty = \frac{\alpha}{K}$$

where *K* is the Kondratjev shape factor, and α is the average heat transfer coefficient. For a cylinder like the ISO 9950 probe, the Kondratjev shape factor *K* is:

$$K = \frac{R^2}{5.783} = 6.75x10^{-6}m^2$$

The generalized Biot Number, Bi_v, is a function of the Kondratjev number[10]:

$$Bi_v = f(Kn)$$

$$log_{10}(Bi_v) = 0.7431 + 3.122log_{10}(Kn) + 1.952log_{10}(Kn)^2 + 0.5469log_{10}(Kn)^3$$

This is shown in Figure 5.

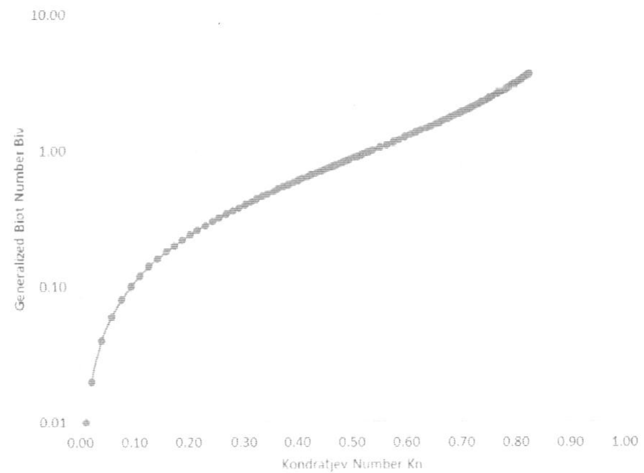

Figure 5. Universal Correlation between the Generalized Biot Number Bi_v and the Kondratjev Number, Kn.

$$Bi_v = \frac{\alpha S}{\lambda V} K$$

Rearranging:

$$\alpha = \frac{Bi_v \lambda V}{KS}$$

For a cylinder:

$$\frac{V}{S} = \frac{R}{2}$$

And substituting:

$$\alpha = \frac{Bi_v \lambda R}{2K}$$

where α is the average heat transfer coefficient (W/m^2 K) over the range of 705°C and the media temperature of 60°C; R is the radius of the cooling curve probe (0.006 m); K is the Kondratjev Shape factor for the probe (6.75 x 10^{-6} m^2); and λ is the thermal conductivity of the steel (W/m K). Substituting the average heat transfer coefficient in the definition of the Quench Severity Factor, H:

$$H = \frac{\alpha}{2\lambda}$$

the Quench Severity Factor H based on the cooling curve can be determined. However, the units of H determined above are in m-1, and the units from the original Grossman paper are in in-1, so conversion is necessary. The variation of H as a function of the cooling rate at 705°C is shown in Figure 6.

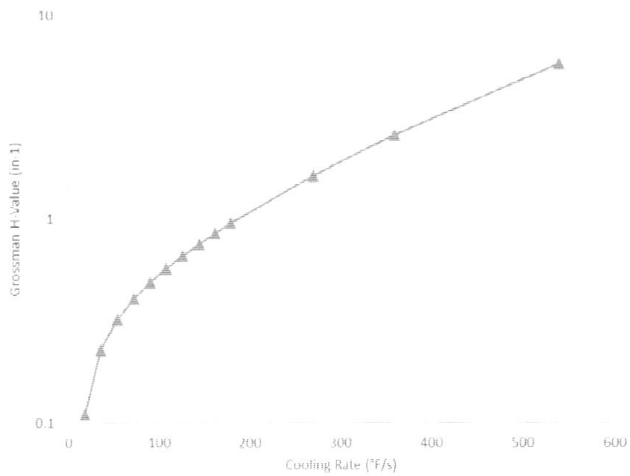

Figure 6. Grossman H-Value as a function of cooling rate (°F/s) at 1300°F.

Typical values for several common quench oils are shown in.

Table 5. Calculated Grossman H-values from cooling curves of commonly used quenchants.

Quenchant	Cooling Rate at 705°C	Cooling Rate at 1300°F	H in^{-1}
Water	117	211	1.30
Houghto-Quench K	60	108	0.65
Houghto-Quench G	37	67	0.43
Houghto-Quench 105	24	43	0.30
Houghto-Quench 100	22	40	0.28

The calculations show a good similarity to the classical H-values of Grossman. Further testing is needed to verify that the calculated results correspond to Jominy data and predicted hardness distributions in quenched bars[11].

To achieve equivalence to the data presented in Jominy End Quench testing, the H-Value only refers to the quench rate at a very narrow temperature range of 705°C (1300°F). To insure that the proper cooling rate and microstructure is achieved, it is necessary to refer to Continuous Cooling Diagrams (CCT) for the steel of interest.

This method makes possible the calculation of the critical size in terms of a standardized quench, and calculation of the critical size from a single test. Using charts , it is possible to predict how a known steel with a specific Jominy curve would behave, and predict the hardness distribution within the bar. Using the concept of equivalent rounds (Table 2), the hardness distribution can be predicted. An example of charts to predict hardness distribution based on hardness correlation to round bars is shown in Figure 7.

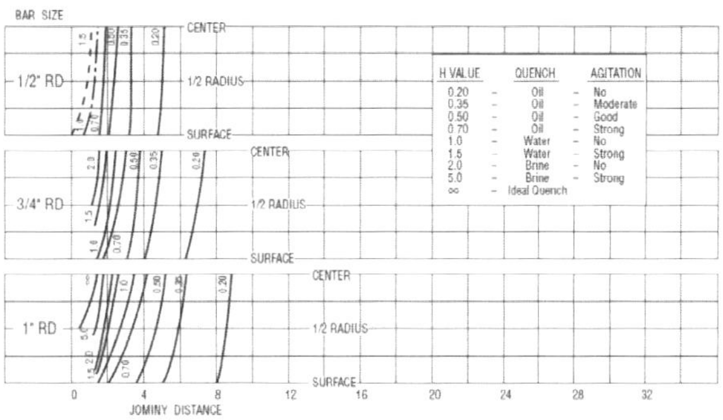

Figure 7. Chart for predicting the approximate cross section hardness of quench round bars using Jominy test results[12].

However, similar physical properties and quenching characteristics will not mean that the oil will perform over time in a similar fashion. The thermal stability of the oil, measured by its oxidative stability must also be considered.

Realistically, a method is needed to quantify the entire quench system (quenchant and agitation), and not just the quench media. A method, proposed by Guisbert[13] offers an excellent method of precisely determining the H-Value of a quench system (the quench oil and agitation system) to achieve repeatable results. In this method, small pins approximately 19 mm in diameter of AISI 1045 steel are placed anywhere within a tray or basket to evaluate quenching performance. This method is a modification of the method recommended by AGMA[14].

Since the pin diameter is fixed, the Grossman data and the Jominy Hardness position can be plotted together (Figure 8 and Figure 9). When the core hardness of the test pin is determined, the actual Jominy hardenability data (Figure 10) can be used to find the equivalent Jominy Hardness Position. The intersection of the Jominy hardness position with the test pin size determines the H-Value.

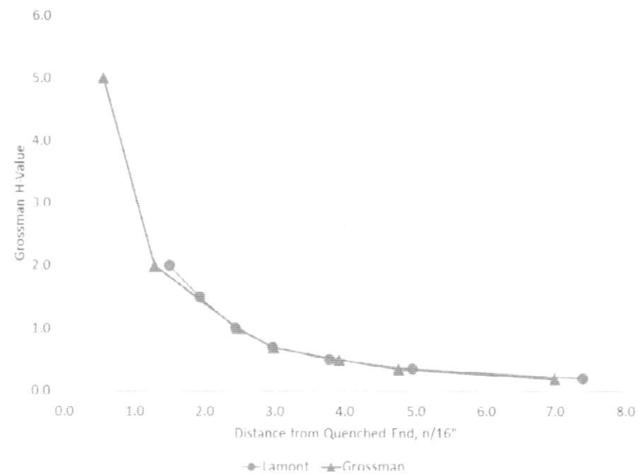

Figure 8. Grossman H-Value as a Function of Jominy End Quench.

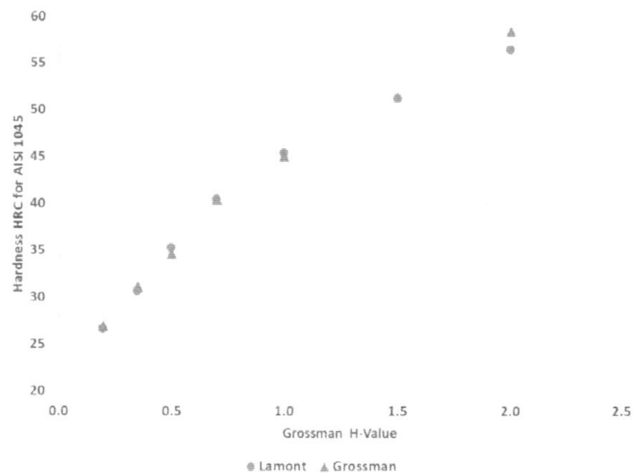

Figure 9. Center hardness as a function of Grossman H-Value for 19mm diameter AISI 1045 Steel.

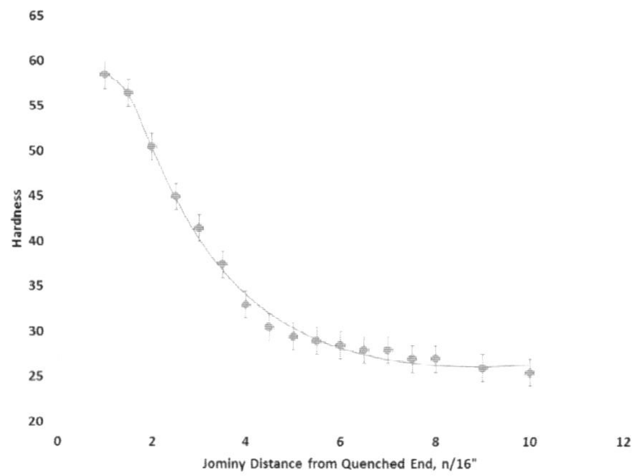

Figure 10. Jominy End Quench data of typical AISI 1045 Steel.

This alternative method is useful for determining the H-Value for specific quenchants, and would allow more precise determination of hardness in a particular quench system (oil and agitation). It would also allow two different quench systems to be compared. This method can also be used to determine the quench uniformity of a quench system by placing pins at eight corners of a quench tank, and one located at the center of the quench tank (assuming a rectangular quench tank). System monitoring of the quench system could also be achieved in this manner.

From an oil selection standpoint, the H-Value could be determined using no agitation, in a similar fashion as the cooling curve, and the H-Value determined. Multiple quench oils could be quenched, and rated according to the H-Value.

Summary

There is not one perfect method to the selection of quench oils. It often involves experience, and an understanding of the entire heat treatment process. This requires a deep knowledge of the alloys used; the part geometry; and the entire quench system. Each of the methods presented have merit in the selection of quenchants used in the heat treatment facility. However, these methods only consider the mechanical properties (hardness) and do not consider the important considerations of distortion. Further, the initial cost, operating cost and disposal cost of specific quench oils have not been considered. It is incumbent on the supplier and the customer to work together as a team to achieve optimal quenching.

Acknowledgements

The author would like to thank the management of Houghton International for permitting me to present this papers at the 2016 ASM/IFHTSE Thermal Processing Congress.

References

[1] E. S. Rowland, S. R. Lyle, Trans. ASM 37, 1946, p 27.

[2] S. Segerburg, "Classification of Quench Oils: A method of Comparison", Heat Treating, December 1988, pp 30-32.

[3] H. Kristoffersen, S. Segerberg, "Designed Experiments to Study Distortion, Hardness and their Variation in hot-oil Quench Carburizing", Heat Treatment of Metals, 2, 1995, pp. 39-42.

[4] L.C.F. Canale, X. Luo, X. Yao, and G. Totten, "Quenchant Characterization by Cooling Curve Analysis", J. ASTM, 6 2, 2009 pp. 1-29.

[5] Chen, C., and Zhou J., "Discussions on Segerberg Formula Concerning the Hardening Power of Quenchants," J. Shanghai Jiaotong Univ., Vol. 36, No. 1, 2002, pp. 78–82.

[6] Grossman, M. A., and Asimow, M., "Hardenability and Quenching," Iron Age, Vol. 145, No. 17, 1940, pp. 25–29.

[7] Grossman, M. A., and Asimov, M., "Hardenability and Quenching," Iron Age, Vol. 145, No. 18, 1940, pp. 39–47.

[8] Reddy, A. V., Akers, D. A., Chuzoy, L., Pershing, M. A., and Woldow, R. A., "A Simple Method Evaluates Quenches," Heat Treating Progress, Vol. 1, 2001, pp. 40–42.

[9] Rosa Otero, "Calculation of Kobasko's Simplified Heat Transfer Coefficients from Cooling Curve Data Obtained with Small Probes", Journal of ASTM International, Vol. 9, No. 4 (2012).

[10] N. Kobasko, E. Souza, L. Canale, G. Totten, "Vegetable Oil Quenchants: Calculation and Comparison of The Cooling Properties of a Series of Vegetable Oils,", J. Mech Eng (Strojniski vestnik) 56 (2010) 2 131-142.

[11] J.L. Lamont, "How to Estimate Hardening Depth in Bars," Iron Age, 1943, October, p. 64-70.

[12] "Timken Practical Data for Metallurgists", 14th Edition, Timken Company, Canton OH.

[13] D. A Guisbert, "Control of Quenching Systems in the Heat Treatment of Gear Products", presented at the ASM Quenching Seminar, Milwaukee, September 2008.

[14] ANSI/AGMA 2004-B89 "Gear Materials and Heat Treatment Manual"

International Federation of Heat Treatment and Surface Engineering 2016
Proceedings of the 23rd IFHTSE Congress
April 18–21, 2016, Savannah, Georgia, USA

Why Fine Filtering for quench oil systems is a must!

Jimmy Pfaffenberger *C.C. Jensen Inc., Newnan, GA USA*
jimmy@ccjensen.com, (770) 683-6810

Abstract

One of, if not the biggest variable in the heat treatment of components is achieving a consistent and repeatable cooling process. New quenching oil does a great job of providing the needed cooling at a known rate. The challenge with oil in general is that it breaks down over time and its properties change. The biggest killer of oil is HEAT! For quench oil, these changes affect the oil's cooling curve, which change the performance of the hardening and strength of the part. Surface finish can also be affected because contaminates can be deposited on the treated part. The high heat from the process produces a bunch of contaminates, which gum up the whole system. This in turn, requires lots of maintenance and increases the risk of oil self-ignition. A properly designed fine filtration system can significantly reduce these effects, while improving quality and reducing operating cost.

This paper will focus on what happens to quenching oil over time. How does it change the cooling curve of the oil and the impact on the quality of the heat treatment? The types of contaminates that are produced and their impact on the process will be explained. The shortcomings of a standard filter system will be explained. The benefits of adding a fine filtration system that keeps the oil as clean as possible will be shown. Finally we will show how to design a great fine quenching oil filtration system.

Introduction

One of the biggest variable in the heat treatment of components is achieving a consistent and repeatable cooling process. New quenching oil does a great job and providing the need and cooling at a known rate. It's critical to the operation to have the right oil for the application, which will deliver the right performance and have a long life. Quench oil serves two primary functions. It facilitates hardening of metal by refrigeration, cooling the surface at a known and controlled rate. It also enhances wetting of the metal during quenching to minimize the formation of un-desirable thermal and transformational gradients, which may lead to increased distortion and cracking. To get the performance and quality that you need does not come cheaply. Quality quench oil can be expensive, but it's worth the investment. So once you have invested in your equipment and you have invested in the right premium quench oily you may think that you have done all that is necessary. Well if you're depending on in-line filter equipment to provide the necessary cleaning for your system, you would be wrong. You not doing all that you can to maximize your ROI. An off-line fine filtration will provide tremendous benefits in the process along with significant reductions in maintenance and repair costs.

Figure 1. Typical heat treating application.

Background on quenching oil

To understand the need for very fine off-line filtration system on your quench oil systems, let's start at the beginning on what happens to the oil during the heat-treating process. The worst thing that you can do to oil is to expose it to high temperatures. Yes, the biggest killer of oil is HEAT! This is the major cause of oxidation in the oil. One of the general rules about oil is that as the operating temperature increases by 10° C you reduce the life of the oil by half. Based on this, quench oil will have a high level of thermal degradation from heat. The base oil and the additive package will change from the extreme forces that are put on it. This results in the formation of insoluble products that will cause deposits on the parts, and sludge, varnish and other soft contaminates will be produced in the quench tank. One of the greatest problems with quench oils is sludge formation. The presence of sludge & varnish could cause non-uniform heat transfer, increased thermal gradients, increased cracking and distortion. Sludge may also plug on-line filters and foul heat-exchanger surfaces.

The loss of heat-exchanger efficiency can cause overheating, excessive foaming and fires. The sludge and varnish will also attach itself to all of the surfaces inside the tank and piping. This requires extensive cleaning to be done to maintain an effect operation and a safe one.

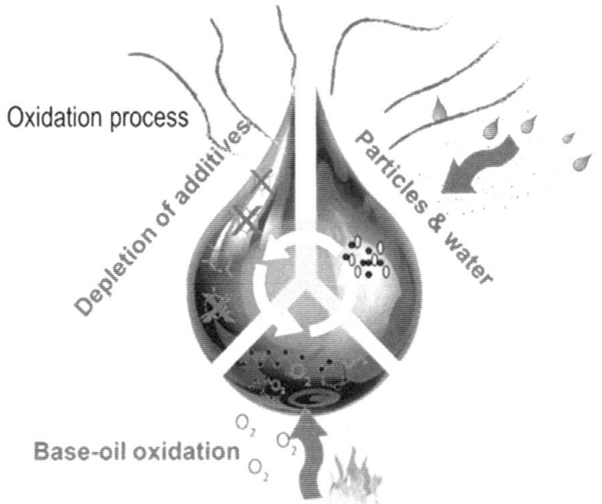

Figure 2. Schematic of oxidation process of quench oil – Illustration source: OelDoc GmbH

In addition to the effects from the heat there are other things that will impact the life of the oil. Heat treating is a messy process and there are many contaminates that are introduced to the system from the outside. This can come from many sources. You have dust and other forms of dirt that enter through the air. Then there are particles and debris that fall off of treated parts. With the high heat you might think that moisture is not a problem but it is.

A small amount of water can have a big impact on the oil and operation. Water is the most common and most harmful contaminant and should be avoided. If the concentration of water is higher than 0.1%, it can lead to granular hardness, distortion, spots and furthermore it is also a potential harm to staff health. At a concentration of 0.5% or above the quenching oil may start to foam thereby creating a risk for fire and explosion. The sources or causes of water contamination can be:

- **Bearing, refrigerated with water**
- **Fans**
- **Heat exchangers: oil/water**
- **Condensation in humid environments caused by high temperature changes**

The quenchant base and additives determine the quality and performance of the oil. Additive depletion is normal and expected when the oil is being used. But when the oil becomes more contaminated, the additives are used up faster than if they were not there. One of the jobs of additives is to control and neutralize contaminates so they don't affect the performance of the oil. But with so many particles, water and sludge, the additive package get depleted very quickly.

Impact on increased degradation and contamination of the quench oil:

The increased degradation and contamination has an impact on the operation in many ways:

1. Black deposits of varnish and lacquer on quenched parts, coolers, circulating systems
2. Sludge accumulation in the tanks, deposits on parts' surface
3. Change the cooling curve of the quench oil: resulting is even longer un-predictable cooling times. Unevenly extended vapor phase, foam formation, risk of splashing.
4. Reduced oil life by 2-4x. The varnish, sludge, and hard particles continue to break down the oil faster. The additives get consumed because they are fighting contaminates.
5. Increase presence of organic acids and increased Total Acid Number. As oil degrades, it forms acidic by-products
6. Changes in oil viscosity. As the oil degrades and particles increase and the viscosity will be increased. This results in reduced efficiency, higher energy consumption, and higher oil consumption.
7. Increased maintenance and downtime. Tanks have to scrubbed, and cleaned. Heat exchangers, circulating equipment and other system parts have shorter lives and unplanned repairs happen more frequently.
8. Increased part cost. Additional cleaning is often required, high dirt load is transferred to the washing baths. Sometimes blasting and grinding is required because of deposits and overheating.
9. Reduced part quality. Cooling curve changes add unpredictability into the quenching process. Visible and performance of the heat treatment
10. Safety: Increase danger for operators and the facility. Hard levels of sludge and water increase risk of steam explosions and fires

These will have an impact on the efficiency of the system, increase operational cost and safety.

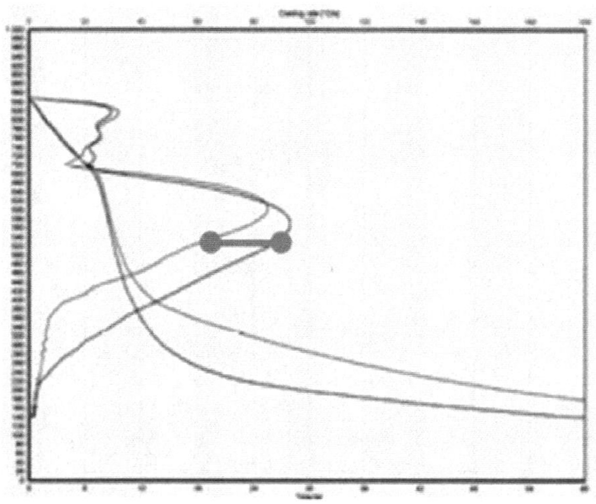

Figure 3. This graph shows the cooling curve of new quench oil and one that has degraded over time. The distance shown in red, is the change in cooling rate.

Why Fine Filtration?

Why do I need a fine filtration system on my quench system? The system already has filter(s) to remove contaminates - it should be good enough. However, that is not the case. Unless the filter system can remove particles down to 3μm or less, has a very high load capacity, able to capture hard particles, soft contaminates (sludge & varnish) and remove water; you are wasting time and money. You might think that particles below 10μm cannot cause any harm, but they build up in the system creating sludge and carbon deposits. Contaminated oil will produce sludge and carbon deposits at a faster rate. Fine filtering will reduce the amount of sludge and carbon produced up to 90%. If you analyze the particulate in the oil, a progressive size distribution will be seen (Figure 4. Typical particle distribution in quench oil.Figure 4). Only 10% of the particles are larger than 10μm. If you are using a 10μm or larger filter, you are only able to retain 10% or less of the particulate. Approximately 70-80% of the particles are between 1-5 microns.

There is a common concern that if the oil is filtered too fine that it will remove additives. The truth is that oil additives cannot be filtered out, because they are fully blended, fully dissolved and active in the base oil (even 1 micron filtration cannot). Additives are dissolved like sugar in coffee. There are a few exceptions: Decomposed and "dead" additives can be filtered out (additives will decompose over time performing their intended function). Polar additives will become attached to particles and be filtered out (this is the job of these AW/EP additives, rust inhibitors etc. so this is normal). Anti-foam additives will be sheared in use and some will be filtered out by means of polar attraction by any filter in the system. Experience has shown that a decrease in the anti-foam level will not give any problems if particles, water and oil degradation products are removed from the oil).

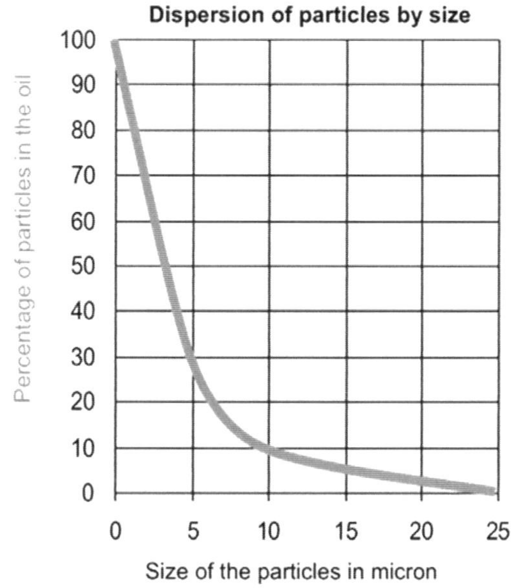

Figure 4. Typical particle distribution in quench oil.

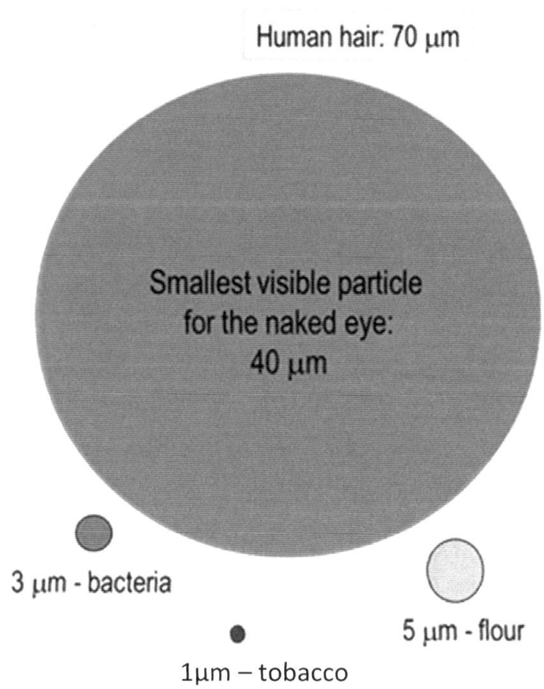

Figure 5. The size of the particles that need to be filtrated is extremely small

Benefits of fine oil filtration

It is important to know what you can expect if you add an off-line fine filtrations system to your quenching operation. These benefits will help justify the upfront cost of the system and help you quantify the savings and other advantages over time.

1. Reducing unplanned down-time (typically by 50 – 70 %)
2. Longer service intervals and lower maintenance costs (typically 50 %)
3. Stable quench, slower change in the cooling curve
4. Reduced costs for system inline filters (longer life of in-line filters and/or less cleaning of filter mesh)
5. Less or no tank cleaning
6. Significantly reduce or eliminated: Sludge, carbon, varnish, hard particles and water
7. Eliminating/reducing cost for oil changes (2-5x the life)
8. Improve surface quality (happy clients)
9. Less or no extra cleaning of quenched parts
10. Reduced risk of fire
11. Reduced environmental impact (less oil changes)

Figure 6. Part on the left is from a system before a fine filtration system has added. You can see the deposits of carbon. The picture on the right was from the same quenching system after fine filtration was installed.

Figure 7. Filter inserts that have captured what the in-line filter could not.

What does a great fine filtration system look like?

Now that we know that we need to have great filtration on our quench system, what does that look like? The system must be off-line, meaning that it is not in the flow of the oil going to the system's pump or the heat exchanger. The required flow and the importance of unrestricted pathway for the oil through these components is vital and this limits how much the filter can do.

A good filter system must be able to capture all contaminates: hard particles, and soft contaminates (sludge, varnish, carbon and water). Depending on the manufacturer, the filter system might have require multiple types of filters to remove all contaminates. There are systems that can accomplish this with one filter media. Be careful when evaluating solutions, though the initial cost is important, the cost of the filter media will determine the true ROI on the investment. The fine filtration system must be able to capture contaminates down to at least 5µm, with the finer the better. All contaminate count, no matter what the size.

The filtration system must be continuous, and running all of the time. The goal is not to just clean contaminate to a specific level and then wait for it to get dirty again. By continuously cleaning the oil, you minimize the degradation of the oil and the opportunity to create new contaminates. There might be a desire to use a mobile filtration system that can be moved from one tank to another. In theory this sounds great, but in practice it significantly reduces that effectiveness of the filtration system. The problem is that you do not totally clean up both the oil and the system surfaces. The contamination that is left in the oil during times that the mobile system is not being used will do harm to the all of the benefits that you are able to achieve with a continuous system. This reduces the effectiveness to less than 50%.

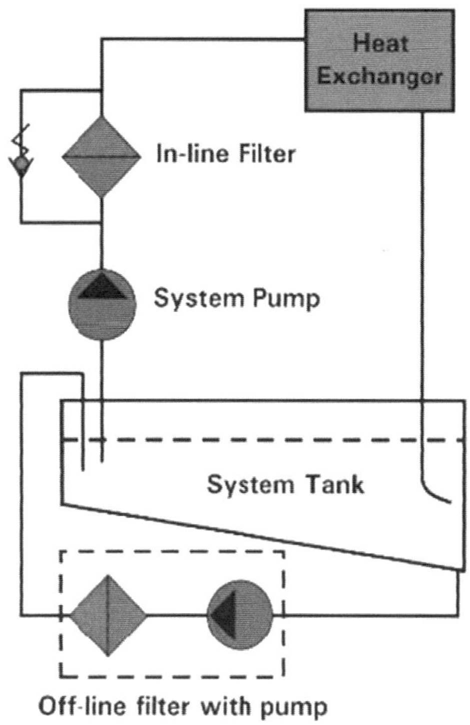

Figure 8. Configuration of an off-line filter system.

One of the great properties of very clean oil is that it acts like a cleaner. Meaning it will clean all of the surfaces of the system over time. Don't expect the clean up to happen overnight. If the system was run without a fine filtration system for years, it could take months for clean oil to scrub all of the surfaces clean.

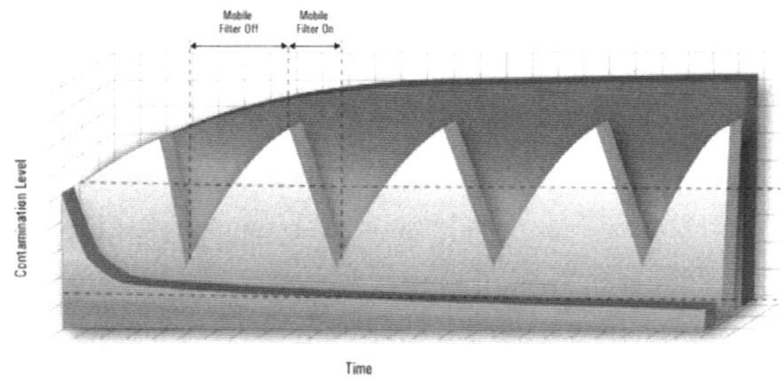

Figure 9. Filtering using a mobile system will have limited impact.

Now let's turn our focus to the filter element itself. This is the replaceable part that captures contaminates. It must be able to hold high levels of contaminates. The most common type of filter element is a pressure filter. They can come in different sizes and made from different materials. These are great for inline filters that capture large particles that could damage pumps or clog heat exchangers. The down side of pressure filters is that they have limited dirt-holding capacity and they are not a good solution for the off-line fine filtration system.

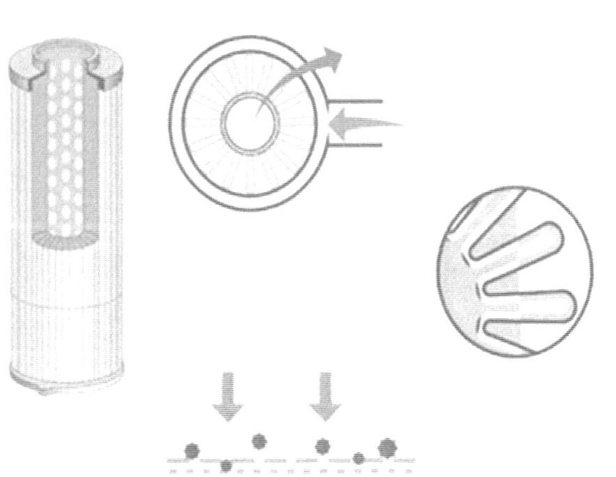

Figure 10. Pressure style filter. There is limited surface area to capture particles and cannot absorb many of the soft contaminates.

A depth type filter is a much better choice because it can hold lots of dirt and has layers that can capture different size particles and soft contaminates. The filter media should be able to last months if not a year. If not, you are trading maintenance on the quenching system for maintenance on the filter system.

This is asking a lot from this filter system, and it has to be very cost effective. The ROI on the system should be 12 months or less. The system must also have a low cost of operation going forward. It may sound too good to be true, but there are fine filter systems in the marketplace that can meet all of these requirement. A fine filtration system should never be looked at as an additional cost or a nice option. Instead it generates you money by lower costs, higher quality parts, less labor and a safer operation.

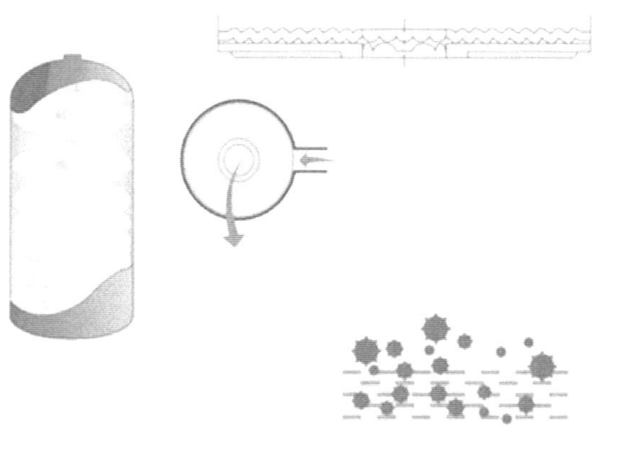

Figure 11. Depth style filter. There is significantly more surface area to capture particles. Depending on the material, it may also be able to absorb many of the soft contaminates.

Conclusions

As discussed in the beginning of this paper, there are going to be conditions in the oil quenching process that will destroy the oil, no matter how good the oil is or how well it is maintained. The heat from the quenching process will slowly kill both the base oil and the additives. There is the dirt that is added from the heat treat shop environment. These contaminates are smaller in size that most people realize and they cannot be remove/captured by standard means. By installing an off-line filtration system that is fine enough to capture the smallest of hard particles; can absorb the soft contaminates and remove water; you can significantly increase the life of the oil, reduce maintenance costs and improve safety.

References

[1] D. Scott MacKenzie, PhD, FASM, "The Mechanism of Quench Oil Oxidation", Heat Treat 2013, Proceedings of the 27th Heat Treating Society Conference (ASM International), 2013 pp. 255-258.

[2] Steffen Dalsgaard Nyman., Clean Oil Guide, CC Jensen A/S (Svendborg, 2015), pp. 9-31.

International Federation of Heat Treatment and Surface Engineering 2016
Proceedings of the 23rd IFHTSE Congress
April 18–21, 2016, Savannah, Georgia, USA

Innovative Quenching Process for Pipe and Bar

Pietro della Putta
SMS Meer SpA, Tarcento, UD, Italy
pietro.dellaputta@sms-group.com

Mario Fabro
SMS U.S.A. LLC, Pittsburgh, PA, U.S.A.
mario.fabro@sms-group.com

Krich Sawamiphakdi, Dan Bader, Sunil J. Nalawadi
TimkenSteel, Canton, OH, U.S.A.
krich.sawamiphakdi@timkensteel.com, dan.bader@timkensteel.com, neil.nalawadi@timkensteel.com

The State-of-the-Art for pipe and bar quench

The increasingly demanding market for hardened steel bars and heavy tubes for OCTG and industrial applications requires the developing of more efficient quenching devices.
SMS Group undertook a significant research effort to expand the original concept of tangential water spraying with the aim of a performance increase and to make it suitable for effectively treating large bars.
Computational fluid dynamics added the capability of predicting the quenching results and allowed to skip most the time-consuming prototype realization phase and to optimize the process parameters.

Introduction

State of the art in continuous Q&T plants prior to the induction and tangential sprayers technologies.
To improve the resistance of carbon steel pipes and tubes, continuous quenching and tempering process lines have been implemented in the industry since the middle of 20th century.
Most of these plants are single-strand lines where the stock material is manipulated in a continuous way to withstand a first heating process ("Hardening" or "Austenitizing"), a water Quenching and finally a second reheating ("Tempering") to finalize its mechanical properties.
The first heating is typically performed through a furnace capable to raise the material above the Ac3 temperature, where the steel microstructure goes through a phase transformation from α-iron (Ferrite) to γ-iron (Austenite).
The Ac3 temperature varies in accordance to the steel chemical composition, in particular with the quantity of Carbon and the presence of alloying that can have antagonistic effects (e.g. Carbon, Nickel, Manganese broaden the γ-iron region whilst Silicon, Chromium, Molybdenum and Tungsten broaden the α-iron).
To ensure that the transformation is uniform and complete, the material temperature is normally exceeding the theoretical Ac3 points by tenths of degrees.
Usual austenitizing temperatures are in the range of 800 to 950°C.

Heating
In a continuous process, depending on size of the stocks to be treated and on the overall capacity of the plant, typical methods to reheat the material included tunnel furnaces or walking beam furnaces, where the heating power comes from combustion burners fed by a gaseous fuel.

The piece reheating through a combustion source necessarily involves heat transmission by radiation, convection and conduction.
By far the predominant heat transfer develops through radiation, where the energy flux between the two bodies 1 (furnace chamber) and 2 (product to be reheated) is regulated by the Stefan-Boltzmann law:

$$Q = [\sigma \cdot (T_1^4 - T_2^4)] / [(1-\varepsilon_1)/(A_1 \cdot \varepsilon_1) + 1/(A_1 \cdot F_{1\to 2}) + (1-\varepsilon_2)/(A_2 \cdot \varepsilon_2)]$$

Where:
ε = emissivity
σ = Stefan's constant
$F_{1\to 2}$ = view factor

Convection from the hot fumes running into the pieces is also contributing quite significantly, while the conduction from the furnace hearth, rollers or supporting skids to the heated piece can be considered as negligible.

The time required for the heat transfer may vary as a function of the variables expressed in the formula above but it is, in general, in the range of 1.5 to 3 s/°F for any kind of combustion furnace.
To the above, a considerable amount of "soaking" time may be required to stabilize the temperature to ensure the uniformity of metallurgical characteristics throughout the piece, resulting in typical residence times of hours inside a combustion furnace.

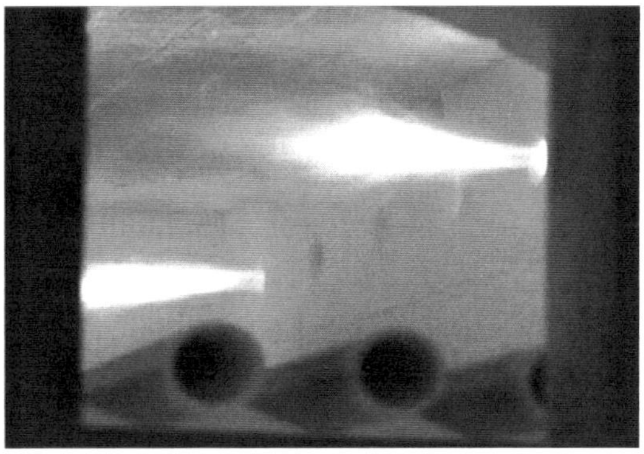

Figure 1: The interior of a combustion furnace for hardening the pipes

The production rate required by the plant determines the choice towards either walking beam furnaces (high throughput) or to tunnel furnaces (low productivity).
In both scenarios the combustion heating intrinsically low process speed requires a furnace capable to accommodate a large number of pieces.

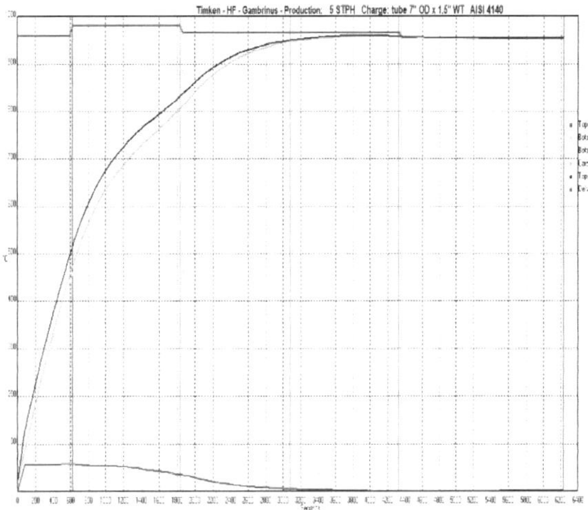

Figure 2: Typical combustion furnace heating curve, showing an almost two-hours long process

Cooling

To obtain the hard quenching necessary to transform the material structure into the tougher Martensitic phase, it is necessary to cool down entirely the workpiece with a cooling media which is typically cold water.

Several quenching systems have been adopted in Q&T plants either consisting in passive devices, i.e. water tanks where the pieces are plunged, or active devices, i.e. sprayers delivering the water into the piece.

In modern continuous lines, the pieces are typically handled and delivered to the quenching system one by one, since this allows a far better control of the cooling uniformity.
If the pieces are led to the cooling media in layers or in bundles, the result is in fact an uneven quenching, because the surfaces do not receive homogeneous contact with the cooling media.
For the quenching of long products such as tubes and bars, it was soon realized that external sprayers could represent an advantage for the cooling process both in terms of homogeneity and in terms of quenching severity compared to the static quench in a water tank. Even inducing agitation of the water inside the vessel couldn't help but a little to increase the heat dissipation rate, and therefore to improve the phase transformation.
The efficiency of OD sprayers depends mainly on the capacity of distributing a large water flow evenly on the pipe surface, therefore systems having a big number of nozzles are preferred over devices with a little number of hoses.
The nozzles are therefore arranged on the circumference in a tight pattern to encircle completely the piece, and in a sufficient number to distribute the cooling media uniformly along the main axis.

Market needs pushing the technology further

Improvements in quenching efficiency

Starting from the late 1980s, it became evident that it was possible to further increase the cooling efficiency of OD water sprayers by impinging the surface of the piece almost tangentially with the water jets.

Figure 3: An early model of tangential spraying device, circa 1985, displaying 16 lines of nozzles in the circumference. The nozzle headers are manually steerable and can be orientated tangentially to the pipe surface.

The physical explanation, although not immediately evident, resides in the capacity of the tangential component of speed of the water jet to remove the layer of vapor generated on the hot surface when in contact with the cold media.
In particular, there is a wide region of temperatures (temperature differential among piece surface and cooling media) where the film boiling effect, creating bubbles over the surface, tends to decrease the heat transfer coefficient. Through the application of tangential water jets it is found that the HTC reduction is largely dampened.

Modern generation of devices for quenching

In the recent years and pushed by a bustling market for new steel pipe production plants, SMS Meer kept on refining the idea of a tangential sprayer for OD cooling, ending up with a very successful design industrialized under the brand name of SMS Quenching Head (QH).

The base concept is to have a modular machine, suitable for being adapted to increasingly productive plants by simply adding spraying modules and increasing the pipe walk-thru speed. The pieces are conveyed longitudinally one by one from the austenitizing furnace upon a roller conveyor and travel inside and through the machine rings.
The pipes are therefore processed in a continuous or "progressive" flow, without interrupting their travel.
At the time of writing, the QH has been implemented in a configuration varying from 2 to 6 modules, reaching a throughput rate of up to about 60 t/h.
The processed tube range goes from diameter of 2 3/8" to 11" (OD 60 to 279 mm).

Figure 4: Quenching Head preassembled in the SMS Meer SpA workshop in Tarcento, UD, Italy.

We can measure the machine quenching capacity in lb/s (or kg/s) as the capability to cool the mass of one test piece from Ac3 down to Mf (Martensite Finish) temperature, without intersecting the Bainite formation region in the relevant CCT diagram.
In this case the steel transformation occurs almost completely from Austenite to Martensite, without formation of intermediate phases (Ms= 99%).

With the appropriate steel grade and tube wall thickness (less than 20 mm when OD only quenching is applied), quenching capacity of up to 36.8 lb/s (16.7 kg/s) has been obtained.
An important feature of the SMS Quenching Head consists in the possibility to change automatically the machine set up parameters in accordance to the different pipes that have to be processed, and in particular:

1. the roller conveyor speed, determined from the required production rate
2. the level of the machine supporting platform, which is adjusted in order to match the machine and the tube centerlines
3. the nozzles angle inclination, in accordance with the tube diameter and the quench severity required.

It is essential for the cooling uniformity that the machine construction and precision of assembly allows to generate a perfectly annular and stable water ring.

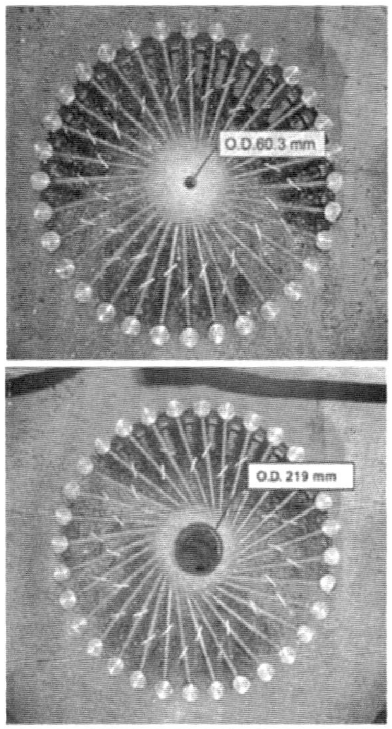

Figure 5: Quenching Head can automatically adjust the spraying angle in accordance with the tube diameter.

A previously unexpressed limitation of the QH progressive quenching is the minimum process speed: whereas the tube mass is brought under the cooling media too slowly, the internal heat conduction through the tube main axis prevails over the water cooling, thus preventing the accomplishment of the necessary cooling rate, or leading to an undesired re-tempering of the material after that the martensitic phase has been already generated.

Different heating solutions

The combustion furnace heating as described under the previous paragraph shows limitation when it comes to reheat:

1. small lots in a discontinuous operation, where the big inertia of the furnace represents a potential slowdown of the whole line
2. at low capacity rate, since the typical tunnel furnace configuration is not very effective and occupies a large layout space.

Especially in case of frequent shut down and consequent furnace start-up, the combustion units cannot be very flexible in catching up with the desired working temperature set points.

The heating times are long and consequently there is the need to have several pieces on heating at the same time, therefore the furnaces tend to be large.

As an alternative heating method, it is also possible to reheat the steel part through Joule heat generated by eddy currents.

An eddy current is an electric current induced in the steel body by a rapidly alternating magnetic field.

This magnetic field can be generated by an induction heater basically consisting of a set of coils through which a high-frequency AC current is sent.

The workpiece is transported longitudinally thorough the coils and progressively heats up.

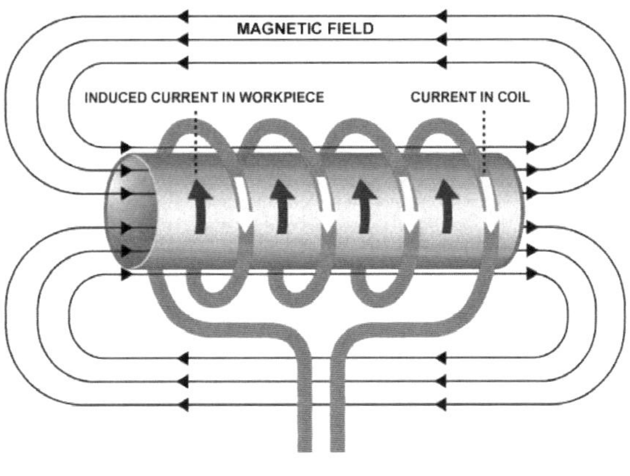

Figure 6: Magnetic field producing an eddy current inside the workpiece according to Faraday's law

The big advantage is that induction heating is capable to generate high power densities which allow to shorten the interaction times to reach the target temperature: this turns into the possibility of having one single workpiece being processed at the time, with significant saving of room and increase in operation flexibility.

The depth of penetration of the parasite currents can also be controlled through the frequency, therefore it is possible to heat up the piece on the very kernel or closer to the surface.

SMS Elotherm developed industrially the process for the quenching and tempering of tubes that resulted in the successfully commissioning of several heat treatment lines fully based on induction heating.

Both the hardening and the tempering of the tubes are performed by a set of inductors working in accordance to the aforementioned principle, and consisting of water-cooled copper coils sunken in a refractory container box.

Figure 7: Induction Furnace consisting on a set of 6 inductors, alternated with transport rollers

The power is distributed to the coils by frequency converters IGBT type, of Elotherm design and built.

The advantage of this configuration certainly resides in the limited amount of layout space, although a drawback of the induction heating is the very short time available for the workpiece to get homogeneous, and to stabilize the structural properties.

This is particularly evident during the annealing process, which normally requires the material to settle for a while at high temperature.

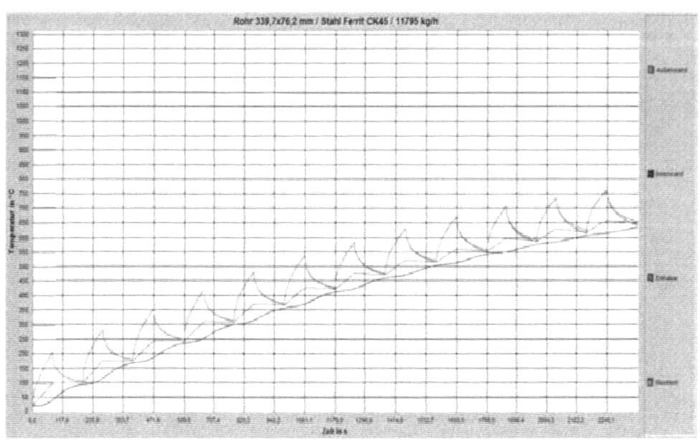

Figure 8: Typical step-heating by induction: as the piece goes through the subsequent coils, it receives peaks of heat which progressively bring the temperature to the target.
As the development of Joule heat is very localized in the material thickness, the temperature is equalized through internal conduction.

The Hybrid revolution

Induction + Combustion as the obvious solution

The combustion furnaces having limitations in terms of room and flexibility and the induction based lines being unreliable when it comes to confer strictly homogeneous properties, the natural evolution became putting together the two technologies trying to get the best of both.

An "hybrid" line making use of both an induction and a combustion heating system can be composed in several ways, but SMS approach to find the most efficient technique considered at least the following standpoints:

1. the induction heating process should be used to confer to the products as much energy as possible, in order to contain the space utilization necessary for large combustion furnaces
2. the induction should be used prominently at low temperature: induction heating is more efficient while the material is ferromagnetic, i.e. below the Curie point; above that temperature the behavior of the steel becomes paramagnetic and the magnetic field resultant eddy currents loose intensity.
3. combustion chambers are required to homogenize the temperature, both during hardening and even more in the tempering process.

Although a general arrangement cannot be found and considered valid for all the possible heat treatment applications, since there are several variable influencing the design (productivity and dimension of workpieces above all), the principles above drive to development of a quite typical layout configuration that can be summarized as a U-turn quench and temper line where, for both the heating process (hardening/austenitizing and tempering/annealing) an induction pre-heater is placed in front of a combustion furnace.

The first is used to deliver a rough and massive heat quantity to the workpiece, while the latter for closing the final temperature gap and for soaking the piece in order to stabilize the temperature within the narrow tolerances required by the process, which can be in the ± 10F for austenitizing and ± 5F for tempering.

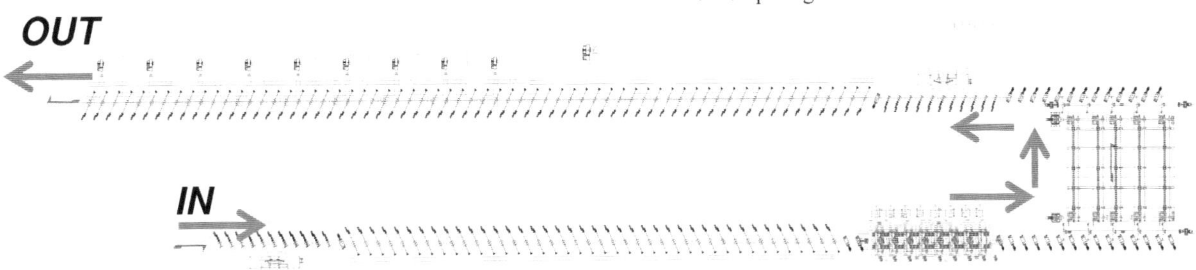

Figure 9: U-turn plant layout, where an in-line austenitizing and quenching (bottom) is followed by a reversal and a tempering line (top) long enough to allow for settling of material mechanical properties.

Both the austenitizing and the tempering systems are then composed by an induction pre-heating unit and a combustion furnace. The combustion furnaces are basically roller hearth type, with single or multiple strand, where the first configuration has to be preferred for the austenitizing process (for the AF), to handle easily one piece at the time toward the quenching system, while the tempering furnace (TF) can be single or multiple in accordance with the layout space availability, the required production rate, the necessary resident time of the workpieces at high temperature etc.

Temperature homogeneity requirement for the austenitizing and the tempering processes are intrinsically different, and this may also lead to the choice of radically different combustion systems. While the end temperature tolerances in the AF definitely allows for the choice of traditional direct flame burners (the burners release the fire directly inside the furnace chamber, where the workpiece travels), the TF urgency of controlling very narrow temperature gaps and keeping the same steady enough for relatively long periods, may force to choose an indirect flame heating method, either with muffle, radiant tubes or with fumes convective chambers.

In SMS modular design, fumes convective furnaces consist in several modules each composed by a lower chamber, where the piece travels supported by rollers, and an upper chamber, equipped with burners, valves and an air blower. The two chambers are separated by a screen but connected by fume baffles which distribute the hot exhausts from top to bottom and all around the workpiece, which is wrapped by a uniform gas flow but never sees the hot flames directly.

Figure 10: Convective chamber module for a TF in SMS shop

The Quenching Shell™ concept

Development of the next generation design

As the necessity of combining in one single line the capability to heat treat a vast assortment of different workpieces arose, SMS developed a brand new device capable to fulfill the task.

The aim was to cover the quench and temper of carbon and alloyed steel tubes and bars, ranging from 4" to 13" of OD and with a wall thickness (for the tubes) of 0.4" to 3".

Length of the workpieces stretched from 12' to 45'.

The request was to obtain high martensitic structure of up to 90 or even 95%, as required by the market and the main standards (e.g. the API-5CT PSL3), and starting with typical high tensile grades like the AISI4140 low alloy steel, having Cr and Mn as main alloying elements.

The target production rate of the line was set as 10 STPH, therefore in a district where an all alone induction furnace would have been too powerful and expensive and a traditional walking beam combustion furnace too underused to be effective.

The choice for the heating system therefore went immediately for a combination of induction preheaters and combustion furnaces, the hybrid system previously described.

More challenging was the identification of the right design for the quenching system, which had to ensure, at the same time, all the following features:

1) high quench severity to reach the desired martensitic transformation
2) extreme temperature uniformity during cooling, both for ensuring uniform structural characteristics and to avoid stresses and bending
3) capability to handle without distinction the wide assortment 4" to 13"
4) possibility to spray the tubes also from the internal side, in order to accelerate the cooling and improve even more the uniformity
5) possibility to harmlessly evacuate shards that may detach from very crack-prone materials during quenching.

A traditional and conservative approach would have suggested to use two different cooling units to be able to perform all these challenging tasks: a progressive sprayer like the aforementioned QH plus a static cooling system, i.e. a tank, to be used for the thicker and heavier stocks.

Combining all the above characteristics in one single machine had been the driver for SMS new R&D project, which ended up in the design of a brand new device called Quenching Shell[(1)]™.

The Quenching Shell shares with the QH a similar tangential OD spraying system, with a very large number of nozzles, in order to guarantee severity 1) and uniformity 2).

The main design difference and revolutionary concept is that the QS, unlike the QH, is not a "progressive" quenching, in the sense that the workpiece instead of continuously and progressively travelling through the machine while being cooled, is brought inside it quickly after the AF discharge, and it is then quenched all at the same time. This is why the method of quenching proper of the QS can be better described as "synchronous".

This aspect confers to the QS the characteristic 3), which is the possibility to manage both small and light tubes (OD 4" x WT 0.4") and the very large hollows or solid bars of 13", where the internal heat conduction would jeopardize the effect of a continuous, "progressive", machine.

Regardless of the diameter of the piece to be quenched, the transport system in service to the machine delivers it quickly from the austenitizing furnace exit to the machine main axis, where the fast cooling can start.

Moreover the QS can be easily equipped with an internal spraying nozzle 4), and it is open on the bottom, as to fulfill 5).

Figure 11: Quenching Shell 3-d modeling, showing the tangential nozzles forming an open annulus on the bottom

Being open on the bottom, the QS offers a convenient slot to insert the workpiece from underneath, and a safe pathway for releasing unwelcome broken shards with no damage.

As a drawback, compared to the traditional QH, the new Quenching Shell design foresees a machine length comparable to the maximum workpiece length, that obviously has to be completely enveloped during quenching.

The OD spraying system shares a similar angle-adjustment system as in the QH, where all the nozzles can be rotated simultaneously in order to keep the jet precisely orientated towards the tangent to the piece surface, regardless of its external diameter.

Key for obtaining a good water distribution and even cooling, despite the nozzle crown being open and not covering the full 360° circle, is to keep the workpiece in constant and fast rotation during the quenching.

The rotation is given thanks to a set of double rotary rollers, having also the function of supporting the piece and lifting it exactly from the transport roller conveyor underneath to the QS centerline level.

---
[1] Patent Pending

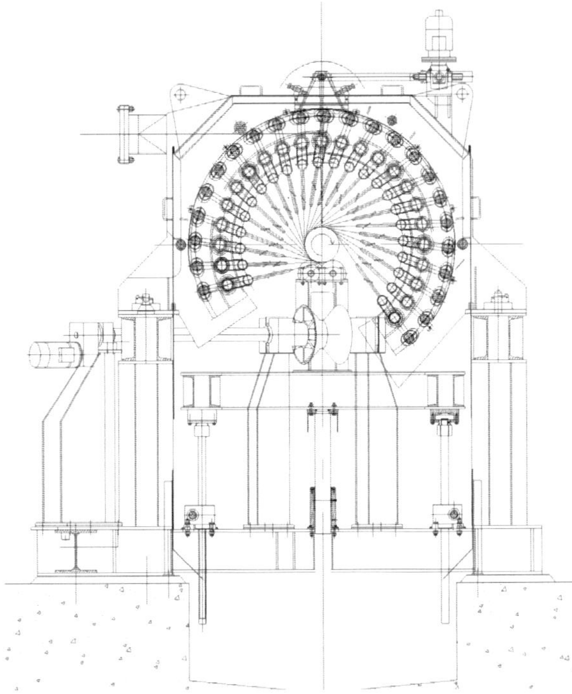

Figure 12: Quenching Shell in the working position

In order to retain the pieces, having different dimension, from jumping off the pocket between the rollers, the same can be also set apart sideways to increase or decrease the gap.

The internal nozzle is installed on a separate skid with adjustable stroke, and can be brought in the spraying position close to one of the tube's edge, as shown on Fig. 11.

Advanced CFD modelling

Computer aided modelling for Q&T simulations: the SMS-Quench

Computational Fluid Dynamics has become a very sophisticated tool not only in the hands of the mechanical and process designers, but also an invaluable help for the metallurgists, whereas through new software routines we are now capable to predict with great accuracy the behavior of the steel workpiece under quench.

To condensate the enormous physical and chemical knowledge necessary to built-up a working model, SMS process engineers made use, as a starter, of two powerful commercial software, that have been heavily customized through the years in order to obtain an instrument tailor made for the own design concept of Quenching Head, first, and then Quenching Shell.

From the 3-d mechanical to the cooling model

At the early stage of this development it was necessary to find a powerful tool to create a reliable computational model of the quenching device both from the mechanical and the fluid dynamic points of view: the choice felt on ANSYS Fluent™, which already contained in-built routines for three-phase cooling (air/water/solid).

At the beginning a plain 2-d archetypal of the QH was created, in order to understand the coolant flow in a cross section.

The planar geometry was subdivided into a mesh grid of triangular and rectangular elements.

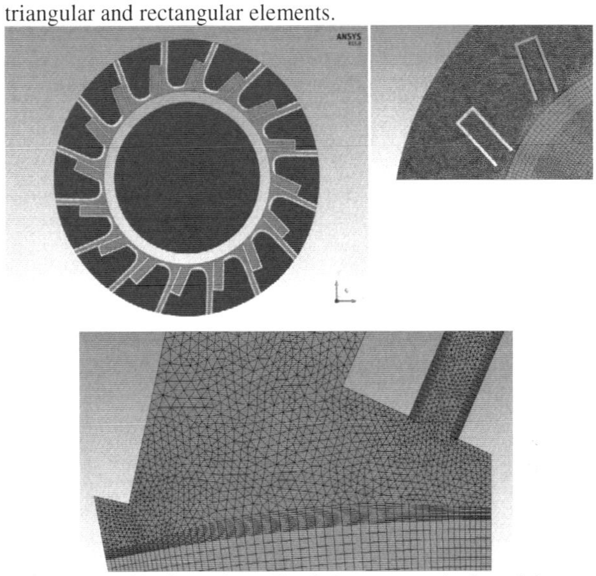

Figure 13: Quenching Head ring geometrical model

The next step had been the implementation of a 3-d model, inclusive of the spatial region around one single water nozzle. By making good use of the longitudinal symmetry and the circular periodicity, the single sector could then be extended to the desired complexity.

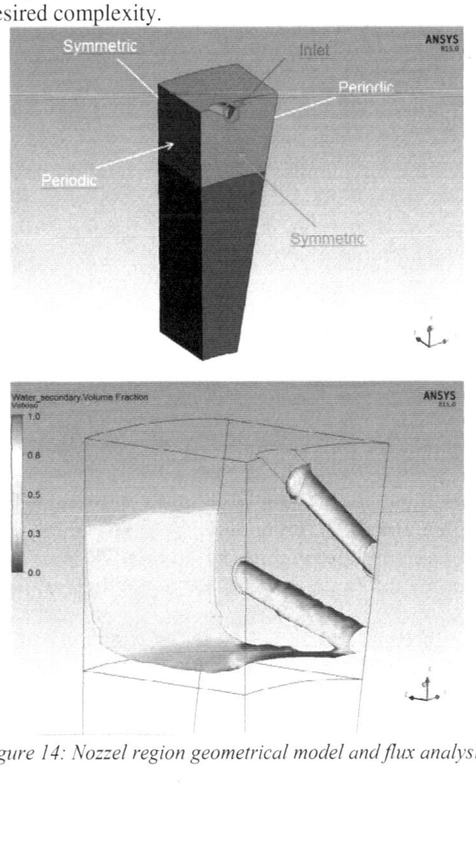

Figure 14: Nozzel region geometrical model and flux analysis

For the purpose of obtaining a streamlined dynamical algorithm for the solution of the cooling simulation, it was necessary to reduce the mesh weight: an optimization of tetragonal and hexagonal elements was worked out until the correct balance between precision on computational speed could be found.

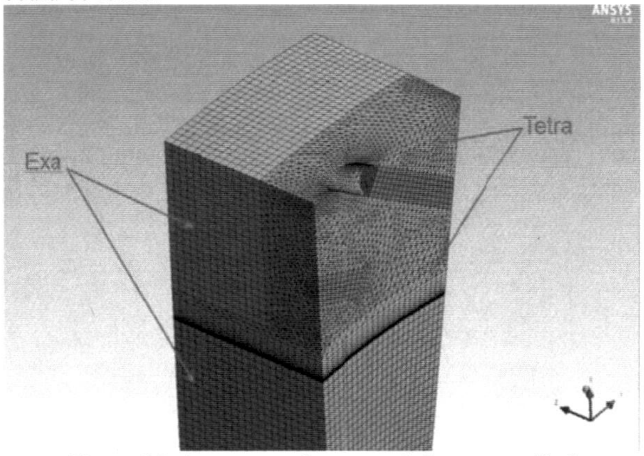

Figure 15: Area around one nozzle hose magnified

The ANSYS Fluent™ contained already some typical fluid/solid and thermal interaction routines, such us the film boiling effect which, although customized, represented the base algorithms applied for SMS modelling.

One typical time simulation run, i.e. the complete cooling analysis of one workpiece processed by the Q.H. model, can be subdivided in 5,000 to 10,000 Time-Steps, each taking as much as 1 second for being calculated by a proper workstation.

The output given by the Fluent software is typically a diagram representing the cooling of different layers of the workpiece while exposed to the water stream, as shown on Fig.16 hereby:

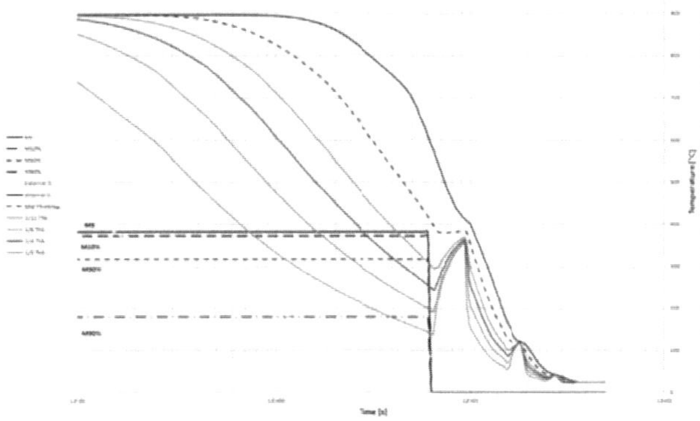

Figure 16: Cooling curve of a steel pipe processed by SMS Quenching Head: the evident spikes shows the transit of the progressing workpiece in the areas between the QH modules, where the inner layers, still hot, tend to reheat the external

Steel characterization and phase transformation analysis

The Fluent model above doesn't yet tell us anything about what phase transformation the material ran through during the cooling process, that all in all is the final aim for the overall CFD modelling.

In order to obtain this, we needed to get a fine characterization of the properties of any steel, such as knowing the physical parameters (specific heat, thermal conductivity, emissivity etc.) as functions of temperature, as well as mastering also the phase transformation dynamics, through the knowledge of TTT (temperature time transformation) and CCT (continuous cooling transformation) diagrams.

These impressive data amount is partially available through the professional software library of JMat Pro™, and has been partly subject of laboratory characterization conducted by SMS Group, necessary to fill in the library gaps and gain the complete information necessary for the transformation analysis. Interpolation models allow to calculate the necessary properties for any steel chemical composition, including those not necessarily present in the software library or already subject of laboratory characterization.

As a final result it is not only possible to correct the cooling diagrams taking into account of the exact physical properties of any specific steel, but also to predict very accurately the phase transformation during the quenching.

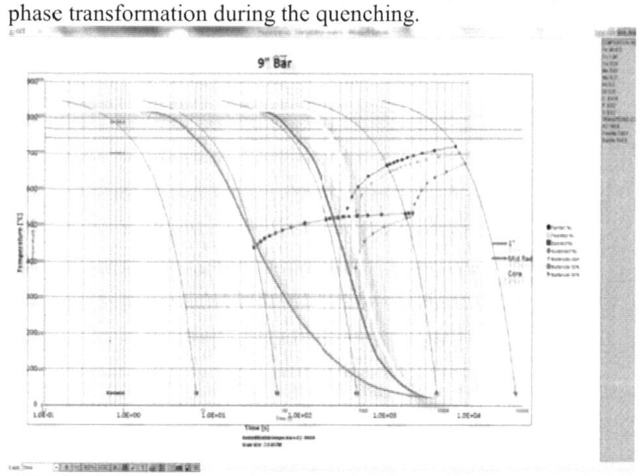

Figure 17: Overlapping of a CCT diagram with the relevant steel tube cooling diagram (in °C over log(time))

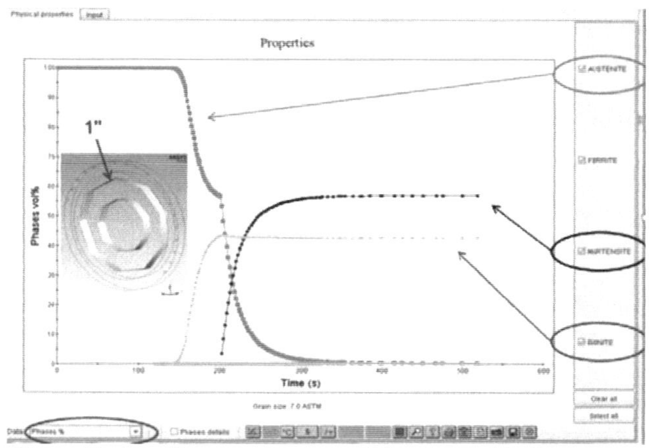

Figure 18: As a direct consequence of the above, this diagram shows the different metallographic phases formation over time, from 100% Austenite down to Martensite and Bainite.

The theoretical predictions have been compared with plant results in order to check the findings and input some correction parameters to the physical formulas.

From circular to open tangential model
The passage from the 360° circular model, proper of the Quenching Head geometry, to the partial annulus (~220°), opened on the bottom, of the Quenching Shell, forced to introduce additional variables in the calculation.

Figure 19: Fully periodical 360° circle becomes an open annular geometry

In particular, the following effects, previously negligible, had to be added the model:

- Gravity
- Rotation of the piece and relevant centrifugal forces
- Surface or Coandă effect (the tendency of a fluid jet to adhere to a convex surface)

Although computationally heavier, the enhanced model suited for the Quenching Shell is still capable to produce reliable outputs that allows to predict the metallurgical result of a quench and temper process cycle.

In addition to the above described models, valid for OD cooling, as per the time of writing in December 2015, an ID nozzle modelling is already at an advanced development stage, and its publication and consequent validation is expected within the second quarter of 2016.

Conclusions

A model with ample potential
The reach of the described methodology is vast: you can not only use the SMS-Quench CFD model as a design helper during the engineering development of a new Q&T line, but also and more profitably as a tool to create real working recipes to be adopted for the plant operation.
Through the SMS-Quench model it is possible to design a tailor made quenching solution, through the selection of the correct project criteria, such as:

- number of quenching modules (for the QH)
- number of nozzles (rows, lines)
- shape of spraying jet
- borehole diameter (influencing the water flow)

Once the machine is sized and designed, it is in any case possible to choose the correct working parameter:

- water flow
- workpiece rotation / travelling speed
- nozzle angle inclination (affecting the quench severity)

And it is also possible to estimate in advance different workpiece quenchability, to suggest steel grade variations to obtain specific target results, and to prepare optimized sets of recipes for the different tubes and bars to be processed.

International Federation of Heat Treatment and Surface Engineering 2016
Proceedings of the 23rd IFHTSE Congress
April 18–21, 2016, Savannah, Georgia, USA

A New Probe to Measure and Evaluate the Real Quenching Intensity Based on Temperature Gradients

***Bozidar Liscic, Bozidar Matijevic*[1]**
[1]Faculty for Mech. Engineering, University of Zagreb, Zagreb, Croatia

***George E. Totten*[2]**
[2]Dept. Mech. and Materials Engineering, Portland State University, Portland, OR, USA

***Lauralice de C.F. Canale*[3]**
[3]Departamento de Engenharia de Materials, Universidade de Sao Paulo, Sao Carlos, SP, Brazil

Abstract

A new Liscic/Petrofer probe is designed to measure and record the quenching intensity in real industrial conditions. It is a cylinder of 50 mm diameter and 200 mm length, made of INCONEL 600, instrumented with 3 thermocouples (TC_s) at the same radius of the cross-section: at 1 mm below surface, at 4.5 mm below surface and at the center, at half the length of the probe. The probe measures the *dynamic of heat extraction*, which is best represented by the *change of temperature gradients*. The results of each test are presented in the form of the following 4 diagrams: The recorded cooling time-temperature curves; The calculated heat transfer coefficient (α) as function of time; The calculated α as function of surface temperature; The calculated surface temperature. A series of 26 tests recently performed at the Quenching Research Center (QRC) of the Faculty for Mech. Engineering, University of Zagreb, Croatia, has shown that by using this probe and relevant criteria, distinctive data are obtainable, essential for selection of quenchants and quenching conditions which cannot be obtained by small laboratory test probes. These data can e.g. document the differences between a mineral oil and a vegetable oil, as well as to show the influence of quenchant temperature and agitation

Key words: Quenching intensity; Temperature Gradient Method; Heat transfer coefficient.

Introduction

Cooling characteristics of a liquid quenchant are usually determined by cooling curve analysis using a laboratory test, in most cases by ISO 9950 or ASTM D6200 [1] standards, using small cylindrical test specimens of 12.5 mm dia. × 60 mm, having one thermocouple (TC) at its geometrical center. These tests are invaluable for checking a quenchant sample or for regular monitoring of a quenching bath but they are of limited value for measurement and recording real quenching intensity of workpieces in workshop conditions.

The Liscic/Petrofer probe is designed to measure and record the cooling intensity of all types of liquid quenchants using different quenching techniques (Immersion quenching; Intensive quenching; Delayed quenching; Martempering; Austempering) to characterize quenching processes in industrial practice and for comparing their quenching intensity.

Because of its cylindrical shape, the probe is applicable for heat transfer calculations of axially symmetric workpieces with diameters between 20 and 100 mm. The probe is a cylinder of 50 mm diameter and 200 mm length. Fig. 1a is the schematic of the probe with the handle and Fig. 1b is a photo of the probe itself. The ratio between length and diameter L/D = 4 : 1 which assures that the heat dissipation through both ends of the probe is negligible so that in the cross-section at half length of the probe where the TC_s are positioned only radial heat flow exists. This is a prerequisite for 1-D heat transfer calculations. The probe is constructed of INCONEL 600, a Ni-Cr austenitic alloy, which during heating and cooling does not undergo structural transformation and is oxidation resistant. The probe is instrumented with 3 sheathed and grounded TC_s: one placed 1 mm below surface, the second one 4.5 mm below surface, and the third one at the center of the cross-section at half length of the probe.

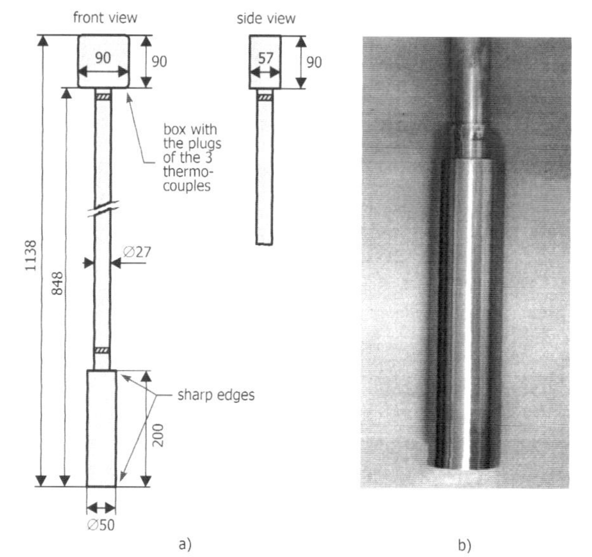

Fig. 1 Sketch of the Liscic/Petrofer probe with the handle (a), Photo of the Liscic/Petrofer probe (b)

All TC_s are placed along the same radius. The size of the probe and its mass (3.3 kg without handle) ensures sufficient heat capacity and radially symmetric heat flow in the cross-sectional plane where the TC_s are located. By placing the outer TC (the measured data of which are used as input for heat transfer calculations) at 1 mm below surface, minimum damping effect of transient surface temperature is assured. Also the lagging effect (because of thermal diffusivity) is minimized, due to small distance to surface. The response time (which depends on the outer diameter of the TC) is relatively short because the diameter of the TC at 1 mm below surface is only 1 mm.

The procedure of a test is as follows: The probe is heated to 850 °C until the central TC reaches this value, than it is quickly transferred to the quenching bath and immersed vertically. It is of utmost importance that the transfer from the furnace to the quenching bath is always performed within the same short period of time, as well as immersing the probe with the same speed, strictly vertically, so that equal cooling conditions can be guaranteed around the perimeter of the probe. The probe is connected to a temperature data acquisition system which contains 3 A/D converters and amplifiers, and a personal computer. The acquisition software enables recording of all 3 TC signal outputs with a frequency of 0.02 s (50 measurements per second) during the entire quenching process and simultaneously drawing the 3 cooling curves in real time.

Temperature Gradient Method: The main feature, when comparing the quenching intensity among different quenching processes, is the ability to extract heat from a heated workpiece i.e. the *dynamic of heat extraction*. This implies that the real quenching intensity in workshop practice cannot be estimated by a single number only (as is the case with Grossmann's H-value) and neither can it be measured and recorded by the small test specimen used for laboratory tests.

Physically the most correct way to measure and record the heat extraction dynamic during quenching is by the heat flux density (W/m^2) on the surface of the quenched workpiece, from which the heat transfer coefficient (HTC) is calculated. The heat flux density (q) on the workpiece surface is inherently connected with the *changing temperature gradients* inside the workpiece by the known physical rule that "the heat flux on the surface of a body is directly proportional to the temperature gradient at the surface multiplied by the thermal conductivity of the material of the body being cooled".

$$q = \lambda \times \frac{\partial T}{\partial x} \tag{1}$$

where: q is the heat flux density (W/m^2) i.e. the quantity of heat transferred through a surface unit per unit of time; λ is thermal conductivity of the body material (W/mK) and $\frac{\partial T}{\partial x}$ is the temperature gradient inside the body perpendicular to the surface (K/m).

When measuring the quenching intensity by a multithermocouple probe, the change of the heat flux inside the probe at point of time can be determined as the temperature difference (ΔT) between the points where the TC_s are placed. This method is known in literature as the Temperature Gradient Method.

Calculation of the heat transfer coefficient when using the Liscic/Petrofer probe: Every quenching process in vaporizable liquids is a complex thermodynamic and fluid-dynamic problem, described in [2], encompassing different phases (vapor film, nucleate boiling and convection). To describe the heat transfer from a heated metallic workpiece to the quenchant, determination of the temperature dependent Heat Transfer Coefficient (HTC) is used. To calculate the HTC, a cooling curve measured near the surface of the probe is necessary. In this case the probe is considered as a long radially symmetric body of a given radius "r". In a simplified one-dimensional (1-D) model, the temperature distribution $T(x, t)$ inside the cylinder, for times $t > 0$, depends on the radial distance $x(0, r)$ from the center of the cylinder and is determined by the 1-D heat conduction equation:

$$\rho c \frac{\partial T}{\partial t} = \mathrm{div}(\lambda \,\mathrm{grad}\, T) \tag{2}$$

All physical properties: ρ (density), c (specific heat capacity), and λ (heat conductivity) of the probe material are temperature dependent so the entire problem is nonlinear. The initial condition $T(x, 0)$ is known (measured), and the problem is to determine the surface heat transfer coefficient (HTC), designated "α" in the boundary condition for $x = r$.

$$\lambda \frac{\partial T}{\partial x} = \alpha \left(T - T_{ex(t)}\right) \tag{3}$$

where: $T_{ex(t)}$ is the measured external temperature of the quenchant. To determine α, an additional cooling curve $T(t)$ is measured at a point $x = r_1$ near the surface (in this case 1 mm below surface). The inverse problem of computation α is solved by the following numerical procedure:

1. Solve the heat conduction equation (2) within the range $(0, r_1)$ with measured $T(t)$, the Dirichlet boundary condition.
2. If $r_1 < r$, extend the solution towards the boundary from $(0, r_1)$ to $(0, r)$ and
3. Calculate α from equation (3) with measured $T_{ex(t)}$ by using numerical differentiation.

Since temperatures are measured at discrete times, they must to be smoothed before use. This is done by a cubic spine least-squares approximation to obtain a sufficiently smooth global approximation over the entire time range.

Numerical solution of the heat conduction equation (2) is performed by the nonlinear implicit method with a simple iteration per time step to adjust all physical properties to the new temperatures.

The solution extension in step 2 is computed by local extrapolation based on a low degree polynomial least-squares approximation. The same approximation is also used for the numerical differentiation needed to compute α in step 3.

How realistic and exact will be the determination of the HTC depends also on the input data from the point where the temperature inside the probe is measured.

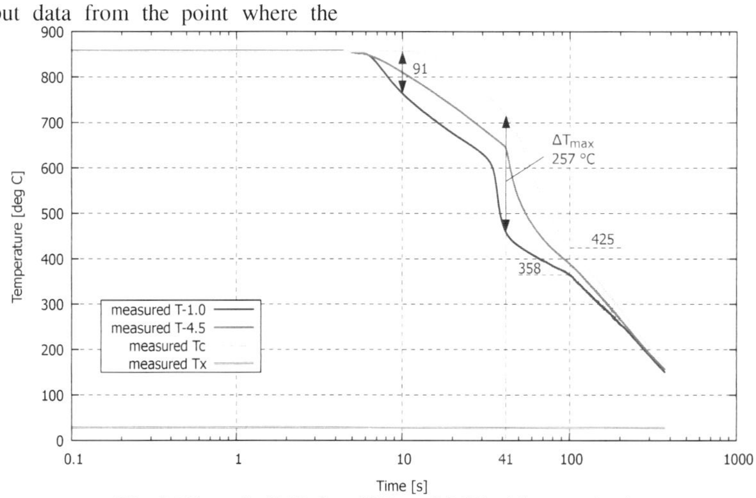

Fig. 2 Mineral oil (Kalenol32), at 28 °C without agitation

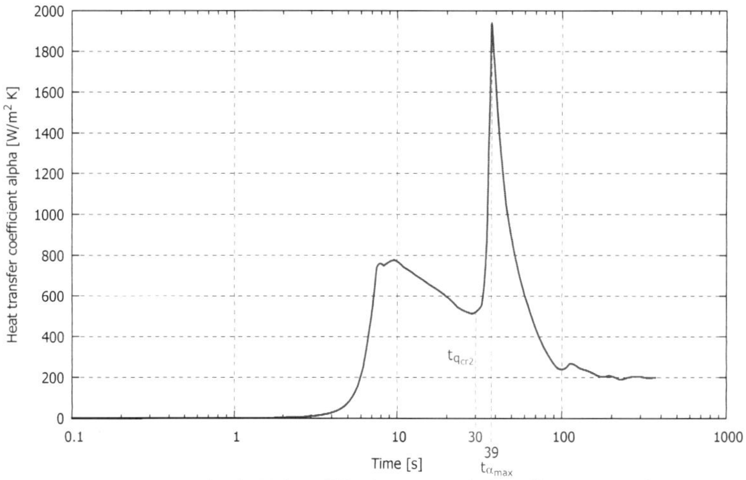

Fig. 3 Mineral oil (Kalenol32), heat transfer coefficient $\alpha = f(t)$

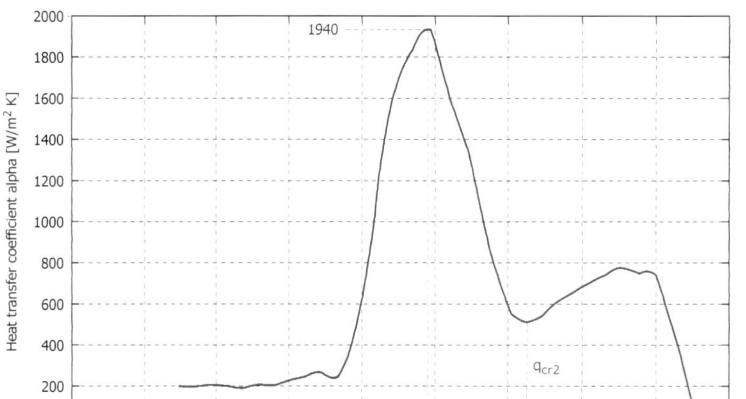

Fig. 4 Mineral oil (Kalenol32), heat transfer coefficient $\alpha = f(T_s)$

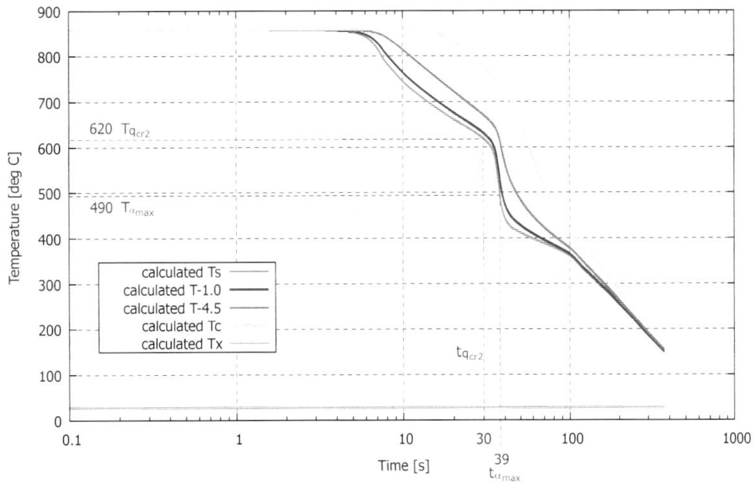

Fig. 5 Mineral oil (Kalenol32), calculated surface temperature

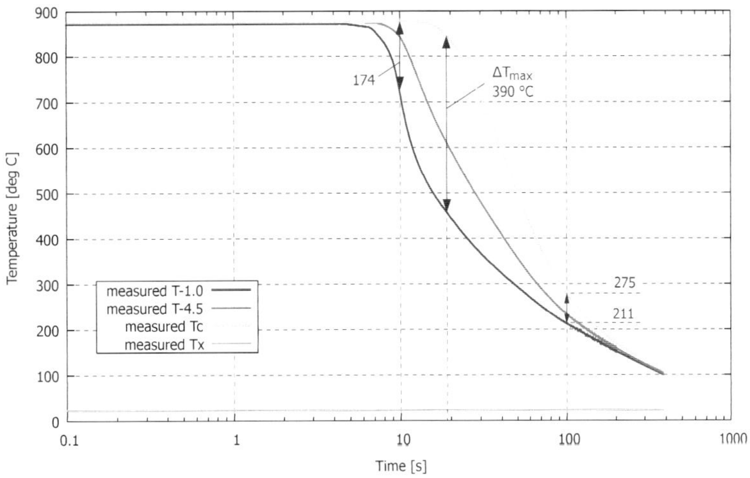

Fig. 6 Vegetable oil (Palm-oil), at 23 °C without agitation

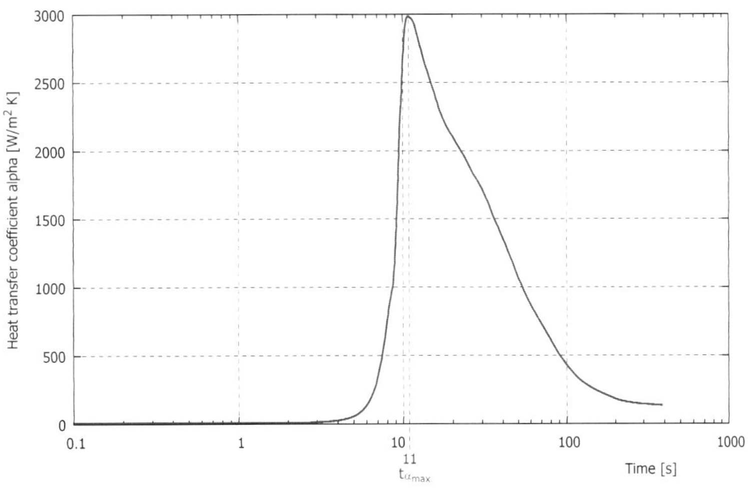

Fig. 7 Vegetable oil (Palm-oil), at 23 °C without agitation, heat transfer coefficient $\alpha = f(t)$

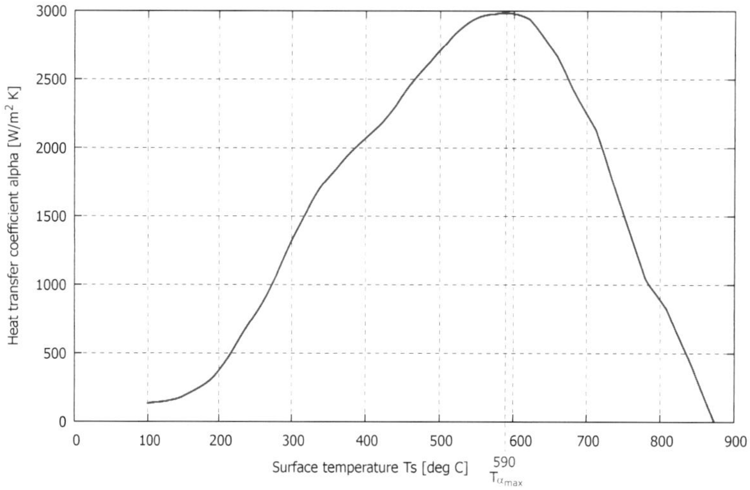

Fig. 8 Vegetable oil (Palm-oil), at 23 °C without agitation, heat transfer coefficient $\alpha = f(T_s)$

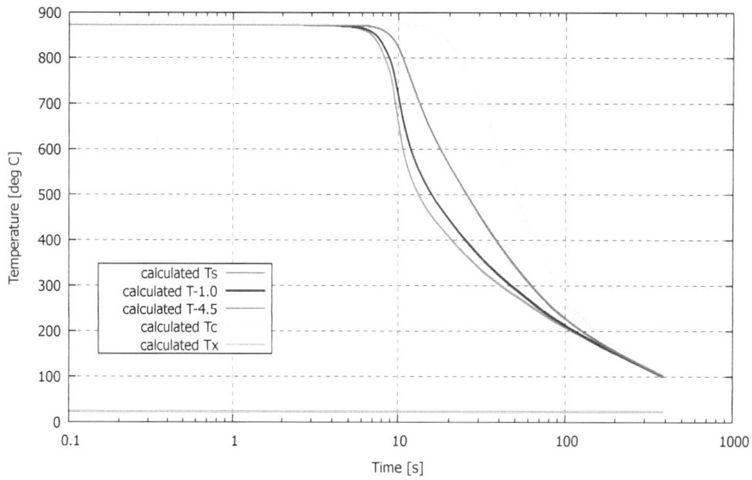

Fig. 9 Vegetable oil (Palm-oil), calculated surface temperature

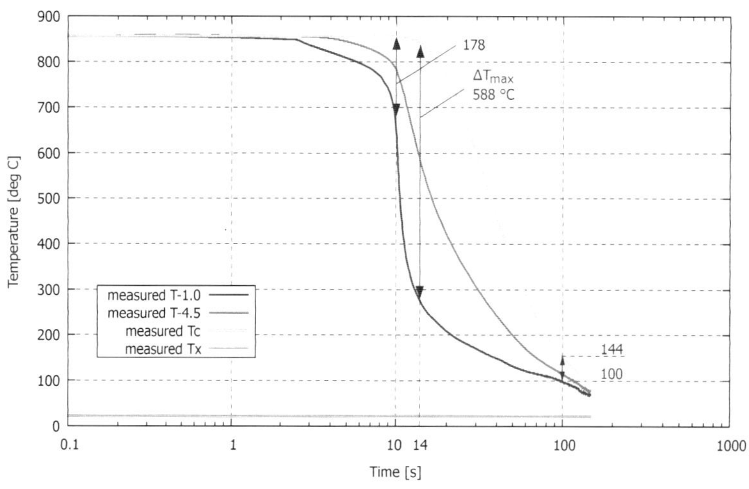

Fig. 10 Cooling curves recorded by the Liscic/Petrofer probe when quenched in still water at 22 °C

Distinctive Characteristics of Cooling Curves and Calculated Heat Transfer Data

When performing tests using the Liscic/Petrofer probe, the results are presented by the 4 following diagrams:

a) The recorded cooling curves
b) The calculated heat transfer coefficient (HTC) as function of time
c) The calculated HTC as function of surface temperature
d) The calculated surface temperature, as shown in Figs. 2 to 5 for a mineral oil.

In order to impartially analyze and use the experimental results, the following *characteristic criteria* from those diagrams must be applied:

1. ΔT_{10} Temperature gradient between the center and the point 1 mm below surface at 10th s after immersion
2. $t_{\alpha\,max}$ Time from immersion to the moment when α_{max} occurs
3. ΔT_{max} Maximal temperature gradient
4. α_{max} Maximal value of the heat transfer coefficient
5. $T_{\alpha\,max}$ Temperature at which α_{max} occurs
6. ΔT_{100} Temperature gradient between center and the point 1 mm below surface at 100th s after immersion.

For quenchants having a clearly distinguishable film boiling phase (as is the case with the oil, see Figs. 2 to 5), the transition point between film boiling and nucleate boiling (the second critical heat flux density q_{cr2}) can be detected.

The most important feature of each quenching process is the *dynamic of heat extraction* which is best represented by the *change of temperature gradients*. The above specified criteria include therefore the following 3 temperature gradients: ΔT_{10}, ΔT_{max} and ΔT_{100}. The temperature gradient at 10th s after immersion (ΔT_{10}) shows how quickly a quenchant reacts. The maximal temperature gradient (ΔT_{max}) is of greatest importance showing the maximal thermal stresses developed during the quenching process. The time when ΔT_{max} occurs is also important because the temperature at 1 mm below surface depends on it. If this temperature, at the moment of ΔT_{max} , is lower than the martensite start temperature (M_s) of the relevant steel, structural stresses due to martensite formation may be added to maximal thermal stresses and distortion or cracking may occur. The temperature gradient ΔT_{100} at 100th s after immersion provides a information on the duration of the quenching process.

Differences Between a Mineral Oil and a Vegetable Oil

When performing tests with the Liscic/Petrofer probe and analyzing the results using the above specified criteria, substantial differences are found between a mineral quenching oil and a vegetable oil (in this case a Palm-oil), as shown in Table I. The mineral oil used was Kalenol32 (a product of INA – Maziva Zagreb, Croatia) which exhibits a clearly distinguishable film boiling phase (see Fig. 4) that changes to nucleate boiling phase at 620 °C after 30 seconds (see Fig. 3). The vegetable Palm-oil does not exhibit the film boiling phase as can be seen from its heat transfer coefficient in Fig. 8.

Table I. Comparison between mineral oil Kalenol32 and vegetable Palm-oil at room temperature without agitation

Quenchant	**Kalenol32 at 28 °C**	**Palm-oil at 23 °C**	**Measure**
ΔT_{10}	91	174	°C
$t_{\alpha\,max}$	39	11	s
ΔT_{max}	257 at 41 s	390 at 19 s	°C
α_{max}	1940	3000	W/m^2K
$T_{\alpha\,max}$	490	590	°C
ΔT_{100}	425 - 358 = 67	275 - 211 = 64	°C

The main differences from Table I are the following: The temperature gradient at 10th s after immersion (ΔT_{10}) is 1.9 times greater for Palm-oil (compare Fig. 6 and Fig. 2), and the time to for the maximal heat transfer coefficient ($t_{\alpha\,max}$) is 3.5 times shorter (compare Fig. 7 and Fig. 3), indicating that Palm-oil reacts more quickly because it exhibits no film boiling phase. The maximal temperature gradient (ΔT_{max}) is 52 % greater at Palm-oil and occurs 22 s earlier (compare Fig. 6 and Fig. 2). The maximal value of the heat transfer coefficient (α_{max}) is 55 % greater for Palm-oil. The temperature gradient at 100th s after immersion (ΔT_{100}), although almost equal for both oils, occurs for Palm-oil at much lower temperatures indicating that the duration of the cooling process for Palm-oil will be shorter.

Influence of Quenchant Temperature

Table II. Comparison between mineral oil Kalenol32 at room temperature and at 60 °C without agitation

Quenchant	**Kalenol32 at 28 °C**	**Kalenol32 at 60 °C**	**Measure**
ΔT_{10}	91	50	°C
$t_{\alpha\,max}$	39	40	s
ΔT_{max}	257 at 41 s	257 at 41 s	°C
α_{max}	1940	2055	W/m^2K
$T_{\alpha\,max}$	490	490	°C
ΔT_{100}	425 - 358 = 67	425 - 366 = 59	°C

Results from Table II demonstrate the following: At 60 °C the mineral oil Kalenol32 exhibits a decreased temperature gradient at 10th s after immersion (ΔT_{10}), i.e. it reacts slower than at room temperature probably because of the longer film boiling phase. Time for the maximal heat transfer coefficient ($t_{\alpha\,max}$), as well as the maximal temperature gradient (ΔT_{max}), are equal in both cases. The value of the maximal heat transfer coefficient (α_{max}) at 60 °C is 6 % greater due to its lower viscosity but it occurs at the same temperature in both cases.

The temperature gradient at 100th s after immersion (ΔT_{100}) does not change substantially in both cases.

Influence of agitation

Table III. Comparison between still mineral oil Kalenol32 and the same oil intensively agitated, at room temperature

Quenchant	Kalenol32 still	Kalenol32 agitated	Measure
ΔT_{10}	91	150	°C
$t_{\alpha\ max}$	39	19	s
ΔT_{max}	257 at 41 s	324 at 24 s	°C
α_{max}	1940	2300	W/m^2K
$T_{\alpha\ max}$	490	530	°C
ΔT_{100}	425 - 358 = 67	285 - 185 = 100	°C

Intensive agitation resulted with much earlier reaction of this mineral oil. The temperature gradient at 10th s after immersion (ΔT_{10}) is 65 % greater than still oil. Time for the maximal heat transfer coefficient ($t_{\alpha\ max}$) is 20 seconds shorter. The maximal temperature gradient (ΔT_{max}) is 26 % greater and occurs 17 s earlier when agitated intensively. The maximal value of the heat transfer coefficient (α_{max}) is 19 % greater than for still oil and occurs at a 40 °C higher temperature. The temperature gradient at 100th s after immersion (ΔT_{100}) is much greater than for still oil, and occurs at much lower temperatures indicating a shorter duration of the quenching process. It is interesting to see that for the mineral oil Kalenol32, how agitation influences the transition point between film boiling and nucleate boiling (the second critical heat flux density q_{cr2}). For still oil, this transition point occurs at 620 °C and appears 30 s after immersion. For intensively agitated oil, this transition point occurs at 600 °C but appears at only 17 s after immersion. It can be concluded that agitation does not substantially change the temperature of the transition point but reduces the time when it occurs, i.e. *agitation reduces the film boiling phase.* By using the Liscic/Petrofer probe, this effect can be precisely documented.

Use of the Liscic/Petrofer probe for Quenching in Water and Water-Based Solutions

This probe can be also used for testing the quenching intensity of water and water-based solutions, including different polymer-solutions. Fig. 10 shows cooling curves recorded when quenching in still water of 22 °C. There are striking differences in recorded cooling curves (concerning primarily the maximal temperature gradient ΔT_{max}) for a mineral oil and water, as seen when comparing Fig. 2 and Fig. 10. While in this case, the Kalenol32 oil exhibits a ΔT_{max} of 257 °C which occurs at 41 s after immersion, water exhibits a ΔT_{max} of 588 °C which occurs 14 s after immersion. The values of the maximal heat transfer coefficient (α_{max}) are in this case also very different: for Kalenol32: 1940 W/m^2K and for water: 13270 W/m^2K.

Conclusions

The series of experiments described here provide clear evidence that testing the quenching intensity of liquid quenchants by the Liscic/Petrofer probe in industrial conditions can provide information not attainable with the small laboratory probes. This evaluation is based on *dynamic of heat extraction* from the quenched workpiece which is best represented by the *changing temperature gradients.* As result of each test, 3 experimentally recorded cooling curves and diagrams of calculated: surface temperature, heat transfer coefficient as function of time and as function of surface temperature are obtained. In order to impartially and exactly analyze each test, characteristic criteria from these diagrams must to be applied. The analysis of all of the tests performed confirm the following important facts:

- Contrary to the information obtainable when testing liquid quenchants by small laboratory probes, the analysis of tests when using the Liscic/Petrofer probe may be used for the selection of optimal quenchant and quenching parameters under industrial conditions.
- The testing procedure using the Liscic/Petrofer probe is sufficiently sensitive to clearly differentiate when changes of the quenching parameters (quenchant temperature or agitation rate) occur.
- The Liscic/Petrofer probe can be used for different types of liquid quenchants under different quenching conditions in industrial practice.

References

[1] ASTM D6200-97: *Standard Test Method for Determination of Cooling Characteristics of Quench Oils by Cooling Curve Analysis*, Annual Book of ASTM Standards, ASTM International, West Conshohocken, PA, 2001.

[2] Liščić, B. Tensi, H.M. and Luty, W, Eds 1992. *Theory and Technology of Quenching*, Springer –Verlag, Berlin, pp. 41-72.

International Federation of Heat Treatment and Surface Engineering 2016
Proceedings of the 23rd IFHTSE Congress
April 18–21, 2016, Savannah, Georgia, USA

Pre-oxidation treatment to increase the service life of heat treatment furnace conveyor belts made of AISI 330Cb Industrial application

Combe C., Cardey PF., Centre Technique des Industries de la Mécanique (Technical Center of Mechanical Industries)
CETIM, SAINT-ETIENNE, Metallic Materials and Surfaces division– FRANCE
cecile.combe@cetim.fr, pierre-francois.cardey@cetim.fr – +33 (0)4 77 79 40 51

Buscail H., Issartel C., Riffard F., Rolland R., Laboratoire Vellave sur l'Elaboration et l'Etude des Matériaux (Materials Development and Research Laboratory)
LVEEM, LE PUY EN VELAY - FRANCE
henri.buscail@udamail.fr - +33 (0)4 71 09 90 71

Abstract

Heat treatment furnace conveyor belts are simultaneously exposed to high temperatures and chemically reactive gaseous species. Besides the economic and technical characteristics (forming, weldability, etc.) which condition the choice of the material used in the conveyor belts, other constraints need to be taken into account, such as: creep resistance (microstructural stability), high temperature strength and high temperature corrosion resistance. The main high temperature corrosion processes which furnace conveyor belts are exposed to are oxidation and carburisation. With regard to carburisation and oxidation resistance, previous laboratory studies have demonstrated the positive effect of the presence of a silica surface layer (obtained with a pre-oxidation treatment) on AISI330Cb heat resisting stainless steels [1] [2].

Pre-oxidation treatments were carried out in 2014 on three industrial conveyor belts. This paper presents the microstructural evolution of one of these conveyor belts.

Introduction

CETIM's Attachments and Heat Treatments commissions have been working for ten years on improving the service life of conveyor belts used in heat treatment furnaces. Due to their operating conditions and their design (material grades, dimensions, etc.), furnace conveyor belts have rather variable service lives [2] [3].

There are many models of conveyor belts available, due to the wide variety of applications. The industrial conveyor belt considered in this paper is installed on a continuous furnace and it is made of spirals, rods and side-plates (Fig. 1)
Only the spirals and rods are made of AISI330Cb alloy, while the side-plates are made of AISI310 alloy (material grade without niobium).

The AISI330Cb stainless steel grade (X10NiCrSiNb35-22 as per Standard NF EN 10095 [4]) is generally used in high temperature applications in carburising atmospheres. The chemical composition of the base material of the studied conveyor belt is presented in Table 1.

Table 1 – Chemical composition of the studied alloy (spirals and rods).

Chemical composition (weight %)	X10NiCrSiNb35-22 as per EN 10095:1999	X10NiCrSiNb35-22	
		Spirals	Rods
% Carbon (C)	≤0.15	0.021	0.013
% Manganese (Mn)	≤2.00	0.45	0.26
% Silicon (Si)	1.00-2.00	1.47	1.80
% Sulphur (S)	≤0.015	≤0.005	≤0.005
% Phosphorus (P)	≤0.030	0.015	0.015
% Chromium (Cr)	20.0-23.0	20.3	20.1
% Nickel (Ni)	33.0-37.0	35.2	35.1
% Nitrogen (N)	≤0.11	Not determined	Not determined
% Niobium (Nb)	1.00-1.50	1.20	1.15
% Molybdenum (Mo)	-	0.056	0.016
% Copper (Cu)	-	0.044	0.027
% Titanium (Ti)	-	≤.020	≤.020
% Iron (Fe)	Base	Complement to 100%	Complement to 100%

Figure 1 – Structure of a conveyor belt with alternate pitch spirals
(a- as received; b- after pre-oxidation treatment)

Interest of a pre-oxidation treatment

Oxidation and carburisation are the two main high temperature corrosion processes (dry corrosion) to which the alloys used in conveyor belts are exposed.

High temperature corrosion resistance is provided by the presence of stable, fine, dense and adhering protective oxides. The oxides which provide the best level of protection are alumina Al_2O_3, silica SiO_2 and chromium (III) oxide Cr_2O_3. In these oxides, the diffusion of the metal, oxygen and carbon elements is sufficiently low to guarantee an acceptable corrosion rate. Therefore, aluminium, silicon and chromium are extremely interesting alloying elements to make heat resisting material grades.

In the case of the AISI330Cb alloy, the presence of chromium (20.0-23.0%) makes it possible to obtain a protective surface layer of chromium (III) oxide. A layer of manganese chromite is often found on the chromium (III) oxide layer. This layer has no corrosion protection properties, as it is porous and poorly adhering, however its presence can limit the evaporation of the chromium oxides [5] [6].

The chromium (III) oxide layer provides protection mainly in isothermal conditions and for temperatures which do not exceed 1100°C (beyond this temperature, the layer is vaporised). However, in thermal cycling conditions (case of the studied conveyor belts), the chromium (III) oxide layer tends to detach, leaving the substrate unprotected. The adhesion of the chromium (III) oxide layer can be significantly improved by adding rare earth elements [7] [8] [9]. In carburising atmospheres, the chromium (III) oxide layer can be destabilised by chromium carbides (M_7C_3, M_3C_2, etc.).

Carburisation adversely affects the ductility and toughness of materials, particularly at room temperature, due to the additional stresses resulting from the increase in volume of the matrix caused by the presence of these carbides [10].
Moreover, the loss of chromium which occurs below the surface during the formation of the chromium (III) oxide layer will constitute a favourable ground for subsequent carburisation in the event of fracture of the oxide layer as a result of a creep, fatigue or thermal cycling process.

In given conditions (temperature and gaseous atmosphere), the addition of silicon (1.00-2.00%) in the AISI330Cb alloy will allow a SiO_2 layer (amorphous or crystallised) to form at the interface between the substrate and the Cr_2O_3 layer. Fig. 2 provides a scanning electron microscope picture which shows the SiO_2 and Cr_2O_3 layers for a ferritic stainless steel with 0.5%Si content [11]. The presence of a silica layer improves the high temperature corrosion protection properties (better adhesion of the oxide layer in thermal cycling conditions and better carburisation resistance) [10].

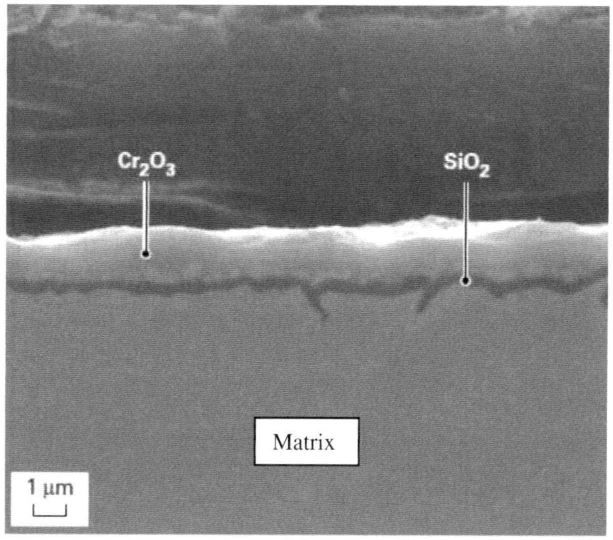

Figure 2 – Protective oxide layer formed at 900°C on a ferritic stainless steel with 17%Cr and 0.5%Si content [12]

Interest of adding sodium hydroxide in small quantities [12]

In its initial state, the microstructure of the AISI330Cb alloy is composed of an austenitic matrix scattered with chromium carbides and niobium carbides. Between 700 and 900°C, the intergranular niobium carbides are not stable and as a result dissolved [13]. Then, carbon combines with the chromium present in substitutional solid solution in the matrix to form new chromium carbides. At the same time, niobium combines with the nickel and silicon present in substitutional solid solution to form the G-phase (nickel and niobium silicide). Finally, the silicon trapped in the G-phase will not be able to diffuse to form the silica layer. Adding sodium allows the niobium to be trapped to form a $NaNbO_3$ type compound (superficial action), and the silicon remains free to form a silica layer.

Conditions of the industrial pre-oxidation treatment

Studies previously carried out on this subject [1] [2] [12] showed that a silica surface layer on the AISI330Cb alloy could be obtained in conditions detailed below:

- Impregnation of the conveyor belt with pure sodium hydroxide (or with sodium hydroxide mixed with calcium chloride) in very low concentration,
- Gradual temperature increase from room temperature to 900°C and holding at 900°C for a long period (the laboratory tests were performed for 48 hours),
- The atmosphere of the furnace, as from room temperature, must be made of pure dinitrogen, or hydrogenated dinitrogen or argon.

These conditions, defined in laboratory on small-size test specimens, were reproduced in August 2014 on a complete conveyor belt. The treatment consisted of immersion in a low-concentration sodium hydroxide solution, followed by gradual temperature increase to 900°C for 36 hours (among which 6 hours at 900°C) in pure dinitrogen.

The industrial furnace considered in this paper is used for quenching, carbonitriding and case-hardening heat treatments at temperatures ranging between 780 and 930°C. Until today, the conveyor belts had rather variable service lives (often less than 12 months).

Experimental methods

We monitored the conveyor belt from its initial condition (before first use), just after the pre-oxidation treatment and then in service after 2 months, 6 months and 15 months of operation (Table 2).

Table 2 – Identification of the samples

Initial state (as received)	t_i
After pre-oxidation treatment	t_0
In service after 2 months of operation	t_0 + 2 months
In service after 6 months of operation	t_0 + 6 months
In service after 15 months of operation	t_0 + 15 months

The characterisation work below was carried out in the laboratory on samples taken from the conveyor belt (spirals and rods):

- Scanning electron microscope examination of the surface oxide layer,
- Monitoring of the material's microstructure with an optical microscope,
- Assessment of the mechanical characteristics by Vickers hardness tests on micrographic sections,
- Checks of the diameter of the spirals.

Results

Formation of the silica layer

The examination of the silica layer was performed after coating the spirals and rods, in longitudinal direction, in hard curing resin. The samples were mechanically polished with 1-µm grade paste.

- The pre-oxidation treatment (t_0) initiated the formation of the silica layer at the interface between the chromium (III) oxide layer and the substrate. However, the surface was only partially covered (Fig. 3).
- The examinations performed after 2 months of operation showed that the silica layer continued to extend during operation in the furnace, but it still featured significant discontinuities (Fig. 4)
- The chromium (III) oxide and manganese chromite layers were also visible.
- After 6 months of operation, the silica layer continued to extend (Fig. 5) but the surface was still not completely covered.
- At 15 months of operation, depending on the examined zones, we noticed that, when there was no oxide layer (due to scaling and/or to a silica discontinuity), the matrix was somehow carburised from the surface. Such carburisation does not occur in the zones where the oxide layer is present (refer to paragraph: "Micrographic examinations").

XRD analyses were carried out at t_0 and at t_0+6 months. Only the Fe(γ) FCC phases, the chromium(III) oxide and the manganese chromite could be detected. The silica layer was not diffracted, which implies either that it was too thin to be detected or that is was not crystallised (amorphous silica).

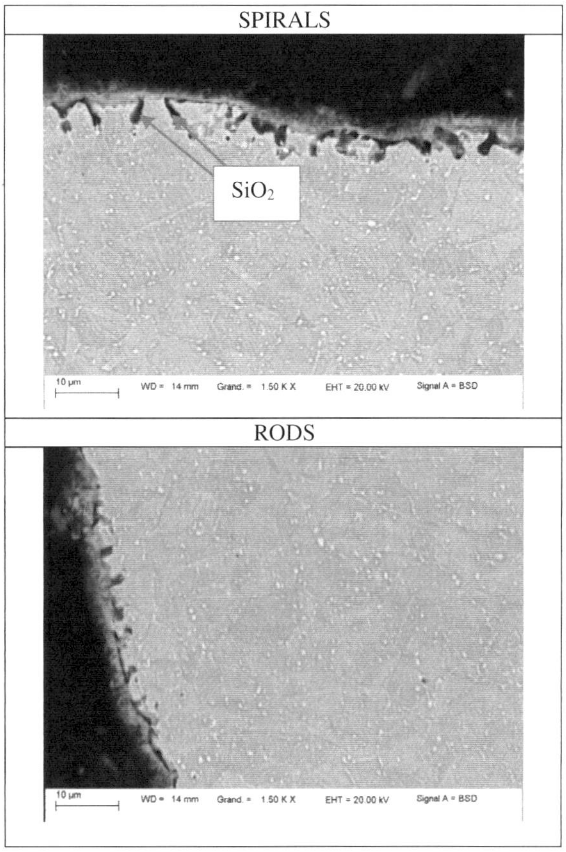

Figure 3 – Scanning electron microscope examination (back-scattered electron detector) at t_0 (etching with nitrohydrochloric acid)

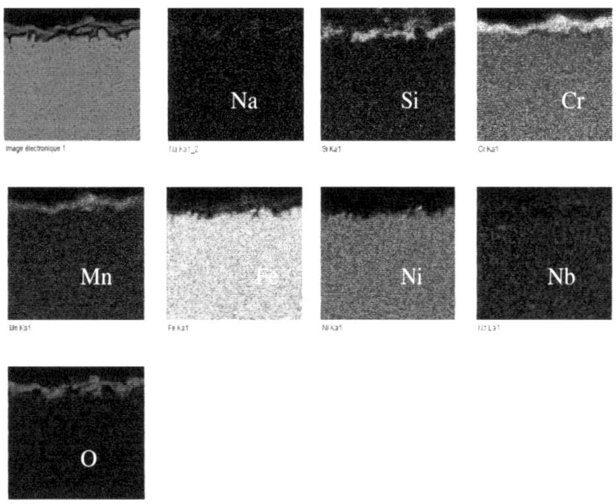

Figure 4 – X-ray mapping of elements on spirals, at t+2 months (etching with nitrohydrochloric acid)

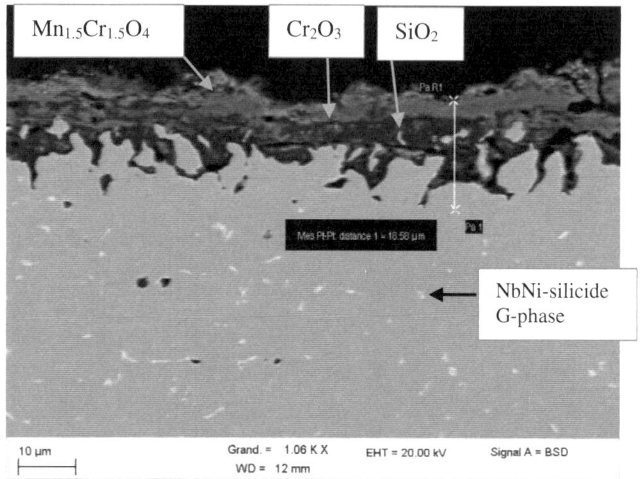

Figure 5 – Scanning electron microscope examination of the spirals at t0+6 months (examination of the oxide layer) without prior chemical etching of the sample

Dimensional checks

The dimensional checks at t_0 + 6 months and t_0 + 15 months were carried out using an optical machine with uncertainty of measurement of 0.02 mm, after ultra-sonic cleaning in ethanol. The results are presented in Table 3.

Table 3 – Dimensional checks

Conveyor belt	Measured characteristic	Measurement results
t_i	Spiral diameter	From 1.20 to 1.21 mm
t_0 + 6 months (ageing for 6 months)		From 1.19 to 1.21 mm
t_0 + 15 months (ageing for 15 months)		From 1.22 to 1.25 mm

Between initial condition and 15 months of operation, the diameter of the spirals slightly increased.

Hardness tests

Vickers hardness tests with 10 kg load were carried out on core micrographic sections. The evolution is presented in Graph 1.

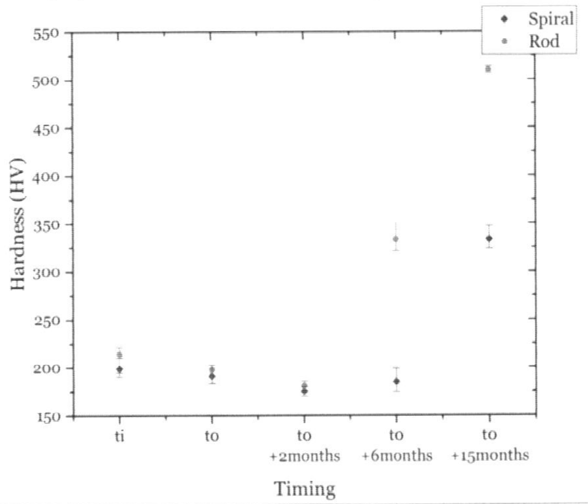

Graph 1 – Evolution of the HV10 Vickers hardness

Between initial condition and t_0 +2 months, the hardness value decreases. After 6 months of operation, the hardness value measured on the rods significantly increases (+150HV). At 15 months, the same increase is visible on the spirals (increase by +150 HV), while the hardness of the rods further increased (+300 HV with respect to initial condition).

Micrographic examinations

In initial condition t_i, the matrix is austenitic and equiaxed with homogeneous grain size. The austenitic matrix is scattered with fine precipitates:

- Intragranular and intergranular chromium carbides ($M_{23}C_6$ type),
- Intragranular and intergranular niobium carbides (MX), finely scattered,
- A few titanium carbonitride spots.

The examinations at t_0 and t_0+2 months revealed the following:

- Slight coalescence of the existing chromium carbides and formation of new chromium carbides,
- Disappearance of niobium carbides (MX) at grain boundaries,
- Appearance of a new phase at grain boundaries (G-phase),
- Reduction of the mechanical characteristics due to relocation of dislocations,
- Constant grain size.

After 6 months, the examinations showed the following:

- Excessive grain growth reaching, in some areas, a depth of 100 μm (Fig. 6). Fig. 7 et Fig. 8 show, respectively, a zone without grain growth and a zone with matrix carburisation (found on a rod at t_0+6 months). The micro-hardness profile clearly shows the increase in the surface hardness in the carburised zone.
- Coalescence and formation of new chromium carbides (mainly $M_{23}C_6$ and M_7C_3 types).
- Coalescence of the new phase located at grain boundaries (Fig. 9).

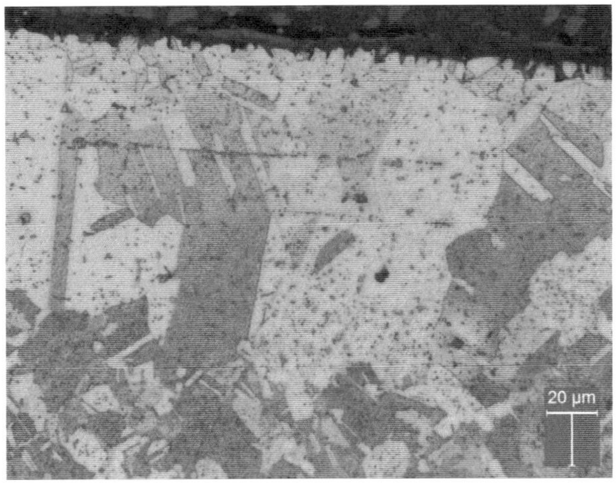

Figure 6 – Optical microscope examination of a spiral at t0+6 months (grain growth on the surface)

Figure 7 – Optical microscope examination on a spiral at t0+6 months (no grain growth)

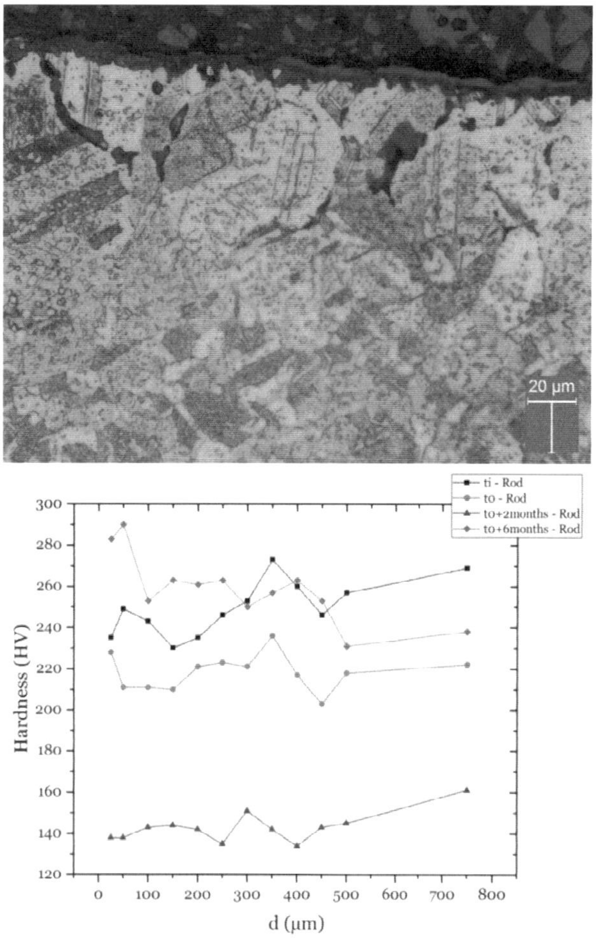

Figure 8 – Optical microscope examination on a rod at t0+6 months and micro-hardness profile (in carburised zone)

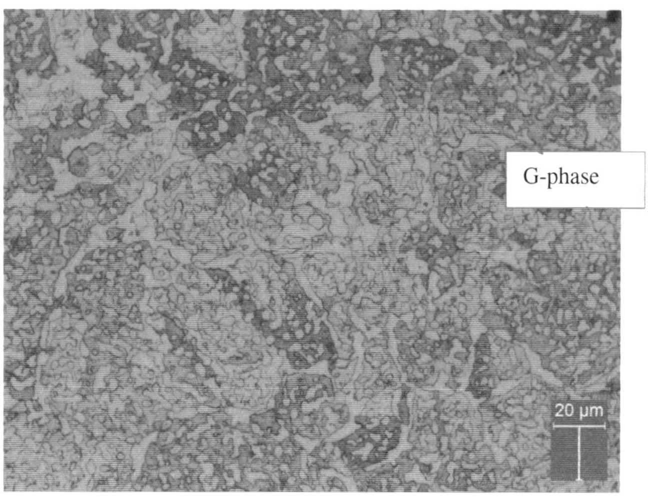

Figure 9 – Optical microscope examination on a spiral at t0+6 months, core structure

Conclusions

The pre-oxidation treatment performed at 900°C for 6 hours led to formation and growth of the silica surface layer. We found that the silica layer continued to expand during belt operation, but it does not cover the entire surface of the spirals and rods. This fact leads to local preferential attack zones for high temperature corrosion (carburisation and oxidation). Therefore, the temperature step at 900°C is not sufficient to obtain a silica layer that covers the entire surface. The laboratory pre-oxidation treatments had been performed for 48 hours continuously on a zone, and in this particular case, the 6 hours are not continuous (but in cyclic condition), which means that each zone "actually" received the heat treatment at 900°C for less than an hour.

The examinations carried out between initial condition and 6 months of operation revealed some damage mainly due to a creep phenomenon:

- Growth of the austenitic grain,
- Dissolution of niobium carbides at grain boundaries and coalescence of a new phase (phase with high content of silicon, nickel and niobium),
- Coalescence of the existing chromium carbides and formation of new, intergranular and intragranular chromium carbides,
- Slight reduction of the mechanical characteristics until 6 months (due to relocation of dislocations)

Between 6 and 15 months of operation, carburisation becomes visible in the zones where the silica did not seem to be present initially and/or where the it was missing due to debonding in operation, leading to slight dimensional increase. The rods became carburised before the spirals, probably because the pre-oxidation treatment was less efficient on the rods due to their containment.

After 15 months of operation, the plastic deformation of the conveyor belt led to some offset in operation. This resulted in rough running points between the side-plates and the bed, quickly leading to their destruction then to the ejection of the treated parts. The conveyor belt was finally scrapped.

The work group is carrying on with its investigations on:

- The application of an industrial pre-oxidation treatment with increased temperature steps at 900°C in order to obtain a silica layer correctly covering the entire surface,
- Understanding the processes which lead to the formation of the silica layer (by studying, among others, the plates made of AISI310 alloy which does not contain niobium),
- Understanding the creep damage processes,
- Improving the adhesion of the oxide layers by adding rare earth elements.

Acknowledgments

We would like to thank all the companies which actively take part in this study: Thermi-Lyon Développement, STTS, Agrati, Lisi Automotive and RD Technologies.

References

[1] Buscail, H. & al "Influence of various gaseous environments on SiO2 formation on the 330Cb alloy at 900°C" (2012) *Corrosion science*, pp. 535-541.

[2] Fleurentin, A. & al "Optimisation de la durée de vie de l'acier réfractaire 330Cb sous atmosphère carburante and réductrice" (2012). *Traitements & Matériaux*, pp. 15-21.

[3] Fleurentin, A., "Optimisation de la durée de vie des réfractaires utilisés dans les fours de traitements thermiques à tapis" (2008). *Traitement thermique*, pp.27-32.

[4] NF EN 10095 (1999) – *Norme AFNOR* Heat resisting steels and nickel alloys. Classification index: A35-584.

[5] Buscail, H. & al "Etude de l'oxydation de l'alliage 330Cb entre 800 et 1000°C" (2009). *Traitement thermique*, pp. 33-41.

[6] Buscail, H. & al "Etude de l'influence de la carburation sur l'oxydation de l'alliage 330Cb à 900°C" (2010). *Matériaux et Techniques*, pp. 209-218.

[7] Buscail, H. & al "Influence of lanthanum coatings on a model 330 alloy Oxidation at High Temperatures" (2014) *Oxidation of metals*, pp. 127-138.

[8] Buscail, H. & al "Effect of various lanthanum sol-gel coatings on the 330Cb oxidation at 900°C" (2011) *Applied surface science*, pp. 678-686.

[9] Issartel, C. & al "Effect of oxidation with lanthanum Sol-Gel coating on the 330Cb carburization at 900°C" (2013) *Oxidation of metals*, pp. 467-478.

[10] Fleurentin, A. & al "Intérêt de la pré-oxydation d'aciers réfractaires austénitiques" (2010) *Traitements & Matériaux*, pp. 31-36.

[11] Antoni, L. & al "Corrosion sèche des métaux - mécanismes" [Techniques de l'Ingénieur, m4221].

[12] Buscail, H. & al "NaOH coatings effects on the 330Cb alloy oxidation at 900°C" (2016 - publication in progress) *Oxidation of metals*.

[13] Bonaccorsi, L. & al "Creep damage and metallurgical characterization of high alloys reformer tubes after long service time" (2014) [ECCC, Rome].

International Federation of Heat Treatment and Surface Engineering 2016
Proceedings of the 23rd IFHTSE Congress
April 18–21, 2016, Savannah, Georgia, USA

Effect of Microstructure on the Fatigue Crack Growth Behaviour In Low Carbon Steel

Daisuke Watanuki
Basic Technology Research Center, NSK Ltd., 1-5-50 Kugenuma-Shinmei, Fujisawa, Kanagawa 251-8501, Japan
Watanuki-d@nsk.com

Hangyue Li, and Paul Bowen
School of Metallurgy and Materials, College of Engineering and Physical Science, The University of Birmingham, Edgbaston, Birmingham B15 2TT, United Kingdom

Abstract

In order to optimize heat treatments of large-scale rolling bearings, the relationships between microstructure and mechanical properties need to be thoroughly understood. In the current study, the effect of microstructure on fatigue life, fatigue crack growth threshold (ΔK_{th}) and fracture toughness were investigated for 18NiCrMo14-6, typically used for large-size bearings. The martensite microstructure was processed by conventional quenching and tempering, and the two-phase microstructures, martensite+bainite, and bainite+austenite were processed by austempering. Fatigue crack propagation properties of two-phase microstructures were compared with three martensitic microstructures tempered at different temperatures to match the tensile properties of those of the two-phase microstructures. Fatigue crack growth tests were carried out and the ΔK_{th} values of both dual-phase microstructures are higher than that of the martensitic counterparts. In contrast, it was found by uniaxial tensile fatigue tests using cylindrical specimens, that the fatigue strength at 10^7 cycles of the two-phase microstructure is lower than that of the martensite single-phase microstructure. It is also postulated from fracture surface observations that the variations in ΔK_{th} seen in different microstructures result from plasticity-induced crack closure and roughness-induced crack closure.

1. Introduction

Rolling bearings are mechanical parts which support loaded and rotating shafts. In order to reduce friction loss, ring elements and rolling elements have to hold very large loads over an extremely small contact area such as point or line. Therefore flaking failure occurs by the effect of cyclic high contact stress on the raceway. Therefore, very high hardness is required on the raceway and is achieved by quenching steel which contains over 0.5% carbon. On the other hand, sometimes unusually large stresses are loaded on rolling bearings which are used in industrial machines. Therefore resistance against fatigue crack propagation and fracture is also important. In order to achieve high surface hardness and high toughness in the core of bearings, carburized bearings are used which are made of 0.2% steel after carburizing to increase surface carbon content.

Toughness of steels for rolling bearing is often compared only using their absorption energy in Charpy impact tests. However, testing impact absorption energy does not correspond to the fact that failures of rolling bearing are commonly considered as a fatigue phenomenon. Fatigue crack growth properties of bearing material are still not understood sufficiently. Therefore, it is important to inspect fatigue crack growth properties of steels used for rolling bearings.

Fatigue crack growth properties are often discussed with respect to hardness. Research concerning microstructural influence on fatigue crack propagation behaviour for bearing steels is limited. Sato *et al.*[1] evaluated the effect of microstructure on fatigue strength of carbon steel and reported that fatigue strength was reduced by the existence of a secondary phase. Other work by Nie *et al.*[2] suggested that fatigue crack growth can be delayed by having a two phase microstructure for a spring steel. However there is lack of understanding of the relationship between fatigue properties and damage tolerance properties, and the role of microstructure.

In this study, the effects of having a two-phase microstructure on near threshold fatigue crack growth were investigated for a low carbon steel 18NiCrMo14-6 used for rolling bearings. The secondary bainite and austenite phases were obtained and controlled by heat treatment.

2. Experimental

A low carbon steel 18NiCrMo14-6 was used in the current study. The composition of alloy is shown in Table 1. The material is a case hardening steel which is used for large rolling bearings since it has high toughness and high hardenability. The as received material was hot rolled and annealed bar of 50 mm diameter. The dimensions of test pieces for fatigue crack growth tests were either 100mm x 20mm x 10mm or 100mm x 15mm x 7mm. For tensile and fatigue tests cylindrical specimens of 4.5mm diameter in the gauge section were utilised. The gauge length was 12 mm for these cylindrical specimens.

Table 1: Chemical composition of 18NiCrMo14-6. (mass%)

C	Si	Mn	P	S	Ni	Cr	Mo
0.17	0.28	0.47	0.007	0.003	3.27	1.36	0.17

Oversized blanks were extracted from position centered at 1/4 diameter of the bar steel to avoid possible variation in microstructure. They were heat treated under the conditions shown in table2 before machining to final dimensions.

All specimens were austenitized at 1093K for 30 minutes. They were then subjected to different cooling and tempering conditions. A martensitic microstructure was gained in M-1, M-2, and M-3 by quenching after austenitizing, to simulate the normal core microstructure of carburized bearings. They were subsequently tempered at 453K, 673K, and 773K to achieve different tensile properties. M+B material was austempered at 723K for 30 seconds, water quenched, and tempered at 453K. B+A material was also austempered at 723K for 30 minutes, water quenched, and tempered at 453K.

Fatigue crack growth tests were conducted under a four point bend loading configuration to compare mode I fatigue crack growth resistance. The test pieces were parallel ground to the final dimensions after heat treatment. A notch of 3mm depth was machined. Fatigue crack growth tests were performed under a constant stress ratio R, of 0.1, using an electromagnetic fatigue test rig. Direct current potential drop (DCPD) method was used to measure crack length under testing, in which 15 A current was applied to the specimen. The crack lengths were calculated using an in-house calibration equation. The initial and final crack length were measured from the fracture surface and incorporated in the calculation of crack lengths.

Fracture toughness tests were also conducted under four point bending using the same samples as for fatigue crack growth tests. The test pieces were fatigue precracked to a length of about 14mm. The test pieces were loaded to failure under a constant crosshead displacement rate. During the testing load versus crack opening displacement (COD) were recorded using a two channel chart recorder. With a fixed precrack length and testpiece geometry fracture properties of different microstructures were interpreted by comparing load versus displacement curves.

Fatigue life tests were carried out on cylindrical specimens under uniaxial tension using a servo-hydraulic mechanical test machine. The surface of the testpieces was carefully polished using 1200 grade SiC paper along the load direction to avoid crack initiation from tool marks. Tensile fatigue tests were performed using a hydraulic test rig, and stress ratio R was 0.1, same as for the fatigue crack growth tests.

Table 2: Heat treatment conditions of testing materials

Mark	Heat treatment condition
M-1	1093K, 30min – Oil Quenching – 453K, 120min
M-2	1093K, 30min – Oil Quenching – 673K, 120min
M-3	1093K, 30min – Oil Quenching – 773K, 120min
M+B	1093K, 30min – 723K, 30sec – water quenching – 453K, 120min
B+A	1093K, 30min – 723K, 30min – water quenching – 453K, 120min

3. Results and discussion

3.1 Microstructure and tensile properties

Figure1 shows SEM micrographs of heat treated materials.

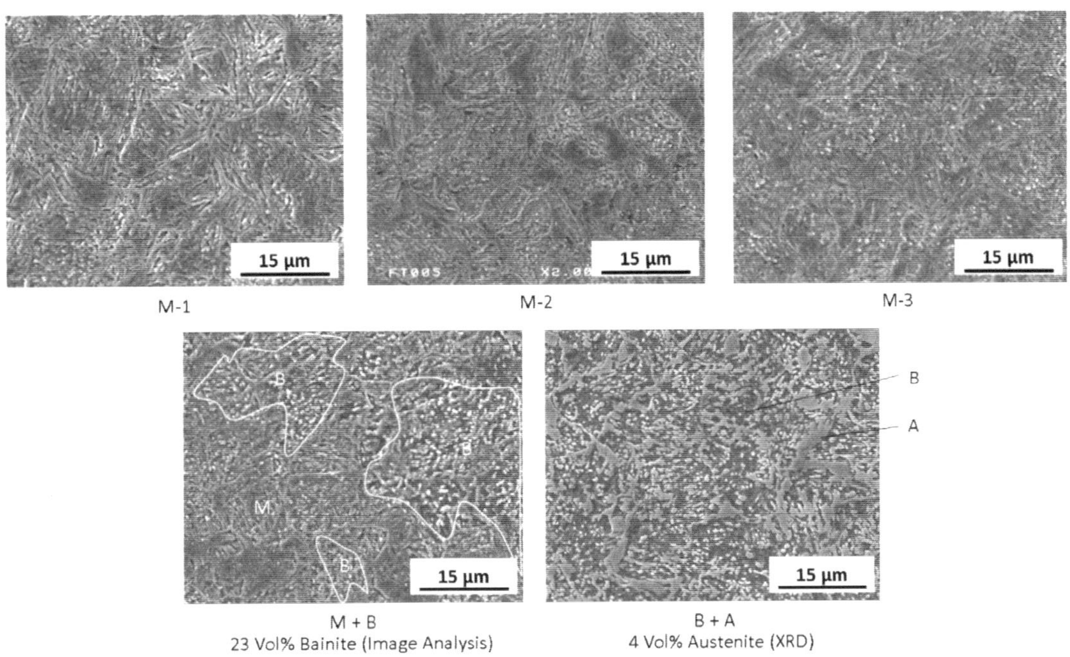

Figure 1: SEM micrographs of tested microstructures: M-1, M-2, and M-3 consist of Martensite. M+B consists of Martensite and Bainite. B+A consists of Bainite and Austenite.

M-1 is a lath martensite microstructure which seems to have a streaky shape, and is almost the same as the normal core microstructure of carburized rolling bearings. The lath structures in M-2 and M-3 are less clear than that of M-1 and precipitation of carbides can be found. Regions of different density of carbides were recognized in M+B which was quenched in the middle of austempering. Areas with many carbides are considered to be bainite, and its volume fraction was measured as around 23% by image processing based on ten micrographs of 1000x magnification. This volume fraction generally agrees with the bainite transformation rate assumed from a TTT-diagram of this alloy. On the other hand, in B+A, most of its microstructure is the same as the microstructure that is regarded as bainite in M+B, and some parts have a flat microstructure. This flat microstructure is considered as austenite, and its volume fraction was measured by XRD to be around 4%. From those observations above, it is confirmed that required microstructures were obtained through all of the heat treatments.

Figure2 compares stress - strain (S-S) curves of each microstructure, and table3 provides tensile strength (UTS), 0.2 % proof stress, and Vickers hardness. In uniaxial tensile testing, all materials shifted from elastic stages to plastic regions not showing an obvious yield point, reached the ultimate tensile strengths, necked at the centre of the parallel part, and broke at the point at which elongations were between 23 % and 27 %.

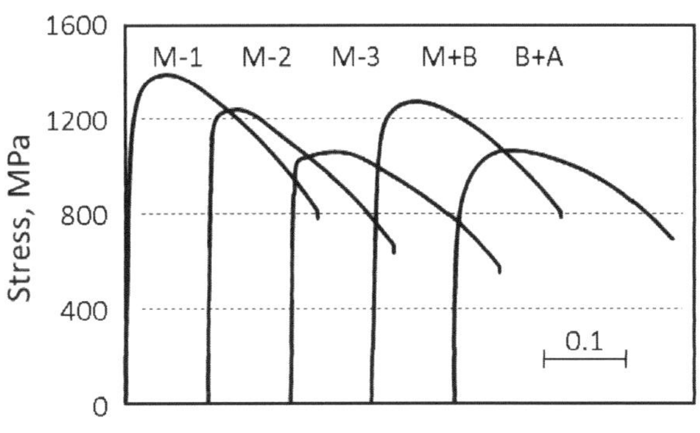

Figure 2: S-S curves of various microstructures

Table3: Heat treatment conditions of testing materials

	UTS (MPa)	$\sigma_{0.2}$ (MPa)	Hv (kgf/mm^2)
M-1	1388	1088	451
M-2	1242	1145	393
M-3	1263	1010	358
M+B	1265	1010	429
B+A	1067	740	360

UTS: ultimate tensile strength, $\sigma_{0.2}$: 0.2% proof stress, Hv: Vickers hardness

3.2 Fatigue crack growth resistance curves and threshold

Figure3 shows representative fatigue crack growth curves obtained for different microstructures. Figure4 shows an example of a fracture surface opened after testing. There was little difference between the five microstructures in the Paris regime where fatigue crack growth rates are relatively fast, over 10^{-5} mm/cycle. This tendency changed toward lower ΔK values It appears that the fatigue crack arrested easily in microstructure having secondary phase, i.e. M+B and B+A, as compared to martensitic single phase structures. Regarding ΔK at crack growth rates under 10^{-7} mm/cycle as ΔK_{th}, noticeably higher threshold values were observed for M+B and M+A microstructures Focusing on crack growth behaviour by fractographic examination, it is observed that the crack grows evenly from the initial notch and a relative straight crack front were observed for most cases. Sometimes the crack front was not straight and it was considered that this might be caused by residual stress induced by heat treatment. So some tests were conducted with reduced thickness by removing the surface layer in which residual stress may exist. This did result in a much more uniform cracks. Note that a load shedding procedure was applied initially to approach ΔK_{th} for fatigue crack growth tests. After reaching fatigue threshold the load was increased and the test was conducted under constant amplitude loading until the final crack length was reached.

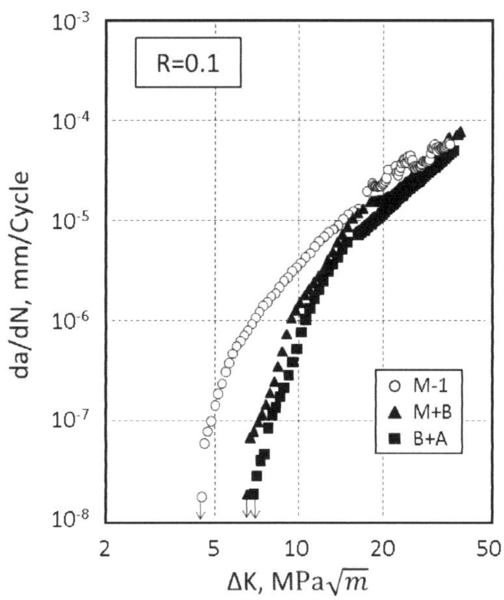

Figure 3: Relationship between fatigue crack growth rate, da/dN, and Mode I stress intensity factor range, ΔK

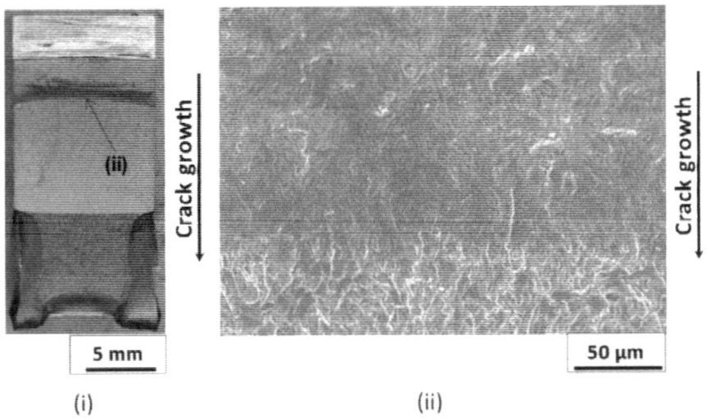

Figure 4: Fracture surface of crack growth test specimen: Red lines indicate crack tips at threshold. (i)Fracture surface. (ii) SEM image at the indicated area in (i).

Therefore, as shown in fig.4 (i), a line is present at which ΔK_{th} was recognized and the load was increased for measuring faster fatigue crack growth rates. In some specimens, oxidation marking which is indicated in fig.4 (i) with an arrow, was found on the fracture surface near to ΔK_{th}. As shown in fig.4 (ii), it was found from SEM micrographs that the fracture surface was flat at this magnification, and fatigue crack grows through a transgranular path.

Figure5 provides a plot of measured ΔK_{th} values for tested materials against tensile strength and 0.2% proof stress, in order to examine the relationship between microstructure and crack growth behaviour. Data points with * indicate specimens which had oxidation markings in the slow crack growth regime. Overall, ΔK_{th} increases with decreasing tensile strength or 0.2% proof stress. These trends agree with the report by Kitsunai[3]. In this figure, two-phase materials are indicated by solid marks. Considering all data, two-phase microstructure tends to have larger ΔK_{th} compared to those of martensite single-phase structures with the same tensile strength, and oxidation markings were found on the fracture surfaces of specimens which showed relatively large ΔK_{th}.

From the points above, the effect of microstructure on ΔK_{th} is examined. It is generally known that ΔK_{th} is strongly related to the crack closure phenomenon. Many types of effect such as plastic deformation, oxides, and bending of crack path, are known as factors of crack closure.[4, 5]

There is obvious evidence indicating the existence of plasticity-induced crack closure in this study, as shown in fig.2 and table3, all of tested materials shows a good ductility of over 20% elongation and necking. However yield stress varies significantly for different microstructures. It is anticipated that larger plastic zones are present for softer microstructural conditions such as M+B, M+A, M-2 and M-3. This could introduce larger plasticity induced crack closure, which is evident. Hence the microstructural influence on near threshold fatigue crack growth and threshold values may be mainly realized through altering yield stress rather than microstructure features due to their very fine size and relative flat fracture surface morphology at threshold. This is supported by the observation that ΔK_{th} is negatively correlated with yield stress. However subtle contributions from increases in fracture surface roughness have been observed with two phase microstructures.

Secondary, oxidation-induced crack closure is also examined. As indicated with * in fig.5, specimens showing large ΔK_{th} tends to have oxidation markings and formation of oxidation and ΔK_{th} value is related strongly. However, a temperature effect at the tested crack tip is also thought as a factor in oxidation. Hence, it is also conceivable that oxidation occurred due to the heat generated at the crack tip near to ΔK_{th} with slow crack growth speed and cyclic deformation. A microstructure effect on the presence or absence of oxidation marking was not recognized, thus, the effect of oxidation on delaying the crack growth is unclear.

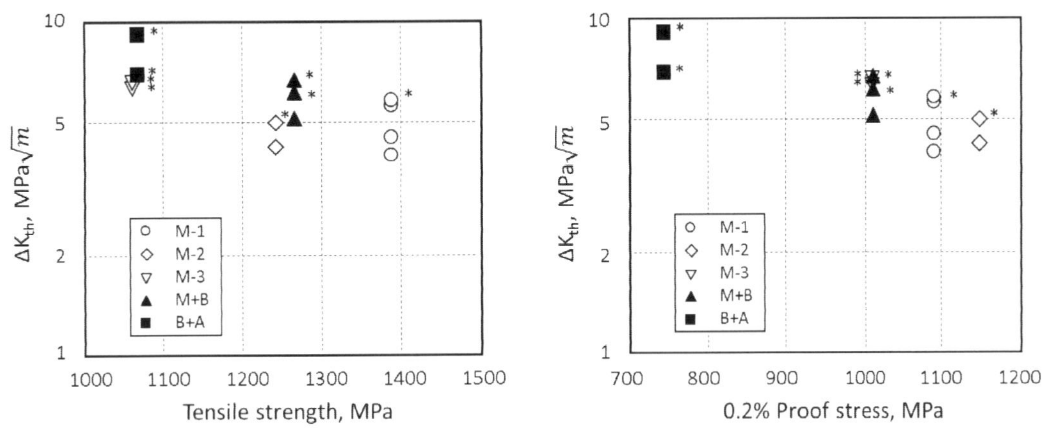

*Figure 5: Threshold stress intensity factor range against tensile strength and 0.2% proof stress. (*Specimens had surface oxidation at threshold.*

Lastly, roughness-induced crack closure is examined. Figure 6 shows SEM micrographs of M-1 and B+A fracture surfaces at ΔK_{th}. Fracture surface morphology of B+A is rougher than that of M-1 at just before ΔK_{th}. Moreover, streaky texture is seen in M-1 which corresponds to the lath microstructure observed in fig.1. Therefore it is considered that crack growth differs at low stress conditions due to the microstructure induced roughness on the fracture surface. Although there has not been a numerical evaluation because of the wear of fracture surface, from the fact that the crack growth path is transgranular, these observation suggest that crack growth rates can be delayed by microstructure control under low stress conditions.

3.3 Fracture toughness

Figure 7 shows the representative curves of load against crack opening displacement of fracture toughness tests. These results are measured using specimens of the same size with the same fatigue pre-crack length, and the load is standardized based on their crack length. The load shown in Fig.7 is calculated by assuming that the crack length of each specimen is exactly same. For all heat treatment microstructures, the load increases monotonically as the crack opens to the maximum load, then it decreases. Maximum load values are almost the same among the various microstructures. However, the area below the load versus COD is very different despite similar slopes of the linear elastic sections. Apparently more energy is absorbed for B+A which has the highest threshold value. For all conditions a valid K_{IC} was not obtained. From an energy point of view, higher toughness could be achieved with B+A and M-3 microstructures. The curve of M+B is very similar to that of M-2 which has a similar tensile strength. Thus, crack propagation behaviour in this fracture toughness test is more affected by tensile strength compared to microstructure.

Summarizing the results of fracture toughness tests and fatigue crack growth tests, the two-phase microstructures have higher ΔK_{th} compared to the single-phase microstructure of the same tensile strength. Both two-phase and single-phase microstructures have almost the same crack growth rate in the Paris regime, and the same maximum load in fracture toughness tests. This suggests that a two-phase microstructure is superior in terms of damage tolerance properties.

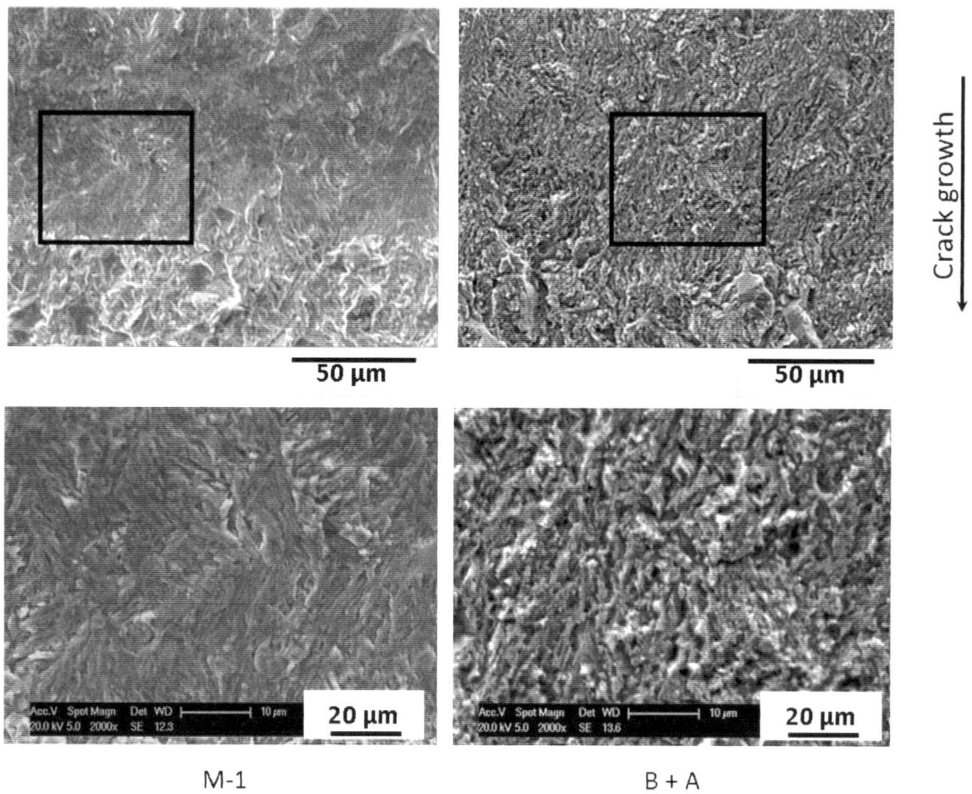

Figure 6: SEM micrographs shows the fracture surfaces near to threshold of the crack growth test specimens.

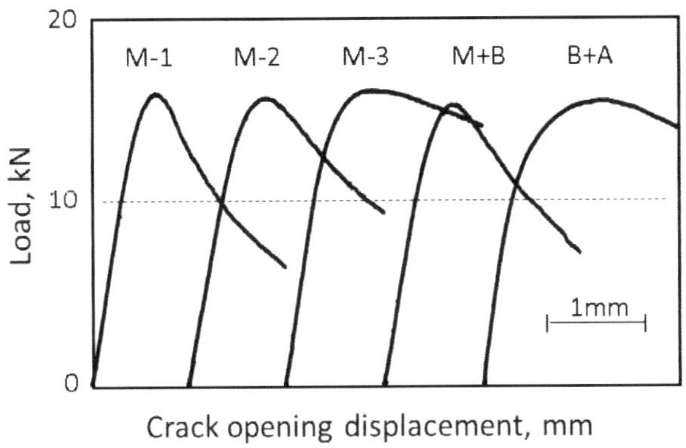

Figure 7: Load plotted against crack opening displacement in various microstructures. Load values are normalized based on their crack length.

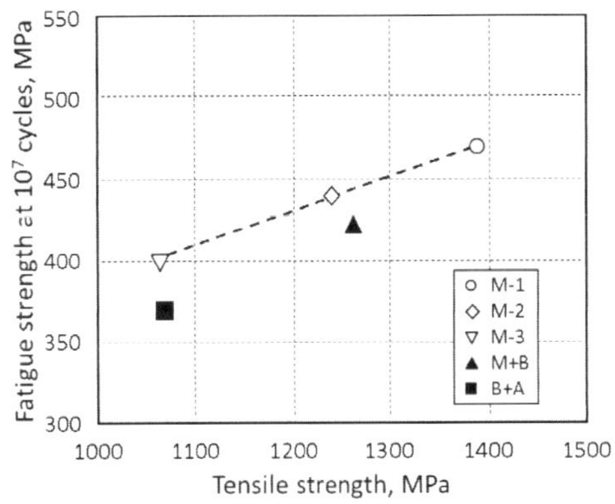

Figure 9: Fatigue strength at 10^7 cycles of each microstructures plotted against tensile strength.

3.4 S-N curves

Fatigue crack growth behaviour using specimens with precrack was examined above.

The effect of microstructure on fatigue strength was examined using uniaxial tensile fatigue tests. Figure 8 shows representative S-N curves of each microstructure. Fatigue strength at 10^7 cycles is plotted against tensile strength in Figure 9. The fatigue S-N curve of martensite single-phase microstructure demonstrates a flat appearance and has the highest fatigue endurance.

Fatigue strength at 10^7 cycles of two-phase microstructures M+B and B+A are lower than that of martensite single-phase structures. The fatigue life of such plain-sided small laboratory test pieces is largely life spent on crack initiation. This is particularly the case with M-1. The transition from a flat S-N curve of M-1 to a slanting appearance indicates an increasing contribution from life spent on crack propagation with decrease of stress level. The difference in slope between B+A and M+B is difficult to interpret. However, conjoint with low fatigue endurance in B+A, it is tentatively suggested that B+A has a lower resistance against crack initiation but higher resistance against crack growth.

Even for the initial crack growth which is less affected by crack closure, microstructure can have an effect on initial crack propagation. For example, non-metal inclusions works as a defect in the matrix, and conversely, a microstructure which has ductility hinders initial crack propagation. Therefore, in order to determine whether or not there were microstructure effects on the propagation of the initial cracks in this study, initial crack size in the fatigue limit regime was estimated. The threshold stress intensity factor of a surface crack on an infinite solid on which uniform stress, σ_0, is applied, can be given as below from the crack length. (Y is a constant)

$$K = Y \cdot \sigma_0 \sqrt{\pi \cdot a} \quad (1)$$

From this equation 1, initial crack length "a" is estimated from ΔK_{th} and fatigue strength at 10^7 cycles which is regarded as the fatigue limit. Figure 10 shows the calculated crack length for each microstructure. The values are from 20μm to 90μm.

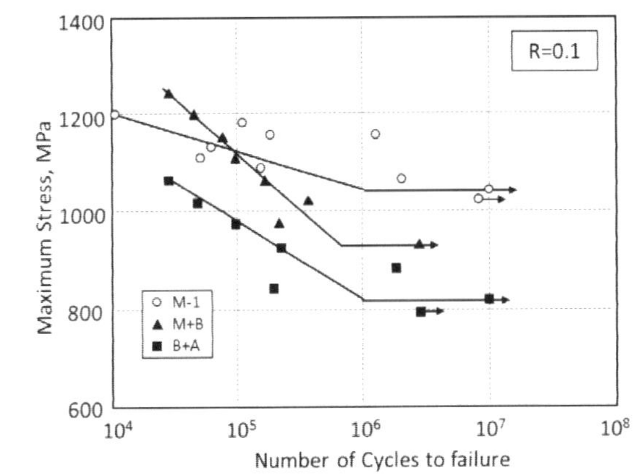

Figure 8: S-N curves of various microstructures

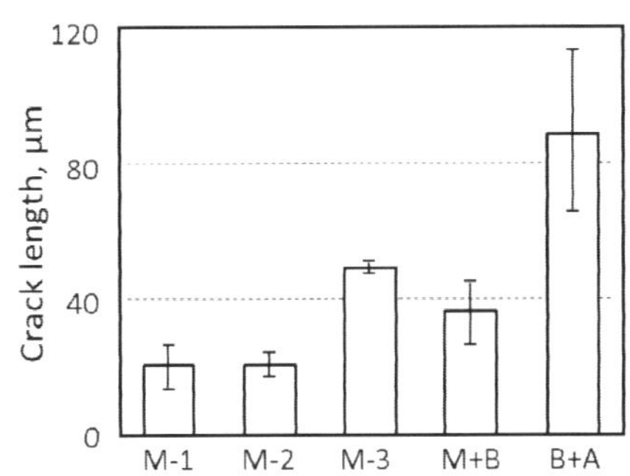

Figure 10: Crack length calculated from ΔK_{th} of crack growth test and fatigue strength at 10^7 cycles of fatigue life test. The error bars show the maximum and minimum values

Compared to this, the primary austenite grain size in this kind of heat treatment is normally from 10μm to 20μm. Packet, block and retained austenite grain sizes which have large angular grain boundaries are smaller than the primary austenite grain size. Thus in the current study, there was little influence of microstructure on initial crack growth resistance. Overall, it is considered that because of the small effect in fatigue crack closure and small microstructural effect on initial crack growth, fatigue limits of two-phase microstructures are low despite their large ΔK_{th} value.

4. Conclusions

In order to evaluate fatigue crack growth properties in low carbon steel and to find a direction for superior performance, fatigue crack growth resistance, fracture toughness, and fatigue life were tested in low carbon steel 18NiCrMo14-6 heat treated as martensite single-phase materials and two-phase materials of M+B and B+A. Conclusions are below.

- ΔK_{th} of M+B and B+A measured by fatigue crack growth testing using samples with a precrack, were higher than those of martensite single-phase structures of the same tensile strength.
- Fatigue strength at 10^7 cycles of two-phase microstructure is lower than that of martenisite single-phase microstructure.
- The maximum value of standardized load in fracture toughness test is constant between two-phase microstructures and martensite single-phase.
- It is estimated from fracture surface observation of crack growth tests that microstructure dependence on ΔK_{th} are due to plasticity-induced crack closure and roughness-induced crack closure.

References

[1] Sato, T. *et al.*, "Effects of Microstructure on the Fatigue Property of High Carbon Steel" Tetsu Hagane, Vol. 66, (1979), pp. 239-243

[2] Nie, Y. H. *et al.*, "Very high cycle fatigue behaviour of 2000-MPa ultra-high-strength spring steel with bainite-martensite duplex microstructure" Fatigue & Fracture of Engineering Materials & Structures, Vol. 32, (2009), pp. 189-196

[3] Kitsunai, Y. "Effect of Microstructure on Fatigue Crack Growth Behavior of Carbon Steels" J Mater Sci Soc Jpn, Vol. 29, (1980), pp. 1018-1023

[4] Kobayashi, H. *et al.*, "Fatigue Crack Growth Characteristics and Crack Closure in A 508-3 Steel at Elevated Temperature" Trans. JSME A, Vol. 50, (1984), pp. 1003-1010

[5] Suresh, S. *et al.*, "A Geometric Model for Fatigue Crack Closure Induced by Fracture Surface Roughness" Metall. Trans. A, Vol. 13A, (1982), pp. 1627-1631

International Federation of Heat Treatment and Surface Engineering 2016
Proceedings of the 23rd IFHTSE Congress
April 18–21, 2016, Savannah, Georgia, USA

Life Extension of Furnace and Fixture Alloys by Surface Engineering in Carburizing Atmospheres

Anbo Wang, R. D. Sisson, Jr.
Center for Heat Treating Excellence, Worcester Polytechnic Institute, Worcester, Massachusetts, 01609, U.S.A

Abstract

The heat-treating industry is in need of heat-treatment furnace materials and fixtures that have a long service life and reduced heat capacity. In this paper, RA330, RA602CA, 304L, Inconel 625 alloys were selected to study their performance in an industrial carburizing furnace for times up to one year. These alloys were exposed in both the as fabricated and aluminized condition. The test samples were exposed to 0.7%C carburizing atmosphere at approximately 900°C for 3 months, 6months, and 12months. In this paper the microstructural development of both aluminized and unaluminized alloys during the prolonged exposure in the carburizing environment will be presented and discussed. It was found that the weight gain for the aluminized alloys was uniformly less than the as-fabricated alloys.

Introduction

Four alloys were selected for the alloy lifetime tests, including RA330, RA602CA, 304L, and INCONEL alloy 625. Samples have been aluminized by Alcoa Howmet in Branford, CT. These samples were divided into 10 groups. Group 1 and Group 6 were kept at WPI for microstructural characterization. The other eight groups were sent to Bluewater Thermal Solutions for extended exposure in the carburizing furnace (T=871°C, $C_p = 0.8\% - 1.1\%$).

RA330, RA602CA, Inconel625 were received from Rolled Alloys, and Stainless steel 304L was purchased from Peterson Steel, Inc.

RA330 is well known as the corrosion-resistant alloy offering an exceptional combination of strength and resistance to carburization, oxidation and thermal shock. RA330 finds wide applications in high temperature industrial environments where good resistance to the combined effects of carburization and thermal cycling is prime requisite.[1,2]

Stainless steel 304 is the original "18-8" stainless steel that have lower price compared to Fe-Ni-Cr base alloys.

The aluminizing was selected to improve the service life of alloys. Aluminizing is widely used to increase the high temperature oxidation and carburization resistance of nickel-based alloys. The thickness of the aluminide layer on the alloys increases with increasing aluminizing time. The intermetallic nickel aluminides that formed during aluminizing act as the reservoir of aluminum to maintain a protective Al_2O_3 layer on the material surface during high temperature services. $Alpha - alumina\ Al_2O_3$ provides effective protection from high temperature oxidation and carburization. Samples have been aluminized to compare the anti-corrosion properties of aluminized alloys and chromia former alloys. RA602CA, Inconel625, 304L, and RA330 were exposed in carburizing furnaces. Aluminizing coating was also conducted on these samples. These samples were exposed side by side in the furnace to compare the performances of these alloys.

After 3, 6, 12months exposure, samples were removed from the furnace. These samples were placed in a batch carburizing furnace and the carbon potential of the atmosphere is approximately 0.7%. Carburizing temperature is set at 871°C to 982°C in this furnace. And the carburizing time for each cycle depends on the size of the products, smaller parts such as gaskets will takes 1-3 hours, and the larger parts takes much longer time.

The weight gain of each sample has been measured. After exposure in the carburizing batch furnace, the weight gains were measured. The results of 6 months exposure are presented below.

Carbon potential: 0.5-1.1% average 0.7%

Temperature: 871°C -982°C average 927°C

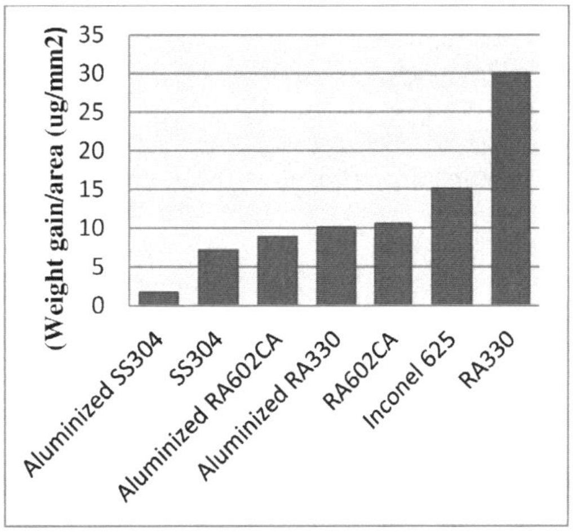

Figure 1: Weight gain per area

As presented in Figure 1 aluminized 304L has the lowest weight gain per area. The weight gain is comprised of both oxidation and carburization. For some alloys a reduced weight gain may be measured due to spalling of the oxides.

In general, aluminized samples have lower weight gain per area. The alumina Al_2O_3 layers coated samples have less weight gain per area than chromium oxide Cr_2O_3 coated samples.

Additionally, the alloys have been exposed in the carburization furnace for 6 months. Carbon continually diffuses through the oxide layer into the alloys substrate. Therefore, weight gain per area as a function of time shows that weight gain increased for 304L, RA330, Inconel625 and RA602CA during the exposure as the expose time increases.

Results and Discussion

304L stainless steel

304L samples were provided by Peterson Steel Corporation, located in 61 West Mountain Street, Worcester, Massachusetts 01606. Stainless steel 304 is the original "18-8" stainless steel which is less expensive compare to Fe-Ni-Cr based alloys. The composition of 304L is listed in Table 1.

	Cr	**Ni**	**Mn**	**Si**	**Cu**
MIN%	18.0	8.0	-	-	-
MAX%	20.0	10.5	2.0	0.75	1.0
	P	**S**	**C**	**Fe**	
MIN%	-	-	0.04	-	
MAX%	0.04	0.03	0.08	Balance	

Table 1: Chemical Composition of 304L

304L rod was 1 foot long and 1 inch diameter cylinder. And it was machined to 1 inch diameter and 0.4 inch thickness disc. Half of the 304L samples were aluminized for metallographic characterizations.

As shown in Figure 2, the microstructure of the exposed 304L reveals a layer of Cr_2O_3 on the surface. EDS analysis was conducted on the sample, the composition of the spectrum 1 has shown in Figure 2. The matrix is Austenite as shown in Spectrum 2. All the samples were etched with 10% oxalic acid by electrical etching, 30 seconds for each sample. The samples were etched with oxalic acid, so internal carbides were revealed by the etching. Single phase austenite was also identified in the core area.

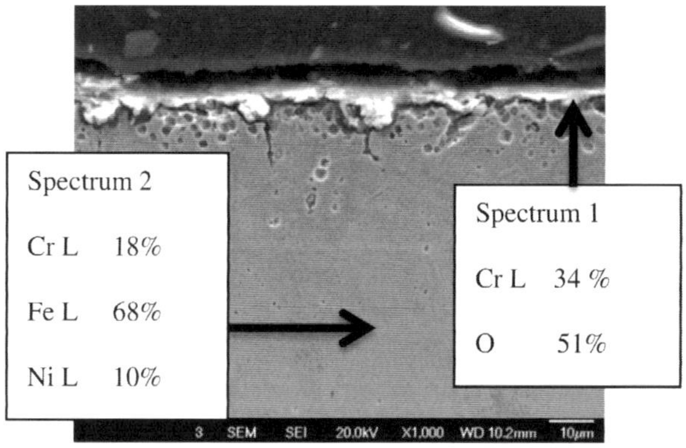

Figure 2: SEM photomicrographs of exposed 304L plate at the edge

Chromium oxide layer was formed on the surface of the 304L alloy due to high chromium content in the alloy. The formation of chromium oxide can provide oxidation and carburization resistance for the 304L. As shown in Figure2, the surface layer was identified as the chromium oxide. The surface region contains high concentration chromium oxide and beneath that is the austenitic Fe-Ni phase. In addition, a layer that rich in Fe, C, O was formed on top of the chromium oxide. Based on the EDS mapping and XRD analysis, the layer could be the combination of Fe_2O_3 and Fe_3O_4.

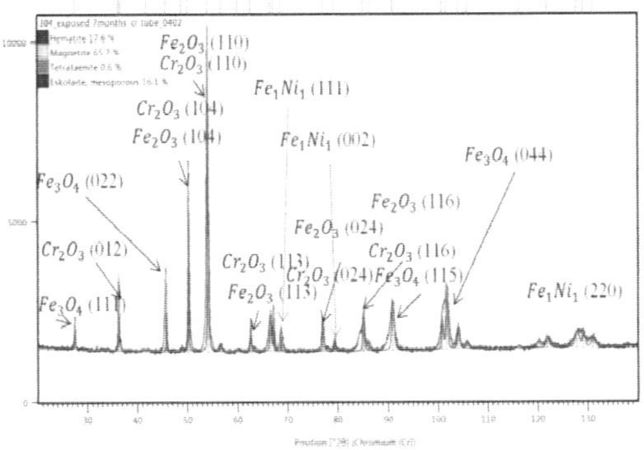

Figure 3: XRD analysis of exposed 304L

According to XRD analysis in Figure 3, hematite and magnetite were formed during the exposure which may due to the relatively high oxygen pressure in the furnace. No chromium carbides were identified by XRD due to the short exposure and insufficient carbon activity in the atmosphere. No excessive carburization was found for 304L after 6 months exposure.

Aluminized 304L stainless steel

Highest carbon content was observed at the surface of exposed sample which is 0.39% of C as seen in Figure 4. Carbon concentration decreases as the depth x increases.

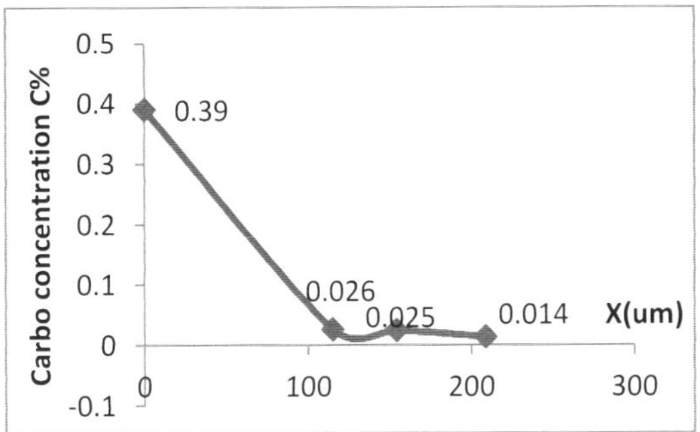

Figure 4: Carbon profile of aluminized 304L exposed for 6 months

The depth of diffusion zone for aluminized 304L is $680\mu m$ from the surface as shown in Figure 5.

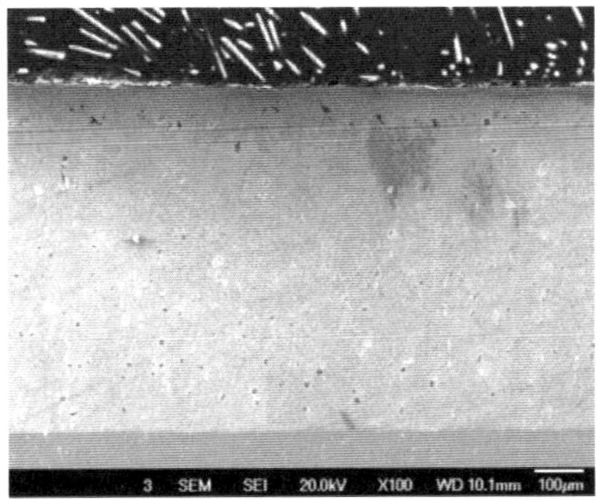

Figure 5: SEM photomicrographs of aluminized 304L exposed for 6 months

Figure 5 shows the cross section SEM photomicrographs of aluminized samples which have been exposed inside the furnace for 6 months. And the samples were aluminized before furnace testing, so a thick layer of aluminum rich diffusion zone was observed in the Figure 6. The thickness of the diffusion zone is about 680 μm. βNi_1Al_1 phase was identified by the XRD that could also be observed in Figure 5 beneath the aluminum oxide layer on the Austenite Fe_1Ni_1 matrix.

a)

b)

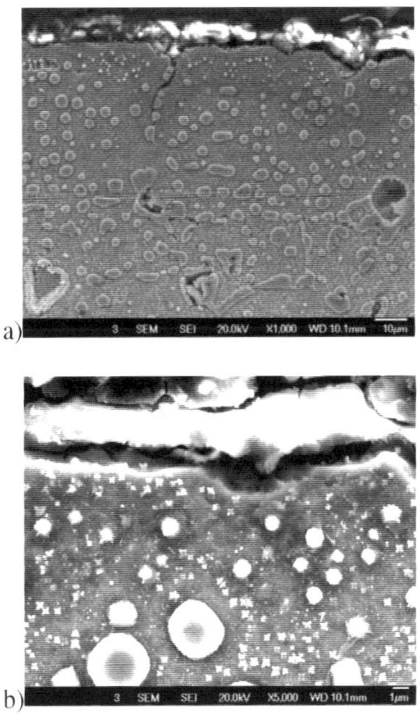

Figure 6: a) SEM photomicrographs of aluminized 304L exposed for 6 months at edge b) Aluminized 304L exposed for 6 months at higher magnification

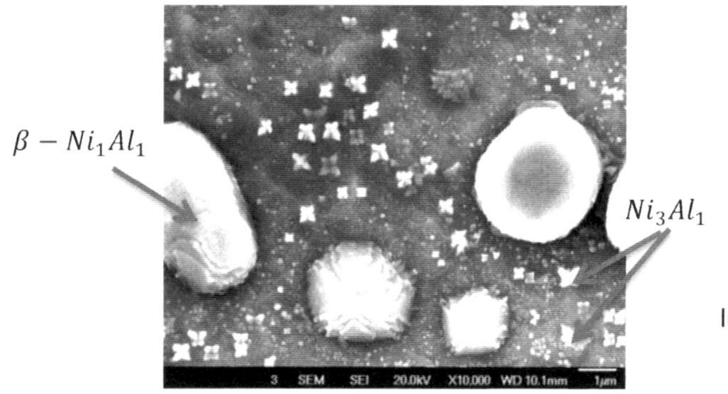

Figure 7: SEM photomicrographs of Nickel aluminide precipitates in the diffusion zone

As shown in Figure 6, a dense layer of alumina was formed on the surface of alloy. EDS line analysis in the Figure 9 shows the aluminum and oxide content decreases as a function of depth that indicated the rich of aluminum oxide on the surface (Right side of photo). In Figure 7, smaller precipitate was identified by EDS phase as $\gamma'-Ni_3Al$ because the smaller $\gamma'-Ni_3Al$ precipitates were rich in Nickel and Aluminum content and lack of chromium and iron content, as shown in Figure 9. Larger white spots were identified at the diffusion zone as $\beta - Ni_1Al_1$ phase as shown in the Figure 7. Similar results were revealed by the XRD in Figure 8.

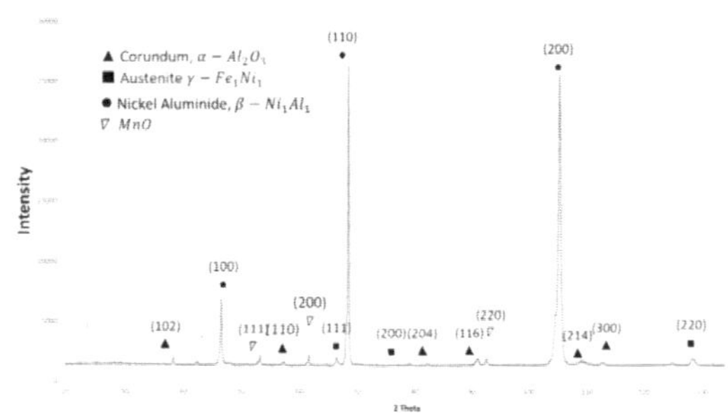

Figure 8: XRD analysis of exposed aluminized 304L Cr tube

Figure 6 shows a continuous layer of aluminum oxide on the surface. $\gamma' - Ni_3Al$ precipitated beneath the oxide layer, on the austenite phase Fe-Ni. $\beta - Ni_1Al_1$ phase was also identified on the austenite matrix. The size distribution of the precipitates is bimodal. The reason is under investigation. The Al_2O_3 layer is effective at reducing carbon uptake from the carburizing atmosphere.

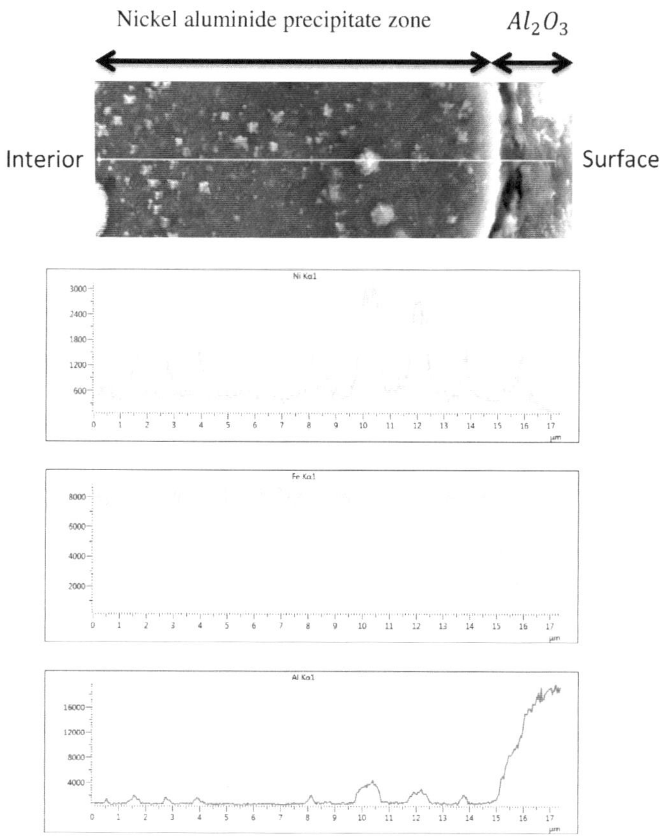

Figure 9: EDS line analysis on aluminized 304L at the cross section in surface region

RA330

RA330 is a widely used alloy for fixtures in the carburizing furnaces. After exposure in the furnace for 3, 6, 12 months, RA330 samples were removed from the furnace.

	Cr	Ni	Mn	Si	Cu
MIN%	18.0	34.0	-	1.0	-
MAX%	20.0	37.0	2.0	1.5	1.0
	P	**S**	**C**	**Fe**	
MIN%	-	-	0.04	-	
MAX%	0.03	0.03	0.08	Balance	

Table2: Chemical Composition of RA330

The optical photomicrographs for the unexposed and exposed cross section have been taken. Figure 10 and Figure 11 presents the single phase microstructure of unexposed (upper) and exposed (lower) RA330. The carburizing case formed at the sample surface after 6 months service in carburizing furnace is seen in Figure 11. Samples were electrolytic etched with 10% Oxalic Acid at 10V for 30s.

The exposed RA330 sample shows in the lower photomicrographs indicated that a thin layer of intergranular oxidation (IGO) was observed at the surface of exposed RA330 sample as seen in Figure 11.

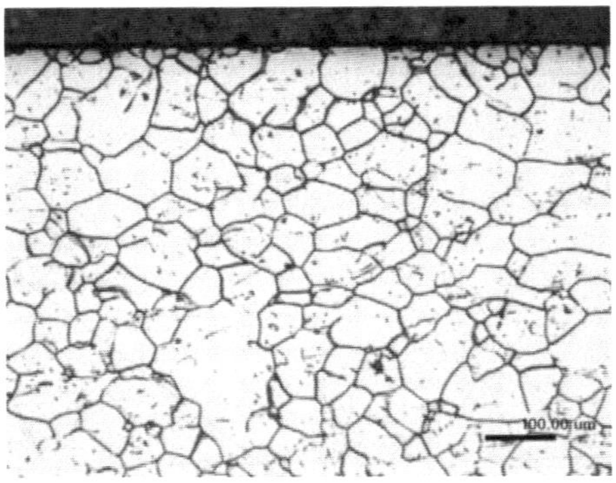

Figure 10: Cross section optical photomicrographs of unexposed RA330

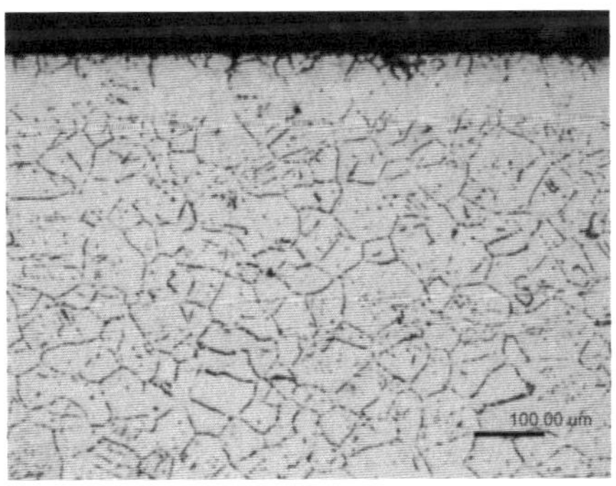

Figure 11: Cross section optical photomicrographs of exposed RA330

Aluminized RA330

Figure 12 shows the aluminized case at the surface of RA330 which is composed of two layers. The XRD pattern presents that the outer layer is mainly nickel aluminides ($\beta - Ni\,Al$) with some aluminum oxides (Corundum, $\alpha - Al_2O_3$). And the EDS analysis presents that the inner layer is discontinuous nickel aluminide ($\beta - Ni\,Al$) on the austenite ($Fe\,Ni$) matrix. The alumina layer is stable and adherent on the substrate, no defect was identified. Figure 13 shows the discontinuous $\beta - Ni_1Al_1$ at depth about $20\mu m$ from the surface, the black areas were voids caused by etching.

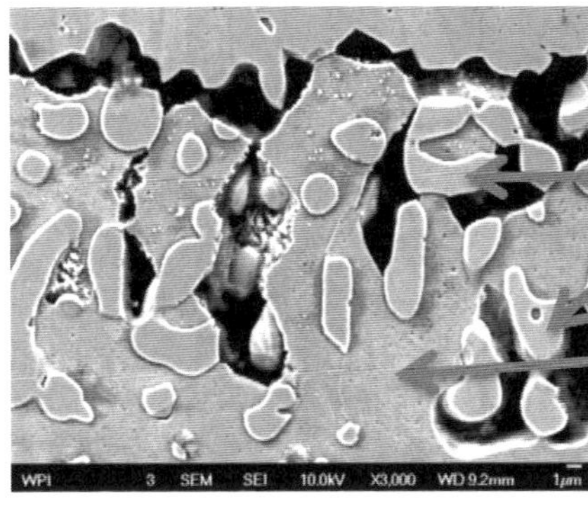

Figure 12: SEM photomicrographs of exposed Aluminized RA330

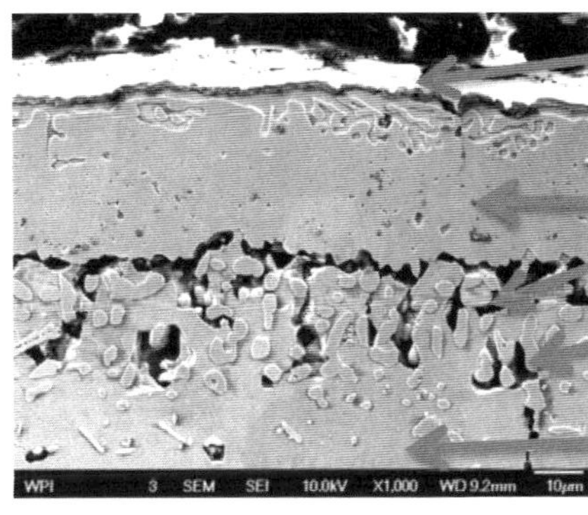

Figure 13: SEM photomicrographs of exposed Aluminized RA330

The $M_{23}C_6$ carbides form in regions with high chromium content. $M_{23}C_6$ strongly influences the properties of these alloys that carbides could lead to metal dusting. Therefore, decreasing the carbon uptake is important. RA330 is Ni-Fe based alloys that Cr% is higher than 19%, so continuous Cr_2O_3 protective layer formed at the surface. And the SiO_2 also protects austenitic matrix from carburizing. According to our weight gain measurement, aluminizing significantly decreased the carbon uptake during carburization. Besides, alumina $\alpha - Al_2O_3$ is well known as better protection coating at high temperature above 980°C. Furthermore, the price for RA330 is $11.63 per pound for ¼" PLATE.[1] Consider the nickel unit price is relatively high and RA330 has lower price than nickel base alloys, aluminized RA330 would be an ideal choice for fixtures in the furnace.

RA602CA

RA602CA is a nickel based alloy that employs high chromium content along with aluminum and yttrium additions that produces a tightly adherent oxide, allowing the alloy to operate at temperature in excess of 1230°C.[3]

	Cr	**Ni**	**Cu**	**P**	**Fe**	**C**
MIN%	24.0	-	-	-	8.0	0.15
MAX%	26.0	Balance	0.1	0.02	11.0	0.25
	Al	**Ti**	**Y**	**Zr**	**Si**	**Mn**
MIN%	1.8	0.1	0.05	0.01	-	-
MAX%	2.4	0.2	0.12	0.1	0.5	0.15

Table3: Chemical Composition of RA602CA

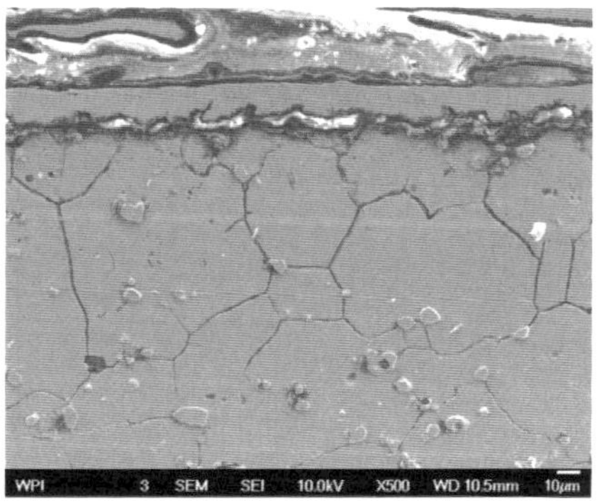

Figure 14: SEM photomicrographs of exposed RA602CA

RA602CA have more than 24% of Cr content which is sufficient for continuous Cr_2O_3 protective layer as shown in Figure 14. $M_{23}C_6$ carbides precipitated in various locations, but typically at grain boundaries.

Aluminized RA602CA

Figure 15 shows Al_2O_3 scale on the surface of aluminized RA602CA. In addition, a 20 μm nickel aluminide intermetallic compound layer was shown in Figure 15. Beneath the intermetallic layer, the interior matrix of RA602CA is uniform austenite iron nickel phase.

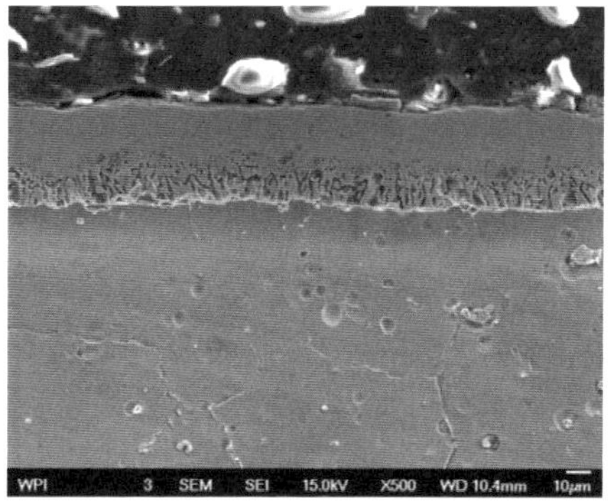

Figure 15: SEM photomicrographs of unexposed Aluminized RA602CA

Due to the exposure in gas carburization furnace, chromium carbides $Cr_{23}C_6$ and Cr_7C_3 precipitated in the grain boundaries at austenite matrix which shows in Figure16.

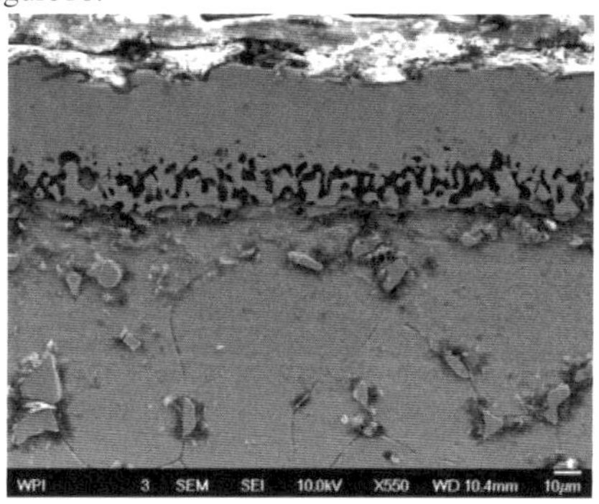

Figure 16: SEM photomicrographs of exposed Aluminized RA602CA

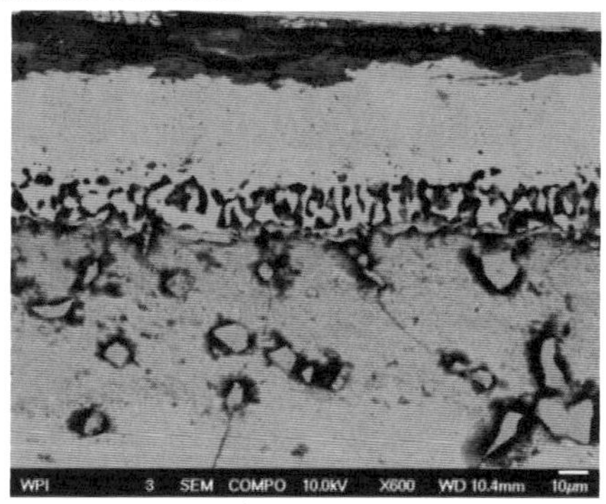

Figure17: Back scattered electron photomicrographs of exposed Aluminized RA602CA

Back scattered electron photomicrographs also indicated the chromium carbides in the grain boundaries. Besides, carbides were also discovered in the core area of cross section after exposure as seen in Figure 18. This is because RA602CA has up to 3% carbon in it. Therefore, after long term high temperature exposure, chromium carbides precipitate in the matrix.

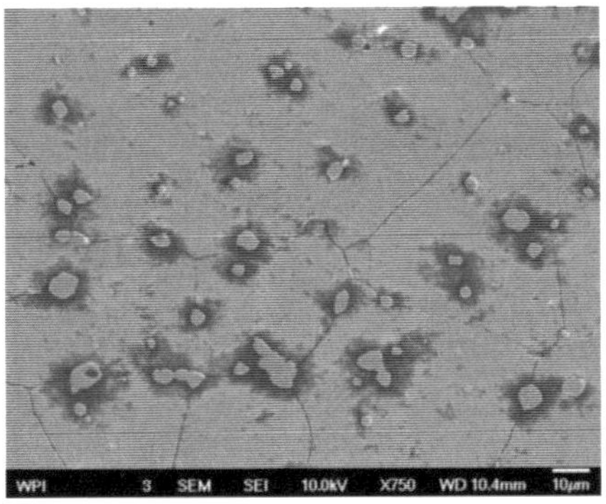

Figure 18: SEM photomicrographs of exposed Aluminized RA602CA plate at the core

Inconel 625

Alloy 625 is a solid-solution matrix-strengthened face-centered-cubic alloy.

	Ni	Cr	Fe	Al	C
MIN%	58.0	20.0	-	-	-
MAX%	-	23.0	5.0	0.4	0.1
	Mg	**Niobium**	**Mo**	**Co**	**Si**
MIN%	-	3.15	8.0	-	-
MAX%	0.5	4.15	10.0	1.0	0.5

Table4: Chemical Composition of Inconel 625

A continuous Cr_2O_3 layer and Ni_3Nb_1 beneath that could be identified in the Figure 19. At higher magnification, Ni_3Nb precipitate could be clearly identified in the Figure 20.

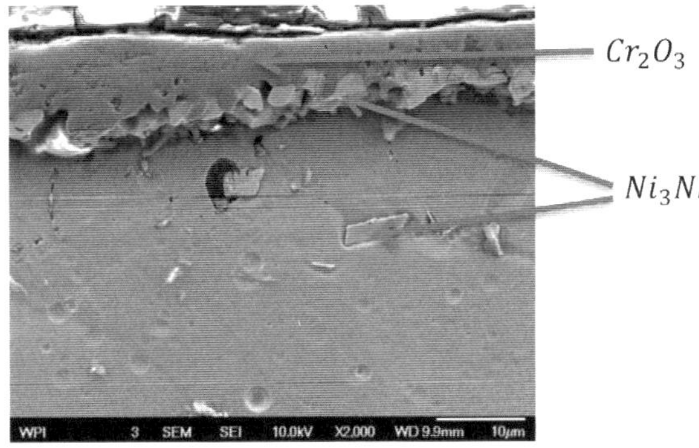

Figure 19: SEM photomicrographs of exposed Inconel 625 at the edge

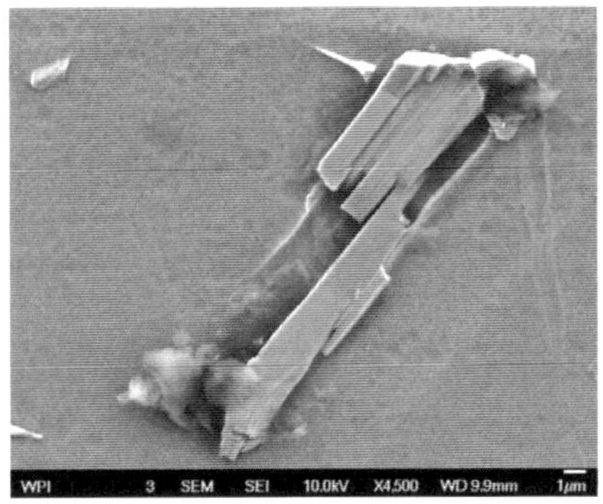

Figure 20: SEM photomicrographs of exposed Inconel 625 precipitates at core

Aluminized Inconel 625

The γ'-phase typically consists of Al and Ti as the major alloying elements, i.e. $\gamma' - Ni_3\,Al$; while the γ''-phase (Ni_3Nb) precipitates tend to be more efficient strengtheners than γ'-phase precipitate. SEM photomicrographs were also conducted on the Alloy 625. At higher magnification, Figure 21 shows Ni_3Nb precipitate which could be clearly identified. According to the Figure 21, Ni-rich intermetallic compounds $\beta - Ni\,Al$ are also found in the matrix.

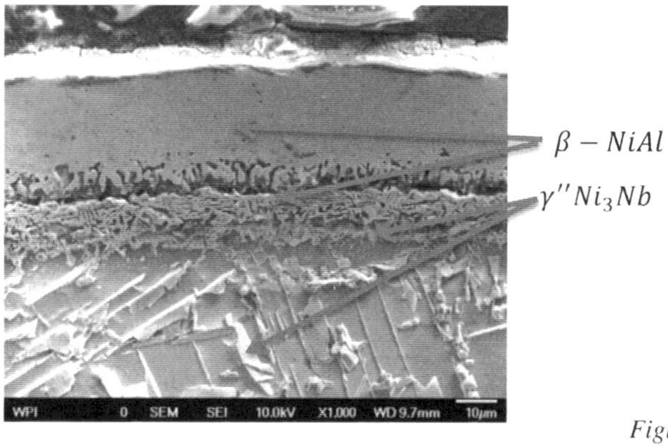

Figure 21: SEM photomicrographs of new Aluminized Inconel 625 at the edge

And by using EDS the possible phases of the Alloy 625 could be estimated. Figure 21 shows the Ni_3Nb precipitate in the matrix. And the EDS presents that the outer layer is mainly Al_2O_3 phases.

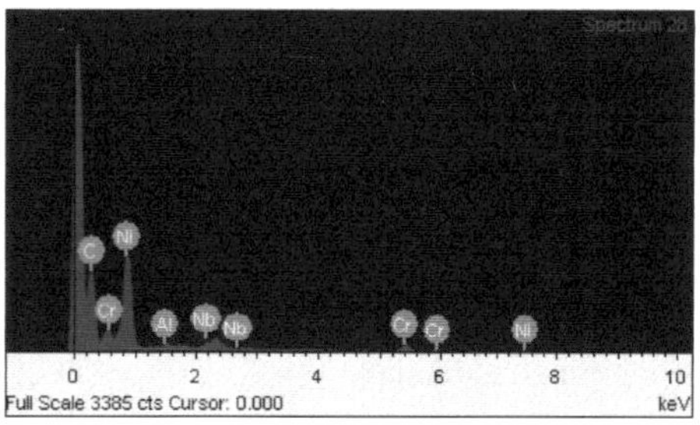

Figure 22: EDS analysis of Inconel 625 precipitates

After exposed in carburization furnace for 12 months, carbides were also identified at grain boundaries. Ni_3Nb carbides also found in the core area of cross section as shown in Figure 24.

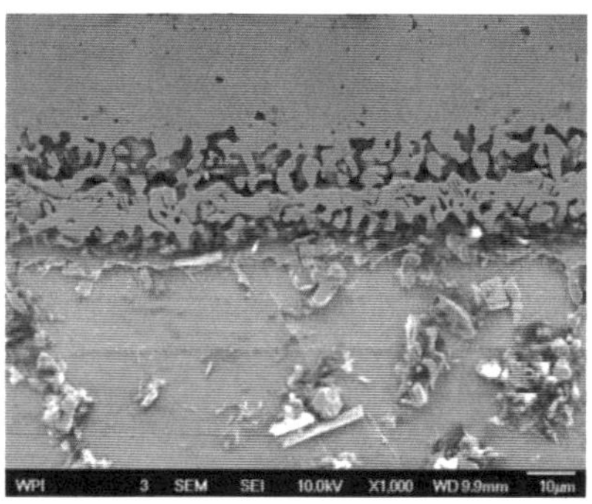

Figure 23: SEM photomicrographs of exposed Aluminized Inconel 625 at edge

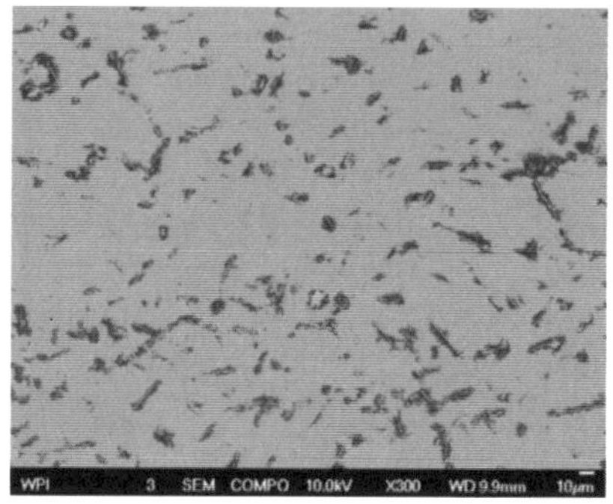

Figure 24: SEM photomicrographs of exposed Aluminized Inconel 625 at the core

Summary

As seen in Figure 1 the alloys that had been aluminized gained significantly less weight than the same alloys without aluminizing.

The aluminizing process formed aluminum rich intermetallic (i.e.$\beta - Ni\ Al$) that provided aluminum to form alumina at the surface. This alumina resulted is a reduced oxidation rate and decreased carbon absorption.

References

[1] Rolled Alloys, Quotation, https://www.rolledalloys.com/ecommerce/#Q1227526

[2] *RA330® vs 309/310 - Rolled Alloys*. Available from: http://content.rolledalloys.com/technical-resources/alloy-comparison/RA330vs309-310_AC_US_EN.pdf.

[3] Marc Glasser, "RA602CA: An Alloy for the Highest Temperature Processes", 28th ASM Heat Treating Society Conference, 2015

International Federation of Heat Treatment and Surface Engineering 2016
Proceedings of the 23rd IFHTSE Congress
April 18–21, 2016, Savannah, Georgia, USA

Design of Advanced Tempering Diagrams of Hot Work Tool Steel Using Multifunctional CNPTB Specimen

B. Podgornik , G. Puš, V. Leskovšek, M. Godec
Institute of Metals and Technology, Ljubljana, Slovenia
bojan.podgornik@imt.si, +386 1 4701930

Abstract

The aim of this research work was to determine the effect of vacuum heat treatment conditions on different properties and their correlations for low Si content AISI H11 type hot work tool steel using single CNPTB test specimen. Work was also focused on the potential of CNPTB test specimen for designing advanced tempering diagrams and optimizing vacuum heat treatment of hot work tool steels. Results show that CNPTB specimens allow simultaneous determination and correlation of multiple properties of hot work tool steels, with compression and bending strength increasing with hardness, and strain hardening exponent and bending strain with fracture toughness. On the other hand, the best machinability and surface quality of hardened hot work tool steel are obtained for hardness values between 46 and 50 HRC and fracture toughness below 60 MPa√m.

Introduction

In hot metal forming applications, like forging, stamping, rolling or die casting, the service life of the tool is limited due to the extreme working conditions, including thermal, mechanical and impact loading, high contact pressures and abrasive flow of work material [1,2]. Under such complex working conditions, which may vary significantly across the tool, tool surface is deteriorated and damaged through different wear mechanisms and thermo-mechanical fatigue processes [3-5].

In recent years requirements on tool resistance and performance are becoming increasingly demanding, which is strongly related to tool design, material selection and heat treatment optimization [6]. On one hand, increase in energy costs [7], the competitiveness of companies from emerging countries, and investments in new technologies dictate increased productivity, efficiency and performance of the existing tools [8]. As die costs represent between 15 and 30% of the total forging process costs, producing more parts leads to a notable cost reduction and increased profitability [1]. Another major factor is coming from the automotive sector. The automotive industry is continuously working on the vehicles weight reduction in order to lower fuel consumption and CO_2 emission, while maintaining or even improving the strength of components in accordance with increasing safety demands [9]. This requires introduction of low-weight high-strength materials, which are very difficult and demanding to form [10,11]. High-strength work material combined with increased complexity of the formed parts results in tools working under extreme stresses.

Increased tool demands mean strengthened property requirements for the tool material, including temper resistance, yield and ultimate tensile strength, wear resistance, fatigue and shock resistance, but mainly ductility and fracture toughness, which are essential for high temperature forming applications [12,13]. Properties of tool steels depend on the balanced chemical composition and the steel processing route, but mainly on the heat-treatment process, which defines final microstructure. In general, required properties, mainly focused on hardness and toughness are generated by a controlled heat treatment process, consisting of an austenitization treatment with a subsequent hardening and a multiple tempering procedure [12,14]. Traditionally, trade-off between high hardness and sufficient toughness is required, where hardening and tempering parameters have to be chosen from quite a narrow range [15]. On the other hand, through optimized vacuum heat treatment procedure fine grained microstructure with homogeneous fine secondary carbides distribution and reduced retained austenite content can be obtained [13], thus satisfying ever greater demands on tool properties in terms of greater fracture toughness and thermal fatigue resistance while maintaining high hardness and strength [16].

The most common properties used for optimizing heat treatment of tool steels are hardness and fracture toughness. High hardness (46-50 HRC) is related to resistance to plastic deformation and wear resistance, and toughness to tool ability to resist crack initiation and propagation [5,16]. However, beside hardness and toughness there are also other tool steel properties, which depending on the application are becoming important as we move toward more and more complex tools. These include thermal conductivity, creep and wear resistance, compressive and bending strength as well as machinability [11]. Although different tool steel properties can be determined using standard test methods, each one requires specific and often quite unique and demanding test specimens. Furthermore, different geometries of standard specimens mean different heat treatment conditions, which result in microstructure deviation [17], making it practically impossible to directly correlate properties after heat treatment. On the other hand, circumferentially notched and fatigue-precracked tensile bar (CNPTB) specimen has been found as the best alternative, which after fracture toughness measurement can be used to determine many other properties [18]. The advantage of the CNPTB specimen lies in the radial symmetry, which makes the specimen particularly suitable for studying the influence of tool steel properties and to optimize its heat treatment [19].

The aim of this research work was to determine the effect of vacuum heat treatment conditions on different properties and their correlations for low Si content AISI H11 type hot work tool steel using single CNPTB test specimen. Furthermore, work is intended to show the potential of CNPTB test specimen for designing advanced tempering diagrams in order to optimize vacuum heat treatment of hot work tool steels.

Experimental

Material and heat treatment

Material used in the investigation was low Si content AISI H11 type hot-work tool steel with the following composition (wt %): 0.36% C, 0.22% Si, 0.25% Mn, 5.02% Cr, 1.25% Mo, and 0.43% V. Material was delivered in the form of forged and soft annealed blocks, from which CNPTB specimens [18] (Fig. 1) were machined in the short transverse direction.

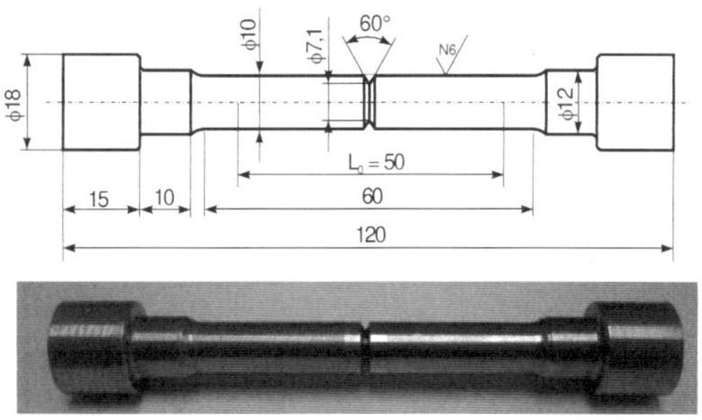

Figure 1: CNPTB specimen.

Heat treatment of the specimens was performed in Ipsen VTC 324-R horizontal vacuum furnace with uniform high-pressure gas quenching system. The specimens were first preheated to 850°C, followed by progressive heating at 10°C/min to the austenitizing temperature of 990°C and 1030°C, respectively. After soaked for 20 min, specimens were quenched in N_2 gas at a quenching speed of 3°C/s and double tempered for 2 h. For each austenitizing temperature first tempering was performed at 540°C, followed by second tempering carried out at six different temperatures, varied between 550 and 630°C. In terms of vacuum heat treatment conditions employed (austenitizing and tempering temperature) twelve sets of specimens, each consisting of at least 12 samples were prepared in order to obtain statistically relevant results. After heat treatment, series of properties were evaluated including fracture toughness, hardness, compression strength and strain hardening exponent, bending strength and machinability, with the results presented in the form of combined tempering diagrams.

Properties measurement

Fracture toughness measurements were performed on non-standard CNPTB specimens [18], with the fatigue pre-crack of about 0.5 mm created prior to the final heat treatment. Pre-cracking was done under rotating-bending loading of 400 N for 4500 cycles. As shown in [20], heat treatment has no negative effect on the crack tip blunting and measured fracture toughness values. After heat treatment CNPTB specimens were subjected to tensile loading until fracture, using Instron 1255 tensile-test machine, room temperature conditions and the cross-head speed of 1.0 mm/min. By measuring load at fracture (P), diameter of the brittle fractured area (d) and knowing diameter of the nominal non-notched specimen diameter (d_0 = 10 mm) fracture toughness can be calculated using Eq. 1 [21,22]:

$$K_{Ic} = \frac{P}{d_0^{3/2}} \cdot \left(-1.27 + 1.72\frac{d_0}{d}\right) \quad (1)$$

Fracture toughness measurements were followed by measurement of ***Rockwell-C hardness***, carried out by Willson-Rockwell B2000 machine. Hardness measurements were performed circumferentially on both halves of each fractured CNPTB specimen and average value calculated.

From one half of the fractured CNPTB specimen, ϕ10x12.5 mm cylinder was cut for ***compression test*** (Fig. 2). Compression tests were carried out at room temperature in Instron 1255 machine according to ASTM E9-09 standard and constant cross-head speed of 2.0 mm/min. Compression test results comprised yield strength, maximum compression strength and strain hardening exponent, determined on a log-log plot of true stress-true strain between yield and maximum compression stress. The other half of the fractured CNPTB specimen was used to machine cylindrical ***4-point bending test*** specimen, with the diameter of 5 mm and length of 60 mm (Fig. 2). After machining by high-speed turning, bending specimens were center-less ground in order to obtain required surface roughness of 0.2 μm. Room temperature 4-point bending tests were performed according to ASTM E290-09 standard using Instron 1255 machine and cross-head speed of 2.0 mm/min. Support span was 40 mm and load span 16 mm. During testing maximum bending strength and deformation at 3.5 kN load were determined.

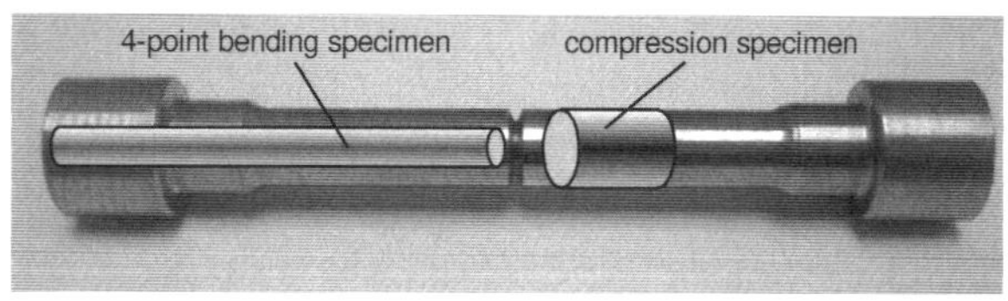

Figure 2: 4-point bending and compression test specimens' extraction.

Finally, effect of heat treatment on the ***machinability*** of hot work tool steel was evaluated by analyzing surface roughness of 4-point bending specimens after turning process. All specimens were first pre-machined using Sandvik-Coromant

DNMG11 R0.4 cutting inserts, feed rate of 0.12 mm/rev, depth of cut 0.3 mm and cutting speed of 100 m/min. Final machining was done with VBMT 16 04 cutting inserts, feed rate of 0.08 mm/rev, depth of cut 0.2 mm and the same cutting speed of 100 m/min. For each specimen new cutting inserts were used and surface roughness analyzed by Alicona InfiniteFocus G4 system in terms of R_a, R_z, R_p, R_v, R_{sk} and R_{ku} roughness parameters, as defined in ISO 4287 standard.

Results

Microstructure

After quenching from 990°C and double tempering microstructure of the investigated AISI H11 type hot work tool steel consists of some undissolved primary carbides and small secondary carbides uniformly distributed in the matrix of fine needles-like tempered martensite (Fig. 3a). By increasing austenitizing temperature to 1030°C a small fraction of bainite has been observed beside typical microstructure of tempered martensite, as well as pronounced dissolution of primary carbides (Fig. 3b). Increased tempering temperature, on the other hand, results in intensified precipitation [12] and increased volume fraction of small secondary carbides (Figs. 4a,b).

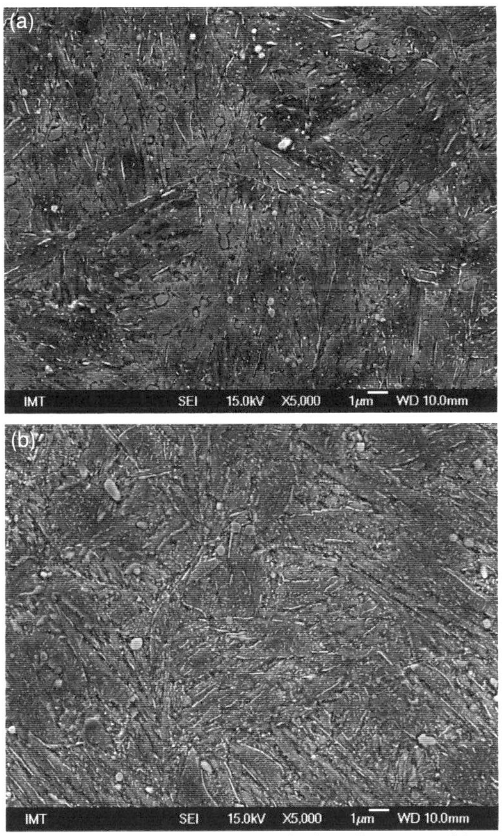

Figure 3: Microstructure of quenched and tempered hot work tool steel, quenched from 990°C and tempered at (a) 550°C and (b) 630°C.

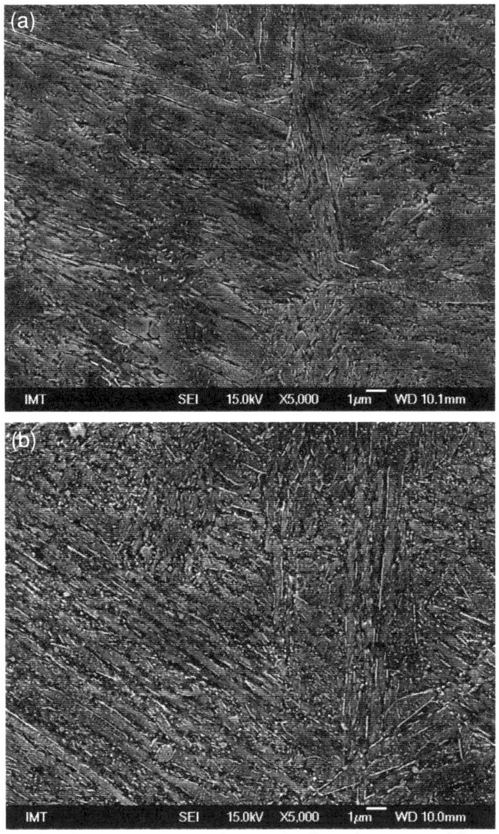

Figure 4: Microstructure of quenched and tempered hot work tool steel, quenched from 1030°C and tempered at (a) 550°C and (b) 630°C.

Hardness and fracture toughness

Tempering diagram displaying hardness and fracture toughness of the investigated AISI H11 type hot work tool steel as a function of austenitizing and tempering temperature is shown in Fig. 5. Hardness obtained after quenching from 990°C and double tempering at 630°C was 40 HRC and increased to 49 HRC as the tempering temperature was reduced to 550°C. Fracture toughness, on the other hand, dropped from 87 MPa√m to less than 30 MPa√m. Further increase in hardness was obtained by increasing austenitizing temperature, with temperature of 1030°C resulting in hardness increase of about 5% and maximum hardness of ~52 HRC. However, contrary to high Si content hot work tool steels [23], higher austenitizing temperature resulted also in elevated fracture toughness, ranging from 45 MPa√m even up to 115 MPa√m, as shown in Fig. 5. Combined properties of increased hardness and fracture toughness for the investigated hot work tool steel can be contributed to small fraction of bainite formation, giving better fracture toughness vs. hardness ratio (0.75 vs. 1.20).

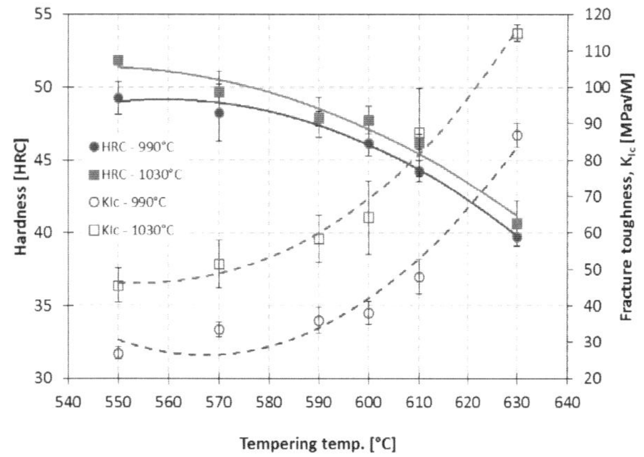

Figure 5: Hardness and fracture toughness tempering diagram.

Compression test results

Yield and maximum compression strength of the investigated hot work tool steel range between 1200 and 1780 MPa, and 1450 and 2050 MPa, respectively. Analogously to hardness, yield and maximum compression strength are increasing with austenitizing temperature (for ~5%), but decreasing with tempering temperature, as shown by tempering diagram in Fig. 6. In terms of strain hardening exponent, representing measure for material ductility, it was found more or less independent on the heat treatment conditions. For low and high austenitizing temperature and tempering temperatures between 550 and 600°C strain hardening exponent shows a constant value of about 0.45. However, as the tempering temperature exceeded 610°C strain hardening exponent for the investigated hot work tool steel increased to about 0.6 (Fig. 6).

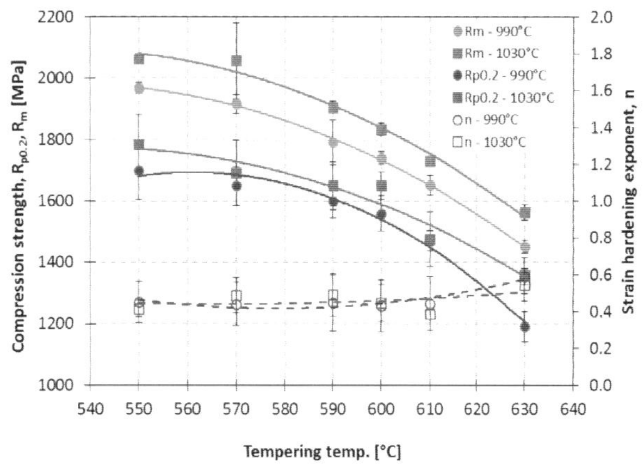

Figure 6: Tempering diagram displaying compression test results.

Bending strength

In the case of 4-point bending test, yield and maximum strength for the investigated steel quenched from 990°C increased from 1860 MPa and 3260 MPa to 2600 MPa and 4550 MPa, respectively, as the tempering temperature was lowered from 630°C to 550°C. By increasing austenitizing temperature to 1030°C, yield and maximum bending strength also increased, reaching peak values of 2700 MPa and 4780 MPa at the tempering temperature of 550°C (Fig. 7). Similarly as fracture toughness, bending strain at the preset stress of 3500 MPa increased with tempering temperature, showing values between 3 and 10%, Furthermore, by increasing austenitizing temperature resistance of the investigated steel to bending deformation was improved, with the deformation being reduced by about 20%, as shown in Fig. 7. Synergistic effect of increased hardness and fracture toughness, obtained through bainite formation is manifested also in improved bending resistance.

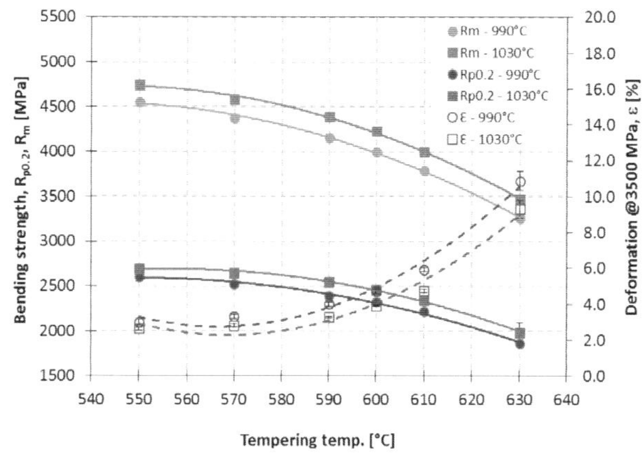

Figure 7: Tempering diagram for 4-point bending test.

Machinability

Results of surface roughness measurements, performed on machined 4-point bending specimens to evaluate hot work tool steel machinability are shown in Fig. 8. As shown in Fig. 8a, surface roughness parameters including average surface roughness R_a, maximum peak to valley height R_z, maximum profile peak height R_p and maximum profile valley depth R_v, they all increase with increased tempering as well as austenitizing temperature, indicating increased tearing component and worse machinability. For the machining conditions used and applied austenitizing and tempering temperatures, R_a value increased from 0.4 to 0.8 μm, and R_z value from 2.5 to 4.5 μm (Fig. 8a). Furthermore, skewness (R_{sk}), being measure of surface profile asymmetry has been reduced with higher tempering temperature from ~0 to -0.5 and kurtosis (R_{ku}), displaying sharpness of the surface profile increased from 2.5 to above 3. Austenitizing temperature, on the other hand, had only marginal effect on skewness and kurtosis values, as shown in Fig. 8b. Kurtosis being above 3 characterizes surfaces with sharp peaks and negative skewness plateau like topography [24]. As shown by surface roughness analysis, machinability and surface quality of hot work tool steel improve by lowering tempering temperature and/or austenitizing temperature (Fig. 8).

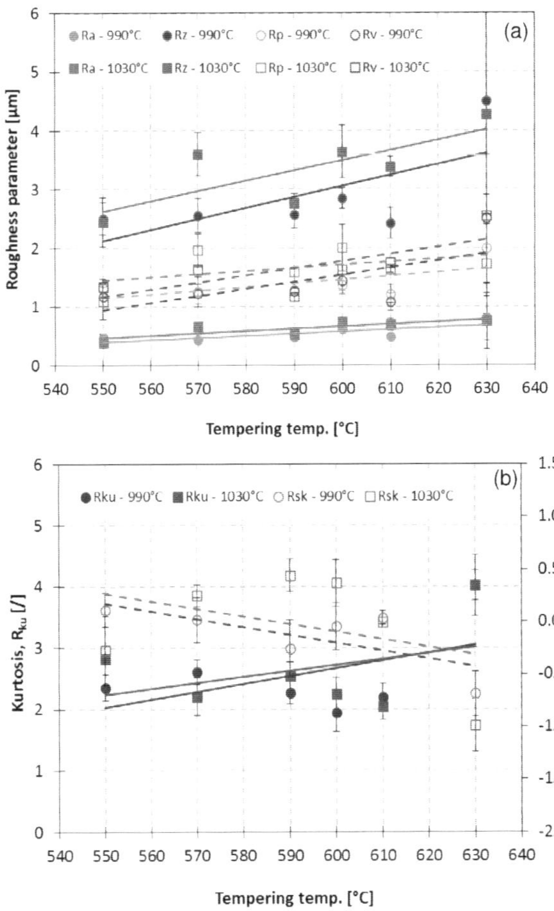

Figure 8: Surface roughness analysis of machined hot work tool steel.

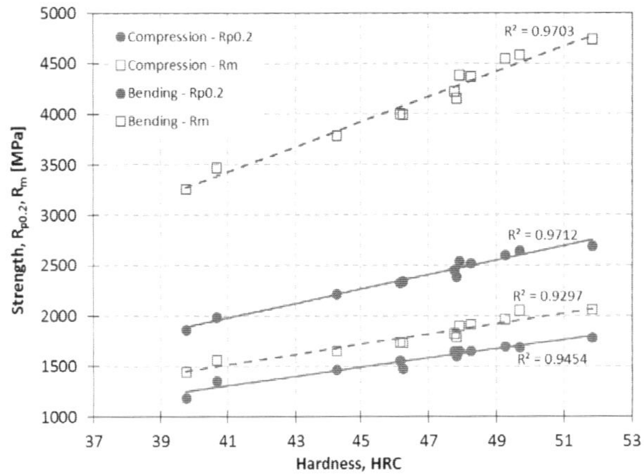

Figure 9: Correlation between hardness and compression and bending strength of hot work tool steel.

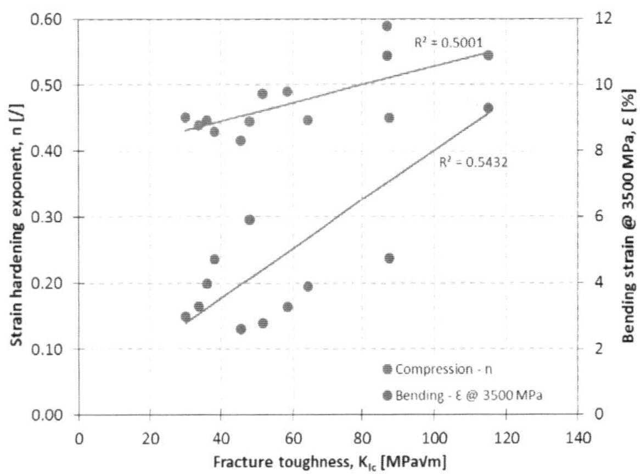

Figure 10: Correlation between fracture toughness and strain hardening exponent and bending strain of hot work tool steel.

Discussion

In the present case and for the heat treatment conditions applied strong correlation ($R^2 > 0.92$) between investigated hot work tool steel hardness and compression and bending strength has been observed. As shown in Fig. 9 compression and bending strength linearly increase with material hardness. Yield and maximum compression strength of the investigated hot work tool steel can be expressed as 33.5·HRC and 38.6·HRC, respectively, and bending yield and maximum stress as 51.0·HRC and 88.3·HRC, respectively. On the other hand, although strain hardening exponent and bending strain show correlation with fracture toughness, both increasing with improved fracture toughness of the material, correlation factor R^2 is rather weak, of only about 0.5 (Fig. 10).

In terms of surface quality and machinability things are not so straight forward. As shown in Fig. 11a roughness parameters (R_a, R_z, R_{sk}, R_{ku}), indicating improved machinability of hot work tool steel improve with increased hardness obtained through increased austenitizing temperature and/or reduced tempering temperature (Fig. 5). However, if material is too hard, above 50 HRC surface quality starts to deteriorate again, displaying sharp asperities and emphasized valleys. Similar observation is valid for fracture toughness, with surface quality and machinability being deteriorated with increased toughness of the material (Fig. 11b). In general, for the investigated AISI H11 type hot work tool steel the best machinability and surface quality is obtained when material hardness is between 46 and 50 HRC and fracture toughness below 60 MPa√m, as shown in Fig. 11.

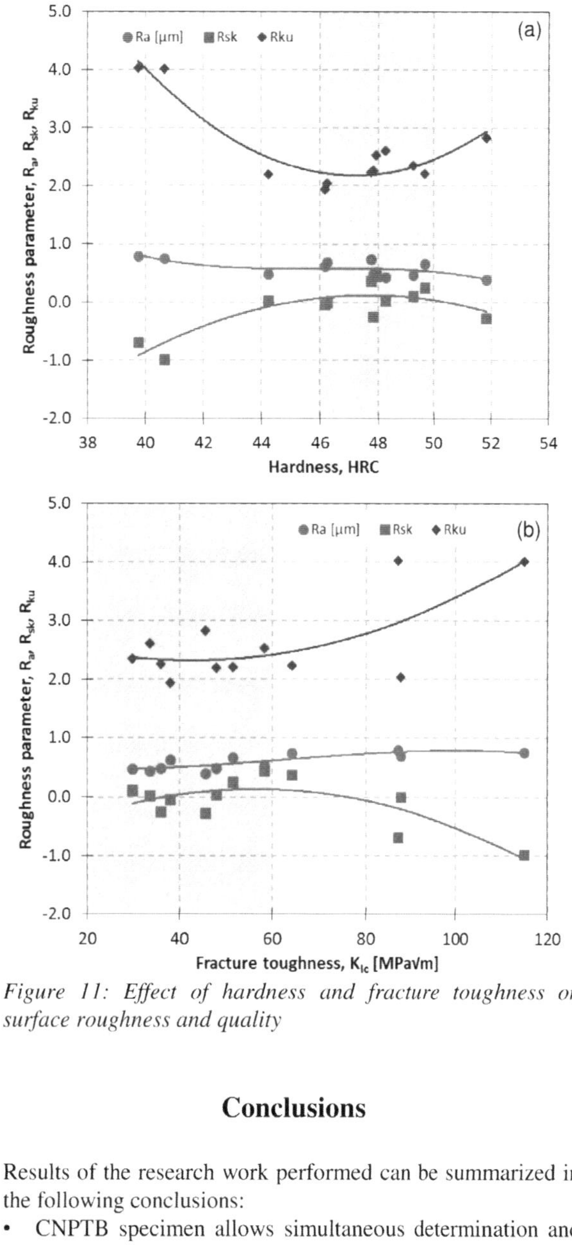

Figure 11: Effect of hardness and fracture toughness on surface roughness and quality

Conclusions

Results of the research work performed can be summarized in the following conclusions:

- CNPTB specimen allows simultaneous determination and correlation of multiple properties of hot work tool steels, ranging from hardness, compression and bending strength, fracture toughness and strain hardening exponent, wear resistance and machinability, which can be directly correlated to microstructure and used to design advanced tempering diagrams.
- In the case of AISI H11 type hot work tool steel with low Si content hardness, strength as well as fracture toughness increase with increased asutenitizing temperature and reduced tempering temperature.
- Yield and maximum compression and bending strength show direct and very stron correlation with hardness. Also strain hardening exponent and bending strain increase with fracture toughness, however, correlation is rather weak.
- Surface quality and machinability of heat treated hot work tool steel mainly depend on hardness but also on fracture toughness. Best results are obtained when hardness is between 46 and 50 HRC and fracture toughness below 60 MPa√m.

References

[1] Pérez, M. and Belzunce, F.J., "The Effect of Deep Cryogenic Treatments on the Mechanical Properties of An AISI H13 Steel," *Materials Science and Engineering: A*, Vol. 624 (2015), pp. 32–40.

[2] Podgornik, B., *et al.*, "Vacuumheat Treatment Optimization for Improved Load Carrying Capacity and Wear Properties of Surface Engineered Hot Work Tool Steel," *Surface and Coatings Technology*, Vol. 261 (2015), pp. 253–261.

[3] Mellouli, D., *et al.*, "Hardness Effect on Thermal Fatigue Damage of Hot-Working Tool Steel," *Engineering Failure Analysis*, Vol. 45 (2014), pp. 85–95.

[4] Malm, S. and Norström, L.A., "Material-Related Model for Thermal Fatigue Applied to Tool Steels in Hot Work Applications," *Metal Sci.*, Vol. 9 (1979), pp. 544–550.

[5] Leskovšek, V., *et al.*, "The Influence of Austenitizing and Tempering Temperature on the Hardness and Fracture Toughness of Hot-Worked H11 Tool Steel," *Journal of Materials Processing Technology*, Vol. 178 (2006), pp. 328–334.

[6] Roberts, G. *et al.*, Tool Steels (fifth ed.), ASM International (Materials Park, 1998).

[7] euroforge.org, http://www.euroforge.org/fileadmin/user_upload/Downloads/Energy_Euroforge.pdf (26.1.2016).

[8] Shirgaokar, M., Technology to Improve Competitiveness in Warm and Hot Forging - Increasing Die Life and Material Utilization, PhD Dissertation, Ohio State University (Columbus, 2008).

[9] Ellera, T.K., *et al.*, "Determination of Strain Hardening Parameters of Tailor Hardened Boron Steel up to High Strains Using Inverse FEM Optimization and Strain Field Matching," *Journal of Materials Processing Technology*, Vol. 228 (2016), pp. 43–58.

[10] Behrens, B.A., *et al.*, "Precision Forging Processes for High-Duty Automotive Components," *Journal of Materials Processing Technology*, Vol. 185 (2007), pp. 139-146.

[11] Podgornik, B. and Leskovšek, V., "Experimental Evaluation of Tool and High-Speed Steel Properties Using Multi-Functional KIc-Test Specimen," *Steel Research International*, Vol. 84 (2013), pp. 1294–1301.

[12] Eser, A, *et al.*, "Multiscale Modeling of Tempering of AISI H13 Hot-Work Tool Steel – Part 1: Prediction of Microstructure Evolution and Coupling with Mechanical Properties," *Computational Materials Science*, Vol. 113 (2016), pp. 280–291.

[13] Li, Jing-Yuan, *et al.*, "Mechanism of Improvement on Strength and Toughness of H13 Die Steel by Nitrogen," *Materials Science and Engineering: A*, Vol. 640 (2015), pp. 16–23.

[14] Ramezani, M., *et al.*, "Evaluation of Carbon Diffusion in Heat Treatment of H13 Tool Steel under Different Atmospheric Conditions," *Journal of Materials Research and Technology*, Vol. 4 (2015), pp. 114–125.
[15] Lerchbacher, C., *et al.*, "Direct or Indirect: Influence of Type of Retained Austenite Decomposition During Tempering on the Toughness of a Hot-Work Tool Steel," *Materials Science and Engineering: A*, Vol. 564 (2013), pp. 163–168.
[16] Telasang, G., *et al.*, "Structure–Property Correlation in Laser Surface Treated AISI H13 Tool Steel for Improved Mechanical Properties," *Materials Science and Engineering: A*, Vol. 599 (2014), pp. 255–267.
[17] Dossett, J.L. and Totten, G.E., ASM Handbook Volume 4A: Steel Heat Treating Fundamental and Processes, ASM International (Materials Park, 1991).
[18] Podgornik, B., *et al.*, "Experimental Evaluation of Tool Steel Fracture Toughness using Circumferentially Notched and Precracked Tension Bar Specimen," *Materials Performance and Characterization*, Vol. 3 (2014), pp. 1-9.
[19] Leskovsek, V., *et al.*, "Development of a Modified Method for Determining Fracture Toughness of HSS With Non-Standard Cylindrical Round-Notched Tension Specimens," *Metall. Alloys Technol.*, Vol. 27 (1993), pp. 195–204.
[20] Ule, B., *et al.*, "Estimation of Plane Strain Fracture Toughness of AISI M2 Steel from Precracked Round-Bar Specimens," *Engineering Fracture Mechanics*, Vol. 65 (2000), pp. 559–572.
[21] Wei, S., *et al.*, "Fracture Toughness Measurement by Cylindrical Specimen With Ring-Shaped Crack," *Engineering Fracture Mechanics*, Vol. 16 (1982), pp. 69–82.
[22] Gdoutos, E.E., Fracture Mechanics Criteria and Applications, Kluwer Academic Publishers (London, 1990).
[23] Mesquita, R.A. and Kestenbach, H.-J., "On the Effect of Silicon on Toughness in Recent High Quality Hot Work Steels," *Materials Science and Engineering: A*, Vol. 528 (2011), pp. 4856-4859.
[24] Sedlaček, M., *et al.*, "Correlation between Standard Roughness Parameters Skewness and Kurtosis and Tribological Behaviour of Contact Surfaces," Tribology International, Vol. 48 (2012), pp. 102-112.

International Federation of Heat Treatment and Surface Engineering 2016
Proceedings of the 23rd IFHTSE Congress
April 18–21, 2016, Savannah, Georgia, USA

Microstructural Evolution in Microalloyed Steels with High-Speed Thermomechanical Bar and Rod Rolling

Robert Cryderman, Blake Whitely, and John Speer
Advanced Steel Processing and Products Research Center, George S. Ansell Department of Metallurgical and Materials Engineering, Colorado School of Mines, Golden Colorado, USA

Abstract

Bars and rods are rolled at high total deformations, high strain rates, and short inter-pass times compared to products such as plates or structural sections where extensive studies have been conducted to understand the effects of microalloying and the limited range of thermomechanical process parameters. Data are presented to illustrate how microalloying and high-speed thermomechanical processing affect the as hot-rolled microstructures for a variety of steel grades and applications. Simulations on a Gleeble®3500 using torsional deformation and controlled time-temperature schedules as well as interrupted quenching have allowed examination of the evolution of prior austenite grain size and morphology. The austenite condition, in combination with the final cooling schedule, influence final hot-rolled microstructures and can lead to significant effects on microstructures and mechanical properties after subsequent heat treatment.

Introduction

There have been many studies leading to models of austenite grain development in flat rolled plates and strip where deformation essentially occurs in two dimensions (plane strain) such that the thickness is reduced and the length is increased. In contrast, rolling to long products such as bars and rods provides deformation in three dimensions and increased total deformation. For example rolling a 250 mm slab to 16 mm plate achieves an elongation of about 16:1 as compared to rolling a 200 mm square billet to a 40 mm round bar for an elongation of 32:1. Smaller finished sizes are also possible using the same semi-finished sections to produce light gauge strip at 1.5 mm for an elongation of 167:1 or to produce 5.5 mm diameter rod for an elongation of 1680:1. Average actual strains in long products are further increased by the redundant deformation that occurs during the bar forming process. A simplified analysis by Lee et al demonstrates that actual strain per pass is 1.7 – 2.0 times the area reduction for oval passes and 2.0 – 2.5 times the area reduction for round passes in the oval-round pass sequences commonly utilized for bar and rod rolling. [1] Consequently, total strains in bar rolling are on the order of 4 times higher than plates and 20 times higher in rod as compared to light gauge strip.

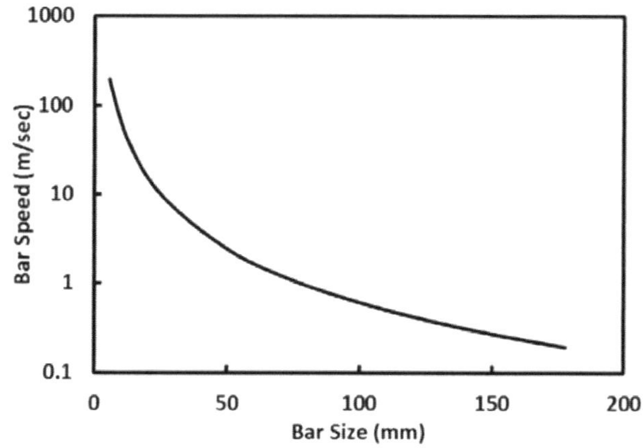

Figure 1: Bar speed increases exponentially as size is reduced through in-line rolling stands. Example shown is for rounds at 150 ton/hour rolling rate.

Modern bar and rod mills are typically designed to utilize in-line rolling stand arrangements that necessitate a constant mass throughput rate for all rolling stands. Consequently, the linear speed through consecutive rolling stands increases as the size becomes smaller as illustrated in Figure 1.

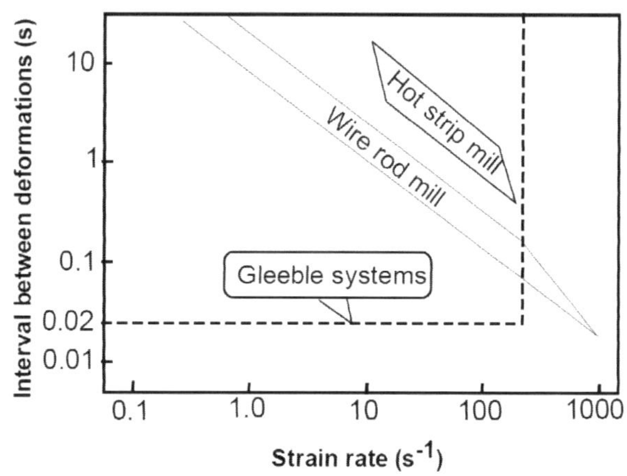

Figure 2: Inter-pass times and strain rates (initial-final passes) for wire rod and thin strip compared to Gleeble® 3800 capability above and left of the dashed line. [2]

As the rolling speed increases, the inter-pass time is reduced and the strain rate is increased as illustrated by Kuziak in Figure 2. [2]

Development of Gleeble® and similar systems equipped with hot torsional deformation capability has allowed simulation of the pass strains, inter-pass times, and strain rates for hot rolling bars and thin plates. However, the limitations of the electro-mechanical torsion equipment prevent replication of the strain rate and interpass times encountered in the high speed rolling of small diameter rods as illustrated in Figure 2. [2]

Strains, strain rates, and inter-pass times determine the characteristics of austenite grains (austenite "condition") at the completion of hot rolling. As shown in Figure 3, large equiaxed austenite grains created during heating prior to rolling are deformed to a distorted shape during deformation at each rolling pass. After a single rolling pass, the distorted grains recrystallize and then begin to grow. The amount of strain, strain rate, and temperature affect the rates of recrystallization and grain growth. [3] At high strains, strain rates, and temperatures recrystallization occurs dynamically (DRx) during the rolling pass. At lower strains, strain rates, and temperatures the recrystallization occurs statically (SRx) after the rolling pass. Recrystallization can be completely suppressed at temperatures below a critical value (T_{nr}) with limited strains and strain rates. [4] T_{nr} can be affected by the strain and strain rates. Short inter-pass times limit the amount of recrystallization and grain growth between stands – or increase T_{nr} as multiple passes are applied before recrystallization occurs. [5] T_{nr} is heavily affected by the addition of small amounts of V, Nb, and Ti as these additions slow grain boundary movement by solute drag and by combining with C and N to form precipitates that limit recrystallization and grain growth kinetics. [6 - 8]

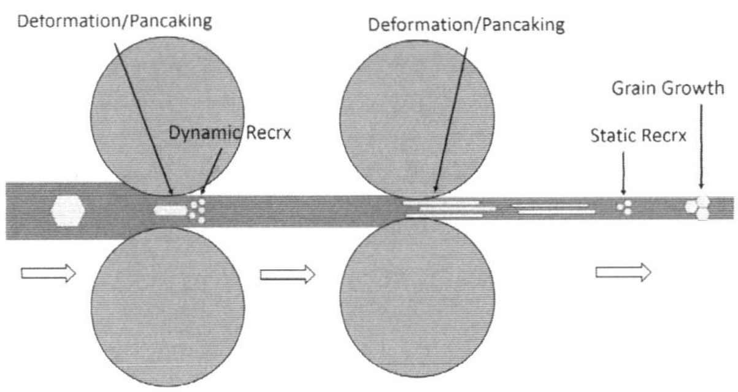

Figure 3: Schematic diagram showing deformation, dynamic recrystallization, static recrystallization, and growth of austenite grains during multi-pass hot rolling.

Steels with different chemical compositions have been tested using the Gleeble or similar equipment to establish the effects of strain and strain rates at various temperatures on recrystallization and grain growth rates. The test data have been assembled into kinetic equations that allow calculation of grain sizes at the entry and exit of each rolling pass. [5, 9, 10]

These equations can be used to show that lowering the temperature during the final few rolling passes decreases the austenite grain size and can change the shape of austenite grains for a given steel chemical composition. [5] Additional investigations have been conducted on low carbon plate steels to show that the prior austenite grain size and shape modifies the microstructure that is formed during subsequent cooling. [11] Generally, the increased grain boundary area associated with smaller deformed austenite grains increases the rate of nucleation of small ferrite grains and reduces the time for ferrite nucleation as illustrated in Figure 4. [11]

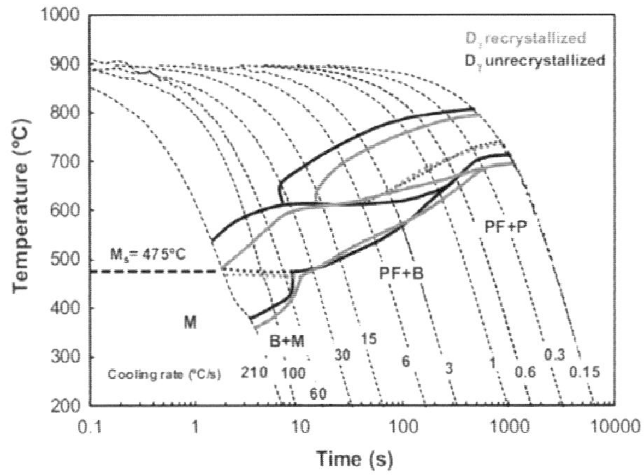

Figure 4: Comparison of CCT diagrams for transformation from recrystallized and un-recrystallized austenite in a 0.06% C- 1.20% Mn-0.0062% Nb-0.053% V plate steel. [11]

Similar effects are anticipated in bar steels with higher carbon and alloy contents compared to the amounts common in plate steels. For example, reducing the austenite grain size has been shown to reduce the size of ferrite grains and increase the amount of ferrite formed in a series of 0.4% C steels containing about 1% Mn and 0.1-0.2% V. [12]

The following sections describe experiments that were conducted to illustrate the effects of thermomechanical rolling of bars on the microstructures, and the influence of these microstructures on subsequent heat treatment.

Thermomechanically Rolled Bars

Bars of chemical compositions listed in Table 1 were hot rolled from 152 mm square billets in a rolling mill described schematically in Figure 5. The rolling mill consists of a reheat furnace, eight roughing stands in H-V arrangement, water cooling boxes for cooling before finish rolling, space to allow temperature equalization in the bar cross section, an eight stand 3-roll type reducing mill, and a three stand 3-roll type precision sizing block. [13] The bars were cooled in air as straight bars on a walking beam cooling bed after rolling. Initial heating temperatures were 1150-1200 °C for the 1045 Steels and 1080-1125 °C for the 16MnCr5 steel. The final rolling temperatures were adjusted to the desired levels by utilizing the water cooling boxes located immediately after the rough rolling stands.

Table 1: Chemical Compositions of Test Steels in wt. pct.

Steel	C	Mn	Si	Ni	Cr	Mo
16MnCr5	0.18	1.13	0.21	0.10	1.05	0.04
1045 Al	0.45	0.72	0.24	0.08	0.12	0.04
10V45	0.45	0.82	0.28	0.07	0.15	0.03
10V45Nb	0.46	0.85	0.27	0.08	0.14	0.03
Steel	**Al**	**V**	**Nb**	**Ti**	**B (ppm)**	**N (ppm)**
16MnCr5	0.029	NA*	NA	NA	NA	
1045 Al	0.021	0.003	0.001	0.001	NA	97
10V45	0.000	0.084	0.001	0.001	NA	127
10V45Nb	0.000	0.092	0.020	0.001	NA	124

*NA = none added

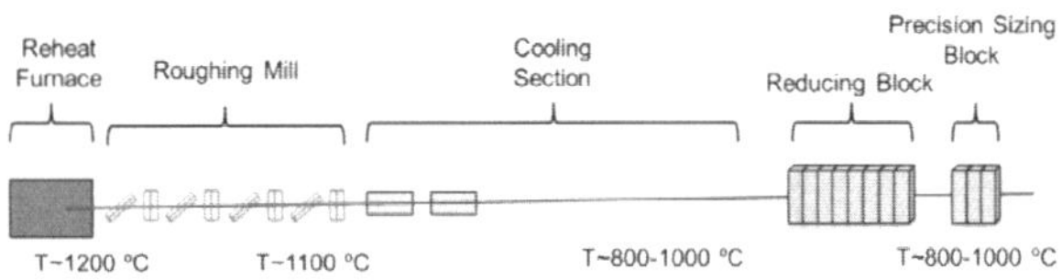

Figure 5: Schematic diagram of rolling mill for evaluating effects of final rolling temperature. [13]

Steel 16MnCr5 is a low carbon hardenable steel commonly utilized to produce driveline components by forging and machining followed by carburizing and hardening. The chemical composition includes controlled amounts of Mn and Cr to achieve the hardenability required for hardening after carburizing. Conventional rolling without intermediate cooling resulted in a final rolling temperature of 1018 °C for 38 mm diameter bars. The addition of intermediate water cooling lowered the final rolling temperature to 886 °C. Calculated T_{nr} for Steel 16MnCr5 according to the Boratto equation is estimated to be about 906 °C. [14] Prior austenite grain size is expected to be considerably smaller with the lower final rolling temperature.

The microstructures from cross sections at the mid radius of the 38 mm diameter bars after air cooling are shown in Figure 6. [15] The higher, conventional finish rolling temperature resulted in a microstructure consisting of proeutectoid ferrite at prior austenite grain boundaries and bainite as shown in Figure 6(a). The lower finish rolling temperature resulted in a microstructure consisting of ferrite and pearlite with well dispersed pearlite colonies as shown in Figure 6(b). The area fraction of free ferrite increased from 16% to 48% with the lower final rolling temperature. Examination by scanning electron microscopy, Figures 6(c, d) confirmed the change from dispersed carbides to fine lamellar pearlite.

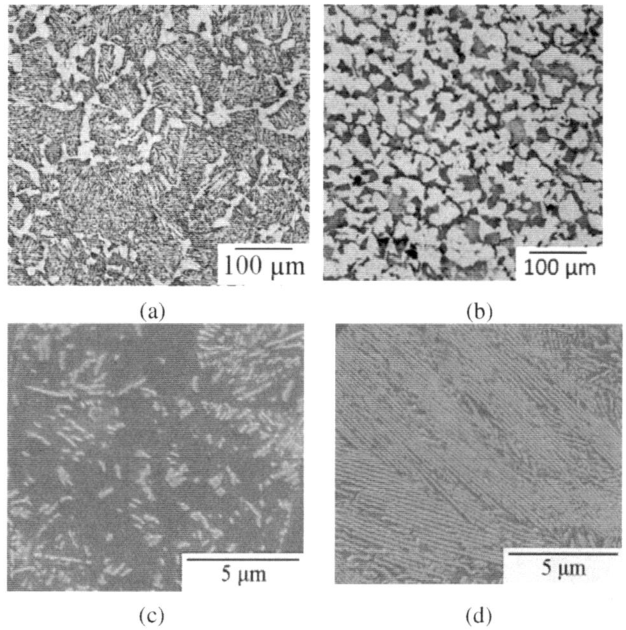

Figure 6: Microstructures of 16MnCr5 hot rolled bars finish rolled at 1018 °C (a, c) and 886 °C (b, d) as revealed by light optical (a, b) and scanning electron(c, d) microscopy. (2% picral etch) [15]

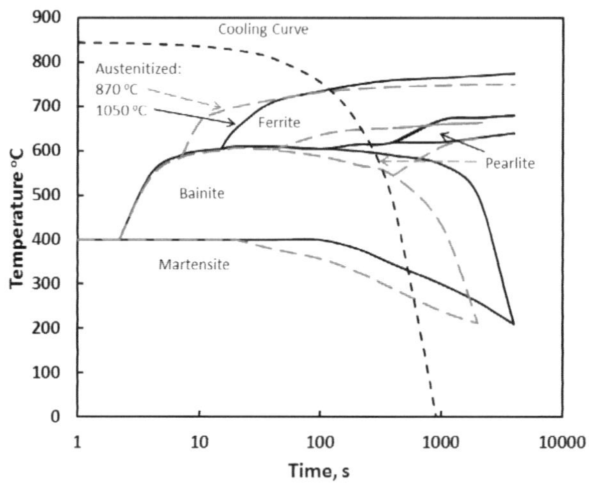

Figure 7: Effect of austenitizing temperature (solid line 1050 °C and dashed line 870 °C) on the continuous cooling transformation of 16MnCr5 steel. [16]

The effects of 16MnCr5 prior austenite grain size, (as determined by austenitizing temperature) on continuous cooling transformation kinetics are illustrated in Figure 7 which was redrawn to provide a direct comparison of the CCT diagrams for austenitizing temperatures of 870 °C and 1050 °C. [16] The cooling curve shown in Figure 7 is for air cooling 38 mm bars after finish rolling. The finer austenite grain size accelerated the formation of grain boundary ferrite, leading to a higher ferrite fraction, and increased the size of the pearlite transformation field so that pearlite was formed rather than bainite.

SAE 1045 steels are commonly used for production of shafts that are induction hardened to resist fatigue at high torsional and bending loads. Attainment of the desired final microstructures and properties with the short reheating times for induction surface hardening is highly influenced by the starting microstructure. It has been shown that a finer ferrite-pearlite microstructure is more readily hardened than coarse ferrite-pearlite microstructures. [17]

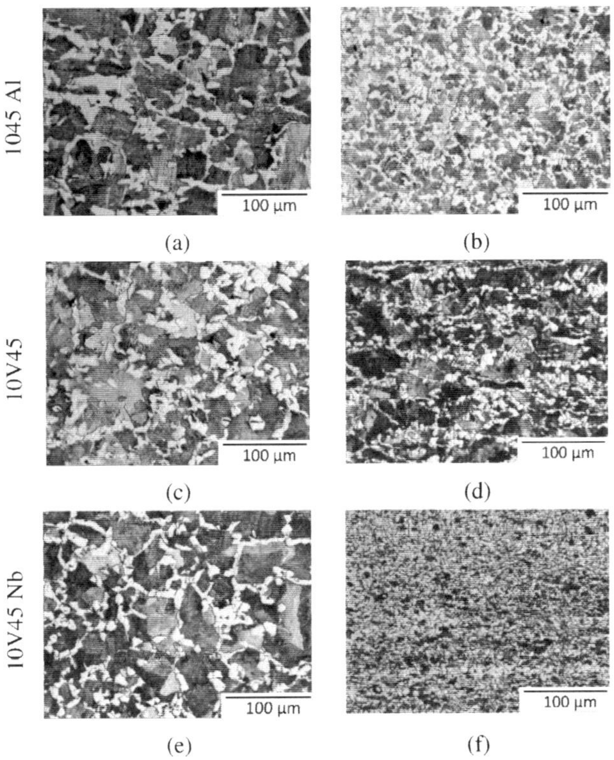

Figure 8: Microstructures of SAE 1045 as- rolled bars after finish rolling at 1000 °C (a, c, e) and 800 °C (b, d, f) as revealed by light optical microscopy (2% nital etch). [13]

The effects of microalloy additions (Table 1) and low temperature thermomechanical rolling (TMR) on the microstructures of 1045 steel are illustrated in Figure 8. In all three 1045 steels, the grain size was substantially reduced by lowering the final rolling temperature from 1000 °C (HR) to 800 °C (TMR). The lower rolling temperature also increased the amount of free ferrite.

Specimens from the three 1045 steels were processed on a Gleeble® 3500 thermomechanical simulator utilizing torsion to simulate the time-temperature-deformation sequence for the industrial rolling cycles. Specimens were quenched after reheating to 1200 °C, and immediately after final deformations at 1000 °C and 800 °C. An example showing the austenite grain shape and partial recrystallization is shown in Figure 9. A tangential sectioning technique was utilized to reveal the prior austenite grain deformation on the shear plane during torsion. [13] The deformed prior austenite grains for the TMP simulations on the three 1045 Steels are shown in Figure 9. Analysis of the microstructures showed the same amount of shear deformation in the unrecrystallized grains for all three steels with 40% recrystallization in Steel 1045 Al, 8% recrystallization in Steel 10V45, and <1% recrystallization in Steel 10V45 Nb. At the higher 1000 °C deformation temperature, all three steels showed full recrystallization. [13]

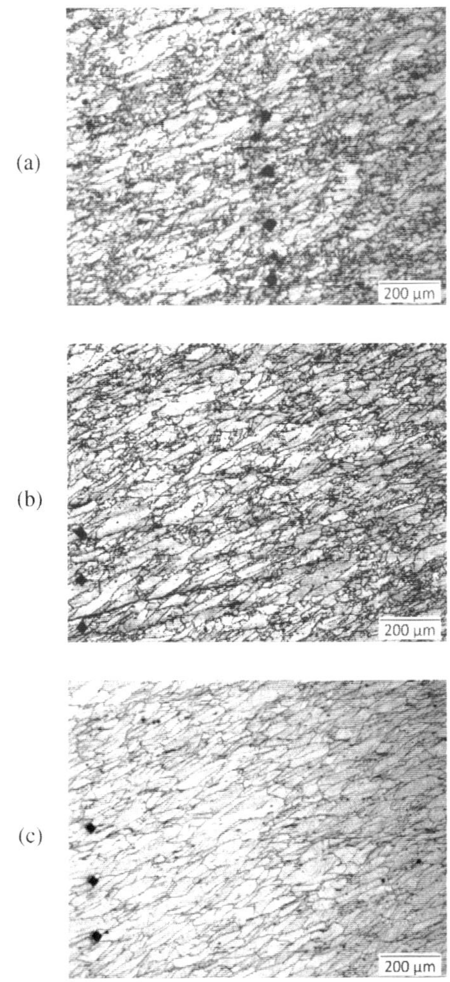

Figure 9: Microstructures of Gleeble® specimens quenched after final deformation at 800 °C showing prior austenite grain sizes, shapes, and amount of recrystallization (saturated picric etch) for (a) Steel 1045 Al, (b) 10V45, and (c) 10V45 Nb. [13]

The quenched Gleeble® specimens were re-sectioned to show the equiaxed cross-sections of the prior austenite grains for all three test conditions and facilitate measurement of the grain diameters. As shown in Figure 10, the large prior austenite grains present after heating to 1200 °C were refined by the HR simulation and further refined by the TMR simulation. After the rolling simulations, Steel 1045 Al consistently exhibited the largest prior austenite grain size and Steel 10V45 Nb exhibited the smallest prior austenite grains. Compared to the HR simulation, the TMR simulation provided reductions in grain size of about 30% for Steel 1045 Al and 50% for the two microalloyed steels.

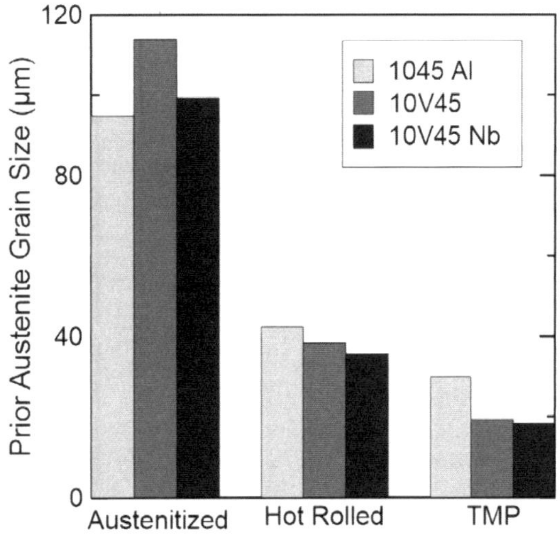

Figure 10: Prior austenite grain sizes measured on Gleeble® torsion simulation specimens quenched after reheating (austenitizing), HR deformation (1000 °C) and TMP deformation (800 °C). [13]

Heat Treatment of Thermomechanically Rolled Bars

The following examples show that the as- rolled microstructure can substantially influence the resulting microstructure after final heat treatment. The 16MnCr5 steels in the two rolling conditions from Figure 6 were sub-critically annealed at 692 °C for 6 hours and the resulting microstructures are shown in Figure 11.

After 6 hours of annealing, the carbides for both prior rolling conditions were fully spheroidized as shown Figure 11 (c, d). However, the distribution of the spheroidized carbides within the overall microstructure was substantially different as shown in Figures 11 (a, b). These microstructural differences could be expected to have an influence on microstructures and properties after the final carburizing and hardening of the finished parts.

The response of the 1045 steels as-rolled microstructures to supercritical heat treatments varied depending on the

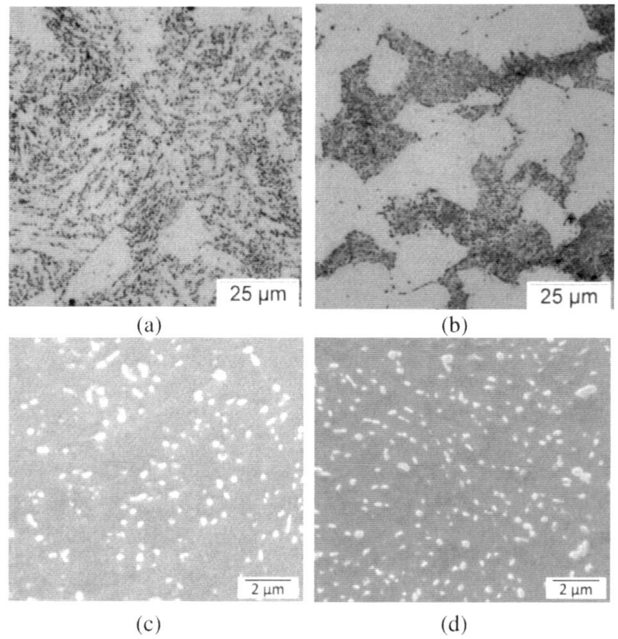

Figure 11: Microstructures of 16MnCr5 specimens annealed at 692 °C for 6 hours; (a, c) finish rolled at 1018 °C, (b, d) finish rolled at 886 °C, (a, b) light optical micrographs, (c, d) scanning electron micrographs. (2% picral etch) [15]

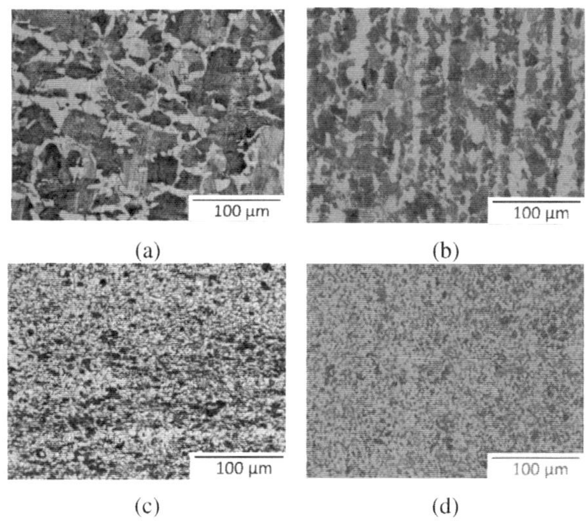

Figure 12: Effects of lamellar pearlite (LP) annealing on the microstructures of 1045 steel specimens: (a) Steel 1045 Al HR at 1000 °C, (b) Steel 1045 Al HR and LP annealed, (c) Steel 10V45 Nb TMR at 800 °C, and (d) Steel 10V45 Nb TMR and LP annealed.

microalloys, the prior rolling conditions, and the type of heat treatment. Two examples are illustrated in Figure 12 for Steel 1045 Al that was initially HR at 1000 °C and for Steel 10V45Nb that was TMR at 800 °C. After rolling, specimens of each steel were lamellar pearlite (LP) annealed by heating to

805 °C (A_{c3} + 50 °C) for 1 hour followed by furnace cooling at 0.06 °C/s (in the range of 800 – 300 °C). Comparison of HR Steel 1045 Al in Figure 12 (a) and (b) shows that the LP anneal reduced the pearlite colony size and introduced a banded structure parallel to the rolling direction (vertical in (b)). LP annealing of TMR Steel 10V45 Nb imparted a limited effect on the microstructure after the LP anneal as shown in Figures 12 (c) and (d). The overall result is that the LP annealed microstructure is substantially modified by the initial microstructure and the addition of micro alloys (Figures 12 (b) and (d).

As noted previously, 1045 steel is commonly utilized to produce induction hardened shafts. Specimens obtained from the three 1045 steels (Table1) were heat treated to simulate an induction hardening thermal cycle on a Gleeble® 3500. The thermal cycle, calculated using Elta 2-D, is shown in Figure 13 and was developed to simulate a 2 mm deep case depth on a 38 mm diameter shaft using a 2-turn coil. [18] The heating portion of the cycle shows an interruption at about 700 °C which reflects the space between the first and second turn during scan hardening. The cooling after heating was first accelerated with forced air to simulate heat transfer to the core and then water quenched at a rate of 3500 °C/s. Gleeble® specimens processed to the HR and TMP conditions were subsequently heat treated to the simulated induction hardening schedule in Figure 13.

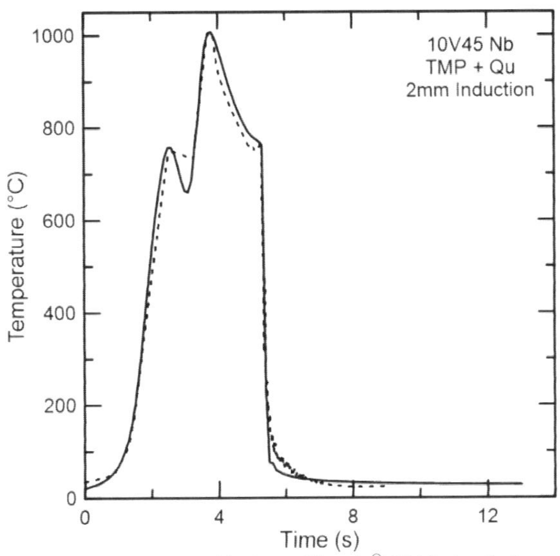

Figure 13: Thermal profile for a Gleenle® 3500 simulation of induction hardening a 38 mm diameter shaft to a 2 mm case depth showing modeled (solid line) and experimental (dashed line) time-temperature profiles.

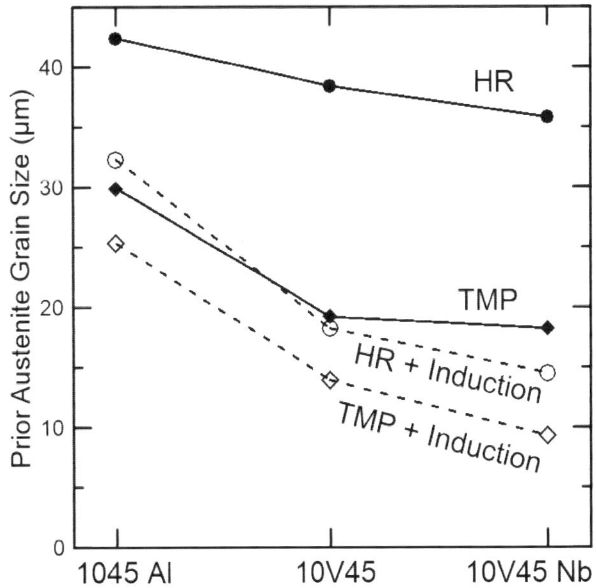

Figure 14: Effects of prior rolling practice (HR vs. TMP) and microalloy content on the prior austenite grain size after induction hardening of 1045 steels to a 2 mm case depth.

The prior austenite grain size after simulated induction hardening was measured for all test specimens and the results are summarized in Figure 14. For the combinations of microalloy contents and rolling conditions evaluated, the prior austenite grain size was reduced by the induction hardening simulation. For Steel 10V45 Nb, the induction hardening simulation reduced the prior austenite grain size by about 50% for the HR and TMP conditions, and consistently attained the smallest prior austenite grain size for all conditions. For Steel 1045 Al the induction hardening simulation reduced the HR and TMP prior austenite grain sizes by about 20% and consistently exhibited the largest grain size for each condition.

Summary and Conclusions

Bars and rods are being thermomechanically rolled to modify the prior austenite grain size and shape. Implementation of the thermomechanical rolling for bars and rods is different than processes developed for flat rolled or structural steel forms due to the higher strains associated with bi-dimensional reduction and the higher rolling speeds. Suitable rolling mill designs are necessary to impart the high strains, high strain rates, and short inter-pass times. Careful selection of microalloys consistent with the specific base chemical compositions can enhance the effectiveness of thermomechanical rolling in achieving refined final microstructures.

Modification of austenite morphology by thermomechanical rolling changes the final microstructure of the as rolled product by affecting the transformation kinetics during subsequent cooling. The finer austenite grain size influences the CCT diagram and can accelerate the nucleation of ferrite. In steels like 16MnCr5 phase fractions can be substantially modified so that bainite is suppressed in favor of pearlite formation. In steels like 1045 that readily transform to ferrite and pearlite, the finer austenite grain size leads to finer ferrite-pearlite microstructures and an increased fraction of free ferrite.

The effects of thermomechanical rolling on microstructures after subsequent heat treatment are not well documented in the literature. For subcritical annealing, the long range (>100-200 microns) microstructural features are relatively unaffected by the heat treatment, even though the short range features (<20 microns) may be modified. The microstructure after supercritical heating such as normalizing or LP annealing is heavily influenced by the microstructure in the bars after hot rolling. A similar effect occurs during the short time higher austenitizing temperatures attained during induction heating. Fine prior austenite grain sizes like the 10 micron sizes observed for the 10V45 Nb after TMP and induction hardening have been shown to increase the fatigue strength of case hardened components. [19]

An understanding of the prior microstructural effects from TMP on the microstructures after final heat treatment is needed to optimize the heat treatment process parameters and the performance of the finished parts.

Acknowledgements

The authors gratefully acknowledge the continued support of the sponsors of the Advanced Steel Processing and Products Research Center, an industry/university cooperative research center at the Colorado School of Mines.

References

[1] Lee, Y. Choi, S., and Hodgson, P.D., "Analytical Model of Pass-by-Pass Strain in Rod (or Bar) Rolling and its Applications to Predictions of Austenite Grain Size," *Materials Science and Engineering*, A336 (2002), pp. 177-189.

[2] Kuziak, R., "Physical Simulation of Thermomechanical Treatment Employing Gleeble 3800 Simulator," *Metal 2006*, Hradek Nad Moravici.

[3] Bianchi, J. H. and Karjalainen, "Modelling of Dynamic and Metadynamic Recrystallization During Bar Rolling of Medium Carbon Bar Steel," *Journal of Materials Processing Technology*, Vol 160 (2005) pp. 267-277.

[4] Kozasu, I., Shimuzu, I. T. and Kubota, H., *Trans. Iron and Steel Institute of Japan*, Vol. 11 (1971), pp. 367.

[5] Maccagno, T.M.,Jonas, J.J., and Hodgson, P.D., "Spreadsheet Modelling of Grain Size Evolution during Rod Rolling," *ISIJ International*, Vol. 36 (1996), No. 6, pp.720-728.

[6] Homsher, C.N. and Van Tyne, C.J., "Empirical Equations for the No-Recrystallization Temperature in Hot Rolled Steel Plates," *Proceedings Materials Science and Technology Conference*, Montreal Quebec, Canada, Oct 27-31, 2013, pp. 1815-1822.

[7] Speer, J. G. and Hansen, S. S. "Austenite Recrystallization and Carbonitride Precipitation in Niobium Microalloyed Steels," *Metall. Trans. A*, vol. 20, (1989), pp. 25–38,.

[8] Vervynckt, S., Verbeken, K., Thibaux, P. and Houbaert Y., "Recrystallization-Precipitation Interaction during Austenite Hot Deformation of a Nb Microalloyed Steel," *Materials Science and Engineering A*, Vol. 528 (2011), pp. 5519-5528.

[9] El Mehtedi, F. et al., "Predicition Models of the Final Properties of Steel Rods Obtained by Thermomechanical Rolling Process," *La Metallurgia Italiana*, n. 3 (2013), pp. 31-37.

[10] Kruse, M., Schwarz, M., Schuck, M., "Kocks Microstructure Simulator (KMS) – New Technical Tool for Process Simulation of Long Products," *La Metallurgia Italiana*, n. 1 (2014), pp. 5-10.

[11] Olasolo, M., Uranga, P., Rodriguez-Ibabe, J.M., and López, B., "Effect of Austenite Microstructure and Cooling Rate on Transformation Characteristics in a low carbon Nb-V Microalloyed Steel," *Materials Science and Engineering*, A 528 (2011), pp. 2559-2569.

[12] Cryderman, R.L. et al. "Effects of Chemical Composition, Heat Treatment, and Microstructure in Splittable Forged Steel Connecting Rods," *SAE Int. J. Mater. Manf.* 8(3):2015, doi:10.4271/2015-01-0522

[13] Whitley, B. M., Easter, C. T., Cryderman, R. L., and Speer, J. G., "Thermomechanical Simulation and Microstructural Analysis of Microalloyed Medium Carbon Bar Steels," *Proceedings of International Conference on Advances in Metallurgy of Long and Forged Products*, AIST, July 12-15, 2015, Vail Colorado, USA.

[14] Barbosa, R., Boratto, F., Yue, S., and Jonas, J.J., "The Influence of Chemical Composition on the Recrystallization Behavior of Microalloyed Steels," *Processing, Microstructure and Properties of HSLA Steels*, (1988), pp. 51-61.

[15] Schaneman, R.A., "The Effects of Prior Microstructure on Spheroidizing Kinetics and Cold Workability in Bar Steels," *M. S. Thesis, Colorado School of Mines*, (2009).

[16] Totten, G.E. and Howes, M.A.H., *Steel Heat Treatment Handbook*, New York, NY: Marcel Dekker, 1997, pp. 540.

[17] Matlock, D.K. "Metallurgy of Induction Hardening of Steel," *ASM Handbook*, Vol. 4C, 2014, pp 45-57.

[18] Goldstein, R, Fluxtrol, Auburn Hills, MI 48326, private communication, July 27, 2015.

[19] Hyde R.S., Matlock, D.K. and Krauss, G "Quench Embrittlement: Intergranular Fracture Due to Cementite and Phosphorous in Quenched Carbon and Alloy Steels", *Proceedings of the 40th Mechanical Working and Steel Processing Conference*, 1998, pp. 921-928.

International Federation of Heat Treatment and Surface Engineering 2016
Proceedings of the 23rd IFHTSE Congress
April 18–21, 2016, Savannah, Georgia, USA

Hot Hydroforging of Lightweight Bimaterial Gears and Hollow Products

Bulent Chavdar
Eaton, Southfield, MI, USA
bulentchavdar@eaton.com

Robert Goldstein
Fluxtrol Inc., Auburn Hills, MI, USA
rcgoldstein@fluxtrol.com

Lynn Ferguson
DANTE Solutions Inc., Cleveland, OH, USA
lynn.ferguson@dante-solutions.com

Abstract

Feasibility of making lightweight powertrain products with hot hydroforging of steel/low density material hybrid billets is explored. A bimaterial billet is designed such that a steel wall encloses a low density core 100%. Furthermore the low density core is selected among the materials that have lower melting or softening temperature than steel such as aluminum and glass. In hot hydroforging the bimaterial billet is heated to 1000-1200 C range similar to the conventional hot forging of steel. However, in hot hydroforging the core is in liquid or viscous state while steel shell is in solid state similar to the conventional hydroforming. During hot hydroforging the viscous/liquid core has negligible resistance to flow thereby providing a uniform hydrostatic pressure inside the steel and enabling a uniform deformation of the solid steel wall.

Steel/aluminum bimetal billets were prepared. Then, the bimetal billets were hot hydroforged in closed dies in one blow. A uniform steel wall thickness was observed all around the forged part upon cross sectioning. However, there was also a large shrinkage void in the aluminum core. The large shrinkage void is formed due to the CTE mismatch between steel and aluminum and the volume increase of aluminum during phase change. The large shrinkage void can be eliminated if aluminum is replaced by glass that has a matching CTE to that of steel. Furthermore, glass doss not have to be fully melted at forging temperatures thereby mitigating the risks of phase change. On the other hand the molten aluminum core can be emptied out of steel shell after forging thereby giving rise to the novel concept of "investment forging". A hollow part with uniform steel shell can be formed for the ultimate weight and cost reductions. For example investment forging of hollow steel valves for engine applications is feasible by hot hydroforging.

Introduction

Bimaterial products can be hot forged from a bimaterial billet where the steel wall encloses the lightweight core fully. A bimaterial forged product can be substantially lighter than a steel forged product. For example a bimaterial composite gear with 50/50 steel/glass composition by volume would reduce the weight of gear by about 35%. Weight reduction is one of the strategies to reduce the energy consumption and CO_2 emission in motor vehicles. Every pound of weight reduction increases the energy efficiency of a midsized car by 0.01% [1]. Furthermore every pound of weight reduction reduces the CO_2 emissions by 6 mg per 100 mile driving distance or by 20 pounds of CO_2 for the life time of a vehicle [2].

A bimaterial billet can be forged in solid state however a better forging quality can be achieved if the core material is viscous thereby providing uniform hydrostatic pressure to steel wall during forging similar to a hydroforming process. However, the similarity only pertains to the hydrostatic pressure developed inside the deforming billet neither to the process temperatures nor to the method of providing the liquid phase. Hence, the forging of a bimaterial billet where the core material is in viscous or liquid state prior to the forging is named as "hydroforging". Furthermore, if the metal outer wall is heated to temperatures greater than its recrystallization temperature while the core is in viscous or liquid state prior to forging then the hydroforging process is named as "hot hydroforging". In other words, hot hydroforging can be defined as hot forging of lightweight products from a hybrid billet of a metal shell and a low melting core. For example, when a billet of a steel shell and glass core is hot hydroforged at temperatures higher than 1000C, the glass core is in a viscous state and the steel is in solid state above the recrystallization temperature.

Hydroforming is a well known process that uses a high pressure hydraulic fluid to press a working material into a die at room temperature [3,4]. While hydroforming is done at room temperatures the hot hydroforging is done at temperatures greater than 1000C enabling deformation of steel into intricate topologies with a streamlined forged microstructure without a fracture. Other differences between the hydroforming and hot hydroforging are that the amount of fluid is constant in hot hydroforging and the fluid may solidify and become an integral part of the product after forging and cooling unlike the case of hydroforming. The hydroforging process described in references [5] and [6] are very similar to the pressure assisted injection forging [7,8] and much different from the hot hydroforging technology described in this article. The lightweight core material will need to have a lower melting or softening temperature than the steel for hot hydroforging. Glass, aluminum and magnesium are such candidate lightweight materials. A literature search did not yield any prior study on the bimaterial billet forging where the core of billet is fully melted prior to the forging except our own paper in Reference [9]. Hence, we continue to discuss the industry first hot hydroforging process in this paper.

Our lightweight bimaterial gear concept includes a lightweight core material that is fully enclosed in a steel shell and formed into the teeth area thereby interlocking with the steel in the interior and contributing to the transfer of torque during the operation of gear. The formation of mechanical interlocking reduces the need for metallurgical bonding between the steel and the lightweight core. The concept gear has all around steel surfaces that engage with the mating gear at the outer diameter and the shaft at the inner diameter. It is envisioned that the steel frame would carry the bulk of torque from one component to the other and also would provide a long fatigue life to the gear. On the other hand the lightweight core would serve as a rigid-elastic continuum to transfer stresses inside the steel thereby providing toughness and reducing risk of plastic deformations to the steel shell under loading.

Behrens and Kosch describe a method of forging bimetal gears where steel does not enclose aluminum 100% and the torque transfer has to go through the steel, and the aluminum in a serial configuration, [11], [12]. Each component has to bear 100% of loads individually and they have to form a strong intermetallic bond. In our concept the loads are shared by the steel and the lightweight core. Furthermore, the torque is transferred from one component to the other mostly by mechanical interlocking. Furthermore, the mismatch of thermal expansion coefficients of steel and aluminum is always a big concern regardless of the forging method. A better material choice with a matching CTE is glass instead of aluminum. Glass and steel gear cannot be forged together by the method of Behrens and Kosch. On the other hand, steel/glass gears can be hot hydroforged together as demonstrated by modeling in this article. Another important distinction of hot hydroforging process is that it enables creating forged hollow products when the core is low melting metal such as aluminum or magnesium. The molten core can be emptied right after the final forging operation by piercing two holes one for draining the fluid and the other is for breaking the vacuum. Forged hollow products combine the benefits of streamlined microstructure with the maximum weight savings.

In Ref. [13] a C15 steel was cladded with 316L stainless steel and forged in solid state. Although the application did not have a lightweighting purpose it stressed the importance of friction condition at the die/billet interface that had a great impact on the final material distribution, forging loads and cracking occurrence. In Ref. [14] two semisolid alloys were mixed at micro level uniformly, then heated the bimaterial mixture between their liquidus and solidus temperature before forging. In the bimaterial hot hydroforging process two materials are not mixed. Hot hydroforging is also fundamentally different from the squeeze casting or liquid forging process where the molten material is poured into the bottom half of die and solidified under pressure [15]. In hot hydroforging the molten material is always contained within the solid steel shell of a billet and never contacts the die.

In this study the feasibility of hot hydroforging was investigated by modeling and also experimentation. Structural finite element analysis was conducted on ANSYS® software to determine the minimum steel wall thickness required to be able to transfer the same order of torque as in the case of all steel gear. Induction heating simulations of bimaterial billets were conducted on ELTA™ (1D) and FLUX™ (2D and 3D) software to establish a heating method with steel aluminum hybrid billets and presented in [9]. The induction heating method will be discussed further on a separate paper in this conference [10]. The stress state, dimensional changes and phase changes that occur during heating and cooling of bi-material billets were simulated using the DANTE® heat treatment software. The hot forging and the hydroforming simulations of steel-aluminum bimetal billets were conducted on the DEFORM® software. The steel/aluminum bimetal billets were prepared. A three piece closed die was designed and made for hot hydroforging experimentation. Hot forging in solid state and hot hydroforging experiments with liquid core were done on bimetal billets. The results were discussed and the future directions were identified.

Enabling Net Shape Gear Forging

Heavy duty manual transmissions require large size spur gears and helical gears made of carburizing steel. Most forged gears are made by two step manufacturing processes, hot forging followed by machining. First a tall cylindrical billet of steel is hot-forged to a flat shape in a series of busting and blocking operations and then a center hole and the teeth are formed by machining. Forty five percent of the starting material is wasted in the machining operations where the most waste is due to the teeth machining. The near net shape forged steel gears with light weight core can enable an 80% reduction in machining, thereby saving 2 to 4 kg waste per an average size gear. The raw material cost of a steel gear with light weight core may be significantly less than that of a gear produced from a solid steel billet because the volumetric cost of the light weight core can be lower than the cost of steel that is replaced. The reduction in waste also has a potential to reduce the carbon footprint of gear manufacturing operations. The objectives of hot hydroforging of gears can be summarized as follows:

- Lightweight gears with net teeth and near net center.
- The hybrid gear has all steel outer structure.
- Investment forging is enabled (molten core can be emptied).
- Press loads are reduced and larger gears can be forged.

- 30% to 50% weight reduction per gear can be achieved.
- Up to 10% weight reduction per transmission is expected.
- 60% to 70% reduction in machining scrap rate due to near net teeth is expected.
- Gear teeth can be through hardened and tempered in one-heat of forging.
- Alternatively, gear teeth can also be induction hardened after forging.

Hot hydroforging concept is an enabler for a near net forging of large gears with the teeth from lightweight bimaterial billets as discussed in [9]. Figure 1 shows the design of a bimaterial billet where the outer structure is made of 100% steel and the inner structure is 100% lightweight core. Figure 2 shows the assembly of the bimaterial billet by fitting the aluminum core in a steel tube and welding two steel endcaps to the steel tube. Figure 3 shows the concept of hot hydroforging a bimaterial gear in a two blow forging operation. In the first blow a flat shape is forged and in the second blow the teeth and the center hole are forged. From safety perspectives all blows are recommended to be done in closed dies. Since the core is in viscous or liquid state, the core can be squeezed out of center in the blocking operations. If the core is in solid state, it is not possible to squeeze out the material from the center. Squeezing the core material out of center and bringing the opposing steel walls in contact at the center will enable the containment of core within steel even after machining the steel center slug out. The resulting all steel walls at the inner diameter of the center hole will enable the assembly of hybrid gear on a steel shaft either by welding or by cutting splines on the steel wall and the steel shaft.

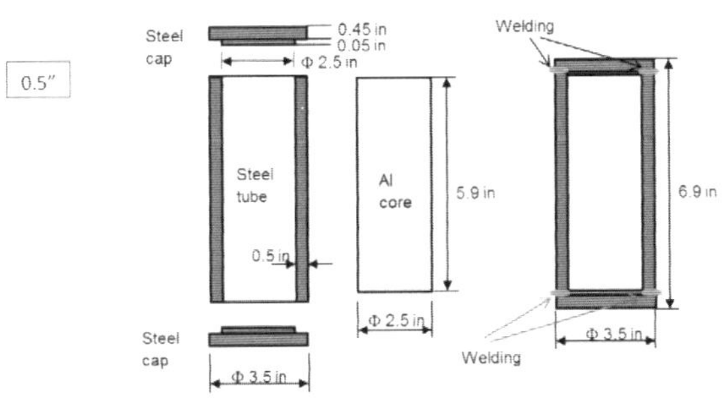

Figure 1: Bimaterial billet. The outer structure is steel, the inner structure is a lightweight material such as aluminum.

Figure 2: Assembly process of steel-aluminum bimetal billet. a) steel tube, Al core and two steel end caps before assembly, b) aluminum core is inserted into steel tube, c) steel end caps are welded to steel tube to complete the assembly.

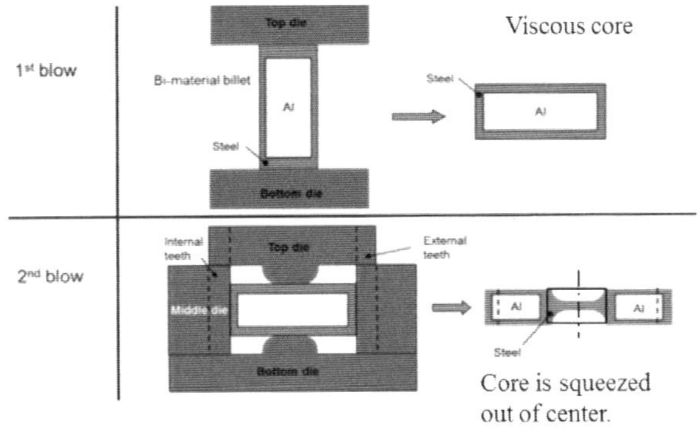

Figure 3: Hot hydroforging concept of net shape gear forging with teeth and the center hole [9].

Structural Finite Element Analysis of Bimaterial Gear

Finite Element Analyses (FEA) were conducted to investigate the load carrying capacity of idealized lightweight gears prior to experimentation. The structural finite element analysis of bimaterial gear has the following objectives: determining the optimum steel wall thickness with alternative core materials, determining the size and load carrying capacity of gear and the deflections of bimaterial gear under load. Lightweight materials such as aluminum and glass have lower elastic modulus than steel does. When the core of a steel gear is replaced by a lightweight material it is expected that the deflections and the bending stresses on the steel casing of a composite tooth will be higher than those on an all steel tooth under the same load. It is clear that the thinner the steel casing is the higher the weight reduction. However, the stresses will also be higher when the steel casing is thinner. Hence, there is an optimum steel wall thickness where the weight reduction is high enough (>25%) to be attractive and the stress increase in the steel casing is either tolerable or manageable with minor design changes (<10%).

A simple spur gear was selected as the basis for structural analysis. The gear has 24 teeth with a coarse pitch of 15.7 mm, outer and inner diameters of 129.0 mm and 68.8 mm and a width of 60.3 mm. Weight reduction, deformation and stress calculations were carried out for three different cases of steel wall thicknesses with an aluminum core. The weight reductions are 30%, 35% and 40% with the steel wall thicknesses of 3.5 mm, 2.5 mm, and 1.5 mm respectively. In these calculations it is assumed that the teeth are all steel and the outer diameter of aluminum core has a uniform distance of 3.5 mm, 2.5 mm, or 1.5 mm from the roots of teeth, from the outer side walls and from the ID of the gear for each thickness case as shown in Figure 4. The interface between the steel and the aluminum is assumed to be bonded. Although it is not considered in the current calculations, if the lightweight core had penetrated deep into the teeth area then additional 9% weight reduction could have been achieved for the case of 1.5 mm uniform steel wall thickness.

Figure 4. The cross section of spur gear shows the idealized thin steel casing in green and the lightweight core in gray.

Finite element analysis was conducted to compare the stresses in the lightweight gears and the all steel gears. The 3D model of gear is created on Pro/Engineer® software and transferred to ANSYS® software. The gear is loaded on the pitch line of 119.8 mm diameter with a point load of 24181 N normal to the tooth surface as shown in Figure 5. The tangential component of 22638 N creates 1356 Nm torque at the gear center and the radial component of 8500 N acts towards the gear center. For the boundary conditions, a fixed cylindrical support was applied on the surface of hole as shown in Figure 6. The whole gear model was meshed with Solid186 elements and analyzed. The number of elements is determined such that the mesh error was less than 5%. The maximum bending stress at the root of tooth is explored by smart meshing. Weight reduction and stress predictions are presented in Table 1 as function of steel wall thickness.

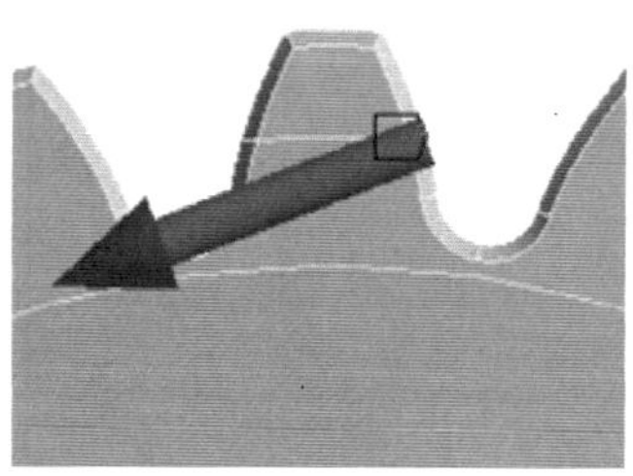

Figure 5. Static point load is applied normal to the surface at the pitch line.

Figure 6. Boundary condition: Fixed cylindrical support at the ID.

The maximum bending stress occurs at the root of a tooth. The von-Mises stress is 128 MPa for all steel gear. The von-Mises stress is 152 MPa, 158 MPa and 172 MPa for 3.5 mm, 2.5 mm and 1.5 mm steel wall thicknesses. Although the von Mises stress is higher with the bimetal gears than with all steel gear all of the calculated values are lower than yield and fatigue strengths of a typical structural steel. The ratio of percentage of stress increase to percentage of weight reduction is defined as the weight reduction penalty factor. The lowest penalty factor is provided by the 3.5 mm steel wall thickness (0.63). A. 2.5 mm steel wall thickness can also be considered as an aggressive weight reduction target with 0.66 penalty factor. However, 1.5 mm steel wall thickness should not be considered due to the increased penalty factor (0.86) as shown graphically in Figure 7.

Table-1

Steel wall Thickness (mm)	Weight reduction (%)	von-Mises stress (MPa)	Max Prin (MPa)	Min Prin (MPa)
All steel		128	125	-141
1.5	40	172	165	-193
2.5	35	158	149	-171
3.5	30	152	139	-165

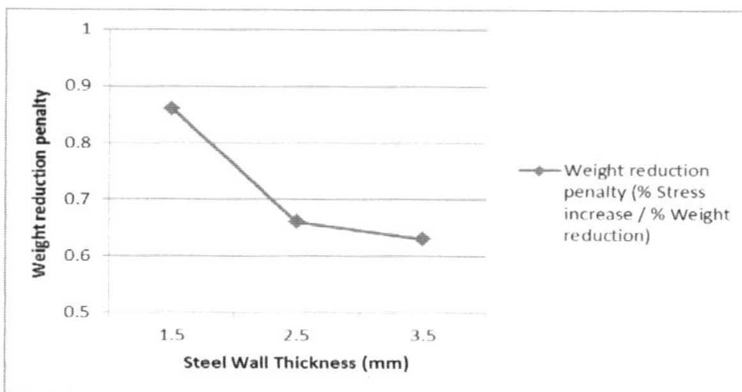

Figure 7. Weight reduction penalty increases with the reduced steel wall thickness

First Hot Hydroforging Experiments of Steel-Aluminum Billet

The simulations and experimentation with the hot hydroforging of steel-aluminum bimetal billets were discussed in detail in [9]. The learnings are summarized here. Simulations showed that if the bimetal billets of steel and aluminum are forged in solid state then the steel wall thickness does not remain uniform during forging. Thinning, thickening, folding and cracking happen all at the same time at various locations on the deforming steel wall. Hence, it was necessary to have the core material melted prior to forging to achieve much more uniform steel wall thickness than the one attainable by solid state forging. The liquid phase inside steel provides a uniform hydrostatic pressure and no resistance to flow that enables the deformation of steel into intricate shapes with a uniform wall thickness as shown by hydroforming simulations in [9]. The uniformity of steel wall thickness is quantified with "non-uniformity index" that is the thickness ratio of the thickest wall to the thinnest wall in a forged bimetal product. A non-uniformity index of 1 means a perfectly uniform wall. It was shown by modeling that the hot hydroforging operation improves the steel wall non-uniformity index significantly as compared to the solid state forging by means of multi-blow indexed forging [9].

Table 2. Design of experiments

Factors	Levels	
Wall thickness	0.25"	0.5"
Steel tube type	Seamless	With seam
Interference fit	0.003"	0.005"
Forging technique	Hot (solid state)	Hot hydroforging

The predictions by modeling were verified by hot hydroforging experimentation with steel-aluminum bimetal billet. The end cap weld zones are identified as the high risk areas for structural failure prior to the experimentation. A closed die design was conceived as shown in Figure 8 in that the end zones of billets are confined in die pockets that prevent the deformation of heat affected zones during busting forging. In the experimentation phase the effects of initial steel wall thickness, the steel tube type, the interference fit as well as the forging technique on the forge-ability of bimetal billets were investigated. Table 2 shows the design of experiments and the levels of factors.

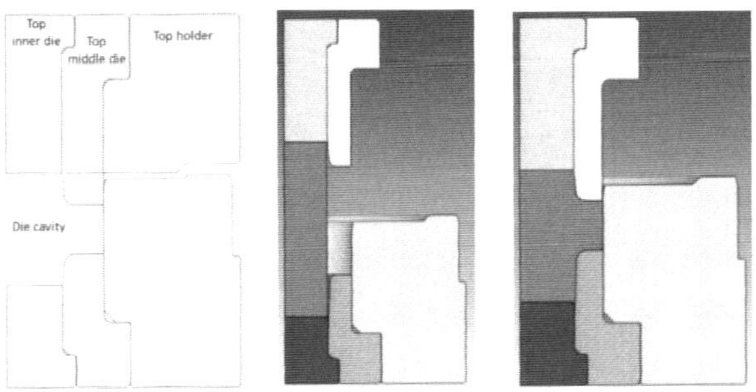

Figure 8: Left: Closed die design for busting. The pockets in the top and the bottom dies keeps the weld zone of end caps under compression during forging. Middle and right: Forging simulations.

The most significant factor affecting the forge quality was the forging technique. Wall thickness and the steel tube type had no effect on the forge quality. Interference fit affected the induction heating phase in that the higher the interference fit is higher the chance of billet cracking during the heating phase prior to forging. Hence, it was concluded that interference fit is not needed in the bimetal billets.

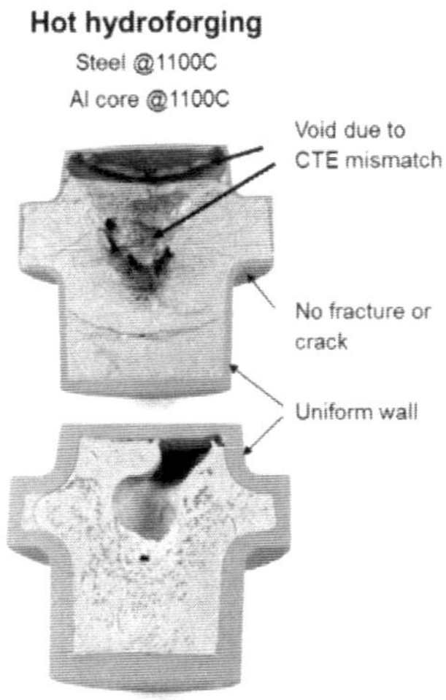

Figure 9. Hot hydroforging experiments of bimetal billets. The steel wall thicknesses of top and bottom billets are 0.25" and 0.5" respectively. The hot hydroforged billets have uniform steel wall thickness without folds or cracks[9].

The steel wall thickness was much more uniform with hot hydroforging (Fig. 9) than the case of solid state forging (Fig. 10) overall. Both the billets with 0.25" and 0.5" steel wall thickness were forged successfully in that the steel wall thickness was uniform all around after the hot hydroforging as compared to solid state forging. While solid state forging exhibited folds, cracks and non-uniform deformation of steel wall, the hot hydroforging did not exhibit any folds or cracks and the steel wall thickness uniformity was much better. The steel wall thickness was thinned around the corner radius of the middle die but did not fracture with hot hydroforging as it did in solid state forging.

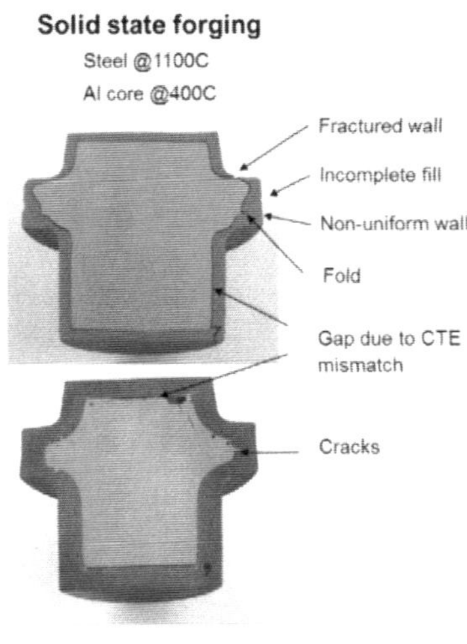

Figure 10. Busting forging experiments of bimetal billets in solid state. The steel wall thicknesses of top and bottom billets are 0.25" and 0.5" respectively. The billets forged in solid state have non uniform steel wall prone to folds and cracking [9].

Figure 11. The macro and micro structure of the weld zone between the steel end cap and the steel tube before (the top pictures) and after (the bottom pictures) the hot hydroforging show that the weld zone healed after hot hydroforging [9].

Large shrinkage voids formed in the aluminum core after hot hydroforging due to the CTE mismatch. Aluminum shrinks much more than steel does during cooling. Furthermore, the volume of aluminum contracts 6% when its phase changes from liquid to solid. These two effects contribute to the voids in the aluminum core. Since the voids in the core are not desirable a different material than aluminum needs to be selected as the core. Experiments with aluminum core were useful to demonstrate the steel wall thickness uniformity in hot hydroforging.

A significant healing of the microstructure of the weld zone was observed after the hot hydroforging. Figure 11 compares the microstructure of the weld zone of the electron beam welded steel end cap and the steel tube before and after the hot hydroforging. The dendritic directional grains of the weld bead and the coarse-grained heat affected zone transformed into a uniform normalized microstructure upon hot hydroforging [9].

Modeling Results of Heating and Cooling Steel-Glass Billet

After demonstrating that a uniform steel wall thickness with uniform microstructure is achieved by hot hydroforging, our attention shifted to eliminating the large shrinkage void in the core. The large void in the core is caused by the mismatch of CTEs of steel (10 ppm) and aluminum (20 ppm) and 6% volume change in aluminum during the phase change. One idea to eliminate the void is to use a low density core material with a CTE that is similar to that of steel. Conventional soda lime glass was identified as a low density (2.5 gcc) core material with matching CTE (10 ppm). Furthermore, glass having amorphous structure neither have abrupt phase change nor abrupt volume expansion unlike aluminum upon heating and cooling. Induction heating and air cooling cycle of steel-glass bimaterial billet were discussed in [9]. It was shown that due to the faster rise of temperature in steel a gap forms between the steel and glass during heating, however, the gap closes upon cooling. An additional advantage of steel-glass billet could be the possibility of furnace heating or conventional low frequency induction heating due to the matching CTEs of steel and glass. In contrast it is necessary to use an expensive high frequency induction heating system to heat steel-aluminum bimetal billets to avoid billet cracking caused by the CTE mismatch as discussed in [9]. The feasibility of furnace heating and air cooling of steel-glass billet is investigated by modeling in the DANTE® heat treatment software in this paper. The model considers the three dimensional displacement of constituents as function of temperature and the allotropic phase changes in steel.

Steel-glass billet with 0.5" steel wall thickness is selected to model the stress-deformation behavior during furnace heating followed by air cooling. The initial microstructure of steel is assumed to be 50% ferrite and 50% pearlite. The objectives of modeling were to investigate the likelihood of billet cracking during furnace heating prior to forging and to predict any gap or void formation between steel and glass during air cooling after forging. It is assumed that glass completely fills steel cylinder at initial room temperature condition. Four sided elements are used for the finite element modeling of both steel and glass. Axisymmetry and vertical symmetry are assumed in order to reduce the computation time resulting in a quarter model. Steel cylinder is modeled with 25,120 elements and 25,563 nodes while glass core has 20,000 elements and 20,331 nodes. Contact surfaces are defined between glass and steel surfaces. Time temperature schedule is made of 1 hour heating at 1050C and 1 hour air cooling to 20C.

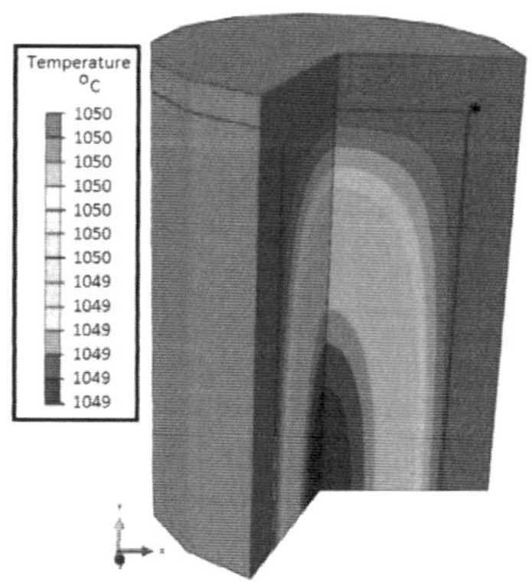

Figure 12. Temperature distribution in the steel-glass billet at the end of 1 hour heating period.

Uniform temperature is achieved in the bimaterial billet at the end of 1 hour heating as shown in Figure 12. The predicted radial and axial displacements are shown in Figure 13 for this cylindrical shape. A radial gap ranging from 1.8 μm to 32.5 μm is predicted, and an axial gap of about 2 μm is predicted. The time history of gap formation in radial direction at mid-height of the cylinder and at 30 mm from the top of glass are shown in Figure 14. The final radial gap is zero at mid-height of cylinder and 32.5 μm at 30 mm from the top of glass. These gap prediction assume no bonding between the steel and glass.

A forged geometry should be investigated as it should behave differently than this simple cylinder.

The steel node 20251 that is marked with an asterisk on Fig. 12 experiences 500 MPa tensile stress due to the volumetric expansion of the steel as the austenite transforms to ferrite and pearlite as shown in Figure 15. Glass may need to be chamfered at the corners in order not to resist the expansion of steel thereby reducing the axial stress. Otherwise, such high stress may compromise the integrity of bimaterial billet prior to forging.

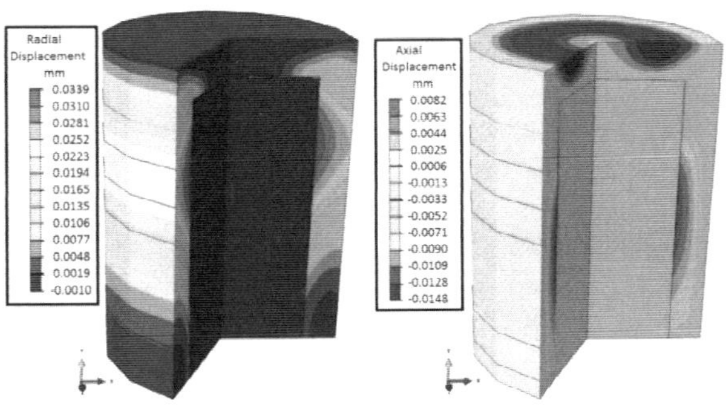

Figure 13. Final radial and axial displacement contours after heating and air cooling.

Final stress states after heating and air cooling are shown in Figure 16. The residual stresses in radial direction are near zero while there is compressive residual stress less than 10 MPa in the axial direction in glass. The compressive residual stress is beneficial for the fracture toughness of glass. The residual stresses are compressive on the inside and tensile on the outside of steel. All residual stresses, radial, axial and hoop are less than 100 MPa in magnitude therefore posing no threat of fracture in steel.

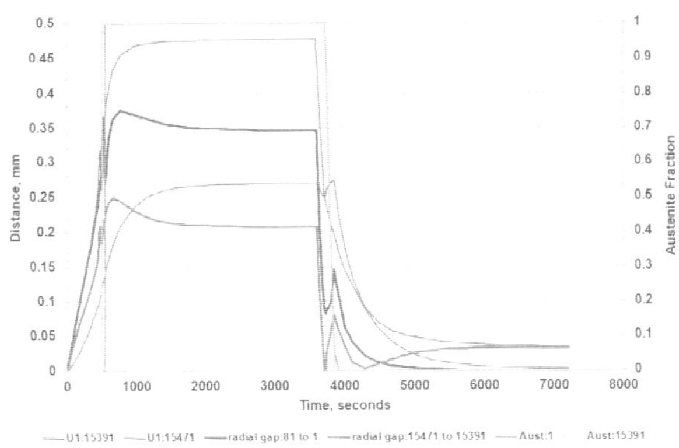

Figure 14.The red curve shows the radial gap between steel and glass at mid-height of cylinder and the green curve shows the radial gap at 30 mm from the top of glass. While there is no gap at mid-height, the maximum gap of 32.5 μm occurs at 30 mm from the top of glass.

Showing the feasibility of furnace heating of bimaterial steel-glass billets is significant in that the bimaterial billets can be heated and forged by conventional methods without requiring new investment in high frequency induction heating.

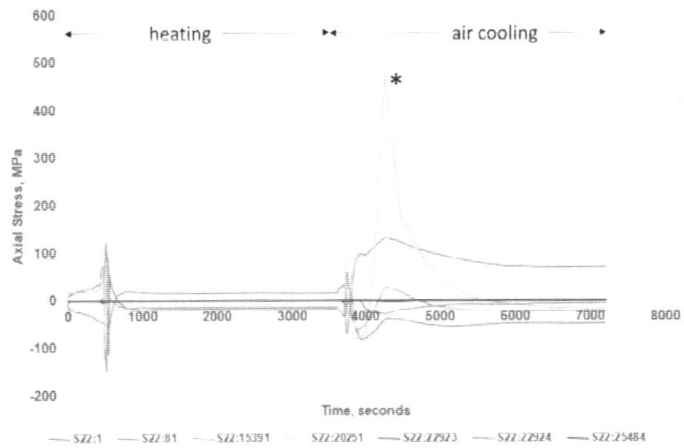

Figure 15. Axial stress histories on various locations on steel cylinder. The highest axial stress occurs at the inner corner of steel end cap and the steel cylinder as marked with an asterisk on Figure 12.

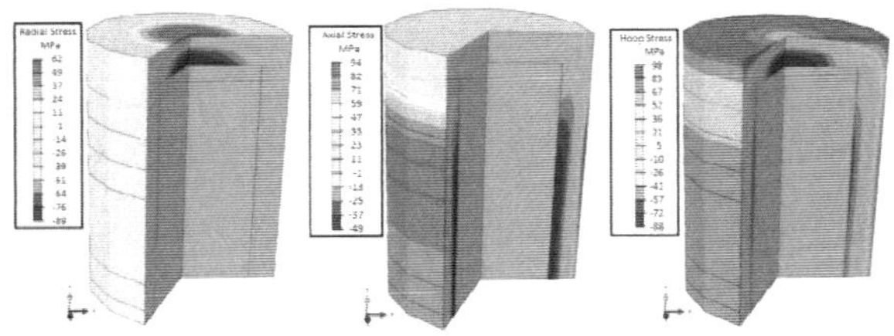

Figure 16. Final stress states after heating and cooling. Left: radial stress, middle: axial stress, right: hoop stress.

Most gears need to be heat treated after forging in order to increase the strength, toughness and wear resistance. Since hot hydroforging of bimaterial gears can enable near net shape gear forging, the gear can be quenched immediately and cooled in air or furnace to get a through hardened steel wall with a relatively soft and elastic glass core. The heat capacity of glass core should be able to temper the steel after quenching. When it is carefully planned such heat treatment method can replace the expensive carburizing and nitriding heat treatments of gears that require additional steps.

Discussions

Lightweighting is an energy saving strategy for the transportation industry. Transmission and driveline components account for 5 -10% weight of a typical class 7/8 tractor. The gears account for 20% weight of a heavy duty manual transmission. Currently 100% of truck gears are made of steel. If the gears are forged from a bimaterial billet such as 50/50 steel/aluminum by volume than 33% weight reduction in gears is possible. This translates into 0.5% reduction in the weight of a truck and 0.2% improvement in the fuel efficiency.

The torque capacity of a gear can be increased by increasing the size of gear and also by forging instead of by machining. Small gears can be forged since their press loads are manageable. However, most big gears are made by machining instead of forging due to big forging press requirements. Hot hydroforging enables forging of large gears on the existing presses due to the reduced press loads.

Feasibility of making lightweight gears with hot hydroforging of steel/low density material hybrid billets is explored for powertrain applications. The functional requirements of gears are determined for transmission applications. An integrated computational engineering approach is followed to determine the theoretical weight reduction targets that meet functional requirements. Design and testing methods are developed to mitigate the risks. Our bimaterial gear concept has a steel wall on the outside and a lightweight core on the inside. The steel wall will carry most of the torque while the lightweight core serves as a rigid-elestic space holder and reduces the weight of gear. Two material options are identified for the low density, low melting core: aluminum and glass. Both materials are relatively low cost and readily available. Aluminum is a lightweight metal and second to the steel in terms of the availability and the cost. However there are many risks and challenges when aluminum and steel are used together in a structural product such as a gear. The biggest technical challenge is due to the difference in the coefficients of thermal expansion (CTE) of steel and aluminum. The CTE of aluminum is twice the CTE of steel. Furthermore, the volume of aluminum increases upon melting. These two factors contribute to the formation of shrinkage voids in the core upon cooling after a hot hydroforging. After conducting intensive digital prototyping efforts, the experimental studies were carried out to prove the concepts.

For the experimental phase, steel/aluminum bi-metal billets were prepared. Then, the bimetal billets were hot hydroforged into a so called "pancake" shape in closed dies in one blow. In hot hydroforging experiments the bimaterial billet was heated to 1000-1200 C range with a high frequency induction heater as guided by the induction heating modeling. Aluminum is allowed to melt inside the steel without fracturing the steel case by a cyclic on/off control of the power of heater. During hot hydroforging the liquid core flows effortlessly and provides a uniform hydrostatic pressure inside the steel thereby enabling a uniform deformation of the solid steel shell. Upon cross sectioning a hot hydroforged part, a uniform steel wall thickness was observed all around the part. However, there were also large voids in the aluminum core as expected. Using glass as the core material instead of aluminum will solve the void issue since glass has a matching CTE to steel. The benefits of matching thermal expansion behavior to eliminate shrinkage void had been demonstrated in [9] by deformation modelling of high frequency induction heating and air cooling of steel-glass bimaterial billets. In the present paper, the deformation studies are extended to the furnace heating and air cooling of steel-glass bimaterial billets. The results demonstrate that the bimaterial billets can be furnace heated before forging and air cooled after forging safely. A very small gap, less than 33 microns, is predicted between steel and glass at certain locations, however, these simulations assumed no adhesion between the steel and glass. In reality, glass and steel can form very strong adhesion and as result may not have any gap.

Heating the steel-glass billets to forging temperatures will be safer, simpler and faster than heating the steel-aluminum billets. It can be also envisioned that hot hydroforged steel-glass bimaterial gears can be quenched right after forging and be tempered by the latent heat of glass during the air cooling. The novel heat treatment procedure can save time and money by replacing the time consuming and costly carburizing and nitriding heat treatment procedures. Hence, the next series of hot hydroforging experimentation can be planned around the steel-glass material combination. Glass is 10% lighter than aluminum thereby providing further weight reduction benefits. These options need to be explored further.

On the other hand, aluminum core enables a novel concept of "investment forging" where the molten aluminum core can be emptied out of the part right after forging analogous to investment casting. A hollow part with uniform steel wall can be formed for ultimate weight savings. For example, investment forging of hollow steel valves for engine applications is feasible by hot hydroforging. In some cases the core can be emptied out later after creating evacuation holes at room temperature and reheating the part if it is difficult to empty the core right after forging. The hollow core can be filled with sodium to increase the heat transfer in exhaust valves as already been practiced today. Any low melting substance that does not go into gas phase at forging temperatures is a candidate as the core in investment forging. Core materials such as aluminum or magnesium that are extracted from the product during investment forging can be recycled.

The experimentation with hot hydroforging addressed several risk factors that had been identified and put into test. One of the concerns was if the degree of interference fit would have any impact on the quality of bimetal billet forging. The experiments indicated that an interference fit is not needed. Another concern was about the vulnerability of the weld seams to failure during forging. The experiments has shown that it is reasonable to expect that the weld zones of billet will heal and have no adverse effect on the performance of the end product upon hot hydroforging [9]. These guidelines lead to significant cost savings for the preparation of billets in a mass production scenario.

Conclusions

The feasibility of hot hydroforging technology to form lightweight structural components from steel-glass bimaterial billets are advanced further in this paper. It is shown by the furnace heating and air cooling simulations of steel-glass bimaterial billet that the huge shrinkage void that forms during the hot hydroforging of steel-aluminum bimetal billets will be mitigated by steel-glass bimaterial billets. Gross reduction or total elimination of shrinkage void is attributable to the matching CTEs of steel and glass. Furthermore, glass neither exhibit abrupt phase change nor abrupt volume increase in the range of forging temperatures. Volume increase upon melting is a huge problem with the aluminum core. Hence it is feasible to make net shape lightweight structural components by hot hydroforging of steel-glass billets. Such components will have 100% steel outer wall and 100% glass core. The continuous steel outer structure will bear the load while glass core will provide toughness due to its rigidity and lower elastic modulus than that of steel.

It is possible to make low cost hollow products by means of investment forging that is a special case of hot hydroforging technology. In investment forging the low melting core is emptied after hot hydroforging. Hollow products of steel, titanium, copper alloys etc. can be created by investment forging. The remaining metal shell will have the strength benefits of a forged microstructure. For example investment forging of hollow steel valves for engine applications is feasible by hot hydroforging and needs to be explored further.

Acknowledgments

This research was funded and conducted at the Corporate Research and Technology Group at Eaton in Southfield, MI. Billet induction heating simulations were provided by Fluxtrol Inc. in Auburn Hills, MI. The high frequency induction power supply was provided by Interpower Induction Services Inc. in Fraser, MI. Forging services were provided by Walker Forge Inc. in Clintonville, WI. Further billet heating, cooling and deformation modeling were provided by DANTE Solutions Inc.in Cleveland, OH. The authors would like to thank following team members for their contributions: Dr. Xi Yang, Mike Killian from Eaton, Jake Butkovich, Dr. Peter Fritz from Eaton.

References

[1] R. Heuss, N. Mueller, W. v. Sintern, A. Starke, A. Tschiesner, "Lightweight, heavy impact", Advanced Industries, Feb. 2012, McKinsey & Company.

[2] "Impact of Vehicle Weight Reduction on Fuel Economy for Various Vehicle Architectures", Research Report Conducted by Ricardo Inc. for The Aluminum Association, 2008-04.

[3] M. Ahmetoglu, T. Altan, "Tube Hydroforming: State-of-the-Art and Future Trends", Journal of Materials Processing Technology, Vol. 98, pp 25-33, 2000.

[4] M. Koç, "Advances in Tube Hydroforming - An Enabling Technology for Low-Mass Vehicle Manufacturing - Material, Lubrication, Loading, Simulation Issues, and Alternatives", Tsinghua Science and Technology, ISSN 1007-0214 04/18, Vol. 9, No 5, pp527-545, Oct. 2004.

[5] M. Roeper, S. Reinsch, "HYDROFORGING: A New Manufacturing Technology for Forged Lightweight Products of Aluminum", Proceedings of IMECE2005, ASME International Mechanical Engineering Congress and Exposition, November 5-11, 2005, Orlando, Florida USA, IMECE2005-80424.

[6] B. Alzahrani, G. Ngaile, "Preliminary Investigation of the Process Capabilities of Hydroforging", Materials, Vol. 9, No. 40; doi:10.3390/ma9010040, 2016.

[7] R. Balendra, Y. Qin, "Pressured-Assisted Injection Forging of Thick-Walled Tubes", International Journal of Machine Tools and Manufacture, Vol. 35, No.11, pp 1481-1492, 1995.

[8] Y. Ma, Y. Qin, R. Balendra, "Forming of Hollow Gear-Shafts with Pressure-Assisted Injection Forging (PAIF)", Journal of Materials Processing Technology, Vol. 167, No. 2, pp. 294-301, 2005.

[9] B. Chavdar, R. Goldstein, X. Yang, J. Butkovich, L. Ferguson, "Hot Hydroforging for Lightweighting", IDE 2015; Bremen, Germany; Sep. 23-25, Vol. 5, pp. 117-128

[10] R. Goldstein, B. Chavdar, L. Ferguson, "Modeling of the Heating Sequences of Lightweight Steel/Aluminum Bimaterial Billets for Hot Forging and Hot Hydroforging", 23rd IFHTSE Congress; Savannah, GA - USA; April 18-22, 2016.

[11] B. –A. Behrens, K. –G. Kosch, "Production of Strong Steel-Aluminum Composites by Formation of Intermetallic Phases in Compound Forging, Steel Research International", Vol. 82, No. 11, 2011, pp. 1261–1265.

[12] B. –A. Behrens, K. G. Kosch, "Development of the Heating and Forming Strategy in Compound Forging of Hybrid Steel-Aluminum Parts", Materialwissenschaft und Werkstofftechnik Vol. 42, No. 11, (2011), pp. 973-978.

[13] Jincai W, Langlois, L., Rafiq, M., Bigot, R., Hao, L. "Experimental &Numerical Study of the Hot Upsetting of Weld Cladded Billets", Key Engineering Materials, Vol. 554-557, No. 1, (2013), pp. 287-299.

[14] Yang, Z., Dong, J., Zhou, L., Shen, L., Zhang, W. "A Study of the method of Manufacturing Bimaterial Composite Parts Through Semisolid Metal Processing", Metallurgical and Materials Transactions A: Physical Metallurgy and Materials Science, Vol. 42, No. 6, (2011), pp. 1709-1716.

[15] ASM Handbook, Vol. 15-Casting, (2010), p. 1153..

International Federation of Heat Treatment and Surface Engineering 2016
Proceedings of the 23rd IFHTSE Congress
April 18–21, 2016, Savannah, Georgia, USA

Effect of processing route on microstructure and impact toughness of duplex stainless steel

Amit Powar[1*], Amol Gujar[1], Niketan Manthani[1], Vinayak Pawar[1], Rajkumar Singh[1]
[1]Kalyani Centre for Technology and Innovation,
Bharat Forge Ltd., Pune, Maharashtra, India.
E-mail: amit.powar@bharatforge.com

Abstract

The main objective of the present work is to study the effect of change in processing route and resulting changes in microstructure and mechanical properties, mainly impact toughness, of F51 duplex stainless steel material. Upsetting of F51 cylindrical bars were done in the temperature range of 1220-1210°C. After the upsetting operation some bars were water quenched and some were air cooled. The samples were further solution treated at 1050°C for 30 min. Optical microscopy, scanning electron microscopy, EBSD, quantitative metallography and impact testing were used to study the microstructure and mechanical properties during the different stages of processing route. It has been observed that microstructures of both routes have a significant difference and air cooled samples after the forging showed good impact toughness than the water quenched sample. XRD was done and revealed phases of austenite and ferrite without any intermetallic phase. A detailed analysis was done and the results suggest that, a small change in the processing route makes an enormous difference in the final achieved properties.

Abbreviation

DSS	Duplex Stainless steel
α	Ferrite
γ	Austenite
σ	Sigma Phase
χ	Chi Phase
π	The nitride π phase
R	Intermetallic compound enriched in molybdenum
M23C6	Carbides
Cr2N	Chromium Nitride
DRV	Dynamic recovery
DRX	Dynamic recrystallization
SFE	Stacking fault energy
LAGBs	Low-angle grain boundaries
SEM	Scanning electron microscope
EBSD	Electron Back Scatter Diffraction
XRD	X-ray Diffractometry

Introduction

Duplex stainless steel (DSS) is a combination of 50% ferrite (α) and 50% austenite (γ). This result is achieved by the controlled addition of alloying elements, each offering specific attributes in respect of strength and ability to resist different environments. The material combines good mechanical strength and ductility with moderate to good corrosion resistance in a variety of environments [1]. The yield strength of these steels is more than twice that of the single-phase stainless steels either of ferritic or austenitic stainless steels. They also have superior toughness and ductility when compared to the ferritic and martensitic types, in addition to superior intergranular and stress corrosion resistance in comparison to the austenitic type. This favorable combination of properties makes this class of stainless steels widely employed in oil and gas, petrochemical, pulp and paper, and pollution control industries. They are frequently used in aqueous solutions containing chlorides, where they have substituted with advantage (major reductions in weight and welding time) the austenitic stainless steels that are more susceptible to stress and pitting corrosion [2].

DSS are prone to the formation of various intermetallic phases (σ, χ, π, R), carbides (M23C6) and nitrides (Cr2N), when exposed to temperatures lower than 1000°C [3]. The sigma phase has particular importance among them, as its small amount causes a considerable reduction in plasticity and impairment in corrosion resistance [4]. Thus, the proper control of the thermo-mechanical processing and subsequent ageing treatment are obligatory, not only for avoiding harmful precipitates, but also for achieving the desired mechanical properties [5]. DSS can be processed by different routes, i.e., casting, forging, extrusion or rolling. These forming operations are usually performed at high temperatures. However, the difference in the mechanical responses of austenite and ferrite under hot working conditions, can lead to the formation of edge cracks or an inappropriate surface finish. It is thus difficult to establish the hot working regime [6].

So far, many studies have focused on the heat treatment process, mechanical properties, and fracture mechanism as well as the fatigue life of 2205 DSS [6-9], but the microstructural changes during the processing requires further systematic investigation. The aim of the present work is to study the effect of change in processing route and resulting changes in microstructure and mechanical properties, mainly impact toughness, of F51 duplex stainless steel material.

Experimental method

The steel studied is F51 duplex stainless steel and the chemical composition is given in Table 1.

Table 1: Chemical composition of F51 (BFL)

Element	C	Mn	P	S	Si	Cr	Ni	Mo	N
Composition	0.019	1.36	0.024	0.0005	0.46	22.19	5.66	3.19	0.194

Cylindrical specimens of 30 mm in diameter and 60 mm in height were machined from as received bars for the upsetting test. A high temperature upsetting test was carried out over the temperature range of 1220-1210°C on a 1-ton hammer. Approximately 70% of reduction ratio was given to all the samples. In order to investigate the microstructural evolution during the processing, two different manufacturing methods were used. In method-A, bars were quenched in water immediately after the upsetting operation. And in method-B, bars were air cooled after the upsetting operation. All the samples were solution treated at 1050°C for 30 min followed by water quenching. After the heat treatment process, metallographic samples were prepared and etched with Beraha reagent (85mL water, 15mL HCL, 1g $K_2S_2O_5$). Carl Zeiss optical microscope and Zeiss (EVO 18) scanning electron microscope (SEM) were used for the microstructural analysis. EBSD scans were also carried out for the identification of phases present and knowing their volume fractions. EBSD scans were conducted in the area of 500* 300 µm at 500X magnification with 0.6 step size. Impact test samples of the 2mm V-notch were extracted from heat treated samples and specimens were made according to ASTM E23 standard. The impact test was performed on a Zwick testing machine at -50°C temperature. X-ray diffractometry was also carried out for the identification of any intermetallic phases present in the material.

Results and discussion

Microstructure evolution during processing:

The microstructures of F51 DSS before hot deformation are shown in fig. 1. The etchant used Beraha reagent, colors the ferrite in the microstructure with blue and then remaining austenite will be in white. It can be clearly seen in fig. 1, that the two phase microstructure consists γ-austenite islands in the α-ferrite matrix.

The microstructure in the processing stages, by method-A and method-B are shown in fig. 2 and fig. 3 respectively. The austenite islands are slightly elongated due to the upsetting operation. The presence of a BCC ferrite phase, which has numerous slip systems and high stacking fault energy (SFE) in the vicinity of an FCC austenite phase with less slip systems and low SFE strongly affects the recrystallization behavior. The kinetics of recrystallization of austenite in the duplex structure is much slower than in single phase austenitic steel. The dominant dynamic restoration mechanism for ferrite and austenite is dynamic recovery (DRV) and dynamic recrystallization (DRX) respectively [10-13]. The coexistence of hard austenite and soft ferrite at high temperatures is found to result in a strain partitioning at the early stages of deformation, when the strain is mostly accommodated by the ferrite phase. At higher strains, the load is transferred from the ferrite to the austenite leading to an increment of dislocation density in the latter till triggering of DRX.

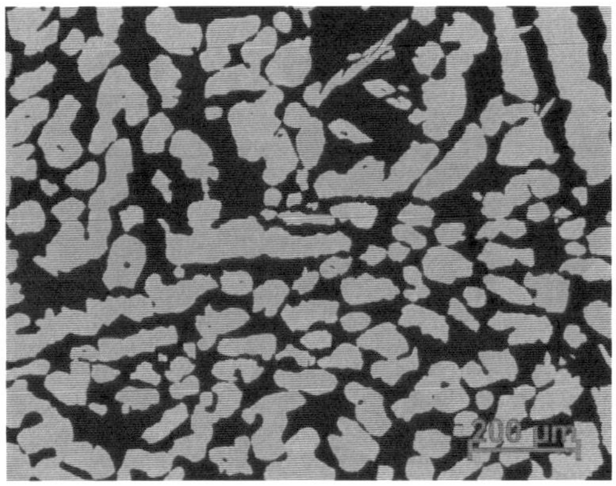

Figure 1: Microstructure of as-received F51 DSS. (Optical, 50X magnification)

It has been observed that grain-boundary serration and bulging are the most important DRX mechanism in duplex stainless steel at the 1220-1210°C working temperature. These are the sites for new DRX grains on the serrated γ/γ boundaries. The same phenomenon can be seen in fig. 2 and fig. 3. Only the notable difference is, in fig. 2, the new DRX grains on the serrated γ/γ boundaries are less as compared to fig. 3. This is because of, water quenched samples haven't had sufficient time to recrystallize the new grains. But in case of air cooled samples, it had got sufficient time to recrystallize the new grains. This will provide homogeneously distributed nucleating sites for the subsequent formation of austenite, which led to a refinement of the austenite grain size by an order of magnitude. It has been also found that, the sites formed after the hot deformation has grown further in the solution treatment process and changed the distribution of austenite completely. The presence of various intermetallic phases (σ, χ, π, R), carbides ($M_{23}C_6$) and nitrides (Cr_2N) were not observed in both the methods as sample sizes are small and cooling rates are higher, which skips the required cooling rates for the formation of intermetallic phases.

The volume fraction of each phase changed during hot deformation by hot upsetting. The phase distributions after the solution treatment from the EBSD are shown in fig. 4(a) and fig. 5(a). The percentages of phases in various stages of processing are shown in the Table 2. The amount of ferrite increased after the hot upsetting at high temperature. Above 1050°C, the amount of γ phase decreases considerably with the deformation temperature, which is attributed to the γ→δ phase transformation.

Table 2: Volume fraction of phases in different stages of processing

Stages		α (%)	γ (%)
As received material		47	53
After upsetting	Air Cool	53	45
	Water Quench	55	45
After Solution treatment	Air Cool	56	44
	Water Quench	53	47

The microstructure evolution during the hot upsetting of the investigated duplex microstructure started with the formation of low-angle grain boundaries (LAGBs) in ferrite without any major influence on austenite. This can be seen in the EBSD images as shown in the fig. 5(b) and fig. 4(b), where low angle grain boundaries marked in the white color. These LABGs are the result of DRV as the dominant restoration mechanism in ferrite. These LABGs evolve and lead to a gradual buildup of higher misorientation between the neighboring subgrains. The increase in ferrite subgrain size appears to be proportional to the increase in the average misorientation of the sub-boundaries in that phase. In contrast to ferrite, some discrete grain boundaries were observed in austenite. These actual grains were mostly formed on the serrated initial grain boundaries indicating DRX. The degree of recrystallization becomes larger with an increase in upsetting temperature.

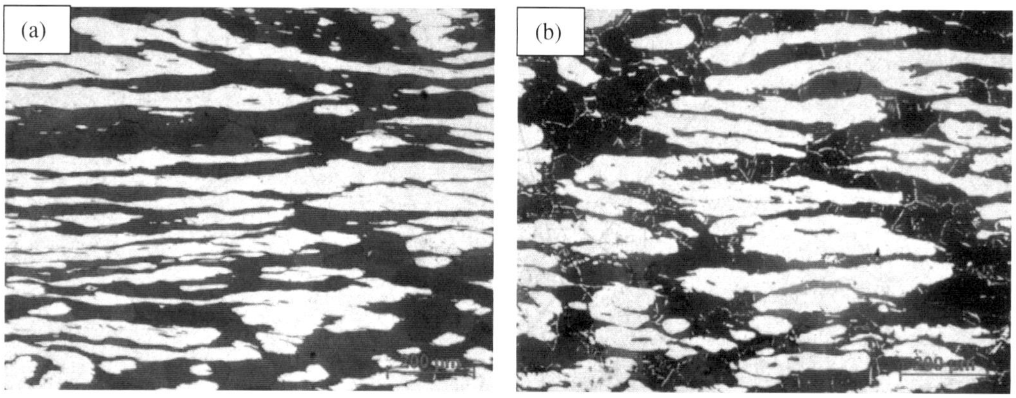

Figure 2: Microstructural evolution of F51 in method-A. (a) after upsetting (b) after solution treatment (Optical, 50 X magnification)

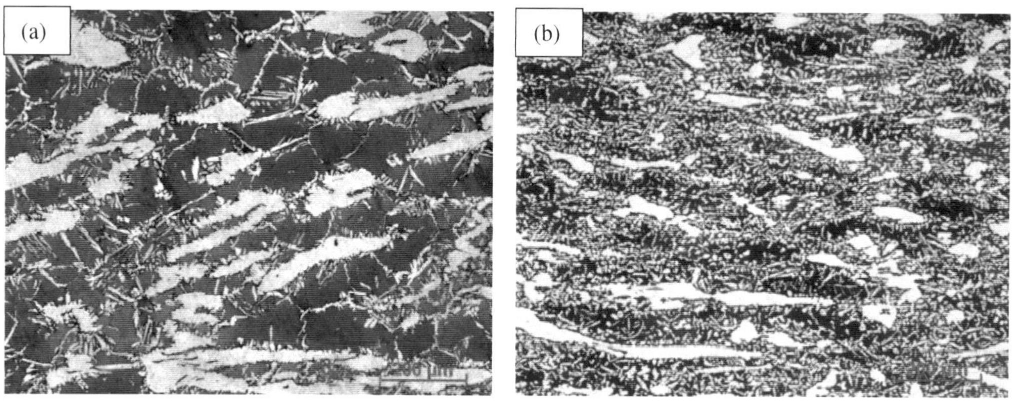

Figure 3: Microstructural evolution of F51 in method B. (a) after upsetting (b) after solution treatment (Optical, 50 X magnifications)

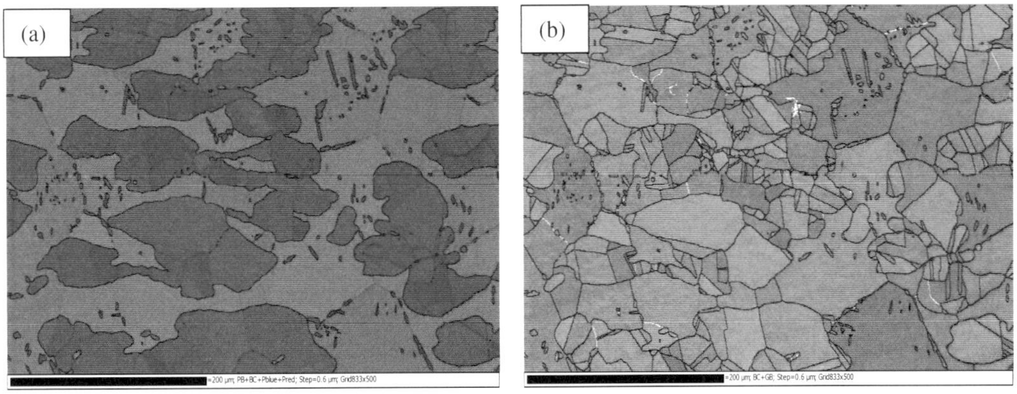

Figure 4: EBSD images of method-A after solution treatment (a) Phase distribution, blue-Austenite, red-Ferrite (b) Grain size distribution.

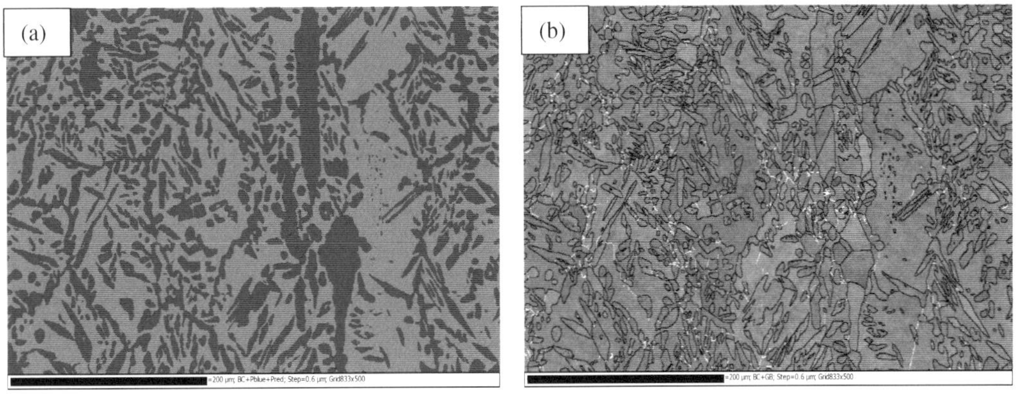

Figure 5: EBSD images of method-B after solution treatment (a) Phase distribution, blue-Austenite, red-Ferrite (b) Grain size distribution.

Change in impact properties:

Charpy v-impact tests were performed at subzero temperature (-50°C). The specimens were taken in the transverse direction of the bar. Fig. 6 shows the impact behavior of the material in the processing stages. From fig. 6, it can be seen there is a large difference in the impact toughness achieved by two methods. In method-A there is a marginal gain in the impact energies after upsetting and subsequent solution treatment (ST) process, instead properties improved lot in case of method-B. This rise in the impact properties of the method-B can be explained with the microstructure, which is having the more DRX grains formed at the grain boundaries and uniform distribution of austenite. The main difference that can be highlighted between method-A and method-B is the grain size. EBSD results shown 13 μm average grain size for method-A and 9 μm for method-B, smaller the diameter, fine the microstructures and higher the impact toughness. Noteworthy this difference can be caused by the longer cooling time with air cooling. Even if α and γ phases having a similar volume fraction, this grain size is having an enormous effect on the impact toughness properties.

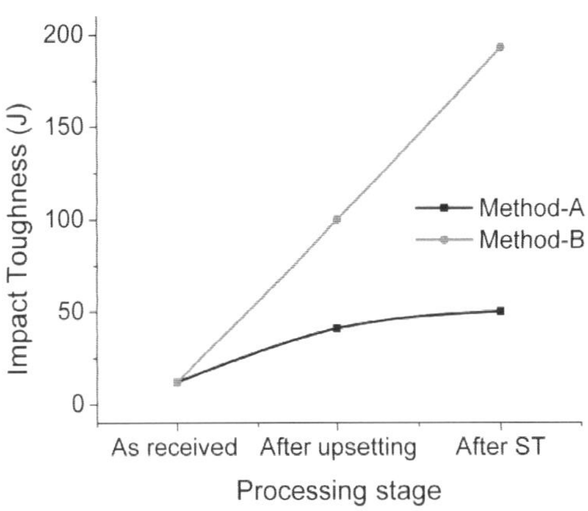

Figure 6: Variation of impact toughness properties in processing stages

XRD results:

Figure 7 shows the X-ray diffraction patterns for as received and solution treated method-A samples. Only samples from method-A were tested as it shows a less impact toughness and checked for intermetallic phase formation during the processing. Figure 7 shows, in both sample peaks corresponding to austenite and ferrites were identified and no other phase precipitates during the solution treatment. In as received material the strongest X-ray peaks were those of austenite, whereas in case of method-A ferrite having the highest peak intensity. The XRD peak intensities of ferrite increased after the solution treatment. This is a good pattern with low background.

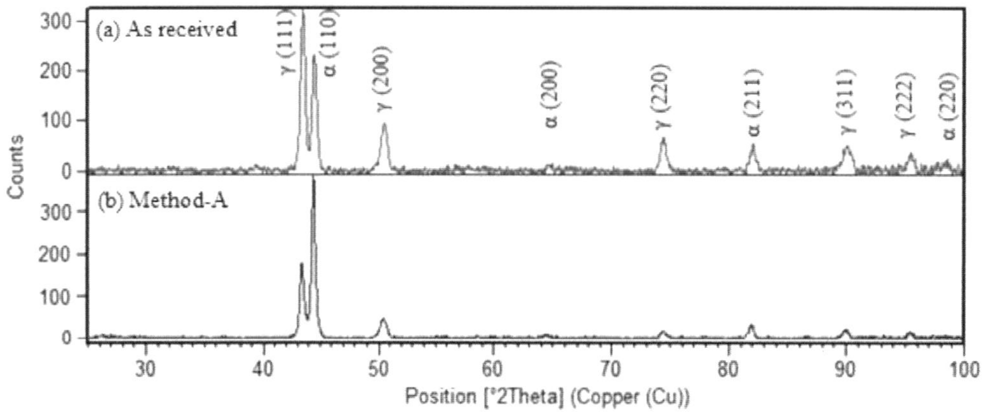

Figure 7: XRD pattern of materials (a) as received (b) after solution treatment method A.

Conclusions

The work aimed to contribute to study the influence of processing routes and heat treatment on microstructural features and subzero impact properties of Duplex Stainless Steel. Cylindrical bars were water quenched and air cooled after the upsetting at 1220-1210°C followed by solution treatment. Very good subzero (-50°C) impact properties were achieved with air cooling after the upsetting operation. From the results of the investigation of the microstructure, it is concluded that, the optimization of the subzero impact properties of this steel strongly depends on the development of the microstructure of the alloy between hot deformation and processing. Grain-boundary serration and bulging are the most important DRX mechanism in duplex stainless steel at the 1220-1210°C working temperature. They are the sites for new DRX grains on the serrated γ/γ boundaries. But water quenching has not given a sufficient time to recrystallize the new grains. In case of air cooling, it has got more time and it leads to the refinement of the austenite grain size by an order of magnitude by creating homogeneously distributed nucleating sites. Low-angle grain boundaries (LAGBs) were observed as the result of the DRV, a dominant restoration mechanism in ferrite. These LABGs evolve and lead to a gradual buildup of higher misorientation between the neighboring subgrains. Presence of intermetallic phases (σ, χ, π, R), carbides (M23C6) and nitrides (Cr2N) were checked by XRD. No intermetallic phases were observed as working temperature is higher and cooling rates skips the range required for the formation.

Acknowledgments

The authors gratefully acknowledge the extended support provided for this work by KCTI (Kalyani Centre for Technology and Innovation) for providing financial funding, laboratory and library facilities. The authors also acknowledge the support provided by Bharat Forge Ltd, Pune and DSIR (Department of Scientific and Industrial Research), Govt. of India. Finally, the authors would like to express special thanks and gratitude to review committee and top management of Bharat Forge Ltd for granting the permission to publish/present the research work.

References

[1] Smuk, O., Microstructure and properties of modern P/M super duplex stainless steels. 2004.

[2] Rosso, M., I. Peter, and D. Suani, About heat treatment and properties of Duplex Stainless Steels. Journal of Achievements in Materials and Manufacturing Engineering, 2013. 59(1).

[3] Alvarez-Armas, I. and S. Degallaix-Moreuil, Duplex stainless steels. 2013: John Wiley & Sons.

[4] Stradomski, Z. and D. Dyja, Sigma phase precipitation in duplex phase stainless steels. Sigma phase precipitation in duplex phase stainless steels, 2009.

[5] Tehovnik, F., et al., Microstructure evolution in SAF 2507 super duplex stainless steel. Materiali in tehnologije, 2011. 45(4): p. 339-345.
[6] Fan, G., et al., Hot ductility and microstructure in casted 2205 duplex stainless steels. Materials Science and Engineering: A, 2009. 515(1): p. 108-112.
[7] Cabrera, J., et al., Hot deformation of duplex stainless steels. Journal of Materials Processing Technology, 2003. 143: p. 321-325.
[8] Chen, T., K. Weng, and J. Yang, The effect of high-temperature exposure on the microstructural stability and toughness property in a 2205 duplex stainless steel. Materials Science and Engineering: A, 2002. 338(1): p. 259-270.
[9] Lo, I.-H. and W.-T. Tsai, Effect of selective dissolution on fatigue crack initiation in 2205 duplex stainless steel. Corrosion science, 2007. 49(4): p. 1847-1861.
[10] Yang, Y. and B. Yan, The microstructure and flow behavior of 2205 duplex stainless steels during high temperature compression deformation. Materials Science and Engineering: A, 2013. 579: p. 194-201.
[11] Momeni, A. and K. Dehghani, Hot working behavior of 2205 austenite–ferrite duplex stainless steel characterized by constitutive equations and processing maps. Materials Science and Engineering: A, 2011. 528(3): p. 1448-1454.
[12] Chen, L., et al., Processing map for hot working characteristics of a wrought 2205 duplex stainless steel. Materials & Design, 2011. 32(3): p. 1292-1297.
[13] Farnoush, H., et al., Hot deformation characteristics of 2205 duplex stainless steel based on the behavior of constituent phases. Materials & Design, 2010. 31(1): p. 220-226.

International Federation of Heat Treatment and Surface Engineering 2016
Proceedings of the 23rd IFHTSE Congress
April 18–21, 2016, Savannah, Georgia, USA

Effect of Electron Beam Welding on Microstructure and Mechanical Properties of Ti-6Al-4V Alloy

Sandeep Thakare[12]*, Prof. N. Prabhu[2], Prakash Kattire[1], Valmik Bhavar[1], Dr. R. K. P. Singh[1]

1. Kalyani center for technology and Innovation, Bharat forge limited, Pune, Maharashtra, India

2. IIT Bombay, Mumbai, Maharashtra, India

Abstract

Several trials were conducted to join two 7 mm thick, mill annealed Ti-6Al-4V alloy plates, by electron beam welding (EBW). Different values of the processing parameters, such as, accelerating voltage, beam current and welding speed, were selected with the objective of obtaining full depth penetration in the weld. The weldments were characterized for their mechanical properties, microstructure and phase composition. While the microstructure of the base metal consisted of slightly elongated α phase and transformed β phase, the microstructure of the fusion zone consisted of dispersed acicular martensite α' phase. The microstructure of the HAZ consisted of two distinct regions, viz., fine grain and coarse grain zone regions. The results indicated that microhardness in the fusion zone was higher than that in the base metal. The UTS and YS of the weld metal were higher than that of the weld joint and base metal. However, the % elongation is lower due to the formation of the acicular martensite α' phase in the weld metal.

Keywords

Electron beam welding, Ti-6Al-4V alloy, Acicular martensite.

1 Introduction

Fusion welding process is one of most widely used process to join two metal plates. Out of all the different fusion welding processes, electron beam welding (EBW), is mostly used in fabrication of titanium alloy components due to its amenability for mass production and minimum damage to parent material [1-2]. The high power density obtained in electron beam makes it possible for joining metals such as titanium, having high melting points and which are extremely reactive when hot or molten[4-5]. Since electrons are easily scattered by air molecules, EBW is carried out under high level of vacuum with little or no contamination in the weld joints [3].

*corresponding author. Tel.: +91 9130084180
E-mail address: sandeepthakare@bharatforge.com

Titanium and most of its alloys are readily welded with the help of electron beam process without the risk of oxidation and subsequent degradation in ductility. Due to this the EBW process is used widely in aero engine industry for welding critical titanium alloy parts safely [6-7].

The aim of the study is to characterize the microstructural changes which take place in various zones during EBW of Ti-6Al-4V and its effects on the mechanical properties of the welded samples.

2 Experimental Procedure

A mill annealed Ti-6Al-4V alloy plate of dimension 200 mm length, 80 mm width and 7 mm thickness was received from Timet UK Ltd. The chemical composition of this material was analyzed by vacuum emission spectrometer which is given in table 1.

Table 1. Chemical composition of received Ti-6Al-4V alloy

Elements	Fe	V	Al	C	O	N
Wt. %	0.18	3.96	6.29	0.007	0.18	0.010

The area of the test plates where the EBW was to be performed was carefully machined to obtain a perfect square butt joint. In order to eliminate any possibility of impurities in fusion zone, the specimen was cleaned with acetone. The welding was performed on a 80kV, 12kW electron beam welding machine. The vacuum in the working chamber was maintained at $5x10^{-4}$ mbar. Two plates of Ti-6Al-4V titanium alloy each having the dimensions 200 mm x 80 mm x 7 mm, were mounted on a self- made clamp. The welding was performed without any filler metal. Several trials were conducted to join the two 7 mm thick plates, by electron beam welding (EBW). Different values of the processing parameters, such as, accelerating voltage, beam current and welding speed, were selected with the objective of obtaining full depth penetration in the weld. The selected values of the processing parameters, finally used are given in Table 2.The actual welded sample is shown in fig. 1.

Table 2 Parameter used for electron beam welding of Ti-6Al-4V butt joint

Voltage (kV)	Beam current (mA)	Welding speed (mm/min.)	Heat Input (kJ/mm)
50	60	800	3.8

A Carl Zeiss Axio vision optical microscope was used to determine the microstructure in different regions of the weldment. A Bruker D8-Discover X-ray diffractometer was used to identify the phases in different regions. XRD analysis was performed using X'pert Highscore plus software. For performing Electron backscattered diffraction (EBSD), specimens were cut into small cubes 10mm x 10 mm x 2 mm and then electropolished and EBSD analysis was performed using orientation Imaging Microscopy (OIM) software. The microhardness was determined using a Clemax CMT microhardness tester under a load of 500 gm and a dwell time of 10s. Microhardness was measured at ten different locations. Tensile samples were obtained in both longitudinal and transverse directions as shown in fig.2. The tensile specimens were prepared according to ASTM E8 specification [8] shown in fig.3. The tensile testing was carried out at room temperature using Zwick/Roell Z250 tensile testing machine.

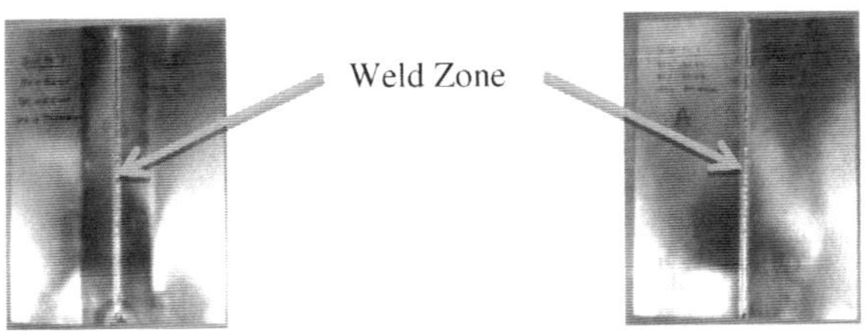

Figure.1 Photograph showing the electron beam welded sample

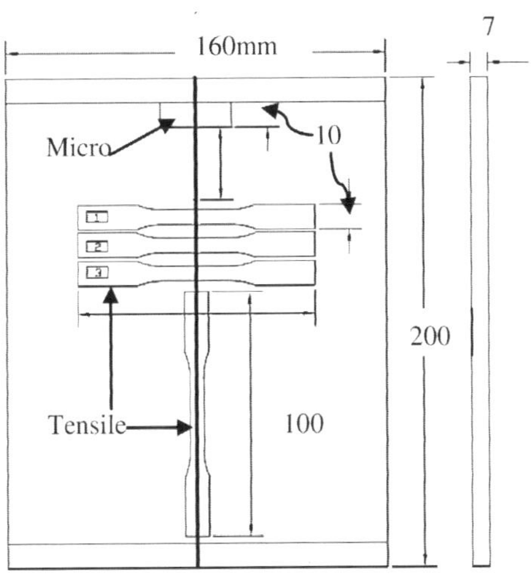

Figure.2 Schematic diagram illustrating orientations of microstructure and tensile specimens

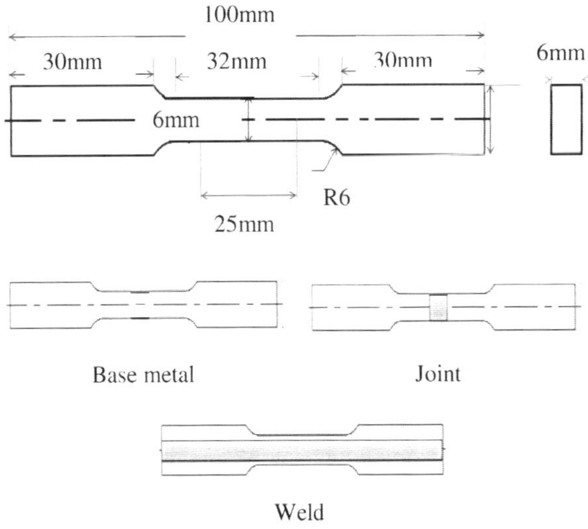

Figure.3 Sub size tensile specimens as per ASTM E8

3 Results and discussion

3.1 Macrostructure of electron beam welded joint

The cross-sectional macrograph of electron beam welded Ti-6Al-4V joint is shown in Fig.4. From the macrostructure of EBW joint three different zones can easily be identified: base metal (BM), fusion zone (FZ) and the heat affected zone (HAZ).

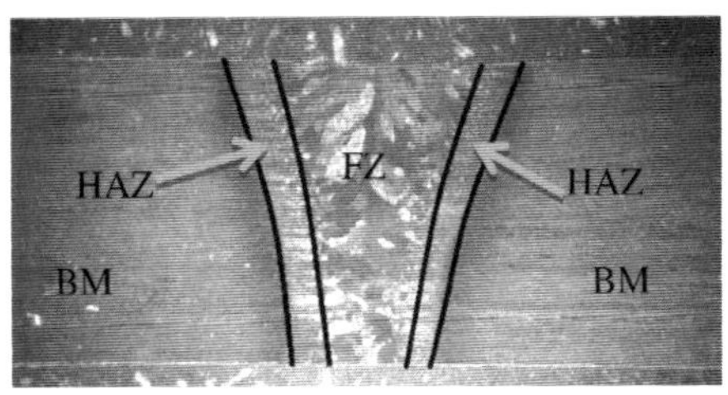

Figure.4 Cross-sectional macrograph of the EBW joint

3.2 Microstructure

3.2.1 Microstructure of base alloy

The microstructure of base metal is shown in Fig. 5(a). The microstructure consists of two phases, slightly elongated grains of primary α-phase (light) and transformed β-phase (dark). The X-ray diffractogram shown in Fig. 5(b) clearly shows the peaks corresponding to α- and β-phases.

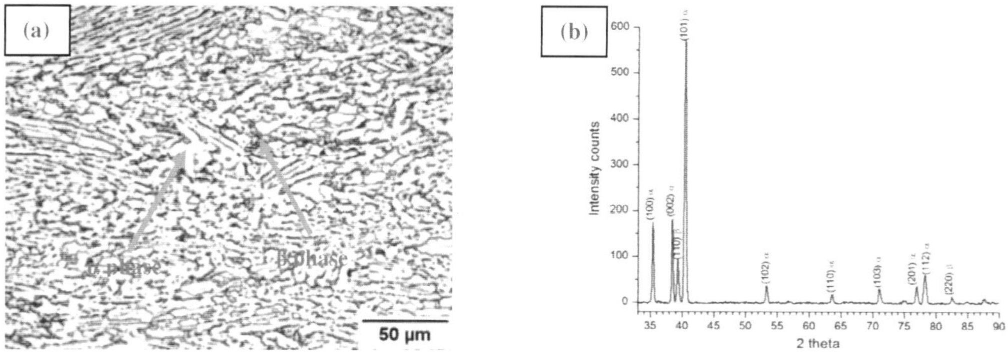

Figure.5 (a) Microstructure of base material, (b) X-ray diffractograms of the base metal

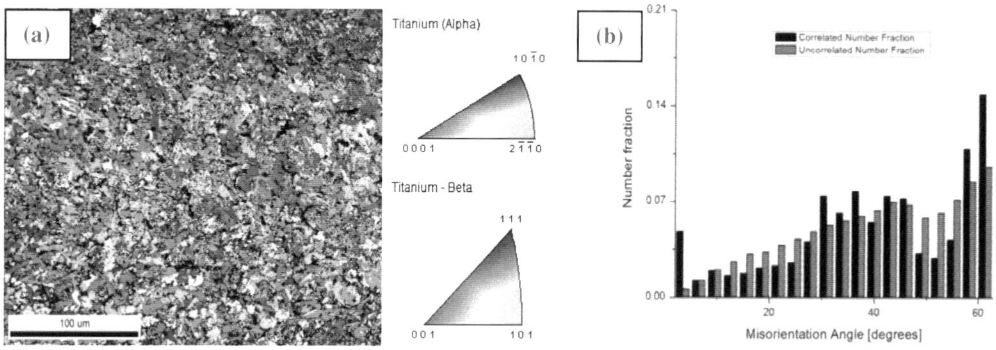

Figure.6 (a) Inverse pole figure map of base alloy Ti-6Al-4V (b) Misorientation angle distributions for base alloy Ti-6Al-4V

From the Inverse Pole Figure (IPF) map shown in Fig. 6(a) it is clear that the grains are randomly oriented. Misorientation histograms for base metal are plotted in Fig.6 (b). Correlated misorientations map display the misorientation between neighboring points and uncorrelated misorientation data shows misorientation between randomly chosen points in the data set. It can be seen from the correlated (black) and uncorrelated (red) misorientations graph that, this sample crystallographic orientation are fully random and is said to have poor texture.

3.2.2 Microstructure of fusion zone

The microstructure of the fusion zone is shown in Fig.7 (a). The microstructure consists of mainly acicular α' martensite plates. This is also confirmed from the X-ray diffractogram shown in Fig. 7 (b).

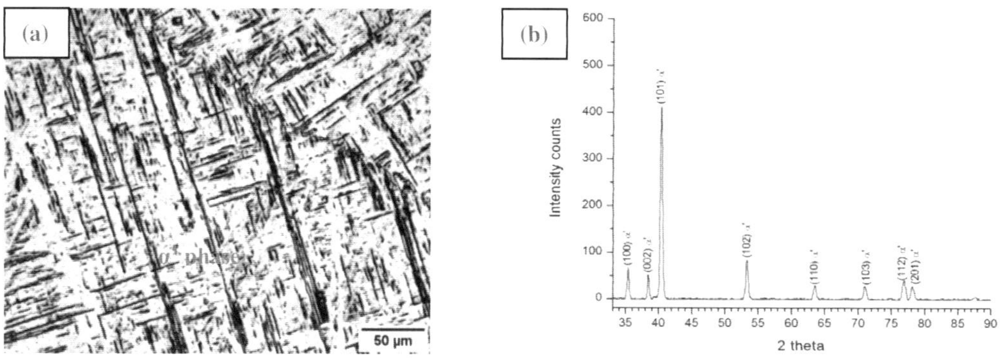

Figure.7 (a) Microstructure of fusion zone, (b) X-ray diffraction patterns for fusion zone

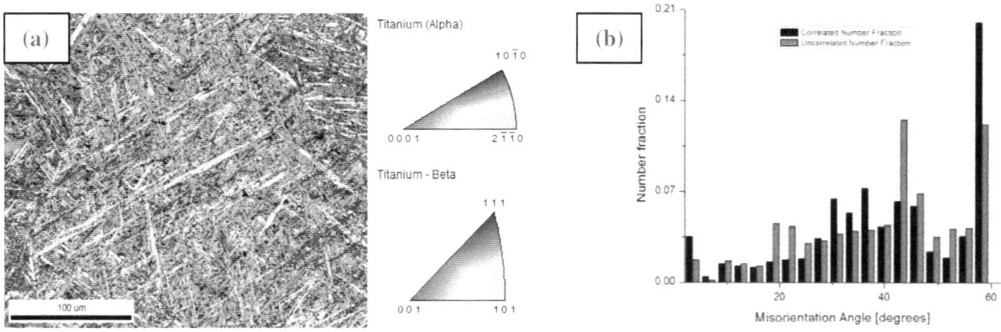

Figure.8 (a) Inverse pole figure map of Fusion zone of electron beam welded Ti-6Al-4V sample, (b) Misorientation angle distribution in the FZ of EB welded Ti-6Al-4V sample

From the IPF map shown in Fig.8 (a) it is clear that the grains are randomly oriented. From the misorientation histograms for base metal are plotted fig.8 (b). It can be seen from the correlated (black) and uncorrelated (red) misorientations graph that, in the fusion zone, the crystallographic orientation of the grains is fully random and is said to have no texture.
The fusion zone undergoes rapid cooling on account of good conductivity. The liquid metal first crystallizes to form β phase, and then subsequently transforms into α phase martensitic. Therefore in weld zone a large volume fraction of dispersed acicular martensite α' phase is formed.

3.2.3 Microstructure of heat affected zone (HAZ)

The microstructure of the HAZ is shown in Fig.9. It can be seen that based on the microstructure HAZ can be divided into two distinct regions: coarse grain region which is near the FZ and fine grained region which is near the BM.

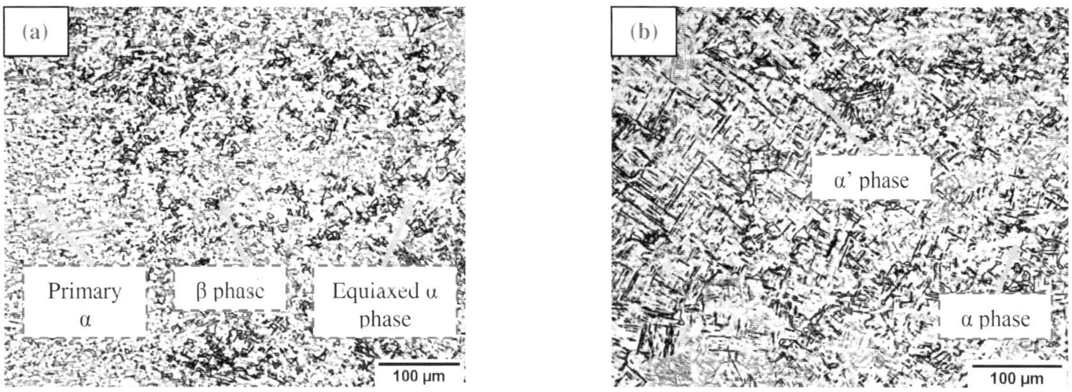

Figure.9 Microstructure of heat affected zone: (a) Fine grain region, (b) Coarse grain region

It can be seen from Fig.9 (a) which shows the microstructure of fine-grained region, it can be found in the fine-grained region that the amount of primary α phase reduces compared with the BM zone and the amount of β phase increases significantly. The transformation of α phase to β phase occurred in the fine grain zone in the welding process due to heating, so the zone reaches phase transition temperature. As this zone is far away from heat source, so the temperature is low and phase transition was not complete. In the subsequent cooling process, the fine α equiaxed grains precipitated from the β phase matrix. Therefore the microstructure of fine-grained region was primary α phase + β phase + equiaxed α phase. It can be seen from Fig.9 (b) the microstructure of coarse-grained region. It can be seen in this zone that the amount of primary α phase significantly reduces, and apparent coarsening happens. The transformation of α phase to β phase occurred due to heating, so the zone reaches phase transition temperature. As this zone is near fusion zone temperature is high and phase transition complete. In the subsequent cooling process transformation from the β phase to α phase by atomic diffusion has no time to complete, because of fast cooling rate. Acicular martensite α′ phase is formed by diffusionless transformation of β phase. Therefore, the microstructure of coarse-grained region was primary α phase + α′phase.

4 Mechanical properties

4.1 Microhardness

The microhardness profile across the weldment is plotted in Fig. 10. Microhardness values in the fusion zone are higher than those in the HAZ and BM. Also, the microhardness values of HAZ are higher near the fusion zone-side and are lower towards the base metal-side.

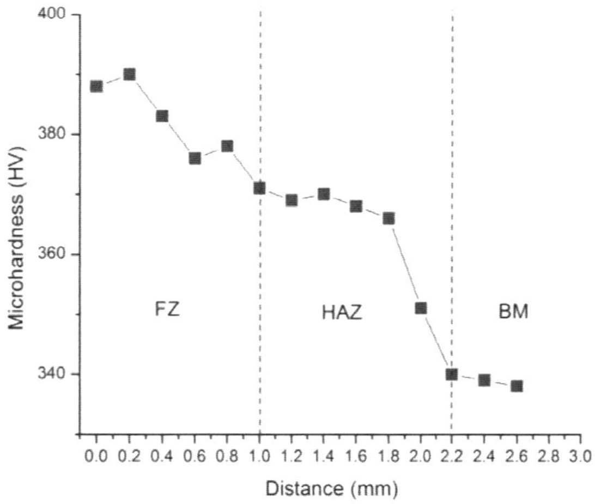

Figure.10 Microhardness across FZ, HAZ and BM of electron beam welded joint

4.2 Tensile properties

The data obtained from the tensile tests of base metal, joint and weld sample are plotted in the form of bar graph as shown in Fig.11. The yield strength and ultimate tensile strength of the weld is higher than that of the joint and base metal and the % elongation of weld is lower than that in the joint and the base metal.

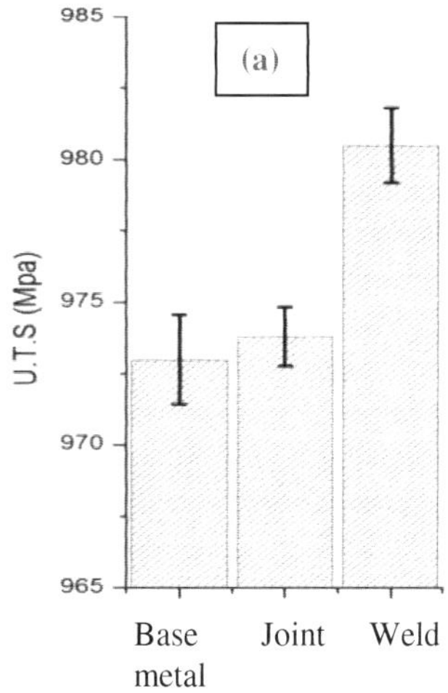

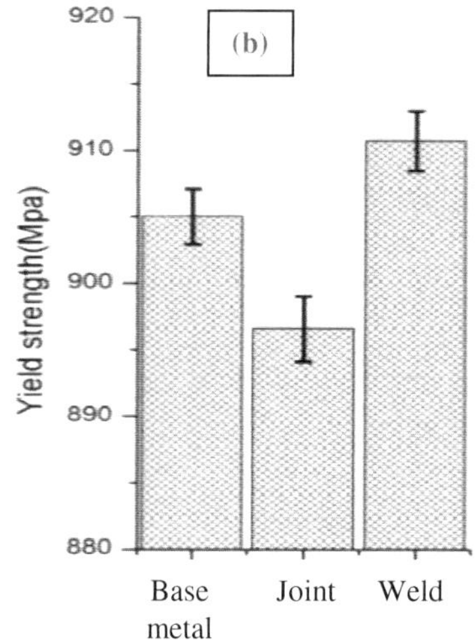

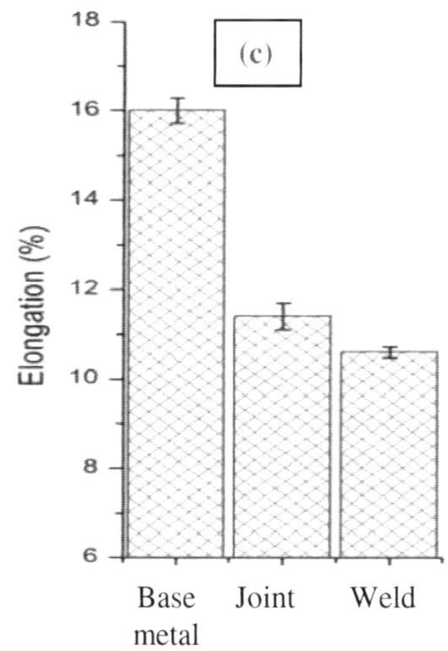

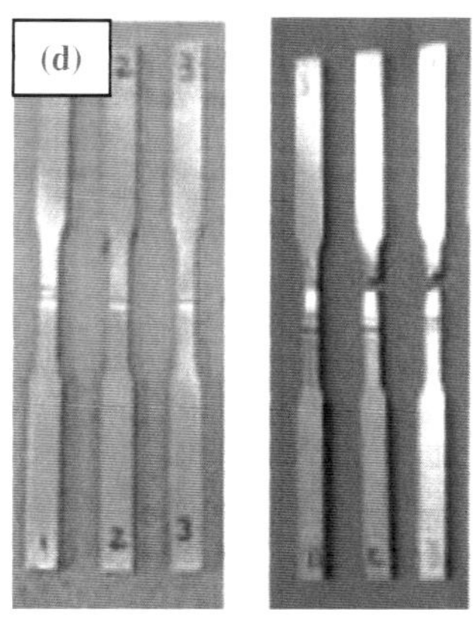

Figure.11 (a)Ultimate tensile strength (b)Yield strength (c) Elongation % (d) Tensile specimen

Also it is clear from Fig. 11 (d) that all welded sample in transverse direction broke in the region of the base metal and not a single sample broken in weld joint area. This indicates that the weld region is comparatively stronger than base alloy region.

Conclusions

- The microstructure of the mill annealed Ti-6Al-4V alloy consists of two phases, elongated grains of primary α-phase and transformed β-phase.
- The microstructure in the fusion zone consists of acicular martensite α' phase.
- The microstructure of the HAZ has two distinct regions: the fine grain zone near base metal having primary α phase + β phase + equiaxed α phase and coarse grain zone, near the fusion zone having α phase + α′ phase.
- Microhardness values of weldment in the fusion zone are higher than those in the heat affected zone and base metal.
- The fracture location of all EBW samples is in base metal. The ultimate tensile strength and yield strength of the weld is higher than that of base metal, but the % elongation of weld is lower than that base metal.

References

[1] Z.Sun, R.Karppi, "The application of electron beam welding for the joining of dissimilar metals: an overview", J.Mater process Tech, Vol.59 (1996), pp. 257-267.

[2] C.L.M Cottrell, "Electron beam welding: A critical review", Materials and Design, Vol.6, (1985), pp. 285-291.

[3] Piotr Lacki, Konrad Adamus, "Numerical simulation of the electron beam welding process", Computer and structures, Vol.89, 2011, pp. 977-985.

[4] K.Y.Benyounis, A.G.Olabi, "Optimization of different welding processes using statistical and numerical approaches-A reference guide", Advances in Engineering Software 39, 2008, pp. 483-496.

[5] Mathew J. Donachie., "Titanium, a Technique Guide". ASM International, 2nd edition, 2000.

[6] Gerd Lutjering and James C. Williams. "Titanium". Springer-Verlag, Berlin Heidelberg New York, 2nd edition, 2007.

[7] H Schultz. "Electron beam welding". Abington publishing, Cambridge, 1993.

[8] ASTM. Standard test method for tensile testing of metallic materials. ASTM. E8M-09; 2009.

International Federation of Heat Treatment and Surface Engineering 2016
Proceedings of the 23rd IFHTSE Congress
April 18–21, 2016, Savannah, Georgia, USA

Induction Hardening Shafts of Shafts and Shaft-like Components

Dr.Valery Rudnev, William West, Aaron Goodwin, Steve Fillip
Inductoheat Inc., An Inductotherm Group Company
32251 N. Avis Dr., Madison Heights, MI 48306 USA
rudnev@inductoheat.com; wwest@inductoheat.com; agoodwin@inductoheat.com; sfillip@inductoheat.com

Abstract

This presentation reviews selected innovations related to induction hardening of various automotive components, including but not limited to induction surface hardening of complex geometry shafts. Thanks to several innovative designs (patented and patent pending), important goals were achieved. Process flexibility in shaft scan hardening has been substantially enhanced thanks to a novel inverter design that allows controlling independently *frequency* and *power* during scan hardening. This innovation optimizes heat source generation and allows improving quality of induction hardened components maximizing production rate and process flexibility.

When single-shot hardening of shaft-like components including output shafts, flanged shafts, yoke shafts, sun shafts, intermediate shafts, drive shafts and others, coil life is often limited due to a necessity of severe copper profiling allowing to "*squeeze*" coil current in a certain area maximizing power density and localizing heat intensity. This seemingly unavoidable feature of the great majority of single-shot inductors represents a "weak link", limiting coil life expectancy. Thanks to innovative design (patent-pending) of a single-shot inductor, its life was increased approximately NINE (9) folds and it was verified by the tool-room tags of the users. Other benefits include measurable improvement in process robustness process sensitivity has been dramatically reduced as well.

Introduction

Quality, price and delivery have traditionally been the three key benchmarks in judging induction heat treating equipment. Because of a number of current developments these three are joined by a fourth that is equally important-flexibility. In the past it may have been acceptable to focus on one or two of these areas but because of the growing increase in customer specifications and requirements it is necessary to focus on all four of these areas.

A changing landscape in the area of automotive supply has led through the years to a different set of requirements for first tier suppliers. In the past, it may have been common for a contract to remain with certain suppliers for a period of many years. With the current conditions it may be common for a contract to move from supplier to supplier on an annual basis. When this is the case it is necessary to win the contract, seek out and source the required equipment to do the processing, purchase the equipment, start the equipment up and complete a Production Part Approval Process (PPAP) to be in production in a very short period of time. In addition, the equipment must be high quality, very reliable and easy to maintain. Set up and maintenance information must be immediately on hand and easily accessible.

Finally, the concern of having a particular contract one year and possibly losing it the next has given rise to the need for greater flexibility, the new fourth benchmark, in capital equipment to allow easy retooling and reprogramming of the equipment if necessary to process different parts in the future.

For capital-equipment manufacturers, the consideration of these types of requirements has led to the development of a new generation of induction heat treating equipment that will meet and exceed the need for quality, price, delivery and flexibility.

Hardening of steels, cast irons and powder metallurgy components represents the most popular applications of induction heat treating. Induction hardening of parts (e.g., machine tools, hand tools, powertrain components, bearings, links, pins, rods, frames, blades, springs, fasteners, gear-like components and many others) may be done for the purpose of obtaining certain properties that include but are not limited to strength, fatigue, and/or wear resistance [1-3]. For example, in torsion and in bending the stress is greatest at the surface, and is zero at the center. For this reason, an induction surface hardening can improve component performance with these two load cases, as it increases the strength at the surface where it is needed the most [1]. In addition, surface hardening normally leaves the surface of the part with a beneficial residual compressive stress of appreciable magnitude, which acts as the crack arrester and increases torsional and bending fatigue life.

Heating inductors provide required three-dimensional (3-D) heat generation during the heating cycle that can produce suitable austenized pattern and upon quenching the desired hardness pattern. Inductors can be of different styles and geometries to accommodate the shape of the workpiece needed to be heat treated and process specifics. In everyday practice, an inductor is also called an induction coil or simply a coil. However, its geometry often does not resemble the shape of a classical coil. Figure 1 shows examples of various inductor designs used in induction hardening.

There are four primary heating modes of induction hardening: *scanning*, *progressive*, *single-shot* and *static* modes. Depending upon the applied method and application specifics, parts may or may not be rotated, moving in a linear motion or be motionless.

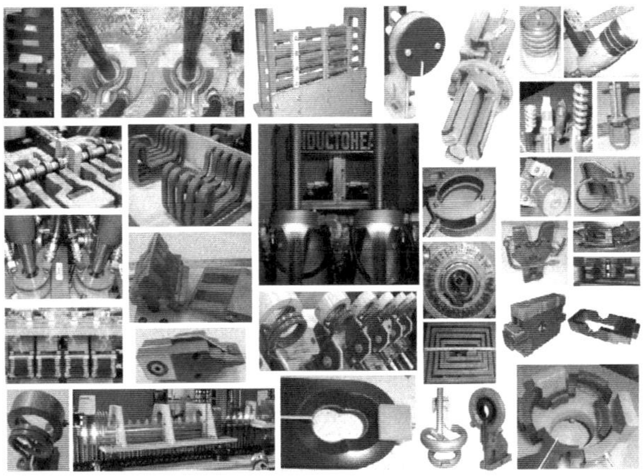

Figure 1: Examples of various inductor designs used in induction hardening. Courtesy of Inductoheat Inc., An Inductotherm Group Company

Steel shafts and shaft-like components are among parts that traditionally undergo induction hardening (Fig.2). Shaft-like components are usually induction hardened using *scanning* or *single-shot* heating modes.

Scan Hardening of Shafts

In scan hardening, the inductor or workpiece or both may move linearly relative to each other during the hardening cycle. Depending on the workflow of parts the system can be built as vertical, horizontal, or even at an angle, though vertical scan hardening is by far the most popular design concept due to a number of reasons including a smaller footprint and natural quench flow to outline a few. According to scan hardening, the coil and/or part move relative to each other. The workpiece generally rotates inside the coil to even out the induction hardened pattern around the circumference.

Scan hardening systems provide noticeable process flexibility with respect to the workpiece length and, to some extent, variations in the part diameter. In addition, this scanning mode provides the ability to vary the speed and power during the process, which controls the amount of heat applied to different areas of the part.

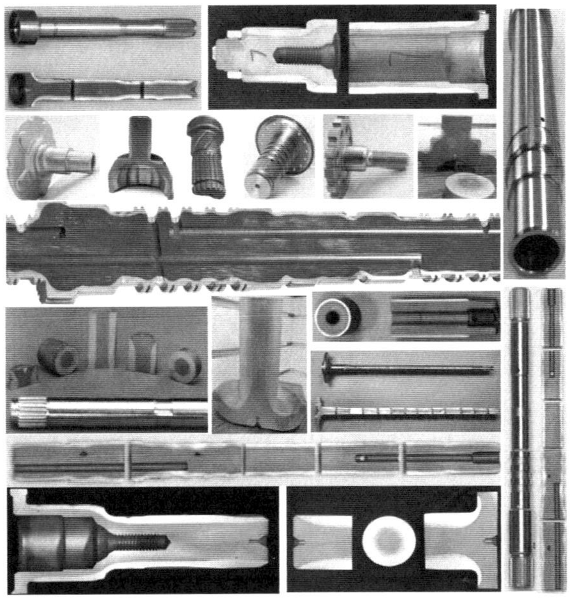

Figure 2: Examples of various shafts and shaft-like components. Courtesy of Inductoheat Inc., An Inductotherm Group Company

Scan hardening can be performed on outside diameters (OD) and/or inside diameters (ID) of cylinder components as well as on flat surfaces. As stated earlier, both horizontal and vertical coil arrangements can be used, with the vertical design being the most popular for hardening of short and moderate-length shafts [1, 2].

When scanning the outside surfaces of solid shafts, the induction coil typically encircles the part. A quench ring is positioned next to the coil in order to spray-quench the area that has been austenized. In other cases, machined integral quench (MIQ) inductors can be used. On some machines the coil remains stationary and the part moves, while on other machines the opposite is true.

The scan hardening process starts with the coil positioned at one end of the part. The part begins to rotate, the power is turned on and the coil may remain stationary for a period of time to drive the heat into a certain region of the part. This is known as the dwell and is typically used to preheat the fillet areas. The part and/or coil then begin to move relative to each other, which is why the term scanning is used.

With the scan hardening of cylindrical parts, using a single-turn or multi-turn coils, induced eddy currents flow predominately circumferentially. Figure 3 shows the results of FEA computer modeling of a vertical scan hardening of a hollow shaft utilizing Inductoheat's proprietary software [3].

Figure 4 shows another example of FEA computer modeling of the sequential dynamics of a horizontal induction scan hardening system for heat treating of a hollow shaft with diameter changes, undercut, and grooves [4]. Because the shaft is symmetrical, only the top half was modeled.

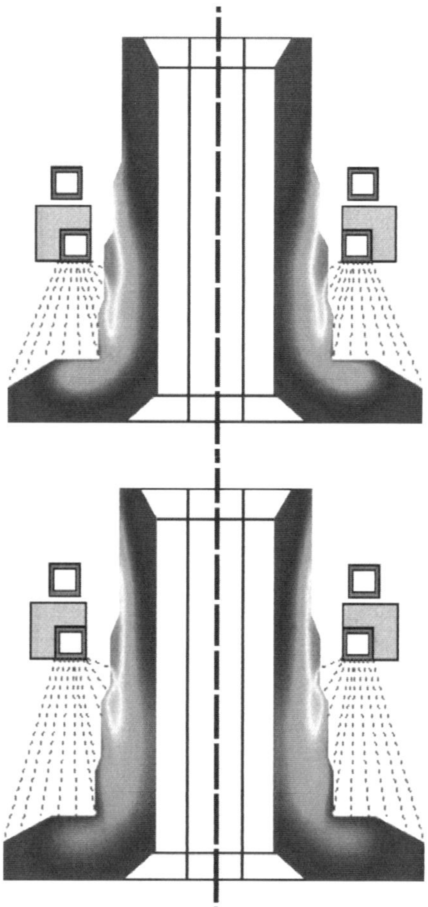

Figure 3: Dynamics of temperature profiles during induction scan hardening of hollow shaft utilizing Inductoheat's proprietary FEA software [3].

As the heat front moves along the length/height of the part the scan rate and/or coil power may be changed to affect the case depth as needed. This is typically done where the diameter or the configuration of the part changes, when there is a combination of the solid and hollow sections of the shaft, or where other geometrical irregularities (e.g., shoulders, fillets, diameter and wall thickness changes, etc.) are present.

The great majority of commercially available power sources for medium and high frequencies are designed to provide certain frequency of output AC current that cannot be instantly changed during the scanning operation. Therefore, the depth of heat source generation (that is related to current penetration depth) might not be an optimal for a particular portion of the heat treated part, its geometrical features and/or localized case depth requirements.

In some cases, available frequency might be substantially lower than optimal, but in other cases it might be significantly higher for obtaining particular hardness case depth in a certain area of the part during scan hardening.

If an available frequency is noticeably higher than optimal, it results in too small current penetration depth δ that might not be sufficient for a proper austenization of the subsurface region at required hardness case depth. Therefore, additional time is needed to allow the thermal conduction to provide a sufficient heat flow from the workpiece's surface toward required depth to ensure its proper austenization. This is commonly accomplished by a reduction of both: the scan rate and power density. Otherwise, the surface might be overheated and undesirable metallurgical structures might be formed (for example, grain boundary liquation, severe grain coarsening, etc.). Unfortunately, this not only adds unnecessary cycle time, but it might still lead to some metallurgical issues related to excessive surface temperatures.

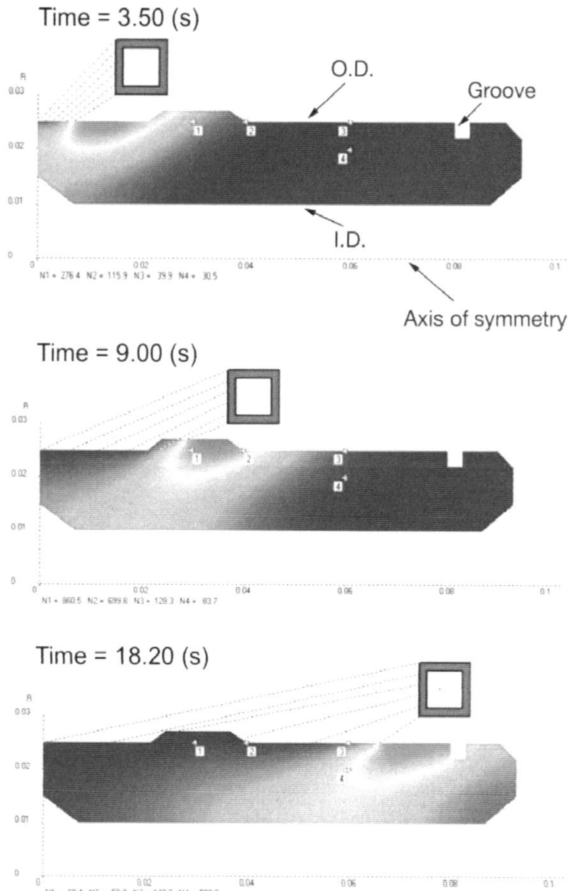

Figure 4: FEA computer modeling of sequential dynamics of horizontal induction scan hardening system for heat treating of hollow shaft with diameter changes, undercut, and grooves utilizing Inductoheat's proprietary software [4]. Because the shaft is symmetrical, only the top half was modeled.

In contrast, if available frequency is noticeably lower than optimal frequency, it might produce a much deeper austenized layer than needed. In order to minimize the negative impact of using lower than desirable frequency, the majority of induction heat treaters are trying to suppress a thermal conduction by increasing both: scan rate and power density.

Thus, conventional scan hardening technologies have to rely upon a compromise between achieved quality, production rate and process capability. Regardless of the fact that those recipe/protocol modifications can help to minimize a negative impact of using other-than-optimal frequencies by relying upon *suppression* or *enhancement* of thermal conduction, they often cannot eliminate it. Depending upon the shaft's geometry, this could limit the ability to meet a pattern specification and might negatively affect the achieved metallurgical quality of heat treated components and distortion requirements.

According to theory of induction heating, applied frequency is the most powerful parameter that directly affects the depth of heat source generation [1, 2]. Obviously, in order to address geometrical subtleties of heat treated components in an optimal manner, it would be much more beneficial to apply various combinations of frequencies, power densities and scan rates at various stages of the scan hardening operation. This would maximize production and improve metallurgical quality of the heat treated components. Unfortunately, the great majority of available inverters do not have such a capability.

The new generation of patented transistorized inverters eliminates this limitation and simplifies achieving required hardness pattern specifications, allowing independently controlled frequency and power (Figure 5) during induction scan hardening. This allows optimizing electromagnetic and thermal conditions when heat treating components with geometrical irregularities and various required hardness case depths. As an example, Figure 6 shows the Statitron-IFP™ inverter that provides Independent Frequency and Power control.

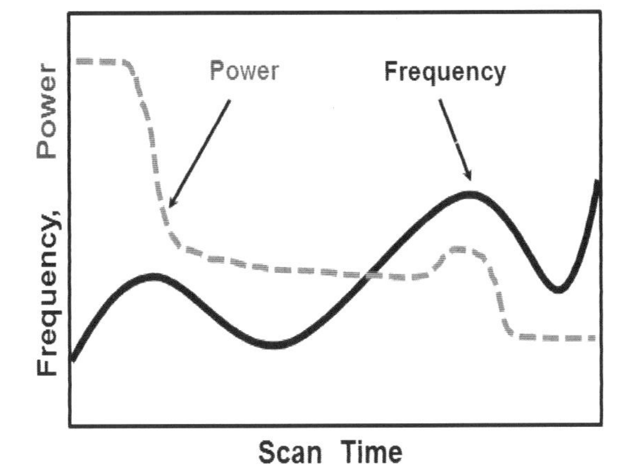

Figure 5. Independent Frequency and Power control capability of new inverters.

Statitron-IFP™ inverter technology is the long-awaited desire of commercial heat treaters, providing the greatest process flexibility, measurably expanding the equipment capabilities and improving the metallurgical quality of induction heat treated components.

Figure 6. Patented Statitron-IFP™ inverter is capable in providing Independent Frequency and Power control. It was specifically designed for induction heat treating applications allowing instantly and independently adjust both: frequency (within 5kHz - 60kHz) and output power (up to 450kW). Courtesy of Inductoheat Inc., An Inductotherm Group Company

Single-Shot Hardening of Shafts

Steel shafts and shaft-like components are among parts that traditionally induction hardened using not only scanning but also single-shot technique. With the single-shot method, neither the shaft nor inductor move linearly relative to each other; the part typically rotates instead.

According to a single-shot hardening method, the entire region to be hardened is heated at the same time. In order to provide the required temperature distribution before quenching, the heat is sometimes applied in several short bursts (pulse heating) with a timed delay between them to allow the thermal conduction towards the areas that might be difficult to heat

Single-shot inductors are made of tubing or CNC machined from solid copper conform to the area of the part to be heated. This type of inductor requires the most care in fabrication because it must operate at very high power densities and workpiece's positioning is critical in respect to coil copper profiling.

A single-shot inductor induces eddy currents that primarily flow along the length of the part. An exception to this rule would be the arc-shape (cross-over) regions of a single-shot inductor, where eddy current flow is circumferential. Normally, the single-shot method is better suited for hardening short and moderate length parts or parts where only a relatively short areas is to be heat treated [1,2]. This method is also better suited to cylindrical parts having axial symmetry and geometrical irregularities, such as diameter changes, fillets, shoulders, and so forth. A single-shot inductor, to some degree, must follow the contour of the entire part or the area required to be hardened. As an example, Figure 7 shows conventional CNC-machined single-shot inductors.

There are several ways to control heat generation at different regions of a heat treated shaft. This includes an inductor copper profiling resulting in a variation of coil-to-shaft electromagnetic coupling and/or machining different widths of the current-carrying faces of the inductor. Besides that, a magnetic flux concentrator can be attached to certain areas of the hardening inductor to enhance localized heat intensity as it is shown on Figure 7.

As a result, certain regions of coil copper might carry extremely high current densities resulting in an intense heat generation. Regardless of an attempt to position water-cooling pockets as close to the current carrying face of an inductor as possible and utilization of high-performance pumps, a coil copper might be still overheated, leading to an accelerated deterioration of the copper surface and eventual premature coil failure.

Designers often attempt to take advantage of electromagnetic proximity effect when applying single-shot inductors to intensify heating of certain areas of the workpiece. Unfortunately, in order to do so, a workpiece-to-inductor distance might be needed to be reduced too drastically.

Dramatically reduced copper face results in the highest current density that is required for generation of sufficient heat sources to be induced in the fillet region. In such applications, it is particularly important to maintain the safe copper-to-part gap in that region, minimizing real-life disturbances associated with an effect of worn bearings and the part wobbling during its rotation, for example.

Because a combination of intense heat generation within the current carrying face of coil copper with intense heat radiation/convection from workpiece's surface (in particularly when relatively small coil-to-shaft air gaps are used) may lead to a substantial copper overheating. This may promote water vaporization and the formation of a steam vapor barrier in that region [5].

Seemingly unavoidable feature of the great majority of conventionally-designed single-shot inductors that is associated with severe copper profiling often represents a "weak link" limiting coil life expectancy, since certain regions of coil copper might carry extremely high current densities resulting in an intense heat generation and localized overheating.

Regardless of an attempt to position water-cooling pockets as close to the current carrying face of an inductor as possible and utilization of high-performance pumps, a coil copper might be still overheated causing accelerated deterioration of the copper surface, which speeds up the onset of inductor copper cracking (due to stress fatigue and stress corrosion, for example) and eventual premature coil failure. As a result, coil life is often shortened to 22,000-24,000 heat cycles (industry average). Therefore, the number of instances where coil current density is increased should be kept to a minimum.

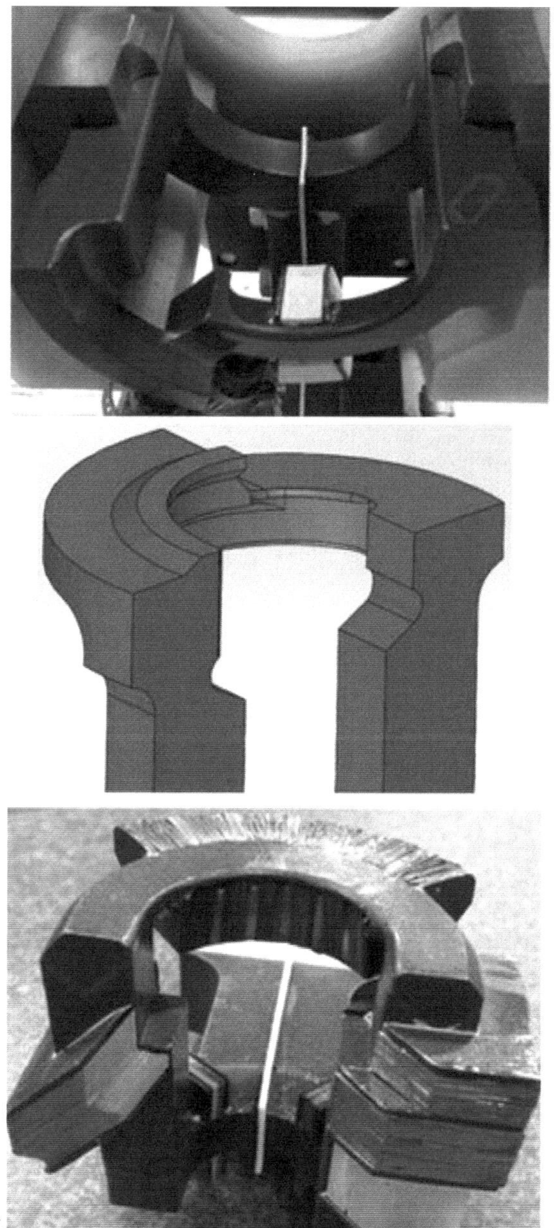

Figure 7. Conventional single-shot inductor for induction surface hardening of shaft-like components having fillets and diameter changes. Courtesy of Inductoheat Inc., An Inductotherm Group Company

An attempt to intensify a heat generation in the fillet forces designers to minimize inductor-to-fillet gaps. This could cause an arcing problem.

Recently developed novel inductor design (Fig.8, patent pending) dramatically reduces localized coil current density in areas prone to overheating and cracking [6]. The presence of a two-collar section reduces coil current by one half, which dramatically reduces correspondent localized heat generation in the most critical region of the inductor copper and significantly extends coil life.

In addition, for a shaft positioned asymmetrically (to some degree) within the inductor or if an appreciable wobbling occurs during a rotation (due to worn bearings), there is a reduced heating effect produced in one of the two half-collar sections that has an increased inductor-to-shaft gap. This is offset by an increased induced heating effect produced in the other half-collar section that has a reduced inductor-to-shaft gap. Consequently, process sensitivity associated with positioning the shaft within the inductor is reduced over that with a conventionally designed single-shot inductor.

In one application of the new inductor (patent-pending), one of the world's largest suppliers of automotive parts achieved a nine-fold increase in a single-shot coil life compared with that for conventional inductors. This is verified by the manufacturer's tool-room tag showing that the inductor (which the customer named "*magic coil*") was still considered in good shape after 225,000 heat cycles (Figure 9).

Other benefits include measurable improvement in process robustness, coil reliability, and maintainability

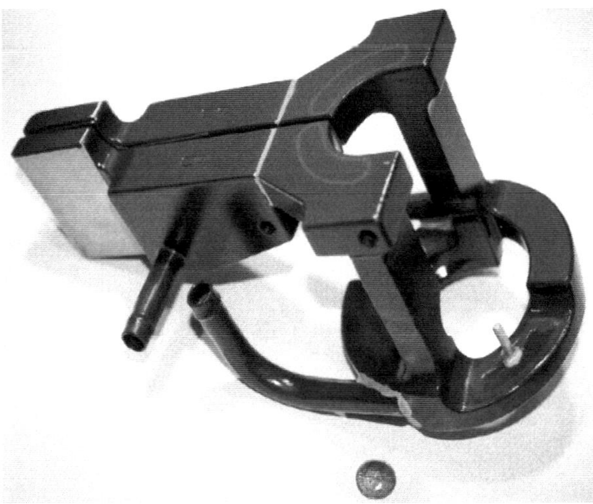

Figure 8. Inductoheat recently developed a novel inductor design that substantially surpasses performance of conventional single-shot inductors. Customer has even named this unique inductor design (patent pending) as a "*magic coil*". *Courtesy of Inductoheat Inc., An Inductotherm Group Company*

Details of this design and explanation of its benefits will be discussed in detail in the PowerPoint presentation.

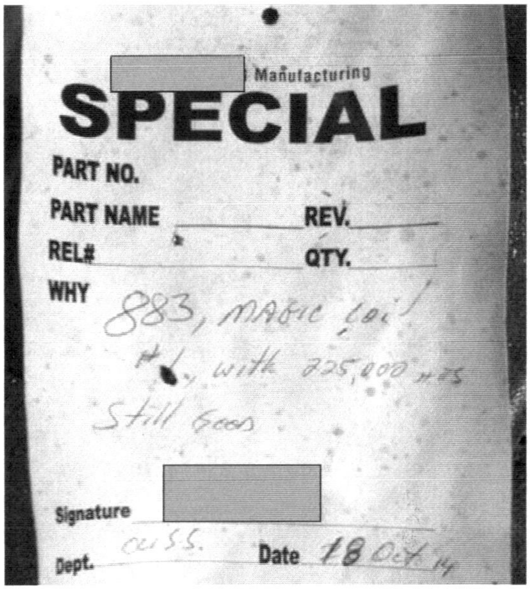

Figure 9. A tool-room tag from one of the world's largest supplier of automotive parts that indicates an achievement of 225,000 hits (nine-fold increase) of a single-shot coil life compared to conventional inductors.

Conclusions

Presentation discusses two innovations related to induction scan hardening and single-shot hardening of shaft like components (including but not limiting to induction surface hardening of output shafts, flanged shafts, yoke shafts, sun shafts, intermediate shafts, drive shafts, etc.). Advanced designs (patented and patent-pending) focus on ensuring metallurgical quality of heat treated components and improving process robustness and flexibility. Customer recorded a nine-fold increase in a single-shot coil life compared to conventional inductors.

References

[1] Rudnev, V., Fett, G.A., Griebel, A., Tartaglia, J., "Principles of Induction Hardening and Inspection," ASM Handbook Vol.4C: Induction Heating and Heat Treating, V.Rudnev and G.Totten (eds.), ASM Int'l, 2014, pp.58-86.

[2] Rudnev, V., *et al.*, Handbook of Induction Heating, CRC Press, 2003, 800p.

[3] Rudnev, V., "Q & A Regarding Quenching and Case Hardening of Solid Shafts vs Hollow Shafts", Professor Induction Series, *Heat Treat. Progress*, ASM Int'l, Sept 2009, p.29-32.

[4] Rudnev, V., "Computer Modeling Helps Prevent Failures of Heat Treated Components," *Advanced Materials and Processes*, ASM Int'l, October 2011, pp. 28-33.

[5] Rudnev, V., "Systematic Analysis of Induction Coil Failures and Prevention," ASM Handbook Vol.4C: Induction Heating and Heat Treating, V.Rudnev and G.Totten (eds.), ASM Int'l, 2014, pp.646-672.

[6] Rudnev,V., Goodwin,A., Fillip,S., West,W., Schwab,J., St.Pierre,S., *Keys to Long-lasting Hardening Inductors: Experience, Materials and Precision*, Advanced Materials and Processes, ASM Int'l, Oct., 2015, p.48-52.

International Federation of Heat Treatment and Surface Engineering 2016
Proceedings of the 23rd IFHTSE Congress
April 18–21, 2016, Savannah, Georgia, USA

Effects of the inductive hardening process on the martensitic structure of a 50CrMo4 steel

Annika Vieweg, Gerald Ressel, Petri Prevedel, Stefan Marsoner, Reinhold Ebner
Materials Center Leoben Forschung GmbH, Roseggerstraße 12, 8700 Leoben

Abstract

This study focusses on the influence of a fast heat-treatment, as occurring with induction heating processes, on the microstructural features and their effect on the mechanical properties of a 50CrMo4 steel.
To this end three different heating rates (1, 10 and 100 K/s) and three different austenitizing temperatures (950, 1000, 1050°C) were applied to a ferritic-pearlitic steel. After quenching the resulting martensitic microstructures were investigated and compared to a conventional furnace heat treating sequence using electron back scattered diffraction (EBSD). Thus, the effect of variable heating rates during austenitization on the resulting martensitic microstructure, including the evaluation of packet and block sizes and orientations, was analyzed.
Through the investigations of this work, it could be shown, that the adaption of the Kurdjumov-Sachs (K-S) variants is changing with the prior heat treatment. The occurrence of all 24 K-S variants in the same intensity increases and the misorientation angle across a martensite block decrease with higher austenitizing temperatures, despite fast heating rates.
Therefore, it is suggested, that the austenitizing temperature has more impact on the martensitic structure than the heating rate.

Introduction

In general induction hardening is a clean and efficient process, which can be easily integrated into an in-line production, which makes it an attractive way for heat treating steel [1, 2]. Through the inductive heating, high heating rates are applied to the workpiece, thus changing transformation kinetics as well as properties of the materials compared to conventional furnace heat treatments. High heating rates retard the austenitic phase transformation to higher temperatures, thus requiring different temperatures for an inductive process, than used within a conventional treatment [3, 4]. Furthermore, it has been reported in literature that through high heating rates and short holding times the austenitic state is not achieved homogeneously [5]. E.g. Sackl et al. [5] could prove inhomogeneity regarding carbon of an inductively heated sample, compared to a conventional heat treated sample using atom probe tomography.
Since it is difficult, to observe properties of the austenitic state of a material during the heat treatment, conclusions on occurring inhomogeneities and stresses among the austenitic structure during and at the end of the heat treatments need to be drawn from the as-quenched, martensitic, structure. In order to do so, knowledge of martensite is important.
In general, martensite can be divided into plate and lath martensite [6]. For this study only the characteristics of lath martensite are of interest, which have been summarized elsewhere [7]. However, important for the understanding of the structural features is the orientation relationship of the fcc to bcc lattice shift during the martensitic phase transformation. The most prominent orientation relationships are the so-called Kurdjumov-Sachs (K-S) and the Nishiyama-Wassermann (N-M) relationship [7, 8].

For the present study it is important to understand the hierarchic structure of laths martensite within the prior austenite grains. The austenite grains transform to packets, which are furthermore divided into blocks, and these blocks are further composed of the martensitic laths. Each packet is transformed from the same habit plane and each block is composed of two so-called variants. The amount of packets within one prior austenite grain varies with the prior austenite grain size. In general theory, one prior austenite grain is constructed of four packets, in correspondence to the four possible habit planes. With decreasing austenite grain size less packets occur per austenite grain, and the packet and block size are decreasing, whereas the lath width was shown to be mostly unaffected [9, 10].

Swarr and Krauss [11], as well as Roberts [12] proposed, that the mechanical properties of martensite can be correlated to the packet size. A decreasing packet size leads to increasing yield strength and a decreasing brittle-ductile transition temperature [11, 12]. Another influence on the mechanical properties can be the occurrence of internal stresses, which can worsen the mechanical properties as well. Qi and Morris [7] stated, that stresses can occur if the 24 K-S variants do not occur with the same volume fraction. If they appear in the same volume fraction over the material this results in pure dilatation and no shear stress within the material [7].

Consequently, this study concentrates on the attributes of the lath martensite structure to evaluate the influences of the austenitizing treatment on the austenite grain sizes, block sizes and inner distortions of the martensite. To this end, different heat treatments are applied and compared to a chosen conventional treatment. The martensitic structures are subsequently evaluated in regards to their transformation kinetics, relationships and microstructure.

Experimental

The steel investigated in this study is a 50CrMo4 with the chemical composition as listed in Table 1.

The initial microstructure, a ferritic pearlitic condition, was produced from the originally rolled 22 mm diameter material according to recommendations from literature [13]. The ferritic pearlitic condition was adjusted through austenitizing the samples at 850°C for 20 min, followed by cooling to 650 °C and a subsequent isothermal treatment at this temperature level for 25 min. Finally the samples were air cooled to room temperature.

Table 1: Chemical composition of the 50CrMo4 steel used in this study. Values are in weight percent and Fe is balanced.

C	Mn	Cr	Mo	Si	P	S	Fe
0.49	0.71	1.05	0.18	0.27	0.016	0.010	bal.

For the dilatometer experiments, samples with a diameter of 4 mm and a length of 10 mm were manufactured at half radius of the original material. Different heat treatments, as listed in Table 2, were conducted using a Bähr DIL850A dilatometer in order to study the phase transformation behavior. All samples were quenched after austenitizing with a cooling rate of 280 K/s (λ of 0.013). All investigations were conducted in the center of the dilatometer sample for good comparison of the individual states.

Phase transition temperatures were defined by determining the temperature when the first derivative of the dilatometer curves is crossing the zero line.

For characterizing the microstructure samples were cut along the long side, grinded and subsequently polished down to 1 µm. For prior austenite grain size evaluation samples were etched with picric acid with sodium dodecyl benzyl acid as a wetting agent [14]. Grain size evaluation was done via optical light microscopy (OLM) and using the intercept method.

Table 2: Nomenclature of the heat treatments and applied heating rates, as well as austenitizing temperatures and holding times.

Name	Austenitizing Temperature [°C]	Heating rate [K/s]	Holding time [s]
Conventional	850	10	1200
1-1	950	1	3
1-2	1000		
1-3	1050		
10-1	950	10	
10-2	1000		
10-3	1050		
100-1	950	100	
100-2	1000		
100-3	1050		

For scanning electron microscopy (SEM) and electron backscattered diffraction (EBSD) samples were ion polished using a Hitachi IM4000plus Ion milling device. EBSD measurements were carried out at a Zeiss EVO MA25 by investigating an area of 70x70 µm at a step size of 50 nm. The data were analyzed using the Orientation Imaging Microscopy (OIM) data analysis software from EDAX and subsequently cleaned using the grain dilatation fit, where an angle of 10.5° was chosen (as seen to be best fit in literature [15]). The analysis of the blocks is done by the phase fraction of certain block sizes using the OIM analysis software. Images of the martensitic structure were prepared by using the Focused Ion Beam (FIB) of the SEM device for etching.

Results

Phase transformation behavior

In a first step the phase transformation temperatures during heating (A_{c1}, A_{c2} and A_{c3}) were determined from the dilatometer curves and are depicted in Figure 1. The focus of the investigations was to study the role of different heating rates on the austenitization behavior. The phase transformation temperatures (A_{c1} and A_{c3}) are shifted to higher temperatures with increasing heating rates. Solely the A_{c2}, the Curie temperature is unaffected with heating rate. For the investigated heating rates the temperature difference between A_{c1} and A_{c3} is not increasing with higher heating rates, thus remaining constant at about 40°C for all three heating rates.

Furthermore, the martensite start temperature was also evaluated from the dilatation curves. The conventional state shows a martensite start temperature of 272°C, whereas the martensite start temperatures for all inductively heated specimens is about 260 ± 2°C, therefore no trend could be observed regarding the different austenitizing treatments.

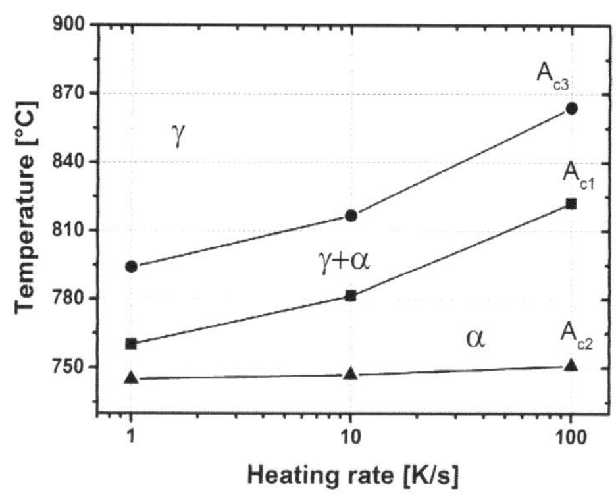

Figure 1: Time temperature austenitizing diagram for the ferritic pearlitic initial microstructure. Circles correspond to A_{c3}, squares to A_{c1} and triangles to A_{c2}.

Characterization of the microstructure

Figure 2 shows SEM-micrographs of the conventional (a), the 10-1 sample (b) and the 100-3 sample (c). Consequently the microstructure from the conventional sample can be compared to the samples heated to 950°C with 10 and 100 K/s. Sample 100-3 exhibits the coarsest microstructure, whereas 10-1 and the conventional sample seem to obtain the same microstructural sizes.

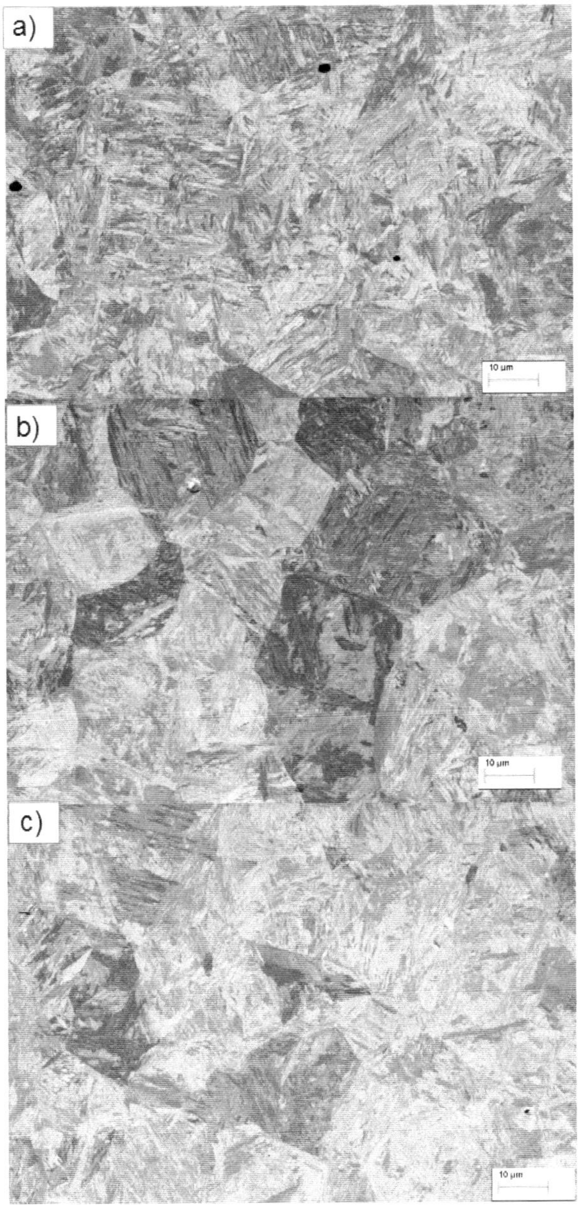

Figure 2: SEM-Micrographs of the conventional (a), the 10-1(b) and the 100-3 sample (c).

To characterize the sizes of microstructural elements in detail the prior austenite grain sizes and the martensite block sizes were evaluated and are depicted in Figure 3.

The conventional sample, depicted by the dotted line, obtains a mean prior austenite grain size of 13.7 µm. No significant difference was found for the heating rates of 10 and 100 K/s for all austenitizing temperatures, whereas the prior austenite grain sizes were higher for the heating rate of 1 K/s. Only 100-1 and 10-1 exhibiting mean grain sizes of 11.9 and 12.2µm, have a smaller grain size than the conventional sample All other inductively heated samples show a higher prior austenite grain size compared to that of the conventional heat treated sample, despite a shorter austenitizing time.

As it is assumed that different prior austenite grain sizes might lead to differences in martensitic block sizes, the block sizes of selected states were evaluated by means of EBSD. The block size, also depicted in Figure 3, follows the trend of the prior austenite grain size. It should be noticed, that 10-1 obtains a smaller block size than specimen 100-3. However, the block sizes follow the same trend as that found for the prior austenite grain sizes. Furthermore, the conventional state obtains a block size between the two selected inductively heated states.

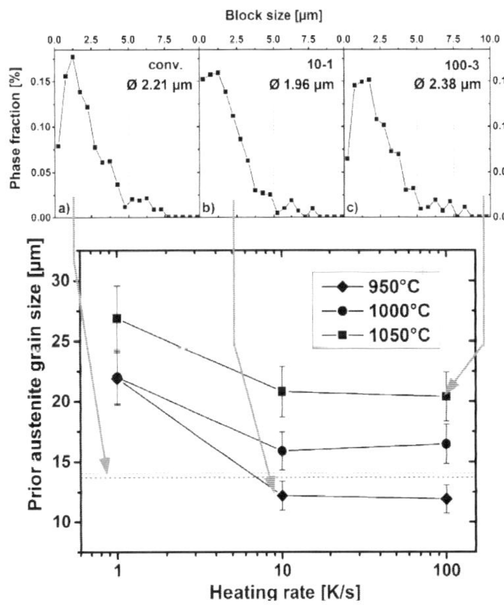

Figure 3: Prior austenite grain sizes depending on the heating rate (bottom). Solid black lines correspond to the inductively heated samples, grey dotted line refers to the conventional treatment. Block sizes for 100-3, 10-1 and conventional are depicted at the top and correlated to the prior austenite grain sizes through gray arrows.

For visualization of the differences in block sizes and the crystallography the inverse pole figures (IPF) of two different states (10-1 and 100-3) are depicted in Figure 4 a) and b), respectively. The thin black lines within Figure 4a) and b) indicate an angle of >10.5° between adjacent areas, thus

encircling blocks [15]. The thick black lines mark a selected prior austenite grain for later analysis regarding the K-S variants.
The IPF images indicate the differences in size of the block and packet structures of the 10-1 and the 100-3 samples. Differences of microstructural sizes, such as coarser structure of 100-3, are as well visible. Furthermore, the sample 10-1 shows more areas of bad indexing (depicted by the black arrows), which may be caused either by small crystal sizes or by higher crystal distortion. The sample 100-3 does show very little areas with microstructural elements, which are badly indexed.

For a more detailed description of the martensitic structures the grain average misorientation was chosen. With this method the misorientation angle within the single blocks, as encircled by thin black lines in Figure 4, was determined. The results are shown in Figure 5. The bars represent the conventional state, which shows that 12 % of the blocks show very low or no inner misorientation. The samples heated to 1050°C (100-3 and 10-3) show the same trend as the conventional sample, peaking at around 1°, but they only show 4% of the blocks with very low or no misorientation. The sample 10-1 shows a shift in the peak to 1.3°, thus indicating a higher amount of misorientation across the blocks in the scan area.
For further investigations of the martensite quality and occurring micro strains and misorientations, pole figures were recorded. To this end selected prior austenite grains were cropped out of the inverse pole figure and the pole figures were subsequently analyzed. For all samples various prior austenite grains were analyzed. Depicted are the variants for the austenite grains with the best agreement to the computed K-S variants. For convenience the pole figures were rotated to obtain a better comparability with the computed K-S pole figures [7].
Figure 6 compares the pole figures of four different heat treatment conditions with a computed one (Figure 6a). Figure 6b shows a pole figure that was determined from a region within one prior austenite grain of the conventional sample. Figure 6d and c show inverse pole figures from the marked austenite grains of the depicted heat treatment conditions 10-1 and 100-3 in Figure 4. From these four pole figures it can be concluded, that the martensitic transformation is occurring with a K-S relationship. Nevertheless, for the conventional heat treatment condition (Figure 6b) all K-S variants are equally pronounced, whereas for 10-1 (Figure 6d) some variants seem to be less pronounced (indicated by the red arrow). Additionally, the intensity of the single K-S points is not as high as for the conventional sample. However, the samples 100-3 and 10-3 (Figure 6c/e) show an increasing intensity regarding the K-S locations, but as well do not show the same intensity for every variant.

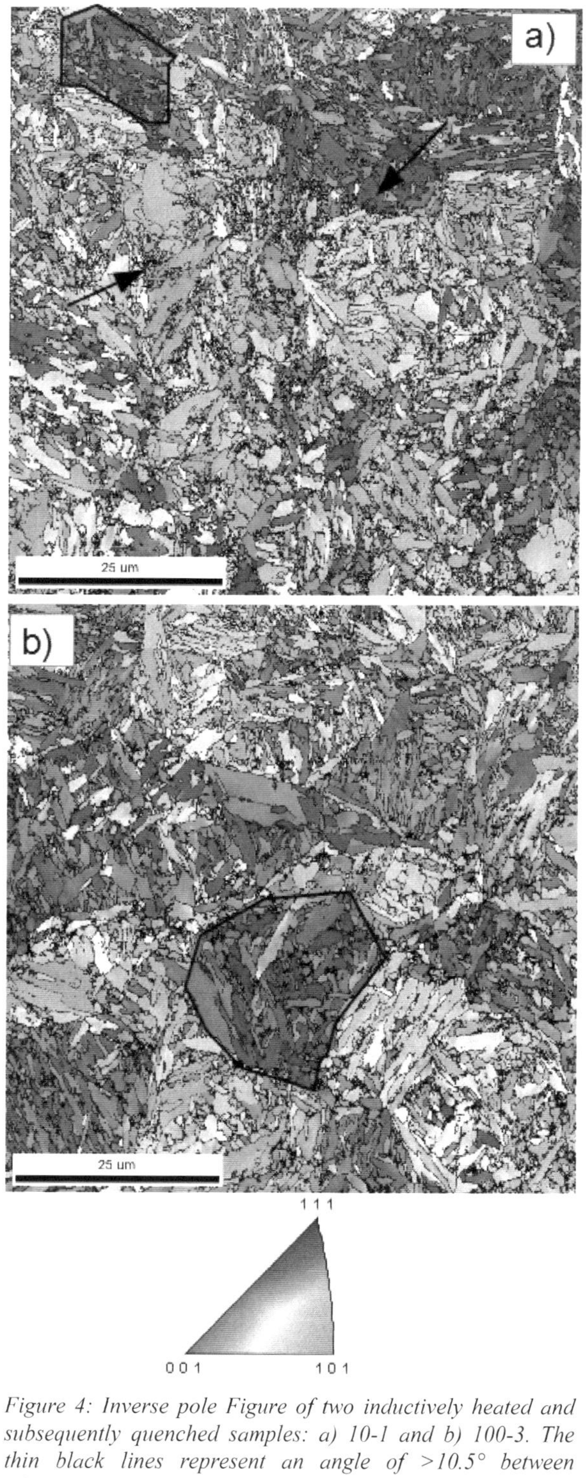

Figure 4: Inverse pole Figure of two inductively heated and subsequently quenched samples: a) 10-1 and b) 100-3. The thin black lines represent an angle of >10.5° between adjacent areas, thus encircling blocks. The thick black lines indicate the selected prior austenite grain for the pole figures shown in Figure 6. Black arrows indicate small microstructural areas with bad indexing.

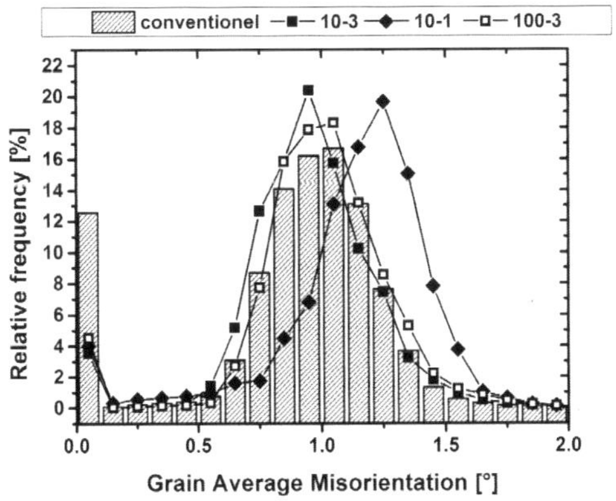

Figure 5: Relative frequency of the grain average misorientation across blocks for different heat treated samples. The bar chart represents the conventional state, whereas the squares correspond to the samples heated to 1050°C (closed 10-3, open with 100-3) and the diamonds represent a heating to 950°C with 10 K/s (10-1)

Discussion

As depicted in Figure 1 the heating rate is largely affecting the transformation kinetics from the ferrite (α) to austenite (γ). In accordance with literature [4, 13], A_{c1} and A_{c3} are shifted to higher temperatures with increasing heating rates. A_{c2}, that indicates the magnetic transformation (Curie-Temperature), is almost unaffected by the heating rate.

Due to this difference of transformation kinetics, the chosen austenitizing temperatures can be assumed to result in different austenitizing conditions, since the temperature difference between the austenitizing temperature to A_{c3} is less distinct for a heating rate of 100 K/s (~80°C for 100-1) than for 10 K/s (~120°C for 10-1) and 140°C for a heating rate of 1 K/s. Nevertheless, these differences seem to have no impact on the austenite grain growth rate, since the prior austenite grains are almost of similar size for 10 K/s and for 100 K/s and not increasing drastically for a heating rate of 1 K/s (Figure 3). However, no conclusions on the homogeneity of the formed austenite can be drawn.

The constant martensite start temperatures (M_s) for the different heat treatment conditions may cause misleading conclusions. Two factors are thought of as the main influences, affecting M_s: (i) an increase in prior austenite grains size and (ii) an increase in carbon dissolved within the matrix is decreasing the M_s [16, 17]. It is not clear how these effects are superimposed, or if either of them is pronounced enough to affect the martensite start temperature. The detailed effect must be studied in more detail, but is not part of this investigation.

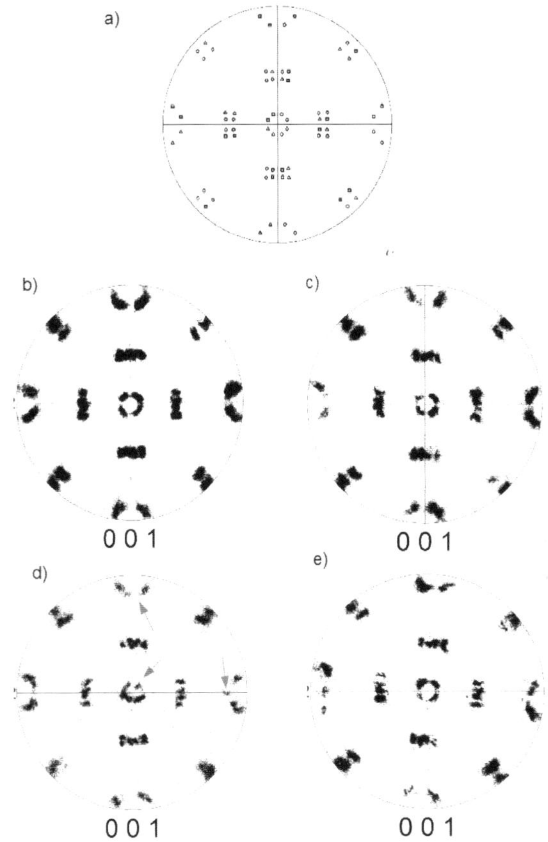

Figure 6: Pole Figures measured within one prior austenite grain: a) computed K-S variants according to [7], b) conventional state, c) pole figure of 100-3 from the prior austenite grain marked in Figure 4b) d) pole figure from 10-1 of the prior austenite grain marked in Figure 4a, and e) pole figure from 10-3. Red arrows are pointing at less pronounced variants.

Comparing the micrographs of the conventionally hardened material and the 100-3 state (Figure 2) it is obvious, that the material heated to 1050°C exhibits a coarser microstructure than the conventionally heated material, despite the shorter heat treatment. This leads to the conclusion that the austenitizing temperature plays a more important role concerning the microstructure development, than the heating rate or the holding time.

As already stated by Furuhara [10] it was observed in this work, that the block size is also correlating with the prior austenite grain size, which leads to a larger block size with increasing prior austenite grain size. Since the block size is correlating with the packet size, it can be expected, that the packet size is increasing as well. Since a small packet size is desirable for good mechanical properties, especially toughness [11], it could be expected that heating to an austenitizing temperature of 1050°C is not beneficial for the toughness of the material, in regards to the microstructural size, since the prior austenite grain size and consequently the packet size are increasing compared to conventional heat treatment.

The inverse pole figures of the different investigated states show the trend of a smaller block and packet size for lower austenitizing temperatures. The structure of the sample heated to 950°C (10-1) is finer than the one heated to 1050°C (100-3), but, as well, shows increasing areas with higher distortions (represented by the small, badly indexed areas as indicated exemplarily by arrows in Figure 4). Despite the fact, that sample 10-1 depicts a very good image quality for the FIB contrast (Figure 2b), a lot of inner distortion became visible using EBSD.
The grain average misorientation of 10-1 shows a higher average misorientation as well, which represents the misorientation along a block. Since the conventional heat treatment condition and the samples heated to higher temperatures show less grain average misorientation, it can be assumed that more imperfect prior austenite grains exist in the inductively heated specimens with the lower maximum temperatures. Hence, this fact would lead to imperfect martensite blocks. Since the block is composed of a variant pair with a low misorientation, an increased misorientation between this variant pair is resulting in a shape strain across the block according to Morito et al. [15]. It is expected, that such a strain across the blocks might cause inferior mechanical properties, such as lower fracture toughness.

Regarding the K-S variants of the prior austenite grains, it can be seen, that the conventional sample follows the K-S relationship and exhibits all K-S variants in a selected prior austenite grain (Figure 6). The sample 10-1 shows a K-S relationship, but it is obvious, that several variants are less pronounced or not existing (depicted by red arrows in Figure 6d) within a prior austenite grain, thus resulting in stress across the austenite grain, which is suggested by Qi and Kinney [7, 18]. The conventionally heat treated sample, however, shows all 24 K-S variants with the same intensity, although the concrete positions of the variants are broadened compared to all inductively heated states. Similar results have been observed in an earlier study [5]. Both to 1050°C inductively heated samples show almost the same tendency regarding the K-S variants, despite the different heating rates. In these variants, some are not as pronounced as others, but the pole figure is closer to the one from the conventional state, regarding the broadness of the variant points, than samples inductively heated to lower maximum temperatures. Consequently, compared to the 10-1 sample it can be expected, that the strain over the austenite grain is less distinct for both samples heated to 1050°C. The pole figures are not as homogeneous as the one of the conventionally heat treated sample, which might be a result of the slightly higher grain average misorientation of these two conditions (Figure 5) compared to the conventional one. Resulting from this information, it seems that the austenitizing temperatures have more impact on the internal strains and stresses of the as-quenched structure than the heating rate.

It has to be mentioned that it is not yet clear how the formerly highlighted facts are affected during a subsequent tempering treatment, which is common for such steels. This is subject of further investigations.

Conclusions

The following conclusions can be drawn from this study:

- Especially for high heating rates (10 and 100 K/s), the heating rate has less impact on the microstructure of the material than the austenitizing temperature. Microstructural investigations show the same prior austenite grain size for different heating rates with the same austenitizing temperatures. Prior austenite grain size increased with increasing austenitizing temperature, despite shorter treatment times through high heating rates.
- As indicated by the measured pole figures, it is assumed, that the austenite formed at low austenitizing temperatures had not sufficient time to form perfectly, thus resulting in lacking K-S variants. Therefore it is assumed, that mechanical properties are negatively affected compared to the conventionally heat treated material, because of local strain in various blocks.
- Furthermore, higher austenitizing temperatures enhance a more even appearance of the K-S variants than an increase in heating rate.
- Transformation kinetics from ferrite to austenite show the expected trend, that the transformation temperatures are shifted to higher temperatures for higher heating rates.
- It could be confirmed, that the martensite block size is correlating with the prior austenite grain size.

Resulting from this study it is recommended to use rather a higher austenitizing temperature than a slower heating rate to achieve less distorted microstructure elements. A reasonable compromise needs to be done regarding the final austenitizing temperature to still obtain a beneficial packet size.

Acknowledgments

Financial support by the Austrian Federal Government (in particular from Bundesministerium für Verkehr, Innovation und Technologie and Bundesministerium für Wissenschaft, Forschung und Wirtschaft) represented by Österreichische Forschungsförderungsgesellschaft mbH and the Styrian and the Tyrolean Provincial Government, represented by Steirische Wirtschaftsförderungsgesellschaft mbH and Standortagentur Tirol, within the framework of the COMET Funding Programme is gratefully acknowledged. The authors want to thank Bernhard Sartory for the help with the EBSD measurements.

References

[1] Rudnev, V. *Handbook of induction heating*. Dekker, (2003).

[2] Zinn, S. and Semiatin, S. *Elements of induction heating: Design, control, and applications*. EPRI, (1988).

[3] Clarke, K., VanTyne, C., Vigil, C., and Hackenberg, R. *Journal of Materials Engineering and Performance* **20** (2), 161–168 (2011).

[4] Cunningham, J., Medlin, D., and Krauss, G. *Journal of Materials Engineering and Performance* **8** (4), 401–408 (1999).

[5] Sackl, S., Leitner, H., Zuber, M., Clemens, H., and Primig, S. *Metallurgical and Materials Transactions A* **45** (12), pp. 5657–5666 (2014).

[6] Krauss, G. *Materials Science and Engineering A* **273-275**, 40–57 (1999).

[7] Qi, L., Khachaturyan, A., and Morris, J. *Acta Materialia* **76** (0), 23–39 (2014).

[8] Iwashita, K., Murata, Y., Tsukada, Y., and Koyama, T. *Philosophical Magazine* **91** (35), 4495–4513 (2011).

[9] Maki, T. and Wayman, C. *Acta Metallurgica* **25** (6), 695–710 (1977).

[10] Furuhara, T., Morito, S., and Maki, T. *Journal De Physique. IV* **112 I** (2003).

[11] Swarr, T. and Krauss, G. *Metallurgical Transactions A* **7**(1), 41–48 (1976).

[12] Roberts, M. *Martensite Fundamentals and Technology*, 119–136 (1970).

[13] Orlich, J. *Atlas zur Wärmebehandlung der Stähle*. Max-Planck-Institut für Eisenforschung, (1973).

[14] Dreyer, G., Austin, D., and Smith, W. *Metal Progress* **16**, 116 (1963).

[15] Morito, S., Yoshida, H., Maki, T., and Huang, X. *Materials Science and Engineering: A* **438-440**, 237 – 240 (2006). Proceedings of the International Conference on Martensitic Transformations.

[16] Kitahara, H., Ueji, R., Tsuji, N., and Minamino, Y. *Acta Materialia* **54**(5), 1279 – 1288 (2006).

[17] Bleck, W. *Werkstoffkunde Stahl für Studium und Praxis*. Mainz, Aachen, 2nd Edition, (2004).

[18] Kinney, C., Pytlewski, K., Khachaturyan, A., and Morris, J. *Acta Materialia* **69**, 372–385 (2014).

International Federation of Heat Treatment and Surface Engineering 2016
Proceedings of the 23rd IFHTSE Congress
April 18–21, 2016, Savannah, Georgia, USA

Minimizing Case / Core Initiated Failures in Induction Hardened Components

Michael A. Pershing
Caterpillar Inc, East Peoria, IL USA

Gary Raab
Caterpillar Inc, Mossville, IL USA

Weizhou Li
Caterpillar Inc, Mossville, IL USA

Abstract

Induction case hardening experiments were performed by heat treat simulation for a large section cylinder of SAE 4140 steel. The simulation is first calibrated to experimental induction hardening results on SAE 4140 steel. Results are presented showing hardness vs. depth and residual stress vs. depth results on prior heat treat and induction processing parameter combinations. The hardness and residual stress levels at the case/core interface and transition region are examined. The goal is to identify material and processing combinations that will prevent case to core transition initiated failures in pins or shafts with high bending or torsional loads. Results are discussed with this in mind. The use of heat treat simulation aids in this thought process and saves laboratory test time and money.

Introduction

Surface hardening using induction is a common and useful heat treatment that provides excellent localized hardness and can provide hardness and strength to a defined depth. The near surface or case hardness normally leaves the part in compressive residual stress state at the surface and throughout the hardened depth area. However, the transition region typically experiences a residual stress that becomes tensile and stays tensile into the core region. X-ray diffraction measurements of residual stress in the axial and circumferential (hoop), and radial direction have shown this on several induction hardened components.

The general residual stress pattern encountered in induction hardened parts and the issue of is explained in several references [1,2,3]. Work by Kristoffersen and Vomacka [4] and also Xu and Kuang [5] show residual stress patterns from varying induction process parameters and the influence those stress patterns will have on fatigue is discussed. Subsurface failure from the case to core transition region is of particular concern. This issue is especially prevalent in heavy bending applications or heavy torsional applications where the applied stress stays high throughout a large portion of the cross-section of the shaft or cylinder. Typically, the surface is hardened to a relatively high surface level and additionally it is also in compressive residual stress. This prevents failures in the near surface region. The problem is that as the hardness drops and the depth of hardening to HRC 50 or HRC 45 is reached, the residual stress changes to a tensile residual. There is a further problem of tempering otherwise strong quench and tempered material in the core if a prior quench and temper heat treat operation has been completed. With the drop in hardness and strength occurring at the same depths as the residual stress is turning tensile, the applied stress has an opportunity to exceed the fatigue limit.

It is important to understand this issue in larger shafts (50-150mm diameter) with hardened depths of 3-12 mm. These will typically be heavily loaded shafts or cylinders where the case to core transition region is prone to low cycle fatigue failure. In order to alleviate this condition, it is desired to determine induction parameters that will help minimize this condition. The influence of further tempering of the prior quench and tempered heat treatment is also a factor to be considered.

This paper first uses an experimental result from an induction hardened SAE 4140 cylinder and then uses that result to be sure the fundamental induction heat treat simulation parameters are correct. It then uses heat treat simulation to examine hardness and residual stress patterns for two different scan speed and power level combinations and the corresponding hardened depth. It also examines an SAE 4140 quench and tempered cylinder and compares it to an as-received SAE 4140 bar.

Experimental Methods

Material

The material used in these studies was 4140 bar stock in the as-rolled condition and then induction hardened. The chemistry of the bar stock is shown in Table 1. The 4140 bar stock was machined to fit a scanning quench induction coil. After machining, the bars were induction hardened to a case depth of

6.5mm at 50HRC, the speed and power were varied to maintain consistent hardened depth.

Table 1: Chemistry of the steel induction hardened in this study.

	SAE 4140
C	0.43
Mn	0.88
P	0.008
S	0.009
Si	0.24
Ni	0.09
Cr	1.04
Mo	0.20
Cu	0.16
V	0.003
Al	0.024

Induction Parameters

The bars were hardened on an Ajax/Tocco induction scanning system, fitted with a 10kHz heating station delivering a maximum power of 400kW. To achieve the desired hardened depth in the bar the induction parameters shown in table 2 were used. Also shown are the parameters later utilized by heat treat simulation. The bar was water quenched during scanning and the quench was left on for 140sec after scanning to cool to handling temperature.

Table 2: Induction parameters used.

Sample	Coupling Distance (mm)	Scanning Speed (mm/sec)	Coil Frequency (kHz)	Power setting (kW)	Material Condition
1	18.6	3.81	10	152	As-rcv'd
2*	18.6	1.27	10	65	As-rcv'd
3*	18.6	3.81	10	152	Q&T
4*	18.6	1.27	10	65	Q&T

**- indicates simulation only*

Residual Stress Measurement Technique

To obtain residual stress measurements, x-ray diffraction was used to determine the stresses present in both the circumferential and axial directions (Figure 1). To gather stress readings the test area was electro-polished and measurements were taken between each short electrolytic polish and repeated until the desired depth was achieved (Figure 2). The bar is periodically ground to remove excess material when measuring deeper case depths (Figure 3). The equipment used was an AST X2002 X-ray stress analyzer, with a 2.65mm collimator. Young's Modulus of 231,840 MPa was used.

Figure 1: Test bar in the AST X2002 ready to be tested.

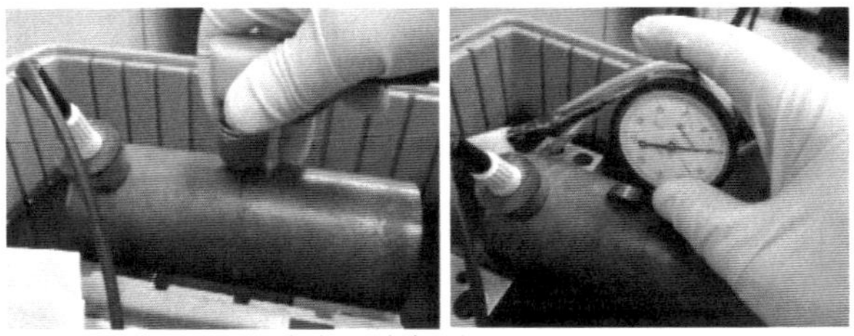

Figure 2: Electropolish and depth measurements

Figure 3: Bar during testing, excess material was removed by grinding to allow room for stress analyzer to operate.

Residual Stress and Hardness Results

The residual stress measurements were taken to a depth where the stress leveled off. In this case measurements were stopped at a depth of 16mm. For comparison, micro-hardness and contour hardness was taken to the same 16mm depth. The results were plotted against depth in Figure 4.

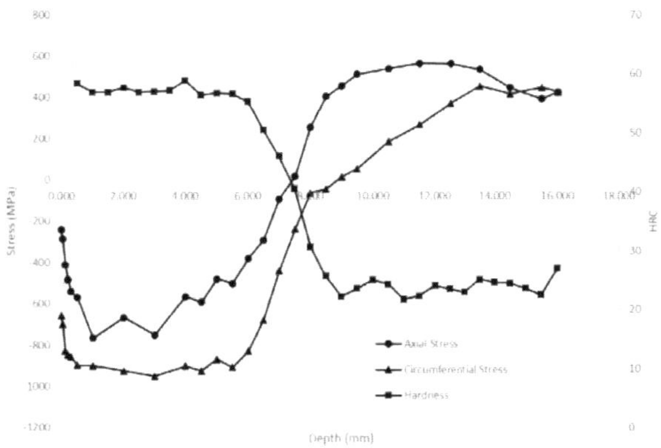

Figure 4: Plot of Hardness and Residual Stress vs depth.

Heat Treat Simulation

Caterpillar Inc. started to develop a heat treat simulation package called QSIM [0 in the late 1980s. This predictive tool has gradually become an integral part of heat treat process development, troubleshooting, and material selection in Caterpillar. As discussed in previous papers [6], it includes thermodynamics data, transient behavior, phase transformation, related transformation strain and plasticity data, and each individual constituent stress-strain curve under a defined temperature range.

The temperature field is governed by a standard heat transfer equation:

$$\nabla \bullet (k\nabla T) + \dot{q} = \rho c_p \dot{T} \tag{1}$$

where ρ, c_p, k and q are density, heat capacity, heat conductivity and internal heat generation, respectively. The latent heat associated with phase transformations is considered in the model by an enhanced specific heat method. All the material properties are defined as functions of temperature and microstructure.

The most significant factor in transient thermal analysis during steel heat treatment is heat extraction rate at a part boundary. This heat extraction rate is taken into form as:

$$\left(\frac{dQ}{dt} \right)\Bigg|_{\Gamma} = h[T(t,x)_s - T_\infty] \tag{2}$$

where Γ is the part boundary, T_∞ is the equivalent ambient temperature, h is the effective heat transfer coefficient, and $T(t,x)_s$ is the time and location dependent surface temperature.

During the quench process, three stages of cooling occur: vapor blanket, boiling, and convection [7[7]]. The three stages of cooling make the heat extraction rate a strong function of part surface temperature, quenchant properties, and the relative velocity of the steel part and the quenchant. The effective heat transfer coefficient is determined by the quench characterization scheme for the given quenchant and part geometry.

Austenite decomposition is a key element in modeling microstructure formation during the quenching of steel. Many research efforts have been made in describing austenite decomposition kinetics [8[8],9,10,11]. There are two types of phase transformation involved during heat treatment: diffusion and diffusionless transformations. For diffusion-based transformations, the fraction of the transformed constituent, f_T, is a function of time t, nucleation rate N, and growth rate G. It can be expressed as [12[12]]:

$$f_T(t) = 1 - \exp[-(\pi/3)NG^3t^4] \tag{3}$$

where nucleation rate and growth rate are dependent on temperature, diffusion coefficient of carbon in austenite, free energy barrier and austenite grain size.

For diffusionless transformation, that is when austenite transforms into martensite, martensite-start temperature is first calculated as a function of chemistry. The percentage of martensite is then calculated according to the transformation database deducted from the measured dilatometric curve for the particular alloy, in this case it is for SAE 4140.

The total strain in this quasi-static elastic-plastic stress model can generally be expressed as:

$$\dot{\varepsilon}_{ij} = \dot{\varepsilon}_{ij}^{e} + \dot{\varepsilon}_{ij}^{p} + \dot{\varepsilon}_{ij}^{tp} + \dot{\varepsilon}_{ij}^{tr} + \dot{\varepsilon}_{ij}^{th} \tag{4}$$

where the superscripts ε, p, tp, tr, and th represent the strain caused by elastic, plastic, transformation plasticity, transformation strain and thermal strain. For multiphase mechanical response, behavior of each individual pure micro-constituent was evaluated from uniaxial tensile tests under various temperatures. For 4140, dilatometric data was measured to determine transformation strain and that information is used when 4140 material is chosen.

Induction

The physical process of induction is a complex process. It is a coupled process of electromagnetic, heat transfer, phase transformation and mechanical stress evolution.

The electromagnetic process is governed by the Maxwell equation:

$$\begin{aligned} \nabla \times \mathbf{H} &= \mathbf{J} + \frac{\partial \mathbf{D}}{\partial t} \\ \nabla \times \mathbf{E} &= -\frac{\partial \mathbf{B}}{\partial t} \\ \nabla \bullet \mathbf{B} &= 0 \\ \nabla \bullet \mathbf{D} &= \rho^{charge} \end{aligned} \tag{7}$$

Here **E** is electric field intensity, **D** is electric flux density, **H** is magnetic field intensity, **B** is magnetic flux density, **J** is conduction current density, and ρ^{charge} is electric charge density.

The constitutive equations are:

$$\begin{aligned} \mathbf{D} &= \varepsilon\varepsilon_0 \mathbf{E} \\ \mathbf{B} &= \mu\mu_0 \mathbf{H} \\ \mathbf{J} &= \sigma_e \mathbf{E} \end{aligned} \tag{8}$$

Here, ε, μ_r, and σ_e are relative permittivity, magnetic permeability and electrical resistivity, respectively.

For ferromagnetic material like steel, the relation between **B** and **H** is highly nonlinear and history dependent (hysteresis effect) (Figure 5). In induction hardening application, the heating is mainly due to Joule heating from inductive eddy current. Hysteresis heating can be neglected [[13]. The B-H relation is also temperature dependent. A significant drop in magnetic permeability happens around the Currie point where steel transfers from ferromagnetic material to paramagnetic.

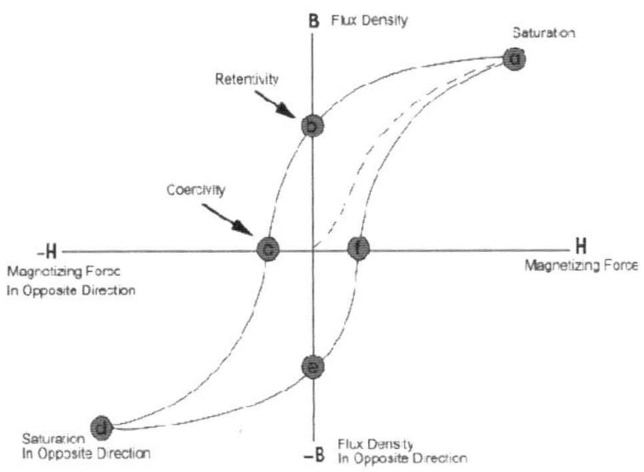

Fig. 5 B-H relation in ferromagnetic material

By neglecting the hysteresis effect, the Maxwell equation can be converted to a time harmonic form. By using magnetic vector potential, the equation can be further reduced to:

$$\frac{1}{\mu_r\mu_0}\nabla^2\mathbf{A} = -\mathbf{J}_s + j\varpi\sigma_e\mathbf{A} \tag{9}$$

Where, **A** is magnetic vector potential, $\mathbf{J}_s$ is source current density, and ω is oscillation period. By applying this form, the effect of time history only comes through the temperature dependency of electrical conductivity and magnetic permeability.

A coupled analysis flowchart of thermo-electromagnetic is shown in Figure 6. The electromagnetic calculation was first performed by assuming an initial magnetic field density. Further iterations by calculating **H** distribution were performed till convergence. Joule heat value then was calculated as internal heat source for heat transfer analysis. Resulting temperature was used for next thermo-electromagnetic analysis.

Here, a sequential coupling scheme was used when calculating thermo-electromagnetic interaction.

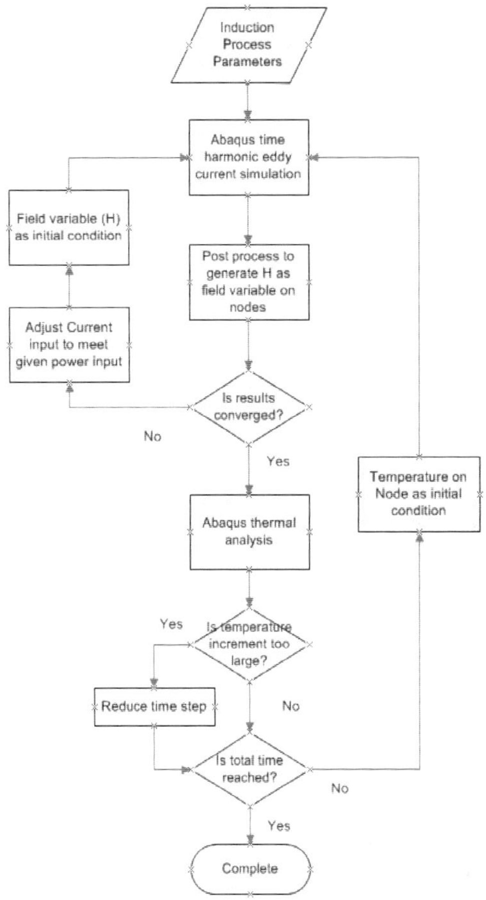

Fig. 6 Flow chart for induction calculation

Once the induction heating process calculation is completed, the heating history is used for phase transformation calculation to determine austenization volume fraction. Similar to phase transformation during quenching, the austenization process is a time history and temperature dependent process. The Johnson-Mehl-Avrami equation is used [14]

$$f = 1 - \exp(-b_k t^{n_k}) \tag{10}$$

Here, f is austenite volume fraction, and b_k and n_k are transformation kinetic constants. [15[15]]

An additive rule is applied to calculated austenite volume fraction. To simplify the model, grain growth was neglected.

Results and Discussion

Two finite element models were built for this induction hardening simulation: 1) Electromagnetic model with cylinder bar, heating coil and surrounding space meshed; In this model, the movement of the coil was accomplished by redefining the coil element set at each time step. 2) Heat transfer, phase transformation and stress evolution model with only the cylinder bar meshed. The bar length in the model is long enough to avoid any end effects.

Induction simulation was first performed on a baseline condition with 3.81 mm/s (0.15 in/s) scanning speed and 152kW power supplier output at 10kHz. Since the modeling domain only includes the coil and part region and, due to power loss, the power on the coil is different from the output coming from the power supply, a calibration process is performed first to identify the system efficiency. Figure 7 shows the results of the calibration.

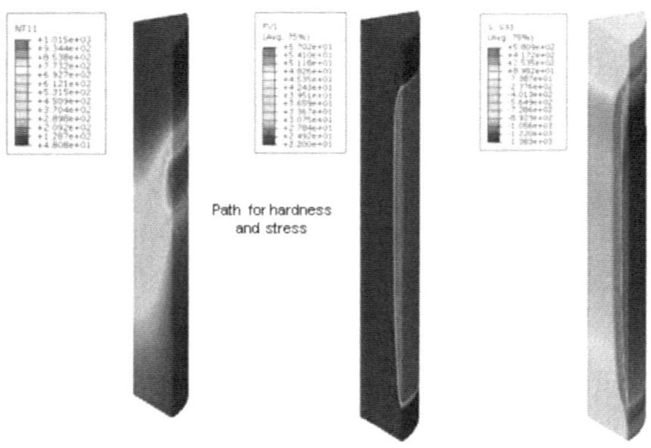

Fig. 7 General induction hardening results: (1) Temperature during scanning; (2) Hardness; (3) Residual stress.

After calibration, the simulation predicted hardness, axial residual stress and circumferential residual stress are compared with the measured results for this bar (sample 1 from Table 2). The comparison is shown in Figure 8.

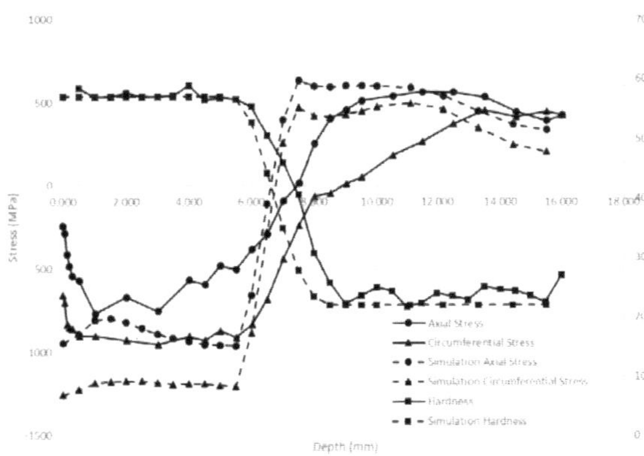

Fig. 8 Hardness and residual stress comparison for baseline case.

From this comparison, hardness roughly agrees with measurements, although the simulation predicted a rather narrow transition zone compared to actual measurements. The transition size is mainly affected by temperature gradient and transformation kinetics. For the residual stress comparison, simulation predicted the proper trend and level of residual stress. However, the very sharp transition of residual stress from compression to tensile has not been observed in measurement. One possible reason is that the simulation did not take self-tempering effects on residual stress relaxation into account.

With a good comparison between experimental results and heat treat simulation completed, further simulations were conducted. Analysis of a quench and tempered starting condition run at the same 3.81 mm/s (0.15 in/s) scan rate was performed followed by a slow 1.27 mm/s (0.05 in/s) scan rate for the two starting material conditions of SAE 4140 (as-received pearlite/ferrite and tempered martensite from a prior quench and temper operation). For the slow scan rate, the power was adjusted to give similar hardened depth. Figure 9 shows the hardness results from these four simulations:

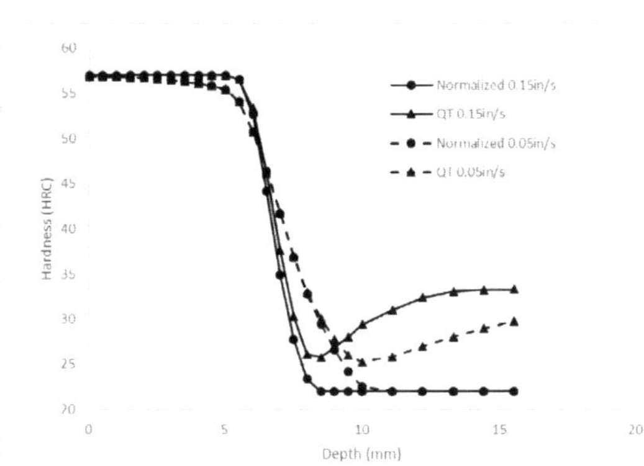

Fig 9. Hardness from different scanning speed and initial microstructure resulting from heat treat simulation.

With a much slower scanning speed, the transition zone turns out to be a lot wider. This is mainly due to more even temperature distribution during heating, which leads to a large area with partially transformed austenite. After cooling, this forms a mixture of martensite and non-transformed pearlite/ferrite structure with the hardness values staying between martensite and pearlite/ferrite through the broader transition zone. For a quench/tempered microstructure, the core hardness is higher than the as-received hardness. For the simulation on quench and tempered pieces, the hardness dropped below the core hardness in the heat affected zone due to modeling of the further tempering coming from the induction heating. The slower scan rate extends the softened zone created by the additional heat from slow scanning.

Figure 10 shows the axial residual stress comparison: The overall trend is that slow scanning speed produces lower tensile residual stress after the transition zone. The peak tensile residual stress is about 200MPa lower for the slower scanning cases. This can be attributed to thermal gradient effect: low power and low scanning speed creates a shallow temperature gradient. Shallow (low) temperature gradients lead to less stress. This gives us a potential option to adjust the induction process with slower scan speed in order to improve the overall fatigue strength of a shaft or pin.

Comparing residual stress profiles predicted by simulation between the two different initial microstructures, there is only a very slight difference. The piece that was initially quench/tempered finishes with a higher tensile peak stress for the fast scanning case. For slow scanning, the difference is even smaller. This could be due to tempering effects reducing a large volume of core material to a similar hardness as in the as-received condition.

The loss of hardness in the transition zone on quench and tempered steels leads to a further complication in design of the product and process. The benefits of holding a strong core in the quench and temper step may be deceptive because of the loss of hardness due to the induction hardening step. This loss in hardness occurs right at the point where residual stress is turning to tensile. Opting for a slower scan rate reduces the tensile residual stress just behind the hardened case but it also extends the zone of lower hardness. Both of these effects must be considered to properly avoid fatigue failures at this point.

Use of as-received 4140 allows a slower scan speed to be used with full confidence that tensile residual stress will be reduced behind the hardened case. Of course, the core is typically of lower strength (lower hardness) and has ferrite grains for easier nucleation of fatigue cracks.

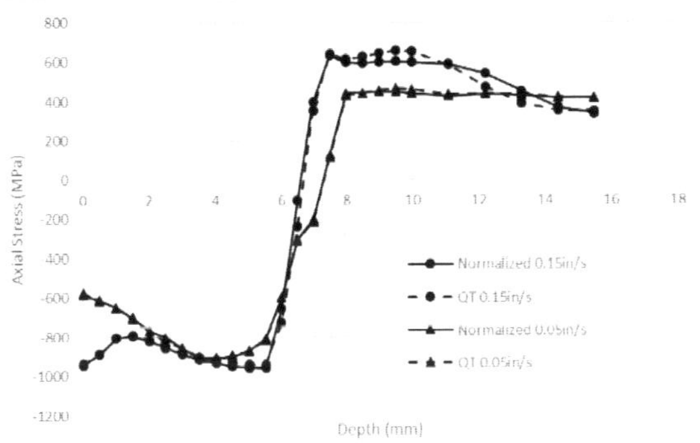

Fig 10 Axial residual stress from different scanning speed and initial microstructure.

Conclusions

1. Experimental hardness and residual stress results from an induction scanned and hardened 4140 steel in the as-received (ferrite/pearlite) state were favorably compared to a heat treat simulation using the same induction parameters and steel material.

2. The transition zone at the case /core interface was predicted to be sharper than what was measured in the heat treated piece, both for hardness and residual stress although the difference in residual stress measurement was more stark. This difference is speculated to be due to the lack of modeling the self tempering effects on residual stress in the simulation model.

3. Heat treat simulations comparing scan rates and prior heat treatment conditions showed that a slow scan rate lowers the tensile residual stress behind the case / core interface for both 4140 steel starting in the as-received condition and in the quench and tempered condition.

4. Heat treat simulations showed 4140 with a prior quench and temper heat treatment suffered a loss of hardness behind the case/core interface that could further negatively affect fatigue strength.

5. Heat treat simulation also indicated that the slow scan speed lengthens the zone of a softened area behind the case/core interface in 4140 steels with a prior quench and temper heat treatment.

References

[1] Haimbaugh, R., Practical Induction Heat Treating, ASM International (Materials Park, OH, 2001), pp. 114-115.

[2] Weiss, K., Rudnev V., Cook, R., Loveless, D., and Black, M., "Induction Tempering of Steel", *Advanced Materials and Processes,* August, 1999.

[3] Rudnev, V., Loveless, D., Cook, R., and Black M., Handbook of Induction Heating, Taylor & Francis (New York, 2003), pp. 300-301.

[4] Kristoffersen H, and Vomacka, P., "Influence of Process parameters for Induction Hardening on Residual Stresses", *Materials and Design,* vol. 22, 2001, pp. 637-644.

[5] Xu D-H, Kuang Z-B, "A Study on the Distribution of Residual Stresses Due to Surface Induction Hardening", *J. Eng. Matls Tech,* Vol. 118, 1996, pp. 571-575.

[6] Chuzhoy L., Clements T., McVicker J., and Burris K., 1993, "Numerical Simulation of Quenching Process at Caterpillar", SAE Transaction 931172.

[7] Bodin J. and Segerberg S., 1992, "Quenching and Distortion Control", G.E. Totten Eds., *ASM International,* Material Park, Ohio.

[8] Li M., Niebuhr D., Meekisho L., and Atteridge D., 1998, "A Computational Model for the Prediction of Steel Hardenability", Metallurgical and Material Transactions B, 29B

[9] Kirkaldy J. and Venugopalan D., 1983, "Phase Transformation in Ferrous Alloys", Marder D. and Goldstein J., eds, *AIME,* New York, NY, pp128-48.

[10] Pont D., Bergheau J., Rochette M., Fortunier R., 1994, "Identification of a kinetic model for anisothermal metallurgical transformations in steels", *Inverse Problems in Engineering Mechanics,* Bui H., Tanaka M. et al, eds, Balkema, Rotterdam, pp151-156.

[11] Leblond J. and Devaux J., 1984, "A new kinetics model for anisothermal metallurgical transformation in steel including effect of austenite grain size", Acta Metalugica, 32, pp137-146.

[12] Brooks C., Principles of the Heat Treatment of Plain Carbon and Low Alloy Steels, ASM International (Materials Park, OH, 1996)
[13] Rudnev, V., Loveless, D., Cook, R., and Black M., Handbook of Induction Heating, Taylor & Francis (New York, 2003), p. 153.
[14] Johnson W. and Mehl R., 1939, Tans. AIME, 135.
[15] Denis S., Farias D. and Simon A., 1992, "Mathematical Model Coupling Phase Transformations and Temperature Evolution in Steels", ISIJ international, 32, pp316-325.

International Federation of Heat Treatment and Surface Engineering 2016
Proceedings of the 23rd IFHTSE Congress
April 18–21, 2016, Savannah, Georgia, USA

Induction Camshaft Hardening Technology with Practically Undetectable Distortion

Gary Doyon, Dr.Valery Rudnev, John Maher, Randy Minnick, Glen Desmier
Inductoheat Inc., An Inductotherm Group Company
32251 N. Avis Dr., Madison Heights, MI 48306 USA
gry@inductoheat.com; rudnev@inductoheat.com; jmaher@inductoheat.com; rjm@inductoheat.com; gcd@inductoheat.com

Abstract

This presentation reviews various induction camshaft hardening technologies presently used in industry. The Pros and Cons of different processes will be analyzed. The causes for hardness pattern deviation and excessive distortion will be assessed. Advanced non-rotational CamPro™ camshaft hardening technology will be discussed. This process lends itself for hardening several camshaft lobes. Therefore, high production rates can be achieved by statically heat treating multiple cam lobes simultaneously.

One of the major advantages of this non-rotational camshaft hardening technology is related to its ability to harden camshafts with practically undetectable distortion, minimizing needed grinding stock and, in some cases, completely eliminating the necessity of the entire subsequent camshaft straightening operation.

Another attractive feature of this process is related to its ability to produce uniform true contour hardness patterns, regardless of the complexity of lobe profile, its size, topology and topography of the camshaft. Heat patterns are "locked in place" and very repeatable since neither the camshaft nor the inductors are moving during the heat treating operation.

Depending upon the specifics of the camshaft's geometry and heat treat specifications, the combined savings on energy consumption may exceed 18% comparted to commonly used processes.

Introduction

Induction hardening using electromagnetic induction is a common method for heat treating different automotive components including camshafts, which belong to the group of critical engine/powertrain components.

A camshaft controls the timing and the speed of the opening and closing of the intake and exhaust valves. It consists of several sets of cam lobes and bearings (Fig.1). Number of lobes, their size, profile, positioning and orientation are dependent upon the camshaft type and engine specifics. Fig. 2 shows some examples of a variety of cam lobe shapes. Gray and nodular cast irons as well as medium carbon steels are used to fabricate camshafts [1].

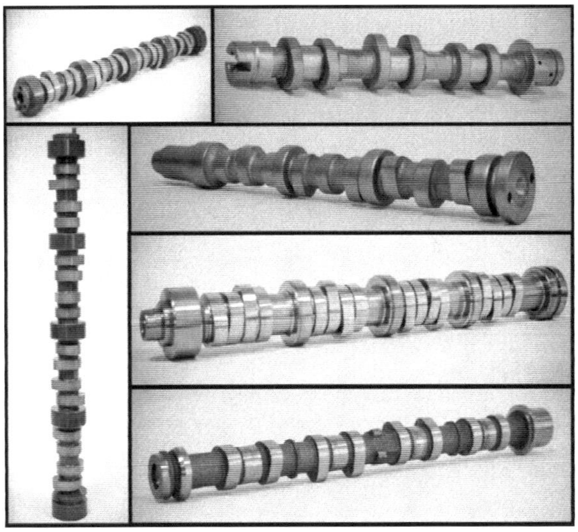

Figure 1: A family of automotive camshafts. Courtesy of Inductoheat Inc., An Inductotherm Group Company

During its operation, the cam may run millions of cycles and experience considerable wear and contact stresses. A good combination of wear resistance and strength is essential for cam lobes which require induction hardening of the working surfaces. But in addition to strength and wear resistance, camshafts need to be straight to offer quiet and vibration free performance.

The shape of the cam lobes can vary appreciably depending upon the specifics of the engine design. Figure 2 shows some examples of cam lobe shapes. The cam profile is the working surface of the cam having contact with the follower.

Figure 2: Examples of cam lobe shapes. Source [1].

Heat treating of the camshafts is followed by grinding/polishing and straightening stages. Grinding is an important finishing operation. Final grinding of the camshaft's bearing surfaces is needed to minimize surface roughness and ensure dimensional accuracy of the camshaft.

The process of grinding can be accomplished using three grinding conditions: gentle, normal or severe (abusive) grinding. A variety of different types/grades of the cutting tools and cutting parameters can be used. The grinding tool (grinding wheel) consists of abrasive particles made of different materials and interconnected with an appropriate binder featuring a different degree of porosity, hardness, wear resistance and abrasive properties.

Complex and interrelated thermo-mechanical and chemical processes take place during surface grinding [4]. Excessive heat generated due to inappropriate grinding conditions can negatively affect the camshaft's performance and wear resistance of bearing surfaces by altering the microstructure, decreasing beneficial compressive residual stresses, and even reversing desirable residual stress distribution.

The amount of grinding stock that is removed from the hardened case directly affects the life of the cutting tool and overall cost effectiveness. Therefore, attempts should be made to minimize crankshaft distortion.

Distortion of Camshafts

An ability to produce straight crankshafts is another critical necessity of modern technologies, which directly affects a process cost-effectiveness and product quality.

There are several factors that affect distortion of heat-treated components including type of material, its prior microstructure, geometry, hardness profile, etc. Camshafts have relatively complex geometry with a lack of symmetry.

One of the critical factors affecting distortion is the amount of heat generation and an ability to form a uniform hardness layer regardless of irregular shape of the lobe. The greater the amount of heated metal, the larger the expansion, which causes greater distortion. Therefore, attempts should be made to minimize the amount of metal being heated while providing sufficient hardness case depth.

Induction Hardening of Camshafts

Depending on the camshaft's geometry and production requirements, the shafts may be induction heat treated using scan hardening of a single lobe or static (single-shot) hardening of a single lobe or multiple lobes.

Both vertical and horizontal induction hardening designs were used by different manufacturers. Heat times are in the 3-8 second range depending on type of material, prior microstructure and geometry. Shorter heat times were required for quenched-and-tempered and normalized prior microstructures. Frequencies of 3 kHz to 40 kHz are typically used, depending on required case depth and camshaft geometry.

Scan hardening is typically used for relatively low production rates and somewhat wide lobes. Depending on the process specifics, lobes can be rotated during heat treating. When lobes are scan hardened without rotation, the inductor heating face can be machined to better accommodate lobe geometry and properly compensate electromagnetic proximity effects. In this case, measures should be taken to provide an appropriate lobe orientation with respect to an inductor.

Scan inductors offer the greatest flexibility by hardening lobes of various lengths/widths applying a minimum amount of power since only a portion of the single lobe is heated using an inductor with a relatively narrow copper face. This process is often used for hardening large lobes (e.g., ship and train cams).

The main limitations of applying scan hardening to automotive camshafts is related to low production rates due to single-lobe processing and some difficulty in obtaining the required hardness pattern of closely positioned lobes. Quench back splash and undesirable localized tempering soak back are additional issues.

Besides that one of the major problems that is associated with conventional induction hardening of camshafts with closely positioned lobes and journals is related to undesirable heating of adjacent areas that have previously been hardened (so-called temper back effect).

The complexity of this problem arises from the fact that, due to external electromagnetic field propagation, the eddy currents are induced not only in the camshaft area that is encircled by an inductor, but in adjacent regions as well.

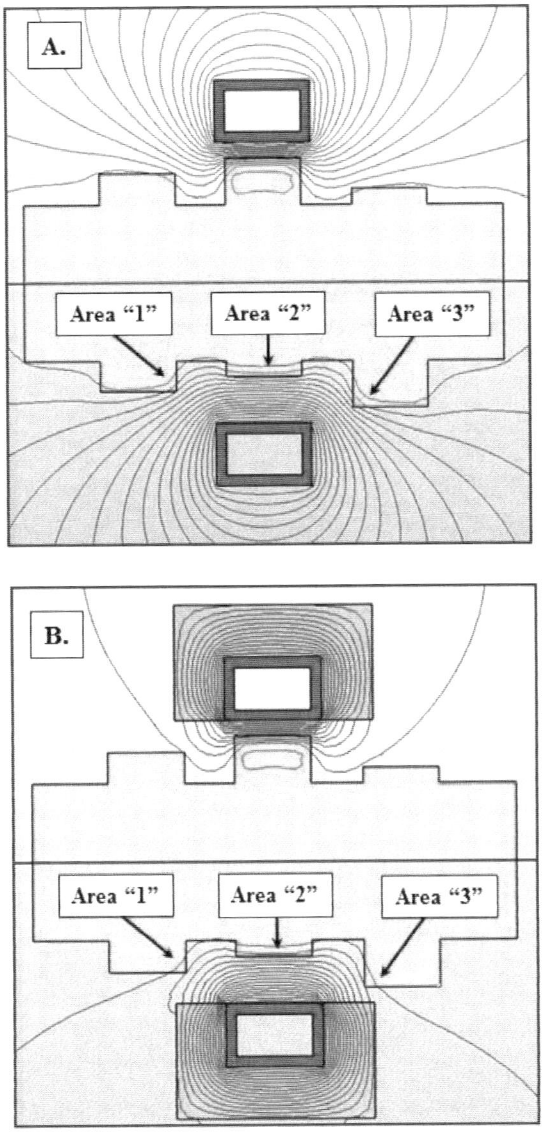

Figure 3. Results of FEA computer modeling of coil field distribution in camshaft single-lobe hardening with closely-positioned lobes. Source: Rudnev V. Invited Lecture on Aspects of Induction Heat Treating, Peoria ASM Chapter meeting, Peoria, IL, January 13, 2015

A magnetic field spreads around an induction coil and links with electrically conductive surroundings, which could include neighboring regions of the camshaft (including adjacent cam lobes and journals) and possibly certain areas of the machine or fixtures that may appreciably deform magnetic field distribution. As a result of induced eddy currents, heat will be produced. This heat can cause undesirable metallurgical changes in edge areas of the camshaft lobes that were hardened in a previous process stage.

To illustrate this, Figure 3A shows the results of FEA computer modeling of coil field distribution in camshaft single-lobe hardening with closely-positioned lobes utilizing a bare coil (Fig.3,A) and coil with "U"-shaped magnetic flux concentrator (Fig.3,B). Due to poor electromagnetic coupling in the heel area of the middle lobe (Area "2") and the necessity to provide sufficient austenization, there will be significant heat generation in neighboring lobes (Area "1" and, in particular, Area "3") [5]. Corners of neighboring lobes are particularly susceptible for undesirable temper-back of previously hardened regions or even their re-hardening (Fig.3A).

Another conclusion can be drawn from Figure 3. Conventional inductor designs are associated with a substantial variation in electromagnetic coupling between the coil inside diameter (coil ID) and different regions of the cam lobes (compare nose-to-inductor coupling vs. heel-to-inductor, for example).

Proximity effect manifests itself in localized redistribution of the electromagnetic field of the coil resulting in respected deviation of the heat source generation. Due to proximity effect, more imaginary lines of the electromagnetic field are closing their loops in air without reaching the surface of the heel of the lobe compared to its nose. This results in an appearance of areas with a temperature deficit or heat surplus that inevitably leads to an appearance of "hot" (the nose) and "cold" (the heel) regions and respected hardness case depth variation.

An attempt to provide a sufficient austenization of the heel region and needed hardness depth upon quenching often results in measurable overheating of the nose of the lobe which can promote some brittleness and sensitivity to intergranular cracking and grain boundary liquation (incipient melting).

An increased use of electric arc furnaces that utilize recycled scrap, which may be rich in *Cu* or other low melting point residuals. Over the last decade, the amount of copper residual has increased in most commercial steel grades. Copper, as a residual element is not readily oxidized and removed from liquid steel in the steelmaking process using electric arc furnaces [6]. Copper has a relatively low melting point (1085 °C).

The rejection of the *Cu* from the iron oxide into the base metal may create a near-surface copper-enriched zone at the *FeO* interface [7]. This takes place because the solubility of copper in iron oxide *FeO* is very low and at high temperatures, where the oxidation rate of iron is quite high the copper is liquid and can penetrate along grain boundaries with ease, dramatically weakening the grain boundaries. Therefore, even in cases where the scale might be removed and a thin surface layer is machined, there might still be steel areas that exhibit a localized copper surplus.

If during camshaft hardening, its surface is overheated then besides the appearance of known unwanted metallurgical conditions and excessive grain growth, a phenomenon of grain boundary liquation may occur. This is the most common cause of cracking, since it weakens grain structure and greatly increases the steel's brittleness and sensitivity to developing intergranular cracking upon quenching.

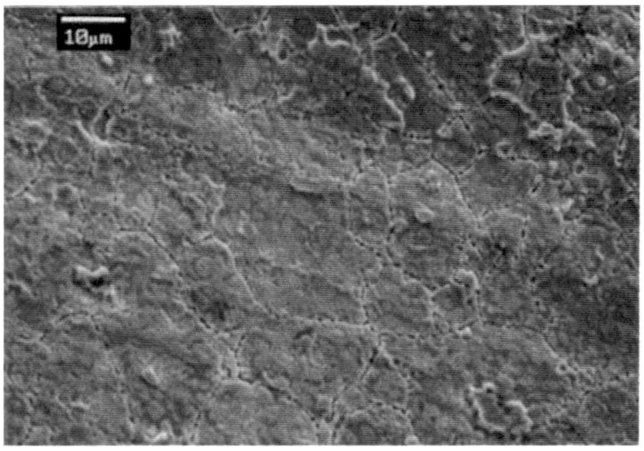

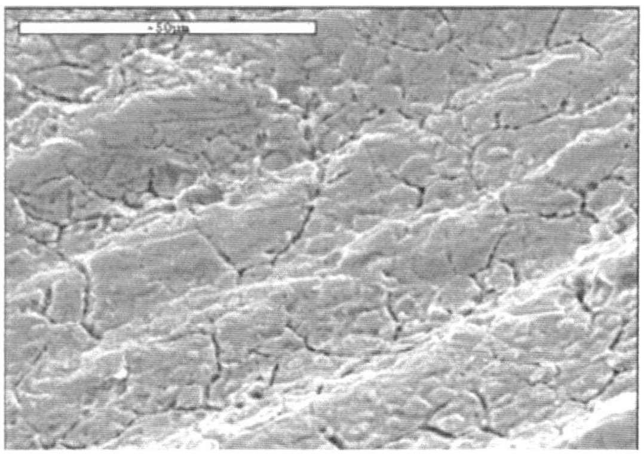

Figure 4. Results of FEA computer modeling of coil field distribution in camshaft single-lobe hardening with closely-positioned lobes. Source: Rudnev V. Invited Lecture on Advanced Case Hardening Technologies, Chrysler Technology Center, Auburn Hills, Michigan, October 16, 2015.

As an example, Figure 4 illustrates steel microstructures of grain boundary liquation. It should be noted that each image has a different magnification indicated by the scales positioned in the top-left corner of each image. Chains of liquated areas located at the grain boundaries are easily visible there.

The phenomenon of grain boundary liquation can be amplified by preferential segregation of manganese, sulfur and some other elements at the austenitic grain boundaries. Both phosphorous and sulfur markedly affect the steel overheating. This lowers the melting point in the grain boundary region compared to the nominal solidus temperature of the steel leading to several undesirable phenomena during hardening (e.g., during cooling) and potentially promoting a temper embrittlement with increased manganese content.

Therefore, undesirable combinations of impurities, residual and trace elements, elements used in the steel making process, as well as non-metallic and intermetallic constituents could increase brittleness and crack sensitivity. Higher than desirable temperatures that occur during steel austenization could cause adverse microstructural changes including severe grain coarsening and incipient melting (grain boundary liquation). Those changes could be associated with the loss of strength, ductility and toughness of steel increasing crack susceptibility and making components prone to intergranular cracking. As example Figure 5 shows an intergranular cracking of the cam nose regions cause by grain boundary liquation.

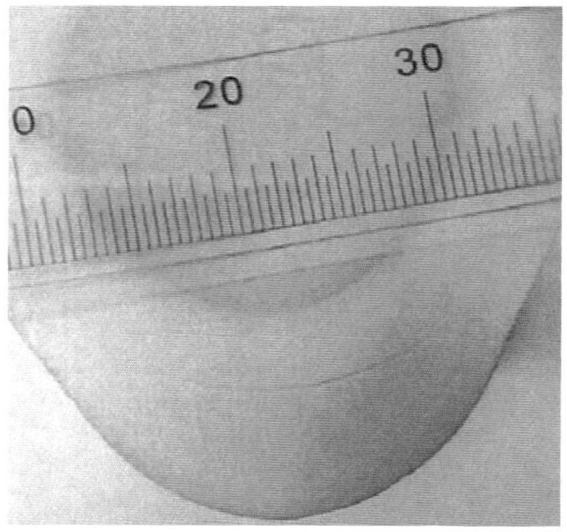

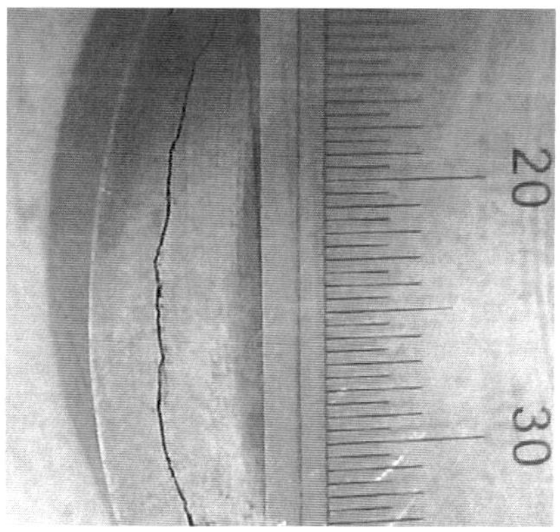

Figure 5. An intergranular cracking of the cam nose regions caused by a grain boundary liquation. Source: Rudnev V. Invited Lecture on Aspects of Induction Heat Treating, Peoria ASM Chapter meeting, Peoria, IL, January 13, 2015.

Examples shown in Figures 4–5 emphasize the importance of obtaining true uniform hardening patterns and avoiding localized heat surplus and severe grain coarsening.

Limitations of Conventional Camshaft Hardening

Single-shot heating of multiple lobes is commonly used when surface hardening small and medium-sized camshafts with lobes of similar size and shape and having the same or very similar axial gaps between them. Camshafts are rotated during induction hardening. In order to increase the output, multiple lobes are heat treated simultaneously. A corresponding number of single-turn inductors are connected electrically in a series, which provides the required heating of multiple lobes.

Coil copper is often profiled in an axial direction to obtain desirable power density distribution. Quenching can be incorporated into an inductor design, or it can be done out of place after completing the heating cycle.

Regardless of camshaft rotation there is still noticeably deeper case depth in the nose compared to a base circle (the heel) area, because during a rotation the nose of the cam lobe has better electromagnetic coupling with an inside diameter of the coil copper.

Besides that, inability to form a contour-like hardened pattern of the lobe and excessive heat generation is inevitably associated with larger and unequal metal expansion /contraction that are inevitably associated with appreciable distortion requiring using a straightening operation.

The necessity of having a camshaft straightening operation results in additional cost and may produce cracking. Besides, it reduces useful compressive residual surface stresses.

Advanced Non-Rotational Camshaft Hardening

Patented non-rotational technology (SHarP-C™) that was developed for hardening crankshafts has recently been effectively expanded for producing true contour hardening of camshafts. The inductor consists of a top (passive) inductor and a bottom (active) inductor (Figure 6).

The bottom inductor (being active and connected to a power supply) is stationary, while the top (passive) inductor can be opened and closed during camshaft loading and unloading. Each inductor has profiled areas where the cam lobes, which need to be heat treated, can be located while the top inductor is in its open position. Due to the active/passive approach, electrical contact issues associated with clamshell coils are not a problem.

Following loading of the camshaft into the heating position, the top inductor pivots into a closed position, and the power is applied from the power supply to the bottom (active) inductor. Electrical current flowing in the bottom coil will instantly induce the eddy currents that start flowing in the top inductor. This is due to a lamination pack that serves as a magnetic flux coupler, which electromagnetically couples top and bottom sections similar to a transformer effect. Therefore, cam lobes "see" the non-rotational inductor as a classical encircling and highly electrically efficient induction system with a uniform lobe-to-inductor gap.

Figure 6. CamPro™ machine utilizes patented non-rotational hardening technology (SHarP-C™ technology). Courtesy of Inductoheat Inc., An Inductotherm Group Company.

As expected, the heating face of such inductors can be easily profiled to achieve the desirable hardness pattern. Specially designed quench slots are used to accomplish the process of quenching as well as coil copper cooling.

Non-rotational induction hardening technology provides several principle benefits such as dramatically reduced distortion, simple operation, better quality, superior reliability and equipment maintainability as well as life-cycle cost reduction. SHarP-C inductors are robust, rigid and repeatable, being CNC machined from solid copper block.

All brazed or banded components in manufacturing inductors are eliminated. This, in turn, dramatically reduces the possibility of inductor distortion during its fabrication and eliminates the associated hardness-pattern drift. This advanced process lends itself for hardening several lobes (Figure 6). Therefore, high production rates can be achieved by statically heat treating multiple cam lobes simultaneously.

A measurable advantage of recent installations that have implemented this novel process is the ability to harden camshafts with practically undetectable distortion. In some cases, this eliminates the entire camshaft straightening operation.

One of the most attractive features of this technology is its ability to produce uniformly distributed true-contour hardening patterns (Figure 7) regardless of the fact that the camshafts belong to a group of geometrically irregular components.

Figure 7. Illustration of obtaining true contour hardening patterns (top) even on closely-positioned camshaft lobes and (bottom) using CamPro™ machine that utilizes patented non-rotational hardening technology (SHarP-C™ technology). Courtesy of Inductoheat Inc., An Inductotherm Group Company.

If required, the heat-affected zone (*HAZ*) can be minimized, which further reduces metal expansion and, obviously, minimizes shape distortion by putting less energy into the camshaft. The core of the journal, acting as a shape stabilizer, remains relatively cold during the entire heating cycle. Heat patterns are locked in place and very repeatable since neither the camshaft nor the inductors are moving during heating. The same pattern is achieved over many cycles.

In contrast to alternative processes, where axial pressure is applied to rotate a camshaft, no axial force is applied during heat treating because the camshaft is being rested on V-shaped blocks (Figure 6).

Conclusions

Compound benefits of this patented technology include the following three major points.

- Neither the part nor the inductor moves during heat treating, preventing misalignment, wear or electrical short circuit potential.
- The inductor is CNC machined, ensuring exact dimensional accuracy.
- Utilizing SHarP-C™ technology in recent installations has proven to produce superior camshaft straightness along with excellent hardening characteristics right off the production line.

References

[1] Doyon,G., Rudnev, V., Maher,J., "Induction Hardening of Crankshafts and Camshafts," ASM Handbook Vol.4C: Induction Heating and Heat Treating, V.Rudnev and G.Totten (eds.), ASM Int'l, 2014, pp.173-186.

[2] Rudnev, V., *et al.*, Handbook of Induction Heating, CRC Press, 2003, 800p.

[3] Doyon,G., Rudnev, V., Maher,J., Minnick,R., Desmier,G., "Elimination of Straightening Operation in Induction Hardening of Automotive Camshafts", *Int. J. Microstructure and materials Properties (IJMMP)*, 2016 (to be published).

[4] Grum, J. (2007a) 'Influence of various induction-heating and grinding conditions on residual stress profiles', *Int. J. Microstructure and Materials Properties*, Vol. 2, No. 3–4, pp.310–325.

[5] Rudnev, V. *Invited Lecture on Aspects of Induction Heat Treating*, Peoria ASM Chapter meeting, Peoria, IL, January 13, 2015.

[6] Krauss, G., *Steels: Processing, Structure and Performance*, ASM Int'l, 2014.

International Federation of Heat Treatment and Surface Engineering 2016
Proceedings of the 23rd IFHTSE Congress
April 18–21, 2016, Savannah, Georgia, USA

Local Hardening of Ductile Cast Iron – The Importance of Temperature

D.E. Guisbert, M.H. Guisbert and D.A. Guisbert
QA Metallurgical Services, LLC, Niles, MI 49120, USA

Abstract

A common problem experienced in the local hardening of ductile (spheroidal graphite) iron is incorrect surface temperature. Many heat treaters experienced in the hardening of steels, do not take into consideration the basic differences between iron and steel. Because iron contains discrete particles of graphite, the composition of the matrix will change with heating temperature and time. This change includes the diffusion of carbon from the graphite into the matrix, which can cause the matrix to become saturated with carbon at higher temperature. In contrast, the matrix composition of steel does not change with heating temperature. Poor temperature control can result in incomplete hardening, excessive levels of austenite, liquation and cracking. This paper reviews the basic difference between iron and steel, and shows examples of problems caused by underheating or overheating.

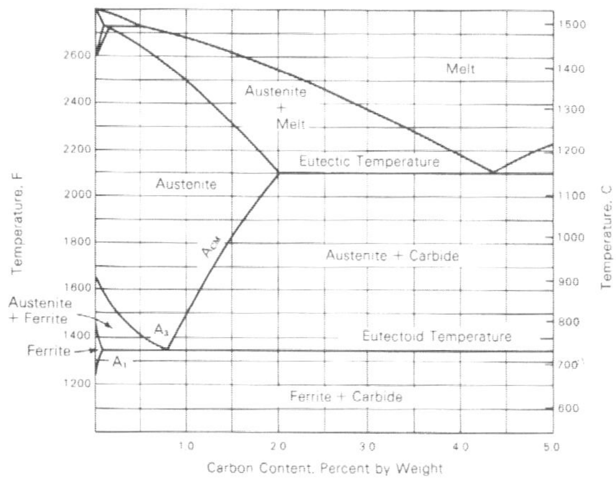

Figure 1. Iron-Iron Carbide binary equilibrium phase diagram, stopping at 5 weight percent carbon (1)

Introduction

Ductile iron castings have many advantages over cast steel. These advantages include the purchased cost of the base casting and the ability to cast more complex shapes with lower levels of shrinkage. In addition, ductile iron tends to exhibit better surface appearance with fewer surface defects. Ductile iron also exhibits good machinability. Like steel however, not all grades respond well to local hardening.

The hardening of ductile iron however, is unlike that of steel. In ductile iron components, the matrix composition changes during heating. This composition change is defined by carbon enrichment of the matrix, where carbon from the graphite nodules diffuses into the surrounding metal. This means that the matrix composition will be a function of the heating temperature and time. Because we are dealing with a changing product, the temperature and time at temperature will be critical.

To illustrate these differences, consider that when heating a carbon steel such as 1050, the matrix may consist of pearlite and ferrite, or tempered martensite, the carbon content of the austenite is simply the nominal 0.50 weight percent of the steel. This is clearly demonstrated from inspection of the typical iron-iron carbide phase diagram, Figure 1. This shows that when a 0.5 weight percent carbon material is heated, austenite begins to form at a temperature over 720°C (1330°F). At a slow heating rate, the first austenite to form will have a carbon content corresponding to that of the pearlite, approximately 0.8 weight percent. At higher temperature, the amount of austenite formed will increase while the average carbon content decreases until what remains is a single phase austenite microstructure with a nominal 0.5 weight percent carbon. This austenite represents a fully dissolved "in-solution" product that is more optimum for hardening.

A typical ductile iron casting contains 3.2 to 4.1 weight percent carbon, so we must consider a more complete view of the iron-iron carbide diagram. But at these higher carbon levels, the carbon is found in the form of free graphite, so there is a need to consult the iron-graphite diagram, Figure 2. Also, these castings will contain 1.8 to 3.0 weight percent silicon for graphite stabilization, so there is a need to consider a slice through the iron-carbon-silicon ternary diagram, Figure 3. Theses diagrams are of very similar shape. The main equilibrium difference is that the matrix of an iron-graphite and iron-graphite-silicon product consists of ferrite and graphite, which is a starting microstructure that is generally impractical for local hardening. In practice we prefer a starting matrix microstructure that is more pearlitic.

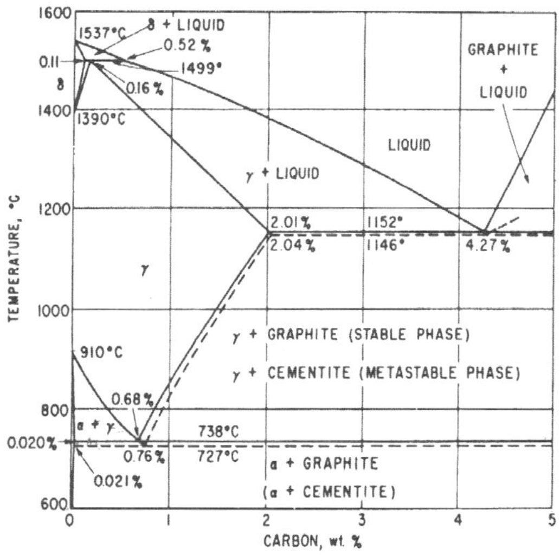

Figure 2. Iron-Graphite binary equilibrium phase diagram, stopping at 5 weight percent carbon (2)

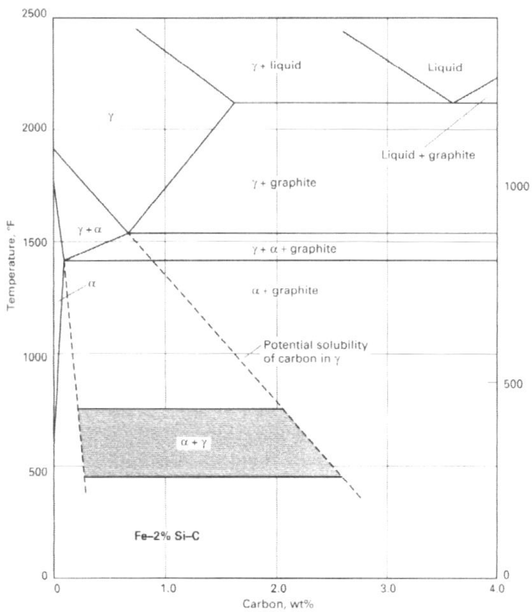

Figure 3. Section through the Iron-Carbon-Silicon ternary phase diagram at 2 weight percent silicon (3)

In production practice for steel components, heating rates are very high. It is not uncommon to see particles of undissolved ferrite and carbide in steels where the starting microstructure was of pearlite and ferrite. These undissolved particles will be seen unless temperatures higher than those used for standard furnace hardening practices are used. When visiting companies that are induction heat treating steels, we often see heat treaters using surface temperatures in excess of 1090°C (2000°F). These high temperatures may cause some grain growth, but if the time is short and the part is free from cross holes and sharp corners, these high temperatures are not only tolerated they are used with great success.

As mentioned earlier, the matrix composition of a typical iron casting will change during hardening. For instance, if the starting matrix microstructure consists of 80 percent pearlite and 10 percent ferrite, this is like having a 1060 carbon steel with graphite nodules. On heating the matrix will first transform to a mixture of austenite at 0.8 percent carbon, with undissolved ferrite and graphite, Heated further, the matrix becomes fully austenitic, but the austenite is now of 1.0 weight percent carbon, because the graphite provides carbon to enrich the matrix. Upon further heating the carbon content of the austenite can reach 1.5 to 2.0 percent at around 1150°C (2100°F), where liquation or melting along the grain boundaries or around graphite nodules occurs.

The time factor in heating must also be taken into account. Typically, heat treating guidelines indicate that it takes several minutes for the carbon content of the austenite to reach near equilibrium levels. But when very high surface temperatures are used during local hardening, changes will occur at a faster rate. So carbon diffusion into the matrix will be higher, along with the increased solubility of carbon in the austenite.

Susceptibility to melting along the boundaries is also influenced by the local composition. Given that boundary regions exhibit higher concentrations of manganese, sulfur and other elements that are associated with lower melting temperatures.

Case Studies

The following case studies represent examples of products that were submitted for analysis because of excessive wear, low hardness, cracking or other surface flaws.

Example 1

Low surface hardness was experienced on a cast sprocket. Surface hardness was 45 to 48 HRC, with specified surface hardness of 50 to 58 HRC. The case microstructure consisted of tempered martensite, with bainite and undissolved ferrite particles, Figure 4. This microstructure reveals that the product was marginally in solution, which prevented the desired hardness from being achieved during the quench.

The solution prescribed in this case was to use a slightly higher heating temperature, so that the product could be effectively quench hardened. It is possible that a longer heating cycle would have produced the same result, but case depth was already sufficient and a deeper case was not desired. Higher heating temperature was achieved by using a higher power percentage. The resulting microstructure, Figure 5, consisted of tempered martensite higher levels of retained austenite surrounding the graphite nodules. Castings heat treated with the higher surface temperature were of the specified hardness.

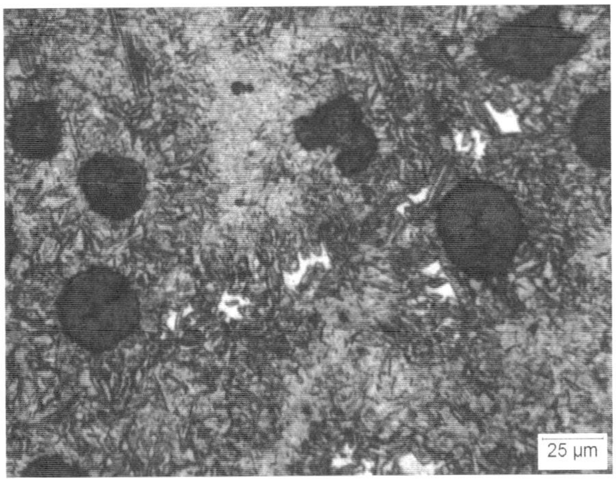

Figure 4. Low hardness case microstructure – Nital Etchant

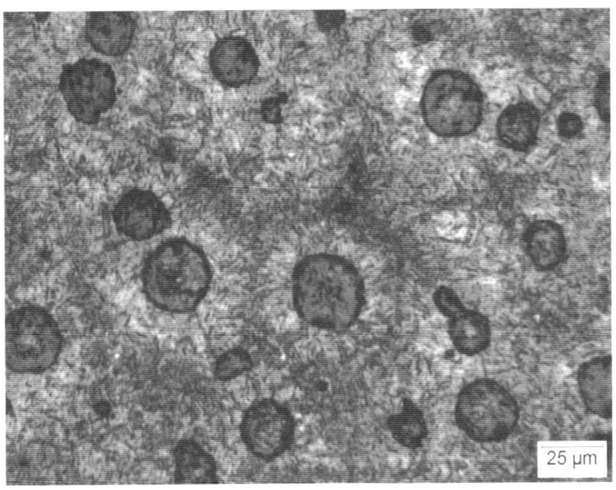

Figure 5. Casting from the same lot as shown in Figure 4, hardened using a higher surface temperature – Nital Etchant

Example 2

Edge cracking occurred on the sharp corner to a tapered hole in a control arm, Figure 6. This product is heat treated using coils off the end face on both sides of the arm. Fracture features reveal that the breakage is due to localized melting of the corners, due to overheating, which is not an uncommon problem for heat treaters, even with steel products.

Melting was verified using metallographic techniques, which revealed high levels of the iron carbon eutectic structure, ledeburite, characteristic of white iron, Figure 7.

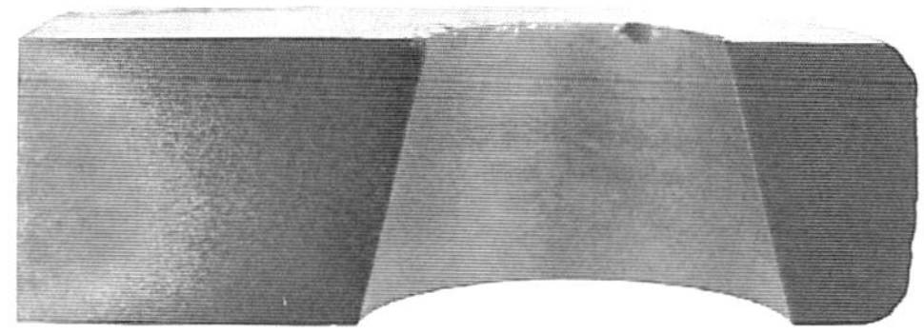

Figure 6. Edge cracking on the sharp corner of the tapered cross hole – Etched in Nital to show the heat pattern through the cross section

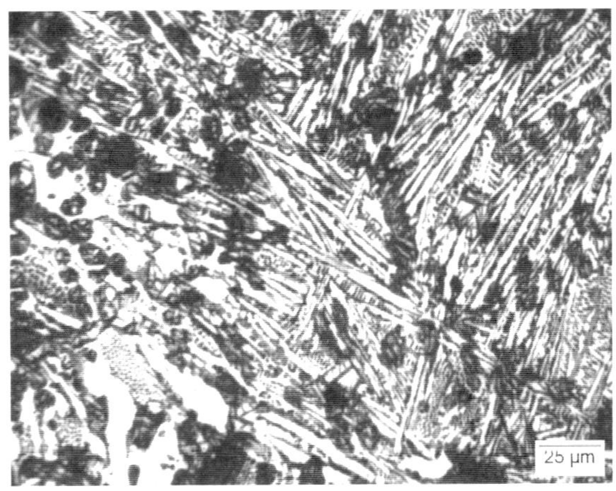

Figure 7. Localized melting at the sharp corner to the tapered cross hole resulted in the formation of the iron-carbon eutectic – Nital Etchant

The solution to this problem involved a significant reduction in the surface heating temperature, or induction power percentage. Improved fixtures were also required in order to ensure that the arm remain in proper position between the coils. After these changes, the final product exhibited suitable hardness and a very good high strength microstructure.

Example 3

Inconsistent surface hardness and edge cracking was experienced at the outer diameter of a roller product. It was found during metallographic examination that the surface was overheated, causing excessive levels of retained austenite, local melting and cracking, Figures 8 and 9.

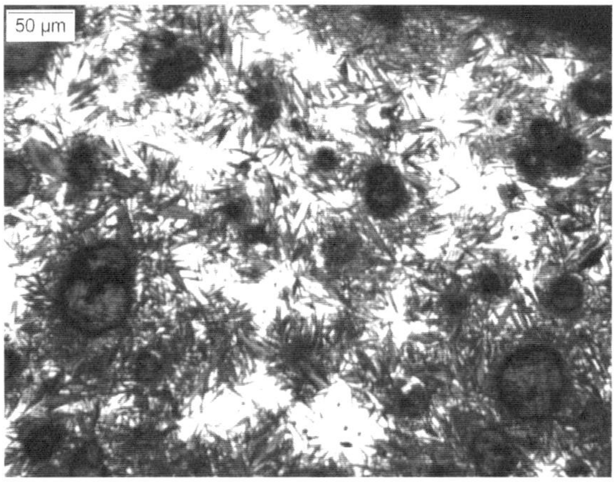

Figure 8. Areas with excessive levels of retained austenite (white) caused by overheating prior to hardening – Nital Etchant

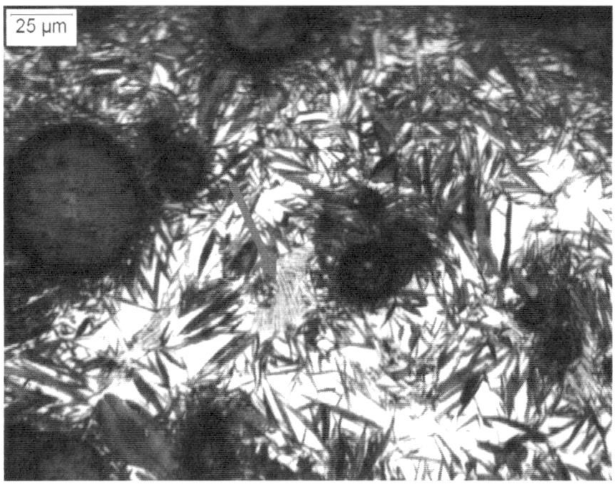

Figure 9. Overheated area exhibiting small pockets of iron-carbon eutectic, caused by local melting – Nital Etchant

For reasons unknown, the part was heat treated using a higher than prescribed power percentage. If this had been a steel component, it would not have caused a problem. But with iron, the matrix became saturated with carbon and the product began to melt adjacent to the graphite nodules that were dissolving into the matrix. The solution was to use the prescribed power percentage.

Example 5

A ductile iron spindle was found to have surface porosity on a portion of the outer diameter after induction hardening, Figures 10 to 12. This porosity was not found in the casting before heat treatment, as several pieces were sectioned and examined prior to hardening. Machined surfaces were also inspected prior to hardening, but no porosity was found.

Figure 10. Portion of a spindle exhibiting porosity on the induction hardened 85 mm diameter

Figure 11. Same as Figure 10 only a higher magnification view of the surface porosity

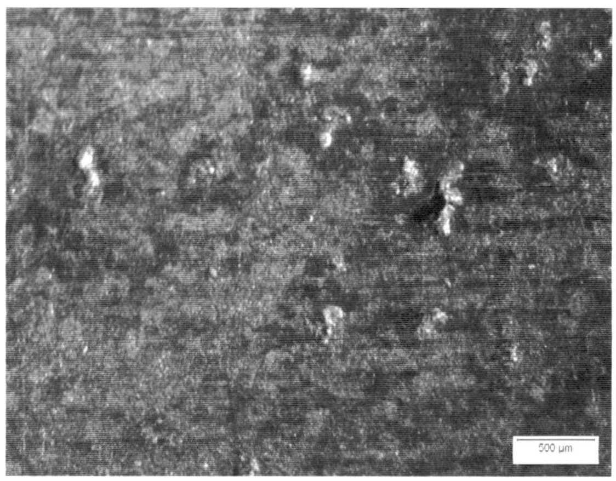

Figure 12. Same as Figures 10 and 11 only a higher magnification view of the surface porosity

A metallographic cross section through the spindle confirmed that the porosity was within the hardened zone, Figure 13. The shape of the pattern is also consistent with the fact that this is the start or dwell location of the scan.

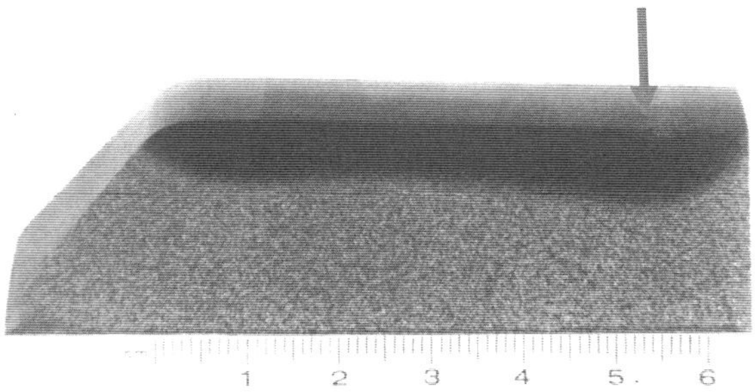

Figure 13. Heat pattern of the induction hardened case, showing the location of the highest concentration of porosity at the dwell or start portion of the scan – Nital Etchant

Additional cross sections in other areas of the spindle confirmed the deeper heat pattern in the area of porosity and that the porosity was primarily in the dwell portion of the pattern length, Figure 14.

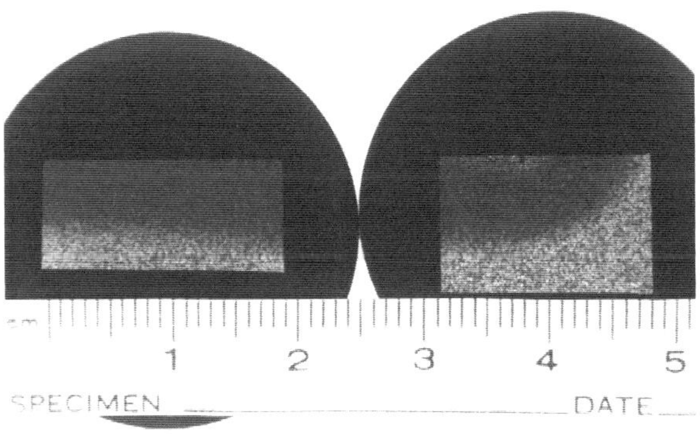

Figure 14. Additional cross sections of hardened case and through one of the largest surface pores – Nital Etchant

At higher magnification it is seen that the large areas where graphite had been present were replaced by zones of iron-carbon eutectic phase, and there were surface voids or porosity, Figure 15. At greater depth within the case, it is apparent that the graphite is being dissolved and surrounded by the iron carbon eutectic phase, Figure 16. These features all indicate that the surface is being overheated and locally melted. The solution to this problem involved reducing the power percent during the dwell portion of the scan, coupled with a smaller reduction in the power percentage during the scan.

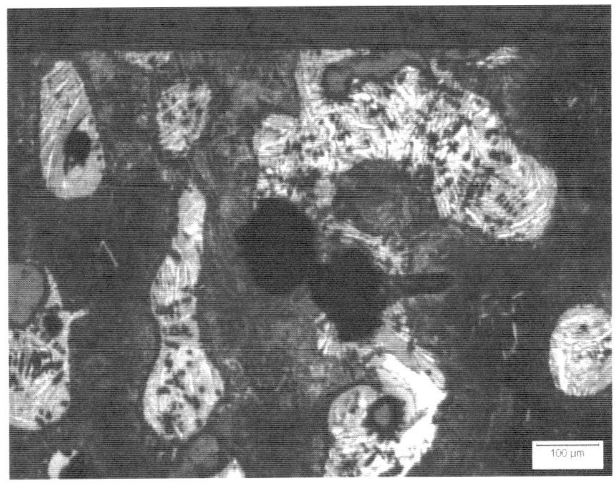

a)

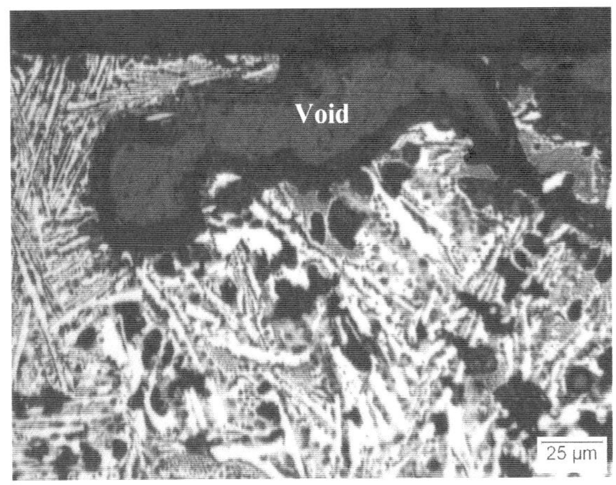

b)

Figure 15. Metallographic images of the same cross section through a large void, showing the iron-carbon eutectic phase (light zones) where local melting occurred – Nital Etchant

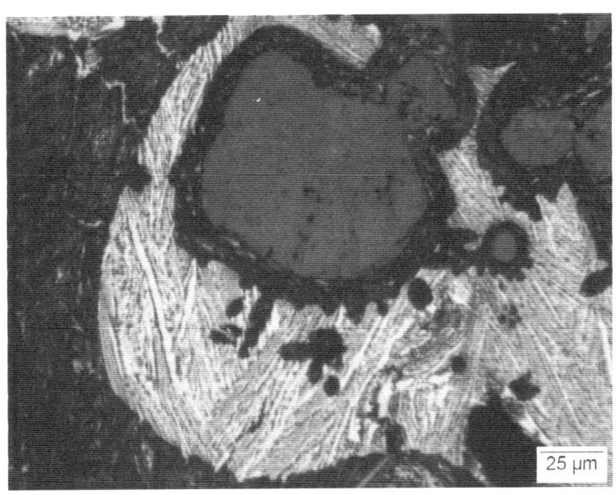

Figure 16. At greater depth, showing a graphite nodule surrounded by molten metal – Nital Etchant

Conclusion

Ductile iron castings are locally hardened every day with great success, but problems can occur when parts are either underheated or overheated. Operators who have successfully hardened steel components do not always understand the differences between steel and iron. High surface temperatures can be acceptable when heat treating steel, but iron is less forgiving as the matrix composition can change during the heating portion of the hardening process.

Some problems we have seen occur because surface temperatures are not monitored. This is made difficult in many cases because there is no clean "line of sight" for monitoring the temperature through use of dedicated pyrometry or simple operator verification.

References

[1] Iron Castings Handbook, Published by Iron Castings Society (1981), p 134

[2] The Making, Shaping and Treating of Steel, 10th Edition, USS, Published by the Association of Iron and Steel Engineers (1985), p 1235

[3] Rundman, K., "Heat Treatment of Cast Irons", Heat Treating, Vol. 4, ASM Handbook, ASM International (1991), p 685

International Federation of Heat Treatment and Surface Engineering 2016
Proceedings of the 23rd IFHTSE Congress
April 18–21, 2016, Savannah, Georgia, USA

Effect of the Surface Morphology and Surface Finish on the Corrosion Resistance of Ti6Al4V via Direct Metal Laser Sintering

Yangzi Xu, Yuan Lu, Jianyu Liang, Richard D. Sisson, Jr.
Worcester Polytechnic Institute, Worcester, MA, USA

Abstract

The effects of surface finish and surface modification on the corrosion behavior of additively manufactured Ti6Al4V in Ringer's solution was investigated. The Ti6Al4V alloy used in this study was fabricated by direct metal laser sintering (DMLS), and post-annealing treatment. The surfaces of as-fabricated and heat treated samples were modified by mechanical grinding, mechanical polishing, mechanical polishing with one scratch and vibropolishing with colloidal Si suspension. Surface morphologies were analyzed with Olympus laser microscope. Roughness analyses were performed to determine the surface roughness (R_a) values and the hybrid roughness parameters: Root Mean Square (RMS) Surface Slope (Sdq) and Developed Interfacial Area Ratio (Sdr) of the investigated samples. The mill annealed commercial Ti6Al4V alloy was also measured for comparison. The microhardness values under mechanical polishing and vibropolishing were also measured. It was found that samples under VP with less surface roughness value possess higher corrosion resistant, and the scratch generated before immersion would make the specimen less pitting resistant.

Introduction

Additive manufacturing is a process to create parts by addition of materials layer by layer [1]. Direct Metal Laser Sintering is an additive manufacturing process that can direct build three-dimensional metallic components from metal powders in accordance with a sliced 3D computer aided design (CAD) model [2]. The huge advantage of this rapid prototyping technique is able to create three-dimensional freeform objects, especially applied in biomedical and aerospace applications. The direct metal laser sintering technique is used in fabricating titanium and its alloys that are widely used in orthopedic and dental industry because of their high strength-to-weight ratio, good mechanical properties, corrosion resistance, chemical inertness and biocompatible [2]. Pure titanium and Ti6Al4V are widely used in total knee replacement, hip replacement and dental implants [3]. However, due to high-localized thermal input and fast cooling rate during direct metal laser sintering, the as-sintered parts contain non-equilibrium phases. In Ti-6Al-4V, the as-fabricated microstructure is composed of acicular martensite with poor ductility [4]. Therefore, proper post treatment, hot isotactic pressing and heat treatments, usually apply to the as-fabricated metallic parts to ensure better performance.

The surface microstructure of orthopedic parts is another important design feature, because rough surface is more osteoconductive than smooth surface by promoting cell growth [2]. However, high friction coefficient and poor wear resistance is a problem for titanium alloys, and rough surface has higher corrosion rate than smooth surface in body fluid [2][5]. Therefore, the corrosion resistances of DMLSed Ti6Al4V as affected by surface finish needs to be studied.

In this study, annealing treatment was conducted on Ti-6Al-4V fabricated by direct metal laser sintering. The effect of surface morphology and surface finish on the corrosion behavior of as-fabricated and annealed samples was experimentally investigated. Surface roughness (Ra) and hybrid parameters (Sdq and Sdr) after different surface finishes were analyzed. The microstructure, phase percentage and micro-hardness values were measured by scanning electron microscope, X-ray diffraction and microindentation. The corrosion behavior of post heat-treated samples was characterized electrochemically in simulated body fluid and the effect of different surface finish was emphasized. The mill annealed Ti6Al4V was also studied for comparison.

Materials and Method

Materials and specimen preparations

Commercial Ti-6Al-4V pre-alloyed powder particles fabricated by gas atomization were purchased from Carpenter Inc. (Wyomissing, PA) and used in this study. The Ti6Al4V specimens were fabricated by E-manufacturing solutions Inc. (Krailling, Germany) via direct metal laser sintering (DMLS). Samples were produced at a scanning speed of 1250 mm/s, a laser power of 340W, a hatch spacing of 120 μm and a layer thickness of 60 μm. The sintering temperature was 700°C. Samples were cut into cubes and degreased in acetone before heat treatment. Half of the as-fabricated samples were heat treated in a vacuum furnace under 799°C for 4 hours, and followed by air-cooling. The mill annealed Ti6Al4V specimens were purchased from McMASTER-CARR Inc. (Robbinsville, NJ).

The specimens were mounted by EpoxiCure™ 2 (Beuhler Inc.). The as-fabricated, annealed and mill annealed specimens were prepared with various surface finishes: mechanical grinding using #1200 silicon carbide paper, mechanical polishing up to 0.03 μm alumina paste, mechanical polishing with one scratch and vibropolishing with 0.5 μm colloidal Si suspension. The applied surface finish processes are listed in Table 1. Kroll's reagent (2ml HF, 5ml HNO_3 and 100ml H_2O) was used on the polished specimens to reveal the microstructure.

Characterization techniques

A Nikon optical microscope (Nikon Instrument Inc.) was used to evaluate the microstructure of specimens under low magnification, and JEOL JSM-7000F scanning electron microscope (SEM) was used for high magnification characterization. Surface morphology tests were performed on an Olympus 3D laser microscope OLS4100 under x20 with three by three stitching of the images, and the MountainsMap® Premium 7.1 (Besançon, France) was used to analyze the generated three-dimensional images. For a specific specimen, ten random profiles were extracted from the measured topography layer. The average surface roughness (R_a value) and hybrid parameters (Sdq and Sdr) were calculated based on the software. Phase identification was conducted on a PANanalytical Empyrean 2 X-ray diffractometer (XRD) with Cu-Kα radiation. The Vickers microindentation hardness values were measured with a CM-400AT Microhardness tester (Deterco, Inc.) under load of 200g.

Table 1: Surface finishes methods and their designations.

Surface finish methods	Designation
Mechanical grinding	MG
Mechanical polishing	MP
Mechanical polishing + one scratch	MP + scratch
Vibropolishing	VP

Corrosion tests

The electrochemical tests were conducted in a three-electrode cell with Ag/AgCl (3M NaCl) as the reference electrode, a Platinum wire (surface area: 3.6 cm^2) as the counter electrode and the tested Ti-6Al-4V specimen as the working electrode. All electrode potentials are referred to the reference Ag/AgCl electrode. The volume of the test cell was 100 ml. Naturally aerated Ringer's solution (6.5 g/l NaCl, 0.14 g/l KCl, 0.12g/l $CaCl_2$, 0.2g/l $NaCO_3$ and 0.4g/l glucose) with a pH of 6.5 was used as the electrolyte and was maintained at 37°C during tests. A Cu wire was attached to the back of the tested specimen by conductive tape, and the whole part was cold mounted by epoxy resin to expose a flat surface of 0.25 cm^2 to the electrolyte. Post curing, the specimens were prepared under different surface finishes as listed in Table 1. A potentiostat (Bio-Logic, Inc.) was used to measure the open circuit potential after immersion of the working electrode into the electrolyte until the potential reached a plateau value, and then generated cyclic polarization curves. A forward scan started from -1000 mV more active (more cathodic) at a constant scan rate of 1 mV/s to higher potential. The scan direction was reversed until the potential reached 3000 mV. Bio-Logic® software was used to analyze the electrochemical parameters values.

Results and Discussion

Microstructure and phase analysis

The microstructure of the as-fabricated, annealed and mill annealed specimens are shown in Figure 1. XRD patterns of the investigated specimens are plotted in Figure 2. The as-fabricated specimen in Figure 1(a) shows predominantly α' martensite microstructure because of the fast cooling rate after sintering process. The martensite phase would lead to increased microindentation hardness in the as-fabricated specimens, and reduced ductility [6]. After proper post-annealing heat treatment, coarsening of α lath can be observed in Figure 1(b). In Figure 2, the XRD pattern of annealed specimen indicates the formation of β phase at 2θ equals 38.5° which represents β (011). Figure 1(c) shows the equiaxed α and β (α/β) phase in mill annealed Ti6Al4V, which generally exhibits superior mechanical properties (higher ultimate tensile strength and ductility) [6]. The XRD pattern of mill annealed specimen possesses higher amount of equilibrium β phase.

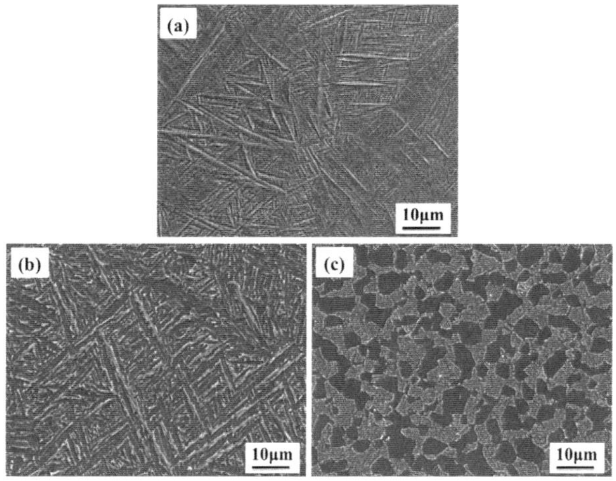

Figure 1: Microstructure of (a) as-fabricated, (b) annealed and (c) commercial mill annealed Ti6Al4V specimens.

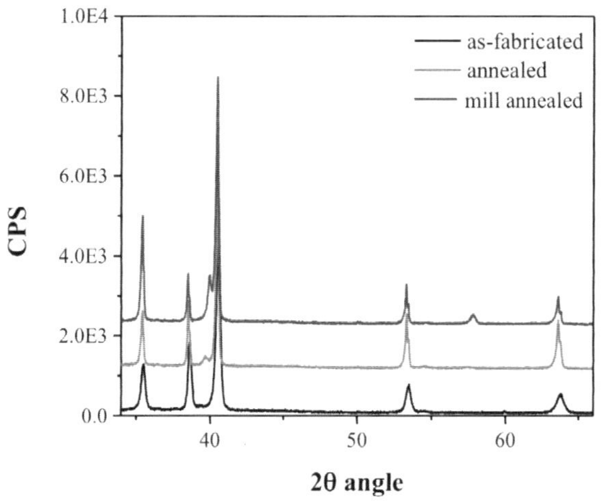

Figure 2: XRD patterns of the investigated specimens.

The effect of various surface finishes

Various surface finishes were applied on the as-fabricated, annealed and mill annealed specimens. The optical images and measured surface morphologies of the as-fabricated specimen under different surface finishes are shown in Figure 3. Due to the manufacturing process, improper sintering and defects of pre-alloyed powder particles may lead to internal pores of the as-fabricated parts [4]. As shown in Figure 3, some micro-pores and unidirectional scratches can be observed in MG specimen. Based on the color scale bar, depth of the scratches was approximately 0.3μm. During mechanical polishing, the applied force may lead to unbinding of some improper sintered powder particles that loosely bonded to the matrix, so the pores are more obvious after mechanical polishing. In Figure 3, the blue dots in the topography image of MP specimen represent the internal pores with depth of approximately 0.5μm. The vibropolished surface was flat and lack of significant pores, VP in Figure 3, because the applied force was lower than MP.

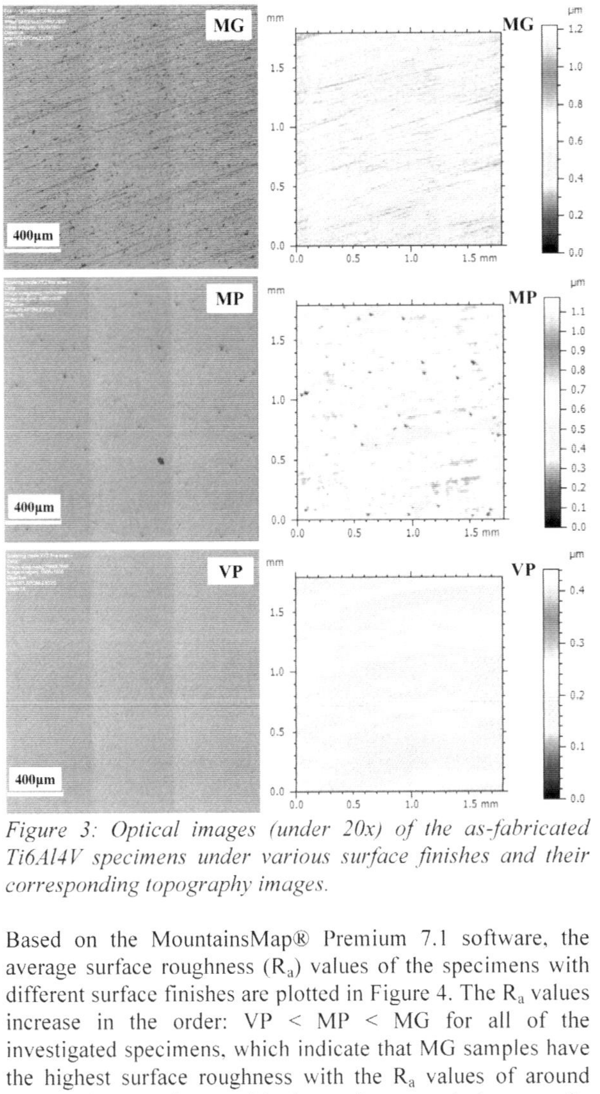

Figure 3: Optical images (under 20x) of the as-fabricated Ti6Al4V specimens under various surface finishes and their corresponding topography images.

Based on the MountainsMap® Premium 7.1 software, the average surface roughness (R_a) values of the specimens with different surface finishes are plotted in Figure 4. The R_a values increase in the order: VP < MP < MG for all of the investigated specimens, which indicate that MG samples have the highest surface roughness with the R_a values of around 100nm. In accordance with the surface morphology results shown in Figure 3, vibropolsihing with lower applied force can achieve a surface with lower R_a values (approximately 10nm) than mechanical polishing (approximately 30nm) for the as-fabricated and annealed specimens. Interestingly, R_a values of the mill annealed specimens under MP (23nm) and VP (13nm) were in a small differences as plotted in Figure 4. This is because the manufacturing process of mill annealed Ti6Al4V generates lower porosity (higher density) and more compact parts compared to DMLS [6]. Since a higher applied force may remove the internal loosely bonded powder particles, VP is able to develop a surface with lower roughness for DMLSed Ti6Al4V specimens.

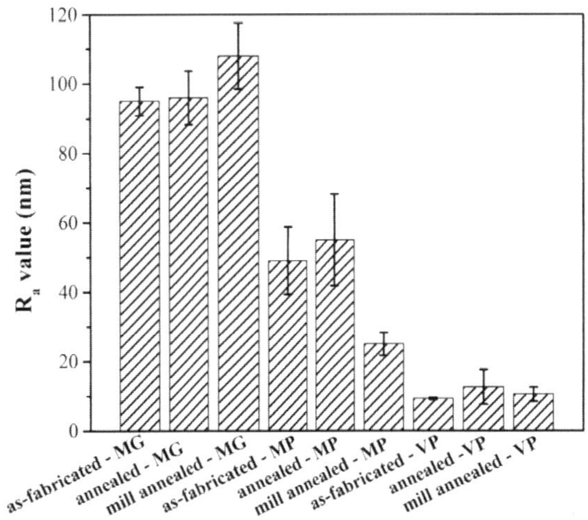

Figure 4: Surface roughness (R_a) values of the investigated specimens.

Apart from the two-dimensional surface roughness (R_a value) parameter, the three-dimensional hybrid parameters were also studied to analyze the surface morphologies. The resulting hybrid parameters (Sdq and Sdr) are listed in Table 2. The root mean square gradient (Sqd), is the RMS-value of the surface slope within the sampling area, and the surfaces area ratio (Sdr) expresses the increment of the interfacial surface area relative to the area of the projected plane [7][8][9]. With higher R_a value of the mill annealed MG sample compared to DMLSed specimens, its Sdq value (0.204) was lower, which indicates a wider spaced texture on the surface. A lower Sdr value (1.87%) of mill annealed MG refers to a less intricate surface. Similarly, the mill annealed VP also have lower Sdq (0.013) and Sdr (0.008%) values. The results of hybrid parameters indicate that DMLSed specimens possess a finer and more intricate texture.

Table 2: Hybrid parameters (three-dimensional surface parameters) of the designated specimens.

Specimens	Sdq	Sdr (%)
as-fabricated - MG	0.226	2.26
annealed - MG	0.251	2.80
mill annealed - MG	0.204	1.87
as-fabricated - MP	0.046	0.098
annealed - MP	0.044	0.096
mill annealed - MP	0.045	0.090
as-fabricated - VP	0.029	0.041
annealed - VP	0.039	0.075
mill annealed - VP	0.013	0.008

Figure 5 shows the cyclic polarization plots for the as-fabricated and annealed specimens under various surface finishes: mechanical grinding (MG), mechanical polishing (MP) and vibropolishing (VP). The corrosion parameters deduced from these curves are listed in Table 3. The corrosion rate (CR) values were calculated based on the corresponding corrosion current density (I_{corr}), specimen weight and density from the software. The resulting I_{corr} and CR decreased in the order: MG > MP > VP in the as-fabricated and mill annealed specimens, which means MG lost more material than the others and VP lost less. In the annealed specimen, MP specimen possessed higher I_{corr} and CR. Annealed VP specimen has relatively lower I_{corr} and CR than MG because of its lower surface roughness as plotted in Figure 4. The corrosion potential (E_{corr}) is an indicator of the stability of surface condition. Thus less variability in E_{corr} values in a specific specimen with different surface finishes indicates a consistent surface processing [5]. Therefore generally, samples under VP with smaller surface roughness values possess higher corrosion resistant. The as-fabricated and annealed specimens under VP have better corrosion resistant than the mill annealed specimen, because it was reported that the acicular microstructure in Ti6Al4V can form denser oxide layer than the equiaxed microstructure [10]. The corrosion behavior were also characterized by the breakdown potential (E_b) and re-passivation potential (E_p), and the results are listed in Table 3. At potential higher than E_b the metal undergoes corrosion. No significant difference in E_b between the investigated samples with various surface treatments. The difference between E_b and E_p is related to the pitting corrosion resistance of the material, a lower (E_b-E_p) value indicates better corrosion resistance. Among the investigated specimens with various surface finishes, no significant difference in (E_b-E_p) values. Compared to the mill annealed sample, the DMLSed specimens have no hysteresis loops, so they have no susceptibility to pitting corrosion. For the as-fabricated MP + scratch, the hysteresis loop is obvious due to pitting corrosion which will be discussed in the next section.

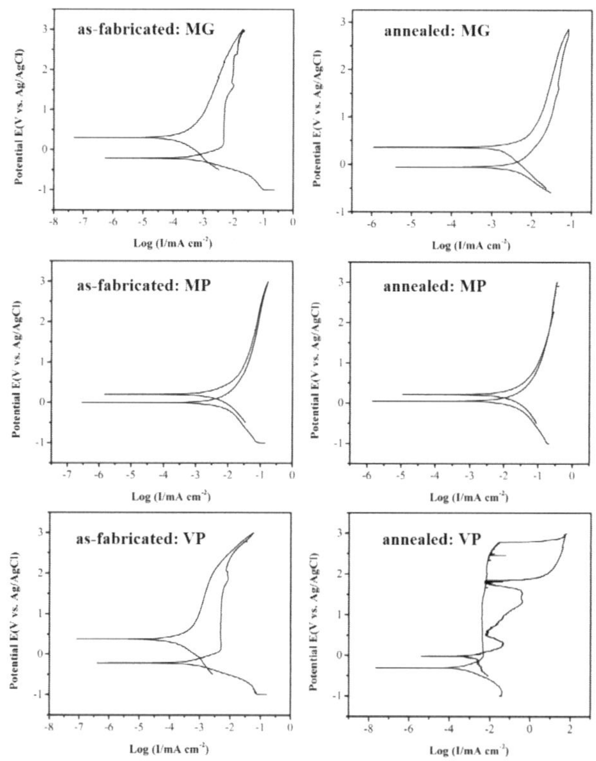

Figure 5: Cyclic polarization plots for the as-fabricated and annealed specimens under various surface finishes: MG, MP and VP.

Table 3: Cyclic polarization test results of the investigated specimens.

Specimen name	I_{corr} (µA/cm^2)	CR (µm/year)	E_{corr} (mV vs. ref *)	E_b (V vs. ref *)	E_p (V vs. ref *)	E_b-E_p (V vs. ref *)
as-fabricated - MG	0.77	0.307	-55.0	~ 1.61	-	-
as-fabricated - MP	0.44	0.176	-8.7	~1.9	-	-
as-fabricated - VP	0.34	0.136	-209.1	2.16	-	-
annealed - MG	0.68	0.267	-213.8	1.47	-	-
annealed - MP	0.17	0.177	-299.5	1.41	-	-
annealed - VP	0.21	0.121	-306.7	1.32	1.84	0.52
mill annealed - MG	0.93	1.105	28.3	1.57	0.10	1.47
mill annealed - MP	0.34	0.399	-300.9	1.25	-	-
mill annealed - VP	0.27	0.305	-80.6	1.69	-	-
as-fabricated - MP + scratch	0.36	0.125	-356.8	1.42	0.76	0.66

* The reference electrode was Ag/AgCl.

The effect of surface scratch

The optical images and measured topography image of the as-fabricated specimen under MP + scratch are shown in Figure 6(a) and (b), respectively. In Figure 6(b), pores on the investigated surface are still obvious with depth of approximately 0.5µm (the blue dots). Based on the color scale bar, depth of the intended scratch was approximately 1.0µm.

The as-fabricated MP + scratch Ti6Al4V specimen was immersed in Ringer's solution at 37°C to reach a plateau value in open circuit potential, and followed by cyclic polarization test. The resulting polarization curve is plotted in Figure 7, and the deduced electrochemical parameters are listed in Table 3. Compared to the MP specimen, more active corrosion potential (E_{corr}) was observed in the as-fabricated specimen under MP + scratch of -356.8 mV. The current density (I_{corr}) of

MP and MP + scratch sample was 0.44 and 0.36μA/cm^2, respectively. The corresponding corrosion rate (CR) are 0.176 and 0.125 μm/year. A slightly lower corrosion current density and corrosion rate was detected in the mechanical polished as-fabricated specimen with a scratch. The breakdown potential (E_b) refers to the onset of pitting formation, and the re-passivation potential (E_p) represents the protection potential below which the pits can no longer propagate are also listed in Table 3 [11]. The lower the difference, the more corrosion resistant the material is [12]. Hysteresis loop can be observed in the MP + scratch specimen as plotted in Figure 7, but absent in the MP sample which indicates a scratch before immersion makes the sample less pitting resistant.

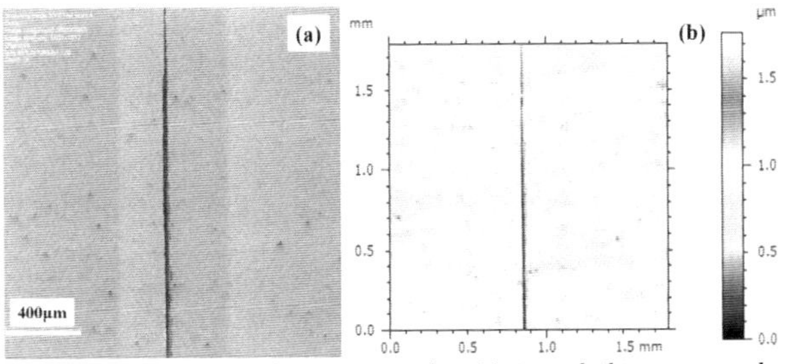

Figure 6: (a) Optical image (under 20x) and (b) topography image of as-fabricated Ti6Al4V with mechanical polishing and one scratch.

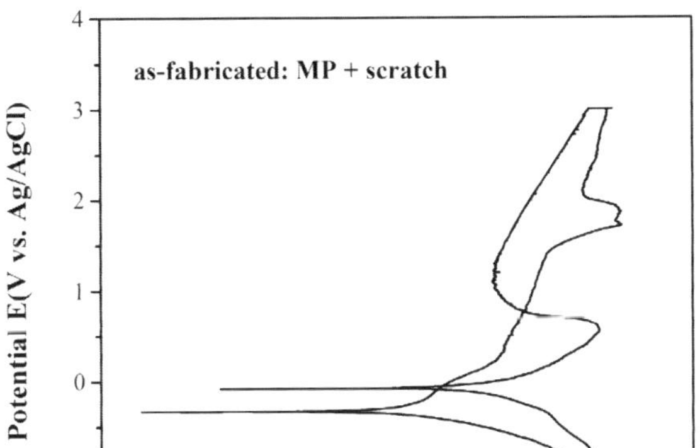

Figure 7: Cyclic polarization curve of the as-fabricated specimen with mechanical polishing and one scratch.

The optical images and topography image of the as-fabricated specimen with MP + scratch after electrochemical test is shown in Figure 8(a) and (b), respectively. Under 50x magnification, it is obvious that pits formed on both sides of the scratch after cyclic polarization. In Figure 8(b), red arrows refer to the developing of pits from the sides of scratch. The red circle shows a large area of pitting formed near the scratch. The three-dimensional surface morphology converted from Figure 8(b) is shown in Figure 9 with the cross-section of the scratch facing outward. The scale bar along the vertical direction indicates that the highest depth of the scratch was 4.0μm as pointed out by the thinner red arrow. The thicker red arrow and red circle refer to an area with significant bumps along the length direction of the scratch. Compared to the surface morphology of as-fabricated MP + scratch before the cyclic polarization test shown in Figure 6(b), numerous small bumps on the un-scratched surface can be observed in Figure 9. Salt deposition may happen on the metallic specimen surface after corrosion. Therefore, the detected depositions on the surface are highly possible of precipitated salt from the electrolyte. According to Revie and Greene, surface scratching is of minor importance in usual surgical implant application [13][14]. An increase in corrosion rate was only observed upon scratching after immersion, so the MP + scratch would have little to no effect on corrosion resistance. Pitting formation near the scratch and salt deposition on the sample surface were observed.

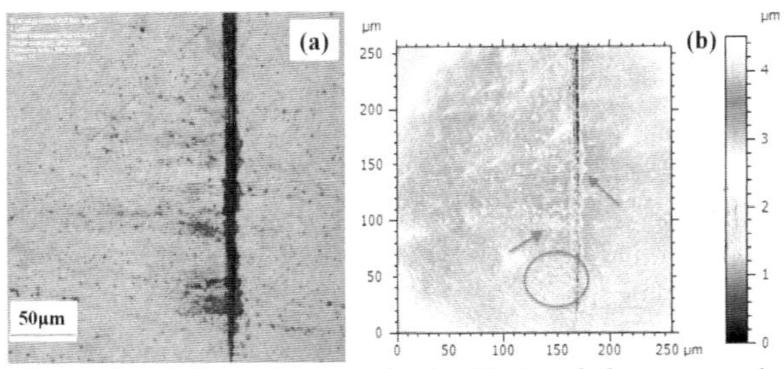

Figure 8: (a) Optical image (under 50x) and (b) topography image of as-fabricated Ti6Al4V with MP + scratch after cyclic polarization test. The red arrows and circle represent pits formed on both sides of the scratch.

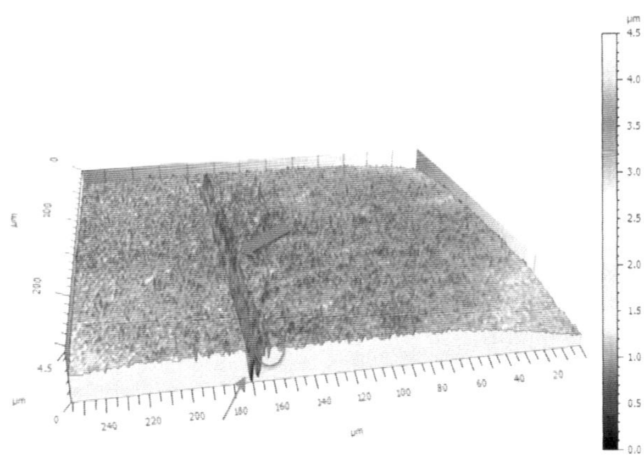

Figure 9: Three-dimensional surface morphology of the as-fabricated specimen with MP + scratch after cyclic polarization test. The thinner arrow refers to the scratch. The thicker arrow and red circle represent salt precipitation from the electrolyte (in warm color) on the surface.

Conclusions

The effect of various surface finishes on the corrosion behavior of direct metal laser sintered Ti6Al4V were examined. The surface roughness parameter (R_a values) increase in the order: VP < MP < MG for all of the investigated specimens, which indicates that MG samples have the highest surface roughness. The three-dimensional surface morphology parameters (Sdq and Sdr) were also calculated based on the software. The results show that Sdq and Sdr values decreased in the order: MG > MP > VP, which indicates VP specimens have less finer and intricate surfaces. Compared to the mill annealed specimens, the DMLSed parts possess finer surface morphology due to the internal pores and printing defects. With decreasing surface roughness (MG > MP > VP), VP specimens have the lowest corrosion current density and corrosion rate, because less interface area will decrease the corrosion susceptible sites.

Acknowledgments

The direct metal laser sintered samples were kindly provided by E-manufacturing solutions Inc. (Krailling, Germany). The help of Professor C. A. Brown with the facilities in surface metrology laboratory is gratefully acknowledged.

References

[1] Parthasarathy, J., Starly, B., & Raman, S. (2011). A design for the additive manufacture of functionally graded porous structures with tailored mechanical properties for biomedical applications. *Journal of Manufacturing Processes*, 13(2), 160-170.

[2] Traini, T., Mangano, C., Sammons, R. L., Mangano, F., Macchi, A., & Piattelli, A. (2008). Direct laser metal sintering as a new approach to fabrication of an isoelastic functionally graded material for manufacture of porous titanium dental implants. *Dental materials*, 24(11), 1525-1533.

[3] Niinomi, M. (1998). Mechanical properties of biomedical titanium alloys. *Materials Science and Engineering: A*, 243(1), 231-236.

[4] Xu, Y. Effect of heat treatment on the microstructure, mechanical properties and corrosion behavior of Ti-6Al-4V via direct metal laser sintering. In *ASM Heat Treat Society 28th Conference and Exposition October 20–22, 2015*. ASM.

[5] Cisse, O., Savadogo, O., Wu, M., & Yahia, L. H. (2002). Effect of surface treatment of NiTi alloy on its corrosion behavior in Hanks' solution. *Journal of biomedical materials research*, 61(3), 339-345.

[6] Murr, L. E., Quinones, S. A., Gaytan, S. et al. (2009). Microstructure and mechanical behavior of Ti–6Al–4V produced by rapid-layer manufacturing, for biomedical applications. *Journal of the mechanical behavior of biomedical materials*, 2(1), 20-32.

[7] Baránborn, I., Czifra, Á., & Kalácska, G. (2011). Height-independent topographic parameters of worn surfaces. *Sustainable Construction and Design*, 2(1), 35.

[8] Stout, K. J. (2000). *Development of methods for the characterisation of roughness in three dimensions*.

[9] Blateyron, F. (2006, September). New 3D parameters and filtration techniques for surface metrology. *In Proceeding of JSPE Annual Congress*.

[10] Cvijović-Alagić, I., Z. Cvijović, S. Mitrović, V. Panić, and M. Rakin. "Wear and corrosion behaviour of Ti–13Nb–13Zr and Ti–6Al–4V alloys in simulated physiological solution." *Corrosion Science* 53, no. 2 (2011): 796-808.

[11] Hryniewicz, T., Rokicki, R., & Rokosz, K. (2009). Corrosion and surface characterization of titanium biomaterial after magnetoelectropolishing. *Surface and Coatings Technology*, 203(10), 1508-1515.

[12] Thierry, B., Tabrizian, M., Savadogo, O., & Yahia, L. H. (2000). Effects of sterilization processes on NiTi alloy: surface characterization. *Journal of biomedical materials research*, 49(1), 88-98.

[13] Johansson, Berit I., and Bo Bergman. "Corrosion of titanium and amalgam couples: Effect of fluoride, area size, surface preparation and fabrication procedures." *Dental Materials* 11, no. 1 (1995): 41-46.

[14] Revie, R. Winston, and Norbert D. Greene. "Corrosion behaviour of surgical implant materials: II. Effects of surface preparation." *Corrosion Science* 9, no. 10 (1969): 763-770.

International Federation of Heat Treatment and Surface Engineering 2016
Proceedings of the 23rd IFHTSE Congress
April 18–21, 2016, Savannah, Georgia, USA

Future Trends of Low Temperature Surface Hardening of Stainless Steel

Andreas Karl[1], *Martina Wägner*[1]
Patrick Jacquot[2]
[1]*Bodycote, Landsberg, Germany*
[2]*Bodycote, Pusignan, France*

Abstract

The wide commercial application of austenitic stainless steels is a result of their excellent corrosion resistance. However, poor tribological behaviour, especially low abrasive / adhesive wear resistance and a tendency to fretting, has prevented the use of these materials in applications where both corrosion and wear resistance are required. Since the 1980s, low temperature carburising or nitriding has offered a solution to enhance mechanical properties without altering the corrosion resistance. These thermo-chemical diffusion processes form carbon or nitrogen S-phase while avoiding carbide or nitride precipitation that causes sensitisation.

This paper shows current applications and limitations of such processes giving consideration to such industries as automotive and medical. Besides improved wear, fatigue, galling, fretting and cavitation resistance, the use of these processes is greatly determined by the corrosion resistance achieved. Due to the fact that the supersaturated microstructure is in para-equilibrium, the precipitation of carbides or nitrides can be exacerbated by an inadequate surface condition. Also, the occurrence of undesirable phases or non-metallic inclusions in stainless steels influences their corrosion properties even more after surface hardening than for non-hardened surfaces.

This is why dedicated alloys for low temperature surface hardening of stainless steels are in focus currently. Compared to nitriding steels the main consideration is to safely avoid undesirable precipitations in the diffusion zone and inadequate surface conditions.

Furthermore, awareness of how certain heat treatments can fundamentally and positively extend the field of applications can offer greater scope for design engineers. This means, as is often the case for classical hardening processes, the engineer takes into account the properties and characteristics of the surface hardened stainless steel when designing a component. For example, uncomplicated slide bearings in corrosive environments have great potential for functional integration. Also, the unique ductility and cavitation resistance in even the smallest bores opens up new applications for high pressure injection. The wear and corrosion resistance of stainless steel can be significantly enhanced by using low temperature surface hardening.

Introduction

The major difference of corrosion resistant alloys to other groups is their outstanding passivation behaviour. Whether stainless steel, Ni-Cr or Co-Cr alloys, the resistance to environmental degradation makes passivated alloys the preferred choice for a diverse range of applications. End users in households come in contact with kitchen sinks, stainless knives or cookware. Other classical applications range from chemical or petrochemical processing to medical engineering to marine applications. In particular, the attribute of low lifecycle cost makes their use attractive.

Although other material groups can have lower purchasing costs, the longevity of stainless steel and low maintenance cost, combined with a favorable recycling rate, makes the alloy fit for the future trend of sustainable use of materials.

However, most passive alloys are characterised by low hardness and thus wear resistance that prevents them from being used for most applications where good tribological properties are desired. Thermochemical low temperature surface hardening has already proven in many fields of application, that it can overcome the disadvantages of stainless materials. Interstitial alloying with carbon or nitrogen can significantly increase the tribological properties and therefore widen the scope of applications. Taking into account that the inherent passive layer complicates surface hardening, the process technology needs to overcome some barriers. Formation of chromium-rich precipitates and thus a loss of corrosion resistance is the major concern when hardening stainless steel. The influences on the risk of the formation of precipitates at given process parameters are diverse. Hardening parameters, such as time, temperature and partial pressure of interstitials, have to be in a range where precipitation in a given alloy is avoided, however, the given alloy might be in a condition that favours precipitations. For example, cold deformed austenitic grades have a much higher tendency to form nitrides during low temperature nitriding than the solution annealed condition [1].

Role of hardening process parameters

Low temperature surface hardening of stainless steel by introducing carbon or nitrogen into the surface has attracted quite extensive research among materials engineers. A good overview of the fields of research is given by Dong [2]. Another work focusing on low temperature carburising has also found its way to the ASM handbook [3]. Since the conventional knowledge during the last century was that austenitic stainless steel cannot be properly hardened without losing the corrosion resistance, these treatments prove the opposite. Low temperature carburising and nitrocarburising allow a hardened case of more than 1000HV while maintaining the corrosion properties.

Applying a treatment temperature below a level where precipitates can form, say 450°C, leads to good corrosion resistance of the hardened zone but provides several challenges. The protective passive layer is quite stable at such low temperatures and thus would prevent the interstitial atoms from diffusing into the surface. A suitable activation or removal of the passive layer is the key for achieving uniform surface hardening. Such low treatment temperatures also prolong the time needed for a sufficiently thick case of 15 to 40μm. But with high enough partial pressures for carbon or nitrogen, a relatively quick supersaturation with interstitials can be achieved. Supersaturation in this case means that the typical solubility for carbon in AISI 316 is <0.015 at.% [4] and the maximum values achieved in carbon S-phase are in excess of 12 at.%.

This also means that the supersaturated layer is in para-equilibrium and thus metastable. It will decompose into more stable phases, nitrides or carbides, when heated above a certain time/temperature regime [5]. For every application it has to be taken into account that the long-term stability of carbon and nitrogen S-phase is at roughly 400°C and 350°C, respectively [6]. Short-term stability reaches much higher values.

Looking at the mechanical properties, the gradual or smooth hardness increase for carbon S-Phase provides fatigue benefits compared to nitrogen S-Phase and results in a significant improvement to the fatigue strength of the material [7]. On the outermost surface these compressive stresses can reach more than -2GPa with a smooth profile to the base material for carburised and more than -6GPa with an abrupt decrease for nitrided 316L stainless steel [8].

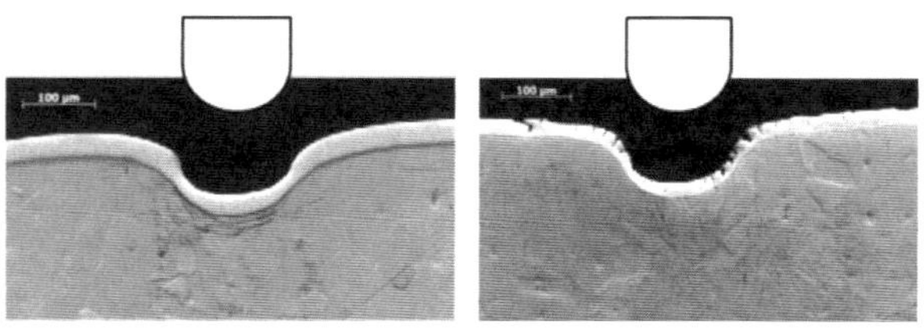

Figure 1. Micrographs after hammer & nail test, left: low temperature carburised S-phase deforms in a ductile manner, right: low temperature nitrocarburised case acts brittle and shows cracks [9].

Compared to low temperature nitriding, carburising has several advantages. These advantages are a function of a lower driving force to create precipitates and a smoother transition from the hardened S-phase to soft, untreated material. The achievable diffusion depth without forming precipitates is improved and the allowable load bearing capacity to failure is significantly increased [5]. Also, the ductility of nitrogen S-phase at rapid loading is quite limited. This is why a combination of both processes known as low temperature nitrocarburising was introduced. By addressing some of the limitations, it combines the beneficial properties, such as a hard surface and a smooth transition to the core. However, the fatigue strength and the high ductility of pure carbon S-phase is still at a higher level. A hammer and nail test, simulating rapid loading with plastic deformation, reveals that carbon S-phase deforms uniformly without forming cracks, Figure 1.

Table 1: Comparison of characteristics of low temperature nitriding and carburising, modified from Dong [2]

Characteristic	Carburising	Nitriding
Treatment temperature	300°C – 550°C	300°C – 450°C
Typical layer thickness	5 – 40 μm	2 – 20 μm
Layer structure	Carbon S-phase	Nitrogen S-phase
Corrosion resistance	Good	Good
Surface hardness	900–1100 HV0.05	1400–1600 HV0.05
Hardness profile	Gradual change	Abrupt change
Wear resistance	Good	Very good
Layer toughness	Good	Poor
Load bearing capacity	High	Low
Fretting fatigue	Good	Very good
Fatigue properties	High	Low

Role of substitutional alloying elements

One solution for improving the corrosion and wear properties of low temperature surface hardened stainless steel would be the development of dedicated alloys for those processes. An approach could be the alloying of standard stainless steels with strong nitride/carbide forming elements (SNCFE). The main goal is to improve the corrosion resistance by ensuring that chromium forms the passive layer and is not bound by precipitates. To achieve this, it is important to have a proper understanding of the role of substitutional elements in the formation of S-phase in austenitic stainless steels, as substitutional elements have a significant influence on S-phase thickness and performance 0. Williamson et al. [10] found that carbide/nitride forming elements such as Ti, V, Nb; Cr, Mn and Mo increase the nitrogen solubility, with V and Ti being the strongest nitride forming elements.

Most recently, Christiansen, et al. [11] discussed the possible addition of Ti or Al to austenitic stainless steels in order to increase process temperature. Therefore, they treated a commercially available austenitic A286 alloy containing ~2% Ti. It is believed that the hardening mechanism not only derives from formation of supersaturated, austenitic S-phase, but also from the formation of hard titanium rich phases.

Furthermore, thermodynamic calculations of possible alloys (AISi316L+Ti and/or Al) were presented and showed that SNCFE contents higher than 4% lead to the formation of ferrite and/or Laves phases. In order to maintain a fully austenitic structure additional alloying of Ni may be necessary.

Nanocrystalline materials

Low temperature interstitial surface hardening of nano-crystalline stainless steels provides the possibility to improve the diffusion depth. It is possible to achieve such microstructures with severe plastic deformation by surface mechanical attrition treatment (SMAT). Due to possible oxide inclusions after SMA-Treatment, a polishing process before nitriding should be considered.

Chemki et al. presented low temperature plasma nitriding of SMATed AISI 316 at 425°C for 20h achieving a 50% higher diffusion depth compared to non-SMATed AISI 316 [12]. This is of special interest for increased wear resistance and load bearing capacity.

Laleh, et al. showed that plasma nitriding of a pre-SMATed AISI 316 at only 300°C also leads to significantly higher diffusion depth, compared to as-received material [13]. Diffusion at such low temperatures can further decrease the risk of distortion or precipitations for certain alloys and conditions.

Change in magnetic properties

An effect that has not been investigated extensively is a change in the magnetic properties of austenitic stainless steels, although there is no indication that precipitations are present. After formation of S-phase via laser carburisation, the resulting S-phase was investigated by Mössbauer spectroscopy. Due to isomer shifts and quadrupole splitting, the formation of an amorphous and soft ferromagnetic phase was revealed [14].

A similar effect, observed on ion nitrided stainless steel, suggests that the ferromagnetic effect is governed by Cr-N interactions [15].

For highly alloyed austenitic stainless steels with three to seven wt% Molybdenum, in combination with very high carbon supersaturation, a ferrite number (FN) of up to 17 Fe% was found. Besides the peak shift typical for S-phase, a second peak or peak broadening can be seen on such alloys, Figure 2.

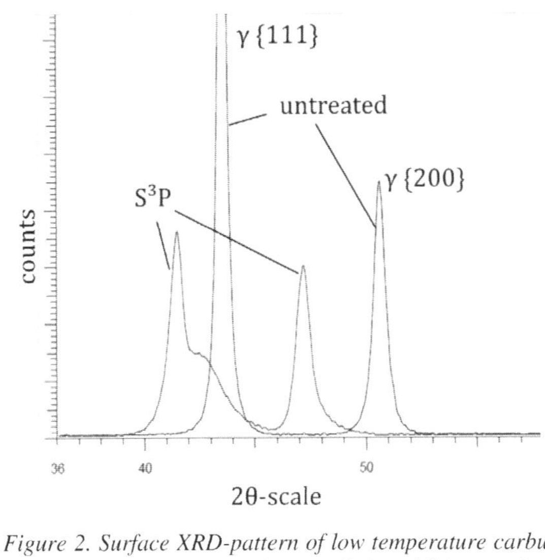

Figure 2. Surface XRD-pattern of low temperature carburised (S^3P treated) 1.4529 stainless steel. In addition to the peak shift that is typical for S-phase compared to the untreated condition, a peak broadening and signs of a second phase appear.

Such a dramatic change in the magnetic properties, without the formation of other phases than austenite, can potentially be used for technological applications. Selective hardening and a locally limited magnetic surface region can be used for monitoring e.g. rotational speed of shafts or transversal speed or position of a slide bearing.

The next step would be to apply a complex pattern of magnetic areas on non magnetic stainless surfaces for part identification. Such an invisible and hard to find feature might make it easier to trace the use of unauthorisised spare parts.

S-phase coatings

Common processes to form S-phase are gas-based or plasma-based thermochemical treatments. Successful results are also reported with ion implantation, plasma immersion ion implantation and magnetron sputtering [2]. The latter PVD coating can apply a relatively flexible combination of different coatings on various materials. Due to the low treatment temperature of around 204°C the S-phase coating can also be applied to certain polymers. Doing so, and combining the hard and chemically resistant S-phase with an active anti-microbial element like Ag or Cu, can achieve a desirable combination of properties [16].

Figure 3. SEM images of S-phase coatings, left: Monolayer doped with low concentrations of copper, right: fracture surface of silver/S-phase multilayer with a bi-layer period of 70nm [16].

Design rules

The general design rules for heat-treatment-suitable part design are not necessarily valid for low temperature processes. Standard case hardening or martensitic hardening of stainless steel requires quenching steps that pose certain challenges. To achieve a uniformly hardened component, the heating and cooling rates of the material have to be uniform. Thus, the wall thickness should be kept constant and the process hardware has to allow for a uniform temperature control. This also influences part distortion or warpage during heat treatment.

For low temperature surface hardening the process temperatures in most cases are low enough to avoid recrystallisation of the bulk material, especially for austenitic stainless steels, Ni-base and Co-Cr alloys. This means that mechanical parts in the final machining condition can be hardened without a change in geometry. Tight tolerances can be achieved without any rework.

For martensitic and PH materials the low temperature heat treatment can cause a change in the base material strength as annealing temperatures are often below 450°C. This means that a final annealing step should be in excess of the process temperature of low temperature surface hardening.
Rules for easy to clean design often apply for low temperature surface hardening. In particular, the fact that hardening processes remove or attack the passive layer during the passivation step makes it necessary to reform the protective layer. Passivation processes are often of aqueous nature and a certain flow of the passivating agent is beneficial, e.g. in blind holes such a flow can hardly take place. Easy to clean design is also mandatory for equipment used in the food and beverage industry [17].Surface scratches can allow bacteria to survive chemical cleaning as the detergents cannot effectively enter small cavities. Because of this, a significant proportion of S-phase treatment demand comes from the food and beverage processing industry. Due to the improvement in wear rate and scratch resistance following S-phase treatment, components are produced which are much easier to clean.

Functional integration

The trend of functional integration, combining many functions in one complex component, is boosting additive manufacturing (AM) technologies. A similar result is possible with low temperature surface hardening.
The combination of low temperature surface hardening and stainless steel can give the engineer the opportunity to change and/or better optimise a component or subassembly design. Many kinematic parts are assembled from different materials with different properties. As an example from the automotive industry, the consideration of the use of S³P allowed for better design.
A connecting rod assembly, Figure 4, has the function of transmitting movement from an actuator to a receiving part. The two bearing zones have to provide a sufficient resistance to fretting wear (1, 3). The connecting rod needs to provide sufficient mechanical strength to not deform at operating forces and a certain ductility for safety reasons. The assembly of the three components can be quite expensive under certain circumstances, e.g. glue cannot be used due to high operating temperatures.

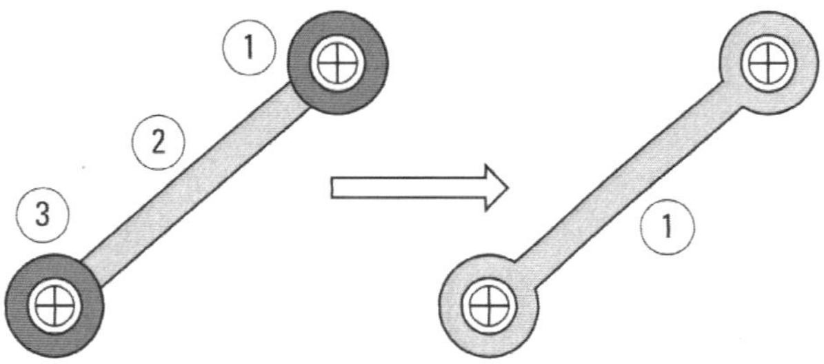

Figure 4. Functional integration example: kinematic connector; left: assembly of three parts, (1) (3) bearing zone material for fretting and wear resistance, (2) connecting rod for mechanical strength and certain ductility; right: (1) cold drawn surface hardened [9]

The function of these three components can be unified in one low temperature carburised part, Figure 4, schematic on the right. The stainless steel base provides sufficient strength and ductility and the carbon S-phase at the bearing zone delivers good fretting and wear resistance. This functional integration provides significant cost savings. Further, problems related to the use of different materials, like galvanic corrosion or deformation due to different coefficients of thermal expansion, can be minimised.
As mentioned, additive manufacturing is becoming more widely used and the ease of manufacturing complex shapes will probably make this technology indispensable in the future. For S-phase technology, the opportunity is to improve these costly parts and prolong their lifetimes. However, AM technology has some disadvantages. Selective laser sintering (SLS) still has a lack of precision and only achieves rough surface conditions. For functional components with tribological loading it is a challenge to withstand the forces under such circumstances. Low temperature surface hardening can be successfully applied, but can have a lack of surface integrity and thus mechanical resistance, Figure 5. Current developments such as laser-remelting, electrochemical polishing or sand blasting of the surface have the potential to solve this problem, but are still not an industry standard.

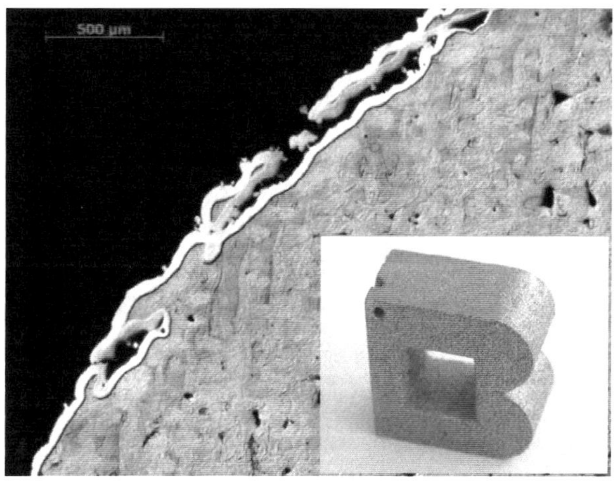

Figure 5. Small component made by means of SLS. In addition to the porous microstructure the surface and thus also the diffusion zone is not uniform in shape. The mechanical properties can be improved with post machining. The S-phase is uniform and is formed on all exposed surfaces.

Combination with other processes

A current obstacle in identifying new applications for S-phase engineering is a lack of knowledge on the part of end users and/or their suppliers. Manufacturing austenitic stainless steel is already challenging, but limited awareness of the scope of use of subsequent low temperature surface hardening provides further limitations. Overly harsh machining or the use of inappropriate material grades or qualities often leaves the impression that low temperature surface hardening is not a solution but an additional problem when using stainless steel.

Deformation martensite is sometimes used to increase the strength of stainless steel. Further, severe manufacturing or dull cutting tools can cause deformation martensite in austenitic stainless steels what is not desirable as it reduces the corrosion resistance. During low temperature carburising or nitriding, Cr-rich precipitates are formed quicker than in non-deformed austenite. To overcome this problem prior solution annealing can be added to the production route. These annealing processes might cause stress relief and thus distortion. This is why it cannot be applied to most precision parts.
Further, the inherent compressive stresses in S-phase can change the surface appearance slightly. Particularly for consumer products a mirror finish can be mandatory and will not be achieved after surface hardening. The slight increase in roughness for highly polished surfaces was measured to be in the range of 0.04μm Ra for the untreated and 0.08μm Ra for the S^{3}P treated condition. To overcome this effect, a subsequent mirror polish can be applied, Figure 6.

Figure 6: Low temperature carburised austenitic stainless steel fork. Due to the high surface hardness the scratch resistance, and thus the shine, is highly improved. To obtain the surface finish a fine polishing is required.

When surface hardening processes leave residues on the surface other post processes, such as pickling or electro-polishing, can be used. These value-adding processes increase the optical homogeneity and the corrosion resistance.
An approach to improve mainly the wear resistance of components is the application of physical vapor deposition (PVD) or low temperature plasma assisted chemical vapor deposition (PaCVD) on S-phase. The diffusion zone with a smooth transition to the base material acts as a support layer for the harder coating. This helps to reduce crack susceptibility and increases the lifetime, Figure 7.

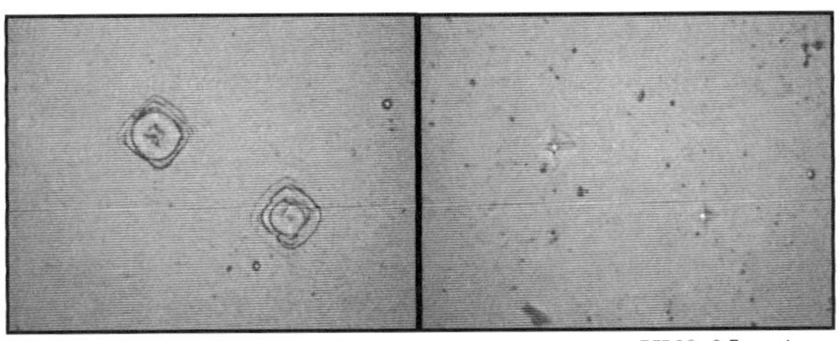

Figure 7: Surface hardness measurements, HV0.05 micro-indentations at 500x magnification; left: diamond-like carbon (DLC) coating, right. S^{3}P treated stainless steel + DLC coating with lower tendency for cracking.

A summary of some important pre- or post-processes for S^{3}P low temperature surface hardening is given in Table 2.

Table 2: Additional processes complementary to low temperature surface hardening

Process	**Pre- or post-process**
Solution annealing	pre
Chemical passivation	post
Electropolishing	pre and/or post
Plasmapolishing	post
Mechanical polishing	pre and/or post
PVD/PaCVD coating	post
DLC coating	post

Engineering applications

Since the first researchers started to investigate the properties of S-phase in the 1980s many successful engineering applications have been in-service. A particular example from the nuclear industry is often referred to: austenitic stainless steel control rods, with a height of about 4m and 9.7mm in diameter, treated by low temperature nitriding. This provides a corrosion resistant, and up to 90% better wear resistant, case prolonging the lifetime of the rod [5].
Another classical application is ferrules for tube fittings. The hardened case allows the ferrule to cut into the tube properly and provide a leak tight seal with a low installation torque due to the improved galling resistance of the material [2].
For dosing pumps in the food or pharmaceutical industry the beneficial tribological properties of carbon S-phase are used to simplify the design, Figure 8. The old design had a polymer seal between the metallic components that had to be replaced frequently. The main advantages of the new design are less maintenance and a much better cleanability of the piston [18]. This principle, namely replacing the polymer seal design by S-phase sliding contact, has also found its way into extensive use in the automotive industry.

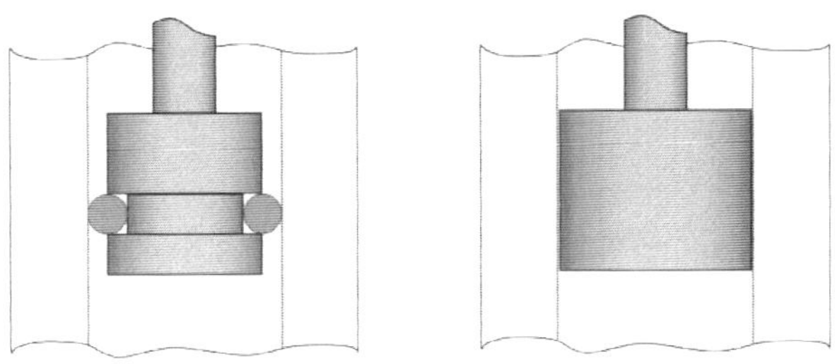

Figure 8: Design change of a dosing pump from conventional design (left) to modified design with an S-phase layer (right) [18].

To date, the stainless steel fastener market, and especially the prevention of galling during installation or disassembly, is still mainly addressed by pastes. These provide lubrication and prevent metallic contact. As soon as liquids are in contact, contamination is possible and thus highly undesirable in health and safety critical installations. Further, the short-term stability of such pastes is not well proven and may not be sufficient in buildings that must last for decades.

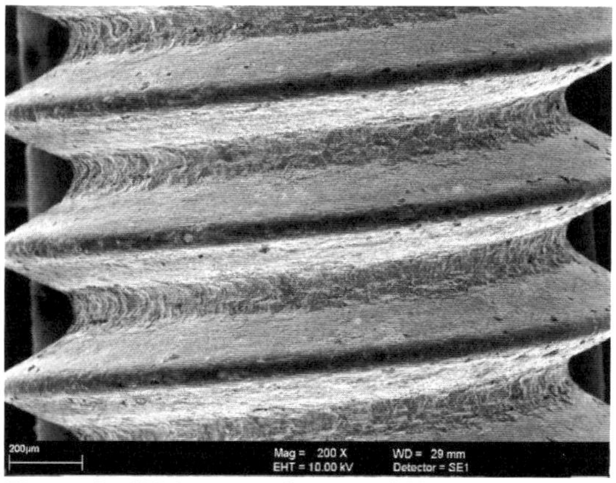

Figure 9: SEM-image of a DIN 912 M2x10 stainless steel screw with a highly deformed and inhomogeneous surface

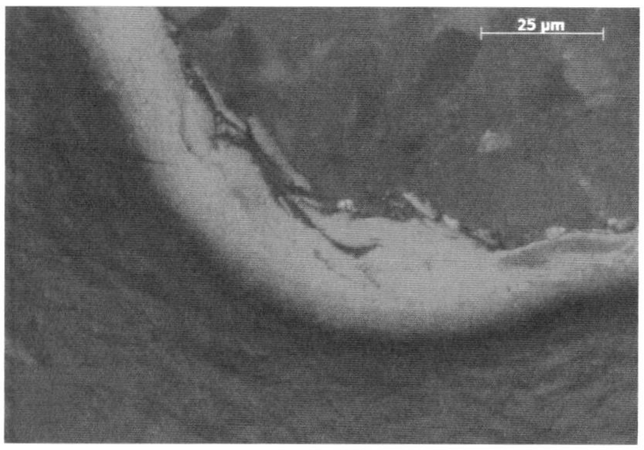

Figure 10: Microstructure thread root of the screw in Figure 9 in the low temperature surface hardened condition. Due to the high degree of deformation and subsequent formation of compressive stresses small crevices form during the diffusion process which can lower the corrosion resistance.

The use of S-phase equipped steel for stainless steel screws, nuts and bolts is increasing but has not yet entered the mass market. Two main limitations can be observed: the additional process cost is a barrier; and technological gaps in terms of materials, have not yet been solved.
In order to allow for cost-effective mass production of screws, production tools are optimised for a long service lifetime, and the surface morphology is not seen as critical, Figure 9.
The surface has inherent weak points that can open up during surface hardening and form unfavorable crevices which lower the corrosion resistance, Figure 10. This technological barrier can be addressed by a different production route or over-alloying of the screw material. Overcoming these barriers would open a huge field of applications.
Besides this, low temperature surface hardening is used for consumables such as watches or tableware, mainly due to its improvement in scratch resistance. A selection of fields of application for S³P treatments is given in Table 3.

Table 3: Selection of some fields of application for low temperature surface hardening incl. main reason for their use

Field of application	Main reason for use
Screws, nuts, bolts	Galling resistance
Spiral or flat springs	Fatigue resistance
Shafts, spindles	Wear and fatigue resistance
Tableware, watches	Scratch resistance
Fluid handling (pumps, valves, propellers)	Galling, wear and cavitation resistance
Medical components	Wear resistance

Conclusions

Low temperature surface hardening of stainless steel allows the improvement of tribological properties without altering the corrosion resistance. Despite 30 years of experience and thousands of successful applications, these technologies are still not considered as standard heat treatments. The relatively thin hardened case and the limitations of temperature stability are the main technological barriers of physical nature.
However, an increasing knowledge among materials and mechanical engineers has led to a widening of application fields to nearly all areas where stainless materials are used.
The merits of low temperature carburising, nitriding and nitrocarburising should be considered for each application. As soon as the specification sheet mentions certain risks concerning in-service temperature in excess of 300°C, welding or laser marking, fatigue loading or rapid impact, carbon S-phase is the preferred choice. Also, when cold deformed parts have to be hardened, carbon processes achieve precipitation free diffusion zones where nitrogen tends to form nitrides. However, nitriding and nitrocarburising have certain advantages, namely higher hardness and better wear resistance. When considering the properties, the more robust and safe technology is carbon diffusion, although with additional processes like solution annealing for austenitic steels, nitrocarburising can produce desirable properties. Further, carbon diffusion has some limitations alloy-wise. Stainless martensites and PH martensites should preferably be nitrocarburised in order to achieve a corrosion resistant, thick and fatigue resistant case.
Design and availability of special alloys for S-phase engineering might be unrealistic from an industrial point of view in the next five or more years, but would offer significant improvements from a temperature stability and diffusion depth point of view. Further research might speed up this trend.
To use S-phase as a coating on various material groups has the potential for niche applications. But end users will have to decide if PDV technology for S-phase technology is the right choice as it has technological restrictions such as geometrical accessibility and coating adhesion.
The effect of a change in magnetic properties on austenitic stainless steels is not yet fully understood but might provide technological advantages for demanding sensor applications or part identification.
Besides these niches, a greater awareness among design engineers will drive the use of surface hardened stainless steel

components. The opportunities within functional integration and possibilities for much easier design are still not well known. One suggestion to close this gap is to implement certain techniques within the surface hardening companies. Beside abrasive processes such as grinding, polishing or blasting, the implementation of chemical or electrochemical pre- or post-processing allows service providers to directly influence the technological properties. Providing this in one facility avoids a more complex supply chain for end-users.

References

[1] F. Bottoli, G. Winther, T. L. Christiansen, and M. A. J. Somers (2015). Influence of Plastic Deformation on Low Temperature Surface Hardening of Austenitic and Precipitation Hardening Stainless Steels by Gaseous Nitriding. HTM Journal of Heat Treatment and Materials: Vol. 70, No. 5, pp. 228-238.

[2] Dong, H., S-Phase surface engineering of Fe-Cr, Co-Cr and Ni-Co alloys, *Int. Mater. Rev.,* Vol 55(No.2), 2010, p 65-98

[3] Collins, S.R., et al.; Low temperature Carburizing of Austenitic Stainless Steel. *ASM Handbook, Volume 4D, Heat treating of Irons and Steels*, 2014

[4] Y. Cao, F. Ernst, G.M. Michal. "Colossal carbon supersaturation in austenitic stainless steels carburized at low temperature." Acta Materialia. 2003, 51, S. 4171-4181.

[5] T. Bell, C. X. Zang. "Low Temperature Nitriding and Carburizing. "Advanced Materials & Processes. June, 2002.

[6] S. Thaiwatthana; X. Y. Li; H. Dong; T. Bell. Comparison studies on properties of nitrogen and carbon S phase on low temperature plasma alloyed AISI 316 stainless steel; Surface Engineering 18(6) pp. 433-437

[7] L.Ceschini, G. Minak. "Fatigue behaviour of low temperature carburised AISI 316L austenitic stainless steel." Surface & Coatings Technology. 2007. doi:10.1016/j.surfcoat.2007.07.066.

[8] T. L. Christiansen, T. S. Hummerlshoj, M. A. J. Somers. "Expanded austenite, crystallography and residual stress." *Surface Engineering*. 2010, Bd. Vol 26, No 4

[9] A. Karl; M. Wägner (2015). Surface hardened stainless steel in automotive applications. SSW conference. Maastricht, Netherlands

[10] Williamson, D. L., Davis, J. A., & Wilbur, P. J. (1998) Effect of austenitic stainless steel composition on low-energy, high-flux, nitrogen ion beam processing. Surf. Coat. Tech. 103 (1), 178-184

[11] Christiansen, T. L.; Dahl K. V.; Somers M. A.J.; New stainless steel alloys for low temperature surface hardening?; Proceedings of 8th European Stainless Steel & Duplex Stainless Steel Conference, Graz, 2015, p. 513-522

[12] Chemki, M., et al.; The effect of surface mechanical attrition treatment on low temperature plasma nitriding of an austenitic stainless steel. Surface and Coatings Technology, Elsevier, 2013, 221, pp.191-195.

[13] Laleh, M., et al.; Low-Temperature Nitriding of Nanocrystalline Stainless Steel and its Effect on Improving Wear and Corrosion Resistance; *Journal of Materials Engineering and Performance*, 2012

[14] P. Schaaf, et al. (2009). Transformation of expanded austenite to an amorphous ferromagnetic surface layer during laser carburization of austenitic stainless steel. HTM J., 64(4), pp. 242-248.

[15] R. L. O. Basso, et al. (2009). Magnetic and structural properties of ion nitrided stainless steel. J. Appl. Phys., 105(124914).

[16] D. Formosa, et al. (2015). Towards novel long-lasting S-Phase based anti-bacterial coatings. *22nd IFHTSE Congress.* Venice, Italy

[17] W. Burggraaf (2015), Don't Spoil your Market. A look at the Cleanliness of Stainless Steels. SSW conference. Maastricht, Netherlands

[18] Bos, M. J. (1998, March). Case hardening of austenitic stainless steel pump components. *World Pumps*, pp. 30-34.

International Federation of Heat Treatment and Surface Engineering 2016
Proceedings of the 23rd IFHTSE Congress
April 18–21, 2016, Savannah, Georgia, USA

COMPUTER SIMULATION OF INDUCTION THERMAL PROCESSING OF FERROUS AND NON-FERROUS ALLOYS

Collin Russell, Dr. Valery Rudnev
Inductoheat Inc., An Inductotherm Group Company
32251 N. Avis Dr., Madison Heights, MI 48306 USA
crussell@inductoheat.com, rudnev@inductoheat.com

Abstract

Computer simulation plays a critical role in the successful design of induction heating and heat treating systems. In fact, in some cases, computer simulation is the only remaining option that is available to an engineer in the design of an induction heating system and the determination of optimal process parameters. Computer simulation is also a valuable tool when searching for root cause of failures; it can help to assess how real-life factors such as deviations in dimensional tolerances and geometric irregularities affect the sensitivity and robustness of a process. Induction heating incorporates a complex combination of electromagnetics, heat transfer, circuit analysis, and metallurgical phenomena that are interrelated and highly non-linear due to material properties' dependence upon magnetic field intensity, temperature, chemical composition, microstructure, etc. Various aspects of numerical computer simulation of different induction heating and heat treating processes are discussed in this presentation and a number of common induction heating and heat treating problems are addressed in the form of real-life case studies.

Introduction

In a rapidly changing global economy, the ability of induction heating equipment manufacturers to minimize time between receiving requests for quotation (RFQs) from customers and preparing quotations based upon accurate computer simulation results is critical. The competitive industrial environment does not offer the luxury of several months, weeks or even days of waiting in order to prove the feasibility of certain process through trial and error; quick and reliable results are a requirement in the modern marketplace.

Induction Scan Hardening

Scanning inductors are often used to induction harden steel components [1-4]. One of the main advantages of scan hardening compared to single-shot hardening or static hardening is associated with its superior flexibility for hardening parts of different lengths and diameters (though some limitations do exist). Scanning inductors consist of one or more turns and its copper is often profiled to better accommodate geometrical complexities of the part such as fillets, undercuts, etc. To illustrate, consider the induction scan hardening of a 64mm (2.5in.) diameter medium carbon steel hollow axle shaft using a two-turn inductor and an electrical frequency of 1 kHz (Figure 1).

The inductor moves along a shaft during scanning. Coil power and scan rate are carefully controlled to accommodate the varying geometry of the shaft, including shoulders, diameter changes, wall thickness variations, etc. To properly harden the shaft's fillet area (above the flange), a short dwell is often used at the beginning of the process; in this stage the inductor is energized, but does not move. Quenching is not applied during the dwell stage.

Achieving a case depth with a minimum required hardness in the fillet without exceeding the maximum case depth in regions next to the fillet often presents challenges. Those challenges can be associated with several phenomena [1,2]:

- *From an electromagnetic perspective,* it is challenging to induce sufficient heat generation (power density) in the fillet without creating excessive temperatures in adjacent regions; fillets with small radii are more challenging due to the increasingly undesirable combination of electromagnetic proximity and ring effects.
- *From a heat transfer perspective,* during heating, the presence of various cold areas adjacent to the fillet creates an unusually pronounced localized cooling effect ("cold sink" effect) in the fillet region compared with other areas of the shaft. Thus, during induction heating, more heat is removed from the fillet area due to thermal conduction than is removed from other areas of the shaft. This requires that the inductor is designed to generate more heat in the fillet to compensate for this effect.
- *From a design perspective,* the shaft geometry may have certain features (e.g., edges, sharp corners, cuts, thin walls, etc.) located near the fillet, which are not permitted to be re-austenitized or through hardened during the fillet hardening cycle.

Figure 1 shows both temperature distribution and magnetic field distribution at the end of the dwell stage. In this application, per the customer's request, a bare two-turn coil (outlined in white) is used.

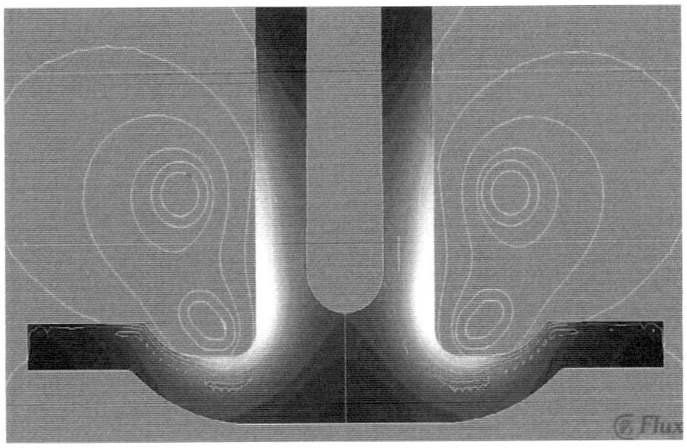

Figure 1. Results of FEA computer simulation of scan hardening a hollow axle shaft [2].

In order to increase power density in the fillet region, the bottom turn is deliberately deformed in the fabrication process. Closer coupling of the bottom turn to the fillet and flange compared to the shaft's body compliments the profiling of the lower turn. Upon completion of the dwell stage, the fillet is sufficiently pre-heated and scanning begins. In many cases, a faster scanning speed is used initially to rapidly move the coil out of the dwell position in order to provide sufficiently quick quenching of the fillet area and also to avoid excessive heating of the mass directly above the fillet.

It is important to remember that induction hardening of steel components is a two-step process involving heating and quenching. When scanning outside surfaces (e.g. shaft's outside diameter), the quenching device is typically positioned next to the coil in order to spray-quench the area that has been properly austenitized. In other cases, a machine integral quench (MIQ) inductor can be used. In either case, the quenching device consists of a quench chamber with numerous holes (orifices) that allow the quenchant to impinge on the part at a specific angle and distance. It is imperative to be able to simulate both the heating and quenching stage of an induction hardening process.

During scanning, a pre-heating and post-heating effect is created due to the propagation of magnetic field lines around the inductor. At any specific point along the surface that is to be hardened, an induced heat source is created both before and after a current-carrying turn is directly adjacent to this point. These effects, practical consequences of electromagnetic end effects, must be taken into consideration when designing induction scan hardening equipment and processes.

The so-called "comet-tail effect" manifests itself as apparent heat accumulation in the subsurface regions of the part below the inductor. Upon quenching, the temperature of the surface is cooled sufficiently below the M_s temperature; however due to subsurface heat accumulation, tempering back of the as-quenched surface regions (resulting in potential soft spots within the case depth) can occur if adequate quench intensity is not provided at the surface. Sufficient quench-out is essential to prevent this undesirable phenomenon.

Selective Induction Heating

Many applications (induction brazing, soldering, hardening, tempering, stress relieving, annealing, shrink fitting, localized warm and hot forming, etc.) require induction heating of only select areas of the workpiece. Compared to processes involving heating of an entire workpiece, selective heating processes inherently involve more complex and non-uniform electromagnetic and thermal condition; consequently the design of the inductor(s), selection of frequency, and determination of electrical parameters are typically much more involved.

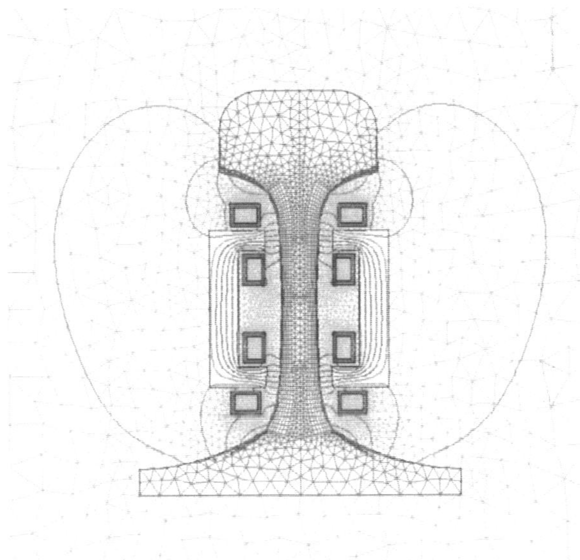

Figure 2. Results of FEA computer simulation: Magnetic field distribution when of induction heating selected areas of steel rail prior to sizing [1].

Finite element analysis (FEA) computer simulation of selective induction heating of a steel rail from ambient to hot forming temperatures is shown in the Figures 2 and 3. Butterfly inductors (one on each side of the rail) are utilized to provide the desired selective heating effect.

Current flow in the center turns of the inductors is in the same direction, generating the main heating effect. The top and bottom turns of the inductors, the so-called "return legs" of the inductors, conduct current in the direction opposing the current flow in the center turns. The top and bottom turns induce sufficient heat sources to compensate for the thermal energy conducted away from the center of the rail due to the cold sink effect – both during heating and during transportation to the subsequent operation in the manufacturing process.

U-shape lamination stacks positioned around the center turns have a two-fold impact on performance of butterfly-type inductors:

- Serving as magnetic flux-concentrators to boost the heating intensity of center turns.
- Providing electromagnetic decoupling of coil turns with currents flowing in opposite directions; this also increases overall coil electrical efficiency.

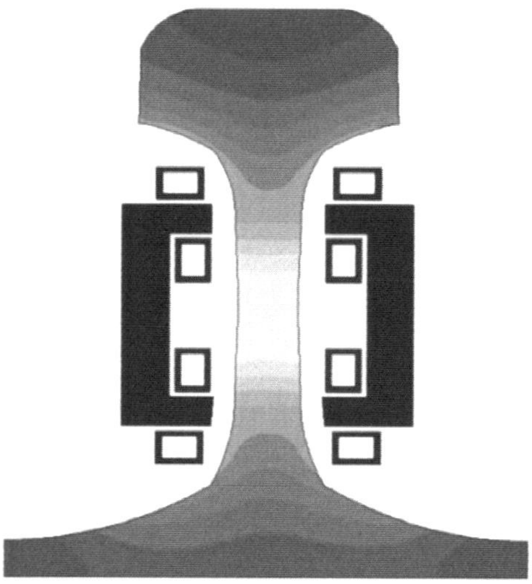

Figure 3. Results of FEA computer simulation: Temperature distribution at the end of induction heating selected areas of steel rail prior to sizing [1].

3-D Inductor Design Challenges

In many cases, it is possible to take advantage of axisymmetric symmetry when simulating the induction heating of cylindrical bodies. This permits the use of an approximate 2D simulation to be conducted as opposed to a 3D simulation. There are cases, however, in which the geometry of the inductor does not provide valid geometric and physical axisymmetry; in these cases 3D simulation is required. A typical single-shot hardening process is an example of a case in which 3D simulation is required to obtain a valid representation of the real induction heating process.

It is imperative to remember that the accuracy and validity of numerical computer simulation results hinges on the correct characterization of the physical and mathematical problem, appropriate selection of boundary conditions and initial conditions, suitable meshing of the representative model, and accurate definition of applicable material properties. At the conclusion of a simulation, most modern 3-D simulation software does not usually provide any feedback information regarding the accuracy of the obtained results. Experience shows that proper FEA meshing is one of the most crucial factors affecting the accuracy of numerical simulations. Regions of high current concentration and areas where the electromagnetic field has substantial gradients must be properly accounted for using a sufficient number of elements. Determining the validity of the results is dependent upon the software user's combination of theoretical knowledge, practical experience, and numerical simulation expertise.

Figures 4-7 show an example of 3-D FEA computer simulation. The simulated process is the single-shot surface hardening of a carbon steel flanged shaft [4]. The heating time is approximately 5.5 seconds. One of the most challenging aspects of this process is attaining sufficient hardened case depth in the fillet (at the shaft-flange joint).

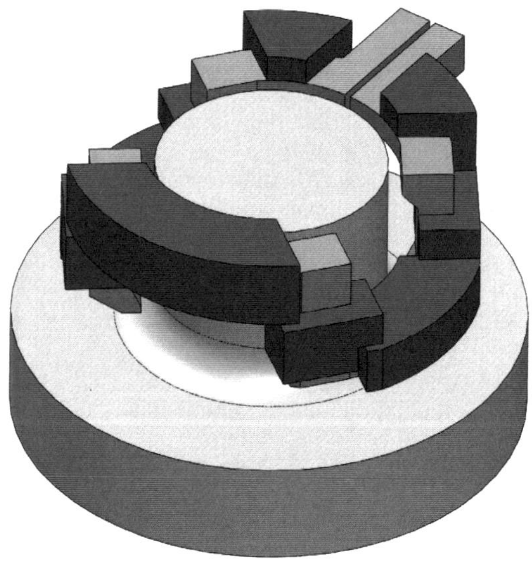

Figure 4. Machined single-shot inductor, including copper (orange) and flux concentrator (black), for hardening a carbon steel (AISI 1040) flanged shaft (grey) [4].

The design of single-shot inductors is challenging for a variety of reasons. Relative to a conventional multi-turn (solenoidal) inductor, the geometry of single-shot inductors (Figure 4) inherently results in a multi-directional and highly non-uniform 3-D distribution of current flow in the inductor and workpiece (Figure 5). This complexity, combined with the rotation of the workpiece that is required in single-shot designs, makes intuitive projection of the resultant 3-D temperature distribution (Figure 6) exceedingly difficult.

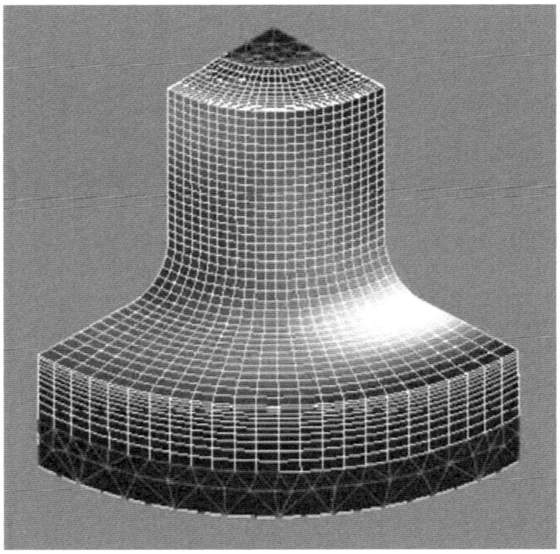

Figure 5. Non-uniform circumferential distribution of current density (A/m^2) during the single-shot hardening of a flanged shaft utilizing a frequency of 10 kHz [4].

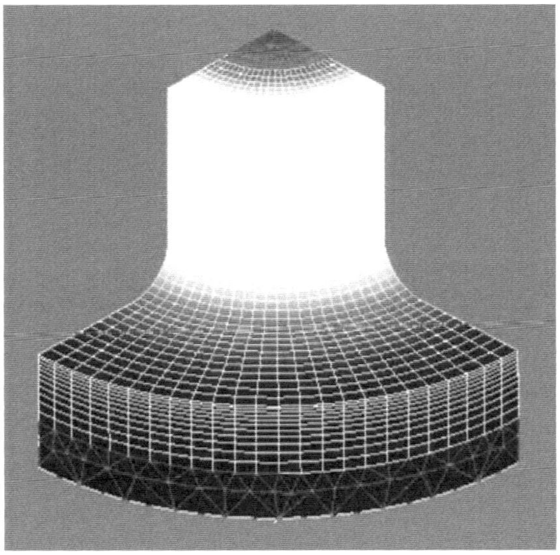

Figure 6. Nearly uniform circumferential distribution of temperature (°C) during the single-shot hardening a flanged shaft utilizing a frequency of 10 kHz [4].

Is the FEA the Ultimate Modeling Technique for All Induction Heating Applications?

In the last decade the finite element method (FEM) has become a dominant numerical simulation tool for a variety of engineering applications. Though FEM (the numerical method utilized in FEA) is a highly effective technique, it cannot be considered to be the ideal computational tool for all induction heating applications. In some cases, a combination of different numerical methods can be more effective.

All numerical computer simulation techniques require a network mesh of the modeled area. In the simulation of a typical induction heating process, that area includes the induction coil(s), heated workpiece, and fixtures and other electrically conductive bodies located in close proximity to the induction coil. Mesh generation affects the accuracy of simulation results, required time for numerical computation, and required time for pre- and post-processing. As an example, Figures 7 and 8 show meshes for two popular numerical computer modeling techniques: finite element method (FEM) and boundary element (BEM) respectively.

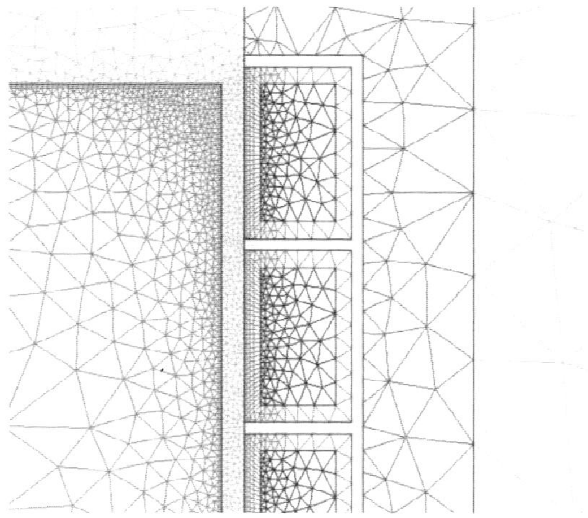

Figure 7. Example of finite element method (FEM) computer simulation mesh.

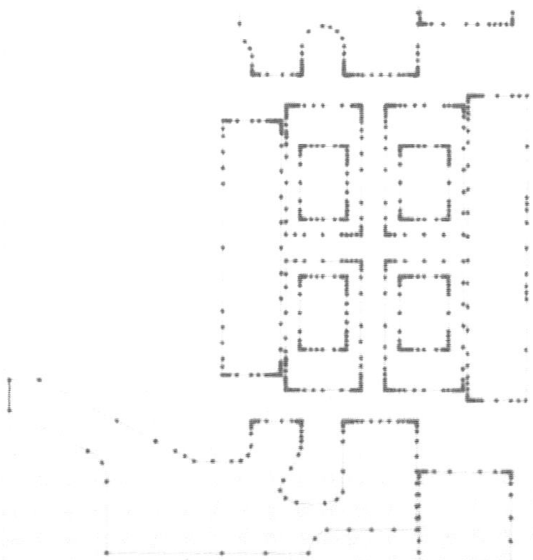

Figure 8. Example of boundary element method (BEM) computer simulation mesh.

Even a cursory look at the comparison of network meshes reveals substantial differences. Depending on the specifics of a given induction process, one particular type of mesh can be more preferable than another. Selection of an appropriate numerical technique is a critical part of the computer simulation process.

As an example, the BEM only requires discretization of the boundaries of electrically conductive components of the system (as shown in Figure 8). This substantially simplifies one of the most time-consuming parts of preparing the numerical model, especially compared to FEM models. With the BEM, unknowns of the electromagnetic field (i.e., magnetic vector potential) are expressed in terms of an integral over the boundary of the area of interest. In this case, the mathematical problem can be sub-divided into two smaller problems: external and internal electromagnetic problems. The internal problem describes the electromagnetic field distribution within the body of the workpiece. The external problem describes the field distribution in external regions. Using an iterative procedure, both problems can be solved.

FEA requires generating a network mesh of the entire modeling area, not just along boundaries. That network includes induction coils, the workpiece, electrically conductive tooling, flux concentrators, and so on. Unfortunately, to suit the conditions of smoothness criteria and continuity of the governing differential equation, it is also necessary to generate a mesh within electrically nonconductive areas, such as air regions.

Electromagnetic field distribution in the air, in most cases of coil design and process development, can be considered useless information (though such information can be of interest during the final design stage when evaluating electromagnetic field exposure). The need to always carry out computation of the electromagnetic field in air regions can be considered a disadvantage of the FEM, particularly when considering large systems in which air represents a substantial portion of the model.

Continuous and progressive multi-stage horizontal heating are two popular induction heating technologies used to heat small- and medium-size bars and billets. The workpieces (i.e. billets, blanks, bars, rods, etc.) are moved (via pinch roll(s), pusher, indexing mechanism, etc.) through a single coil or multi-coil induction heater. Different coils can be powered by different power supplies having unique frequencies and power levels. Components are sequentially heated at certain predetermined rate. Figure 9, for example, shows an induction system consisting of twelve inline 0.8m long induction coils.

Quite often, the length of multi-stage systems exceeds 4m and, in some cases, can be upwards of 20m long. In long systems, because the FEM requires meshing and evaluation of not only the heated workpiece and induction coil(s) but also the air around them, computation time can be exceedingly (and in some cases impractically) long.

Figure 9. A multi-coil continuous induction heater.

Many of the commercial codes used for computer simulation of induction heating processes are "all-purpose" programs developed originally for modeling electro-thermal processes taking place in common electrical machines/components (e.g. AC motors, transformers, etc.). The need to sell their products to as many customers as possible forced early software developers to produce more universal simulation tools that could be used within a broad industrial base. These programs were later adapted to provide capability to simulate of induction heating.

Consequently, certain process subtleties related to induction heating were either overlooked or substantially simplified by commercial software developers. As a result, leading induction heating and heat treating manufacturers developed proprietary subject-oriented software capable of more robust, yet process-specific, simulation. For example, Inductoheat's proprietary **i**HAZ™ temperature profile modeling software enables advanced temperature monitoring for induction bar and billet heating processes. This numerical software not only simulates the electromagnetic and thermal phenomena inherent to induction heating, but it also takes into account the circuitry and load matching capabilities of different types of modern semi-conductor power supplies. This software gives users (Inductoheat's engineers as well as our customers) the ability to obtain customized process parameters for obtaining specific workpiece temperature profiles and maximizing efficiency. The software, which is provided to customers as part of equipment purchases, also features built-in automated PLC communication capability.

In addition to the development of independent proprietary simulation programs, some induction manufacturers have also extended the capabilities of commercial software. For example, Inductoheat has developed a number of subroutines and procedures to better predict metallurgical results of induction heat treatment processes (e.g. predicting case patterns in surface hardening applications and calculating expected hardness reductions in tempering applications).

Based upon experience, there is not a single universal computational method that optimally fits all induction heating processes. As a result, our software designers utilize and integrate both commercial and proprietary computer simulation software. This allows engineers and analysts to select the technique that is most appropriate for each individual application.

Conclusions

In the fast-paced manufacturing industry, the efficiency and reliability that computer simulation can provide in the design of induction heating equipment and processes is extremely valuable. It is important, however, to remember that all computer simulation programs, both commercially-developed and proprietary, have limitations. The implications of these limitations must be thoroughly understood in order to fully take advantage of the benefits that computer simulation has to offer.

References

[1] V. Rudnev, Computer Modeling Helps Prevent Failures of Heat Treated Components, *Advanced Materials & Processes,* October 2011, p.6-11.

[2] V. Rudnev, C.Russell, G.Doyon, Computer modeling, induction heat treating, are here to stay, *Industrial Heating,* September, 2013, p.71-74.

[3] V.Rudnev, Computer Modeling of Induction Heating: Things to be Aware of, Things to Avoid, *Industrial Heating*, May 2011, 41-45p.

[4] C.Russell, Design and Validation of Induction Heat Treating Processes Utilizing 2D & 3D Computer Simulation, Proceedings of ASM Heat Treating Conference, Detroit, September, 2015.

International Federation of Heat Treatment and Surface Engineering 2016
Proceedings of the 23rd IFHTSE Congress
April 18–21, 2016, Savannah, Georgia, USA

Modeling of the Heating Sequences of Lightweight Steel/Aluminum Bimaterial Billets for Hot Forging and Hot Hydroforging

Robert Goldstein
Fluxtrol Inc., Auburn Hills, MI, USA
rcgoldstein@fluxtrol.com

Bulent Chavdar
Eaton, Southfield, MI, USA
bulentchavdar@eaton.com

Lynn Ferguson
DANTE Solutions Inc., Cleveland, OH, USA
lynn.ferguson@dante-solutions.com

Abstract

Recently, a concept to produce lightweight products by hot forging a steel shell that had a lightweight core was presented that could lead to component weight savings up to 50%. Some targeted products are gears, valves, and flanges. The steel shell is envisioned to carry most of the load in a target application while the lightweight core serves as a space holder during the forming process. After forming, the lightweight material may either remain in the component and contribute to the load carrying capacity, or be emptied out to achieve the ultimate weight reduction.

In this paper, the concept studied is hot forged from a bimetal billet, which is a steel tube press fit with a solid aluminum core and welded shut with steel end caps. For the experimental part of the studies Al 7075 was selected as the core material due to its high strength to weight ratio and 1020 steel was selected because of its availability as a tube. Induction heating was selected as the heating method for bimetal forging. This is due to the ability of induction heating to rapidly heat the steel layer. Successful bimetal forging of a closed vessel requires the steel layer to be in the austenite phase prior to the aluminum reaching high temperatures to prevent compromising the weld seams. Modeling of the induction heating process is complex due to the dimensional movement of components during the process. A method was developed to accurately model the induction heating process and predict power requirements. The method will be described and the results of the models will be compared to experimental findings. The forming process will be discussed in another paper at the conference. The simulation presented is for solid state forging of a steel aluminum billet, but the method for modeling the process is the same for hot hydroforging or other material combinations.

Introduction

Vehicular manufacturers and their suppliers are constantly being challenged to improve fuel efficiency, without sacrificing performance or increasing the cost. One of the ways to achieve fuel efficiency improvements is to reduce the weight of components. Removing material from non-load bearing areas, changing materials and changing heat treatment specifications are common methods being employed in the field to try to achieve these goals. Oftentimes, seemingly minor changes can lead to major complications to the manufacturing process, leading to a substantial increase in the component cost for a relatively minor weight reduction.

Bimaterial products are of interest to manufacturers, because they have the potential to offer significantly higher strength to weight ratios than components made of only one material. The challenge is how to produce them economically. One concept being explored is to produce bimaterial products using a hot forging process. Theoretical studies using computer simulation and experimental trials have been conducted to develop a process for manufacturing bimaterial gears using hot forging or a combination of hot forging plus machining. The bimaterial billets in this work consist of a lightweight core fully enclosed within a steel shell.

A bimaterial billet can be forged in the solid state, however better forging quality can be achieved if the core material is allowed to melt or flow, thereby providing uniform hydrostatic pressure to the steel shell during forging similar to a hydroforming process. However, the similarity only pertains to the hydrostatic pressure developed inside the deforming billet, not to the process temperatures. While hydroforming is done at room temperature, the hot hydroforging is done at temperatures greater than 1000C, enabling deformation of steel into intricate topologies without a fracture. Other differences between hydroforming and hot hydroforging are that the volume of fluid is constant in hot hydroforging and the fluid solidifies and becomes an integral part of the product after forging and cooling. The lightweight core material will need to have a lower melting or softening temperature than the steel for hot hydroforging. Aluminum, magnesium, and glass are such candidate lightweight materials [1].

Our lightweight bimaterial gear concept includes a lightweight core material that is fully enclosed in a steel shell and formed into the teeth area thereby interlocking with the steel in the interior and contributing to the transfer of torque during the operation of gear. The formation of mechanical interlocking is complementary to a strong metallurgical bond needed between the steel and the lightweight core. The exposed surfaces of the concept gear are all steel surfaces that engage with the other components of a transmission system, such as the mating gear(s) at the outer diameter and a shaft at the inner diameter. It is envisioned that the steel frame of the bimaterial gear would carry the bulk of torque from one component to the other and also would provide a long fatigue life to the gear. Furthermore, steel surfaces could be induction hardened to withstand very high Hertzian contact stresses between the mating components. On the other hand, the lightweight core would serve as a continuum to transfer stresses inside the steel shell, thereby providing rigidity and reducing risk of plastic deformation of the steel shell under loading. A comparison of this concept to other competitive bimaterial components is contained in the following papers [1, 2].

In this study the feasibility of hot forging and hot hydroforging was investigated by modeling and also experimentation. Induction heating simulations of bimaterial billets were conducted using ELTA™ and Flux™ 2D software. The hot forging and the hot hydroforging simulations of bimaterial billets were conducted initially in simplified models using DEFORM® and later in more complete models using DANTE® software. The steel/aluminum bimaterial billets were designed and prepared. A three piece closed die was designed and made for hot hydroforging experimentation. Hot forging and hot hydroforging experiments were done on bimaterial billets.

The goal of this paper is to present a method of modeling the heating process. The proper thermal, metallurgical and dimensional results are necessary in order to correctly design the forming process. The forming process and results are discussed and the future directions are identified and presented more in depth in another paper at this conference [2].

Bimaterial Gear Concept

Heavy duty manual transmissions require large size spur gears and helical gears made of carburizing steel. Most forged gears are made by two step manufacturing processes, hot forging followed by machining. First a tall cylindrical billet of steel is hot-forged to a flat pancake shape and then a center hole and the teeth are formed by machining. Forty five percent of the starting material is wasted in the machining operations where the most waste is due to the teeth machining. The near net shape forged steel gears with light weight core can enable an 80% reduction in scrap, thereby saving 2 to 4 kg waste per an average size gear. The raw material cost of a steel gear with light weight core may be significantly less than that of a gear produced from a solid steel billet because the volumetric cost of the light weight core can be lower than the cost of steel that is replaced. The reduction in scrap also has a potential to reduce the carbon foot print of gear manufacturing operations.

Hot forging and hydroforging concepts are an enabler for near net forging of large gears with the teeth from lightweight bimaterial billets. Figure 1 shows the components of the bimaterial billet design. Figure 2 shows the assembly of the bimaterial billet by fitting the aluminum core in the steel tube and welding the two steel endcaps to the steel tube. Figure 3 shows the initial concept of hot hydroforging a bimaterial gear in a two blow forging operation. Figure 4 shows the concept of bimaterial gear where the outer structure is made of 100% steel and the inner structure is 100% lightweight core. A more detailed description of the manufacturing process is contained in another paper at this conference [2].

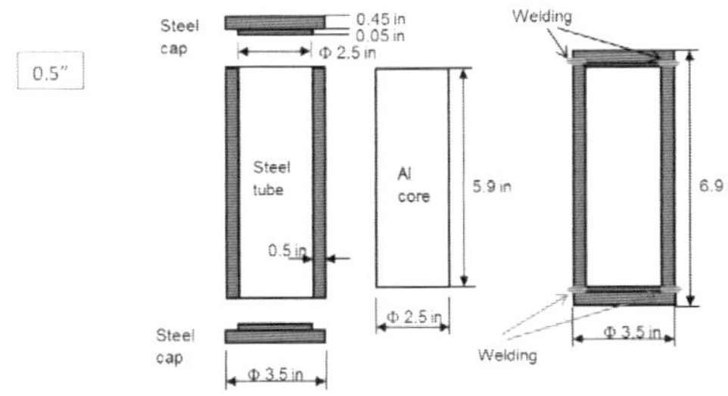

Figure 1: Bimaterial gear concept. The outer structure is steel, the inner structure is a lightweight material.

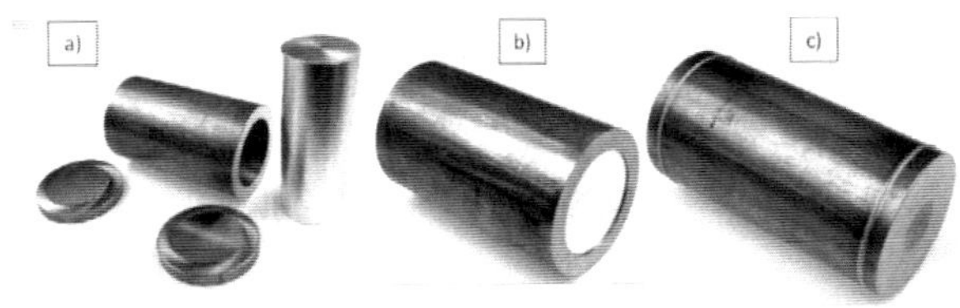

Figure 2: Assembly process of steel-aluminum bimetal billet. a) steel tube, Al core and two steel end caps before assembly, b) aluminum core is inserted into steel tube, c) steel end caps are welded to steel tube to complete the assembly.

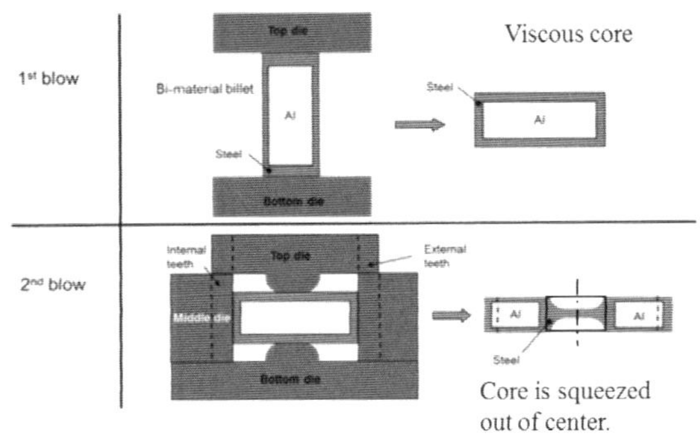

Figure 3: Hot hydroforging concept of net shape gear forging with teeth and the center hole [1].

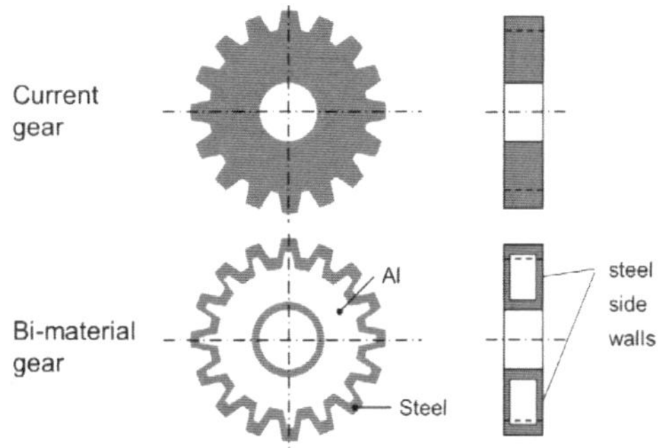

Figure 4. Bi material gear concept where the outer structure of gear is 100% steel and the inner structure is 100% lightweight material such as aluminum [1].

Initial heating trials were conducted in a furnace for the steel-aluminum billets. With the slow, uniform heating process, the welds holding the caps on the die fractured during the heating process due to the thermal expansion of the aluminum core during the heating process (Figure 5).

Figure 5. Cracking and leaking of the Al after heating to 600C, 0.5" billet

In order to prevent the cracking of the bimetal billet, the steel wall needs to be heated and expanded away from the aluminum core before aluminum is heated. The steel needs to reach above recrystallization temperatures before aluminum expands and catches up with the steel wall to prevent any cracking in the steel. When the steel is well above the recrystallization temperature even if aluminum catches up with the steel and exerts pressure on the steel it does not cause a problem since then steel can deform outwards easily to accommodate the expansion of aluminum without any fracturing. Based upon this concept, induction heating was explored.

Modeling of the Heating Process

For safety reasons, initially the hot forging concept was pursued. The goals for the induction heating modeling were to determine if it was possible to heat the steel casing and then to maintain a temperature at the mid-radius of greater than or equal to 750 C in the steel billet for a period of 10 s (transportation time from induction heater to forging press) without melting the aluminum core (Temperature below 600 C). The upper limit on surface temperature of the steel was set to 1350 C. The billet OD was 88.9 mm (3.5 inch). The wall thickness of the steel was 12.7 mm (0.5 inch). The radial gap and the top gap between the aluminum core and the steel shell was 0.076 mm (0.003 inch).

One of the big unknowns for the induction heating simulation is the thermal contact between the steel casing and the aluminum billet during the process. If the steel casing is heated substantially faster than the aluminum, a gap should grow between the steel and the aluminum bodies. If there is a gap, then heat transfer from the steel to the aluminum will move from conduction to radiation, resulting in much slower heating of the aluminum core. Calculations were made for the temperature of steel and aluminum that would result in the same thermal expansion (Figure 6). If the aluminum temperature is above the red line at a corresponding steel temperature, then the aluminum will be pressing up against the steel. If the aluminum temperature is below the red line at a given steel temperature, then the two bodies will be separated.

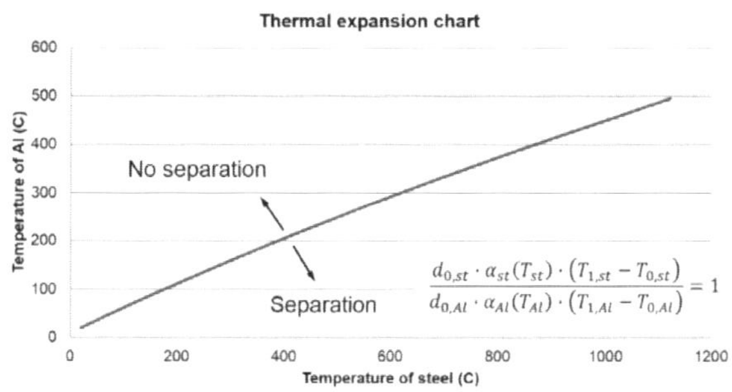

Figure 6. Temperature when aluminum and steel have the same thermal expansion

Preliminary computer models were made using ELTA™ software for a straight bimaterial tube of the same dimensions comparing the two threshold conditions – ideal physical contact and no physical contact. It was found that with ideal physical (thermal) contact, it was highly questionable even with a surface temperature of 1350 C if it would be possible to maintain the necessary temperature in the steel shell (750 C) without melting the aluminum core (600 C) through the transportation phase (5- 10 s). Figure 7 shows the temperature results for the ½" steel wall billet with the induction equipment available (up to 600 kW at 10 kHz) for the tests (Figure 7). For the ¼" billet, the process did not appear to be feasible with the ideal thermal contact assumption.

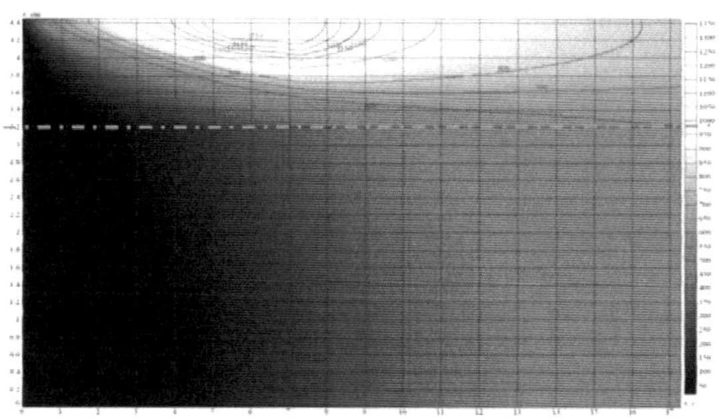

Figure 7. Temperature distribution in the ½" wall bimetal billet assuming ideal physical contact – dashed green line represents steel aluminum interface.

With no physical contact between the components, the results looked much more promising. Even with no physical contact between the components, there would still be heat transfer between components by radiation and convection. To simulate the heat transfer, a thin layer with very low thermal conductivity was used. Under this scenario, the ¼" billet was more difficult to maintain temperature in the steel compared to the ½" billet due to the higher surface area to mass ratio. Figure 8 shows the temperature distribution for a ¼" billet with no physical contact between components. It is clear that even with a 1200 C maximum temperature, the process appears feasible.

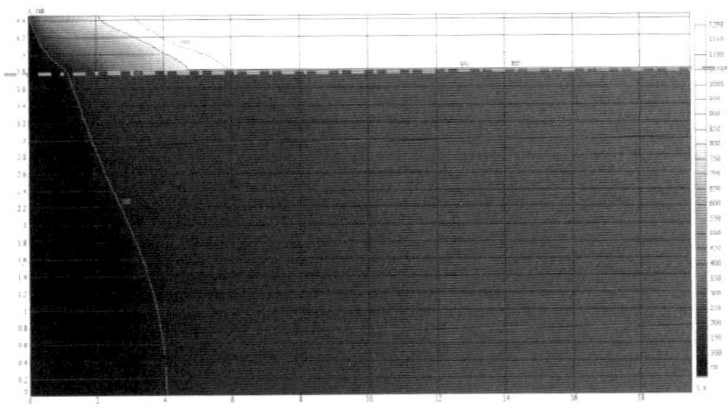

Figure 8. Temperature distribution in the 1/4" wall bimetal billet assuming no physical contact – dashed green line represents steel aluminum interface.

Based upon the promising results from the ELTA™ software, more complicated, multi-software 2D modeling was conducted. 2D coupled electromagnetic plus thermal modeling of the process was done using Flux2D to develop a thermal cycle with the assumption of physical contact on the top and bottom surface of the billet. The power density distribution was then mapped and exported to Microsoft Excel. These results were then inputted into the DANTE® software for thermal, metallurgical, stress and dimensional movement models to validate the proposed cycle. If thermal contact was not as expected, Fluxtrol would then update the thermal cycle to achieve the desired results and the revised models would be sent to the DANTE® software. This procedure has been used successfully for induction heat treating of tubes [3] and axle shafts [4].

At this point, it was determined that a 500 kW, 10 kHz induction heating power supply from Interpower Induction (www.interpowerinduction.com) would be used for the tests. One of the limitations of this unit was that it was only possible to utilize two stages of power in a given cycle. A number of different potential cycles were developed for both the ¼" billet and ½" billets using 2D modeling. Only one selected cycle for the ½" 2D billet heating results is shown in this paper to demonstrate the concept.

For this proposed cycle, the goal was to get the core as soft as possible without melting to prevent rapid quenching of the steel shell (and resulting tensile stresses) during the forming process where physical contact would occur between the steel and aluminum. Based upon these criteria, a 26 second heating cycle was developed. The cycle was designed such that the first stage would use high power (90%) for 4.2 s to rapidly heat the surface up to the maximum temperature and ensure that the steel shell separated from the aluminum core (Figure 9). After this initial, high power stage, a much lower power level (20 – 25%) would be used for the second stage (21.8 s) while the temperature equalized in the thickness of the steel shell (Figure 10). After heating, a 5 second transportation stage without power was modeled (Figure 11).

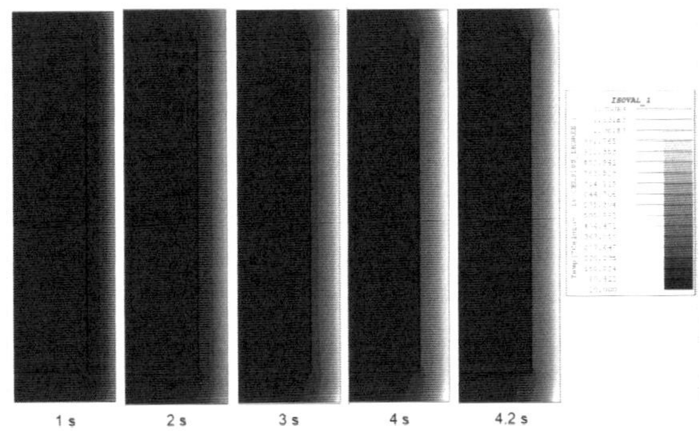

Figure 9. Temperature distribution in the 1/2" wall bimetal billet at the end of the 4.2 s, 90% power preheating stage

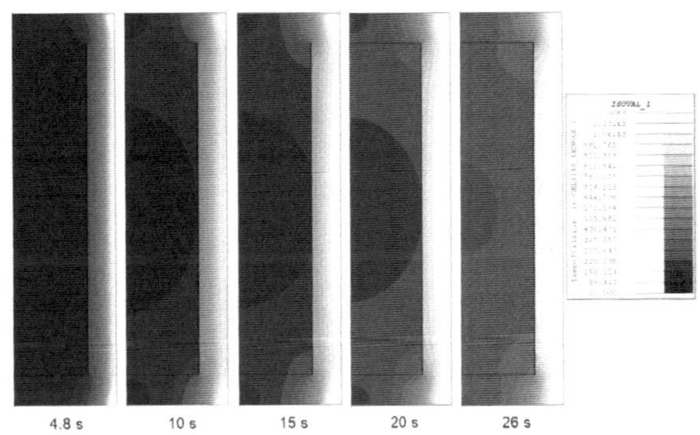

Figure 10. Temperature distribution in the 1/2" wall bimetal billet at the end of the 21.8 s, 20-25% power holding stage

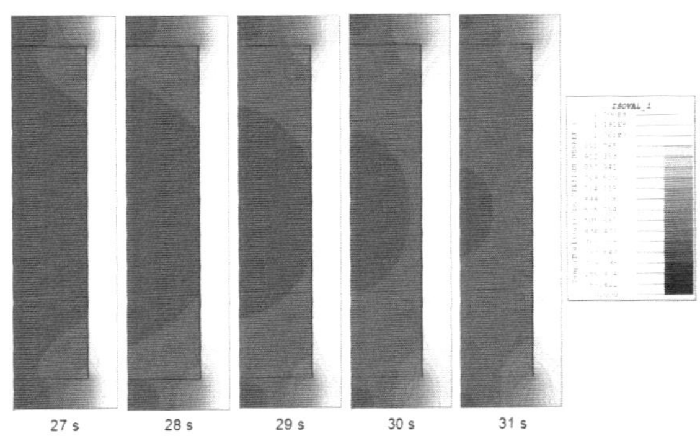

Figure 11. Temperature distribution in the 1/2" wall bimetal billet at the end of the 5 s, transportation stage

Due to the relatively short process and lack of thermal contact with the aluminum, there is a large temperature gradient between the hot, outer diameter of the billet and the center of the top and bottom steel caps due to the comparatively poor thermal conductivity of steel. If outer walls were to be in intimate contact with the aluminum (which would be molten at these temperature), the entire outer surface temperature would be close to uniform due to the much higher thermal diffusivity of aluminum.

The power densities at the different time steps were then exported into the DANTE® model. One difference between the Flux2D model and the DANTE® model was in the physical contact between the aluminum and the steel. In the DANTE® model, physical contact was assumed only for the bottom surface. This is the more appropriate assumption as it was expected that the steel will grow both axially and radially during the heating process.

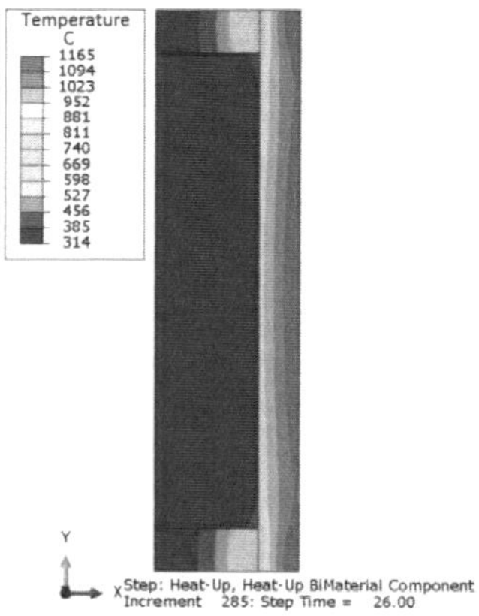

Figure 12. Temperature distribution at the end of 26 second heating cycle (DANTE® software)

The temperature distribution at the end of the heating cycle in the DANTE® models is shown in Figure 12. The temperature distribution is similar to the one from Flux 2D, but the magnitude of temperature in DANTE® model is slightly lower. This is due to differences in the surface heat transfer coefficients (convective and emissive) used in the respective models. This could be resolved by adjusting the thermal boundary conditions. The aluminum core temperatures are quite similar (the lower start point on the temperature scale makes the DANTE® software model aluminum core appear cooler).

Figure 13 shows the radial dimensional movement for the steel ID and aluminum OD that occurs during the process. It is clear that at the very beginning of the heating process, the steel grows rapidly and the gap between the steel and the aluminum widens. In the middle of the process, the gap is nearly constant as the steel and aluminum grow at similar rates. Near the end of the heating process the gap begins to decline as the steel expansion slows and the aluminum continues to grow. During the transportation phase, the steel contracts as it cools, while the aluminum continues to grow as it is still being heated by radiation from the hot steel.

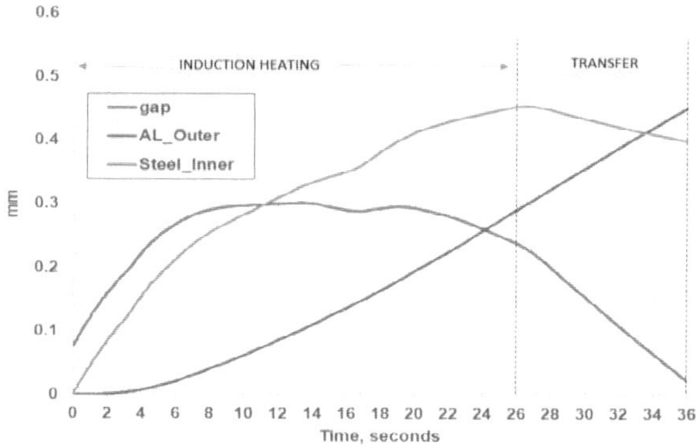

Figure 13. Dimensional movement predicted during the heating cycle

Experimental Validation of the Heating Process

To verify the concept, it was decided to test a billet with a small hole drilled in the top of a ¼" wall billet to monitor the temperature of the aluminum in the Interpower lab. The tests were monitored using a camera mounted on a tripod. Figure 14 shows the billet before and a few seconds after a 4 second heat pulse at 400-500 kW, 10 kHz. It is clear that the billet maintains temperature in the center of the OD and also that the billet has grown by several millimeters. This much energy inputted into the billet would be insufficient to maintain this temperature if the billet were in good thermal contact with the aluminum core. In addition, further evidence of the cold core is that the upper and lower surfaces had much lower temperatures.

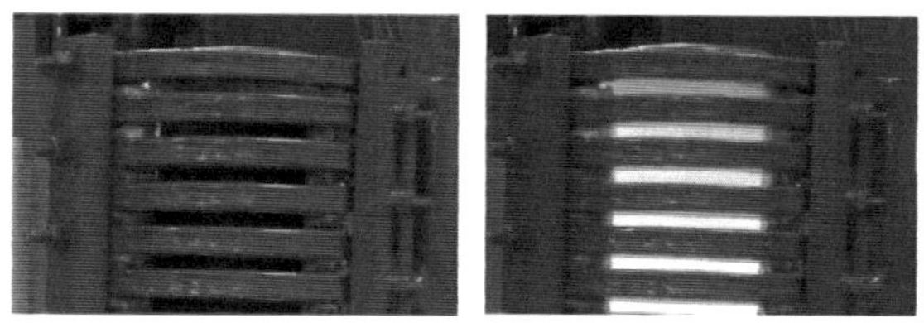

Figure 14. 1/4" wall bimetal billet before and several seconds after a 4 second heat pulse at 400 – 500 kW, 10 kHz.

Initial heating trial results showed agreement with the models, verifying that with the proposed thermal cycle there would be separation of the materials. Based upon this, industrial tests were planned. The 500 kW, 10 kHz generator was moved from Interpower to Walker Forge for the forging tests. The forging installation at walker forge had a lift to remove the billet from the coil and a robot to transport the billet from the heating coil to the forging die. It was estimated that this process would take approximately 10 seconds.

A series of trials were performed building sequential up to the desired cycle. A two stage heating cycle 6 seconds high power (90% was programmed into the machine, but power dropped somewhat as the Curie point was reached), 20 seconds 20% According to the pyrometer on the machine, the temperature in the center of the billet at the end of the heating cycle was approximately 1250 C, very close to the calculated temperature. The temperature in the billet 2 seconds after the end of heating as the part is being lifted out of the coil is shown in Figure 15. It is clear that on the outer surface of the billet, temperature is relatively uniform. Due to the high temperature, it was difficult to get the camera to focus to discern the exact distribution. This also agrees quite well with models (Figures 10-12).

Approximately 2 s after heating **In Forge Press**

Figure 15. ½" wall bimetal billet before 2 seconds after a 26 second, two stage heating cycle

As the billet is placed in the die, the surface temperature of the billet has declined, but it remains high. In the image, it can be clearly seen that the top cap of the billet is substantially cooler than the sides. This also agrees well with the models and offers further proof that the aluminum core is not in contact with the steel shell.

Discussions

Lightweighting is an energy saving strategy for the transportation industry. Transmission and driveline components account for 5-10% weight of a typical class 7/8 tractor. The gears account for 20% weight of a heavy duty manual transmission. Currently 100% of truck gears are made of steel. If the gears are forged from a bimaterial billet such as 50/50 steel/aluminum by volume than 33% weight reduction in gears is possible. This translates into 0.5% reduction in the weight of a truck and 0.2% improvement in the fuel efficiency.
The torque capacity of a gear can be increased by increasing the size of gear and also by forging instead of by machining. Small gears can be forged since their press loads are manageable. However, most big gears are made by machining instead of forging due to forging press tonnage requirements. Hot hydroforging enables forging of large gears on the existing presses due to the reduced press loads. Feasibility of making lightweight gears with hot hydroforging of steel/low density material hybrid billets for powertrain applications is discussed in detail in another paper at this conference [2].

Our bimaterial gear concept has a steel shell on the outside and a lightweight core on the inside. The steel shell will carry most of the torque while the lightweight core serves as a space holder and reduces the weight of gear. Two material options are identified for the low density, readily formable core at temperatures equal to or below the forging temperature of steel: aluminum and glass. Both materials are relatively low cost and readily available.

For the initial trials, steel/aluminum bi-metal billets were studied. Aluminum is a lightweight metal and second to the steel in terms of the availability and the cost. However there are many risks and challenges when aluminum and steel are used together in a structural product such as a gear. The biggest technical challenge is due to the difference in the coefficients of thermal expansion (CTE) of steel and aluminum. The CTE of aluminum is approximately twice the CTE of steel. Furthermore, the volume of aluminum increases upon melting. To avoid billet fracture on heating, it is critical that the weld seam on the top and bottom of the billets is in the austenitic phase before the aluminum core expands and exerts pressure on the welds. Based upon this, induction heating was viewed as the best available technology for this process.

For the initial tests, hot forging of the steel shell with a warm, solid aluminum core was viewed as the lowest risk option. Extensive virtual prototyping of the induction heating process was conducted using a combination of ELTA™, Flux2D™ and DANTE® software to model the various electromagnetic, thermal, structural transformation, stress and dimensional movement processes. A number of potential thermal cycles were developed for billet wall thicknesses of ¼" and ½" with varying levels of core temperature prior to forming. In this paper, we present the simulation results for one of these cases, but all of the models showed good agreement with the experimental results. This means that the models are scalable and it is possible to use the same virtual prototyping process to develop thermal cycles for a variety of shell/core material/geometry combinations.

Then, a set of heating and forging trials were made using the bimetal billets in a so called "pancake" shape in closed dies in one blow. Initial hot forging trials using aluminum cores showed mixed results. The samples did not crack upon heating and they were able to be successfully formed. However, post cooling sectioning of the samples showed varying levels of internal cracking/tearing and also gaps between the aluminum core and steel shell.

The internal defects were likely from a combination of forging folds and post forming temperature equalization. During the forming process, the hot steel is put into direct contact with cooler aluminum core. Once in direct contact, the aluminum temperature rises and the steel rapidly cools as the aluminum core "quenches" the shell. As the aluminum heats and the steel cools, the aluminum expands while the steel tries to contract. This puts significant tensile stresses on the internal surface of the steel while the steel shell falls below the AC1 temperature. Figure 16 shows the DANTE® software predictions for dimensional movement in the billet for the 26 second heating cycle for half inch billets if forming did not occur and the sample was allowed to air cool. The models show that the gap would go to zero. Once in contact, both the aluminum and steel contract in concert as temperature equalizes between the two components. The steel and aluminum dimensions grow again slightly as temperature equalizes in the aluminum. Finally, as the components cool the aluminum contracts at a greater rate than the steel and a significant gap develops (0.2 mm).

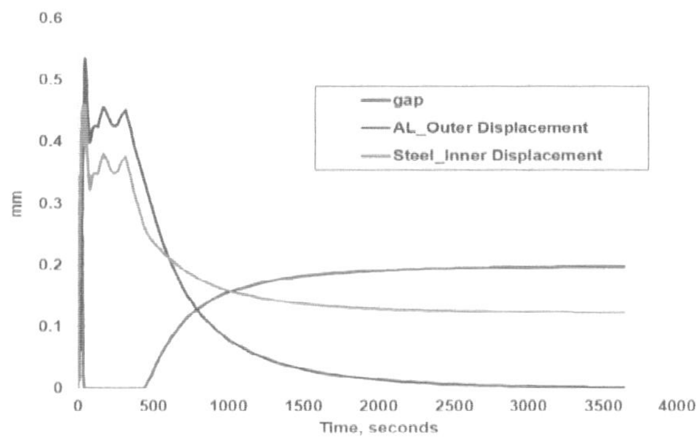

Figure 16. Dimensional movement predicted during the heating cycle with post heating air cooling, no forming

One can predict that the greater the temperature differential between the two components, the greater the post forming dimensional movement and stress level would be. Also, the thinner the steel wall, the larger temperature drop in the steel would be after forming. For the ¼" steel billets, this resulted in crack formation in the radius [1]. The cracks would form sometime after the forming process and the cracks continued to grow while the billets sat and cooled. On the converse, the lower the equalized temperature, the lower the post equalization contraction will be resulting in smaller shell core gaps. This was also evident in the trials [1].

After the hot forging trials, hot hydroforging experiments were conducted to resolve the issues related to temperature equalization. The bimaterial billet was heated to 1000-1250 C range with a high frequency induction heater as guided by the induction heating modeling. Aluminum was allowed to melt inside the steel without fracturing the steel case by a cyclic on/off control of the power of heater. During hot hydroforging the liquid core flowed effortlessly and provides a uniform hydrostatic pressure inside the steel thereby enabling a uniform deformation of the solid steel shell. Upon cross sectioning a hot hydroforged part, a uniform steel wall thickness was observed all around the part. However, there were also large voids in the aluminum core as expected. Using glass as the core material instead of aluminum will solve the void issue since glass has a matching CTE to steel. The benefits of matching thermal expansion behavior to eliminate shrinkage void had been demonstrated [1] by deformation modelling of induction heating and air cooling of steel-glass bimaterial billets.

Heating the steel-glass billets to forging temperatures will be simpler and faster than heating the steel-aluminum billets. It can be also envisioned that hot hydroforged steel-glass bimaterial gears can be quenched right after forging and be tempered by the latent heat of glass during the air cooling. The novel heat treatment procedure can save time and money by replacing the time consuming and costly carburizing and nitriding heat treatment procedures. Hence, the next series of hot hydroforging experimentation can be planned around the steel-glass material combination. Glass is 10% lighter than aluminum thereby providing further weight reduction benefits. These options need to be explored further.

On the other hand, aluminum core enables a novel concept of "investment forging" where the molten aluminum core can be emptied out of the part right after forging analogous to investment casting. A hollow part with uniform steel shell can be formed for ultimate weight savings. For example, investment forging of hollow steel valves for engine applications is feasible by hot hydroforging. In some cases the core can be emptied out later after creating evacuation holes at room temperature and reheating the part if it is difficult to empty the core right after forging. Any low melting substance that does not go into gas phase at forging temperatures is a candidate as the core in investment forging. Expensive core materials such as aluminum or magnesium that are extracted from the product during investment forging can be recycled.

The experimentation with hot hydroforging addressed several risk factors that had been identified and put into test. One of the concerns was if the degree of interference fit would have any impact on the quality of bimetal billet forging. The experiments indicated that an interference fit is not needed. Another concern was about the vulnerability of the weld seams to failure during forging. The experiments has shown that it is reasonable to expect that the weld zones of billet will heal and have no adverse effect on the performance of the end product upon hot hydroforging [1, 2]. These guidelines lead to significant cost savings for the preparation of billets in a mass production scenario.

Conclusions

Bimaterial forging is a new and exciting technology that has the potential to substantially reduce the weight of steel components used for many automotive powertrain components. Currently 100% of truck gears are made of steel. If the gears are forged from a bimaterial billet such as 50/50 steel/aluminum by volume than 33% weight reduction in gears is possible.

Our bimaterial gear concept has a steel shell on the outside and a lightweight core on the inside. The steel shell will carry most of the torque while the lightweight core serves as a space holder and reduces the weight of gear. The steel shell would have welded caps on top and bottom.

The initial lightweight core studied was aluminum. Due to the mismatch in CTE, it is necessary to get the weld seam in the austenitic phase quickly, so that it will be able to easily expand when the aluminum expands. This means that the heating process must be fast, so induction heating was selected as the heating method.

The method for modeling the induction heating process for steel aluminum bimaterial billets was demonstrated. The modeling method involved coupling the electromagnetic and thermal program Flux2D™ with the thermal, structural, stress and dimensional movement program DANTE®. The results showed that using induction heating, it was possible to successfully heat the components without risking cracks to the weld seam during heating. Experimental validation of the heating process was conducted and the experimental results agreed very well with the modeling predictions.

One selected case was demonstrated in this paper; however, a number of different heating scenarios were both modeled and tested. These models all showed good agreement with the experimental predictions. This means that the modeling is scalable and can be used to predict results with different core materials or core phase (i.e., hot hydroforging).

In the future, it is anticipated that the results of the Flux2D™ and DANTE® software models should be coupled with additional modeling packages for subsequent manufacturing steps. The DANTE® software results could be used as an input for forming simulation. Forming simulation would be followed by cooling simulation. Cooling simulation could be followed by modeling of the subsequent machining, heat treatment and final loading to determine the component performance.

Acknowledgments

This research was funded and conducted at the Corporate Research and Technology Group at Eaton in Southfield, MI. Billet induction heating simulations, thermal profile design and validation support were provided by Fluxtrol Inc. in Auburn Hills, MI. The high frequency induction power supply was provided by Interpower Induction Services Inc. Fraser, MI. Forging services were provided by Walker Forge Inc. in Clintonville, WI. Further billet heating, cooling and deformation modeling were provided by DANTE Solutions Inc. in Cleveland, OH. The authors would like to thank following team members for their contributions: Dr. Xi Yang, Mike Killian from Eaton, Jake Butkovich, Dr. Peter Fritz from Eaton.

References

[1] B. Chavdar, R. Goldstein, X. Yang, J. Butkovich, L. Ferguson, "Hot Hydroforging for Lightweighting", IDE 2015; Bremen, Germany; Sep. 23-25, Vol. 5, pp. 117-128

[2] B. Chavdar, R. Goldstein, L. Ferguson, "Hot Hydroforging for Lightweight Bimaterial Gears and Hollow Products", 23rd IFHTSE Congress; Savannah, GA - USA; April 18-22, 2016.

[3] V. Nemkov, et.al, "Stress and Distortion Evolution during Induction Case Hardening of Tube", 26th ASM International Heat Treating Society Conference, Cincinnati, OH USA; October 31 – November 2, 2011.

[4] Li, et.al, "Effect of Spray Quenching Rate on Distortion and Residual Stresses during Induction Hardening of a Full-Float Truck Axle", 27th ASM International Heat Treating Society Conference, Indianapolis, IN USA; September 16 - 18, 2013.

International Federation of Heat Treatment and Surface Engineering 2016
Proceedings of the 23rd IFHTSE Congress
April 18–21, 2016, Savannah, Georgia, USA

Influence of Synthesis on the Mechanical Properties of Sintered and Nanoporous Copper

Alex Moore, David Carroll, Rakesh Behera, Chaitanya Deo, and Antonia Antoniou
Georgia Institute of Technology, 770 State Street, Atlanta, GA 30332, USA
amoore31@gatech.edu

Keywords: Nanoindentation, Nanoporous Copper Foam, Spark Plasma Sintering, Fine Grained Copper

Introduction

Indentation is a widely used, marginally destructive testing method to estimate mechanical properties of materials. Nanoindentation is a special indentation technique which can evaluate mechanical properties of materials at smaller dimensions, which traditional indentation tests fail to address. Nanoindentation can also accurately probe local mechanical property variation in the nanoscale.

In this study we have used the nanoindentation technique to examine mechanical properties of different forms of copper. Using a Berkovich indenter we have obtained the hardness and modulus values of copper synthesized using Spark Plasma Sintering (SPS). The powder used in the SPS process is in the nanometer range, which has the potential to result in grains from the nanometer to bulk range. The grain size is tied to the mechanical properties via the barriers of the indentation plastic zone expansion and preferred method of plastic deformation. We will explain the combination of sintering variables (pressure and temperature) used to synthesize dense copper pellets using SPS. In addition to spark plasma sintered bulk copper, we have evaluated the mechanical properties of nanoporous (NP) copper foam. The NP Cu films are synthesized on a silicon substrate by electrochemical dealloying. The NP foam behaves much differently than the bulk material due to the high porosity volume fraction. The high porosity can cause issues during nanoindentation leading to fluctuations in the data, and therefore the average mechanical properties are obtained through a series of indentations.

The forms of copper chosen to be examined tend to show unique mechanical and other physical properties. Fine grain copper, if in the nanocrystalline (NC) region, have been seen to regain some of their ductility due to the inverse Hall-Petch effect at nanosized grains, where some of the plasticity is due to grain shifting and reorienting. However, if the fine grain copper is in the ultrafine grain (UFG) region, we can expect a strengthening of the material properties due to the Hall-Petch relation. NP foams like gold have been seen to have a high yield strength (*1-3*). However NP copper, unlike NP gold, is seen to have a complex internal microstructure inside the nanofoam struts (*4*). This complex microstructure could lead to alternative mechanisms for plastic deformation (e.g. twinning and grain boundary shifting), causing a decrease in strength (*4, 5*) when compared to NP gold.

Materials and Experimental Details

Sample Production
The spark plasma sintered sample was created using copper nanopowder with particle sizes between 40 and 60 nm, and was sintered into a 10 mm diameter, 2 mm thick copper pellet using a Dr. Sinter SPS-211Lx. Initially, 1.41 grams of the copper powder was measured into the graphite die and two punches were used to keep the powder in place. The graphite die and punches were placed into the SPS device and a thermocouple was placed in a hole bored into the die to monitor process temperature. The chamber was evacuated of air, and a pressure of 50 MPa was applied to the punches and held constant throughout the process.

The temperature was then raised to 500 °C over five minutes and then slowly increased to 550 °C over another minute. This was the final temperature and it was held for five minutes. Temperature versus time and the displacement over time is shown in Figure 1. Once the SPS process was complete, the sample was polished, and the relative densification compared to theoretical density (8.96 g/cm^3) was measured using the Archimedes' principle. The relative density was measured to be 94.19%.

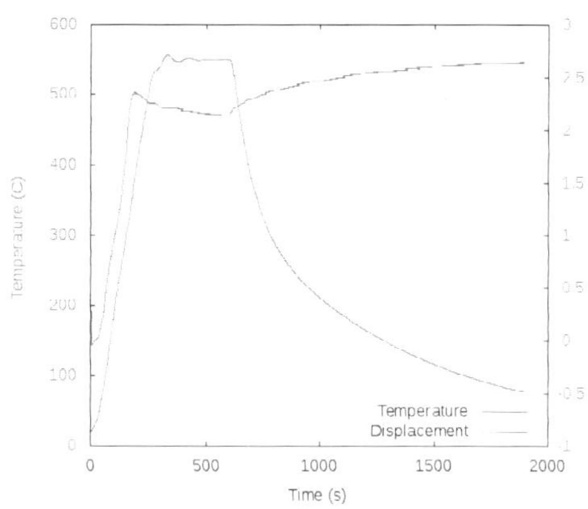

Figure 1: Temperature and sintered displacement versus time profile of the spark plasma sintering process of copper.

In order to visualize the grain structure, the sample was etched with a 22% nitric acid solution for 15 seconds, and then rinsed. The resulting surface after etching can be seen in the optical brightfield and darkfield images in Figure 2 below. The etching allows the grain size to become apparent under a scanning electron microscope (SEM), as seen in Figure 3. The resulting grain size from the SEM images was determined to be between 300 nm to 2 µm with an average of a little less than 1 µm.

There have been only a few previous spark plasma sintered copper publications, starting with different particle sizes. Ritasalo et al. (*6*) used 410 nm copper particles sintered at various temperatures between 600-800 °C and holding times between 1-30 min, resulting in grain sizes between 0.46-1.35 µm. However, Zhang et al. (*7*) used a larger copper particle size of 1 µm, and found that temperature was the main deciding factor on grain size growth. Below temperatures of 750 °C, Zhang et al. (*7*) found the average grain size to be approximately 2.2 µm, but between 750 and 800 °C, they found that grain size growth rates increased significantly, with an average grain size of approximately 6.6 µm.

The resulting grain size was larger than expected. Our grain size is between the values reported by Ritasalo et al. (*6*) and Zhang et al. (*7*), even though our starting copper particle size was smaller. We believe this to be a result of early Ostwald ripening, where small nanoparticles, when exposed to air, deposit onto much larger nanoparticles, resulting in a larger initial particle size before sintering.

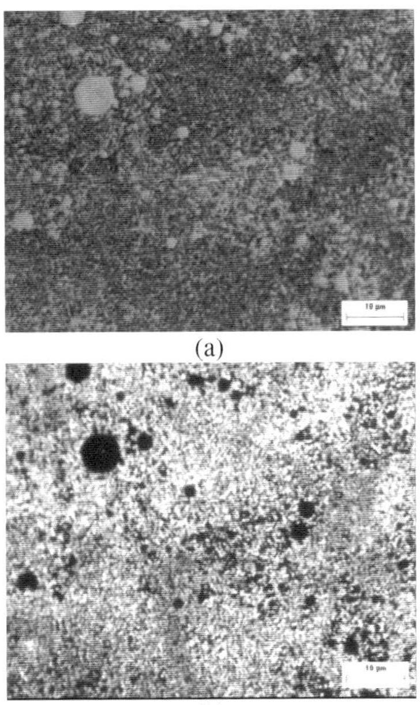

Figure 2: (a) Brightfield and (b) darkfield optical images of etched spark plasma sintered copper.

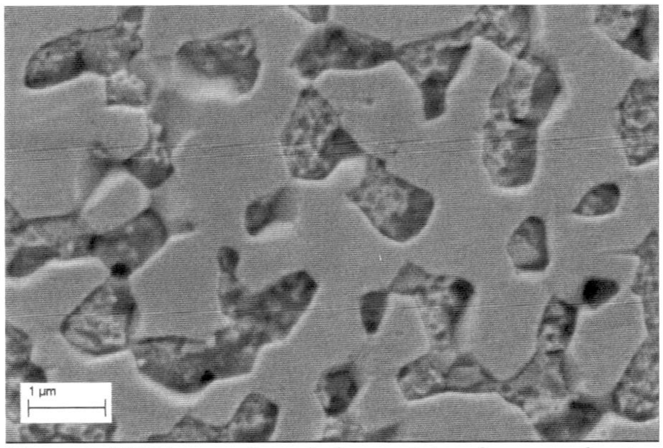

Figure 3: SEM image of spark plasma sintered copper, showing some of the grains after etching.

The NP copper foam was synthesized via a similar process as reported by Liu et al. (*4*). An amorphous copper silicide sample $Cu_{0.41}Si_{0.59}$ was co-sputtered on a silicon substrate. After which the NP copper foam was constructed by dealloying the amorphous copper silicide, where the silicon is preferentially dissolved by an aqueous solution of 3% HF and an externally applied voltage in the range of 0.1–0.25 V. The resulting NP copper foam was approximately 50% of the theoretical copper density, and had approximately 30 nm strut thickness. In addition, the overall thickness of the NP copper foam on the silicon substrate was determined to be about 890 nm.

Indentation Theory

The nanoindentation experiments were conducted on a NanoTest Vantage micromaterials nanoindenter. The experiments were conducted using a load controlled depth versus time hysteresis (seen in the Figure 4), where the load was ramped to the maximum defined load during 25 seconds, then a 25 second hold at the max was performed, allowing residual stresses to equilibrate, then a 25 second unload time was performed, and lastly a 60 second hold was performed at 10% of the maximum load to account for thermal drift at room temperature (which is generally negligible).

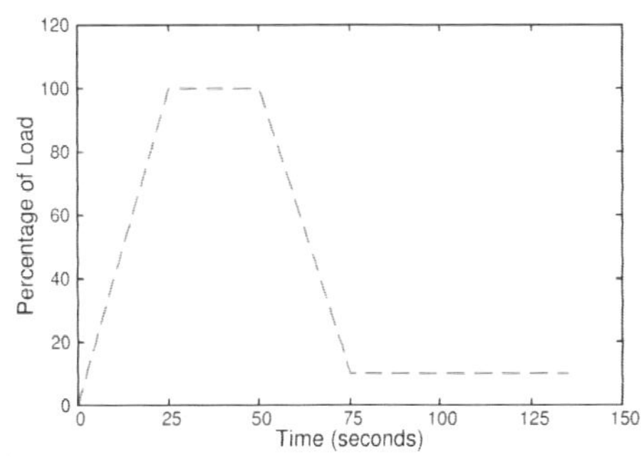

Figure 4: Percentage of maximum load versus time of the indentations performed using the Berkovich indenter.

The resulting data from the indenter are the indentation load, depth, and time throughout the experiment. However, some of the resulting depth measured is due to the instrument and not the sample. In order to account for this phenomenon known as the instrument compliance (C_f), a depth correction is applied to the data. The C_f is determined using equation (1) on indentations of various depths on samples with known material properties, and the thermal drift is determined by the depth change at the 60 second hold time at the end of the experiment. After the C_f was determined, the indentation depth (h) was modified using equation (2).

$$C_{total} = \frac{dh}{dP} = \frac{\sqrt{\pi}}{2\sqrt{A_p}}\frac{1}{\beta E_r} + C_f \quad (1)$$

$$h'' = h' - C_f P \quad (2)$$

In the equations above, C is the stiffness, A_p is the projected area, E_r is the reduced modulus, P is the load, h' is the corrected depth of the initial contact, and h'' is the corrected depth after instrument compliance is taken into account. The corrected depth h'' is the depth used for all hardness and modulus calculations.

The hardness, is determined from the loading curve, while the modulus, is determined from the unloading curve with a protocol summarized below. The unloading curve is distinctly curved and well approximated using a power law fit (*8*). The power law fit is applied to the upper 80% of the unloading data in accordance with the ISO-14577 (*9*) by fitting the constants "a" and "m" to the equation:

$$P(h) = a\left(h - h_f\right)^m \quad (3)$$

where P(h) is the power law fit function, h is the depth, and h_f is the final indentation depth.

The elastic recovery of the material during unloading has some deviations from ideality that are corrected for by the use of ε and β terms. The correction factors used are the standard correction values of ε=0.75 and β=1.034, and are applied using the equations below.

$$\frac{dP}{dh} = \frac{1}{\beta}\frac{dP}{dh}_{measured} \quad (4)$$

$$h_c = h_{max} - \varepsilon\frac{P_{max}}{dP/dh} \quad (5)$$

The diamond area function (DAF) is the function which provides the true contact area as a function of the contact depth (h_c). The DAF was determined indirectly by performing indentations into fused silica and tungsten, which have known Young's modulus and hardness values, and by using the known modulus to determine the true contact area by applying the analysis procedures in reverse. The created DAF was fit to a constant Young's modulus of 71.7 GPa for fused silica and 411 GPa for tungsten, across the load range of 0.5-200 mN. The DAF is in the form of equation (6) where the constants C_1 and C_2 are the fitted parameters.

$$A_p = A(h_c) = C_1 h_c^2 + C_2 h_c \quad (6)$$

One of the useful mechanical properties to be obtained from nanoindentation is the indentation hardness, or the mean contact pressure given a particular load. The indentation hardness is obtained by taking the maximum indentation load and dividing it by the projected area of contact at the maximum load, as seen the equation (7) below.

$$H_{IT} = \frac{P_{max}}{A(h_c)} \quad (7)$$

Where h_c is the contact depth of penetration, P_{max} is the maximum load, and A(hc) is the diamond area function.

Another useful mechanical property obtained from nanoindentation is the reduced modulus (E_r) of the system. The reduced modulus can be obtained using the slope of the elastic unloading curve, as seen in equation (8). Using information about the indentation system, E_r can be converted to the Young's modulus of the sample, as seen in equations (9) and (10) below.

$$E_r = \frac{1}{2}\frac{dP}{dh}\frac{\sqrt{\pi}}{\sqrt{A}} \quad (8)$$

$$\frac{1}{E_r} = \frac{(1-v_s^2)}{E_s} + \frac{(1-v_i^2)}{E_i} \quad (9)$$

$$E_s = \frac{E_r E_i(1-v_s^2)}{E_i + (v_i^2 - 1)E_r} \quad (10)$$

Where v_s is the Poisson's ratio of the sample, v_i is the poissons ratio of the diamond indentor (0.07), E_s is the Young's modulus of the sample, E_r is the reduced indentation modulus, and E_i is the Young's modulus of the diamond indentor (1141 GPa).

Not only does indentation allow the average hardness and modulus values to be obtained, but their depth dependence can give additional information as well (*10, 11*). When the indentation plastic zone is being hindered by grain boundaries or other barriers, the hardening increases, and when the plastic zone increases beyond the barriers, the hardening decreases.

Results and Discussion

Bulk Copper

In order to develop confidence in our methods, we began by performing indentions on a bulk copper sample. The bulk copper was obtained from a cut tensile test specimen that had been polished using mechanical polishing methods down to a 0.05 micron colloidal silica solution. Bulk copper has several issues that have been noted (*12, 13*) when performing indentation experiments. Bulk copper has a tendency to pile-up or sink-in during an indentation, causing changes to the indentation contact area and thus leading to errors. The pile-up or sink-in phenomenon has been observed to increase with indentation strain (*8, 13, 14*) and can be seen in Figure 5. In addition, the indentation size effect (ISE) is an issue with copper, creating drastic differences in hardness values (*8, 10*). In order to minimize pile-up/sink-in effects, as well as indentation size effects, surface oxidation, and surface polishing effects (*11*), a 10 mN load was used for the indentations, which resulted in an approximately 500 nm maximum depth. The resulting Young's modulus and hardness values were found to be 122.4 ± 6.1 GPa and 1.68 ± 0.09 GPa respectively. The obtained bulk modulus agrees well with previous publications (*15-19*), and the hardness value is within the range of previously published values (*6, 10, 11, 13, 15, 19*).

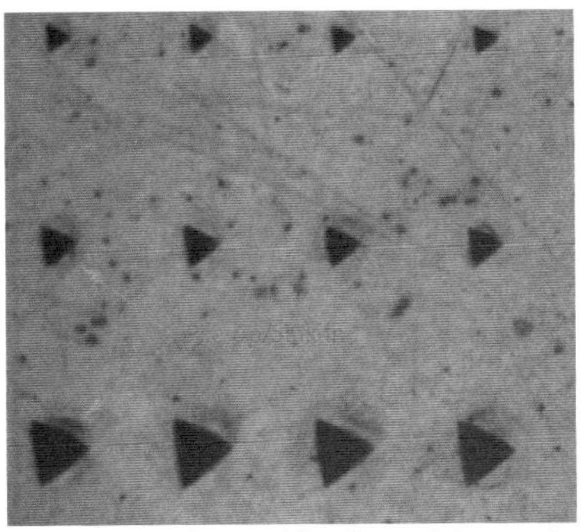

Figure 5: 40x optical magnification of increasing size of indentations in copper showing evidence of pile-up/sink-in phenomena. Pile-up regions can be seen at the edges of the indents and become larger with increasing size.

Spark Plasma Sintered Copper

The SPS copper was indented with a series of increasing loads, enabling a view of the material properties versus indentation depth as seen in Figure 6. The results obtained show clear indications of ISE phenomena indicated by the increasing modulus and hardness as the depth decreases. In addition to ISE phenomena, it is likely that surface oxide layers also contributed partially to the higher hardness values. However, as the indentation depth increases, the modulus and hardness begin to level out around values of 110 GPa for Young's modulus and 1.7 GPa for hardness. The obtained Young's modulus and hardness values are within the range reported by Chang and Chang in 2007 (*19*), who performed indentation on different sized grains of copper and obtained hardness values between 0.9-2.5 GPa and modulus values between 75-175 GPa. The modulus and hardness are also in the range reported by Ritasalo et al. (*6*), who published values between 80-160 GPa for the modulus and 2-3 GPa for hardness.

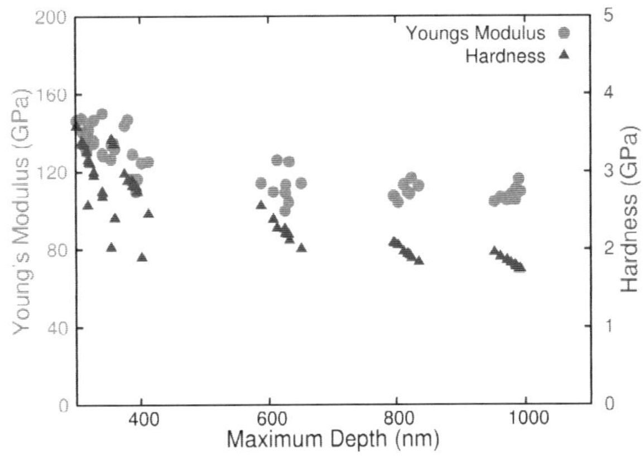

Figure 6: The Young's modulus and hardness of the spark plasma sintered copper sample versus depth, illustrating the depth dependence of the properties.

In addition to traditional Berkovich analysis, a spherical indenter with 100 μm radius was used to conduct a spherical Field and Swain indentation experiment on the SPS copper. The sample was loaded and then partially unloaded with incrementally increasing loads, which were then used to determine the flow stress and flow strain of the sample upon loading. The procedure used is documented by Field and Swain (*20*), and is a commonly used technique.

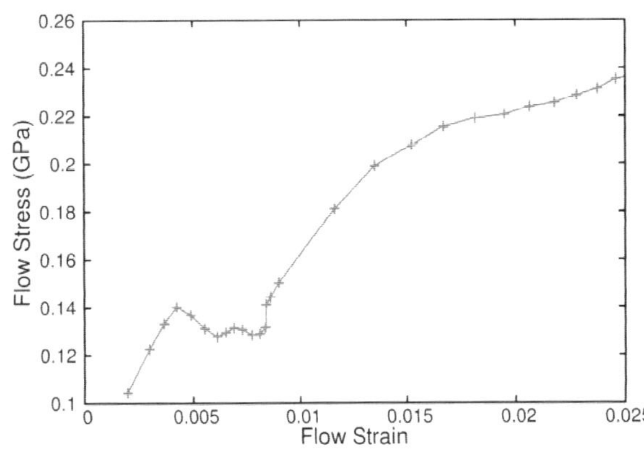

Figure 7: Flow stress versus flow strain using the nanoindentation Field and Swain technique on the spark plasma sintered copper.

The indentation flow stress-strain plot shows that there are multiple regions of plasticity in the NC copper. The flow stress-strain curve begins with a linearly increasing slope indicating elasticity, but then starts to decrease indicating a yield point for a drastic plastic transition, after which there are slight oscillations in the curve, and lastly as the curve continues the response transitions into a more traditional plastic response.

We found a yield strength of approximately 0.14 GPa or 140 N/mm^2 for our sample using the Field and Swain flow stress-strain curve, which is about a third of the yield strength found by Ritasalo et al. (*6*) using hardness values in the Tabor relation (*21*).

Nanoporous Copper Foam

Indentation of NP Cu is a useful tool for measuring the mechanical properties of the intricate system, but has many complexities and difficulties along the way. The NP Cu has a complex system of interconnected struts that collapse upon indentation of the surface. This permanent deformation of the surface is seen to happen during initial contact with the surface, before the indentation experiment even begins. Therefore, a correction is applied to the depth to account for this initial contact indentation. In addition, NP Cu with its large amount of surface area, allows a more rapid and penetrating depth of oxide formation to occur. The results of the NP Cu below are therefore a mixture of copper and copper oxide, causing an increase in hardness and modulus over the pure metallic nanoporous copper as seen in Cheng and Hodge (*22*) and Kreuzeder et al. (*23*).

The nanoindentations on the NP Cu foam were conducted using a 5 μN initial load, and a 100 μN maximum load resulting in an average depth of approximately 32.13 nm. The average indentation depth of 32.13 nm is approximately 3.6% of the overall foam thickness, indicating the likelihood that substrate effects are negligible. The Young's modulus decreases as a function of porosity (*16*). A series of indentations resulted in a Young's modulus of 28.5 ± 4.32 GPa and a hardness of 1.63 ± 0.39 GPa. The nanoindentation reduced modulus is similar to those obtained by Kreuzeder et al. (*23*), while the hardness values are a little high. The higher hardness values are most likely due to the oxidation of the NP Cu foam (*22*) as well as the decreased strut thickness (*4*). The hardness values can be used in a modified Tabor relation for porous metals (*1, 24, 25*), where the yield strength is related to the hardness. The resulting yield strength of NP copper foam and bulk copper [approximately 500 MPa (*26*)] can then be related by their relative densities and a proportionality constant, as shown by Liu et al. (*4*). The relation can be seen in equation (11), where H_f is the hardness of the NP copper foam, σ_s is the yield strength of the solid strut, ρ_f/ρ_s is the relative density, and C is the proportionality constant. The proportionality constant is set equal to 0.6, which was found to be reasonable by Liu et al. (*4*). This results in an average strut strength of approximately 5.43 GPa.

$$\frac{H_f}{\sigma_s} = C\left(\frac{\rho_f}{\rho_s}\right) \quad (11)$$

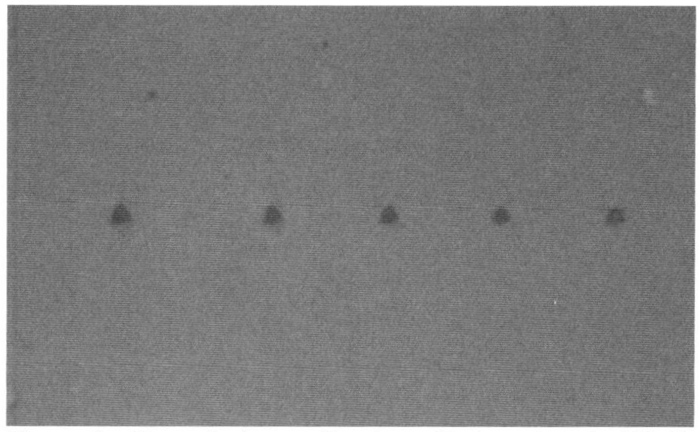

Figure 8: 40x optical magnification of indentations in NP Cu foam.

Conclusions

Spark plasma sintering of nanopowder allows for the manipulation of grain size. Our spark plasma sintering procedure allowed the sample to be 94.19% of the theoretical density, and an average grain size of approximately 1 μm. In order to explore ultrafine grain and nanocrystalline copper, more work is needed to refine the spark plasma sintering process by lowering the temperature and possibly the pressure as well.

Nanoindentation allowed the SPS copper Young's modulus and hardness values to be obtained. The modulus values obtained were in good agreement with previously published values, while the hardness values were a little higher than previous publications. This increased hardness could be due to oxides, but may also be due to the presence of nanotwins, work hardening from numerous dislocations, or other such barriers to the indentation plastic zone. More research is needed to unravel the contributions to the recorded hardness. Lastly, spherical Field and Swain indentation was performed on the SPS sample, giving the flow stress and flow strain plot, allowing the yield strength of approximately 0.14 GPa to be obtained.

The mechanical properties of NP Cu foam were also obtained using nanoindentation. The indentations of NP Cu foam resulted in a Young's modulus of 28.5 ± 4.32 GPa and a hardness of 1.63 ± 0.39 GPa, which is similar to the reported values by Kreuzeder et al. (*23*). The average strut yield strength for the NP Cu foam of 5.43 GPa, obtained from the modified Tabor hardness relation (*1, 4*), is well above the bulk copper value of approximately 500 MPa (*26*). This increase in yield strength demonstrates how NP Cu foam is a high yield strength material, despite its complex internal structure.

References

1. J. Biener, A. M. Hodge, A. V. Hamza, L. M. Hsiung, J. H. Satcher Jr, Nanoporous Au: A high yield

strength material. *Journal of Applied Physics* **97**, 024301 (2005).
2. J. Biener *et al.*, Size effects on the mechanical behavior of nanoporous Au. *Nano letters* **6**, 2379-2382 (2006).
3. M. Hakamada, M. Mabuchi, Mechanical strength of nanoporous gold fabricated by dealloying. *Scripta Materialia* **56**, 1003-1006 (2007).
4. R. Liu *et al.*, Synthesis and mechanical behavior of nanoporous nanotwinned copper. *Applied Physics Letters* **103**, 241907 (2013).
5. I. C. Cheng, A. M. Hodge, Strength scale behavior of nanoporous Ag, Pd and Cu foams. *Scripta Materialia* **69**, 295-298 (2013).
6. R. Ritasalo *et al.*, Spark plasma sintering of submicron-sized Cu-powder—Influence of processing parameters and powder oxidization on microstructure and mechanical properties. *Materials Science and Engineering: A* **527**, 2733-2737 (2010).
7. Z.-H. Zhang, F.-C. Wang, L. Wang, S.-K. Li, Ultrafine-grained copper prepared by spark plasma sintering process. *Materials Science and Engineering: A* **476**, 201-205 (2008).
8. A. C. Fischer-Cripps, *Nanoindentation*. (Springer, 2011), vol. 1.
9. I. C. Secretariat, in *ISO 14577*. (1 rue de Varembé, 1211 Geneva 20 Switzerland).
10. A. H. Almasri, G. Z. Voyiadjis, Nano-indentation in FCC metals: experimental study. *Acta mechanica* **209**, 1-9 (2010).
11. Y. Liu, A. Ngan, Depth dependence of hardness in copper single crystals measured by nanoindentation. *Scripta Materialia* **44**, 237-241 (2001).
12. Y. Wang, D. Raabe, C. Klüber, F. Roters, Orientation dependence of nanoindentation pile-up patterns and of nanoindentation microtextures in copper single crystals. *Acta Materialia* **52**, 2229-2238 (2004).
13. K. Kese, Z. C. Li, Semi-ellipse method for accounting for the pile-up contact area during nanoindentation with the Berkovich indenter. *Scripta Materialia* **55**, 699-702 (2006).
14. K. McElhaney, J. Vlassak, W. Nix, Determination of indenter tip geometry and indentation contact area for depth-sensing indentation experiments. *Journal of Materials Research* **13**, 1300-1306 (1998).
15. X. Deng, N. Chawla, K. Chawla, M. Koopman, Deformation behavior of (Cu, Ag)–Sn intermetallics by nanoindentation. *Acta materialia* **52**, 4291-4303 (2004).
16. P. G. Sanders, J. A. Eastman, J. R. Weertman, Elastic and tensile behavior of nanocrystalline copper and palladium. *Acta Materialia* **45**, 4019-4025 (1997).
17. D. E. Armstrong, A. J. Wilkinson, S. G. Roberts, Measuring anisotropy in Young's modulus of copper using microcantilever testing. *Journal of Materials Research* **24**, 3268-3276 (2009).
18. X. Deng, M. Koopman, N. Chawla, K. Chawla, Young's modulus of (Cu, Ag)–Sn intermetallics measured by nanoindentation. *Materials Science and Engineering: A* **364**, 240-243 (2004).
19. S.-Y. Chang, T.-K. Chang, Grain size effect on nanomechanical properties and deformation behavior of copper under nanoindentation test. *Journal of applied physics* **101**, 033507 (2007).
20. J. Field, M. Swain, A simple predictive model for spherical indentation. *Journal of Materials Research* **8**, 297-306 (1993).
21. D. Tabor, *The Hardness of Metals, Clarendon*. (Oxford, 1951).
22. I. C. Cheng, A. M. Hodge, Morphology, oxidation, and mechanical behavior of nanoporous Cu foams. *Advanced engineering materials* **14**, 219-226 (2012).
23. M. Kreuzeder *et al.*, Fabrication and thermo-mechanical behavior of ultra-fine porous copper. *Journal of materials science* **50**, 634-643 (2015).
24. S. Corcoran, R. Colton, E. Lilleodden, W. Gerberich, Anomalous plastic deformation at surfaces: nanoindentation of gold single crystals. *Physical Review B* **55**, R16057 (1997).
25. L. J. Gibson, M. F. Ashby, *Cellular solids: structure and properties*. (Cambridge university press, 1997).
26. K. Lu, L. Lu, S. Suresh, Strengthening materials by engineering coherent internal boundaries at the nanoscale. *Science* **324**, 349-352 (2009).

International Federation of Heat Treatment and Surface Engineering 2016
Proceedings of the 23rd IFHTSE Congress
April 18–21, 2016, Savannah, Georgia, USA

Grain refinement strategies in high strength cast steel

David C. Van Aken[1], Terrell O. Webb, and Joshua D. Green

Department of Materials Science and Engineering,
Missouri University of Science and Technology, Rolla, MO, 65409
1. e-mail: dcva@mst.edu

Abstract

This paper reviews the physical metallurgy of grain refinement produced by standard heat treatment practices. Grain refinement by normalization is shown to be dependent upon the starting microstructure. High alloy steels with as-cast microstructures that are bainitic are shown to be problematic with respect to obtaining grain refinement by standard normalization practices. These as-cast bainitic microstructures contain remnant austenite that is stable during heating, which subsequently grows during austenitization, and upon impingement reproduces the original parent austenite grain. The addition of a tempering heat treatment for these high alloy steels is shown to eliminate the austenite, increase grain multiplication upon heating, and produce grain refinement.

Introduction

It is well established that reduction of the prior austenite grain size improves the performance of tempered martensitic steel. Swarr and Krauss [1] have shown that the martensitic lath packet size decreases as the prior austenite grain size is reduced and Grange [2] has shown a Hall-Petch behavior for the yield strength with respect to prior austenite grain size for quenched and tempered 8650 and 4340 grades. It should be pointed out that more recent literature reveals that the yield strength of lath martensites correlate best with the lath block size rather than the packet size. [3] Regardless, lath packet size or lath block size or both might be expected to decrease as the prior austenite grain size is reduced and an increase in strength is expected. A complete description of lath martensitic terminology may be found in the work of Morito et al. [4]. In addition to increased yield strength a refinement in prior austenite grain size is expected to improve low temperature notch toughness as shown by Porter et al. for a 5%Ni Cr Mo V steel. [5] Thus, refinement of the prior austenite grain size is of great importance in the development of high strength and high toughness steel.

Manipulation of prior austenite grain size and crystallographic texture in wrought products can be obtained by thermomechanical processing where recrystallization of deformed austenite is manipulated to provide both desired texture and grain size control. [6] In wrought alloys grain refinement can be produced by recrystallization and grain pinning with AlN, NbC, VN, etc. Smaller grain diameters are produced when plastic deformation exceeds 20-30% with the grain pinning agents precipitating on the dislocation substructure. Newly formed or recrystallized grains then grow to the pinning diameter. In contrast, grain refinement of cast steels must rely upon grain multiplication during polymorphic phase transformation. Grain refinement in cast steel in the absence of grain pinning agents is the focus of this paper.

To fully understand the evolution of grain structure when steel is austenitized, the heat treater must know how the prior microstructure affects grain multiplication during phase transformation, the kinetics of grain growth, and how second phase particles pin the growing austenite grains. Grain refinement in plain carbon and low alloy steels is obtained predominately by the practice of normalization in the foundry industry. Polymorphism in iron is responsible for grain refinement when steels are normalized or heat treated by quench hardening. The following paragraphs will review some basic concepts related to grain refinement and build simple models that can be used to estimate grain multiplication factors associated with phase transformation upon heating and cooling.

Grain multiplication during a polymorphic transformation ($\alpha \rightarrow \gamma$) will occur in pure iron as the steel is heated above 912°C (1674°F). Nucleation of new grains (γ) will typically occur at grain corners and grow until impingement. Grain coarsening will occur after complete transformation and the growth of the average grain diameter (L_3) follows a parabolic time (t) dependence, i.e. $L_3 \propto \sqrt{t}$. Here we use the measure of mean linear size or L_3 as determined by the Heyn intercept method, to describe the grain diameter, see ASTM E112. [7] A simple model of grain multiplication can be built using a grain structure of space filling Kelvin tetrakaidecohedrons as shown in Fig. 1. Each individual tetrakaidecohedron has 14 faces, 36 edges, and 24 corners. The volume (V) of a single tetrakaidecohedron can be calculated [8] based upon L_3 as

$$V = 1.767(L_3)^3 \quad (1)$$

Close examination of Fig. 1(a) reveals 4 grains share each grain corner; thus, there are 6 corners for each grain and a grain multiplication factor of 6 may be inferred during polymorphic phase transformation. It should be noted that grain multiplication should occur upon both heating and cooling. Using this same argument of grain multiplication a single heating and cooling cycle will transform the original parent grain into 36 new grains assuming grain coarsening at the elevated temperature is ignored. Fig. 1(b) shows the effect of two heating and cooling cycles to produce 1296 new grains from a single parent grain and a final grain diameter (L_3) of 9.3μm will result if the starting grain diameter is taken as 100μm.

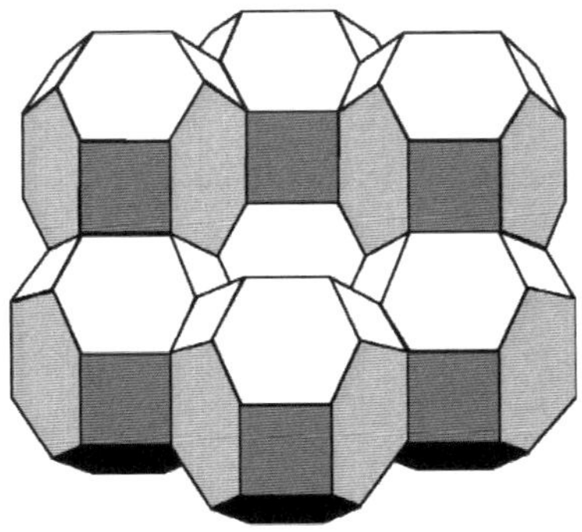

(a)

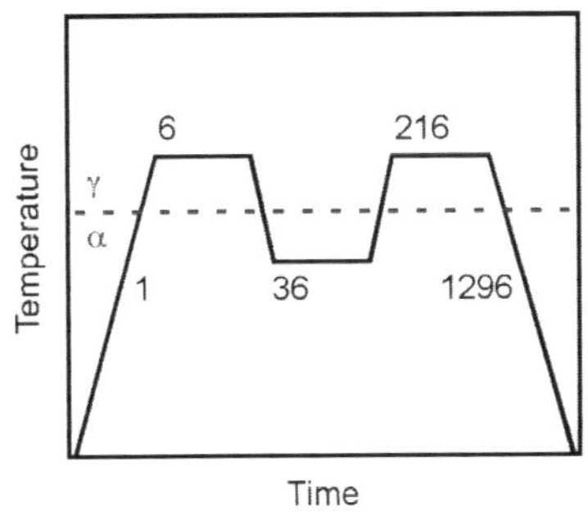

(b)

Figure 1. (a) A cluster of tetrakaidecahedron shaped grains. (b) An illustration of grain multiplication assuming that each polymorphic phase transformation produces a grain multiplication of 6.

Refinement by grain multiplication is more complicated in mixed microstructures of ferrite, carbide, and potentially austenite. Grain multiplication during heating of multiphase microstructures will be dependent upon the distribution of ferrite and the carbon source to obtain austenite of the appropriate composition. Work of Vilella, as reported by Grossman and Bain, shows nucleation of austenite at spheroidal Fe_3C carbides. [9] In plain carbon steels, the austenite forms first in the pearlitic microstructure and these new austenite grains grow into the remaining ferrite. Upon cooling, proeutectoid ferrite is expected to form at austenite grain corners and grain edges. Thus, multiplication factors greater than 6 might be expected. An example of grain multiplication is shown in Fig. 2 for cast 1524. The initial ferrite grain size (L_3) is approximately 200 μm and upon a single normalization cycle the ferrite grain structure is reduced to approximately 30 μm. The resultant grain multiplication (n) for this heating and cooling cycle can be calculated using eq. (1) as

$$1.767(200\mu m)^3 = n1.767(30\mu m)^3$$

$$n = \frac{(200\mu m)^3}{(30\mu m)^3} = 296$$

where 296 new grains occupy the volume of the original 200μm diameter grain. The average multiplication factor for each transformation (heating then cooling) would be approximately 17.

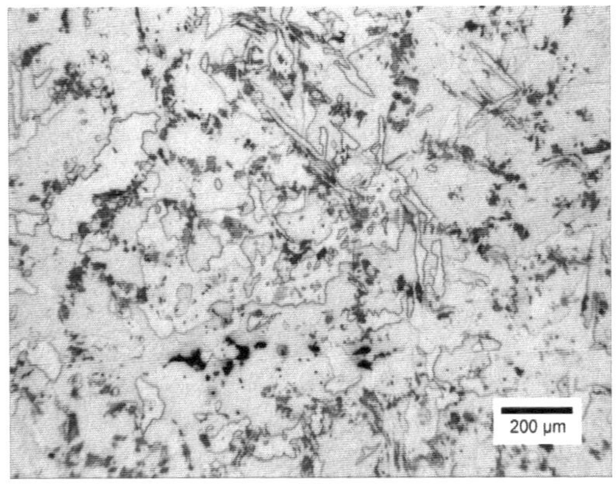

(a)

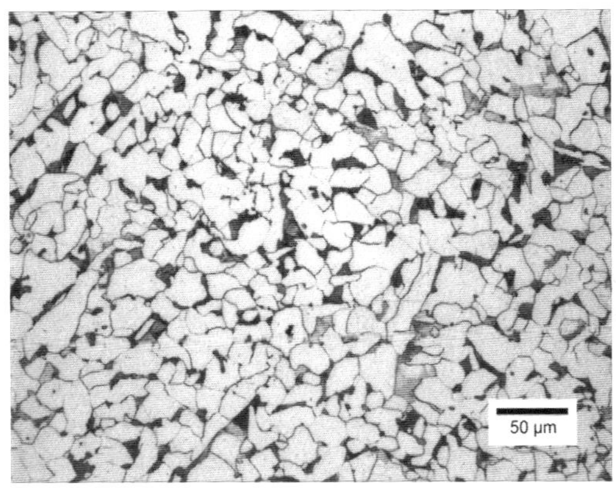

(b)

Figure 2. Microstructures of 1524 (a) as cast and (b) normalized at 925°C (1700°F). Each microstructure consists of ferrite grains and pearlite. Etched with 2% nital.

Grain multiplication can be reduced by crystallographic factors during the polymorphic phase transformation and by grain growth during the elevated temperature hold. Quite often the nucleated product grain will have a specific orientation relationship to one of the adjacent parent grains. A random boundary is established with the other parent grains. The Kurjumov-Sachs orientation relationship describes the most commonly observed relationship between ferrite and austenite in steel where close packed planes and close packed directions are parallel. The Kurjumov-Sachs orientation relationship is written as follows using the standard Miller index notation:

$$\langle 1\bar{1}0\rangle_\gamma // \langle 1\bar{1}1\rangle_\alpha \ and \ \{111\}_\gamma // \{110\}_\alpha \qquad (2)$$

If two adjacent corners to the same parent grain nucleate with the same orientation relationship to the parent grain, then the two new grains will coalesce upon impingement to a single grain and the multiplication factor will be reduced. As a result it is often easier to determine the grain multiplication factors by experiment as shown above for the normalization of the 1524 steel.

Grain growth at elevated temperature will also reduce the extent of grain refinement by increasing the volume of the average grain and thus reducing the number of grain corners per unit volume upon cooling. Fig. 3 shows a schematic double normalization process where grain growth is incorporated and the multiplication factors for heating are

$$n_1 = \left(\frac{L_3(t_0)}{L_3(t_1)}\right)^3 \text{ and } n_2 = \left(\frac{L_3(t_2)}{L_3(t_3)}\right)^3 \qquad (3)$$

The prior austenite grain diameters $L_3(t_2)$ and $L_3(t_4)$ can be experimentally measured after each heating and cooling cycle. However, to determine $L_3(t_1)$, $L_3(t_3)$, and the grain multiplication factors n_1 and n_2 requires knowledge about the grain growth kinetics of austenite.

Oil quenching and introduction of a tempering step after the first austenitization cycle, as shown in Fig. 3, has been shown to reduce the prior austenite grain size in high strength steel castings [10, 11] and carburized steel. [12] Webb and Van Aken have shown improved tensile ductility as measured by either elongation to failure or reduction in area and higher Charpy V-notch impact energy using the intermediate tempering or subcritical annealing step. [10] The resulting quenched and tempered steel exhibited a prior austenite grain diameter of 56 μm and a refined lath martensitic microstructure with increased frequency of martensite lath boundary misorientation angles. A comparison of the mechanical properties from eference 10 is shown in Table I. For carburized and hardened steel, Grange introduced a Stage II temper at 371°C (700°F) after oil quenching from the carburization cycle to transform retained austenite to bainite. Grange hypothesized that the refined ferritic microstructure after tempering, when combined with rapid heating to the hardening temperature, produced the observed grain refinement and the increased case hardness.

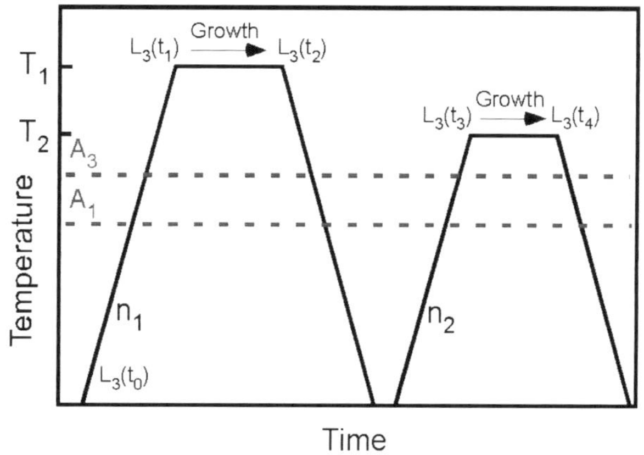

Figure 3. Schematic showing a double normalization heat treatment. The prior austenite grain size of the casting is $L_3(t_0)$. At the end of the first and second normalization cycles the prior austenite grain diameters would be measured as $L_3(t_2)$ and $L_3(t_4)$. Calculation of the grain multiplying factors, n_1 and n_2, requires that the austenite grain diameter at the beginning of the austenitization cycle be determined by estimating the grain growth at temperature.

Table I. Mechanical properties for W-steel with respect to grain size

Prior austenite grain size	0.2% offset yield strength MPa (ksi)	Ultimate tensile strength MPa (ksi)	Elongation to fracture	Reduction in area	V-notch Charpy impact at energy (-40°C)	Dynamic fracture toughness K_{ID}
265±88 μm	1207 (175)	1269 (184)	2.7%	6.7%	48J (36 ft-lb)	111 MPa√m
56±12 μm	1210 (176)	1586 (230)	8.3%	21%	58J (43 ft-lb)	110 MPa√m

Experimental procedure

Two tungsten-containing steels have been examined in the course of this investigation and the steel chemistries are listed in Table II. The W-steel was supplied as 50 mm (2 inch) thick weld plates. The W-steel from Foundry 1 is the same steel previously studied in the paper by Webb and Van Aken. [10] This same steel was used here to study grain growth and grain multiplication. American Cast Iron Pipe Co. (ACIPCo) provided optical emission spectrometry results for the W-steels. Carbon and sulfur were measured using a LECO CS6000, and, oxygen and nitrogen were measured using a LECO TC 500. LECO measurements were performed by Missouri University of Science & Technology.

Table II: Steel chemistry and foundry designation.

Alloy	C	Mn	Si	Cr	Ni	Mo	W	V	Cu	P	S	N	Al	O
W-Steel Foundry 1	0.25	0.77	0.88	2.64	1.03	0.425	0.92	0.097	0.12	0.006	0.004	0.009	0.008	0.022
W-Steel Foundry 2	0.26	0.63	1.02	2.72	1.03	0.45	0.75	0.092	0.07	0.009	0.004	0.013	0.022	0.015

The kinetics of austenite grain growth was studied using the W-steel from Foundry 1. The steel was first grain refined by heating to T_1, holding for 4 hours to dissolve alloy carbides, and quenched. A second austenitization was conducted at T_2, held for one hour and quenched. A prior austenite grain diameter of 100 μm was measured after heat treatment and this heat treatment appears later in the paper as method 3 in Table III. Test coupons were then heated to 1093°C (2000°F), 1121°C (2050°F), 1149°C (2100°F), and 1177°C (2150°F) for varying lengths of time and then subsequently water quenched to produce martensite. Grain growth coupons were double bagged in chemically treated stainless steel to inhibit decarburization. Each metallographic specimen was etched with a Marbles reagent modified by adding an equal volume of glycerin. The modified Marbles reagent sufficiently highlighted the prior austenite grain boundaries to permit grain size measurement. Grain diameters were measured in accordance with ASTM E112 [7] using the Heyn intercept method and diameters are reported as the mean linear distance, L_3, and reported with a 95% confidence level for the individual test measurement.

Studies involving the formation of austenite were conducted by heating the W-steel (Foundry 2) into the intercritical temperature range at 800°C (1472°F) and holding for varying times to determine the rate of austenite formation. Each experiment was conducted using 6 specimens cut to a thickness of approximately 10 mm (3/8 inch) and welding a thermocouple to one of the coupons. Specimens were stood upright in the furnace to allow free convection between each specimen. Specimens were removed at 0, 5, 10, 20, 30, and 40 minutes starting as soon as the temperature of the thermocoupled specimen was within 2 C° of the set temperature and then quenched in water. Rockwell C-scale hardness measurements were taken on each coupon after surface grinding to remove potential decarburization and to make the surfaces flat and parallel. A total of ten measurements were taken on each specimen and the reported uncertainty represents the 95% confidence level of the mean value.

Standard metallographic techniques were used to prepare specimens for optical microscopy. Specimens used for orientation image mapping using electron backscatter diffraction (OIM-EBSD) were vibratory polished using 0.02 μm colloidal silica. OIM-EBSD was performed on a FEI Quanta 200F Scanning Electron Microscope and images were subject to a cleanup procedure, setting the minimal confidence index to 0.1 and matching orientation of data points that are within 5 degrees of 4 of its closest neighboring data points. The electron beam was operated at an accelerating voltage of 20.0 kV and an emission current of 11 nA with a 0.025 μm step size during mapping. Quantification of retained austenite was performed using X-ray diffraction and calculated utilizing the Rietveld refinement described by Martin et al. [13]

Results

A typical microstructure of the cast W-steel is shown in Fig. 4. The as-cast prior austenite grain size was typically on the order of 1-2 millimeters and the microstructure was characterized as being bainitic. Austenite grain growth was studied using the W-steel supplied by Foundry 1. Grain growth data was assumed to follow parabolic time dependence (normal grain growth) and described by the following kinetic equations:

$$\left[L_3(t)\right]^2 - \left[L_3(0)\right]^2 = kt \qquad (4)$$

$$k = k_0 \exp\left(\frac{-Q}{RT}\right) \qquad (5)$$

where $L_3(t)$ is the grain diameter measured after time t, $L_3(0)$ is the initial grain diameter, k is the rate constant determined from eq. 5 at absolute temperature T, R is the universal gas constant, k_0 and Q are kinetic constants for the W-steel. A plot of the square of the measured grain diameters versus time is shown in Fig. 5. Continuous grain growth was observed at 1149°C (2100°F) and 1177°C (2150°F) and the data could be fit to a linear trend line indicating normal grain growth as shown in Fig. 5(a). At temperatures below 1149°C (2100°F) the linear portion of the grain coarsening was restricted to short times. After 2 hours at 1121°C (2050°F) the grain diameter remained constant at approximately 230 μm. After 6 hours at 1093°C (2000°F) the grain diameter reached 230 μm and no further grain growth was observed. The 230 μm grain diameter is consistent with the secondary dendrite arm spacing of the W-steel casting and the presence of persistent carbides in the interdendritic regions may explain grain pinning. An initial austenite grain size of 100 μm upon heating the W-steel to the coarsening temperature was assumed when determining the rate constants. Rate constants, k in mm^2/hr, were determined from the slopes of the linear trend lines. An Arrhenius plot of the rate constants is shown in Fig. 5 (b). An activation energy (Q) of 537 kJ/mole and k_0 equal to 2.96x 10^{18} mm^2/hr were determined from the Arrhenius plot. Grain coarsening behavior can now be calculated using the kinetic data.

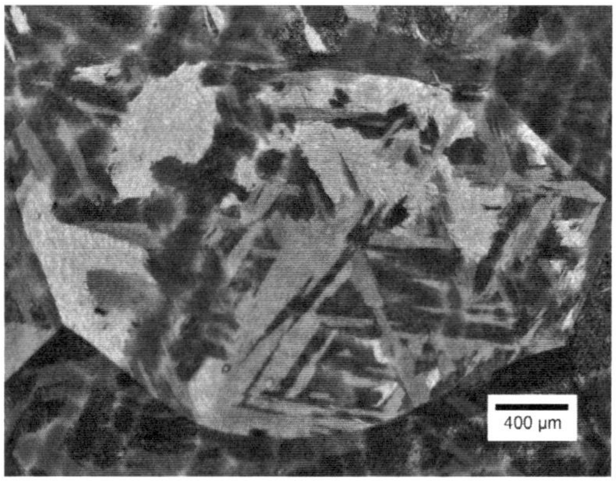

Figure 4. As-cast microstructure of the W-steel showing bainitic laths of ferrite. Etched with modified Marbles reagent.

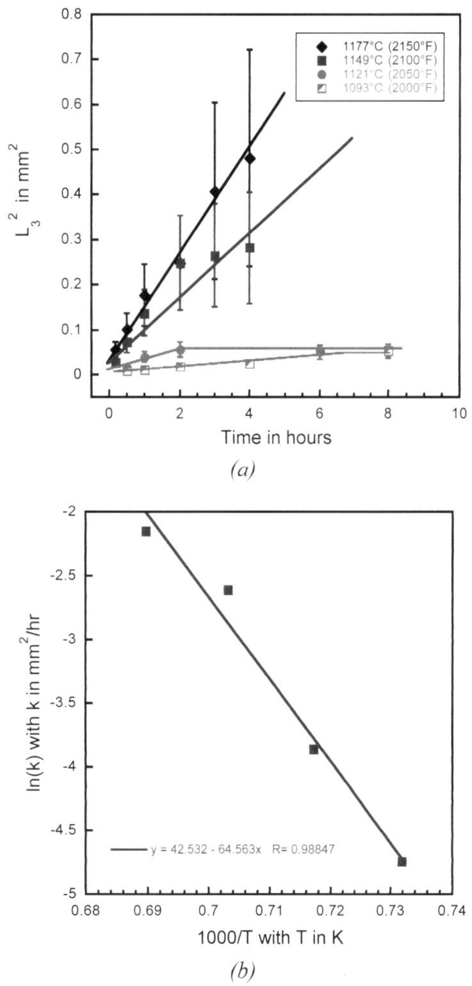

Figure 5. (a) Normal grain growth is observed for the W-steel in 3 of the 4 (b) Arrhenius plot of rate constants determined from the slope of the lines shown in (a). The slope of the Arrhenius plot is equal to -Q/R and the y-intercept is ln(k_0).

Grain multiplication factors (n_1 and n_2) as shown in Fig. 3 can now be calculated for various heat treatments (see Table III) conducted on the W-steel. An OIM-EBSD study of the previous heat treatment (labeled method 2 in Table III) reported by Webb and Van Aken [10] is shown in Fig. 6. The measured grain diameters (L_3) of the as-cast (1800 µm) and that resulting from the first austenitization cycle (1454 µm) indicate little to no grain refinement. Austenite grain growth may now be calculated to determine the size of the austenite after the initial heating using the kinetic results above. The measured prior austenite diameter of 1454 µm would have grown from an initial austenite grain diameter of 1414 µm and the grain multiplication factor of 2.1 for n_1 is calculated from the ratio of the grain diameters cubed $(1800/1414)^3$. Grain growth and multiplication factors are summarized in Table IV for the heat treatments listed in Table III. It should be noted that the uncertainty in the measured grain size is quite large for grain diameters greater than 200 µm; however, order of magnitude changes in the multiplication factor may indicate significant changes in the mechanism of austenite formation.

Differences in multiplication factors appear to be influenced by subcritical annealing for both the as-cast and steels quenched directly after the first heating cycle. Initial heating of the as-cast W-steel has a much smaller grain multiplication factor ($n_1 = 2.1$) upon heating than the same steel that has been subjected to an 8-hour subcritical anneal, as for method 4, where n_1 is 670. A comparison of methods 2 and 3 also indicate that the multiplication factor, n_2, increases from 4000 to 66,000 when a Stage 3 temper of 4 hours is added prior to austenitization to harden the steel. It should be noted that the prior austenite grain size after normalization in method 4 is the pinned diameter as reported in the grain growth study. Also apparent is that the multiplying factors (n_2) are quite different between the air-cooled bainitic (method 1) and quenched (method 3) martensitic microstructures.

An OIM-EBSD examination of a normalized W-steel is shown in Fig. 7 and reveals that the steel transforms to bainitic laths of ferrite when air-cooled, see Fig. 7(a) and 7(b), with remnant austenite between the ferrite laths as shown in Fig. 7(c). The crystallography of the remnant austenite indicates a single parent austenite grain as shown by the FCC crystal orientation distribution shown in Fig. 7(e).

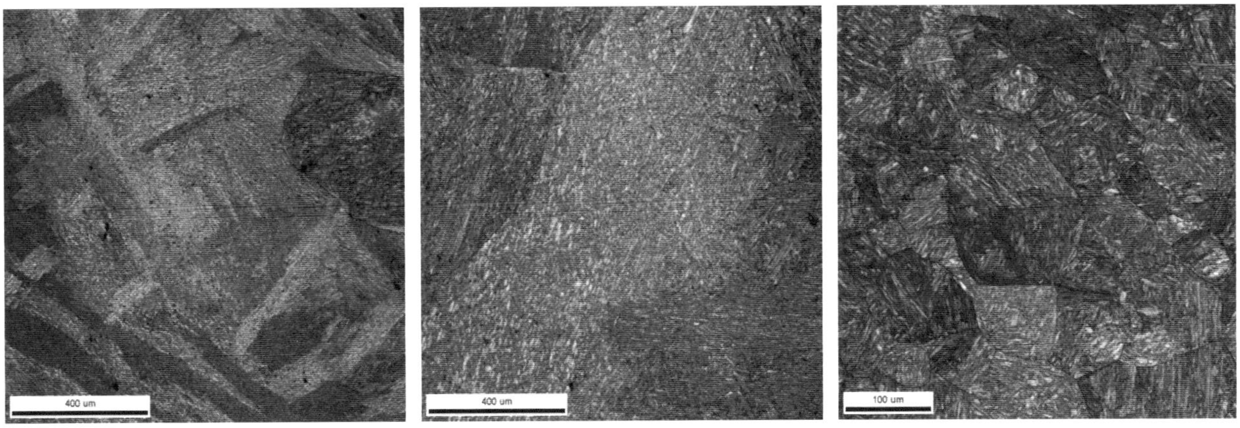

(a) (b) (c)

Figure 6. OIM-EBSD images from various stages of heat treatment from the same W-steel specimens as reported by Webb and Van Aken. [10] (a) As-cast microstructure with starting grain diameter of 1800μm. (b) Oil quenched microstructure of the steel austenitized for 2 four hours at T_1 with measured grain diameter of 1454μm. (c) Water quenched microstructure of the W-steel after hardening from T_2 and holding one hour producing a prior austenite grain diameter of 50μm.

Table III: Heat treat schedules for determining grain multiplication factors

Heat treat method	1	2	3	4
Temper, $T<A_1$	None	None	None	8 hours
1st austenitization	8 hours at T_1+25°C	2 hours at T_1	4 hours at T1	4 hours at T1
Cooling	Air	Quench	Quench	Air
Temper $T<A_1$	None	4 hours	None	4 hours
2nd austenititization	4 hrs. at T_2 +50°C	1 hour at T_2	1 hour at T_2	1 hour at T_2
Cooling	Quench	Quench	Quench	Quench

Table IV: Prior austenite grain diameters and calculated multiplication factors

Heat treat method	Measured $L_3(t_0)$ in μm	Calculated $L_3(t_1)$ in μm	Grain multiplier n_1	Measured $L_3(t_2)$ in μm	Calculated $L_3(t_3)$ in μm	Grain multiplier n_2	Measured $L_3(t_4)$ in μm
1	1,800*	1,414*	2.1	1,756*	259	311	283
2	1,800	1,414	2.1	1,454	36	65,900	50
3	1,800*	1,414*	2.1	1,492*	94	4000	100
4	1,979	226	671	226	87	17.5	94

*Calculated assuming similar processing as method 2, but grain growth at either different temperatures or different times or both.

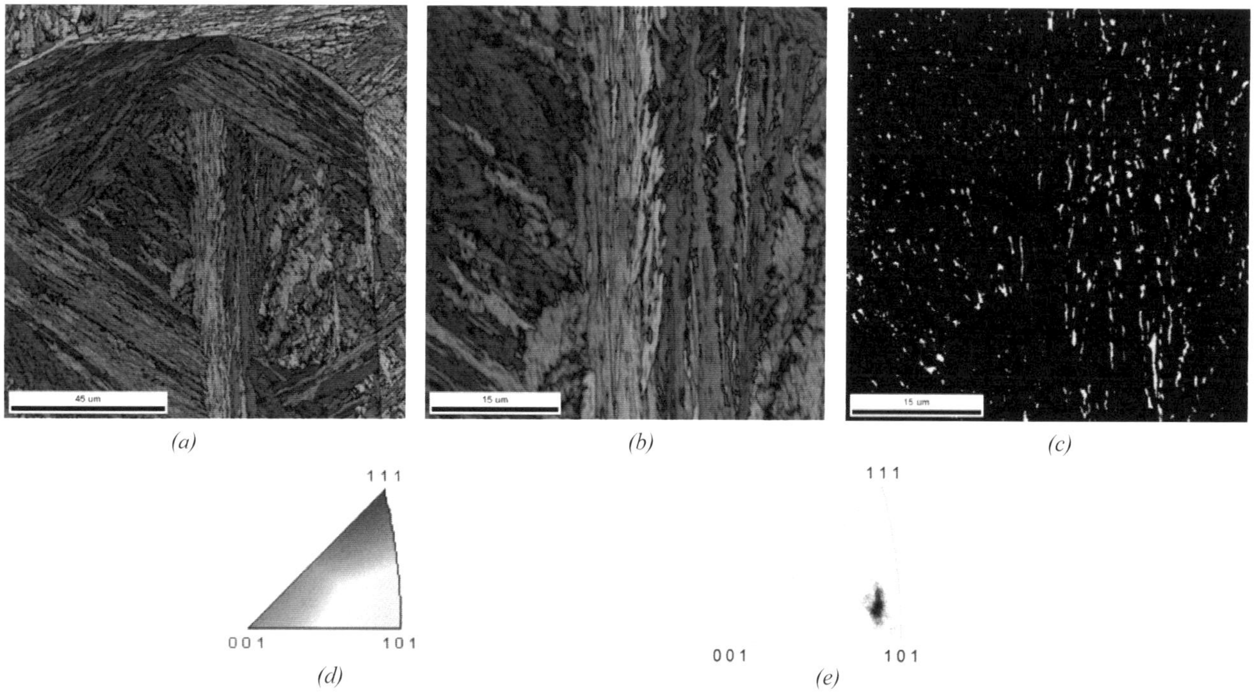

Figure 7. OIM-EBSD of normalized (4 hours at T_1 and air cooled) W-steel showing a bainitic microstructure of ferrite in (a) and (b) and remnant austenite in (c). Crystallographic variants of ferrite are highlighted using (d) It should be noted that each highlighted austenite portion shown in (c) has the exact same crystallographic orientation and reflects the crystal orientation of the parent austenite grain from which the laths of ferrite formed.

X-ray diffraction patterns for the bainitic microstructure produced by a first stage normalization and then after a subcritical anneal are shown in Fig. 8. Quantification of the austenite in the normalized steel was performed utilizing the Rietveld refinement described by Martin et al. [13] The normalized steel contains 13.2% austenite by volume, and 0.6% austenite was detectable after tempering for four hours. This may suggest a significant time dependence of austenite transformation during tempering. Some carbide is apparent in both conditions but cannot be identified from the single diffraction peak at approximately 2θ equal to 42.2°.

Intercritical annealing experiments were conducted at 800°C in an effort to understand the role of starting microstructure (as-cast with a bainitic microstructure versus martensite) and the role of the tempering prior to heating in the formation of austenite. The martensitic microstructures were produced by first heating to T_1, holding 4 hours, and quenching in water. Each of the six specimens were heated to 800°C with a thermocouple attached to one of the specimens. Upon reaching 798°C the first specimen was removed and quenched in water, i.e. time zero. Additional specimens were removed and quenched in water after an additional 5, 10, 20, 30, and 40 minutes. The thermocoupled specimen was quenched last. Quenched hardness using a C-scale Rockwell hardness test as a function of time is shown in Fig. 9. In general, the hardness increased with time at temperature indicating that the volume of martensite (austenite formed during intercritical anneal) was increasing with time. It should be noted that the as-cast bainitic microstructure appears to reach the maximum hardness in the time zero specimen whereas the other three conditions show that the hardness increases with time indicating a slower rate of austenite formation.

A comparison of the as-cast bainitic microstructures with and without the prior tempering is shown in Fig. 10 for the intercritically annealed steel. A nearly fully martensitic microstructure is observed after 10 minutes at 800°C when heating the as-cast bainitic microstructure as shown in Fig. 10(a). The dark etching features shown in Fig. 10(a) are more easily discerned in Fig. 10(b), which shows a martensite, carbide, and ferrite microstructure. The ferrite appears as thin remnant laths (white contrast) between the regions of martensite (brown contrast). The convex nature of the martensite interface indicates a diffusional growth of austenite at the intercritical temperature into the ferrite. Some carbide precipitation is evident at these boundaries and within the ferrite as might be expected for a 3-phase intercritical region for the W-steel. The tempered bainitic structure is shown in Figs. 10(c) and 10(d). Carbide precipitation is evident through out the microstructure and the martensite appears as small islands. Growth of the austenite is crenulated as the austenite grows through the carbides rather than being lenticular and parallel to the previous ferrite laths as shown in Figs. 10(a) and 10(b). No significant microstructural differences were observed when the starting microstructure was martensitic or tempered martensite as might be expected based upon the hardness versus time results shown in Fig. 9(b).

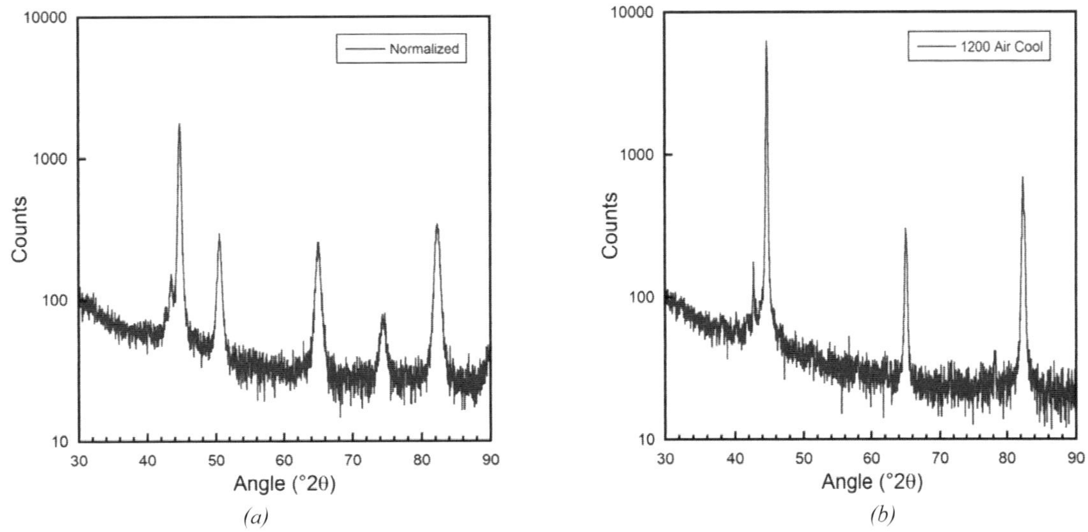

Figure 8. Diffraction patterns for normalized W-steel. (a) Air cooled from T_l with microstructure as shown in Fig. 9. (b) Same steel as in (a) but after a 4 hour temper. Diffraction peaks for austenite are found in (a) at 43° for {111}, 50.8° for {200} and 74.3° for {220}. Variation in the background counts produced a value of 0.6% austenite after the 4 hour temper for (b).

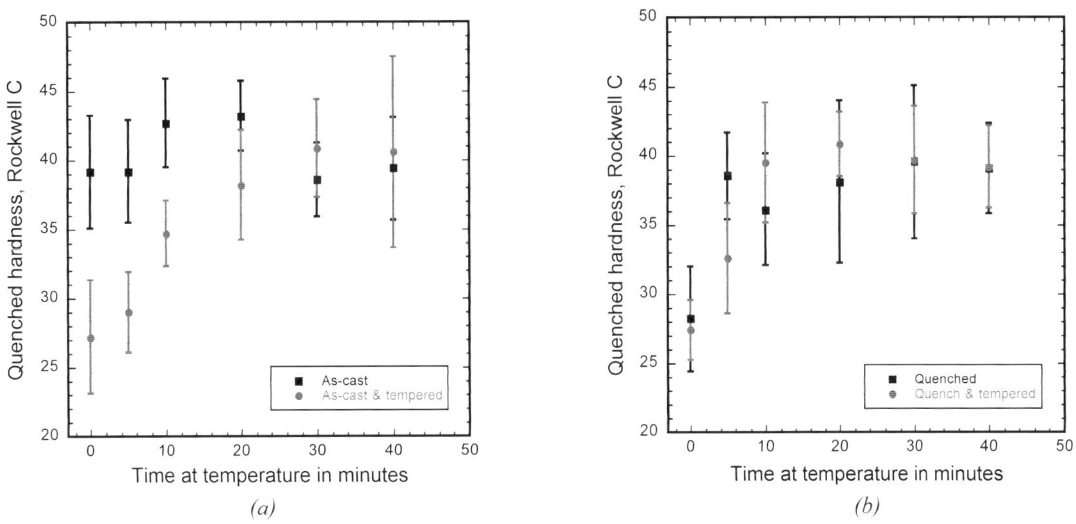

Figure 9. As-quenched hardness for W-steel (Foundry 2) heated to and held at 800°C (1472°F) for varying lengths of time to nucleate and grow austenite. (a) A comparison of an as-cast microstructure of bainite versus tempered bainite. (b) A comparison of quenched martensite versus tempered martensite.

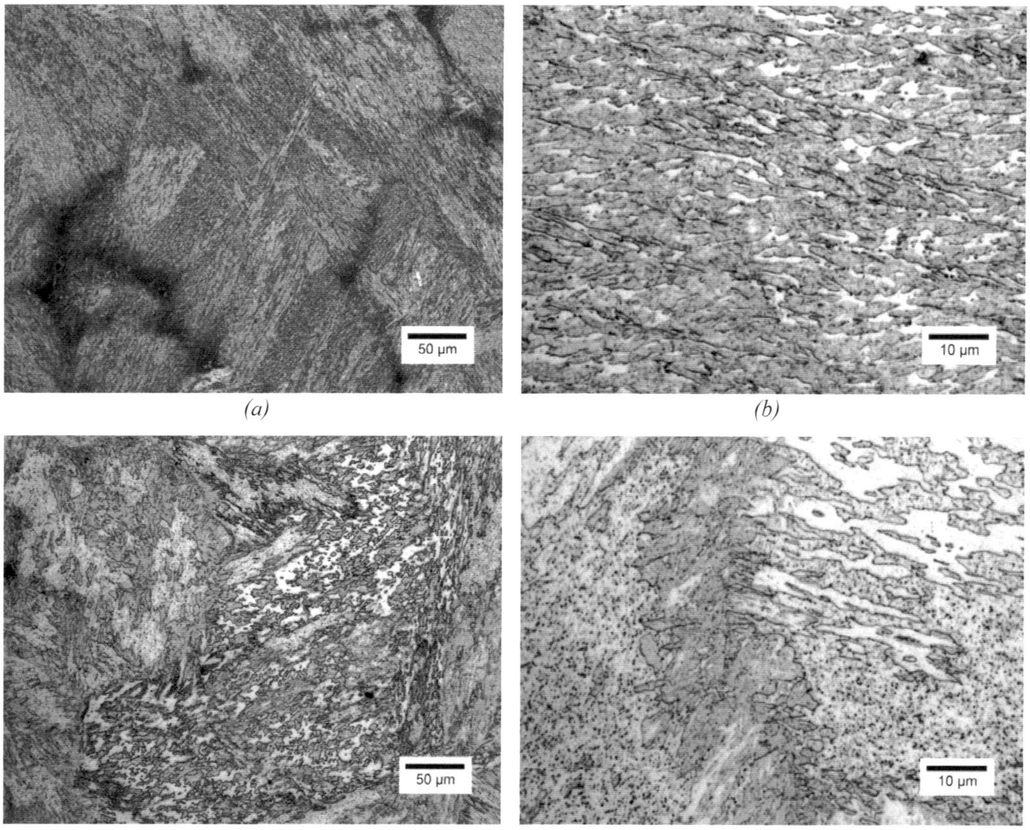

(a) (b)

(c) (d)

Figure 10. Austenite formation at 800°C (1472°F) in the as-cast microstructure of bainite is shown in (a) and (b) and for the tempered cast material in (c) and (d). Each was held for 10 minutes at 800°C (1472°F) and quenched into water. Nearly complete transformation to austenite was observed in the as-cast bainitic microstructure. Figure (b) shows the austenite formed in long laths with remnant ferrite in between these laths. Persistent carbides are discernible in the ferrite. The ferrite appears as dark etching features in (a). In contrast, substantial carbide precipitation has occurred in the tempered specimen and the regions of transformed austenite appear more irregular as shown in the central area of figure (d). Etched with 2% nital.

Discussion

Any discussion of phase transformation in steel must start with an understanding of the Kurdjumov-Sachs orientation relationship, see eq. (2). Manifestation of this crystallographic relationship are often displayed in both bainitic and martensitic microstructures. This relationship will produce distinct spatial relationships between lath packet variants. For example, the image of the as-cast W-steel shown in Fig. 4 shows three ferrite lath packets forming equilateral triangles within a single prior austenite grain. These shapes result from the basic crystallography of a cubic parent phase where there are 4 independent {111} planes. Thus, a maximum of 4 lath packets will be observed in any single prior austenite grain for either bainite or lath martensite. Each lath packet (bainite or martensite) has a common {111}, but there can be up to six crystallographic variants with two different <111> of the ferrite aligned with three different austenite <110>. This difference is best demonstrated using the OIM-EBSD shown for bainite in Fig. 7(a). Parallel ferrite laths with different colors within a packet represent different variants of the Kurdjumov-Sachs orientation relationship. Upon transformation back to austenite these different variants are capable of transforming to different austenite grains producing grain multiplication or by transforming to the same austenite grain with no contribution to the multiplication factor.

Large volumes of remnant austenite in the bainitic structure, as shown in Fig. 7(c), suggest another explanation for low grain multiplication factors when these microstructures are heated. Upon heating there will be no incubation time to nucleate austenite, since the austenite is already present and this may explain the rapid hardening shown in Fig. 9(a) for the as-cast microstructure of the W-steel. Growth of this remnant austenite upon heating may simply grow to consume the ferrite laths and reform the original parent austenite grain. This mechanism of preexisting austenite growth and impingement can explain the low grain multiplier of 2.1 calculated for the method 2 heat treatment shown in Table IV and may also explain the observations of Grange [12] for carburized steels. It needs to be emphasized that the remnant austenite within these bainitic microstructures is exactly the same in each of the different lath packets as shown in Fig. 7(c) and 7(e).

Time dependent nucleation and growth of austenite is observed when the starting microstructure is martensitic or bainitic after tempering prior to heating to austenitize the steel as shown in Fig. 9. For the tempered bainite, the grain multiplier, n_1, changed from 2.1 to 671 for the first austenitization cycle of the method 4 heat treatment. The second heating cycle of the method 4 heat treatment appears to be much less effective (n_2= 17.5) and to explain these differences in n_1 we must consider how these bainitic microstructures develop during cooling.

In Cr-Mo steels there is a phenomenon of incomplete transformation [14], which is responsible for the development of an austenite bay in the time-temperature transformation diagrams of these steels. Metallurgists erroneously divide the transformation of the steel into a pearlitic transformation above the bay and a bainitic transformation below the bay. [15] In fact, the development of the austenite bay results from a phenomenon of solute drag at the ferrite and austenite interface. [14] Chromium, molybdenum and to a lesser extent manganese accumulate at the ferrite growth front and slow the transformation. Carbon is partitioned to the austenite during the transformation and growth of the ferrite eventually comes to a halt producing a condition of incomplete transformation or "growth stasis." Precipitation of carbide may break the stasis and allow further transformation of the austenite. The rate of transformation will be dependent upon the diffusivity and diffusion distance of the alloy to grow the M_xC_y carbide. Tempering may serve the purpose of carbide precipitation and depletion of carbon from the austenite. This may suggest that the scale of the starting microstructure might influence the effectiveness of tempering in removing the remnant austenite. In general, the dimension of the bainitic ferrite lath should be proportional to the prior austenite grain size and cooling rate; and as a result, the dimension of the remnant austenite should also vary. The consequence of austenite scale may manifest itself as a time dependence in complete removal of the remnant austenite during tempering, i.e. coarse bainitic microstructures as observed in cast microstructures may require longer subcritical annealing cycles. A bainitic microstructure is also expected in the normalized microstructures for steels with high alloy content and the scale of the remnant austenite will be dependent upon casting size or cooling rate and prior austenite grain size. A fast cooling rate will decrease the size of the remnant austenite and may allow for shorter times during the subcritical anneal. Also, a slow heating rate upon the second austenitization cycle may be sufficient to remove the remnant austenite and eliminate the need for tempering. Induction heating might be expected to enhance the retention of austenite and further reduce the grain refinement. The fine scale of the remnant austenite combined with a slow heating rate may explain the larger grain multiplication factor (n_2) of 311 in method 1 heat treatment. Retained austenite in martensitic microstructures exists as sub-micron layers between the laths of martensite. The thin retained austenite in martensite may thus be consumed during heating or during tempering and this may explain the high grain multiplication factors, n_2, for the method 1 and method 3 heat treatments.

The order of magnitude increase in the multiplication factor with the addition of tempering in the martensitic microstructure is both difficult to ignore and explain. In the previous paper by Webb and Van Aken [10] it was suggested that recrystallization of the martensitic microstructure during tempering may enhance the nucleation of additional grains, but a significant difference in how austenite nucleates and grows in the two martensitic microstructures is not obvious. There is some evidence that the forming austenite is more globular in the tempered microstructure. This may indicate a deviation from the Kurdjumov-Sachs orientation relationship and the early development of more equiax austenite grains. However, the majority of the new austenite forms as laths with remnant ferrite in between indicating that the Kurjumov-Sachs orientation relationship constrains the growth of the new austenite. Further work will be required to fully understand these details; however, the results presented here can be used to develop robust heat treatments to develop improved grain refinement for cast steel, e.g. method 4.

Conclusions

The inability to produce grain refinement during normalization in high alloy steels is associated with a bainitic microstructure containing remnant austenite. Upon heating this remnant austenite grows to reform the original parent austenite grain and little to no grain refinement occurs during the normalization heat treatment. Tempering can be used to remove the remnant austenite and enhance grain refinement during subsequent austenitization. The addition of tempering prior to austenitization for the purpose of grain refinement should find good application to higher alloy steels such as 4320 and 4340 to improve mechanical properties in quenched and tempered castings.

Acknowledgements

This work has been supported under the Steel Founders' Society of America (SFSA) program on Casting and Advanced Steel Technology that has been funded by the Army Research Laboratory under the provisions of cooperative agreement W911NF-12-2-0033. The authors are grateful for steel castings supplied by Bradken-Atlas LP, and Bradken-Tacoma. The authors wish to thank Mr. Don Denney at American Cast Iron Pipe Co. (ACIPCo) for optical emission spectroscopy of the W-steels. Richard Harris and Rachel Abrahams, Air Force Research Laboratory at Eglin AFB, are gratefully acknowledged in providing support for conducting Electron Backscattered Electron Diffraction Orientation Image Mapping. The authors also acknowledge the support of the Materials Research Center and in particular Dr. Eric Bohannan for guidance and training in using the FIB and performing the XRD work.

References

[1] T.E. Swarr and G. Krauss, "Boundaries and the Strength of Low Carbon Ferrous Martensites," in Grain Boundaries in Engineering Materials, eds. J.L. Walter, J.H. Westbrook, and D.A. Woodford, Claitors Pub. Div. , Baton Rouge, LA, (1975) pp. 127-138.

[2] R.A. Grange, "Strengthening Steel by Austenite Grain Refinement," Trans. Am. Soc. Met. **59** (1966) p. 26-48.

[3] S. Morito and T. Ohba: "Crystallographic analysis of characteristic sizes of lath martensite morphology", in Fundamentals of Martensite and Bainite toward Future Steels with High Performance T. Furuhara and K. Tsuzaki, eds., ISIJ (2007), p. 57-63.

[4] S. Morito, H. Tanaka, R. Konishi, T. Furuhara, and T. Maki, "The morphology and crystallography of lath martensite in Fe-C alloys," Acta Materialia vol. 51 (2003) pp. 1789–1799.

[5] L.F. Porter, S.J. Mananello, D.S. Dabkowski, and J. H. Gross, "Ultraservice steels with yield strengths of 130 to 200 ksi," Metals Engineering Quarterly, American Society of Metals August (1966) pp.17-32.

[6] H. Hu, G. R. Speich, and R.L. Miller, "Crystallographic Texture, Retained Austenite, and Austenite Grain Size on Mechanical and Ballistic Properties of Steel Armor Plates," DTIC AMMRC CTR 74-39 (1976).

[7] ASTM E112-13, Standard Test Methods for Determining Average Grain Size.

[8] W.F. Hosford, Physical Metallurgy, CRC Taylor & Francis, NY (2005) pp. 476-477.

[9] M.A. Grossman and E.C. Bain, Principles of Heat Treatment, 5th ed. ASM, Metals Park, OH (1971) pp. 38-39.

[10] T.O. Webb and D.C. Van Aken, "An Innovative Heat Treatment to Improve the Ductility of High Strength Steels," Steel Founders' of America Technical and Operating Conference paper 2.1, Chicago, IL, December 13, 2014.

[11] D.C. Van Aken, T.O. Webb and J. D. Green, "Grain refinement strategies in high strength cast steel," Steel Founders' of America Technical and Operating Conference paper 5.6, Chicago, IL, December 12, 2015.

[12] R.A. Grange, "Method of Hardening Case -carburized Steels," SAE Trans., **82** (1973) p. 556.

[13] S. Martin, C. Ullrich, D. Simek, U. Martin, and D. Rafaja, "Stacking fault model of e-martensite and its DIFFaX implementation," *J. Appl. Crystallogr.,* vol. 44, (2011) pp. 779-787.

[14] H.I. Aaronson, G. Spanos, and W.T. Reynolds, "A progress report on the definitions of bainite," Scripta Mater. Vol. 47 (2002) pp. 139-144.

[15] H.I. Aaronson, W.T. Reynolds, and G.R. Purdy, "The Incomplete Transformation Phenonmenon in Steel," Metall. Mater. Trans. A, vol. 37A, (2006) pp. 1731-1745.

International Federation of Heat Treatment and Surface Engineering 2016
Proceedings of the 23rd IFHTSE Congress
April 18–21, 2016, Savannah, Georgia, USA

Modeling the mechanical properties of quenched and tempered martensitic steels

Xiaoqing Cai, Yuan Lu, Haixuan Yu, R. D. Sisson Jr.
Worcester Polytechnic Institute, Worcester, MA, USA

Abstract

This study focuses on a physical model to predict the mechanical properties more accurately during the tempering process. It is proposed to determine the mechanical properties in quenched and tempered martensitic steels. The model determines the hardness of the as-quenched martensite as a function of carbon concentration. The effects of tempering are determined by the reduction of carbon with martensite as carbides nucleation and growth. This model work is mostly based on Hollomon-Jaffe equation.

Introduction

AISI 4140 steel is a medium-carbon alloy steels that contains Mn, Cr and Mo. The high hardenability AISI 4140 alloy steel is a widely used alloy steel in engineering applications. To enhance mechanical properties, this steel is quenched and tempered to gain higher toughness and strength before industrial applications [1-2].

Tempering is conducted after quenching to balance hardness and toughness. It has a wide impact on mechanical properties include reducing brittleness and increasing toughness, and microstructure transformation from lath martensite to ferrite and carbides. Improved toughness of high strength lath martensitic structures is a major concern in industry applications. So quench and temper are the major heat treatment to achieve a high combination of toughness and strength [3-6]. Although the transformation process is complicated, Hollomon and Jaffe developed an empirical tempering parameter to predict approximate hardness of quenched and tempered steels [8]. The parameter is

$$H_{HJ}=T(c+\log t) \qquad \text{Equation 1}$$

Where T is temperature in degrees Kelvin, t is time in seconds, and c is a constant that depends on the carbon content of steels [10]. The physical explanation of c was not clear. H is proportional to steel hardness when the tempering time is longer than 15min [8].

Re-Calculate Hollomon and Jaffe Parameters

In the original Hollomon and Jaffe's paper [4], the tests for steel with 0.75, 0.89 and 0.96wt% carbon were conducted. Figure 1 presents the original data from Hollomon's paper. In this paper, the regression equation between hardness and Hollomon-Jaffe parameter is calculated again.

TABLE 6.—*Steel* T (0.75 *Per Cent* C)

Brine quenched after 1 hour at 760°C.; cooled to −70°C. before tempering. As-quenched hardness 66.5 Rc

	Rockwell C Hardness Obtained by Tempering at Indicated Time and Temperature								
Nominal Time of Tempering	100°C.	150°C.	200°C.	250°C.	300°C.	400°C.	500°C.	600°C.	700°C.
10 sec	67.0	67.5	67.5	67.0	65.5	61.5	51.5	42.5	36.5
90 sec	66.5	67.5	65.0	61.5	58.5	51.5	44.5	38.5	29.0
900 sec. (15 min.)	67.5	66.5	62.5	59.5	56.5	49.5	42.5	34.5	22.5
9,000 sec. (2½ hr.)	66.5	64.5	59.5	58.0	53.5	47.5	38.5	28.5	16.5
86,400 sec. (24 hr.)	65.5	61.5	58.5	55.0	51.5	44.5	34.5	21.5	7.5

TABLE 5.—*Steel* W (0.89 *Per Cent* C)

Brine quenched after 1 hour at 840°C.; cooled to −70°C. before tempering. As-quenched hardness 67.0 Rc

	Rockwell C Hardness Obtained by Tempering at Indicated Time and Temperature								
Nominal Time of Tempering	100°C.	150°C.	200°C.	250°C.	300°C.	400°C.	500°C.	600°C.	700°C.
10 sec	67.0	67.5	67.5	67.5	64.5	58.5	48.5	43.0	37.5
90 sec	67.5	67.5	65.5	62.5	59.5	52.5	45.5	38.0	31.5
900 sec. (15 min.)	68.0	66.5	64.5	60.5	57.5	50.5	42.5	34.0	25.5
9,000 sec. (2½ hr.)	68.5	63.5	61.5	58.5	54.0	46.5	38.0	28.5	18.0
86,400 sec. (24 hr.)	67.5	62.5	59.5	55.5	52.5	44.5	33.0	22.5	12.0

TABLE 3.—*Steel* R (0.96 *Per Cent* C)

Brine quenched after 1 hour at 870°C.; cooled to −70°C. before tempering. As-quenched hardness, 66.5 Rc

	Rockwell C Hardness Obtained by Tempering at Indicated Time and Temperature								
Nominal Time of Tempering	100°C.	150°C.	200°C.	250°C.	300°C.	406°C.	500°C.	602°C.	700°C.
10 sec	66.5	67.0	68.0	66.5	65.0	60.5	52.5	43.5	35.5
90 sec	65.5	66.5	66.5	62.5	61.0	52.5	44.5	35.5	27.5
900 sec. (15 min.)	67.0	65.5	63.5	60.0	56.5	49.5	41.5	28.5	22.5
9,000 sec. (2½ hr.)	67.5	64.5	62.5	57.5	55.5	45.5	35.5	23.5	14.5
86,400 sec. (24 hr.)	66.5	63.5	60.5	56.0	53.5	43.5	30.5	4.5	4.5

Figure 1 The original data from Hollomon's paper [4]

The test results are shown in Table 1. HJ stands for the Hollomon-Jaffe parameter.

Table 1 Regression equation for Hollomon-Jaffe parameter

Carbon content wt%	HJ parameter	R^2	Regression equation
0.75	12	0.98	Hardness=98.6-0.005*HJ
0.89	12	0.98	Hardness=98.7-0.005*HJ
0.96	12	0.96	Hardness=99.8-0.005*HJ

Discussion on Hollomon and Jaffe Parameters

Figure 2 presents the typical relation among hardness, tempering temperature and time. Two stages can be designated to the tempering process. Stage I is from the beginning of tempering experiments to 60s. At stage I, the hardness decreases quickly. Stage II is from 60s to the end of tempering process. In stage II, the carbides are growing and clustering. The rate of hardness decreasing becomes constant, apparently smaller than the rate in stage I. For stage I, the hardness reduction at higher tempering temperature is larger compared to at lower tempering temperature. For stage II, the slightly larger rate of hardness decreasing at higher tempering temperature is seen compared to at lower tempering temperature.

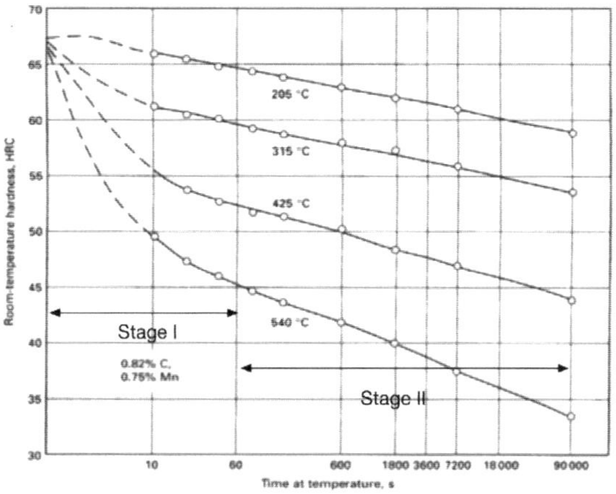

Figure 2 Effect of time at four tempering temperatures on room-temperature hardness of quenched 0.82% C steel. Note nearly straight lines on logarithmic time scale [7].

In order to describe the hardness after 60s under different tempering temperature, Equation 1 can be assumed as:

$$Hardness = H_{aq} - H_{stageI} - H_{stageII} \qquad \text{Equation 2}$$

H_{aq} is the as-quenched hardness. H_{stageI} is the hardness reduction in stage Ⅰ. $H_{stageII}$ is the hardness reduction is stage Ⅱ. Based on the previous discussion, hardness reduction in stage I seems to be only dominated by temperature. The martensite will not decompose under specific temperature T_1 [10,8]. Under T_1, the carbon in martensite will cluster and increase the stress in martensite matrix, which causes the hardness to increase a little [10]. Then the H_{stageI} can be written as:

$$H_{stageI} = x_1(T - T_1) \qquad \text{Equation 3}$$

x_1 is the coefficient for stage Ⅰ. Hardness reduction in stage Ⅱ is dominated by both temperature and time. At very low temperature T_2, the carbides growing rate can be seen as 0. For steel with 0.31 wt% carbon, after 86400s at 373K tempering temperature, the hardness still keeps the same [8]. The $H_{stageII}$ can be written as:

$$H_{stageII} = x_2(T - T_2)(logt - log60) \qquad \text{Equation 4}$$

x_2 is the coefficient for Stage Ⅱ. Substituting equation3 and equation4 to equation2, hardness after tempering can be described as:

$$Hardness = H_{aq} - x_1(T - T_1) - x_2(T - T_2)(logt - log60) \qquad \text{Equation 5}$$

Equation5 can be rewritten as:

$$Hardness = H_{aq} + x_1T_1 + x_2T_2(logt - log60)x_2Tlog60 - x_2T\left(\frac{x_1}{x_2} + logt\right) \qquad \text{Equation 6}$$

Letting $c = \frac{x_1}{x_2}$, equation6 can be written as:

$$Hardness = H_{aq} + x_1T_1 + x_2T_2(logt - log60)x_2Tlog60 - x_2T(c + logt) \qquad \text{Equation 7}$$

Considering the typical tempering process, the range of tempering time is from 3600s to 86400s and the range of tempering temperature is from 373K to 873K. In order to simply equation7, the tempering time is chosen as 10800s and tempering temperature as 623K. T_1 and T_2 assume as 373K. Equation7 can be rewritten as:

$$Hardness = H_{aq} + 373x_1 + 1949x_2 - x_2T(c + logt) \qquad \text{Equation 8}$$

After comparing equation 8 with regression equations from Table 1,

$$x_2 = 0.005$$
$$x_1 = cx_2 = 0.005 \times 12 = 0.06$$

Substituting x_1 and x_2 into equation8:

$$Hardness = H_{aq} + 32 - 0.05HJ \qquad \text{Equation 9}$$

The as-quenched hardnes for 0.75, 0.89 and 0.96wt% carbon can be seen as 67HRC, substituting H_{aq} into equation9:

$$Hardness = 99 - 0.05HJ \qquad \text{Equation 10}$$

Equation 10 is the same as the calculated regression equation in Table 1, which demonstrates the assumption from equation 2 to equation 10 is reasonable.
Recalling $c = \frac{x_1}{x_2}$, x_1 is the coefficient for stage I and x_2 is the coefficient for stage II. The constant c strands for the hardness reduction mechanism in stage I and stage II. If c is the same for two different steels, it requires x_1 and x_2 are all the same. The physical meaning of x_1 and x_2 will be discussed below.

Experimental Procedures

Materials
The AISI4140 steel that was purchased from Peterson Steel Co. as 304.8mm long by 12.7mm diameter rods. The analysis of alloy compositions from the supplier and measured by OES at WPI are presented in Table 2.

Table 2 Analysis of chemical composition of AISI4140 rods in weight percent

	Peterson	WPI
C	0.41	0.397
Mn	0.86	0.85
S	0.024	0.019
Si	0.25	0.229
Ni	0.08	0.122
Cr	0.93	0.89
Cu	0.23	0.226
Mo	0.17	0.19
Al	0	0.028
Fe	-	-

Thermal process

Before the tempering experiments, all the samples were austenitized at 850°C held for one hour in Endothermic gas at 0.4% carbon potential, followed by quenching in oil at Bodycote Co. in Worcester. After quenching, the furnace tempering procedure was conducted at WPI and induction tempering procedure was conducted at Thermatool Co. in East Haven, Connecticut. Table 3 shows the matrix of tempering experiments.

Table 3 Tempering experiments matrix

	Furnace Tempering				
Tempered Temperature (°C)	250	350	450	550	650
Tempered Time (hr)	1	1	1	1	1
	3	3	3	3	3
	7	7	7	7	7
	10	10	10	10	10
	15	15	15	15	15

(a) Furnace tempering experiments matrix

	Induction Tempering				
Tempered Temperature (°C)	250	350	450	550	650
Tempered Time (min)	1	1	1	1	1
	3	3	3	3	3
	5	5	5	5	5
	10	10	10	10	10
	15	15	15	15	15

(b) Induction tempering experiments matrix

Test Methods

The selected samples were mechanically polished and etched with 2% nital solution, and the microstructures were observed by optical microscopy and scanning electron microscopy (SEM).

The Rockwell C hardness was measured on the surface of each rod and the reported average hardness was from measurements over 40 locations on each rod.

The X-Ray diffraction pattern was measured by Panalytical Empyrean X-ray diffractometers. The tube was Cr.

Results

Hardness

The average of surface hardness of tempered AISI4140 was presented in Figure 3. The data shows that the surface hardness of both furnace and induction tempered AISI4140 decreases with the increasing tempering temperature and time. It can be also seen that the surface hardness is more sensitive to tempering temperature than time for both furnace and induction tempering. A comparison of the hardness results of furnace and induction tempering process, shows that at the same tempering temperature, the hardness of furnace tempering is lower than induction tempering due to the longer times. The hardness of as-quenched sample is the highest at 53HRc.

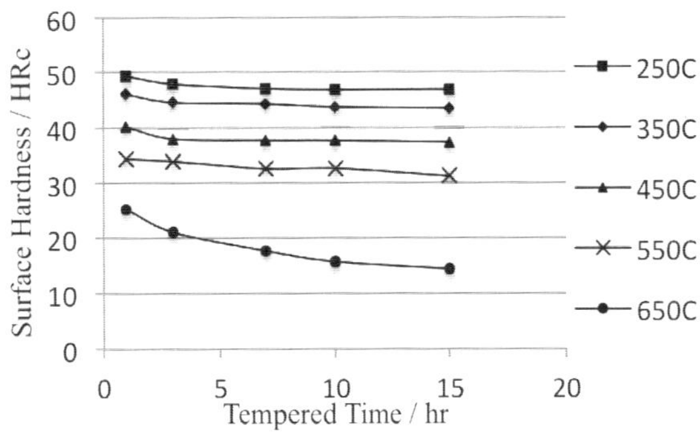

(a) Furnace tempering

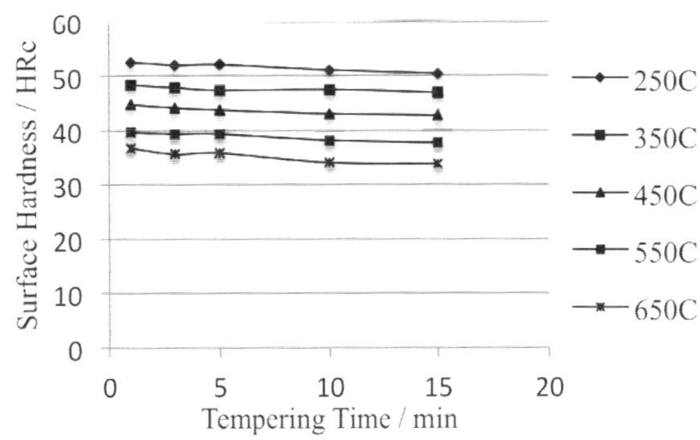

(b) Induction tempering

Fig. 3 Variation of surface hardness with tempering temperature and time

Microstructure

The microstructure of tempered AISI 4140 tempered at 250 ℃ - 650 ℃ is shown in Figure 4. These scanning electron micrographs reveal that AISI 4140 tempered at 250 ℃ mainly consists of lath martensite and some carbides which maybe epsilon carbide ($Fe_{2.4}C$). With the increasing tempering temperature, lath martensite transform to ferrite and carbides.

It can be observed that the sample tempered at 450 ℃ consists lath martensite and ferrite and rod-shape carbide. When the tempering temperature increased to 650 ℃, some spheroidal shape carbide precipitate. These spheroidal shape cabides are cemente[3].

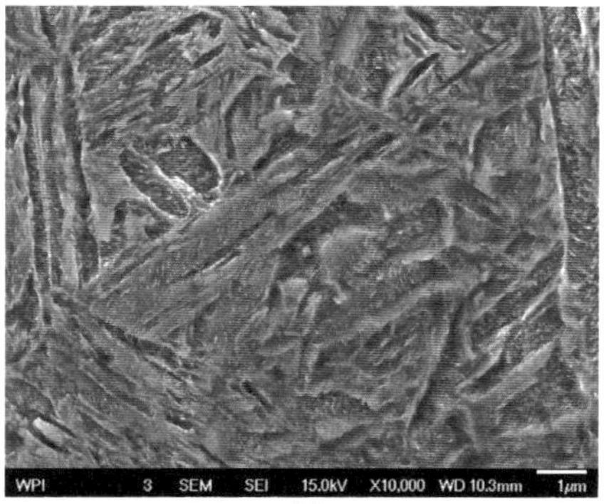

(a)

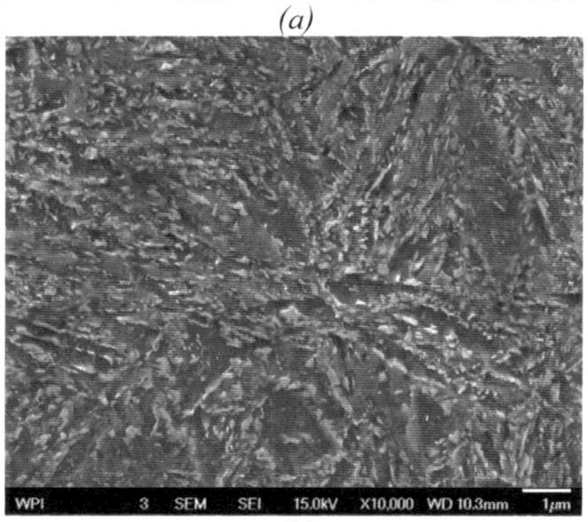

(b)

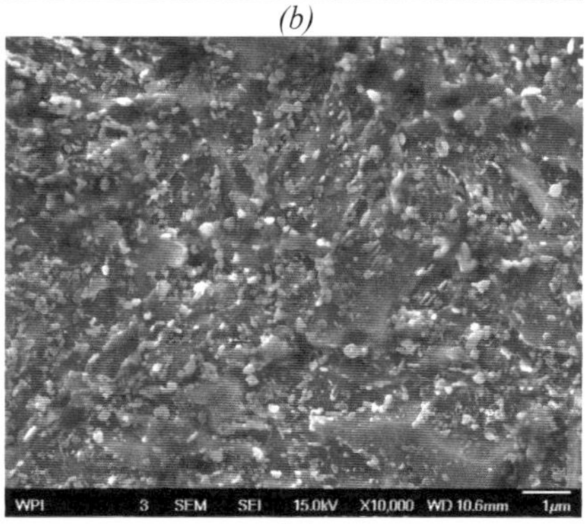

(c)

Fig. 4 SEM micrographs of tempered samples:
(a) Tempered at 250°C and held for 1hr
(b) Tempered at 450°C and held for 1hr
(c) *Tempered at 650°C and held for 1hr*

X-Ray Diffraction

Figure 5, Figure 6 and Figure 7 presents the X-ray diffraction pattern for AISI 4140 steels..

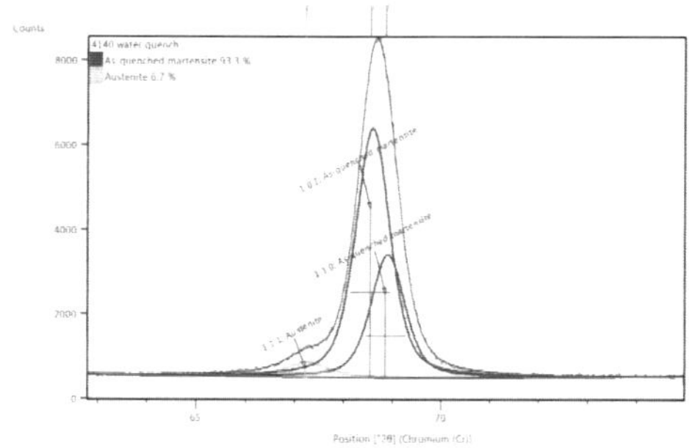

Figure 5 AISI 4140 oil quench X-ray diffraction pattern

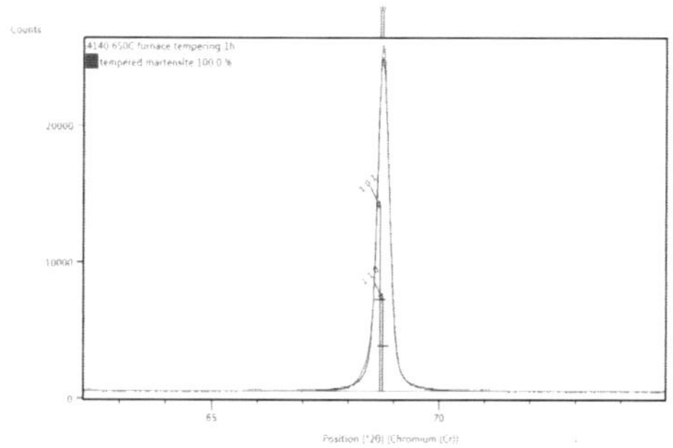

Figure 6 AISI 4140 650 ℃ tempering 1h X-ray diffraction pattern.

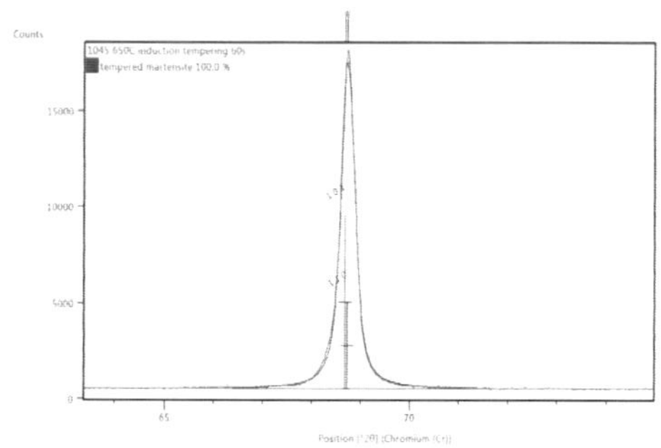

Figure 7 AISI 4140 650 ℃ induction tempering 60s X-ray diffraction pattern

Discussion

Mechanical properties

It can be seen from Figs. 3, the tempering temperature has a significant impact on surface hardness. Compared with the tempered samples, the as quenched sample has the highest hardness. That can be explained with the phase transformation, during the quenching process. Austenite transforms to martensite and the crystal structure transforms from face center cubic to body center tetragonal. It forms high dislocation density in the martensite that increases the hardness. During the tempering process, with increasing the tempering temperature lath martensite transform to ferrite and carbides. Carbon content of martensite decreases as carbon diffuses into carbides. At the same time the internal stress will be relieved as the atoms rearrange and dislocation density is reduced. These factors cause the reduced hardness and strength and increased toughness [3].

Based on the previous discussion, the hardness reduction from oil quench (53HRC) to 650℃ induction tempering for 1min (36.7HRC) can be designated as stage I. the hardness reduction from 650℃ induction tempering for 1min to 650℃ for 1h (25.3) can be designated as stage II.

Two reason may exist for hardness reduction in tempering process: (1) the carbon diffuses from martensite structure and form martensite with lower carbon or ferrite; (2) the carbides clusters with the increasing of tempering time. The assumption is that the main contribution of hardness reduction at stage I is that carbon diffuses from the martensite structure. Because the tempering time is only 60s at stage I, the carbides clustering, probably taking longer time, does not have significant effect on hardness reduction.

X-Ray diffraction pattern

Based on the figure 5 to figure 7, the Rietveld refinement method is used to get accurate lattice parameter. Under each X-ray diffraction pattern, it is the difference between experimental result and simulation result. The Rwp (weighted R profile) is less than 6% for all these three patterns.

Figure 8 presents the lattice parameters of AISI 4140 steel under different heat treatment conditions. The c/a ratio quickly changes from 1.008 to 1.002 within 60s tempering. It should be noted that the induction heating time to desired tempering temperature is less than 30s. The significant c/a ratio change demonstrates the assumption, which carbon diffused from the martensite structure.

If assuming after tempering 60s, the structure of tempered martensite is still body center tetragonal and carbides does not have enough time to transform and cluster, the weitht percent carbon in tempered martensite can be calculated.

$$a = 2.876 - 0.013x \qquad \text{Equation 11}$$
$$c = 2.867 + 0.116x \qquad \text{Equation 12}$$

Equation 11 and equation 12 are the equations to reveal the relation between lattice parameter and carbon content for as-quenched martensite [11]. x stands for the weight percent carbon. Although these equations were generated based on as-quenched martensite, it should also be accurate enough to predict the carbon content in tempered martensite for a very short tempering time condition, such as 60s. When the c/a ratio is 1.003, the weight percent carbon in tempered martensite was calculated as 0.07%. Considering the original weight percent carbon in as-quenched martensite is 0.4%, 82.5% carbon atoms in as-quenched martensite diffuses out of the structure cell and leaves only 0.07% carbon in tempered martensite structure, a body center tetragonal structure with 1.003 c/a ratio.

From the X-ray diffraction, hardness analysis and optical microstructure, the reason for hardness reduction in stage I is that carbon diffused from martensite structure and form martensite with lower carbon. For stage II, the hardness reduction is mainly due to carbides clustering. For this precipitate weaken mechanism, the decrease of dislocation may have direct relationship with hardness reduction.

It was found that c=16 is suitable for steels containing 0.15-0.45% carbon.

Hardness Prediction

According to Hollomon-Jaffe equation [8], it is known that hardness is affected by tempering temperature and time. The constant C in the Hollomon-Jaffe equation is determined by statistical analysis of the fitted data. The constant C shown was selected based on the r^2 value. The experimental hardness data presented in Figure 3 are plotted with the Hollomon-Jaffe relationship. The results are shown in Figure 8. C equals to 13 for furnace tempered AISI4140 and C equals to 15 for induction tempered AISI4140. The hardness of AISI4140 can be predicted with Hollomon-Jaffe equation for different tempering processes.

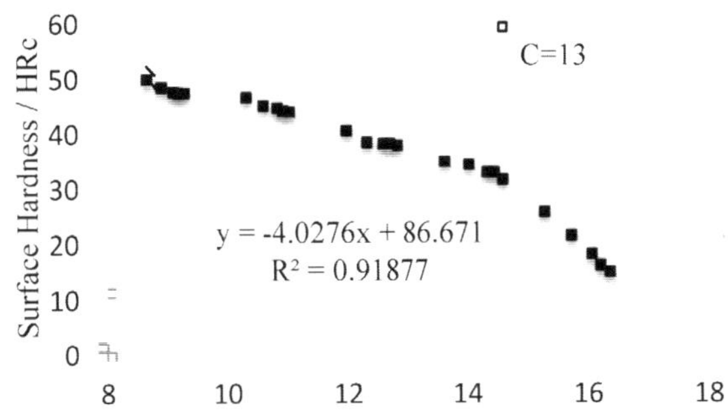

(a) Furnace tempering

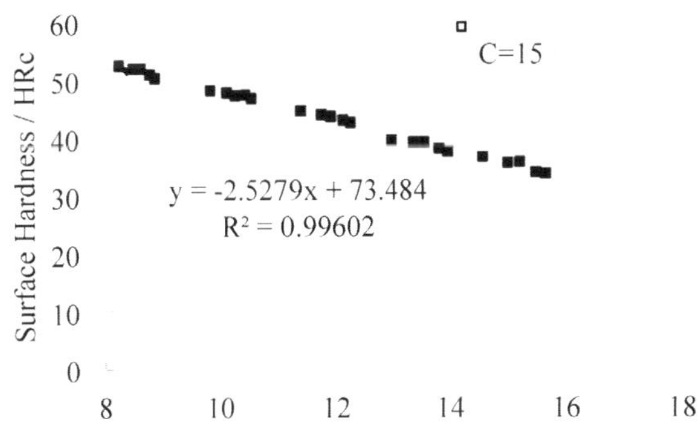

(b) Induction tempering

Fig. 8 The relationship between surface hardness and Hollomon-Jaffe parameter

Conclusions

1) Lath martensite transforms to ferrite and carbides during tempering. During this process, the shape and size of carbides also change.
2) Surface hardness decrease with the increasing tempering temperature and time. This phenomenon is well modeled with Hollomon-Jaffe analysis.

3) Comparing surface hardness of induction and furnace tempering procedures, it is found that samples can reach the same hardness with high tempering temperature short tempering time and low tempering temperature long tempering time.
4) For AISI 4140 steels quenching and tempering tests, the significant c/a ratio decreasing was found in stage I, which demonstrates that carbon diffuses from martensite structure is the main reason for hardness reduction in stage I.
5) $c = \frac{x_1}{x_2}$, x_1 is the coefficient for stage I and x_2 is the coefficient for stage II. The constant c stands for the hardness reduction mechanism in stage I and stage II.
6) It was found that c=16 is suitable for steels containing 0.15-0.45% carbon. However, based on the experimental data, c=13 for furnace tempered AISI 4140 and c=15 for induction tempered AISI 4140. This should be figured out in the future work.

Acknowledgments

The support of the Center for Heat Treating Excellence (CHTE) at Worcester Polytechnic Institute and the member companies is acknowledged. Special thanks to Dr. Lesley D. Frame of Thermatool CO. in Connecticut for their support.

References

[1] ASM Metals Handbook. Heat Treating, Metals Park, vol. 4. Ohio: American Society for Metals; 1991.

[2] Chandler H. Heat treating guides. Metals Park, Ohio: American Society for Metals; 1996

[3] George Krauss, Steel: heat treatment and processing principles (2nd Edition), ASM International.

[4] D. H. Huang, G. "Thomas, Structure and mechanical properties of tempered martensite and lower bainite in Fe-Ni-Mn-C steels", Metall. Trans. 2A (1971) 1587-1596.

[5] Woei-Shyan Lee, Tzay-Tian Su, "Mechanical properties and microstructural features of AISI 4340 high-strength alloy steel under quenched and tempered conditions", Journal of Materials Processing Technology 87 (1999) 198-206.

[6] M. H. Khani sanij, S.S. Ghasemi Banadkouki, A.R. Mashreghi, M. Moshrefifar, "The effect of single and double quenching and tempering heat treatments on the microstructure and mechanical properties of AISI 4140 steel", Materials and Design 42 (2012) 339-346.

[7] E. C. Bain and H.W. Paxton, *Alloying Elements in Steel*, American Society for Metals, 1966, p185, 197

[8] J.H. Hollomon, L.D. Jaffe, "Time-temperature Relations in Tempering Steel," Metals Technology, Vol 162, 1945, p223-249.

[9] D.K. Bullens, "Steel and its heat treatments", Wiley, New York, 1948.

[10] ASM Handbook Vol 4 Heat Treating, ASM International, ISBN 0-87170-379-3, p122

[11] B. D. Cullity, *Elements of X-ray Diffraction*, Addison-Wesley Publishing Company, inc.

International Federation of Heat Treatment and Surface Engineering 2016
Proceedings of the 23rd IFHTSE Congress
April 18–21, 2016, Savannah, Georgia, USA

Thermodynamic and Kinetic Simulation and Experimental Results Homogenizing Advanced Alloys

Paul D. Jablonski, Joseph J. Licavoli and Jeffrey A. Hawk,
National Energy Technology Laboratory, United States Department of Energy,
Albany, OR USA

Abstract

Segregation of solute elements occurs in nearly all metal alloys during solidification. The resultant elemental partitioning can severely degrade as-cast material properties and lead to difficulties during post processing (e.g. hot shorts, incipient melting, etc). Many cast articles are subjected to a homogenization heat treatment in order to minimize segregation and improve their performance. Traditionally, homogenization heat treatments are based upon past practice or time consuming trial and error experiments. Through the use of thermodynamic and kinetic modeling software, NETL has designed a systematic method to optimize homogenization heat treatments. Use of the method allows engineers and researchers to homogenize casting chemistries to levels appropriate for a given application. The method also allows for the adjustment of heat treatment schedules to fit limitations on in-house equipment (capability, reliability, etc.) while maintaining clear numeric targets for segregation reduction. In this approach, the Scheil module within Thermo-Calc is used to predict the as-cast segregation present within an alloy, and then DICTRA (Diffusion Controlled TRAnsformations) is used to model homogenization kinetics as a function of time and temperature. Examples of computationally designed heat treatments and verification of their effects on segregation and properties of real castings are presented.

Introduction

Solidification is a process whereby liquid metal freezes; however, during solidification some solute elements are rejected into the remaining liquid while other elements preferentially diffuse to the solid [1]. This causes chemical heterogeneity in the solidified microstructure. Solute segregation is governed by the solidifying phase (or phases) which may not necessarily be the phases present at room temperature or in service. With sufficient segregation an alloy may also form non-equilibrium matrix phases such as austenite in ferritic/martensitic steels or delta-ferrite in austenitic steels. The formation of non-equilibrium solid phases further complicates the overall microstructural chemistry since elements partition differently based on the phase(s) that are solidifying.

Ferritic/martensitic steel P91 and austenitic 316 stainless steel are good examples of the intricacies of solute segregation. Under equilibrium conditions, Alloy P91 solidifies in the presence of a ferritic (BCC) matrix while 316 solidifies with an austenitic (FCC) matrix (under equilibrium solidification conditions within certain chemistry restrictions). In reality, while P91 may begin solidifying with primary ferrite eventually solute segregation stabilizes the austenite. As a result, continued solute segregation is governed by austenite growth instead of ferrite growth resulting in distinctly different segregation regions and a primary solid that is solute poor while the last to freeze solid is solute rich. The case of 316ss is more complicated. Within the alloy specification chemistry there are certain compositions that form primary delta ferrite while others form intermediate delta ferrite as a result of segregation. A single alloy chemical specification can thus have multiple non-equilibrium solidification pathways and segregation behavior. Left unaddressed, solute segregation can result in undesirable scatter in performance in such areas as oxidation resistance and long term mechanical performance (e.g., creep) resulting from various levels of chemical and microstructural heterogeneity.

In alloys designed for high temperature service microstructural effects are often the most important since long-term exposure can lead to undesirable phase formation much earlier than expected. After these phases form, concomitant breakdown of microstructural features responsible for strength, creep resistance, grain size control, and oxidation resistance may occur. For example, in Ni-based superalloys, the segregation of Cr and Mo can lead to the formation of embrittling TCP (topologically close packed) phases. In high Cr iron based alloys such as 304ss and 316ss brittle sigma phase may form due to segregation of Cr (either directly from the solidification process or as a result of exposure at temperature). Finally, Z-phase formation in creep resistant ferritic/martensitic steels significantly decreases creep performance. Z-phase can be stabilized by a number of elements such as Cr, Mo, Nb, V and N. Danielson [2] showed that increasing the Cr levels from 9 to12 weight percent (a compositional change commonly observed in a highly segregated structure) increased the Z-phase precipitation rate by a factor of 50 illustrating how sensitive these alloys are to small changes in composition. Furthermore, chemical inhomogeneity, can lead to localized variations in phase solvus and recrystallization temperatures which both produce undesired microstructural non-uniformities. These examples provide very clear evidence of the need to utilize a holistic approach to ensure pathways for alloy degradation are controlled.

To promote chemical homogeneity within the microstructure, and eliminate coring and other undesirable chemistry based solidification artifacts, heat resistant alloys must undergo a thorough homogenization treatment prior to any subsequent heat treatment activity, be it solutionizing for nickel superalloys, austenitizing/normalizing for steels, or aging heat treatments to facilitate strength improvements in other alloy systems. In many cases, the solution heat treatment can be skipped if the homogenization heat treatment temperature and cooling rates are adequate.

It is common practice to homogenize heat resistant castings, especially single crystal and directionally solidified nickel superalloys used in both power generation and aerospace turbines [3-5]. It is less common to homogenize wrought alloys because many believe the thermo-mechanical processing (TMP) involved sufficiently breaks down the cast microstructure. However, newer heat resistant alloys for fossil energy applications (e.g. ferritic/martensitic steel, and solid solution strengthened, low volume γ' fraction nickel superalloys) make significant use of refractory and transition elements to strengthen the matrix and develop complex carbides for grain size control and long-term microstructure stability. The refractory elements, by their nature, segregate radically during solidification by either partitioning to the liquid or diffusing very slowly in the solidified alloy. As a result, solidified structures are characterized by extremely non-uniform matrix composition profiles compared to other systems. TMP operations may do no more than move these regions relative to one another, or in some cases, cause even more severe segregation due to anisotropic deformation. Additionally, reheating and holding steps common in most TMP operations cannot be relied upon for homogenization because there is insufficient time for refractory elements to diffuse. Utilizing a one hour normalizing sequence after TMP for martensitic heat resistant steel, for example, does little to further disperse the refractory elements W, Mo or Co.

Until recently, the process of developing an alloy homogenization has been fairly stagnant. The steps involved typically include identifying the incipient melt temperature of an alloy and estimation of the time at temperature necessary to facilitate solid state diffusion in the alloy. Accurate estimates of incipient melt point will depend on sampling and measurement technique. Methods include differential scanning calorimetry (DSC) (fairly accurate temperatures but small sample sizes) and isothermal exposure (less accurate temperatures but much larger samples). Solid state diffusion time estimates are complicated by the fact that there is chemical diffusion involved rather than simple tracer diffusivity thus the *local* chemistry becomes an important factor which changes throughout the process. As discussed previously, there are many factors that must be considered when assessing local chemistry in cast structures, necessitating a more formal manner by which estimates may be made. Thermodynamic and kinetic simulation software, while not perfect, have evolved quite significantly over the past 10 years and can now facilitate time saving approaches to the developing homogenization heat treatments for heat resistant alloys.

Given the complex nature of heat resistant alloys, computationally guided experiments have become a necessity. Computational tools help alloy developers make reasonably accurate predictions of phase equilibria thermodynamics and diffusion driven kinetics. These predictions provide guidance for alloy processing (including heat treatment) without relying on significant trial and error. In the approach described herein commercially available thermodynamic and kinetic software has been utilized in a carefully designed algorithm to homogenize complex nickel-base superalloys, as well as austenitic, martensitic, and ferritic steels. Specifically, this approach has led to highly homogenized solid solution strengthened cast nickel superalloys (Haynes 230, Inconel 617, and Inconel 625), γ' strengthened nickel superalloys (Nimonic 105, Nimonic 263, Inconel 740, and Haynes 282), austenitic stainless steels (304 and 316), ferritic stainless steels (403, and 441) and ferritic/martensitic alloys (P91, P92, and TAF) and other experimental alloys. Given this success, the homogenization sequence is now being utilized for every heat resistant alloy produced at NETL with very good results.

The computational approach (which may be used to homogenize any alloy) will be discussed in broad terms with some specific examples provided. The examples include many of the previously mentioned alloy systems. Discussion as to effectiveness will be provided through microscopy of the microstructures and mechanical property results, primarily through measurement of the creep rupture life

The present homogenization algorithm utilizes simple microstructural measurements from the as-cast microstructure to scale the degree of segregation in the solidified ingot using the well-known Scheil-Gulliver approach. Overall ingot chemistry information serves as a primary input to DICTRA and ThermoCalc and is subsequently used along with microstructural measurements to predict the chemical profile evolution over time. The chemical profile information is then used to provide an estimate of the new incipient melt temperature after partial homogenization has taken place. This new incipient melt point can then be used to incrementally increase heat treatment temperatures to achieve high chemical uniformity over less time. This approach can also accelerate the development of homogenization heat treatments for newly developed heat resistant alloys since the only model inputs are easily obtained microstructural and chemical information.

Previous efforts had relied on the alloy solidifying from the melt as a single matrix phase [6-7]. However, not all alloys solidify in this manner. For example, many ferrous alloys begin to solidify with a BCC crystal structure and finish with an FCC structure. Other alloys start with an FCC structure and finish with a BCC structure. Still other alloys may form a significant amount of second phase such as carbide or some other ordered phase. In these instances, characterizing only the matrix does not adequately capture the overall nuances of the local chemistries found in these alloys. The present algorithmic computational approach calculates the *incremental* solid chemistry, i.e., the chemistry of each new addition to the solid from the overall amount of liquid and its chemistry, as the alloy progresses from the initial stages of solidification through the point when all liquid is converted to solid. In this way, each incremental solid is assembled together subsequently to form an overall chemistry profile that can then be used to calculate the interdiffusion required to bring about homogenization. The description of the solid chemistry flows from that of the liquid as follows:

Chemistry just before first solid forms: $f_s = 0$: C_0.

Chemistry when first solid forms:

$$C_{s_1} = \frac{(C_o - f_{l_1} C_{l_1})}{inc\ f_l} \quad (3)$$

Chemistry for each additional solid that forms is given by:

$$C_{s_{1+i}} = \frac{\left(C_o - f_{l_{1+i}} C_{l_{1+i}} - \Sigma f_{l_i} C_{l_i}\right)}{inc\ f_l} \quad (4)$$

where C_o is the overall alloy chemistry for each element; C_{s_1} is the chemistry of the first solid to form; $C_{s_{1+i}}$ is the chemistry of the s_{1+i} solid to form; f_{l_n} is the fraction of nth liquid to form; C_{l_n} is the composition of the liquid when the nth liquid forms; and $inc\ f_l$ is the incremental fraction of liquid that forms when the nth liquid forms. To homogenize an alloy to a high degree requires multiple steps based on redefining this "overall chemistry" for each homogenization step (i.e., specific temperature and time at a given temperature increment) until the required degree of homogenization is reached.

At this point, the Scheil Module (ThermoCalc [8]) is utilized to predict the extent of chemical microsegregation present as a result of alloy solidification. Whenever possible, the actual measured chemistry of the solidified cast alloy is used; however, in many cases the nominal or aim chemistry will suffice. Close attention is given to potent melt point suppression elements such as C, Si and Mn. Based upon the chemical segregation predicted, computational kinetic modeling (DICTRA [9]) is then used to design an optimized homogenization heat treatment that flattens the sinusoidal concentration profile of each alloying element within the dendrites of the matrix.

Subsequently, the Scheil Module and relevant database from the thermodynamic software (ThermoCalc) is used to predict the non-equilibrium solidification range of each alloy. Specifically, the Scheil Module in the commercial software package (i.e., ThermoCalc) is an implementation of the Scheil-Gulliver Model [10-11], and in this case assumes that diffusion occurs infinitely fast within the liquid phase and no diffusion occurs in the solid state. The analytical Scheil equation is commonly derived for an idealized system in which the liquidus and solidus are linear with respect to composition. When this is true, $C_s / C_l = k$, where C_s is the composition of the solid and C_l is the composition of the liquid at a given temperature. When an incremental amount of solid (df_s) forms, $C_s df_s$ solute transfers from the liquid to the solid. As a result, the incremental change in the liquid composition is given by:

$$dC_l = (C_l - C_s) df_s / (1 - f_s) \quad (5)$$

This equation can be integrated to describe the composition of the liquid as a function of fraction solid (f_s) formed:

$$C_l = C_o (1 - f_s)^{(k-1)} \quad (6)$$

In the more complicated (and more usual) case in which the liquidus and solidus curves are not linear, C_s and C_l will be described analytically with subsequent integration of Eq. (5). The Scheil Module within the commercial software package (ThermoCalc) calculates Δf_s, ΔC_s and ΔC_l in a user-defined increment of temperature until a predetermined amount of liquid remains. Fast diffusing elements such as B, C, N, and O can be allowed to back-diffuse within the solid, but this was not considered in the present case. The default results obtained from the Scheil simulation for a given alloy are the amount of solid phase formed versus solidification temperature. Not only does this show how the solidification sequence progresses, it also shows the phases that develop (many are non-equilibrium), as well as the solidification temperature range. However, this is not directly useful in defining the homogenization heat treatment sequence beyond establishing the initial incipient melt temperature. A homogenization heat treatment consists of certain times at certain temperatures to establish the level of chemical uniformity desired. After the Scheil simulation is run, the amount of each alloying element in a given phase present can be determined. In the case of the alloys investigated, the weight fraction of each of the alloying elements in the liquid phase is then calculated as a function of temperature. The liquid phase chemistry is used to calculate the overall solid phase chemistry as described by Eqs. (3) and (4). Thus, the chemical segregation that occurs in the solid across a secondary dendrite arm constitutes the undesired microsegregation that needs to be eliminated through an appropriate homogenization heat treatment. Standard metallographic techniques are used to measure the secondary dendrite arm spacing (*sdas*) of the castings. From these measurements, a value of approximately 1/2 the maximum *sdas* was used as a conservative estimate of the required diffusion distance. The weight fraction of matrix phase was scaled to this distance and used in the DICTRA software, along with the chemistry profiles across the matrix phase. Thus, the positional chemistries were established

Experimental Details

Typically, the experimental heats were formulated from high purity raw materials. To date nickel superalloys, austenitic and ferritic-martensitic steels and high entropy alloys have been manufactured using this approach. The starting materials were melted and cast in a Vacuum Induction Furnace (VIM). Typically, molten alloy was cast at 50°C superheat above the ThermoCalc predicted liquidus temperature for each alloy composition into 100 mm diameter round graphite molds [12-13]. When simulating a thick wall casting, the mold was placed in a secondary container with loose-packed sand around the outside. The sand height was carefully controlled in order to be above the final height of the solidified ingot to promote a slower cooling rate. In those instances where the final casting will be subsequently thermo-mechanically processed into a wrought product, the alloy heats were poured into free standing 75 mm diameter graphite molds. Commercial-scale castings have subsequently been made by multiple methods, including VAR (vacuum arc remelting) and sand casting. The scale of the microstructures in those commercial sized castings as well as the NETL laboratory castings was subsequently measured using standard metallographic techniques and the *sdas* in these castings used as the scale of the matrix phase as previously described.

No attempt was made to optimize the final cast article for shrinkage aside from the addition of a small ingot "hot top" riser in the 75 mm castings. As such, the upper surface in the central portion of the solidified 100 mm castings exhibited significant shrinkage (Figure 1). The microstructures of the castings were evaluated through standard metallographic techniques, including sectioning, grinding, polishing, and etching followed by optical microscopy. The *sdas* was measured using a line intercept method. A sample from the casting was then subjected to the homogenization heat treatment for its specific microstructural *sda*s. A side-by-side comparison of the as-cast and homogenized specimens was provided by mounting them together in the same metallographic mount, thus providing identical metallographic preparation conditions. Whenever possible, mechanical property testing, such as room and elevated temperature tensile tests as well as creep rupture tests, was performed on selected samples following the usual ASTM mechanical testing procedures including E21 for elevated tensile and E139 for creep testing.

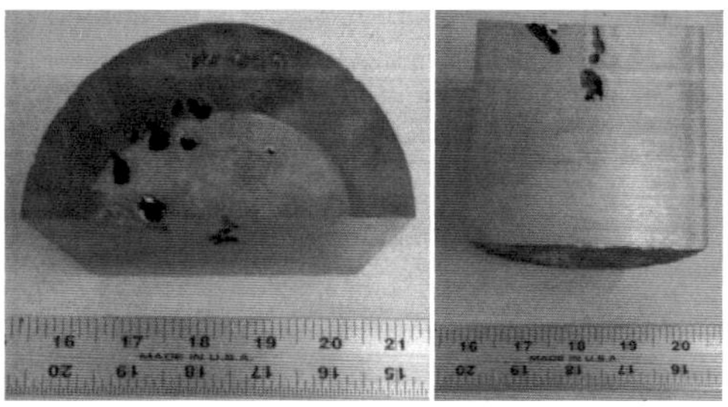

Figure 1: Example of a casting using the mold design in ref. [12-13]. The upper photograph shows the top of the solidified article. The bottom photograph shows the cross-section with extent of shrinkage. This was typical of all materials manufactured in this manner.

Results and Examples

A variety of alloys have been produced and heat treated according to the computationally optimized homogenization algorithm. In some cases multiple examples of the same alloy cast under similar conditions (and thus anticipated to have similar scaled microstructures) have been investigated. Likewise, in some cases the same alloy has been cast at small scale in the laboratory as well as at larger scale in a commercial facility, or perhaps by a different method (or both), thereby providing a range of *sdas* characteristic lengths. Several of these examples follow.

Example 1: Haynes 282

Haynes alloy 282 is a relatively new wrought Ni-based superalloy initially developed for the aerospace industry [14]. The alloy has a nominal composition of Ni-19.5Cr-10Co-8.5Mo-2.1Ti-1.5Al with small amounts of C and B and prescribed limitations on the amounts of Fe, Mn and Si. Haynes alloy 282 has been evaluated for thick section steam turbine casing and valve body components. These components, because of the size requirements, are most economically made by sand casting. Initial trial lots of Haynes 282 were cast into 100 mm diameter molds using the "enhanced" slow-cooling technique to emulate larger castings [12-13]. The as-cast Haynes 282 ingot chemistry was evaluated using the commercial thermodynamic software packages as outlined in the previous section. A critical parameter important in this approach is the incipient melt point. The initial incipient melt point is given by the Scheil module. However, the incipient melt point changes throughout the homogenization process, and thus to optimize alloy homogenization it is necessary to iterate the new virtual incipient melting temperature. Using the prescribed algorithms the homogenization temperature and time at this temperature is adjusted incrementally in order to meet the desired chemical uniformity goals. Metallographic measurements have been used to verify that in typical slow cooled castings the *sdas* varied between 47 μm and 96 μm for those castings with a diameter of 100 mm. From these measurements, for example, a value of 50 μm (or ~ ½ the maximum *sdas* length) was used as a conservative estimate of the required diffusion distance. An example of the technique is shown in Figure 2 which shows the as-solidified profile for Mo along with a schematic representation of a typical dendrite.

As shown, molybdenum segregates to the interdendritic region of this alloy. Note that the inhomogeneity is scaled to half of the *sdas*. Figure 3 shows the Mo profile in the as-cast state and after several simulated time steps at 1100ºC. This temperature was chosen because it was safely below the estimated non-equilibrium incipient melt temperature, ~1167ºC. After running these simulations it was evident that even after 80,000 s (~22.4 h) at 1100ºC, the Mo has not been substantially homogenized. Evaluating the local chemistry after a shorter time step indicated that the incipient melt temperature had risen, and now approached the equilibrium incipient melt temperature. This fact is advantageous in improving the efficiency of the Haynes 282 homogenization cycle.

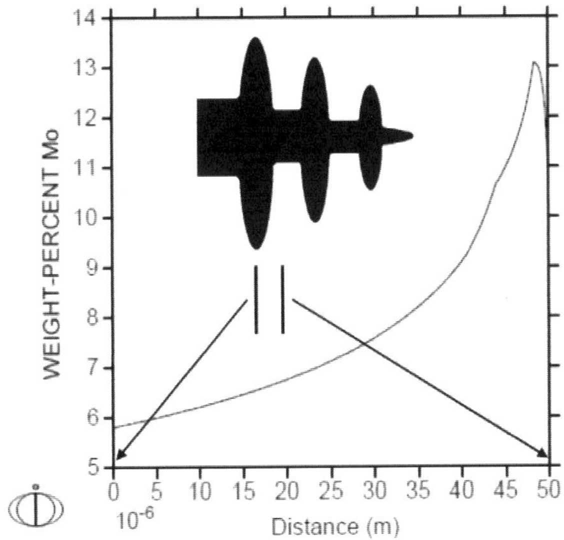

Figure 2: The as-solidified mass change profile of Mo along with a schematic representation of one half the distance between the dendrite arms is shown above.

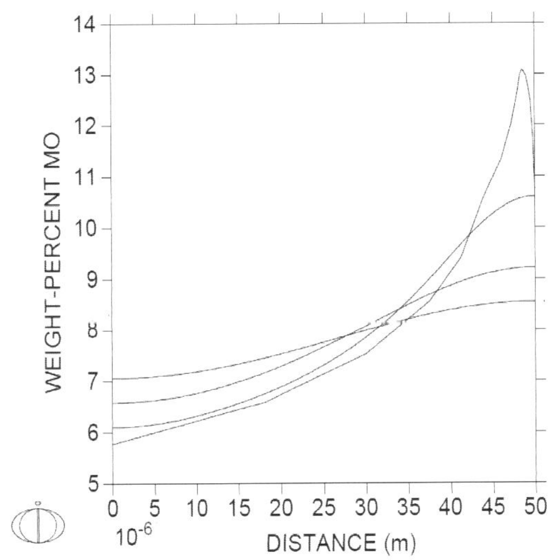

Figure 3: The Mo mass profile in as-solidified condition for one half the dendrite arm spacing and after 10 ks, 40 ks, and 80 ks isothermal exposure at 1100°C is shown above (as time increases the profile approaches horizontal). Note that substantial inhomogeneity remains even after 80 ks at 1100°C.

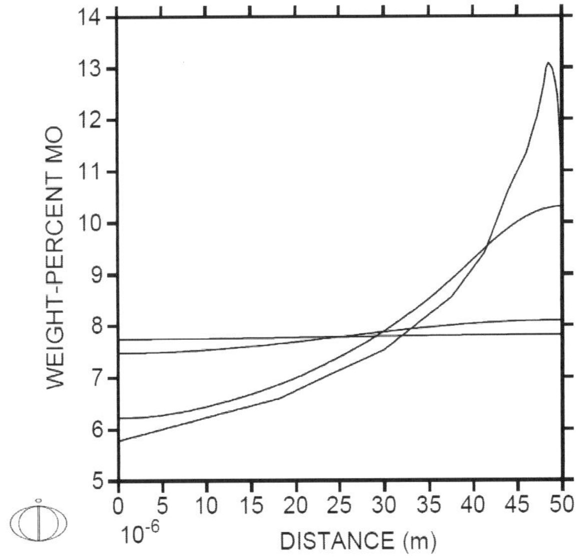

Figure 4: The Mo mass profile in the as-solidified condition for one half the dendrite arm spacing and after 10 ks, 40 ks, and 80 ks total isothermal exposure time is shown above. After 10 ks at 1100°C the simulation temperature was raised to 1200°C. Note the substantial improvement in the Mo uniformity over the single temperature simulation (Figure 3).

For example, after 10,000 s (~2.8 h) at 1100°C the incipient melt temperature has increased to ~1245°C, thus allowing the homogenization temperature to be safely raised as well. A second computational evaluation was made with 10 ks at 1100°C followed by the remaining time at 1200°C (see Figure 4). In this case, it is quite evident that the efficiency of the homogenization cycle has greatly improved by incorporating a second, higher, temperature treatment. It should be noted that the choice of temperature margin between the homogenization cycle and the incipient melt point is somewhat arbitrary. Factors considered in the selection of this temperature include the accuracy of the estimated incipient melt point (mainly dependent on the quality of the thermodynamic and kinetic/diffusion databases and degree of accuracy in the measurement of the *sdas*), furnace temperature control margin, and furnace temperature uniformity (both dependent on available equipment). Other considerations include maximum furnace temperature capability. The impact of the homogenization cycle on the microstructure is shown in Figure 5, which compares the solidified cast structure with a similar representation of the as-cast and homogenized structure under identical imaging conditions. Clearly, the alloy exhibits less segregation after homogenization. Perhaps more important is the comparison between the homogenized and non-homogenized mechanical properties. Figure 6 show creep rupture performance of the cast, cast and homogenized, and wrought product forms. While the as-cast, homogenized, and heat treated (to develop γ' precipitates) castings compared favorably to the wrought material, the non-homogenized cast and heat treated samples generally did not do as well in terms of creep rupture.

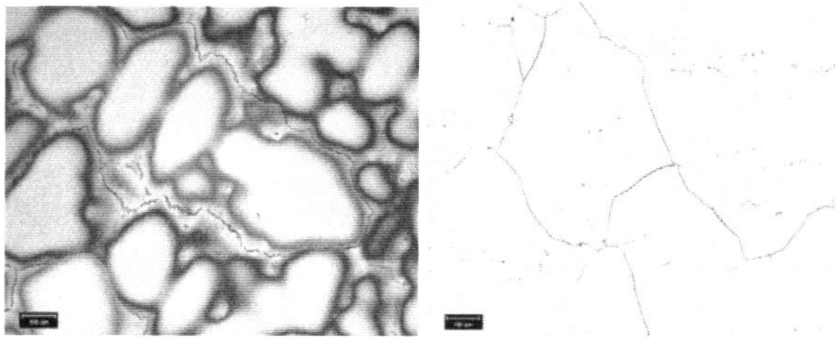

Figure 5: Examples of the as-cast (left) and homogenized (right) microstructure of Haynes 282 are shown above. Dendritic coring (i.e., element segregation) is clearly visible on the left while only grain boundaries and carbides are visible on the right. Micron markers indicate 100 μm.

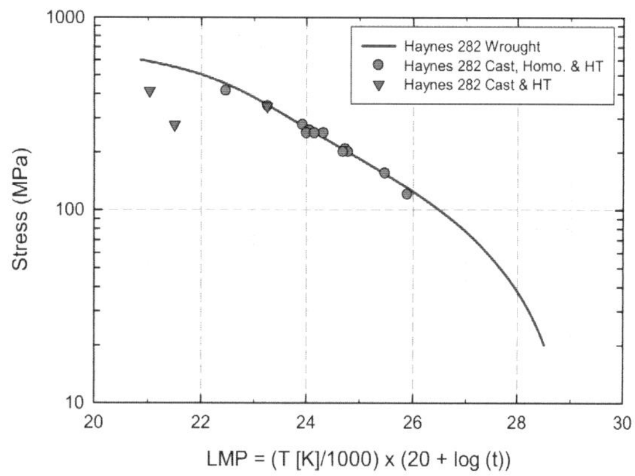

Figure 6: Larson-Miller Parameter plot for Haynes 282 comparing the cast and heat treated (inverted triangles) to cast, homogenized and heat treated (solid circles) to that of the wrought processed and heat treated (solid line) is shown above. Note that the cast, homogenized and heat treated version of Haynes 282 compares favorably to that of the wrought version of Haynes 282 in terms of creep rupture life, while those tests of the cast alloy without homogenization generally do not.

The impact of microstructural scale is significant, if somewhat predictable. Figure 7 shows a plot of the predicted Mo simulation profile after virtual homogenization of as-cast material using *sdas* of 100, 200 and 400 μm. It is clear that while modulation in the 100 μm Mo profile has "flattened" (i.e., significant reduction in segregation), the same cannot be said for the coarser microstructures. Haynes 282 has been cast on a larger scale and the measured *sdas* was on the order of 200 μm. For Haynes 282 an initial virtual simulation of 3 hours at 1100°C raises the incipient melt points to the following: 1245°C, 1219°C and 1219°C for *sdas* lengths of 100, 200 or 400 μm, respectively. Thus, these results clearly show that it gets progressively more difficult to raise the temperature for castings that exhibit coarser microstructures with large representative *sdas* lengths.

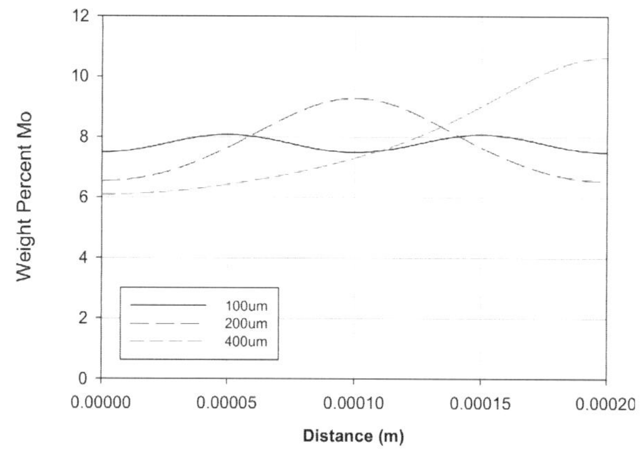

Figure 7: The Mo mass profile (in weight percent) is shown for various sdas lengths for Haynes 282 after homogenizing at 1100°C/3 h + 1200°C/9 h. For this homogenization cycle, the degree of homogenization is dependent of the coarseness of the as-cast dendrite structure.

While this has been fundamentally understood for quite some time, it was not until research on this project discovered the underlying reason, i.e., the computational algorithm is based on thermodynamics and kinetics/diffusion that describes the change in incipient melt point after any increment of heat treatment based on the overall solid chemistry. Some advantage can be gained by going to a higher temperature initially as this raises the incipient melt temperature more rapidly. However, for this alloy the equilibrium incipient melt temperature is only on the order of 1260°C, so the upper furnace temperature can become the limiting homogenization processing factor. Successful homogenization heat treatment cycles based on this approach have been developed for large castings of Haynes 282 using the following homogenization schedules:

a) 1130°C/3 h + 1200°C/3 h + 1210°C/14 h (135 kg step block casting)
b) 1100°C/6 h + 1200°C/48 h (450 kg step block casting)
c) 1133°C/4 h + 1190°C/8 h + 1223°C/30 h (4,500 kg ESR/VAR ingot)
d) 1125°C/1 h + 1160°C/3 h + 1190°C/6 h + 1200°C/10 (635 kg centrifugal casting)

These four heat treatment schedules were designed for commercial heat treating furnaces and subsequent results verified the variability and adaptability of this approach. The homogenization cycles produced good material which was verified from metallographic analysis and mechanical property evaluation.

Example 2: Alloy 263

Alloy 263 is another well-established Ni-base superalloy with a nominal composition of Ni-20Cr-20Co-5.85Mo-2.4Ti-0.3Al with small amounts of C and B as well as composition limitations on Fe, Mn and Si [15]. Similar sized ingots to that of Haynes 282 (both small and large) were cast using alloy 263 chemistry. The length scale of *sdas* in the small castings was once again approximately 100 μm. The larger castings had *sdas* approximately equal to 200 μm. The homogenization cycles developed for alloy 263 were:

a) 1100ºC/3 h + 1200ºC/9 h (laboratory-scale castings)
b) 1100ºC/1 h + 1200ºC/4 h + 1250ºC/18 h
c) 1100ºC/1 h + 1190ºC/44 h
d) 1100ºC/1 h + 1200ºC/4 h + 1232ºC/22 h

with (b), (c) and (d) homogenization cycles for a 450 kg step block but with limitations on furnace capability, (c) and (d). For example, one commercial furnace had an upper temperature of 1190°C while another was able to go to 1232°C. Following the homogenization schedule given for (b), (c) and (d) gave for Mo a variation of: 91.4% - 108.5% using (b) with 23 hours in the furnace; 91.6% - 108.3% using (c) with 45 hours in the furnace; and 91.1% - 108.8% using (d) with 27 hours in the furnace. Each homogenization cycle resulted in successful homogenization as evidenced by metallographic inspection and mechanical property determination. Figure 8 shows the LMP plot for both wrought and heat treated, and laboratory as-cast, homogenized and heat treated versions of alloy 263.

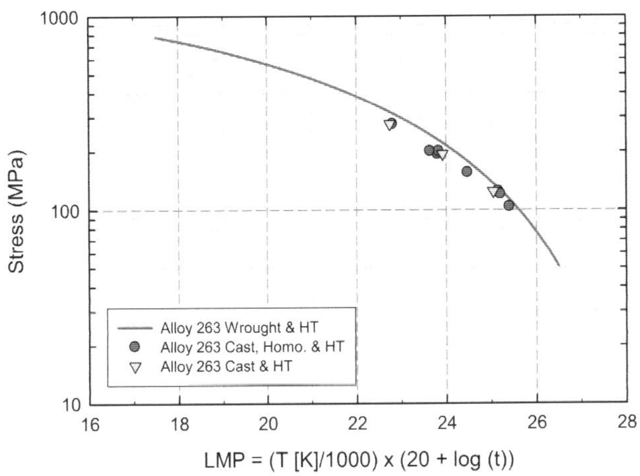

Figure 8: Larson-Miller Parameter plot for alloy 263 comparing the cast and heat treated (inverted triangles) to cast, homogenized and heat treated (solid circles) to that of the wrought processed and heat treated (solid line) is shown above. Note that the cast, homogenized and heat treated version of alloy 263 compares favorably to that of the wrought version of alloy 263 in terms of creep rupture life, while those tests of the as-cast and heat treated version generally do not.

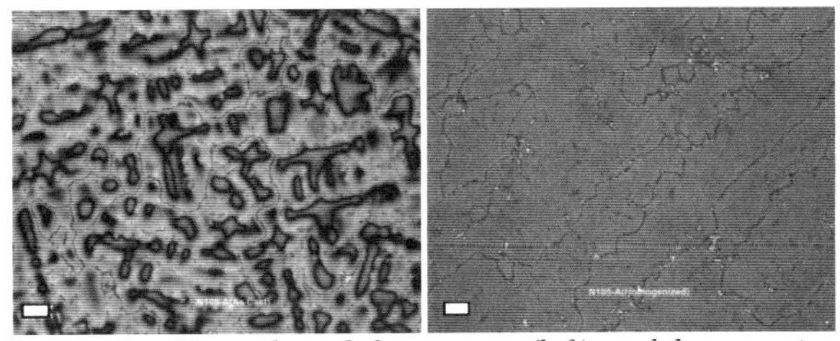

Figure 9: Examples of the as-cast (left) and homogenized (right) microstructure of Nimonic 105 are shown above. Dendritic coring (i.e., element segregation) is clearly visible on the left while only grain boundaries and carbides are visible on the right. Micron markers indicate 50 μm.

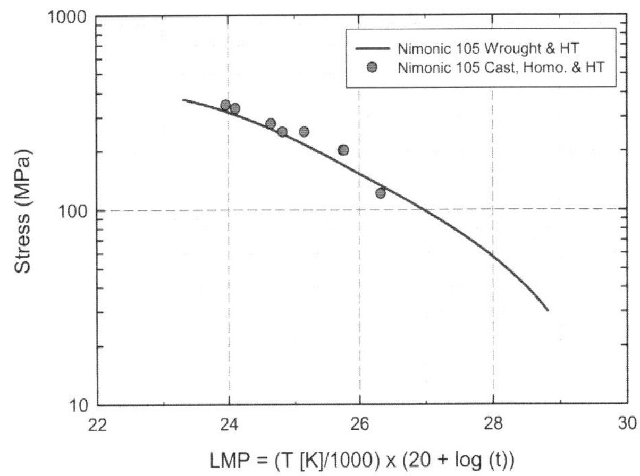

Figure 10: Larson-Miller Parameter plot comparing the cast, homogenized and heat treated (closed circles) versus wrought and heat treated version of Nimonic 105 (solid line). Note that the cast, homogenized and heat treated version on Nimonic 105 compares favorably to that of the wrought and heat treated alloy.

Example 3: Alloy 105

Alloy 105 (or Nimonic 105) is yet another well-established Ni-base superalloy with a nominal composition of Ni-14.8Cr-20.0Co-5.0Mo-1.2Ti-4.7Al with small amounts of C and B and limitations on Cu, Fe, Mn, Si and Zr [16]. Small scale ingots of this alloy were produced with resultant maximum *sdas* ≤ 100 μm. The homogenization cycle used in this case was: 1100ºC/3 h + 1200ºC/9 h. Good homogenization was achieved (see Figure 9). Figure 10 shows the resultant creep testing that was performed on the cast, homogenized and heat treated alloy 105. As with the Haynes 282 alloy there was good correlation between wrought and as-cast/homogenized/heat treated creep rupture lives for stress-temperature combinations.

Examples 4: Stainless and Ferritic/Martensitic Steels

The use of the homogenization algorithm can be applied to stainless steels with equally effective results. For example, ferritic stainless steel 441, which has a nominal composition of Fe- 18.5Cr-0.5Nb-0.3Ti, and also has limits on Mn, Si, Ni, and other minor elements, was evaluated similary to the previous examples. Small-scale laboratory ingots of this alloy were produced with resultant maximum *sdas* ≤ 100 μm. The homogenization cycle used in this case was: 1100°C/3 h + 1200°C/9 h, since the initial incipient melt temperature was about 1300°C. Good homogenization was achieved.

More complicated ferrous alloys such as ferritic/martensitic steels contain more elements (upwards of 12), some of which are hard to diffuse elements such as Mo and W. Additionally, this class of alloys is generally austenitic at the high heat treatment temperature. It is well known that elements diffuse more slowly in the austenitic phase than they do in the ferritic phase, thus alloys having the austenitic phase as the high temperature phase are more difficult to successfully homogenize. Nevertheless, the homogenization approach discussed herein has been applied to small scale ingots of ferritic/martensitic alloys which had resultant maximum *sdas* ≤ 100 μm. In one example, the initial incipient melt temperature was 1193°C. The homogenization cycle used for this case was: 1125°C/1 h + 1250°C/8 h. After the first temperature hold, the new incipient melt temperature was calculated to have risen to slightly over 1300°C, thus allowing the application of the second higher hold temperature to successfully complete alloy homogenization. The effectiveness of this treatment was evident when comparing the creep performance of the homogenized alloy to similar commercial alloys within this class of materials (i.e., those alloys having similar although not exact chemical compositions). At 650°C and 25 ksi applied stress for the creep test, the homogenized alloy had a life of 1618 h while similar commercial alloys had lives ranging between 614 h and 655 h. The improved performance became even more dramatic as the applied stress was reduced to 20 ksi and the time under creep conditions increased. The homogenized alloy had a life of 5,388 h while the commercial alloy lives ranged from 1,783 h to 2,428 h, perhaps demonstrating the benefit of a chemically uniform microstructure.

Discussion

Obtaining good mechanical properties for thick wall castings of nickel superalloys and ferritic/martensitic alloys is very important in the power industry. For land based power plants that will be operated at 760°C and 34.5 MPa, for example, superior heat resistant materials will be needed. Superior creep strength will be needed for these components to reach at least 100,000 hours in service, if not longer. Most commercial gas turbine nickel-based castings are in relatively thin sections compared to the steam turbine high- and intermediate pressure rotor casing or for chests that house the main steam valves into the rotors. As such normal investment casting techniques traditionally used for nickel superalloys in gas turbines are not appropriate. Additionally, the slow cooling in these thick section castings leads to significant segregation that must be addressed if the full potential in terms of strength and creep life are to be realized. Furthermore, because of the size-scale of these machines and the high cost of Ni alloys, the use of Fe alloys will also have to be extended to the highest possible temperatures as well. The homogenization approach described herein allows these goals to be achieved.

NETL examined the technology gap in thick wall nickel superalloy casting and approached it using modified nickel compositions proven to succeed in the wrought form. It was quickly determined that in order to achieve similar mechanical properties segregation in the solidified nickel castings needed to be addressed, especially when solution/age heat treatments failed to sufficiently improve the strength and creep life to levels needed in A-USC steam cycles. NETL envisioned an approach where the best thermodynamic and kinetic property data was used to develop a computationally based heat treatment cycle in which a physical feature, i.e., the *sdas*, was used as a "physical feature of merit" in order to derive the multi-step homogenization cycle. Furthermore, equations (3) and (4) fully describe the incremental solid chemistry in such a way that it is possible to describe the positional chemistry independent of the phases that form during solidification. This is very useful in cases where slow diffusing elements are involved and tend to segregate to second phases (e.g., Mo or Nb in Ni alloys) or in the case of the solidification of a matrix phase that is partially BCC and then FCC as is often the case with ferritic/martensitic alloys. In order to test this approach a number of nickel superalloys proven to perform well in A-USC conditions have been produced as small-scale laboratory castings, but done in such a way as to simulate the cooling in thicker wall castings. These castings were homogenized to the highest degree possible based on the approach described herein before subsequent traditional heat treatments used for these alloys.

It has, therefore, been possible to produce cast versions of each of the following:

a) Nickel solid solution strengthened superalloys (e.g., alloys 230, 617 and 625)
b) Gamma prime strengthened nickel superalloys (e.g., alloys 105, 263, 740, and 282)

Adding the homogenization cycle in the production sequence of each of these as-cast alloys has greatly reduced, and/or eliminated completely, alloy segregation due to thick section cooling from the melt. For the γ' strengthened alloys (i.e., alloys 105, 263, and 282 in particular) subsequent aging heat treatments produced mechanical strength equivalent to the wrought product. This was especially evident in the creep rupture test results where life as measured by the Larson-Miller Parameter was equivalent to the wrought alloy values. On the other hand, solution strengthened castings (i.e., alloys 230, 617 and 625) continue to exhibit a debit in terms of creep capability compared to the wrought version of the alloy and this is most probably due to a lack of process strain from thermo-mechanical processing which can facilitate global microstructural changes in the alloy.

The creep performance of the solution strengthened superalloys, as shown in Figure 11, clearly shows that the creep capability is uniformly less than that of the wrought version of the corresponding alloy. This behavior is consistent with the differences that may result when a solid solution alloy is not thermo-mechanically processed and wherein the stored energy of deformation is not available to induce recrystallization, thereby, refining the grain size in a controlled manner. Since there is little or no γ' precipitates in these alloys, the loss in strength due to thermo-mechanical processing differences cannot be compensated for without chemistry modification.

Although not shown here, tensile mechanical behavior at 800°C for the solution strengthened cast and homogenized and the γ' strengthened cast, homogenized and heat treated superalloys was considered to be very good, especially in terms of ductility measures. Typically, elongation-to-failure for the solution strengthened cast and homogenized superalloys was at least 20%, or greater. For the γ' strengthened alloys elongation to failure for tensile tests performed at 800°C was around 15%, or greater, for alloys 105, 263 and 282.

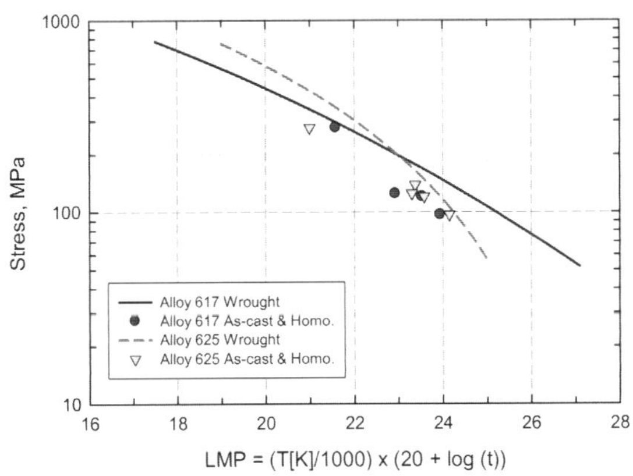

Figure 11: Larson-Miller Parameter plot comparing the cast and homogenized (closed circles) versus wrought and heat treated version of Alloy 617 (solid line). Also included is the Larson-Miller Parameter plot comparing the cast and homogenized (closed inverted triangles) versus wrought and heat treated version of Alloy 625 (dashed line). Note that the cast and homogenized 617 variant deviated from the wrought LMP for 617 at low stress levels, while the cast and homogenized 625 variant deviates from the wrought LMP for 625 at the higher stress levels.

In ferrous alloys the improvement in creep performance of 9-12% Cr ferritic-martensitic steels that have been given a homogenization cycle prior to thermo-mechanical processing is most likely due to greater uniformity in the distribution of elements. This is particularly important for this class of materials since they are susceptible to the formation of strength reducing phases (e.g., Laves and Z-phase) when exposed to high temperatures over extended periods. Regions of possible chemical inhomogeneity, leading to pockets of hard to diffuse elements like W or Mo or Co provide opportunities for these undesirable phases to form relatively early in their creep life (compared to their expected 30 year operating life).

A number of other observations were made during the course of homogenization modeling simulations, experimental manufacturing on the laboratory and commercial scales, and mechanical property verification activities for nickel superalloys. During the computational modeling activities it was found that certain refractory elements, including W, Mo, and Nb, do not homogenize even after ~22 hours at 1100°C. This was the primary motivation for exploring and developing a computational multi-step homogenization heat treatment cycles. Also, it was observed during optical microscopy and SEM of the cast structures that segregation of the second phase strengthening elements, including Al, Nb and Ti, consistently occurred in these alloys, to the point that 1/2 to 2/3 of the casting could be considered "lean". It was also during these early microstructural investigations that Cr-poor regions were discovered in the solidified casting as predicted by the thermodynamic modeling activities. In selected alloys significant Co segregation was also predicted, and as such had to be addressed. Thermodynamic modeling suggested that significant partitioning of Mn and Si to the interdendritic region might also occur. These many factors made apparent the need to homogenize these materials since to one degree or another, some or all of the above factors came into play for each of the nickel superalloy castings. It is apparent from the evidence presented that to produce consistent tensile and creep properties a more uniform microstructure in terms of the chemical distribution of elements is needed. Other manufacturing factors, such as shrinkage, voids and other melt-solidification anomalies, consistent with thick castings, still occurred. While important to note, these manufacturing related anomalies are best addressed through optimization of the melting and casting practices upon scale-up.

Summary and Conclusions

The application of carefully controlled multi-step homogenization cycles, based on thermodynamics, kinetics, and measurement of the physical feature of merit, i.e., *sdas*, were achieved through application of a specific computational algorithm, and was used to significantly reduce, and/or eliminate, visible elemental segregation in slow cooled nickel and steel castings. As a consequence, tensile mechanical properties for these alloys were more consistent and creep rupture life was greatly improved as were particular measures of ductility for both tensile and creep tested specimens.

Creep life for computationally homogenized γ' strengthened nickel superalloy castings has been shown to be equivalent to fully wrought versions of the same alloy. Solution strengthened superalloys (i.e., alloys 230, 617 and 625), while performing well in the cast and homogenized condition, did not perform to quite the same level as their wrought counterpart. This resulted primarily because the grain structure could not be controlled to the same degree as in the wrought variants. Similar behavior has also been observed for the 9-12% Cr ferritic-martensitic steel castings.

Acknowledgments

The authors would like to thank Mr. Edward Argetsinger and Mr. Joseph Mendenhall for assistance in melting and heat treating as well as Mr. Paul Danielson metallographic work.

Disclaimer

"This report was prepared as an account of work sponsored by an agency of the United States Government. Neither the United States Government nor any agency thereof, nor any of their employees, makes any warranty, express or implied, or assumes any legal liability or responsibility for the accuracy, completeness, or usefulness of any information, apparatus, product, or process disclosed, or represents that its use would not infringe privately owned rights. Reference herein to any specific commercial product, process, or service by trade name, trademark, manufacturer, or otherwise does not necessarily constitute or imply its endorsement, recommendation, or favoring by the United States Government or any agency thereof. The views and opinions of authors expressed herein do not necessarily state or reflect those of the United States Government or any agency thereof."

References

[1] D.A. Porter and K.E. Easterling, *Phase Transformation in Metals and Alloys* (New York, NY: Van Nostrand Reinhold Co., 1981) 208.

[2] H.K. Danielsen, P.E.D. Nunzio, J. Hald, "Kinetics of Z-Phase Precipitation in 9 to 12 pct Cr Steels" *Metall. Mater. Trans.* A 44 (2013) 2445.

[3] G.L. Erickson, "The Development and Application of CMSX-10," *Superalloys*, R.D. Kissinger, D.J. Deye, D.L. Anton, A.D. Cetel, M.V. Nathan, T.M. Pollock and D.A. Woodford, Eds. (Warrendale, PA: TMS, 1996) 35-44.

[4] E.C. Caldwell, F.J. Fela and G.E. Fuchs, "Segregation of Elements in High Refractory Content Single Crystal Nickel Based Superalloys," *Superalloys*, K.A. Green, T.M. Pollock, H. Harada, T.E. Howson, R.C. Reed, J.J. Schirra and S. Walston, Eds. (Warrendale, PA: TMS, 2004) 811-818.

[5] S.R. Hedge, R.M. Kearsey and J.C. Beddoes, "Designing Homogenization-Solution Heat Treatments for Single Crystal Superalloys," *Mater. Sci. Eng. A*, 527 (2010) 558-5538.

[6] P. D. Jablonski and C. J. Cowen, "Homogenizing a Nickel-Based Superalloy: Thermodynamic and Kinetic Simulation and Experimental Results," Metall. Mater. Trans. B-Proc. Metall. Mater. Proc. Sci., Vol.40, 2009, 182-186.

[7] P.D. Jablonski and C.J. Cowen, "Computationally Optimized Homogenization Heat Treatment of Metal Alloys," S119,015 and S130,803.

[8] THERMOCALC, Version 3.1 (Stockholm, Sweden: Royal Institute of Technology, 2013).

[9] DICTRA, Version 27 (Stockholm, Sweden: Royal Institute of Technology, 2008).

[10] E. Scheil, "Bemerkungen zur schichtkristallbildung," *Z. Metallkd.*, 34 (1942) 70-72 (in German).

[11] G.H. Gulliver, "The Quantitative Effect of Rapid Cooling Upon the Constitution of Binary Alloys," *J. Inst. Met.*, 9 (1913) 120-57.

[12] P.D. Jablonski, J.A. Hawk, C.J. Cowen and P.J. Maziasz, "Processing of Advanced Alloys for A-USC Steam Turbine Applications," *Proc. 6th International Conference on Advances in Materials Technology for Fossil Power Plants*, D. Gandy, J.P. Shingledecker and R. Viswanathan, Eds. (Palo Alto, CA: Electric Power Research Institute, 2010) 872-885.

[13] P.D. Jablonski, J.A. Hawk, C.J. Cowen and P.J. Maziasz, "Processing of Advanced Cast Alloys for A-USC Steam Turbine Applications," *Journal of Metals*, 64 (2012) 271-279.

[14] Haynes International, Haynes Alloy 282 literature, H-3173A (2008).

[15] Special Metals, Nimonic Alloy 263 literature, SMC-054 (2004).

[16] Special Metals, Nimonic Alloy 105 literature, SMC-081 (2004).

International Federation of Heat Treatment and Surface Engineering 2016
Proceedings of the 23rd IFHTSE Congress
April 18–21, 2016, Savannah, Georgia, USA

Improvement in Creep Rupture Ductility of Creep Strength Enhanced Ferritic Steel by New Heat Treatment Process with Intermediate Tempering

Kazuhiro Kimura and Kota Sawada
National Institute for Materials Science, Tsukuba, Ibaraki, Japan
kimura.kazuhiro@nims.go.jp, +81(0)29-8592229

Abstract

Creep rupture ductility of creep strength enhanced ferritic (CSEF) steels has been investigated. Creep rupture ductility of CSEF steels decreases in the long-term where the stress is lower than 50% of 0.2% offset yield stress at the temperatures which is referred as a half yield. A half yield is regarded to be an elastic limit of the steels. Creep deformation mechanism of the steels is considered to be different in high- and low-stress regimes divided by a half yield. Accuracy of creep rupture life prediction is improved by region splitting analysis in which creep rupture life is analyzed independently for high- and low-stress regimes divided by a half yield. The higher strength steel tends to indicate more significant decrease in creep rupture ductility. High ductility in the high-stress regime above half yield should be provided by a plastic deformation, which easily takes place with the assistant of external stress above elastic limit and it has been speculated that a remarkable drop in ductility in the low-stress regime is derived from a concentration of creep deformation into a tiny recovered region formed at the vicinity of grain boundary. Preferential recovery at the vicinity of grain boundary is caused mainly by lowered creep strength due to plastic deformation during martensitic phase transformation. New heat treatment process with intermediate tempering on partially transformed dual phase microstructure obtained by cooling from normalizing temperature to the temperature between Ms and Mf, retains the potential to improve creep rupture ductility without loss of creep rupture life. Improvement in creep rupture ductility of Grade P92 steel has been obtained by a modified heat treatment process with an intermediate tempering on partially quenched dual phase which reduces residual stress introduced by martensitic phase transformation accompanied with shape change and volume expansion.

Introduction

Creep strength enhanced ferritic (CSEF) steels have been widely used as header, main/reheat steam pipes and heat exchanger tube of Ultra Supercritical (USC) power plant, and heat recovery steam generator of combined cycle power plant. Energy efficiency of USC power plant has been improved by increasing steam condition of temperature and pressure through the introduction of CSEF steels which possess higher creep strength at the elevated temperatures. However, remarkable drop in creep rupture ductility of CSEF steels was recognized in the long-term [1]. Mechanism of ductility drop observed on ferritic steels has been investigated by many researchers [2-10]. Harmful effect of alloying elements and impurities such as aluminum, sulfur, phosphorus, arsenic, antimony, tin, niobium, vanadium and boron on creep rupture ductility has been reported on Cr-Mo-V type ferritic steels [2-5]. Creep cavity formation in association with coarse precipitates and nonmetallic inclusions such as Laves phase, BN, MnS and Al_2O_3 is considered to be a cause of low ductility in the long-term [6, 8-10]. According to a wide range of investigations on creep strength property of CSEF steels, tightened composition range was recommended for Grade 91 as Class 2 steel, in order to increase in the performance of the material with respect to creep rupture ductility [10].

On the other hand, good correspondence between remarkable drop in creep rupture ductility and stress condition was reported [7]. High ductility observed in the high-stress regime indicates significant drop with decrease in stress at 50% of 0.2% offset yield stress (Half Yield) that is considered to be an elastic limit at the temperatures [7]. This observation suggests that not only chemical composition, but also another factor may have an effect on creep rupture ductility of CSEF steels. To clarify another cause of ductility drop will be useful to improve reliability and availability of high temperature components manufactured by CSEF steels.

In this study, mechanism of remarkable drop in creep rupture ductility of ASME Grade P92 steel was investigated and a possibility of improvement in creep rupture ductility of the steel was examined.

Experimental Procedures

The material used in the present study was ASME SA-335/SA-335M Grade P92 steel with a dimension of 315 mm in outside diameter and 38 mm in wall thickness. The chemical composition of the steel examined is listed in Table 1 [11]. The pipe was normalized at 1070°C for 2 h and tempered at 780°C for 2 h. The specimen was taken in parallel to longitudinal direction of the pipe. Tensile test data and creep test data of the P92 steel reported in NIMS Creep Data Sheet, No.48A as MJP heat [11] was used in this study.

Optical micrograph of the P92 steel in the as tempered condition is shown in Figure 1. Tempered martensitic microstructure is observed in the as tempered condition of the steel studied.

Table 1: Chemical composition (mass%) of the Grade P92 steel studied [11].

	C	Si	Mn	P	S
SA-335 P92	0.07 -0.13	≤0.50	0.30 -0.60	≤0.020	≤0.010
MJP	0.110	0.10	0.41	0012	0.0037

Ni	Cr	Mo	W	V	Nb
≤0.40	8.50 -9.50	0.30 -0.60	1.5 -2.00	0.15 -0.25	0.04 -0.09
0.17	9.26	0.42	1.67	0.16	0.057

Al	N	B	Ti	Zr
≤0.02	0.03 -0.07	0.001 -0.006	≤0.01	≤0.01
0.01	0.0462	0.002	0.003	<0.001

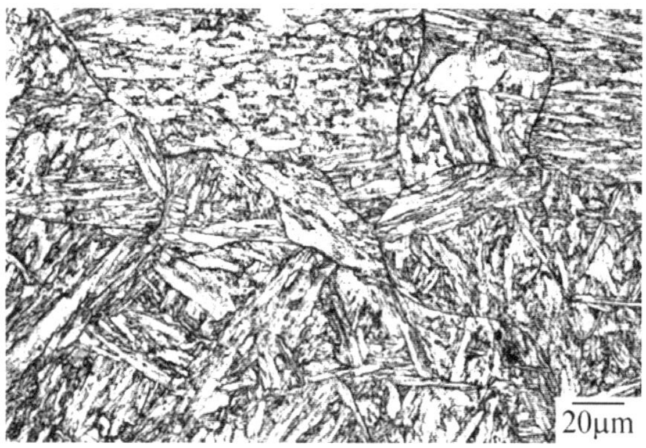

Figure 1: Optical micrograph of the Grade P92 steel studied in the as tempered condition.

Results and Discussion

Creep rupture strength and ductility

Stress versus time to rupture curves over a range of temperatures from 550 to 700°C of the Grade P92 steel are shown in Figure 2. The curves were evaluated by means of region splitting analysis with a boundary condition of half yield and a Larson-Miller parameter. Parameters obtained for high- and low-stress regimes are reported in NIMS Creep Data Sheet, No.48A [11]. Slope of the curve at 550°C is almost constant from short-term to long-term, however, it becomes steeper in the long-term at 600°C and above. Similar change in stress dependence is also observed on the minimum creep rate, however, good linear relationship was recognized on a Monkman-Grant plot over a wide range investigated [11].

Crept specimens of the Grade P92 steel creep ruptured at 650°C and various stresses are shown in Figure 3. Good creep rupture ductility with a large reduction of area is observed on the specimens creep ruptured at 140 and 100MPa, however, creep rupture ductility decreases with decrease in stress. Reduction of area of the creep ruptured specimen decreases with decrease in stress and significantly low creep rupture ductility is recognized on the crept specimens creep ruptured at 70 and 60MPa.

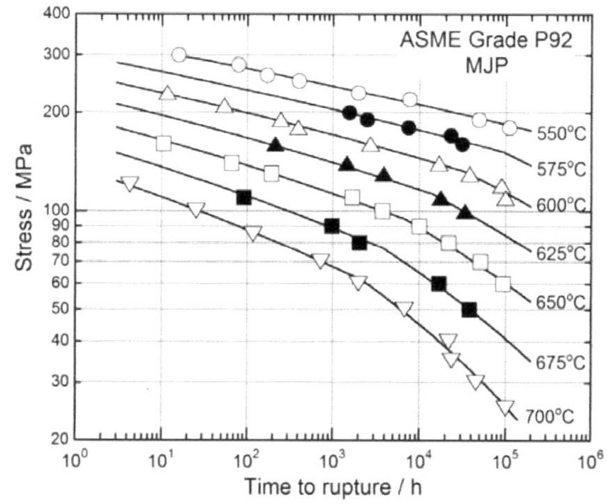

Figure 2: Stress versus time to rupture curves of Grade P92 steel MJP heat [11].

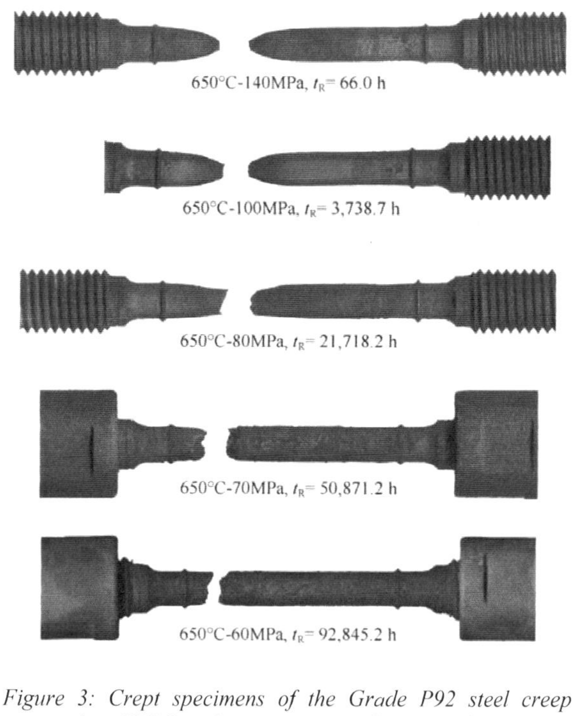

Figure 3: Crept specimens of the Grade P92 steel creep ruptured at 650°C and over a range of stresses from 140 to 60MPa.

Rupture elongation and reduction of area of the Grade P92 steel are plotted against time to rupture and shown in Figure 4 (a) and (b), respectively. In the short-term less than about 1,000 hours, good creep rupture ductility is observed, however, it indicates remarkable drop with increase in creep exposure time. With increase in temperature, large drop in ductility is recognized after shorter creep exposure time.

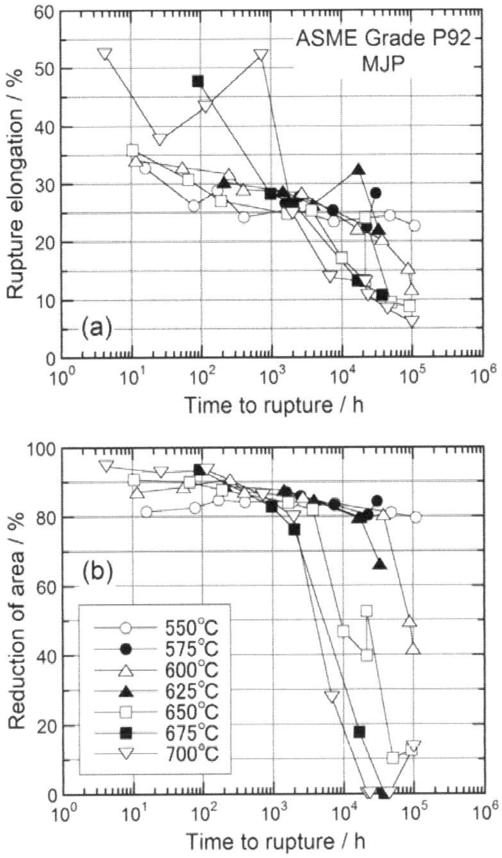

Figure 4: Rupture elongation (a) and reduction of area (b) of Grade P92 steel MJP heat [11].

Mechanism of decrease in creep rupture ductility

Mechanism of ductility drop observed on ferritic steels has been investigated by many researchers [2-10]. Creep rupture ductility of Cr-Mo-V steels is reduced by aluminum in solid solution [2]. Reduction in impurities such as sulfur, phosphorus, arsenic, antimony and tin is beneficial to improve rupture ductility of 1CrMoVTiB steels [3] and to suppress tempering and long-term embrittlement of 3.5%NiCrMoV rotor steels [5]. Alloying elements which increase creep strength tend to reduce creep rupture ductility, in particular, niobium, vanadium and boron strongly increase rupture strength and decrease ductility [4]. Cipolla et al. [6] reported that brittle fracture occurs at the onset of coarsening of Laves particles, which result to be the preferential site for cavities nucleation in the 9% Cr steels. On the other hand, amount of Laves phase precipitation in Grade 92 steels decreases with increase in temperature [13], and remarkable drop in creep rupture ductility is observed independent of temperature. Gu et al. [8] identified several particles of BN, MnS and Al_2O_3 associated with cavities in the crept P92 steels. They observed very small cavities associated with the BN particles in the head-gauge transition area of the crept samples and it was considered to be the onset of creep damage [8]. According to the above results, Parker et al. [9,10] concluded that proper control of the composition is required to increase in the performance of CSEF steels and tightened chemical composition is recommended for Grade 91 as Class 2 steel.

On the other hand, good correlation between significant decrease in creep rupture ductility and stress condition is recognized [7]. Reduction of area of the P92 steel shown in Fig. 4(b) is plotted against a ratio of stress to 0.2% offset yield stress, hereinafter referred as yield ratio and shown in Figure 5. Remarkable drop in reduction of area is observed at half yield with decrease in yield ratio, independent of temperature. This behavior is recognized on not only Grade P92 steel, but also the other type of CSEF steels [7]. Since half yield is considered to be a macroscopic elastic limit where successive plastic deformation of polycrystalline material initiates [12], remarkable drop in creep rupture ductility in the long-term should be derived from change in stress condition from high-stress regime exceed elastic limit to low-stress regime below it.

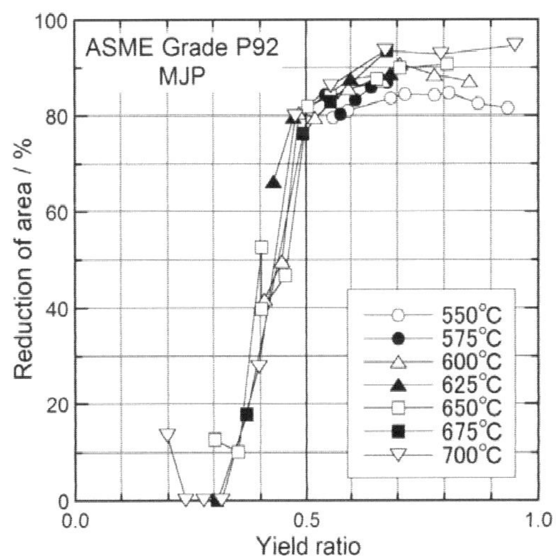

Figure 5: Reduction of area plotted against a yield ratio of the Grade P92 steel MJP heat [11].

Bright field TEM image of the T91 steel creep ruptured after 34,101.0 h at 600°C and 100MPa is shown in Figure 6 [14]. Preferentially recovered region is observed along prior austenite grain boundary (PAGB), however, fine microstructure is still observed in the other area. Although this microstructure is observed on not Grade P92 steel, but Grade T91 steel, however, inhomogeneous progress in recovery is a common feature of CSEF steels crept in the low-stress regime below half yield. Since creep rate in a recovered area is significantly larger than that in grain interior and creep deformation is localized in a preferentially recovered tiny area in the vicinity of PAGB, magnitude of creep deformation of overall material is significantly reduced and results in large drop of creep rupture ductility.

As causes of preferential recovery at the vicinity of PAGB, faster grain boundary diffusion, annihilation site of dislocation, faster coarsening of particles on grain boundary and concentration of internal stress introduced by martensitic transformation are considered. In addition to those factors, it has been supposed that creep strength of certain part, especially at the vicinity of grain boundary is lower than that of the other area. Martensitic transformation results in change in shape and expansion of volume and, therefore, local strain and residual stress are introduced. Such local strain is mainly introduced in austenite matrix in the beginning of martensitic transformation [15, 16], however, plastic deformation may take place in transformed martensite phase at the final stage of martensitic transformation in order to accommodate shape and volume change. Since a certain amounts of plastic deformation reduces creep strength [17, 18], creep strength in the local area plastically deformed during martensitic transformation should be reduced, and such local area tends to be produced at the vicinity of boundaries such as PAGB. Consequently, if the residual stress introduced by martensitic transformation is reduced, magnitude of plastic deformation due to martensitic transformation should also be reduced.

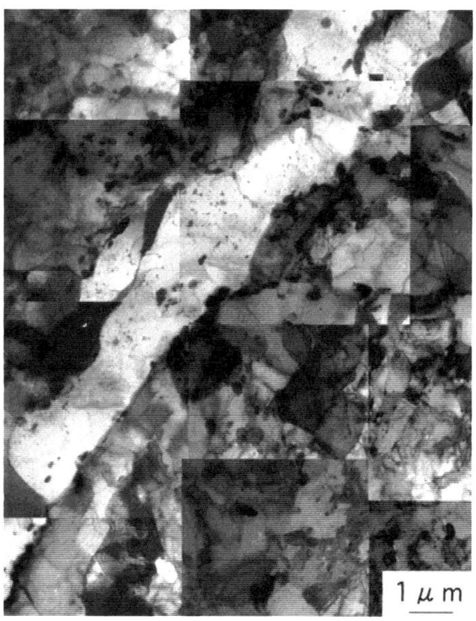

Figure 6: Bright field TEM image of Grade T91 steel MGC heat creep ruptured after 34,141.0 h at 600°C and 100MPa [14].

Influence of intermediate tempering after partial quench

CCT curves of Grade P92 steel, MJP heat are shown in Figure 7. Start and finish temperature of martensitic phase transformation, Ms and Mf temperatures are about 400 and 240°C under fast cooling condition, respectively. Although slight increase in Mf to about 260°C is observed after 50,000 seconds, Ms of 400°C is almost constant independent of cooling rate.

In order to reduce residual stress introduced by martensitic phase transformation, intermediate tempering was employed after partial quenching to the temperature between Ms and Mf during cooling from normalizing temperature to room temperature. Heat treatment conditions are shown in Figure 8 and Table 2. Normalizing heat treatment for 2 h at 1070°C and final tempering for 2 h at 780°C are the same as those of original MJP heat. The test blocks taken from the identical seamless pipe of Grade P92 MJP heat were subjected to four different heat treatment process of DTA, DTB, DTC and DTD as shown in Table 2. Similar heat treatment process has been investigated by Tamura et al. [19] and Paul et al. [20], however, material is heated to the final tempering temperature from the temperature between Ms and Mf in their processes. Moreover, their purpose is to increase creep strength, and the influence of such heat treatment on creep rupture ductility in the long-term has not yet been reported.

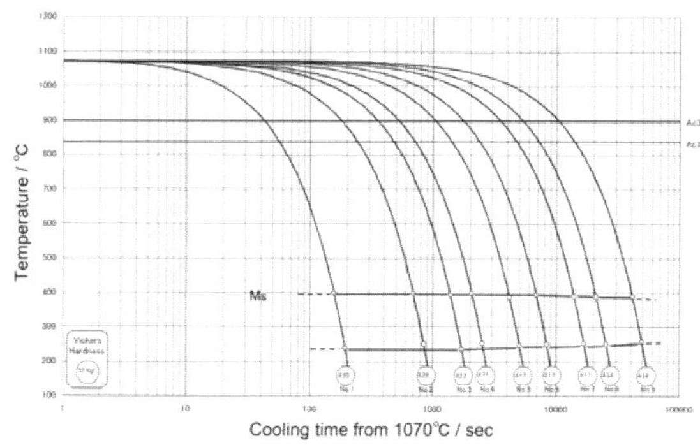

Figure 7: CCT curves of the Grade P92 steel MLP heat.

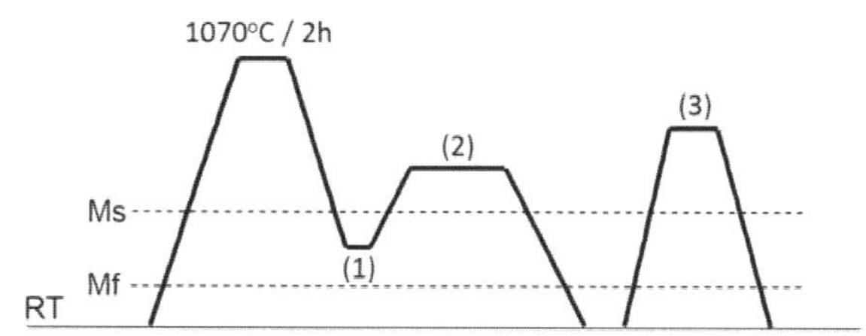

Figure 8: New heat treatment process with an intermediate tempering after partial quench conducted on the Grade P92 steel. (1) partial quenching, (2) intermediate tempering and (3) final tempering.

Table 2: New heat treatment process with an intermediate tempering after partial quench conducted on the Grade P92 steel.

Steels	(1) Partial quenching	(2) Intermediate tempering	(3) Final Tempering
DTA	350°C / 0.5 h	570°C / 24 h	780°C / 2 h
DTB	350°C / 0.5 h	570°C / 24 h	730°C / 2 h
DTC	350°C / 0.5 h	590°C / 24 h	780°C / 2 h
DTD	320°C / 0.5 h	570°C / 24 h	780°C / 2 h

Creep test at 650°C and 90MPa has been conducted on the Grade P92 steel and the steels subjected to the additional heat treatment shown in Figure 8 and Table 2, and creep curves, creep rupture ductility of the steels are shown in Figures 9 and 10, respectively. Moreover, crept specimen of the steels creep ruptured at 650°C and 90MPa are shown in Figure 11. The steels subjected to new heat treatment process indicate higher creep rupture ductility than that of original MJP heat. Beneficial effect to improve creep rupture ductility of the new heat treatment process with an intermediate tempering after partial quenching is clearly recognized. Although DTA and DTB indicate slightly shorter creep rupture life, however, those are within the standard range of Grade P92 steel [21], and creep rupture lives of DTC and DTD are essentially the same as that of original MJP heat. Above results have been obtained on the material taken from the identical seamless pipe of Grade P92 steel, therefore, chemical composition including impurities and nonmetallic inclusions are essentially the same each other. Consequently, creep rupture ductility of CSEF steels can be improved by heat treatment process which control martensitic phase transformation behavior. In order to optimize heat treatment process for both creep rupture strength and ductility of CSEF steels, mastensitic phase transformation behavior and influence of intermediate tempering after partial quenching on the following martensitic phase transformation should be investigated in detail.

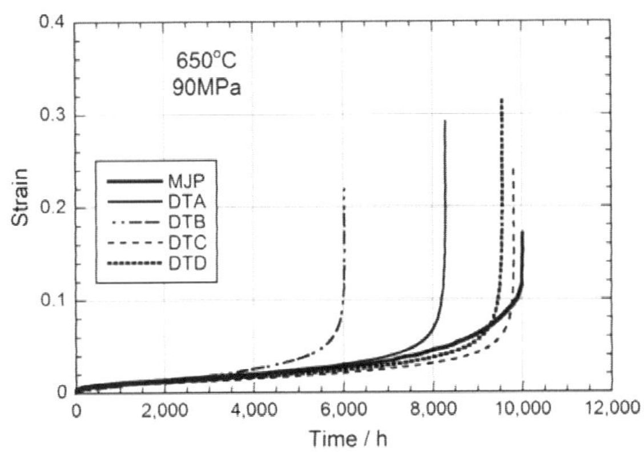

Figure 9: Creep curves at 650°C and 90MPa of the Grade P92 steel MJP heat and the steels subjected to the additional heat treatment shown in Figure 8 and Table 2.

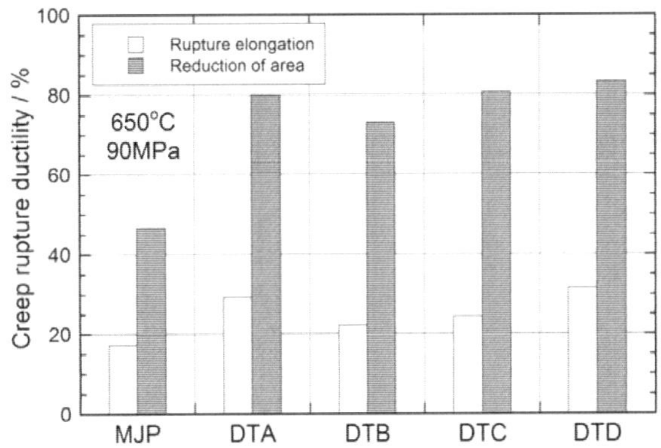

Figure 10: Creep rupture elongation and reduction of area at 650°C and 90MPa of the Grade P92 steel MJP heat and the steels subjected to the additional heat treatment shown in Figure 8 and Table 2.

Figure 11: Crept specimens creep ruptured at 650°C and 90MPa of the Grade P92 steel MJP heat and the steels subjected to the additional heat treatment shown in Figure 8 and Table 2.

Conclusions

In this study, mechanism of remarkable drop in creep rupture ductility of creep strength enhanced ferritic steels was investigated and a possibility of improvement in creep rupture ductility of the steel was examined. The results are summarized as follows.

(1) Detrimental alloying elements including impurities and cavitation associated with coarse hard particles such as Laves phase, BN, MnS and Al_2O_3 are considered to be a cause of low ductility of CSEF steel.

(2) Remarkable drop in creep rupture ductility is recognized at half yield independent of temperature, and that is considered to be caused by concentration of creep deformation in the tiny

area of preferentially recovered region at the vicinity of prior austenite grain boundary.

(3) It has been supposed that creep strength of certain part, especially at the vicinity of grain boundary is reduced by plastic deformation introduced by martensitic phase transformation accompanied with change in shape and expansion.

(4) New heat treatment process with intermediate tempering after cooling from normalizing temperature to the temperature between Ms and Mf is a beneficial to improve creep rupture ductility without loss of creep rupture life.

Acknowledgments

The creep tests were performed as a part of NRIM/NIMS Creep Data Sheet project. The authors are grateful to all the members participated in the NRIM/NIMS Creep Data Sheet project for their sincere efforts.

References

[1] Hald, J., "Creep strength and ductility of 9 to 12% chromium steels", Mater. High Temp., 21 (2004), pp. 41-46.

[2] Viswanathan, R. and Beck, C.G., "Effect of Aluminum on the Stress Rupture Properties of Cr-Mo-V Steels", Metall. Trans. A, 6A (1975), pp. 1997-2003.

[3] Strang, A., and Holdsworth, S.R., "Rupture Ductility of 1CrMoV High Temperature Bolting Steels", Proc. Rupture Ductility of Creep Resistant Steels, York, Dec. 1990, The Inst. Metals, UK (1991).

[4] Greenwell, B.S., and Beech, S.M., "The Rupture Ductility of High Alloy Ferritic Steels", Proc. Rupture Ductility of Creep Resistant Steels, York, Dec. 1990, The Inst. Metals, UK (1991).

[5] Mayer, K.H., and Berger, C., "Influence of Steel Cleanness on the Operational Properties of Turbine Components", Proc. New Steels for Advanced Plant up to 620°C, London, May 1995, EPRI, USA (1995).

[6] Cipolla, L., Di Gianfrancesco, A., Venditti, D., Cumino, G., and Caminada, S., "Microstructural evolution during long term creep tests of 9%Cr steel grades", Proc. CREEP8, San Antonio, Texas, USA, July 22-26, 2007, CREEP2007-26030.

[7] Kimura, K., Sawada, K., and Kushima, H., "Creep Rupture Ductility of Creep Strength Enhanced Ferritic Steels", J. Pressure Vessel Tech., 134 (2012) 031403. [DOI: 10.1115/1.4005876]

[8] Gu, Y., West, G.D., Thomson, R.C., and Parker, J., "Investigation of creep damage and cavitation mechanisms in P92 steels", Proc. 7th Int. Conf. on Advances in Materials Technology for Fossil Power Plants, Waikoloa, Hawaii, USA, Oct. 22-25, 2013, pp. 596-606.

[9] Parker, J., "Creep cavitation in CSEF steels", Proc. 7th Int. Conf. on Advances in Materials Technology for Fossil Power Plants, Waikoloa, Hawaii, USA, Oct. 22-25, 2013, pp. 714-731.

[10] Parker, J., Siefert, J., and Shingledecker, J., "The Benefits of Improved Control of Composition of Creep-Strength-Enhanced Ferritic Steel Grade 91", Electric Power Research Institute, 2014, 3002003472.

[11] NIMS Creep Data Sheet, No.48A, National Institute for Materials Science, 2012.

[12] Kimura, K., Sawada, K., Kubo, K., and Kushima, H., "Influence of stress on degradation and life prediction of high strength ferritic steels", Proc. 2004 ASME/JSME PVP Conf., San Diego, California, USA, July 25-29, 2004, PVP2004-2566.

[13] Hald, J., "Metallurgy and creep properties of new 9-12% Cr steels", Steel Research, 67 (1996), pp. 369-374.

[14] Kushima, H., Kimura, K., and Abe, F., "Degradation of Mod.9Cr-1Mo steel during long-term creep deformation", Tetsu-to-Hagane, 85 (1999), pp. 841-847.

[15] Miyamoto, G., Shibata, A., Maki, T., and Furuhara, T., "Precise measurement of strain accommodation in austenite matrix surrounding martensite in ferrous alloys by electron backscatter diffraction analysis", Acta Materialia, 57 (2009), pp. 1120-1131.

[16] Zhang, S., Terasaki, H., and Komizo, Y., "In-situ observation of martensite transformation and remained austenite in supermartensitic stainless steel", Tetsu-to-Hagane, 96 (2010), pp. 691-697.

[17] Arav, F., Lentferink, H.J.M., Etienne, C.F., and Wortel, van J.C., "Effect of Fabrication Processes on the Creep Behavior of 9-12% Chromium Steels", Proc. Fifth Int. Conf. Creep of Metals, Lake Buena Vista, Florida, USA, May 1992, ASM.

[18] Case 2179-8, Case of ASME Boiler and Pressure Vessel Code, 2012.

[19] Tamura, M., Kumagai, T., Miura, N., Kondo, Y., Shinozuka, K., and Esaka, H., "Effect of martensitizing temperature on creep strength of modified 9Cr steel", Materials Transactions, 52 (2011), pp. 691-698.

[20] Paul, S., Zeiler, G., Putschoegl, A., and Plesiutschnig, E., "Development of high creep resistant MARBN steels for forged components in high efficient fossil-fired power plants", Proc. 19th Int. Forgemasters Meeting, Makuhari, Chiba, Japan, Sep. 29-Oct. 3, 2014, pp. 338-342.

[21] Kimura, K., and Takahashi, Y., "Evaluation of long-term creep strength of ASME Grades 91, 92 and 122 type steels", Proc. ASME 2012 PVP Conf., Toronto, Ontario, Canada, July 15-19, 2012, PVP2012-78323.

International Federation of Heat Treatment and Surface Engineering 2016
Proceedings of the 23rd IFHTSE Congress
April 18–21, 2016, Savannah, Georgia, USA

Evaluation of mechanical and fatigue properties of weld deposit and its comparison with that of die steel

Sachin S. Patil, Shreyas S. Kirwai, Santosh Kumar, Atul R. Patil and Dr. Rajkumar P Singh
Kalyani Centre for Technology & Innovation, Bharat Forge Ltd., Pune, Maharashtra, India

Jayant B. Mane
Bharat Forge Ltd., Pune, Maharashtra, India

Abstract

In hot forging process, a billet is heated to a high temperature and then pressed between a set of dies in order to achieve the required shape. During production, dies are continuously under thermal and mechanical cyclic loading. The combination of these two phenomena may lead to die failure due to Low Cycle Fatigue (LCF) cracking. A cracked die cannot be replaced completely due to high costs and long turnaround time associated with procurement of the die steels. Hence, the cracked dies are repaired using the process of gouging – welding – machining. The weld deposited material becomes an integral part of forging dies and the strength of the repaired die depends on the quality and strength of the welding. Present work evaluates the mechanical, metallurgical and LCF properties of the weld deposit and compare it with that of die steel, L6. Generally, Mailam 535 is employed for welding L6. Tensile properties i.e. Yield Strength and Ultimate Tensile Strength of weld deposit were found to be 30% and 4% less than that of die steel, respectively. Charpy impact properties of weld deposit were observed to be 62% and 57% less than that of die steel in longitudinal and transverse direction, respectively. LCF strength of weld deposit weld material was found to be 50% less than that of die steel. Fractography and optical microstructure were studied for failed specimen in mechanical testing to study failure mode of welding material. The difference in properties can be attributed to the different microstructure observed in weld deposit and the die steel.

Introduction

During the hot forging process, dies are continuously under thermal and mechanical cyclic loading. Heating due to contact with high temperature billet and cooling due to lubrication provided on the die surface, lead to thermal fatigue of dies. Moreover, during the production process, the dies have to face impact loads which lead to mechanical fatigue. The combination of these two phenomena may lead to die failure by cracking [1-3].

A cracked die cannot be replaced completely due to high costs and long turnaround time associated with procurement of the die steels [1, 4]. Hence, welding is widely used to repair cracked forging dies. The crack in the dies is removed by gouging or scarfing. This gouged area is then filled with desired filler metal (welding). A multi-layer weld deposit method is used during die weld repair process. The inherent characteristics of melting and solidification of the filler material leads to formation of elongated grain structure also known as "as cast microstructure". Commonly, Metal Arc Welding (MAW) is used for this process. Finally, the welded die is machined to its final shape on CNC machine. Thus, welding becomes an integral part of the forging dies.

A forging die cracks at a location which faces the highest value of stresses during its operation. Generally, cracks on the die surface appear after 3000 to 5000 numbers of forging cycles or strokes. This comes under the regime of Low Cycle Fatigue (LCF) [5]. After weld-repair, the highest stresses in the die have to be borne by the weld material. Hence, the life of the die is governed by the properties of the weld deposit after the weld repair cycle. Thus, it is important to study the LCF strength of weld material which governs the life cycle of forging dies.

The aim of the present work is to evaluate the mechanical, metallurgical and low cycle fatigue properties of the Mailam 535 welding material and compare it with that of die material i.e. AISI L6 (DIN 1.2714).

Experimental details

Materials

AISI L6 (die material) and Mailam 535 (welding material) are selected for the present study. Chemical composition of L6 and Mailam 535 is shown in Table 1.

Table 1 : Chemical composition of L6 & Mailam 535 material

Material	Chemical composition wt. %						
	C	Mn	Si	Cr	Ni	Mo	V
L6	0.5	0.8	0.2	1.0	1.7	0.4	0.1
Mailam535	0.09	1.35	0.42	2.96	1.81	1.17	0.25

Welding process

During experimental work, a Shielded Metal Arc Welding (SMAW) method was employed. Schematic of the welding process is shown in Fig. 1. In this method, the welding is deposited layer by layer such that two successive layers are parallel and in the same direction to each other. Direction of welding is referred as longitudinal direction whereas direction perpendicular to welding direction is referred as transverse direction. Before welding, base metal was scarfed by using 10 mm carbon electrodes to make 'U' shaped groove, which was then filled with welding material. During welding, base metal is preheated to 500 °C and inter-pass welding temperature was maintained as 300 °C minimum.

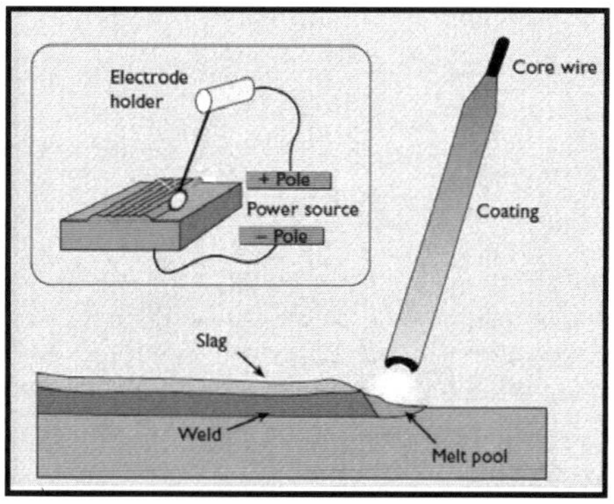

Figure 1: Schematic of welding process [6]

Mechanical testing

To compare L6 and Mailam 535 material, following mechanical tests were carried out.

Mechanical testing: Tensile testing

Tensile specimen for L6 was directly extracted from as received (quench and tempered) material. For Mailam 535, first, a volume of approximately 40 mm width/breadth and 30 mm depth was gouged from a block of material made of L6 (300 x 100 x 60 mm) as shown in Figure 2. The gouged surface was, then, filled by weld material Mailam 535. The tensile test piece was taken out in such a way that its gauge length is made up of welding material (Refer Fig. 2). This was done to minimize the amount of welding required. Tensile test specimens were prepared from L6 and Mailam 535 material as per standard ASTM E8. The testing was done on Universal Testing Machine of ZWICK/ROELL make (Z250).

Three specimens were tested for both L6 and Mailam 535 and their average values were taken as tensile strength of that material. For tensile test, the specimens were extracted only in transverse direction as it replicated actual working condition of forging dies. Tensile test results, furthermore, were used for Elasto-plastic CAE analysis in ANSYS to calculate the LCF strength

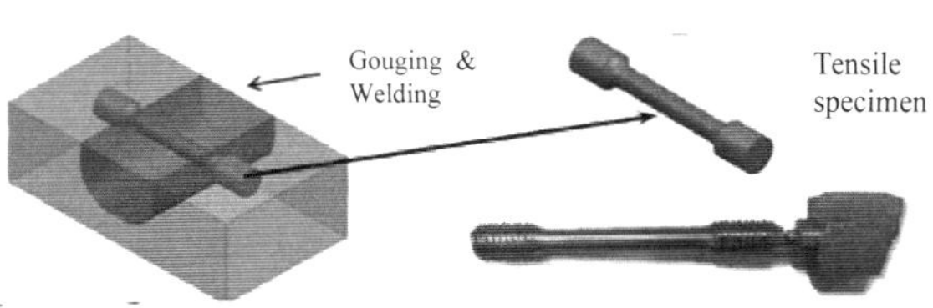

Figure 2: Extraction of tensile specimens from welded block

Mechanical testing: Charpy Testing

Charpy specimens for L6 were directly extracted from as received material. For the Mailam 535 specimens, a block of L6 with size (200 x 150 x 120 mm) was taken. On its top surface welding layers were deposited with approximate height of 30 mm. The specimens were extracted from welding as shown in Fig. 3.

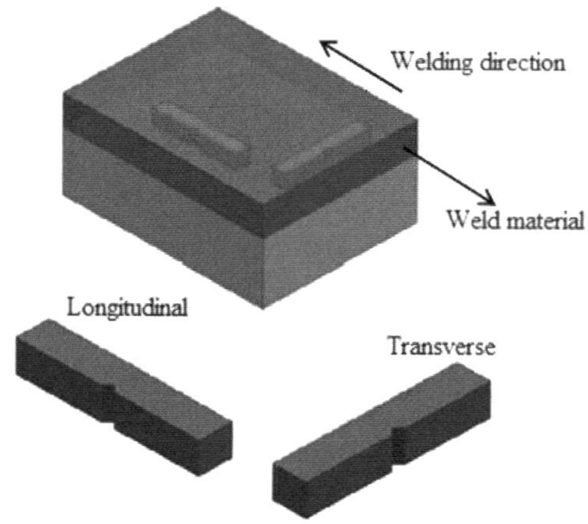

Figure 3: Extraction of Charpy impact specimens from welded block

Three specimens were tested for both materials in longitudinal and transverse direction and their average value is taken as Charpy impact energy of that material. Charpy impact test specimens were prepared as per standard ASTM A370-13. The testing was done on ZWICK/ROELL make machine (450 J capacity).

Mechanical testing: Low cycle flexural fatigue (LCF) test

LCF test was performed on 250 KN Servo-hydraulic machine of INSTRON make at room temperature for both materials i.e. L6 and Mailam 535. The geometry of LCF specimen is shown in Fig. 4. LCF specimen for L6 was directly extracted from as received material. But, for Mailam 535, a volume of approximately 40 mm width/breadth and 30 mm depth was gouged from a block of material L6 (300 x 150 x 100 mm). The gouged surface was filled by weld material Mailam 535. The specimens were extracted from welded blocks such that its highest stressed zone (middle portion) was made up of welding as shown in Fig. 5.

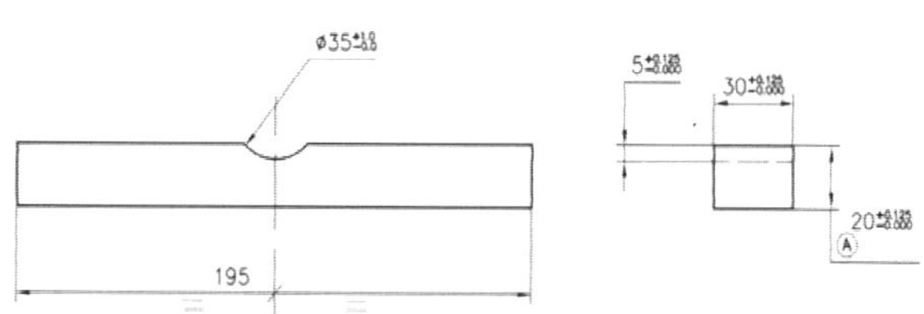

Figure 4: Flexural fatigue specimen drawing

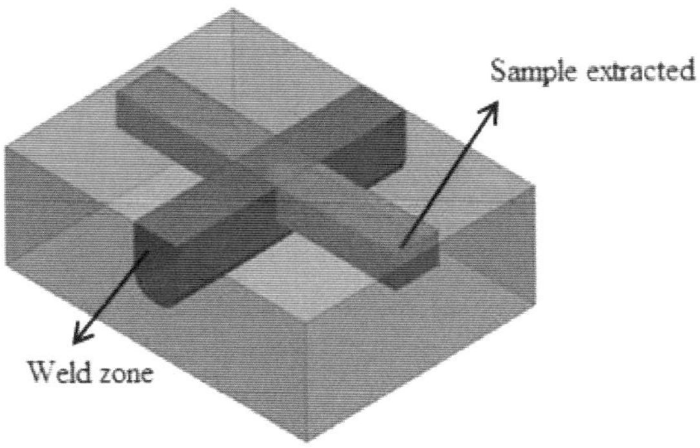

Figure 5: Flexural fatigue specimen extraction from welded block

For LCF test, stress-strain calibration was done using a strain gauge. A repeated cyclic (sinusoidal) load was applied to the test piece as shown in Fig. 6. This cyclic load was applied in such a way that throughout the test the bottom face of the test piece is continuously under tensile stresses while the top face is under compressive stresses. The load cycle was such that the stresses at the bottom face of the test piece underwent minimum stress of 100 MPa and maximum stress as per setting. The minimum and maximum stresses were both tensile in nature. This was done to avoid slipping of the test piece while testing.

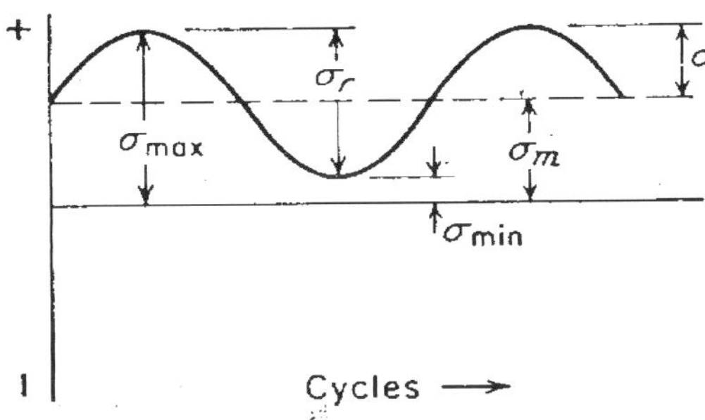

Figure 6: Repeated loading cycle applied during LCF testing [7]

For both L6 and Mailam 535 material, three specimens were tested. The fatigue testing was started at stress level approximately 80% of yield strength. This stress level was applied in cyclic form as explained earlier. The test was run for 3000 cycles at 5 Hz frequency. If the test piece did not fail then the stress level was increased by a step of 200 MPa. This process was repeated till the stress level at which the test piece failed below 3000 cycle was reached. The next test piece was then tested at the stress level 100 MPa less than the failure stress level of previous piece. For 2nd and 3rd test piece, the stress increment of 100 MPa was taken. The flexural fatigue strength value is taken as average of these three tests. Here, Fatigue strength is the stress value above which specimen fails in less than 3000 cycles.

The test set up was as shown in Fig. 7. LCF specimen rested on roller 1 and 2, whereas load is applied on specimen through roller 3.

Figure 7: Low Cycle Flexural Fatigue test setup

The stress-strain calibration was done in the elastic region. The failure stresses, calculated using this stress-strain calibration, turned out to be higher than the yield strength of the material. Hence, it was concluded that the failure stresses calculated, using the linear extrapolation of calibration curve, were wrong and required plasticity correction. The linear extrapolation predicted the value of stress along the brown line in Fig. 8.while the stress strain relation followed the curve shown by blue color in the same figure. The plasticity correction can be done using the Neuber's plasticity correction formula. Using this formula, any point on the pure elastic tensile curve (i.e. point A in Fig. 8) can be projected on the plastic curve (i.e. point B in Fig. 8). This was done using CAE analysis in ANSYS software.

For CAE analysis, the Engineering Stress vs. Engineering Strain curve was obtained from the tensile test results. This engineering stress-strain curve was converted to True Stress-Strain curve using the following formulas:

$$\varepsilon = \ln(1 + e) \quad (1)$$

where,

ε	:	True Strain and
e	:	Engineering Strain.

$$\sigma = s(1 + e) \quad (2)$$

where,

σ	:	True Stress,
s	:	Engineering Stress and
e	:	Engineering Strain.

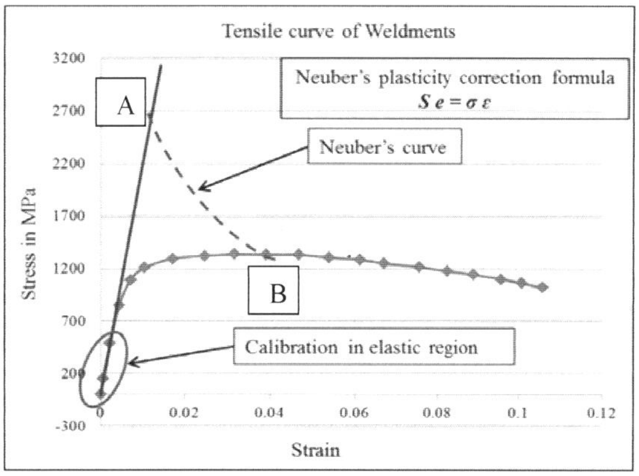

Figure 8: Neuber's plasticity correction curve

The True stress strain values obtained, thus, were used for the non-linear stress analysis in ANSYS. During the setting of simulation, the material was defined as a non-linear material. The value of young's modulus was defined as 1.9 E5 N/mm^2. The stress vs. strain values were entered manually. The first point entered was the point after which the curve was departing from its linear nature. After this all the stress and strain values were entered as per the data generated from tensile curve. The boundary conditions applied were in line with the actual condition.

Figure 9 shows the boundary conditions applied during the CAE analysis of the LCF specimen.

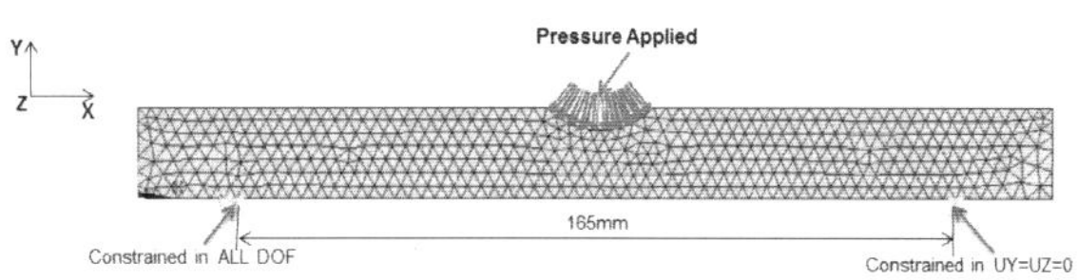

Figure 9: Boundary conditions applied during CAE analysis

Metallurgical Analysis

Metallurgical analysis was done for both L6 and Mailam 535 to compare their microstructure. Fractography was also conducted to study the mode of failure during mechanical testing.

Metallurgical Analysis: Microstructural analysis

1 inch cubical samples were prepared for both L6 and Mailam 535. Mounting of all the samples was done on STRUERS CITOPRESS-10 machine. A CARL ZEISS optical microscope with Image Analyzer software Z.2m was used for microstructure analysis using Electron Backscattered diffraction (EBSD). The microstructural analysis was performed in all the three directions on the 1 inch cubical samples.

Metallurgical Analysis: Fractography

Fractography was performed using SEM of CARL-ZEISS MA EVO18 make, equipped with an X-ray energy dispersion system and operated at 5-20 kV. Fractography was performed on the failed test pieces of the following tests:

1. Tensile test,
2. Charpy test and
3. Low cycle flexure fatigue test.

Experimental results

Tensile test results

Tensile test results are presented in Table 2, Fig. 10 and Fig. 11.

Table 2: Tensile test results of L6 and Mailam 535

Material	YS (MPa)	UTS (MPa)	% RA	% EA
L6	1216	1310	39.40	10.60
Mailam 535	843	1256	22.50	7.26

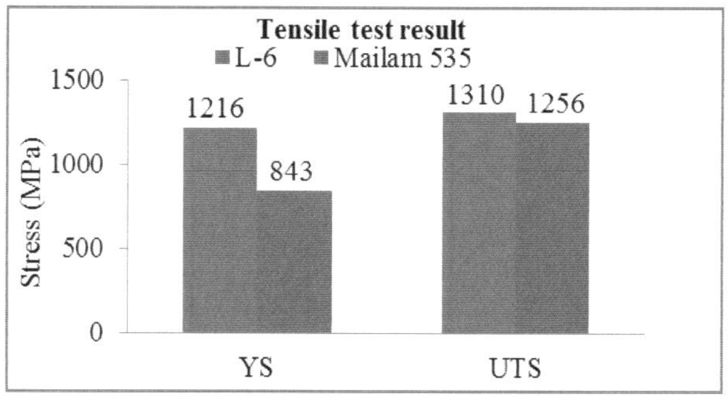

Figure 10: Comparative bar chart for YS and UTS

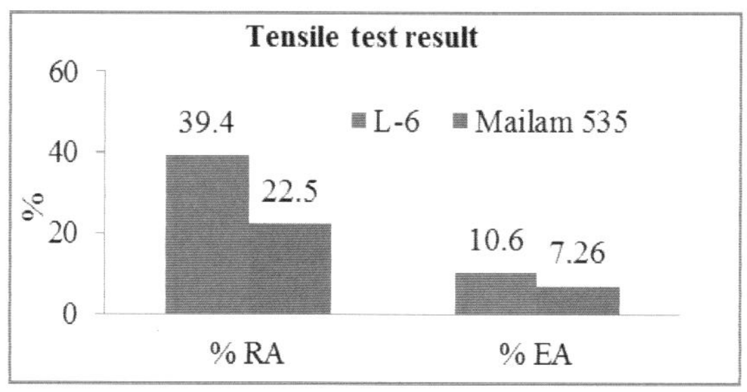

Figure 11: Comparative bar chart for %El and %RA

It can be observed from the results that L6 material has YS, UTS, %RA and %El is 30%, 4%, 43% and 31% higher than that of Mailam 535.

Charpy impact testing results

Charpy impact test results are presented in Table 3 and Fig. 12.

Table 3: Charpy impact test results (V-Notch)

Material	Direction	Avg. Impact Value (J)
L6	Longitudinal	58.70
	Transverse	36.50
Mailam 535	Longitudinal	22.10
	Transverse	15.73

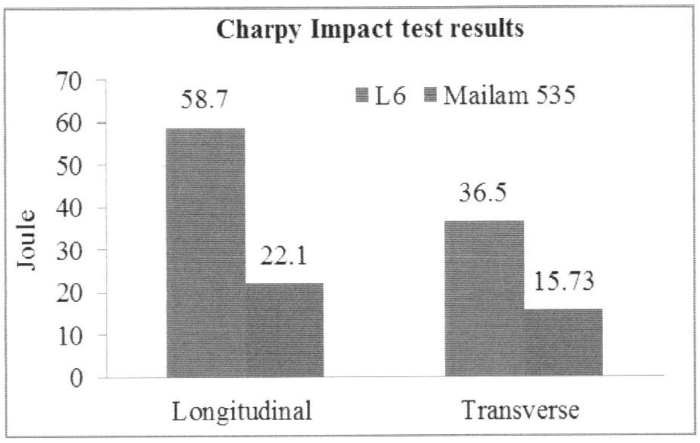

Figure 12: Comparative bar chart for Charpy Impact values

The impact value of the L6 material are 62.3% & 56.9% greater than that of the Mailam 535 welding material in longitudinal & transverse direction, respectively.

Results of Low cycle flexural fatigue test

LCF test results are presented in Table 4 and Fig. 13.

Table 4: Results of LCF flexural fatigue test – Pure elastic

Material	Pure elastic fatigue strength (MPa) @ 3000 cycles	Max. load applied (kN)
L6	2267	69.2
Mailam 535	1233	33.5

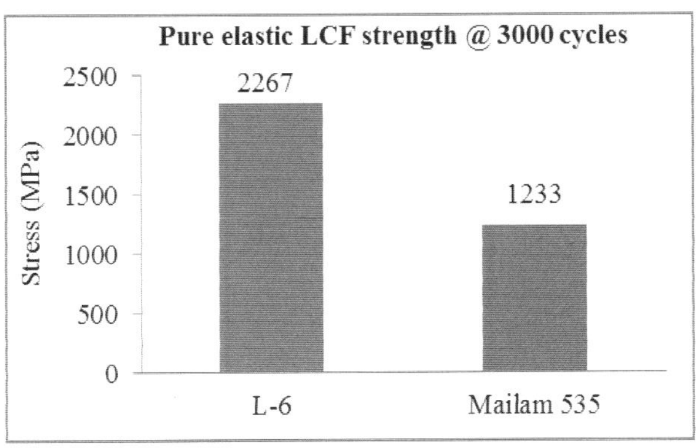

Figure 13: Comparative bar chart of flexural fatigue test – Pure elastic

Maximum load data were recorded for each specimen and average value for the same is presented in Table 4. As mentioned earlier, the calibration for stress strain was done in pure elastic region. As explained earlier, the pure elastic stresses obtained from linear extrapolation of calibration curve were corrected using elasto-plastic analysis done in ANSYS.

Results of CAE analysis for plasticity correction

Plasticity correction for L6

For this purpose the tensile test result was used. In CAE analysis, the 1st principle stress distribution was obtained as shown in Fig. 14. The results obtained from the CAE analysis were in the form of values of maximum 1st principle stress and load applied for each time step. These values are plotted in Fig. 15.

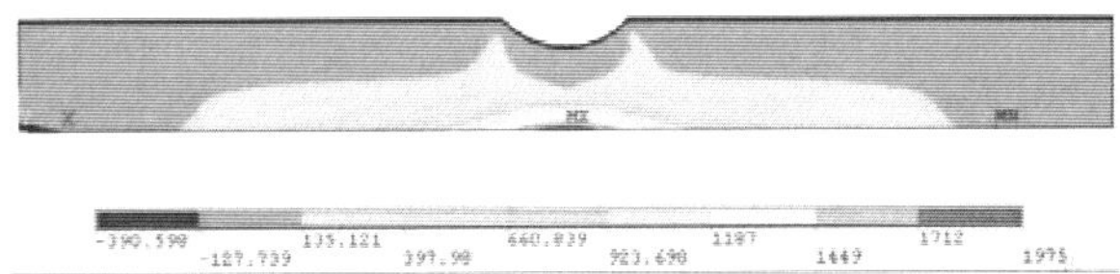

Figure 14: 1st principle stress distribution in the LCF test piece

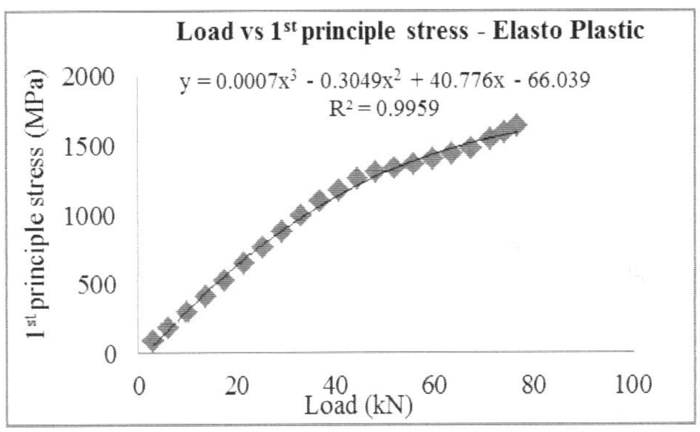

Figure 15: Load vs. Max 1st principle stress for L6

The load vs. 1st principle stress data was fitted using a 2nd order polynomial equation. The equation is as given below:

$$y = 0.0007 \times x^3 - 0.3049 \times x^2 + 40.776 \times x - 66.039 \quad .... (1)$$

where,

y: Maximum 1st principle stress at that FEM time step

x: Load applied at that FEM time step.

Plasticity correction for Mailam 535 welding

For this purpose the tensile results were used. The results obtained from the CAE analysis are plotted in Fig. 16.

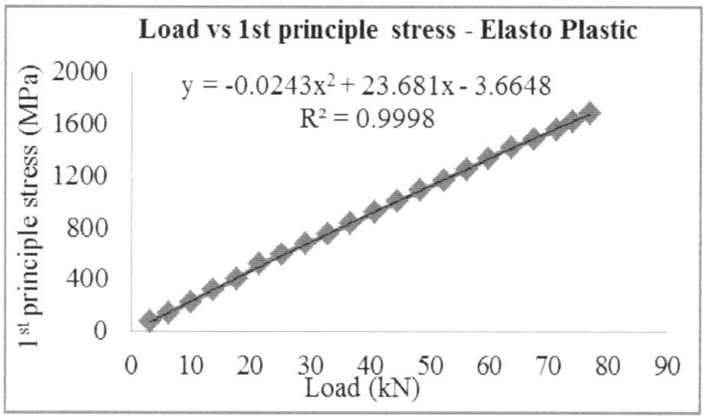

Figure 16: Load vs. Max 1st principle stress for Mailam 535

The load vs. 1st principle stress data was fitted using a 2nd order polynomial equation. The equation is as given below:

$$y = -0.0243 \times x^2 + 23.681 \times x - 3.6664 \ldots\ldots\ldots\ldots (2)$$

where,

y: Maximum 1st principle stress at that FEM time step

x: Load applied at that FEM time step.

Using the equation (1) and (2), the Elasto-plastic stresses generated during the flexure fatigue testing were calculated. The calculated results are given in Table 5 and shown in Fig. 17. After CAE analysis, it was observed that flexure fatigue strength @ 3000 cycles for die material is 50 % more as compared to that of Mailam 535.

Table 5: Results of LCF flexural test – Plasticity corrected

Material	Elasto-plastic Fatigue strength (MPa)@ 3000 cycles
L6 (Using eq. 1)	1525
Mailam 535 (Using eq. 2)	763

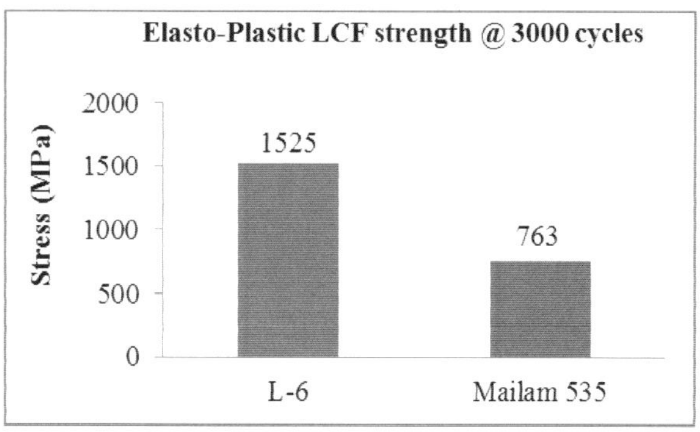

Figure 17: Comparative bar chart of flexural fatigue test – Plasticity corrected

Microscopic characterization

Figure 18 shows the microstructure of quenched and tempered L6 material, which shows fine, tempered martensitic microstructure. Fig. 19 and Fig. 20 present the microstructure of Mailam 535 welding material. Since, the features of the microstructure are very large (due to its cast structure); a mosaic image at 100X for a matrix of 5 x 4 was taken in (shown in Fig. 20) for weld material.

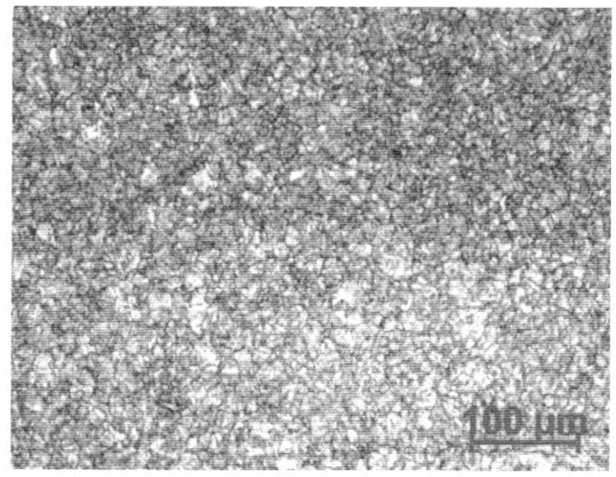

Figure 18: Microstructure of L6 material

The microstructure of the welding is essentially made up of ferrite flakes. From Fig. 19, it is observed that the welding has elongated grain structure. Mosaic image of welding is as shown in Fig. 20. It is clearly seen that the microstructure has only elongated grains. The micro analysis was performed in all three directions. This analysis did not reveal any variation in results. EBSD of welded samples was carried out which confirms the elongated grains present (refer Fig. 21).

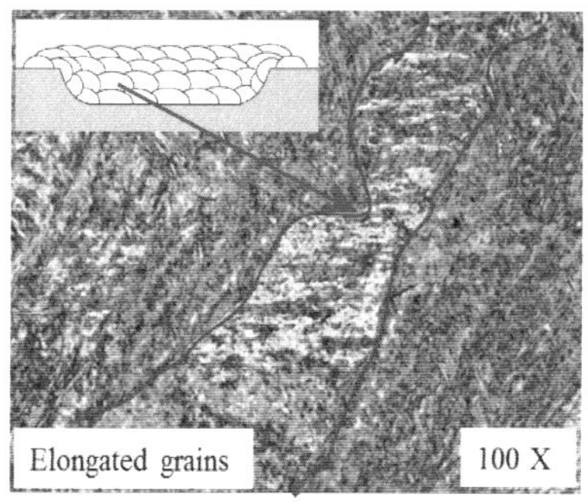

Figure 19: Microstructure of Mailam 535

Figure 20: Mosaic image of Mailam 535

Figure 21: EBSD IPF map of Mailam 535

Fractography

Fractography of failed specimen during mechanical testing for both L6 and Mailam 535 was carried out to study the difference in failure mode of the specimen.

Fractography of tensile test failed specimen

SEM images of failed tensile test specimen for Mailam 535 and L6 are given in Fig. 22 and Fig. 23, respectively. It can be seen that the failed sample of the Mailam 535 material shows a brittle fracture (shiny faceted surface) (refer Fig. 22 (A) & (B)), whereas for L6 material dimples were observed on the fractured surface which shows that the sample failed in ductile mode (refer Fig. 23 (A) & (B)).

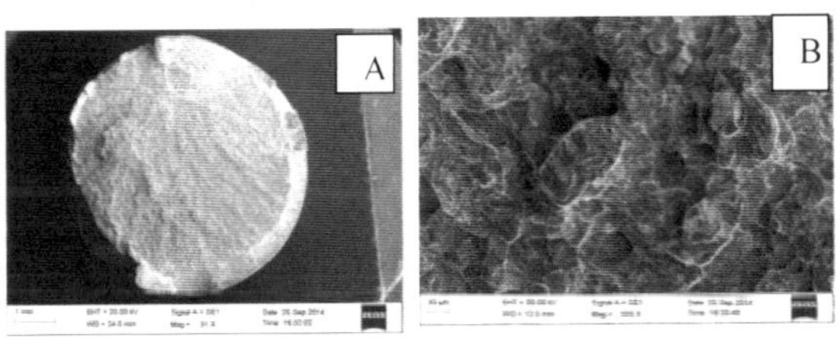

Figure 22: SEM of failed tensile test sample of Mailam 535

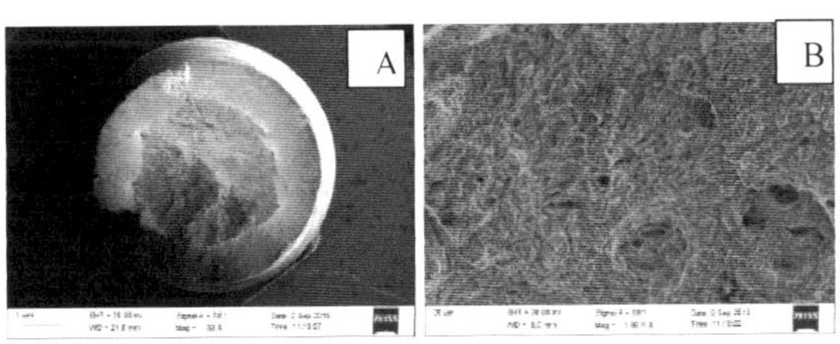

Figure 23: SEM of failed tensile test sample of L6

Fractography of Charpy impact test failed specimen

A SEM image of Charpy impact test specimen for Mailam 535 is shown in Fig. 24. For comparison purpose the images were taken near v notch and away from v notch. Fig. 24 shows that the fractured surface near the V notch (i.e. from where the fracture initiates in the Charpy test) is cleavage (brittle mode) type, while the same shows dimples (ductile mode) away from the V notch for Mailam 535 weld material. But for L6 material images taken near v notch and away from v notch; both show dimples on fractured surface i.e. material failed in ductile mode as shown in Fig. 25. The images were taken from longitudinal as well as transverse direction samples.

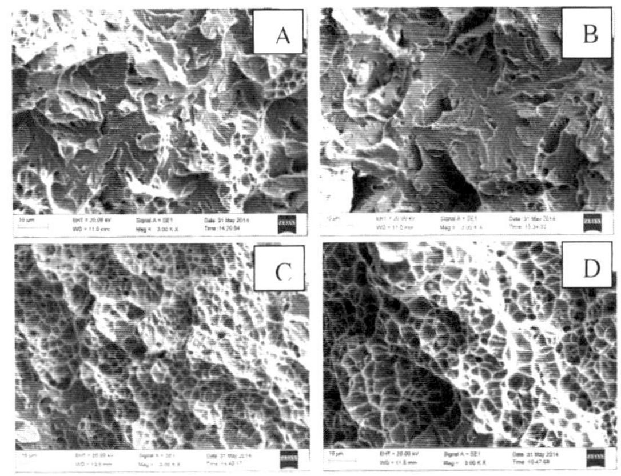

Figure 24: SEM image for Mailam 535 near V notch (A) Transverse, (B) Longitudinal and away from notch for (C) Transverse, (D) Longitudinal

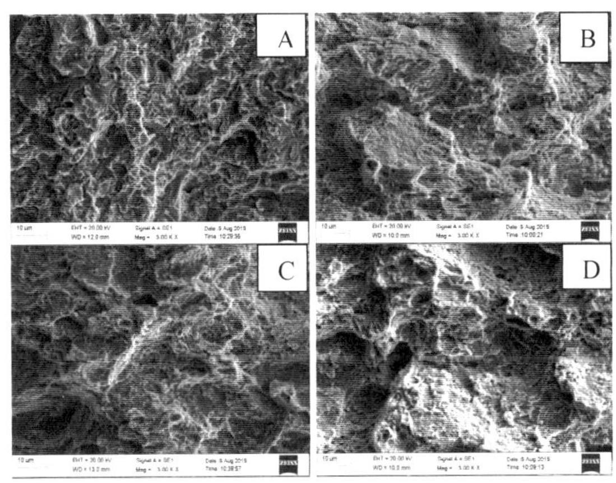

Figure 25: SEM image for L6, near V notch (A) Transverse, (B) Longitudinal and away from notch for (C) Transverse, (D) Longitudinal

Fractography of low cycle fatigue test failed specimen

Fractography of flexural fatigue specimen was also carried out. The SEM images were taken at crack initiation locations (refer Fig. 26 (A) & (B)) in order to find difference between the failure modes. L6 material shows a distinct crack initiation point and the material failed in ductile manner i.e. fractured surface shows less waviness. EDX analysis was carried out at crack initiation point which shows the peaks of MgO, SiO_2 & Al_2O_3 inclusions (refer Fig. 27). For weld specimen no crack initiation point was observed. The specimen failed in brittle manner i.e. fractured surface shows more waviness and shiny facets.

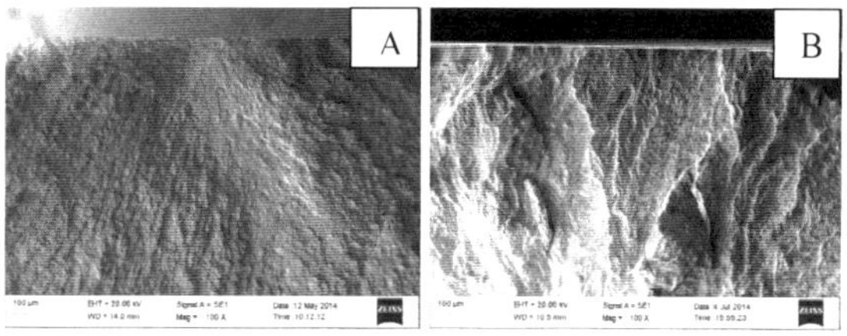

Figure 26: SEM of failed LCF test sample of (A) L6 and (B) Mailam 535

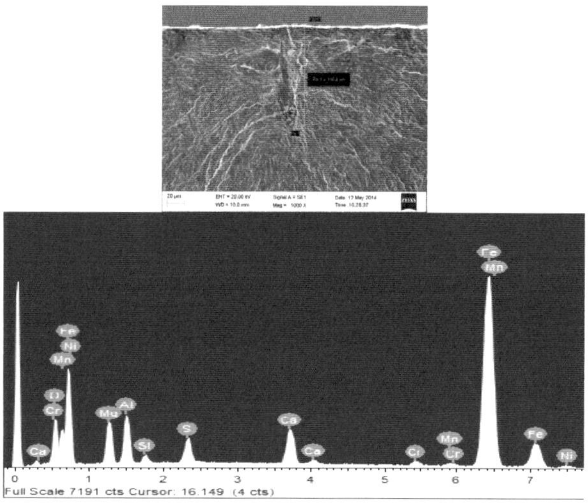

Figure 27: EDX of failed LCF test sample of L6

Discussions

As is explained in the "Introduction" section, if a hot forging die fails due to cracking then it has to be repaired using the 'gouging – welding – machining' process. The forging die is normally made up of die steel (or tool steel) which is in the form of a forged and heat treated material. On the other hand, the weld deposit is a cast structure and no heat treatment is done for it (to avoid any side effects on the die material).

The microstructure of material has a significant effect on its properties. The microstructure of die material has tempered martensite (refer Fig. 18) having very fine prior austenitic (equiaxed) grain size. As opposed to this the weld deposit has large elongated dendritic grains (refer Fig. 19 and Fig. 21). To counter the effect of coarse grains and absence of heat treatment, the weld metal is highly alloyed (refer Table 1). Even with this high alloying content the microstructure difference significantly degrades the mechanical properties of the weld deposit (refer Fig. 10, Fig. 11 and Fig. 12).

The alloying of the weld material is designed in such a way that even in cast condition and in absence of heat treatment, it achieves same hardness as that of die material. This is reflected from the fact that the UTS of both the materials is similar (refer Fig. 10). But, the difference in microstructure reduces the YS, %RA and %El. This means that although the strength of the materials is same but the resistance to plastic deformation and ductility of the weld deposit is less.

The reduction in ductility/toughness is also confirmed by the Charpy impact test results (refer Fig. 12).

As far as LCF is concerned, the literature survey shows that the fatigue failure basically consists of 3 phenomena's. These are crack initiation (i.e. formation of micro cracks having crack length up to 1 μm), stable crack propagation and then final failure. These phenomena's are visible on the fractured surface as the crack initiation zone, the stable crack growth zone and sudden failure zone [8-11].

The crack initiation zone (normally showing intra granular failure) is normally the area around the point at which the crack has initiated. This zone is very small in size (few microns). Previous studies done on poly crystalline materials have shown that under fatigue loading the micro cracks are produced simultaneously at multiple locations. The number of micro cracks generated goes on increasing with increase in the number of cycles. Out of these micro cracks only one (or very few) crack grows depending on the stress-strain fields around it [5, 8-11]. In case of L6 die material samples, the micro crack formed near inclusions propagates preferably to other micro cracks. This gives a prominent crack initiation site on such samples (refer Fig. 27). But, for the Mailam 535 welded samples, analyzed in this study, no crack initiation zone is visible (Refer Fig. 26 (B)). This can be explained as follows. The phenomenon of LCF is observed in very highly stressed zone. The maximum stresses present are around the YS of the material which is close to the plastic regime of the stress strain curve (evident from Table 2 & Fig. 8). In this regime, the resistance to fatigue failure is dependent more on the ductility/toughness of the material. The dendritic microstructure present in the case of the welded sample is brittle in nature hence; it easily allows crack nucleation and propagation. Hence, the fatigue performance of the L6 specimen is better than the Mailam 535.

The ratio of fatigue strength and UTS is also an important parameter. The said ratio obtained for L6 is 1.22 as against 0.60 obtained for Mailam 535. Similarly tensile and Charpy properties of L6 are higher than that of Mailam 535 due to microstructural difference.

Conclusions

1. L6 shows higher values of YS, UTS, %El & %RA than that of Mailam 535 by 30%, 4%, 42% & 31% respectively.
2. L6 shows higher impact value in longitudinal and transverse direction than that of Mailam 535 by 62% & 57% respectively.
3. LCF strength of L6 is higher than that of Mailam 535 by 50% after plasticity correction.

4. L6 material shows fine tempered martensitic microstructure, whereas Mailam 535 material shows dendritic cast structure i.e. very large elongated grains.
5. L6 material shows crack initiation point and material failed in ductile manner, whereas no crack initiation point was observed for Mailam 535 and material failed in brittle manner

Acknowledgement

The authors gratefully acknowledge the extended support provided to this work by KCTI (Kalyani Centre for Technology and Innovation) for providing financial funding, laboratory and library facilities. The authors also acknowledge the support provided by Bharat Forge Ltd, Pune and DSIR (Department of Scientific and Industrial Research), Govt. of India. Finally, the authors would like to express special thanks and gratitude to review committee and top management of Bharat Forge Ltd for granting the permission to publish/present the research work.

The help of Girish Shegavi, Abhijit Patil and CAE team of CDFD & KCTI for their support regarding the elasto plastic simulation of the flexure fatigue test is gratefully acknowledged.

References

[1] Taylan, A. *et al.*, Cold and Hot Forging Fundamentals and Application, ASM International (Materials Park, Ohio, 2005), pp. 295-311.

[2] Deshpande, M. *et al.*, "Selection of die materials and surface treatments for increasing die life in hot and warm forging," 644-FIA Tech Conference, Chicago (2011) pp.1-32.

[3] Shivpuri, R. *et al.*, ASM Handbook-Volume 14A Metalworking: Bulk Forming, ASM International (Ohio, 2005), pp. 62-83.

[4] Mazurkiewicz, A. *et al.*, "Comparative analysis of wear mechanism of different types of forging dies," Archives of materials Science and Engineering, Vol. 49, No. 1 (2011), pp. 40-45.

[5] Mellouli, D. *et al.*, "Hardness effect on thermal fatigue damage of hot-working tool steel," Engineering Failure Analysis, Vol. 45 (2014), pp. 85-95.

[6] "Welding of Tool Steel", www.uddeholm.com.

[7] Dieter, E. *et al.*, Mechanical Metallurgy, McGraw-Hill Book Company (London, 1988) pp. 375-432.

[8] Pang, W. *et al.*, "Hardness and microstructural gradients in the heat affected zone of welded low-carbon quenched and tempered steels," Australasian Welding Journal, Vol. 56, No.2 (2011), pp. 36-48.

[9] Patil, Sachin S. *et al.*, "Study of low cycle flexural fatigue performance of different hot forging tool steels," 7th International Conference on Creep, Fatigue and Creep-Fatigue Interaction, IGCAR, Kalpakkam, India, January 2016.

[10] Patil, Sachin S. *et al.*, "Comparison of low cycle fatigue behavior of AISI H13 treated using different surface hardening processes," MISE 2015, Brisbane, November 2015.

[11] Lampman, S. *et al.*, Heat Treating, ASM Handbook vol. 4, ASM International (Ohio, 2003), pp. 607-25.

International Federation of Heat Treatment and Surface Engineering 2016
Proceedings of the 23rd IFHTSE Congress
April 18–21, 2016, Savannah, Georgia, USA

Sub-zero Celsius treatment: a promising option for future martensitic stainless steels

Matteo Villa, Thomas L. Christiansen, Marcel A.J. Somers
Technical University of Denmark, 2800 Kongens Lyngby, Denmark
matv@mek.dtu.dk, tch@mek.dtu.dk, somers@mek.dtu.dk

Abstract

A series of samples of (in wt.%) 11.5Cr-0.67C martensitic stainless steel grade were austenitized in Argon for 1 hour at temperatures ranging from 1010°C to 1190°C. Additionally, a series of samples of (in wt.%) 15.0Cr-5.8Ni-1.0Mo-0.03C (EN 1.4418) martensitic stainless steel grade were solution nitrided at 1110°C for 2 hours in N_2 at a pressure of 1 atm. Some of the samples were subjected to a sub-zero Celsius treatment. Four types of sub-zero Celsius treatments were investigated: (i) isothermal holding at -78°C; (ii) immersion in boiling nitrogen followed by immersion in water; (iii) immersion in boiling nitrogen followed by isothermal holding at -78°C; (iv) isothermal holding at -78°C followed by immersion in boiling nitrogen and subsequently in water. The effectiveness of the different sub-zero Celsius treatments in promoting an additional hardening response was investigated by reflected-light microscopy and Vickers micro-hardness indentation. Complementary electron back-scatter diffraction was applied for determining the phase fractions of austenite and martensite. Data shows that sub-zero Celsius treatment yields an additional hardening response when austenite is retained in the material. The relevance of the additional response was observed to be a function of the sub-zero Celsius treatment applied, as well as of the stability of the retained austenite.

Introduction

Martensitic stainless steels are hard, high strength, sufficiently tough steels, that show resistance to corrosion in aqueous environment at low temperatures [1]. These properties are obtained with a microstructure consisting of a fine martensite matrix containing a sufficiently high Cr content, coupled with finely dispersed precipitates [1]. The thermal processing for achieving this microstructure contains three steps: austenitization, to form austenite; quenching, to completely transform the austenite into martensite; tempering, to form finely dispersed precipitates within the martensite. The chemical composition of martensitic stainless steels has to fulfill a series of constraints.

Firstly, the chemical composition must be tailored to provide the passive surface that yields the stainless characteristic. In Ref. [2-6] this condition is handled by imposing the constraint that the Cr content in martensite after tempering must exceed 12%. This constraint can be relaxed under significant alloying with Co; according to Ref. [1], a Cr content of 9% can suffice for stainless properties if Co-content is at least 15%.

Secondly, the chemical composition must be tailored so that austenite is the only thermodynamically stable phase in a large temperature interval within the range 800°C-1300°C. In Refs. [2-6], this condition is approached with a commercially available software, Thermocalc©, imposing the constraint that the fraction of austenite in thermodynamic equilibrium must exceed 99% in the temperature interval of interest for austenitization. In general terms, this constraint implies that alloying with ferrite stabilizing elements, such as Cr, V, Mo, Ti and Al, must be balanced by alloying with austenite stabilizing elements, such as C, N, Ni, Cu and Co.

Thirdly, the chemical composition must be tailored so that the austenite completely transforms into martensite during quenching. In Ref. [1-6] this condition is handled by introducing the constraint that the martensite start temperature, M_s, must be higher than 150°C (or, more conservatively, 200°C). In this case, complete transformation of austenite into martensite occurs during quenching to room temperature [1]. M_s can be calculated applying equations available in the literature [7], among which the one in Ref. [8] is the most applied. The constraint on M_s implies that alloying with C, N, Ni, Cr and Cu must be limited, because these elements all lower M_s. Alternatively it must be balanced by sufficient alloying with Co, which is the only element known to significantly increase M_s.

Lastly, the chemical composition must be tailored to obtain the formation of favorable precipitates during tempering. A series of possible precipitates are listed in Ref. [1-6,9] and their effectiveness in maximizing the mechanical properties of martensitic steels is evaluated in Ref. [3]. Examples of favorable types of precipitate are MX and M_2X carbides, Cu particles and $Ni_3(Al/Ti)$ intermetallics. These are obtained by alloying the steel with C+V/Mo, Cu and Ni+Al/Ti, respectively.

In the light of the above constraints, it is evident that conflicting objectives exist in the design of new martensitic stainless steel grades. Targeting a simultaneous improvement of all properties is generally not possible. For example, enhanced corrosion resistance can be promoted by increasing the Cr content above its minimum value for stainless characteristics [1]; however, Cr can prevent the formation of martensite by decreasing M_s, as well as by destabilizing austenite at high temperature. Increasing the C content is the most effective method to increase the hardness and the strength of steels [1,9]; however, alloying with C significantly decreases M_s and imposes additional constraints (i.e. the presence of Mo and / or V to improve toughness and prevent the formation of Cr carbides, which are detrimental for corrosion resistance). Alloying with Cu, and/or Ni+Al/Ti are other possible solutions for high strength materials [2-6,9]; however, both these solutions are similarly associated with a lowering of M_s. Evidently, the properties of martensitic stainless steels, when considered individually, can all be improved by increasing the fraction of alloying elements. Generally, such improvement is impeded by enhanced austenite stability by increasing the contents of alloying elements.

In Refs. [1-6], two possibilities have been considered to improve the performances of new martensitic stainless steels, without relaxing the constraint M_s>150°C. In the first solution, Co is added to the steel to extend the alloying possibilities. In addition to increasing M_s, Co is an austenite stabilizer and it can improve the corrosion resistance of the material at constant Cr content [1]. This solution is evidently applied in Ref. [1], which suggests a new grade containing (in wt.%) 15%Co-9%Cr-1.5%Ni-0.2%C as bearing materials with superior performance. Unfortunately, Co is expensive and significant alloying with this element is not suited for the majority of cases. In the second solution, the chemistry and heat treatment parameters are tailored to maximize the properties of the material by optimizing its microstructural features, among which the type, distribution and size of the precipitates formed during tempering. This solution is applied in Refs. [2-6], to obtain new ultra high strength martensitic stainless steels.

An alternative strategy is to relax the constraint M_s>150°C, including an additional treatment step at subzero Celsius temperature prior to tempering. This solution was extensively applied in the past for the development of grades with outstanding performances, as demonstrated by the case of 17-7 PH and 15-7 PH semi-austenitic stainless steel grades [10-12], and Nanoflex© maraging stainless steel [13-14]. In these materials, austenite partially transforms into martensite upon quenching to room temperature. Complete transformation is obtained applying an additional treatment step at sub-zero Celsius temperatures [10-16].

The present work aims to illustrate that sub-zero Celsius treatment is a very promising solution to significantly extend the alloying possibilities for martensitic stainless steels, and consequently to improve their performances.

The investigation consisted of the two following steps. Firstly, the fraction of alloying elements dissolved in the austenite of two martensitic stainless steel grades was systematically increased following two different approaches (see next section). Thereafter the materials were characterized both after conventional quenching to room temperature, as well as after a series of additional treatment steps at sub-zero Celsius temperatures.

Experimental methods

Materials and Heat Treatment

The alloys investigated in this work were a Fe-Cr-C model alloy with composition Fe-11.5Cr-0.67C (in wt-%) and a commercial EN 1.4418 with composition 15.0Cr-5.8Ni-1.0Mo-0.03C (in wt.%).

In the first approach, the fraction of alloying elements dissolved in the austenite phase of the Fe-Cr-C alloy was systematically varied by variation of the austenitization temperature. In the second approach the fraction of alloying elements dissolved in austenite was systematically varied by applying a nitrogen gradient in EN 1.4418.

Fe-Cr-C samples were disks 2mm in diameter, 1 mm thick, which were austenitized at temperatures in the range from 1010°C to 1190°C with 20°C intervals applying a continuous flow of 0.2 l/min Argon in a horizontal tube furnace. After austenitization, samples were quenched to room temperature at an average cooling rate of 40°C/min. EN 1.4418 samples were squared plates, 5mm x 5mm, 2mm thick which were solution nitrided at 1110°C for 2 hours in a horizontal tube furnace, applying a continuous flow of 0.5 l/min nitrogen at a pressure of 1 bar. After solution nitriding, the samples were quenched to room temperature at an average cooling rate of 40°C/min. Samples in this condition are the reference state and designated RT, to benchmark the effect of sub-zero Celsius treatment.

Sub-zero Celsius treatment

After quenching to room temperature, the samples were stored at room temperature for three days and thereafter subjected to four different sub-zero Celsius treatments: (i) samples DI were kept at the sublimation temperature of dry ice (i.e. -78°C) for 24 hours; (ii) samples BN were immersed in boiling nitrogen for 60 s and thereafter upquenched to room temperature in water; (iii) samples named BN+DI were immersed in boiling nitrogen for 60 s and thereafter treated at -78°C for 24 hours; (iv) samples DI+BN were stored at -78°C for 24 hours and thereafter immersed in boiling nitrogen for 60 s and upquenched to room temperature in water.

Metallography

After heat and cold treatment, samples were prepared for metallographic investigation. Preparation for hardness measurement and microscopic investigation consisted of standard metallographic techniques. Hot mounting implied heating of the samples to 180°C for 300 s and was identical for all samples. Preparation for electron back-scatter diffraction, EBSD, investigation consisted of polishing with colloidal silica suspension as the final step, without prior mounting. Reflected light microscopy was performed with an Olympus GX 41 microscope equipped with a Leica DFC 450 C camera. Kalling´s II reagent was used to reveal the microstructure of the material. EBSD for quantitative phase fraction determination was only applied to the Fe-Cr-C samples austenitized at 1190°C. To this end, a FEI Helios Nanolab 600 equipped with an EBSD system from EDAX-TSL and Hikari camera was used. Hardness measurements were performed using a Future Tech model FM-700 micro-hardness tester equipped with a Vickers indenter applying a load of 0.5 kg (Fe-Cr-C) or 0.1 kg (EN 1.4418) and a dwell time of 10 s.

Results and discussion

Variation of the austenitizing temperature of Fe-Cr-C

As-quenched samples (RT)

Fig. 1 illustrates the results of metallographic investigation for the 11.5Cr-0.67C steel. All micrographs refer to samples in condition RT (i.e. after austenitization and quenching to room temperature). Four austenitization temperatures, 1010°C, 1070°C, 1130°C and 1190°C, were selected as representative of the entire temperature range under investigation and are reported in Fig. 1 from top to bottom, respectively.

Fig. 1 shows that the microstructure of the material is strongly dependent of the austenitization temperature. Austenitization at the lowest applied temperatures (1010°C and 1070°C in Fig. 1, top) yields the formation of a fine martensitic microstructure during cooling to room temperature, which is decorated with a large number of fine primary (i.e. formed during austenitization) carbides. The primary carbides are observable in Fig. 1 as white ellipsoids uniformly distributed in the material. Gradually increasing the austenitization temperature promotes coarsening of the microstructure together with the dissolution of primary carbides, which are no longer observable in the material treated at 1170°C and at 1190°C (see for example 1190°C in Fig. 1, bottom). The correlation between the presence of primary carbides and a fine microstructure is not surprising, since carbide phases pin the austenite grain boundaries, hence slowing down grain growth during austenitization.

The presence of retained austenite was also observed in Fig. 1 as dark polygonal areas located in-between the light martensite units (see for example 1190°C in Fig. 1, bottom). Retained austenite was observed in the samples austenitized at, and higher than, 1110°C and its fraction was observed to increase with an increase of the austenitization temperature (compare for example 1130°C and 1190°C in Fig. 1, bottom).

Regarding the corrosion resistance of the material, the dissolution during etching of the samples provides qualitative information. It was observed that to reveal the microstructure of the material after austenitization at 1010°C etching in Kalling´s II reagent for 3 s was sufficient. For Fe-Cr-C treated at 1070°C and 1130°C 4 s were necessary, while 5 s were necessary for the material treated at 1190°C. Evidently, the resistance of the material to the chemical attack of Kalling´s II reagent increased with an increase of the austenitization temperature.

Summarizing, the metallography for the RT series of Fe-Cr-C samples are consistent with an enrichment of the austenite in C

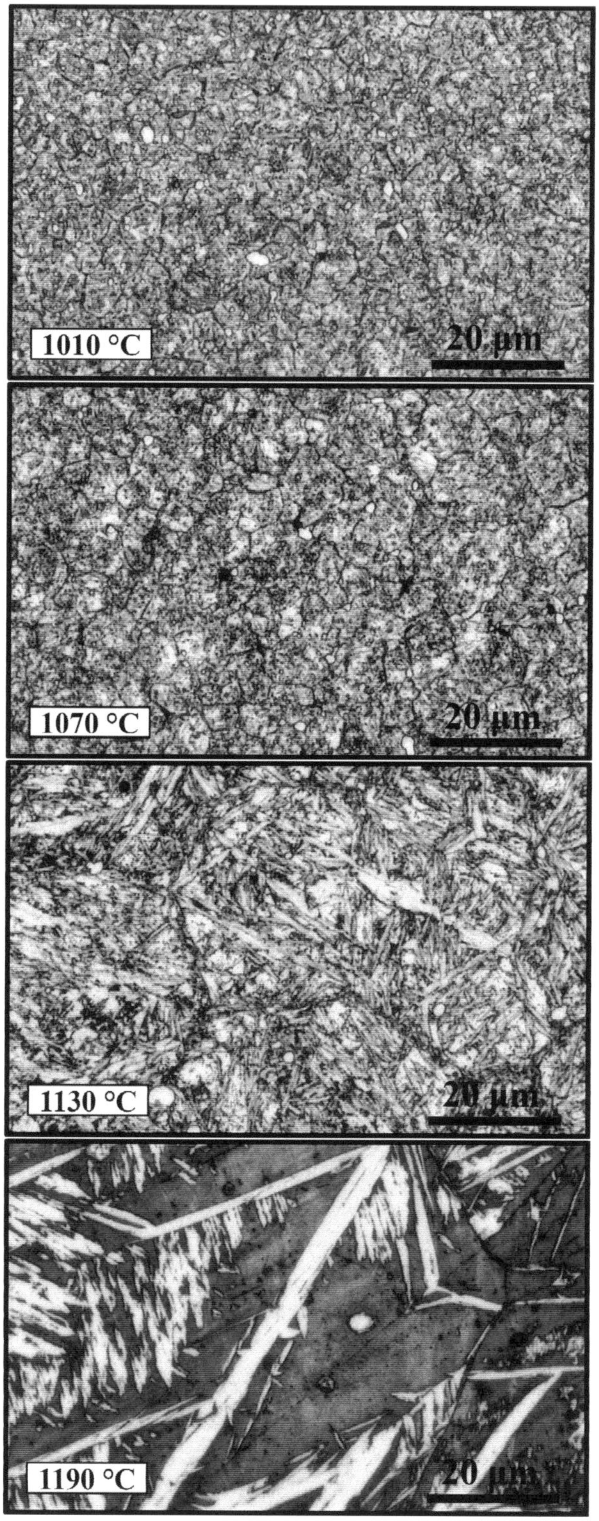

Figure 1: Micrographs showing the microstructure of the samples in RT condition after austenitization at, from top to bottom: 1010°C, 1070°C, 1130°C and 1190°C. Optical microscopy. Etching with Kalling´s II reagent for 3-5 s.

and Cr with increasing austenitizing temperature. Firstly, enrichment of austenite in Cr is expected to improve the resistance of the material to aggressive environmental conditions, as it was observed during etching in Kalling's II reagent. Secondly, enrichment of the austenite in C and Cr with increase of the austenitization temperature is known to stabilize this phase against its transformation into martensite during quenching, eventually promoting its retention, as clearly observed for the material treated at, and at temperatures higher than, 1110°C. Finally, enrichment of the austenite in C and Cr is directly evidenced by the reduction of the fraction of primary carbides in the microstructure with increasing austenitization temperature.

Sub-zero Celsius treated samples

Metallography was also applied to investigate the microstructure of the material after additional treatment at sub-zero Celsius temperatures. Investigation was performed for samples in all the four sub-zero Celsius conditions under investigation, i.e. DI, BN, DI+BN and BN+DI. Micrographs relative to the sub-zero Celsius treated samples were considered redundant for the present discussion and are not reported. The investigation showed that the effect of sub-zero Celsius treatments on the microstructure of the material is merely to reduce the fraction of retained austenite in the sub-zero Celsius treated specimens as compared to the RT ones. A clearly observable difference for the differently sub-zero Celsius treated samples was not reported.

EBSD investigation was applied in order to complement the study of the microstructure with quantitative data. The investigation was performed for the samples treated at 1190°C, where austenite and martensite were the only two phases present in the material. The measured fraction of retained austenite was 58% for condition RT, 40% for conditions DI, DI+BN and BN, and 35% for condition BN+DI, respectively. Evidently, even if limited to the sample treated at 1190°C, the investigation confirms that sub-zero Celsius treatments yield a reduction of the fraction of retained austenite in the samples as compared to conventional quenching to room temperature. Additionally, it suggests that not all the sub-zero Celsius treatments are similarly effective in promoting an additional formation of martensite: for the material austenitized at 1190°C, immersion in boiling nitrogen followed by long isothermal holding at -78°C appears to be the most effective treatment.

To assess the effectiveness of the different sub-zero Celsius treatments versus austenitization temperature, additional investigation was performed by hardness measurements. The results of the investigation are reported in Fig. 2, which shows that the hardness of the material is independent on the condition under investigation when the austenitization treatment is performed at temperatures lower than 1090°C, whereas the condition of the material influences its hardness when the austenitization treatment is performed at higher temperatures. Additionally, the hardness of the material is

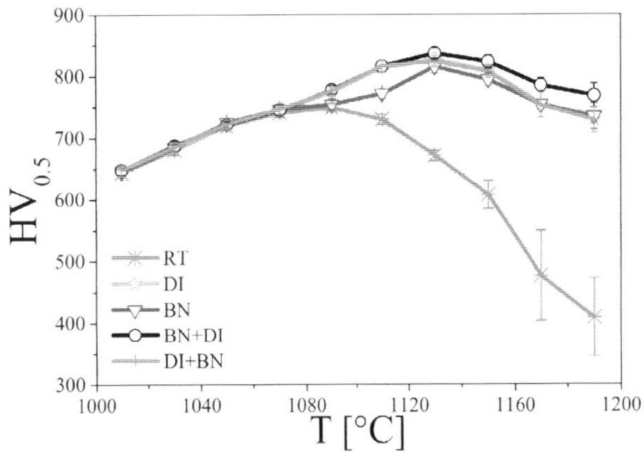

Figure 2: Hardness of the material $HV_{0.5}$ for the different conditions under investigation plotted versus austenitization temperature T. All data are reported in terms of averaged values ± standard deviation of the measurements.

strongly dependent on the austenitization temperature. In all investigated conditions, i.e. RT, DI, BN, BN+DI and DI+BN, the hardness of the material initially increases with increasing of the austenitization temperature, reaching a maximum in correspondence of austenitization temperatures of 1090°C and 1130°C for the samples RT and for the sub-zero Celsius treated, SZT, samples, respectively, and then decreases when the austenitization temperature is further increased. The maximum hardness is 750 HV for samples RT and ranges between 810 HV and 840 HV for the SZT samples, depending on the specific sub-zero Celsius condition under investigation.

Hardness data is discussed in the following.

The first point of the discussion is that hardness measurements yield information on the microstructure of the RT samples. In fact, data in Fig. 2 reveals the presence of retained austenite in the RT samples austenitized at, and higher than 1090°C. The presence of retained austenite is revealed by the higher hardness measured for the SZT samples as compared to the corresponding RT samples. Higher hardness indicates an additional conversion of austenite into martensite during sub-zero Celsius treatment, which is only possible when the austenitization temperature is sufficiently high to promote the presence of retained austenite in the RT samples. On the other hand, austenite appears fully transformed into martensite during cooling to room temperature when austenitization is performed at temperatures below 1090°C, as shown by hardness values that are independent of whether an additional sub-zero Celsius treatment is applied or not.

Data appears consistent with metallographic investigation. The fact that metallography does not reveal the presence of retained austenite after austenitization at 1090°C can be interpreted as an indication that the regions of retained austenite in the material treated at 1090°C are too small to be resolved by optical microscopy.

The second point of the discussion is that hardness measurement indicates that an additional treatment step at sub-zero Celsius temperatures is fundamental to maximize hardness. As presented in the following, this conclusion can be generalized for all martensitic stainless steels with a microstructure composed of austenite and primary carbides at the austenitization temperature, like grades AISI 420 and AISI 440, and can be extended to include considerations on the resistance of the material to corrosion. High hardness and high corrosion resistance are the properties tailored in the development of martensitic stainless steels.

Data in Fig. 2 shows the presence of maxima in the hardness as measured versus austenitization temperature for all investigated conditions. The observation of maxima can be interpreted in terms of the counteracting effects on hardness of the enrichment of the austenite in C and Cr with increasing austenitization temperature: on the one hand, enrichment of austenite in C promotes an increase in the hardness of the martensite that forms during quenching, and consequently of the martensitic microstructure; on the other hand, C and Cr both prevent the formation of martensite by stabilizing the austenite, which is a phase significantly softer than martensite, eventually yielding softening of the material.

According to the above reasoning, softening can only take place after austenitization at a temperature somehow higher than the one firstly promoting the formation of retained austenite. In the present case, retained austenite in the RT samples is firstly reported at 1090°C, in correspondence with the maximum hardness, 750 HV, and softening is firstly observed upon austenitization of the material at a temperature 20°C higher, 1110°C.

The effect of sub-zero Celsius treatments is to promote additional conversion of austenite into martensite as compared to conventional quenching to room temperature. Additional conversion of austenite implies that softening, and accordingly the maximum in hardness, can only take place upon an additional increment of the austenitization temperature as compared to the RT condition. Furthermore, the hardness at the point of maximum will be higher for the SZT samples as compared to the hardness at the point of maximum for the RT samples. In the present case, softening of the SZT samples is observed upon austenitization at 1150°C, a temperature 40°C higher than for samples RT, after hardness reached a maximum for the material austenitized at 1130°C. The maximum value is in the range 810-840 HV, 60-90 HV higher than the maximum value obtained for the RT samples (i.e. 750 HV).

Evidently, application of a sub-zero Celsius treatment after austenitization and quenching to room temperature is mandatory practice when hardness maximization is targeted. To maximize hardness it is not sufficient to sub-zero Celsius treat an RT sample austenitized to obtain maximum hardness in the RT condition, but it is required to re-consider the full heat treatment cycle, firstly establishing the austenitization temperature yielding maximum hardness for the sub-zero Celsius treated conditions. This temperature will be significantly higher than the austenitization temperature yielding maximum hardness for the RT condition, 40°C higher in the present case.

Remarkably, an increase in austenitization temperature for the present steel, and similarly for grades AISI 420 and AISI 440, is associated with an enrichment of the austenite in Cr. Consequently, a component austenitized to obtain maximum hardness after sub-zero Celsius treatment will not only show higher hardness as compared to the same component after conventional quenching to room temperature in optimal conditions, but it will also present superior corrosion resistance. In a few words, it will show better properties.

The third point of the discussion is that hardness measurement indicates that not all the sub-zero Celsius treatments are equally effective in promoting an additional hardening response as compared to conventional quenching to room temperature, and that their effectiveness is dependent on the austenitization temperature. In general, the most effective treatments are those relying on the isothermal formation of martensite during a long holding step of 24 hours at -78°C.

No differences are reported for treatments DI and DI+ BN, as an indication that when the material is kept 24 hours at -78°C, no additional phenomena take place during subsequent cooling to boiling nitrogen temperature (i.e. -196°C). Moreover, austenitization at 1090°C and 1110°C, yields maximum hardness for the DI, DI+BN and BN+DI conditions. For these conditions, hardness is identical within experimental accuracy, and it is 20-40 HV higher than for the BN samples. Evidently, 24 hours isothermal holding at -78°C is more effective than immersion in boiling nitrogen for 1 minute to maximize hardness. Finally, austenitization at, and higher than 1130°C, always yields maximum hardness for the BN+DI samples only, as an indication that an additional holding step of 24 hours at -78°C after immersion in boiling nitrogen promotes an increase in hardness.

Based on the hypothesis that hardness reflects the progress of the austenite to martensite transformation, data indicates that martensite forms isothermally at -78°C, and isothermal formation of martensite takes place both when the material is directly cooled to -78°C, as well as when the material is firstly cooled to -196°C, by immersion in boiling nitrogen prior to isothermal holding at -78°C. This hypothesis is indeed validated by EBSD data, at least concerning the material treated at 1190°C.

Isothermal martensite formation in martensitic stainless steels at sub-zero Celsius temperatures was reported several time in the literature [13,15-21] and it is now a well-known, but also largely not understood, phenomenon. The present investigation shows that this phenomenon cannot be neglected in heat treating practice, because it may have a significant effect on the properties of the material (hardness in the present case).

Solution nitriding of EN 1.4418

Solution nitriding [22-24] was applied to obtain a graded material by the presence of a nitrogen concentration profile. In the presence of an N gradient, a gradient occurs in the stability of austenite against martensite formation, and thus a gradient in M_s, which is expected to be lowest for the highest nitrogen content. Thus, the effectiveness of the various SZTs can be investigated. It was calculated using Thermo-Calc version 4.1 with the Thermo-Calc TCFE7.0 database [25] that the equilibrium nitrogen content in EN 1.4418 under the applied solution nitriding conditions is 0.46wt.%. The average nitrogen content in the sample as determined with combustion analysis is 0.26wt.%.

Metallography

Fig. 3, illustrates the results of metallographic investigation of solution nitrided EN 1.4418 martensitic stainless steel. The microstructure of the material is shown versus depth below the surface of the case. The left sides of the micrographs show the microstructure of the material at the surface, whereas the right side refers to the center of the samples. The conditions under investigation were RT, DI, BN, and BN+DI, showed from top to bottom in Fig. 3, respectively.

The microstructure at the surface of the case (Fig. 3, left) consists of austenite, which appears as a zone extending to 0.1-0.15 mm below the surface. At larger depths, the microstructure evolves into martensite (Fig. 3, right). From the case to the core the martensite morphology changes from lenticular to lath type. Metallographic investigation does not clarify whether the center of the samples is fully martensitic, or whether small regions of retained austenite are present. The extent of the austenite layer is observed to be a function of the sub-zero Celsius history, and it is the thickest in the RT sample and the thinnest in the BN+DI sample.

The presence of austenite in the near surface region of EN 1.4418 martensitic grade is consistent with a significant uptake of strongly austenite-stabilizing N during solution nitriding The enrichment in N is known to stabilize austenite against its transformation into martensite during quenching by lowering M_s [7]. When M_s is lowered to below the quenching temperature, N stabilized austenite is retained in the microstructure.

Hardness measurements

Additional information on the microstructure and properties of the material versus gradual enrichment of the austenite in N was obtained with hardness measurements. Data are reported in Fig. 4 versus depth below the surface of the case.

Fig. 4 shows that close to the center of the case, the hardness of the material is about 360 HV for sample RT and within the range 370-380 HV for the sub-zero Celsius treated samples. Moving towards the surface of the case, the hardness of all samples firstly increases up to a maximum hardness, which value and position depend of the condition of the material.

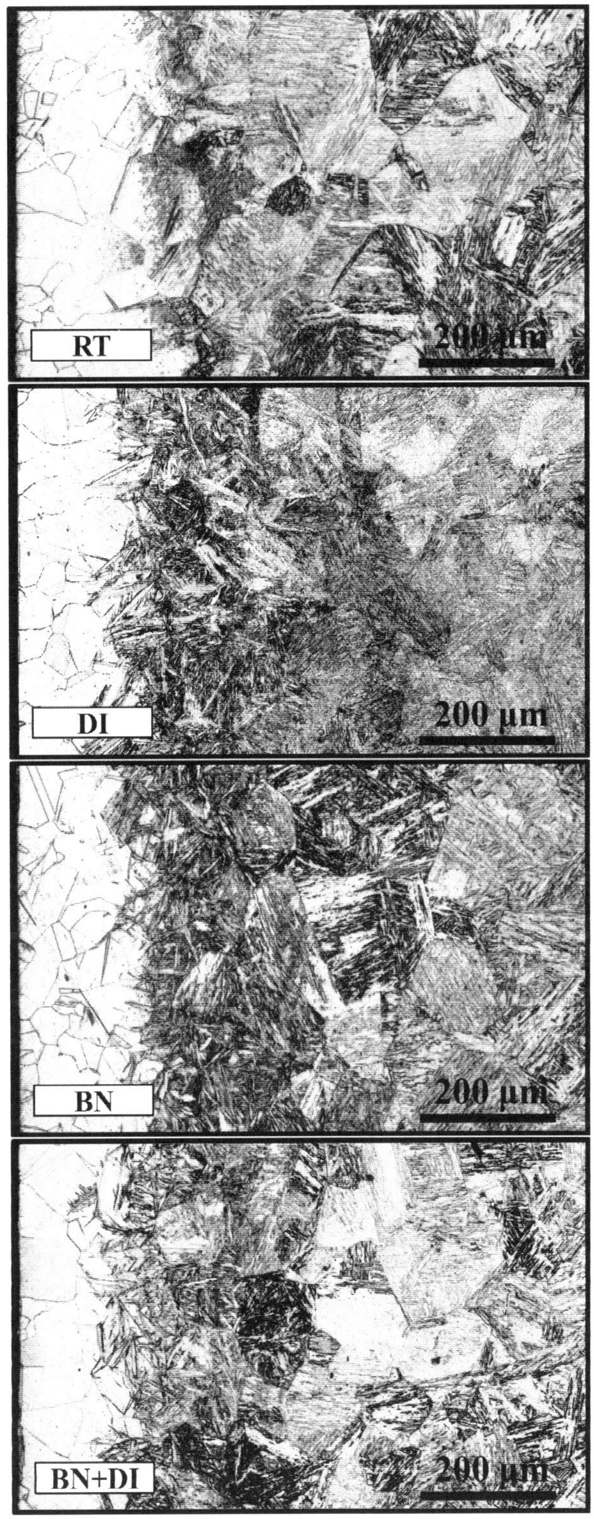

Figure 3: Micrographs showing the microstructure EN 1.4418 case nitride 2h at 1110°C versus depth for all the investigated condition. From top to bottom: sample RT; DI; BN; BN+DI. Optical microscopy. Etching in Kalling's II reagent for 10 s.

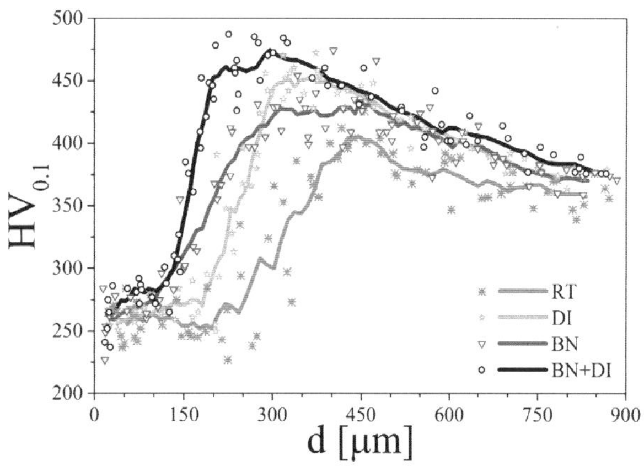

Figure 4: Micro-hardness measured on the cross section of the material versus depth below the surface of the case and material condition. Points refers to single measurements, lines to data averaged over 7 neighboring points.

Beyond the position of the maximum, the hardness decreases sharply and finally reaches a plateau at 250-270 HV, at a depth of 0.1-0.2 mm. The maximum hardness values equal about 400 HV, 430 HV, 450 HV and 470 HV, for the samples RT, BN, DI and BN+DI respectively, and are reported at a depth below the surface of about 0.45 mm for the sample RT, 0.35 mm for the samples DI and BN and 0.3 mm for the sample BN+DI.

Hardness data is further discussed in the following.

Firstly, hardness measurement suggests the presence of retained austenite over the entire thickness of the RT sample, because the SZT samples all show a higher hardness in the core as compared to the RT.

Secondly, hardness measurement indicates that not all the sub-zero Celsius treatments are equally effective in promoting an additional martensite formation. Along the N gradient, the most effective treatment is BN+DI. The DI treatment is as effective as BN+DI for small N-enrichment of austenite. The BN treatment instead is not so effective for small degrees of enrichment, but becomes more effective than treatment DI for a large N-enrichment of austenite.

Interestingly, the effectiveness of the different sub-zero Celsius treatments versus enrichment of the austenite in alloying elements follows the same pattern for both the investigations presented in this manuscript (see Fig. 2, left to right, and Fig. 4, right to left). Evidently, the austenitization temperature is not the key parameter that determines the effectiveness of the different sub-zero Celsius treatments. Rather, the stability of austenite versus martensite formation is decisive.

Lastly, hardness measurement indicates that an additional treatment step at sub-zero Celsius temperatures is fundamental to maximize the properties of martensitic stainless steels, providing their chemical composition is correctly tailored to exploit the possibilities offered by sub-zero Celsius treating.

Fig. 4 shows the evolution of hardness versus depth. Data can be interpreted considering two factors: the counteracting effects on hardness of the enrichment of the austenite in N, and the effect of sub-zero Celsius treatments on the microstructure.

The effect of the enrichment of the austenite in N on hardness is twofold. On the one hand, enrichment of austenite with N promotes an increase in the hardness of the martensite that forms during quenching, because of solid solution hardening. On the other hand, enrichment of austenite with N promotes the retention of a gradually larger fraction of this phase after quenching, and consequently softening of the material. Hardening is dominant close to the center of the samples, where the microstructure is mainly martensitic. Softening is dominant in the near-surface region, where the fraction of retained austenite becomes significant, and continues until austenite becomes the only phase present. At this point, the hardness of the material reaches a plateau value corresponding to the hardness of the austenite. The balance between the two effects yields maximum hardness for a certain fraction of N, i.e. at a certain depth below the surface.

The effect of sub-zero Celsius treatments is to promote an additional conversion of austenite into martensite as compared to conventional quenching to room temperature. Consequently, maximum hardness is observed for the SZT samples at a lower depth, i.e. at a higher N content, as compared to the sample RT. Furthermore, the hardness at the point of maximum is higher, 40-70 HV in the present example, for the SZT samples as compared to the hardness at the point of maximum for the RT one, as directly follows form a higher N content at the point of maximum.

Evidently, by balancing the N content in the steel to maximize its hardness only after sub-zero Celsius treatment of the material, a new grade significantly improved in respect to this property could be obtained. Similarly, tailoring the chemical composition of the base material with additional alloying with, for example Cr to increase corrosion resistance, Cu to increase strength, ... is expected to yield a new grade with superior performance upon sub-zero Celsius treatment of the steel.

Conclusions

In the present work, two approaches were followed to investigate how sub-zero Celsius treatments can be effective solutions to increase the alloying possibilities of martensitic stainless steels and consequently to improve their properties.

In the first example, the material under investigation was a (in wt-%) Fe-11.5Cr-0.67C martensitic steel. The contents of Cr and C dissolved in austenite were systematically varied by varying the austenitization temperature. The application of a sub-zero Celsius treatment allowed to obtain a (mainly) martensitic microstructure enriched in Cr and C as compared to conventional quenching to room temperature. Additional alloying with Cr and C simultaneously yielded an increase in hardness up to 90 HV and an improvement in the resistance of the material to aggressive environmental conditions as compared to conventional quenching to room temperature.

In the second example, the material under investigation was a (in wt-%) Fe-15.0Cr-5.8Ni-1.0Mo-0.03C martensitic stainless steel. A nitrogen concentration profile in austenite was established by high temperature gas nitriding. The application of a sub-zero Celsius treatment allowed to obtain a (mainly) martensitic microstructure enriched in N as compared to conventional quenching to room temperature. Additional alloying with N yielded an increase in hardness up to 70 HV as compared to the conventional quenching to room temperature.

In both examples, optimal results were obtained by immersion of the sample in boiling nitrogen followed by isothermal holding 24 hours at -78°C. Inverting the two steps for the Fe-Cr-C steel yielded less effective results, equivalent to isothermal holding 24 hours at -78°C alone. Neither was only immersion in boiling nitrogen as effective as immersion in boiling nitrogen followed by isothermal holding 24 hours at -78°C.

Data indicate that sub-zero Celsius treatment should be included as a mandatory thermal step in the design of new martensitic stainless steels, because it expands the number of possibilities to design new martensitic stainless steel grades with improved properties. Additionally, isothermal formation of martensite at sub-zero Celsius temperatures deserves further investigation, because the phenomenon can potentially yield significant improvements of the properties of the steel upon sub-zero Celsius treatment.

Acknowledgments

This work was supported by the Danish Council for Independent Research (G.R. grant: DFF-4005-00223). Dr. Kristian Vinter Dahl is acknowledged for the Thermocalc simulation.

References

[1] Campbell, C.E. and Olson, G.B., "Systems design of high performance stainless steels I. Conceptual and computational design", *J. Computer-Aided Mater. Des.*, Vol. 7, No. 3 (2001), pp. 145-170.

[2] Xu, W., *et al.*, "Designing nanoprecipitation strengthened UHS stainless steels combining genetic algorithms and thermodynamics", *J. Comput. Mater. Sci.*, Vol. 44, No. 2 (2008) pp. 678-689.

[3] Xu, W., *et al.*, "A combined optimization of alloy composition and aging temperature in designing new UHS precipitation hardenable stainless steels", *J. Comput. Mater. Sci.*, Vol. 45, No. 2 (2009), pp. 467-473.

[4] Xu, W., *et al.*, "Computational design of UHS maraging stainless steels incorporating composition as well as austenitization and aging temperatures as optimization parameters", *Phil. Mag.*, Vol. 89, No. 20 (2009), pp. 1647-1661.

[5] Xu, W., *et al.*, "A new ultrahigh-strength stainless steel strengthened by various coexisting nanoprecipitates", *Acta Mater.*, Vol. 58, No. 11 (2010), pp. 4067-4075.

[6] Xu, W., *et al.*, "Genetic design and characterization of novel ultra-high-strength stainless steels strengthened by Ni3Ti intermetallic nanoprecipitates", *Acta Mater.*, Vol. 58, No. 10 (2010), pp. 3582-3593.

[7] Lo, K.H., *et al.*, "Recent Developments in Stainless Steels." *Mater. Sci. Eng. R*, Vol. 65, No. 4-6 (2009), pp. 39–104.

[8] Ishida, K. "Calculation of the Effect of Alloying Elements on the Ms Temperature in Steels", *J. Alloy. Compd.*, Vol. 220, No 1-2 (1995), pp. 126–131.

[9] Saha, A and Olson, G.B., "Computer-aided design of transformation toughened blast resistant naval hull steels: Part I", *J. Computer-Aided Mater. Des.*, Vol. 14, No. 2 (2007), 177-200.

[10] "Armco 17-7PH and PH 15-7Mo Stainless Steel and Strip", Arco steel corporation, Middletown, Ohio, 1966.

[11] Slunder, C.J., *et al.*, "Thermal and Mechanical Treatment for Precipitation-Hardening Stainless Steel", NASA, Washington, D.C., 1967.

[12] Ludwigson, D.C. and Hall, A.M., "The Physical Metallurgy of Precipitation-Hardenable Stainless Steel", Defence Metals information center, Report III, 1959.

[13] Holmquist, M., *et al.*, "Isothermal Formation of Martensite in a 12Cr-9Ni-4Mo Maraging Stainless Steel", *Scr. Metall. Mater.*, Vol 33, No. 9 (1995), pp. 1367–1373.

[14] Olson, G.B., "Advances in theory: Martensite by design", *Mater. Sci. Eng. A*, Vol. 438-440, No. 165 (2006), pp. 48-54.

[15] San Martin, D., *et al.*, "Real-Time Martensitic Transformation Kinetics in Maraging Steel under High Magnetic Fields", *Mater. Sci. Eng. A*, Vol. 527, No. 20 (2010), pp. 5241–5245.

[16] Villa, M., *et al.*, "The Sub-Zero Celsius Treatment of Precipitation Hardenable Semi-Austenitic Stainless Steel" *Heat Treating 2015: Proceedings of the 28th ASM Heat Treating Society Conference*, Detroit, MI, October 2015, pp. 431–435.

[17] Villa, M., *et al.*, "Anomalous kinetics of lath martensite formation in stainless steel", *Mater. Sci. Technol.*, Vol. 31, No. 11 (2015), pp. 1355-1361.

[18] DasGupta, S.C. and Lement, B.S.. "Isothermal Formation of Martensite at Subzero Temperatures in High Chromium Steel" *Trans. AIME – J. of Metals*, Vol. 3, No. 9 (1951), pp. 727–731.

[19] Kurdjumov, G.V., "Nature of Martensitic Transformations" *J. of Metals*, Vol. 11, No. 7 (1959), pp. 449–453.

[20] Villa, M., *et al.*, "Investigation of martensite formation in Fe based alloys during heating from boiling nitrogen

temperature", *Metall. Ital.*, Vol. 11., No. 12 (2015), pp. 39–46.
[21] Lobodyuk, V.A., and Estrin, E.I., "Isothermal Martensitic Transformations", *Phys. Usp,* Vol. 48, No. 7 (2005), pp. 713–732.
[22] Berns, H. "Stainless Steels Suited for Solution Nitriding", *Mater. Werks.*, Vol. 33, No. 1 (2002), pp. 5–11.
[23] Berns, H., *et al.*, "Case Hardening of Stainless Bearings and Tools Using Nitrogen", *Mater. Werks.*, Vol. 27, No.1 (1996), pp. 25–36.
[24] Berns, H. and Siebert, S., "High Nitrogen Steels. High Nitrogen Austenitic Cases in Stainless Steels", *Isij Int.*, Vol. 36, No. 7 (1996), pp. 927–931.
[25] Andersson, J.O., *et al.*, (2002) Thermo-Calc and DICTRA, Computational tools for materials science. Calphad, Vol. 26, No.2 (2002), pp. 273-312.

International Federation of Heat Treatment and Surface Engineering 2016
Proceedings of the 23rd IFHTSE Congress
April 18–21, 2016, Savannah, Georgia, USA

Modeling and Measurement of Residual Stresses and Machining Distortions in Aircraft Engine Disks

M.G. Glavicic and Robert Ress III
Rolls-Royce Corporation, Indianapolis, Indiana, USA
michael.glavicic@rolls-royce.com, robert.a.ress@rolls-royce.com

Shesh Srivatsa
GE Aviation, Cincinnati, Ohio, USA (Retired)
shesh.srivatsa@gmail.com

James Castle
Boeing Research and Technology, St Louis, Missouri, USA
james.b.castle@boeing.com

Byung Kwan Chun, Jin Yong Oh and Wei-Tsu Wu
Scientific Forming Technologies Corporation, Columbus, Ohio, USA
bchun@deform.com, joh@deform.com, wwu@deform.com

T. M. Holden
Northern Stress Technologies, Deep River, Ontario, Canada
holdent@magma.ca

B. Clausen, D. W. Brown and T. Sisneros
Los Alamos National Laboratory, Los Alamos, New Mexico, USA
clausen@lanl.gov, dbrown@lanl.gov, tsisneros@lanl.gov, marko.knezevic@unh.edu

M. Knezevic
University of New Hampshire, Durham, New Hampshire, USA
marko.knezevic@unh.edu

Jeff Simmons
Air Force Research Laboratories, WPAFB, Dayton, OH, USA
jeff.simmons@afrl.af.mil

Abstract

A study was undertaken to quantify the effects of residual stresses on machining distortion. Distortion can be caused by material bulk stresses resulting from heat-treating operations and/or by local near-surface machining induced stresses. There is a strong need to understand the effects of heat treating and machining on distortion and to predict, minimize, and control these distortions. A 3-D finite element model (FEM) was developed and validated to predict machining distortions of aircraft engine and airframe structural forgings. An automated procedure was developed to simulate the distortion due to material removal in multiple operations in the presence of bulk residual stresses. Residual stress measurements using neutron diffraction, hole-drilling and contour methods were performed for model validation. The residual stresses measured by the different methods were qualitatively in agreement with one another. The model was first validated on 3-D shaped Alloy 718 parts similar to production engine forgings and then on a production Alloy 718 disk. Good agreement was obtained between the measured distortions and the model predictions. The predicted residual stresses agreed with the measurements qualitatively in the regions where the different measurement methods showed low variability.

Introduction

Aircraft engine and airframe components that are machined from forgings or plate stock have bulk residual stresses resulting from forging and heat treatment processes. Heat treatment involves rapid quenching from a high temperature. Temperature gradients during quenching cause thermal stresses which drive localized plastic deformation and residual stress build-up which leads to distortion during quench, during machining and also during service. Part distortion can be caused by material bulk stresses resulting from heat-treating operations and/or by local near-surface machining-induced stresses.

A major contributor to the cost of forged components in both military and commercial aircraft is the large buy-to-fly weight ratio, which is the ratio of the weight of the forging to that of the finished part. Typically, this is between 4 and 10 for aircraft engine components and much higher for airframe components. The excess material is removed by various machining operations. Due to the internal residual stresses induced during heat treatment, the component reaches a new equilibrium state by deflection when a portion of the material is removed from the component during the machining process. Using trial-and-error approaches, it can be very challenging to plan a sequence of machining operations to ensure that the final component is produced within the very tight dimensional tolerances (typically as low as ~0.002") required for assembly and satisfactory performance in service.

There is a strong need to understand the effects of heat treating and machining on distortion and to predict, minimize, and control these distortions. Using the methods developed in this work, part distortions during machining can be predicted in order to determine more aggressive machining passes. Typical results achieved were the prediction of dimensional tolerances to within 20% or 0.005" absolute tolerance, with the correct trends in distortion.

To reduce the material and manufacturing costs and to facilitate the machining process planning, a procedure was initially implemented within the Finite Element Method (FEM) based commercial code DEFORM™ and successfully validated for two-dimensional (2-D) axisymmetric components [1]. Subsequent efforts were directed towards developing and validating the method for prediction of distortions from three-dimensional (3-D) features such as holes and dovetail shaped retention slots.

In this paper, the overall 3-D modeling methodology is described first. This is followed by details of the measurements of machining distortions and model validation using the measured data. Next the measurement of residual stresses is described using several different methods (neutron diffraction, hole-drilling and contour methods). A comparison of the results from the different measurement methods and model validation are presented. Finally, the paper concludes with a summary of the accomplishments and future work to be done. Remember that you should not re-state material that is readily available in the archival literature. Simply summarize it and then add a reference or two.

MODELING METHODOLOGY

The overall modeling methodology consists of two parts: first, modeling of the heat treatment process to compute the residual stresses in the part, and second, modeling of the machining or material removal process to compute the distortions due to the residual stresses.

During the heat treatment process a part continuously deforms as the temperature evolves under the specific thermal exposure (e.g., solution treatment, quenching, annealing, or aging). The focus here was to investigate the machining distortion due to the bulk residual stress resulting from heat treatment for engine disks and not by local near-surface machining-induced stresses. A coupled thermo-mechanical analysis was performed to predict the temperature and resulting residual stress fields subject to the specified thermal boundary conditions. The analysis consisted of heating the part from room temperature to the heat treat temperature, holding at temperature, transfer to the quench media, and cooling back to room temperature. The commercially available finite element software DEFORMTM (available from Scientific Forming Technologies Corporation) was used. This software predicts the transient temperature field, the evolution of stresses and strains and the resulting part distortion. Typically, finite element meshes consisting of up to 4000 elements for 2-D problems and up to 200,000 elements for 3-D problems were used for the simulations. It was established that with this mesh size, the results were mesh size independent.

In the machining process material is removed from the heat-treated part in order to produce a machined component with the specified final dimensions. A machining process typically involves a series of cutting operations with multiple passes per operation. Different cutting tools and fixtures may be introduced between operations. A simulation procedure was developed to automatically carry out multiple operation machining distortion simulations starting from an initial work piece with bulk residual stress distribution from the end of heat treatment.

Within one operation (or one set of fixtures), boolean, remeshing, interpolation, and stress re-equilibrium are carried out for each machining pass. The machined configuration is obtained by subtracting the volume-to-be-removed (Boolean operation) from the finite element mesh of the work piece. The material to be removed is defined using computer-aided design (CAD) and computer-aided manufacturing (CAM) programs. The programs and commands needed to operate a particular machine are typically in G-code format. G-code translators were developed to facilitate the direct import of machining path data into the finite element model. This enables a multi-pass machining operation to be set up quickly. Local remeshing is used to create a new mesh for the machined component and the state variables such as stress, strain and displacements are interpolated from the old mesh (before material removal) to the new mesh (after material removal). As the shape continuously evolves during the machining process, the internal stress is redistributed and the part distorts when reaching a new equilibrium state.

At the end of each operation (or the beginning of a new operation) a free-state analysis is done by removing the clamping loads before the component is placed into the next set of fixtures. The process is repeated for each machining operation until the final finished part shape is reached. There are two states to represent the work piece during the machining condition: the clamping state when the part is under the constraint of fixtures, and the free-state when the part is removed from the fixture and is ready for the next operation. The contact condition between the work piece and the fixtures, clamping force, or prescribed displacement can be specified at the clamping state. An automatic assignment of the boundary conditions to eliminate the rigid body motion was developed for the free-state analysis. The elimination of rigid body motion is necessary in order to obtain a unique solution.

To ensure easy and error-free setup of multiple machining operations a 3-D graphics display and a preview of the model at different machining operations is generated. The distortion results were converted to the same coordinate reference frame as used in the measurements so that the two can be compared easily. The procedure described above was applied to model the machining distortion of an engine disk. The results are presented and discussed in subsequent sections.

The following material, geometry and processing data are needed for residual stress and machining distortion modeling.

Material Data (all as a function of temperature, deformation conditions and microstructural evolution that occurs during processing): constitutive data (stress-strain in plastic region), elastic modulus, Poisson's ratio, creep and stress relaxation data, thermo-physical properties (thermal expansion, specific heat and thermal conductivity). Standardized tests were used to generate these data. The material property data are proprietary to the various engine companies involved in this study and therefore, cannot be presented here.

Geometry Data: The geometry of the component, machining fixtures and the material removed during each machining operation in IGES, UG or similar CAD format.

Processing Data: The heat treat temperature, transfer time to quench media and the heat transfer coefficients during heatup, transfer and quench. The processing data are proprietary to the forging suppliers and therefore, cannot be presented in detail.

MACHINING DISTORTION MEASUREMENTS

Six disks were machined from proprietary shaped forgings of nickel based 718 super-alloy. The disks were heat treated prior to the machining process using production process parameters. Temperature data was gathered during the heat treatment of a thermo-coupled disk. These data were used with an inverse method for obtaining the heat transfer coefficients during quench. Using the modeling procedure described above, the residual stresses after heat treatment were computed.

Six 2-D turning operations, followed by 3-D milling of three slots in three passes and a final 3-D turning operation were simulated. The overall 2-D material removal plan (i.e., machining operations) for the axisymmetric disk is shown in Figure 1. During machining, two nodes in the radial direction and two nodes in the axial direction were constrained to account for the fixtures. The locations of the constrained nodes are shown in Figure 1 by the solid triangles. In the free state distortion simulations only one node in the axial direction must be constrained to remove the rigid body motion in an axi-symmetric simulation.

The initial 3-D model was created by a 360° revolution of the 2-D model. The residual stresses were interpolated to the 3-D tetrahedral mesh consisting of about 200,000 elements. In order to reduce the interpolation error resulting from this 2-D to 3-D transformation, a finer mesh was used in the regions of steep gradients and finer geometrical features. The geometries of the slots to be machined by milling are shown in the Figure 2. The work piece was held by the fixtures shown in Figure 3. A clamping pressure of 100 psi was applied on the top face of the disk by four chucks (90° apart from each other). The locating faces were on the opposite side of the chucks. The locating faces were modeled by fixed velocity boundary conditions. Figure 4 shows the finite element model of the machining setup for this operation. The three slots were removed in three passes within the same operation, followed by remeshing, interpolation and stress re-equilibrium. The free state distortion simulation was also performed.

In the final turning operation the workpiece was held in three chuck jaws, 120° apart, applied to the outer diameter (OD) surface. The locating fixtures are on the bottom side. The procedure of the boolean operation followed by remeshing, interpolation and stress re-equilibrium was performed. The last stress re-equilibrium iteration, determining the final shape of the disk, was performed in the free state.

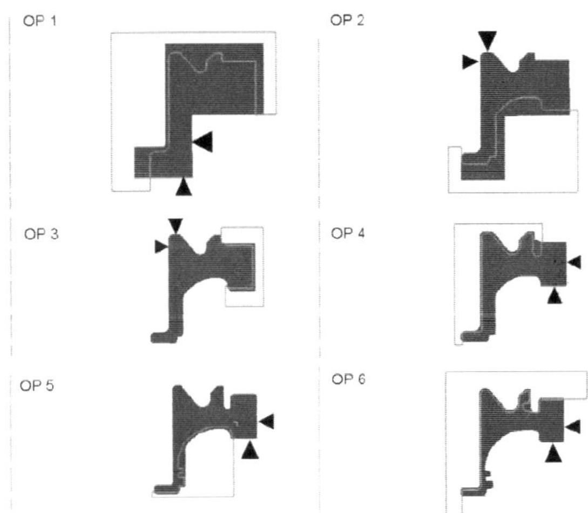

Figure 1: Modeling of six turning operations. The light blue outline shows the cutter path and the material removed during each machining operation. The black triangles indicate clamping locations where the finite element nodes are constrained.

MACHINING DISTORTION VALIDATON

In this section, the validation of the modeling predictions is presented. The predicted displacement in the hoop direction suggests that residual hoop stress is in compression since the slots are closing in. In the axial direction, the disk distorts downward 90° from the slots and upward 0° and 180° from the slots, and is slightly warped. In the radial direction, the disk expands 90° counterclockwise (CCW) and clockwise (CW) from the slots. The ring is slightly out of round.

Figures 5 - 8 show a comparison of the measured and predicted disk distortion for four surfaces: outer diameter (OD), inner diameter (ID), top, and bottom. Radial distortions were measured at the OD and ID faces at several vertical locations and at 20 clock positions (plotted as angle in the figures, 0 to 360°). Figures 5 and 6 show the comparison for the OD and ID, respectively along two vertical locations: the top most location and the bottom most location. Axial distortions were measured at several horizontal locations and at 20 clock positions. Figures 7 and 8 show the comparison for the top surface and the bottom surface, respectively along two horizontal locations: close to OD location and close to ID location. The measurements were made after the milling the three slots and after the final turning operation. The axial and the radial distortions are plotted as a function of the angular position. The error bars show the prescribed dimensional tolerances of the feature and not the error in the measured CMM data.

The modeling predictions are in a very good agreement with the measurements. As mentioned above, the part was inspected multiple times during the production and compared the predicted and measured distortion. Only the comparison after the last turning operation is shown. The other comparisons were also in very good agreement.

Figure 2: FEM model showing the slots to be removed by a boolean operation

RESIDUAL STRESS MEASUREMENT

All residual stress measurement methods are indirect and rely on converting a measured strain or displacement to a stress. Measurement techniques differ in respect of the stress components measured; depth (near-surface vs. through-thickness); mapping dimensionality (1-D, 2-D, 3-D); spatial resolution; sensitivity at low stress levels; destructive vs. non-destructive; near-surface resolution; difficulty of calibration; experimental and computational cost and effort. No single technique is applicable in all cases and validation requires a combination of measurements depending on part geometry and surface vs. internal measurements.

Two forged and heat treated Alloy 718 pancake forgings (18.1" diameter, 2.75" thick) were also examined. The forgings were not aged in an attempt to retain the high residual stresses so as to yield measurable machining distortions. Residual stresses in one of the two forgings were measured using the contour method (Figure 9). The second forging was machined to an engine disk like shape and the residual stresses were measured first with the non-destructive neutron diffraction method followed by hole-drilling and then the destructive contour method.

.

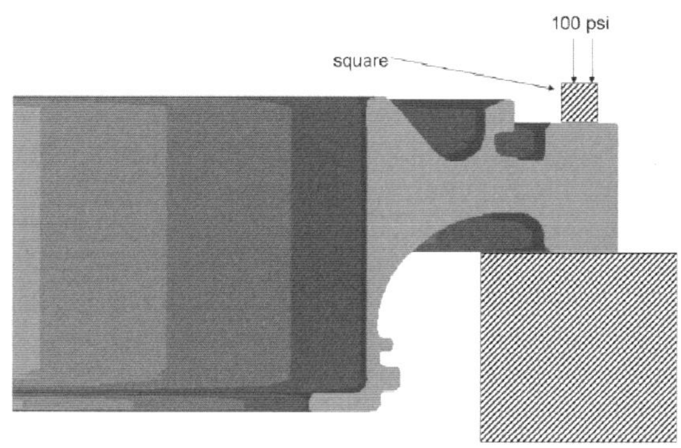

Figure 3: Holding of the workpiece during the milling operation.

Figure 4: A pressure of 100 psi was applied on the top surface while a fixed velocity boundary condition was applied to the bottom surface below the pressure definition.

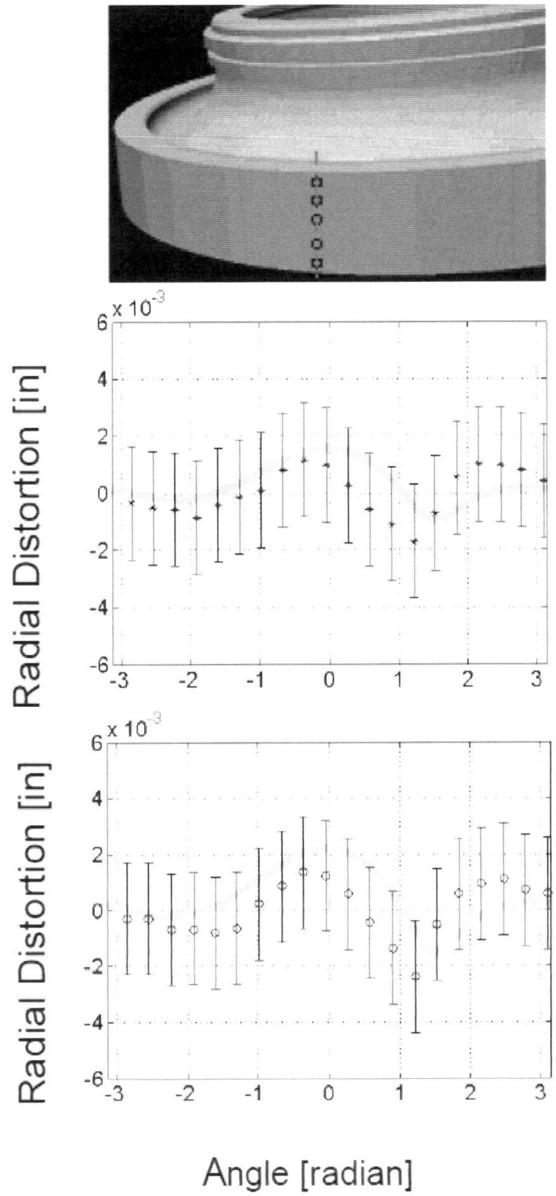

Figure 5: OD CMM inspection: 5 vertical and 20 clock locations. Comparison of measured (symbols) and predicted (solid lines) OD radial distortion at two vertical locations: the most top and the most bottom. Error bars indicate the prescribed machining tolerances (± 0.002") of the feature.

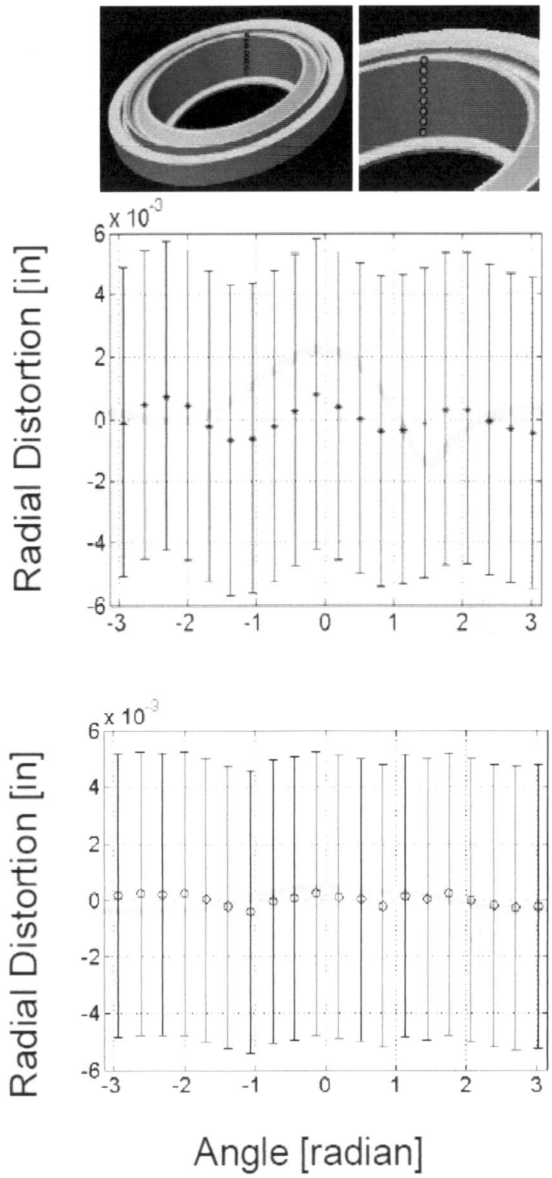

Figure 6: ID CMM inspection: 8 vertical and 20 clock locations. Comparison of measured (symbols) and predicted (solid lines) ID radial distortion at two vertical locations: the most top and the most bottom. Error bars indicate the prescribed machining tolerances (± 0.005") of the feature.

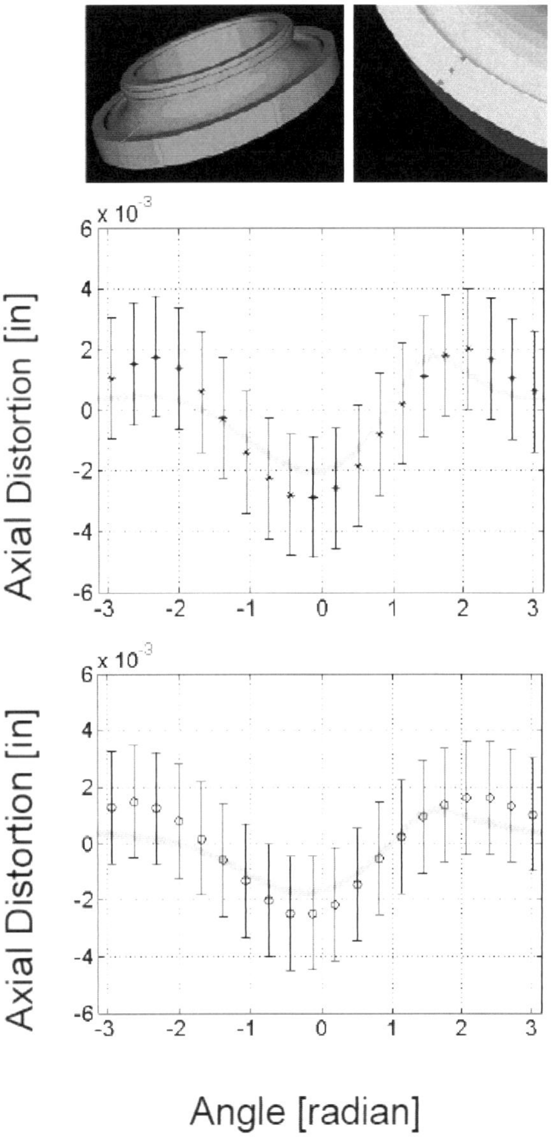

Figure 7: CMM inspection of the bottom: 3 horizontal and 20 clock locations. Comparison of measured (symbols) and predicted (solid lines) axial distortion at the bottom of the OD rim at two horizontal locations: close to OD and close to ID. Error bars indicate the prescribed machining tolerances (± 0.002") of the feature.

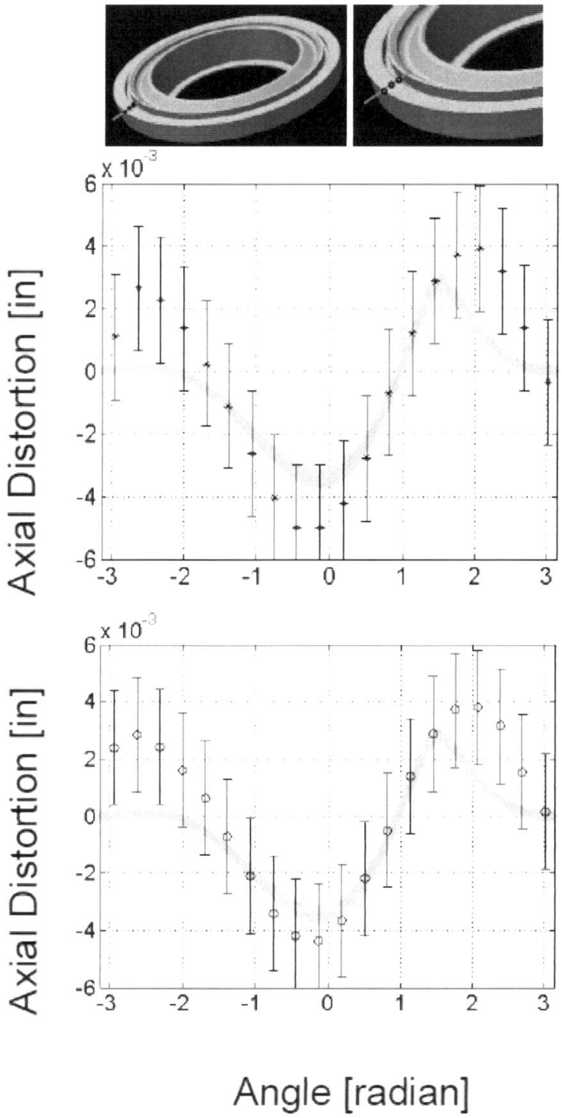

Figure 8: CMM inspection of the top: 3 horizontal and 20 clock locations. Comparison of measured (symbols) and predicted (solid lines) axial distortion at the top of the OD rim at two horizontal locations: close to OD and close to ID. Error bars indicate the prescribed machining tolerances (± 0.002") of the feature.

Contour Method

In this method [9] a part is sectioned along a plane of interest (usually by wire EDM). Strains released by the cut are measured by strain gages. The contour of the cut surface is measured. Using a finite element model the topography measured on the plane of the cut (usually averaged from the two sides of the cut) is forced back into plane to obtain the original residual stress within the component. This avoids the use of ill-posed inverse methods to obtain stresses from measured strains. The contour method can provide 2-D stress distributions over the complete cross-section of the cut face.

Following contour sectioning of both the pancake and the disc component, an Elastic modulus = 216 GPa and Poisons ratio = 0.3 were used in the processing of the contour data for the calculation of a residual stress values.

A plot of the measured residual hoop stress in the pancake forging using the contour method is shown in Figure 10a. The model predicted residual stress is shown in Figure 10b. Tensile residual stresses exist in the interior of the forging, with the highest levels of tensile residual stress located subsurface near the outer rim. Generally, a high magnitude compressive stress exists around the edges of the forging. The general shapes of the contours that were measured and modeled were similar, but there are some differences in the magnitudes.

Line plots at selected locations (Figure 11) were generated from the contour method data to quantitatively describe the stress distribution over the measurement plane. Comparisons of the modeled versus measured data (Figure 12Figure) reveals that the modeled residual stress data is higher than the measured residual stress data at all locations in the interior of the pancake.

A plot of the measured residual hoop stress in the machined forging using the contour method is shown in Figure 13. The model predicted residual stress is shown in Figure 14. The general shapes of the contours that were measured and modeled were similar, but there are some differences in the magnitudes.

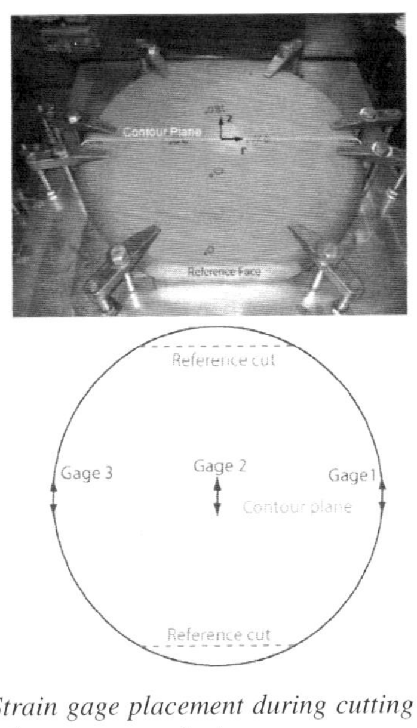

Figure 9: Strain gage placement during cutting of reference surfaces for the contour method.

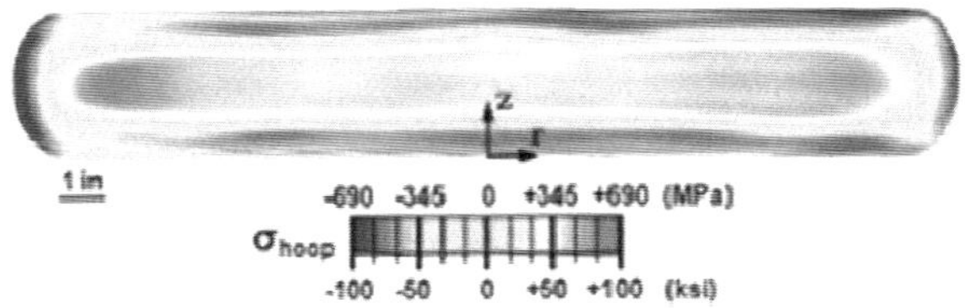

(a)

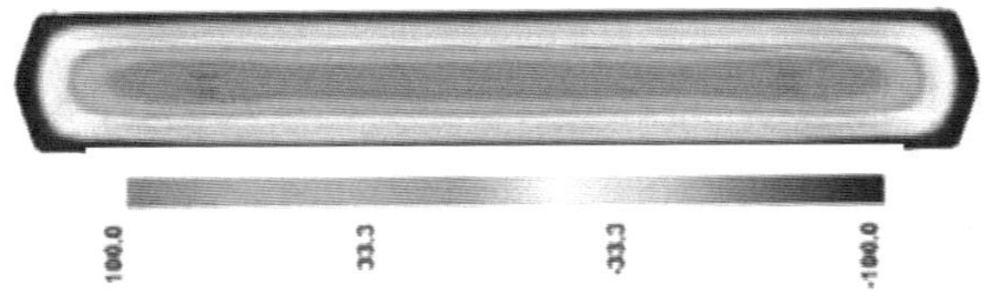

(b)

Figure 10: Contour map of the (a) measured and (b) predicted residual hoop stress over the contour measurement plane.

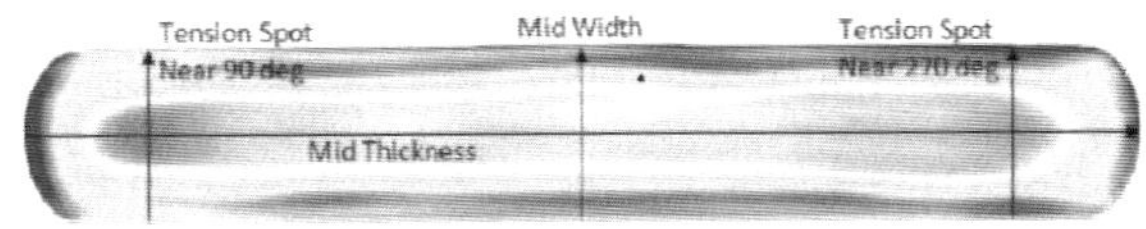

Figure 11: Line plot locations superimposed on contour results.

Neutron Diffraction Method

Neutron diffraction utilizes thermalized neutrons to measure the strains in the lattice of crystalline materials and then stresses are inferred. Neutron diffraction can be used to measure in-depth residual stresses non-destructively since the penetration of neutrons is much greater than that of x-rays, typically centimeters instead of microns.

Stress measurements using neutron diffraction require an accurate knowledge of the unstressed lattice spacing as a function of location in the component. A comb shaped specimen was extracted from the pancake that had undergone residual stress measurements using the contour method (Figure 15). The intent of the comb design was to take into account chemical segregation affects within the pancake and to determine the unstressed lattice spacing of the gamma nickel phase as a function of radial location within the pancake. The dimension of the individual tines on the comb was selected to be 2.5 mm x 2.5 mm x 2.5 mm so that the gage volume interrogated in the residual stress measurements (2 mm x 2 mm x 2 mm) could be entirely contained within a single tine of the comb.

Each tine in the comb then had its unstressed lattice spacing determined in the hoop, axial and radial directions via Rietveld analysis of the entire diffraction pattern collected. The measured lattice parameter data was fit to a third order polynomial so that the variability observed in the experimental lattice parameter data could be used in a consistent fashion in the correction of the data. At any given radial location, the measured unstressed lattice spacings were within 0.01% in the hoop, axial and radial directions which is within the range of experimental limitations. All residual stress data were corrected as a function of radial location in an unbiased fashion using the third order polynomial fits.

In the neutron diffraction work, the residual strains in each direction were calculated from the lattice parameters measured as a function of location within the component relative to an appropriate reference unstressed lattice parameter. Having measured the strains in the hoop, radial and axial direction, the principal elements of the stress tensor (assuming cubic symmetry) were calculated using standard stress-strain relationships. The diffraction elastic constants which relate the macroscopic stress in the sample to the macroscopic strain in the crystal lattice were taken to be equal to the macroscopic elastic constants (Elastic modulus = 216 GPa and Poisons ratio = 0.3). The principal axes were assumed to be the hoop, radial and axial directions.

Finally during the neutron diffraction residual stress measurements, the sample was positioned optically with the aid of two computerized LeicaTM theodolites that, by triangulation, located fiducial marks on the outside diameter of the machined forging at the desired position of measurement with an accuracy of ± 0.1mm. The sample was moved into the gage volume by a computer driven translator table (maximum translations of ± 30 cm in x and y, ± 60 cm in z, 360° of rotation) so that the lattice parameter could be mapped as a function of position.

Hole drilling Method

In this method a strain gage rosette is used to measure the strains when a hole is drilled. In hole drilling, the strain gage is around the hole. Stresses are computed from the measured strains. For a through hole in a thin plate with no depth gradient a closed form solution is available. A more complicated solution accounts for the area of the strain gages. For more general cases, numerical solutions have been developed.

Hole drilling was performed at 8 locations (Figures 16 and 17) to a depth of 0.04" with 4 duplicate residual stress measurements at axially symmetric locations along the perimeter defined by planes A, B, C and D (Figure 18) which were selected using the following criteria:

1. The angle between the two opposing planes was selected (20°) so that the hole-drilling measurements on one plane would not affect the measurements at the symmetrical location on the other side.
2. The four planes were kept as far away from the plane to be used for the contour measurements (70°), in order to ensure that the hole-drilling did not affect the contour data that was collected subsequently on the plane which had been measured using neutron diffraction.

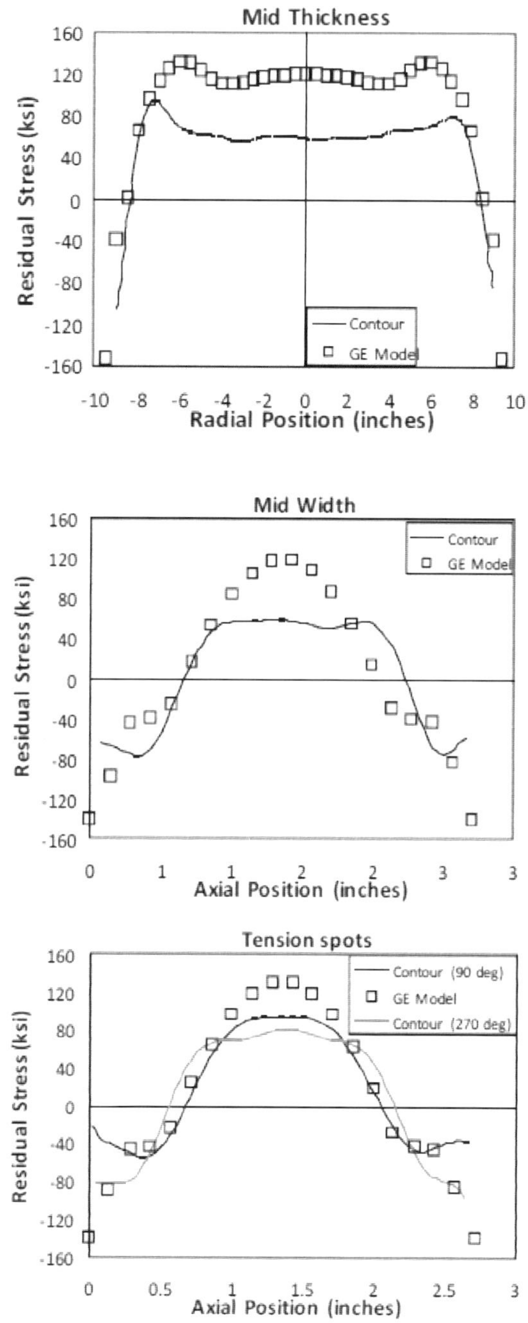

Figure 12: Modeled versus measured residual stress data at a) the mid thickness, b) mid width, and c) tension spots of the pancake forging.

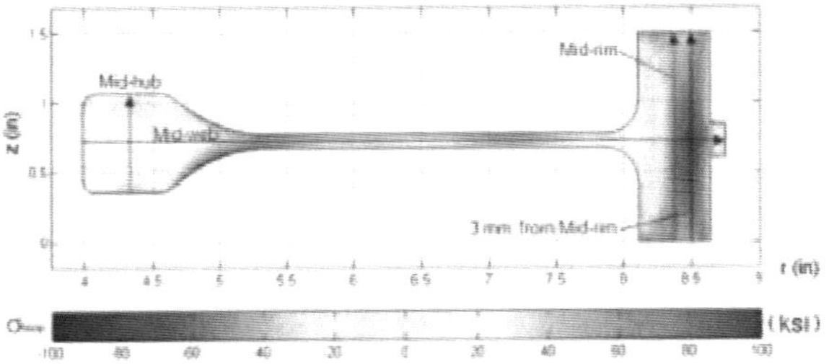

Figure13: Contour bulk residual stress results in machined forging and locations of data strips extracted for comparisons with other data.

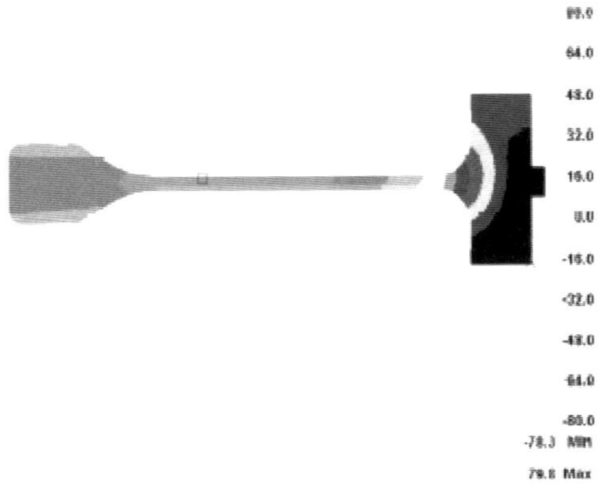

Figure 14: Modeled residual hoop stresses in machined forging.

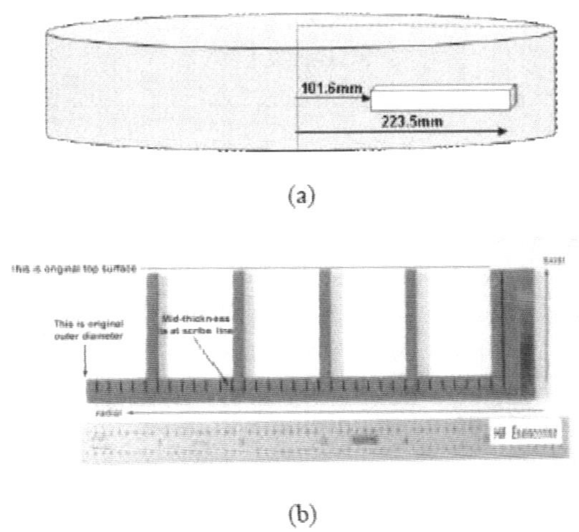

Figure 15: Unstressed lattice spacing comb used in diffraction experiment. a) extraction location in relation to pancake centerline; b) photograph of comb extracted from pancake.

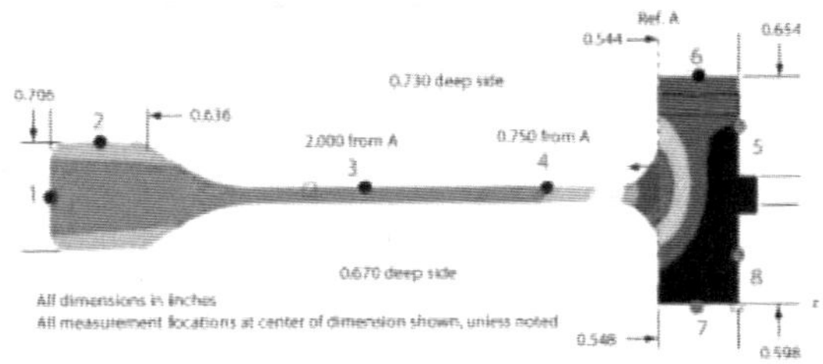

Figure 16: Positions of 8 hole drilling residual stress measurements along the perimeter of the machined component.

Figure 17: Photograph of the hole-drilling strain gauge layouts on the machined component

RESIDUAL STRESS VALIDATION

Mid-Web Strip (Figure 19)

All the residual stress data measured by the different methods followed similar trends. Taking the calculated error bars for hole-drilling data into account (+ 1 standard deviation), the results from hole-drilling and neutron diffraction are within the experimental uncertainty of one another in the hoop, radial and axial directions. The contour data also followed this trend, but had an oscillatory behavior to it which is not uncommon and is caused when the wire used to EDM (electro discharge machining) the disk stalls during the cutting process. Moreover, for the direction in which the component was sectioned in half during the contour work (0° degree to 180° in Figure 18) oscillatory behavior would be expected for the stress components in this direction (Figure 19) and not in the stress component perpendicular to the cutting direction (Figure 20 and Figure 21). In spite of these shortcomings, the contour data trends are consistent with the other measured data, and at many locations are within the experimental uncertainty of the other measurement techniques.

The modeled data show the same general trends as the measured data but with some significant deviations. The cause of these deviations is unknown at the current time.

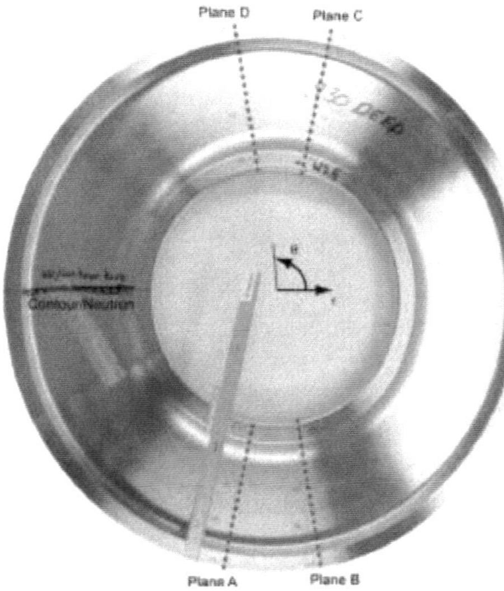

Figure 18: Orientation of the planes A, B, C and D in relation to the plane used in the contour and neutron diffraction residual stress measurements

Mid-Rib Strip (Figure 20)

All of the measured data follow the same trend and are for the most part within the experimental uncertainty of one another. As expected, no oscillatory behavior in the contour data in this direction are observed. A comparison of the modeled and the measured data reveals significant discrepancies at both the surface and the interior of the component. Again the reasons for these discrepancies are unknown.

3 mm from the Mid-Rib Strip (Figure 21)

All of the measured data follow the same trend and are for the most part within the experimental uncertainty of one another. As expected no oscillatory behavior in the contour data in this direction are observed. A comparison of the modeled and the measured data reveals significant discrepancies at both the surface and the interior of the component. Again the reasons for these discrepancies are unknown.

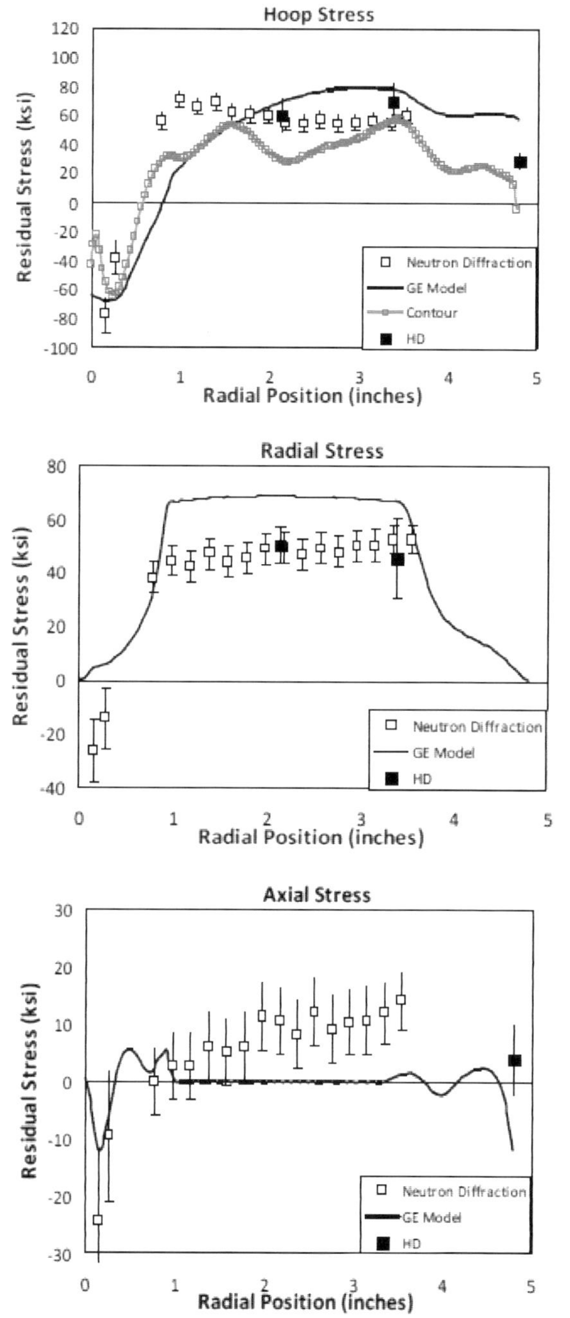

Figure 19: Measured and modeled residual stresses along the Mid-Web strip of the machined component (HD=Hole Drilling).

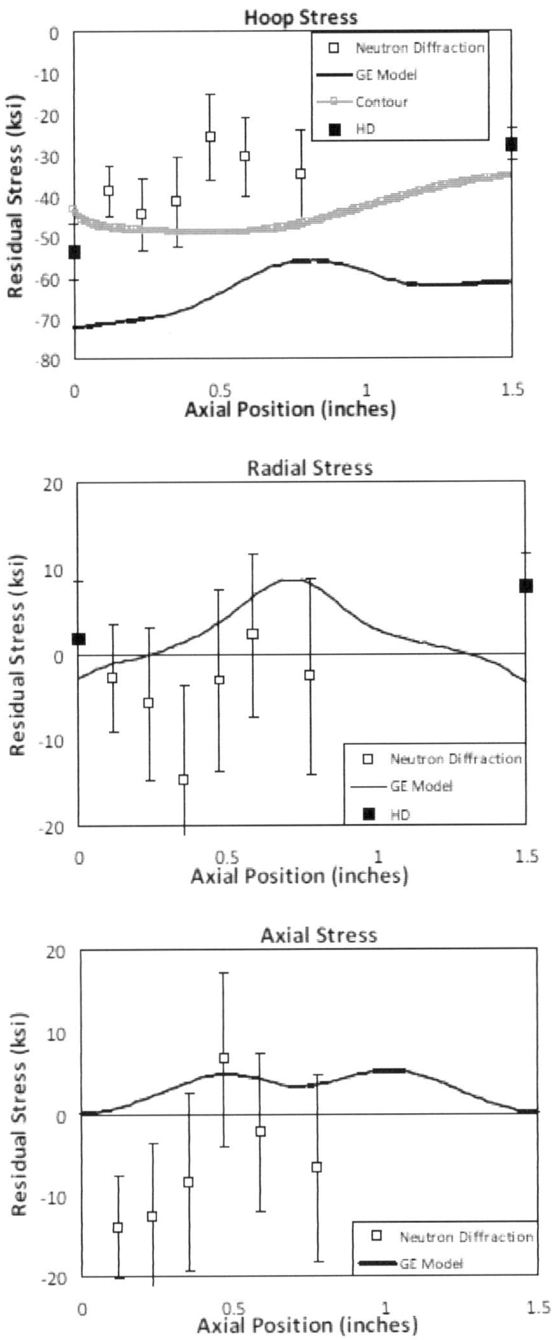

Figure 20: Measured and modeled residual stresses along the Mid-Rib strip of the machined component (HD=Hole Drilling).

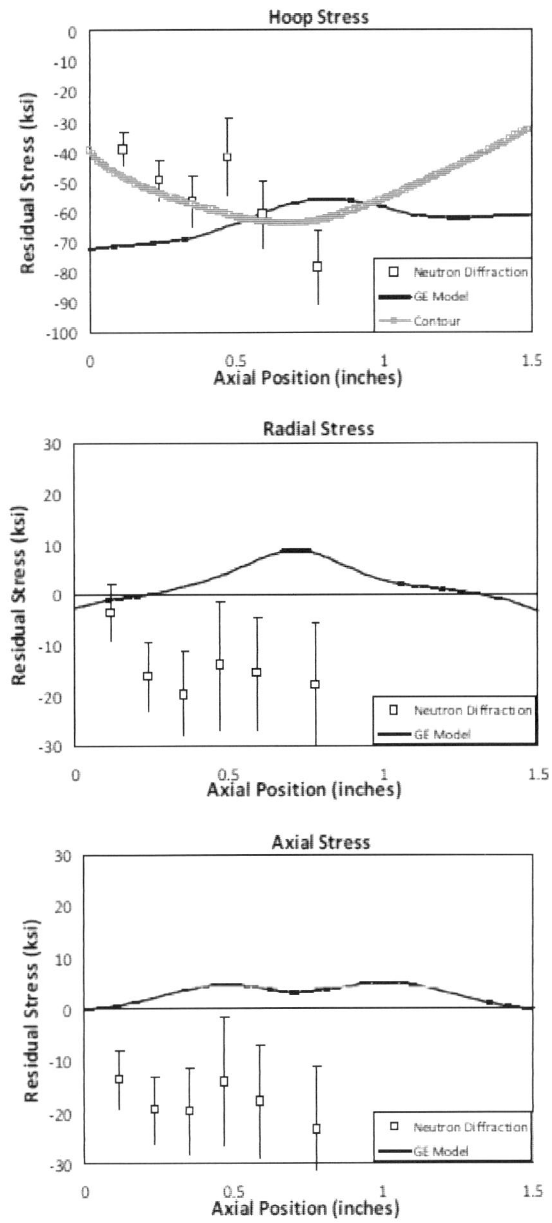

Figure 21: Measured and modeled residual stresses along the 3mm from the mid-rib strip of the machined component.

Conclusions

A detailed understanding of the processing - induced residual stress field in flight critical components like turbine disks can have a significant impact on engine design. The general concept proposed here - multiple residual stress determination experiments combined with a finite element model of the processing of the metal is an excellent way for making real progress on this important problem.

An automated multiple operation material removal procedure was developed and implemented in the commercially available software DEFORM™. The procedure is capable of performing multiple machining operations and multiple passes per operation. Material is removed by Boolean operations followed by remeshing to the as-machined shape and interpolation of the state variables. The stress equilibrium iterations are performed in the clamped as well as in the free state. The machining distortion behavior of an engine disk was modeled per production practices. The case study consisted of six 2-D turning operations, one 3-D milling operation of three slots, and one 3-D turning operation, modeled in a fully automated mode. Comparisons of the measured and predicted distortions demonstrate the accuracy and robustness of the developed methodology.

The results of this work revealed that all experimental results were qualitatively in agreement with one another, while the modeling results qualitatively agreed with the measured results in most regions but not all. In some regions the modeling results were found to be in disagreement with the measured values of up to three independent measurement techniques (neutron, hole drilling and contour). The source of these discrepancies is currently unknown at this time.

Based upon the results presented it can be concluded that the ability to predict the correct distortion in a component to a given accuracy using the modeling techniques employed does not infer that the modeled bulk stress has been accurately modeled at all locations. The modeled bulk residual stress state can be assumed to follow the same general trends predicted by the model, but the true stress state, as measured experimentally at specific locations, can differ substantially within the core sections and near the surface of a component. Hence it can be concluded that an agreement in modeled and measured distortion on a component can be considered as an indication that the bulk stress state has been modeled accurately in a semi-quantitative fashion only.

This study demonstrated that the residual stress induced during manufacturing of aircraft engine components can be predicted semi-quantitatively with confidence. The accuracy of input material data and heat transfer coefficients has a significant influence on the modeling results. Most of the residual stress is accumulated during the quenching process, caused by fast cooling and steep thermal gradients. The residual stress decreases after the aging process, which is a stress relaxation process. Models for this process have also been successfully used for many production components.

For engine parts, the bulk heat treat residual stresses are the significant distortion driver. Machining induced stresses can generally be neglected except in very thin parts. The complete heat treatment process (solution quench and stress relief) must be modeled to obtain the initial residual stress distribution for machining distortion analysis. Stresses from processes prior to heat treatment can generally be neglected since the heatup and hold at the heat treat temperature wipes out most or all of any prior stress. The material data used for modeling the heat treatment process must be generated on material with the same thermo-mechanical processing history as the actual parts. The data must be representative of the actual process. The use of on-cooling tests is recommended. Heat transfer coefficients should be generated on representative part geometry. Sufficient thermo-couples should be placed all round the part to capture the local HTC variation accurately. Fine finite element meshes should be used in thin-walled regions and in areas where stress gradients are steep. Measured and modeled distortion data must be reduced to a common reference frame for comparison.

For these reasons a sensitivity study is recommended to establish which inputs most strongly impact the modeling outputs of interest. Variations in the critical inputs should be quantified to assess the accuracy of the modeling outputs. Efforts can then be focused on reducing the impact of variability by focusing on the most critical variables. The sensitivity analysis can also define the resolution needed in the input material property data. In general, for a given part geometry and material, residual stresses increase with: solution temperature, elastic modulus, thermal expansion coefficient, and quench heat transfer coefficients.

Acknowledgments

This work was funded by the USAF MAI (Metals Affordability Initiative) programs GE6 and GE7. The authors would like to acknowledge the staff of LANL for their help with the neutron diffraction experiments and Hill Engineering for their help with the contour measurements. The Lujan Center at the Los Alamos Neutron Science Center at LANSCE is funded by the U.S. DOE, Office of Basic Energy Sciences. Los Alamos National Laboratory is operated by Los Alamos National Security, LLC, for the National Nuclear Security Administration of the U.S. Department of Energy under contract DE-AC52-06NA25396.

References

[1] Yin YL, Wu WT, Srivatsa S, Semiatin L, Gayda J, Modeling (2004) Machining Distortion of Aircraft-Engine Disk Forging. 8th International Conference on Numerical Methods in Industrial Forming Processes NUMIFORM 2004, Columbus, OH, June 13-17, 2004.

[2] Prime M, http://www.lanl.gov/contour/. The Contour Method for Measuring Residual Stress.

International Federation of Heat Treatment and Surface Engineering 2016
Proceedings of the 23rd IFHTSE Congress
April 18–21, 2016, Savannah, Georgia, USA

Distortion Potential of Cold Forming Prior to Induction Heat Treatment Processes

Dawid Nadolski; Alwin Schulz; Hans-Werner Zoch
Stiftung Institut für Werkstofftechnik (IWT) and MAPEX Center for Materials and Processes, University of Bremen, Germany
zoch@iwt-bremen.de

Abstract

Cold forming of steel is known as a process of near net shape production with significant economic advantages compared to machining. The degree of cold working of the material however introduces high residual stresses into the workpiece which are released during subsequent heating processes e. g. induction hardening. Those stress relief effects combined with plastic deformations during rapid heating and fast quenching as well as volume changes by transformation of austenite into martensite in the heated zones lead to distortion which causes losses in production costs due to straightening and/or grinding operations.
The influencing parameters of the forming process (degree of deformation), the tool design (opening angle of extrusion tool), the lubrication conditions and the hardening parameters (surface hardening depth) on the distortion behavior after inductive case hardening have been analyzed by Design of Experiments (DoE) methods. Diameter and length changes and changes in curvature and inclination angle of shouldered shafts made of SAE 1055 have been investigated for two different case depths.
The test results revealed the important distortion potential of the extrusion process and its parameters in the overall distortion after hardening. The significant influence of the induction hardening process on the size changes, not the shape changes, was proven.
How the initial microstructure of the raw material influences the surface hardness and hardness depth as well as the residual stress distribution in inductive surface hardened samples is demonstrated in a second investigation. The hardness response of pearlitic, spheroidize annealed, bainitic and martensitic quenched and tempered microstructure depends very much on the carbon solution in the very fast induction heat treatment process. This is of importance e. g. for contour hardening processes where only very short heating cycles have to be applied.

Introduction

Especially in the automotive industry mass production combined with high competitiveness in production costs are main boundary conditions for the process planning. The production system itself also can significantly influence the productivity. Of course the prior and main requirements are given by the component and its loading and stress situation. In modern lightweight design surface hardening of steel parts offers the desired high fatigue strengths and surface hardness for wear resistance. Compared to the also common case carburizing, induction case hardening offers similar local properties as surface hardness and hardening depth, beneficial compressive residual stresses in the surface region and low crack propagation rates in the softer core. Some advantages of induction heat treatment however are the possibility of partial hardening of surface areas, the short process cycle times, low overall energy consumption, and the easy integration capability into inline production.
To fulfil the requirements of modern production inductive surface hardening can provide significant contributions to it.
The selection of the proper raw material and its initial condition for induction hardening is of importance. First the core hardness or yield limit influences the required case depth to achieve the necessary part strength in cyclic loading [1], the microstructure before heating also affects the austenitising and finally the yield strength of the core is influencing the plastic deformations taking place during rapid heating and quenching.

Especially the high temperature gradients during heating and quenching interact in a complex way which will end up in a specific residual stress pattern and size and shape change. First, the fast heating for austenitising, which very often is in the range of a few seconds, leads to an expansion of the surface layer, which gets under strong compressive strain and some yielding, at the same time the core gets under tension with also possible yielding. In the second phase a rapid quenching leads to contraction of the austerities' surface with tensile stresses and possible elongation while the core sees compressive stresses. In the third phase, the martensitic transformation of the case takes place under simultaneous tensile stresses with some transformation plasticity and leads to a volume increase of martensite against austenite. Those steps summarize in the final know residual stress distribution with high compressive stresses in the surface, dropping into the tension region of the core almost at the depth of the hardened layer [2].

The final distortion of the part is ruled by those mechanisms, too. Yet the resulting distortion behavior of a component sees much more influences as distortion is known as an overall outcome of the entire process chain, where also the raw materials and forming or machining steps are contributing their distortion participation.
This is especially valid for components which see intensive cold working before induction heat treatment. The cold extrusion process is known to supply quite precise near-net-shape parts, however the high - and not necessary homogeneous - cold working process introduces high residual stresses into the component, which are released during heating and interact in a complex way with the mechanism described

above. As distortion is equivalent with an increase in production costs due to straightening and/or hard machining operations, the main influences on the final distortion should be investigated in this research, after which possible countermeasures can be derived.

Investigations and Results

Induction hardening of cold formed products

The challenge in controlling dimensional changes of induction heat treated cold formed parts is given in the initial strength of the raw material and the triaxial residual stress distribution after cold extrusion. Both influence the final residual stress pattern and the size and shape change, compared to the initial state, very much.

The degree of cold strengthening furthermore has a big influence as well as the die design of the cold extrusion tool. If it is possible to predict the resulting distortion behaviour, soft measures could be adapted regarding the final required dimensions. Yet the mechanisms which would allow this approach are known too little up to now. Practice therefore uses to run expensive trial and error experiments to adjust the ruling dimensions of raw material and die. The main task of this research was to establish fundamental knowledge on the influence of cold working and induction heat treating parameters on the final distortion of components. Representing typical industrial parts the investigations were performed on shouldered shafts, for which cold forming is a common process.

An additional evaluation reflects on the influence of the prior microstructure on the hardness and residual stress patterns after induction hardening.

Process chain and manufacturing of samples

For the experiments the steel C53G (1.1213; AISI 1055) with an initial diameter of 30 mm was used. The chemical composition of the materials was determined by spark optical emission spectroscopy (S-OES) as given in Table 1.

Table 1: Chemical composition of the used steel, determined by S-OES

Element in mass-%				
C	**Si**	**Mn**	**P**	**S**
0,55	0,26	0,70	0,015	0,026

The material was used in a spheroidized initial microstructure. Due to the relatively low hardness (182 HV1), this microstructure is predestined for cold forging processes.

The shafts were cold formed by cold forward-extrusion of coated cylindrical blanks in one single step to shouldered shafts. The cold worked parts were released from the die by means of an ejector, pressing in opposite direction to the forward-extrusion. After the cold forging process the shafts were induction hardened. For all experiments three parallel samples were used. Before and after the hardening process coordinate measurements were carried out to determine the size and shape changes. The entire process chain is shown in Figure 1.

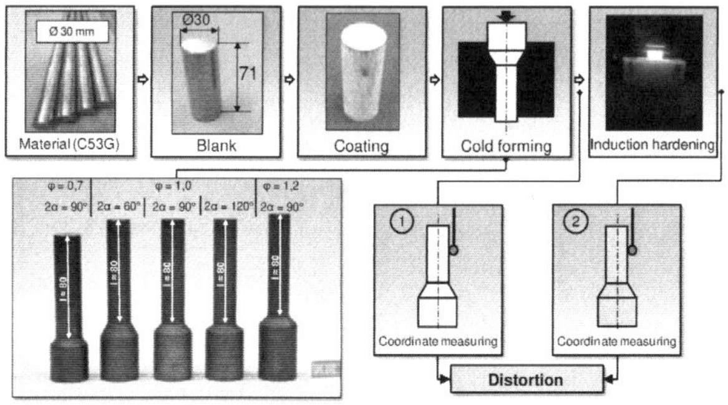

Figure 1: Process chain of cold extruded and induction heat treated shafts

Before extrusion machined cylindrical blanks (d_0=30 mm; h_0=71 mm) were Zn phosphatized and coated with zinc soap (ZWEZ Lube Z; coefficient of friction $\mu = 0.04$) resp. a solid lubricant (ZWEZ Lube MD; $\mu = 0.08$). After coating the blanks were extruded at room temperature with different forging dies on an extrusion press Schuler SMG HZPUI 260/100 at the Institute of Forming Technology and Lightweight Construction (IUL) at Technical University Dortmund. In a first step the smaller die diameter was varied (d_1 = 21.15 mm; 18.20 mm; 16.45 mm). The geometrical change of a formed workpiece can be described by the degree of deformation, which is defined by the equation:

$$\varphi = \ln\frac{A_0}{A_1} = \ln\frac{d_0^2}{d_1^2}$$

From the logarithmic ratio of the initial area A_0 to the deformed cross-sectional areas A_1, three different degrees of deformation (φ = 0.7; 1.0; 1.2) are resulting. For all three degrees of deformation an opening angle of the die of $2\alpha = 90°$ was chosen. Additionally for the mid degree of deformation (φ = 1.0) two more die geometries with different opening angles (2α = 60°; 120°) were used. For all shaft geometries the length of the deformed part was kept constant at 80 mm. Since all extruded shafts originally had the same blank geometry, the remaining length of the non-extruded part of the shaft varies for the different dies parameters.

After the cold forming process the lubricant was removed and the geometry of the shouldered shafts was measured by a coordinate measuring equipment (Leitz PMM 654). As main geometry factors the diameters of the extruded and non-extruded part of the shaft, the total length as well as the curvature of the extruded part of the shaft and the inclination between both sections of the shaft were determined.

After the measurement all shaft geometries were induction hardened by the universal hardening machine VL1000 SINAC 200/300 S MFC (EFD Induction). For all shafts a scanning hardening process was adjusted using an adapted coil. Starting from the end face of the extruded shaft part passing the shoulder zone down to the non-extruded shaft part the whole surface layer was hardened by polymer quenching. By varying feed rate and power an approximately constant austenitization temperature at the surface was adjusted over the entire length of the component. In order to assess the influence of the hardening depth on the distortion two different depth ranges of hardened layer of the shafts were aimed at by varying the hardening frequency (high frequency HF = 200 kHz or medium frequency MF = 12 kHz). The difference between the two process variants is exemplarily shown in Figure 2 on micrographs and hardness depth profiles for one extruded shaft variant (φ = 1.0; 2α = 90°; solid lubricant).

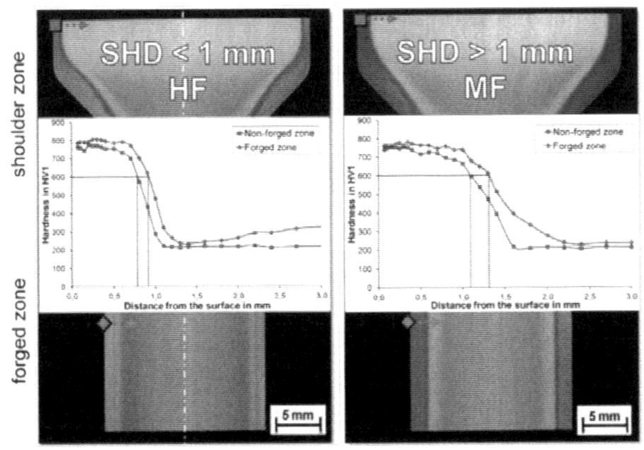

Figure 2: Overview micrographs and hardness depth profiles of both hardening variations at the example of one shaft variant (φ=1.0; 2α=90°, solid lubricant)

A martensitic surface layer can be observed for both of the hardness variants covering the whole length. The high frequency hardened shaft exhibits a smaller depth of hardening in the forged and non-forged area than the shaft hardened by medium frequency.

Experimental design and results

A statistically based set up of experiments was applied including all forming and heat treatment processes and their varying influencing factors in order to be able to estimate their significance on the size and shape change. The Design of Experiments (DoE) method was used, supported by the software Cornerstone®.

Table 2: Experimental design

Factor	Denomination	Level		
A	Degree of deformation (φ)	0,7	1,0	1,2
B	Opening angle (2α)	60°	90°	120°
C	Lubricant	Zinc soap (μ=0,04)	Solid lubricant (μ=0,08)	
D	Surface hardness depth (SHD)	< 1 mm	> 1mm	

All three influencing factors of forming (degree of deformation; die opening angle; lubricant) and the variation of heat treatment (SHD < 1mm; SHD > 1 mm) were combined and statistically evaluated (Table 2).

Due to the software support the influence of those factors as well as their interdependency on certain characteristics could easily be quantified.

Regarding the *size changes* of the shafts after cold extrusion and induction surface hardening an influence of degree of deformation (factor A), of surface hardening depth resp. hardening frequency (D) and lubricant (C) was detected. The influence of the die opening angle (B) was negligible (Table 3).

Table 3: Effects of the main factors and interactions on the size changes of cold extruded shafts as a result of induction hardening

Measured variable	Pareto chart
Change in diameter (non-extruded zone) 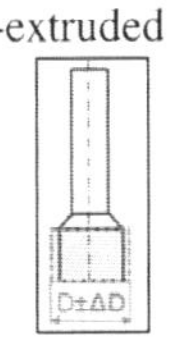	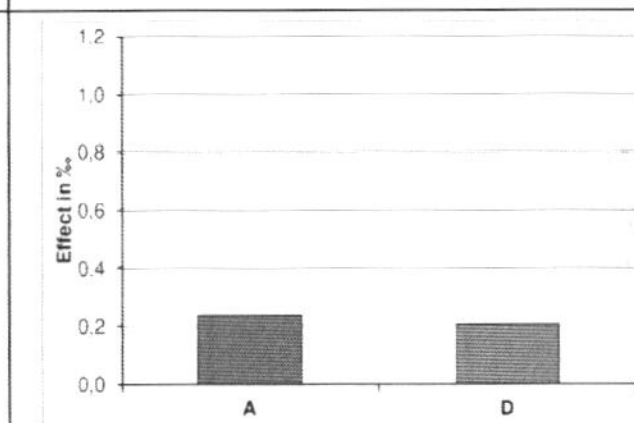
Change in diameter (extruded zone) 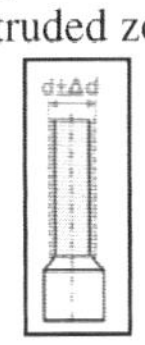	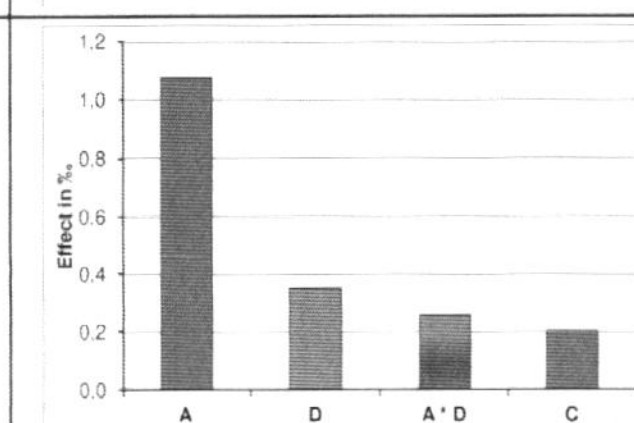
Change in length 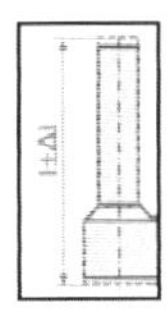	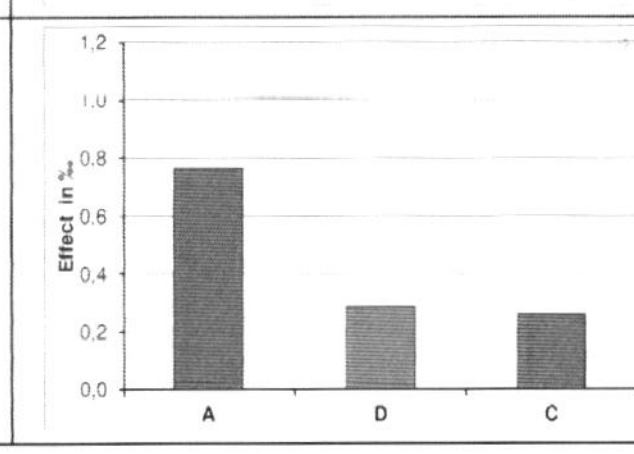

The diameter increase of the non-extruded zone was measured to be small in general (average + 0.08 ‰), but a not very strong relationship to the deformation degree and hardening depth was noticed (effect 0.2 ‰). The diameter increase reflects the compression of this zone due to the extrusion process and the martensite formation in the hardened surface.

In the extruded zone there was a strong effect of the degree of deformation (A), smaller effects of SHD (D) and lubricant (C) as well as an detected interdependency between deformation degree and hardening depth (A*D). The average diameter increase of 0.89 ‰ due to residual stress relief in the cold worked zone and volume increase in the hardened layer was about 10 times bigger than in the non-extruded area. The influence of deformation degree was of dominant influence.

The length change showed a mean decrease in length of about -0.94 ‰, where the influence of the deformation degree (A) was strongest with lower significance of hardening depth (D) and lubricant (C). Interdependencies did not occur. Also in length change the size deviation results from a stress relief by induction heating and quenching and martensite formation effects. The visualization of all size changes in dependence of the varied influencing parameters is given in Figure 5 in enlarged depiction.

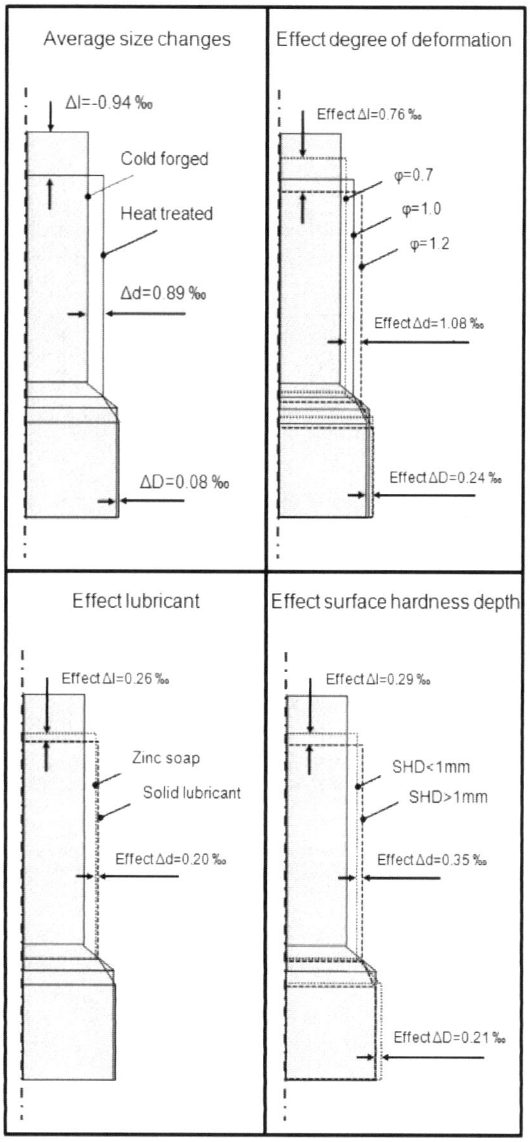

Figure 3: Visualized size changes of shafts after cold extrusion and induction hardening (average values and varied degree of deformation; effects enlarged)

Similar investigations were performed regarding the influences of cold forming and induction hardening on the *shape changes* of extruded shafts (Table 4).

The changes of curvature of the extruded zone showed a certain influence of degree of deformation (A) and of die opening angle (B). The effect on curvature was not very strong; the average shift ΔK was about 19 µm.

With the same dependencies on deformation degree and opening angle a stronger effect was detected for the inclination between the extruded and non-extruded zone of the shafts. The average shift here was about 77 µm.

Table 4: Effects of the main factors on the shape changes of cold extruded shafts as a result of induction hardening

Measured variable	**Pareto chart**
Curvature change 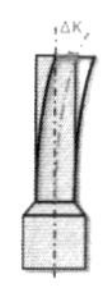	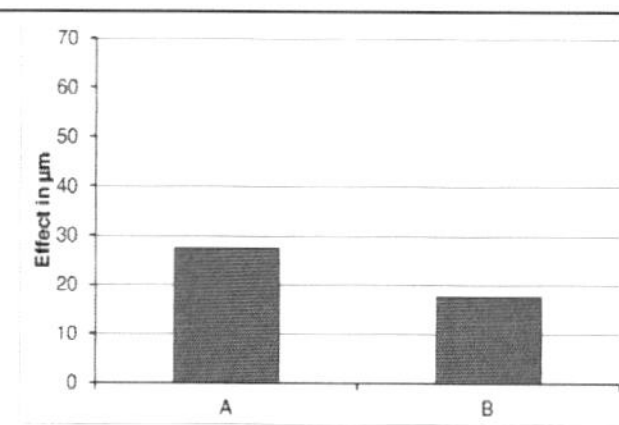
Change of inclination (extruded zone) 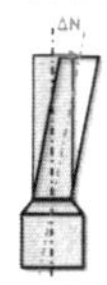	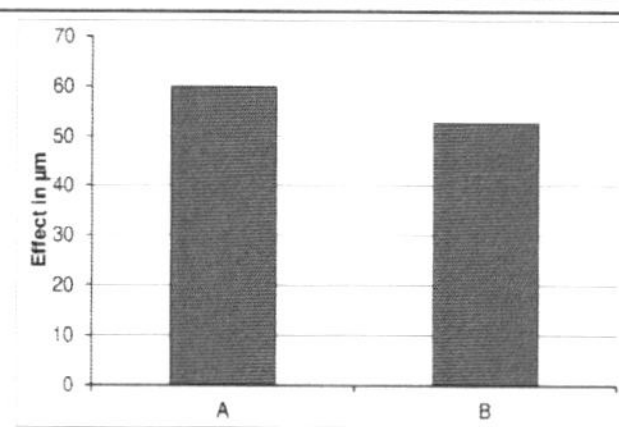

As the deformation degree and the hardening depth were supposed to have the biggest influences, an additional test was performed to eliminate the cold extrusion process and the introduction of cold working and residual stresses, yet keeping the geometry of the shaft and the hardening pattern constant.

Shafts of the same geometry of two extrusion variants with the deformation degree of 0.7 and 1.2 were soft machined and subsequently induction hardened. As soft machining will not introduce residual stress and cold working besides some negligible surface near reactions, the influence of those parameters on the final dimensional changes after induction hardening could be separated. The results are given in Table 5.

As can be seen, the length decrease is much less for the turned shafts compared to the cold extruded ones. The diameter change in the zone representing the extruded area also is less for the turned shafts, even more significant for the higher deformed shaft (φ = 1.2) compared to the less deformed area (φ = 0.7). Of course the ratio between surface hardening depth and shaft diameter is different also in both variants.

The interplay between cold working process and its introducing of residual stresses and the hardening process which releases stresses and creates new strains and volume increases determines the final size change of the shaft.

Table 5: Size changes of induction hardened shafts (SHD > 1mm) with different pre-processing: comparison of soft machining and cold extrusion before hardening

Geometry/ Degree of Deformation	Relative size change after induction hardening	Cold extrusion	Soft machining (turning)
0.7	Δl/l	- 0.80 ‰	- 0.14 ‰
	Δd/d	0.49 ‰	0.36 ‰
	ΔD/D	0.03 ‰	- 0.22 ‰
1.2	Δl/l	- 1.56 ‰	- 0.40 ‰
	Δd/d	1.77 ‰	0.88 ‰
	ΔD/D	0.07 ‰	0.06 ‰

Changes in shape e. g. curvature and inclination very often are supposed to result from inhomogeneous residual stress distributions. In order to investigate the cause of the measured shape deviations the surface near residual stresses of extruded shafts were measured by X-ray diffraction in circumferential direction (Figure 4). The residual stress patterns showed some inhomogeneous distribution as well as different signs (tension and compression) dependant on the degree of deformation.

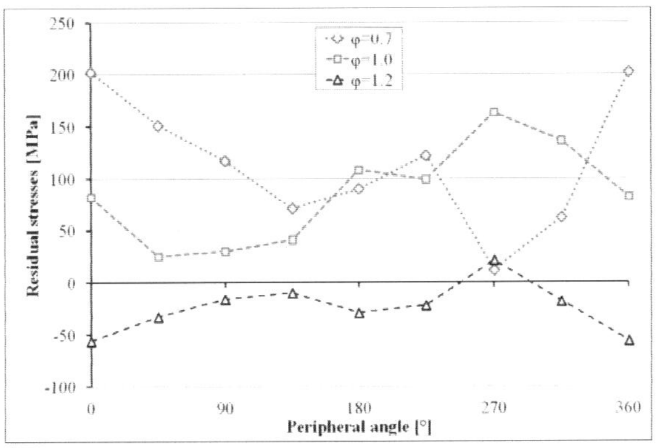

Figure 4: Circumferential near surface residual stresses after cold extrusion, measured in the extruded zone

Similar results in general have been presented by Tekkaya [3] on cold formed components. The finally resulting stresses are influenced by the extrusion process and its stress conditions and of the friction between workpiece and die.

As inhomogeneous distributed residual stresses act as internal forces during stress relief, the difference between minimum and maximum stress along the circumference (absolute values) was calculated and compared with the change in curvature after the induction hardening process. As shown in Figure 5 there is a good correlation between both measures which indicates, that the inhomogeneous residual stresses is the reason for the resulting curvature change. Further investigations showed that the direction of curvature change correlates well with the angle of maximum resp. minimum residual stress levels. The inductive heating of the shafts acts as stress relief in this case.

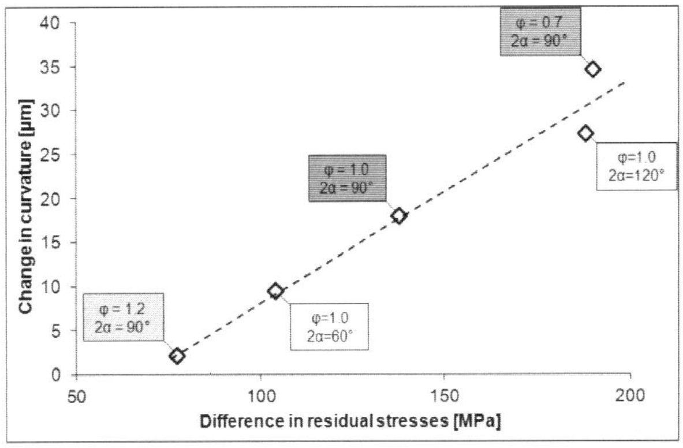

Figure 5: Change in curvature of the extruded zone of the shafts in correlation with the difference in residual stresses, measured in circumferential direction after induction hardening

Influence of initial microstructure on induction hardening results

As shown before, the introduction of residual stresses and cold working by cold extrusion creates a high distortion potential, which is transferred and changed along the entire process chain. Minimizing finally resulting distortion requires to consider all production steps from raw material until finish heat treated or grinded components - the Distortion Engineering approach [4].

The microstructure (and segregation, residual stress condition, etc.) of the raw material can be of great influence on the distortion behavior. For high productivity and precision of the shaped part, the initial microstructure before cold extrusion should offer low yield strength and high ductility. In the investigation reported before therefore a spheroidize annealed microstructure was used. Regarding the following process of induction hardening a different microstructure however would perform better in the kinetics of austenite formation and carbon distribution. Possible alternative microstructures have been investigated in another project to determine the influence of initial microstructure on the case depth and residual stress distribution after induction case hardening.

The material used for this investigation was the quench and temper grade 42CrMo4 (1.7225; AISI 4140) in three different initial heat treatment conditions: normalized and quenched and tempered with lower (550°C) and higher tempering temperature (650°C). Micrographs and hardness values are given in Figure 6, the ultimate tensile strength having been calculated from hardness.

Normalized
850 °C 1 h / cooling in air

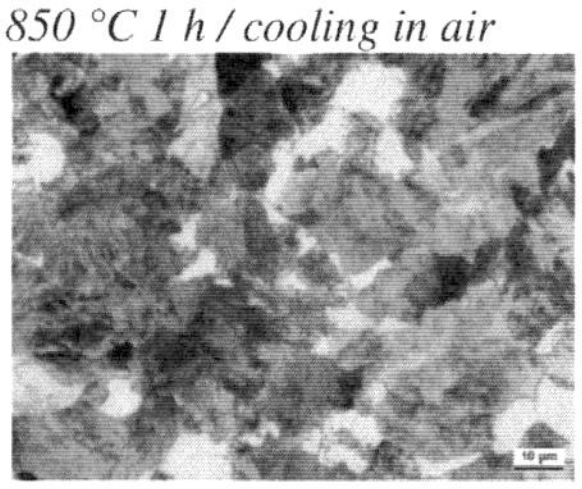

hardness 230 HV
tensile strength ≈ 740 MPa

Q&T (high temperature)
850 °C 1 h / oil / 650 °C 2 h

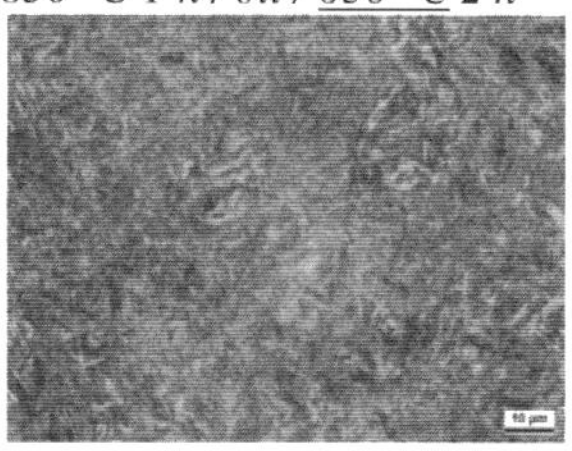

hardness 277 HV
tensile strength ≈ 900 MPa

Q&T (low temperature)
850 °C 1 h / oil/ 550 °C 2 h

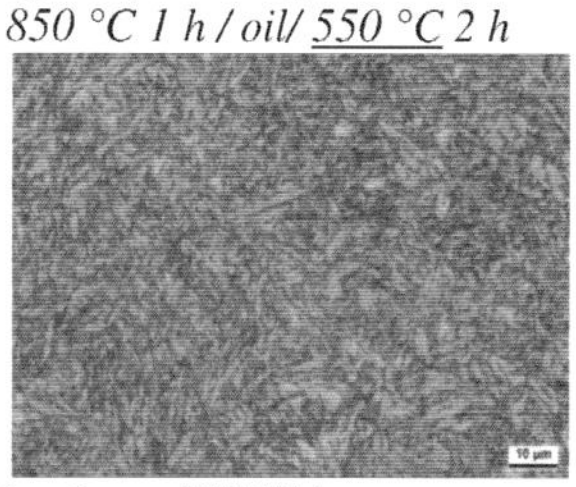

hardness 358 HV
tensile strength ≈ 1.150 MP

Figure 6: Initial microstructure of 42CrMo4 (AISI 4140) before induction hardening. (Sample size: 60 mm diameter × 10 mm height)

After induction surface hardening with medium frequency heating (induction unit as before) the hardness response of those microstructure was different regarding surface hardness and surface hardening depth as expected (Figure 7). Higher surface hardness as well as deeper case depth were found for the low tempered Q&T condition due to a more homogeneous structure and better pre-distribution of carbon within small carbides of the microstructure. The difference between the normalized and high tempered Q&T material was quite small.

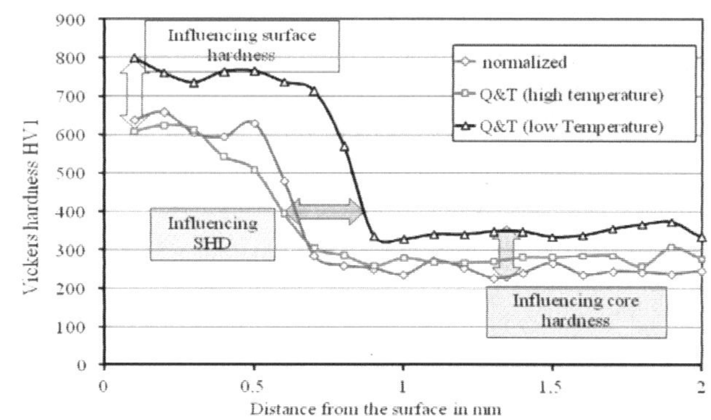

Figure 7: Surface hardness of induction surface hardened 42CrMo4 (AISI 4140) at different initial microstructure

To illustrate the influence of initial microstructure on the kinetics of austenite formation and carbon dissolution an early publication of Orlich et al. was redrawn for a similar steel 50CrMo4 (1.7228; AISI 4174) in Figure 8 [5].

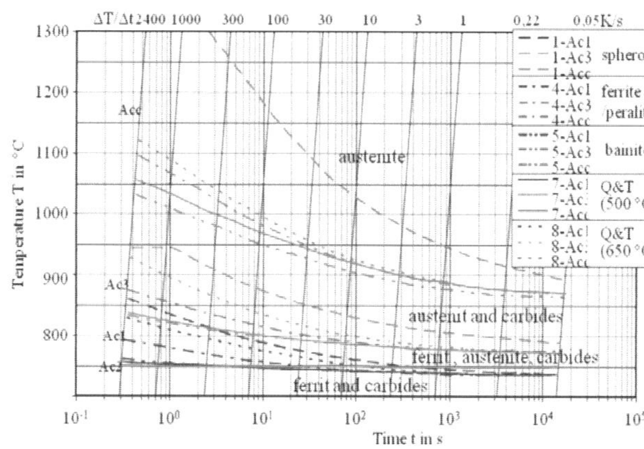

Figure 8: Transition temperatures in continuous TTA-diagram of the steel 50CrMo4 (AISI 4174) at different initial microstructures (according to [6])

The TTA diagram shows the influence of initial microstructure on the transformation temperatures A1 and A3. The accelerated heating, which is evident for induction hardening, leads to a further shift to higher temperatures. Therefore the initial microstructure is of importance for the set up of parameters in induction hardening, which of course is well known.

From the same publication the hardness response of different initial microstructures and maximum austenitizing temperatures again shows the influence of the initial condition (Figure 9). The temperature range for optimum hardness response for the different microstructures is remarkably high. As hardness is not a proper measure to describe fatigue properties, the increasing grain size of high austenitizing temperatures also have to be taken into account.

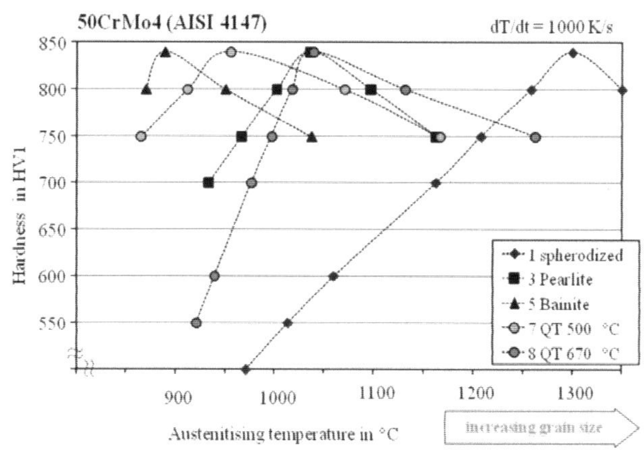

Figure 9: Influence of initial microstructure on hardness (data taken from [5])

Very important for fatigue properties are the surface residual stresses. How those are influenced by the initial microstructure before induction hardening, the samples of Figure 6 were measured by X-rax diffraction, the tangential residual stresses are given in Figure 10.

The surface residual stresses reveal high compressive values as desired for properly induction hardened parts. Yet obviously the initial microstructure is of almost negligible influence on the value of the compressive stresses. A small indication is given for some higher compressive stresses for the low tempered Q&T condition, which again confirms the superior behavior of a fine pre-distribution of carbon resp. carbides. More significant in this case is the depth effect of different microstructures on the residual stress pattern. The zone where compressive residual stresses drop into the tension zone is deeper for the low tempered Q&T condition. As described before, the generation of residual stresses during induction surface hardening is the result of interaction of some mechanisms with own influencing parameters. Figure 10 shows e. g. the difference in core hardness of the microstructures, which reflect the resistance of the core material against yielding, when tensile strain occurs during the rapid heating of the surface near areas.

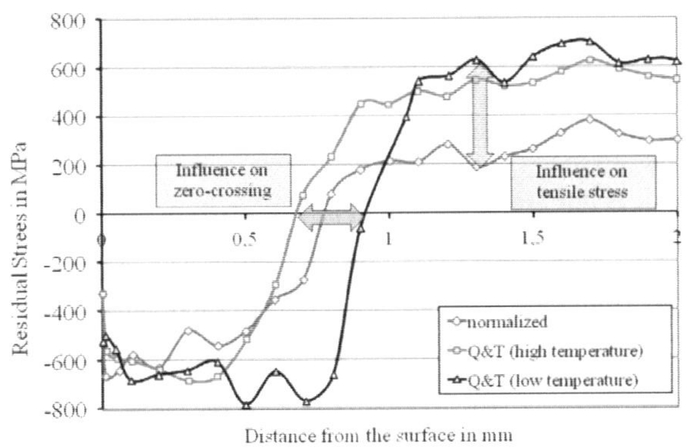

Figure 10: Tangential residual stresses of induction surface hardened 42CrMo4 (AISI 4140) at different initial microstructure (Sample size Ø 60 × 10)

Conclusions

Cold forming processes are of economical importance due to high productivity rates and the possibility to achieve near-net-shape geometries which require less stock removal. Yet, processes like cold extrusion introduce high cold working and residual stresses into the workpiece. Those form a so-called distortion potential which is activated e. g. by a heat treatment process like induction surface hardening.

How different process parameters of cold extruded single-shouldered shafts influence the size and shape changes after a subsequent induction hardening process was investigated using the method of Design of Experiments (DoE). For the given set-up of parameters of geometry of the raw material and tool design as well as different degrees of deformation, dry lubricants and surface hardening depths it was found that the size changes dominantly are controlled by the degree of deformation and the surface hardening depth, less by the lubrication coating of the blank before extrusion.

Regarding shape changes, e. g. for the shafts the changes in curvature and inclination of the shouldered shaft, are influenced by the degree of deformation and the opening angle of the extrusion tool. With this basic knowledge, process descriptions can be improved in order to minimize distortion effects or to apply a pre-correction of dimensions in the planning state already.

As the microstructure of the raw material is on the one side influencing the extrusion behavior due to strengthening effects and introduction of residual stresses, on the other side simultaneously affects the hardening behavior of the steel, some comparisons were performed between different initial microstructures. The microstructure requirements of the hardening process (homogeneous microstructure, good carbon resp. carbide distribution) are contrary to the requirements of the cold extrusion process (soft condition, low hardness, high ductility, ferritic-carbidic microstructure with larger carbides). So, to achieve an optimum process design, a suitable compromise has to be found for best fulfillment of both requirements.

Acknowledgments

The project was funded by the German Federal Ministry for Economic Affairs and Energy via the German Federation of Industrial Research Associations (AiF) regarding a resolution of the Budget Committee of the Federal Parliament, which is gratefully acknowledged. We wish to express our gratitude to our project partner, Institute of Forming Technology and Lightweight Construction (IUL) at Technical University Dortmund for the excellent collaboration as well as to our industry partners in the project working group.

References

[1] Kloos, K. H. / Braisch, P.: Schwingfestigkeit induktiv randschichtgehärteter bauteilähnlicher Proben. Forschungsbericht für die Deutsche Forschungsgemeinschaft im Rahmen des DFG-Schwerpunktprogramms "Fertigung und Bauteilverhalten". Darmstadt, 1988

[2] Fahry, H., Zoch, H.-W. u. Schlicht, H.: Entstehung von Eigenspannungen und Verzügen bei der induktiven Randschichthärtung von Bauteilen. HTM 44 (1989) 3; p. 149 - 156

[3] Tekkaya, A. E.: Ermittlung von Eigenspannungen in der Kaltmassivumformung. Berichte aus dem Institut für Umformtechnik, Universitat Stuttgart, Nr. 83, Springer-Verlag, Berlin, 1986

[4] Zoch, H.-W.: From single production step to entire process chain – the global approach of Distortion Engineering. Mat.-wiss. u. Werkstofftech. 37 (2006) 1, S. 6-10

[5] Orlich, J.; Rose, A. u. Wiest, P.: Atlas zur Wärmebehandlung der Stähle - Band 3. Verlag Stahleisen, Düsseldorf, 1973

International Federation of Heat Treatment and Surface Engineering 2016
Proceedings of the 23rd IFHTSE Congress
April 18–21, 2016, Savannah, Georgia, USA

Prediction of quench-induced residual stresses in Al-Cu-Mg impeller forgings

Denis Carron
Université Bretagne Sud, FRE CNRS 3744, IRDL, Lorient, France
denis.carron@univ-ubs.fr

Nicolas Chobaut, Jean-Marie Drezet
Ecole Polytechnique Fédérale de Lausanne, Laboratoire de Simulation des Matériaux, Lausanne, Switzerland
nicolas.chobaut@epfl.ch , jean-marie.drezet@epfl.ch

Peter Saelzle
ABB Turbo Systems Ltd, Baden, Switzerland
peter.saelzle@ch.abb.com

Abstract

In this work a finite element analysis is used for the calculation of residual stresses in AA2618 forgings for large turbocharger impellers. Heat transfer during quenching is determined from temperature measurements on forgings equipped with thermocouples. Rather than a complex and time-consuming precipitation characterization and modelling task, a simple but realistic approach to account for precipitation upon cooling is chosen. An elasto-viscoplastic constitutive model with additive hardening is used with an uniform (not space dependent) and temperature dependent yield strength. A limited number of thermomechanical Gleeble tests have been used to identify the model parameters and to validate it. Calculated strains and stresses are compared to neutron diffraction measurements.

Introduction

High strength aluminum alloys (AA) which are widely used for aerospace and engineering applications owe their superior mechanical properties to the heat treatment process where quenching is a key step. Fast quenching is needed to avoid coarse precipitation which is detrimental to the final yield strength. From a metallurgical point of view, an ideal quench must lead to a supersaturated solid solution in the entire component. However fast quenching cannot be achieved in the center of large components that experience a too low cooling rate to prevent any precipitation. On the other hand, fast quenching induces also residual stresses (RS) that - even reduced by stress relief [1] - can cause unacceptable distortions during machining [2] and unfavorable stresses in service. Control and prediction of quench-induced RS is thus of particular interest for the industry to prevent such negative impacts. The problem is complex since these RS do not depend only on the magnitude of the cooling rate and on the component size and shape but are also influenced by possible precipitation during quenching.

This is the case in forged AA2618 large parts used for the fabrication of the compressor impellers in turbochargers. In this heat treatable Al-Cu-Mg based alloy [3], two types of precipitation may take place during quenching and affect the formation of internal stresses. A first precipitation occurs at intermediate temperature (415-300°C [4]) for low cooling rates. These large S/S' precipitates are undesirable since they do not harden the material significantly while reducing the amount of elements in solid solution and thus the hardening potential. A second precipitation of Cu-Mg hardening clusters occurs at low temperature (below 300-250°C [5,6]) even for high cooling rates. The resulting effect of these two types of precipitation is a possible yield strength increase at the surface of large quenched components [7]. Heat transfer during quenching, thermally induced deformations and precipitation phenomena are thus interacting and therefore need to be taken into account to predict RS after quenching.

In this work a finite element analysis is used for the calculation of residual stresses in axisymmetric AA2618 forgings. Heat transfer during quenching is determined from temperature measurements on forgings equipped with thermocouples. The general approach to account for precipitation is to use a thermo-metallurgical-mechanical model based on a yield stress model coupled to a precipitation model [7]. This kind of model requires a lot of effort to characterize and model the precipitation of metastable non-stoichiometric phases and clusters during quenching [8,9]. In the present work, a thermomechanical model accounting for precipitation in a simple but realistic way is chosen [9-11]. Two elasto-viscoplastic constitutive models with additive hardening are used to achieve simulations. One of the thermomechanical models (TM model) ignores precipitation. The other model (TMG model) takes into account precipitation in a simple way through uniform (not space dependent) and temperature dependent yield strength values. Using a Gleeble machine, a limited number of interrupted quench-tests (IQ) have been performed to characterize the mechanical behavior of AA2618 upon cooling and determine the numerical model parameters. For the sake of validation, the identified model parameters are used to simulate non-isothermal Gleeble constrained cooling experiments. Strains and stresses in axisymmetric forgings are then calculated by finite element analysis and compared to measurements with neutron diffraction.

Experimental procedure

Materials

Two identical axisymmetric AA2618 (Al-2.4Cu-1.55Mg-1.1Ni-1.1Fe wt.%) forgings of 0.36 m in length and 0.53 m in diameter with a central axial through-hole were used for temperature and RS measurements. To determine the heat transfer during quenching, one forging has been equipped with 7 thermocouples of 1.5 mm in diameter some of them located only 3 mm beneath the surface. The other forging, which has experienced the same heat treatment, was non-destructively tested by neutron diffraction, first in the as-quenched state and then in T6 state. For the Gleeble tests, specimens were cut from AA2618 forging material, presolutionized in a furnace for 4h at 530°C, quenched in cold-water and stored in a freezer at ca. -10°C to retard any precipitation. More details on the experimental procedure can be found in Refs. [11] and [12].

Temperature measurements during quenching

After solutionizing the forgings were quenched by vertical immersion in a boiling water basin. A vertical water jet directed to the bottom side of the forging is activated to increase the cooling rate in the temperature range of. 400-300°C (Fig. 1).

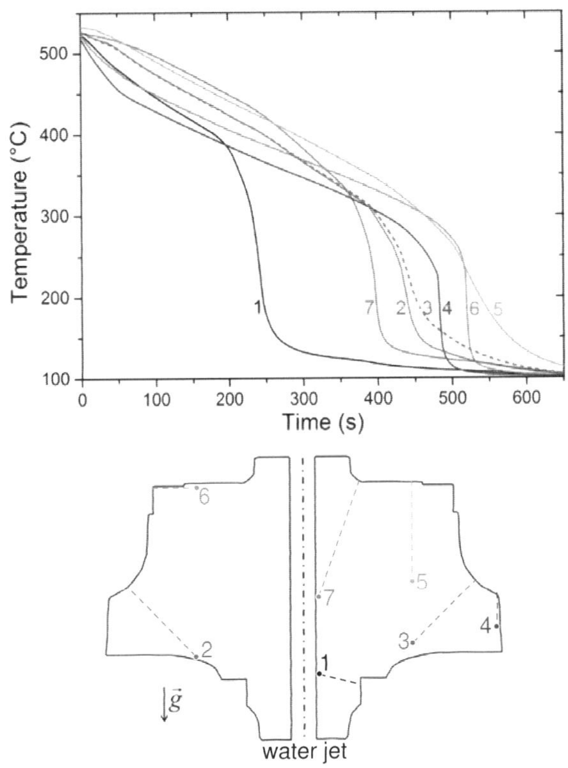

Figure 1: Measured cooling curves in boiling-water quenched forging and position of thermocouples with drillings in dashed lines.

From the solutionizing temperature, temperature begins to decrease almost linearly with time but is not uniform within the forging. The bore (TC1, TC7) and the bottom side experience the fastest cooling since they are exposed to the water jet. The center of the massive cross section (TC5) cools more slowly and with very little variation in the cooling rate. The subsequent huge temperature gradients induces internal thermal stresses. After ca. 650 seconds temperature becomes almost uniform and as a result internal stresses will no longer evolve.

Residual strain and stress measurements

Residual strain measurements have been carried out in the forging in as-quenched state by neutron diffraction (ND) using the SALSA diffractometer located at the Institut Laue-Langevin (IIL) in Grenoble, France. The same forging was then artificially aged to T6 state and then measured at the STRESS-SPEC diffractometer located at FRM II in Garching, Germany. Reference measurements have been made on a stress-free probe material cut from the forging. Using the (311) aluminum reflection and reference interplanar spacing, strain measurements in three mutually orthogonal directions (radial, axial and hoop) were performed on three radial scan lines (Fig. 2).

Residual elastic strains are converted into stresses using the generalized Hooke's law with the diffraction peak-specific elastic constants for texture-free aluminum $E_{(311)}$ = 70.2 GPa and $\nu_{(311)}$ = 0.35 [13]. The measurement method is fully detailed in Ref. [9].

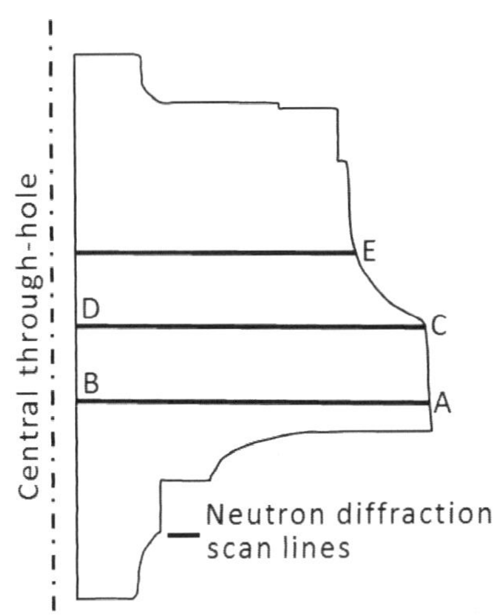

Figure 2: Positions of neutron diffraction scan lines [AB], [CD] and [EF].

Gleeble IQ-tests and constrained cooling experiments

A Gleeble 3500 machine is used to achieve the thermomechanical tests. For all the tests, specimens are first heated by Joule effect at 30 K/s to the solutionizing temperature while the force is maintained near zero to allow free dilatation. Then they are cooled at zero force control to the tensile testing temperature (IQ-tests) or are cooled while keeping both extremities blocked by fixed jaws (constrained coolings). An optimized geometry was designed to localize deformation at the mid-length of the specimen where the cross section is minimal [10]. Specimens are held by water-cooled grips. A diametral extensometer is used to monitor strain during the test.

The IQ-tests are used to characterize the mechanical behavior of AA2618 upon cooling and identify the parameters of the constitutive law used for the TM and TMG models. After the in-situ solutionizing treatment, specimens are cooled with:

- a fast interrupted quench "IQ1" with a cooling rate greater than 100K/sec above 300°C in order to characterize the solid solution without precipitates,

- or an interrupted quench "IQ2" with an almost constant cooling rate of 20 K/s. This cooling avoids coarse S/S' precipitation above 300°C [4] and it represents the surface cooling measured in forgings between 250°C and 150°C where precipitation hardening occurs [5,6] (Fig. 3).

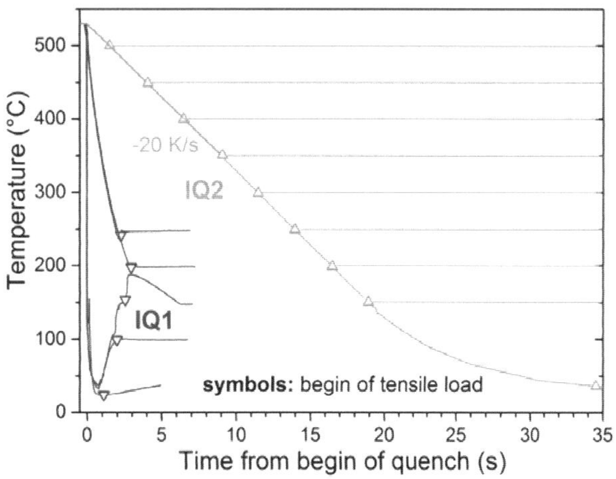

Figure 3: Interrupted quench-test thermal cycles.

For the IQ1-quenches, cooling by copper grips and additional air or water cooling were used. Since water quench (WQ) interrupted at 100°C and 150°C caused too high overshoots, WQ to 25°C, followed by a fast reheat was rather performed. For the IQ2-quenches, cooling by copper grips with or without additional air cooling was applied.

The coolings were interrupted at different temperatures to perform isothermal tensile loads at constant displacement rates. Experimental stress-strain curves can be found in Refs [9] and [14].

The constrained cooling experiments are used to test and validate the TM and TMG model. Three different cooling conditions from the solutionizing temperature are imposed in the Gleeble machine (Fig. 4):

- continuous cooling "CC1" representative of the slowest temperature evolution measured at the surface of a large forging quenched in boiling water, with a typical temperature plateau around 100°C,

- continuous cooling "CC2" faster than 20 K/s above 300°C and fast below 150°C without "CC2a" or with "CC2b" air cooling,

- continuous cooling "CC3" slightly faster than "CC1" above 150°C and as fast as possible without additional air/water below 150°C.

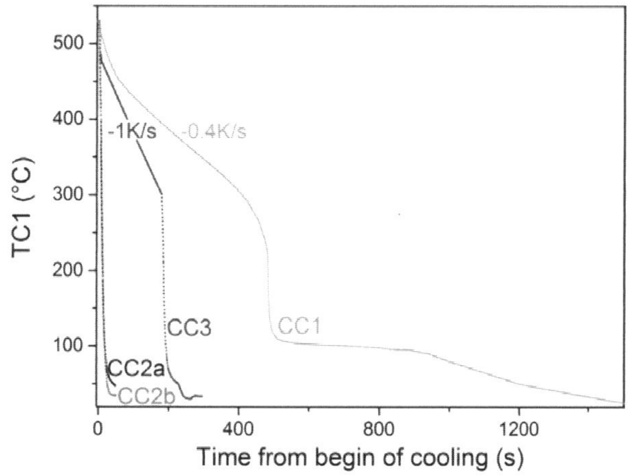

Figure 4: Constrained cooling experiment thermal cycles.

Above 100°C "CC1" and "CC3" coolings are representative of the slowest (respectively fastest) temperature evolutions measured at the surface of large forgings quenched in boiling water [14]. Instead of holding the temperature at 100°C as in "CC1", "CC3" cools rapidly to limit precipitation hardening.

"CC2a" was chosen to avoid the formation of S/S' phases at high temperature and limit precipitation hardening at low temperature. "CC2b" is close to "IQ2" used to feed the TMG model.

During cooling, the specimens undergo thermal contraction but are constrained, thus leading to the build-up of tensile axial stresses (Fig. 5). Differences between stress evolutions are attributed to [11]:

- positive strain-rate effect at temperature higher than 300°C,

- solute loss during cooling "CC1" and "CC3" in the 230-300°C temperature range which decreases solution strengthening and/or precipitation hardening potential,

- precipitation hardening by cluster formation at low temperature.

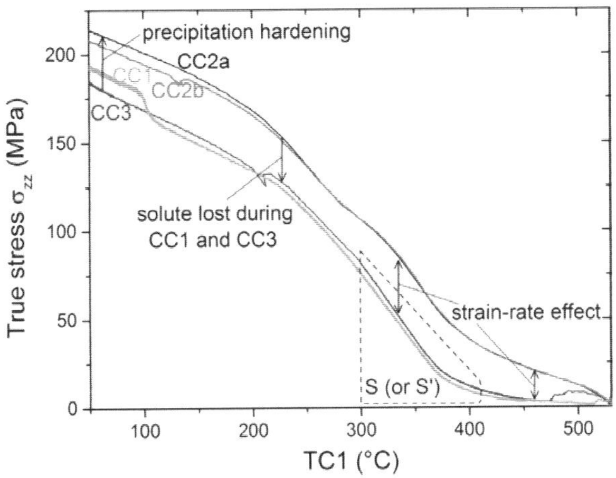

Figure 5: Measured axial stress vs temperature during constrained cooling experiments.

Modelling

Thermo-mechanical behavior

To fit the experimental results, an elasto-viscoplastic constitutive law with additive hardening is chosen [14]. The uniaxial tensile load at constant temperature is defined by:

$$\sigma_{zz} = \sigma_y + H\left(p_{cum}\right)^n + K\left(\dot{p}\right)^m$$
$$\text{with } \dot{p} = \left|\dot{\varepsilon}_{zz}^{in}\right| \text{ and } p_{cum} = \int_{T<T_{cum}} \dot{p}\,\mathrm{d}t \qquad (1)$$

where σ_{zz} is the axial flow stress, $\dot{\varepsilon}_{zz}^{in}$ is the inelastic strain-rate and p_{cum} is the inelastic deformation accumulated below T_{cum} = 350°C. This temperature corresponds to the temperature above which inelastic deformation has no effect on the subsequent low temperature behavior. It represents a simple way to consider plastic strain recovery at high temperature. It is further assumed that the Bauschinger effect is negligible during quenching as this is the case for water quenched plates [10]. In Eq. 1, σ_Y is the yield strength at 0% strain offset, H and n are hardening parameters, K is the consistency and m is the strain-rate sensitivity.

The temperature-dependent parameters (H, n, K, m), given in Ref. [11], are considered independent of precipitation. Precipitation mainly affects σ_Y, which depends on the nature, size and volume fraction of precipitates as well as on solute concentration in the matrix. The same parameter values are thus used for both TM and TMG models which differ only by σ_Y values.

Using the stress-strain curves achieved after IQ-tests [9,14], σ_Y yield strength values are determined in order to be used in the TM and TMG models (Fig. 6).

The TM model ignoring precipitation is fed by σ_Y values after fast "IQ1" coolings that correspond to a state as close as possible to supersaturated solid solution.

The TMG model uses σ_Y values obtained through tensile tests after slower "IQ2" coolings where precipitation hardening by cluster formation occurs below 300°C [5].

At high temperature σ_Y is considered little affected by S/S' precipitation due to the short time spent during quenching. Thus the same values, obtained after "IQ2" coolings, are used by the two models.

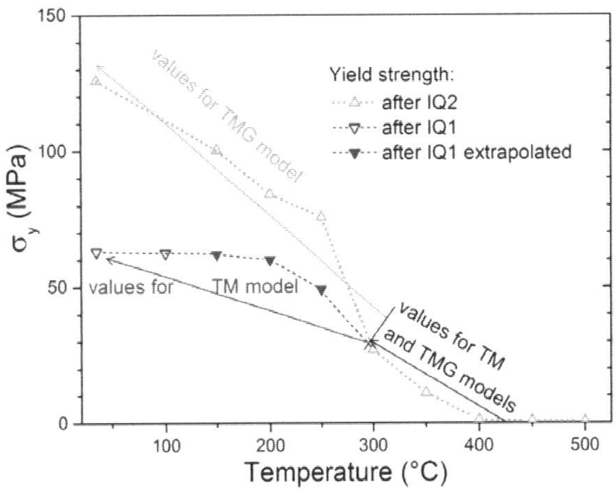

Figure 6: Measured yield strength vs. temperature during interrupted quench-tests.

Constrained cooling simulations

Constrained cooling of Gleeble specimens is simulated with the Abaqus software. Radial thermal gradient is neglected and an axial temperature profile temperatures measured by thermocouples located at specimen surface is imposed for the stress analysis. The specimen is free to expand during heating but is constrained during cooling using a boundary condition that simulates the blocked-jaw condition experimentally imposed.

For the three cooling conditions considered in this work, the experimental results of blocked-jaw tests are shown in Fig. 7, together with the results by finite element simulation using the TM and TMG models.

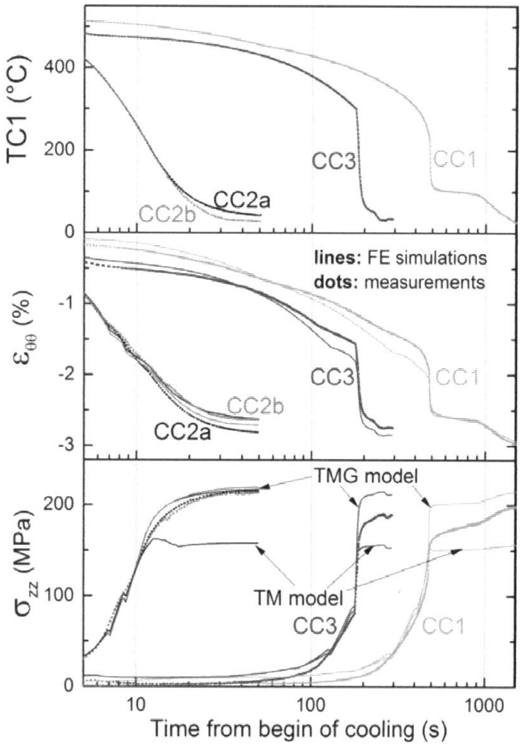

Figure 7: (a) Cooling cycles and comparison between measured (short dots) and simulated (solid lines) (b) hoop strain and (c) axial stress evolutions.

An excellent agreement is found between calculated and experimental thermal cycles. The measured hoop strain evolution are qualitatively well reproduced by the simulation even for long times. The TM and TMG models give identical strains but different stresses [11].

At high temperature, the stress predicted by the simulations using TM or TMG models is close ("CC2a" and "CC2b" coolings) or slightly higher ("CC1" and "CC3" coolings) than the measured one. At lower temperature (< 200°C), the agreement between measurements and simulation is good with the TMG model provided that the constrained cooling and the interrupted quenches have a similar cooling rate with an identical thermal history at high temperature ("CC2a" and "CC2b" coolings). By comparison the TM model underestimates stresses.

For coolings "CC1" and "CC3" similar to boiling water quenching of forgings, the simulation using the TMG model (resp. TM model) overestimates (resp. underestimates) the stresses measured during constrained cooling. For unconstrained coolings as in boiling water quenched forgings, and assuming that the thermal field is well known, it can thus be inferred that the TM model will give a lower bound for RS predictions. An upper bound would be given by the TMG model.

Finite element computation of quenching of forgings

In order to simulate the strongly position-dependent heat flow during quenching, heat transfer coefficients were determined on the different surfaces by inverse methods using an axisymmetric model of quenching in boiling water [12]. As shown in Fig. 8, the simulated temperature profile matches very well the measured temperatures. This is essential for a reliable stress prediction.

Then an uncoupled heat transfer and subsequent thermal-stress analysis is performed. The temperature field is recorded in the Abaqus results file during the heat transfer analysis. This file is then used as input to the thermal-stress analysis of quenching. The thermal-stress simulations have been executed with both material models, i.e. the TM and the TMG model. The artificial ageing step is not simulated.

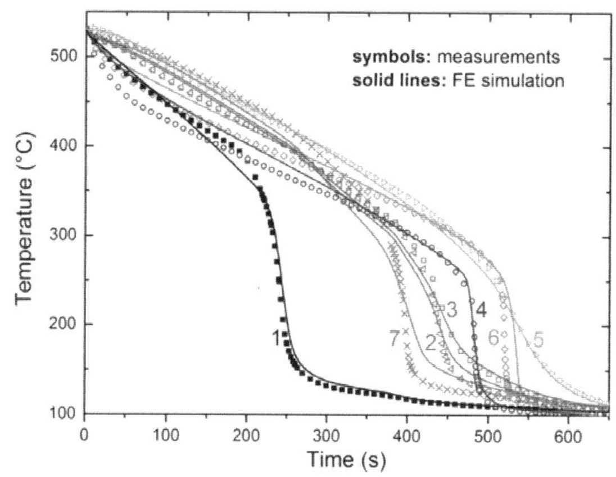

Figure 8: Comparison between measured and simulated temperature in the forging.

In this model, the stress tensor in cylindrical coordinates writes:

$$\underset{\sim}{\sigma} = \begin{pmatrix} \sigma_{rr} & \sigma_{rz} & 0 \\ \sigma_{zr} & \sigma_{zz} & 0 \\ 0 & 0 & \sigma_{\theta\theta} \end{pmatrix} \qquad (2)$$

where r, θ and z are the radial, hoop and axial directions respectively. Shear components $\sigma_{rz} = \sigma_{zr}$ are small but different from zero due to the forging shape.

Fig. 9 shows in direct comparison the simulated as-quenched residual strain profiles using the TMG model and the residual strains measured by neutron diffraction in as-quenched and T6 state.

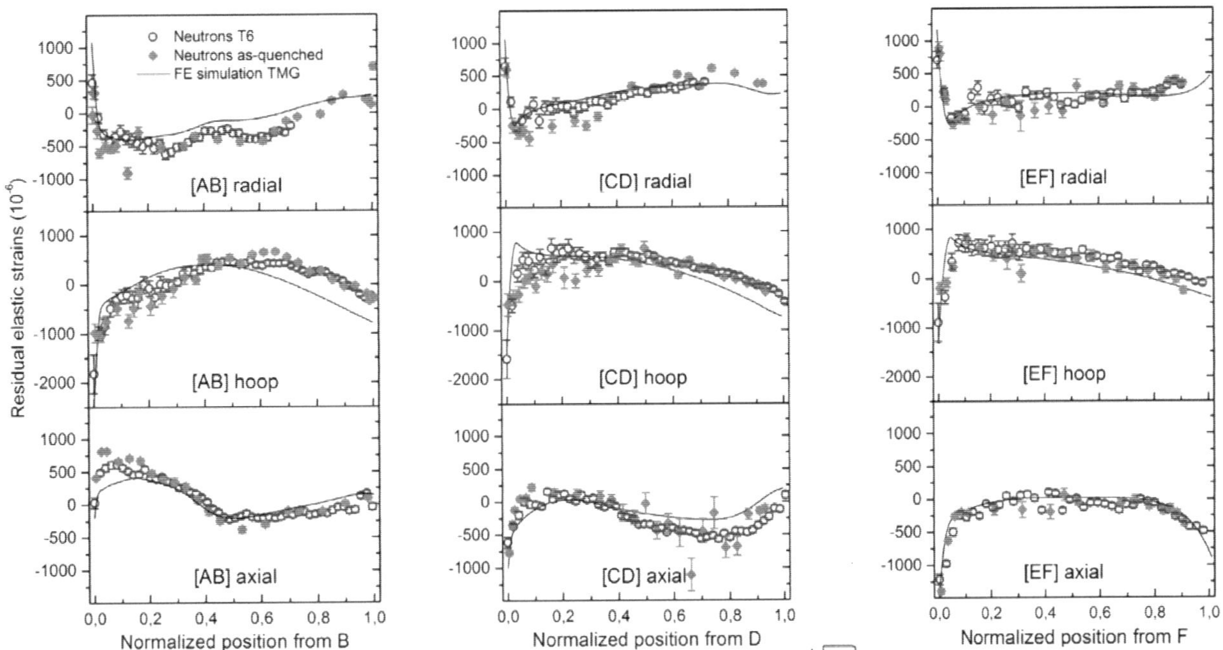

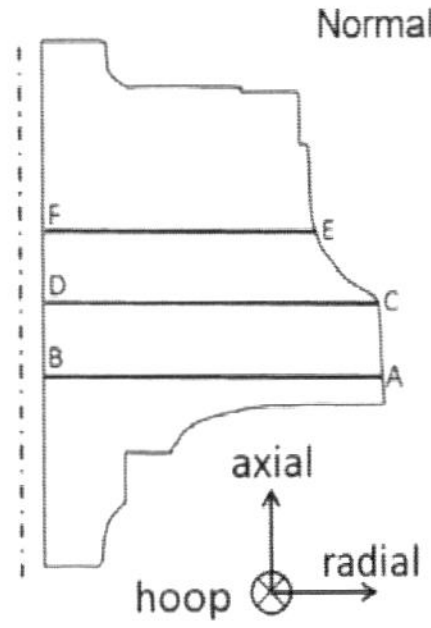

Figure 9: Comparison between measured and simulated residual elastic strains along [AB], [CD] and [EF] in forgings. Measurements were achieved in as-quenched and T6 condition, simulation only consider the as-quenched condition. The same legend applies for all the graphs.

The measured residual strain profiles from the two heat treatment conditions are very close to each other. In other words the peak age T6 heat treatment of AA2618 does not influence the strains and stresses induced during quenching because of the rather low temperature of this treatment, and the high thermal stability of the AA2618.

Overall, the strain measurement results are well reproduced by the FE quenching simulation. This is due to the quality of both the calculated thermal field evolution and the material model.

RS calculated from the measured strains are compared to simulated RS with the TMG model in Fig 10. It shows that the FE quenching simulation using the TMG model predicts relatively well the measured RS inside the forging. In general stresses seem slightly less well predicted than strains but this should be viewed in relation to the larger differences between measured RS profiles from the two heat treatment conditions compared to those between strain profiles. It is worth recalling that all stress components are dependent on all measured strain components by the use of Hooke's law. Thus uncertainties in one strain direction, for example due to long penetration paths and thus lower measurement statistics, are visible in all three stress directions.

The as-quenched residual stress distributions simulated with the TMG model are shown in Fig. 11. As already shown by the measurements, the forging is in a bi-axial compression state close to the surfaces and in a tri-axial tension state at the center. The maximal tensile RS is found close to the central hole for the hoop component.

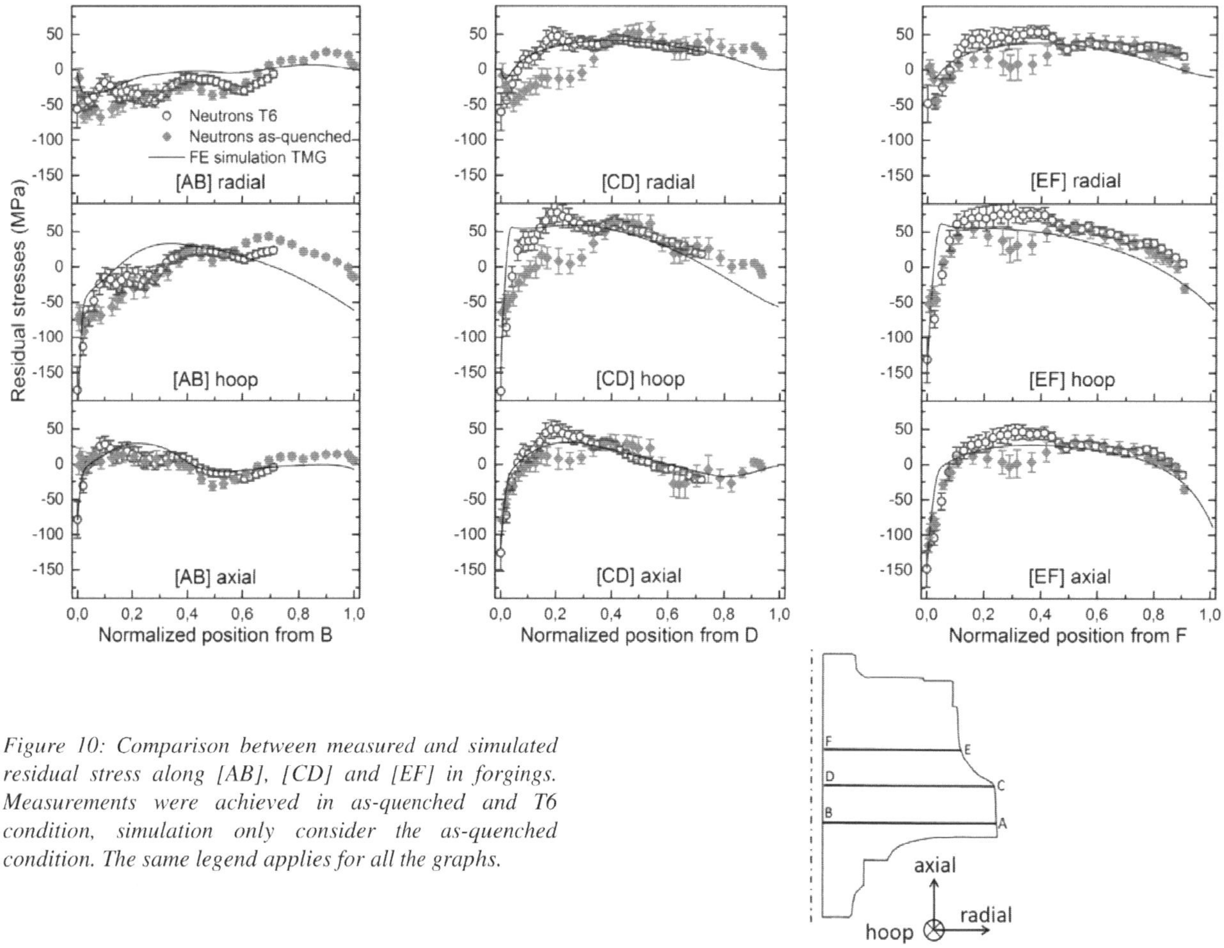

Figure 10: Comparison between measured and simulated residual stress along [AB], [CD] and [EF] in forgings. Measurements were achieved in as-quenched and T6 condition, simulation only consider the as-quenched condition. The same legend applies for all the graphs.

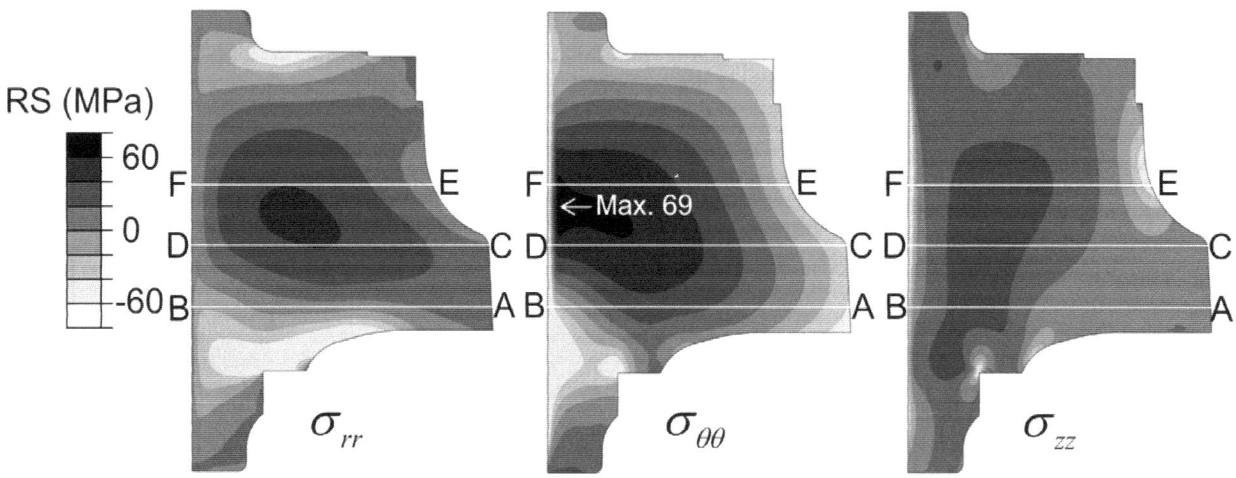

Figure 11: Calculated residual radial, hoop and axial stress components using the TMG model.

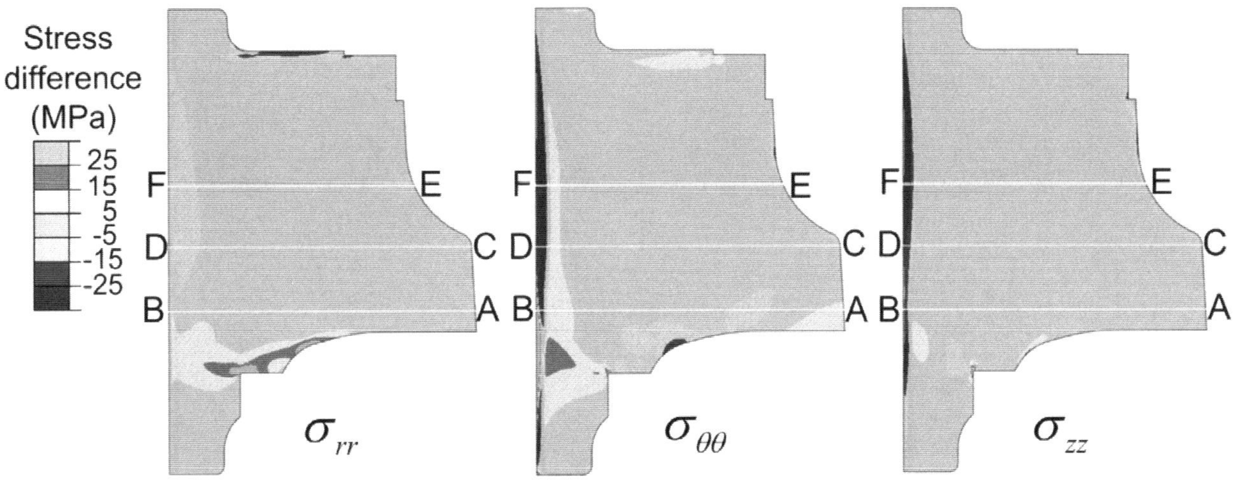

Figure 12: Subtraction of stresses in the forging predicted by the TM model to those predicted by the TMG model.

A direct comparison of the predicted stresses by the usage of the TM model and the TMG model is shown in Fig. 12, where the stress differences are depicted.

The TM model, where precipitation is neglected, gives in the most regions and especially in the center of the forging RS very close to those predicted by the TMG model. The differences are lower than +/- 5 MPa. Bigger differences occur along the surface regions and mainly along and close to the central bore for hoop and axial stress. To explain these local discrepancies in the comparison of stresses some assumptions can be made:
- close to the central hole, plastic deformation occurs below 300°C and since yield strength values used for TMG model are higher than those used for TM model, calculated RS are also higher in absolute values,
- due to the complex cooling conditions and high temperature gradients, the simulated RS close to the central hole are very sensitive to the boundary conditions and the meshing density. As a consequence, calculated RS values in these areas may be judged cautiously.

Nevertheless, the low difference in the large-area center region between RS predicted by the TM and the TMG models indicates that precipitation occurs in the forgings with a central hole, but has a relatively small impact on RS. The impact is more pronounced in water quenched thick AA7xxx plates [10] where surface precipitation hardening occurs by cluster formation during quenching. In the absence of a central hole, differences between TM and TMG predictions are also more pronounced [9]. In all cases, precipitation being accounted for through uniform (not space dependent) but temperature dependent yield strength values determined after cooling representative of the component surface, the relatively good prediction of RS by the model means the surface - i.e. the coldest area with the highest σ_y value - dictates the RS distribution.

Conclusions

Interrupted quench-tests have been performed to determine the mechanical behavior of AA2618 at different temperatures below its solutionizing temperature and to identify the parameters of the constitutive law used for two material models – TM, ignoring precipitation and TMG, accounting for precipitation in a simple way. The identified model parameters were used to simulate Gleeble constrained coolings in a blocked jaws configuration. At high temperature (> 300°C), the agreement between constrained cooling measurements and simulation is relatively good whether the TM or the TMG model is used. At lower temperature, the agreement between measurements and simulation is good with the TMG model provided that the constrained cooling and the interrupted quenches have a similar cooling rate with an identical thermal history at high temperature.

The TM and the TMG model are used in a FE analysis for the calculation of RS after quenching in large AA2618 forgings. The thermal field evolution in the quenched forging is well modelled which is essential to a reliable stress prediction. The residual strain profiles in as-quenched and T6 state appear to be very similar meaning that the peak age treatment does not affect the internal stresses. The neutron diffraction measurements are well reproduced by the FE quenching simulation using TM or TMG model. Hence considering or ignoring the impact of possible precipitation on the yield strength in the material model appears to be of second order in the present case.

Acknowledgments

This work is funded by the Competence Center for Materials Science and Technology in the frame of the project entitled "Measurements and modeling of residual stress during quenching of thick heat treatable aluminum components in relation to their microstructure" involving EPF Lausanne, PSI Villigen, Univ. Bretagne-Sud Lorient, Constellium and ABB Turbo Systems Ltd. The Gleeble 3500 machine of Univ. de Bretagne Sud was co-financed by European Regional Development Fund. The authors are grateful to Prof. P. Pilvin and W. Berckmans (Univ. Bretagne Sud) for providing SiDoLo and for the instrumentation of the Gleeble specimens respectively. The authors are indebted to W. Erb (ABB Turbo Systems Ltd) for thermocouple and strain gauge instrumentation of the forging. The authors are also grateful to the international neutron source beam line scientists, Th. Pirling at the Institut Laue Langevin and M. Hofmann at FRM-II for the provision of beam time.

References

[1] Boyet, J.C., Boivin, M., "Numerical calculations of residual–stress relaxation in quenched plates", *Materials Science and Technology*, Vol. 1 (1985) pp. 786-792.

[2] Dubost, B., Bouet-Griffon, M., Jeanmart, P.H., Hommette, O., "Prediction and minimization of residual stresses in quenched aluminium alloy die forgings", *Proc. Int'l Conference on Residual Stresses*, Nancy, France, November 1988, pp. 581-586.

[3] Pouget, G., Sigli, C., "Thermal Stability of Al-Cu-Mg Alloys", *Materials Science Forum*, Vols. 794-796 (2014) pp. 691-696.

[4] Chobaut, N., Carron, D., Drezet, J.M., "Characterisation of precipitation upon cooling of an AA2618 Al-Cu-Mg alloy", *Journal of Alloys and Compounds*, Vol. 654 (2016) pp. 56-62.

[5] Schloth, P., Menzel, A., Fife, J.L., Wagner, J.N., Van Swygenhoven, H., Drezet J.M., "Early cluster formation during rapid cooling of an Al-Cu-Mg alloy: in situ small-angle X-ray scattering", *Scripta Materialia*, 2015, vol. 108, pp. 56-59.

[6] Schloth, P., Precipitation in the high strength AA7449 aluminium alloy: implication on internal stresses on different length scales, PhD thesis n°6525, EPFL, (Lausanne, 2015).

[7] Godard, D., Archambault, P., Denis S., Gautier E., Heymes, F., "Modelling of heat treatment residual stresses. Application to high strength aluminum alloys including precipitation effects", *Proc. 7th Int'l Seminar on Heat Treatment and Surface Engineering of Light Alloys*, Budapest, Hungary, September 1999, pp. 249-257.

[8] Godard, D., Archambault, P., Houin, J.P., Gautier E., Heymes, F., "Mechanical softening kinetics at high temperatures in AlMgZnCu alloy: experimental characterization and microstructural interpretation", *Proc. 6th Int'l Conf on Aluminum Alloys ICAA6*, Toyohashi, Japan, July 1998, pp. 249-257.

[9] Chobaut, N., Measurements and modelling of residual stresses during quenching of thick heat treatable aluminium components in relation to their precipitation state, PhD thesis n°6559, EPFL (Lausanne, 2015).

[10] Chobaut, N., Carron, D., Arsène, S., Schloth, P., Drezet, J.M., "Quench induced residual stress prediction in heat treatable 7xxx aluminium alloy thick plates using Gleeble interrupted quench tests", *Journal of Materials Processing Technology*, Vol. 222 (2015) pp. 373-380.

[11] Chobaut, N., Carron, D., Saelzle, P., Wagner, J.N., Pirling, T., Hofmann, M., Drezet, J.M., "Stress generation during the quenching of large AA2618 forgings part I: Determination of the mechanical behavior using Gleeble interrupted quench tests and validation against constrained coolings", submitted to *Metallurgical and Materials Transactions A*, (2016).

[12] Chobaut, N., Carron, D., Saelzle, P., Wagner, J.N., Pirling, T., Hofmann, M., Drezet, J.M., "Stress generation during the quenching of large AA2618 forgings part II: finite element stress-strain computations and validation against neutron diffraction measurements", submitted to *Metallurgical and Materials Transactions A*, (2016).

[13] Hutchings, M. T., Withers, P. J., Holden, T. M., and Lorentzen, T., Introduction to the characterization of residual stress by neutron diffraction, CRC Press, (Boca Raton, 2005), pp. 213.

[14] Chobaut, N., Saelzle, P., Michel, G., Carron, D., Drezet, J.M., "Quench-induced stresses in AA2618 forgings for impellers: a multiphysics and multiscale problem", *JOM*, 2015, vol. 67, pp. 984-990.

International Federation of Heat Treatment and Surface Engineering 2016
Proceedings of the 23rd IFHTSE Congress
April 18–21, 2016, Savannah, Georgia, USA

Study Of Residual Stresses After Steel Quenching With Vegetable Oils

Carlotta Civera
Siderval Spa, Talamona (SO), Italy
carlotta.civera@siderval.it

Barbara Rivolta
Politecnico di Milano, Dipartimento di Meccanica, Milano, Italy, 20156
barbara.rivolta@polimi.it

Rosa L. Simencio Otero, João Gilberto Lúcio, Lauralice C. F. Canale
Universidade de Sao Paulo, Escola de Engenharia de Sao Carlos, Sao Carlos, SP, 13566-590, Brazil
rosa_simencio@yahoo.com.br, jgilberto@rassini-nhk.com.br, lfcanale@sc.usp.br

George E. Totten
Department of Mechanical and Materials Engineering, Portland State Univ., Portland, OR 97201
getotten@gmail.com

Abstract

Although quenching is necessary to increase the mechanical properties of steel, this process is often accompanied by high thermal stresses and transformation stresses which may lead to high residual stresses and distortions and possible cracking. Therefore, it is desirable to develop quenching processes that maximize the material properties while minimizing the residual stresses and distortions. This means that there is often a compromise between the factors involved to assure overall optimal process performance. The objective of this work was to measure and compare the experimental results of residual stresses that arise after quenching AISI 4140 steel grade into four different quenching fluids. Quenching was performed with 3 soybean derived oils and one fully formulated commercially available petroleum-based oil. Residual stresses were experimentally determined on cylindrical samples by two different methods: X-ray analysis and hole-drilling method according to ASTM E837-08. A detailed analysis of the residual stresses was carried out. The obtained results were related to the experimental heat transfer coefficient and viscosity of each fluid.

Introduction

Quenching involves a solution treating (austenitizing) operation, typically in the range of 750-1100°C, and subsequent cooling in a suitable quenching medium to room temperature. The heat treatment is performed to prevent ferrite and pearlite formation at high temperatures and to allow the formation of bainite and martensite at low temperature. Typically, for carbon and alloy steels, martensite is the desired microstructure due to its relatively high strength and hardness [1, 2, 3].

In addition to strength and hardness, quenching inevitably contributes to distortion and residual stresses in the components which the heat treatment industry is strongly interested in minimizing. Experimental investigation of the residual stresses has been studied since the early 1930s and it still receives much attention, especially with computer modeling and simulation of stress-producing processes such as heat treatment and quenching [4]. Unacceptable residual stresses and distortions may occur during manufacturing of heat treated components [5]. Therefore, it is desirable to develop quenching processes able to maximize the material strength while minimizing the residual stresses and distortions.

The function of a quenching medium is to enhance heat-transfer rate extraction from the hot-metal during the cooling process and to produce the desired metallurgical transformation [6]. Typical quenching media include: air, pressurized gases, fluidized beds, molten metals or salts, water, aqueous polymer solutions, and specially formulated petroleum oils [7, 8]. The quenching media selected depends on the quenching severity required to obtain the desired properties while at the same time providing optimal residual stresses and distortion control.

Petroleum based oils are the most commonly used quenchants for industrial heat treatment [9, 10]. Many components are oil quenched to achieve consistent and repeatable mechanical and metallurgical properties, and optimum distortion. Mineral oils exhibit good thermal-oxidative stability over a broad range of operating conditions. However, petroleum based products are non-renewable, toxic, and contribute to air and water pollution [11, 12]. With the increasing awareness and importance of environmental issues such as global warming, air and water pollution, more environmentally friendly quench media are of interest as alternatives to petroleum oil-based quenchants [6, 9].

Vegetable and animal oils have been used as quenchants for steel for hundreds of years, if not longer, until they were generally replaced by petroleum oil quenchants in the early 1900s [13]. However, in view of the current industrial demands for more environmentally friendly quenchants, there is a renewed interest in vegetable oils as quenchants. Currently, vegetable oils are one of the most commonly identified alternative basestocks to petroleum oils primarily because they are biodegradable and obtained from renewable sources such as soybean, sunflower, palm, canola, and hazelnut [14, 15, 16]. Vegetable oils are abundantly available, biodegradable, and eliminate or reduce harmful emissions.

Although soybean oil has shown excellent performance in various commercial applications, its oxidation instability and its narrow viscosity range potentially limit its use as a quenchant [14, 17, 18]. Adhvariu et al. [19] reported that soybean oil oxidizes at a rate that is at least an order of magnitude greater than petroleum oil.

An increasing number of studies have been performed on vegetable oils to study quenching characteristics such as: the cooling mechanism, heat transfer, wetting behavior, spread activation energy, quench severity, oxidation stability, antioxidants, and identification of potential corrosion inhibitor additives. Some computational studies have also been performed to simulate residual stresses that arise after quenching steels in petroleum oil and vegetable oil quenchants. Results for distortions and the geometrical variations in C-ring samples were published by the authors in [20] after quenching in several vegetable oils with special reference to the heat transfer coefficients. In this work, the authors carried out measurements of residual stresses on cylindrical samples by two different methods: the X-ray technique and the hole drilling method. The results obtained were compared after quenching in three different vegetable oils and one traditional mineral oil.

Materials and Methods

Cylindrical samples of 38.1mm (1.5") diameter and 25.4mm (1") height SAE/AISI 4140 were machined and quenched in four different quenchants. The samples were austenitized at 900°C and quenched at 60°C. Quenching was performed using the three renewable fluids and one fully-formulated and commercially available petroleum-based quenching oils shown in Table 1.

Table 1: Vegetable oils quenchants and a petroleum derived oil.

Quenchants
SO (Soybean Oil)
ESBO (Epoxidized soybean oil)
EF60 (40% ESBO + 60% FAME)
Houghto-Quench® KB

The renewable fluids considered all come from soybean oil: soybean oil (designated as SO in this work), epoxidized soybean oil (designated as ESBO in this work), and a blend of ESBO and FAME (Fatty Acid Methyl Ester: designated as FAME in this work) with 60% blend of FAME in ESBO (designated as EF60 in this work). The soybean oil (SO) was purchased in the supermarket of São Carlos (Brazil). The epoxidized soybean oil was donated by Inbra (Indústrias Químicas Ltda) and was of interest because a substantial portion of the oxidative unstable double bonds were converted to epoxide functionality. The FAME derivative of soybean soapstock, designated as FAME 3P, was obtained from Cognis do Brasil Ltda and used with ESBO to produce the blend EF60. (Soybean oil soapstock is a by-product of the caustic refining process of soybean oil). [20]

The fully formulated commercially available petroleum-based quench oil used was Houghto-Quench KB, an accelerated, cold quenching oil. The petroleum oil-based quenchant used is suitable for use up to 80°C.

The residual stress analysis was performed by two different methods: a non-destructive method and a semi-destructive method.

X-ray diffraction Rigaku machine (model MSF 3) has been used to non-destructively test residual stress in the specimens under consideration according to ASTM E2860-12: Standard Test Method for Residual Stress Measurement by X-Ray Diffraction for Bearing Steels [21] and ASTM E915-10: Standard Test Method for Verifying the Alignment of X-Ray Diffraction Instrumentation for Residual Stress Measurement [22]. The X-ray technique measures the residual stresses on the surface of the specimen according to the distortion of crystallographic parameters. The experimental value for Young modulus was chosen according to ASTM E111-04: Standard Test Method for Young's Modulus, Tangent Modulus, and Chord Modulus [23]. On the external circumference of the samples the residual stresses have been measured in the longitudinal (a) and circumferential (c) direction of the specimen in correspondence of the middle of the height of the probes (Fig. 1).

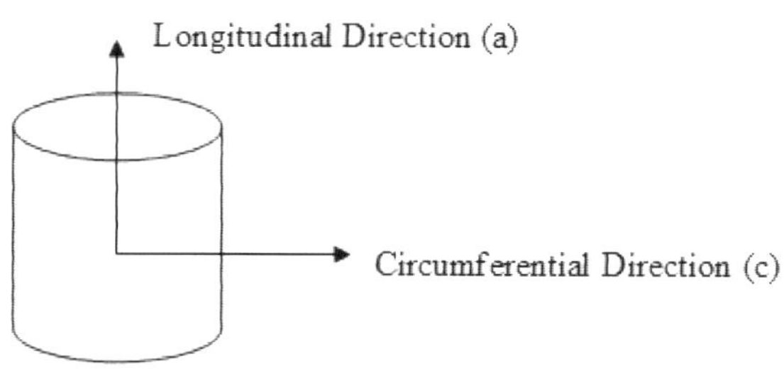

Figure 1: Sketch of the longitudinal and transversal direction on a general sample.

The measurement of the residual stresses in the longitudinal direction (a) of the circular surface of the specimen has been measured in the middle of the height of the probe taking the specimen centre and the two faces of the probe as a reference (Fig. 2 and Fig. 3). The arrow shows the direction of the goniometer rotation (Fig. 2).

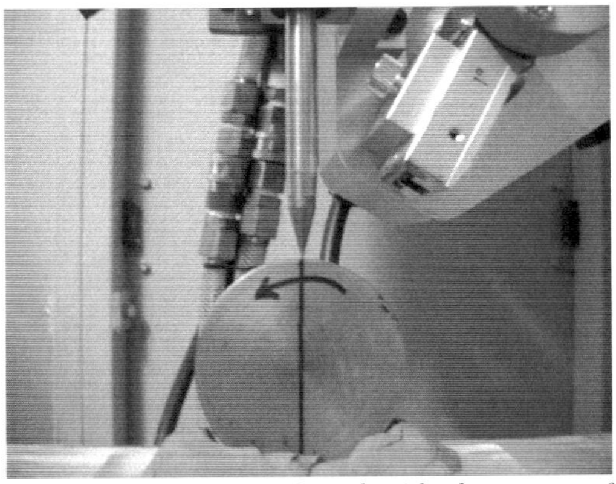

Figure 2: Instrument tip aligned with the centre of the specimen. The blue harrow shows the direction of the goniometry rotation.

Figure 3: Instrument tip positioned in the middle of the height of the specimen.

The hole-drilling technique was used as a semi-destructive technique according to ASTM E837-08 "Standard Test Method for Determining Residual Stresses by the Hole–Drilling Strain-Gage Method" [24]. The residual stress results were used in comparison with the values obtained using the X-ray non-destructive technique.

The hole-drilling strain gage method determines residual stresses near the surface of the component. In this work, a Sint Technology Restan MTS3000 was used to produce a hole of 1 mm or 2 mm depth on the surface of the specimen (Fig. 4). The hole drilling measurement procedure involves the positioning of a strain gage rosette with three radial grids and the drilling of a blind hole through the geometric center of the rosette.

The test parameters are the following:

- B-type extensometers with 5.13 mm diameter;
- Nominal diameter of the hole: 1.8 mm;
- Speed 0.2 mm/min;
- 40 drilling steps with 0.05 mm depth per each step (final depth of the hole: 2mm).

Readings of the strains produced are made by relieving the residual stresses. The residual stresses are calculated from the measured strains using special calculation algorithms. To drill the hole, a drill bit of TiAlN-coated tungsten carbide end mills (1-SINTCTT/1) inverted cone milling cutter has been used (Fig. 5). A strain gage rosette with three radial grids attached on the sample under analyses measures the deformations arising during the advancement of the hole. The deformations are acquired by a compatible computer. According to the ASTM reference calculations, special software calculates the principal stresses up to 1 mm depth.

Residual stress analyses on samples of dimensions 38.1mm x 25.4mm (1.5" x 1") was performed on the external cylindrical surface of the samples. The hole drilling measurements have been performed in the middle of the height on the external circumference of the cylinder like the X-ray measurements.

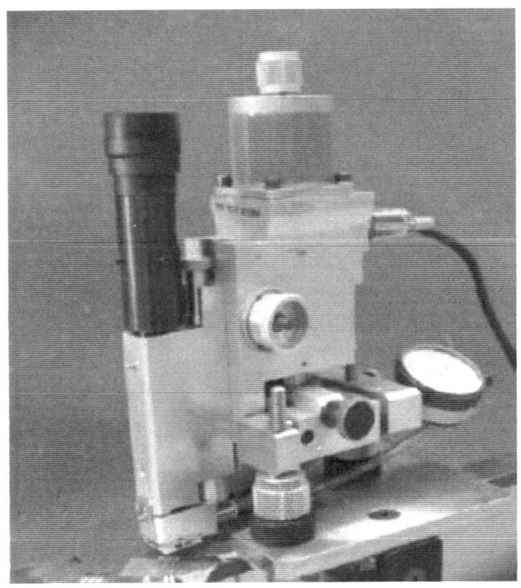

Figure 4: The MTS3000 or RESTAN (Residual Stress Analyzer) is an instrument designed for measuring residual stresses by the blind hole method developed and patented by SINT Technology.

Figure 5: Strain gage rosette with three radial grids drill bit of TiAlN-coated tungsten carbide end mills (1-SINTCTT/1) inverted cone milling cutter.

The hole-drilling strain-gauge method was performed on four samples with dimensions 38.1mm x 25.4mm (1.5" x 1"). The specimens analysed with the hole drilling method included the three samples quenched with vegetable oils and one sample quenched with the accelerated mineral oil. The samples selected for the hole drilling measurements were quenched in SO (soybean oil), ESBO (epoxydized soybean oil), EF60 (40% ESBO + 60% FAME) and the accelerated mineral oil HKB (Houghto-Quench® KB).

Discussion

The residual stresses measured with the hole-drilling strain gauge method were plotted as a function of hole depth. Figs. 6-9 show the results obtained from the hole-drilling after quenching the sample in the following quenchants: SO (soybean oil), ESBO (epoxidized soybean oil), EF60 (40% ESBO + 60% FAME) and HKB (accelerated mineral oil).The results are expressed in terms of stresses and shear stresses along the longitudinal (a) and circumferential directions (c).

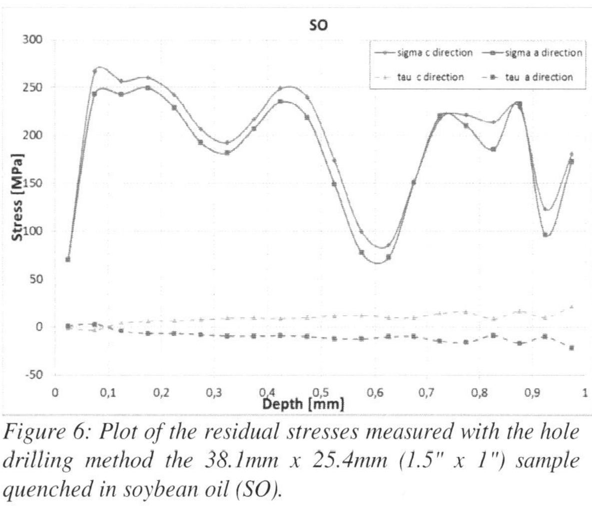

Figure 6: Plot of the residual stresses measured with the hole drilling method the 38.1mm x 25.4mm (1.5" x 1") sample quenched in soybean oil (SO).

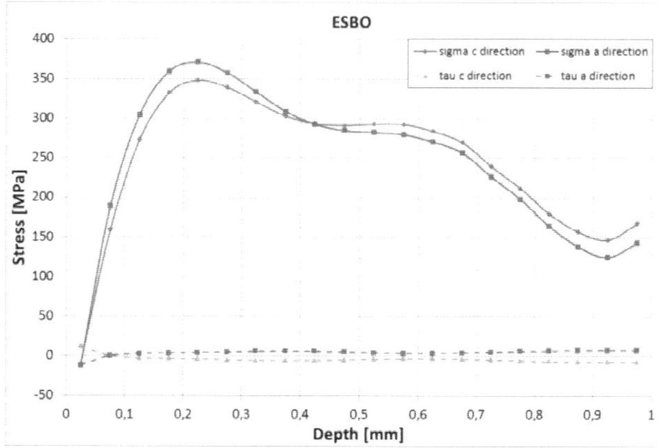

Figure 7: Plot of the residual stresses measured with the hole drilling method the 38.1mm x 25.4mm (1.5" x 1") sample quenched in epoxidized soybean oil (ESBO).

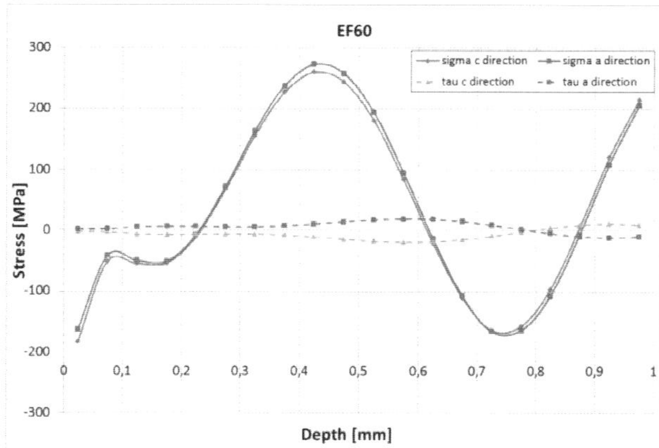

Figure 8: Plot of the residual stresses measured with the hole drilling method the 38.1mm x 25.4mm (1.5" x 1") sample quenched in EF60 (40% ESBO + 60% FAME).

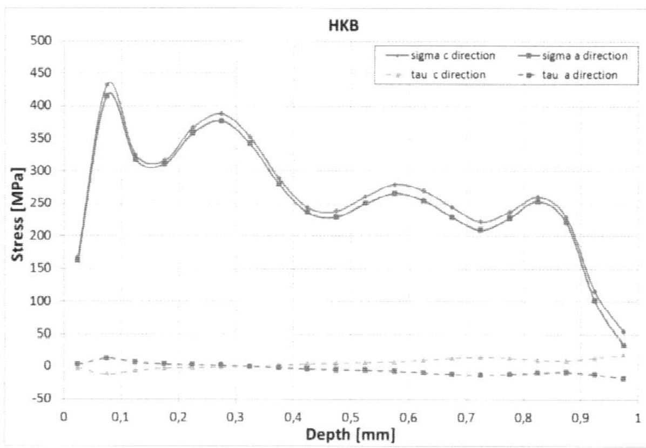

Figure 9: Plot of the residual stresses measured with the hole drilling method the 38.1mm x 25.4mm (1.5" x 1") sample quenched in Houghto-Quench® KB (HKB).

Figs. 10-13 show the values of the equivalent residual stresses (calculated with the longitudinal (a) and circumferential (c) residual stresses obtained from the x-ray analysis results and the hole drilling results on the AISI 4140 samples of dimensions 38.1mm x 25.4mm (1.5" x 1") according to:.

$$\sigma_{VonMises} = \sqrt{\sigma^2_{longitudinal} + \sigma^2_{circumferential} - \sigma_{longitudinal}\sigma_{circumferential}}$$

Apart from the cylinder quenched in EF60, for each of the hole drilling measurements, the values related to the equivalent residual stresses in the first 0.1-0.2mm below the surface increase up to the highest peak. The curve of the equivalent stresses after 0.1-0.2 mm up to 1 mm shows other peaks in decreasing order of magnitude as the hole gets deeper.

The highest peak of the equivalent residual stresses has been measured on the sample quenched in the accelerated mineral oil HKB (439 MPa at 0.075mm depth), followed by the two vegetable oils: ESBO (360 MPa at 0.225mm depth), and SO (256 MPa at 0.075mm depth).

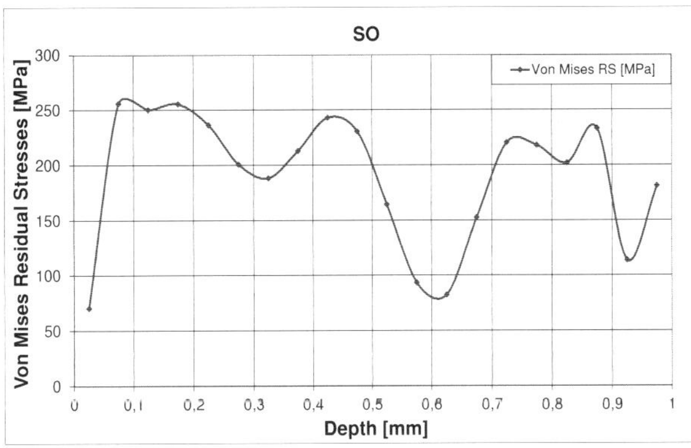

Figure 10: Plot of the equivalent residual stresses calculated with the results the hole drilling measurements of the 38.1mm x 25.4mm (1.5" x 1") sample quenched in soybean oil (SO).

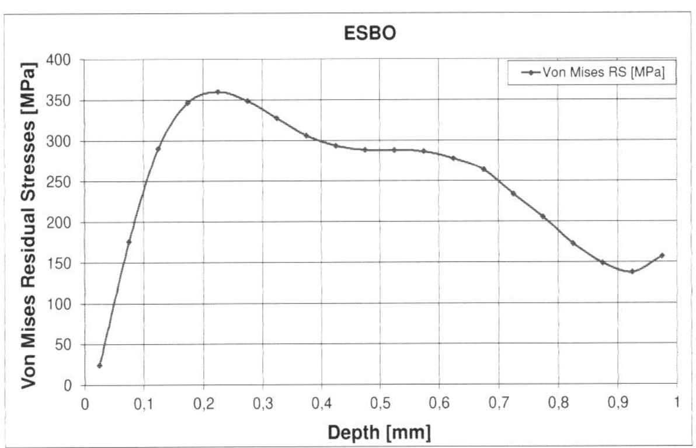

Figure 11: Plot of the equivalent residual stresses calculated with the results the hole drilling measurements of the 38.1mm x 25.4mm (1.5" x 1") sample quenched in epoxidized soybean oil (ESBO).

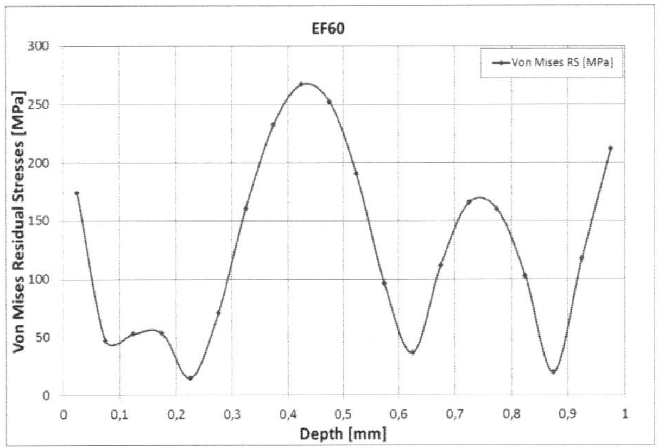

Figure 12: Plot of the equivalent residual stresses calculated with the results the hole drilling measurements of the 38.1mm x 25.4mm (1.5" x 1") sample quenched in EF60 (40% ESBO + 60% FAME).

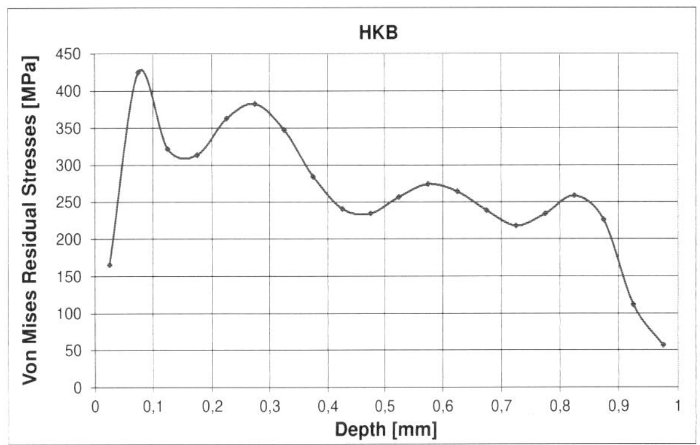

Figure 13: Plot of the equivalent residual stresses calculated with the results the hole drilling measurements of the 38.1mm x 25.4mm (1.5" x 1") sample quenched in Houghto-Quench® KB (HKB).

Fig. 14 shows a comparison between the results obtained with the X-ray technique and the equivalent stresses from the hole drilling. The values of the stresses measured using hole drilling were taken at the depth of 0.225 mm from the surface [20]. The measurements performed with x-rays and hole drilling on the cylinder quenched in the same coolant are comparable.

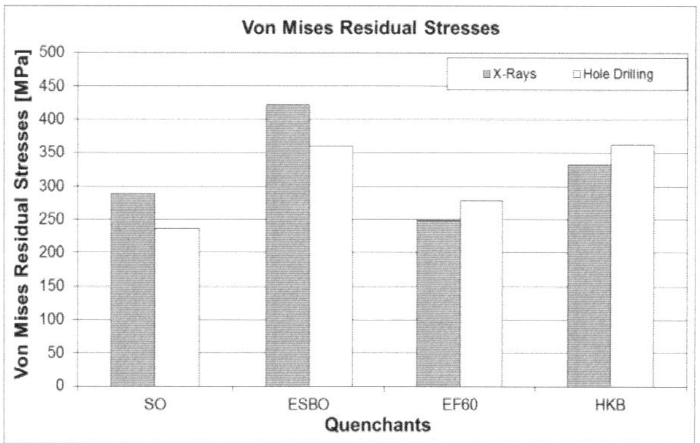

Figure 14: Plot of the equivalent residual stresses calculated with the results the hole drilling measurements of the 38.1mm x 25.4mm (1.5" x 1") sample quenched in Houghto-Quench® KB (HKB).

Conclusions

The experimental study was performed to investigate the effect of different oils on the residual stresses after quenching cylindrical specimens. The residual stresses measured on the AISI 4140 specimens of 38.1mm of diameter and 25.4mm of height are higher for samples quenched into ESBO, SO and the mineral oil. Lower values are measured on samples quenched in the blend. With probes of 38.1mm of diameter and 25.4mm of height, high heat transfer at high temperatures seems to have a strong effect on the residual stresses together with high heat transfer in the martensitic temperature interval. In fact, ESBO measures the higher values of residual stresses among the specimens of the same size. It has been also detected that the measurements performed with X-rays and the hole drilling method on the cylinder quenched in the same coolant can be considered comparable. Moreover, from the results obtained with vegetable oils and the petroleum derived oil do not lead to very different values of residual stresses.

References

[1] Totten, G. E., C. E. Bates, and N. Clinton, Handbook of quenchants and quenching technology, ASM international, 1993

[2] Canale l. C. F., Totten, G. E., "Chapter 1 – Hardening of Steel" in Liscic, B., H. M. Tensi, L. C. F. Canale, and G. E. Totten, Quenching theory and technology. CRC, (2nd Edition) , p. 1-42. 2009

[3] Liscic B. , H.M. Tensi, G.E. Totten, G.M. Webster, "Non-Lubricating Process Fluids: Steel Quenching Technology" in Totten, G. E., S. R. Westbrook, and R. J. Shah, Fuels and lubricants handbook: technology, properties, performance, and testing (Vol. 37). ASTM International, 2003

[4] Sedighi, M., C. McMahon, The influence of quenchant agitation on the heat transfer coefficient and residual stress development in the quenching of steels: Proceedings of the Institution of Mechanical Engineers, Part B: Journal of Engineering Manufacture, v. 214, p. 555-567, 2000

[5] Canale, L. C. F., Totten, G. E., Overview of distortion and residual stress due to quench processing. Part I: factors affecting quench distortion, International Journal of Materials and Product Technology, 24(1), 4-52, 2005

[6] Prabhu, K. N., P. Fernandes, Determination of wetting behavior, spread activation energy, and quench severity of bioquenchants: Metallurgical and Materials Transactions B, v. 38, p. 631-640, 2007

[7] Totten G.E., H.M. Tensi and L.C.F. Canale, "Chemistry of Quenching: Part I – Fundamental Interfacial Chemical Processes Involved in Quenching", In Proc. of 22nd Heat Treating Society Conference and the 2nd Surface Engineering Congress, Eds. N.B. Dahotre, R.J. Gaster, R.A. Hill, O.O. Popoola, 15-17 September 2003, Indianapolis, IN, ASM International, Materials Park, OH, p. 141-147.

[8] Totten G.E. , H.M. Tensi and L.C.F. Canale, "Chemistry of Quenching: Part 2 – Fundamental Thermophysical Processes Involved in Quenching", Proc. of 22nd Heat Treating Society Conference and the 2nd Surface Engineering Congress, 15-17 September 2003, Indianapolis, IN, ASM International, Materials Park, OH, p. 148-155.

[9] Fernandes, P., and K. N. Prabhu, Comparative study of heat transfer and wetting behaviour of conventional and bioquenchants for industrial heat treatment: International Journal of Heat and Mass Transfer, v. 51, p. 526-538, 2008

[10] Totten G.E. , J. L. Dossett, N.I. Kobasko, "Quenching of Steel", in ASM Handbook, Vol. 4A – Steel Heat Treating Fundamentals and Processes, Eds. J. Dossett and G. Totten, ASM International, Materials Park, OH, p. 89-157, 2013

[11] Denis J. , J. Briant, J-C. Hipeaux, "Chapter 1 – Analysis of Oil Constituents", in Lubricating Properties, Analysis and Testing, Editions Technip,Paris, France p. 1-106, 2000

[12] Simpson, A. T., & Ellwood, P. A., Polycyclic aromatic hydrocarbons in quench oils, Annals of Occupational Hygiene, 40(5), 531-537., 1996

[13] Otero, R. L. S., Canale, L. C., & Totten, G. E.,. Use of Vegetable Oils and Animal Oils as Steel Quenchants: A Historical Review-1850-2010, Journal of ASTM International, 9(1). , 2011

[14] Said, D., G. Belinato, G. S. Sarmiento, R. L. S. Otero, G. E. Totten, A. Gastón, and L. C. Canale, Comparison of Oxidation Stability and Quenchant Cooling Curve Performance of Soybean Oil and Palm Oil, Journal of Materials Engineering and Performance, p. 1-8., 2013

[15] de Souza, E. C., Fernandes, M. R., Augustinho, S. C. M., Canale, L., & Totten, G. E.. Comparison of structure and quenching performance of vegetable oils, Journal of ASTM International (JAI), 6(9), 2009

[16] Canale, L. C. F., & Narazaki, M., Quenching and cooling, residual stress and distortion control, International Heat Treatment and Surface Engineering, 4(3), 96-97, 2010
[17] Belinato, G, L. C. F. Canale, and G. E.Totten, Effect of Antioxidants on Oxidative Stability and Quenching Performance of Soybean Oil and Palm Oil Based Quenchants, Journal of ASTM International, Vol. 8, No. 9, Paper ID JAI103376, 2011
[18] de Souza, E. C., G. Belinato, R. L. S. Otero, É. C. A. Simêncio, S. C.M. Augustinho, W. Capelupi, C. Conconi, L. C. F. Canale, G. E. Totten, Thermal Oxidative Stability of Vegetable Oils as Metal Quenchants", Journal of ASTM International, 9(1), Paper ID JAI 103817, 2011
[19] Adhvaryu,A., S. Z. Erhan, Z. S. Liu, J. M. Perez, Thermochim Acta, 364, 87-97, 2000
[20] Civera, C., Rivolta, B., Simencio-Otero, R. L., Lu´cio, J. G., Totten, G. E., and Canale, L. C. F., "Vegetable Oils as Quenchants for Steels: Residual Stresses and Dimensional Changes," Materials Performance and Characterization, Vol. 3, No. 4, 2014, pp. 1–20, doi:10.1520/MPC20140039. ISSN 2165-3992
[21] ASTM E2860-12:Standard Test Method for Residual Stress Measurement by X-Ray Diffraction for Bearing Steels, Annual Book of ASTM Standards, ASTM International, West Conshohocken, PA, 2012.
[22] ASTM E915-10: Standard Test Method for Verifying the Alignment of X-Ray Diffraction Instrumentation for Residual Stress Measurement, Annual Book of ASTM Standards, ASTM International, West Conshohocken, PA, 2010.
[23] ASTM E111-04: Standard Test Method for Young's Modulus, Tangent Modulus, and Chord Modulus, Annual Book of ASTM Standards, ASTM International, West Conshohocken, PA, 2010.
[24] ASTM E837-08: Standard Test Method for Determining Residual Stresses by the Hole-Drilling Strain-Gage method, Annual Book of ASTM Standards, ASTM International, West Conshohocken, PA, 2008.

International Federation of Heat Treatment and Surface Engineering 2016
Proceedings of the 23rd IFHTSE Congress
April 18–21, 2016, Savannah, Georgia, USA

Parallel PSO method for estimation Heat Transfer Coefficients

Imre Felde, Sándor Szénási, Gergő Pintér
Óbuda University, Budapest, Hungary
felde.imre@nik.uni-obuda.hu

Wei Shi
Department of Mechanical Engineering, Tsinghua University, Beijing, P. R. China

Rafael Colas, Oscar Zapata-Hernández
Facultad de Ingeniería Mecánica y Eléctrica, Universidad Autónoma de Nuevo León, México

Abstract

The methodology based on the Particle Swarm Optimization (PSO) method, as a recent stochastic optimization technique to solve complex inverse heat transfer problems is outlined. The temporospatial Heat Transfer coefficient obtained on the surfaces of a cylindrical work piece is reconstructed by solving the inverse heat conduction problem. The fitness function to be minimized by the PSO approach is defined by the deviation of the measurements and the calculated temperatures is minimized. The PSO algorithm has been parallelized and implemented on a GPGPU architecture. Numerical results are demonstrated that the determination of Heat Transfer Coefficient functions can be performed by using the PSO method, as well as, the GPU implementation; provide a less time consuming and accurate estimation.

Introduction

Inverse Heat Conduction Problems (IHCP) are known as "reverse engineering" problems, due to the reversal of a cause-effect sequence, in the field of heat transfer analysis. An inverse problem means that some of the initial, boundary conditions or material properties are not fully specified as determined from the measured temperature profiles at some specific locations. The inverse problems in most situations are likely to be ill-posed [1]. Solutions of the inverse problem are very sensitive to measurement errors, i.e. small errors in the measured data values can produce very large errors in solutions. In general, the exclusivity and stability of an inverse problem solution is not guaranteed. In recent years, the inverse problems have been studied extensively due to their applications in various engineering disciplines.

The most of the methods approach the inverse heat conduction problem, as an optimization problem, i.e. the problem is defined as the minimization of a cost function or a fitness function measuring the distance between measurements and predictions [2,3]. With the improvement of computer capability, a variety of numerical techniques and computational methods have been developed to provide accurate solutions for IHCP in the last decade. Among these methods, stochastic optimization methods have become a popular means of solving inverse problems, due to their capability of finding the global optimal result without computing the complicated gradient of the objective function

The introduction contains the opening text for the main body of the manuscript. It should include a statement of the problem or situation and the approach taken to resolve it. This first section may also contain a summary of the past developments and background of what is already known and/or published elsewhere. This is best summarized in your own paper, with references to other publications containing more-extensive discussions of this background information. All references are placed at the end of the paper.

Genetic algorithms [4,5] are applied successfully for solving many types inverse heat transfer problem. The quantitative evaluation of different numerical optimization techniques showed [6] that stochastic methods could serve more accurate results for IHCP than gradient approaches in recovering complex thermal boundary conditions. The Particle Swarm Optimization (PSO) algorithm became popular in the recent years due to its ability of maintaining a good balance between the convergence and diversity. Applications of PSO algorithms in the field of heat transfer are still limited. An inverse application of boundary elements method to estimate the thermal conductivity and the shape of an inclusion was implemented [7]. Qi et al. [8] applied the multi-phase PSO method to solve the inverse radiation problem. The proposed method applied the benefits of both two-group PSO and multi-start PSO algorithms. The effectiveness and efficiency of Particle Swarm Optimization technique in inverse heat conduction analysis were analysed by Vakili and Gadala [9]. Three variations of the PSO method, i.e. basic PSO, repulsive PSO, and complete repulsive PSO, were performed to solve the boundary inverse heat conduction problem, in one, two and three dimensions. The results showed that PSO can reduce the stability problems of the classical methods, for solving the inverse heat conduction problems.

In this work, an inverse analysis for the reconstruction of local coordinate and a time-varying Heat Transfer Coefficient, in two-dimensional cylindrical coordinates is investigated. The inverse heat conduction analysis is based on the application of a PSO technique. Transient temperature measurements at multi-locations in the body of the work piece, obtained by the solution of the direct heat transfer problem, served as the virtual experimental data required to solve the inverse analysis. The fitness function which is defined by the quadratic residual between the measurements and the calculated temperatures is minimized. The PSO algorithm has been parallelized and implemented on a GPU architecture. Numerical results are demonstrated that the determination of Heat Transfer Coefficient functions can be performed by using the PSO method, as well as, the GPU implementation; provide a less time consuming and accurate estimation.

The Heat Transfer Model

The determination of the Heat Transfer Coefficient is an important issue of the IHCP and has been extensively studied. Improvements and adaptations of the numerical algorithms on the applications are still an active area of research for obtaining stable and reliable results. A two-dimensional axis-symmetrical heat conduction model is considered to estimate the temperature distribution in a cylindrical work piece (the radius and length of the cylinder is noted by R and Z). The cylinder is subjected to a longitudinal local coordinate and time varying Heat Transfer Coefficient HTC(z,t) on all its surfaces. Both the thermal conductivity and the heat capacity are varying with the temperature, k(T) and $C_p(T)$. The dimensional mathematical formulation of this nonlinear transient heat conduction problem can be described as follows:

$$\frac{\partial}{\partial r}\left(k\frac{\partial T}{\partial r}\right)+\frac{k}{r}\frac{\partial T}{\partial r}+\frac{\partial}{\partial z}\left(k\frac{\partial T}{\partial z}\right)+q_v=\rho C_p\frac{\partial T}{\partial t} \tag{1}$$

with the initial and the boundary conditions

$$T(r,z,0)=T_0 \tag{2}$$

$$k\frac{\partial T}{\partial z}\bigg|_{\substack{0\le z\le Z\\ r=R}}=HTC(z,t)\left[T_q-T(r,z,t)\right] \tag{3}$$

where r and z is the local coordinate, t is the time, ρ is the density, T_0 is the initial temperature and Tq is the temperature of the cooling medium. In this work, the weighted Schmidt explicit finite difference method is used to discretize the Eqs. (1-3) and solve the direct problem.

The Inverse Heat Transfer Model

Assuming that the temperature inside the work piece and/or on its surface is measured during the heat transfer process, it is possible to solve the inverse heat conduction problem by determining the time/or temperature variations of the thermal boundary conditions [1,2,3]. The temperature at different times is given by measurements at n points in the solid region, located at r_k, (k=1…n). On calling T_k^m, the measured temperatures, and T_k^c, the calculated temperature at those points, the solution of the present inverse problem can be obtained by minimizing the following fitness function

$$S=\sum_{k=1}^{n}\left(T_k^m-T_k^c\right)^2=\min \tag{4}$$

where n is the total number of measured temperatures, i.e., the number of points multiplied the number of measurements at each point. The inverse problem is recast as an optimization problem. A variety of numerical and analytical techniques have been developed to solve the optimization problems.

The particle swarm optimization algorithm

The Particle Swarm Optimization (PSO) algorithm introduced by Kennedy and Eberhart [10] in 1995 is a stochastic optimization technique which draws inspiration from the social behavior of a flock of birds or the collective intelligence of a group of social insects with limited individual capabilities. The basic PSO model consists of a swarm of M particles moving in a problem search space. Each particle is a potential solution of the global optimum over a given domain D. For a N-dimensional search space, the position of the ith particle is represented as $X_i = (x_{i1},x_{i2}, . . . ,x_{iN})$. At each generation, the new particle position is found by adding a displacement to the current position where the displacement is the particle velocity multiplied by a time step of one as shown in Eq. (5)

$$X_i^{n+1} = X_i^n + V_i^{n+1} \tag{5}$$

In Eq. (5), X_i^n and X_i^{n+1} represent the current and previous positions of particle i, V_i^{n+1} i is the current velocity of particle i and is represented as $V_i^{n+1}= (v_{i1},v_{i2}, . . . ,v_{iN})$. The velocity of each particle is also updated at each generation and is given by:

$$V_i^{n+1} = V_i^n + c_1r_1(P_{best,i} - X_i^n) + c_2r_2(G_{best} - X_i^n) \tag{6}$$

where V_i^n and V_i^{n+1} are the current and previous velocities of the particle i, respectively. Each particle maintains a memory of its previous best position, say $P_{best,i} = (p_{i1}, p_{i2}, . . . ,p_{iN})$, where the position giving the best fitness function value. The best one among all the particles in the swarm is represented as the global best position, say $G_{best} = (p_{g1},p_{g2}, . . . ,p_{gN})$. The new velocity in Eq. (6) can be seen as the sum of three parts. The first part of Eq. (6) represents the previous velocity and is called the momentum part. The second part of the Eq. (6) represents the tracking of best position for individual particle and is called the cognition part. The third part of the Eq. (6) represents the cooperation among particles in the swarm and is called the social component. The cognitive learning coefficient c_1 and social learning coefficient c_2 are known as accelerating factors, and r_1 and r_2 are two random numbers generated by the uniform distribution within 0 and 1. The relative sizes of these components determine their contribution to the new particle velocity. The most common setting for c_1 and c_2 are 2.0 for the standard PSO algorithm.

One difficulty encountered in the standard PSO algorithm is that it could easily fall into local optimum in many optimization problems. Better selection of the inertial weight provides a balance between global and local exploration and exploitation. Therefore, less iteration is needed to find the optimal solution and improve the performance of the algorithm. A constriction factor was incorporated into the PSO algorithm by Clerc's suggestion [11] to insure the convergence of the algorithm (PSOC). The velocity term was update as follows:

$$V_i^{n+1} = C_3(V_i^n + c_1 r_1(P_{best,i} - X_i^n) + c_2 r_2(G_{best} - X_i^n)) \quad (7)$$

When Clerc's constriction method is used, the constant multiplier C_3 is approximately equal to 0.7298 and the two coefficients c_1 and c_2 are 2.05. The Clerc's approach has been used for the determination of Heat Transfer Coefficients.

Computation procedure for the PSO algorithm

The aim of the inverse analysis is to iteratively estimate the unknown HTCs using the PSO procedure which results a negligible difference between measurements taken at the given locations of the work piece and temperatures computed from the numerical model. The fitness functions value of each particle at the nth iteration is given by the difference between the measured and calculated temperature curves, Eq (4) at the position X_i^n. The computational steps of the PSO algorithm described above are given as follows:

- Step 1: Generate the initial particles in a swarm by randomly generating the position and velocity for each particle.
- Step 2:Evaluate the fitness function of each particle.
- Step 3: Update the $P_{best,i}$ for each particle, if its fitness is smaller than the fitness of its previous best position ($P_{best,i}$).
- Step 4: Update the G_{best}, if the fitness function of a particle is smaller than the fitness of the best position of all particles (G_{best}).
- Step 5: Update each particle according to Eqs. (5) and (7).
- Step 6: Repeat the loop until the stopping criteria or a predefined number of generations is reached.

It is strongly advised to parallelize the computational jobs in Step 2 due to the fact that there are no interferences between the iterations as well as there is no communication between the particles in a given iteration. Therefore, these parts are executable in a data parallel fashion, which is ideal for GPU implementation [12,13]. We used the following configurations for the tests:

- Graphics accelerator: NVIDIA Tesla K40c
- Architecture: Kepler (GK110B)
- Number of shades: 2880
- SMX Count: 15
- Memory: 12GB GDDR5

Verification of the proposed method

The Heat Transfer Coefficients obtained during immersion quenching of a cylindrical bar have been estimated by using the proposed approach.

A cylindrical rod was produced from X8NiCrS18-9 austenitic steel 1.4305 (chemical composition is in Table 1, while physical properties is summarized in Table 2) in the IWT - Stiftung Institut für Werkstofftechnik (Bremen, Germany). The work piece have been equipped with 10 TCs which were located various distances from the top surface and 1 mm under the vertical surface of the cylinder (Fig 1.). The diameter of the cylinder was 50 mm while the length was 100 mm.

C%	Si%	Mn%	P%	S%	Cr%	N%	Ni%	Cu%
<0.1	<1.0	2.0	0.045	0.2	0.18	0.11	8.0	1.0

Table 1. Chemical composition of X8NiCrS18-9 steel grade

Temperature, °C	Specific heat, J/kgK	Conductivity, W/mK	Density, Kg/m³
0	443.3	14.6	7920
200	530.9	17.14	7832
400	568.5	19.68	7865
600	591.5	22.22	7660
800	627.1	24.76	7570
1000	985.3	27.30	7480

Table 2. Physical properties of X8NiCrS18-9 steel grade

The rod has been heated up to 860°C and immersed in oil (Isorapid 277) without agitation at 80°C temperature. The signals of the TCs have been recorded during the cooling process. Fig 2. represents the cooling curves acquisited during the quenching process.

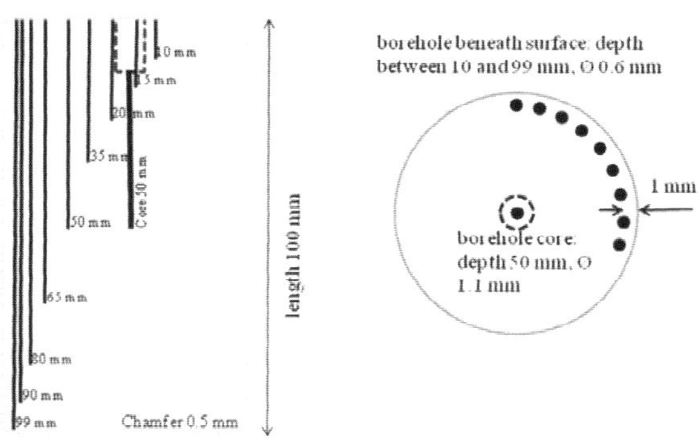

Figure 1. Locations of TCs mounted in the cylindrical work piece (produced in the IWT - Stiftung Institut für Werkstofftechnik)

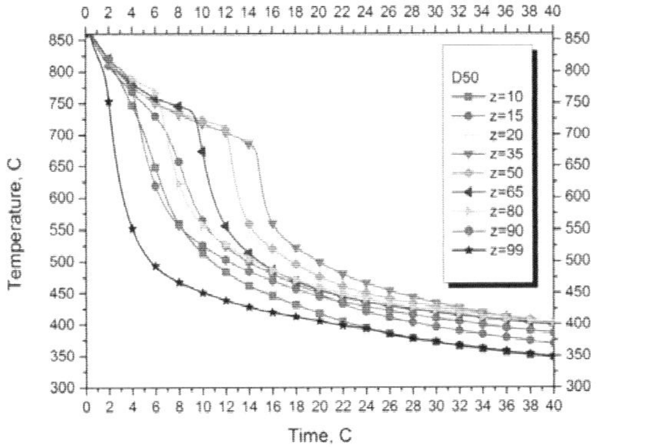

Figure 2. The cooling curves recorded during quenching of the cylindrical probe

A 2D axis-symmetric heat transfer model was applied to calculate the temperature distribution during the cooling process. The physical properties of stainless steel (Table 2) were was assigned to the virtual work piece.

Inverse computations have been carried out by including PSO algorithm, in order to predict the HTC(z,t) functions. The Heat Transfer Coefficient function was defined by a matrix where the rows stood for the local coordinates "z" given by the locations of the thermos couples. The columns of the matrix included the time instances. In the recent investigation 10 vectors (HTC(t)) consisted the matrix. The exact values of the HTC(z,t) function was calculated by the bilinear interpolation procedure.

A hybrid approach is applied to estimate the Heat Transfer Coefficient function: four calculation cycles of PSO and one calculation cycle of Levenberg-Marquardt Method [14] (LMM, a gradient optimization technique) has been used. A swarm size was set to 3000 particles for each computational steps. The concept of the sequential hybrid estimation is based on the following assumptions:

1. In PSO Step 1, the HTC(t) vectors for each TC locations have been determined by 5 time instances (Fig 3, "PSO Step 1" chart).
2. Extra 2 time instances are given to the HTC(t) vectors in PSO Step 2 ((Fig 3, "PSO Step 2" chart). The HTC(z,t) matrix included 10x7=70 elements.
3. In PSO Step 3, 3 new time instances added again to the HTC(t) vectors. The total number of elements in the HTC(z,t) matrix was 100.
4. In PSO Step 3, the number of time instances was increased to 14. The HTC(z,t) matrix included 10x14=140 elements.
5. In the last calculation step, the LMM was applied.

The calculation process in each PSO and LMM Steps stopped when the relative deviation of the fitness function between the former and the recent iteration step was less than 10%.

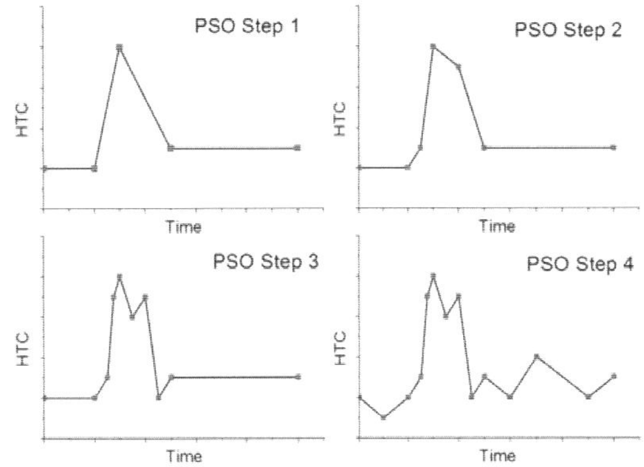

Figure 3. The concept of the sequential hybrid estimation process

The measured T_k^m, and the reconstructed T_k^c cooling curves at the locations of the TCs (k=1,2...10) are shown in Fig 4.

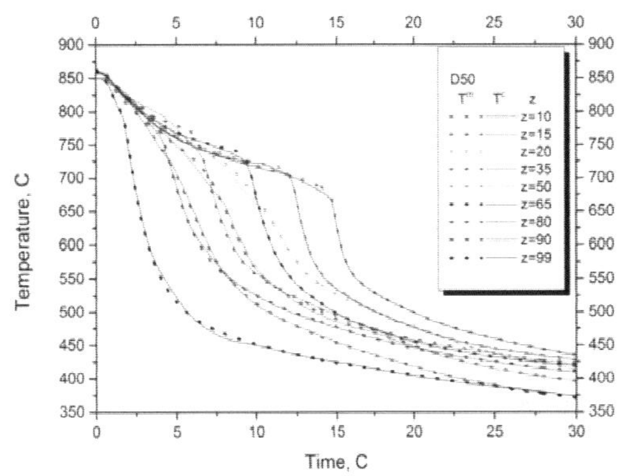

Figure 4. The measured T_k^m, and the reconstructed T_k^c cooling curves at the locations of the TCs (k=1,2,..10)

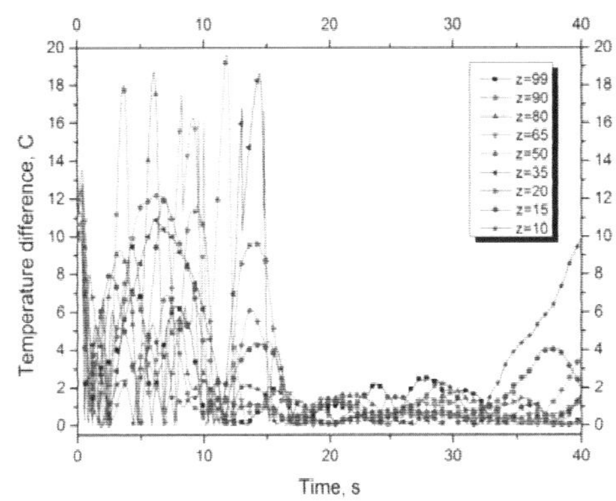

Figure 5. Difference between measured and calculated cooling curves at the locations of the TCs (k=1,2,..10) as a function of time

Satisfactory agreement of original and predicted cooling curves can be observed. The difference between the measured and estimated samples as a function of time for each TC positions is shown in Fig. 5.

In order to quantify the magnitude of deviation between the measured and the recovered temperature samples the mean, standard deviation and maximum value of the difference of cooling curves in each positions were calculated (Table 3). The highest value of temperature difference (19.60°C) was given at the middle of cylinder's length (z = 50 mm). The standard deviation were given in the range of 1.88 to 3.39 while the mean of value the differences was between 2.15 and 4.39. The reason of relatively high maximum deviations could be originated to the fact that the proper time instances needed to reconstruct the Heat Transfer Coefficient have not been found exactly by PSO algorithm. Due to the low value of differences the hybrid PSO approach applied to estimate the complex Heat Transfer Coefficient in a two dimensional axis-symmetrical model seems to be a feasible approach providing an acceptable accuracy.

TC location, mm	Mean, °C	Standard deviation, °C	Maximum deviation, °C
10	2.22	2.29	10.60
15	2.65	2.60	11.96
20	2.74	3.32	11.65
35	3.34	4.39	18.58
50	3.39	4.31	19.60
65	2.06	3.55	17.68
80	2.22	3.30	18.71
90	1.88	2.88	18.09
99	1.91	2.15	10.18

Table 3. The statistical information of deviations between measured T_k^m, and the reconstructed T_k^c cooling curves

The predicted Heat Transfer Coefficient function obtained by the PSO technique is shown in Fig. 6.

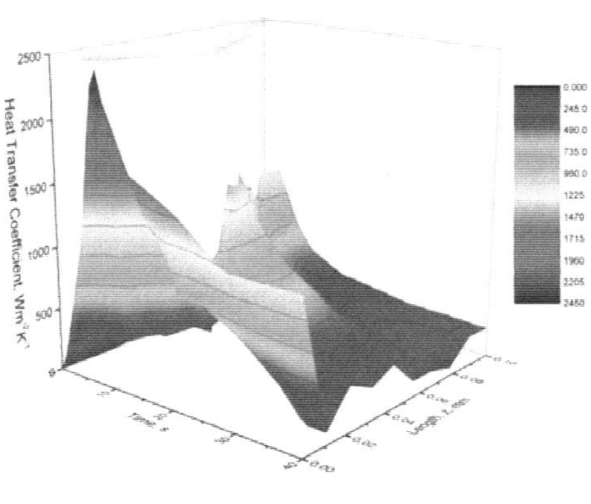

Figure 6. The estimated HTC(z,t) function

It is important to report the results of the computational time required for PSO based estimation. The calculations have been made in:

1. Sequential processing (the fitness value of each particle has been estimated one after another in the swarm in one generation)
2. Multi-tier processing (8 parallel computation of particles at the same time)
3. Full parallelization on GPU architecture

The implementation of version 1 and 2 a PC hardware setup (Processor: Intelr CoreTM i7-2600, Architecture: Sandy Bridge, Number of cores: 4, Memory: 16GB DDR2) has been applied while for version 3 the GPU setup is used mentioned earlier. The time measured during performing the inverse estimations are summarized in Table 4. The parameters show the solution performed on the GPU architecture provides a significant acceleration of calculation time.

PSO Implementation	Calculation time, h
Sequential processing	27.01
Multi-tier processing	11.73
GPU processing	0.81

Table 4. The calculation times (in hours) required for HTC estimations

Conclusions

An inverse analysis using a sequential approach of Particle Swarm Optimization algorithm and of Levenberg-Marquardt Method has been presented to reconstruct the Heat Transfer Coefficient in a two dimensional heat conduction problem. The Heat Transfer Coefficient obtained on the surfaces of a cylindrical probe was considered as functions of local coordinates and time. The temperature samples applied as measured cooling curves were acquisited during immersion quenching experiment. The obtained results underline the feasibility of the procedure and the capabilities for the PSO technique to reconstruct a complex surface Heat Transfer Coefficient without using any prior information of the unknown transient functions. The PSO algorithm has been carried out in high performance GPU configuration using the CUDA environment. The GPU implementation of the inverse heat conduction problem provides significant acceleration of the prediction compared to sequential or multi-tier computation used on a personal computer.

Acknowledgments

We acknowledge the financial support of this work by the Mexican-Hungarian S&T bilateral project TÉT MX -1-2013-0001 and „Cooperative research on control of heat treatment stress and distortion for high value added machinery products" (2014DFG72020) projects. The authors would like to express the acknowledgement for Dr. Thomas Lübben (IWT - Stiftung Institut für Werkstofftechnik) for supporting the experimental part of this work.

References

[1] Beck J. V; Blackwell B; St Clair Jr. C.R: Inverse Heat Conduction, Wiley, New York, 1985.

[2] Alifanov O.M.: Inverse Heat Transfer Problems, Springer, Berlin/Heidelberg, 1994.

[3] Özisik M.N.; Orlande H.R.B.: Inverse Heat Transfer: Fundamentals and Applications, Taylor & Francis, New York, 2000

[4] Verma S.; Balaji C: Multi-parameter estimation in combined conductione radiation from a plane parallel participating medium using genetic algorithms, International Journal of Heat and Mass Transfer 50, 1706-1714., 2007

[5] Kim K.W.; S.W. Baek: Inverse surface radiation analysis in an axisymmetric cylindrical enclosure using a hybrid genetic algorithm, Numerical Heat Transfer Part A e Applications 46 (4) 367-381., 2004

[6] Felde I.: Estimation of Thermal Boundary Conditions by Gradient Based and Genetic Algorythms, MATERIALS SCIENCE FORUM 729: pp. 144-149. (2012)

[7] Ardakani M.D.; Khodadad M.: Identification of thermal conductivity and the shape of an inclusion using the boundary elements method and the particle swarm optimization algorithm, Inverse Problems in Science and Engineering 17 (7), 855-870., 2009

[8] Qi H.; Ruan L.M; Shi ; An W.; Tan H.P.: Application of multi-phase particle Swarm Optimization Technique to inverse radiation problem, Journal of Quantitative Spectroscopy & Radiative Heat Transfer 109, 476–493. 2008

[9] Vakili S.; Gadala M.S.: Effectiveness and efficiency of Particle Swarm Optimization technique in inverse heat conduction analysis, Numerical Heat Transfer Part B: Fundamentals 56 (2), 119–141., 2009

[10] Kennedy J.; Eberhart R.C.: Particle Swarm Optimization, in: Proceedings of the IEEE International Conference on Neural Networks, 1942–1948. , 1995

[11] Clerc M.: The swarm and the queen: Towards a deterministic and adaptive Particle Swarm Optimization, in: Proceedings of the Congress on Evolutionary Computation, 1951–1957., 1999

[12] Szénási S.; Felde I.; Kovács I.: Solving One-dimensional IHCP with Particle Swarm Optimization using Graphics Accelerators, Proceedings of the 10th IEEE International Symposium on Applied Computational Intelligence and Informatics, Timisoara, Románia, 365-369., 2015

[13] Kirk D. B.; Hwu W. W.: Programming Massively Parallel Processors: A Hands-on Approach, 1st ed. San Francisco, CA, USA: Morgan Kaufmann Publishers Inc., 2010.

[14] M.N. Özisik, H.R.B. Orlande, Inverse Heat Transfer: Fundamentals and Applications, Taylor & Francis, New York, 2000Downey, D. F. *et al.*, Ion Implantation Technology, Prentice-Hall (New York, 1993), pp. 65-67. [A book reference …]

International Federation of Heat Treatment and Surface Engineering 20
Proceedings of the 23rd IFHTSE Congress
April 18–21, 2016, Savannah, Georgia, USA

Residual Stress and Bending Fatigue Strength in Carburized and Quench Hardened Pyrowear 53 Steel Gears

B. Lynn Ferguson, Andrew M. Freborg and Zhichao Li
DANTE Solutions, Cleveland, OH, USA
lynn.ferguson@dante-solutions.com (440)234-8477

Abstract

Carburization of alloy steels promotes the formation of compressive residual surface stress upon quenching, and that compressive surface stress enhances fatigue life. To further investigate the role of residual stress on fatigue strength, a project was undertaken to assess the role of residual stress magnitude on bending fatigue life of a spur gear through innovative quenching and the achievement of deeper compressive surface stress. DANTE Solutions demonstrated the feasibility of improving the bending fatigue life of Pyrowear 53 steel gears by achieving deeper compressive residual stress in carburized and quench hardened parts. At the same time, concepts of integrated computational engineering (ICME) were employed for simulation of the steel heat treatment and then of the gear service stresses.

Computer simulations of two different quenching processes, conventional oil quenching and intensive quenching, were conducted using the DANTE heat treatment simulation software to predict differences in final residual stress state. Although similar hardness profiles were predicted for both processes, the predicted surface stresses at the center of the gear root were -600 MPa and -300 MPa, with the intensive quenching producing higher compression. These predictions agreed with XRD measurements. The stresses predicted at the tooth fillet were even more compressive and maintained similar separation between the two quenching methods. The simulations showed that timing and sequence of martensite formation that occurred during the quenching process was related directly to the magnitude of compressive residual surface stress.

Tooth bending fatigue tests, conducted by Gear Research Institute, showed an endurance limit difference of 15% between the two quenching methods, with higher surface compression yielding higher fatigue life. Scatter in the data was significant, even with surface conditions within product specification. Isotropic surface finishing increased the endurance limit, and the difference between quenching methods was maintained.

Introduction

Material and Process

Both the automotive and aerospace industries have a need to reduce weight and improve vehicle performance. For powertrain gearing this means increasing the power density of the transmission as weight is removed and performance is increased. Importantly, this must be done without degrading the service life of the components, and for gears that means that bending fatigue life and contact fatigue life must be improved. Several "new" gear steels have been developed toward this goal, each with improved strength and resistance to softening as operating temperatures have increased.[1,2] In helicopter transmissions, Pyrowear 53 steel, a patented alloy by Carpenter Technology Corp., has replaced AISI 9310 steel in many applications.

Pyrowear 53 is a carburizing grade of steel that was developed in the 1970's.[2] Its chemistry is given in Table I. Major characteristics of this steel are its high hardenability, its combination of high strength and toughness, and its ability to retain strength under loss of lubricant conditions. Pyrowear 53 provides an enhanced safety margin in comparison to carburized 9310 steel. As a result, it is used in many military applications where performance and safety outweigh cost considerations.

Table I. Typical Pyrowear 53 Steel Chemistry

Fe	C	Mn	Si	Cr	Ni	Mo	Cu	V
bal.	0.1	0.35	1.0	1.0	2.0	3.25	2.0	0.1

One issue with powertrain improvement is the time and testing required to substitute a new material and/or geometry into the drivetrain. The testing of a new component is costly and the time period can be excessively long. For this reason, many potential enhancements are shelved until entire new systems are being designed. Existing systems tend to remain as-is, even though potential improvements may be available. However, process modifications that use existing materials and geometries offer a path for improvement of existing systems at minimal time and cost. Heat treatment is one such process, since all gears rely on thermal processing to meet performance requirements. If a change in the heat treatment process can result in performance enhancement of the transmission components, then improved transmission performance can be realized without an expensive redesign and testing program. Since gear life is dependent on the nature and magnitude of surface residual stress, a specific goal of this study was improvement in the level of surface compressive residual stress. The technology applied was the use of intensive quenching in place of conventional immersion quenching in agitated oil to produce enhanced residual compressive surface stresses.

Prior research showed that a change in quench hardening practice from immersion quenching to intensive quenching could improve bending fatigue life.[3] Notched bars of Pyrowear 53 steel were carburized on one face and hardened conventionally or by intensive quenching, and then subjected to three point bending fatigue testing. The result was an improvement in cyclic bending life for the intensively quenched samples, and this was related to greater surface compressive stress as measured by X-RAY diffraction and also predicted by computer simulation.

Process Simulation

Before embarking on an experimental program of manufacturing and testing gears, computer simulation was used to investigate gas carburization and quench hardening processes in terms of residual stress. The DANTE software package was used for this task. Parameters specific for Pyrowear 53 steel were developed previously, so both carbon diffusion and phase transformations were modeled. The DANTE software is a set of user subroutines that link with commercial finite element solvers ABAQUS/STANDARD[1] and ANSYS Mechanical[2]. A database of mechanical, thermal and phase transformation properties for many carburizing and hardenable steels is part of DANTE, and this includes the Pyrowear 53 steel addressed in this study.

The structure of the DANTE software is shown in Figure 1. Inputs consist of component geometry, process schedule, and material data. The simulation calculations include carbon diffusion to get the carbon distribution in the part, and thermal, mechanical and metallurgical calculations to determine the phase fractions, temperatures, stress state, and dimensional changes in the part as heating and quenching takes place. The results include the local histories for stress, displacement, phase fractions and temperature, and the final hardness profile, residual stress state, and part dimensions.

Material data is critical to the accuracy of heat treatment simulation of steels, especially since the crystal structure and properties change significantly during the heating and cooling steps. The carbon gradient from carburization also results in significant property changes. For Pyrowear 53, the high hardenability results in austenite transforming to martensite during quenching, even at slow cooling rates. Figure 2 reports the martensite start temperature for different carbon levels as determined from dilatometric tests. Knowledge of when martensite forms is critical since the plastic properties for

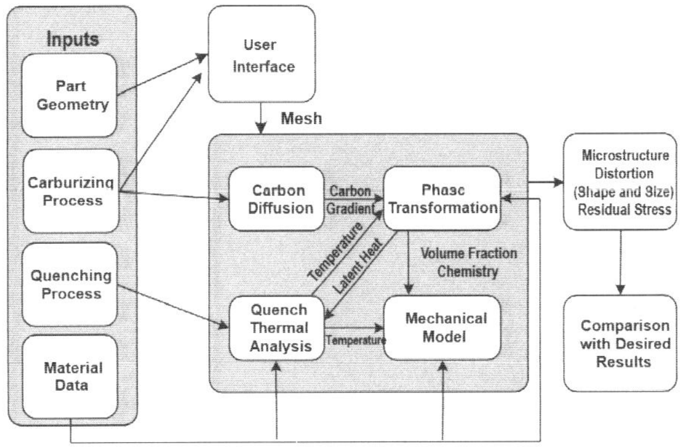

Figure 1: Flowchart showing the structure of DANTE calculations.

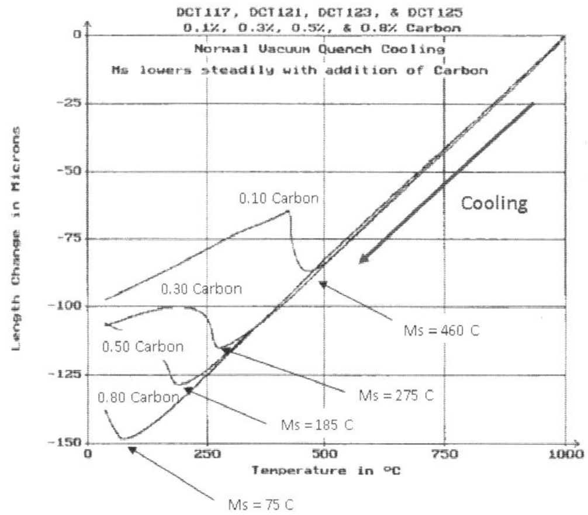

Figure 2: Martensite start temperature for carburized Pyrowear 53 steel.

martensite are so different from those of austenite. For accurate simulation of a quench hardening process, these data must be available.

Process data are also important to the accuracy of the heat treatment simulation. Especially important are the thermal boundary conditions such as the surface heat transfer during quenching. Because oil boils as heat is extracted from the racked parts, heat transfer changes as the part cools. The local agitation or flow of oil also varies around the part, so cooling is not uniform. Specific tests were not conducted to determine the heat transfer data for the gears used this study, so 'typical' values for oil quenching were used.[5] Figure 3 shows such a 'typical' data set for oil quenching.

For intensive quenching, the aim is to extract heat from the parts by using extremely fast flowing water so that no film boiling occurs, and even nucleate boiling may be absent. IQ Technologies, Inc. supplied heat transfer data for their process. A constant value of 30 kW/(m^2*C) was used. More recent work has reported values of 25 kW/(m^2*C) as being typical for intensive quenching.[6]

[1] ABAQUS/STANDARD is a product of Dassault Systemes Simulia Corporation, Providence, RI, USA.

[2] ANSYS Mechanical is a product of Ansys, Inc., Canonsburg, PA, USA.

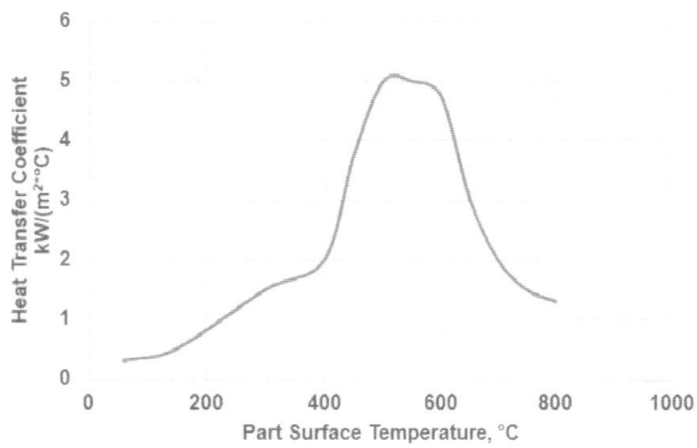

Figure 3: Surface heat transfer used for oil quenching of gears.[5]

Experimental Procedures

A total of thirty two gears were purchased from a helicopter OEM in the as-machined condition. These gears were machined from Pyrowear 53 bar stock and the geometry was specifically designed for single tooth bend testing. The gear has 40 teeth, a module of 2.54, and a face width of 6.35 mm.

Once received at our office, the gears were heat treated and then finish ground. Two quenching practices were followed while all other heat treatment steps were the same. The purpose of this was to produce different residual stress states in gears having the similar microstructures. The heat treatment steps are given in Table 2. All 32 gears were vacuum carburized in one batch; prior to carburization, the gears were copper plated so that only the tooth flanks and root were carburized. Figure 4 shows a view of one spur gear at a stage during its heat treatment processing. The copper plating was used to prevent carburization of some surfaces, and also to prevent decarburization during certain process steps. Sixteen gears were then hardened and tempered following the baseline schedule which used oil quenching, and the remaining gears were hardened and tempered following the intensive quenching schedule, see Table 2. After tempering, approximately 0.13 mm of stock was ground from the tooth faces of the oil quenched gears, and 0.18 mm of stock was ground from the intensively quenched gears. The aim was for the only process difference to be the quench step, so the effect of grind stock differences was investigated using simulation prior to proceeding with finish grinding.

Table 2: Heat Treat Schedules for the Two Sets of Gears.

	Baseline OQ Process	IQ Process
Vacuum Carburize	8 hours at 927° C	8 hours at 927° C
Subcritical Anneal	2 hours at 635° C	2 hours at 635° C
Austenitize & Quench	913° C, quench in oil at 66° C	913° C, intensive quench to 20° C
Deep Freeze	1 hour at -73° C	1 hour at -73° C
Double Temper	2 hours at 232° C	2 hours at 232° C

Figure 4: Spur gear used for bending fatigue test.

Heat Treat Process Simulations

Heat treat process simulations were conducted to predict the levels of residual stress and dimensional changes that could be expected from both processes before green gears were procured,. This was a necessary task as the amount of growth due to heat treatment was not known, the difference in growth between the two quenching methods was not known, and there was concern that differences in finish grinding could remove any residual stress differences between the two processes. These simulations proved to be extremely useful, as explained below.

A single tooth sector of the gear was meshed using hexahdral elements, also known as brick elements. Figure 5 shows the meshed tooth sector, with a manified view of the tooth and root cross section. The mesh density is higher along surfaces in order to capture the carbon gradient and the steep thermal and stress gradients that develop during quenching.

The predicted carbon profiles at the center of the root, the fillet or corner of the root, and at the pitch diameter are shown in Figure 6. These predicted data sets were extracted from a 3D model of the gear. While the green dimensions were unknown to us, the final dimensions and the typical amount of finish grind stock, 0.005 inches (0.13 mm), was provided to us. The models predicted the amount of growth expected for the two process routes, and the residual stress states after hardening and before finish grinding could be compared. It was also necessary to compare growth differences between the baseline and IQ processes in order to assess how the metal removal would affect the differences in residual stress. In other words, could differences in growth due to heat treatment plus the differences in stock removal to achieve identical final gear dimensions remove any differences in the residual stress state that might exist after quench hardening? This was critical

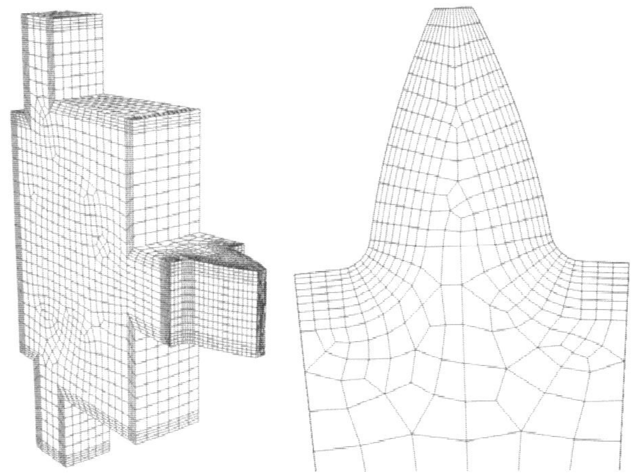

(a) (b)

Figure 5: (a) Hexahedral mesh developed for a section of the spur gear. (b) Magnified view of the mesh in the tooth and root.

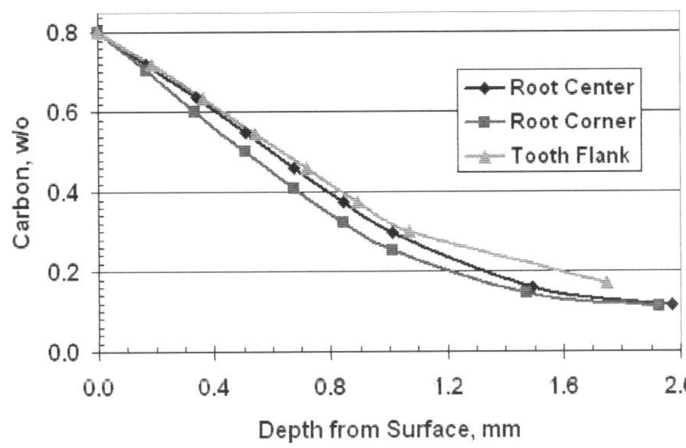

Figure 6: Carbon profiles at the indicated locations.[4]

information to the study, and simulation provided the method to answer this question.

Using the DANTE software, it was determined that a baseline gear would grow approximately 0.033 mm (0.0013 inches) per side and the IQ processed gears would grow approximately 0.076 mm (0.003 inches) per side, or 0.043 mm (0.0017 inches) more than the baseline gears, see Figure 7. This meant that the standard (0.005 inches) would be ground from the baseline gears and that 0.18 mm (0.007 inches) of stock would be ground from the intensively quenched gears.

The same simulations predicted the residual stress states that would be produced by the two quench hardening processes. These are shown in Figure 8, as are the depths of stock to be removed. From Figure 8, it is clear that the simulations predicted that the IQ route would produce a significant increase in residual compression over the baseline process.

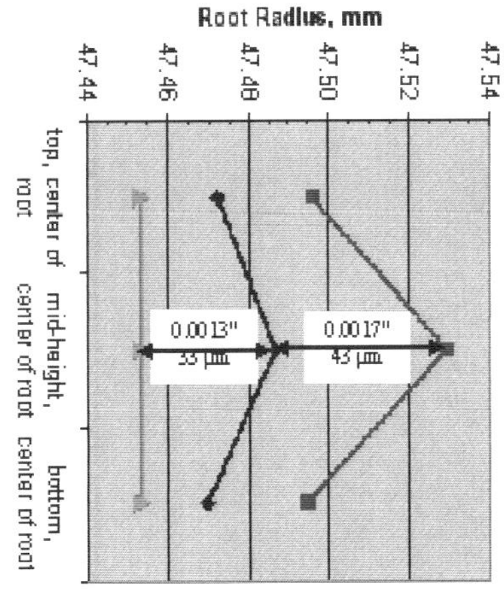

Figure 7: Radial location of the gear root, showing the initial or green location(▸), and the heat treated locations for the baseline (◆) and IQ (■) processes.[4]

The compressive field in the IQ processed gear was predicted to extend deeper into the gear body than that of the baseline gear. These predictions showed that even with the anticipated difference in stock removal that a significant difference in residual compression would exist after finish grinding.

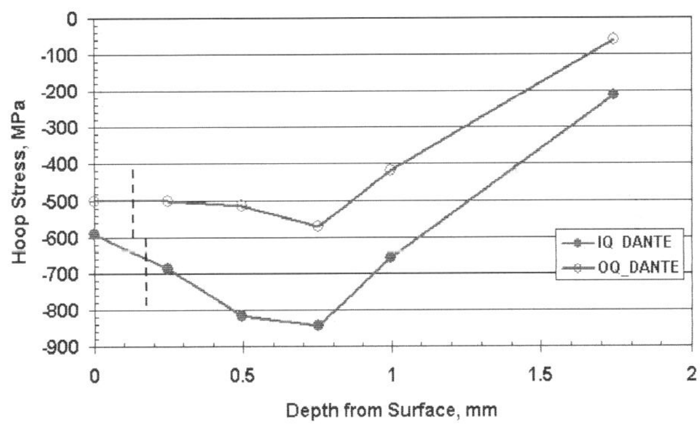

Figure 8: Predicted residual hoop stress at center of tooth root for the baseline and IQ processes. The grinding stock removal is indicated by the dashed lines.[4]

Figures 9 and 10 examine the model predictions in more detail by looking at data at specific locations over the course of the quench. Figure 9(a) identifies two locations associated with the root fillet, one at the surface and the other one at 1.25 mm from the surface. The carbon level at the surface is 0.65% and in the sub-surface location is 0.20%. Figure 9(b) shows that the surface cooling rate is slightly faster than that at the sub-surface point. There is a predicted temperature drop during the air transfer from the furnace to the quench tank, as indicated by the graph in (b) for the15 second period before the quench starts. The sub-surface location starts to form martensite at less than 10 seconds of quenching, which is over 30 seconds earlier than the surface location, and that sub-surface martensite formation is quickly completed in ~20 seconds. At 150 seconds of quenching, the surface martensite transformation has been active for over 100 seconds but is still only 70% completed.

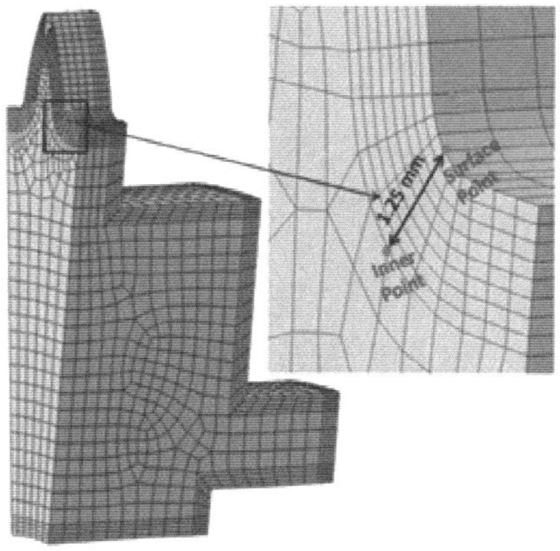

(a)

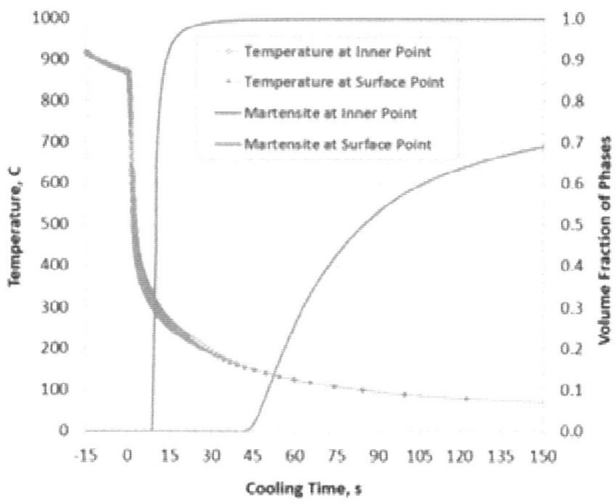

(b)

Figure 9: Response of gear root fillet during quenching: a) surface and inner points selected, and b) relations between temperature and martensite transformation at the two selected points.

This difference in the timing of the phase transformation has a significant effect on stress evolution. Figure 10 shows the stress histories for these same locations. It is clear from both (a) and (b) graphs that several stress reversals take place. The first stress reversal is due to the thermal gradient between the surface and sub-surface, and it occurs while both locations are fully austenitic. The next stress reversal occurs when the sub-surface is transforming to martensite and the surface remains austenite. The final stress reversal occurs when the sub-surface has completed its transformation to martensite and thermal contraction of the gear body pulls both the sub-surface martensite and surface austenite in compression. As the gear body continues to cool and transform to martensite, the sub-surface stress rebounds as the core expands, but the surface stress continues to become more compressive. The volumetric expansion of the surface as martensite forms is restricted by the already transformed sub-layers, and this is the contributor to surface compression. At the end of quenching process, high magnitude of residual compression is predicted at the surface point, and low magnitude of compression is predicted at the specific subsurface point. In regions deeper than this specific sub-surface point, tensile residual stress is predicted.

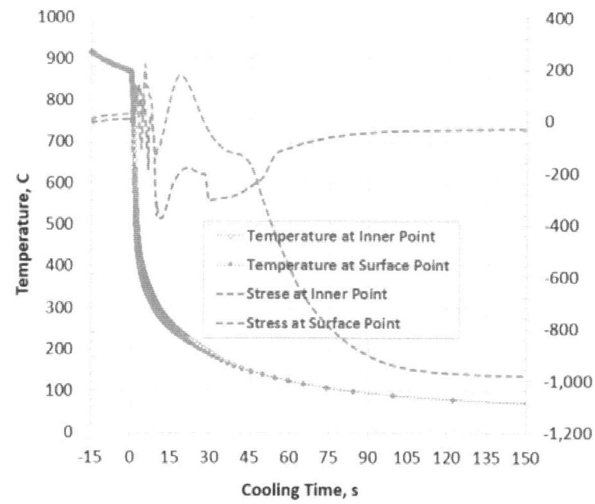

(a)

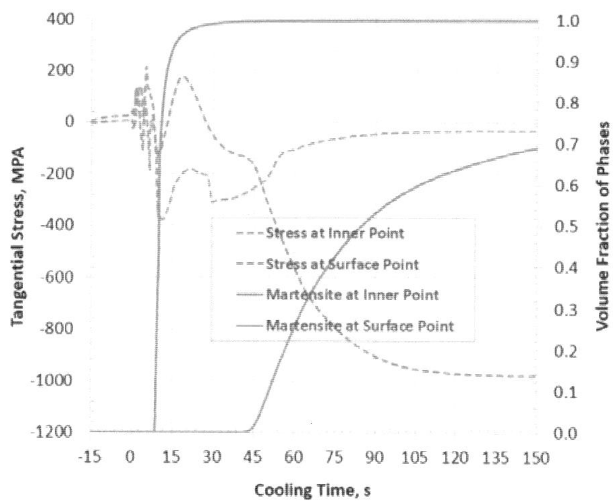

(b)

Figure 10: Response of gear root fillet during quenching: a) relations between temperature and tangential stress, and b) relations between temperature and martensite transformation at the two selected points.

Gear Characteristics

After heat treatment and final grinding, one gear from each batch was characterized in terms of microstructure, microhardness of the case and core, and residual stress state.

With the high hardenability of Pyrowear 53, the microstructures for both quenching methods were tempered martensite in the carburizied case and core of the gear tooth. The measured microhardness of the intensively quenched gears was slightly higher than that of the oil quenched gears in the carburizcd casc, i.c HRc 58 to 60 for IQ gears vs. HRc 58 to 59 for OQ gears, and these values were measured after grinding. These values are lower than expected for this carburizing schedule, even accounting for surface grinding; values $\geq$ HRc 60 were measured on witness bars. The core hardnesses were HRc 42 for both gear sets.

X-RAY diffraction was performed on a baseline (OQ) gear and an intensively quenched (IQ) gear to compare residual stress values. Similar to the simulations, hoop stresses calculated from the X-RAY diffraction (XRD) measurements showed that the intensive quenching process produced deeper compressive surface stress in the tooth root, see Figure 11. The agreement between the predicted hoop stress and measured hoop stress is very good for the IQ gear, while the predicted hoop stress profile for the OQ gear is slightly more compressive than that from XRD measurements. The XRD data demonstrates the difficulty of stress depth profiling for complicated geometric locations such as a gear root. The scatter band for both processes approach 100 MPa (14.5 ksi). The two XRD data sets for each gear are measured from opposite root positions.

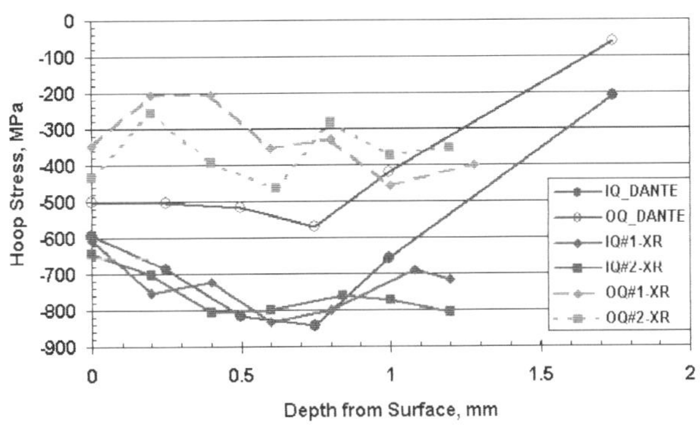

Figure 11: Hoop stress profiles at the center of the root for OQ and IQ gears as predicted by computer simulation and calculated from XRD measurements.[4]

From the computer simulations, the minimum principal stress profile for the root fillet was predicted to be more compressive than at the center of the root for both sets of gears. The direction of the minimum or most compressive principal stress was tangential to the gear tooth/root surface. This is significant because this is the typical cracking location for bending fatigue failures. For the OQ gear, the minimum principal stress at the surface of the root fillet location was predicted to be -600 MPa (-87 ksi), while the predicted minimum principal stress for the IQ gear was -820 MPa (-118.9 ksi). The maximum principal stress at this location was predicted to be -30 MPa (-4.4 ksi) for the OQ gear and -60 MPa (-8.7 ksi) for the IQ gear.

Single Tooth Bending Fatigue Testing

The gears were tested at Gear Research Institute which is located at Pennsylvania State University in State College, PA, using a servohydraulic testing machine with a specially designed fixture to apply a cyclic bending load to two teeth. A schematic of the fixture is shown in Figure 12. One tooth was loaded by a tup connected to the ram of the machine which cycled at 40 Hz, while a second tooth was seated against a shaped anvil. The upper and lower anvils were shaped such that no teeth needed to be removed for load application. The ratio of minimum to maximum load was fixed at 1:10 so that one root fillet of each contacted tooth was always loaded in tension. The machine turned off automatically when excessive ram displacement due to tooth cracking occurred. Runout was defined as 10^7 cycles without failure of either tooth. The location of the applied load was the point of first contact with a mating gear, which corresponds to the condition that generates the maximum stress in the gear root fillet. This is the typical failure location due to bending fatigue.

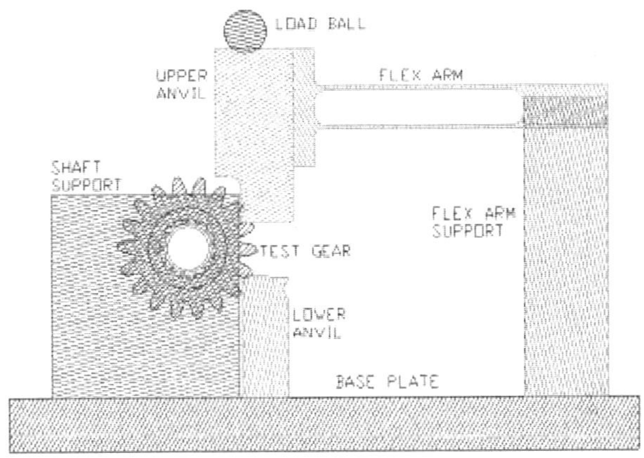

Figure 12: Schematic of tooth bending fatigue fixture at Gear Research Institute.

The fatigue test data for the OQ and IQ gears are shown in Figure 13. Qualitatively, the IQ gears have higher tooth bending fatigue strength than the baseline OQ gears. The number of runout tests (10^7 cycles) at a given load are indicated along the right side of the figure. Since each test loads two teeth, each runout case produced two data points. A failure produced only one data point, that for the cracked tooth root. The IQ gears have approximately a 15 % improvement in life over the baseline OQ gears.

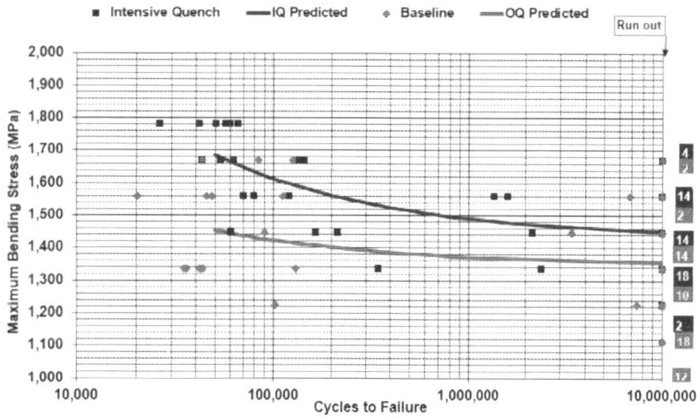

Figure 13: Single tooth bending fatigue data.

Service Stress

The study showed that increased residual surface compression in the gear root resulted in improved bending fatigue life. In this case, intensive quenching was used to produce a higher magnitude of compression than the baseline oil quenching. A higher magnitude of residual surface compression produced higher fatigue strength for parts having similar microstructures and hardness levels. Questions exist as to whether this translates to better rolling contact performance for these gears. Gear failure includes more than just bending failure, with spalling also being a major issue. To investigate stress states further, a dynamic model was developed where gears were contacted under load while being rotated. The residual stress state from quench hardening was the starting stress state. This approach to modeling is an example of ICME where the effects of material processing on microstructure and properties are carried forward to applications so that service stress state can correctly account for a non-neutral starting condition.

The root experiences cyclic stress during the single tooth bending test, and since the test was conducted with a stress ratio of 0.1, tension is always present. During loading under rotation, the same root location goes through a somewhat different stress cycle, with tensile loading being the high magnitude as in the bending test. In both cases, the benefit of initial residual compression on the local root stress is apparent. The initial compressive stress decreases the stress realized during service and this is the reason for improved bending fatigue life.

The bending stress assuming an initial stress free condition was calculated using the finite element method and is shown in Figure 14. The position of tooth loading is indicated to be near the tooth tip. As expected, the maximum tensile stress of almost 1600 MPa (232 ksi) is predicted to occur on the surface of the root fillet. Figure 15 is for the same finite element model except the residual stress state from the OQ heat treatment is included as the initial stress state. Now, the maximum tensile stress at the root fillet position is reduced to 1110 MPa (161 ksi) because of the initial compressive residual stress. Because a subsurface crack is more difficult to generate than a crack that starts on the free surface, fatigue resistance is improved. The maximum principal stress under bending load

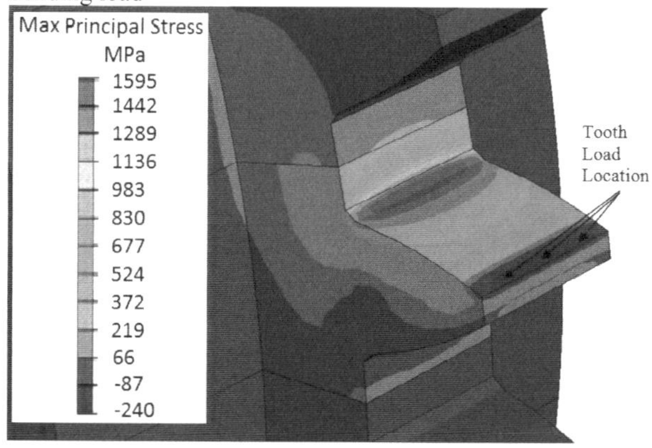

Figure 14: Maximum principal stress plot for tooth bending model with a stress free initial state.

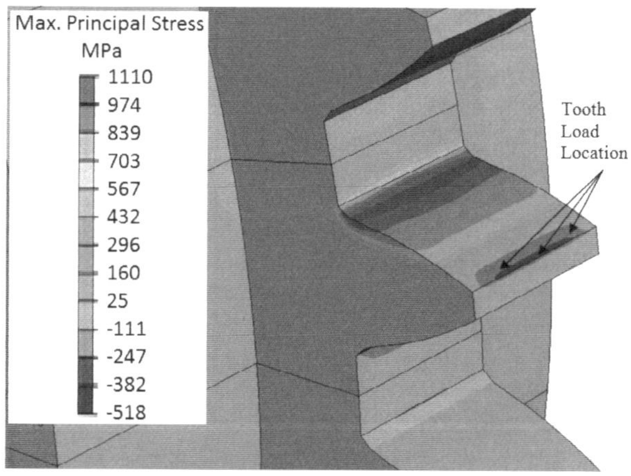

Figure 15: Maximum principal stress plot for tooth bending model accounting for initial residual compressive surface stress state.

is even less for the IQ gear because of even deeper compressive residual stress at the root fillet. This is shown more quantitatively in Figure 16(a) where maximum principal stress is plotted as a function of depth from the fillet surface. Figure 16(b) shows the calculated minimum principal stress for these three cases of different initial residual stress. For the stress free condition, not only is the maximum stress the highest, but the minimum principal stress is also tensile. Clearly the benefit of the residual compressive surface stress in the carburized case on the bending stress is evident. This residual surface compression allows the gear to carry a bending stress that approaches the yield strength of the alloy without actual yielding occurring.

The flank area at the pitch diameter experiences compressive loading from the mating tooth, but also frictional loading due to the rotation as sliding occurs between the teeth. The direction of the shear stress due to friction and sliding is opposite on each side of the contact point. Now the role of surface compression is conflicted as it does not affect the level of shear stress. However, the magnitude of surface

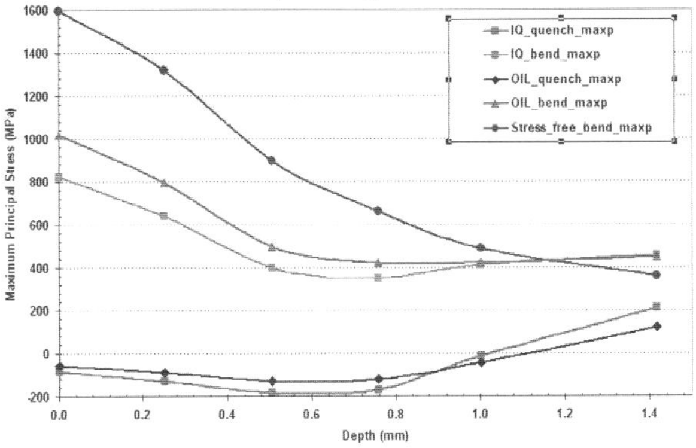

(a) Maximum Principal Stress

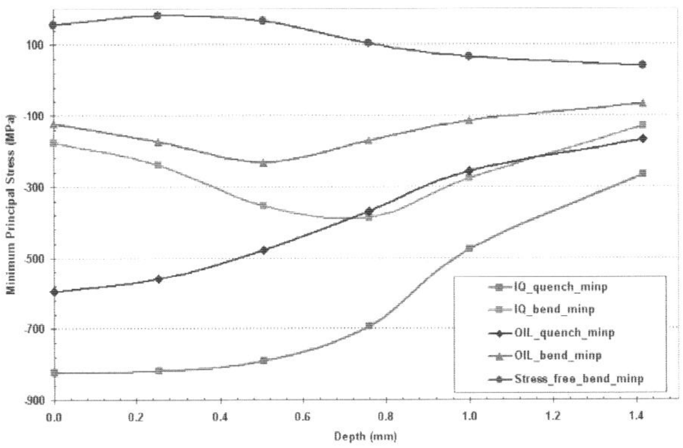

(b) Minimum Principal Stress

Figure 16: (a) Maximum principal stress and (b) minimum principal stress vs depth from the root fillet surface for finite element models with and without residual compressive surface stress due to carburizing and quench hardening.

compressive stress does retard cracking, so it should benefit resistance to spalling. While this was not directly tested in this study, another study on rolling contact of other carburized steels did show that surface compressive stress magnitude related to resistance of spalling.[7]

Summary

Gear tooth bending fatigue life is certainly improved by achieving higher magnitudes of surface compressive stress. For the carburized Pyrowear 53 steel and this gear geometry, intensive quenching produced deeper surface compressive stress than oil quenching and higher resistance to bending fatigue.

Accurate process modeling was used in this study to demonstrate that oil quenching and intensive quenching would produce the desired difference in residual stress states. Accurate material data were available in the DANTE material database for Pyrowear 53 steel so that this modeling could be confidently conducted.

Simulation of gears in service requires that the stress state resulting from prior processing, including heat treatment, be taken into account.

Further study of dynamically loaded gears is recommended to investigate the role of initial stress state on service stress and subsequent life. If bending fatigue is a limiting condition, then increased surface compression will definitely help gear life.

Acknowledgments

Part of this work was funded by the US Army AATD through SBIR Contract #W011W6-05-C-0017. The authors would like to thank B. Smith and E. Ames of AATD, J. Powell and M. Aronov of IQ Technologies, Inc., B. Hansen of Sikorsky Aircraft Corporation, S. Rao of the Gear Research Institute, and D. Schwam of Case Western Reserve University.

References

[1] C.P. Kern, et al., "Manufacturing and Processing of a New Class of Vacuum-Carburized Gear Steels with Very High Hardenability,", AGMA Technical Paper 11FTM27, Oct 2011.

[2] Committee on Helicopter Transmission Gear Materials, "Materials for Helicopter Gears," Final Report No. NMAB-351, National Materials Advisory Board, October 1979.

[3] Freborg, Andrew, et al., "Bending Fatigue Strength Improvement of Carburized Aerospace Gears", Proceedings of the 23rd ASM HTS Conference, Sept. 25-28, 2005, pp 186-195.

[4] B.L. Ferguson, A.M. Freborg, and Zhichao Li, "Residual Stress and Heat Treatment – Process Design for Bending Fatigue Strength Improvement of Carburized Aerospace Gears," Zeitschrift für Werkstoffe Wärmebehandlung Fertigung, HTM v62, pp 279-284, June 2007.

[5] "Predictive Model and Methodology for Heat Treatment Distortion," NCMS Report 0383RE97, 1997.

[6] B. Lynn Ferguson, Zhichao Li and Andrew M. Freborg, "Characterizing Water Quenching Systems with a Quench Probe," Journal of Materials Engineering and Performance, 23(12), pp 4197-4201, 2014.

[7] Andrew Freborg, B. Lynn Ferguson and Zhichao Li, "Thermal Process Optimization for Gear Shafts," Final Report US ARMY Contract # W911W6-09-D-0016, January 12, 2015.

International Federation of Heat Treatment and Surface Engineering 2016
Proceedings of the 23rd IFHTSE Congress
April 18–21, 2016, Savannah, Georgia, USA

The Development of a Process-Structure-Properties-Performance (PSPP) Map for Aluminum-Zinc-Magnesium-Copper Alloys Used In Aircraft Applications

A. N. Goulding, R. W. Neu
George W. Woodruff School of Mechanical Engineering, Georgia Institute of Technology, Atlanta, GA, 30332
School of Materials Science and Engineering, Georgia Institute of Technology, Atlanta, GA, 30332
T. H. Sanders, Jr.
School of Materials Science and Engineering, Georgia Institute of Technology, Atlanta, GA, 30332

Abstract

In 2011, the White House announced the Materials Genome Initiative, which aims to cut down the cost and time required to discover, develop, and optimize a material and deploy it in a commercial system. One method of achieving this is by integrating experimental test methods, computational tools, and known knowledge databases. While the potential for computational models to speed and assist process-structure-property optimization is not a new concept, one of the key issues in the implementation of models in this space is the lack of a standard method of communication regarding the physical and chemical mechanisms that drive a materials system, and the amount and type of data required to accurately characterize that materials system in its entirety. This work focuses on developing process-structure-property-performance (PSPP) maps as this missing standard communication method. Here we detail the steps that should be taken to generate a map for any system, and then briefly discuss how those steps were applied to high-strength Al-Zn-Mg-Cu alloys. These alloys are highly utilized in aircraft applications and are extremely complicated and commercially viable materials which are repeatedly subject to re-optimization, and are thus a good example of the value of these maps and how they can be used.

Introduction

In 2011, the White House announced the Materials Genome Initiative (MGI), which is based on the idea that advanced materials systems are foundational and transformative technologies. One of the major goals of the MGI is to encourage researchers to integrate experimental test methods, computational tools, and known knowledge databases to find ways to decrease the time and cost the currently goes in to discovering, developing, and optimizing a material system for deployment in a commercial system[1].

One of the potential applications for this kind of integration is in the development and property optimization stage of material development. These stages are often cost and time consuming, largely because of the strong reliance in materials research and development on scientific intuition and trial and error experimentation. Much of the design and testing of new materials systems is currently performed through time-consuming and repetitive experiment and characterization loops. However, once enough experimental work has been done to develop the material system, including understanding the chemical and physical mechanisms that drive the process-structure-property-performance relationships, then computational tools can be used to more quickly and efficiently work towards optimization.

While the potential for computational tools to assist with property optimization is not a new concept, it is surprising that more accurate computational models are not used in many materials systems which could benefit from them. On closer examination, there are still two key issues that are limiting the development and integration of these tools. The first, is that there is currently no standard method to communication critical information about the material system from those who develop the system, to those who are interested in modeling it, and beyond. Oftentimes these materials systems are developed by people with a deep knowledge of the system in question, but the computational models are developed by people with a deep knowledge of computational modeling, and who are usually knew to the material system they are working in. Therefore, it is critical that a standard exists to communication information about the mechanisms driving the system. The second key issue is that there is often a lack of understanding about what information must be obtained to accurately describe the material system in its entirety. Having a standard method of communicating the physical and chemical mechanisms which drive the system could be useful in resolving this issue as well. This work focuses on developing process-structure-property-performance (PSPP) maps as this missing standard communication method. Here we detail the steps that should be taken to generate a map for any system, and then briefly discuss how those steps were applied to high-strength Al-Zn-Mg-Cu alloys. These alloys are typically utilized in aircraft applications and are extremely complicated and commercially viable materials which are repeatedly subject to re-optimization, and are thus a good example of the value of these PSPP maps and how they can be used.

Generating a PSPP Map

Process-Structure-Property-Performance maps as they are described here are meant to be a standard method of communication for the physical and chemical mechanisms that are involved in the system, as well as the data or information required to fully characterize it. Here, we lay out some general guidelines for how to make these maps for any materials system.

When developing a PSPP map for a material, it is important to consider how developed the materials system is. Although the method detailed here is flexible enough to account for both more established systems as well as newer materials that are still in development, the overall process or result may be altered depending on how much is currently known about the mechanisms that drive the system being mapped. Just as the development of a materials system is an inherently iterative process, the mapping of the system will also be iterative, with new maps being developed as new information becomes available. Since these maps are meant to serve as an allegorical tool and help researchers and modelers navigate these often complex and interconnected materials systems, it is unsurprising that they share many of the same characteristics observed in geographical maps. Although the underlying relationships that govern a material system to do not change depending on the application, it is possible that maps with various levels of specificity may exist for different applications or defined materials spaces. Much like geographical maps, PSPP maps are visual standards, but can be customized depending on the features of interest and the level of detail or total size required.

Given the various levels of specificity that are available, an important aspect to consider when building a PSPP map for a material system is the scope, or total size, of the materials design space that is desired. As a general guideline, maps should be large enough to capture all of the features, processing options, or properties of interest, but specific enough to be useful to the application or project they are built for. Ultimately, the scope and level of detail that is captured in the "Structure" column will set the space.
Even though the "Structure" column is pivotal to defining the space included in the map, it is not considered the best place to begin the mapping process. Rather, users who are interested in developing a new map should start with the "Process" and "Properties" columns as a first step. Under the "Properties" column, the user should begin by listing the various properties of interest. If the user has a particular set of "Performance" metrics, sometimes called a "performance profile", these may be useful in determining what properties need to be included in the map. It is recommended that the properties initially be listed and boxed individually as much as possible. After the first iteration of the map has been detailed, properties that share similar driving mechanisms can be grouped for convenience if desired.

Under the "Process" column, it is recommended that the user begin by listing all of the available or currently known processing steps involved in the production of the material defined by the system. Each processing step should be boxed, and within each box the variable parameters involved in that step can be listed and grouped together (e.g. a box containing the "Annealing" step might include variable parameters such as "annealing time" and "annealing temperature"). When the processing route being detailed contains two possible steps that are mutually exclusive, they must either be redefined so that they can be included together in one box, or one must be discarded and the system being mapped will be constrained (e.g. a metal alloy which may be either extruded or hot-rolled can have either a boxed step labeled "Extrusion", a boxed step labeled "Hot-Rolling", or these two processing steps could be simultaneously combined in a more general boxed step labeled "Plastic Deformation"). The "Process" column should generally consist of a set of linear steps that constitute the entire processing route. However, other parameters, such as the parent composition of the material, can also be included. Ultimately, the "Process" column should contain all directly controllable parameters involved in the making of the material being characterized.

Once the "Process" and "Properties" columns have been detailed, the various structural features of the material, which are indirectly controlled through the processing, can be generated. To do this, the user should begin by examining the elements listed in the "Process" column and, using what is already known about the material, generate a list of structural features that are influenced by the different processing parameters. The structural features can be boxed either individually or in relevant groupings depending on the level of detail that is desired. Recall that the scope and detail of the "Structure" column will ultimately define the design space encompassed. Similar to the "Properties" column, if the user is unsure of which structural elements would be well suited to grouping, it is best to leave them boxed individually until the first iteration of the map has been detailed. Afterwards, structural features that are of similar length scales and have similar effects on the processing parameters may be considered for grouping.

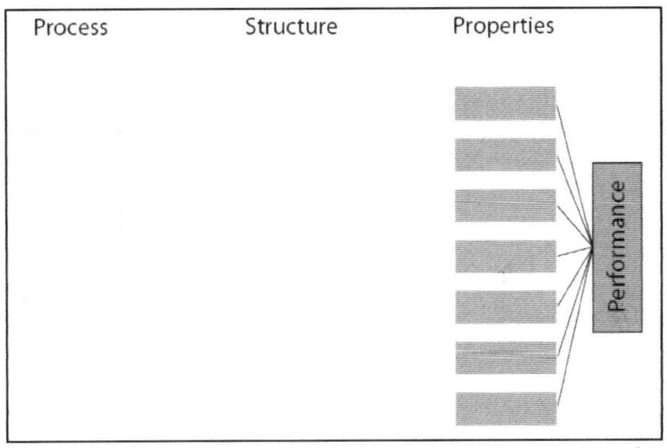

Figure 1: The user should begin the mapping process by generating all the necessary boxes contained in the "Process" and "Properties" columns.

Once the structural elements have been boxed, it is important to map the relationships between the two columns. If a relationship between an element in a processing box and an element in a structural box is known, indicate this relationship by connecting the boxes with a solid line. If a relationship is expected, but not confirmed, use a dashed line. The presence of a dashed line is useful to indicate that a relationship between these boxes is expected, but that further work must be done by the materials science community in order to confirm the presence or absence of this relationship. It is important that the lack of a line connecting two boxes indicates that the two elements in question have been confirmed to *not* share a relationship. Therefore, if the effect of a processing step on the various structure elements is not fully or adequately known, indicate this by dashing the processing box in question. Even if the processing step in question does have some confirmed or expected relationships, this can still be a useful way to indicate to the community that this step may have effects on the structure that are not fully understood. As a point of clarification, while the presence of a solid line should indicate that there is an influence of a processing box on an associated structure box, it is not an indication that the exact relationship is known. Even in situations where the two parameters are known to be, for example, linearly or inversely related, this information should not necessarily be indicated as a part of this map.

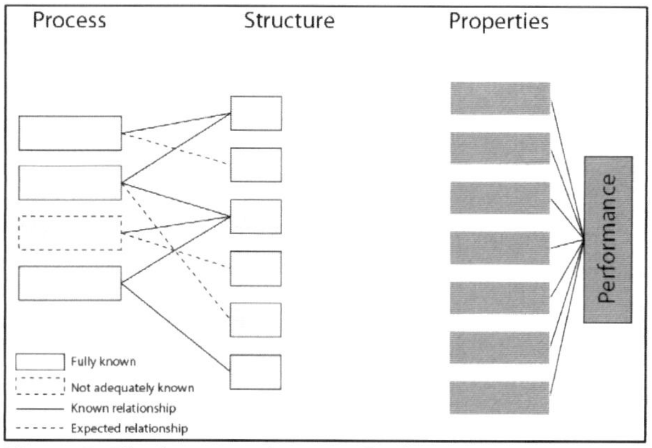

Figure 2: The known and expected relationships between the processing parameters and the structural features should be indicated by a series of solid or dashed lines and boxes.

Once the relationships between the "Process" column and the "Structure" column have been mapped, similar steps can be taken to map the "Structure" column to the "Properties" column. Here, the user should generate a second list of structural features that drive or affect the properties that are listed under the "Properties" column. Once again, if the features that drive a property are unknown or not fully explored, this should be indicated by dashing the box containing that property, as shown in Figure 3.

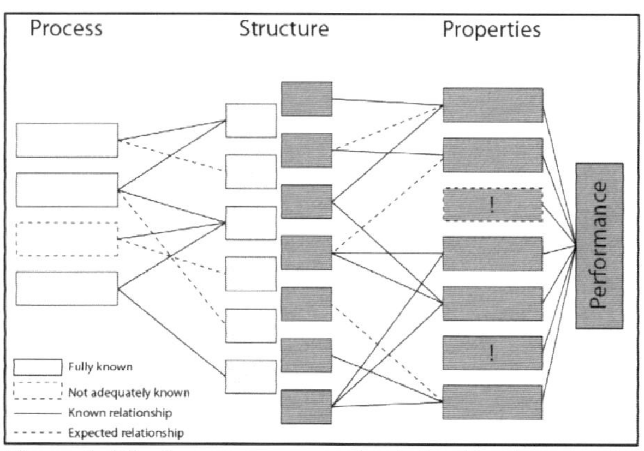

Figure 3: The known or expected relationships between the properties and the structural features that drive them should be indicated with alternatively solid or dashed lines. If the features that drive a property are unknown, this should be indicated with a dashed box around the property in question.

At this point, it is possible that one or more property boxes may be unconnected to the rest of the map because it is unknown what those relationships with the various structural elements may be. However, if a property box is floating and is contained by a solid box, this is an indication that either (1) the user has not included all of the necessary structural elements required to characterize the material, or (2) that the "property" in question is not a true property. For example, yield strength, electrical conductivity, and density are all material properties, which are dictated *only* by the structure of the material. Cost, however, which is driven by many factors not captured in the map, is not a true material property and should not be included in this column.

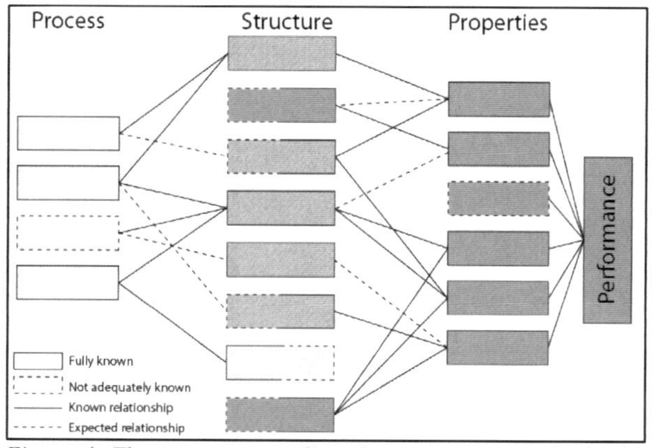

Figure 4: The two structure lists that were generated, should be merged. If the relationship between a structural feature and either the "Process" or "Properties" column is not fully known, this should be indicated by half of a dashed box, as shown here.

Once these two sets of relationships have been mapped, and any misidentified properties have been resolved, the user can proceed with merging the two independently generated "Structure" lists. Here, if a box appears in one list, but not the other, then dash one half of the box to indicate that the relationship between this feature and the opposite column is potentially unknown. It is also possible that, since the structure is indirectly controlled through one or more elements listed under the "Process" column, while one set of processing parameters may have been identified to affect the feature, others are still thought to be missing. Since this indications that more work needs to be done by the community in this area, this would again be indicated by a half dashed box.

For boxes that appeared in only one of the two "Structure" lists, these half-dashed boxes should be solidified *if and only if* it is established that no relationship is missing between the feature in question and the opposite column. If a solidly boxed element in the "Structure" column is not connected to anything in the "Property" column, then it is likely an indication that the structural feature in question, while indirectly controllable by the processing steps, is extraneous to the material system as it has been currently defined. Unless it is likely to be linked to a property of interest at some point in the future, it is suggested that it be removed from the map.

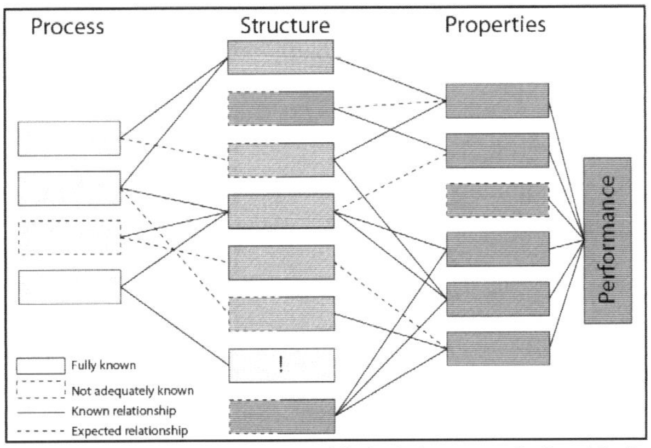

Figure 5: If and only it is established that no relationship is missing between the feature in question and the opposite column, should a half dashed "Structure" box be replaced with a fully solid one.

If a solidly boxed element in the "Structure" column is not linked to the "Process" column then this is an indication that the processing step or steps that control this feature are not included in the existing processing list and it is therefore incomplete. If the processing step that controls the feature in question cannot be added to the map, perhaps because it is not known, then it is still possible to use the map to characterize the material system. But to do so, it will be critical to measure the structural feature in question, since it cannot be indirectly measured by controlling the processing steps that drive it. This information should be communicated on the map by double boxing the structural feature in question.

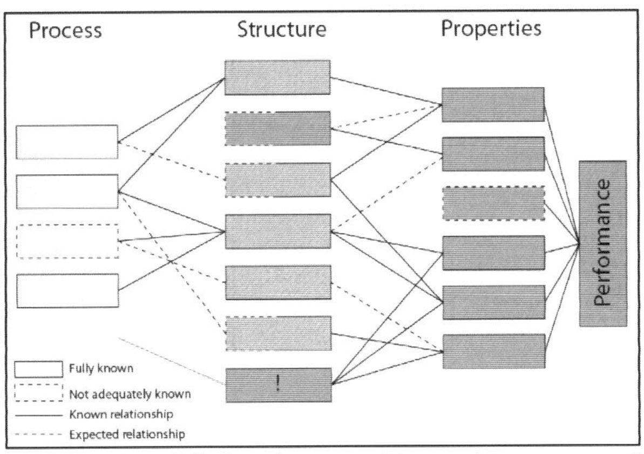

Figure 6: If a solidly boxed structural feature is not connected to an element in the "Process" column, then it is an indication that the directly controllable processing steps listed are incomplete.

Once all of the known and unexplored relationships are indicated on the map by following the above steps, the first iteration of the map should be complete. At this point, if the user is unsatisfied with the size of the materials design space that has been defined, or with the level of detail included in the map, they can iterate through the steps detailed above: adding, combining, or dividing boxes as necessary until the desired level of specificity is obtained. Additionally, once the map has been detailed, targeted exploration can be used to solidify those unconfirmed relationships as indicated by dashed lines and boxes.

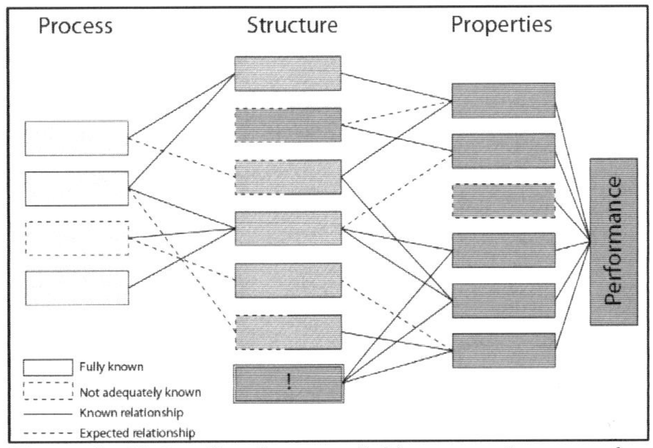

Figure 7: If the directly controllable processing step that drives the 'floating' structural feature in question cannot be included, perhaps because it is unknown, then this information must be included by double boxing the structural feature.

Al-Zn-Mg-Cu System

Now that some general guidelines for how to make these maps for any materials system have been detailed, these steps can be used to create a PSPP map for high-strength Al-Zn-Mg-Cu alloys used in aircraft components. This class of Al-Zn-Mg-Cu alloys, which includes AA7075, AA7050, AA7055, AA7079, AA7449, and many others, has been used in aircraft applications for over 75 years and is well established in the industry. Because these alloys are constantly being developed to provide a wide range of performance profiles for specific applications, they are continually developed. However, the extremely complex relationships that exist between the many processing routes, the complicated microstructural features that extend over multiple length scales, and the variety of properties of interest, make property optimization through computational modeling techniques challenging. In fact, many of these relationships cannot currently be predicted computationally with any real accuracy in this system, even after 75+ years of development[2]. Some initial successes, however, have been made in this space – such as relationships which predict yield strength as a function of a variety of microstructural features[2,3,4,5]. The next step in this materials system is to enable the modeling of multiple elements within this space, possibly through machine learning methods such as artificial neural networks. To do this successfully, a PSPP map for this alloy system must first be developed. Here the authors have used Al-Zn-Mg-Cu alloys as an example to illustrate the steps used to create the map.

Processing Steps and Properties of Interest

One advantage of this material class is that is, to date, quite well developed. As a result, the processing steps and their variable parameters, as well as the effect of changing those parameters on the microstructure are well established. Industrially in these systems, the material is first solidified through direct chill casting methods before being homogenized[6]. While this class of aircraft aluminum alloys is not commonly forged, they are regularly extruded or hot-rolled into stock pieces such as plates, sheets, and bars depending on the application[6]. Since material can only be extruded or hot-rolled, but not both, these steps are contained in one box titled "Plastic Deformation". After they are formed in to stock pieces, they are typically solution heat treated, quenched, and aged. Occasionally, they can be pre-strained or cold-rolled for stress-relief or to increase the dislocation density prior to aging[7].

Although many properties are typically of interest in this material system, the most commonly optimized and developed properties are the yield and ultimate strength, the work hardenability, the ductility, fracture toughness, and the stress corrosion cracking susceptibility. Other properties such as exfoliation corrosion resistance or density could also be considered, and the performance profile is left general.

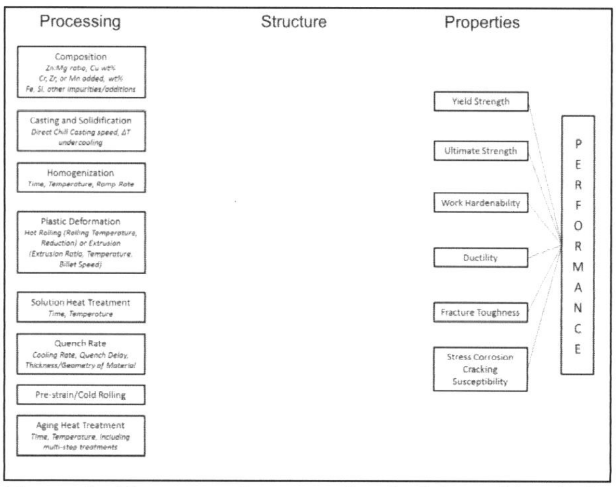

Figure 8: The processing steps involved in making high-strength Al-Zn-Mg-Cu alloys, and some of the most common properties of interest in developing materials for aircraft.

Process-Structure Relationships

Since the material is formed through direct chill (DC) casting, both the casting speed and the degree of undercooling are important and can affect the solidification kinetics[6-8]. Similar to gravity cast systems, semi-continuous casting methods such as DC casting form regions of columnar dendritic growth near the walls of the container, with more equiaxed structures appearing in the middle of the cast. The as-cast material is highly cored with intermetallics in the interdendritic regions[8]. Changing the parameters during this step can affect the solidification kinetics of the casting process, which can lead to changes in the grain size and the size, spacing, and volume fraction of the resulting constituent particles.

Other directly controllable parameters used during this step include additions of Al-Ti-B refiners to change the grain size[6,7,9]. However, care must be taken since small excess amounts of Boron can lead to constituent particles, which have detrimental effects on many properties – including fracture toughness, machinability, and appearance[9]. Cr, Zr, and less often Mn are also added as dispersoid forming elements when the material is formed. The dispersoid particles that are formed using these elements are added to inhibit the recrystallization of the material, but sometimes these additions can form large primary particles during casting instead – leading to more constituent particles throughout the material[6,7,9].

Once the material is cast it is typically homogenized, by heating the material to an elevated temperature for an extended period of time[6,7,10]. The homogenization serves multiple purposes. In addition to redistributing the Zn, Mg, and Cu required to form well dispersed strengthening and grain boundary precipitates, it also serves to redistribute and optimize the precipitation of the Cr- or Zr-containing dispersoid elements which are intended to inhibit recrystallization[6,7,9,10]. The elevated temperature is also used to try and melt as many of the coarse constituent particles as possible. However, care must be taken in setting the homogenization temperature, since many of these constituent particles have relatively low melting temperatures[7,10]. As a result, the time, temperature, and the heating ramp rate, can all affect the dispersion of constituent and dispersoid particles, as well as indirectly affecting the extent of recrystallization through the formation and optimization of the dispersoid particles.

The controllable parameters during the plastic deformation step are obviously different depending on whether the material is hot-rolled or extruded. Regardless, this step is known to have significant effects on the final microstructure, including the texture, the grain characteristics, and the uniformity of the grain characteristics throughout the component[6,7]. Even though this material does not dynamically recrystallize, but rather statically recrystallizes during the solution heat treatment, this step does affect the extent of recrystallization[7,11-13]. As the material experiences deformation, the energy that is added is redistributed inhomogeneously and is often concentrated around coarse constituent particles that remain after homogenization. The more energy that is stored locally in the material, the less thermal energy required during solution heat treatment to begin the recrystallization process. Therefore, the size, spacing, and volume fraction of the constituent particles will also affect the extent of recrystallization, although indirectly, by serving as locations for this added energy to localize. This effect is known as "Particle Stimulated Nucleation" and therefore any processing step that affects the constituent particles, could also have unforeseen consequences on the extent of recrystallization[7].

After the material has been shaped in to a stock piece, it is solution heat treated and then quenched. This step primarily serves to increase the number of vacancies in the system and also increase the amount of primary alloying elements Zn, Mg, and Cu in solution[6,14,15]. Therefore, both of these steps will increase the aging kinetics of the system and have significant impacts on the type, size, spacing, composition, and total volume fraction of precipitates that form both in the interior of the grain and at the grain boundaries. Additionally, since the solution heat treatment is usually conducted at temperatures similar to those used in the homogenization step, it can also melt some of the remaining constituent particles, and as previously discussed the added thermal energy triggers static recrystallization in the material[6,7,14,15].

Prior to aging, the material is occasionally stretched or cold-rolled[7,16]. While further deformation obviously affects the texture of the material, it also can affect the interior of the grain by significantly increasing the dislocation density in these regions. This, in turn, affects the size, spacing, and type of strengthening precipitates that are formed in these regions, since the dislocations serve as heterogeneous nucleation sites[8,16]. During aging, the material can be heated to any one of a range of low to moderately high temperatures for any length of time to encourage the precipitates of strengthening particles in the grain interior and the growth of equilibrium particles at the grain boundary. While aging heat treatments can consist of only one heating step, it is far more common to combine multiple steps together in a single aging process.

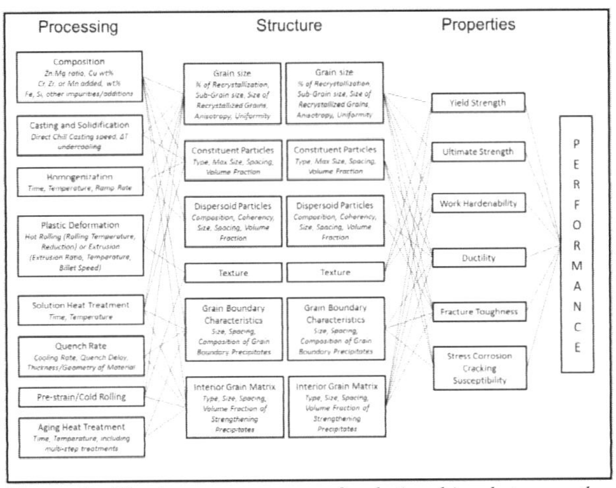

Figure 9: The known or expected relationships between the processing steps and the structural features the influence, as well as the properties and the structural features that drive them are shown here. Since this material system is well developed, no dashed lines or boxes are required.

Structure-Property Relationships

When generating the list of properties that are of interest, we listed each of them individually, as shown in Figure 8. However, sometimes information about the structural features that drive the properties will not be as detailed. For example, in this and many other materials, distinctions between the features that drive the yield strength versus those that drive only the ultimate strength of the material are not always made. And depending on the application and level of detail that is desired in the map, this designation is not always necessary. The main purpose of initially splitting these properties is: (1) to encourage the user to find out if distinctions between the driving mechanisms for each of these properties are known, and (2) to prevent the user from accidentally grouping properties together that are inappropriate. For example, since information about the ductility is also collected as a part of a standard tensile test, users might be tempted to group this property together with the yield strength, the ultimate strength, and the hardening behavior. However, the mechanisms that drive ductility in this system are much more similar to those that drive the fracture toughness. Had the ductility been grouped with the strength and hardening properties, it would have been impossible to distinguish which mechanisms drove which properties. In this material, as in many materials, distinctions between the mechanisms driving the yield strength and the ultimate strength are uncommon. However, as they are very closely related in this material, they are both taken to have the same mechanisms unless shown otherwise.

Although high-strength Al-Zn-Mg-Cu alloys are known for being heat-treatable and precipitation-hardened, there are actually multiple mechanisms responsible for their strength properties[2-6,17]. The most overlooked of these is perhaps the strengthening effect due to the presence of grain boundaries. While the Hall-Petch relationship does hold in these materials for grain sizes at about above approximately 20μm, this effect becomes even more pronounced for smaller grains[17]. Since these materials are often partially recrystallized, the presence of small sub-grain cell structures typically in the range of 2-5μm means that the grain size, the extent of recrystallization, and the sub-grain cell size will all significantly impact the yield and ultimate strength of the alloy[4,5,17-19]. Additionally, the grain size and the shift from large recrystallized grains to unrecrystallized regions of smaller cell structures will also greatly impact the hardening behavior of the material. When dislocations accumulate in fine-grained material, the dislocations are more able to rearrange and annihilate, since they are near grain boundaries. When these dislocations accumulate in larger grains, the distance the dislocation must travel to reach a grain boundary becomes larger than the free slip distance, and so the dislocations become accumulated in the bulk and the material is less able to dynamically recover[17].

In addition to the size and the extent of recrystallization, the texture and the misorientation of the grains and the sub-grain cell structures also affects the yield strength, ultimate strength, and the hardening behavior. As the material is strained, one method of accommodating this excess energy is through the rotation of grains or the polyginzation of the sub-grain cell structures. Therefore, the texture, as expected, will also affect these properties[17,18].

Similar to the effect of grain size, the precipitation-hardening effect is also caused by an increased ability to trap and accumulate dislocations in the material. Here, the formation of coherent solute clusters stabilizes dislocations, since their presence reduces the strain caused by the volume or lattice mismatch that occurs in the matrix. Since small particles are cut by the moving dislocations, their strength contribution is proportional to their average size and volume fraction. When precipitates become large and increasingly incoherent, they become instead physical barriers to dislocation movement, since the only way for a dislocation to circumvent the particles is by bowing into a roughly semi-circular shape between them[6,8]. When this bypassing is done without the aid of cross-slip, it is referred to as Orowan looping, and the strength contribution becomes inversely proportional to the size of the particles[2,6,7,20,21]. It is expected then, that the maximum strengthening effect would be reached when this critical size of the shearing-bypass transition is reached. Indeed the optimization of strengthening precipitate size has been the focus of much of the property optimization work in these material systems to date.

After precipitation hardening has occurred, there are still small amounts of solute that are left in solid solution in the matrix in the interior of the grain boundary. The presence of these solutes contributes to solid solution strengthening. Although this can be a meaningful contribution to the overall strength of the alloy, it is much smaller than some of the others discussed[4]. The amount of strength the remaining solute would be able to impart to the alloy if it was incorporated in to additional precipitates is actually much higher than the contribution from remaining in solid solution. Indeed, new processing methods have recently been developed which increase the total volume fraction of the precipitates in the grain interior, and thereby increase the total strength of the alloy[22].

Fracture toughness is a major design parameter in this class of aluminum alloys, especially as component thicknesses increase. Typically, strength and fracture toughness are considered to be trade-off properties in these materials[6,23]. This view largely stems from the original processing methods that were developed for increasing fracture toughness and aluminum alloys. In these one-step aging treatments, the material has been observed to slowly lose strength as the material is overaged, while the corrosion properties and the fracture toughness are both observed to increase with the increased aging. However, rather than being an inherent trade-off, this change in the properties is driven by a change in the size and spacing of both the strengthening precipitates in the interior of the grain and the equilibrium precipitates at the grain boundary. Here, the slightly elevated temperature these over-aging treatments utilize cause both sets of precipitates to continue to grow. More recently, much work has been done to develop processing methods that allow for the optimization of the size and spacing of these two separate precipitate types – which has in turn led to better strength and fracture toughness or corrosion properties[22-28]. Once again, this demonstrates that it is critically important to understand and be able to distinguish between the structural and physical mechanisms that drive different properties of interest.

Fracture toughness is generally agreed to occur in three distinct phases in these alloys, including (1) void nucleation, (2) void growth, which is also considered to be ductile intergranular fracture since it usually occurs through the growth of voids nucleated near grain boundary regions, and (3) void coalescence, which either occurs by the coalescence of voids and the formation of a void sheet, or through the necking and subsequent failure of the remaining transgranular ligaments[13,14,29]. Because of this transition from intergranular to transgranular fracture at different points of the process, it has been reported that the overall fracture toughness is controlled by this transition. However, a clearer perspective can be gained from considering that an increase in the overall toughness of the material will be gained by prolonging the void growth phase, or through delaying void coalescence[29].

Void nucleation typically occurs through the cracking of coarse constituent particles, which are usually between 1-10μm in size. Since these particles are often quite large and typically appear in clusters, the stress required to break them is quite low, and in many cases they will crack during the multiple thermal and mechanical processing steps required to make the component. Therefore decreasing the size and volume fraction of these constituent particles can significant impact the fracture toughness of the material. However, as the stress increases, voids are also formed at the smaller dispersoid particles, which are typically 0.1-1.0-μm. Thus, it is possible that decreasing the size and increasing the coherency of these particles may also help to increase toughness.

Once nucleated, these voids grow through plastic deformation. As the strain increases, the growth of larger voids can happen concurrently with the nucleation and growth of smaller voids, such as those that form at dispersoids. Although the growth rate of voids that are the same size is independent of void spacing, when small voids are located near larger voids, the growth rate of the smaller voids is significantly accelerated due to strain concentration effects. Decreasing the strain hardening ability can also increase the void growth rate[29]. If the grain interior is very strong compared to the precipitate free zone (PFZ) area near the grain boundary, then void growth is halted, and any additional energy is dispersed in the material through void coalescence along the grain boundaries. This mechanism is responsible for the observed dependence of K_{IC} on the difference in strength between the grain interior and the PFZ. This also explains the severe increase in the amount of intergranular fracture and the overall decrease in the toughness with increasing solute content in Al-Zn-Mg alloys. The increasing solute content increases the strength of the material through the formation of fine particles in the grain interior, thus halting void growth[24]. Therefore the fracture surface goes from being largely transgranular to occurring mainly through intergranular fraction and void coalescence along grain boundaries[24,29,30]. This also helps explain the tradeoff between strength and toughness that is observed with over-aging discussed previously.

Once void growth is halted, the voids will begin to coalesce, and the material will fail. This process can occur either intergranularly through the formation of large void sheets, or transgranularly through the necking and failure of the remaining ligaments. The void distribution, while not observed to affect the growth rate, plays a dominant role in the onset of void coalescence[29]. It has been well observed that the area fraction of grain boundaries plays an important role in determining the overall fracture toughness of an alloy[13,24]. This is due to the acceleration of the void coalescence stage with an increasing areal fraction of the grain boundary, or a decreasing void distribution[24,29]. There is also evidence that supports the theory that the equilibrium particles at the grain boundary facilitate this void coalescence, rather than playing a part in the direct nucleation and growth of voids. Since increasing the size and spacing of the grain boundary particles serves to increase the overall toughness of the material, rather than decrease it. In the over-aged materials discussed earlier, the increase could be accounted for merely by the loss of strength in the grain interior due to the coarsening of strengthening precipitates. But other, more recently developed, processes have shown that this increase is observed even when the interior of the grain is unchanged and the material still possess a high strength [14,23,27,28].

Stress corrosion cracking (SCC) is environmentally assisted fracture, which requires both a corrosive environment and the application of an applied stress to occur. Where typical fracture occurs in three steps consisting of void nucleation, growth, and coalescence, SCC consists of two distinct mechanisms: (1) pitting, which causes a stress concentration and allows for the formation of a crack, and (2) crack growth, which is often measured by the SCC plateau velocity. The stress concentration that must be reached in the initial pitting process before the crack will begin to propagate is given by K_{ISCC}, the threshold stress intensity. Since SCC is effectively accelerated intergranular fracture in the presence of a corrosive environment, it is unsurprising that many of the microstructure parameters that accelerate or affect the fracture toughness and ductility, also affect the SCC resistance[31].

Just as increasing the extent of recrystallization in the material will decrease the overall toughness; increasing amounts of recrystallization will also increase the susceptibility of the material to SCC[13,32]. One mechanism responsible for this is that unrecrystallized sub-grain boundaries, and the dispersoids that are often present at them, are very effective at trapping hydrogen, and therefore their presence can decrease the amount of hydrogen present in the grain boundary regions where SCC is most likely to occur[4,10]. This contributes to an increase in the SCC cracking plateau velocity with an increase in the extent of recrystallization.

Some constituent particles are cathodic with respect to the surrounding the matrix material and can dissolve the material around them through the formation of microgalvanic cells[33]. This decohesion between the particle and the matrix causes pitting, and therefore constituent particles play a role in initiating SCC through the formation of pits that is roughly equivalent to the role they play in void nucleation in the fracture and ductility processes. It was also previously discussed that, although grain boundary particles do not play a role in void nucleation and growth, they are important in the facilitation of void coalescence. Therefore, increasing the size of these particles does not have a detrimental effect on the fracture toughness, and increasing the discrete spacing between them actually has a positive effect. Here, similarly increasing both the size and the spacing of these grain boundary precipitates increases both their ability to effectively trap hydrogen, and increases the amount of time it takes for them to anodically dissolve. [15,22,24,25].

However, changing the chemistry of the microgalvanic cell that is formed between the grain boundary precipitate and the surround PFZ can be more influential on the SCC resistance than changes in size and spacing. It is well known in this material system, that increasing the amount of Cu in the grain boundary precipitates enobles the precipitate, or makes it less electrochemically active[34-36]. Therefore, changing the parent composition of the alloy, or changing the aging process to increase the amount of Cu locally incorporated into grain boundary precipitates will greatly affect the corrosion properties of the material[35,36].

Merge Structure Lists and Adjust for Detail

Once all of the relationships between the three columns have been mapped, the two structures lists have been merged together. Here, the advanced state of this material system is again an advantage, since the two structure columns match and there are no unknown or unexplored relationships that need to be identified. However, this does not mean that the map that has been generated here is static. As discussed in the known structure-property relationships, much work continues to be done to better understand the chemical and physical mechanisms that drive these properties.

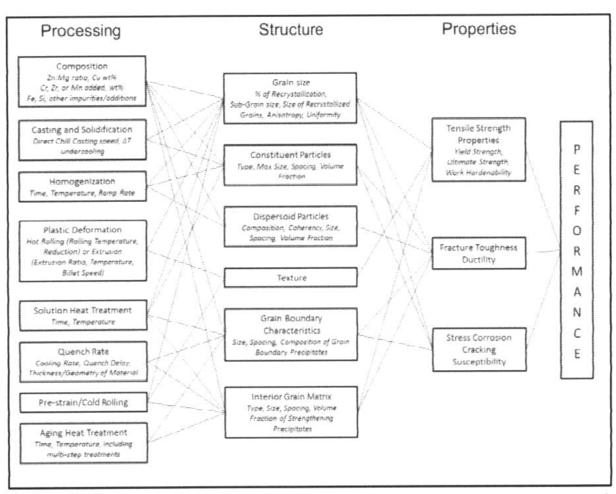

Figure 10: Here, the two structure lists that were generated are merged, and some of the properties have been grouped.

We can also adjust the level of detail by grouping some of the properties we are interested in based on the mechanisms that drive them. Here, we can easily group together yield and ultimate strength, as well as work hardenability. Similarly, we can group fracture toughness and ductility together if desired. Once the PSPP map for a materials system has been developed, it can be used to communicate important information about the physical and chemical mechanisms that drive the system, and about what information must be captured to adequately characterize the material in its entirety.

Discussion

One of the most important things in modeling process-structure-property-performance relationships is making sure that all of the factors that influence the element being predicted are being accounted for. Some of the only successful predictive modeling work in this material system has been in predicting the yield and ultimate strength as a function of various aspects of the microstructure. In each of these cases, this success was due to the ability to individually calculate the different strengthening contributions such as precipitate strengthening, solution strengthening, the strengthening effect due to the presence of grain boundaries, and the strengthening effect due to dislocation-dislocation interactions, and to sum them together using various super-positioning methods.

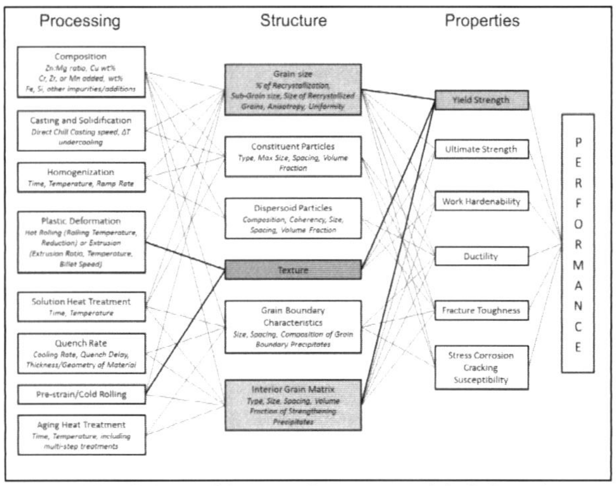

Figure 11: If the yield strength is not modeled directly as a function of the texture of the material, then the model must either account for it indirectly through the processing steps that affect the texture, or the modeling tool will be accurate only for other materials with that texture profile.

The PSPP map, therefore, can be quite useful in making it clear which of these mechanisms needs to be accounted for in such an attempt. But clearly laying out all of the mechanisms that are involved, these individual mechanisms can be more easily decoupled and predicted individually. The map is also useful in understanding which mechanisms may be missing and therefore constraining the usefulness of the model that has been built. This is also true in the case of predicting the yield or ultimate strength as a function of the various mechanisms that were previously mentioned. Using the map, it is easy to see that unless the texture of the material is accounted for, the relationship that was developed may not successfully extend and the predictive model is constrained. In these strengthening equations, the texture is often accounted for with a Taylor factor, which makes rough approximations about the textures that can be accounted for. In this cases understanding that the texture of the material does affect the strength can help users of these predictive modeling tools fully understand any assumptions being made about the material, and adjust their expectations accordingly.

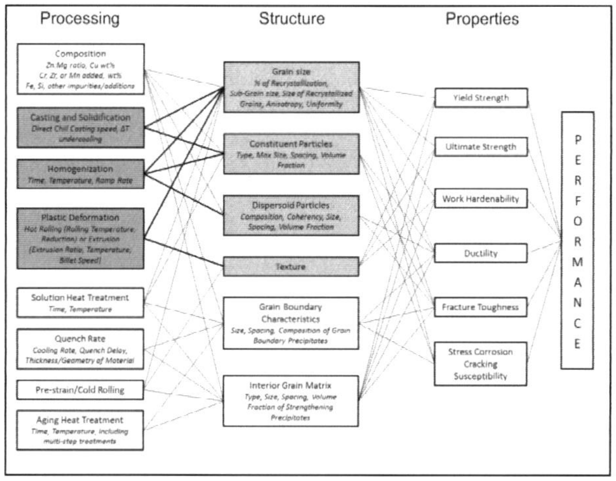

Figure 12: If processing information is unknown, then it is still possible to completely define the material by characterizing the structural features those steps influence.

Another important ability of these PSPP maps is to clearly identify what information must be captured to fully characterize or describe the material. This is especially important in materials systems where some or all of the processing steps are executed by a commercial company. In Al-Zn-Mg-Cu alloys, the casting, homogenization, and formation of a material in to a stock piece is not commonly conducted as a part of academic studies, both in part due to the complexity of these steps as well and the danger than can be involved. Since these materials are made quite commonly industrially, it is far more common to order material that has already been partially or completely processed and simply re-solution heat treat, quench, and age the material for study. Additionally, the casting and homogenization steps in particular are often highly proprietary, and therefore this processing information is rarely if ever reported to the end user. By using the PSPP map for this system, however, it is possible for a user to still fully characterize or describe the material in question, even if these processing details are unknown. Here, we can see that even if the casting, solidification, homogenization, and plastic deformation processes are unreported, the material can still be adequately described by fully characterizing the grain size information, the constituent and dispersoid particle profiles, and the texture. This application of the PSPP map is especially useful when gathering literature databases in a material system, since computational models are only as accurate as the data that is used to develop them. If the data sets that are used to develop and train a computational tool are incomplete, or limited, then the model will be similarly limited at best, or at worst inaccurate.

Acknowledgements

The authors would like to thank Dr. Justin Lamb at Universal Alloys Corporation for his extensive knowledge of casting, solidification, homogenization, and plastic deformation processes as they are currently conducted in commercial aluminum alloy processing today. They would also like to thank Adrienne Muth for the design of these clear and beautiful graphics, and Dr. David McDowell for useful discussions related to the Materials Genome Initiative, and its goals in bridging modeling and experimentation.

References

[1] The National Science and Technology Council, "Materials Genome Initiative for Global Competitiveness", June 2011.

[2] H. R. Shercliff and M. F. Ashby, "A Process Model for Age Hardening of Aluminum Alloys - I. The Model", *Acta. Metall. Mater.*, Vol. 38, (1990), pp. 1789-1990.

[3] A. Deschamps and Y. Bretchet, "Influence of Predeformation and Ageing of An Al-Zn-Mg Alloy - II. Modeling of Precipitation Kinetics and Yield Strength", *Acta mater.*, Vol. 47, (1999), pp. 293-305.

[4] M. Dixit *et al.*, "Structure-Property Correlations in Al 7050 and Al 7055 high-strength aluminum alloys", *Mat. Sci. and Eng. A*, Vol. 478, (2008), pp.163-172.

[5] M. J. Starink and S. C. Wang, "A model for the yield strength of overaed Al-Zn-Mg-Cu alloys", *Acta mater.*, Vol. 51, (2003), pp. 5131-5150.

[6] J. R. Davis, ed., Aluminum and Aluminum Alloys, (ASM International, 1993).

[7] J. Lamb, "Title of his thesis", Ph.D. Thesis, Georgia Institute of Technology, December 2015.

[8] D. A. Porter, K. E. Easterling,and M. Y. Sherif, Phase Transformations in Metals and Alloys, (CRC Press, 3rd ed., 2009).

[9] J. B. Hess, "Physical Metallurgy of Recycling Wrought Aluminum Alloys", *Metall. Trans.*, Vol. 14, (1983), pp.323-327.

[10] J. D. Robson and P. B. Prangnell, "Dispersoid precipitation and process modelling in zirconium containing commercial aluminum alloys",*Acta mater.*, Vol. 49, (2001), pp.599-613.

[11] X. Zhang *et al.*, "Effect of processing parameters on quench sensitivity of an AA7050 sheet", *Mat. Sci. and Eng. A,* Vol. 528, (2011), pp.795-802.

[12] H. Zhang et al., "Hot deformation behavior of Al-Zn-Mg-Cu-Zr aluminum alloys during compression at elevated temperature", *Trans. Nonferrous Met. Soc. China*, Vol. 21, (2011), pp.437-442.

[13] N. U. Deshpande *et al.*, "Relationship between fracture toughness, fracture path, and microstructure of 7050 aluminum alloy: part 1. quantitative characterization", *Metall. Trans. A*, Vol.29, (1998), pp.1191-1201.

[14] N. M. Han *et al.*, "Effect of solution treatment on the strength and fracture toughness of aluminum alloy 7050", *J. Alloys and Com.*, Vol. 509, (2011), pp.4138-4145.

[15] J. Beddoes, "Design of solution heat treatments for aerospace alloys", *Can. Metall. Quar.*, Vol. 50, (2011), pp.215-221.

[16] N. M. Han *et al.*, ""Effect of solution treatment on the strength and fracture toughness of aluminum alloy 7050", *Mat. Sci. and Eng. A*, Vol. 528, (2011), pp.3714-3721.

[17] R. Mahmudi, "Grain Boundary Strengthening in a Fine Grained Aluminum Alloy", *Scipta Metall.*, Vol. 31, (1995), pp.781-786.

[18] S. Zhang *et al.*, "Microstructure evolution and deformation behavior of ultrafine-grained Al-Zn-Mg alloys with fine eta' precipitates", *Acta mater.*, Vol. 58, (2010), pp.6695-6705.

[19] S. Chen *et al.*, "Effect of recrystallization and heat treatment on strength and SCC of an Al-Zn-Mg-Cu alloy", *J. Alloys and Com.*, Vol. 581, (2013), pp.705-709.

[20] R. B. Nicholson *et al.*, "The Interaction of Dislocations and Precipitates", *Acta metal. mater.*, Vol. 8, (1959), pp.172-176.

[21] W. M. Lee and M. A. Zikry, "Modeling the interfacial plastic strain incompatibilities assocaited with dispersed particles in high strength aluminum alloys", *Acta mater.*, Vol. 60, (2012), pp.1669-1679.

[22] J. F. Li *et al.*, "Mechanical properties, corrosion behaviors and microstructures of 7075 aluminum alloy with various aging treatments", *Trans. Nonferrour. Met. Soc. China*, Vol. 18, (2008), pp.755-762.

[23] N. Kamp *et al.*, "Toughness-strength relations in the overaged 7449 Al-based alloy", *Metall. Trans A*, Vol. 33, (2002), pp.1125-1136.

[24] G. M. Ludtka and D. E. Laughlin, "The influence of microstructure and strength on the fracture mode and toughness of 7xxx series aluminum alloys", *Metall. Trans. A*, Vol. 13, (1982), pp.411-425.

[25] J. F. Li *et al.*, "Influence of retrogression temperature and time on the mechanical properties and exfoliation corrosion behavior of aluminum alloy AA7150", *Mat. Char.*, Vol. 60, (2009), pp.1334-1341.

[26] A. Karaaslan *et al.*, "Effect of aging temperature and of retrogression treatment time on the microstructure and mechanical properties of alloy AA7075", *Met. Sci. and Heat Trt.*, Vol. 49, (2007), pp. 443-447.

[27] R. N Lumley *et al.*, "Interrupted aging and secondary precipitation in aluminum alloys", *Mat Sci. and Tech.*, Vol. 19, (2003), pp.1483-1490.

[28] R. N. Lumley *et al.*, "Development of mechanical properties during secondary aging in aluminum alloys", *Mat. Sci. and Tech.*, Vol. 21, (2005), pp.1025-1032.

[29] T. Pardoen *et al.*, "Multiscale modeling of ductile failure in metallic alloys", *Comptes Rendus Phys.*, Vol. 11, (2010), pp.326-345.

[30] T. H. Sanders Jr. and E. A. Starke, "The relationship of microstructure to monotonic and cyclic straining of two age hardening aluminum alloys", *Met. Trans. A*, Vol. 7A, (1976), pp. 1407-1418.

[31] J. R. Davis, ed., Corrosion of Aluminum and Aluminum Alloys, (ASM International, 1999).

[32] A. S. El-Amoush, "Intergranular corrosion behavior of the 7075-T6 aluminum alloy under different annealing conditions", *Mat. Chem. and Phys.*, Vol. 126, (2011), pp. 607-613.

[33] N. Birbilis *et al.*, "Electrochemical behavior and localized corrosion associated with Al7Cu2Fe particles in aluminum alloy 7075-T651", *Corr. Science*, Vol. 48, (2006), pp. 4202-4215.

[34] N. Birbilis *et al.*, "Electrochemical characteristics of intermetallic phases and in aluminum alloys", *J. Electro. Chem. Soc.*, Vol. 152, (2005), pp.B140-B151.

[35] G. Sha *et al.*, "Segregation of solute elements at grain boundaries in an ultrafine grained Al-Zn-Mg-Cu alloy", *Ultramicroscopy*, Vol.111, (2011), pp.500-505.

[36] R. K. Gupta *et al.*, "Relating the early evolution of microstructure with the electrochemical response and mechanical performance of a Cu-rich and Cu-lean 7xxx aluminum alloy", *J. Electro. Chem. Soc.*, Vol. 159, (2012), pp.492-502.

International Federation of Heat Treatment and Surface Engineering 2016
Proceedings of the 23rd IFHTSE Congress
April 18–21, 2016, Savannah, Georgia, USA

Metallurgical Aspects of Distortion and Residual Stresses in Heat Treated Parts

D. Scott MacKenzie, PhD, FASM
Houghton International, Inc., Valley Forge PA 19426
smackenzie@houghtonintl.com

Abstract

While modeling of quenchant behavior and microstructural response has matured, the sensitivity to variation of common metallurgical and production practices are often not considered. As an example, the variation in hardenability of the purchased material is generally not considered when predicting microstructure and residual stress. Machining practices, including worn tools and use of tooling beyond the expected life can have dramatic influence on distortion and residual stress. The quench oil chosen, and the quench part are all important variables to consider. Finally the presence of adjacent parts and the racking method can also effect the residual stress state and distortion. The effects of common metallurgical variables on the residual stress and distortion of heat treated parts are reviewed. Factors such as hardenability, segregation, and rack and fixture design are discussed.

Introduction

The problem of distortion is universal across industry. A study by Thoben (1) indicated that the 1995 losses from heat treatment alone in the German machine, automotive and transmission industry exceeded 850M € (1 Billion USD). This does not include the rest of Europe, Asia or the Americas. The problem is truly immense.

Distortion is not limited to heat treatment, but each step in the manufacturing process contributes to distortion 0. The sources of distortion follow the manufacturing process chain. Looking at distortion, the primary sources are volume changes from phase transformations and precipitation, or from deformation from either plastic or elastic deformation (Figure 1).

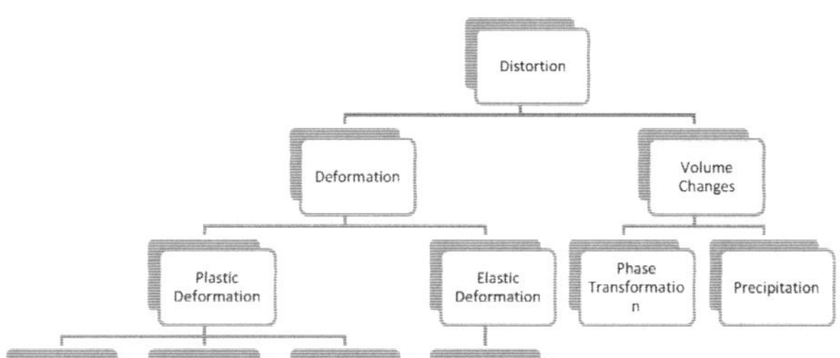

Figure 1. Global sources of distortion (2).

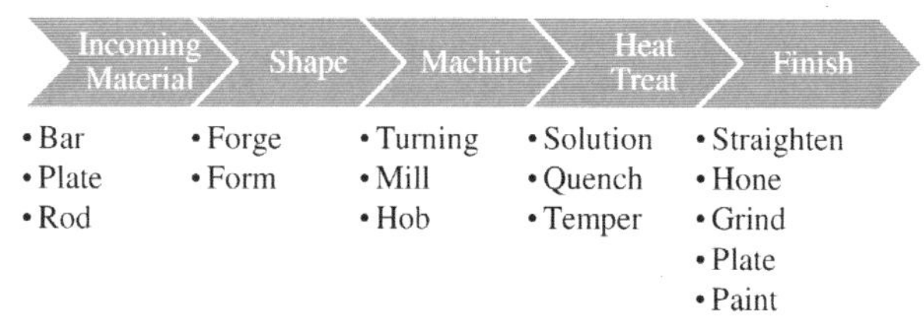

Figure 2. Typical fabrication sequence or process chain of a generic steel or aluminum component.

The literature is filled with papers trying to understand and model the distortion of components. Modeling has been used to understand the heat treatment process and the entire process chain. However, modeling is usually used to model the component under perfect conditions – new tooling, homogeneous steel, uniform heat treating conditions and quenchant. Rarely are sensitivity studies performed to look at the expected variation of distortion due to manufacturing variation.

It is the intent of this paper to review some of the possible variation due to metallurgical and manufacturing processes.

Literature Review

In the typical manufacturing process to manufacture a steel or aluminum part, incoming material is purchased; shaped by forming, forging or casting; machined; heat treated; and finish processed (final machining, plating or painting). This is illustrated in Figure 2.

There are many sources of distortion that have been identified. A fishbone chart illustrating some of the common sources of distortion is shown in Figure 3. It is not the purpose of this paper to review all the possible sources and permutations of distortion, but to highlight some of the important factors.

Prior Microstructure

The incoming raw material plays a critical role in the quality of a finished part. The prior microstructure and hardenability of the alloy grade selected, as well as the cleanliness of the incoming material affect distortion and the mechanical properties (fatigue and impact properties). This is illustrated in Figure 4.

Hippenstiel (3) evaluated the effect of prior grain size on distortion and found that finer grain size contributed to reduced distortion. He also found that by reducing the amount of grain size variation, further reductions in distortion could be achieved.

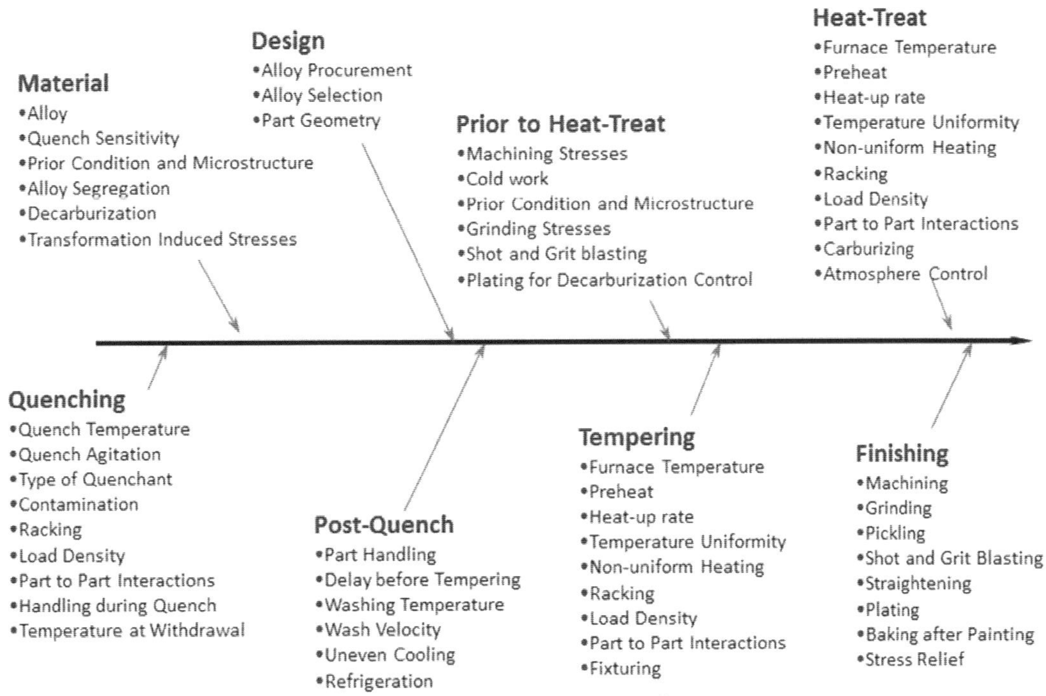

Figure 3. Fishbone diagram showing some of the many factors affecting distortion.

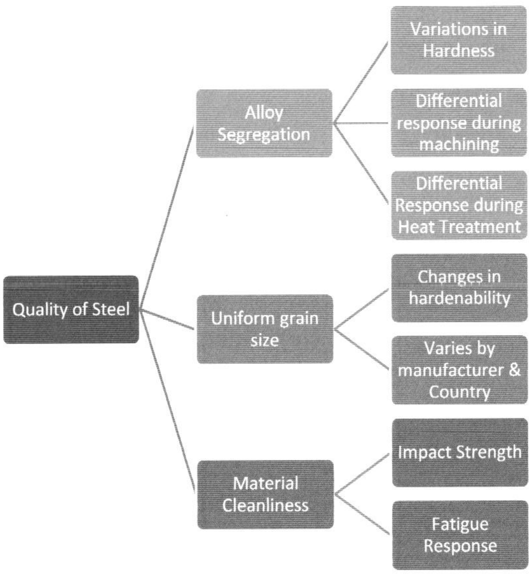

Figure 4. Quality of steel and impact on distortion and property response.

In another study (4), electro-slag remelted and vacuum melted and remelted X38CrMoV5-1 hot-work steels were examined for distortion, residual stresses, microstructure and properties. The vacuum heat treatment were varied to develop a model of phase volume fraction, residual stress and hardness. Using a Gleeble, the transformation kinetics and flow curves for the alloys were measured. A hardness after tempering was developed based on the volume fraction and hardness of each phase. An additional difficulty was the calculation after tempering was overcome by the inclusion of a coarse secondary hardening carbide. The validation of the model showed good correspondence with experiment.

The examination of the prior microstructure on the distortion of press-quenched gears was conducted by Reardon (5). In this study, 464 mm diameter AISI 8620 H steel automotive transmission gears were processed in two separate furnace loads. For the first set of gears, which were designated Series 1, the normalizing temperature that was used was 926°C. These gears received a single normalizing cycle. This is illustrated in Figure 5.

The second set of gears, labeled Series 2, was normalized twice in succession. The first normalizing cycle they received was identical to that used for the Series 1 gears. For the second normalizing cycle, the temperature was raised to 954°C. Each group of 24 gears was subsequently processed through a gas carburizing furnace using an endothermic atmosphere at a temperature of 926°C to generate effective case depths in the range of 0.0016 to 0.0022 m. The gears were all cooled to room temperature after carburizing was completed. They were then reheated to the austenitizing temperature of 854°C, and individually press quenched.

The resulting out of round distortion was measured (Figure 5). It was found that the double normalized gears exhibited substantially lower distortion, and much reduced distortion scatter than the single normalized gears. The double normalized gears had a mean out of roundness of 100μm, with a standard deviation of 25.4 μm. The single normalized gears had a mean out of roundness of 330 microns, with a standard deviation of 154μm.

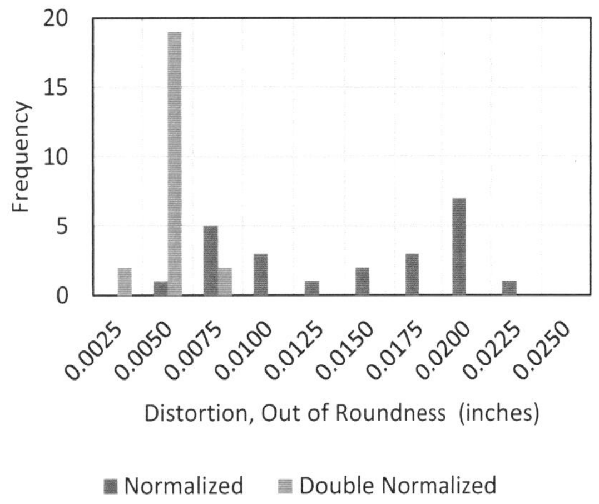

Figure 5. Effect of normalization cycle on the out of roundness of two different groups of carburized AISI 8620H gears after press quenching and normalization.

An examination of the effect of the microstructure prior to heat treatment was examined by Prinz (6). In this investigation, the microstructure of bar material was varied by normalizing, annealing or hardening and tempering. The resultant microstructures were primarily ferrite and pearlite in the normalized "as-delivered" condition, with some bainite and martensite due to the high hardenability. The annealed structure was ferrite and pearlite, while the hardened and tempered structure was tempered martensite. The extent of the distortion was evaluated by measurements before and after heat treatment of machined shafts. The effect of hardenability on the distortion of the heat treated shafts were also examined.

The results of the study showed a strong influence of the prior microstructure on the resulting distortion (Figure 6). Increasing hardenability reduced distortion. However the authors noted that it is likely that the reduced grain size from the prior heat treatments likely changed the hardenability, so the effects are compounded with prior heat treatment effects.

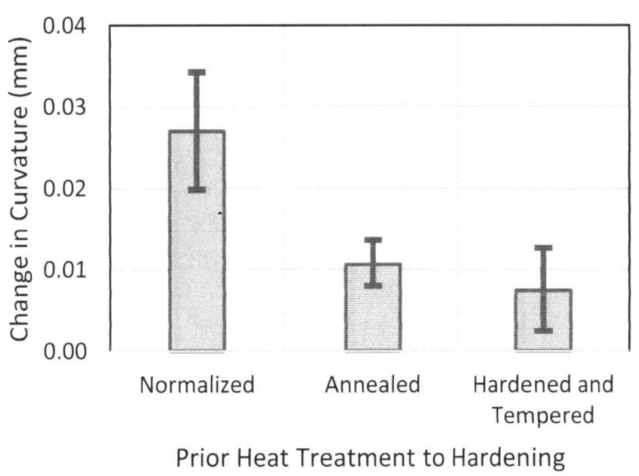

Figure 6. Variation of distortion with microstructure prior to heat treating (6).

Banding or segregation can produce a wide variation in the local hardness, even within the carburized layer (Table 1). This segregation can cause non-uniform response to heat treatment and result in uncontrolled distortion (Figure 7 & Figure 8).

These studies are all consistent, in that it shows that the prior thermal processing prior to machining and heat treatment, play a critical role in controlling the distortion of a heat treated component.

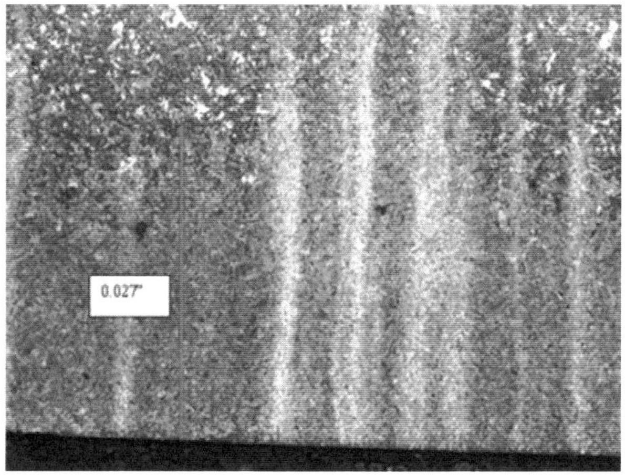

Figure 7. Segregation and banding in AISIS 4320 carburized gear showing non-uniform effective case depth (Courtesy of L. Ferguson DANTE Solutions).

Hardenability

The hardenability of a steel is influenced by its chemical composition and grainsize. The chemistry further affects the martensite start and finish temperatures, M_s. Because of variations in chemistry within an alloy specification. Even within a specific heat lot of material, the chemistry and hardenability can vary. Since higher hardenability steels are more deeply hardened than a lower hardenability steel within the same grade, it can be expected that the higher hardenability chemistry will have a higher core hardness and effective case depth. This would also indicate that the distortion would increase as hardenability increases.

Funatani (7) showed that the distortion of SCr.420 steel (a Cr case hardening grade) increased with a small variation in the Jominy End Quench hardenability. The change in over ball diameter changed from 20 μm to 80 μm as the hardness at 12.5 mm from the quenched end on the Jominy End Quench Test changed from 24 to 30 HRC (Figure 9). Hypoid ring gears of SCM430H showed that the maximum pitch angle increased as a function of the hardenability of the steel increased (Figure 10).

In one study (8), a customer was looking to increase performance of an automotive pinion gear by changing alloy from AISI 8620 to AISI 4320. the distortion of an AISI 8620 and 4320 automotive pinion gear was compared, using identical heat treating conditions. The core requirements were the same (35-40 HRC).

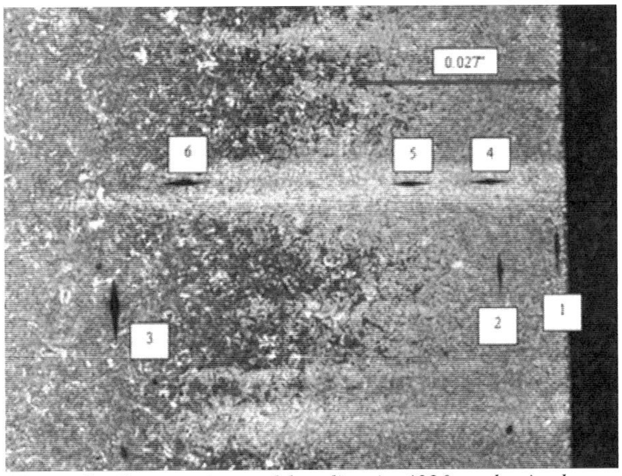

Figure 8. Segregation and banding in 4320 carburized gear showing hardness variation within case (Courtesy L. Ferguson, DANTE Solutions, Inc.). See Table 1 for hardness at locations 1-6.

Table 1. Hardness variation noted within case of carburized AISI 4320 gear (Courtesy L. Ferguson, DANTE Solutions, Inc.).

Location	Hardness (HRC)
1	56.3
2	60.3
3	33.3
4	66.3
5	60.9
6	54.9

The results of the modeling, show that the distortion of the AISI 4320 carburized pinion gear is approximately 0.15 mm – approximately one order of magnitude greater than the bow of the carburized AISI 8620 gear. This behavior was explained by the differences in the Continuous Cooling Curves of the two materials. The AISI 8620 material, with a short martensite shelf, has a significantly lower hardenability than AISI 4320. Fast cooling rates are required to achieve a fully martensitic structure. AISI 8620 also has a significant bainite bay at the expected cooling rates expected during quenching. AISI 4320 has a much higher hardenability, with a large martensite shelf. Slower cooling rates can still achieve a fully martensitic structure.

The cooling rates on opposite sides of the pinion were determined to be approximately 800°C/min and 400°C/min respectively.

For AISI 8620, the CCT diagram predicts that the microstructure would be predominately bainite with small portions of martensite and pearlite for both cooling rates. For AISI 4320, the CCT diagram predicts approximately equal portions of martensite and bainite. However, for the faster cooling rate, martensite would form earlier than for the slower cooling rate.

For 8620, during quenching, the core transforms almost completely to bainite. While some asymmetry of transformation occurs, equal amounts of bainite are formed. There is less volume expansion due to bainitic transformation, resulting in less distortion.

For AISI 4320, the core of the pinion shaft transforms to approximately equal portions of martensite and bainite. However, because of the differential cooling rates, one side of the pinion shaft transforms to martensite much earlier than the opposite side. This asymmetry of martensite formation and the accompanying volume change results in an initial bow of the pinion. Once the opposite side of the pinion transforms to martensite, a stress reversal occurs, with the final bow of 0.15 mm.

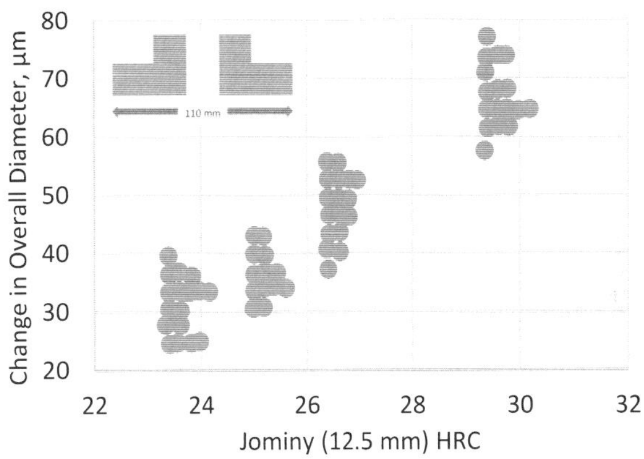

Figure 9. Influence of steel hardenability on the over ball diameter of Carburized Module 2.8 Spur gears. Gears were carburized at 920°C and quenched in 90°C oil (7).

Recommendations included increasing the agitation or modifying the racking to achieve uniform heat transfer around the circumference of the shaft.

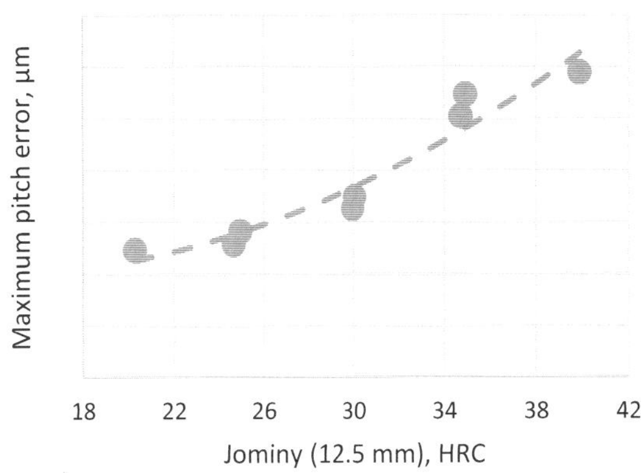

Figure 10. Influence of steel hardenability on the maximum pitch angle of hypoid ring gears (SCM420H) (7).

An investigation into the effect of variation of hardenability within heats of 16NiCrS4 carburizing steel was conducted by Stormvinter, *et al* (9). Three different heats of material (Table 2) were obtained as tubes of 71 mm diameter and a wall thickness of 17 mm. The tubes were sectioned and machined into rings of 70 mm diameter, with a wall thickness of 16 mm and a height of 19 mm. The calculated ideal diameter, DI in mm, for Heats A, B and C were 80.47 mm, 91.73 mm and 100.82 mm. All heats satisfied the requirements for 16NiCrS4.

Table 2. Heat examined by Stormvinter et al (9).

Steel heat	C	Si	Mn	P	S	Cr	Ni	Mo	Cu
Heat A	0.2	0.07	1.00	0.12	0.41	1.01	1.03	0.13	0.17
Heat B	0.2	0.07	1.03	0.16	0.41	1.04	1.05	0.17	0.17
Heat C	0.2	0.07	1.02	0.08	0.41	1.03	1.39	0.16	0.18

The rings were first machined then stress relieved at 600°C for 5 hours. Rings were austenitized at 850°C for 1 hour with no carburizing. The rings were then quenched in an accelerated quench oil at 50°C. The rings were measured for distortion before and after heat treatment. The results of the measurements at the inner and outer diameters are shown in Figure 11 and Figure 12. The hardness of the rings were identical within the error of the hardness measurement.

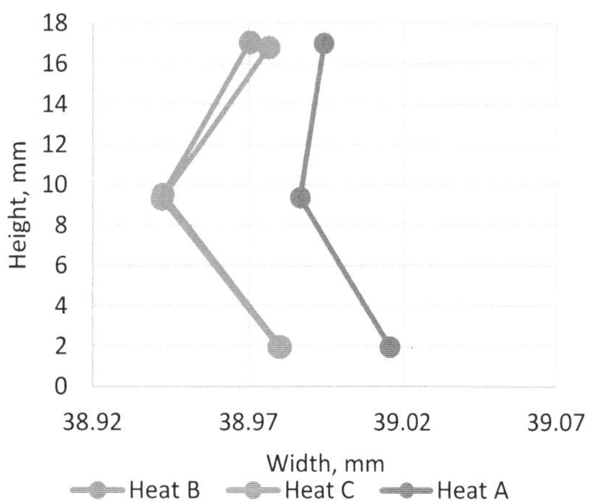

Figure 11. Radial distortion of the inner surface of the heat treated rings. The inner surface becomes convex. The inner diameter shrinks for steel heats B and C, but grows for A. The nominal inner diameter is 38.98 mm.

For ring components, a lower hardenability alloy would tend to cause the ring to become thinner and have a larger overall diameter. For higher hardenability alloys, the ring would become thicker and tend to shrink.

The effect of distortion was also studied (9) on production vehicle crown wheels. In this study the crown wheels were manufactured from forgings of 17NiCrMoS6-4 using a steel from a single steel mill. All crown wheels were machined and heat treated at the same time to minimize unknown influences. The distortion of the crown wheels was then measured using a coordinate measuring machine (Figure 13).

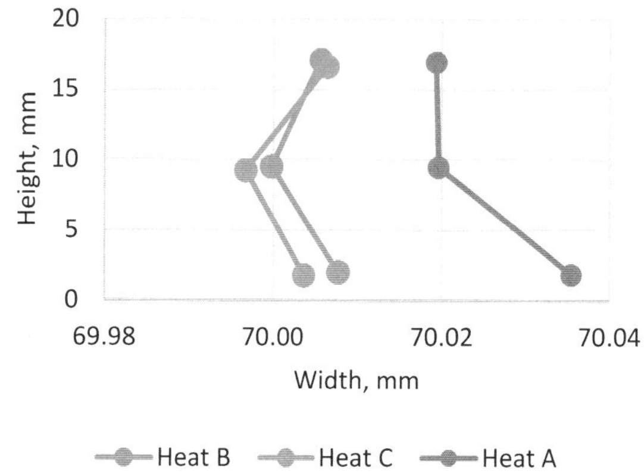

Figure 12. Radial distortion of the outer surface of the heat treated rings. The nominal outer diameter is 70.01 mm. The surfaces become concave.

Four mill heats of steel were examined, and the results show that there is a correlation between the measured hardenability of the steel, and the resultant distortion of the wheel. Different heats, with different hardenability, affected the distortion of the crown wheel.

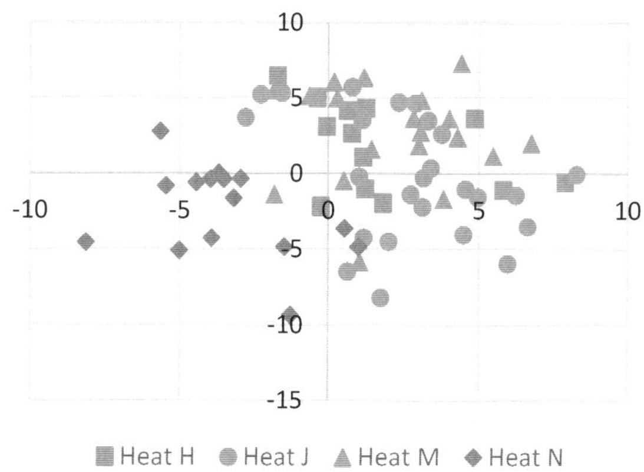

Figure 13. Score scatter plot from principal component analysis. The first PCA component and the Second PCA component are represented by the x-axis and y-axis respectively. Each dot represents the relative distortion of a crown wheel. The distortions are examined corresponding to different steel heats.

Machining

Prior operations such as machining create residual stresses at the surface. This stress is relieved during heat treatment, resulting in distortion.

During turning or milling operations, the work piece is held fixed by a chuck or jaws. The method of holding the work piece can result in residual stresses. In one study (10), 100Cr6 (SAE 52100) bearing rings were evaluated on the influence of cutting speed, depth of cut, and feed rate on the roundness. Two different types of clamping mechanisms were used: a mandrel clamp and segmented jaws.

The study showed (Figure 14) that the residual stresses in the tangential direction of the two rings had a mean residual stress for the different clamping methods were nearly identical with a mean residual stress of 600 MPa. The segmented jaws show a periodicity of three around the circumference, with the tangential residual stress varying between 500 and 700 MPa. The mandrel chuck showed a slight periodicity of 6, but varied to a much less greater degree, with the variation between approximately 590 and 620 MPa.

The segmented jaws had a segment of approximately 120°, however, the real contact was limited to a single point in the middle of each segment. Therefore the clamping forces are induced at three locations 120° apart. This results in the ring bulging at these locations. The mandrel supports the bearing uniformly around the circumference, resulting in a uniform residual stress.

Further work was conducted by Grote (11), comparing the residual stresses generated by different methods of clamping a work piece during internal turning.

The influence of cutting parameters on residual stresses were also investigated for rings that were clamped by a mandrel. Four cutting parameters were examined: cutting speed, nose radius, depth of cut and feed rate. The influence of the different parameters is shown in Figure 15. The parameters used for the DOE is shown in Table 3.

Heat Treatment

Heat treating and quenching is a complex business. The configuration of parts is endless, as is the types of furnaces available for heat-treating. Numerous variables in the quenching process alone govern the ability of a part to meet distortion requirements (Figure 3). Heat-treating is a constant balancing process. It is important to balance the ability of the material to achieve properties, while at the same time control distortion.

Multiple heat transfer regimes occur during the quenching process at the same time. These heat transfer mechanisms change as a function of the quench path. Further changes occur as the quenchant oxidizes, complicating the analysis. Agitation and racking interact, changing heat transfer at different locations on the part. There is no quantitative method to understand the interaction of fluid flow and the part, with the accompanying changes in time dependent heat transfer.

Phase transformations result in volume changes and transformational strains. Diffusional processes (pearlite + ferrite, and bainite) and non-diffusional processes (martensite) are time and temperature dependent.

Sources of stress and strain include thermal strain, creep and strains from phase transformation further complicate the business of heat treating (Figure 16).

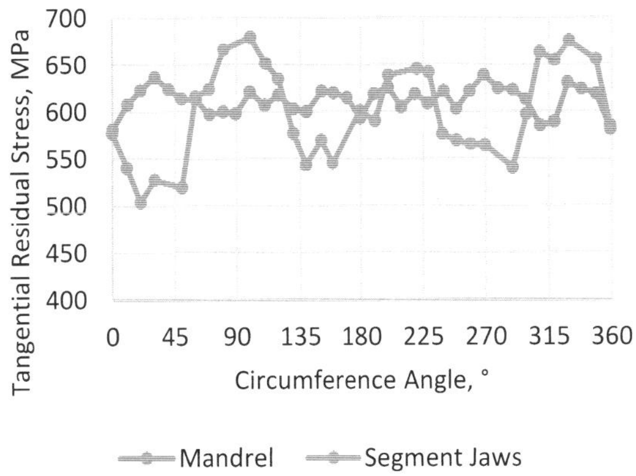

Figure 14. Residual stress distribution around a ring circumference depending on the clamping technique used.

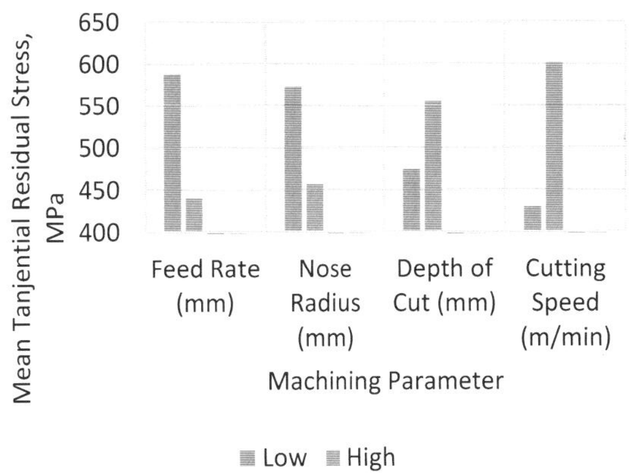

Figure 15. Results of DOE conducted by Nowag, et al, on 100Cr6, using a TiN tool and a 3% emulsion coolant.

Table 3. Levels used by Nowag et al, examining the effects of machining parameters on the external turning of 100Cr6 (SAE 52100).

	Feed Rate	Nose Radius (mm)	Depth of Cut (mm)	Cutting Speed (m/min)
Levels	0.3	0.8	0.7	200
	0.6	1.2	1.2	300

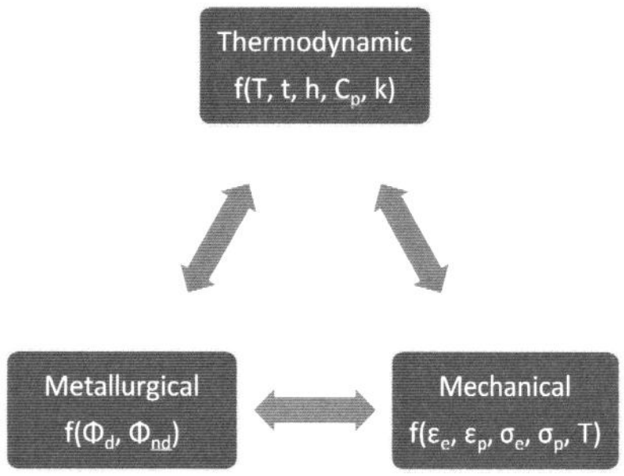

Figure 16. Interrelationships of primary variables on a heat treated part and resulting residual stress and distortion.

Thermal stresses result from thermal gradients: surface to center and surface to surface. Surface to surface thermal gradients result from differential heat transfer around the surface of the part. This could be from inadequate agitation during quenching, or from thinner sections (gear tooth) attached to a thicker or heavier body. Surface to center thermal gradients are due to the surface cooling faster than the interior, limited by the thermal diffusivity of the part. As a general rule, the faster the quench, the greater the thermal gradient.

Transformation stresses are complex in heat treating. This is a coupled stress related to phase changes that occur during heat treatment. As a part is heated, the volume expands due to thermal expansion. As the temperature increases thru the first critical temperature, the volume decreases as ferrite transforms to austenite. The volume then expands due to thermal expansion of austenite until the steady state temperature is reached.

During quenching, there is a volumetric contraction as the work piece cools. As austenite transforms, there is a volume expansion. Martensite expands the most, as a function of the carbon content. Other phases can form, such as bainite or pearlite, which have much smaller volumetric expansion. Generally outer fibers transform first, then inner fibers. This is further complicated by the effect of martensitic transformation start temperatures. Depending on the relative M_s temperatures and the thickness of a part, it is often possible that the core will transform first due to a much higher M_s temperature. As the remaining part transforms, a stress reversal occurs.

An investigation into the influence of carburizing parameter on the distortion behavior of stepped shafts was conducted by Bahnsen *et al* (12). In this study, the influence of prior microstructure, carburizing depth, surface carbon content, process sequence, carburizing temperature and the secondary hardening temperature on distortion were evaluated in a DOE. The results are shown in Figure 17 and Figure 18.

The largest dominating effect was found to be the carburizing depth. All geometrical parameters exhibited a significant sensitivity to the depth of carburizing. Other parameters showed an effect, but not necessarily statistically significant for this investigation.

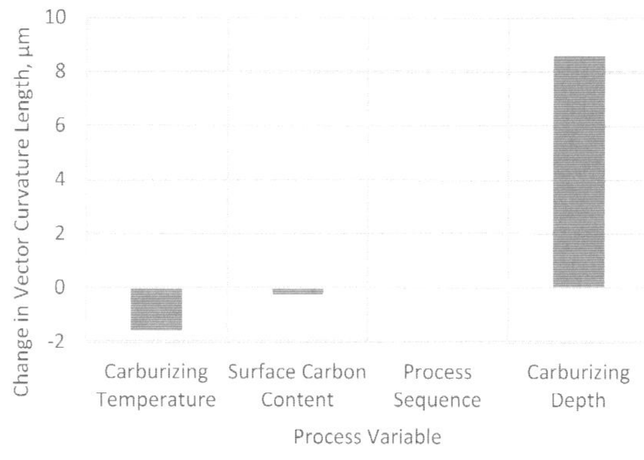

Figure 17. Effects carburizing parameters on curvature vector length of an AISI 5120 carburized shaft 14 mm in diameter.

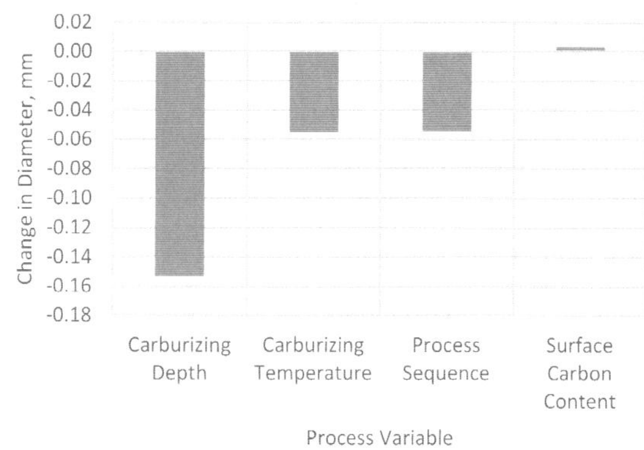

Figure 18. Effects of carburizing parameters on the change in diameter of a AISI 5120 carburized 14 mm shaft.

An investigation into the distortion behavior at elevated carburizing temperatures was conducted by Kleff *et al* (13). Heavy duty transmission components with deep case requirements require long furnaces times to satisfy case depth requirements. The use of elevated carburizing temperatures would significantly reduce the time required for carburizing, and increase production rates. Using a selection of various parts, the distortion was measured after carburizing at 930°C, 980°C and 1050°C. Alternative heating and cooling profiles were also investigated.

Two different grades of steel were used in this study: either 18CrNiMo7-6 or 20MnCrB5. Both grades were processed as a conventional heat with Al or N added. A modified grade with additional Ti (0.002%) and Nb (0.03-0.04%) for grain size control. Case hardening was performed by gas carburizing at the carburizing temperature. The parts were cooled to the hardening temperature and oil quenched. An iso-thermal step at 800°C was added to achieve more homogeneous temperature distribution in the part.

Examination of the radial run-out of the helical bevel gears, as well as the outer and inner diameters was conducted. The radial run-out was found to be small for all temperatures processed.

The inner and out diameters showed a slight increase after carburizing at 940°C and 980°C (≤ 200 μm), or stayed nearly constant when carburizing at 1050°C (Figure 19).

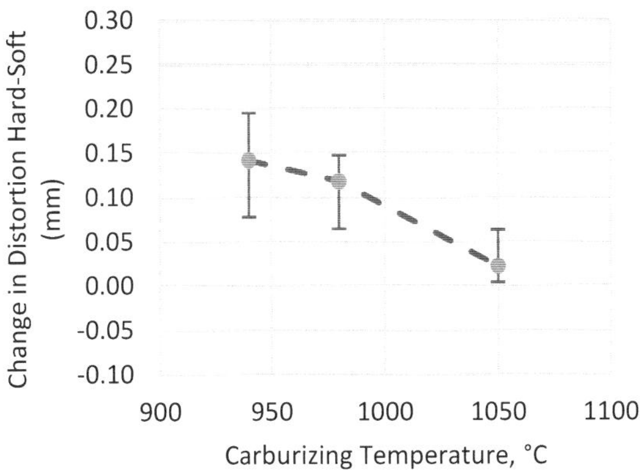

Figure 19. Results of carburizing temperature on the bore diameter of a 191.85mm diameter H6 helical bevel gear.

No negative influence on higher carburization temperatures were observed, provided that an intermediate heating step to insure homogeneous temperature distribution is included.

Since the primary focus was to investigate the use of higher temperatures to shorten production time, quenching parts directly from the carburizing temperature instead of the hardening temperature of 850°C was investigated. Helical gears were carburized at 930C, 980C and 1050C and quenched directly from the carburizing temperature. The results of bore diameter distortion quenched from the hardening temperature of 850°C is shown in Figure 20. Bore diameter distortion quenched directly from the carburizing temperature is shown in Figure 21.

The tooth sections and the bore of the gear showed significantly more scatter of distortion critical dimensions when quenched directly from the carburizing temperature. The tooth flank deviation increased significantly so that more stock would have to be removed. However, this increases the cost of grinding. The results showed that cooling down to the hardening temperature was necessary to achieve good distortion control and process capability.

Quenching

The literature of quenching and control of distortion is as immense as the problem of distortion in industry. The author will try and cover the high points.

A comparison of the distortion resulting from different quenching methods (traditional oil and high-pressure gas quenching) to the same hardness, and the racking method was conducted by Jurci (14). Helical gears and mating pinions made from low carbon structural steels were gas or low pressure carburized to an effective case of 0.7 mm. The dimensional distortion (shrinkage & growth) and the shape distortion (roundness, flatness and tooth deviations) were measured. They also measured the distortion as a function of position in the rack.

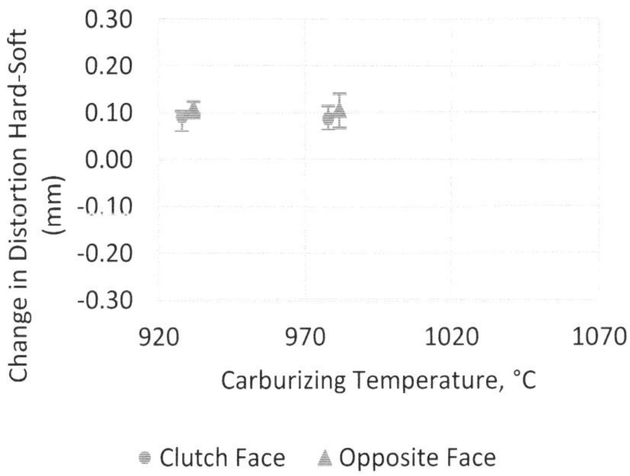

Figure 20. Influence of cooling down from the carburizing temperature to the hardening temperature of 850C on the distortion of AISI 5120 helical gears.

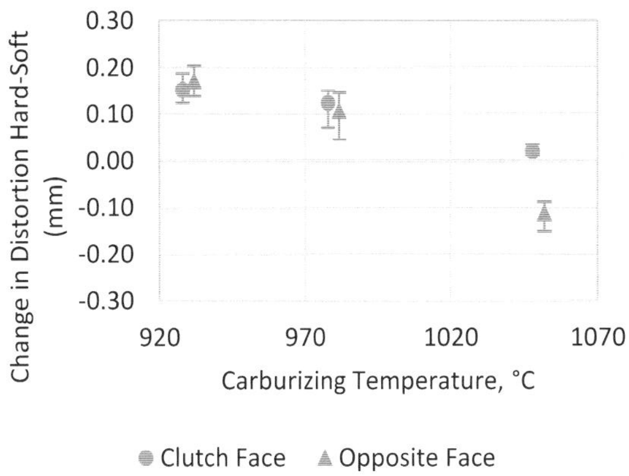

Figure 21. Influence of quenching directly from the carburizing temperature on the distortion of AISI 5120 helical gears.

The authors found significant reduction in distortion using gas-pressure quenching than the traditional oil quenching. However, they found that the use of nitrogen gas as a quenchant lead to higher out-of-flatness than oil quenching. They also observed a greater scatter in the results with gas quenching. This was attributed to inhomogeneity of gas flow with in the quenching chamber.

Braun (15) reviewed the effects of racking and condition of the oil on the resultant distortion of steel components. He found that as an oil aged, greater distortion resulted.

In another study (16), the viscosity and cooling curve was evaluated. A single viscosity base oil, and different quantities of a high viscosity (450 cSt at 40°C) were mixed. A standard speed improver quantity was added to the mixtures. Cooling curves were measured using the JIS K2242 silver probe. They showed that the speed increased from 90°C/s to 100°C as the viscosity increased to approximately 100 cSt. However, they found that the surface hardness of S45C decreased from HRC 55 to HRC 29 as the kinematic viscosity increased. They attributed the improved quenching performance at low viscosities to the composition of the base oils having a low boiling point fraction.

A comparative study comparing the residual stresses and distortion in cylindrical components of AISI 5160H spring steel quenched at various temperatures of 25% poly-alkylene glycol (PAG) and compared to an accelerated quench oil was conducted by Sarmiento (17).

The results showed that no residual stresses were measured in cylinders quenched in either PAG or quench oil for diameters of 13.5 and 20 mm.

In a sensitivity study of quenching variables (Orientation, Slenderness ratio, M_s temperature, Quench Path and Immersion Rate), MacKenzie (18) examined the effect of the primary quenching variables on the hoop and axial stresses in simple cylinders. This study found that the orientation (horizontal or vertical) and the M_s temperature were the primary main effects for increasing radial and hoop stresses.

Racking in a vertical fashion increased the hoop stress, as did decreasing the martensite start temperature. Radial stresses were decreased by racking vertically, while decreasing the martensite start temperature increased radial stresses. The immersion rate was shown to have little effect on the residual stresses as the part cooled. The quench path, specifically the stability of the vapor phase, caused increased residual stresses. A more stable vapor phase resulted in reduced residual radial and hoop stresses. This was thought to be due to decreased thermal gradients present. This result was unexpected, and was thought to be associated with the alloy selected (41XX). This alloy has a large hardenability, so decreased initial thermal gradients would show less resultant residual stresses.

Arimoto (19) reviewed the status of quenching Fe-Ni alloys, and modeling residual stresses. His work showed that transformation plastic strain is generally similar to plastic strain during heat treatment. Plastic strain is mainly produced during rapid cooling during quenching, while transformation plastic strain is due to martensitic transformation. Both the strain distributions show a tendency of the mirror symmetry especially in circumferential and axial components.

Liscic (20) indicated that there are three simultaneous process occurring during the heat treatment process. He also proposed a new quenching technology of utilizing both a volume of gas, and dispersed liquid nitrogen, to specifically control heat extraction and heat transfer to minimize distortion and residual stresses. This method would be amenable to automated control.

Narazaki (21) predicted the heat transfer coefficients and applied them to a steel helical gear. Using the Japanese silver probe, an estimate of the heat transfer coefficient was determined by the lumped method. They found that the difference in material between the silver probe and the steel (SCr420H) gear affects the minimum temperature of the vapor blanket phase. The high conductivity of the silver rod probe resulted in stabilizing the vapor phase, with the result of lowering the minimum temperature of vapor blanket.

Funatani (7) reviewed the effects of quenchant and quenchant temperature on the distortion of SCM420 bearings 80 mm in diameter by 44 mm tall. As the heat extraction speed increased, hardness increased, but elliptical and cylindrical distortion also increased (Figure 22). As the temperature of the quenchant was increased, hardness and distortion also decreased (Figure 23).

Racking and Fixturing.

The impact of racking and fixturing on 120 mm diameter 20MnCr5 (SAE 5120) gear blanks was investigated by Clausen (22). In this study, gear blanks were carburized and quenched. During carburizing, the disks supported by 2-point loading or three point loading. Parts were preheated at 850°C and carburized at 940°C to obtain a carburized depth of 0.8 mm. Quenching was accomplished high pressure nitrogen. Measurements were carried out before and after heat treatment. It was determined that the method of racking was the most critical factor affecting gear blank distortion (Figure 24 and Figure 25).

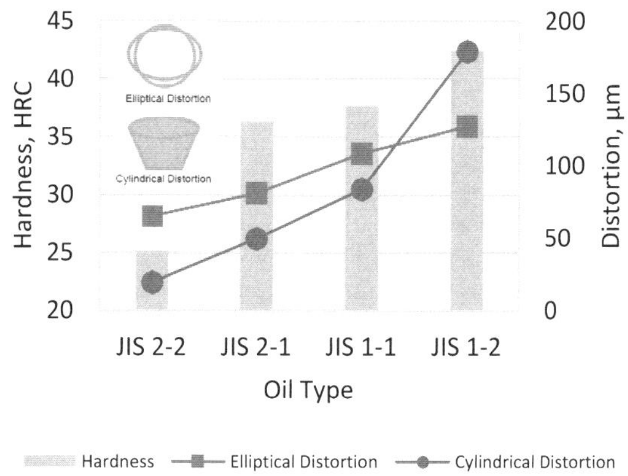

Figure 22. Comparison of quench oil selection for distortion control. SCM420 bearings (80 mm in diameter by 44 mm thick), quenched into oils of increasing speed.

Base grids used to support work are expensive, and must be replaced at routine intervals. To save money, maintenance of these grids is delayed or neglected. As these grids are used, distortion of the grid occurs. Takahashi (23) investigated the distortion of SCM420H helical gears with controlled hardenability. He found that flatness of the gears decreased as a function of time of use of the base grid (Figure 26).

In another study, MacKenzie (24) used CFD and FEA to minimize the distortion of SAE 8620 pinion gears. The gears were originally rack in a pyramid on a grid and quenched in a cold, medium speed oil. Examination of the distortion showed that the pinions would bow along the shaft up to 2 mm. An alternative method of racking and quenching in mar-tempering oil was proposed and modeled. A CFD model was created for the entire quenching system and velocity vectors established. These velocity vectors were applied as to the FEA model of each pinion gear as a boundary condition. A look-up table of heat transfer coefficient as a function of velocity and surface temperature was applied to the FEA model.

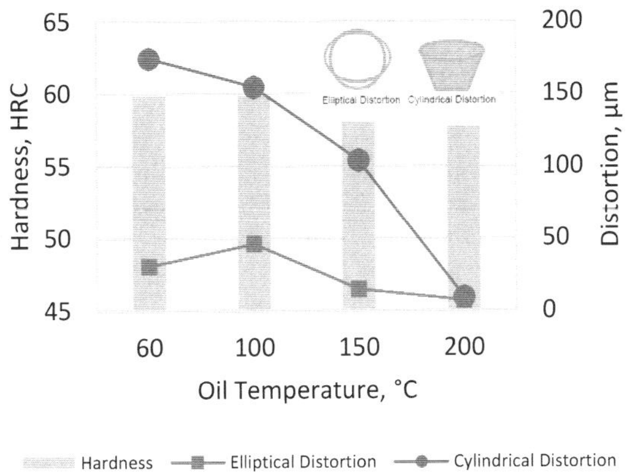

Figure 23. Comparison of quench oil temperature for distortion control. SCM420 bearings (80 mm in diameter by 44 mm thick), quenched into oils at different temperature.

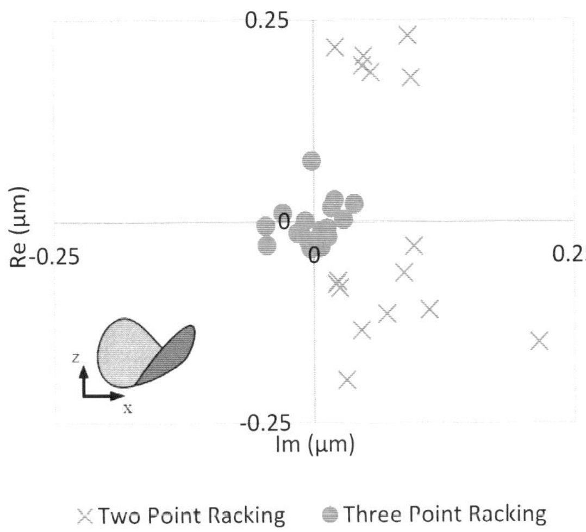

Figure 24. Change in the amplitude of the outer radius dependent on the method of loading.

The predicted bowing of the pinion was predicted to be less than 0.20 mm. Validation of the model showed that the model over-predicted the distortion, with the actual bowing of the pinion shaft averaging 0.12 mm. This reduction in distortion on this single part resulted in a documented $17,000,000 savings in rework, scrap, straightening and warranty issues.

Conclusions

The control of distortion and residual stress is a complex undertaking. Great strides have been made in the modeling and simulation of machining, heat treatment and carburizing. However the best modeling efforts can be undermined by variations in process. Dull tools, feeds and speeds and even clamping methods result in changes and variations in residual stress and distortion.

The steel making process and forming processes, with the variables of grain size, grain flow, prior microstructure and segregation can vastly effect ease of machining, as well as response to heat treatment.

While a great deal of attention has been made to control of residual stress and distortion in heat treating and quenching, these can be undermined by improper attention to the details of grid and fixture life or the changes in quenching due to oxidation of the quenchant.

It is an understanding of the entire process, and specific operations and complexities that control residual stress and distortion. These specific variables can change from machine to machine, or from vendor to vendor. Only by extreme attention to detail and attendant process variables can true distortion control be achieved.

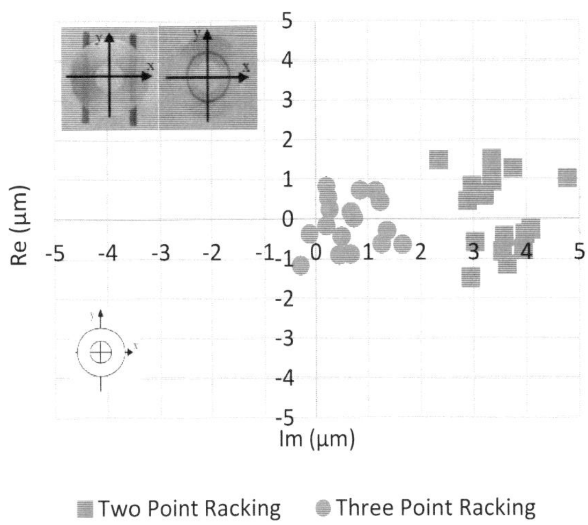

Figure 25. Change in flatness (amplitude of dish slopes) dependent on the type of racking support.

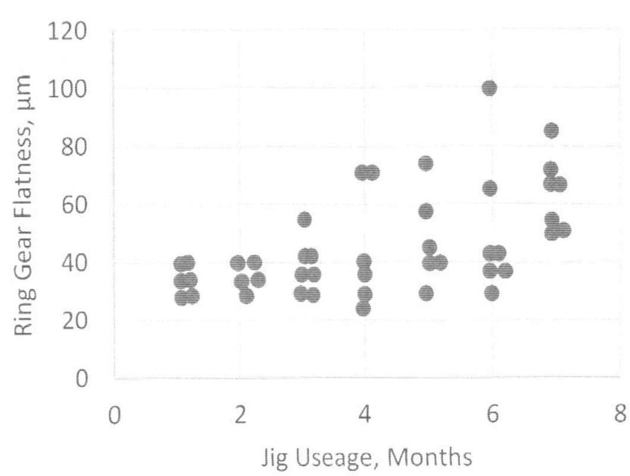

Figure 26. Effect of time-in-use of base grids on the flatness of SCM420H helical gears. Increased distortion and scatter increased as the in-use grid time increased.

Acknowledgments

I would like to thank the management of Houghton International, Inc., for allowing me to present this modest review at the 2016 ASM IFHTSE Advanced Thermal Processing World Congress.

References

1. *Distortion Engineering.* **Thoben, K.-D., et al.** 4, 2002, HTM, Vol. 57, pp. 276-282.

2. *From Single Production Step to Entire Process Chain - the Global Approach of Distortion Engineering.* **Zoch, H-W.** Bremen, Germany : s.n., 2005. IDE 2005. p. 3.

3. *Metallurgical and production-related protocols to reduce heat-treatment distortion in the manufacture of gear components.* **Hippenstiel, F.** [ed.] H-W. Zoch. Bremen, Germany : s.n., 2005. IDE 2005. pp. 57-64.

4. *Comparison of different remelted hot-work tool steels and their distortion, residual stresses and other properties. Experiments and Simulations.* **Schützenhöfer, W., et al.** Bremen, Germany : s.n., 2005. IDE 2005. pp. 65-72.

5. *Controlling Distortion in Heat Treatment Through Press Quenching.* **Reardon, A.** April 2015, Thermal Processing, pp. 24-29.

6. *Metallurgical influence on distortion of the case-hardening steel 20MnCr5.* **Prinz, C., et al.** [ed.] T. Lubben H-W. Zoch. Bremen, Germany : s.n., 2005. IDE 2005. pp. 75-82.

7. *Distortion control via optimization of cooling process and improvement of quench oils.* **Funatani, K.** Bremen, Germany : s.n., 2008. IDE 2008. pp. 265-274.

8. *Effect of Alloy on the Distortion of Oil Quenched Automotive Pinion Gears.* **MacKenzie, D. S. and Ferguson, B. L.** Shanghai, China : s.n., 2010. International Conference on Thermal Process Modeling and Computer Simulation.

9. *Effect of Hardenability and Quenching on Distortion of Steel Components.* **Stormvinter, A., et al.** Bremen, Germany : s.n., 2015. IDE 2015. pp. 39-46.

10. *Effect of Machining Parameters and Clamping Technique on the Residual Stresses and Distortion of Bearing Rings.* **Nowag, L., et al.** Bremen, Germany : s.n., 2005. IDE 2005.

11. *Distortion Engineering in Turning Processes with Standard Clamping Systems.* **Grote, C., Brinksmeier, E. and Garbrecht, M.** Bremen, Germany : s.n., 2008. IDE 2008.

12. *Influence of Carburizing on Distortion Behaviour.* **Bahnsen, C., et al.** Bremen, Germany : s.n., 2005. IDE 2005. pp. 235-242.

13. *High Temperature Carurization - Influences on Distortion Behaviour of Heavy-Duty Transmission Components.* **Kleff, J., et al.** Bremen, Germany : s.n., 2005. IDE 2005. pp. 227-234.

14. *Investigation of Distortion Behaviour of Machine Components due to Carburizing and Quenching.* **Jurci, P., et al.** Berlin, Germany : s.n., 2007. IFHTSE 5th International Quenching and Control of Distortion Conference. p. 39.

15. *Influence of the Quenching Process on the Microstructure and Distortion of Heat Treated Parts with Particular Emphasis on Quench Oils.* **Braun, R.** Bremen, Germany : s.n., 2007. IFHTSE 5th International Conference on Quenching and Control of Distortion. p. 47.

16. *The Influence of Cooling Characteristics and Viscosity on the Hardening Behaviour of Steels.* **Tomita, Y., et al.** Berlin, Germany : s.n., 2007. IFHTSE 5th International Conference on QUenching and Control of Distortion. p. 54.

17. *Modeling Residual Stresses in Spring Steel Quenching.* **Sarmiento, S., et al.** Indianapolis, IN : ASM International, 2001. Heat Treating 2001 - Including Quenching and Distortion.

18. *Effect of Quenching Variables on Distortion and Residual Stresses.* **MacKenzie, D. S. and Lambert, D.** Indianapolis, IN : ASM International, 2003. Proceedings of the 22nd Heat Treating Society Conference and the 2nd International Surface Engineering Congress.

19. *Historical Review of Resdual Stress in Quenched Fe-Ni Alloy Cylinders and Explaination of Its Origin Using Computer Simulation.* **Arimoto, K. and Funatani, K.** Detroit, MI : ASM International, 2007. Proceedings of the 24th ASM Heat Treating Society Conference.

20. *History and Perspective of Controllable Heat Extration During Quenching.* **Liscic, B.** Indianapolis, IN : ASM International, 2001. ASM 2001 Heat Treating Society Conference.

21. *Validation of Estimated Heat Transfer Coefficients during Quenching of Steel Gear.* **Narazaki, M., et al.** Berlin, Germany : s.n., 2007. IFHTSE 5th International Quenching and Control of Distortion Conference.

22. *Identification of Process Parameters affecting Distortion of Disks for Gear Manufacture - Part II: Heating, Carburizing and Quenching.* **Clausen, B., et al.** Bremen, Germany : s.n., 2008. IDE 2008. pp. 41-50.

23. ***Takahashi, A., Morishima, T. and Yamada, H.** 6, 1990, Journal of the Japanses Heat Treating Society, Vol. 30, p. 301.*

24. *Control of Residual Stress and Distortion. **MacKenzie, D. S.** s.l. : ASM International, July 2007, Heat Treating Progress, p. 47.*

International Federation of Heat Treatment and Surface Engineering 2016
Proceedings of the 23rd IFHTSE Congress
April 18–21, 2016, Savannah, Georgia, USA

Surface Structuring as a Method to Influence on Rewetting during Liquid Quenching

Nikolay Kozlov[1], Olaf Kessler
Chair of Materials Science, University of Rostock, Rostock, Germany
[1] nikolay.kozlov@uni-rostock.de

Abstract

Quenching is an important process in the heat treatment of metallic components. This is usually performed as immersion quenching in vaporizing liquids, such as water or oil. Quenching in water or oil is characterized by relatively small effort and high average quenching rate. Disadvantage of this method is the occurrence of the Leidenfrost effect. The Leidenfrost effect manifests itself in the fact that a thin vapor film is formed on the hot component surface, which greatly reduces the heat flow between the component and the quenching medium. Depending on the component geometry and quenching conditions the vapor film collapses unevenly at different time and location. This causes component distortion, residual stresses and possibly unevenness of metal structure. Many studies on the Leidenfrost effect are dealing with influences of quenching medium and quenching conditions. In this work, the influence of component surface structure on the quenching kinetic has been investigated as a novel approach in the surface engineering. The focus of the work is on the investigation of surface structures, which can be produced during the turning process of cylindrical components without much additional effort. The investigation was made with water or oil as a quenching medium by means of steel components with defined groove surface structures. Geometry of the surface structures was varied. The evaluation was based on time-temperature measurements, and rewetting front video observations. It was found that the purposefully structured surface can affect quenching kinetic considerably. The possibility to control and homogenize the quenching process by surface structuring adapted to component geometry was confirmed. This indicates a high potential for homogenizing of immersion quenching in vaporizing liquids and therefore reduction of residual stresses and distortion in the quenched components. The integration of the surface structuring in a manufacturing process of components is relatively easy, because in most cases it only requires an adaptation of turning parameters.

Introduction

In many heat treatment processes of metallic materials, for example martensitic hardening of steels, a rapid quenching has great importance. The aims of quenching are supersaturated solid solutions, which show high solid solution hardening or represent initial states for subsequent precipitation hardening. Quenching should be done as quickly as necessary to set the desired structures and properties. On the other hand it should be done as slowly and evenly as possible in order to reduce component distortion and residual stresses. Quenching can be performed in very different media. Due to low cost and high average quenching rate an immersion quenching in evaporating liquids like water or oil is very often used. Disadvantage of evaporating liquids as a quenching medium is the unevenness due to the Leidenfrost effect [1]. The Leidenfrost effect leads to the formation of a vapor film between hot component surface and quenching medium. The vapor film has an insulating effect and reduces the heat transfer. The collapse of the vapor film often occurs usually at different time and location. This is illustrated in Figure 1 on the example of a metal cylinder quenched in water. In the first moment (t0) after immersion of the hot cylinder in water, the entire surface is covered by vapor film (film boiling). Over time, the cylinder cools down and the vapor film collapses at the lower edge of cylinder (t1) causing very high heat transfer (nucleate boiling). The formed rewetting front moves from bottom to the top of the cylinder. While the upper portions of the samples cool down relative slowly (film boiling), the lower portion of the sample cool down very quickly (nucleate boiling), see t1 and t2. At the end of quenching (t3) the entire cylinder is completely under convection. This time and location difference in the cooling rates along the cylinder can cause unevenness in microstructure as well as residual stresses and component distortion

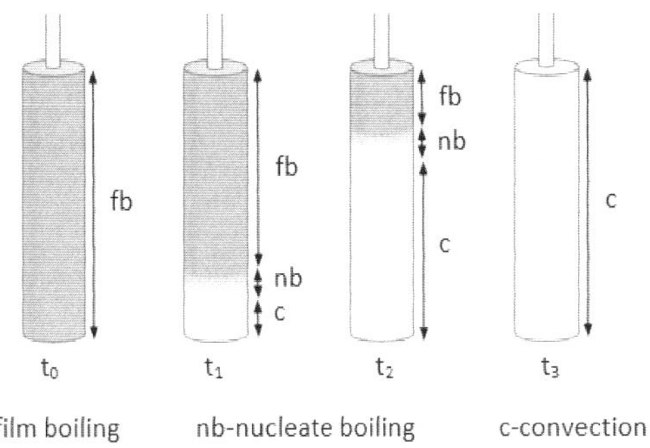

Figure 1: Schematic illustration of immersion quenching in water.

The Leidenfrost effect in evaporating liquids was very often studied and various approaches for reducing its influence exist [2-11]. All these approaches concentrate on the quenching medium and quenching condition but not on the surface structure. The influence of surface structure and sample geometry depends on many parameters and therefore exhibits a complex behavior. Different roughness, material properties, quenching conditions and sample geometries lead to different effects [12-17]. The increasing of roughness can reduce as well as encourage the Leidenfrost effect.

In this work, the influence of the sample surface in combination with sharp and rounded sample edges on the Leidenfrost effect will be investigated. The focus of the investigation is on the influence of grooved surface structures, which can be manufactured during the turning process of cylindrical components without much additional effort. The idea is to destabilize the vapor film by means of the irregularities (groove tips) on the surface. Theoretically, sufficiently high groove tips should pierce the vapor film (see Figure 2) and shorten the film boiling phase.

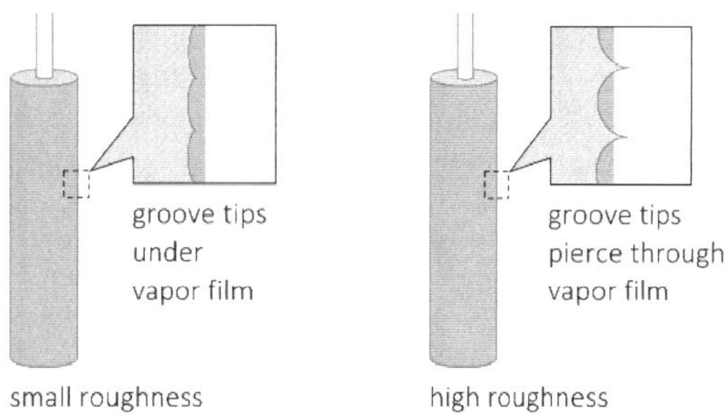

Figure 2: Possible Influence of groove depth on vapor film.

The experiments with different groove depths (roughness) are applied on steel samples in order to derive the relationships between surface structures and quenching kinetics. Based on the obtained results, an attempt will be undertaken to homogenize the quenching process using grooved surface structures. For this purpose, samples with inhomogeneous surface structures will be used in order to control vapor film collapse by means of targeted surface structuring. In case of successfully targeted modulation of the Leidenfrost effect, the economics of manufacturing could be substantially increased. In the ideal case the post-processing of distorted components after quenching could be shortened or completely eliminated.

Materials and methods

Experimental setup

The experimental setup used in this work is illustrated in Figure 3. For details of the experimental setup see [20]. The cylindrical sample is heated in a furnace to a determined temperature and after that quenched in the liquid tank. The tank is a thermostat LAUDA PV 36 with an observation window and has a volume of 36 liters. As quenching medium, tap water and quenching oil ISOMAX 166 were used. Tap water had been purified with activated charcoal filter. To immerse the sample into quenching liquid, a semiautomatic kneeling system was used based on a guide rail and a brake unit. This allowed uniform and reproducible immersion of all samples in the quenching tank and the influence of uneven immersion could be minimized. After immersion, the distance from liquid surface to the upper end of sample is about 40 mm. The distance from the tank bottom to lower end of sample is about 110 mm.

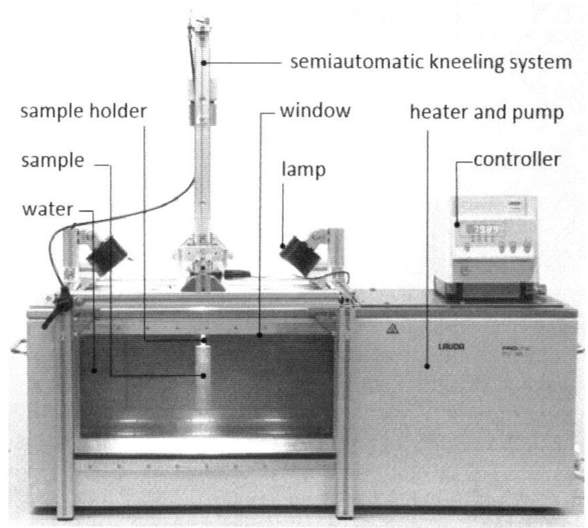

Figure 3: Experimental setup with tap water [20].

As samples, cylinders with diameters of 29 mm and lengths of 120 mm were used. Cylinder materials of austenitic stainless steel X5CrNi18-10 were investigated. This steel is not martensitic hardening, but it is a suitable model for martensitic hardening steels, because cooling curves can be recorded without superposition of transformation heat. Samples with sharp and rounded edges were used (Figure 4). Curvature radii at the rounded samples are 14.5 mm on the bottom and 4 mm on the top of cylinders. The upper and lower halves of samples with uneven surface were structured with two different groove depths (roughness) R_{t1} and R_{t2}.

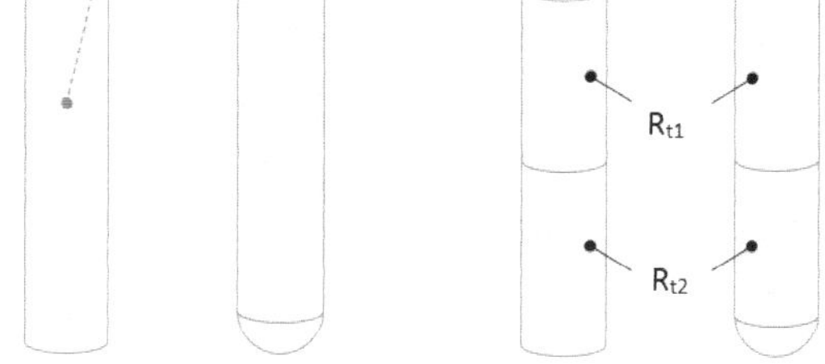

Figure 4: Types of used steel samples.

For fastening on the sample holder, all samples have threaded bores M5×15. Samples with even surface and sharp edges were equipped with thermocouples (type K, diameter 1 mm), which were introduced into the core of sample through an inclined drill bore 1.2×36 mm (see Figure 4, red dashed line). Thermocouple tip lies on the cylinder axis about 30 mm from the upper sample end. The required surface structures (grooves) were manufactured by turning with defined parameters. The geometry of grooves and roughness can be controlled by changing the turning tool geometry (corner radius, corner angle) and varying the turning parameters (feed, cutting depth). The relationships between these parameters are illustrated in the Figure 5

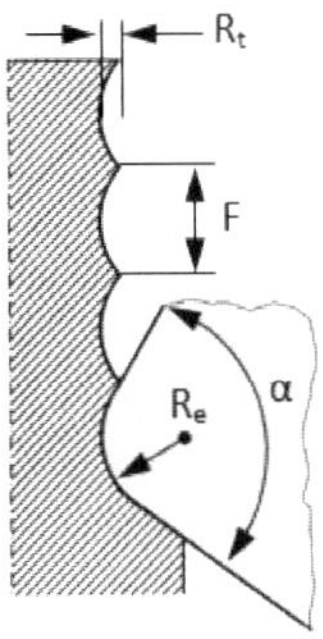

R_t groove depth (roughness)
F feed
R_e corner radius
α corner angle

If $F \leq 2 \cdot R_e \cdot \sin\left(90 - \frac{\alpha}{2}\right)$

then $R_t = R_e - \sqrt{R_e^2 - \frac{F^2}{4}}$

else $R_t = \frac{0.5 \cdot F}{tg(\frac{\alpha}{2})} - \frac{R_e}{\sin(\frac{\alpha}{2})} + R_e$

Figure 5: Creating of groove structure by turning, based on [18].

According to [1] the thickness of vapor film during steel quenching in water (20 °C) amounts from about 100 µm to about 300 µm. The thickness of the vapor film depends on water temperature and component surface temperature amongst others. Based on the assumption that high groove tips pierce through the vapor film, the grooves should be higher than the thickness of vapor film. However, previous investigations [1, 13, 19] show that in the area of small roughness also a significant change of quenching kinetic can happen. Different groove depths were elected for investigation in this work: 1, 10, 30, 80, 150, 300, 500 µm (nominal). The tool corner radii for samples with groove depths of 1, 300 and 500 µm were 0.8 mm. For other groove depths, tools with corner radius 0.2 mm were used. The optimal process parameters were determined experimentally and adapted for investigated materials. The quality of the turning process is verified on micrographs and confocal microscope measurements. Selected micrographs are shown in Figure 6.

The desired groove structures could be successfully manufactured by turning. Only the samples with groove depths of 500 µm (nominal) show modified grooves. Here the tips are not properly formed. This comes about the relatively high strength and ductility of the used stainless steel. In the turning process the material flows aside and the residual material stays on surface. As a result, true groove depth in these samples is higher than nominal 500 µm, which is an exception. For all other samples, true groove depth is slightly smaller than nominal. The comparison of nominal and measured (confocal microscope) groove depths is shown in Table 1. The actual geometry of grooves is further influenced by other factors such as cutting speed, material properties etc. Following, the real groove depths will be considered.

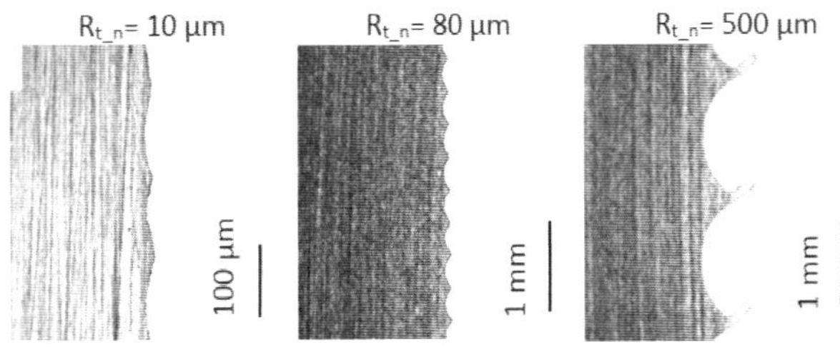

Figure 6: Selected micrographs of groove structures with different depth (nominal).

Since this work is dedicated to the investigation of surface structure, it is important to minimize influence of other factors. Especially the cleanliness of surface plays an important role. Samples were cleaned 10 minutes in ultrasonic bath with 1% cleaning concentrate TICKOPUR R33 and then rinsed with ethanol. Each sample was quenched only once. Although this causes high effort, it is prerequisite for studying the influence of targeted surface structuring. Otherwise, the influence of surface change with repeated use could superimpose the influence of grooved surface structure.

Table 1: Nominal R_{t_n} and true R_{t_t} values of the groove depths in µm.

R_{t_n}	1	10	30	80	150	300	500
R_{t_t}	1	10	27	76	127	270	630

Experimental procedure

Heating of the samples was carried out in an air circulation furnace at 850 °C whereby a uniform temperature distribution within the furnace was guaranteed. The sample was placed vertically in the furnace. Heating of samples was performed until the temperature of 835 °C (± 5 K) in core of the sample has been reached. This was controlled based on the sample thermocouple measurement. As not all samples were equipped with thermocouples (see Figure 4), the heating process was partially controlled by time. The heating time for samples with different surface structures has been determined in advance by means of thermocouple measurements on test samples. After heating, the sample was taken out from the furnace and immersed into the quenching medium by means of semi-automatic kneeling system. The time until immersion was kept less than 10 seconds resulting in a core temperature decrease of maximum 10 K. Usually three samples with identical parameters were quenched. As quenching medium tap water at 30, 60 and 80°C (± 0.1 K) purified with activated charcoal filter and quenching oil ISOMAX 166 at 60°C (± 0.1 K) were used. Since the temperature of quenching medium, especially at temperatures near boiling point, has strong influence on quenching process [1], a forced bath convection with an integrated pump were used during bath heating. Thus, a uniform temperature distribution throughout the quenching tank was guaranteed. Quenching has been performed either with bath agitation (≈20 l/min and ≈32 l/min) or without. In experiments without bath agitation the pump was deactivated about 40 seconds before sample immersion. The increasing of quenching medium temperature after sample immersion amounts maximum 2 K. During experiments the water in the quenching tank was replaced after 30÷40 quenches. A change of the quenching kinetic depending on numbers of quenches was not observed. For documentation and analysis of experiments the whole quenching process was video recorded with camera Canon EOS 60D (60 frames per second) through the window of quenching tank. By means of video recordings the heights of rewetting fronts were measured depending on the duration of quenching process. To improve the contrast and the recording quality, blue background was applied to the back of the quenching tank and special lighting were used. For even samples with sharp edges temperature curves in sample core (Figure 4) were recorded using a temperature measurement system National Instruments NI-cDAQ9174. The temperature was measured with a frequency of 50 Hz.

Results and discussion

Quenching of cylinders with sharp edges

The quenching process of samples with sharp edges in water at 80°C without bath agitation is shown in Figure 7, using the example of the sample with groove depth of 27 µm. In the first moment after immersion, the entire sample is covered by a vapor film. The sample cools down slowly but evenly until the vapor film collapses. The times before the collapse begins are different depending on grove depth. In the example represented in Figure 7 it is about 15 seconds until the vapor film collapses at the lower edge and begins to move upwards. Sometimes later, a second rewetting front forms at the top edge of the sample. This second rewetting front moves downwards and can be seen in Figure 7, at 60 seconds. In the end of quenching process these two fronts meet.

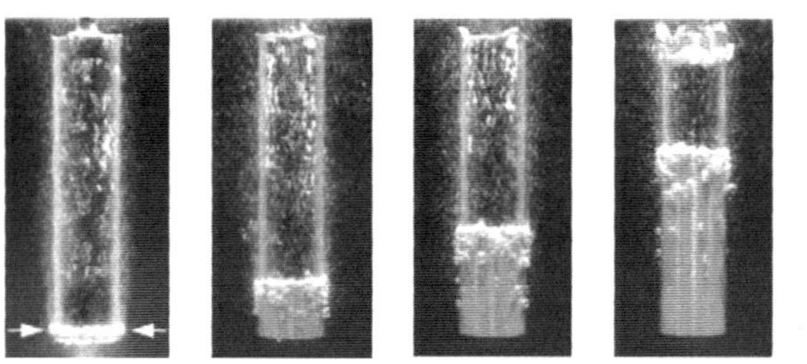

Figure 7: Quenching process of steel cylinders with groove depth of 27 µm at different times after immersion in water (without bath agitation) at 80°C; arrows mark rewetting begin.

The quenching process of samples in water at 60°C and oil at 60°C is faster but in general similar to the process shown in Figure 7. In case of quenching in water at 30°C the quenching process is much faster and more uneven. As shown in Figure 8 the vapor film collapses almost immediately after immersion simultaneously at different locations.

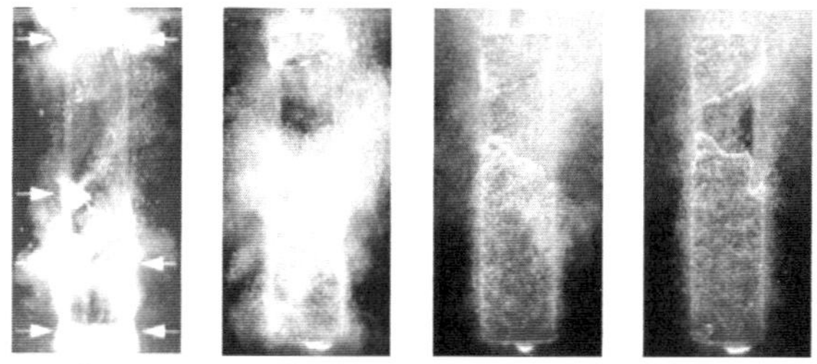

Figure 8: Quenching process of steel cylinders with groove depth of 27 µm at different times after immersion in water (without bath agitation) at 30°C; arrows mark rewetting begin.

The behavior of rewetting can be described with height of rewetting front [1]. Distances from the bottom of the sample to the rewetting front are measured at various times and plotted in a graph (see [20]). Although the height of rewetting front is an appropriate way for visualization of experimental results it can not be used for all experiments carried out in this work. The measurement of height of rewetting front in experiments with water temperature of 30°C is not possible as the rewetting front forms simultaneously at different locations (Figure 8). Besides height of rewetting front, the influence of surface structure on quenching process can be determined based on temperature measurements. The temperature curves of steel samples with sharp edge and different groove depths during quenching in water at 80 °C are shown in Figure 9.

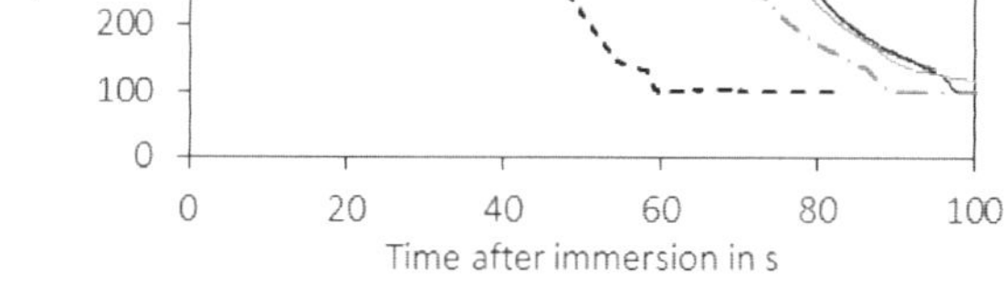

Figure 9: Curves of core temperature during quenching of steel samples with sharp edges in water (without bath agitation) at 80°C.

Because the measurements are performed in the sample core and only in one point (see Figure 4, left) these cooling curves are not appropriate for the whole sample. For a comparative analysis of experimental results, a further evaluation based on rewetting videos was used. The quenching processes were divided in three stages: pure film boiling, rewetting and pure convection. The rewetting stage is defined as the time between first vapor film collapse until the entire sample is under convection. Figure 10 illustrates the dependence of quenching stages on groove depth and bath agitation during quenching of samples with sharp edges in water at 30°C, 60°C, 80°C and in oil at 60°C.

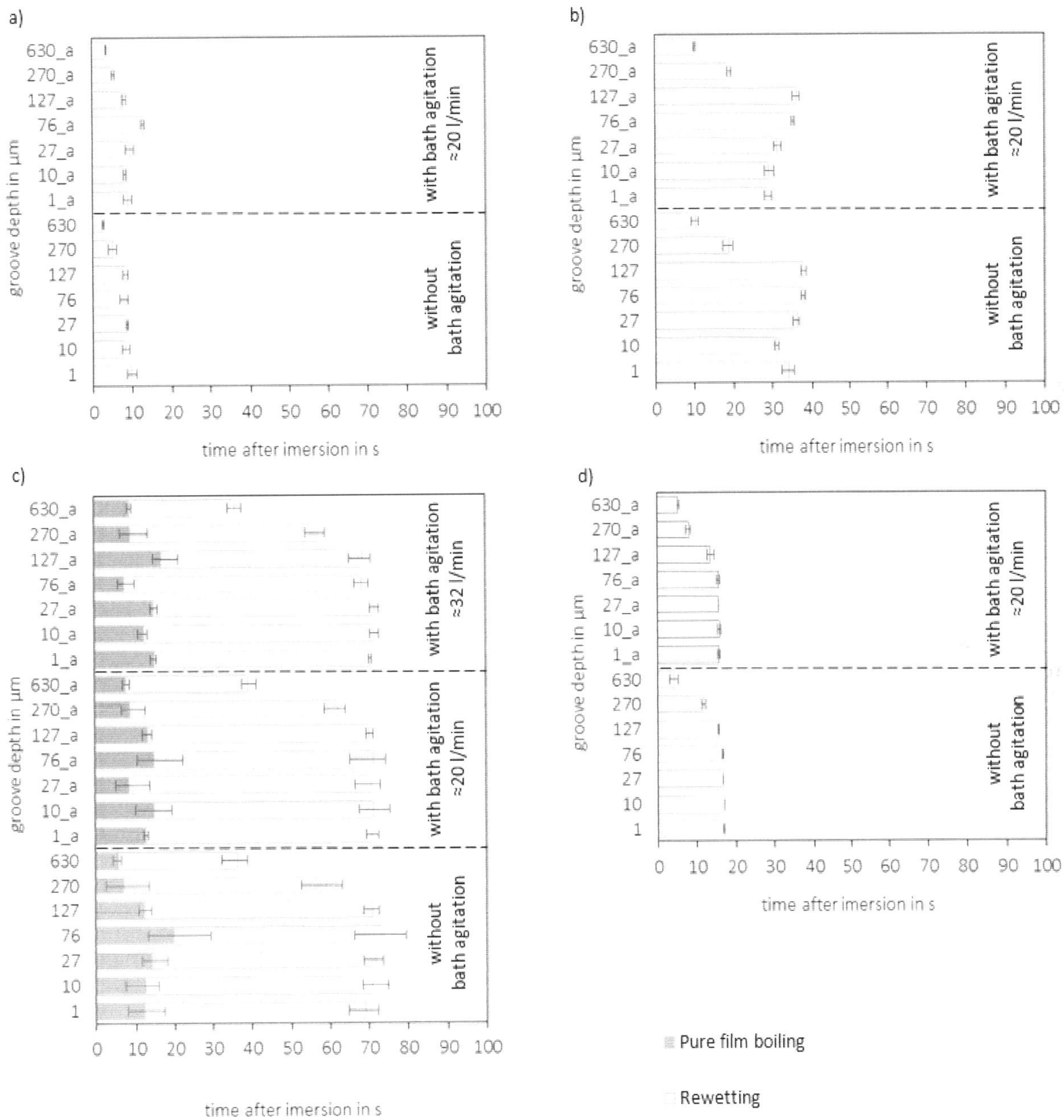

Figure 10: Dependence of quenching stages on groove depth and bath agitation during quenching of samples with sharp edges in water at 30°C (a), 60°C (b), 80°C (c) and in oil at 60°C (d).

In these experiments the influence of bath agitation on quenching process is relative small. A comparative summary of the experimental results presented in Figure 10 shows that the increasing of water temperature leads to a longer rewetting duration. During quenching in water at 80°C the pure film boiling stage occurs (Figure 10, c). In all experiments in water and in oil with groove depth up to 127 µm the differences in the duration of quenching stages are relative small. With a further increasing of groove depth a significant decrease of the quenching duration is possible. Particularly by quenching in water at 80°C but also at 60° C and 30° a decrease of the pure film boiling duration and quenching duration can be achieved. These trends match the results arising from the temperature measurements during quenching in water at 80°C (compare Figure 9 and Figure 10, c). The presented experimental results indicate that surface structuring has a substantial influence on rewetting and can be successfully used for targeted influence on the quenching process.

Quenching of cylinders with rounded edges

The macroscopic geometry of the sample has a great influence on the quenching kinetic and especially on start of rewetting [1, 21, 22]. Sharp edges of the cylindrical sample cool down faster than other areas of the sample and they cause earlier local collapse of vapor film. In order to separate the combined influence of edges and surface structure the experiments with samples with rounded edges (see Figure 4) were carried out in water without bath agitation at 60°C and 80°C. The quenching process in water without bath agitation at 80°C is shown in Figure 11 on the example of sample with groove depth of 27 µm.

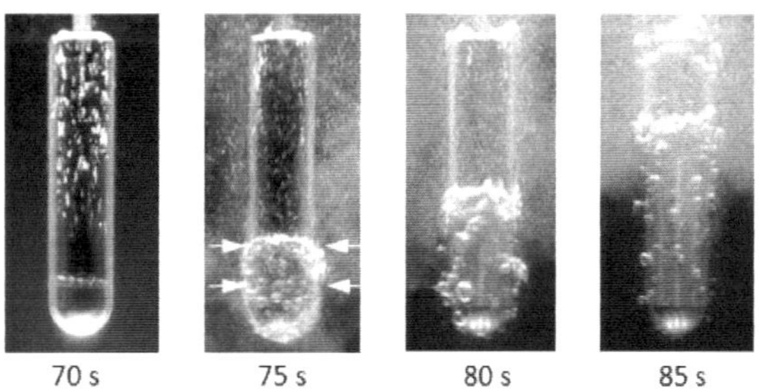

Figure 11: Quenching process of rounded steel cylinders with groove depth of 27 µm at different times after immersion in water (without bath agitation) at 80°C; arrows mark rewetting begin.

In comparison to quenching of samples with sharp edges (see Figure 7) for rounded edges the whole surface of the sample stays much longer covered by the insolating vapor film. After local collapse of the vapor film the rewetting of the rounded sample occurs faster as by samples with sharp edges. The duration of the whole quenching process by experiments with rounded samples is longer (Figure 12).

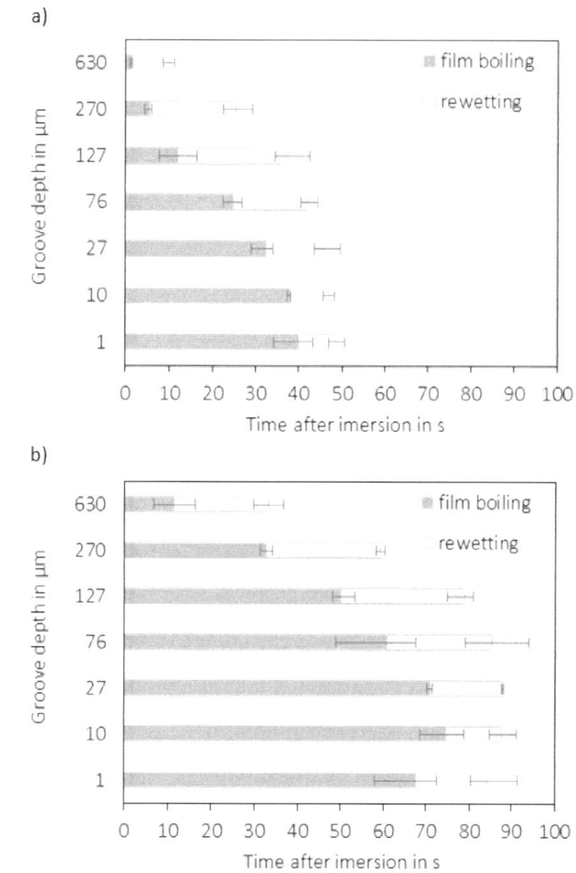

Figure 12: Dependence of quenching stages on groove depth during quenching of samples with rounded edges in water without bath agitation at 60°C (a) and 80°C (b).

The increasing of water temperature leads to a longer rewetting duration. The influence of groove depth on quenching kinetic and especially on rewetting begin by rounded samples is well pronounced. A significant reduction of pure film boiling stage is possible by increasing of groove depth. This points again to high potential of vapor film destabilization by locale surface structuring using samples with possible no sharp edges but rounded edges.

Quenching of inhomogeneous structured cylinders

Based on the experimental results an attempt will be taken to homogenize quenching process by surface structuring. In order to control the collapse of vapor film by means of grooved structures samples with inhomogeneous surface structuring were used. Two different structures with groove depths R_{t1} and R_{t2} (Figure 4) were turned on each sample. The quenching of sharp samples with groove depths 630 µm (upper half of cylinder) and 1 µm (lower half of cylinder) in water without bath agitation at 80°C is shown in Figure 13.

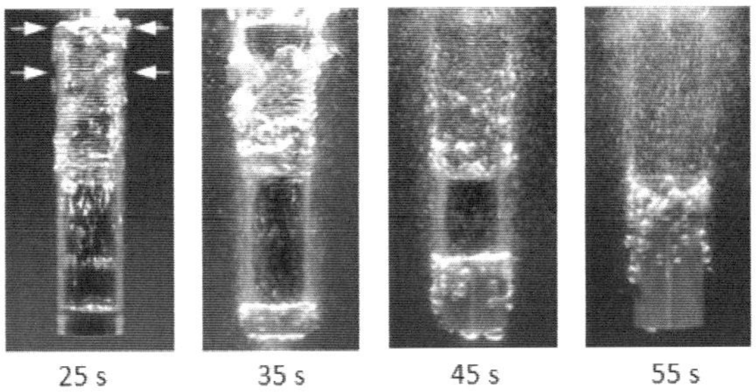

Figure 13: Quenching process of uneven steel samples (sharp edges) with groove depth of 630 µm (upper half) and 1 µm (lower half) at different times after immersion in water by 80°C (without bath agitation). Arrows mark rewetting begin.

The quenching process by such structured samples differs completely from the quenching of even samples. The rewetting now starts not at the lower half (comp. Figure 7) but at the upper half of cylindrical sample were the groove depth is higher. The collapse of vapor film occurs first in areas near to upper end of the sample. Afterwards the rewetting front moves top-down. At that time where the rewetting of lower half of the sample starts, the upper half of the sample is already rewetted for the most part (Figure 13, at 35 s). The quenching of such structured uneven samples (top half 630 µm + bottom half 1 µm) with sharp edges in water at other temperatures of 30°C and 60°C leads to similar results. Similar results were also obtained during quenching in oil without agitation at 60°C (Figure 14).

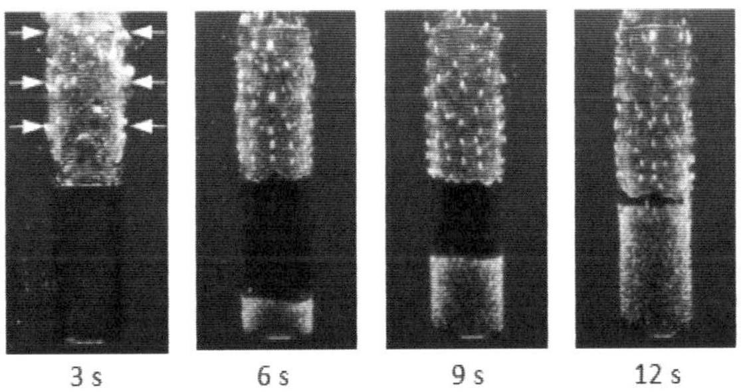

Figure 14: Quenching process of uneven steel samples (sharp edges) with groove depth of 630 µm (upper half) and 1 µm (lower half) at different times after immersion in oil by 60°C (without bath agitation). Arrows mark rewetting begin.

Furthermore rounded samples with uneven structured surfaces were investigated. The quenching process of rounded samples with uneven surface structure (top half 127 µm + bottom half 10 µm) in water without agitation at 80°C is shown in Figure 15. The rewetting starts at the upper half of the samples and then the rewetting front moves top-down. When the rewetting of lower half of the sample starts, the upper half of the sample is already completely rewetted (Figure 15, at 75 s).

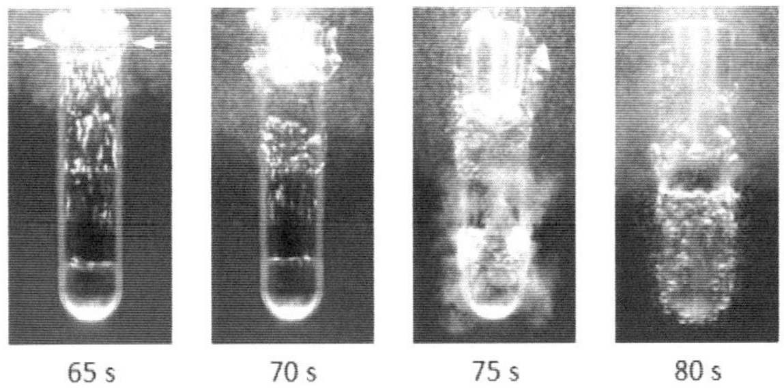

Figure 15: Quenching process of uneven steel samples (rounded edges) with groove depth of 127 µm (upper half) and 10 µm (lower half) at different times after immersion in water by 80°C (without bath agitation). Arrows mark rewetting begin.

With this targeted inhomogeneous surface structuring of the upper and lower half of the sample a strong influence on the rewetting process during quenching in water and oil was reached. For sharp edged cylinders relative large surface structure depths of several hundred micrometer seem to be necessary to cause the desired effect. This is due to the overlapping strong influence of the sharp edge on cooling. But for cylinders with rounded edges, smaller surface structure depths of about one hundred micrometer seem to be sufficient.

Summary

Quenching as part of the heat treatment is a very important step in the manufacturing chain of metallic components. Mostly quenching is done with immersion in evaporating liquids due to low cost and high average quenching speed. The problem of using evaporating liquids, is the occurrence of Leidenfrost effect, which means formation of insulating vapor film surrounding a hot component. By the vapor film, the heat transfer from a component to quenching medium is greatly reduced. After the vapor film is collapsed, a nucleate boiling starts and the cooling rate increases sharply. In general, the collapse of vapor film does not occur at the same time on the entire surface of the component. This results in inhomogeneous residual stresses as well as in component distortion. In this paper, a new approach for a targeted control of quenching process in water and oil by setting defined groove surface structures was investigated. Groove structures were successfully produced by turning with defined parameters on steel cylinders. The influence of bath temperature and agitation together with groove depth on the quenching process was investigated by means of cylindrical samples with sharp and rounded edges. Temperature measurements and video recordings were used for evaluation. The experiments showed that the influence of bath agitation on quenching process is relative small. The theoretical approach, that large groove tips pierce through vapor film and thereby destabilize film boiling could be confirmed. Steel samples with an uneven surface consisting of two areas with different groove depths were quenched. The experiments delivered, that the quenching process could be changed intentionally in selected areas of the samples. This proves that control of the quenching process by surface structuring is possible. To sum up, the surface structure adjustment by chipping (e.g. turning) processes seem to have a high potential to homogenize immersion quenching and therefore reduce residual stresses and distortion in the quenched components. The integration of the surface structuring in a manufacturing process of components is relatively easy, because in most cases it only requires an adaptation of chipping parameters.

Acknowledgements

The authors gratefully acknowledge funding of this work by the German Research Foundation, project KE616/21-1

References

[1] Liscic, B. *et al.*, Quenching Theory and Technology, 2. Ed., CRC Press, (New York, 2010).

[2] Aronov, M. *et al.*, *Proc. Heat Treating Conference*, Indianapolis, Indiana, 2001

[3] Rath J. *et al.*, *HTM J. Heat Treatm. Mat.*, Vol. 64, No. 6 (2009), pp. 338-350

[4] Bach, Fr.-W. *et al.*, *HTM Z. Werkst. Wärmebeh. Fertigung,* Vol. 61, No. 3 (2006), pp. 142-147

[5] Bernard, M. *et al.*, *HTM Z. Werkst. Wärmebeh. Fertigung,* Vol. 60, No. 3 (2005), pp. 150-157

[6] Krause, C. *et al.*, *HTM Z. Werkst. Wärmebeh. Fertigung,* Vol. 63, No. 1 (2008), pp. 22-26

[7] Kohlmann, R., *HTM Härterei-Techn. Mitt.*, Vol. 50, No. 2 (1995), pp. 126

[8] Beck, M. *et al.*, "The Ideal Quenching Medium – Characterisation of Ionic Liquids for Heat Treatment of Metallic Components" *HTM Z. Werkst. Wärmebeh. Fertigung,* Vol. 68, No. 5 (2013), pp. 214-223

[9] Redman, R. *et al.*, "Ultrasonic assisted water quenching of aluminium and steel cylinders" *International Heat Treatment and Surface Engineering,* Vol. 6, No. 3 (2012), pp. 115-121

[10] Redman, R. *et al.*, „Charakterisierung der Abschreckwirkung einer ultraschallunterstützten Wasserabschreckung von Aluminiumzylindern" *HTM Z. Werkst. Wärmebeh. Fertigung,* Vol. 66, No. 5 (2011), pp. 281-289

[11] Lauralice C. *et al.*, "Eliminate Quench Cracking with Uniform Agitation" *Heat treating progress,* No. 7-8 (2004), pp. 27-30

[12] Maaß R., „Untersuchung zur Wärmeübertragung beim Abschreckhärten" *Dissertation,* (TU Clausthal, 1988)

[13] Bernardin, J.D. *et al.*, *Journal of heat transfer-transactions of the asme,* Vol. 124, (2002), pp. 864-874

[14] Baumeister, K. J. *et al.*, *American society of mechanical engineers, winter annual meeting,* New York, NY, November 1970, pp. 91-101

[15] Chandra, S. *et al.*, *Journal of heat transfer- transactions of the asme,* Vol. 116, No. 4 (1994), pp. 999-1006

[16] Nagai, N. *et al.*, *Experimental thermal and fluid science,* Vol. 12, No. 3 (1996), pp. 373-379

[17] Prabhu N. *et al.*, "Effect of surface roughness on metal/quenchant interfacial heat transfer and evolution of microstructure" *Materials and Design,* Vol. 28, (2007) pp. 544-550

[18] Klocke, F. *et al.*, Fertigungsverfahren 1. Drehen, Fräsen, Bohren., Springer (Berlin, 2008)

[19] Lee J., "Role of Surface Roughness in Water Spray Cooling Heat Transfer of Hot Steel Plate" *ISIJ International,* Vol. 49, No. 12 (2009) pp. 1920-1925

[20] Kozlov N. *et al.*, "Influencing on liquid quenching by surface structuring" *International Journal of Thermal Sciences,* Vol. 101 (2016), pp. 133–142

[21] Vergara-Hernández H. J. *et. al.*, "A novel probe design to study wetting front kinematics during forced convective quenching" *Experimental Thermal and Fluid Science,* Vol. 33 (2009), pp. 797–807

[22] Frerichs F. *et. al.*, "The Influence of Geometrical Features on Rewetting Behavior of Cylindrical Components" *Int. J. Microstructure and Materials Properties,* Vol. 7 (2012), Nos. 2/3

International Federation of Heat Treatment and Surface Engineering 2016
Proceedings of the 23rd IFHTSE Congress
April 18–21, 2016, Savannah, Georgia, USA

Mathematical Modeling and Computer Simulation of Non-monotonic Quenching

Božo Smoljan, Dario Iljkić, Lovro Štic
Department of Materials Science and Engineering, Faculty of Engineering, University of Rijeka, Rijeka, Croatia
smoljan@riteh.hr, darioi@riteh.hr, lstic@riteh.hr

Abstract

Numerical modeling of phase transformations and hardness distribution in non-monotonic quenched steel specimen was developed based on the results of simple experimental test i.e. Jominy test. Differences between the cooling curves of investigated steel specimen and cooling curves of Jominy specimen were taken in the account in the model. The hardness in specimen points was estimated by the conversion of cooling time results to hardness by using both, the relation between cooling time and distance from the quenched end of Jominy specimen, and by using the Jominy hardenability curve. The cooling curves at the specimen points was predicted by numerical modeling of cooling by using the finite volume method. Besides characteristic cooling times, $t_{8/5}$, it takes into account the history of cooling. Developed method was compared with methods of simulation of as-quenched hardness based on characteristic cooling times from 800 to 500 °C, $t_{8/5}$ without taking in account history of cooling. Moreover developed model was compared by methods that are based on continuous cooling transformation (CCT) diagram and isothermal transformation (IT) diagram by using Scheil's additivity rule. Developed numerical model for computer simulation of quenching was also experimentally verified. Limitations of proposed numerical model were found out as well. It has been shown that proposed numerical model can be successfully applied for purposes of simulation of interrupted quenching of carbon and low alloyed steel specimens.

Introduction

Simulation of steel quenching is a complex problem. Research of numerical simulation of hardening degree distribution in quenched steel specimen is one of the highest priority researches in simulation of phenomena of steel quenching [1, 2, 3, 4].

Generally, the mathematical modeling of steel quenching can be divided in two parts, numerical simulation of specimen cooling and numerical simulation of specimen hardening. Rate of steel specimen cooling essentially depends on specimen geometry and characteristic physical properties of quenchant and quenched steel. Main physical properties on which cooling rate depends are specific heat capacity of steel, heat conductivity coefficient of steel, steel density, heat transfer coefficient of quenchant. For precise mathematical modeling, these variables must be estimated with high precision. These variables can be predicted experimentally for the concrete quenchant and steel [1]. Another way of estimating data involved in mathematical model of transient temperature field can be done by inversion method based on relation between cooling curves and heat conductivity coefficients [1]. More generally applied method of estimation of needed physical properties for modeling of quenching is method based on achieved results and qualitative analysis of cooling curve. In this way, physical properties are adjusted or calibrated with the results of quenching. Physical properties predicted by this method are usable for large spectra of quenchants with the same intensity of cooling, and for large spectra of steels [5]. Hardness in quenched specimen can be successfully estimated by using thermo-kinetic equations or by isothermal transformation (IT) diagram. In both of these methods Scheil's additivity rule can be used [6]. Also, hardness can be estimated by using continuous cooling transformation (CCT) diagrams [6]. Finally, hardness can be estimated based on characteristic cooling time, $t_{8/5}$ [7, 8]. For every mentioned method, problems appear when modeling the so-called interrupted quenching. Calculation of transient temperature field is necessary in all of these methods.

Mathematical Modeling of Cooling

Transient temperature field in an isotropic rigid body without heat sources can be described by Fourier's law of heat conduction:

$$\frac{\delta(c\rho T)}{\delta t} = div\lambda\, gradT \tag{1}$$

where λ/Wm^{-1}K^{-1} is the coefficient of heat conductivity, ρ/kgm^{-3} is the density, c/Jkg^{-1}K^{-1} is the specific heat capacity. Characteristic initial condition is:

$$-\lambda\frac{\delta T}{\delta n}\bigg|_{s} = \alpha(T_s - T_f) \tag{2}$$

where T_s/K is the surface temperature, T_f/K is the quenchant temperature, α/Wm^{-2}K^{-1} is the heat transfer coefficient.

Eq. 1 can be solved by the finite volume method [9]. In this case domain could be divided in N control volumes. Discretization system has N linear algebraic equations with N unknown temperatures of control volumes, where N is total number of control volumes. Time of cooling from T_a to specific temperature in particular point is determined as sum of time steps, and in this way, the diagram of cooling curve in every grid-point of a specimen is possible to find out [5].

Before setting of model of temperature field change in an isotropic rigid body input data, i.e., specific heat capacity of steel, c, heat conductivity coefficient of steel, λ, steel density, ρ, heat transfer coefficient of quenchant, α must be consistent with the achieved results of microstructure and mechanical properties. Optimization of input data should be done according to achieved results.

Variables ρ and c can be accepted from the literature [6]. If the variables ρ and c were accepted, variable λ and variable α must be estimated, i.e., calibrated according to variables ρ and c. The input values of heat transfer and heat conductivity coefficients have been optimized using Crafts-Lamont diagrams for large spectra of a specimen bar diameter [10]. Heat transfer coefficient, α/Wm^{-2}K^{-1} for cylindrical specimen is approximately equal to:

$$\alpha \approx -\frac{c\rho V}{F(T_s - T_f)}\frac{dT}{dt} = \frac{K}{T_s - T_f} v \quad (3)$$

where V/m^3 is volume of specimen, F/m^2 is lateral area of cylindrical specimen, v/Ks^{-1} cooling rate and K/Jm^{-2}K^{-1} is a coefficient. Coefficient K and cooling rate, v depends on temperature. For an accurate estimation of heat transfer coefficient, α, the dependence of cooling rate, v and coefficient K on temperature must be known. If the dependence of coefficient K on temperature is neglected and coefficient K is known, heat transfer coefficient, α can be estimated directly from diagram of cooling rate vs. temperature in Fig. 1 by Eq 3. Since the value of coefficient K is not known, the characteristic values of heat transfer coefficient, α must be additionally calibrated. Calibration of heat transfer coefficient was provided by varying of heat transfer coefficient values in the established model of cooling of steel bar. Dependences of heat transfer coefficients on temperature for different quenchants with different Grossmann severity of quenching and different bar diameters were defined. Transient temperature field of specimen was calculated for radius position r/R=0, r/R=0.5 and r/R=0.9. Relative differences between distances from a quenched ends of Jominy specimen, estimated by numerical modeling and by Crafts-Lamont diagrams were negligible, relative errors were less than 10% [5].

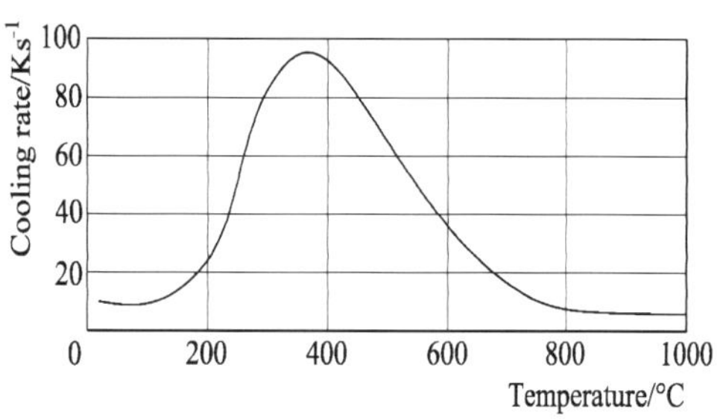

Figure 1: Typical cooling rate for water quenching.

Calibrated values of heat transfer coefficient, α for water are shown in Table 1.

In similar way, heat transfer coefficient, α for air was found out. Calibrated values of heat transfer coefficient, α for air are shown in Table 2.

Table 1. Calibrated values of heat transfer coefficient for water

Temperature, ϑ [°C]	20	116	316	703	1000
Heat transfer coefficient, α [Wm^{-2}K^{-1}]	1636	1722	14523	1417	1250

Table 2. Calibrated values of heat transfer coefficient for air

Temperature, ϑ [°C]	20	100	200	400	600	800	1000
Heat transfer coefficient, α [Wm^{-2}K^{-1}]	12	15	21	33	50	84	113

Mathematical Modeling of Hardness

When using IT diagrams, in accordance to the Scheil's additivity rule, characteristic microstructure transformation is completed when transformed part of microstructure, X is equal to one [6, 8]:

$$\int_0^t \frac{dt}{\tau(X_0, T)} = 1 \quad (4)$$

where $\tau(X_0,T)$ represent the isothermal transformation time for X=X_0 at a temperature T, and t is the total transformation time. Time of isothermal transformation could be calculated by thermo-kinetic equations or could be found out by using IT diagram [8].

Transformed part of microstructure, X can be calculated by Avrami's isothermal equation:

$$X = 1 - \exp(-k \cdot t^n) \quad (5)$$

For purpose of numerical analysis by computer, it is convenient when kinetics of austenite decomposition is defined in an incremental form of Avrami's isothermal equation. By differentiating the Avrami's equation it follows that

$$\frac{dX}{dt} = \exp(-k \cdot t^n) \cdot n \cdot k \cdot t^{n-1} \quad (6)$$

and by extracting the time component from Eq. 6 it follows that

$$\frac{dX}{dt} = n \cdot k^{\frac{1}{n}} \cdot \left(\ln \frac{1}{1-X}\right)^{1-\frac{1}{n}} \cdot (1 - X). \quad (7)$$

Eq. 7 can be written in an incremental form and the volume fraction ΔX of austenite transformed in the time interval Δt_i at temperature T_i can be calculated as follows:

$$\Delta X_{(N)} = n \cdot k^{\frac{1}{n}} \cdot \left(\ln \frac{1}{1 - X_{(N-1)}} \right)^{1-\frac{1}{n}} \cdot \left(1 - X_{(N-1)}\right) \cdot \Delta t_{(N)}. \tag{8}$$

Kinetic parameters k and n from Eq. 8 can be determined inversely by using data from IT diagrams.

According to additivity rule, the non-isothermal transformation kinetics can be described as the sum of a series of the small isothermal transformations.

Figure 2 shows the scheme for microstructure prediction based on cooling curve and IT diagram. In Fig. 2, the temperature range is divided into a series of small finite steps. Maintaining the time interval, Δt_i to sufficiently short times permits the assumption that the conditions are isothermal over each time step. It was assumed that each time step produces such a transformation as occurs in the IT diagram at the same temperature and microstructure composition.

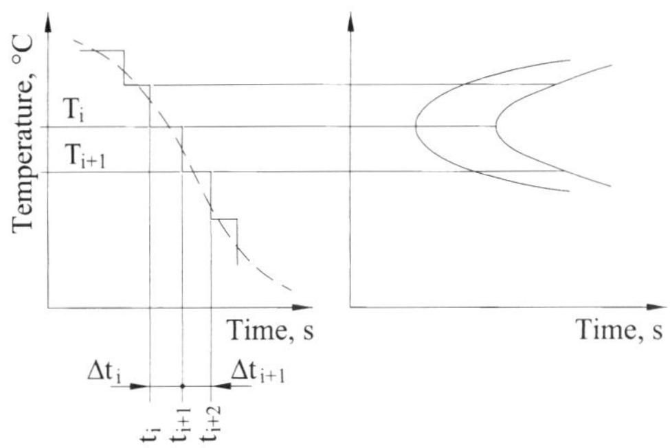

Figure 2: Prediction of microstructure composition and mechanical properties from cooling curve and IT diagram.

Isothermal transformation (IT) diagram for steel EN C45E is shown on Fig. 3.

When using CCT diagrams, hardness can be estimated by drawing the cooling curve in the CCT diagram [6]. This is a very simple method which is often used.
Also, hardness at different specimen points can be estimated by the conversion of the calculated cooling time $t_{8/5}$ to the hardness by using both, the relation between cooling time, $t_{8/5}$ and Jominy distance and the Jominy hardenability curve (Fig. 4) [7].

This method is applicable in cases when there is similarity between cooling curves at certain locations of specimen and cooling curves at certain distances from quenched end of Jominy specimen. This condition is often not satisfied for modern quenchants, as well as for interrupted quenching. Similar problems appear in application of CCT diagrams. A new method of prediction of as-quenched hardness has been developed. In this method, equivalent cooling time $t_{8/5e}$ was used instead of cooling time $t_{8/5}$. Equivalent cooling time $t_{8/5e}$ is function of cooling time, $t_{8/5}$ and history of cooling.

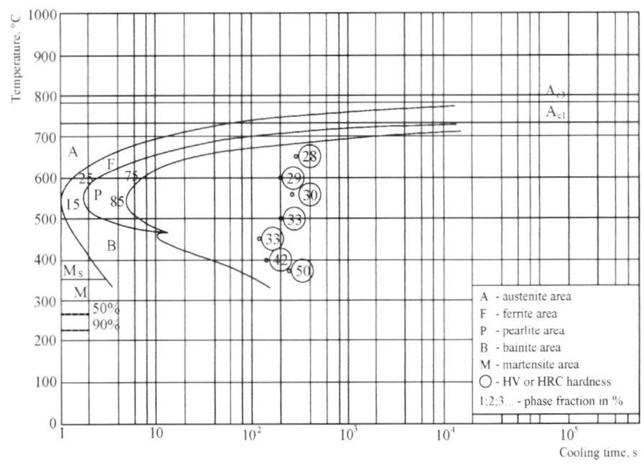

Figure 3: IT diagram for steel EN C45E.

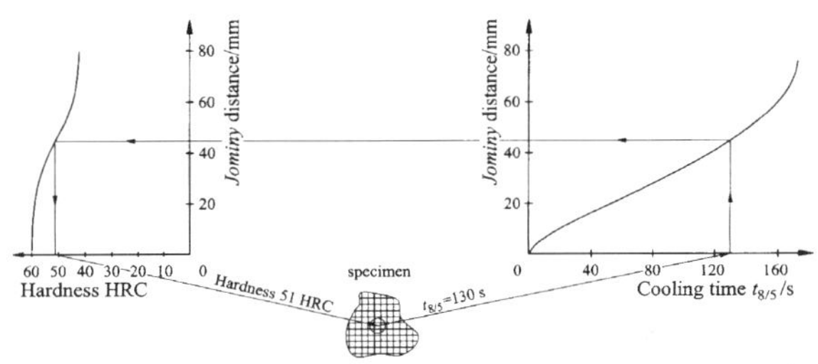

Figure 4: Conversion of the cooling time $t_{8/5}$ to the hardness.

Experimental Procedure

The experiment was performed on specimen made of steel EN C45E. The chemical composition of steel EN C45E is: 0.44 %C, 0.22 %Si, 0.66 %Mn, 0.022 %P, 0.029 %S, 0.15 %Cr, 0.05 %V. Geometry of specimen is shown in Fig. 5. Jominy test results of the steel EN C45E are shown in Table 3.

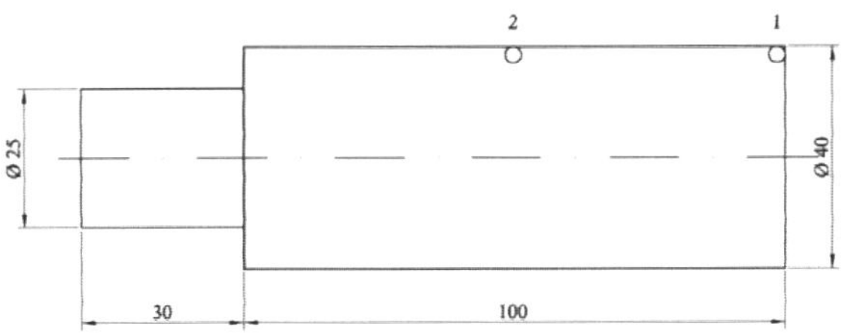

Figure 5: Geometry of the specimen.

Table 3. Jominy test results of steel EN C45E

Jominy distance, [mm]	1.5	3	5	7	9	11	13	15	20	25	30
Hardness HRC	57	55	42	32	30	29	28	27	25	23	21
Jominy distance, [mm]	35	40	45	50	55	60	65	70	75	80	-
Hardness HRC	20	18	17	16	15	15	14	14	14	14	

Specimen was quenched in three different ways. Treatments were defined by measuring temperature and hardness in location 2. During the quenching, temperature was measured by Dostmann electronic HiTemp 2400 infrared thermometer with data acquisition system. Temperatures and cooling times were measured at the same time. In experimental procedure, hardness was tested in locations 1 and 2 (Fig. 5) by HRC method. Achieved as-quenched hardnesses were determined by HRC method and compared with simulation results.

Cooling curves of each treatment in location 2 are shown in Fig 6.

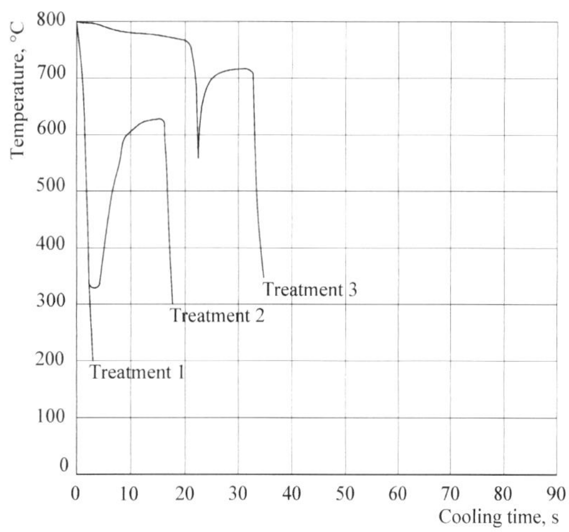

Figure 6: Treatment cooling curves in location 2.

In all treatments austenitizing temperature was equal to 850 °C/30 min. In treatment 1 specimen was quenched from austenitizing temperature to the temperature of quenching medium (Fig. 6). In treatment 2, specimen was interruptedly quenched in water from austenitizing temperature to the desired temperature of 330 °C. After achieving the desired temperature, steel specimen was taken out of water and held in air for 10 s and again cooled in water to the temperature of quenching medium (Fig. 6). In treatment 3, specimen was also interruptedly quenched. First, it was heated to austenitizing temperature and cooled in air to the temperature of 780 °C. After achieving the desired temperature, it was quenched in water to the temperature of 550 °C, taken out of the water and held in air for 10 s and again cooled in water (Fig. 6).

Cooling curves of each treatment in location 1 are shown in Fig 7.

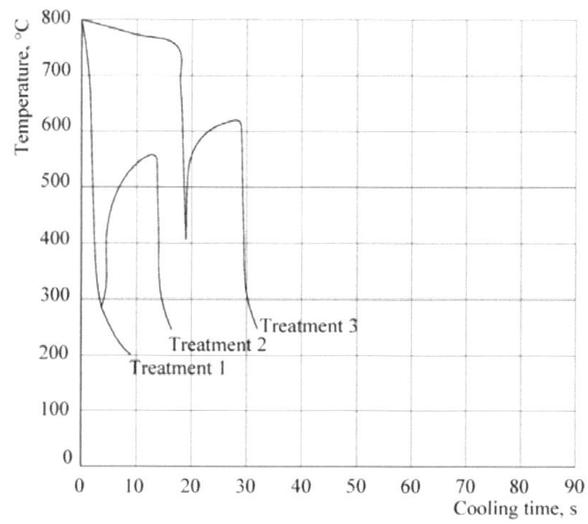

Figure 7: Treatment cooling curves in location 1.

Verification

Results of simulation of as-quenched hardness in locations 1 and 2 by IT diagram (Scheil's additivity rule) method are shown in Table 4. Results of simulation of as-quenched hardness in locations 1 and 2 by using CCT diagram are shown in Table 4 and Fig. 8. Results of simulation of as-quenched hardness by original $t_{8/5}$ method are shown in Table 4 and Fig. 9. Results of simulation of as-quenched hardness by modified $t_{8/5}$ method are shown in Table 4 and Fig. 10.

Results of verification of hardness estimation by certain methods are shown in Table 4. All methods were verified by testing in locations 1 and 2 of cylindrical specimen. Differences between experimental and simulated results of original $t_{8/5}$ method for treatment 1 in locations 1 and 2 are satisfactory. Since this method is originally designed for continuous cooling, it can be assumed that main part of transformations happen during the continuous cooling of specimens. For treatment 2, verification showed large differences of 17 and 19.6 HRC between simulation and experiment results, so this method is not recommended for interrupted quenching. Modified $t_{8/5}$ method and method which uses IT diagrams give similar results, but comparing these two methods, modified $t_{8/5}$ method gives slightly better results. Since the largest difference for treatment 2 is equal to 5 HRC when using the modified $t_{8/5}$ method, it is evident that this method should be further improved. Measurements have shown that modified $t_{8/5}$ method gives the best results in computer simulation of as-quenched hardness.

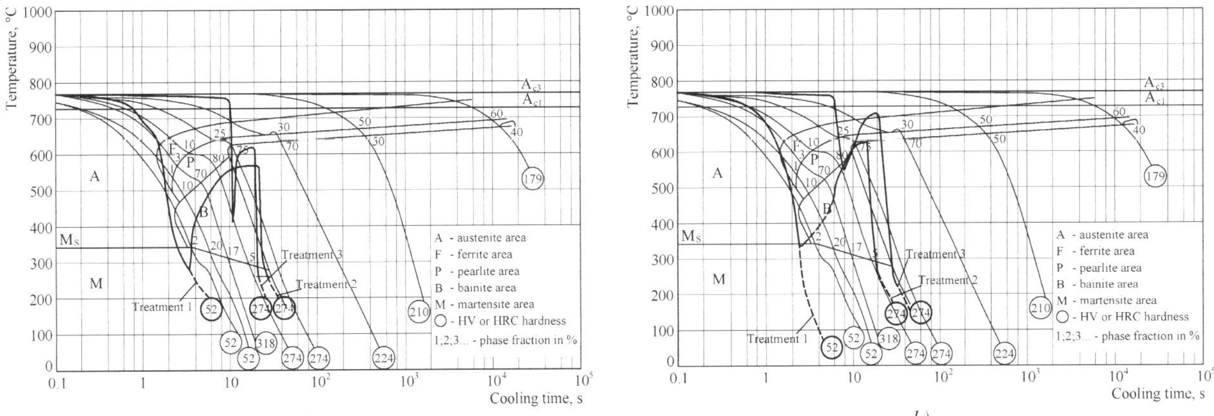

Figure 8: Results of simulation of as-quenched hardness by using CCT diagrams, a) location 1, b) location 2.

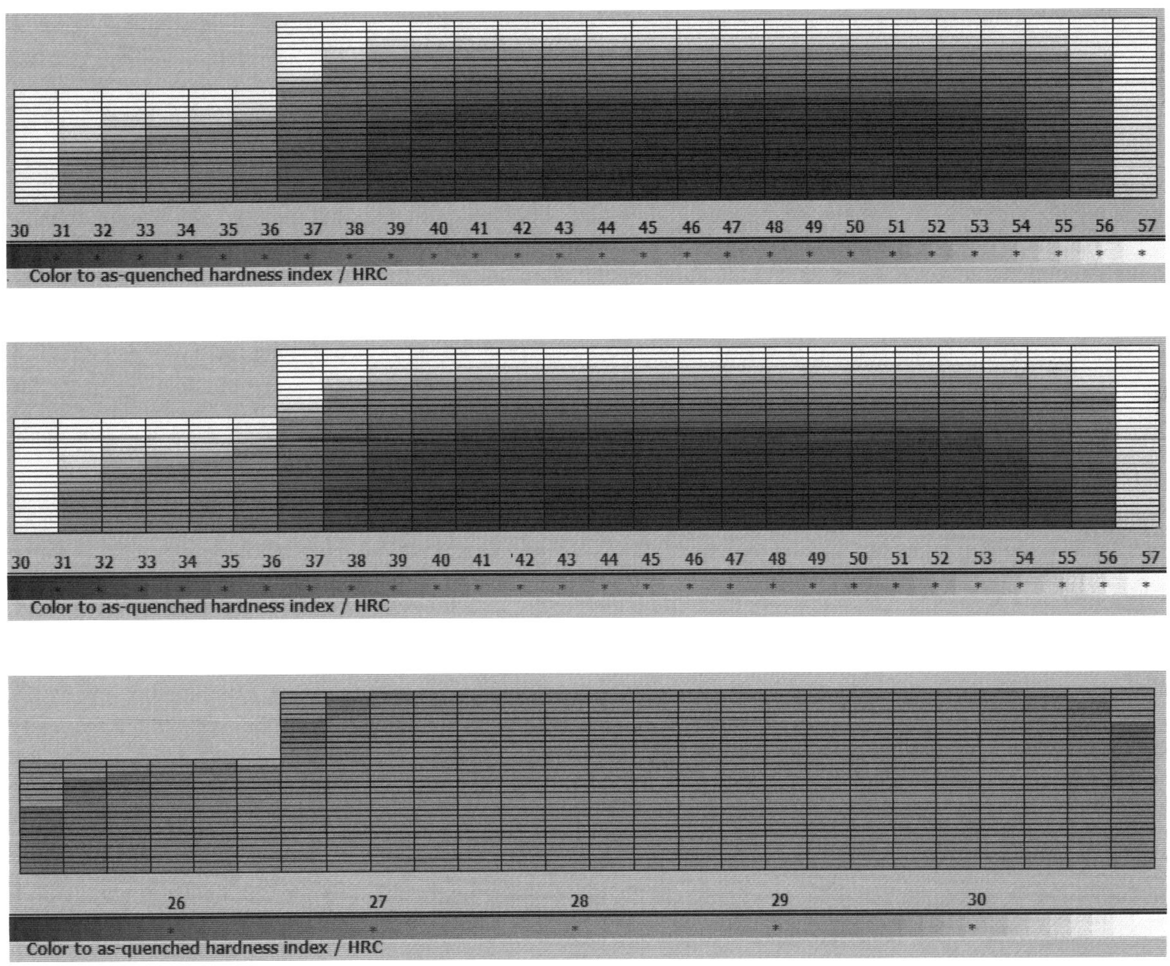

Figure 9: Results of simulation of as-quenched hardness by original $t_{8/5}$ method, a) treatment 1, b) treatment 2, c) treatment 3.

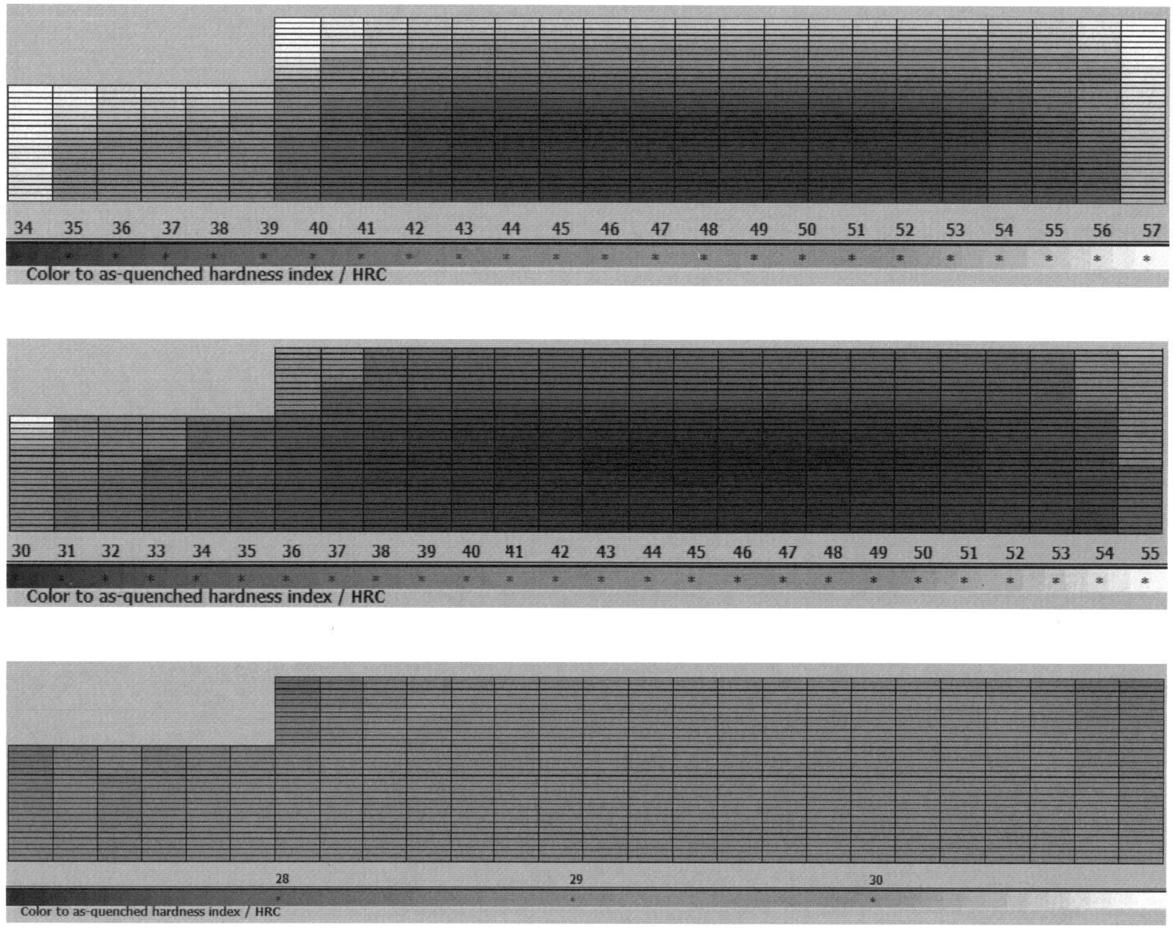

Figure 10: Results of simulation of as-quenched hardness by modified $t_{8/5}$ method, a) treatment 1, b) treatment 2, c) treatment 3.

Table 4. Results of simulation of as-quenched hardness

Method	Location	1			2		
	Treatment	Simulation Hardness HRC	Experiment Hardness HRC	Difference Hardness HRC	Simulation Hardness HRC	Experiment Hardness HRC	Difference Hardness HRC
IT diagram method	1	55.1	55	0.1	51.2	54	-2.8
	2	37.2	40	-2.8	35.6	37	-1.4
	3	30.2	25	5.2	35.3	32	3.3
CCT diagram method	1	52	55	-3	52	54	-2
	2	28	40	-12	28	37	-9
	3	28	25	-3	28	32	-4
Original $t_{8/5}$ method	1	57	55	2	56.6	54	2.6
	2	57	40	17	56.6	37	19.6
	3	28	25	3	27.7	32	-4.3
Modified $t_{8/5}$ method	1	56.2	55	1.2	54.7	54	0.7
	2	45	40	5	36	37	-1
	3	28.3	25	3.3	29	32	-3

Conclusions

Computer simulation of continuous and non-monotonic, i.e., interrupted quenching in water was analyzed. A method has been set up for calculation of transient temperature field in case of interrupted quenching which consisted of quenching in water and holding in air during interruption. Heat transfer coefficients for air and for water have been determined by calibration methods.

Simulation of as-quenched hardness was performed by using CCT diagrams, IT diagrams by using Scheil's additivity rule and by method for estimation of hardness based on characteristic cooling times from 800 to 500 °C, $t_{8/5}$. Moreover, a new method for estimation of as-quenched hardness has been set up. Besides characteristic cooling times, $t_{8/5}$, it takes into account the history of cooling from austenitization temperature, t_A to the final temperature. By analyzing possibilities of simulation with every method, it could be concluded that all methods are applicable in cases of continuous cooling, but in case of interrupted quenching, best results were achieved using the modified $t_{8/5}$ method.

Acknowledgments

This work has been fully supported by Croatian Science Foundation under the project 5371.

References

[1] Liščić, B., "System for Process Analysis and Hardness Prediction When Quenching Axially-Symmetrical Workpieces of Any Shape in Liquid Quenchants," *Materials Science Forum*, Vol. 638-642 (2010), pp. 3966-3974.

[2] Gur, C. H., Pan, J., Handbook of Thermal Process Modeling Steels, CRC Press, (2008)

[3] Smoljan, B., Iljkić, D., Totten, George, E., "Mathematical Modeling and Simulation of Hardness of Quenched and Tempered Steel," *Metallurgical and Materials Transactions B*, Vol. 46, No. 6 (2015), pp. 2666-2673, DOI:10.1007/s11663-015-0451-6

[4] Reti, T., Felde, I., Guerrero, M., and Sarmiento, S., "Using Generalized Time-Temperature Parameters for Predicting the Hardness Change Occurring during Tempering," *Proc International Conference on New Challenges in Heat Treatment and Surface Engineering (Conference in honour of Prof. Božidar Liščić)*, B. Smoljan, B. Matijević, Eds., Dubrovnik-Cavtat, Croatia, 9-12 June 2009, pp. 333-342.

[5] Smoljan, B., "The Calibration of the Mathematical Model of Steel Quenching," *Proc 5th World Seminar on Heat Treatment and Surface Engineering*, M. Salehi Ed., ISSST and IFHT, Vol.1, Isfahan, 1995, pp. 709-715.

[6] ..., Heat Treating, Vol. 4, ASM Handbook, ASM International, 1991

[7] Rose A., Wever, F., Atlas zur Wärmebehandlung der Stähle I, Verlag Stahleisen, (Düsseldorf, 1954)

[8] Smokvina Hanza, S., Mathematical modeling and computer simulation of microstructure transformations and mechanical properties during steel quenching, Doctoral Thesis, Department of Materials Science and Engineering, Faculty of Engineering, University of Rijeka, (2011) (in Croatian)

[9] Patankar, S., Numerical Heat Transfer and Fluid Flow, McGraw Hill Book Company, (New York, 1980), pp. 60, ISBN 0-07-048740-5

[10] Crafts, W., Lamont, J., Hartbarkeit und Auswahl von Stahlen, Springer Verlag, (1954)

International Federation of Heat Treatment and Surface Engineering 2016
Proceedings of the 23rd IFHTSE Congress
April 18–21, 2016, Savannah, Georgia, USA

Using Heat Treat Simulation to Characterize Sensitivity of Quench Hardening Response in Hot Mill Steel Work Rolls

A. M. Freborg, Z. Li and B. L. Ferguson
DANTE Solutions, Inc.
7261 Engle Road, Suite 105, Cleveland, OH 44130
andy.freborg@dante-solutions.com

Abstract

Quench hardening of work rolls for steel rolling is used to impart both residual compression and hardness to the roll surface to promote durability, wear resistance and life. Minimizing required roll re-dressing and mill down time are important cost drivers for improved work roll surface life. Developing the necessary metallurgical and mechanical response of these rolls in heat treatment is complex, often involving a heating and quenching sequence in which thermal gradients are controlled at both the outer surface and by application of internal cooling through the roll's hollow center. Depth of hardness, along with surface and internal residual stresses, are challenging to control. Typically, methods for developing such heat treating practices are based heavily on in-plant experimental trials, as well as "experience."

The advent of accurate heat treatment simulation allows for analytically based process engineering to be used to examine roll heat treatments to improve depth of hardening and overall metallurgical response. In this paper, virtual experiments are conducted to characterize practical heating, cooling and quenching regimes to study how varied practices may be used to control residual stress, hardness and metallurgical structure. Because of heat treatment's highly non-linear nature, simulation is a valuable tool for examining part response sensitivities. The heat treatment response of a typical steel mill work roll is examined under a variety of heating and quenching conditions. The relationship between residual stress and hardness is not straightforward, as will be shown. Therefore, the use of heat treatment modeling, which allows examination of the in-process thermal and transformation stresses in conjunction with microstructural changes, is an important tool for understanding the heat treatment of these rolls.

Introduction

Control of surface residual compressive stresses has long been established as a key factor in promoting increased wear resistance and operating life in steel work rolls for hot mills. [1 - 3] Typically produced from high alloy steels (e.g. Cr > 3.0%), the rolls are designed to take advantage of both Cr carbide and surface compressive stresses to resist wear, thermal fatigue, and cracking from load excursions typically encountered in normal production. [4] Surface compressive stresses are important because they act to both reduce the effective stress at the roll surface by countering surface tension from loading, and thus better resist crack initiation and wear. The benefits of surface compression for reducing both wear and crack initiation propensity are well documented.

For hot mill work rolls, residual surface compression is most typically introduced by heat treatment.[5] The usual heat treating practice for these rolls includes controlled heating to a predetermined austenitizing depth on the roll body, controlled quenching, and subsequent tempering operations, sometimes including cryogenic treatment. Greater control of the austenitizing and hardening response, as well as reduction in core tensile stresses is often achieved by removing the roll core by center boring prior to heat treatment. Through a combination of controlled heating/center cooling, followed by quenching, the heat treater thus has great flexibility in adjusting the depth of heating, hardening depth, and the roll internal thermal gradients during the heat treatment process. Consequently, the resulting surface and internal residual stress profiles possible from such heat treatments can have significant variation.

Quantifying and predicting these thermal, metallurgical and residual stress responses remain a substantial challenge to the manufacturing engineer. Recent advancements in quantitative process simulation (modeling) have made it possible to study in-situ the combined effects of heat treatment thermal and transformation strains on resulting part residual stresses. DANTE® is a finite element based software tool that calculates the residual stress, dimensional change, hardness and metallurgical phase volume fractions of steel parts as a result of heat treatment.[6] The DANTE database includes mechanical and thermal property data for steel microstructural phases as functions of temperature and rate, as well as the necessary phase transformation kinetics parameters to address both heating and cooling transformations.[7] In this study, the DANTE software was used to examine the residual stress response sensitivity of a typical alloy steel, hollow, hot mill work roll. Baseline hardening response using a nominal 203mm depth of austenitizing was first compared with industry reported stress, hardness and retained austenite values for establishing the validity of the model. Subsequently, the depth of austenitizing was adjusted to three (3) additional shallower depths to predict how the depth of heating, surface hardness, surface residual stress, and core residual stress vary with changing austenitizing depth.

Approach

The hot mill work roll chosen for this study was of design typical for rolling hot strip products. The roll chemistry was 3.25%Cr, 0.50%Mo, 0.73%C, with balance of iron. A continuous cooling transformation (CCT) diagram for the steel is shown in Figure 1.

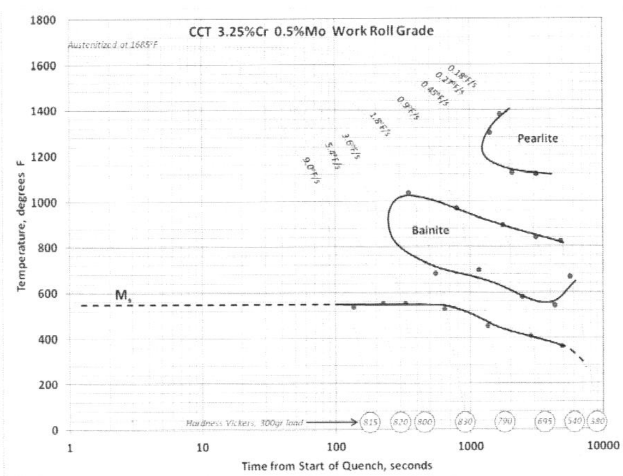

Figure 1: Experimentally determined continuous cooling transformation diagram for the 3.25% Cr alloy roll steel.

The roll chemistry is designed such that a specific austenitized depth on the roll body can be hardened to martensite. The roll core is not austenitized during heat treating, and remains pearlitic throughout the heat treatment.

The roll shape is shown schematically in Figure 2. The overall length is 4400mm, with the roll body itself being 2080mm in length. The diameter of the roll body is 612mm, and there is an 88mm diameter bore extending the length of the roll.

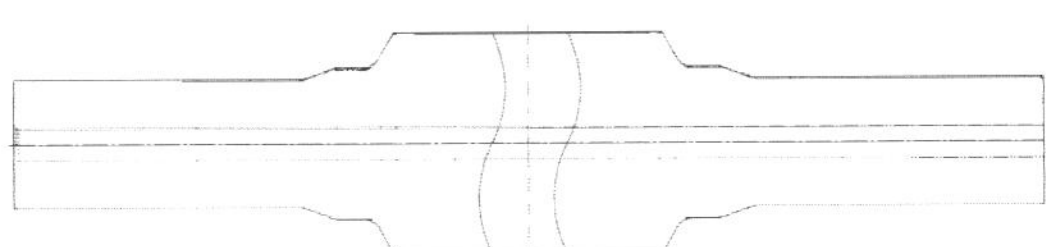

Figure 2: Schematic of the work roll shape.

During heat treatment, the entire roll is preheated to 480°C, and subsequently the roll body alone is flash heated at 965°C to achieve the desired austenitizing depth. Internal cooling is often applied through the roll bore to control the depth of austenitizing to the designed amount.

Upon austenitization of the roll body, the roll is then typically quenched in a high velocity water quench tank. Water quenching is applied to the entire roll surface, including the bore which typically sees somewhat reduced cooling power relative to the roll outer diameter.

For the study presented in this paper, the roll body austenitizing depth was varied between 127 – 203 mm. Subsequently, a high velocity water quench was applied to the roll outer surface, with a reduced water quench in the roll bore. The models were simplified to a 2-D axisymmetric slice though the roll body, as shown in Figure 3.

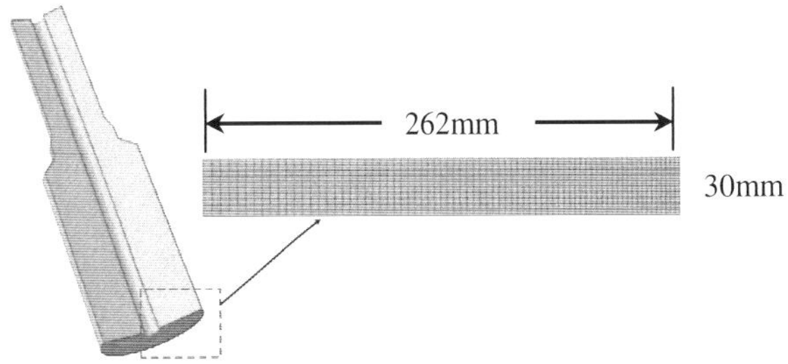

Figure 3: 2-D mesh slice through the roll body for the DANTE models.

Table 1 summarizes the general heat treating practice for the roll.

Table 1: Summary of General Heat Treatment Process Sequence for the Work Roll.

STEP	PROCESS
1	Preheat 480°C
2	Flash Heat Roll Body at 965°C to Achieve Desired Austenitizing Depths of 203mm, 178mm, 152mm and 127mm
3	Water Quench

For this study, four (4) austenitizing depths for Step 2 were examined as indicated in Table 1. Using DANTE, the austenitizing depths were modeled by applying varied internal convection in the roll bore, as per industry practice. The water quench practice was consistent for all cases. Tables 2 and 3 describe the convection heat transfer boundary conditions applied for the bore cooling during heating (to control austenitizing depth), and the convection cooling applied to the roll outer body and bore during the water quench. As industry practice typically holds the rolls for a set austenitizing time, and use internal cooling to control depth, this was the approach taken for the models in this paper.

Table 2: Internal Convection Cooling Heat Transfer Coefficients Applied to Roll Bore to Achieve Desired Austenitizing Depth during Flash Heating.

Austenitizing Depth	Roll Bore HTC
203mm	64 W/m^2 K
178 mm	120 W/m^2 K
152 mm	220 W/m^2 K
127 mm	480 W/m^2 K

Table 3: Water Quench Convection Cooling Heat Transfer Coefficients Applied to the Roll Body during Quenching.

Location	Water Quench HTC
Outer Surface	14,000 W/m^2 K
Inner Bore Surface	4,000 W/m^2 K

The heat treatment model for the 203mm austenitizing depth was executed first, and compared with industry data provided to DANTE Solutions for validation purposes.[8] The favorable agreement, described in the Results section of this paper, prompted the additional model runs for 178mm, 152mm, and 127mm austenitizing depths. The model results were then examined for comparative metallurgical (martensite and hardness profiles), as well as residual stress response.

Results

The predicted roll surface residual stress, hardness and retained austenite values for the 203mm austenitizing depth are shown in Table 4, and compared with industrial reported results for this same roll material and heat treatment. [8]

Table 4: Comparison of DANTE Roll Model with 203mm Austenitizing Depth with Comparable Result Reported to DANTE Solutions from Industry for the Same Roll Material and Heat Treatment.

Attribute	DANTE Model	Reported Industrial Result
Surface Residual Stress	-960 MPa	-890 – -1000 MPa
Surface Retained Austenite	26%	22% - 26%
Surface Hardness	HRC 60	HRC 57 - 61

The favorable results comparison in Table 4 prompted the additional model runs for the 178mm, 152mm and 127mm austenitizing depths. Surface residual stress, surface retained austenite and surface hardness comparisons are presented in Table 5.

Table 5: Summary Results for DANTE Roll Models with Four Different Austenitizing Depths.

ROLL AUST. DEPTH	ATTRIBUTE			
	Outer Surface Residual Stress	Bore Surface Residual Stress	Outer Surface Retained Austenite	Surface Hardness
203mm	-960 MPa	-319 MPa	26%	HRC 60
178mm	-1014 MPa	-412 MPa	26%	HRC 60
152mm	-1065 MPa	-470 MPa	26%	HRC 60
127mm	-1096 MPa	-500 MPa	26%	HRC 60

While the results shown in Table 5 indicate definite variation in surface and roll bore residual stress with varying the austenitizing depth, little variation is seen in surface hardness and surface retained austenite. Therefore, the utility of using surface hardness data and even estimates of surface retained austenite as a means of assessing residual stress response and sensitivity in actual industrial practice is of little use. Furthermore, reaction tensile stresses are present within the roll. For large parts such as this 612mm diameter roll, these stresses can be quite significant. At first glance, using the 127 mm austenitizing depth may seem desirable for achieving greater surface compression. However, internal metallurgical response and the internal stresses must also be assessed. Heat treatment modeling readily allows for this assessment.

Figure 4 shows comparative axial stress profiles for the roll section from the outer surface to the inner bore surface. The profiles reveal that while decreasing the depth of hardness increases the surface compressive stresses on the roll body, the internal reaction tensile stresses also increase dramatically. In fact, the internal stresses in excess of ~900 MPa may have a propensity to produce internal cracking or possible roll failure. It should also be noted that decreasing the austenitizing depth not only increases the internal tensile stresses, but also shifts these higher tensile stresses closer toward the roll mid-radius.

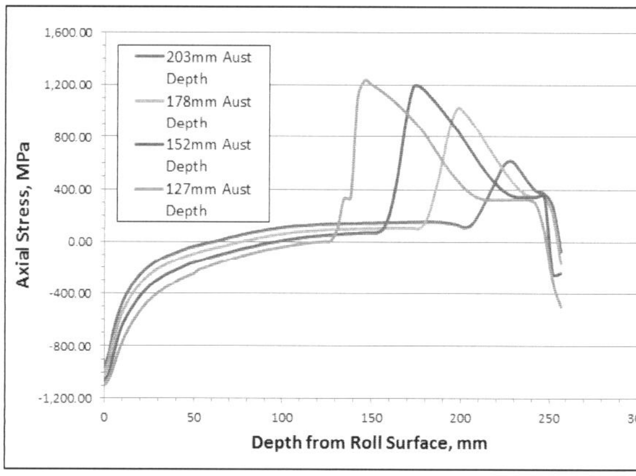

Figure 4: Comparative axial stress profiles for the roll body, showing the effect of austenitizing depth.

The model reveals that the internal tensile stresses occur just below the martensite transformation transition location. Compare Figure 5, which shows the predicted martensite profiles for each model, with Figure 4. The spikes in the internal tensile reaction stresses occur within 15 mm of the martensite transformation boundary.

Examination of the model contours provides graphic illustration of the relationship. Figures 6 – 9 show composite contour plots of the flash heating austenitizing depth, post quench axial stress, and post quench martensite phase fraction. The relationship between shallower austenitizing depth, greater surface compression, and higher internal tension moving toward the roll mid-radius is clearly evident.

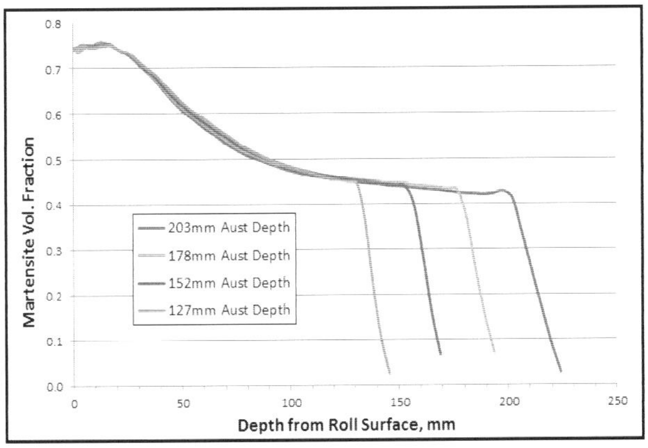

Figure 5: Comparative martensite volume fraction profiles for the roll body, showing the effect of austenitizing depth. Compare with the stress plot in Figure 4.

Figure 6 shows composite contour plots of the austenitized layer after heating, and the axial stress and martensite after quenching.

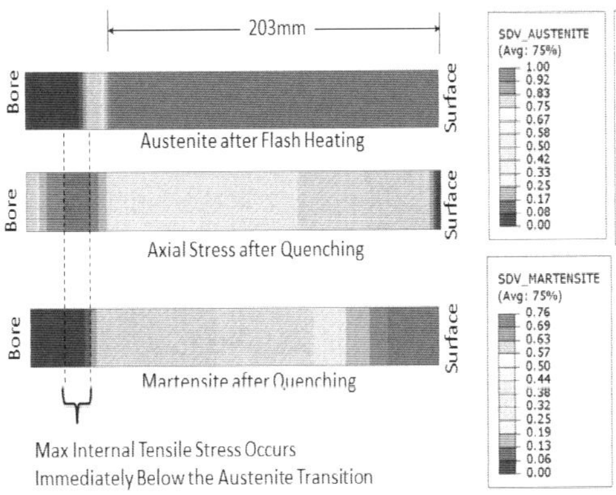

Figure 6: Composite contour plot showing flash heated austenitizing depth relative to the post quenching axial stress and martensite phase fraction profiles for the roll austenitized to 203mm depth.

The contour map composites are valuable in providing a visual cue to the relationship between the location of the reaction tensile stress and the depth of austenitizing. As seen also in Figure 4, and shown further in Figures 7 – 9, as the austenitizing depth decreases, the increased surface compression is accompanied by an increased internal tensile reaction stress which shifts inward towards the roll mid-radius.

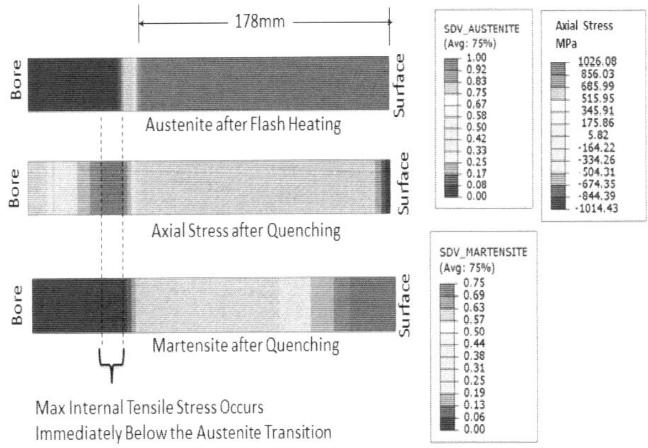

Figure 7: Composite contour plot showing flash heated austenitizing depth relative to the post quenching axial stress and martensite phase fraction profiles for the roll austenitized to 178mm depth.

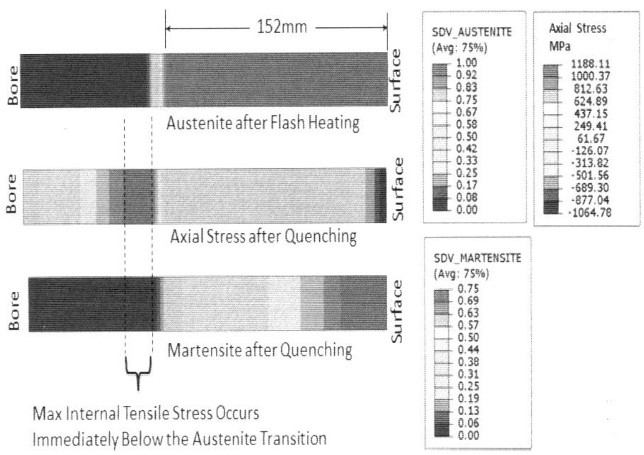

Figure 8: Composite contour plot showing flash heated austenitizing depth relative to the post quenching axial stress and martensite phase fraction profiles for the roll austenitized to 152mm depth.

Examining the residual stress results both numerically and qualitatively through the comparative contour maps shows key correlations between surface compressive stress magnitude, magnitude and depth of the internal reaction tensile stress, and the applied austenitizing depth. The heat treatment models allow both for virtual process design to assess these responses, as well as a tool for understanding the material physics behind the responses. For the subject roll and heat treatment in this paper, these will be described further in the next section.

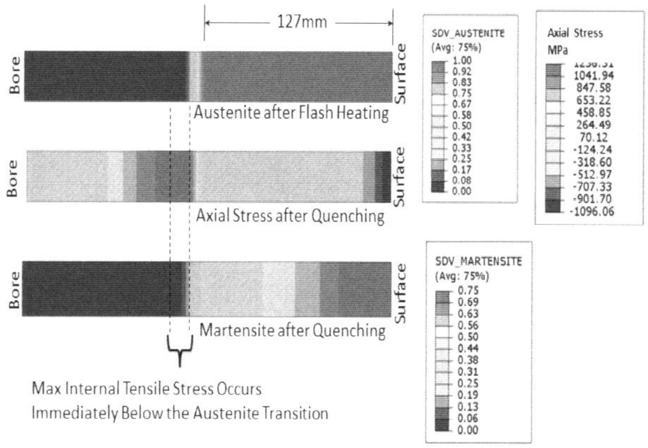

Figure 9: Composite contour plot showing flash heated austenitizing depth relative to the post quenching axial stress and martensite phase fraction profiles for the roll austenitized to 127mm depth.

Discussion

The model results showed a correlation between the applied austenitizing depth, the magnitudes of both the resulting surface compression and internal tensile residual stresses, and the depth at which the maximum internal tensile stress occurs. This data is tabulated from the models in Table 6, and graphically summarized in the plot in Figure 10.

Table 6: Tabulated Stress Results for DANTE Roll Models with Four Different Austenitizing Depths.

	Austenitizing Depth			
	207mm	178mm	152mm	127mm
Surface Compression, Axial	-960MPa	-1014 MPa	-1065 MPa	-1096 MPa
Max Internal Tensile Stress, Axial	616 MPa	1026 MPa	1188 MPa	1236 MPa
Depth from Surface of Max. Internal Tensile Stress	228mm	198mm	173mm	150mm

From both Table 6 and the Figure 10 graphic, the model shows the following:

- Surface residual compression increases (becomes more compressive) with decreasing austenitizing depth
- The internal reaction tensile stress increases (becomes more tensile) with decreasing austenitizing depth
- The location of the internal maximum tensile stress shifts closer to the roll surface with decreasing austenitizing depth

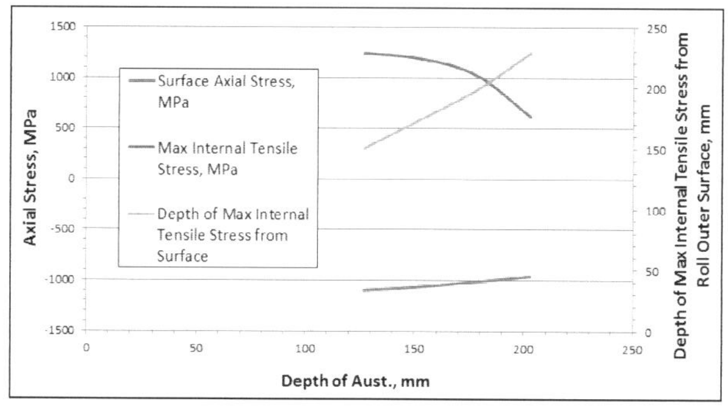

Figure 10: Composite contour plot showing flash heated austenitizing depth relative to the post quenching axial stress and martensite phase fraction profiles for the roll austenitized to 127mm depth.

The driving force behind the residual stresses is a combination of both thermal and transformation strains. As the roll cools, both the austenitized shell and heated pearlitic core will shrink. Transformation of the austenitic shell will produce a combination of martensite, bainite and pearlite. The 3.25% Cr steel hardenability is such that the 50% martensite depth remains a constant 90 mm for all of the austenitizing depths examined in this study (see Figures 5 – 9). Hardness to this depth can be expected to be uniform between all of the austenitized conditions.

The martensite transformation causes a volumetric expansion of the transforming steel. This will place the roll surface region in compression. Simultaneously the roll pearlitic core will shrink under thermal contraction. However, the volume of this contracting core will vary more directly with the austenitizing depth (see again Figures 5 – 9). The shrinking core will act to pull the outer shell inward and also act to increase surface compression. As deeper layers of the shell partially transform to martensite (depth range 127mm – 203mm for this study), the surface compression will be become more dispersed and drop. A more shallow overall martensitic shell means a greater volume of the internal shrinking pearlitic core acting to place the surface in compression. The higher local surface compression will created higher subsurface reaction tensile stress near the transformation boundary, as seen in the stress profile predictions in Figure 4. Conversely, a deeper overall martensitic shell will have less pearlitic core shrinkage, and the deeper martensitic transformation will act to reverse the surface compression response.

Examining the local temperature, axial stress and martensite evolution during the process at key locations in the roll can add additional insight into the mechanism controlling the residual stress magnitudes. In Figure 11, the temperature, and stress evolution is compared at the surface and at the austenitizing depth for the 127mm and 203mm heated rolls. Also plotted is the surface martensite formation. What can be clearly seen in the plots is the effect that greater austenitizing depth has in reducing the surface compression in the 203mm austenitized roll. The expanding martensite transforming at the greater depths acts to reverse the initially induced surface compression. For the 127mm depth austenitized roll, the internally transforming martensite volume is insufficient to produce the stress reversal, and higher surface compression is maintained. Consequently in the 127mm austenitized roll, the subsurface reaction tensile stresses are also higher.

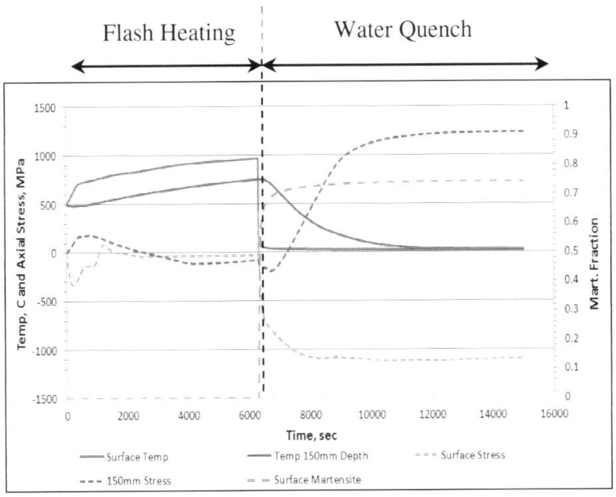

a) Roll austenitized to 127mm depth

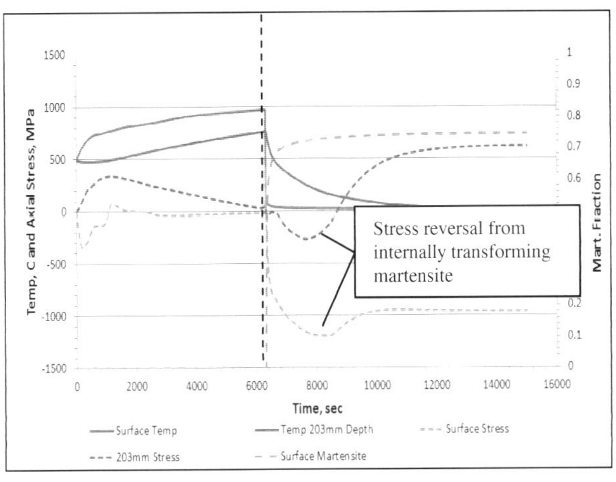

b) Roll austenitized to 203mm depth

Figure 11: Temperature, stress and martensite evolution comparison between the 127mm and 203mm depth austenitized rolls. Deeper martensite transformation in the 203mm austenitized roll produces a stress reversal at the surface.

Conclusions

A DANTE heat treatment process model was validated against provided industrial data for a quench hardened 3.25% Cr hot mill work roll. The model was subsequently utilized to characterize the differing residual stress responses arising from varying the austenitizing depth in the roll prior to quench hardening. The results showed important relationships between the austenitizing depth, surface compressive stress magnitudes, and the depth and magnitude of the internal reaction tensile stresses. The heat treatment simulations showed that increasing the austenitizing depth resulted in a decrease in the surface compressive residual stresses, and also reduced the internal tensile stress. The stresses are controlled by a combination of thermal and transformation strains acting simultaneously. For a given part geometry and heat treating schedule, process modeling incorporating the thermal and transformation strain metallurgical and mechanical effects is necessary to understand key process sensitivities.

References

[1] Kapadia, B. M., and Marsden, K. W., "Failure Analysis of Hot Mill Work Rolls," *Proc 40th MWSP Conference*, ISS, 1998.

[2] Liu, Y. Zhang, H., Somers., B. Marsden, K., and Freborg, A., "Fracture Mechanics-Based Acceptance Criteria of CVC Rolls for a Hot Strip Mill," AIST Iron and Steel Technology, 2013/11.

[3] Lohe, D., Lang, K. H., and Vohringer, O. *Residual Stresses and Fatigue Behavior*, Handbook of Residual Stress and Deformation of Steel, ASM International, 2002, pp. 27-53.

[4] Hashimoto, M., et al., "Development of High Toughness Rolls," Nippon Steel Technical Report, No. 48, Jan 1991.

[5] Scoular, T., and McPate, W. A., *Manufacture of Cast Steel Rolls*, Rolls for the Metalworking Industries, R.B. Corbettt, Ed., Iron and Steel Society, 1990, pp. 95-112.

[6] Ferguson, B. L., and Freborg, A. M., Software to Predict Distortion of Heat Treated Components, USA AMCOM Report TR 02-D-46, 2002.

[7] Ferguson, B. L., Z. Li and A. M. Freborg, "Modeling Heat Treatment of Steel Parts," *Computational Materials Science*, 34 (2005), pp. 274 – 281.

[8] Lehigh Heavy Forge, private communication.

International Federation of Heat Treatment and Surface Engineering 2016
Proceedings of the 23rd IFHTSE Congress
April 18–21, 2016, Savannah, Georgia, USA

Martensite Lattice Parameter Measured by Modern X-ray Diffraction in Fe-C alloy

Yuan Lu, Haixuan Yu, Xiaoqing Cai, Yiming Rong, Richard D. Sisson*

Center for Heat Treating Excellence,
Worcester Polytechnic Institute, 100 Institute Road, Worcester, MA, USA, 01609

Abstract

Martensite crystal structure is widely accepted to be body-centered tetragonal (BCT). With the increase of carbon content, the c/a ratio increases, following the linear relationship. However, it is pointed out by Sherby that the steel with less than 0.6 wt% carbon has body-center cubic (BCC) structure and c/a ratio equals to 1. AISI9310, AISI4140, AISI4150, AISI4161, AISI1080 and AISI52100 were quenched to form fresh martensite. The lattice parameter was determined by X-ray diffraction and Rietveld refinement. The microstructure was observed by Scanning Electron Microscope and hardness was measured. The result shows that for the steel with less than 0.6 wt% carbon, the crystal structure is still BCT structure, however, c/a=1+0.028wt%C. The slope is smaller compared to classical c/a model, c/a=1+0.045wt%C.

Introduction

Martensite is one of the most important microstructures in steels, which provides ultimate strength and wear resistance. From Adolf Karl Gottfried Martens [1][2],intensive work has been conducted to focus on the crystal structures and morphologies by researchers, including George Krauss [3], Harry Bhadeshia [4] and S. Morito [5].

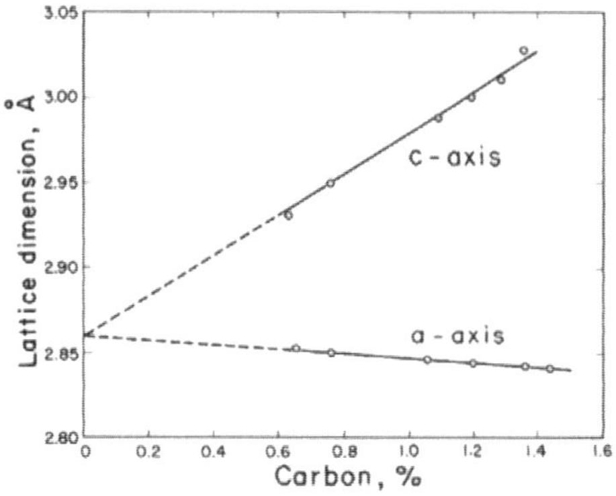

Figure 1 Lattice parameters versus carbon content of quenched martensite in Fe-C steels from E.C.Bain [6]

Figure 1 presents the lattice parameters versus carbon content of quenched martensite in Fe-C steels [6]. The tetragonality of the martensite, measure by X-ray diffraction, c/a, increases with carbon percentage:

$$c/a = 1 + 0.045 \text{ wt\% } C \text{ [4]}$$

However, it seems this equation lacks data when the carbon content in steel is less than 0.6 wt %. The dotted line in *Figure 1* was the assumption, but widely accepted. Sherby [7] claimed after searching references, the c/a ratio relationship was not linear .

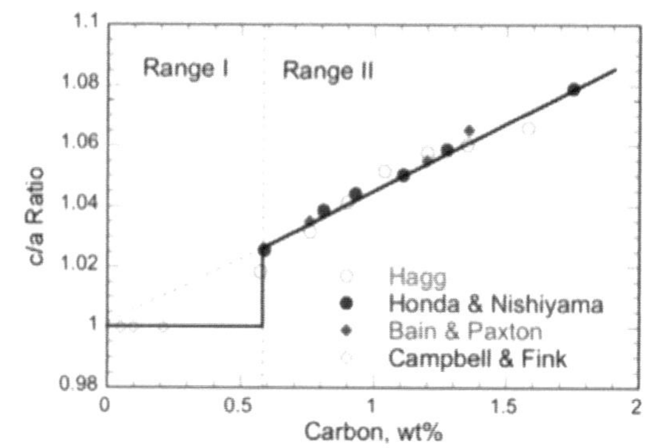

Figure 2 c/a ratio vs. carbon content of quenched martensite in Fe-C steels from Sherby [7]

Based on this phenomena, Sherby later proposed BCC to HCP to BCT martensite transformation process, different than widely accepted BCC to BCT martensite transformation for carbon steels and low or medium alloy steels [8]. This hypothesis is based on BCC martensite structure when the carbon content is less than 0.6 wt%. The author checked the original reference by Campbell, Fink and Kurdumoff, G. [9]. It reported that for steels with less than 0.6 wt%, they could not identify the peak split and tetragonal structure. It should be noted that during 1930s, the X-ray diffraction technology was much less developed that today and the resolution was low. The modern PANalytical Empyrean X-ray diffraction sets have ultra high resolution, with a FWHM of 0.026degrees 2θ for the first reflection of NIST SRM660a LaB_6 [10].

In order to get better understanding of martensite crystal structure, steels with different carbon contents from 0 wt% to 1.5 wt% should be investigated by modern high resolution X-ray diffraction method. AISI9310 (0.1 wt% C), AISI4140 (0.4 wt %C), AISI4150 (0.5 wt % C), AISI4161 (0.6 wt % C), AISI1080 (0.8 wt % C) and AISI52100 (1.0 wt% C) were selected to re-measure the c/a ratio of as-quenched martensite.

Materials and Experimental

AISI9310 (0.1 wt% C), AISI4140 (0.4 wt %C), AISI4150 (0.5 wt % C), AISI4161 (0.6 wt% C), AISI1080 (0.8 wt % C) and AISI52100 (1.0 wt% C) were collected from the Timken company. Each sample was cut into disc with 1 inch in dimeter and 2 inch in length. Before water quenching, all the samples were normalized. The austenitizing temperature and holding time for different steels are listed in *Table 1*.

Table 1 Heat treating parameters for steels

Steel	Austenitizing Temperature [C]	Holding Time [h]
9310	850/950	1
4140	850/950	1
4150	950/1050	1
4161	950/1050	1
1080	950/1050	1
52100	1050/1150	1

Two austenitizing temperatures were selected for each steel grade, in order to investigate the austenitizing temperature effect on martensite crystal structure. The sample was put into the furnace after the furnace reaches the specific austenitizing temperature. After 1h (holding time), the sample was taken out of the furnace and quenched in 20C water within 5s. To prevent oxidation and decarburization effect, high-temperature protective coatings from Advanced Technical Products Supply Co. were used. Then the sample was cut by Mark V CS600 specimen saw. The core of the sample was used for further characterization.

Chemical compositions were measured by Optical Emissions Spectrometer (OES) on the SPECTROMAXx. Phase identification and lattice parameter measurement were conducted by PANanalytical Empyrean 2 X-ray diffractometer (XRD) with Cr-Kα radiation. Rockwell C hardness was measured using equipment from Stanley P. Rockwell Co.

Then the samples were ground, polished and etched with 4% nital solution. A Nikon optical microscope was used to obtain optical microscopy and JEOL JSM-7000F Scanning Electron Microscope (SEM) was for higher magnification characterizations.

Results and Discussions

Chemical composition: OES

The OES experiments were conducted on the core of all the samples, in order to eliminate unnecessary oxidation and decarburization, and accurately measure the carbon content of the steels.

Table 2 presents the chemical compositions of steels. For the same steel grade, such as AISI9310, the chemical compositions are all the same when the austenitizing temperature is 850C or 950C, which demonstrates no oxidation and decarburization were happened to affect the accuracy of carbon content and c/a ratio measurement.

It is clear that the accurate carbon content is more or less compared to the labeled steel grade. For AISI52100, the labeled carbon content is 1%, however, the measured carbon content is 1.24%, much higher than 1%. When drawing the relationship between carbon content and c/a ratio, measured carbon content should be used. Considering the relationship as shown in *Figure 1* and *Figure 2*, the less accurate carbon content may also cause offset of the curve.

Table 2 Chemical compositions of steels

%	C	Si	Mn	Cr	Ni
9310 850C	0.124	0.254	0.63	1.16	3.07
9310 950C	0.123	0.248	0.63	1.15	3.07
4140 850C	0.396	0.225	0.92	0.86	0.09
4140 950C	0.406	0.225	0.93	0.86	0.09
4150 950C	0.449	0.228	0.92	1.02	0.152
4150 1050C	0.449	0.229	0.93	1.03	0.153
4161 950C	0.56	0.169	0.98	0.81	0.191
4161 1050C	0.57	0.169	0.97	0.81	0.189
1080 950C	0.76	0.271	0.83	0.183	0.069
1080 1050C	0.77	0.272	0.82	0.185	0.068
52100 1050C	1.26	0.243	0.397	1.59	0.052
52100 1150C	1.24	0.243	0.399	1.59	0.053

Martensite crystal structure and c/a ratio: XRD

As quenched martensite crystal structure was measure by XRD as shown in *Figure 3* and *Figure 4*.

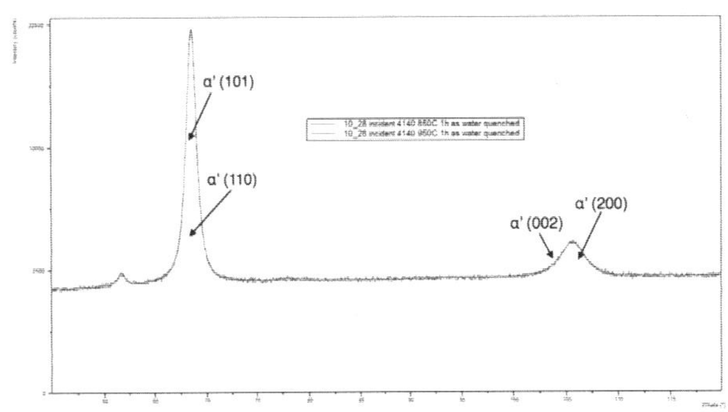

Figure 3 XRD pattern of AISI4140 quenched from 850C and 950C

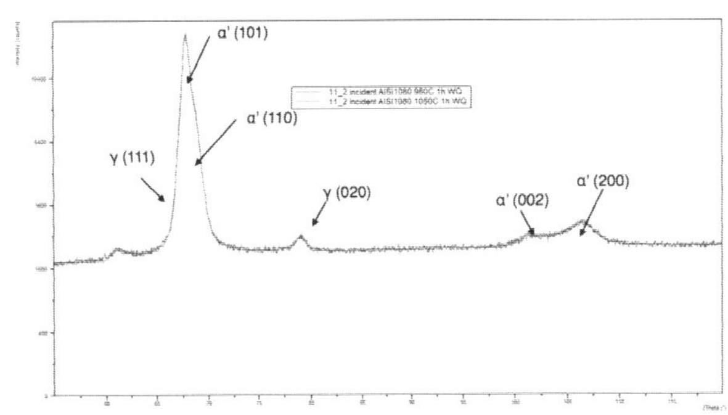

Figure 4 XRD pattern of AISI1080 quenched from 950C and 1050C

From *Figure 3*, α' (002) and α' (200) two peaks can be clearly observed from XRD pattern, which demonstrates that even for steels with less than 0.6 wt% carbon content, the as-quenched martensite crystal structure is still BCT, not BCC. For AISI4140 as shown in *Figure 3*, austenite is hardly noticed. For ASI1080 as shown in *Figure 4*, γ (020) can be clearly observed, which demonstrates retained austenite forms during water quenching process.

It is interesting to notice that although AISI4140 was quenched from different austenitizing temperature, 850C or 950C, the XRD patterns are the same. Similar conditions happened for AISI1080 and all the other steels in this paper. These phenomena can only occur when all the alloying elements already dissolve into austenite at lower austenitizing temperature. After the austenite is homogeneous, continuing increasing austenitizing temperature will only increase austenite grain size, but not affect martensite crystal structure. The Rockwell C hardness of AISI4140 at 950C quenching is lower than at 850C, which also demonstrates that grain size increase causes hardness drops.

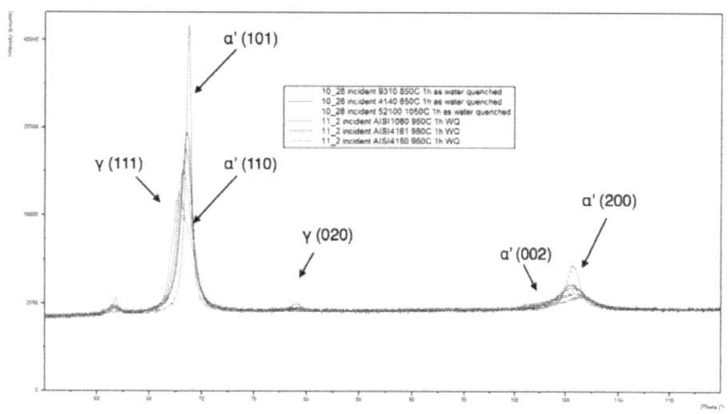

Figure 5 Martensite XRD patterns of steels with different carbon contents

Figure 5 presents martensite XRD patterns of steels with different carbon contents. From AISI9310 to ASI52100 steels, the martensite crystal structure changes gradually. α' (002) and α' (200) two peaks split away when the carbon content increases. γ (111) peaks drifts to lower angle and the intensity increases. It indicates the amount of retained austenite is increasing and the carbon content in retained austenite is also increasing. With the increase of carbon content, the α' (101) martensite peaks also drifts to lower angle, indicating larger lattice parameter and higher carbon content. The decreasing intensity of martensite peaks represents higher distortion order in martensite microstructure.

Although martensite has BCT structure, the slightly distorted c/a is not easy to measure. From *Figure 5*, for low carbon steel, the α' (002) and α' (200) two peaks are overlapped. For high carbon steel, α' (002), α' (200) and γ (111), three peaks are overlapped, which gives great difficulty to accurately measure and calculate c/a ratio. It is another error may have happened in previous experiments.

To accurate measure c/a ratio, Rietveld refinement can be used [11]. Using this technology devised by Hugo Rietveld can obtain the most accurate lattice parameters for martensite. HighScore Plus 4.1 developed by PANalytical was used to conduct Rietveld refinement analysis for martensite crystal structure.

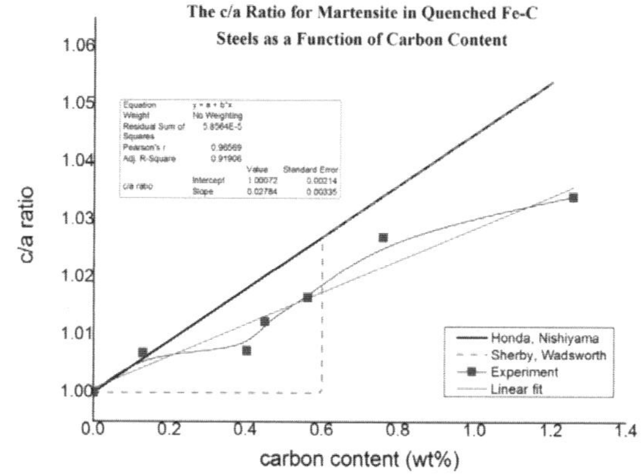

Figure 6 The c/a ratio for martensite in quenched Fe-C steels as a function of carbon content

Figure 6 presents the c/a ratio for martensite in quenched Fe-C steels as a function of carbon content. The black solid line represents classic model developed by Honda and Nishiyama. The red dotted line represents the model developed by Sherby and Wadsworth. The blue dots were measured c/a results by authors. The red solid line is the linear fit for measure c/a results by Origin.

Compared to classic liner model, the authors model has also the linear relationship between carbon content and c/a ratio with smaller slope. The equation for authors model is:

$$c/a = 1 + 0.028 \text{ wt\% } C$$

Considering the classic liner model, the slope 0.028 wt% is 38% smaller than the slope 0.045 wt%. The difference is from chemical composition error and c/a error. With modern OES machine and X-ray diffractometer, 0.028 wt% slope is considered more accurate.

Comparing to Sherby's model, the c/a ratio is clearly not 1 when the carbon content is less than 0.6 wt%. The reason that Campbell, Fink and Kurdumoff, G.[9] considered c/a ratio as 1is lack of more accurate X-ray diffractometer and Rietveld refinement with related software. The error bar of c/a ratio for low carbon steel is larger compared to for high carbon steel. However, the gradual peak shift and peak intensity decreasing of as-quenched martensite indicate the linear relationship between carbon content and c/a ratio.

Tempering at room temperature is another issue to be considered when accurately measuring martensite c/a ratio. The XRD patterns of AISI52100 simply water quenching and water quenching + liquid nitrogen cryogenic treating were measured and compared as shown in *Figure 7*. The XRD experiments were conduct within 30mins after the sample were water quenched or cryogenic treated. From the XRD patterns, after cryogenic treatment, the amount of retained austenite drops dramatically and the the amount of martensite keeps the same. The c/a ratio of martensite, as seen from α' (002) and α' (200) keeps the same, which demonstrates that the cryogenic treatment will not affect the martensite c/a ratio. Although the steels may be auto tempered during quenching, the measured c/a ratio is as close as the "real" as-quenched martensite crystal structure.

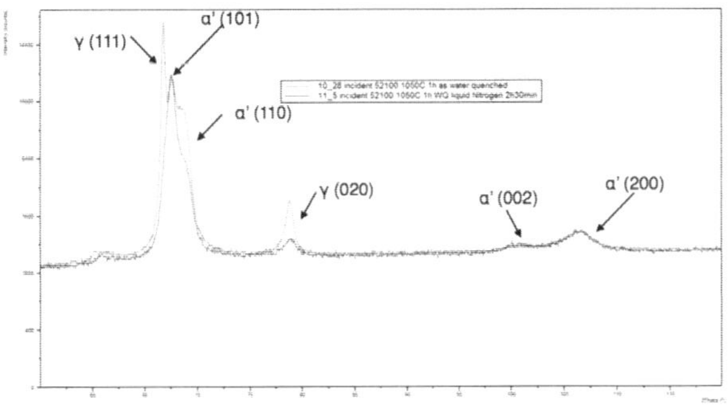

Figure 7 XRD patterns of AISI52100 steels before and after cryogenic treatment

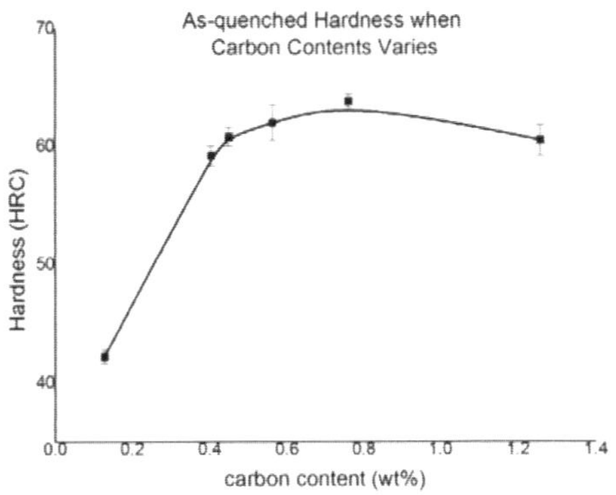

Figure 8 Steel as-quenched hardness when carbon contents varies

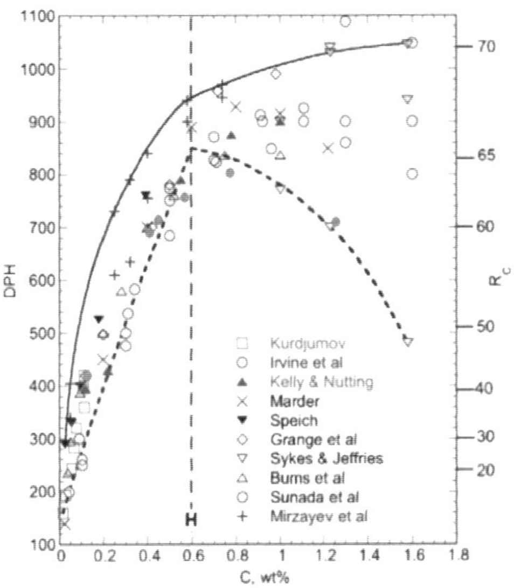

Figure 9 Measured as-quenched hardness by authors compared with the references [8]

Martensite hardness: Rockwell C test

Rockwell C hardness test were conducted on all the samples. The relationship between carbon contents and as-quenched hardness is shown in *Figure 8* and *Figure 9*. The relationship is linear until the carbon content exceeds 0.6 wt%. *Figure 9* is redrawn from Sherby's article [8]. The red dots are author's measured hardness data, which lies in the overall range.

Martensite morphology: optical microscope and SEM

Optical microscope

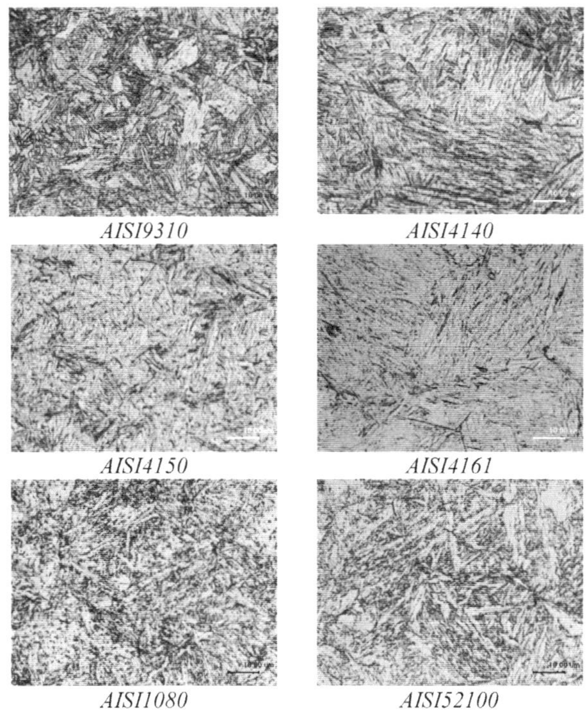

Figure 10 Optical microscopes for steels

Figure 10 presents the optical microscopies for different steel grades. Low carbon steels, including AISI9310, AISI4140, AISI4150 and AISI4161, present lath morphology. Higher carbon steels, including ASI1080 and AISI52100, present plate and lath morphology.

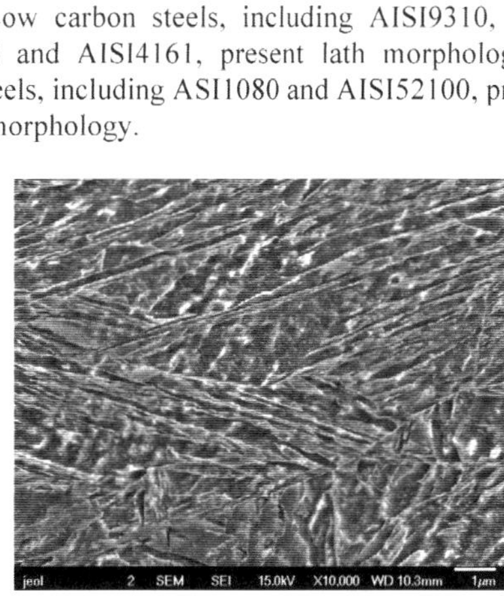

Figure 11 AISI9310 Magnification×10000

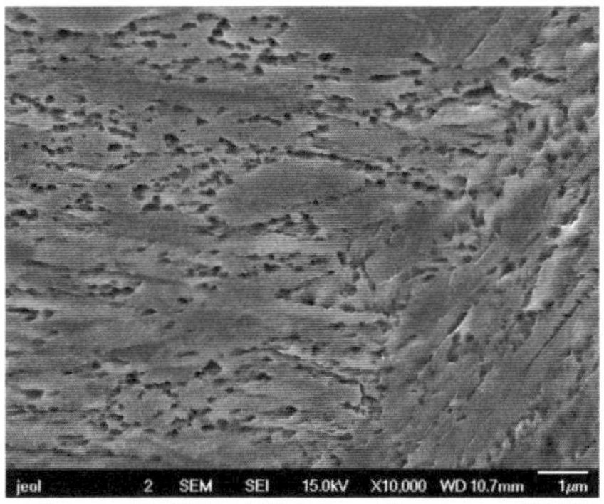

Figure 12 AISI4140 Magnification×10000

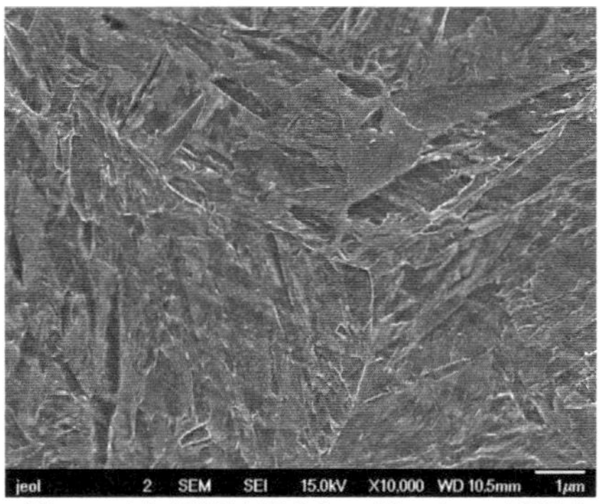

Figure 13 AISI4150 Magnification×10000

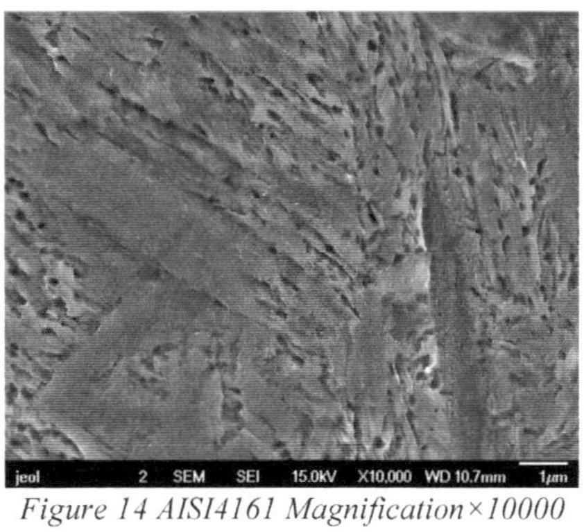

Figure 14 AISI4161 Magnification×10000

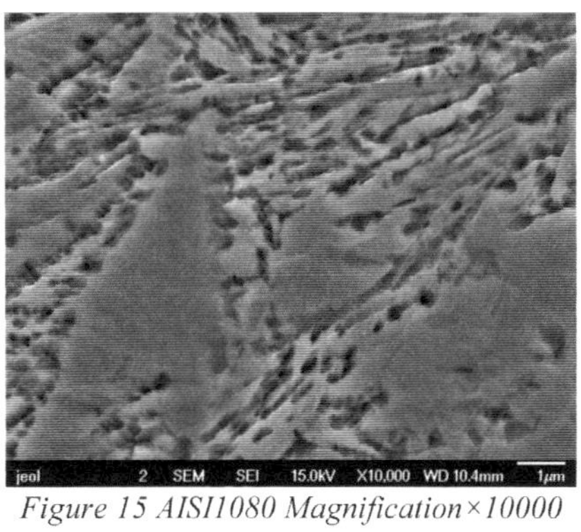

Figure 15 AISI1080 Magnification×10000

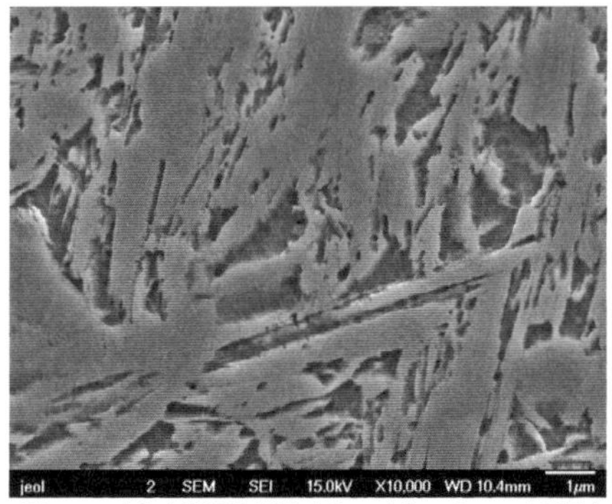

Figure 16 AISI52100 Magnification×10000

SEM

The martensite morphologies under high magnification were shown from *Figure 11* to *Figure 16*. With the increase of carbon content, the amount of lath martensite decreases and the amount of plate martensite increases.

Discussion

The new equation of martensite c/a ratio can be used to accurately estimate carbon content in as-quenched martensite and tempered martensite by knowing the c/a ratio. It could also help to develop more accurate hardness and other mechanical properties model based on martensite crystal structure.
Although the the relationship between carbon content and c/a ratio is linear, carbon atom segregation still exists in martensite as shown in *Figure 17* [12]. It should also be noted that for the steel with less than 1.5 wt% C, every martensite crystal cell does not contain even one carbon atom in the cell. The slightly distorted tetragonal structure, say c/a ratio, is an average expression of more than 40 Iron atoms and 1 or 2 Carbon atoms. A carbon cluster may describe the structure precisely.

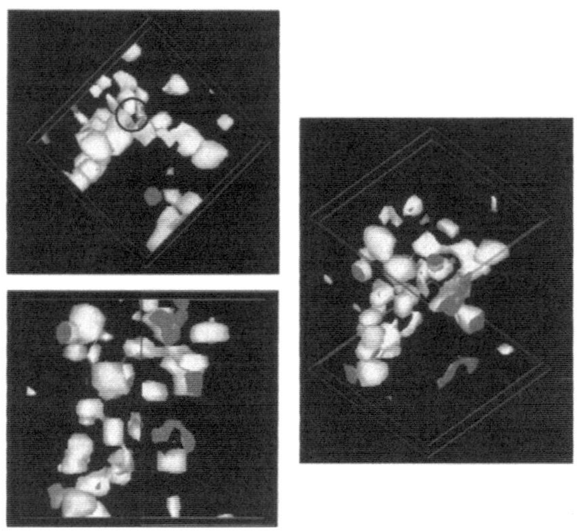

Figure 17 Carbon atom distribution around dislocation in martensite of an Fe-0.18 wt. % C alloy. Sensitive Atom Probe analysis: various reconstructions from a region of analysis 10×10×4nm in size [12]

Conclusions

In this paper, AISI9310, AISI4140, AISI4150, AISI4161, AISI1080 and AISI52100 were quenched to form fresh martensite. The XRD result shows that for the steel with less than 0.6 wt% carbon, the crystal structure is still BCT structure and it follows the relationship listed as below.

$$c/a=1+0.028wt\%C$$

The slope is smaller compared to classical c/a model developed before 1940s. The inaccuracy of previous well accepted model may be caused by less advanced X-ray diffractometer and less advanced OES machine.

This new model could be used to accurately estimate carbon content and hardness of as-quenched martensite and tempered martensite by only knowing c/a ratio of martensite.

Acknowledgments

The authors would like to thank John Imundo, Carl Ribaudo and Christopher Akey from the Timken Company in North Canton, Ohio to provide steels. The authors appreciated Dr. Boquan Li's help on X-ray diffraction data acquisition. Center for Heat Treating Excellence (CHTE) at Worcester Polytechnic Institute provided financial support. This research is under "Gas quenching steel hardenability standard" project, funded by CHTE, WPI.

References

[1] Portella, P. D., and R. Helmerich. "Adolf Martens and his contribution to materials engineering." Engineering Failure Analysis 43 (2014): 2-10.

[2] Edmonds, David V., and Adolf Karl Gottfried Martens. "The "Silicon Age" of Steel: How Alloying with Silicon Is Playing a Crucial Role in Modern Steel Developments."

[3] Krauss, George. "Martensite in steel: strength and structure." Materials science and engineering: A 273 (1999): 40-57.

[4] Bhadeshia, Harry, and Robert Honeycombe. Steels: microstructure and properties: microstructure and properties. Butterworth-Heinemann, 2011.

[5] Morito, S., et al. "The morphology and crystallography of lath martensite in Fe-C alloys." *Acta Materialia* 51.6 (2003): 1789-1799.

[6] Bain, Edgar Collins, and Harold W. Paxton. "Alloying elements in steel." 1966, 291 P. AMERICAN SOCIETY FOR METALS, METALS PARK, OHIO (1966).

[7] Sherby, Oleg D., et al. "The c/a ratio in quenched Fe-C and Fe-N steels-a Heuristic story." Materials science forum. Vol. 539. 2007.

[8] Sherby, Oleg D., et al. "Revisiting the structure of martensite in iron-carbon steels." Materials transactions 49.9 (2008): 2016-2027.

[9] Kurdumoff, G., and E. Kaminsky. "X-ray studies of the structure of quenched carbon steel." Nature 122 (1928): 475-476.

[10] PANalytical Empyrean brochure, http://www.panalytical.com/Empyrean/Specifications.htm

[11] Rietveld, HaM. "A profile refinement method for nuclear and magnetic structures." Journal of applied Crystallography 2.2 (1969): 65-71.

[12] J.Wilde, "M. Eng. Thesis", Department of Materials, Oxford University, 1996

International Federation of Heat Treatment and Surface Engineering 2016
Proceedings of the 23rd IFHTSE Congress
April 18–21, 2016, Savannah, Georgia, USA

In situ X-ray diffraction investigations of martensitic transformation and austenite modifications during quenching of low alloy steels

Jeremy Epp
Stiftung Institut fuer Werkstofftechnik, University of Bremen, Bremen, Germany
epp@iwt-bremen.de

Abstract

In situ X-ray diffraction measurements were performed at the European Synchrotron Radiation Facility in Grenoble, France during quenching of two low alloy steel grades: the ball bearing steel AISI 52100 (100Cr6) and the case hardening steel AISI 5120 (20MnCr5). Variations of the austenitizing temperature were used to achieve varying amounts of carbon in solid solution in the 100Cr6 steel. Two-dimensional diffraction frames were recorded during the complete heat treatment cycles. The integrated frames were analyzed in order to determine the temperature- and time-dependent evolution of phase contents and lattice parameters. With the help of the determined temperature dependent lattice parameters, different effects were investigated during quenching. It appeared that the austenite lattice parameter does not follow only a linear contraction during cooling but that a change of slope occurs systematically at temperatures below Ms. In the case of the ball bearing steel with austenitizing at high temperature, an accelerated decrease takes place, which could be attributed to the generation of high compressive hydrostatic stresses. On the other hand the case hardening steel exhibits a slower decrease of the austenite lattice parameter below Ms. A carbon enrichment of the austenite could be identified. For the ball bearing steel austenitized at low temperature, a competition of both effects is expected to take place.

Introduction

Martensitic transformation in steels has now been investigated for more than 100 years [1, 2] The interest in martensitic transformations is still very high as numerous industrial applications use this transformation to improve wear, mechanical and fatigue properties of parts in engineering components [3]. Moreover, new interest on fundamentals of martensitic transformations appeared in the last decades with the development of computer simulation, where kinetics, distortions and other phenomena have to be well described to reach reliable simulation results [4, 5].
In the context of control and reduction of distortion, FEM-simulation has become an unavoidable tool for the optimization of heat treatment processes. In order to perform reliable simulations, the knowledge about ongoing phenomena as well as reliable material data and models are required. However, in particular in industrial applications, simplified models and old material data are often used, leading to unreliable results.
In situ X-ray diffraction analysis has become a powerful method of materials characterization stimulated by constant advances in instrumentation and data processing. This method allows, contrarily to dilatometry or resistivity measurements, to obtain time-resolved quantitative information about every single phase present in the investigated material [6].
In the present study, in situ X-ray diffraction experiments were performed with synchrotron X-ray radiation at ESRF on beamline ID11 during heat treatment of two low alloy steels with varying parameters. By variations of the austenitizing temperature, different carbon contents in solution were set in hypereutectoid steel, leading to different behaviors during quenching. Two-dimensional diffraction frames were recorded with a time resolution of 0.8 seconds per measurement during the complete heat treatment cycles. The integrated diffraction frames were analyzed in order to determine the temperature- and time-dependent evolution of phase contents and lattice parameters. The martensitic phase transformation kinetics could be precisely described for both steel grades. With the help of the determined temperature dependent lattice parameters, different effects were investigated during quenching.

Experimental methods

Material

A typical ball bearing steel 100Cr6 (AISI 52100) and a carburizing steel grade 20MnCr5 (AISI 5120) were investigated. Their respective chemical compositions are given in Table 1.

Table 1: Chemical composition of both steels in Mass-%

Steel	C	Si	Mn	P	S	Cr	Mo	Ni
52100	0.95	0.21	0.44	0.012	0.005	1.45	0.052	0.11
5120	0.204	0.23	1.35	0.011	0.020	1.02	0.03	0.10

The initial microstructure of the 100Cr6 steel exhibits a ferritic matrix with globular carbides. An optical micrograph of one sample in its initial state after Nital etching is shown in Fig. 1a. The steel grade 20MnCr5 presents a ferrite/pearlite microstructure with a banded morphology (Fig. 1b). The width of the ferrite/pearlite bands is about 25 µm. Electron microprobe analyses showed that carbon, manganese and chromium concentrations present variations with the period of the microstructure for this alloy (about 25µm) [7].

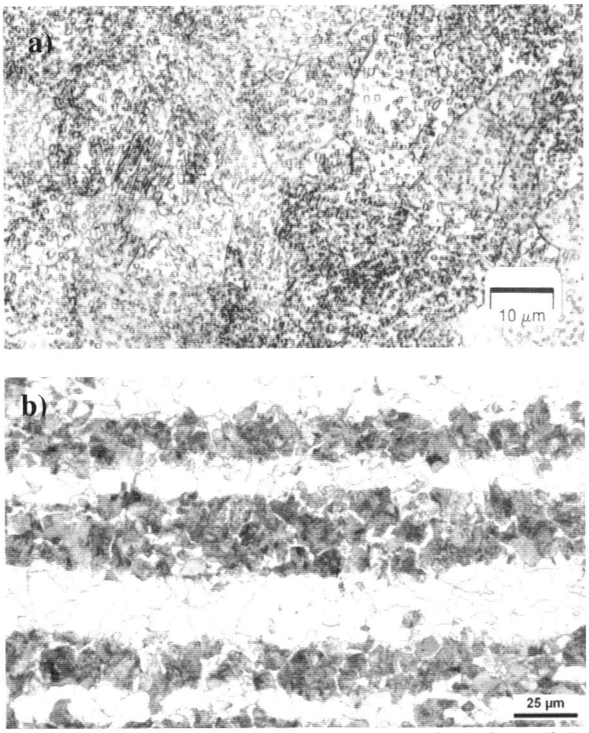

Figure 1: Microstructure of the investigated steels in their initial state: a) 100Cr6 (AISI 52100); b) 20MnCr5 (AISI 5120)

In situ XRD experiments

In situ X-ray diffraction experiments were performed at the European Synchrotron Radiation Facility (ESRF) in Grenoble, France, on beamline ID11. The experiments were executed with a heating device (ETMT, Instron) allowing a controlled heating of samples with different cross section of 1.5 mm with a length of 40 mm by resistivity.

Following heat treatment cycle was used for both steel grades: heating with a rate of 5 K/s up to different austenitizing temperature (Ta) between 800 °C and 940 °C followed by a soaking time of 15 min and quenching to room temperature (RT). The clamping tools huge mass allowed quenching of samples without use of any other quenching media. The different samples cross section led to slightly varying cooling rates with $t_{8/5}$ time around 2.5s. Constant argon purging was used to avoid oxidation of the surface. The temperature was controlled by a type K thermocouple welded on the surface of the samples.

During the entire heat treatment cycle, diffraction frames were recorded in transmission mode with a FRELON camera [8] using an exposition time of 0.7 seconds for each frame during quenching with additional time for read-out of 0.1 s. The beam energy was 71 keV and the primary beam size was set at maximum (about 100µm high and 300 µm width) in order to get the best statistical conditions in terms of diffracting domains. A standard material (LaB_6 powder) was measured to describe the instrumental contribution to the diffraction patterns. The recorded frames were integrated with the program Fit2d developed at ESRF after background correction for further analysis [9]. The analysis of diffraction patterns after integration has been performed with the Rietveld refinement software TOPAS© from Bruker AXS [10]. X-ray analysis of a fine polycrystalline material will result in a convolution of different microstructural features [11]. Different phases are present here and each phase may contribute in different way to the diffraction pattern according to local textures, anisotropic size and strain broadening as well as stacking faults. Microstrain and size broadening was introduced to model the peak widths. Several refinement parameters obtained were fixed and used for the analysis of the temperature dependent diffraction patterns obtained during quenching.

Results and discussion

Phase transformations during quenching

The 2d-diffraction frames recorded at different temperatures during quenching of a 100Cr6 sample austenitized at 915 °C is presented in Fig. 2. It can be observed that at high temperature after 15 min soaking, the diffraction rings are not completely homogeneous but that spots are present. This is due to grain coarsening leading to a decreasing sampling statistic at high temperature. During quenching, no significant changes of the diffraction rings can be observed until Ms is reached (170 °C). Then, diffraction rings belonging to martensite appear, and the diffraction rings of austenite become more and more closed due to a plastic deformation and related increasing number of crystallites.

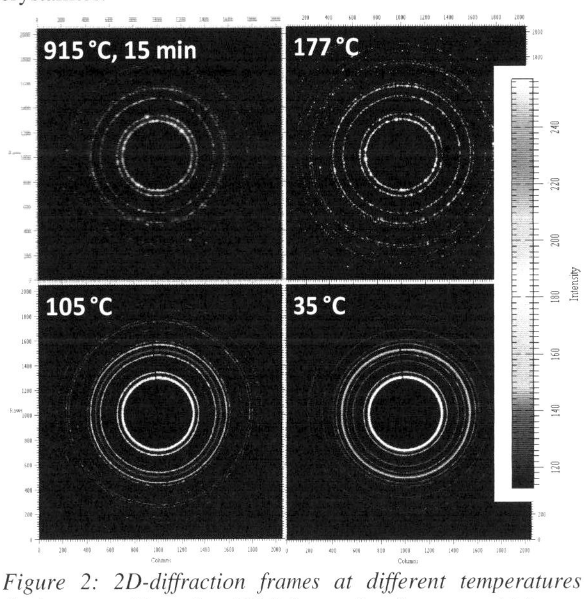

Figure 2: 2D-diffraction frames at different temperatures during quenching of a 100Cr6 sample after austenitizing at 915°C

After integration of the 2d-diffraction frames and analysis of the standard intensity-vs 2Theta diffraction patterns by the Rietveld method, the evolution of phase content could be precisely described. Fig. 3 shows the evolution of austenite content during quenching for different initial austenitizing temperatures of the 100Cr6 material and for one 20MnCr5 sample austenitized at 900 °C. In the case of the 100Cr6, changing amount of undissolved carbides (Fe_3C) remains in the microstructure after austenitizing, as expected. This explains the fact that the initial austenite amount is not at 100 % and that the initial austenite content is increasing with increasing austenitizing temperature. The kinetics of the martensitic transformation is well described. After quenching, the amount of retained austenite present in the 100Cr6 samples is 9, 14, 21 and 30 Mass-% for the austenitizing temperatures 825, 870, 905 and 940 °C respectively, while the 20MnCr5 steel exhibits about 5 Mass-% retained austenite. In the case of the 20MnCr5, small quantity of bainite formed above Ms (about 10 Mass-%).

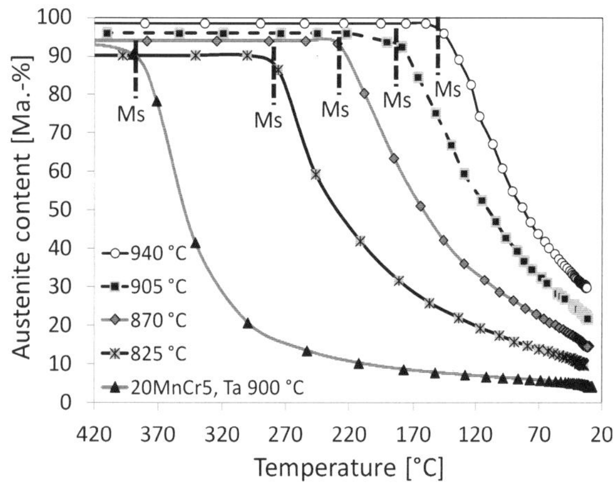

Figure 3: Evolution of austenite content during quenching of four100Cr6 samples for different austenitizing temperatures and of one 20MnCr5 sample

In order to model the austenite → martensite transformation kinetics, the model of Wildau was used (Eq. 1) [12]. It is based on a modification of the Koistinen-Marburger model by an additional exponent n. VM is the volume of martensite formed at the temperature Tq below the martensite start temperature (Ms) in °C, X is the fraction of austenite available just above Ms and b and n are variable parameters. The parameters b and n were determined by fit of the experimental curves in order to achieve a precise modeling of the phase transformation kinetics.

$$V_M = X - \exp\left[\frac{1}{b}(Ms - Tq)^n\right] \quad (1)$$

The determined carbon dependency of both parameters of the Wildau model for average cooling rates of 25 to 15 K/s are given in Eq. 2 and Eq. 3. For low carbon contents in solution (< 0.3 Mass-%), the asymptotic evolution of the carbon dependence of the parameter b gives slightly lower values than obtained experimentally. This can lead to uncertainties in the description of the transformation kinetics. Therefore, for 20MnCr5, the mean experimentally determined value of b should be used: b= -13.

$$b = -3.193 \times \exp(4.292 \times \%C) - 6 \quad (2)$$

$$n = 0.62919 - 0.01489 \times C\% + 0.58094 \times (\%C)^2 \quad (3)$$

Evolution of lattice parameters

From the Rietveld refinements, the lattice parameters of all present phases could be extracted. For martensite a nearly linear contraction with decreasing temperature could be observed. On the other hand, the evolution of austenite lattice parameter during cooling exhibits non-linear evolution for all experiments. The austenite lattice parameters are plotted in Fig. 4 as a function of the temperature during quenching of the considered four 100Cr6 samples austenitized at different temperatures between 825 and 940°C and for one 20MnCr5 sample. With increasing carbon content in solution (with increasing austenitizing temperature for 100Cr6), the lattice parameters are shifted to higher values, as expected. The thermal contraction with decreasing temperature is linear and almost parallel for all experiments until Ms is reached. During further cooling, it can be observed that for the 100Cr6 samples, a change of slope takes place with an accelerated decrease towards room temperature. On the other hand, the evolution of lattice parameter in 20MnCr5 sample exhibits the opposite effect: a non-linear decrease with continuously decreasing slope takes place below Ms.

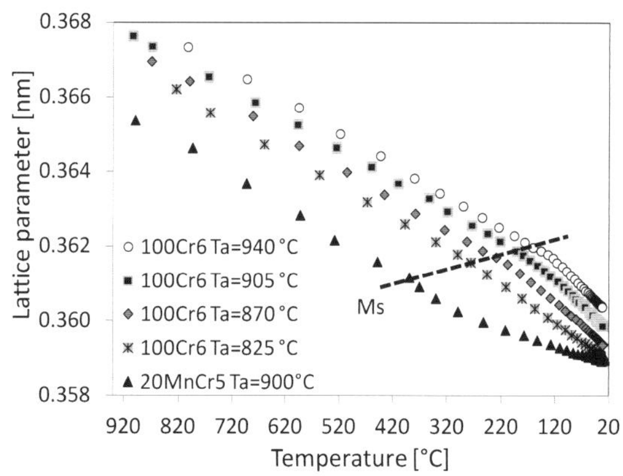

Figure 4: Evolution of austenite lattice parameter during quenching of 100Cr6 and 20MnCr5 samples

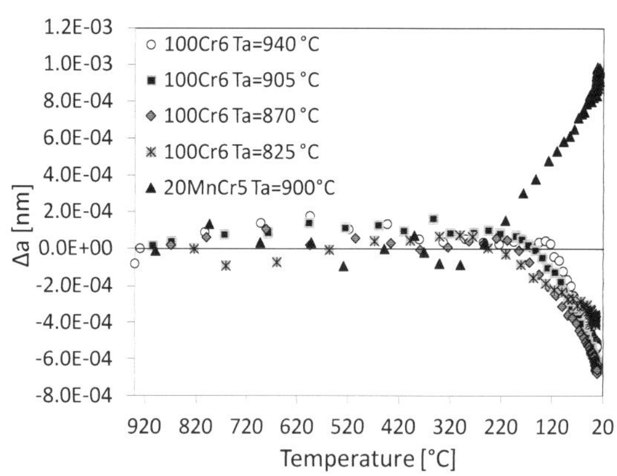

Figure 5: Deviation of the measured austenite lattice parameter from the extrapolated linear thermal contraction during cooling

In order to observe these changes of slope during cooling, the presented data were corrected by the respective linear contraction taking place above Ms ($\Delta a = a^{measured} - a^{thermal}$) and plotted again as a function of the temperature (Fig. 5). All values are close to 0 with small scattering until Ms is reached. During further cooling, the continuous decrease previously observed for the 100Cr6 samples can be clearly observed, while the 20MnCr5 sample exhibits a strong increase.

Carbon content in solution in 20MnCr5

The carbon content in solution in austenite during cooling can be obtained based on the measured lattice parameter evolution (Fig. 4). In order to determine the evolution of the carbon content in solution, the experimentally determined equation of Onink was used (Equation 4) [13]. In that equation, the effect of the carbon content (f_c in At.-%) on the thermal expansion coefficient is taken into account with T in Kelvin.

$$a_\gamma(f_C,T) = (0.363067 + 0.000783 \times f_C) \times \left(1 + (24.92 - 0.51 \times f_C) \times 10^{-6}[T - 1000]\right) \quad (4)$$

After conversion from At.-% to Mass-%, results were plotted in Fig. 6. The straight decreasing line represents the theoretical lattice parameter for 0% of carbon calculated with equation 4. It can be seen that the measured lattice parameter evolution is roughly parallel to the theoretical line from 920 °C to 520 °C with a constant shift. The average calculated carbon content in that temperature range is around 0.21 Mass-%. This fits to the nominal carbon content of this steel batch (0.204 Mass-%). From 450 °C until M_s (range of formation of slight amount of bainite), the calculated carbon content increases slightly. The slowed decrease of austenite lattice parameter in Fig. 4 leading to a positive deviation from the assumed linear contraction as seen in Fig. 5 during cooling below M_s (390 °C) leads to continuously increasing calculated carbon content in solution. Values of about 0.5 Mass-% are then reached at room temperature. Similar effects of carbon enrichment of austenite during quenching were already reported in the literature for low to medium carbon steels [1, 7, 14]. As a consequence of the increasing carbon content in solution in the remaining austenite, a dynamic stabilization of the austenite takes place, leading to an elevated amount of retained austenite after quenching [7].

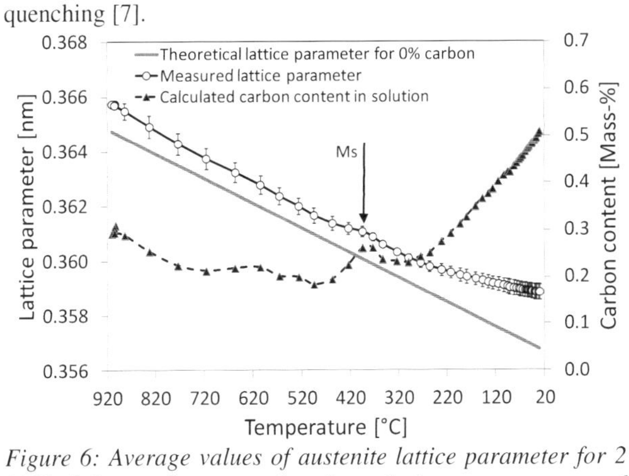

Figure 6: Average values of austenite lattice parameter for 2 20MnCr5 samples as a function of the temperature and calculated carbon content in solution

The same analysis of martensite lattice parameters was performed during quenching. The carbon content in solution in martensite during cooling can be calculated with Equation 5 [15]. For this, the ratio of tetragonal (BCT) martensite lattice parameters c/a is calculated and from the known linear dependence over carbon content, the carbon content in solution can be obtained at each temperature. The advantage of this method is that effects associated with temperature, thermal expansion correction or stresses are strongly reduced as only the tetragonality is considered and no reference values are needed.

$$\%C_M = \frac{\left(\frac{c_M}{a_M} - 1\right)}{0.0443} \quad (5)$$

The average evolution of measured lattice parameters as well as the calculated carbon content from two experiments is presented in Fig. 7 as a function of temperature. In the first stages of quenching (T > Ms) few % of ferrite are detected due to slight surface decarburizing and then about 10 % of bainite is formed, as already shown in [7]. The measured lattice parameters present large error bars but it can be seen that the lattice parameters "a" and "c" have almost the same value leading to a c/a ratio close to one and a calculated carbon content close to 0 as expected during the creation of ferrite/bainite with a BCC structure.

When martensite start temperature is reached, lattice parameter "c" increases quickly to reach a maximum value where it then starts to decrease linearly during further cooling. The lattice parameter "a" decreases continuously during the whole quenching process. The resulting carbon content calculated from the c/a ratio increases rapidly to a value of 0.19 Mass-% at Ms and then increases slightly to reach 0.20 Mass-% at room temperature. When Ms is reached, error bars become very small indicating a very good agreement between both experiments.

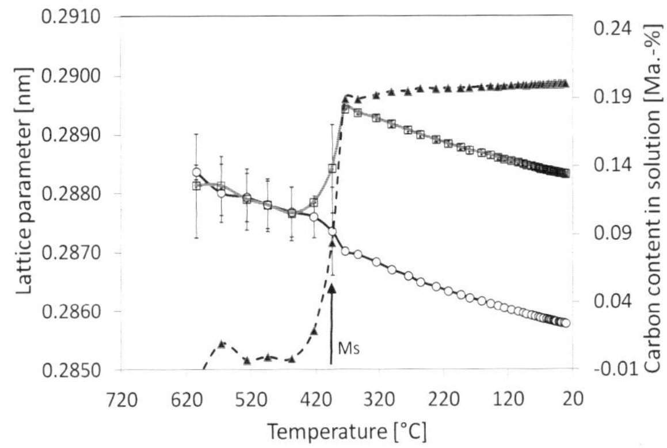

Figure 7: Average values of lattice parameters of ferrite/martensite for two 20MnCr5 samples as a function of temperature and calculated carbon content in solution

Generation of hydrostatic stresses in austenite

In the case of the steel 100Cr6 austenitized at different temperatures, the same analysis of austenite lattice parameter as performed for the steel 20MnCr5 was done. The determined evolution of carbon content in solid solution in austenite during quenching of the 4 samples already presented in Fig .4 are shown in Fig 8. For all samples a similar behavior can be observed: the determined carbon content in solution is almost constant at values between 0.6 and 1.0 Mass-% C according to the austenitizing temperature until Ms. At slight undercooling below Ms, the calculated carbon content in solution exhibits a continuous decrease until room temperature is reached as a result of the negative deviation observed in Fig. 5. The total decrease is different for the four austenitizing temperatures. As the cooling rate is high and Ms is low (in particular for austenitizing above 900 °C), it appears improbable that such strong carbon loss due to self-tempering can occur in austenite during quenching. Therefore, the only plausible explanation for the strong decreasing austenite lattice parameter below Ms is the generation of strong hydrostatic compressive stresses.

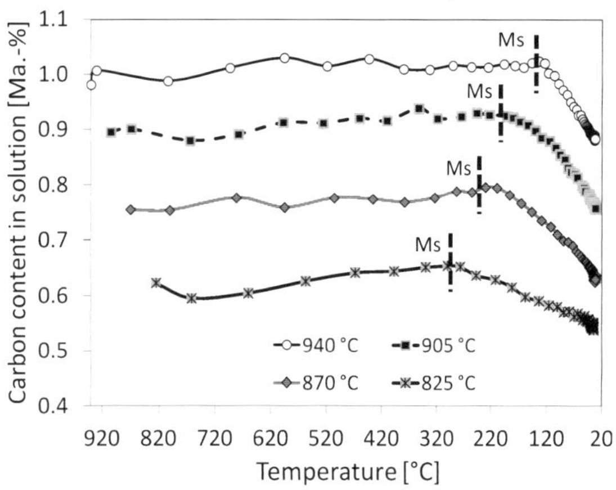

Figure 8: Evolution of carbon content in solution in four 100Cr6 samples for different austenitizing temperatures (the decreasing carbon content below Ms is not a real effect but is attributed to the generation of hydrostatic stresses)

Few studies can be found in the literature about residual stresses in retained austenite during or after quenching of steel. Several authors describe the residual stress state within retained austenite existing at room temperature as a hydrostatic residual stress state under high compressive stresses [15- 17]. The reason for this would be the very large volume expansion associated with the martensitic transformation (> 3%). Of course, due to shear-processes during the martensitic transformation, local residual stresses will not be purely hydrostatic in a single austenite region. However, X-ray diffraction techniques give average information of thousands of "grains" which in general are randomly oriented, and therefore the average information contained in the measured volume might be predominantly hydrostatic.

In the literature, two different sets of data concerning the evolution of austenite lattice parameter at room temperature as a function of the carbon content can be found: the one based on room temperature measurements after quenching (retained austenite) and the one based on high temperature measurements of austenite with thermal expansion correction. From the literature data, equations were developed based on the room temperature measurements (a^{RT}=0.3556 + 0.00443 × % C) and on the high temperature measurements ($a^{stress\text{-}free}$=0.3573 + 0.00327 × % C) with %C in Mass.-% [15]. It can be assumed that the main difference between the high temperature data and the measurements of retained austenite at room temperature is the generation of hydrostatic residual stresses as a consequence of the large transformation strain and the different thermal expansion coefficients of austenite and martensite.

Both equations from the literature can be used to quantify the possible existing hydrostatic residual stress state in retained austenite depending on the carbon content in solution. If it is assumed that the high temperature lattice parameters describe the stress-free state while the equation based on room temperature measurements represents the retained austenite under stress after quenching, equation 6 can be used to calculate the corresponding strain (ε^{RA}) in retained austenite.

$$\varepsilon^{RA} = \frac{\Delta a}{a^{stress-free}} = \frac{a^{RT} - a^{stress-free}}{a^{stress-free}} \quad (6)$$

The resulting hydrostatic residual stresses at room temperature can then be calculated by equation 7 with E^{RA}=207 GPa and ν^{RA}=0.28.

$$\sigma^{RA} = \varepsilon^{RA} \times \frac{E^{RA}}{\left(1 - 2\nu^{RA}\right)} \quad (7)$$

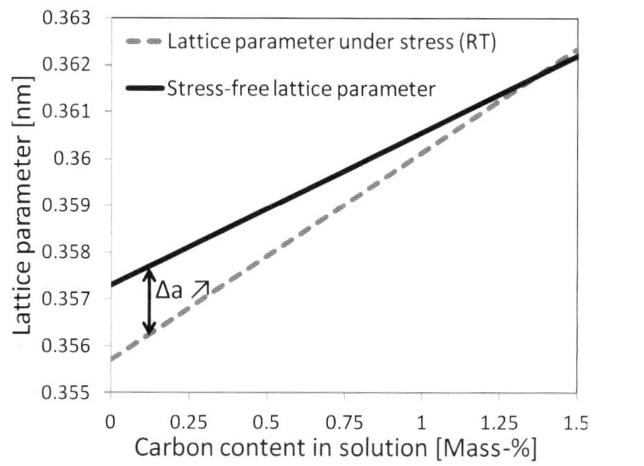

Figure 9: Evolution of the stress-free ($a^{stress\text{-}free}$) and the RT austenite lattice parameters (a^{RT}) as a function of the carbon content in solution (data from the literature)

The evolution of the literature data of the stress-free lattice parameter of retained austenite ($a^{stress\text{-}free}$) and of the lattice parameter of retained austenite after quenching (a^{RT}) depending on the carbon content in solution is presented in Fig. 9. From these theoretical evolutions, it can be observed that at values close to 1.4 Mass-% C both lattice parameters are almost similar and with decreasing carbon content, the gap is increasing continuously. From these data and together with equations 6 and 7, resulting theoretical hydrostatic residual stresses were calculated and plotted in Fig. 10. At 1.4 Mass-% C, the theoretical residual stresses are close to 0 as Δa is very small. With decreasing carbon content in solution, the theoretical values go to always increasing compressive residual stresses. For carbon-free steel (0 % C) a theoretical value of -2200 MPa is obtained. Of course it is thermodynamically almost impossible to have retained

austenite in low alloy steels without carbon. However, it appears from these calculations that according to the data of the literature, retained austenite might be under very large hydrostatic compressive residual stresses after quenching.
The experimental data collected at RT within the frame of the present study were then used to calculate hydrostatic residual stresses in retained austenite based on equations 6 and 7 using the stress-free lattice parameter from the literature, and plotted in Fig. 10. It can be observed that the experimental values can be divided into two distinct zones. For %C < 0.76 Mass.-% the values are close to the theoretical line obtained from the literature data for carbon content in solution with values between -750 MPa to -300 MPa, but with a slight shift. Discrepancies between the experimental values (triangles) and the theoretical evolution (black line) in this range of carbon content can either result from an effect of alloying elements on lattice parameters which can lead to slight variations of the stress-free references, or from a systematic shift in the lattice parameter determination due to the experimental setup.

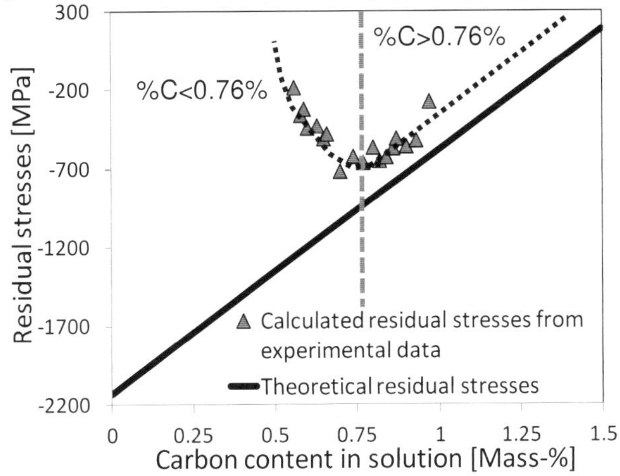

Figure 10: Theoretical hydrostatic residual stresses in austenite calculated based on literature data (black line) and values calculated from lattice parameters measured in the present in situ experiments using stress-free lattice parameters from the literature

When the carbon content in solution decreases below 0.76 Mass-% a continuous deviation of the experimental values from the theoretical line can be observed. In the 20MnCr5 steel, a pronounced carbon enrichment of the austenite could be identified during the martensitic transformation due to carbon partitioning (Fig. 6). With decreasing carbon content in solution in the 100Cr6 steel, Ms increases and therefore carbon diffusion is accelerated. As a consequence, short distance thermally activated carbon partioning from martensite to austenite could also take place and lead to an increasing deviation from the theoretical line for decreasing carbon content in solution above Ms in the case of the 100Cr6 steel. This shift of the stress-free lattice parameter would then lead to "pseudo shifts" of the stress values to tensile which are not real. However, in order to take this into account further investigations have to be performed.

Beside the analysis of the room temperature values, it is possible to evaluate the development of residual stresses during the whole quenching process with the help of the in situ XRD data. As observed previously, the measured lattice parameters of retained austenite during quenching show a non-linear change of the slope (continuous decrease) below Ms for the 100Cr6 samples (Fig. 4 and 5). The change of slope of austenite lattice parameter at temperatures below Ms can be explained by the generation of large compressive hydrostatic stresses. In order to obtain the temperature-dependent stress-free lattice parameter, a linear extrapolation of the austenite lattice parameter measured in the present experiments at temperatures above Ms down to RT was used (evolution presented in Fig. 4). For the further evaluations, only the experiments with initial carbon content in solution above 0.76 Mass.-%C will be considered, as the effect of possible carbon enrichment for lower carbon contents in solution cannot be taken into account at this point of the investigations. The stress calculations were performed by using Eq. 6 and 7 at each temperature during cooling, taking into account the temperature-dependent changes of elastic properties as given in [18].
The calculated evolutions of hydrostatic compressive stresses in austenite during cooling of three experiments with 100Cr6 steel during are presented in Fig. 11. It can be observed that the increase of compressive stresses is low at the very beginning of the transformation. After certain undercooling the compressive stresses increase continuously with decreasing temperature until room temperature is reached. Towards room temperature small cooling steps lead to a large increase of the compressive stresses.
The generation of large hydrostatic compressive stresses in austenite during the martensitic transformation is assumed to be related to the volume change due to martensitic transformation. Therefore, the plot of the evolution of compressive stresses in austenite as a function of the amount of retained austenite (Fig. 12) can give additional information. Here, it can be noticed that for all experiments, only a slow increase of residual stresses takes place in the first stage of the transformation. For increasing amounts of transformed austenite, the increase of compressive stresses follows an almost exponential behavior. The present measurements give average information from a volume of 100 × 300 × 1500 μm^3. When only a small amount of austenite transformed into martensite, the austenite regions that are unaffected by the transformation strain are much larger than the affected region. The average peak shift which is measured is therefore moderate. When the amount of austenite affected by the formed martensitic plates is increasing, the recorded peak shifts (due to stresses) is getting larger. This effect is expected to be superimposed to the proper effect of increasing strains in austenite when an increasing part of the surrounding regions transforms to martensite. Due to both effects, the detected increasing compressive stress exhibits a strong dependence on the amount of transformed austenite.

The residual stress values that are reached at RT after cooling are in the range of -900 and -750 MPa. The end values for these evaluations using the stress free lattice spacing extrapolated from high temperature values correspond to the residual stress values obtained by using the literature data as presented in Fig. 10.

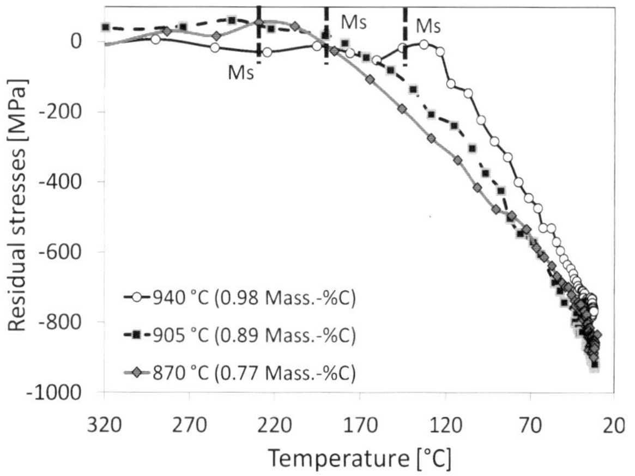

Figure 11: Development of hydrostatic stresses in austenite during quenching of 100Cr6 samples with different austenitizing temperatures as a function of the temperature

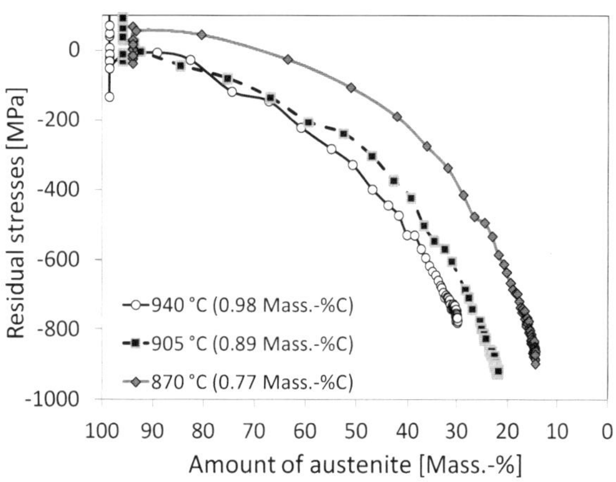

Figure 12: Development of hydrostatic stresses in austenite during quenching of 100Cr6 samples with different austenitizing temperatures as a function of the austenite content

It should be kept in mind that a part of the results obtained here is based on assumptions concerning the stress free lattice parameter of austenite. Indeed, diffraction methods are sensitive to residual stresses as well as to changes of chemical composition resulting in different lattice parameters. A proper separation of both effects is therefore difficult. However, with the help of the in situ experiments, clear evidences were obtained. In the case of the 20MnCr5 steel, carbon enrichment of the austenite could be observed. In the case of the 100Cr6 steel for carbon contents higher than 0.76 Mass-% C no carbon enrichment of the austenite is expected to occur as Ms is low (< 220 °C) and so the diffusion of carbon at long distances is strongly inhibited. In this range of carbon contents (> 0.76 Mass-% C), the experimentally determined high compressive residual stresses are close values calculated on the base of literature data. This confirms that the presented results are reliable. Additionally, full residual stress tensors measured in retained austenite at 100Cr6 samples were presented in a previous paper [19] and the hydrostatic nature of the present stresses could be demonstrated, validating the present results.

Conclusions

In situ XRD measurements were performed at the European Synchrotron Radiation Facility in Grenoble, France on ID11 during quenching of two low alloy steel grades: the ball bearing steel AISI 52100 (100Cr6) and the case hardening steel AISI 5120 (20MnCr5). Variations of the austenitizing temperature were used to achieve varying amounts of carbon in solid solution in the hypereutectoid steel. Two-dimensional diffraction frames were recorded with a time resolution of 0.8 seconds per measurement during the complete heat treatment cycles. The integrated diffraction frames were analyzed by using the Rietveld-method in order to determine the temperature- and time-dependent evolution of phase contents and lattice parameters.

By investigating the changes of austenite lattice parameter during the quenching process, it could be observed that both steel grades exhibit a non-linear evolution below Ms but in opposite direction: a slowed decrease in the case of the 20MnCr5 and an accelerated decrease for the 100Cr6 steel. In the case of the 20MnCr5 steel, a carbon enrichment of the austenite has been identified to be the responsible mechanism. This results in a carbon enriched austenite (0.5 Mass-% C or even more). The carbon enrichment of austenite explains the "high" amount of retained austenite at room temperature due to a dynamic stabilization. In the case of the 100Cr6, the generation of almost hydrostatic compressive residual stresses in the austenitic phase is the dominant effect. This assumption is supported by calculated hydrostatic residual stresses based on data from the literature for high-temperature austenite and tensor measurements at room temperature. With the present in situ experiments, the evolution of hydrostatic stresses in austenite could be followed during the whole quenching process and showed the generation of large compressive residual stresses. For 100Cr6 samples with low austenitizing temperature, the two competing effects are expected to take place. The consideration of the combined influence of carbon content variation and generation of compression stresses in the austenite will require further investigation in the future in order to characterize both effects.

Acknowledgments

The authors gratefully acknowledge the Deutsche Forschungsgemeinschaft (DFG) for financial support of the project HU-905/5-1 as well as Loredana Erra and the ESRF for support and provision of synchrotron radiation facilities.

References

[1] Olsen, G.B.; Owen, W.S., Martensite. First edition. ASM International (1992).

[2] Nishiyama, Z.; Fine, M.E.; Meshii, M.; Wayman, C.M., Martensitic Transformation, Academic Press (New York, 1978).
[3] Metals Handbook, Vol. 1: Properties and Selection: Irons, Steels, and High-Performance Alloys, tenth edition, ASM International (1990).
[4] Acht, C.; Clausen, B.; Hoffmann, F.; Zoch, H.-W., "Simulation of the distortion of 20MnCr5 parts after asymmetrical carburization," *Proc. 1st Int. Conf. on Distortion Engineering*, Bremen, September 2005, pp.251-258.
[5] Frerichs, F.; Lübben, T.; Hoffmann, F.; Zoch, H.-W., "Numerical analysis of distortion due to inhomogeneous distribution of martensite start temperature within SAE 52100 bearing rings," *Steel Research Int.*, (2007), pp. 558-563.
[6] Epp, J.; Surm, H.; Kessler, O.; Hirsch, T., "In-Situ X-Ray Investigations and Computer Simulation during Continuous Heating of a Ball Bearing Steel," *Metal Mater Trans A.*, 2007, pp. 2371-2378.
[7] Epp, J.; Hirsch, T.; Curfs, C.: "In situ X-Ray Diffraction analysis of carbon partitioning during quenching of low carbon steel," *Metallurgical and Materials Transaction*, Vol. 43A (2012), pp. 2210 – 2217.
[8] Labiche, J.-C.; Mathon, O.; Pascarelli, S.; Newton, M.A.; Guilera Ferre, G.; Curfs, C.; Vaughan, G.; Homs, A., "The fast readout low noise camera as a versatile x-ray detector for time resolved dispersive extended x-ray absorption fine structure and diffraction studies of dynamic problems in materials science, chemistry, and catalysis," *Chemistry and catalysis*, Review Sci. Inst, (2007).
[9] Hammersley, A.; Svennson, S.O.; Thompson, A., "Calibration and correction of spatial distortions in 2D detector systems," *Nucl. Instrum. Methods A.*, (1994).
[10] Young, R. A., The Rietveld Method, IUCr Book Series, Oxford University Press (1993), pp. 1-39.
[11] Warren, B.E., X-Ray Diffraction, Addison-Weseley (Reading, Mass., 1969).
[12] Wildau, M., Zum Einfluß der Werkstoffeigenschaften auf Spannungen, Eigenspannungen und Maßänderungen von Werkstücken aus Stahl, PhD-Thesis, Technische Hochschule Aachen (Aachen, 1986).
[13] Onink, M.; Brakman, C. M.; Tichelaar, F. D.; Mittemeijer, E. J.; van der Zwaag, S., "The lattice parameters of austenite and ferrite in FeC alloys as functions of carbon concentration and temperature," *Acta Metall et Mat.*, Vol. 29 (1993), pp. 1011-1016.
[14] Sherman, D. H.; Cross, S. M.; Kim, S.; Grandjean, F.; Long, G. J.; Miller, M. K., "Characterization of the carbon and retained austenite distribution in martensitic medium carbon, high silicon steel," *Metall. Mater. Trans.*, A. 38 (2007), pp. 1698-1711.
[15] Cheng, L.; Böttger, A.; de Keijser, Th.H.; Mittemeijer, E. J., „Lattice parameters of iron-carbon and iron-nitrogen martensites and austenites," *Scripta metallurgica et materialia*, Vol. 24, (1990), pp. 509-514
[16] Golovchiner, K. Y.,"Changes in the austenite lattice parameter during the martensitic transformation in steel," *Physics of Metals and Metallography*, Vol. 37-2 (1974), pp. 126-130.
[17] Villa, M.; Pantleon, K. Somers, M. A. J., "Evolution of compressive strains in retained austenite during sub-zero Celsius martensite formation and tempering," *Acta Materialia*, Vol.65 (2014), pp. 383–392.
[18] Richter, F., Physikalische Eigenschaften von Stählen und ihre Temperaturabhängigkeit, Verlag Stahleisen GmbH (Düsseldorf, 1983).
[19] Epp, J., "Investigation of triaxial stress state in retained austenite during quenching of a low alloy steel by in situ X-ray diffraction," *Advanced Materials Research*, Vol. 996 (2014), pp. 525-531.

International Federation of Heat Treatment and Surface Engineering 2016
Proceedings of the 23rd IFHTSE Congress
April 18–21, 2016, Savannah, Georgia, USA

Hardness-toughness relationship of a boron alloyed quench and tempering steel after different heat treatment cycles

Reinhold S. E. Schneider, Katharina Steineder
Univ. of Appl. Sciences Upper Austria, Wels, Austria

Yuri Toshima, Masahiro Okumiya
Toyota Technological Institute, Nagoya, Japan

Abstract

The hardness-toughness relationship for the steel 42MnNiCrB4-3 was investigated by a dilatometer. This included standard heat treatment (850 °C, 30 min), short-cycle heat treatment (950 °C, 16 s) and double short-cycle heat treatment (2 × 950 °C, 16 s). The tempering process lasted one hour for all samples and temperatures up to 500 °C were examined. Furthermore, isothermal transformation treatment (austempering) with 320 / 360 / 400 °C were performed. To characterize these heat treatment conditions, three notched impact samples were used. After the dilatometer experiments, impact testing and Vickers micro-hardness testing was performed. Additional investigation on the tested samples comprised XRD, LOM as well as SEM. Regarding the hardness-toughness relationship, there are no significant differences between standard HT and short-cycle HT. The results showed a severe reduction in toughness properties due to the effect of 300 °C - brittleness (blue brittleness). Tempering at 200 °C provides a good combination of this relationship for each heat treatment condition. The best hardness-toughness relationship was achieved by isothermal treatment at 320 °C and 360 °C. There is no cleavage fracture at any conditions, but a significant change in fracture surface appearance could be found at the isothermal transformation treatment at 400 °C. For all heat treatment conditions tempering at 250 °C or higher removes the retained austenite.

Introduction

Rapid heat treatment such as induction or laser hardening, has found widespread attention for the heat treatment of machinery components as well as for some tools due to the possibility of the in-line integration of the heat treatment into standard production routes and/or the limitation of the heat treatment to small areas of a work piece such as the surface or edges e.g. [1-5]. Despite the wide application of this technology, there are still only few material data on toughness properties. For the lower toughness range of cold-work tool steels there are some studies available using unnotched impact specimens [6-9]. Expanding the data range to higher impact values the usage of notched impact specimen was applied to characterize the properties of multi-layer welding and post-weld heat treated materials as well as engineering steels [10-13].

In this study, the hardness-toughness relationship of quenched and tempered 42MnNiCrB4-3 was investigated by this established testing procedure. To overcome the reported toughness loss caused by 300 °C - embrittlement [12,13] in the temperature range between 200 °C and 300 °C, the mechanical properties resulting from an alternative isothermal heat treatment were additionally investigated. Especially for sharp edges or drill holes, which are in danger of cracking, austempering is preferable to quenching and tempering because of the residual compressive stresses at the surface [14,15]. Furthermore, a superior overall hardness-toughness relationship is reported of the bainitic microstructure compared to tempered martensite for certain tempering conditions [16,17].

Experimental

Material
The material investigated in this study was the boron alloyed quench and tempering steel 42MnNiCrB4-3. The chemical composition is given in Table 1. Notched micro impact samples (3 mm × 4 mm × 27 mm), according to DIN 50115 but with a notch radius of 0.2 mm, were spark-eroded from hot-rolled strips.

Table 1: Chemical composition of the steel 42MnNiCrB4-3 in wt.-[%]

C	Si	Mn	Al	Cr	Ni	Ti	B
0.42	0.251	0.85	0.036	0.27	0.7	0.039	0.001

Heat Treatment
All heat treatment experiments were performed on a dilatometer (Bähr Dil 506 A/D) using a modified induction system to generate the heat in the material. The program included standard heat treatment cycles as well as single and double short-cycle heat treatments, all of them quench and tempering cycles including a furnace tempering procedure. Furthermore, austempering with an integrated isothermal transformation was also performed.
The various heat treatment parameters are summarized in Table 2 and were similar to those reported in [12,13]. For the standard heat treatment (standard-HT) the samples were heated to the austenitizing temperature of 850 °C in 10 min. for a holding time of 30 min. In comparison the short-cycle heat

treatment (short-cycle-HT) used a 4 s. linear heating from room temperature to austenitizing temperature of 950 °C with a successive holding period of 16 s. The quenching step after austenitizing was performed using nitrogen gas. In difference to [12,13], the 1 hour tempering was performed in a chamber furnace. Different tempering temperatures up to 500 °C were examined.

Table 2: Heat treatment parameters (T_A ... Austenitizing temperature, HT ... Heat Treatment)

HT	**heating**	**holding**	**T_A**	**Tempering**
Standard	10 min	30 min	850 °C	60 min
Short-cycle	4 s	16 s	950 °C	at 150, 200, 250, 300, 400, (500) °C
				Isothermal transformation
Austempering	10 min	30 min	850 °C	5 min at 320, 360 400 °C

For the austempering process, quenching was interrupted at three different temperatures above the martensite start temperature Ms, followed by a 5 min. holding period and successive gas-quenching to room temperature. The necessary holding time at the isothermal transformation temperature was determined by prior dilatometric experiments to guarantee a complete bainite transformation. Figure 1 shows these different heat treatment cycles.

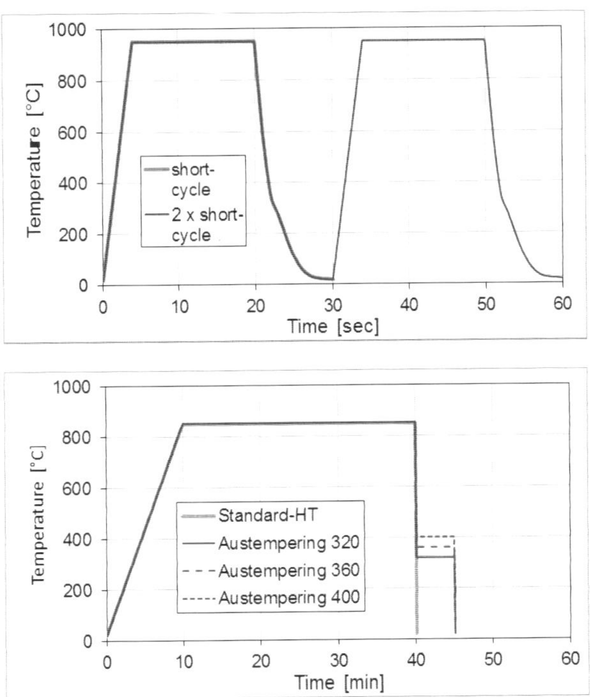

Figure 1: Time-temperature-chars for the different heat treatment cycles; top: short-cycle heat treatment, bottom: standard heat treatment and austempering

To characterize one heat treatment condition at least three notched impact specimens were used. More detailed information on the experimental procedures can be found in [9,10].

Testing Methods

After dilatometric experiments, the samples were tested regarding their impact toughness and Vickers hardness. Impact testing was performed with a 15 J hammer and 22 mm opening width between the counter bearings. Vickers micro-hardness testing with a load of 9.807 N (HV1) was applied. Additional investigation on the tested samples comprised light optical (LOM) and electron optical microscopy (SEM) of the microstructure and the fracture surface as well as X-ray diffraction with Mo-Kα-radiation for the determination of the retained austenite content according to ASTM E 975-03.

Results

Hardness-toughness-relationship

Figure 2 shows the hardness-toughness relationship of the steel 42MnNiCrB4-3 for the three conditions, standard-HT, short-cycle-HT and double short-cycle-HT. For the standard-HT with an austenizing at 850 °C and successive tempering, the curve exhibits in general the expected increase in toughness from about 20 J/cm² to 60 J/cm² with rising tempering temperatures, while the hardness decreases from about 650 HV1 to 350 HV1. Only for the range between 200 °C and 300 °C the results show a severe reduction in toughness due to the effect of 300 °C - brittleness (blue brittleness), while hardness is dropping to about 500 HV1.

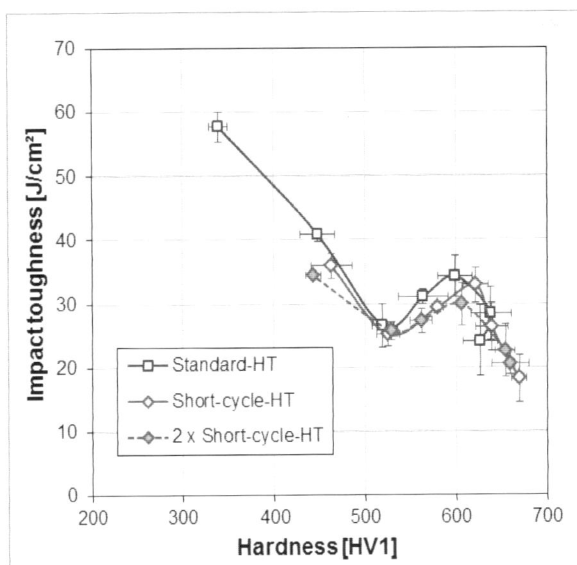

Figure 2: Hardness-toughness relationship of the steel 42MnNiCrB4-3 ; Standard heat treatment compared to short-cycle heat treatment

Short-cycle-HT, with an austenitizing temperature of 950 °C, leads to a similar hardness-toughness relationship as the standard-HT with slightly higher hardness values directly after quenching. Again, tempering at 200 °C provides a favorable property combination. Double short-cycle-HT showed a slight drop in toughness but with similar result within the scatter band of the others. Austempering at 320 °C and 360 °C leads to a significantly better hardness-toughness relationship at hardness levels below 600 HV1 than standard-HT (Fig. 3). The results almost follow the trend of the quenched and tempered samples with tempering temperatures up to 200 °C. Austempering at 400 °C, which results in the lowest hardness of all isothermal treated samples, exhibits a lower toughness level similar to those of quenched and tempered samples at higher tempering temperatures.

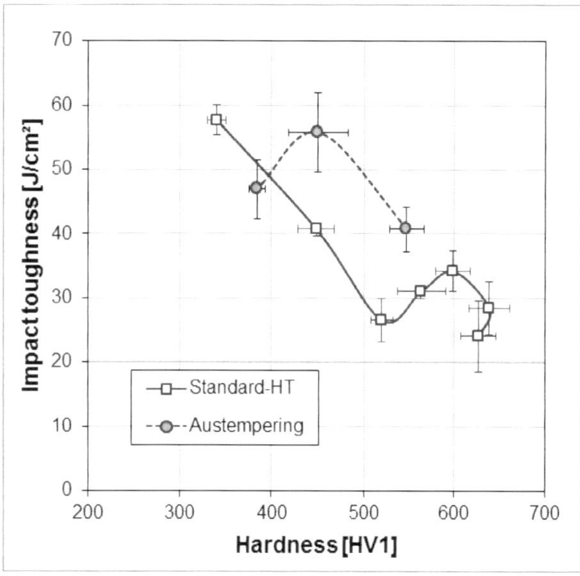

Fiugure 3: Hardness-toughness relationship of the steel 42MnNiCrB4-3; Quenched and tempering compared to austempering

Fracture surfaces

After the impact testing the fracture surfaces of the samples were investigated by SEM. Figure 4 shows fracture surfaces for selected heat treatment conditions. Independent from the austenitizing conditions, there was no cleavage fracture at any of the quenched and tempered samples due to the generally high toughness level (Fig. 4a). Micro dimples, which are typical characteristics of ductile fracture, appear at all quenched and tempered samples. A similar fracture behavior could be found for the austempered samples with austempering temperatures of 320 °C and 360 °C (Fig. 4b).

A significant change in the appearance of the fracture surface could only be found for the isothermal transformation at 400 °C (Fig. 4c). A less ductile fracture leads to a quasi-cleavage fracture. Due to this behavior the material shows a lower level of the impact toughness compared to the other austempered samples, but with similar properties as the standard-HT, tempered between 400 °C and 500 °C. Furthermore, some small TiN were also visible at all fracture surfaces.

a)

b)

c)

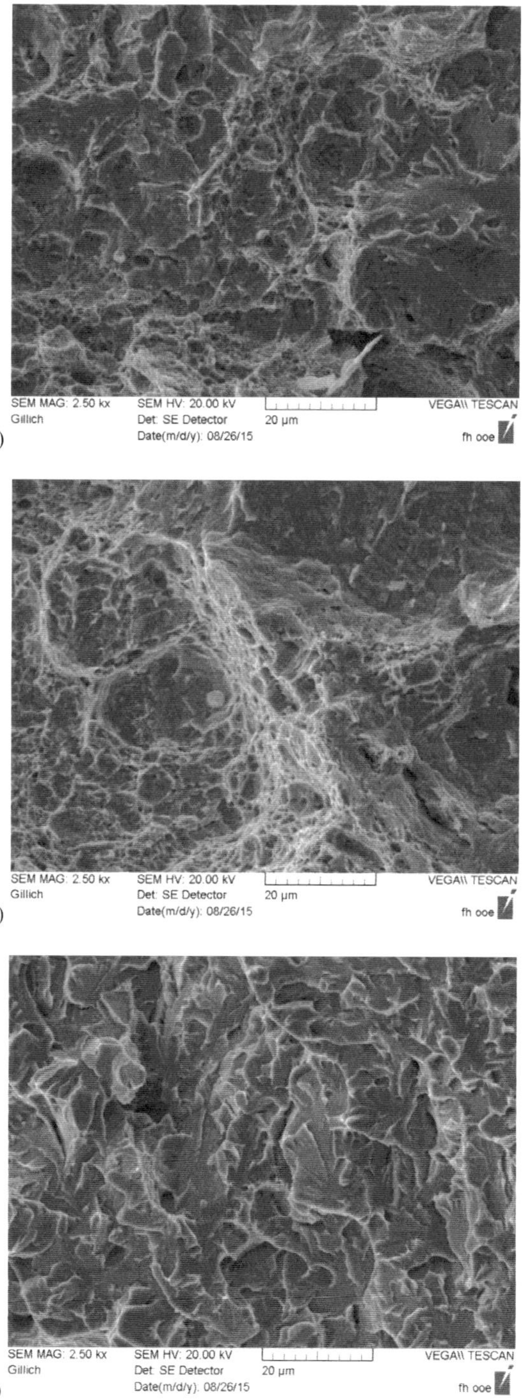

Figure 4: Fracture surface of the steel 42MnNiCrB4-3

(a) 850°C, 30 min + 400°C, 60 min

(b) isothermal with 360°C

(c) isothermal with 400°C

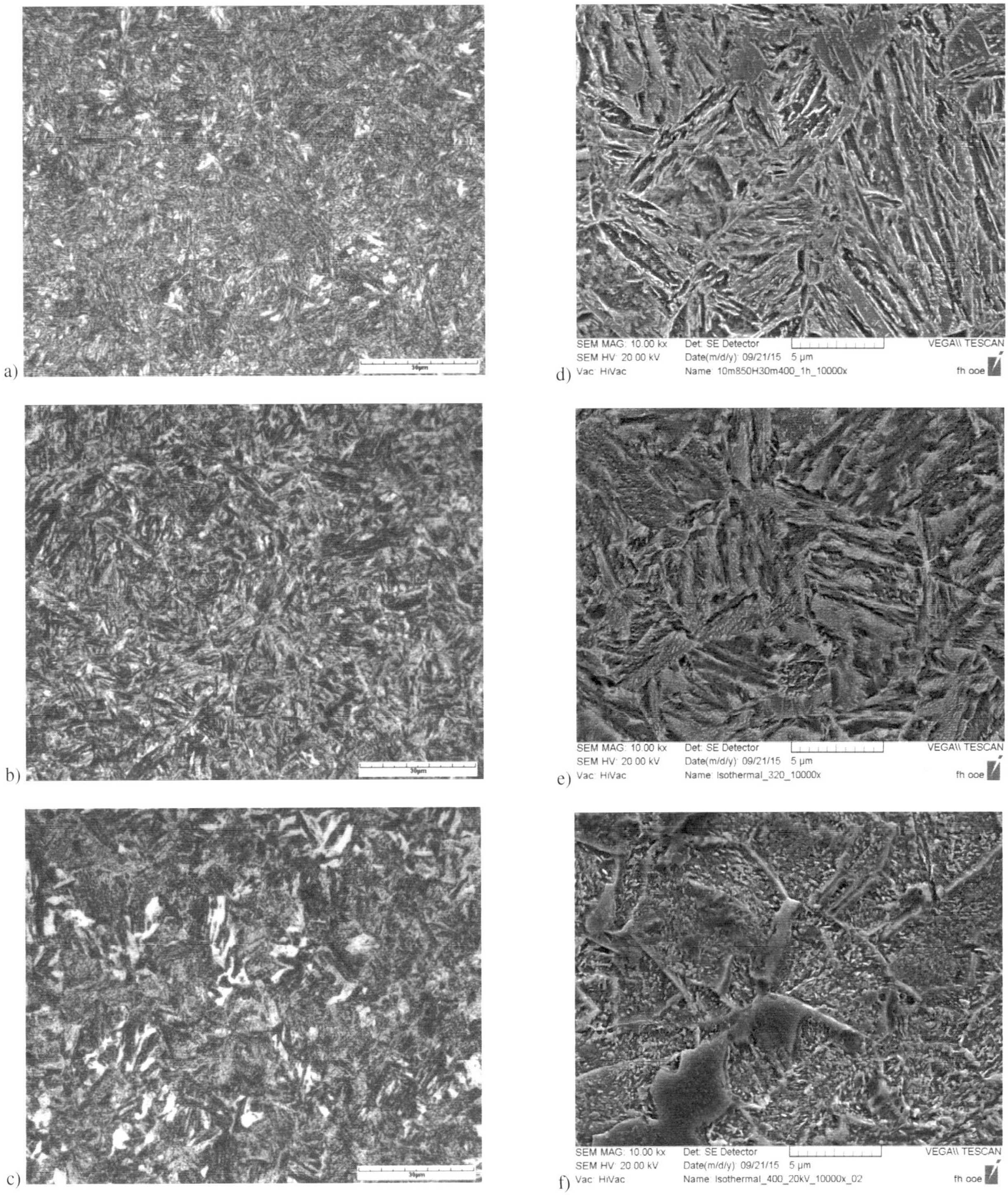

Figure 5: Microstructures of the steel 42MnNiCrB4-3
(a, d) 850°C, 30 min + 400°C, 60 min
(b, e) austempered at 320°C
(c, f) austempered at 400°C

Microstructures

To get a better understanding on the effect of various heat treatment conditions, LOM and SEM investigation on the microstructure were performed. In the Figures 5a and 5d the microstructures of the steel 42MnNiCrB4-3 for the standard HT and tempering at and 400 °C is shown. A needle like martensitic microstructure with few fine and uniformly dispersed temper-carbides can be seen. Similar microstructures could be found at the short-cycle heat treated samples.

Figures 5b and 5e show the microstructure after an isothermal transformation at 320 °C. In comparison to the quenched and tempered microstructure, there are coarser less sharp needle shaped packages with a significantly higher number and density of finely dispersed carbides, typical for lower bainite.

Rising the isothermal transformation temperature to 400 °C (Fig. 5c and 5f) resulted in a significantly coarser, plate like and more globular microstructure still with a dense population of carbides. These carbides have a slightly coarser, plate like appearance and are larger than their counterparts in the other samples.

Retained Austenite

The decomposition of the retained austenite for various tempering temperatures is shown in Figure 6. The retained austenite content after quenching is in the range of only 2 % for all heat treatment conditions and is constantly reduced with rising tempering temperatures. After tempering at 250 °C no more retained austenite could be detected.

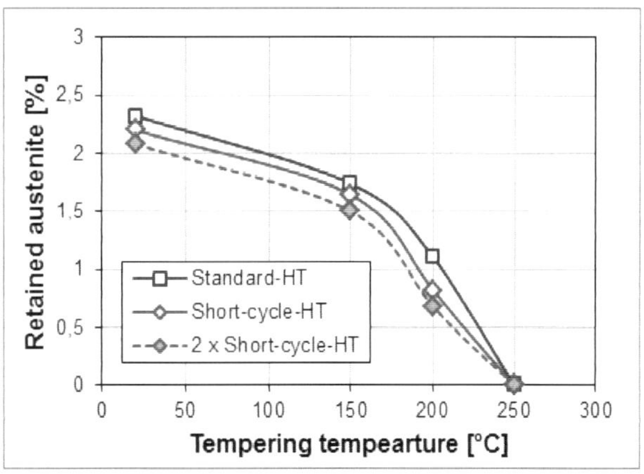

Figure 6: Content of retained austenite for standard-HT, short-cycle HT and double short-cycle HT

Figure 7 shows the retained austenite content after isothermal transformation in comparison to the standard heat-treatment. Similar to the three QT-heat treatments, the relatively low carbon content leads to a low amount of retained austenite (~ 1.5 %) for all austempering temperatures, but the diagram indicates that these contents occurred at temperatures were all retained austenite is transformed after normal tempering.

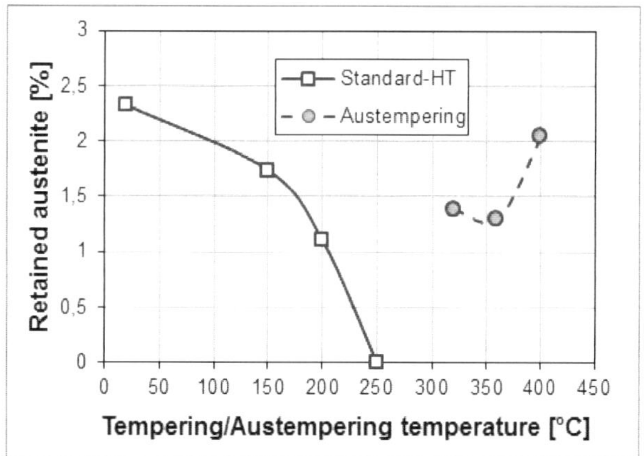

Figure 7: Content of retained austenite for isothermal transformation.

Discussion

In comparison to the previously published results of the steel 51CrV4 [12,13] (Fig. 8), the steel 42MnNiCrB4-3 exhibit significantly higher toughness at all hardness levels. This can be explained by the lower carbon content [18]. In agreement with all steels investigated in [12,13], and independent from the austenitizing conditions (standard-HT, one of two times short-cycle-HT), the best combination of hardness and toughness after quenching and tempering can be found at the tempering temperature of 200 °C.

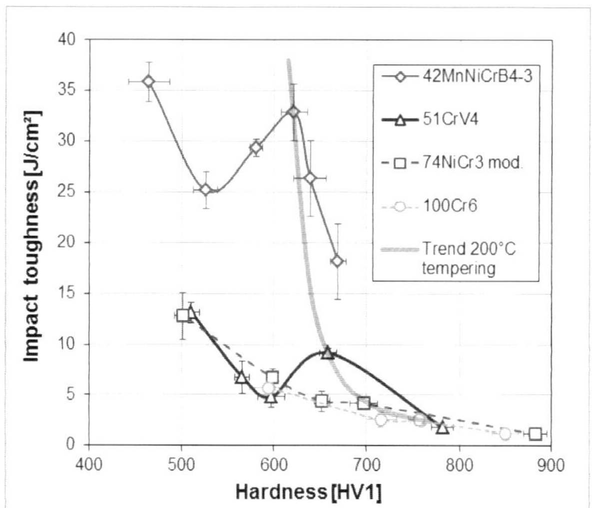

Figure 8: Hardness-toughness relationship of the steel 42MnNiCrB4-3 in the Short-cycle-HT-condition compared to other low alloy steels from [12,13]

Above this temperature all steels are affected by 300 °C - embrittlement (blue brittleness) in a similar way (Fig. 8), which, according to [19-22], can be explained by the diffusion of carbon to dislocation and the precipitation of Fe_3C at grain boundaries. In this steel nitrogen is stabilized by titanium, therefor it is unlikely that nitrogen contributes to this phenomenon.

Austempering offers an alternative way to achieve good material properties. Similar to results in [16, 17], austempering at low isothermal transformation temperatures just above the martensite start temperature Ms result in a superior combination of toughness and hardness copared to QT-heat treatment. As shown in [17] too high transformation temperatures lead to a reduction in toughness, providing combinations of material properties that can be inferior to quenching and tempering.

As shown in earlier works [9-13], there is good agreement between the appearance of the fracture surfaces and the toughness levels. Also in the case of austempered samples, the drop in toughness at an isothermal transformation temperature of 400 °C correlates well with a change in the fracture appearance. Nevertheless, despite this different fracture behavior, the steel possesses still a similar toughness level as quenched and tempered samples with a fully cleavage free fracture surface.

The microstructures of the quenched and tempered state, independent from the austenitising conditions, show typical patterns of tempered marteniste with fine precipitated carbides. After isothermal transformation at 320 °C, and slightly coarser at 360 °C, lower bainite is formed. In contrast, isothermal transformation at 400 °C leads to a different microstructure with coarser, plate like carbides, which might be addressed as upper bainite. This change in the microstructure seems to be the main reason for the reduction in the hardness-toughness combination.

The low amount of retained austenite is a result of the lower carbon content compared to the steels investigated in [12,13], but otherwise shows a similar trend.

Conclusions

- The hardness-toughness relationship shows, within the scatter bands, no significant differences between standard-HT and both variants of short-cycle-HT.
- Tempering at 200 °C provides a favorable property-relationship for all austenitizing conditions.
- In comparison to previously published results for steels with higher carbon contents, all heat treatment conditions exhibit a higher toughness level.
- A very good hardness-toughness relationship for hardness values below 600 HV1, was achieved by austempering at 320 °C and 360 °C respectively. This advantage gets lost when the isothermal transformation temperature is increased to 400 °C.
- There is no cleavage fracture at any conditions, but quasi-cleavage fracture could be found at the isothermal transformation treatment at 400 °C.
- The microstructures consist of tempered martensite with finely disperse carbides, lower bainite, for austempering at 320 °C and 360 °C, and a coarser upper bainite like structure for austempering at 400 °C.
- The lower carbon content leads to low amounts of retained austenite which gets fully removed after tempering at 250 °C or higher. Isothermal transformations also lead to a similar small content of retained austenite.

Acknowledgments

The financial support of the Toyota Technological Institute, for Mrs. Yuri Toshima to conduct the work at the Univ. of Appl. Sciences Upper Austria is gratefully appreciated.

References

[1] Bonß, S., „Laserstrahlhärten, Integration in die Fertigung ermöglicht schlanke Prozesse“, *HTM Z. Werkst. Wärmebeh. Fertigung*, Vol. 63, No. 3 (2008), pp. 147-154

[2] Hanisch, G., „Qualität von Induktionshärteprozessen und –anlagen in der Serienfertigung“, *HTM Z. Werkst. Wärmebeh. Fertigung*, Vol. 61, No. 2 (2006), pp. 71-75

[3] Opitz, T. and Vibrans T., „Induktive Schnellerwärmung von Formplatinen für die Warmumformung im Karosseriebau“, HTM J. Heat Treatm. Mat., Vol. 70, No. 6 (2015), pp. 261-266

[4] Leitner, H. *et al.*, „Kurzzeitwärmebehandlung von Schnellarbeitsstählen“, *BHM Berg- und Hüttenmännische Monatshefte*, Vol. 149, No. 5 (2004), pp. 172-175

[5] Henke, T. and Escher, C., “Innovative surface hardening of large tools using modern simultaneous dual frequency (SFD) generators”, *Proc 7th Int. Tooling Conf.*, Torino, Italy, *May* 2006, pp. 419–426.

[6] Schneider, R. *et al.*, “Short cycle hardening and tempering behaviour of low alloy tool steel 60WCrV7”, *Int. Heat Treatment and Surface Engineering* Vol. 2 No. 3/4 (2008), pp. 121-125

[7] Schneider, R. *et al.*, “Induction Hardening and Tempering of Cold Work Tool Steels”, *Proc 8th Int. Tooling Conf.*, Aachen, Germany, June 2009 pp. 573-583

[8] Schneider, R. *et al.*, “Induction hardening and tempering behaviour of X153CrMoV12-1 (~ D2) cold work tool steel”, *J. Japan Society for Heat Treatment* Vol. 49, No. 1 (2009), pp. 93-96

[9] Schneider, R. *et al.*, „Ermittlung der Härte-/Zähigkeitsbeziehung für kurze Wärmebehandlungszyklen“, *HTM J. Heat. Treatm. Mat.* Vol. 67, No. 1 (2012), pp. 31-38

[10] Schlagradl, T. *et al.*, „Entwicklung einer Methode zur Simulation von Kurzzeitwärmebehandlungen an Mikrokerbschlagproben für die Prüfung der Zähigkeit von hochfesten Schweißzusätzen“, *BHM Berg- und Hüttenmännische Monatshefte*, Vol. 156, No. 3 (2011), pp. 93-98

[11] Schlagradl, T. *et al.*, “Investigation of the hardness–toughness relationship of a welded joint after different

heat treatment cycles", *Weld. World.* Vol. 57, No. 1 (2013), pp. 113–121

[12] Steineder, K. and Schneider, R., "Determination of the hardness-toughness-relationship for different short-cycle heat-treatment conditions on the steels 51CrV4, 74NiCr2 and 100Cr6", *Proc. 20th IFHTSE-Congress*, Beijing, China, October 2012, pp. 93-98

[13] Steineder, K. and Schneider, R., "Influence of short-cycle heat treatment conditions on the hardness-toughness-relationship of the steels 51CrV4, 74NiCr2 mod. and 100Cr6", *HTM J. Heat Treatm. Mat.,* Vol. 68, No. 2 (2013), pp. 77-84

[14] Trojahn, W., „Einfluss der Wärmebehandlung auf das Verschleißverhalten und die mechanischen Eigenschaften von Werkzeugstählen und Werkzeugen", *Swiss Materials* Vol. 3, No. 4 (1991), pp. 15-19

[15] Ölund, P., *et al.*, "Properties of Bainite Hardened SAE 52100 Steel, *Heat Treating - Proc. 18th Conf.*, Rosemont, USA, October 1998, pp. 305–309.

[16] Löwisch, G., *et al.*, „Mechanische Eigenschaften eines nach zweifacher Austenitisierung gehärteten 100Cr6", *HTM Härterei-techn. Mitt.*, Vol. 53, No. 4 (1998), pp. 266-272

[17] Rossow, E., „Zwischenstufenvergüten", *HTM Härterei-techn. Mitt.,* Vol. 3 (1944), pp. 143-158

[18] Krauss, G., „Verformung und Bruch martensitischer Stähle", *HTM Härterei-Techn. Mitt.* Vol. 46, No. 1 (1991), pp.7-15

[19] Roos, K. and Maile, K., Werkstoffkunde für Ingenieure, Springer Verlag (Berlin/Heidelberg, 2008)

[20] Bargel, H.-J. and Schulze, G., Werkstoffkunde, Springer Verlag, (Berlin/Heidelberg, 2004)

[21] Dengel, D., „Kurzzeitanlassen von Stahl", *HTM Härterei-Techn. Mitt.* Vol. 39, No. 5 (1984), pp.182-194

[22] Bleck W, Werkstoffkunde Stahl für Studium und Praxis, Verlag Mainz, (Aachen, 2004)

International Federation of Heat Treatment and Surface Engineering 2016
Proceedings of the 23rd IFHTSE Congress
April 18–21, 2016, Savannah, Georgia, USA

Close relations between metallic materials, heat treatment and thermochemical treatment on the microstructure: contribution of metallographic examinations performed on industrial parts through several case studies

Patrick Jacquot, Bruno Stauder, Nicole Jacquot, Guillaume Fürst
Bodycote, Pusignan, France

Abstract

Metallurgical analysis is widely used in the field of heat treatment of metals to ensure that the quality of the treatments carried out on a wide variety of industrial parts and treatments is properly done. It is therefore a valuable tool for subcontractors of the heat treatment industry to monitor their production and treatment response on various parts. Through different examples, we will present actual cases that involve parts of treated areas of mechanical engineering, automotive, machinery, aerospace. We show that there is a strong interdependence between material and heat treatment, one being related to each other, with successful treatment depending both on the metallurgical quality of the material and the production quality of the heat treatment process. We will discuss some types of treatment: vacuum core-hardening, gas and low pressure carburizing, gas and plasma nitriding and vacuum brazing.
This paper shows current applications of metallurgical investigations carried out on several types of mechanical parts which have been heat treated and exhibit after heat treatment some defects, damages, cracks, failures, anomalies, discolored surface, and others problems. The goal of such investigations is to determine, the precise origin of the defects with more or less certainty. This is done in order to: avoid reoccurrence of the problem; improve the overall level of heat treatment quality; decrease the scrap rates; or improve general knowledge and know-how.
Furthermore, we show that there is strong interaction between: the material to be treated, its structure, machining or forming mode, design of the shapes, surface quality, heat treatment parameters, and finally how it is use in service.

Introduction

Metallurgical inspection, metallographic examination, failure analysis, fractographic analysis, both visual inspection and scanning electron microscopy with energy dispersive X-ray spectroscopy (SEM-EDS) are commonly used techniques to carry out precise investigations on industrial parts.
Damages or defects generated on a part and produced during the heat treatment (HT) operations, may be the result of a defective heat treatment or caused by other means than the heat treatment. These are nevertheless visible after heat treatment. In this case, the heat treatment step may be a way to reveal the problem, but does not reveal the root-cause of the problem [3][4].

The different steps to achieve a well-structured metallurgical analysis of a treated mechanical part having a defect, must be carried out in a logical way, be uncompromising and be only factual. This is really important if you want to achieve an indisputable analysis.
The first step begins with a detailed description of: the part, shape, type and chemical composition of material, manufacturing process (machining, stamping, blanking, rolling, molding, extrusion, cold and hot forging, hydro-forming, sintering, additive manufacturing, grinding, EDM-electro discharge machining, polishing, sand blasting); heat treatment type and parameters achieved and conditions of use in service. This step is very important to collect all possible information on the history of the manufacture and heat treatment performed on parts. The search for such information is sometimes difficult because it involves different people and potentially different suppliers in the manufacture of the part. This step is essential for further investigations and often allows for further interpretation.

Secondly, it is necessary to define, by visual examination, as precisely as possible, the type of fault or type of problem appearing on the part. These defects can be cracks or a complete mechanical failure of the part; improper hardness; improper case or nitriding depth; Microstructure inhomogeneity, surface staining or discoloration, or presence of carburized or decarburized surfaces. Porosity, corrosion, distortion or other factors that result in customer rejections, should also be investigated.

The third step is to provide a first hypothesis on the possible origin of the rejection or failure. This initial hypothesis may or may not be confirmed on further closer examination.

The fourth step is certainly the largest and most tedious. It is to conduct a proper investigation, including metallurgical examinations and analysis required to determine the nature of the defect and finally its origin. Depending on the type of defect identified, samples are collected by cutting and precise sectioning the part where the defect is present. The first operation in preparing a specimen for metallographic analysis is to locate the area of interest. Sectioning is the most common technique for revealing the area where the defect is located. After abrasive cutting, specimen mounting with a resin enables the abrasive grinding and the final polishing steps before the metallographic chemical etching. Chemical etching selectively attacks specific micro-structural features, such as grain size, phase identification, inclusions, oxides, cracks, undesirable compounds that are now visible with an optical microscope.

The fifth step is to achieve a synthesis of all the analysis's, to, provide the reasons for the occurrence of the fault and provide remedies to avoid recreating the defects. All of these elements allow us to establish a detailed written report including all of these steps.

We will now hereinafter presented some examples of concrete case studies from the mechanical industry. We do not present all the details of each of the case studies but only a condensed summary.

Role of case Hardening microstructure on distortion

Case hardening consists of carburizing a steel part in a carbon-rich atmosphere at a high temperature in order to increase the carbon content at the surface of the part. This part is then quenched, forming a high hardness gradient on the surface due to formation of carbon-enriched martensite phase. Case hardening is one of the most used heat treatment methods and is mainly applied on highly loaded parts, like several transmission parts such as: gears, shafts, bearings.
A hub impeller converter in 20CrMo4 steel with an external gear zone was gas carburized to increase the wear and fatigue resistance (Fig.1).

Figure 1: View of the carburized hub impeller converter

The carburizing parameters were: carburizing temperature: 930°C, time: 4h30, quenching temperature: 880°C, Oil quenching at 75°C, tempering at 150°C – 2h. After the case hardening operation, during the dimensional control of the parts, the diameter with the gear zone exceeded the distortion allowed by an inspection gauge, on about 30% of parts. The surface hardness of the case is 62 HRC and the case depth is 1,1mm (Fig.4).

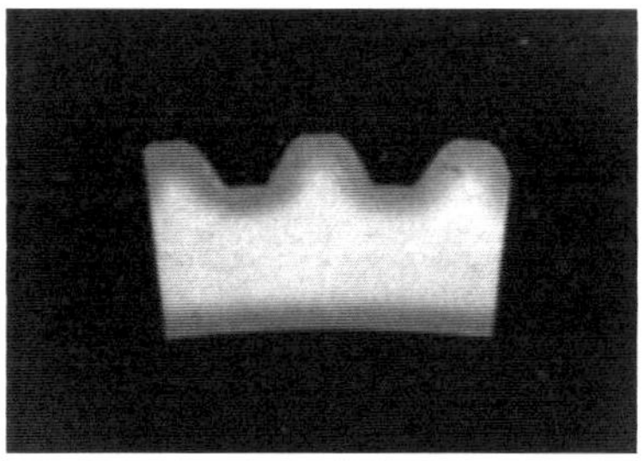

Figure 2: Macrograph of the carburized layer

A check of the case microstructure shows a presence of excessive retained austenite at 0.2 to 0.3 mm depth (Figures 2, 3). The retained austenite content can be estimated on the microscope at 35 - 40%. The microhardness profile shows a normal decrease of microhardness on the first 0,3mm from the surface (Fig.4).

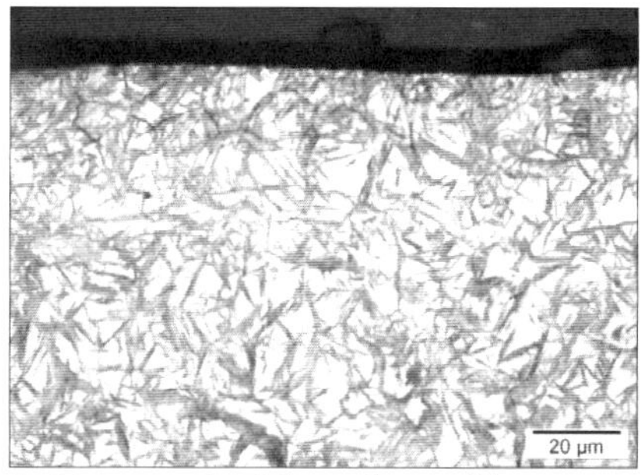

Figure 3: Micrograph of the carburized layer composed of martensite plates and a high content of retained austenite.

The amount of retained austenite in the case increases as the carbon content increases. When dimensional stability is critical, the retained austenite content should be low. The main problem in gas carburizing parts is shape and size distortion. Shape distortion can be reduced by stress relieving between machining as they are mostly due to residual stresses. Whereas, size distortion is due to phase transformation in steel from the austenite to martensite phase transformation. When austenite transforms into martensite there is an expansion in volume, while for tempering there is a contraction due to carbide formation. In the present case, a lower carbon content in the case must be precisely adjusted and a stress relieving treatment after semi-final machining will be a possible solution.

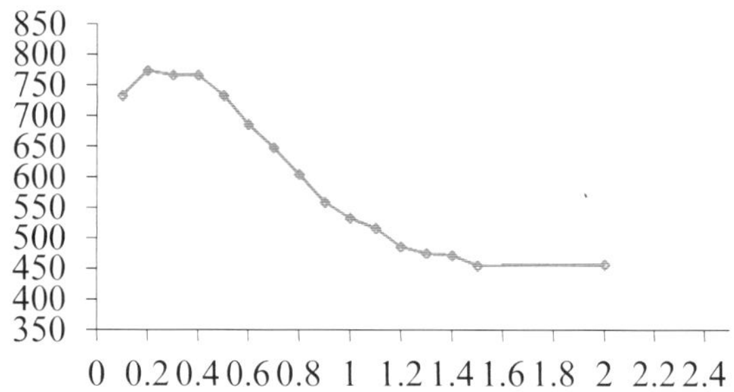

Figure 4: Microhardness profile.

Role of surface cleanliness on case Hardening

Specialists in heat treatment, thermochemical treatments, vacuum coatings and brazing very often encounter processing problems which come from the surface of the treated parts [1]. The causes are of various types: chemical contamination of the surface, passive layers, particular roughness, remelting, porosity, residual stress, work hardening, cold deformation, transfer of metals, surface spots, blisters, scratches, cracking, local oxidation, peeling, discolouration, to name just a few of the main types. As an example on the important effect of a contaminant on a carburizing layer, we can present the following case. A steel pressure plate in 16MnCr5 steel presents a lack of carburization in some areas corresponding to spots present on the surface (Fig 5). The carburizing parameters were: degreasing, low pressure carburizing at 920°C during 3 h followed by a gas quenching with nitrogen at 15 Bars pressure and tempering at 150°C for 2 h.

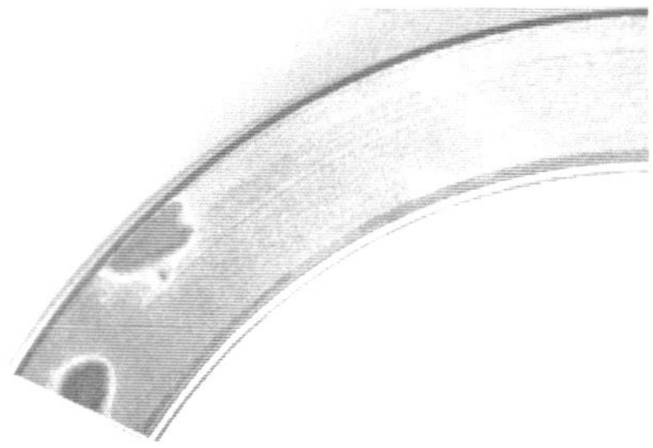

Figure 5: View of the carburized pressure plate in 16MnCr5

The carburized layer was examined at the stained areas, and away from the stained areas. It was found that the carburized layer was absent at the stained areas (Fig. 6). To understand the origin of this carbon diffusion barrier at high temperature, EDS analysis was performed on surface stains. The presence of a very fine thin layer of thickness less than 1 micron and mainly composed of silicon (Fig.7) was identified. One can conclude that the original contaminant might be a silicone lubricant that could not be removed during the cleaning stage prior to carburizing.

Figure 6: View of the lack of carburized layer in some spots

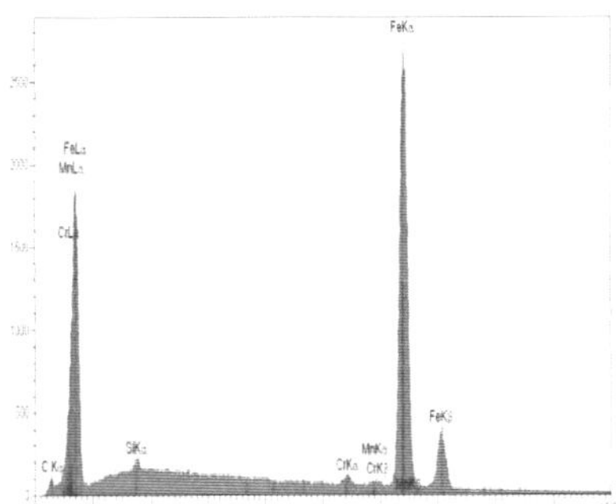

Figure 7: EDS spectra showing a small silicon peak

Role of surface defects on case Hardening

A carbonitrided transmission shaft of a car, manufactured by hot forming of a tube in 25CrMo4 steel, broke in service after a very short life of 50 km. The parameters of the gas carbonitriding were: temperature 870°C, time: 2h30, oil quenching at 100°C (Fig.8). The surface hardness was approximately 890 HV20 and the case depth was 0,4 mm.

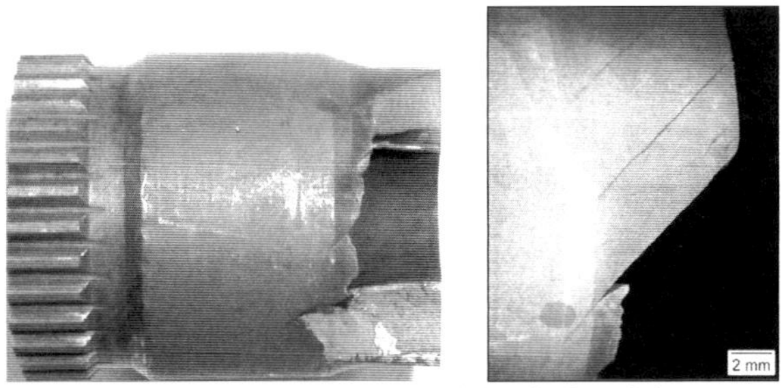

Figure 8: View of the rupture of the shaft and the torsion breaking with cracks oriented at 45 °.

The rupture occurred in torsion as shown with the cracks oriented at 45°. The micrographic observations show some oxidized cracks or defects that are located near the surface and can initiate a sudden rupture. The oxidized cracks originate from the hot forming of the shaft and contributed to initiate the rupture of the shaft in conjunction with a sudden shock in service (Fig. 9).

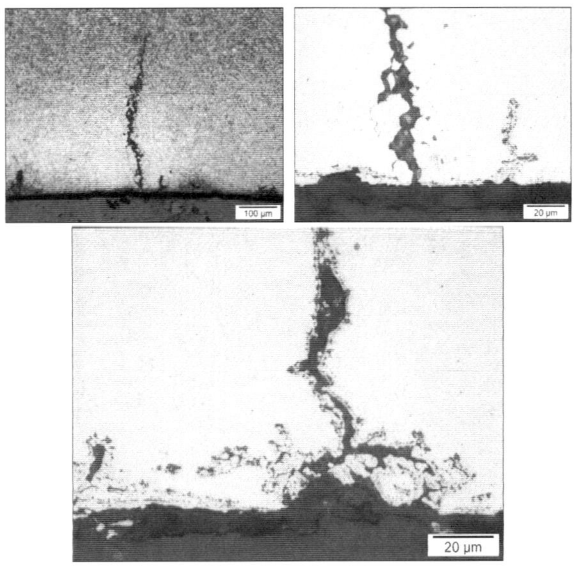

Figure 9: View of oxidized cracks and defects with a depth of 100 µm, generated during hot forming of the shaft.

Role of high cold working on gas nitriding

A fine blanked unison ring in X2CrNiMo17-12-2 stainless steel sheet of 2 mm thick was gas nitrided to a 100 µm diffusion depth to obtain a hardness of about 1200 HV. During the fine blanking, the sheet metal is cut by a cutting tool and the metal is strongly cold formed in its thickness. This provides a high stress level on the cut surface. A gas nitriding cycle was tested on such parts without success on the side faces that have been strongly cold worked (Figs. 10, 11, 12).

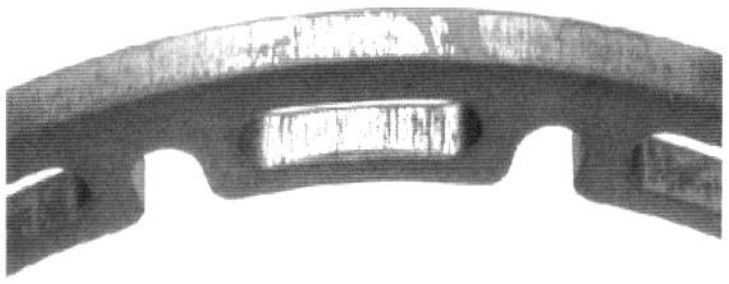

Figure 10: View of the un-nitrided cold worked surface

The other faces of the part that have not been cold worked, exhibited no problems being nitrided. Before nitriding, the surface of the parts are chemically etched and so activated in order to suppress the passive layer that is a diffusion barrier.

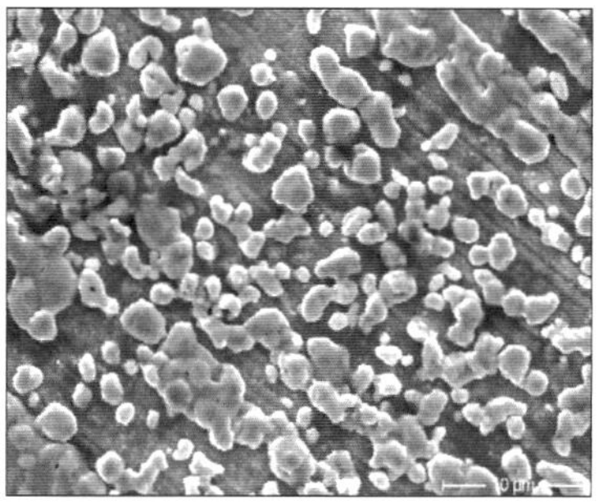

Figure 11: View of the partial growth of the nitriding layer

Despite this, the cold worked surface was not able to permit the nitriding. After an adapted stress relieving treatment, the diffusion layer was totally homogeneous all around the part (Fig. 13).

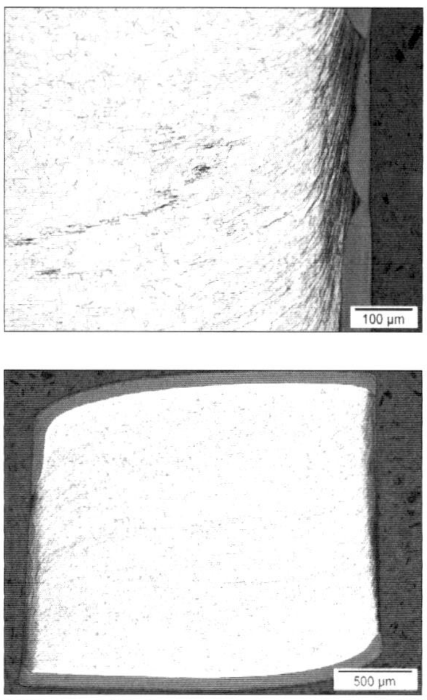

Figure 12: View of the inhomogeneity of the diffusion layer

Figure 13: View of the perfect homogeneity of the nitrogen diffusion layer

Role of surface post-machining on case Hardening

A mechanical part, like a cam in 35NiCrMo16 steel, is low pressure carburized at 900°C and gas quenched, cryogenically treated at -80°C and tempered at 150°C (Fig. 14).

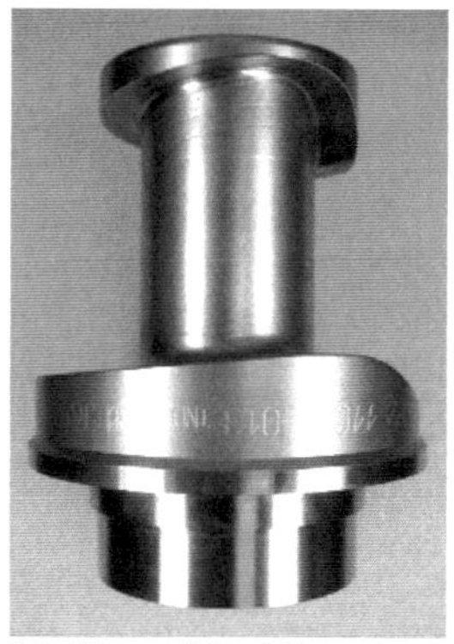

Figure 14: View of the LP carburized cam

After heat treatment, the track of the cam on the parts are ground and polished by honing. After these operations, the track is examined by non-destructive testing using dye penetrant inspection. During this testing, numerous parallel cracks appeared on the track (Fig. 15).

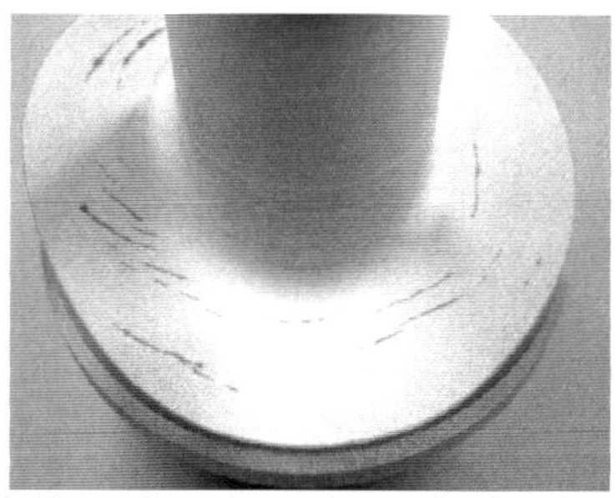

Figure 15: View of cracks on the track after dye penetrant inspection

Metallographic investigations showed that these cracks were created by excessive grinding step that has reheated and re-quenched the surface to a depth of 160 µm producing a hard and very brittle white layer, followed by a softer layer, and generating cracks similar to quench cracks (Fig. 16). This hard layer, re-treated and self re-quenched, is due to too severe grinding conditions (lack of cutting fluid, nature of the grinding wheel, speed ...) which led to significant heating of the carburized surface and allowed the re-quenching of the surface.

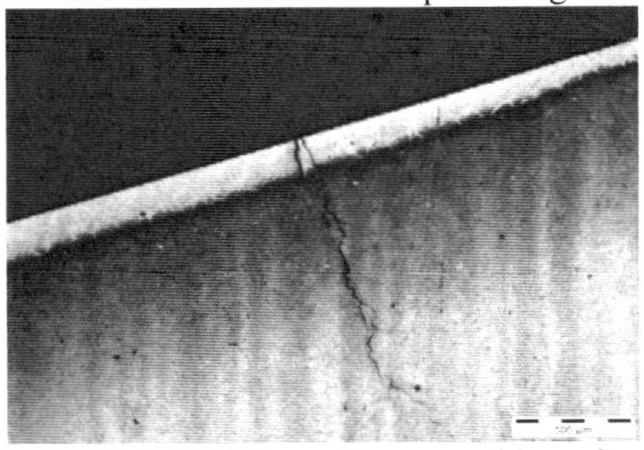

Figure 16: Micrograph of the grinded and honed carburized surface with a crack initiated on the re-treated white layer

This resulted in an increase of the surface hardness followed by a strong decrease of the hardness due to the effect of tempering or thermal affect zone of the under-layer (Heat Affected Zone). The carburizing process is not in question, but rather post-finishing operations of the track surface.

Role of HT conditions or alloy on surface integrity

An extruded and cold formed 6082-T6 aluminium alloy is solution heat treated at 540°C, water quenched and tempered at 175°C. After heat treatment, blisters were formed on the surface of the parts. Swelling of the surface in some areas was also observed. After some metallographic investigations on swelling areas, it was demonstrated that these areas were composed of pores that are grouped mainly under the surface to form a parallel network of cracks or void areas (Fig. 17).

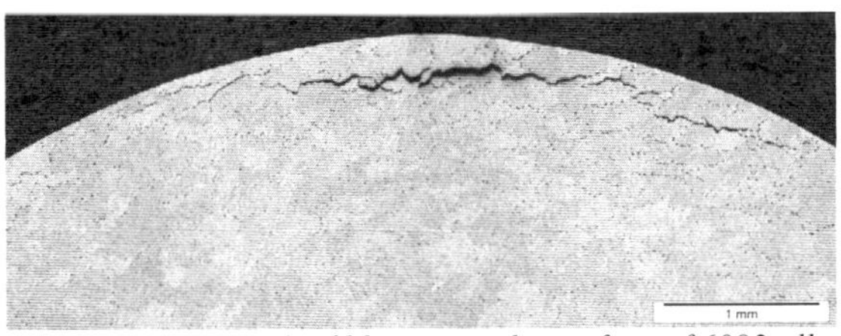

Figure 17: Micrograph of blisters on the surface of 6082 alloy.

We also observed some smaller spheroidal porosity located at grain boundary and throughout the structure of the alloy (Fig. 18). At the surface, the presence of these branched porosities networks caused swelling of the surface by deformation during quenching, in a region of approximately 0.1 to 0.2 mm thick. The origin of this porosity is likely either a degasing failure of the melt or in an inappropriate heat treatment conditions of the metal before or during processing. Excessive moisture level in the air around the parts during solution heat treatment can also produce this type of blister formation. This moisture can be a source of hydrogen, which can diffuse into the aluminum alloy.

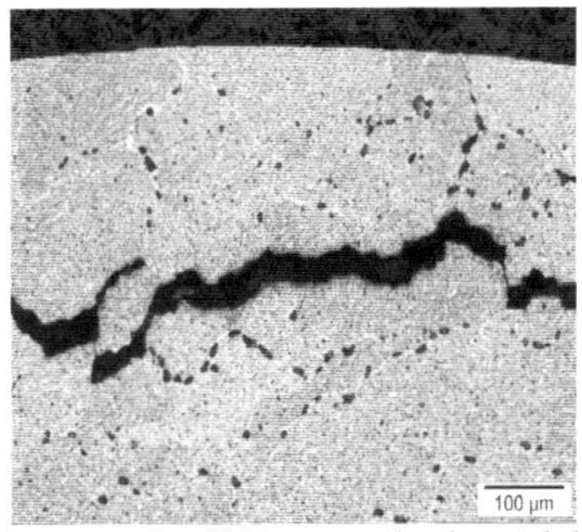

Figure 18: Micrograph of blisters and spheroidal porosities

The hydrogen gas collects at inclusions or grain boundaries and can form blisters on the surface of the part. Extruded parts are the most prone to blistering [2]. Use of various treatments such as anodizing can prevent blistering from occurring.

Role of tool design on heat treatment response

A plastic mold (sizes: 300 x 300 x 36 mm) made of X40CrV14 has been vacuum hardened at 1050°C, gas quenched at a pressure of 2 Bars, and followed by double tempering at 220°C. The final hardness was 53 HRC. After heat treatment, it was observed that the mold was cracked and initiated in an area where the section is reduced (Fig. 19).

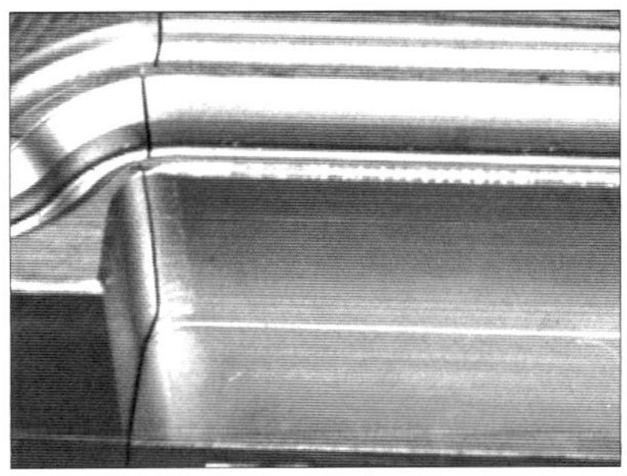

Figure 19: View of a crack on a plastic mold

The micrographic examination showed that the crack initiated at a sharp angle (Figs. 20, 21). It is well know that the risk of cracking increases during quenching at stress concentrations. A precise observation of the microstructure revealed improper structure, with carbide precipitation along the grain boundaries due to a too slow cooling rate.

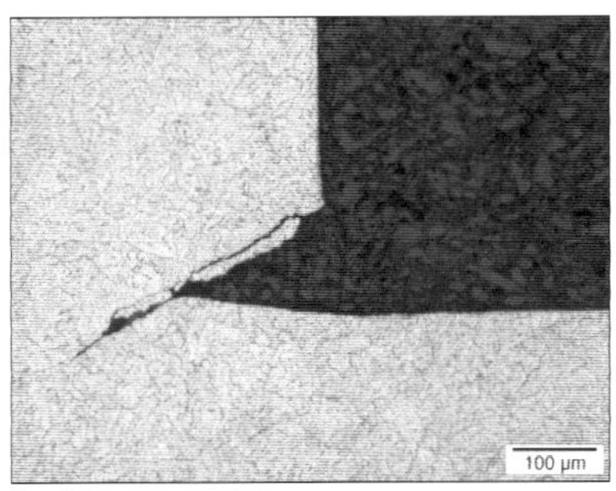

Figure 20: View of a crack on a sharp angle

This leads to low toughness and is conducive to sudden failure in case of constraints (Fig. 22). This is likely what happened in the sharp corner where there is a high concentration of stresses and high constraint. In the end, it can be said that cracking has two distinct origins, but the primary fault is one of design, with a fragile structure and unsuitable design of the mold.

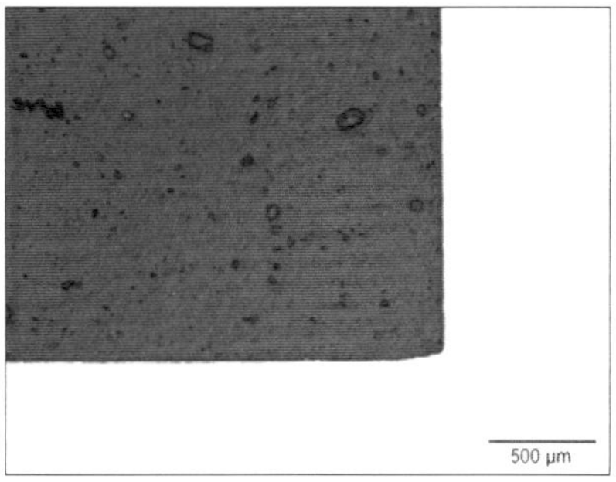

Figure 21: View of a sharp angle of the mold

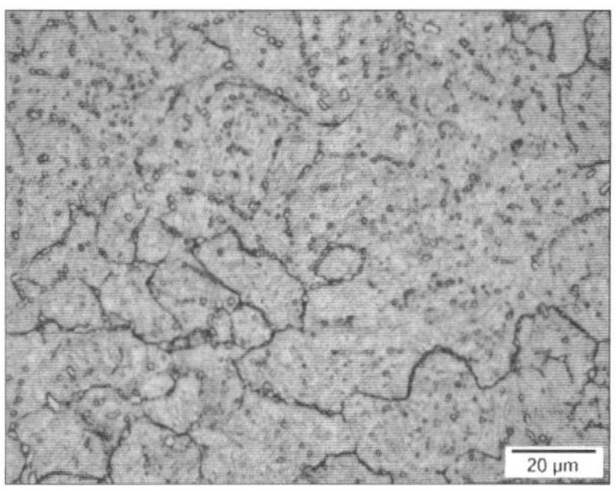

Figure 22: Micrograph of the brittle structure with carbides precipitation along the grain boundaries.

Role of initial structure of a steel on heat treatment

A part called a neck ring made in X40Cr13 martensitic stainless steel was gas quenched at 1020°C, gas cooled at a pressure of 3 Bars, and tempered two times at 230°C to obtain a final hardness of 49 HRC (Fig. 23).

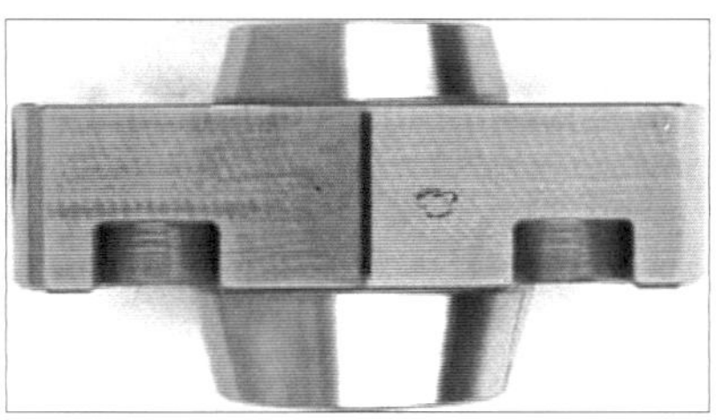

Figure 23: View of the neck ring in X40Cr13 steel.

After heat treatment the parts are ground. It is during this step that a defect appeared visible on the surface (Fig. 26). A section of the part shows that the defect was a 7 mm wide and 45 mm long (Fig. 25) aggregate of inclusions. The microstructure of this defect is shown Figure 27. In order to identify the nature of the defect, EDS analysis was performed and X ray mapping of each elements was obtained. Inclusions of aluminates, silicates, lime and oxides with the detection of Al, Ca, Si, Mg, O elements (Fig. 24) were observed.

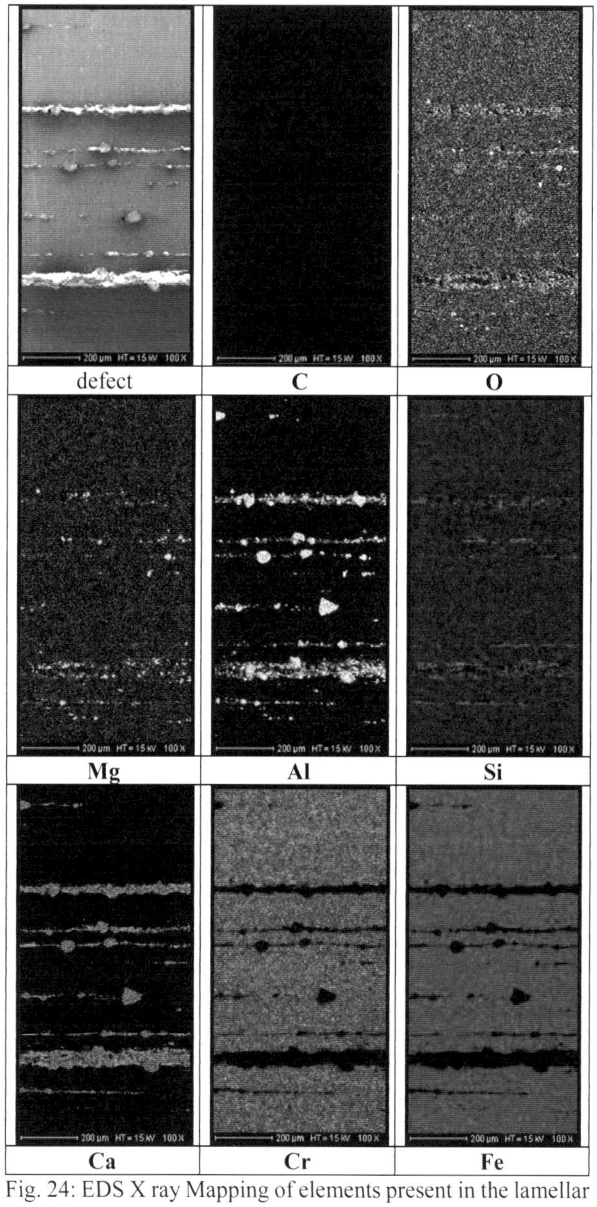

Fig. 24: EDS X ray Mapping of elements present in the lamellar inclusions.

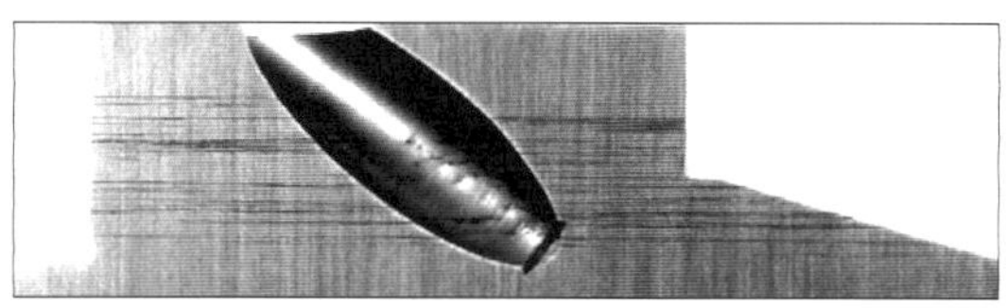

Figure 25: View of the defect like several lines, 7 mm wide.

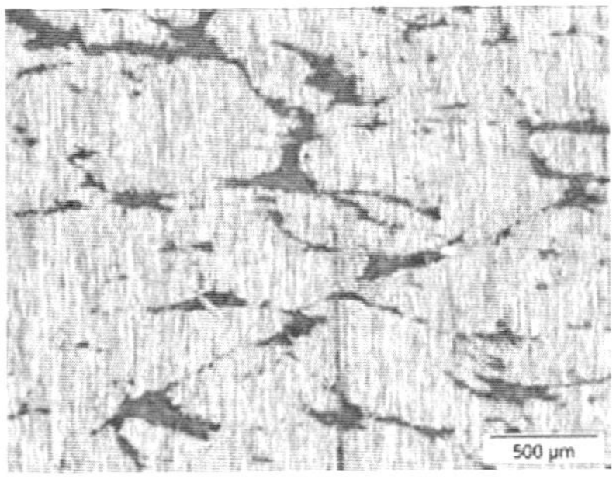

Figure 26: Macrograph of the defect on the ground surface

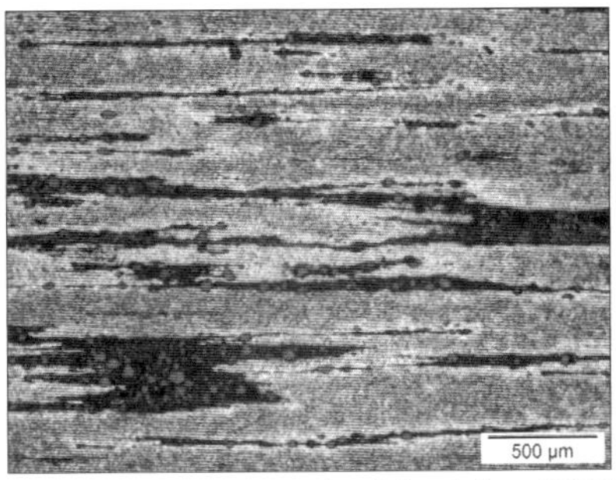

Figure 27: Micrograph of the defect (X25)

Role of HT conditions on the surface quality of the parts

Cast parts of Ti6Al4V titanium alloy was heat treated in a vacuum furnace at 1050°C for 2h under vacuum (V) or under argon (A) atmosphere, followed by Argon gas quenching at 5 Bars. The goal was to check the surface contamination after such treatment. The color of the surface is grey-black. The microstructure of the part treated under vacuum is shown figure 28.

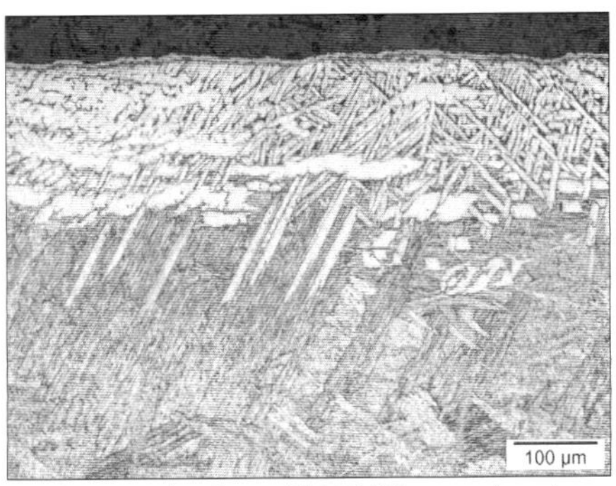

Fig 28: Micrograph of cast Ti6Al4V treated under vacuum

The top surface is composed by a TiC layer of 5 – 8 µm, followed by an alpha case of 190 µm thick, and finally a core structure of lamellar α and β. The microstructure of the part treated under argon is shown figure 29. The top surface is composed by a TiC layer of 4 – 6 µm (Fig. 30), followed by an alpha case of 245 µm thick, and finally a core structure of lamellar α and β. These trials show clearly that the type and technology of vacuum furnace plays an important role on the surface contamination of Ti6Al4V and also the atmosphere used at high temperature.

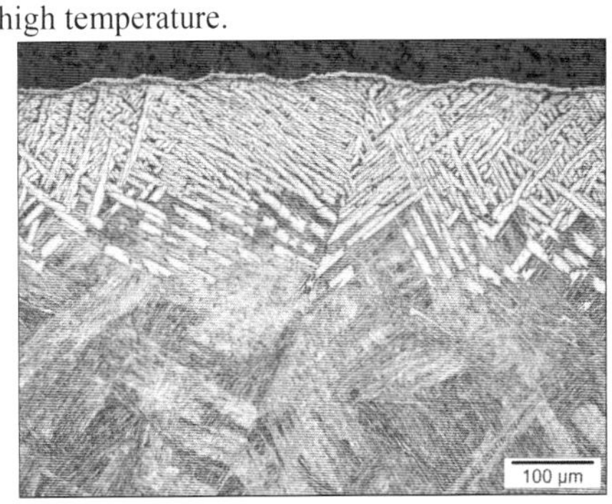

Fig 29: Micrograph of cast Ti6Al4V treated under argon.

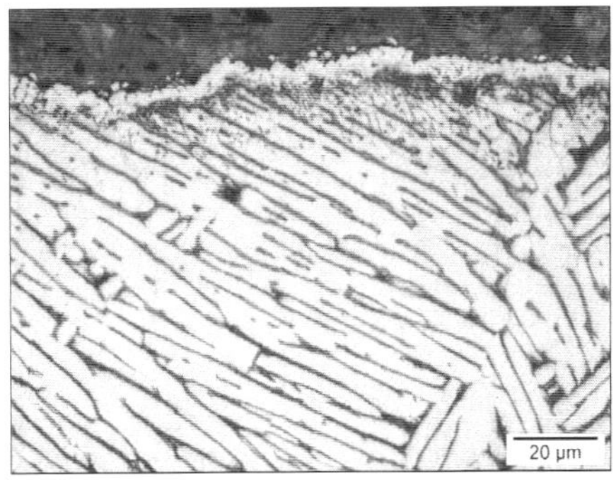

Fig 30: Micrograph of top layer of TiC on cast Ti6Al4V.

Role of HT conditions on the surface quality of the parts

Some shafts in X15CrNi17-3 steel were plasma nitrided at 550°C for 9 h with some specific areas protected by special paint to avoid nitriding. The stop off paint without solvent was deposited with a thickness of about 0.3 – 0.5 mm thick and allowed to dry for 30 min. After nitriding, the protected area was brushed and showed some defects, like arcing crater, on the protected zone (Fig 31).

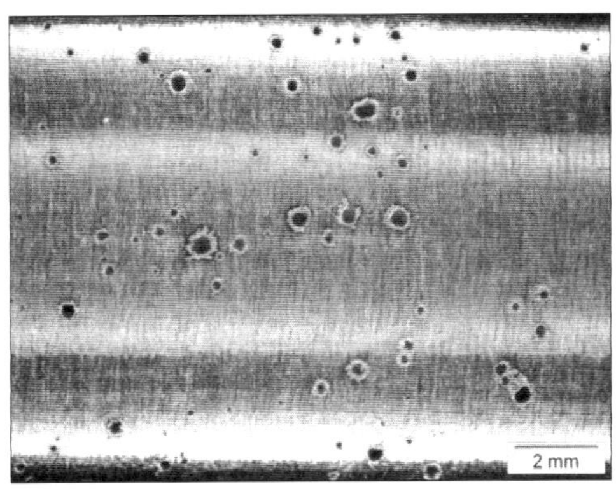

Fig 31: View of arcing craters on the protected surface

Micrographic examinations showed a arc crater, composed of melted steel with some copper from the paint (Fig. 32). The surface is damaged to a depth of 0.1 mm. The origin of arcing was probably from too short drying time and or a too thick paint. This resulted in excessive water present, and resulted in an arc from plasma bombardment. The best solution in plasma nitriding, if possible, is to use preferably a metallic shroud, that doesn't require drying.

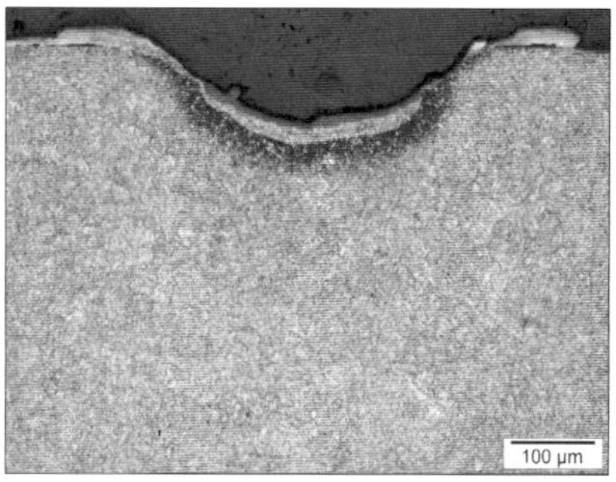

Fig 32: Micrograph of a arcing crater on the protected surface

Role of the surface cleanliness on the quality of treatment
Some ring gears in 31CrMo12 steel were cleaned by solvent and then gas nitrided. Some defects appeared after nitriding and are perfectly visible: matt grey color for nitrided area and a shiny surface for non-nitrided surface (Fig. 33).

Fig 33: View of nitrided areas (toothing and bore) and non-nitrided areas (turned surface).

The non-nitrided shiny surface is shown in figure 34 and a nitrided tooth is shown in figure 35. EDS analysis on shiny surface revealed nothing suspicious. The surface contaminant has a strong diffusion barrier effect and is not easy detected by conventional surface analysis method.

Fig 34: Micrograph of an non-nitrided shiny surface.

Some trials using mechanical surface treatment before nitriding, like sand blasting, show an interesting result to nitriding, like sand blasting, show an interesting result to eliminate the contaminant from the surface (Fig. 36).

Fig 35: Micrograph of a nitrided tooth.

The pre-oxidizing step in this case has been used, but without sufficient efficiency. It is clear that some mineral compounds used as additive in the cutting lubricant were deposited on the surface with or without the help of the cutting stresses, or also transformed during the nitriding step to a form very stable compound presenting a total diffusion barrier for nitrogen (Fig.37).

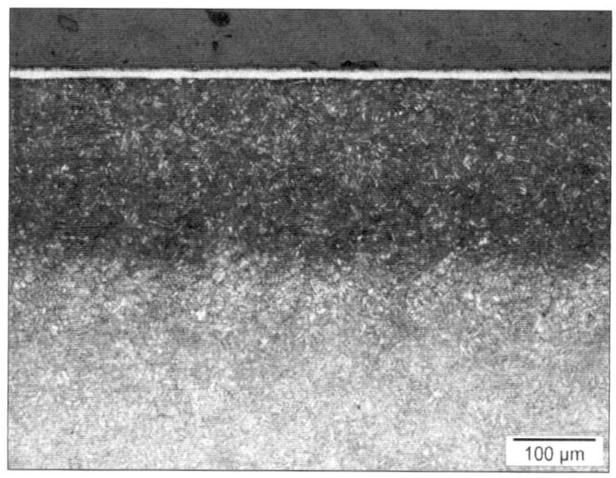

Fig 36: Micrograph of a nitrided layer using a pre-cleaning step of sand blasting.

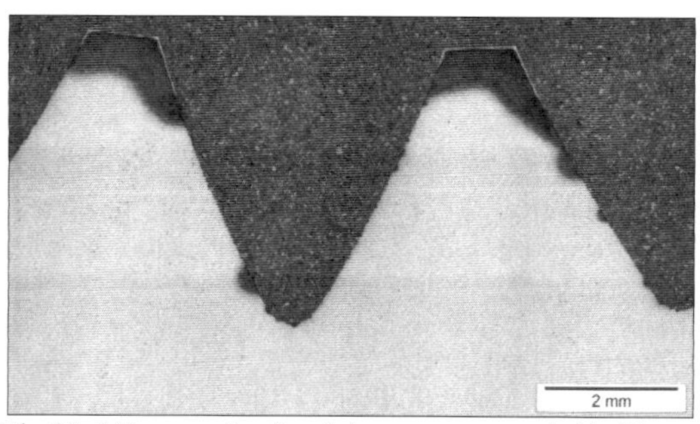

Fig 37: Micrograph of an inhomogeneous nitrided layer

Role of the HT conditions on the quality of vacuum brazed joints.

Some bypass flaps are manufactured by using two simple parts in AISI 316L stainless steel, which are assembled by vacuum brazing process. The brazing step is carried out at 1030°C by using a brazing alloy of NiCr7Si4Fe3B3. The two parts are pre-assembled by spot arc welding and then brazing alloy paste is deposited between the two parts. Metallurgical investigations to check the quality of brazed joint revealed the presence of an inter-diffusion of elements in each interfaces, and the absence of brittle precipitation and porosities (Figs. 38, 39, 40). The thickness of the joint is approximately 20 – 50 µm. Micrographs of the brazed joints showed no brittle precipitation inside the brazed joint. The brazing cycle and the brazing conditions are well suited.

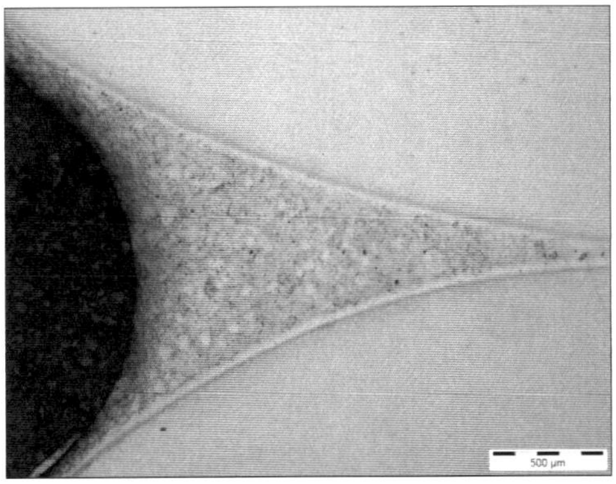

Fig 38: Micrograph of the brazed joint

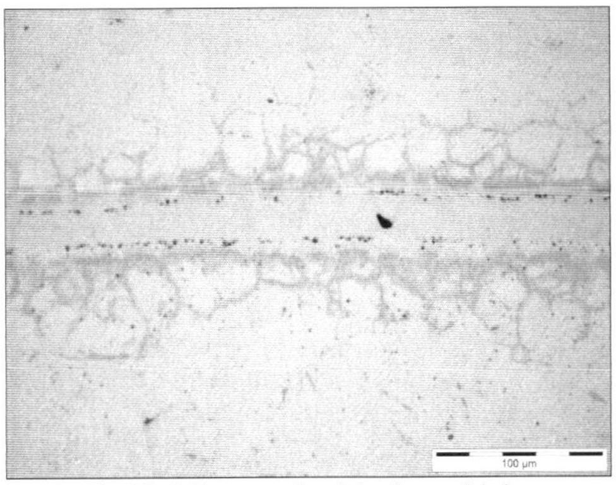

Fig 39: Micrograph of the brazed joint

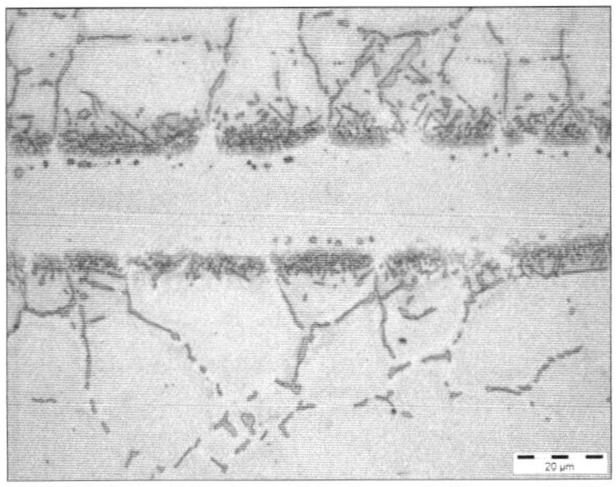

Fig 40: Interdiffusion of the brazed alloy on 316L without brittle phase inside the joint.

Conclusions

The presented examples show perfectly that there is a very close relationship between the metallic materials, heat treatments, thermochemical treatments with their microstructure. Metallurgical investigations allow to better understand the complex phenomena occurring on industrial parts during heat treatments.

It is clear that the interactions between materials, manufacturing, heat treatment processing and service use are important.

A comprehensive and global approach to materials science is required to advance in this field, which is still in full evolution.

Acknowledgments

I would like to especially thank Scott MacKenzie for his help for taking the time to review this paper in English.

References

[1] P. Jacquot, C. Andreux, B. Stauder. "Effect of industrial surface finishes on heat treatment and surface engineering of steel". International Heat Treatment and Surface Engineering, 2010, Vol. 4, N°3, pp. 110-116.

[2] D.S. MacKenzie. "Heat treating Aluminium". Hot Topics in Heat Treatment & Metallurgy, July 2004, Vol.2, N°7, pp. 4-13.

[3] D. Aliya, "How is failure analysis done?". Industrial Heating Magazine, October 2013.

[4] P. Sommer, "Wärmebehandlung-Felher, Schäden und Ursachen". (Heat Treatment-Faults, Damages and Cause) HTM, J. Heat Treatment. Mat. 70 (2015), pp.97- 107.

International Federation of Heat Treatment and Surface Engineering 2016
Proceedings of the 23rd IFHTSE Congress
April 18–21, 2016, Savannah, Georgia, USA

Microstructural Analysis of Carbides in Steel by Electrochemical Extraction Technique

***Haixuan Yu*[1], *Yuan Lu*[1], *Xiaoqing Cai*[1], *Richard D. Sisson, Jr.*[1*]**
Center of Heat Treating Excellence, Material Science and Engineering
[1]Worcester Polytechnic Institute, Worcester, MA, USA

Abstract

The tempering process is used to increase the ductility and toughness of steel after quenching process [1]. During tempering, the as-quenched Martensite can decompose into lower carbon Martensite or ferrite and the structure, size and distribution of carbides have a great effect on the mechanical property of steel. In this paper an electrochemical extraction technique is used to remove the carbides. X-ray diffraction (XRD) is used to determine the average crystallite sizes and crystal structure of carbides. Scanning electron microscopy (SEM) was used to characterize the shape and size distribution of carbides. A variety of carbides in AISI 4140 steel and UNS S42000 subjected to different heat treating processes. Cementite formed in AISI 4140 steel during tempering while $Cr_{23}C_7$ was found in UNS S42000. The crystals of carbides in AISI 4140 steel by induction tempering are finer than the same steel tempered by furnace tempering. The crystals of carbides with lower tempering temperature in AISI 4140 steel are finer than carbides with higher tempering temperature. The carbides in samples with the same hardness but from different tempering process were compared by XRD. It was found that the crystals of carbides from lower temperature with longer time were finer than carbides from higher temperature with shorter time.

Introduction

In order to balance hardness and toughness of steel, the tempering process is used after quenching process. The carbides precipitate during tempering process and the structure, size and distribution of carbides have a great effect on the mechanical properties of steel. Different microscopic methods, such as optical microscopy(OM), scanning electron microscopy(SEM) and transmission electron microscopy(TEM) are to understand the size, shape and distribution of the carbides as well as the strengthening mechanisms [2]. XRD is the traditional way to determine the crystal structure of steel. But when dealing with carbides within the steel, because of the low volume fraction of carbides and influence of steel matrix, it's difficult to identify the carbides by XRD. Extraction technique, combined with XRD and SEM is a method to fully characterize the carbides in steel [3]. Both furnace tempering and induction tempering are used in industry [4]. In this paper, an effective technique for carbide extraction was introduced. The effect of tempering temperature and time on the distribution and size of carbides in AISI 4140 steel and UNS 42000 was studied.

Experimental procedure

Materials

In this work, AISI 4140 steel was purchased from Peterson Steel Co. The dimension of the rod steel is 304.8mm long by 12.7mm diameter. The Optical Emission Spectroscopy (OES) result for the composition of the AISI 4140 steel is listed in Table 1.

Thermal Process

All the AISI 4140 steel samples were Austenitized at 850°C and held for one hour in a neutral atmosphere at 0.4% carbon potential, followed by quenching in oil at Bodycote Co. in Worcester before tempering process. After quenching, the induction tempering procedure was conducted at Thermaltool Co. in Connecticut and the furnace tempering procedure was conducted at WPI. Table 2 shows the matrix of tempering experiments. UNS 42000 were tempered at 400F and 600F for one hour.

Test Methods

The samples for testing were ground, polished and etched with 4% nital solution, and the microstructures were observed by JEOL 7000F SEM.
The Rc hardness was measured on all samples.

Effective Method on Carbides Extraction

Electrolytic extraction can be used to extract carbides [5], which can then be qualitatively or quantitatively analyzed by X-ray diffraction or microanalysis. One of the traditional methods for residue collection is pouring electrolyte through the polycarbonate (PCTE) membrane filter after electrolyze and washing the residue several times with ethanol. This method has many drawbacks. First, filtering electrolyte is time consuming. Second, air pump filtration can break the membrane filter paper easily. Third, the residue may stick into the holes of membrane filter paper. A faster way to extract carbides out was proposed and conducted in this paper.

Table 1: OES result of AISI 52100 steel and AISI 4140 steel in wt.%

Specimen	C	Si	Mn	P	S	Cr	Ni	Mo	Cu
AISI 4140	0.396	0.225	0.920	0.011	0.019	0.860	0.090	0.169	0.215

(1) Sample Preparation
Cut the sample into rectangular plate. Grind all the surfaces of the sample to 120 grit to avoid contamination. Corners of sample should be smoothed to avoid localized high-current density.
(2) Determine Current Density to be Applied
Measure the dimension of each face of the sample and calculate the total area. Current density is about 0.1 A/cm^2.
(3) Electrolyze
Set power supply to the predetermined current. Proceed electrolysis. The proper time for this electrolyze method is about 1 hour.
(4) Residue Collection
After electrolyze, there is a heavy corrosive coating on the sample. That is the carbides remaining on the surface of the sample when steel matrix is dissolved into electrolyte. Use tweezers clamp the sample and dry the sample by blower. Gently swipe the surface and collect the carbides and by-product such as $FeCl_2 \cdot 2H_2O$ on the sample surface. Centrifuge the powder with ethanol. The by-product would dissolve into ethanol and the carbides will remain in the bottom of centrifuge tube. Comparing with traditional residue collection using membrane filter, this method can get the same result but much faster.
(5) Residue Identification
Phase identification was conducted on PANanalytical Empyrean 2 X-ray diffractometer with Cr-K_α radiation at 30 keV and 55 mA. HighScorePlus was used to proceed Rietveld refinement [6] to identify phase percentage and peak parameter of the residue.

Table 2: Matrix of tempering experiments

Tempering Temperature (°C)	250	350	450	550	650
Tempered Time (hour)	1	1	1	1	1
	3	3	3	3	3
	7	7	7	7	7
	10	10	10	10	10
	15	15	15	15	15

(a) Furnace tempering process matrix

Tempered Temperature (°C)	250	350	450	550	650
Tempered Time (min)	1	1	1	1	1
	3	3	3	3	3
	5	5	5	5	5
	10	10	10	10	10
	15	15	15	15	15

(b) Induction tempering process matrix

Result and Discussion

Tempering temperature and time effects on microstructure of AISI 4140 steel

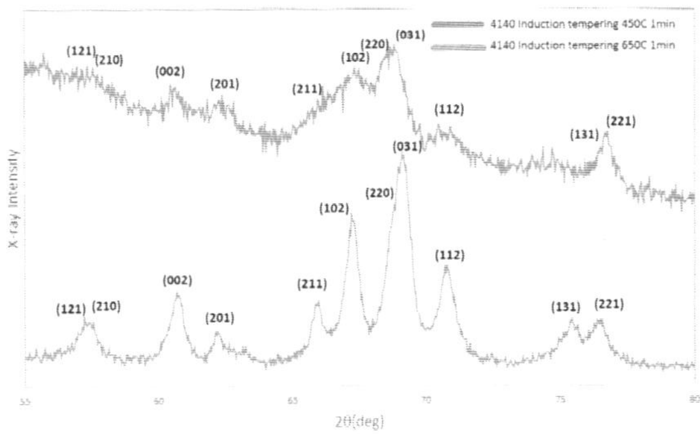

Figure 1: XRD patterns of 4140 steel induction tempering 450°C 1 minute vs 650°C 1 minute

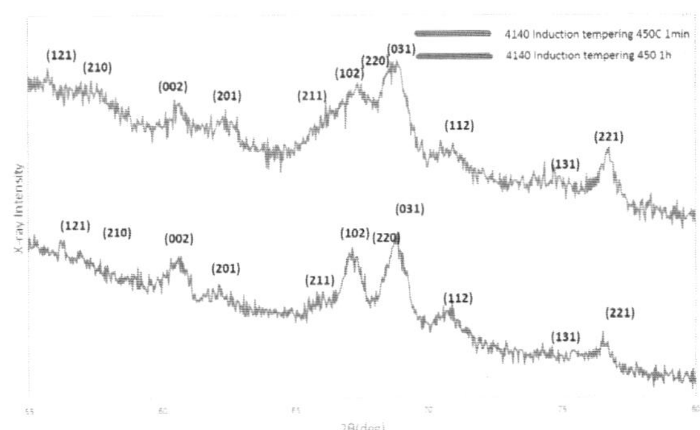

Figure 2: XRD patterns of 4140 steel induction tempering 450°C 1 minute vs 450°C 1 hour

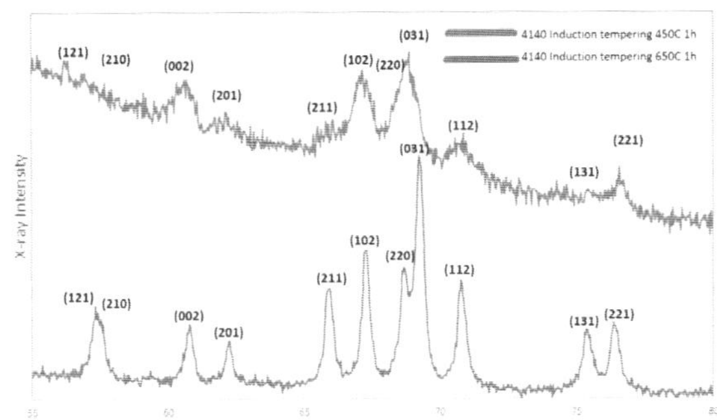

Figure 3: XRD patterns of 4140 steel induction tempering 450°C 1 hour vs 650°C 1 hour

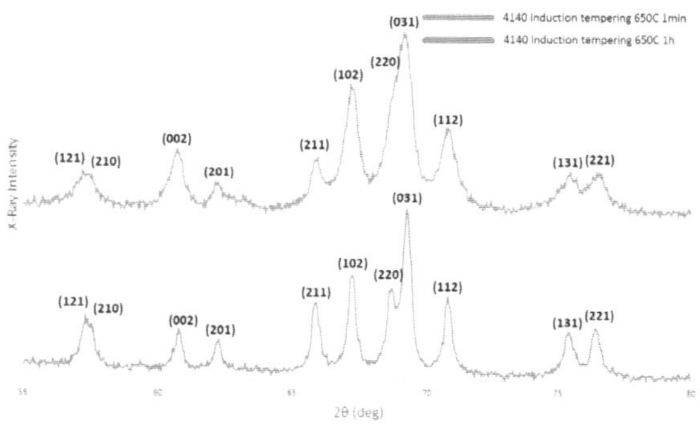

Figure 4: XRD patterns of 4140 steel induction tempering 450°C 1 minute vs 450°C 1 hour

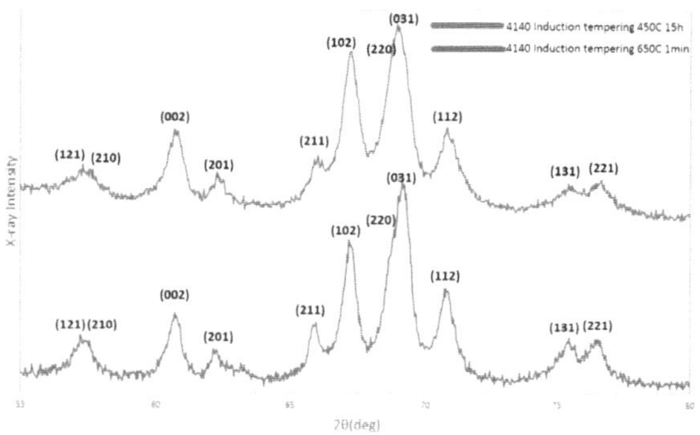

Figure 5: XRD patterns of 4140 steel induction tempering 450°C 15 hours vs 650°C 1 minute

As shown in Figures 1-5, with different tempering temperature and time, the XRD patterns of cementite have the same peak position but different shapes. In Figure 1, the peaks from AISI 4140 steel induction tempering at 450°C for 1 minute are much boarder than the peaks from AISI 4140 steel induction tempering at 650°C for 1 minute. In Figure 2, the peaks from AISI 4140 steel induction tempering at 450°C for 1 minute are boarder than AISI 4140 furnace tempering for 1 hour. In Figure 3, the peaks from AISI 4140 steel induction tempering at 450°C for 1 hour are much boarder than the peaks from AISI 4140 steel furnace tempering at 650°C for 1 hour. In Figure 4, the peaks from AISI 4140 steel induction tempering at 650°C for 1 minute are boarder than the peaks from AISI 4140 steel induction tempering at the same temperature at the same temperature. In Figure 5, AISI 4140 steel furnace tempering for 15 hours and AISI 4140 induction tempering for 1 minute were chosen because they have the same micro hardness (31.275HRC). The shape of these two set of peaks are nearly the same but the peaks from AISI 4140 furnace tempering for 15 hours are still a little bit boarder.

The Scherrer equation, in X-ray diffraction and crystallography [7], is a formula that relates the size of sub-micrometer particles, or crystallites, in a solid to the broadening of a peak in a diffraction pattern. The Scherrer equation is written as:

$$\tau = \frac{K\lambda}{\beta cos\theta}$$

τ is the mean size of the crystalline domains. K is the dimensionless shape factor, with a value from 0.9 to 1. β is the full width at half maximum (FWHM). θ is the Bragg angle. With mean size of the crystalline domain decrease (τ), peaks get broadened.

As we can see in XRD patterns from lower tempering temperature, the peaks are all very boarder. Some of the peaks even overlap. So it is relatively hard to measure the broadening at half the maximum intensity. Rietveld refinement method was used to measure the FWHM for selected peaks.

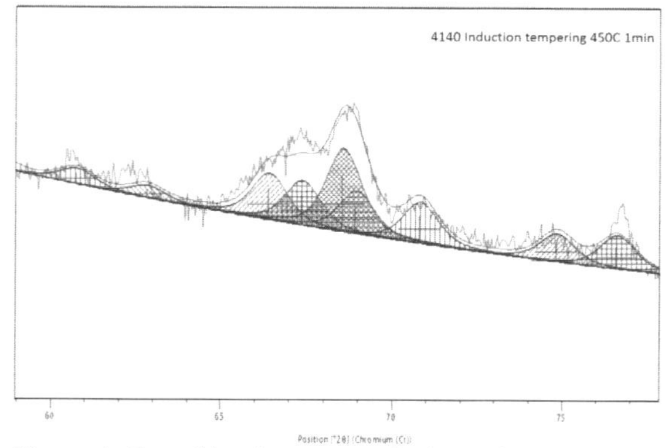

Figure 6: Rietveld refinement method to calculate FWHM of peaks in XRD pattern of 4140 steel induction tempering for 450°C 1 minute

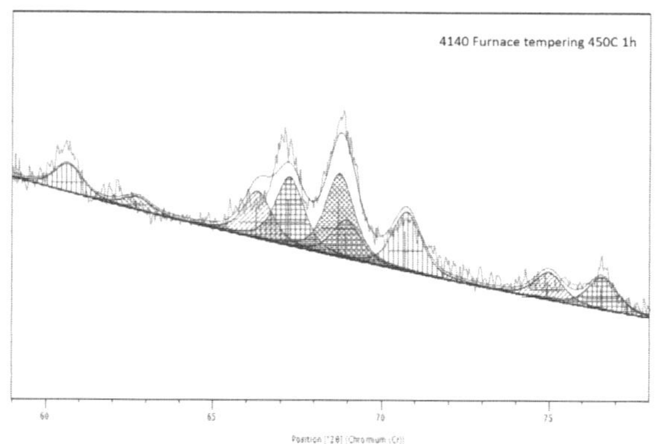

Figure 7: Rietveld refinement method to calculate FWHM of peaks in XRD pattern of 4140 steel induction tempering for 450°C 1 hour

As listed in Table 3, one example of calculation of the average crystalline domains for carbides in AISI 4140 steel furnace tempering 650°C 1 hour was shown. The average crystalline domains size is around 72.5nm. Different crystalline domains of carbides for different tempering temperature and time was listed in Table 4. With increasing tempering time, the crystalline domains increase a little bit if the tempering temperature is 450°C. But the crystalline domains of carbides doubled when the temperature time goes up from 1 minute to 1 hour in 650°C. Even though the sample from furnace tempering at 450°C for 15 hours has the same hardness with the sample from induction tempering at 650°C for 1 minute, they have carbides with a different crystalline domains size.

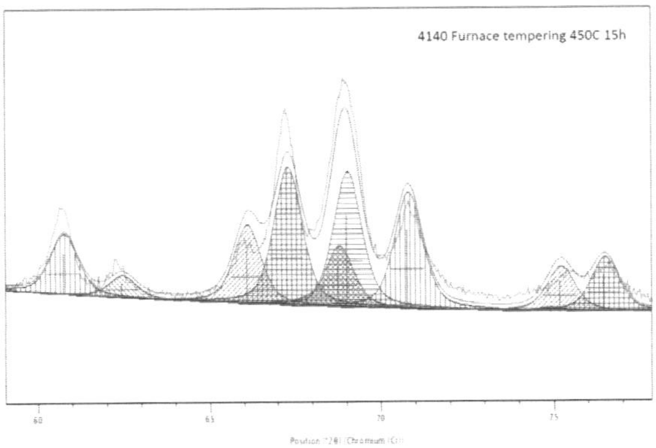

Figure 8: Rietveld refinement method to calculate FWHM of peaks in XRD pattern of 4140 steel furnace tempering for 450°C 15 hours

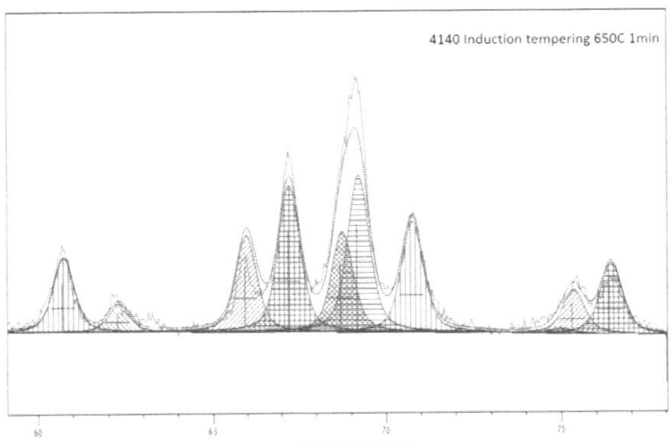

Figure 9: Rietveld refinement method to calculate FWHM of peaks in XRD pattern of 4140 steel induction tempering for 650°C 1 minute

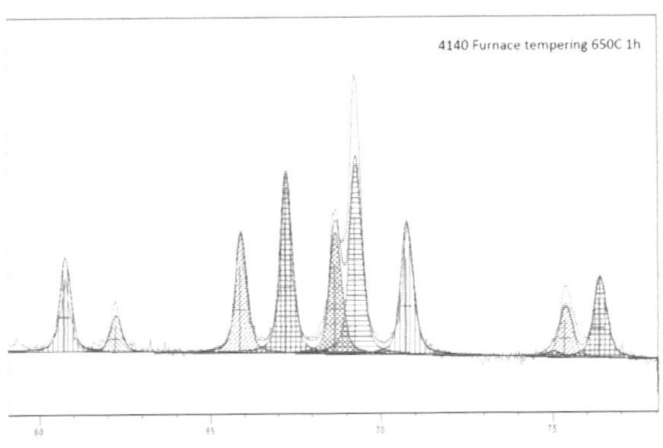

Figure 10: Rietveld refinement method to calculate FWHM of peaks in XRD pattern of 4140 steel furnace tempering for 650°C 1 hour

Table 3: Calculation of crystalline domains of carbides in AISI 4140 steel 650°C furnace tempering 1 hour

No.	Pos. [°2θ]	FWHM [°2θ]	h	k	l	Crystallite Size [nm]
1	60.7365	0.3365	0	0	2	73.6
2	62.2127	0.3395	2	0	1	73.4
3	65.8778	0.3474	2	1	1	72.9
4	67.212	0.3505	1	0	2	72.7
5	68.6564	0.3539	2	2	0	72.6
6	69.2279	0.3552	0	3	1	72.5
7	70.7423	0.359	1	1	2	72.3
8	75.001	0.3703	0	2	2	71.8
9	75.3361	0.3713	1	3	1	71.8
10	76.3396	0.3741	2	2	1	71.7
Average crystalline domains of carbides						72.5

Table 4: Calculation of crystalline domains of carbides with different tempering conditions

	Average crystalline domains of carbides[nm]
AISI 4140 induction tempering 450°C 1 minute	16.5
AISI 4140 furnace tempering 450°C 1 hour	19.8
AISI 4140 furnace tempering 450°C 15 hours	21.8
AISI 4140 induction tempering 650 °C 1minute	32.0
AISI 4140 furnace tempering 650 °C 1 hour	72.5

Microstructural Analysis

After grinding, mechanical polishing and etching with 4% nital, the microstructural analysis was conducted using JEOL F7000 SEM at 20kV and 10mm WD. The microstructure of tempered AISI 4140 steel was shown in Figure 11. With increasing tempering time, the carbides got coarsen. The shape of the carbides for tempering temperature at 450°C is a rod. But if the tempering time goes to 15 hours, the shape of the carbides goes to a little bit rounded. AISI 4140 steel furnace tempering for 15 hours and AISI 4140 induction tempering for 1 minute have the same hardness but the shape of carbides of these two samples are different. The trend for average crystalline size and carbides size with different tempering conditions are similar. As we can see the carbides size is larger than the crystalline domains, which means cementite in samples were not monocrystalline.

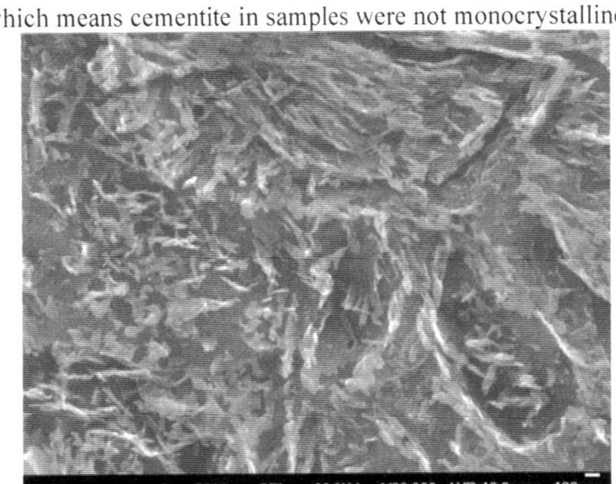

(a)

(b)

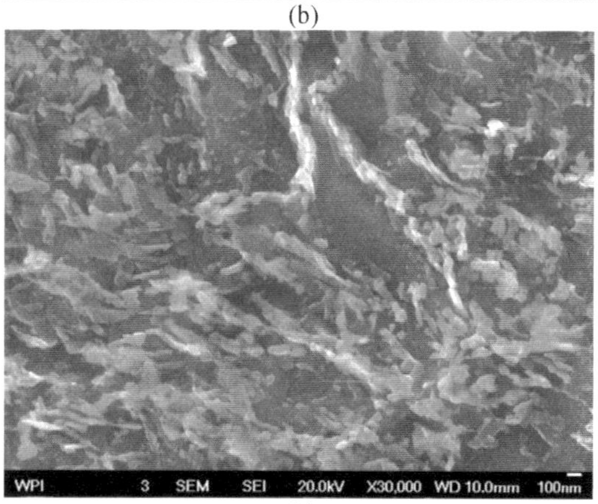

(c)

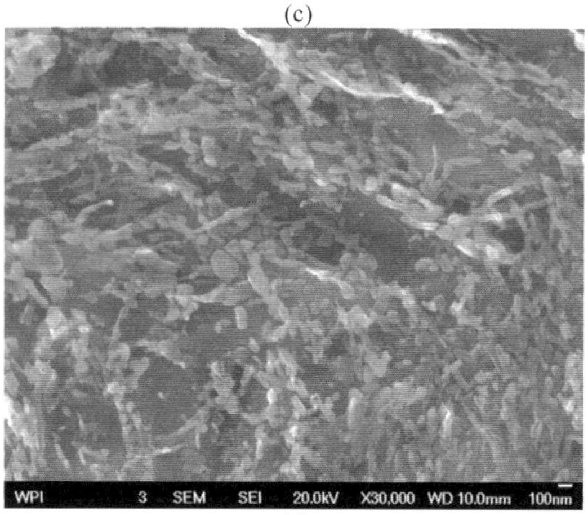

(d)

(e)

Figure 11: SEM micrographs with different tempering conditions: (a) AISI 4140 steel induction tempering for 1 minute (b) AISI 4140 steel furnace tempering for 1 hour (c) AISI 4140 steel furnace tempering for 15 hour (d) AISI 4140 steel induction tempering for 1 minute (e) AISI 4140 steel furnace tempering for 1 hour

Tempering temperature and time effects on microstructure of UNS S42000

The same extraction technique was used on 420 stainless steel. Figure 11 shows the XRD patterns of 420 stainless steel with different tempering condition. $Cr_{23}C_7$ were found in 420 stainless steel. The electrolyze procedure last 1 hour for every sample. Same carbides were found out in as quenched samples, which means not all carbides dissolve in matrix during austenitizing. But with the same electrolyze time, more peak intensity was get with higher tempering temperature, which means more carbides precipitate out with higher tempering temperature.

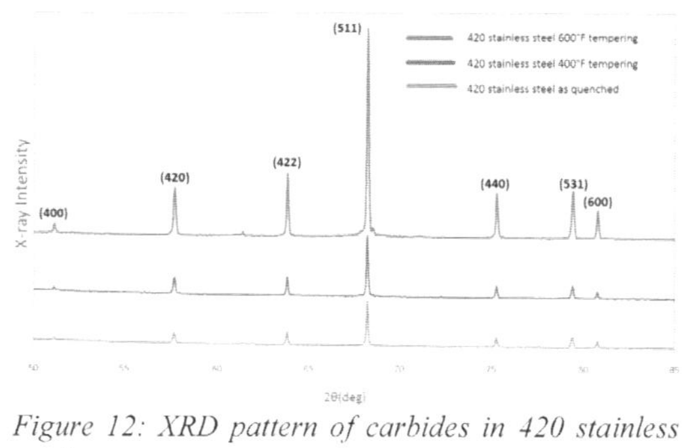

Figure 12: XRD pattern of carbides in 420 stainless steel

Conclusion

Effects of tempering temperature and time for carbides was studied. With higher tempering temperature and longer tempering time, more rounded shape of carbides with larger crystalline domains were formed in AISI 4140 steel. With higher tempering temperature and longer tempering time, more carbides were form in UNS S42000. An effect way for carbides extraction was developed.

Acknowledgments

The authors want to acknowledge members of the Center for Heat Treating Excellence (CHTE) at Worcester Polytechnic Institute (WPI) for their support to this project.

Reference

[1] Rao, P. Prasad, and Susil K. Putatunda. "Influence of microstructure on fracture toughness of austempered ductile iron." *Metallurgical and Materials Transactions A* 28.7 (1997): 1457-1470.

[2] Lagerpusch, Ulrike, Boris Anczykowski, and Eckhard Nembach. "Comparison of atomic force microscopy and transmission electron microscopy of second-phase particles." *Philosophical magazine A* 81.11 (2001): 2613-2628.

[3] Lu, J., D. G. Ivey, and H. Henein. "A Review of Methods to Quantify Nanoscale Precipitates in Microalloyed Steels, Part 1." *Iron & Steel Technology* 10.1 (2013): 232-244.

[4] Krauss, George, Marcus Aurelius Grossmann, and E. C. Bain. "Principles of heat treatment of steel." (1980).

[5] E04 Committee. (2010) "Practice for Electrolytic Extraction of Phases from Ni and Ni-Fe Base Super alloys using a Hydrochloric-Methanol Electrolyte (2014), pp. 1-4.

[6] Rietveld H M "A profile refinement method for nuclear and magnetic structures." Journal of applied Crystallography, 2(2) (1969), 65-71.

[7] Patterson A L. The Scherrer formula for X-ray particle size determination [J]. Physical review, (1939), 56(10):978.

International Federation of Heat Treatment and Surface Engineering 2016
Proceedings of the 23rd IFHTSE Congress
April 18–21, 2016, Savannah, Georgia, USA

"Study of phase transformations during tempering of AISI L6 tool steel by means of dilatometry and high temperature X-ray diffraction"

Douglas I. Quiñones-Salinas*, Rafael D. Mercado-Solís, Luis A. Leduc-Lezama,
Universidad Autónoma de Nuevo León, México

Florentino Fernandez-Guzman, Rene Cerda-Rojas.
Frisa Forjados S.A. de C.V., Nuevo León, México

Abstract

The phase transformation and the influence of heating rate during tempering (100°C to 700°C range) for as quenched AISI L6 tool steel are presented in this work. Phase transformations during tempering of the steel were analyzed by means of dilatometry curves recorded during heating from as quenched state. Heating to 700°C was made at different rates (5, 50 and 150°C/s). High temperature X-ray diffraction (XRD) was used to identify the existing phases in the material and to confirm the presence of carbide precipitate and retained austenite. It is possible that changing the rate reduce probability of the occurrence of some phase transformations (transition carbides) in favor of other (cementite carbides). The results of X-ray diffraction showed that is possible observe a precipitation of carbide and cementite into retained austenite.

Introduction

This project studies phase transformations in AISI L6 low alloy tool steel under tempering temperature range [1-2]. This experiment has been studied mainly by means of dilatometry and high temperature X-ray diffraction. These techniques are very powerful and allows the identification of kinetics occurring during tempering.

Some authors report during tempering stages of martensitic steels, the simultaneous transformation of retained austenite and the precipitation of cementite produce strong signals in the dilatometer [2-3].

Researchers as J. Pacyna *et al*, describes the kinetics of phase transformations during continuous heating from hardened state and subsequent cooling by dilatometry investigations [4]. Also illustrating phase transformations occurring during heating from hardened state at specific rate.

P. Bala *et al*, design CHT diagram, illustrating the kinetics of phase transformations during continuous heating (tempering) from as-quenched state of investigated steel, elaborated using dilatometer. Heating of the investigated steel from the as-quenched state resulted in the occurrence of 3 primary transformations: precipitation of ε carbide, M_3C precipitation and transformation of retained austenite. The differences in hardness and microstructure of samples of the investigated steel in relationship to the heat rate were evaluated [5].

P. V. Morra *et al*, studied the decomposition processes of a carburized cast alloys by means of differential scanning calorimetry and dilatometry. The combination of these two experimental techniques was very useful and allows the identification of the stages occurring during tempering. Activation energies were obtained by performing a Kissinger-like analysis and were used to infer the rate-determining step for each stage of decomposition [6].

For all these above, for this project precipitation processes during tempering can be summarized in the following three stages:

"Stage I": The formation of transition carbides and lowering of the carbon content of the martensite. Consist of the segregation of carbon atoms to lattice defects and the formation of clusters of carbon atoms in the iron matrix ***(100 to 200 °C)***.

"Stage II": The transformation of retained austenite into ferrite and clustered carbon atoms transform into the more stable cementite and carbides ***(200 to 450 °C)***.

"Stage III": Dissolution of cementite particles and the formation of more stable alloy carbides occurs only if strong carbide formers such as chromium, molybdenum, tungsten, vanadium are present ***(500–650°C)*** [1-8].

X-ray is the other technique used by several authors to study the kinetics during tempering. Cusenza *et al*, observe the carbide formation upon isothermal heat treatments by means X-ray diffraction. Cr and Mo rich carbides (M3C, Mr23C6 and Mr7C3) are analyzed as transformed phases from cementite [9]. M. Wiessner *et al* employed the diffraction method to identify austenitic and the ferritic/bainitic/martensitic phases, or the tetragonal lattice distortion and the interstitial carbon content in martensite. Also presented is an important study of the transformation of austenite through tempering heat treatment [10].

Mechanical properties have been studied at different heating rates and temperatures of tempered steels. Researchers, report the influence of temperature on hardness of steel, up to which sample was heated after quenching. They observed the highest increase of hardness takes place in a temperature range. At a certain temperature range, precipitation slows down and matrix recovery occurs causing hardness decrease [11-12].

Methodology

The investigations were performed on one N-Cr-Mo-V tool steel whose chemical composition is presented in Table 1. Before the experiments was necessary to apply a heat treatment with quench in order to obtain the martensitic structure. Therefore for the tool steel was proposed the normalizing and quench heat treatments. The normalizing temperature for the quench heat treatment was determinate using the equation for Ac3 transformation temperature [13].

Table 1: Chemical composition (wt. %) of AISI L6 tool steel.

C	Si	Mn	P	Si	Cr	Mo	V	Ni
0.5	0.2	0.8	0.03	0.02	0.9	0.5	0.1	1.6

Dilatometry

R.I.T.A. horizontal dilatometer was employed for the dimensional change experiments. Cylindrical samples of 3 mm diameter and 10 mm length were machined. The sample was mounted between a quartz scanner connected to a linear variable differential transformer (LVDT). It was kept in a vacuum at all testing times to minimize oxidation and decarburization of the surface. The temperature of the sample was measured with K thermocouples welded at the surface. Dimensional changes were recorded by means of a LVDT. Tempering tests were performed on hardened samples at 100°C to 700°C using different heating rates (5, 50 and 150 °C/sec). Helium flow was used as quenching media.

High temperature X-ray diffraction

The equipment for high temperature X-ray experiments was a PANatycal Empyrean model, using the PIXcel^{3D} as area detector and EVA V2.0 software for peaks acquisition. The samples for high temperature X-ray diffraction were 10 x 20 x 3 mm^3 polished. The experiments were heating from room temperature to 700°C before reaching the overall austenite phase content. The heating rate used was 10 °C/min in a vacuum atmosphere. Diffraction scanning was analyzed around the tempering temperatures every 100 °C. Phase transformations were analyzed after cooling. Double tempering was made to study the precipitation during tempering.

Microstructure

The experimental samples cubes of size 2.5 cm^3 were austenized in a laboratory furnace for 1 hour at 900 °C and oil quenched. Then samples were tempered for 1 hour at 100, 200, 300, 400, 500, 600 and 700 °C. Then double tempered for each temperature. After heat treatment the samples were prepared for metallography on the final dimension of 1 cm^3. Saturated picral etchant was used for microstructure analysis. The photos of the microstructure of quenched and tempered samples were analyzed in a scanning electron microscopy Jeol model JSM 6510 LV.

Micro hardness

Every quenched and tempered sample was measured by micro hardness Vickers. The hardness tests were carried out under a load of 5 Newton and 15 seconds of indentation holding. Ten indentations were then used to calculate the Vickers average value for each tempering temperature. Finally for a better assessment of the hardness value, the micro hardness Vickers was converted to hardness Rockwell C (HRC).

Results

Dilatometry

Figure 1 shows the dilatometric curves with their respective derivative curve at heating at rates of 5, 50 and 150 °C/sec for samples previously oil quenched. The red curves define the heating step and blue one is the cooling step. Delta L is represented in μm unites at the left side of the thermograms, while the derivate curve is in the right side at μm/°C.

a)

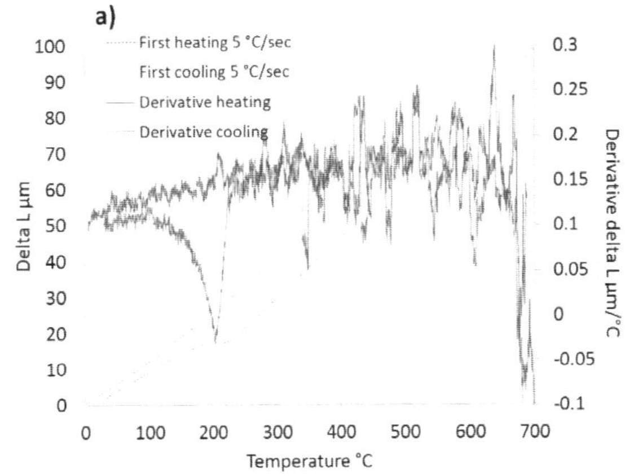

b)

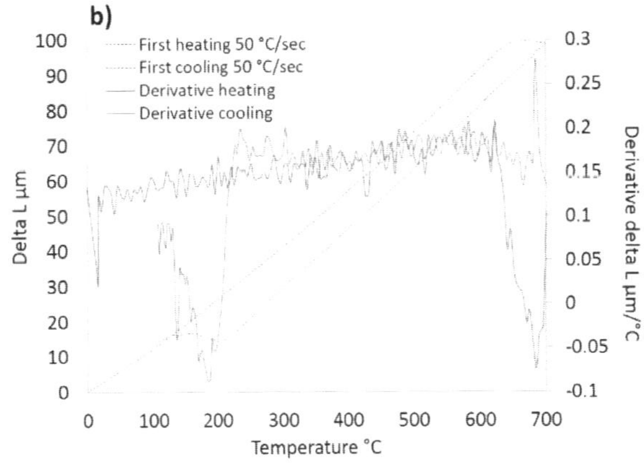

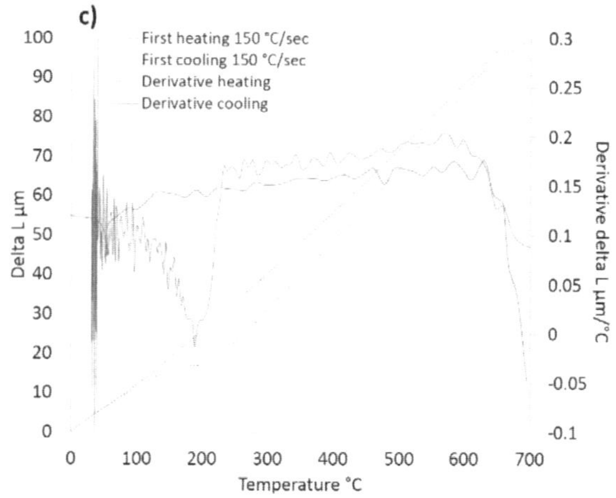

Figure 1. Dilatometry curves with its corresponding derivative of heating with the rate of a) 5 °C/sec, b) 50 °C/sec and c) 150 °C/sec for samples previously oil quenched from 900°C.

High temperature X-ray diffraction

To detect the phase transformation during the tempering, high temperature X-ray diffraction was employed too. Resultant diffractograms are shown in figure 2. Martensite-ferrite, retained austenite, and some carbides (Fe3C, Cr23C6 and Cr7C3) are some of the phases detected and transformed during the heating. Retained austenite is a phase not detected for the EVA V2.0 software. However, some authors present the austenite phase around different heat treatments and identify 2Θ around 118° [14]. Carbide phases as Fe3C precipitate since 500°C and is constant until 25°C after cooling. Two peaks of Cr7C3 were detected at 600°C and Cr23C6 precipitate for a single peak only at 700°C.

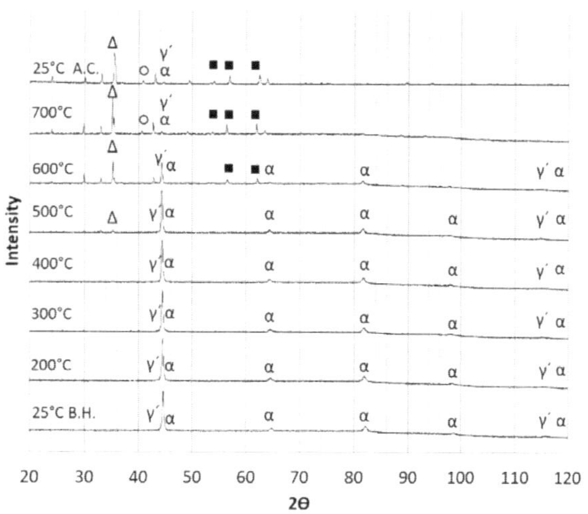

Figure 2. Phases detected during the tempering heating (200°C to 700°C) were carbides, martensite and probably residual austenite at high temperature X-ray diffraction. Fe3C (Δ) , Cr23C6 (○) , Cr7C3 (□) , Martensite – Ferrite (α) and retained austenite (γ΄).

Also diffractions measurements were obtained before heating (B.H.) and after cooling (A.C.) at room temperature. This comparative diffraction pattern shows mostly martensite phase and possibly retained austenite before heating to tempering. While after cooling presents precipitated carbides and martensite were.

Microstructure

Acicular martensite is show in Fig. 3 a) for quenched (900°C) tool steel. The microstructure of all tempering heat treatments are very similar to the quenched condition as is observed in fig. 3.

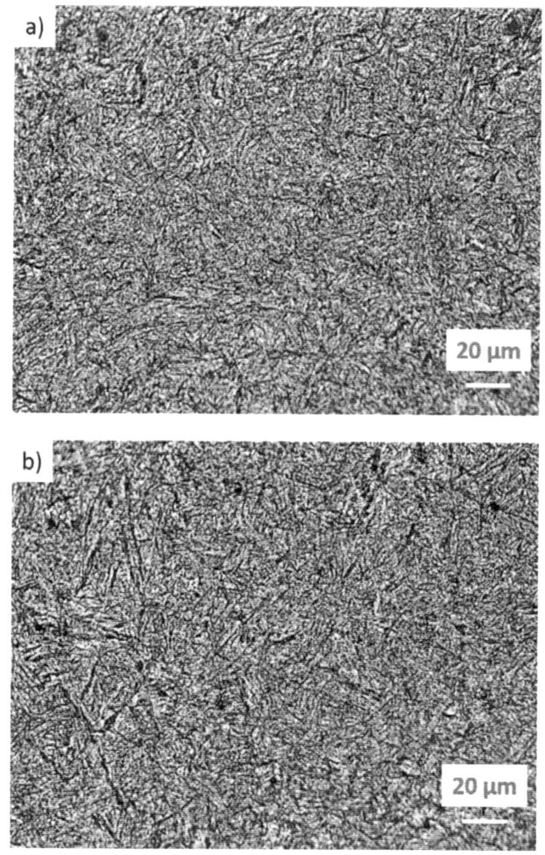

Figure 3. Microstructure from a) quenched and b) tempered condition analyzed by optical microscopy.

SEM was used to study the microstructure of the tool steel. Fig. 4 the quenched microstructure while Fig. 5 is the tempered at 500 °C microstructure. Spherical particles were observed between martensite laths for both conditions.

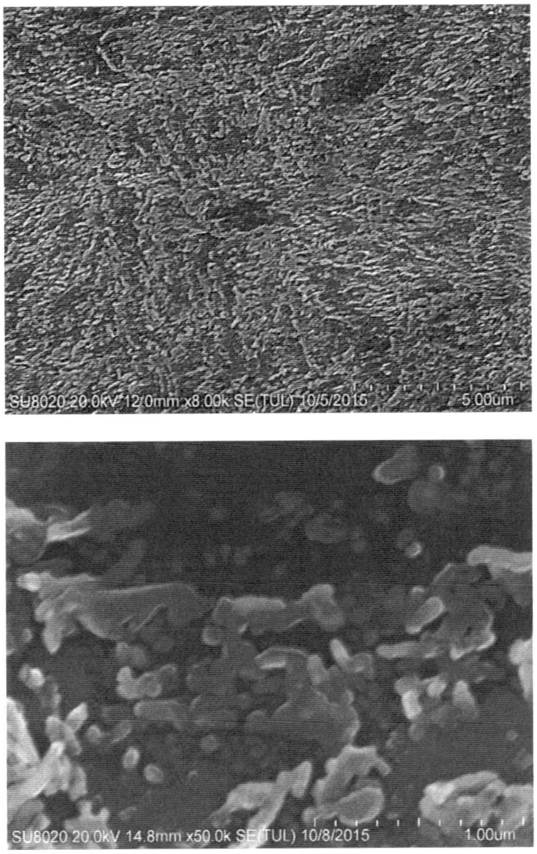

Figure 4. Microstructure of the tool steel heated at 900 °C and oil quenched.

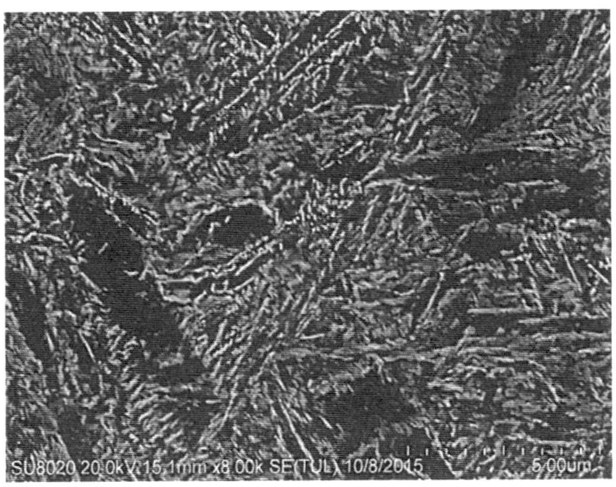

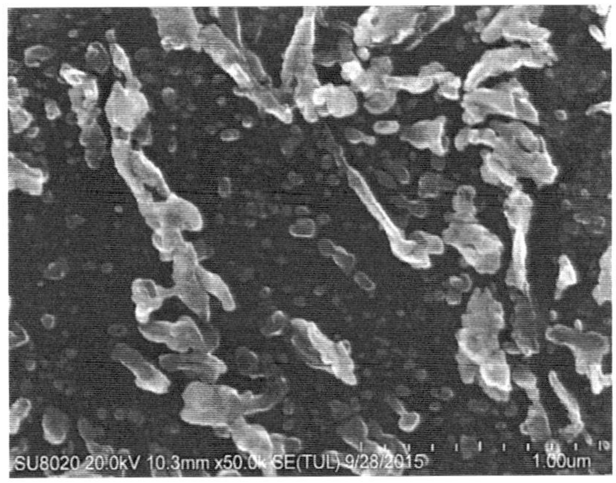

Figure 5. Microstructure of the tool steel oil quenched and then tempered at 500 °C.

Micro hardness

Fig. 6 show the influence of tempering on hardness for the tool steel. Since the quench condition presenting 62 HRC and final tempered sample at 700°C show 31 HRC. Hardness has a decrease in relation of the tempering heating up. Only 500°C and 600°C temperings remains with the same hardness level.

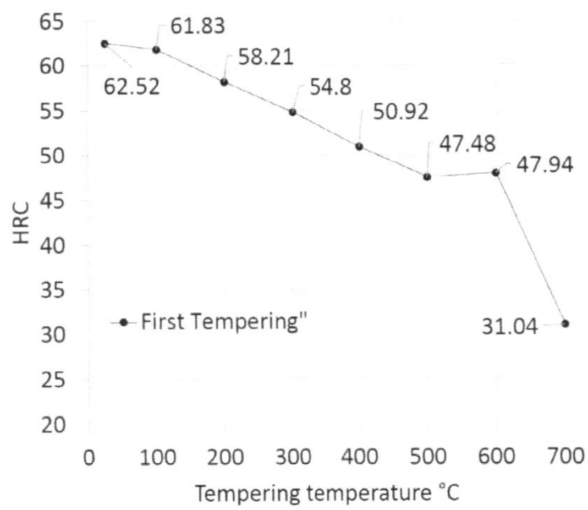

Figure 6. Influence of tempering on hardness for AISI L6 tool steel samples quenched in oil from 900 °C.

Discussion

Dilatometry

The three rates do not show an important volumetric change during the tempering heating. However, during cooling after tempering a contraction effect happens for the three rates around the 200 °C. Between 600°C to 700°C, derivative curves present a contraction effect. However this is related with a heating rate decrease and not a phase transformation. This occurs for each rate and could be related with a possibly martensitic transformation [4]. It is possible that the martensite did not ends to transform after quenching and cooling after tempering works as other quenching heat treatment.

High temperature X-ray diffraction

The diffraction results during tempering shows the nucleation of some precipitates and probably austenite turns to cementite. Austenite peak was referenced with the heat treated steel investigated by Salvatore Cusenza [10].

Two peaks of austenite occurs since room temperature diffractogram. Main precipitate carbide were detected since 500°C: *Fe3C, Cr23C6, and Cr7C3*. As the tempering temperature goes up, austenite peaks seems to decrease and almost disappear while cementite and carbides precipitate. Then the precipitation of carbide and cementite into retained austenite phase apparently occurs, remaining in the cooling after final tempering temperature.

Microstructure

For quenching and different tempering conditions it wasn't possible to observe any microstructural change. The microstructure looks acicular and highly finer. Spherical particles, apparently cementite and carbides, precipitated between the martensite laths. The spherical particles only could be observed at high magnifications and apparently have similar dimensions.

Hardness

The tempering temperature from quench (25°C) condition to 400°C have an expected hardness decrease influence. However the remaining hardness level at high tempering temperatures (500 and 600°C) can be related with a precipitation hardening. Probably related with dissolution of cementite particles and the formation of more stable alloy carbides.

Conclusions

The kinetics of phase transformation on the tempering process of steel was investigated at different heating rates by means of dilatometry, high temperature X-ray diffraction, microstructure and hardness.

- 5, 50 and 150 °C/sec heating rates results to fast for kinetic phase transformation studies. It wasn´t possible to measure a significant volumetric change that represent any carbide precipitation or phase precipitation.
- Possible martensite transformation is observed after tempering.
- By high temperature X-ray diffraction it was possible to observe a precipitation of carbide and cementite into retained austenite.
- Main precipitate carbide were detected since 500°C, *Fe3C, Cr23C6, Cr7C3*.
- For quenching and tempering conditions results with an acicular and highly finer microstructure. Spherical particles precipitated between the martensite laths.
- Hardness test exhibited a probable precipitation hardening occurring at high tempering temperatures (600°C).

References

[1] ASM, Low alloy special purpose tool steel (L Series), "L6 Technical Data Sheet", International Heat Treaters Guide, p. 576 - 579.

[2] Charlie R. Brooks *et al.*, ASM International, 1996, "Principles of the heat treatment of plain carbon and low alloy steels"

[3] ASM, Metals Handbook Vol. 04, Heat Treating.

[4] J. Pacyna, Journal of Achievements in Materials and Manufacturing Engineering, "Dilatometric investigations of phase transformations at heating and cooling of hardened, unalloyed, high-carbon steels", Vol. 46, May 2011.

[5] P. Bala, J. Pacyna, Journal of Achievements in Materials and Manufacturing Engineering, "The kinetics of phase transformations during tempering of low alloy medium carbon steel", Vol. 28, May 2007.

[6] P. V. Morra, Journal of thermal analysis and Calorimetry, "Decomposition of iron-based martensite, A kinetic analysis by means differential scanning calorimetry and dilatometry", Vol. 64, 2001, pages 905-914.

[7] M. Goji, Journal of thermal analysis and Calorimetry, "Thermal analysis of low alloy Cr–Mo steel", Vol. 75, 2004, pages 947-956.

[8] Rafal Dziurka, METAL, "Effect of heating rate on phase transformations during tempering of Cr-Mn-Mo alloy steels", 2013

[9] Salvatore Cusenza *et al*, Appl Phys A, "Amorphous stainless steel coatings prepared by reactive magnetron-sputtering from austenitic stainless steel targets", Springerlink.com, 2009.

[10] M. Wiessner *et al*, "Advanced X-ray Diffraction Techniques for Quantitative Phase Content and Lattice Defect Characterization during Heat Treatment of High Speed Steels", BHM, Vol. 159, 2014.

[11] P. Bala, METAL, "kinetics of phase transformations during continuous heating from as-quenched state of 17-4ph steel", 2015.

[12] R. D. Browski, R. Dziurka, Archives of Metallurgy and Materials, "Tempering Temperature Effects on Hardness and Impact Toughness of 56NiCrMo7 Steel", Volume 56, 2011.

[13] M.A. Alvarado-Meza *et al.*, Journal of Materials Engineering and Performance, "Effect of the High-Temperature Deformation on the Ms Temperature in a Low C Martensitic Stainless Steel", Volume 22, February 2013.

[14] D.L. Johannsen et al., Metallurgical and Materials Transactions A, "Influence of Annealing Treatment on the Formation of Nano/Submicron Grain Size AISI 301 Austenitic Stainless Steels", Volume 37, August 2006.

International Federation of Heat Treatment and Surface Engineering 2016
Proceedings of the 23rd IFHTSE Congress
April 18–21, 2016, Savannah, Georgia, USA

Hardenability Characterization and Microstructural Evolution in Heavy Gauge Plate Steels

Igor Vieira and Emmanuel De Moor
Advanced Steel Processing and Products Research Center, Colorado School of Mines, Golden, CO, USA

Abstract

High strength heavy gauge plate steels for structural applications are used in the construction of buildings, bridges, sport stadia, oil rigs, power plants, *etc*. In order to reach the desired mechanical properties, these steels are subjected to a quenching process followed by a tempering treatment. However, different cooling rates are obtained throughout the thickness of the material during quenching and several grades do not fully harden to martensite. In the present study, three alloys containing 0.17 wt pct carbon and different amounts of molybdenum and vanadium were used. The objective was to characterize hardenability by evaluating the influence of alloying on hardness and microstructure. The Jominy End-Quench test was used to assess hardenability and scanning electron microscopy was conducted for microstructural characterization. The effects of grain size and alloying content on hardenability are discussed for each alloy.

Introduction

High strength heavy gauge plate steels for structural applications are quenched and tempered in order to achieve required mechanical properties. During the quenching process, a temperature profile is obtained along the plate thickness associated with non-uniform cooling rates through thickness. From the surface to the core, the cooling rate decreases resulting in different microstructures. Thus, the material hardenability, defined as the ability of the steel to obtain satisfactory hardening to some desired depth when cooled under prescribed conditions [1], becomes an important characteristic.

Two main factors govern the hardenability of a material: austenitizing conditions and chemical composition. The former parameter, through changes in temperature and time, determines the solubility of alloying elements, coarsening of precipitates, and the prior austenite grain size, which impacts the amount of available nucleation sites. Alloying additions may be present in solid solution, segregated to grain boundaries and present as carbides, nitrides, etc [2].

The present work was designed to investigate the hardenability of plate steels and the alloying impact caused by vanadium and molybdenum additions.

Experimental Procedures

Three alloys were investigated and their compositions are shown in Table 1. The material was provided by Tata Steel as laboratory produced hot rolled plate. A heat size of 50 kg was used which generated two 25 kg ingots with different chemical compositions. Ingots with a cross section of 140 mm^2 were reheated at 1200 °C, and hot rolled to plates with reductions of about 10 pct per pass. The final plate dimensions were 140 mm in width and 35 mm in thickness.

Table 1 - Chemical Compositions of Studied Alloys in wt pct

Designation	C	V	Mo
Base	0.165	<0.001	0.010
Base-V	0.170	0.051	<0.005
Base-Mo-V	0.165	0.051	0.510

Figure 1 schematically describes the two analysis sequences proposed, namely testing of the as-received material, and normalization for one hour followed by Jominy testing. The latter route was planned based on the standard for determining hardenability which includes a normalization step prior to austenitization [3].

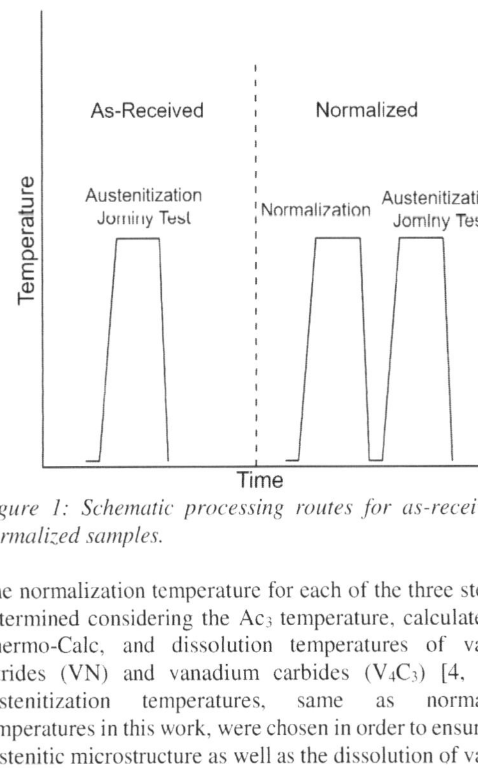

Figure 1: Schematic processing routes for as-received and normalized samples.

The normalization temperature for each of the three steels was determined considering the Ac_3 temperature, calculated using Thermo-Calc, and dissolution temperatures of vanadium nitrides (VN) and vanadium carbides (V_4C_3) [4, 5]. The austenitization temperatures, same as normalization temperatures in this work, were chosen in order to ensure a fully austenitic microstructure as well as the dissolution of vanadium carbides and nitrides. Table 2 presents the transformation temperatures Ac_3, normalization temperatures, and finish rolling temperatures. The finish rolling temperatures were measured with a pyrometer and used as a reference regarding the as-received condition.

Prior austenite grain size (PAGS) was measured for each normalized sample. Cubic specimens measuring approximately 1 cm^3 were prepared from each steel, normalized at the temperatures shown in Table 2 for one hour, and quenched in water. During heat treatment, samples were placed in stainless steel bags with titanium coupons in order to avoid the effects of oxidation and decarburization. Table 3 shows the etching techniques used for each material. Micrographs of 10 fields of each sample were used to evaluate grain sizes. The lengths of the grain intercepts were measured with ImageJ® software, and a distribution was plotted.

Table 2 – Transformation Temperatures Ac_3, Normalization Temperatures, and Finish Rolling Temperatures

Temperatures (°C)	Ac_3	Normalization	Finish Rolling
Base	830	997	1001
Base-V	831	998	1008
Base-Mo-V	843	1010	998

Table 3 - Etching Techniques Used to Reveal Prior Austenite Grain Boundaries

Material	Solution	Heat Treatment
Base and Base-V	2.6g Picric acid, 2 mL HCl, 6mL Teepol, 200 mL, at 65°C	48 hours at 400°C
Base-Mo-V	2.6g Picric acid, 2 mL HCl, 6mL Teepol, 200 mL, at 65°C	24 hours at 400°C

For the Jominy test, specimens were machined from each material measuring 25.4 mm in diameter and 98.4 mm in height, including a section with a reduced diameter on the top to enable the use of washers to hold the samples in place during testing. Samples were placed in a preheated furnace for 35 minutes (isothermal time) at the austenitizing temperatures shown in Table 2, followed by transfer and holding for 15 minutes in the quenching apparatus. Four flats at 90° to each other were ground to a depth of 0.508 mm (0.020 in). It can be noted that the minimum grinding depth required by the standard is 0.381 mm (0.015 in) [3]. Finally, Rockwell hardness measurements were taken every 1.6 mm (1/16 in) from the quenched end for the first sixteen measurements, every 3.2 mm (1/8 in) for the next four measurements, and every 6.4 mm (1/4 in) for the last two measurements. A calibration block certified at 30.3±1.0 HRC was measured to have a hardness of 29.7 HRC, which confirms the Rockwell hardness indenter calibration. Finally, the hardness values from the four flats were averaged and plotted as a function of distance from the quenched end. Microstructural characterization along the length of the Jominy bars was conducted using scanning electron microscopy (SEM). Prior to evaluation, samples were etched with 2pct Nital.

Results and Discussion

Figure 2 contains light optical microscopy (LOM) micrographs of the prior austenite grains from the different materials under normalized conditions. The grain size distribution obtained for each steel is shown in Figure 3. All materials presented a similar distribution, with the majority of the grain sizes ranging from 20 to 60 μm. Such similarity is most likely related to the same difference between normalization and Ac_3 temperatures for all the materials. However, the range of grain sizes was significantly different. Even though all steels presented abnormal grain growth, which takes place at the expense of arrays of fine grains, the Base and Base-V alloys presented wider ranges.

Figures 4 through 6 show the hardness profiles obtained for each of the three materials as a function of the distance from the quenched end in the as-received and normalized conditions. Hardness values below 20 HRC were not included in the plots as such values are not accurate according to the Rockwell C scale [3].

The hardness profiles were in general fairly similar for the as-received and normalized conditions. Such similarity could be related to the small difference between the normalization temperature and the finish rolling temperature, as observed in Table 2. Although the Base material (Figure 4) showed a difference in hardness for similar positions in the bar, the hardness profiles obtained are fairly similar for both conditions.

The hardenability of the normalized alloys was compared and the Base-Mo-V presented the highest hardenability, followed by the Base-V and Base material. The latter two steels showed similar results. The similarity between Base and Base-V, which presented an almost constant drop in hardness, may relate to the modest amount of vanadium present in the Base-V alloy. The present results are in agreement with the work performed by Grange [6] with steels containing 0.19 wt pct carbon austenitized at 900 °C, which shows that 0.05 wt pct vanadium increased the ideal diameter by about 1.7 mm. Since Mo is known to present a strong effect on hardenability [6, 7], the result obtained for the Base-Mo-V was as expected. A plateau is observed in the hardness profile of the Base-Mo-V alloy up to a distance of about 1.6 cm from the quenched end. It is followed by a relatively shallow slope, indicating high hardenability.

In terms of grain boundaries, increased austenite grain sizes are expected to retard the ferrite and pearlite nucleation by decreasing the number of nucleation sites, thereby increasing hardenability. However, an increase in austenitizing temperature, which is responsible for grain growth, also affects the solubility of carbon and alloying elements. Thus, the dissolution effect can play a more pronounced role in hardenability than grain size [2]. Such behavior was observed

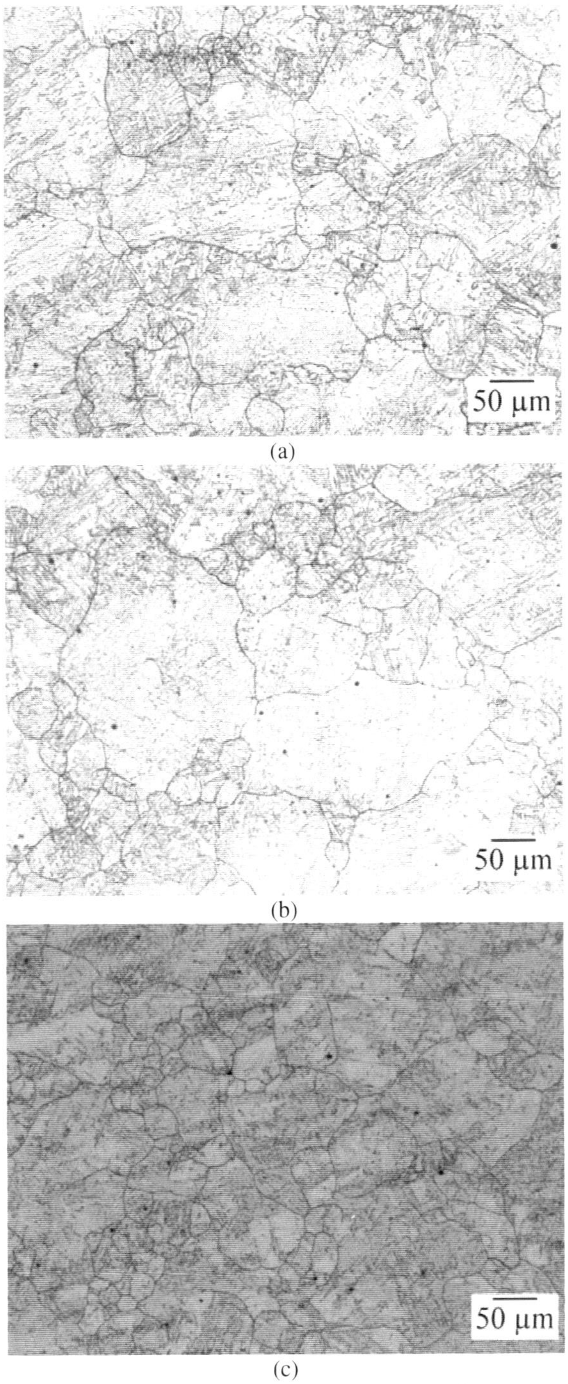

Figure 2: Light optical microscopy micrographs showing prior austenite grains from (a) Base, (b) Base-V, and (c) Base-Mo-V alloys.

in the presented data. The alloying factor seemed to be more significant than the amount of nucleation sites at grain boundaries. So the Base and Base-V alloys, materials with greater grain size, presented lower hardenability than the Base-Mo-V.

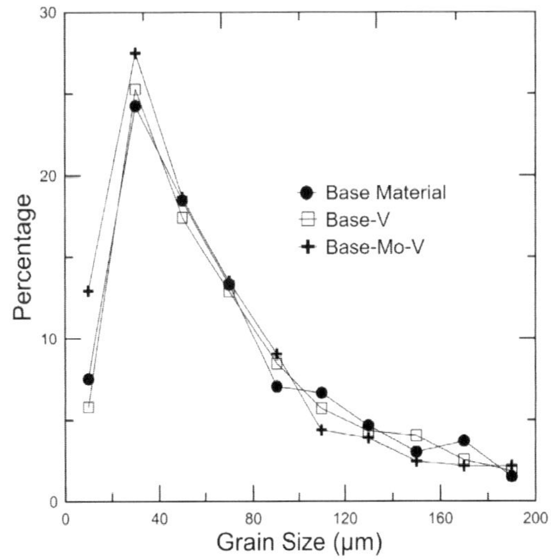

Figure 3: Grain size distributions of the Base, Base-V, and Base-Mo-V alloys, excluding grain sizes greater than 200 μm.

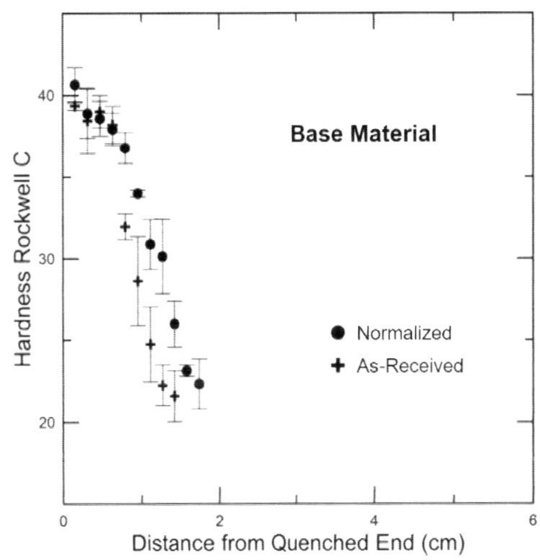

Figure 4: *Hardness as a function of distance from quenched end along Jominy bar plotted for Base material. Error bars represent the standard deviation of the measurements.*

The hardness profiles obtained from normalized and as-received materials were combined and compared to a calculated hardness profile, according to ASTM-A255 [3]. Figure 7 shows the comparison between the experimental and calculated hardness profiles for the three alloys considered. The Base and Base-V steels again presented a similar behavior, with the experimental data slightly lower than the calculated for the first measurements and higher after the intersection between the curves. Base-Mo-V was the exception, presenting very similar results for the first measurements and starting to differ from the calculated value at the J5 position (8.0 mm from the quenched end). The experimental data scatter, represented by the standard deviation in the plots, plays a role in the encountered difference. In addition, the proposed calculation assumes an austenitic grain size ASTM 7 [3], which represents an average grain size

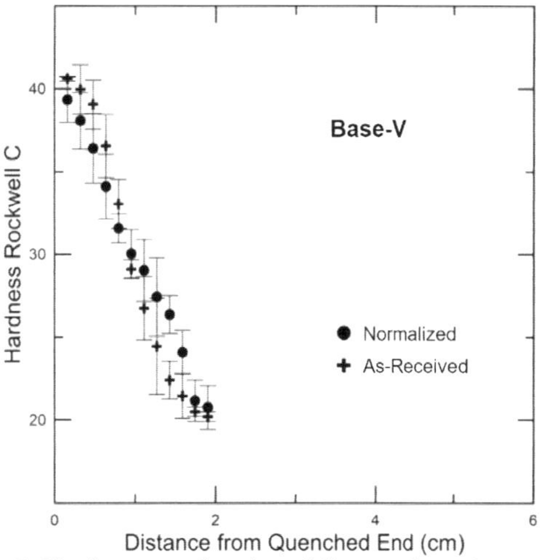

Figure 5: *Hardness as a function of distance from the quenched end along the Jominy bar plotted for Base-V. Error bars represent the standard deviation of the measurements.*

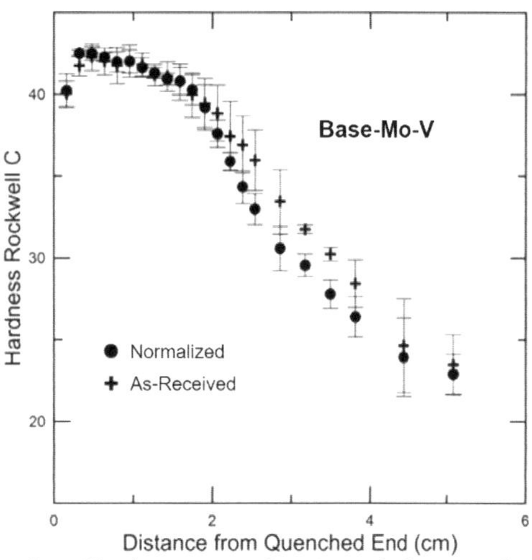

Figure 6 – Hardness as a function of distance from the quenched end along the Jominy bar plotted for Base-Mo-V. Error bars represent the standard deviation of the measurements.

of 31.8 μm [8], smaller than found on average for the considered material. So, with less grain boundaries available for nucleation of the different phases, the material tested should present a higher hardenability, as found. This suggests that the grain size might also be affecting hardenability. Another plausible explanation for the difference found arises from the experimental procedure. The samples were submitted to a water quench at one end of the bar for 15 minutes and were then soaked in water for the final quench. In case some untransformed austenite was still present, the final quench would result in more martensite than if longer times for the completion of transformation were given. As a consequence, the greater amount of martensite would increase the expected hardness. The calculated and experimental results seem to be in

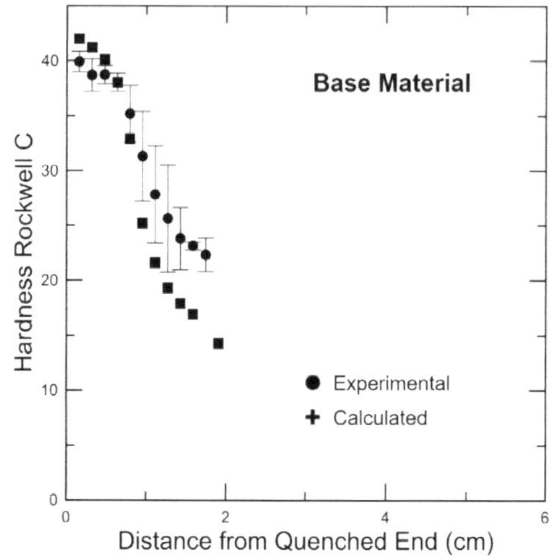

(a)

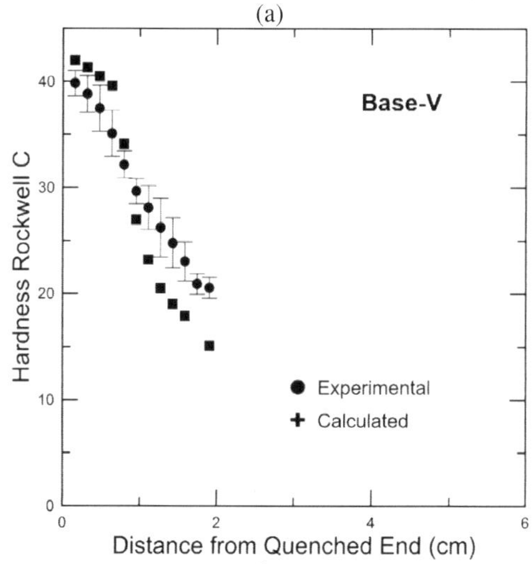

(b)

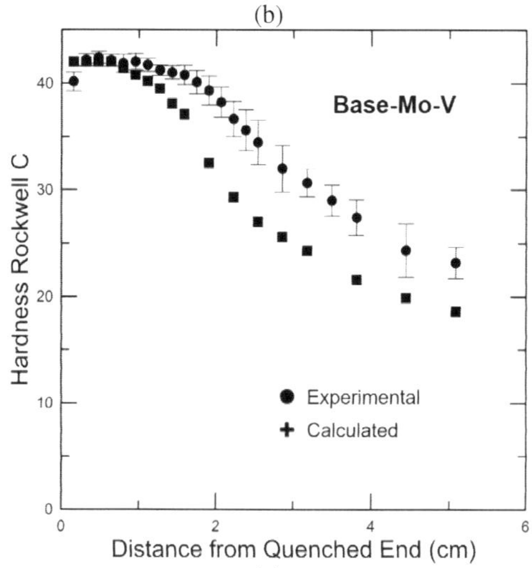

(c)

Figure 7: Experimental and calculated hardness profiles for the (a) Base, (b) Base-V, and (c) Base-Mo-V alloys. Error bars represent the standard deviation of the measurements.

agreement, especially considering the aforementioned sources of discrepancy.

Micrographs were obtained using SEM in order to evaluate the microstructural evolution along the Jominy bars. Figure 8 shows the microstructures obtained from three different positions for each material. The positions under evaluation were chosen due to the similarity in hardness and expected significant microstructural differences in the same material.

The first set of micrographs, position J1 for Base and Base-V, and position J2 for Base-Mo-V, presented a fully martensitic microstructure. The martensite laths are evident in all images. In the second set of micrographs, position J9 for Base, position J8 for Base-V, and position J22 for Base-Mo-V, prior austenite grain boundaries are evident in all microstructures. Parallel features, indicated by arrows in the Base and Base-V micrographs, suggest the presence of bainite, while surrounding features arrows suggest the presence of martensite. The Base-Mo-V alloy, as well as the last set of micrographs (position J11 for Base, position J12 for Base-V, and position J32 for Base-Mo-V), seem to present a mixture of tempered martensite and bainite. Due to the position in the bar, those parts were subjected to slower cooling rates, possibly allowing the material to be tempered during cooling, which would produce tempered martensite. It is challenging to distinguish tempered martensite and bainite in the SEM micrographs. Color etching using LePera [9] is commonly used to differentiate such

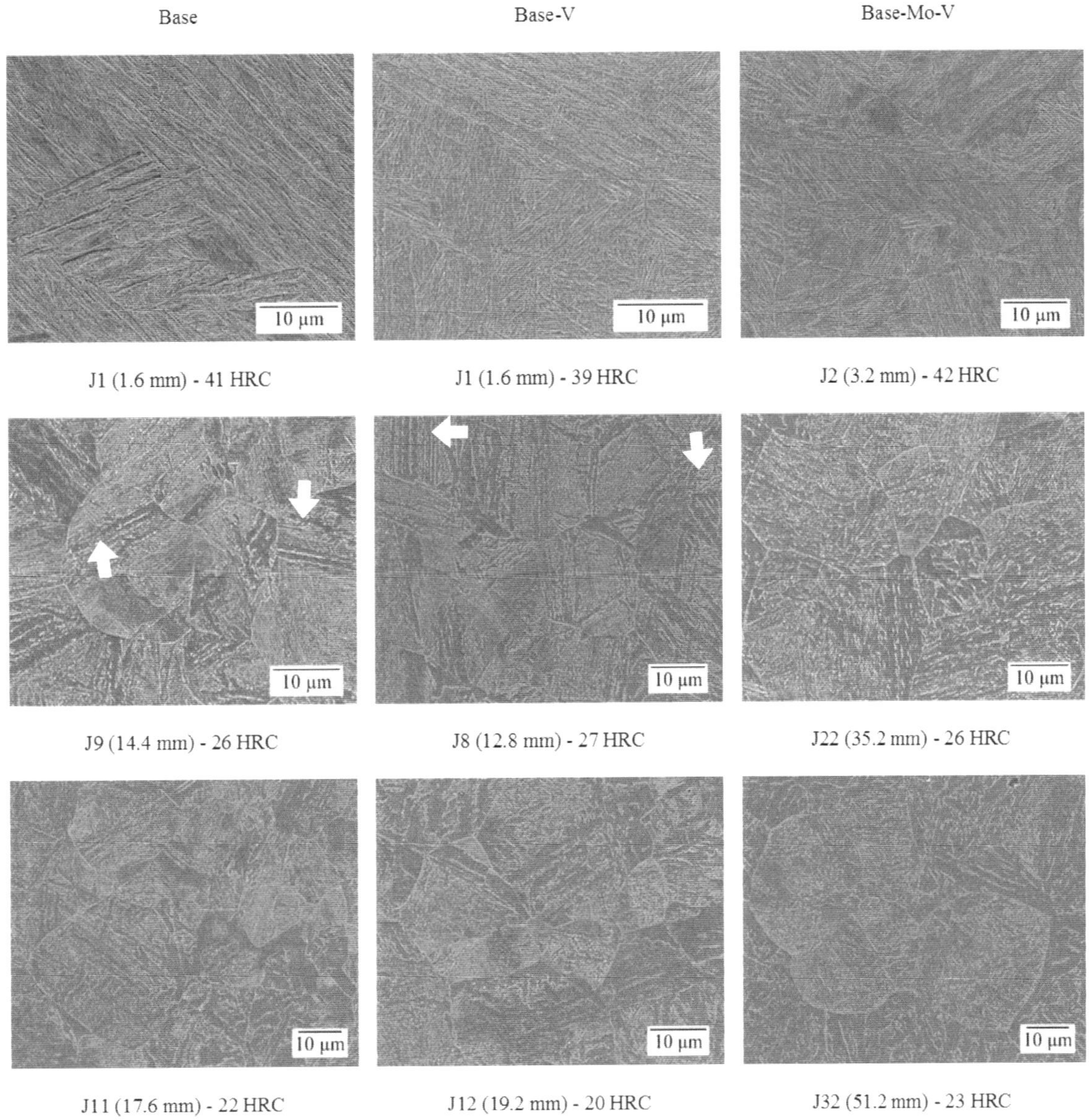

Figure 8: Scanning electron microscopy micrographs for Base material (positions J1, J9, and J11), Base-V (positions J1, J8, and J12), and Base-Mo-V (positions J2, J22, and J32). White arrows indicate features with possible bainitic characteristics while black arrows indicate features with possible as-quenched martensitic characteristics.

microstructures in light optical microscopy, but the technique was not successful in the present work.

Conclusions

Hardenability characterization was performed for three steels with different amounts of vanadium and molybdenum in the normalized and as-received conditions. Results suggested a limited effect from normalization.

The Base-Mo-V steel presented the highest hardenability followed by Base-V and Base material. Comparison among the three materials in terms of hardness profiles and prior austenite grain size distribution suggest a more significant influence on hardenability from alloying elements than grain size.

The hardenability curves obtained from Jominy tests were compared to calculated curves and the experimental results were considered reasonable.

Acknowledgements

The support of the sponsors of the Advanced Steel Processing and Products Research Center, an industry-university cooperative research center at the Colorado School of Mines, and Tata Steel, for providing the material, are gratefully acknowledged.

References

[1] Dossett, J. L. *et al.*, ASM Handbook, Volume 04A - Steel Heat Treating Fundamentals and Processes, ASM International (2013), pp. 26.

[2] Bain, E. C., Functions of the Alloying Elements in Steels, American Society for Metals (1939), pp. 48-58.

[3] ASTM A255-10, "Standard Test Methods for Determining Hardenability of Steel," *ASTM International*, 2010, pp. 1-26.

[4] Turkdogan, E., "Causes and Effects of Nitride and Carbonitride Precipitation During Continuous Casting," *Iron and Steelmaker*, Vol. 16 (1987), pp. 61-75.

[5] Leslie, W., The Physical Metallurgy of Steels, Hemisphere Publishing Corporation (1981), pp. 134-141.

[6] Grange, R., "Estimating the Hardenability of Carbon Steels," *Metallurgical Transactions*, Vol. 4 (1973), pp. 2231-2244.

[7] Krauss, G., Steels – Processing, Structure, and Performance, ASM International (2005), pp. 297-323.

[8] ASTM E112-13, "Standard Test Methods for Determining Average Grain Size," *ASTM International* (2014), pp. 1-28.

[9] LePera, F., "Improved Etching Technique for the Determination of Percent Martensite in High-Strength Dual Phase Steels," *Metallography*, Vol. 268 (1979), pp. 263-268.

Influence of cooling and heating rates on the thermal properties of Ni-Ti shape memory alloy

Boutheina BEN FRAJ[1] and Pr. Zoubeir TOURKI[2]
Laboratoire Mécanique de Sousse LMS, Ecole Nationale d'Ingénieurs de Sousse ENISo, Université de Sousse, Sousse, 4023, Tunisia.

[1] bfb.ingmec@gmail.com, [2] zbrtourki@gmail.com

Abstract

The near-equiatomic NiTi shape memory alloy (SMA) has many unusual thermomechanical properties which are dependent of various parameters such as heat treatment conditions, manufacturing process, chemical composition... The determination of these unique properties is indispensable for studying the behaviour of this alloy under different thermal, mechanical or thermo-mechanical stresses. In this paper, the study is focused on the effect of the cooling and heating rates on the thermal properties of the NiTi alloy during its phase transformation cycle at zero stress. This experimental work is based on the differential scanning calorimetry (DSC) as a thermal analysis method. Calorimetric results show that the temperature, which characterizes the end of the direct martensitic transformation, is significantly influenced by the increasing of the cooling/heating rates from 5°C/min to 20°C/min. Thus, it leads to deduce the influence of this parameter on the thermal hysteresis of this material under these conditions. The evolution of absorbed and released heats as of the total energy dissipation as function of the cooling/heating rate during phase transformation were also discussed in this study. The obtained results are in a good agreement with previous studies.

Introduction

The NiTi shape memory alloys are the most widely used shape memory materials. These alloys have received much scientific and technological attention because of their unusual thermo-mechanical properties such as: one way shape memory effect, two way shape memory effect, super-elasticity and damping effect. These unique properties are closely linked to a thermoelastic phase transformation called "martensitic transformation".

Among the key characteristics of this martensitic transformation, we can cite the transformation temperatures M_s, M_f, A_s and A_f which indicate, respectively, the start and the end of both forward (austenite to martensite) and reverse (martensite to austenite) transformations. These transformation temperatures depend mainly on the alloy's composition and processing. Whatever the parent phase (austenitic or martensitic), the microstructural defects, the degree of order and the grain size can also alter the transformation temperatures by several degrees [1].

The effect of these factors on the transformation temperatures has been widely discussed in literature. However, the influence of the cooling and heating rates on the thermal characteristics of martensitic transformation during the transformation process has not yet been investigated. It has been proved that the finish transformation temperatures are strongly dependent on the cooling/heating rate [2-3-4]. Moreover, it has been reported that the absorbed heat during austenitic transformation as well as the released heat during martensitic transformation and the energy dissipation in a full cycle increase with the increase of the cooling/heating rate [5]. Through microstructural studies, it has been observed that this evolution in the transformation temperatures is attributed to the changes in size and spacing of the Ni_4Ti_3 precipitates formed during the heat treatment process [5-6-7]. On the other hand, the evolution in the absorbed and released heats has been explained as an obvious consequence of the characteristics of thermoelastic transformation mechanism [8-9].

Therefore, this study has been developed to determine the relationship between cooling/heating rate and the thermal properties of a near-equiatomic NiTi SMA, under the phase transformation cycle at zero stress. Thereafter, the effect of the cooling/heating rate on the transformation heats and on the total energy dissipation during the full transformation process, has been also discussed.

Material and experimental procedures

The NiTi SMA was purchased by Baoji Rare Titanium-Nickel Co. Ltd., China. The chemical composition of this material is given in Table 1. All samples are cut from the same NiTi bar and the weight of each one is about 12 mg. The samples were heat treated at 650°C for 1h, followed by cooling in air. After annealing, samples were polished using a waterproof sand paper and then rinsed with ethanol. The measurements were performed with a differential scanning calorimeter DSC 4000/Perkin Elmer with an accuracy of ±0.1°C. The samples were encapsulated in an Aluminium specimen pans. A matching empty specimen pan with crimped lid was used as a reference. The NiTi specimens were treated at various heating and cooling rates, from 5°C/min to 20°C/min, under a constant nitrogen flow rate of 20 ml/min. All the DSC tests consisted of a complete thermal cycle (cooling and heating). This thermal cycle was repeated twice to check the repeatability of the material response.

Table 1 : The chemical composition of the NiTi alloy (wt.%)

Ni	55.89
H	0.001
N	0.002
O	0.048
C	0.041
Ti	Balance

Results and discussion

Fig. 1(a-d) and 2(a-d) illustrate the multiple DSC curves measured with different cooling and heating rates, from 5°C/min to 20°C/min with a step of 5°C/min. As apparent in Fig. 1, cooling the sample have created an exothermic peak. However, heating have produced an endothermic peak as shown on Fig. 2. It appears that the scanning rate has a significant effect on both forward (austenite-martensite) and reverse (martensite-austenite) transformations during cooling and heating processes, respectively.

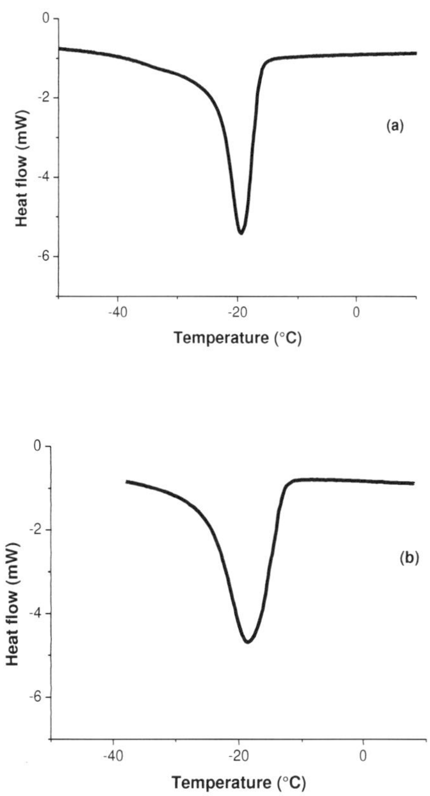

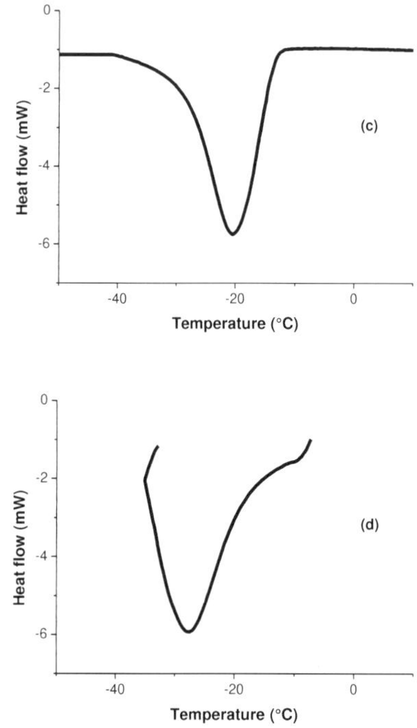

Figure 1: The DSC curves of the NiTi specimen at different cooling rates: (a) 5°C/min, (b) 10°C/min, (c) 15°C/min and (d) 20°C/min.

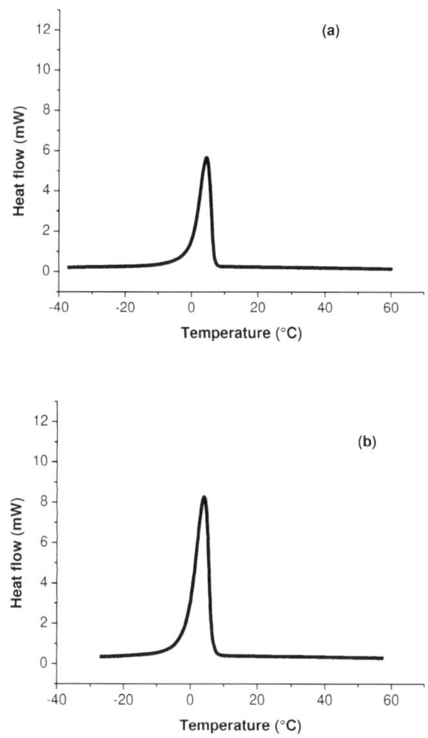

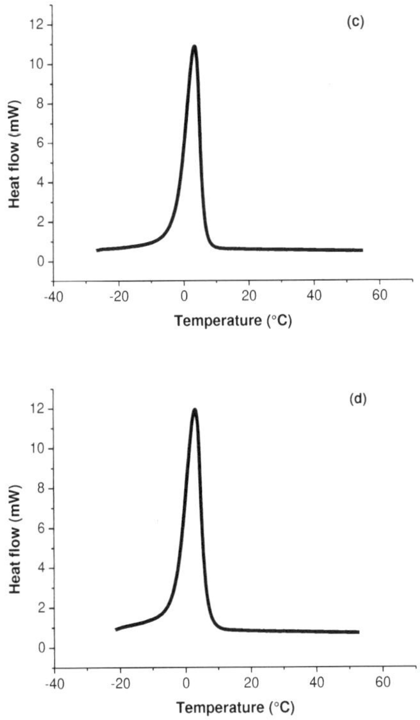

Figure 2: The DSC curves of the NiTi specimen at different heating rates : (a) 5°C/min, (b) 10°C/min, (c) 15°C/min and (d) 20°C/min.

From these calorimetric thermogrammes, the austenitic and martensitic start and finish temperatures have been determined by the intersection of a base line and the tangent to a peak. Furthermore, the areas bounded by the peak and the base line from each curve lead to determine the corresponding transformation heats : released and absorbed heats during forward and reverse transformations, respectively. Wherefore, the discussion of these results will be divided on two sections in order to analyze the effect of the cooling/heating rate on the transformation temperatures then on the transformation heats.

Transformation temperatures

Based on these calorimetric results, the transformation temperatures, indicating the beginning and the end of both the austenitic and the martensitic transformations, are assessed and summarized in the Table 2.

From the collected results, it can be reported that the onset transformation temperatures are not remarkably influenced by the increase of the cooling/heating rate : A_s decreases slightly by almost 3.5°C and M_s increases then decreases at 20°C/min to back finally to its first value measured at 5°C/min.

For the evolution of the finish transformation temperatures, plotted in Fig. 3 as functions of cooling/heating rate, the calorimetric results show that, during heating process, A_f is almost stable. However, it is clear that M_f decreases with the increase of the cooling rate. This significant relation between the cooling/heating rate and the M_f temperature is similar to those proved by previous studies [2-3-4-5].

Table 2 : The transformation temperatures of the NiTi at different cooling and heating rates

Heat treatment conditions	Cooling/heating rate (°C/min)	Transformation temperatures (°C)			
		A_s	A_f	M_s	M_f
650°C for 1 hour	5	-0.96	6.7	-16.23	-23.18
	10	-2.43	6.82	-14.20	-26.52
	15	-3.42	6.56	-13.85	-29.26
	20	-4.25	6.97	-16.91	-35.18

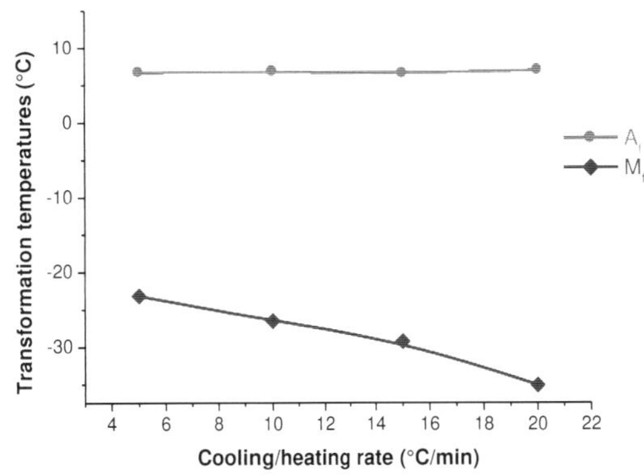

Figure 3: The transformation temperatures as functions of the cooling/heating rate of the NiTi specimen

Consequently, it can be deduced that the difference A_f-M_f, which represents the transformation hysteresis [10], increases with the increase of the cooling/heating rate. In this case, it increases by around 12°C. The evolution of the hysteresis width as function of cooling/heating rate is shown in Fig. 4.

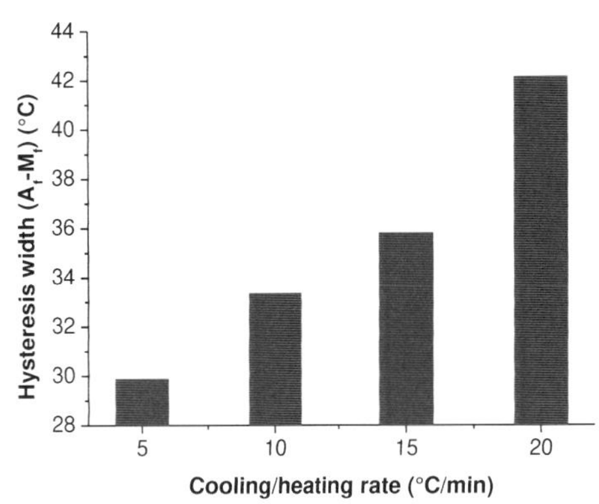

Figure 4: The influence of the cooling/heating rate on the hysteresis width of the NiTi alloy.

From microstructural analysis, it is well known that, during heat treatment, Ni-rich NiTi alloys are capable of developing Ni_4Ti_3 precipitates. These so called intermetallic compounds have an influence on the features of the martensitic transformation by decreasing the nickel-content in NiTi matrix. This behavior leads an increase of the transformation temperatures [11-12]. In addition to this, it has been proved that both size and spacing of these particles have a strong influence on the NiTi behavior during transformation process [6]. In fact, it has been shown that these precipitates are coherent with the martensitic matrix and they act as barrier for dislocation motion [11].

The microstructural behavior of the nearly 50:50 at% NiTi SMA annealed at 450°C for 30 minutes has been studied at various scanning rates [5]. From the optical micrographs illustrated in Fig. 5, it has been well observed that both size and spacing of the Ni_4Ti_3 precipitates have been affected by the variation of the cooling/heating rate. With the increase in scanning rate, these precipitates became more fine and more spaced. Moreover, these microstructural variations lead to an increase in both dislocation density and resistance to martensitic transformation.

Therefore, we can deduce that this reduction of Ni_4Ti_3 precipitates with the increase of cooling/heating rate favors the augmentation of the Ni-content in the NiTi matrix, which subsequently results in the decrease of martensitic transformation temperatures, which has been clearly shown in this present study.

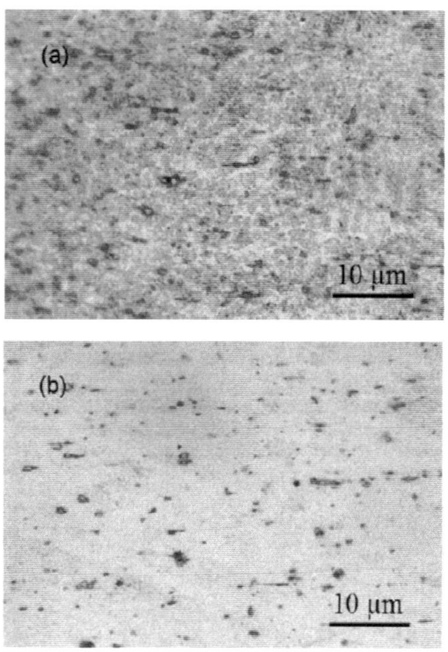

Figure 5: Optical micrographs of the NiTi SMA after heating/cooling processes. (a) The microstructure which includes intensive Ni4Ti3 precipitates obtained at the heating/cooling rate of 5°C/min. (b) A micrograph obtained at the heating/cooling rate of 20°C/min with fine precipitates thinly dispersed in the microstructure. [5]

Absorbed and released heats

When the martensitic transformation takes place, numerous physical properties are modified. During the transformation, a latent heat associated with the transformation is absorbed or released based on the transformation direction. The forward, austenite-to-martensite transformation is accompanied by the release of heat corresponding to a change in the transformation enthalpy (exothermic phase transformation). The reverse, martensite-to-austenite transformation is an endothermic phase transformation accompanied by absorption of thermal energy. For a given temperature, the amount of heat is proportional to the volume fraction of the transformed material [9].

From the thermogrammes obtained on Fig. 1 and Fig. 2, each released and absorbed transformation heat was determined from the areas bounded by the peak and the base line. The values corresponding to each cooling/heating rate are illustrated in Table 3.

Table 3 : The transformation heats (released and absorbed) at different cooling and heating rates

Heat treatment conditions	Cooling/heating rate (°C/min)	Released heat (mJ) (A-M)	Absorbed heat (mJ) (M-A)
650 °C for 1 hour	5	-113.68	49.49
	10	-91.62	75.85
	15	-157.78	119.42
	20	-165.26	159.86

From the data of the Table 3, the released and the absorbed heats were plotted in Fig. 6 and Fig. 7, respectively, as functions of the cooling and the heating rates. The absolute values of these two transformation heats increase with the increase of the scanning rate.

In order to analyze these results, it has been reported that the dislocations associated with high heating/cooling rates generate an internal stress state which restricts the transformations [5].

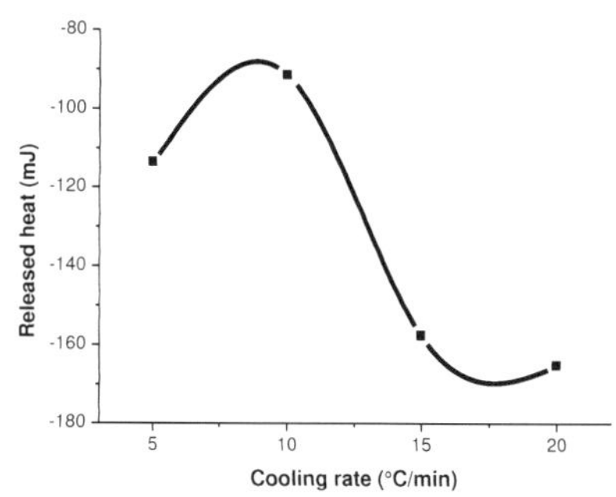

Figure 6: The released heat as function of the cooling rate

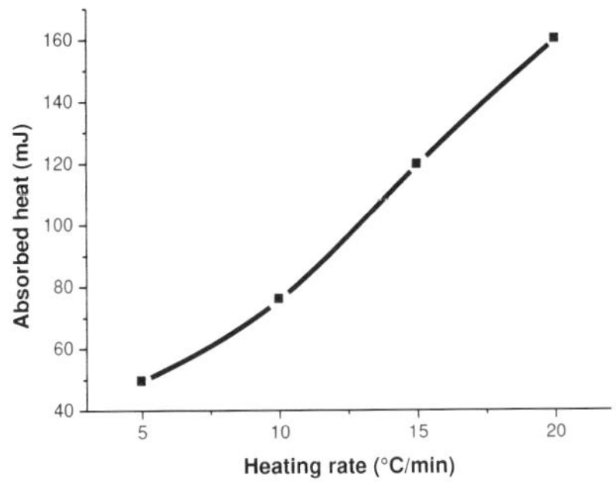

Figure 7: The absorbed heat as function of the heating rate

Furthermore, as the thermoelastic martensitic transformation occurs by the way that the martensite plates grow continuously on cooling and shrink on heating, the rise of irreversible frictional work leads to an increase in the transformation hysteresis [13], as explained in the previous paragraph. This evolution in the transformation hysteresis reflects the energy dissipated during the transformation cycle [9].

The difference between the absorbed and the released heats leads to calculate the total energy dissipation in a full cycle. It increases with the increase of the cooling/heating rate as shown in Fig. 8.

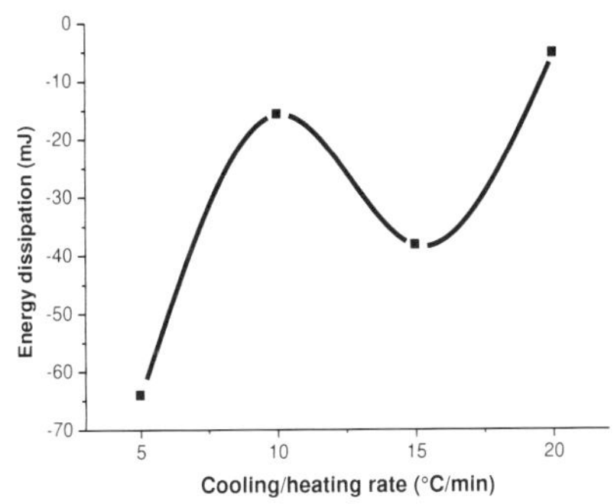

Figure 8: Energy dissipation as function of the cooling/heating rate

Conclusions

Using the differential scanning calorimetry method, the dependence of transformation properties of the NiTi shape memory alloy on the cooling/heating rate, during a full martensitic transformation cycle, has been studied in this paper. It has been reported that by increasing the cooling/heating rate, M_f decreases significantly whereas the other transformation temperatures are not so sensitive to this variation. Consequently, the width of the transformation hysteresis increase simultaneously.

Because of the thermoelastic behavior of the martensitic transformation, it has been concluded that the released heat of the forward transformation (in absolute value), the absorbed heat of the reverse transformation as well as the total dissipated energy during the full cycle transformation, increase with the increase of the cooling/heating rate.

Acknowledgments

This experimental work was been performed in Energy-Materials Laboratory (LabEM) of the high school of science and technology of Hammam Sousse "ESSTHS", Université de Sousse. The authors are thankful to Dr. Slim Zghal for useful discussions and helpful suggestions.

References

[1] J.L. MacQueron et al., "Atomic ordering and martensitic transition in a Cu-Zn-Al shape memory alloy", Journal de physique IV, Vol.1 C4 (1991), pp. C4-259-C4-263.

[2] Z.G. Wang et al., "Effect of heating/cooling rate on the transformation temperatures in TiNiCu shape memory alloys", Thermochimica Acta, Vol. 436 (2005), pp. 153-155.

[3] S.H. Chang and S.K. Wu, "Effect of cooling rate on transformation temperature measurements of Ti50Ni50 alloy by differential scanning calorimetry and dynamic mechanical analysis", Materials characterization, Vol. 59 (2008), pp. 987-990.

[4] Hiroyuki Kato and Kazuaki Sasaki, "Avoiding error of determining the martensite finish temperature due to thermal inertia in differential scanning calorimetry: model and experiment of Ni-Ti and Cu-Al-Ni shape memory alloys", J Mater Sci, Vol. 47 (2012), pp. 1399-1410.

[5] K. Nurveren et al., "Evolution of transformation characteristics with heating/cooling rate in NiTi shape memory alloys", Journal of materials processing technology, Vol. 196 (2008), pp. 129-134.

[6] H. Sehitoglu et al., "Compressive response of NiTi single crystals", Acta Materialia, Vol.48 (2000), pp.3311-3326.

[7] K. Gall and H.J. Maier, "Cyclic deformation mechanisms in precipitated NiTi shape memory alloys", Vol. 50 (2002), pp. 4643-4657.

[8] K. Otsuka and C.M. Wayman, Shape memory materials, Cambridge university, Cambridge press (1999)

[9] Etienne Patoor et al., "Shape memory alloys, Part I: General properties and modeling of single crystals", Mechanics of Materials, Vol. 38 (2006), pp. 391-429.

[10] Zhang Yan-qiu et al., " Influence of cooling rate on phase transformation and microstructure of Ti−50.9%Ni shape memory alloy", Trans. Nonferrous Met. Soc. China, Vol. 22 (2012), pp. 2685-2690.

[11] Jaroslav Capek and Jiri Kubasek, "Influence of temperature of the short-period heat treatment on mechanical properties of the NiTi alloy", Materials Engineering, Vol. 21 (2014), pp. 11-15.

[12] Jafar Khalil-Allafi et al., "Ni_4Ti_3-precipitation during aging of NiTi shape memory alloys and its influence on martensitic phase transformations", Acta Materialia, Vol.50 (2002), pp. 4255-4274.

[13] Z.G. Wang et al., "Design of NiTi alloy two-shape memory coil extension spring", Materials Science and Engineering, Vol. A345 (2003), pp. 249-254.

International Federation of Heat Treatment and Surface Engineering 2016
Proceedings of the 23rd IFHTSE Congress
April 18–21, 2016, Savannah, Georgia, USA

Thermal diffusivity in cast aluminum alloys

Humberto Arenas-García, Alma G. Esmeralda, Rafael Colás
Universidad Autónoma de Nuevo León, Mexico.
humberto.arenas@gmail.com, alma_k56@hotmail.com, colas.rafael@gmail.com

Andrés F. Rodríguez, José Talamantes-Silva
Nemak México, S.A. de C.V., Mexico
andres.rodriguez@nemak.com, jose.talamantes@nemak.com

Abstract

The continuous requirement for the reduction in emissions of internal combustion engines implies the increase in their efficiency, which requires the increase in temperature within the combustion chamber and, with it, the increment in thermal and mechanical demands of the alloy from which the engine is made. Aluminum alloys are used in the manufacture of engine blocks and cylinder heads due to the reduction in weight, and the higher thermal conductivity that these alloys exhibit in comparison to traditional gray iron parts. Combustion engines for the next decades are being designed to deliver higher power output while being reduced in size; therefore the alloys have to be designed to allow for the dissipation of the heat generated within the combustion chamber. This work presents the results of a series of analyses conducted on a wide range of cast aluminum alloys with the aim to evaluate the dependence of the thermal diffusivity with the chemical composition, microstructure and heat treatment imparted. The results show that aging increases the thermal diffusivity, but its magnitude depends on the characteristics of the alloy.

Introduction

The engine block in automobiles converts the chemical energy of fuel into mechanical by burning it within the combustion chambers. Most of the energy produced by the explosion of the air-fuel mixture is converted into heat, which has to be extracted through the walls of the chambers to the proper block and to the refrigerating fluids. Heat extraction has to be efficient otherwise the engine block may be subjected to excessive thermal loads that will modify the mechanical properties of the alloy from which the block is made, leading to its catastrophic failure [1-3]. Effort is put in the evaluation of the conditions existing within the combustion chamber to determine the heat transfer by different means [1-6], but there is a gap of knowledge on the effect that the presence of different components present within the chamber exert on heat flow, and on the thermophysical properties of the materials from which the different components are made.

Engine blocks and cylinder heads in automobiles have been migrating from cast iron to aluminum alloys to take advantage in the reduction in weight to improve fuel consumption and to comply with environmental regulations imposed in different countries [7]. Aluminum alloys exhibit higher thermal conductivity than cast irons, which allows for the reduction in size and weight of the refrigerating systems, but these alloys are sensitive to damage by overheating [8,9].

Aluminum alloys may be produced from commercial primary ingots or from scrap, due to the strong policy in reducing costs of the automotive industry. This problem is of particular interest for Al-Si-Cu alloys, which may have high amounts of residual elements (mainly Fe, Mn and Zn). Al-Si-Mg and Al-Cu alloys are normally prepared from primary ingots and are cast with fewer amounts of residual elements. Heat treating is conducted after solution and cooling the pieces to room temperature, which can be on air, forced steam or immersion in water. The aging condition will also depend on the type of alloy; T7 treatment is preferred in Al-Si-Cu and Al-Cu alloys, as this condition allows achieving the strength requirements in powertrain applications, whereas Al-Si-Mg alloys have to be treated to a T6 condition to achieve similar values of strength [10,11].

Casting is carried out by pouring the liquid metal into a mold made from materials ranging from sand to high strength tool steel, which modify the solidification and cooling rates that affect the microstructure of the solid alloy. The microstructure in silicon containing alloys is made of primary aluminum dendrites surrounded by the Al-Si eutectic aggregate. Mg is added to some alloys to strengthen by precipitation of Mg_2Si particles, but should be avoided in castings hardened by Al_2Cu precipitates, as these elements tend to form a low melting point eutectic that takes place at temperatures below 500°C. Modifying elements are added to change the aspect of the Al-Si eutectic by promoting the formation of fibrous silicon branches; titanium and boron are added to refine grain size by the nucleation of new grains in front of the solidifying front; the presence of tramp elements from scrap promotes the formation of different intermetallic phases [8-14].

Al-Cu alloys are developed for their enhanced mechanical properties that are able to withstand the higher pressures that advanced engines have to sustain; their solidification proceeds in a cellular way and can be aged at room temperature to achieve the required levels of strength [15].

The aim of this work is to present the results of a series of experimental trials conducted in a different aluminum alloys to evaluate their thermal diffusivity. The alloys were cast within an experimental mold that allows for a wide range of microstructures; samples from the various alloys were heat treated following different cycles.

Experimental procedure

The experimental alloys were prepared in a 250 kg gas-fired furnace. The melt was degassed with nitrogen before adding master alloys of titanium and boron, to refine the grain size, and strontium to modify the Al-Si eutectic aggregate, Table I. Liquid metal was poured at 740°C into molds designed to impose a one-direction thermal gradient by a grey iron chill insert at their bottom [15,16]. The samples used in the present study were sectioned from the ingot at three different positions to obtain material with that exhibited either a fine, medium or coarse microstructure, which were characterized by the secondary dendrite arm spacing (DAS) in the Al-Si alloys, and varied within the ranges of 20 to 25, 30 to 40 and 50 to 60 μm.

The material was heat treated following the cycles shown in Table II. The Al5Cu alloy was treated for 2 h at 495°Cs followed by 2 h at 520°C and another 12 h at 535°C. The Al7SiCu samples were solution treated for 5 h at 495°C, whereas the Al7SiMg and Al7SiMgCu were held for 5 h at 530°C. All the samples were cooled down to room temperature by immersing them in water at 90°C or leaving them in still air. Aging was carried out to obtain either T6 or T7 conditions.

Table I. Chemical composition of the alloys.

Element	Alloy			
	Al5Cu	Al7SiCu	Al7SiMg	Al7SiMgCu
Si	0.02	8.1	7.4	7.2
Cu	4.9	3.1	0.01	0.47
Mg	0.31	0.32	0.41	0.32
Fe	0.15	0.69	0.15	0.18
Mn	0.2	0.4	0.06	0.9
Ti	0.05	0.18	0.18	0.14
Sr	0	140	140	140
B	30	30	20	20

Contents in wt.%, except for Sr and B that are in ppm.

Table II. Heat treating of the alloys.

Alloy	Solution	Aging	Temper
Al5Cu	2 hours at 505°C and 3 hours at 525°C	4.5 hours at 160°C	T6
Al7SiCu	5 hours at 495°C	4.5 hours at 235°C	T7
Al7SiMg	5 hours at 530°C	4.5 hours at 190°C	T6
Al7SiMgCu	5 hours at 530°C	4.5 hours at 190°C	T6

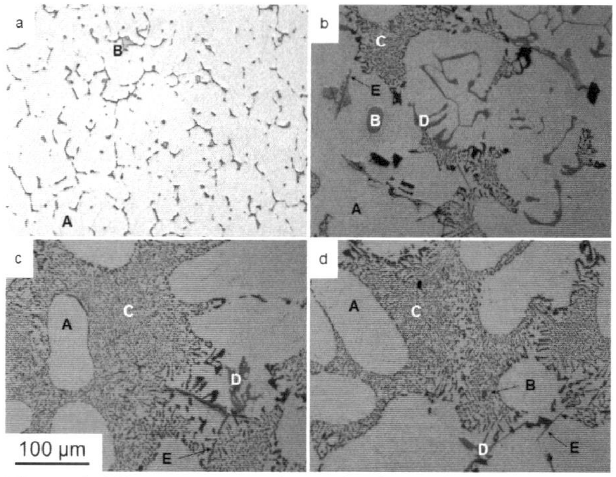

Figure 1. Microstructures of samples from Al5Cu (a), Al7SiCu (b), Al7SiMg (c) and Al7SiMgCu (d) in the as-cast condition. Aluminum dendrites or grains are marked as **A**, *the Cu-rich particles as* **B**, *the Al-Si eutectic aggregate as* **C**, *Fe-rich intermetallics of the α- and β-types by* **E** *and* **F**.

Table III. Microstructural characteristics of the alloys.

Alloy		DAS (μm)	d (μm)	P (%)	P_m (μm)
Al5Cu	F	-	58	0.08	72
	M	-	92	0.21	161
	C	-	122	0.23	232
Al7SiCu	F	21	442	0.08	72
	M	32	472	0.21	161
	C	52	475	0.23	232
Al7SiMg	F	23	403	0.03	64
	M	41	588	0.02	66
	C	62	593	0.22	201
Al7SiMgCu	F	22	337	0.01	20
	M	34	586	0.01	34.3
	C	65	585	0.07	93

d: grain size, P: porosity; P_m: biggest pore
F: fine, M: medium, C: coarse

The specimens were preconditioned to simulate the working conditions in an internal combustion engine; such treatment consisted in holding the specimens for 200 hours at temperatures of 25, 100, 150, 200, 250 and 300°C before measuring their thermal diffusivity.

Thermal diffusivity was measured by the flash laser method [17-20], which is based on the temperature increase on one side of a thin sample after its other side was subjected to a short pulse of concentrated energy provided by a laser lamp. The increase of temperature in the face opposing that impinged by the laser pulse is measured by an infrared detector. The samples used for these measurements were 10 mm in diameter and 2 mm in thickness. Their surface was coated with a graphite spray and mounted within the device. The tests were conducted in the range of 20 to 300°C.

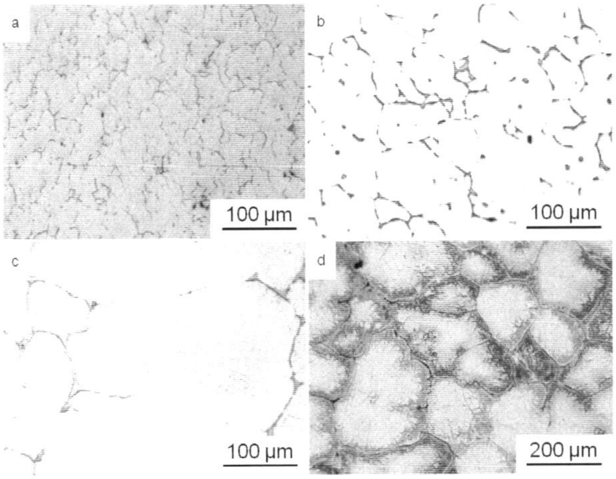

Figure 2. As-cast microstructures of Al5Cu alloy at the fine (a), medium (b) and coarse (c) positions. The grain size at the coarse position is shown in (d).

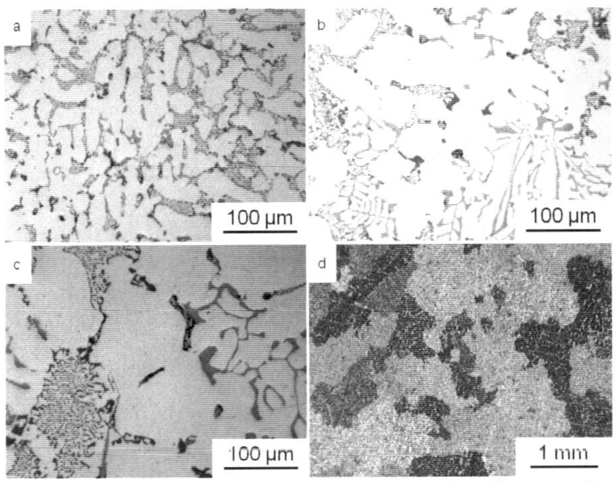

Figure 3. As-cast microstructures of Al7SiCu alloy; similar conditions as in Figure 2.

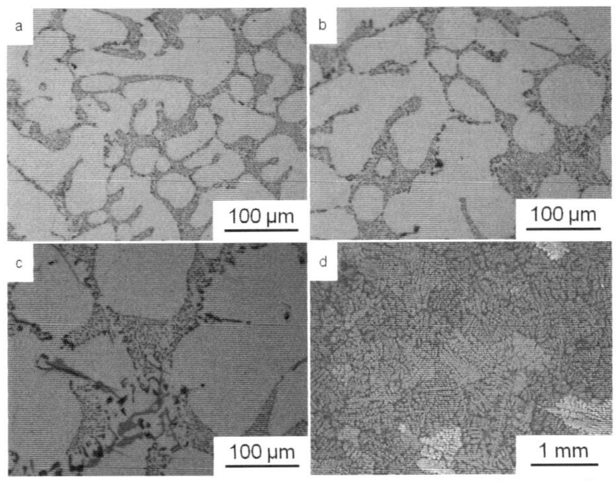

Figure 4. As-cast microstructures of Al7SiMg alloy; similar conditions as in Figure 2.

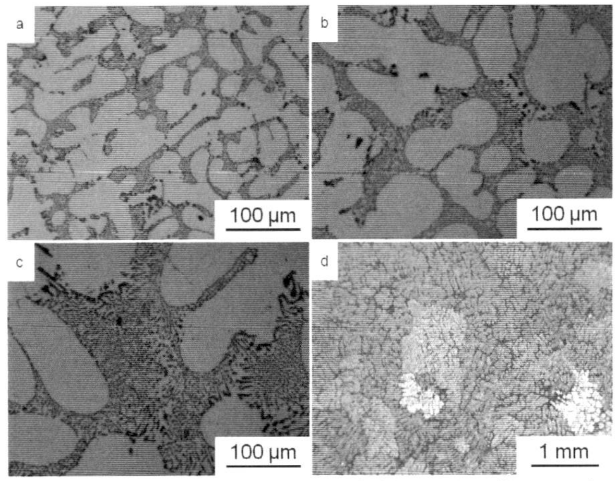

Figure 5. As-cast microstructures of Al7SiMgCu alloy; similar conditions as in Figure 2.

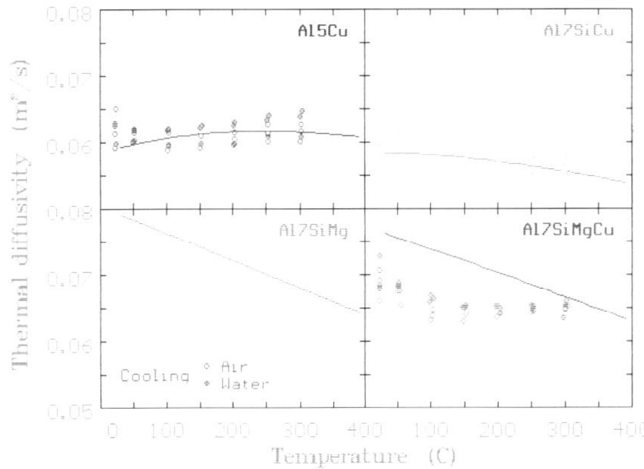

Figure 6. Thermal diffusivity of specimens preconditioned at room temperature. The full lines are computed by the JMatPro software [21].

Results and discussion

Figure 1 shows the as-cast microstructure of the different alloys; the images correspond to samples from the coarse position. The images show the aluminum dendrites, in Al-Si alloys, or grains, in Al5Cu, as **A**, Cu-rich particles as **B**, Al-Si eutectic aggregate as **C**, Fe intermetallics of the α- and β-types by **D** and **E**. Figures 2 to 5 show the effect that the solidification and cooling rate exerts on the microstructure of the alloys. The increase in these rates do not only deduce the dendrite arm spacing (DAS), as the grain size, and the size and amount of pores are also affected, Table III.

Figure 6 shows the changes in thermal diffusivity of the four alloys in the study as a function of temperature for samples of different microstructure and cooled in air and water and preconditioned at room temperature for 200 hours. The full lines were computed by the JMatPro software [21] based on the chemical composition of the alloys.

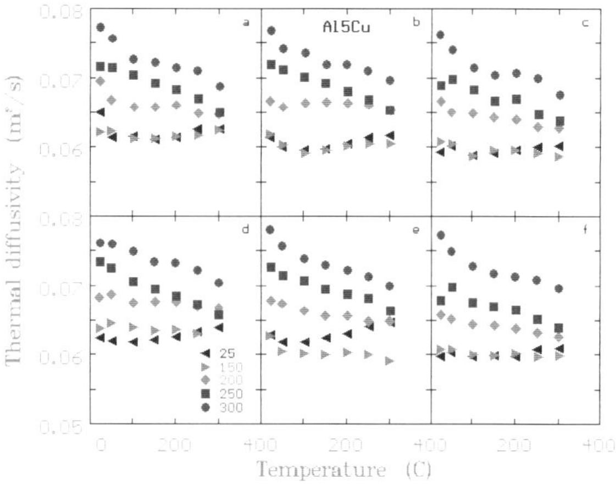

*Figure 7. Thermal diffusivity as of samples of Al5Cu alloy preconditioned at different temperatures. Sample with fine (**a** and **d**), medium (**b** and **e**) and coarse (**c** and **f**) microstructure cooled in air (**a**, **b** and **c**) and in water (**d**, **e** and **f**).*

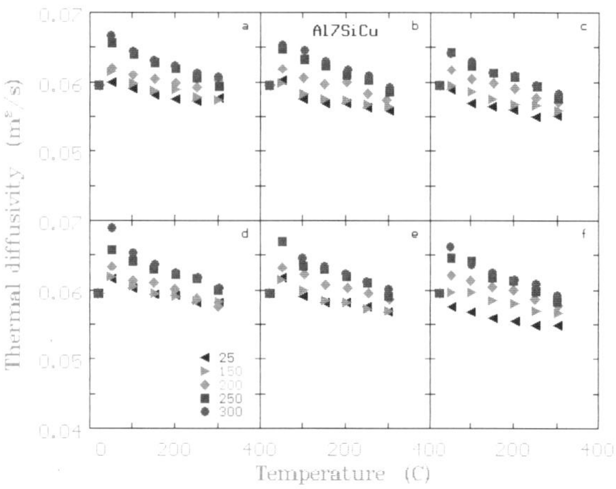

Figure 8. Thermal diffusivity of Al7SiCu alloy; similar conditions as in Figure 8.

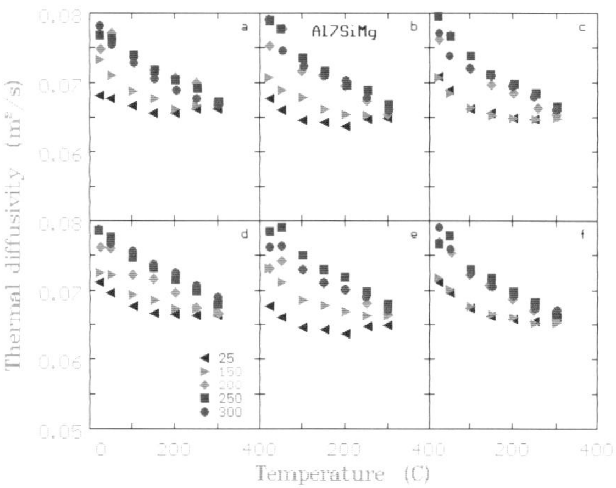

Figure 9. Thermal diffusivity of Al7SiMg alloy; similar conditions as in Figure 7.

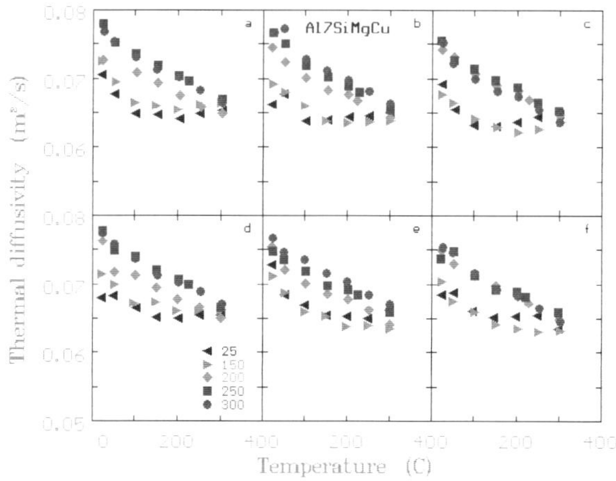

Figure 10. Thermal diffusivity of Al7SiMg alloy; similar conditions as in Figure 8.

Figures 7 to 10 show the effect that the testing temperature exerts on the thermal diffusivity of the different alloys. The samples were preconditioned for 200 hours at the temperatures indicated; the data correspond to measurements carried out on samples with different initial microstructure and cooled in air and in water. It is of interest that preconditioning exerts a very strong effect, as the diffusivity increases with the temperature of the treatment, although this vary in each alloy.

Microstructural refining exerts a small positive effect on the thermal diffusivity of the four alloys, Figures 11 to 14. The measurements were obtained at the temperatures indicated in the figures from samples that were cooled in air and water and were plotted as a function of the dendritic arm spacing (the values for the Al5Cu were assigned as 20, 40 and 60 μm). The values that are represented correspond to preconditioning at room temperature and at 300°C. This behavior coincides with that reported for an Al-Si-Cu alloy [22].

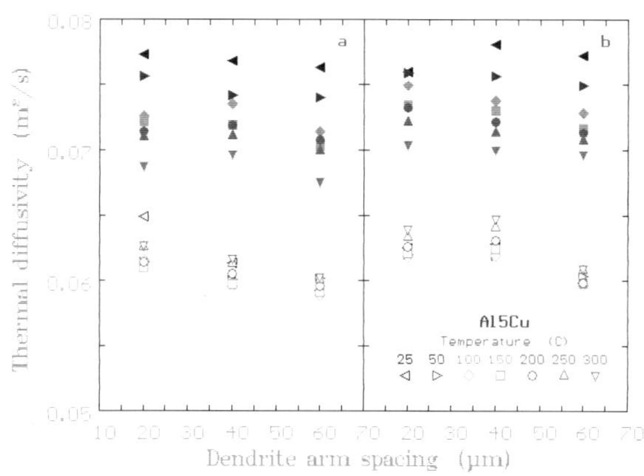

*Figure 11. Thermal diffusivity as a function of DAS at various temperatures in the Al5Cu on samples cooled in air (**a**) and water (**b**) and preconditioned at 25 (open symbols) and 300°C (full symbols).*

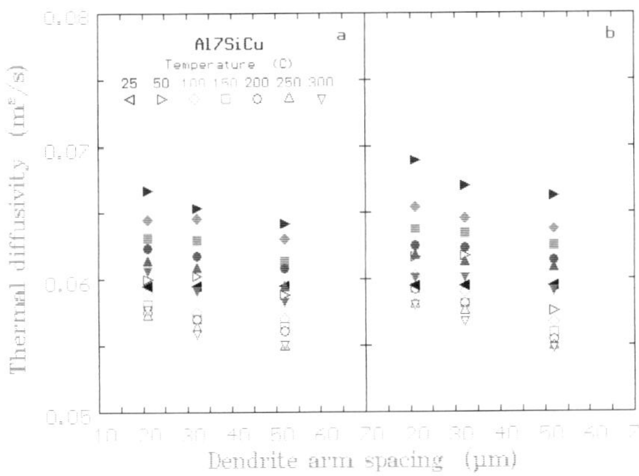

Figure 12. Thermal diffusivity of Al7SiCu alloy; similar conditions as in Figure 11.

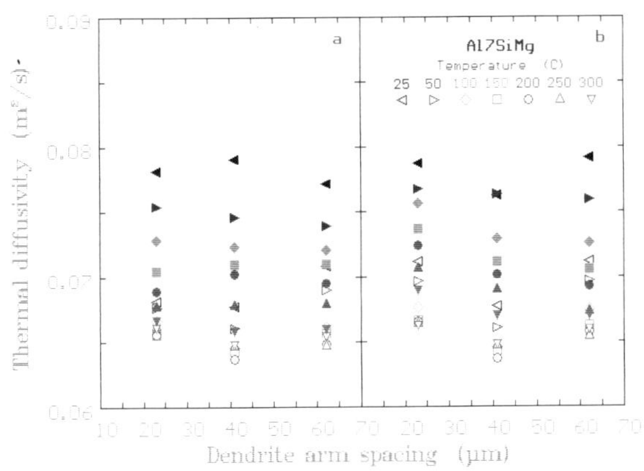

Figure 13. Thermal diffusivity of Al7SiMg alloy; similar conditions as in Figure 11.

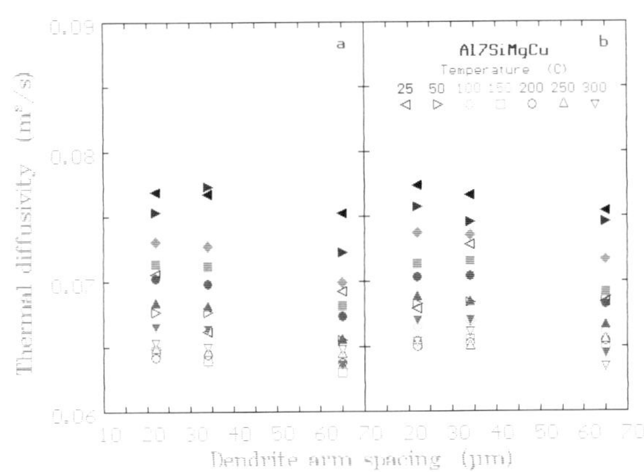

Figure 14. Thermal diffusivity of Al7SiMgCu alloy; similar conditions as in Figure 11.

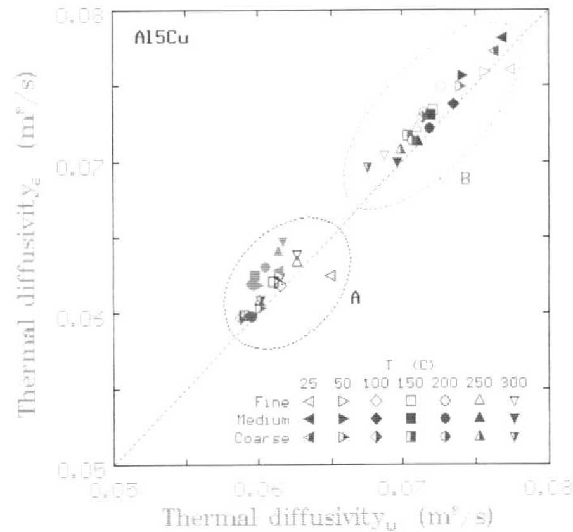

*Figure 15. Thermal diffusivity at various testing temperatures in samples of the Al5Cu alloy cooled in air and in water. Data for samples preconditioned at 25 (**A**) and 300°C (**B**).*

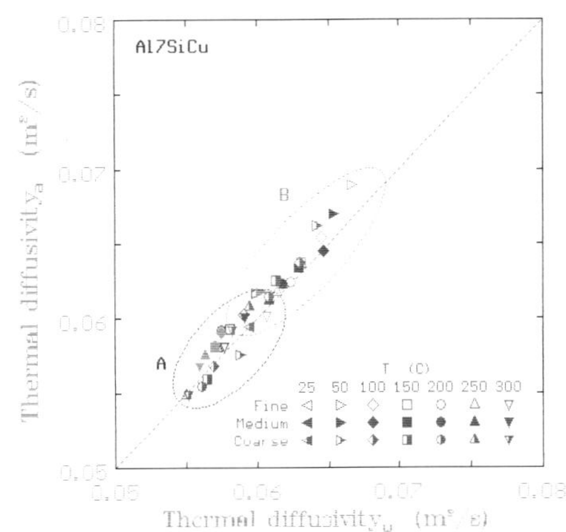

Figure 16. Thermal diffusivity of Al7SiCu alloy; similar conditions as in Figure 15.

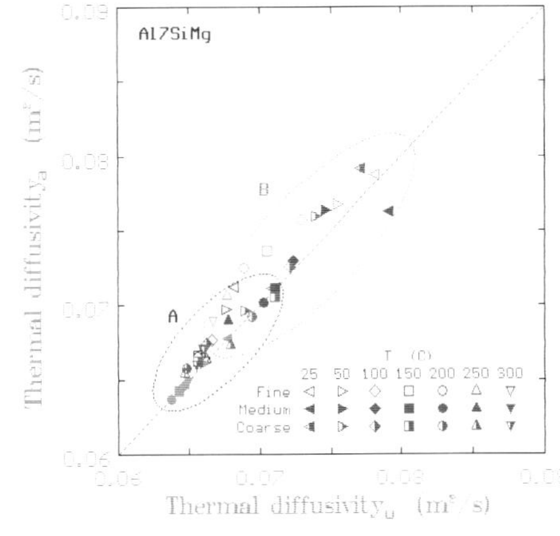

Figure 17. Thermal diffusivity of Al7SiMg alloy; similar conditions as in Figure 15.

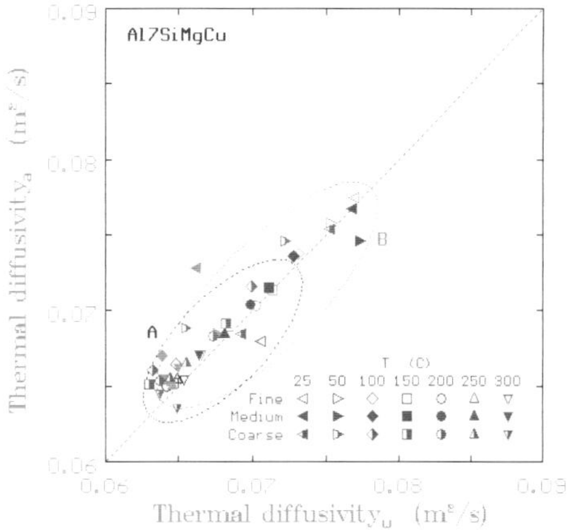

Figure 18. Thermal diffusivity of Al7SiMgCu alloy; similar conditions as in Figure 15.

Figures 16 to 18 compare the thermal diffusivity measured in samples cooled in air (sub index a) against that obtained in samples immersed in water (sub index w) in the various alloys. Only the values of samples preconditioned at 25 and 300°C, identified respectively as **A** and **B**, are shown for the sake of clarity. It can be observed that, despite the spread of the experimental data, the values of thermal diffusivity of samples cooled in air tend to be higher than those recorded on the material cooled at the faster rate. The parameter that exerts the higher effect on the diffusivity is that of the temperature of preconditioning, and this effect is higher in the alloys that age by precipitation of Al_2Cu (Al5Cu and Al7SiCu), rather than Mg_2Si (Al7SiMg and Al7SiMgCu).

Precipitate coarsening is the microstructural phenomenon that takes place during preconditioning at temperatures above aging. Such effect is normally associated with changes on mechanical properties [9-12,23]. Reduction in the cooling rate after solution treating contributes to the incipient precipitation of particles, which will reduce the potential for achieving higher strength in aluminum alloys [24-26]. Both phenomena will enhance thermal diffusivity by reducing the amount of solute elements in the aluminum matrix and increasing the mean free path between the precipitates.

Conclusions

A study to evaluate the thermal diffusivity in four different alloys was conducted. The parameters considered in the study for each alloy were microstructural refining (associated to the dendritic arm spacing in the Al-Si alloys and assigned equivalent values in the Al-Cu alloy), cooling after solution treating either in water or in air, and preconditioning for 200 hours room and at temperatures ranging from 150 to 300°C. The measurements were conducted at the range of 25 to 300°C.

The results show that microstructural refining and slow cooling (air) contribute to the increase of the thermal diffusivity. This property also increases at lower temperatures. Preconditioning at high temperatures enhances the values of the thermal diffusivity.

The results point out that the mechanism responsible for the increment of the thermal diffusivity with preconditioning is that of coarsening of the precipitates, with those of Al_2Cu showing the greatest effect, as coarsening allows for fewer and more spaced precipitates, and the reduction of alloying elements dissolved in the aluminum matrix.

Acknowledgments

The authors would like to thank the support provided by the Graduate and Bilateral Programs of the Mexican National Council for Science and Technology (CONACYT).

References

[1] Heywood, J.B., Internal Combustion Engine Fundamentals, McGraw-Hill, (New York, 1989).
[2] Norris, P.M., Hastings M.C., and Wepfer, W.J. 'An experimental investigation of liquid coolant heat transfer in a diesel engine', J. Exp. Heat Trans., vol. 7, (1994) pp. 43–51.
[3] Hoag, K., Vehicular engine design, Springer Science & Business Media, Wien, New York, 2007.
[4] Borman, G., and Nishiwaki, K., 'Internal combustion engine heat transfer', Prog. Energy Comb. Sc., vol. 13, (1987), pp. 1-46.
[5] Finol, C.A., and Robinson, K., 'Thermal modelling of modern diesel engines: Proposal of a new heat transfer coefficient correlation', J. Autom. Eng., vol. 225, (2011), pp. 1544-1560.
[6] González, J.A., Talamantes-Silva, J., Morales, F., Garza-Montes-de-Oca, N.F. and Colás, R., 'Analyses of heat conduction within combustion chambers in engine blocks', to be published in Int. J. Micros. Mat. Prop. (2016).
[7] Miller, J.D, and Façanha, C., 'The state of clean transport policy. A 2014 synthesis of vehicle and fuel policy developments'. International Council on Clean Transportation (New York, 2014). www.theicct.org/state-of-clean-transport-policy-2014
[8] Campbell, J., Castings, 2nd Ed., Butterworth-Heinemann, (Oxford, 2003).
[9] Colás, R., Velasco, E., and Valtierra, S. 'Castings', in Totten, G.E. and MacKenzie, D.S. (Eds.): Handbook of Aluminum, Vol. 1: Physical Metallurgy and Processes, M. Dekker, (New York, 2003), pp. 591-641.
[10] Rooy, E.L., 'Aluminum and aluminum alloys', in ASM Handbook, Vol. 15: Casting, ASM International, (Materials Park, 1988), pp. 743-770.

[11] Kim, N.J., 'Designing with aluminum alloys' in Totten, G.E., Xie, L. and Funatani, K. (Eds.): Handbook of Mechanical Alloy Design, M. Dekker Inc., (New York, 2004), pp. 441-486.
[12] Polmear, I.J., Light Alloys, Metallurgy of the Light Metals, Arnold (London, 1980).
[13] Bäckerud, L., Chai G., and Tamminen, J., Solidification Characteristics of Aluminum Alloys, Vol. 2: Foundry Alloys, AFS/Skanaluminium, (Des Plains, 1990.
[14] Arnbert, L., Bäckerud, L. and Chai, G., Solidification Characteristics of Aluminum Alloys, Vol. 3: Dendritic Coherency, AFS, (Des Plains, 1996).
[15] Talamantes-Silva, M.A., Rodríguez, A., Talamantes-Silva, J., Valtierra S., and R. Colás, 'Effect of solidification rate and heat treating on the microstructure and tensile behavior of an aluminum-copper alloy', Metall. Mat. Trans. B, vol. 39B, (2008), pp. 911-919.
[16] Carrera, E., González, J.A., Talamantes-Silva, J., and Colás R., Metall. 'Effect of the delay in time between cooling and aging in heat-treated cast aluminum alloys', Mat. Trans. B, vol. 42B, (2011), pp. 1023-1030.
[17] Parker, W.J., Jenkins, R.J., Butler, C.P., and Abbott, G.L., 'Flash method of determining thermal diffusivity, heat capacity and thermal conductivity', J. Appl. Phys., vol. 32 (1961), pp. 1679-1684.
[18] Rudkin, R.L., Jenkins, R.J., and Parker, W.J., 'Thermal diffusivity measurements on metals at high temperatures', vol. 33 (1962), pp. 21-24.
[19] Baba, T., and Ono, A., 'Improvement of the laser flash method to reduce uncertainty in thermal diffusivity measurements', Meas., Sci., Technol., vol. 12 (2001), pp. 2046-2057.
[20] Akoshima M., Hay, B., and Grelard, M., 'Experimental verification to obtain intrinsic thermal diffusivity by laser-flash method', Int. J. Thermophys., vol. 34 (2013), pp. 778-791.
[21] http://www.sentesoftware.co.uk/jmatpro.aspx
[22] Vázquez-López, C., Calderón, A., Rodríguez, M.E., Velasco, E., Cano, S., Colás, R., and Valtierra, S., 'Influence of dendrite arm spacing on the thermal conductivity of an aluminum-silicon casting alloy', J. Mat. Res., vol. 15 (2000), pp. 85-91.
[23] Esmeralda, A.G., Rodríguez, A., Talamantes-Silva, J., and Colás, R. 'Temperature dependent properties of heat treated aluminium alloys', Int. Heat Treat. Surf. Eng., vol. 8 (2014), pp. 65-69.
[24] García-Celis, A.I., Velasco, E., Valtierra, S., Mojica, J.F., and Colás, R., 'Cooling effects on aging in a cast aluminum alloy', in S. Das (ed.), Automotive Alloys II, TMS, Warrendale, 1998, pp. 135-143.
[25] Cavazos, J.L., and Colás, R., 'Aging of an aluminum alloy resulting from variations in the cooling rate', J. Mat. Eng. Perform., vol. 8 (1999), pp. 509-512.
[26] Cavazos, J.L., and Colás, R., 'Quench sensitivity of a heat treatable aluminum alloy', Mat. Sc. Eng., vol. A363 (2003), pp. 171-178.

"Calorimetry for the study of phase transformations during tempering of AISI L6 low alloy tool steel"

Douglas I. Quiñones-Salinas*, Rafael D. Mercado-Solís, Luis A. Leduc-Lezama,
Universidad Autonoma de Nuevo León, México

Florentino Fernandez-Guzman, Rene Cerda-Rojas.
Frisa Forjados S.A. de C.V., Nuevo León, México

Abstract

The present work shows the phase transformation and the influence of heating rate during tempering for as quenched AISI L6 tool steel. The tempering was made at the whole tempering temperature range (100°C to 700°C). The investigations of phase transformations during tempering of the steel were made by the means of calorimetry curves recorded during heating from as quenched state. Heating to 700°C was made it at different rates (0.01, 0.25, 0.35, 0.55 and 1 C/sec). Activation energies have been obtained by performing Kissinger analysis and were used to deduce the rate for each stage of decomposition. Results show that the temperatures of starting and ending of individual phase transformations increase as the heating rate increase. It is possible to reduce probability of the occurrence of some phase transformations (cementite transformation) in favor of other (retained austenite). The different tempering temperature has effect on mechanical properties of the steels like toughness and hardness.

Introduction

This project studies phase transformations in AISI L6 low alloy tool steel under tempering temperature range [1-2]. This experiment has been studied meanly by means of differential scanning calorimetry. This technique is very powerful and allows the identification of kinetic occurring during tempering. Activation energies have been obtained by performing a Kissinger and JMAK analysis and were used to deduce the rate for each stage of decomposition.

Some authors report during tempering, stages of martensitic steels, the simultaneous transformation of retained austenite and the precipitation of cementite produce strong signals as well in the differential scanning calorimeter as in the dilatometer [2-3].

Researchers as J. Pacyna describes the kinetics of phase transformations during continuous heating from hardened state and subsequent cooling by dilatometry investigations [4-5]. Also illustrating phase transformations occurring during heating from hardened state at specific rate.

P. Bala, design CHT diagram, illustrating the kinetics of phase transformations during continuous heating (tempering) from as-quenched state of investigated steel, was elaborated using dilatometer. Heating of the investigated steel from the as-quenched state resulted in the occurrence of 3 primary transformations: precipitation of ε carbide, M_3C precipitation and transformation of retained austenite. The differences in hardness and microstructure of samples of the investigated steel in relationship to the heat treatment were evaluated [6].

P. V. Morra, studied the decomposition processes of a carburized cast alloys by means of differential scanning calorimetry and dilatometry. The combination of these two experimental techniques was very useful and allows the identification of the stages occurring during tempering. Activation energies were obtained by performing a Kissinger-like analysis and were used to infer the rate-determining step for each stage of decomposition [7].

Precipitation processes during the tempering can be summary in the following three stages:

"Stage I": The formation of transition carbides and lowering of the carbon content of the martensite. Consist of the segregation of carbon atoms to lattice defects and the formation of clusters of carbon atoms in the iron matrix ***(100 to 200 °C)***.

"Stage II": The transformation of retained austenite into ferrite and clustered carbon atoms transform into the more stable cementite and carbides ***(200 to 450 °C)***.

"Stage III": Dissolution of cementite particles and the formation of more stable alloy carbides occurs only if strong carbide formers as chromium, molybdenum, tungsten, vanadium ***(500–650°C)*** [1-6].

The main aim of the present work is to build a continuous heating transformation (CHT) diagram of AISI L6 tool steel. The CHT diagram is presented in other work made it by the same authors.

Methodology

The investigations were performed on L6 tool steel whose chemical composition is presented in Table 1. Before experiments, was necessary to apply a heat treatment of quench in order to obtain the martensitic structure. Therefore for the tool steel was proposed the normalizing and quench heat treatments. The normalizing temperature for the quench heat treatment was determinate by JMatPro, using the chemical composition from table 1 and 8 as grain size [8].

Table 1: Chemical composition (wt. %) of the investigated alloy.

C	Si	Mn	P	Si	Cr	Mo	V	Ni
0.5	0.2	0.8	0.03	0.02	0.9	0.5	0.1	1.6

The tempering heat treatments for calorimetry and furnace samples was performed twice with the purpose of study the kinetic precipitation for double tempering.

Calorimetry

Thermal analysis investigations were performed employing a differential scanning calorimeter Netzsch DSC 404 F3 model. The samples were discs with a 6 mm diameter and 25 mg of mass, electrolytically polished into 60% perchloric acid in alcoholic solution. The conditions for elecropolished were 10 seconds under 10 Volts of voltage. Argon was used to provide a protective atmosphere. The test were done from room temperature to 700°C at five heating rates (5, 15, 20, 35 and 50°C/min) in a closed alumina pan. Double tempering was performed for every heating rate. Proteus software was used to analyze the temperature transformations during the heating. Kissinger-like analysis were performed to determine the phase kinetics during tempering. Choosing the maximum peak of the measured curve as the point where phase transformation happen.

Microstructure

The experimental samples were cubes of size 2.5 cm^3 normalized in a laboratory furnace for 1 hour at 900 °C and oil quenched. Then samples were tempered for 1 hour at 100, 200, 300, 400, 500, 600 and 700 °C. Then double tempered for each temperature. After heat treatment the samples were prepared for metallography on the final dimension of 1 cm^3. Saturated picral etchant was used for microstructure analysis. The photos of the microstructure of quenched and tempered samples were analyzed in a scanning electron microscopy Jeol model JSM 6510 LV.

Micro hardness and Charpy impact

Every quenched and tempered sample was measured by micro hardness Vickers. The hardness tests were carried out under a load of 5 Newton and 15 seconds of indentation holding. Ten indentations were then used to calculate the Vickers average value for each tempering temperature. Finally Vickers units were converted to Rockwell C. For impact test bars of 2.5x2.5x10 cm were heat treated by quenched and tempered. The impact tests were performed according to ASTM A370.

Results

Calorimetry

Fig 1 shows the thermograms of first and second tempering of the tool steel at different rates. Red curve represents the heating while the blue one represents the cooling. All the graphics are plotted with experimental heating and cooling curve versus the derivative. Derivate curves define better the reactions during the experiment. Table 2 shows temperatures and energy reaction of peaks detected during heating of first tempering experimental curves.

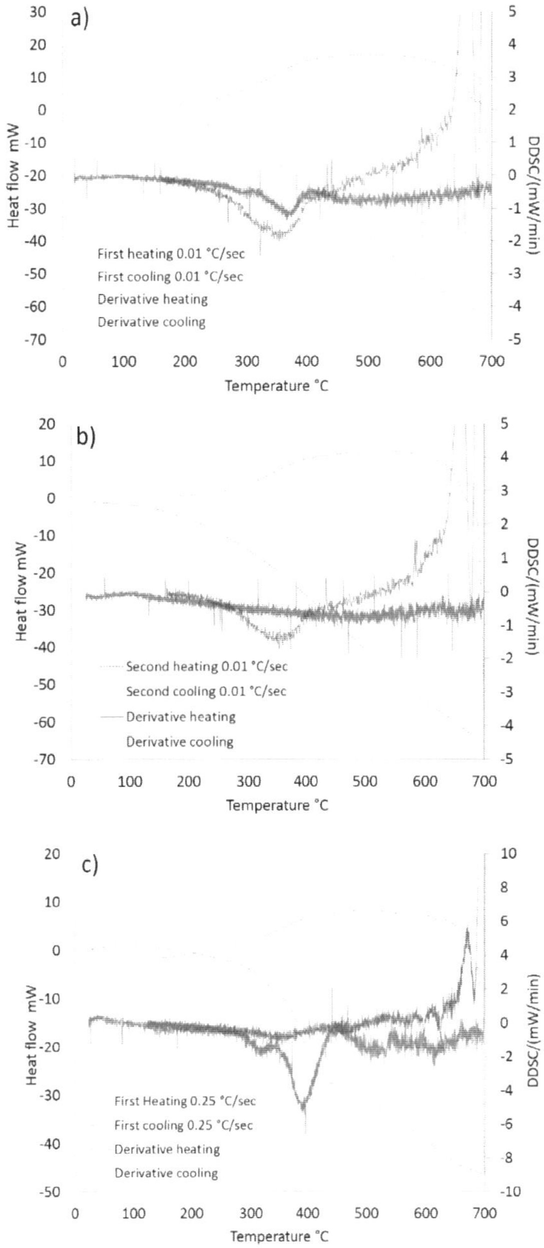

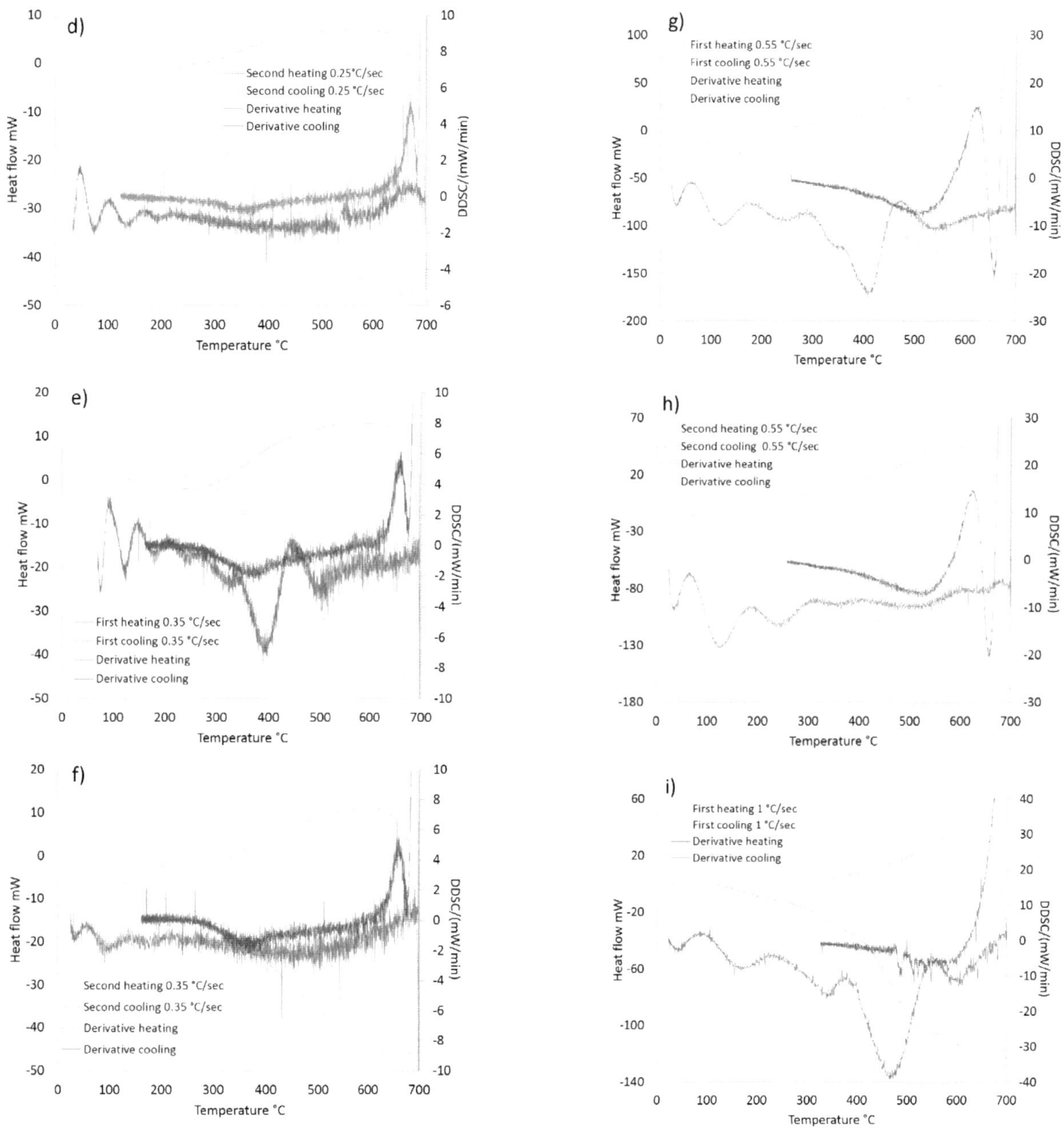

d)
Second heating 0.25°C/sec
Second cooling 0.25 °C/sec
Derivative heating
Derivative cooling
Heat flow mW
DDSC/(mW/min)
Temperature °C
e)
First heating 0.35 °C/sec
First cooling 0.35 °C/sec
Derivative heating
Derivative cooling
Heat flow mW
DDSC/(mW/min)
Temperature °C
f)
Second heating 0.35 °C/sec
Second cooling 0.35 °C/sec
Derivative heating
Derivative cooling
Heat flow mW
DDSC/(mW/min)
Temperature °C
g)
First heating 0.55 °C/sec
First cooling 0.55 °C/sec
Derivative heating
Derivative cooling
Heat flow mW
DDSC/(mW/min)
Temperature °C
h)
Second heating 0.55 °C/sec
Second cooling 0.55 °C/sec
Derivative heating
Derivative cooling
Heat flow mW
DDSC/(mW/min)
Temperature °C
i)
First heating 1 °C/sec
First cooling 1 °C/sec
Derivative heating
Derivative cooling
Heat flow mW
DDSC/(mW/min)
Temperature °C

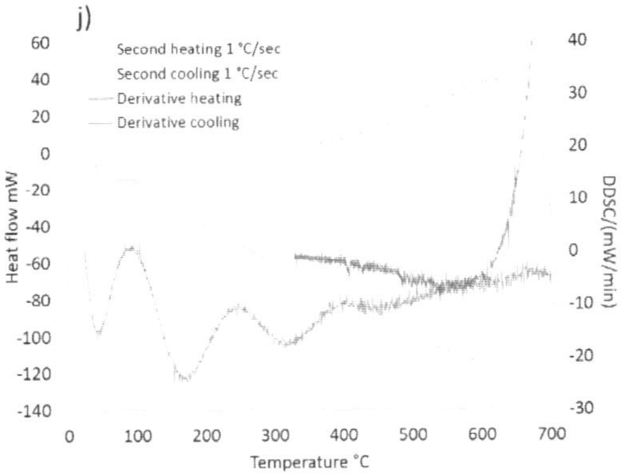

Fig. 1. Measurement of energy reactions during heating of AISI L6 tool steel at a)-b) 0.01, c)-d) 0.25, e)-f) 0.35, g)-h) 0.55 and i)-j) 1 °C/sec for first and second tempering respectively.

Table 2. Temperatures of the initial (Ti), maximum (Tm) and final (Tf) peak during the heating of the tempering and enthalpy for the detected peaks.

°C/sec	Ti (°C)	Tm (°C)	Tf (°C)	J/g
0.001	344	416	441	-37.66
0.25	367	440	462	-37.44
0.35	366	426	464	-35.79
0.55	387	443	503	-42.4
1	462	539	565	-29.48

Microstructure

The microstructures of tested steel after quenching from 900°C and tempered are shown on Fig. 2 a). After quenching from 900°C microstructure consists of martensite. Apparently after all tempering heat treatments at 1 hr. the microstructure not presents change. Fig. 2 b) show the micrograph of tempered sample at 500 °C. Fig. 3 and 4 show the quenched and tempered (500 °C) at low (x8000) and high (x50000) magnifications. Spherical particles were observed between martensite laths for both conditions.

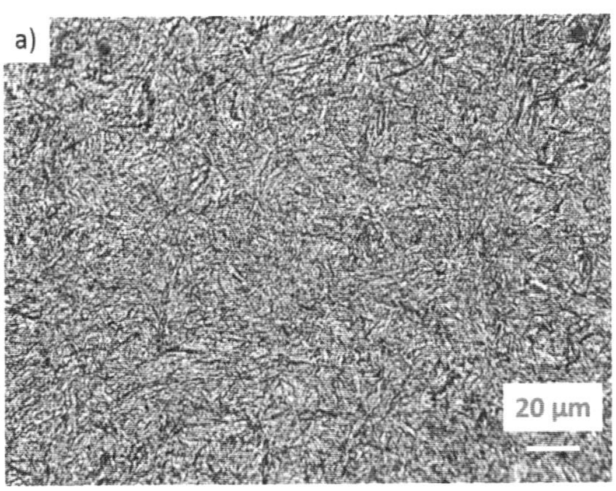

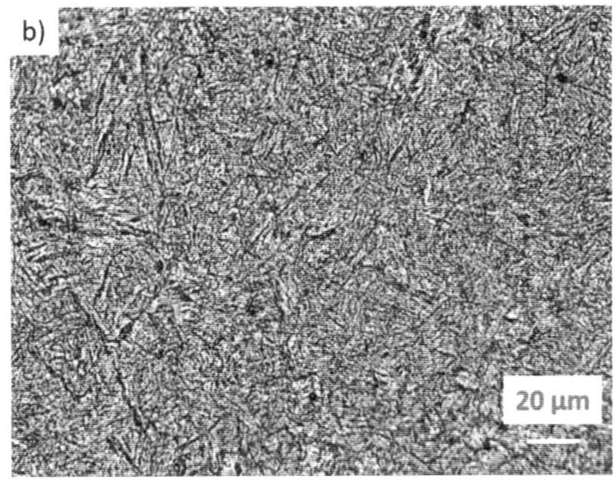

Fig. 2. Microstructure from a) quenched and b) tempered condition analyzed by optical microscopy.

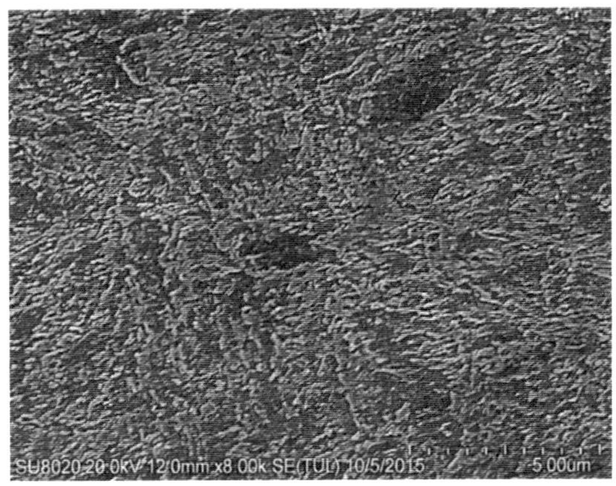

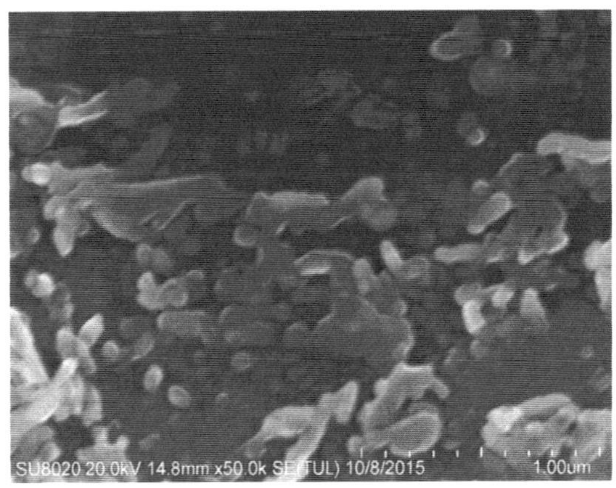

Figure 3. Microstructure from oil quenched condition analyzed by SEM at low and high magnifications.

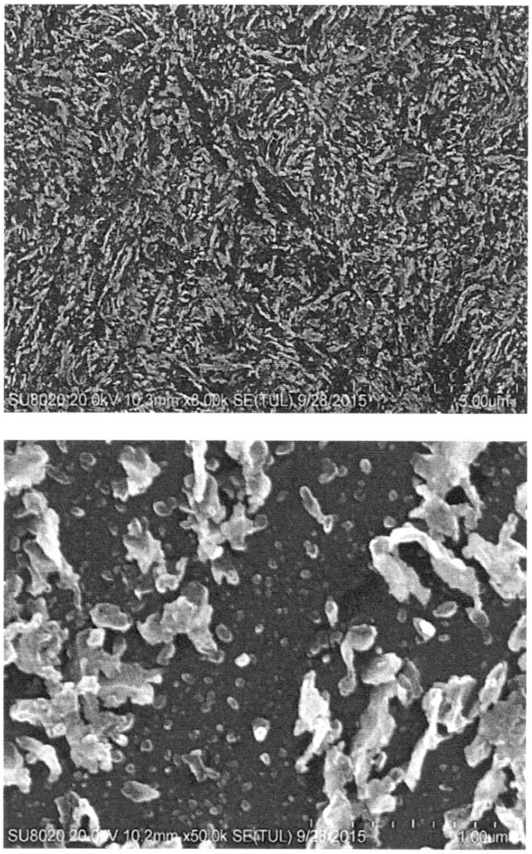

Figure 4. Microstructure from tempered at 500 °C analyzed by SEM.

Hardness and Impact Charpy

As shown in Fig. 5 the hardness of oil quenched sample is 62 HRC. Hardness has a dependence of tempering heat treatment due to the decreasing profile of 61 (100°C) to 31 (700°C) HRC. The grey curve in Fig. 5 present results of Charpy impact test, where energy increasing with tempering temperature, showing a profile from 4 (100 °C) to 52 (700°C) Joules. Only 200 °C Charpy impact sample results with a 45 joules after 4 (100°C) joules and before 10 (300°C) joules outgoing from the trend line curve.

Impact Charpy fracture photographs are presented in fig. 6. Samples a) only quenched, d) 300°C and e) 400°C photos show fragile fracture, while the rest are ductile morphology.

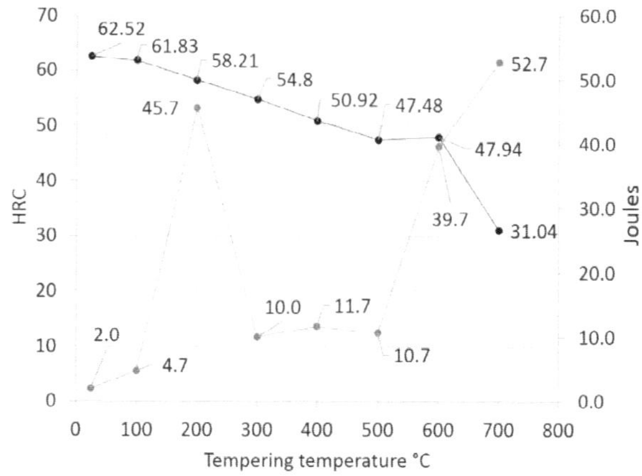

Fig. 5 Influence of the tempering temperature on impact energy and hardness of AISI L6 tool steel for samples oil quenched (900°C) and tempered once (100 °C – 700°C).

Fig 6. Photographs of the AISI L6 tool steel sample fractures: a) oil quenched from 900 °C and then tempered at: b) 100, c) 200, d) 300, e) 400, f) 500, g) 600, h) 700 °C.

Discussion

The results obtained of calorimetry, microstructure analysis and mechanical properties of AISI L6 tool steel are now discussed.

Calorimetry

Calorimetry thermograms show exothermic reactions at 340°C to 460°C. These reactions corresponding to precipitation process at tempering temperatures. The first tempering in the curves show two exothermic reactions. During heating happen one first reaction around 280°C to 350°C and the second one around the temperature of 380°C to 460°C. First reaction can correspond with the "stage II", the nucleation of stable cementite from the carbides and the clustered carbon atoms. The calorimetry experiment with 0.01 °C/sec as heating-cooling rate just present one peak around the 370 °C. However the others rates experiments show two remarkable peaks in derivate curves. This could be the overlapping of nucleation of cementite and the transformation of retained austenite [10]. The second peak is due to formation of alloy carbides from the dissolution of cementite particles.

First tempering show peaks for every heating rate. Apparently kinetic precipitation only happen on the first tempering heating and there are no measurable calorimetry energy changes for a second heating. For figure 1, only the first tempering with a rate of 0.001 °C/sec didn't not present double peak. It is possible that phase precipitation overlap into the same reacted peak. Derivative curves show that heating rates of 0.55 and 1 °C/sec present an increasing of intensity reaction and divide the two apparently overlapped phase transformations. But the energy of the transformation during heating only could be calculated for one peak over the experimental curve. This energy resulted in -42 and -29 J/g for the rates of 0.55 and 1 °C/sec respectively. The slower rates: 0.001, 0.25 and 0.35 °C/sec, shown a very similar energy reaction around -37 J/g.

A possible martensite start peak was detected during the cooling of the derivative curves. These derivative curve cooling peaks was observed for the 0.001, 0.25 and 0.35 °C/sec cooling rates. J. Pacyna is other researcher that present cooling after tempering peaks for one unalloyed high carbon steel, related with martensite transformation [4].

At 0.55 and 1 °C/sec rates, several peaks was observed before 300°C. The above cannot be related with phase precipitation but does with thermal instability of the calorimeter system and the experimental sample. At higher rates, peaks shifts to higher temperatures with high intensity.

Microstructure

Laths of martensite and spherical small particles of possible carbides appear for all the quenched and double tempered samples. The microstructure did not present changes for quenched and tempered samples. For this work it wasn't possible to measure the size and the distribution of the spherical particles, but probably these dimensions can differ with the tempered temperature.

Hardness and Charpy impact tests

The AISI L6 tool steel softens in a slow way and keep hardness of around 48 HRC after tempering at 600 °C. Also this good hardness corresponds to a high toughness, keeping good mechanical properties. High hardness level at high tempering temperatures (600°C) can be related to the precipitation hardening transformation. It should be noted, that 400 and 500 °C tempering temperatures do not change the hardness level. Quenched sample, 100, 300 and 400 °C tempered samples as well show fragile fracture. The 200 °C tempered sample increase energy impact at 10 times of 100 °C energy. This can be due to transition temperature at this tempering temperature range. However for 300 °C tempered sample impact decrease to 10 Joules and keep an expected energy increase behavior until 500 °C. The 600 °C tempered sample reach 39 Joules and 52 Joules at 700 °C, showing other transition temperature. It is possible that at 200°C the stress relief increase energy impact that higher due to carbon atoms release from martensite matrix. Also being possible that this energy decrease due to the cementite precipitation between 300 °C to 450 °C. Thus the dissolution of cementite particles and the formation of more stable alloy carbides precipitate, increasing energy impact from 500 °C tempering temperature.

Conclusions

The kinetics of phase transformation for a tempered steel was investigated at different heating rates by calorimetry, hardness, and energy impact and microstructure analysis. Then the following results were obtained:

- Calorimetry is a powerful technique for phase transformation analysis at tempering temperatures at the range of 300°C to 650°C.
- For calorimetry technique the increase in heating rate shifts highly the possible transformation of retained austenite and cementite precipitation.
- It is possible to measure kinetics transformation during first tempering of AISI L6 tool steel. However second tempering didn't present any calorimetry measurement during heating.
- Slow rates present peaks at the cooling after tempering, which can be related with martensite transformations.
- There is no morphological changes for microstructures of quenched and tempered samples. Quantitative chemical composition is a good option to be sure if the spherical particles are carbides and what type of alloyed carbides could be.
- Good mechanical properties were observed during increasing of the tempering temperature.
- The type of the impact Charpy fracture can be related with the transition temperature and the phase precipitation as well.
- It is possible to have a relation of stress relief with the increasing energy impact.

References

[1] ASM, Low alloy special purpose tool steel (L Series), "L6 Technical Data Sheet", International Heat Treaters Guide, p. 576 - 579.

[2] Charlie R. Brooks *et al.*, ASM International, 1996, "Principles of the heat treatment of plain carbon and low alloy steels"

[3] ASM, Metals Handbook Vol. 04, Heat Treating.

[4] J. Pacyna, Journal of Achievements in Materials and Manufacturing Engineering, "Dilatometric investigations of phase transformations at heating and cooling of hardened, unalloyed, high-carbon steels", Vol. 46, May 2011.

[5] P. Bala, J. Pacyna, Journal of Achievements in Materials and Manufacturing Engineering, "The kinetics of phase transformations during tempering of low alloy medium carbon steel", Vol. 28, May 2007.

[6] P. Bala, J. Pacyna, J. Krawczyk, Achievements in Materials Science and Engineering, "The kinetics of phase transformations during tempering of low alloy medium carbon steel", Vol. 28, Feb 2007, pages 98-104.

[7] P. V. Morra, Journal of thermal analysis and Calorimetry, "DECOMPOSITION OF IRON-BASED MARTENSITE, A kinetic analysis by means differential scanning calorimetry and dilatometry", Vol. 64, 2001, pages 905-914.

[8] JMatPro software.

[9] M. Goji, Journal of thermal analysis and Calorimetry, "Thermal analysis of low alloy Cr–Mo steel", Vol. 75, 2004, pages 947-956.

[10] Rafal Dziurka, METAL, "Effect of heating rate on phase transformations during tempering of Cr-Mn-Mo alloy steels", 2013

[11] S. Primig, Sep. 2011, Thermochimica Acta, El Sevier, "Separation of overlapping retained austenite decomposition and cementite precipitation reactions during tempering of martensitic steel by means of thermal analysis", pages 111-117.

[12] R. D. Browski, R. Dziurka, Archives of Metallurgy and Materials, "Tempering Temperature Effects on Hardness and Impact Toughness of 56NiCrMo7 Steel", Volume 56, 2011.

International Federation of Heat Treatment and Surface Engineering 2016
Proceedings of the 23rd IFHTSE Congress
April 18–21, 2016, Savannah, Georgia, USA

Influence of UNSM Treatment Temperature on Microstructure and Mechanical Properties of Inconel 600

Jun-Hyong Kim, Auezhan Amanov, Young-Sik Pyun and Hak-Doo Kim
Sun Moon University, Asan 31460, Rep. of Korea
pyoun@sunmoon.ac.kr

Abstract

In this study, an ultrasonic nanocrystal surface modification (UNSM) technique was applied to Inconel 600 at temperatures of 25 and 500 ^{0}C. The effects of UNSM treatment temperature on the microstructure and wear behavior were investigated. The hardness with respect to depth from the top surface was measured. Also, the wear behavior of UNSM-treated Inconel 600 specimen at temperatures of 25 and 500 ^{0}C was compared to that of the untreated specimen. The increase in wear resistance by UNSM technique was discussed in terms of increased hardness.

Introduction

Inconel 600, a Ni-based alloy, has been used in steam generator tubines for pressurized water reactors in nuclear power plants, because of its mechanical strength and high resistance to corrosion. Also, with good oxidation resistance at higher temperatures, with good resistance in carburizing and chloride containing environments [1, 2].

In this study, a relatively new ultrasonic nanocrystalline surface modification (UNSM) technique that increases the strength and improves the wear resistance of metallic materials was employed to increase the hardness and to improve the friction and wear behavior of Inconel 600. The UNSM technique strikes the specimen surface up to 20,000 times per second with a tungsten carbide (WC) and/or a silicon nitride (Si_3N_4) ball with a diameter range from 1.0 to 6.0 mm at a frequency of 20 or 40 kHz that modifies the coarse grains into nano-sized grains till the certain depth from the top surface. The most important advantage of UNSM technique over other surface modification techniques is that the controllable static and dynamic loading provides a uniform and homogenous treatment, which leads to the increasingly wide applications of UNSM technique [3].

The main purpose of this study is to evaluate the effectiveness of UNSM treatment temperature on the hardness, and friction and wear behavior under dry conditions. It is expected that the newly developed high-temperature (HT)-UNSM technique can be employed to further increase the hardness and to improve the friction and wear behavior of Inconel 600.

Experimental Procedures

In this study, the specimens made of Inconel 600 with a diameter of 24.8 mm and a length of 5 mm were employed. The specimens were polished using a sand paper for microstructural characterization and friction and wear tests. The specimens subjected to UNSM technique at room and high temperatures under the parameters as listed in Table 1. More details of the UNSM technique can be found in our previous study [4].

Table 1: UNSM treatment parameters at room and high temperatures.

Frequency, kHz	20
Amplitude, μm	40
Horn speed, m/min	1.5
Load, N	20
Interval, μm	30
Ball diameter, mm	2.38

A commercially available ball-on-disk tribometer (CSM, Switzerland) was used to access the friction and wear behavior of the specimens under dry conditions at room temperature at a load of 50 N, with a stroke of 1 mm for 30 min of sliding at a reciprocating speed of 3 cm/s. A bearing steel SAE52100 ball with a diameter of 10.0 mm was used as a counter surface.

Results

The surface microstructure of the polished and UNSM-treated Inconel 600 specimens is presented in Fig. 1. It can be seen from Fig. 1(a) that the microstructure of the polished specimen is not homogeneous with some defects and abrasive grooves on the surface. Clearly, regular UNSM treatment tracks with no abrasive grooves can be observed in Figs. 1(b and c). The average surface roughness of the polished, UNSM-treated at room and high temperatures specimens were found to be about 0.20, 0.17 and 0.28 μm. It is obvious that the specimen exhibited at high temperature exhibited a high surface roughness compared to that of the specimen treated at room temperature. Moreover, as shown in Fig. 2, a little difference in hardness was found between the polished and room temperature UNSM-treated speciment, but the hardness of Inconel 600 increased from about 215 to 474 HV after high-temperature UNSM treatment.

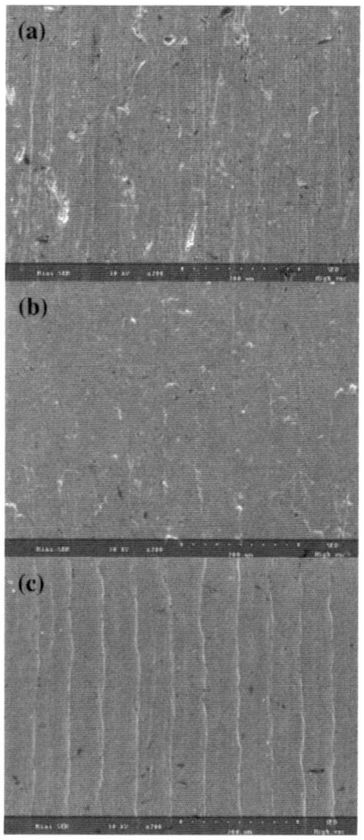

Figure 1: SEM images of the polished (a), UNSM-treated at room temperature (b) and UNSM-treated at high temperature (c) specimens.

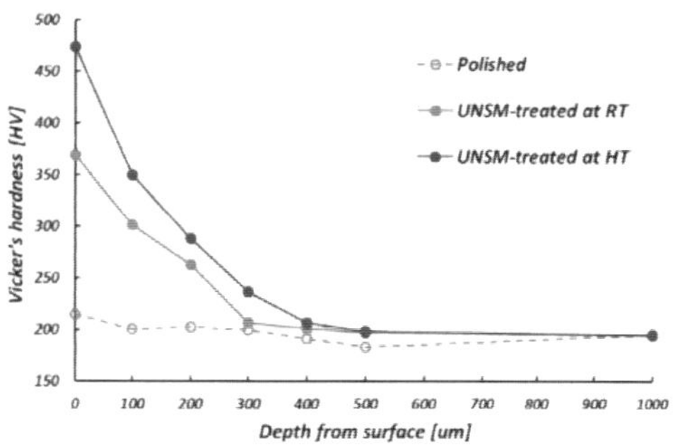

Figure 2: Comparison in Vicker's hardness of Inconel 600 with respect to depth from the top surface.

A comparison in friction coefficient of the polished, UNSM-treated specimens at room and high temperatures treated as a function of sliding time is presented in Fig. 3. It can be seen that the friction coefficient of the polished and room temperature UNSM-treated specimens increased gradually at the initial stage of friction test due to the surface contamination such as absorbed water from atmosphere and initial true contact between the substrate and counter surface. And with continuing sliding it increased gradually, and after passing some time the friction behavior got stabled to a friction coefficient value in the range from 0.20 to 0.17 for the polished and room temperature UNSM-treated specimens, respectively. In case of high temperature UNSM-treated specimens, the friction coefficient was so rough which may be attributed to the high surface roughness.

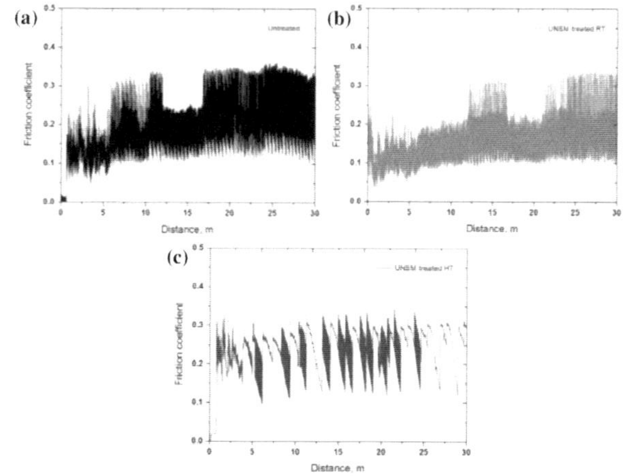

Figure 3: Comparison in friction coefficient for the polished (a), UNSM-treated at room temperature (b) and UNSM-treated at high temperature (c) specimens under dry conditions.

Conclusions

In this study, the hardness, and friction and wear behavior of polished and UNSM-treated Inconel 600 specimens were investigated. It was found that the average surface roughness reduced from 0.20 μm to 0.17 μm for the room temperature UNSM-treated specimen, and increased from 0.20 μm to 0.28 μm for the high-temperature UNSM-treated specimen. The hardness of specimens at the very top surface increased from 215, 369 to 474 HV for the polished, room and high temperatures UNSM-treated specimen, respectively. A deterioration in friction and wear behavior of the high-temperature UNSM-treated specimen compared to that of the untreated specimen may be attributed to the high initial surface roughness and unsuitable UNSM treatment parameters.

References

[1] Kai, J.J., *et al.*, "The effects of heat treatment on the sensitization and SCC behavior of INCONEL 600 alloy," *Metall. Trans.*, Vol. 20A (1989), pp. 1077-1088.

[2] Wang, J.D., *et al.*, "Effect of grain boundary carbides on the mechanical properties of Inconel 600," *Mater. Chem. Phys.*, Vol. 70 (2001), pp. 124-128.

[3] Amanov, A., *et al.*, "Micro-dimpled surface by ultrasonic nanocrystal surface modification and its tribological effects," *Wear*, Vol. 286-287 (2012), pp. 136-144.

[4] Amanov, A., *et al.*, "Effects of ultrasonic nanocrystalline surface modification (UNSM) technique on the tribological behavior of sintered Cu-based alloy," *Tribol. Int.*, Vol. 72 (2014), pp. 187-197.

International Federation of Heat Treatment and Surface Engineering 2016
Proceedings of the 23rd IFHTSE Congress
April 18–21, 2016, Savannah, Georgia, USA

Influence of heat treatment on fatigue and fracture toughness behavior of AISI 4140 steel

Pravin Jadhav
Kalyani Centre for Technology & Innovation, Bharat Forge Ltd, Pune 411036, India
pravin.jadhav@bharatforge.com

R.C. Prasad
Indian Institute of Technology Bombay, Mumbai 400076, India

Rajesh Mane, Manoj Ukhande and Suresh Arangi
Bharat Forge Ltd, Pune 411036, India

Abstract

The present study is concerned with the effect of tempering temperature and time on the microstructure and its correlation with the mechanical properties, endurance strength and fracture toughness of AISI 4140 steel. In this study, the endurance strength of AISI 4140 steel in as received and heat treated conditions (hardening and tempering at different temperatures and time) were determined using Rotary Bending Fatigue tests at a constant stress ratio of R= -1. The fatigue strength was found to increase with decrease in tempering temperature. The fractured surfaces of the failed samples were examined under SEM and the micro mechanisms of fracture are used to explain the variations in the fatigue strength. An attempt has been made to evaluate the fracture toughness i.e. K_{IC} and J_{IC} on the CT specimen for as received material and heat treated samples, respectively. Fracture toughness values significantly improve after heat treatment as compared to the as received material. This is attributed to the change in microstructure from pearlite-ferrite to tempered martensite after heat treatment.

KEYWORDS: *AISI 4140, Fatigue, fracture toughness, Heat treatment.*

Introduction

Fatigue failure is generally considered as the foremost problem affecting any component facing dynamic loading conditions. Almost 90% of failures in mechanical components are due to fatigue [1, 2]. Fatigue failures occur due to the application of fluctuating stresses that are much lower than the stress required to cause failure during a single application of stress. Fatigue is the progressive and localized structural damage that occurs when material is subjected to repetitive or fluctuating stress or cyclic loading [1]. It refers to the time delayed fracture of materials subjected to cyclic stresses below those causing plastic yielding and/or tensile failure. Some investigations state that bending stresses are much more potent in causing fatigue failures as compared to torsional stresses. Therefore, bending stresses are more often focused on, while torsion stresses are often neglected [3, 4].

The fracture toughness values serve as a basis for material characterization, performance evaluation and quality assurance for typical engineering structures including nuclear pressure vessels, piping, oil & gas pipelines, automotive, ship and aircraft structures [5]. The fracture toughness of material denotes the resistance to propagation of crack. Fracture Toughness is an indication of the amount of stress required to propagate a pre-existing crack. If a material has high fracture toughness it will probably undergo ductile fracture, while brittle fracture is characteristic of material with less fracture toughness. Fracture toughness of a material depends upon the volume of the material which deforms plastically prior to fracture [6, 7].This property along with fatigue strength of the material denominates the resistance of material to failure.

Previous studies have revealed that the fatigue strength of a material is directly related to its tensile strength [2]. On the other hand it has also been reported that the fracture toughness of the material is affected negatively by increase in strength of the material as its ductility is reduced [8].

The microstructure of any material is dependent on the manufacturing method used. In steels, the heat treatment plays an important role in determining the final microstructure of the material. The microstructure in turn affects the properties of the material [9]. As described above, while optimizing the heat treatment for a material, both fatigue and fracture toughness properties should be considered.

AISI 4140 is considered one of the most suitable materials for dynamically loaded machine parts of engine and other machines, because of its high tensile strength after quenching and tempering [6]. This type of steel is categorized as quenched-tempered steel as to achieve good mechanical and fatigue properties, the manufacturing process often includes heat treatment (quenching and tempering). The dominant phase present in this material after heat treatment is tempered martensite [10, 11].

Numerous studies have been performed on the effects of microstructure on the fracture toughness of a material. However, those studies used largely different materials to compare results from various microstructural alterations [12]. In order to investigate the effects of microstructures on fracture behavior, while excluding chemistry effects, it is necessary to use samples with identical compositions but different microstructures.

In the present work, the effect of different microstructures, obtained from different tempering temperatures and time, on the fatigue strength, hardness and tensile strength of AISI 4140 has been studied. The fracture toughness of AISI4140 steel with two different microstructures is evaluated at room temperature.

Experimental Work

Material

Medium carbon low alloy AISI 4140 steel produced by cogging followed by rolling was used as base material. Table 1 shows the typical chemical composition of the steel. Round specimens having 20 mm diameter and 120 mm length were extracted for heat treatment from the as received 235 RCS raw material in longitudinal direction.

Table 1: Chemical composition of AISI 4140 steel (weight %)

C	Mn	Si	S	P	Cr	Ni	Mo
0.40	0.89	0.21	0.019	0.010	1.01	0.12	0.20

Heat treatment

The samples were then heat treated in an electrically heated furnace of Therelek make. The heat treatment consisted of hardening at 860°C followed by tempering. The tempering was conducted at different temperatures and tempering time. The details of this process are given in Fig. 1 and Table 2.

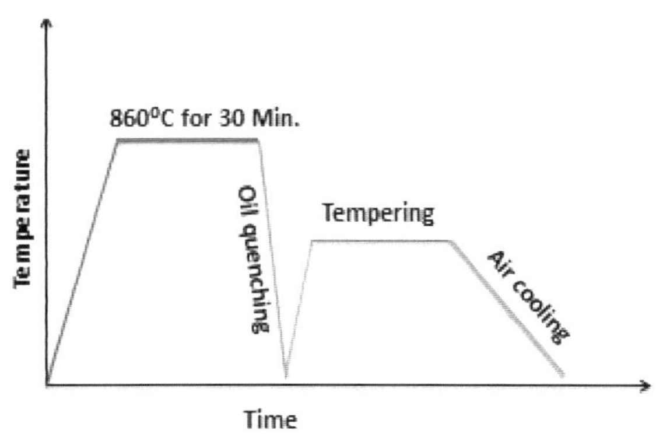

Figure 1: Heat treatment process

Table 2: Heat treatment process and parameters

Batches	Hardening	Tempering
Batch 1	860°C for 30 Min	560°C for 1Hr.
Batch 2		680°C for 1Hr.
Batch 3		560°C for 8Hr.
Cooling	Metaquench 43 Oil.	Air cool

Mechanical testing

The mechanical properties of as received and heat treated materials were determined by conducting hardness and tensile tests (as per ASTM E8), at room temperature. Tensile testing was done on universal testing machine of make Zwick / Roell. The hardness tests were conducted on digital Rockwell hardness tester.

Rotating bending fatigue tests

RBF specimens were prepared according to DIN 50113 as shown in Fig. 2. From the raw material and the heat treated cylindrical specimen. Fatigue tests were carried out on RBF machine of Instron make at constant stress ratio of R = -1. The frequency of cyclic loading was selected to be 4500 rpm.

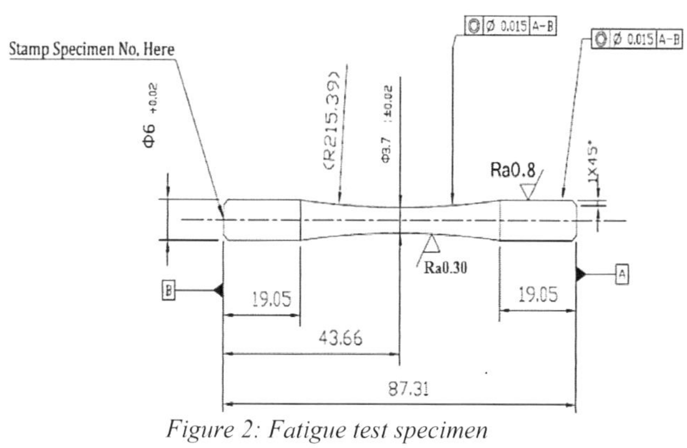

Figure 2: Fatigue test specimen

Fracture Toughness tests

Fracture toughness of as received and heat treated (Batch 3) material was also determined. Fracture toughness test K_{IC} was carried out on compact test (CT) specimen as per ASTM E399 [13] for as received samples. For heat treated samples, the use of linear elastic fracture mechanics (K_{IC}) was found to be not suitable due to extensive plastic zones occurring at the crack tip. Hence, for heat treated samples, elastic plastic fracture mechanics (J_{IC}) tests were carried out as per ASTM 1820 [14]. The orientation of samples was in TL direction, which means that the loading was in transverse direction while the crack propagation was in longitudinal direction as shown in Fig. 3. The test was carried out on servohydraulic MTS machine of capacity 250 kN. The fatigue pre- cracking was done at room temperature under force controlled mode. Test frequency was selected as 20 H_Z and final stress intensity (K_{final}) was selected as 20 MPa $m^{0.5}$.

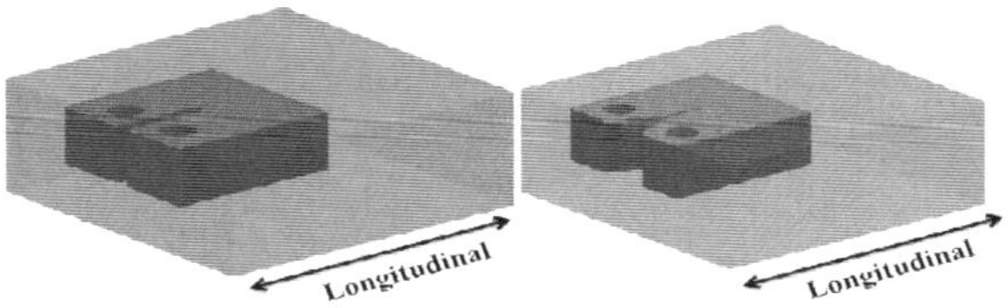

(a) (b)

Figure 3: CT specimen location (a) K_{IC} Specimen. (b) J_{IC} Specimen.

Fractography

The fractured surfaces of the failed tensile, rotating bending fatigue and fracture toughness samples were studied using scanning electron microscopy in order to reveal the fracture mechanisms presented by the various material conditions.

Results and Discussion

Microstructures and mechanical properties

The mechanical test results are displayed in Table 3. It is found that the heat treatment performed on the material has significant effect on its UTS. The reasons for this variation were analysed using microstructural analysis.

Table 3: Mechanical properties and Rotating bending fatigue test Result

Material condition	UTS (MPa)	ES (MPa)	Fatigue Ratio (%)	Hardness (HRc)
Base material	981	415	42.30	25
Batch I	1170	644	55.00	39
Batch II	835	425	50.80	26
Batch III	808	375	46.41	24

The microstructures obtained for different material conditions are shown in Fig. 4. The UTS and hardness of base material was found to be 981 MPa and 25 HRc respectively due to the soft ferritic and pearlitic microstructure as shown in Figure 4(a). The increase in properties in Batch 1 as compared to base material is attributed to the tempered martensitic microstructure after quenching and tempering as shown in Figure 4(b). The proportion of tempered carbon is found to increase at higher tempering temperature as shown in Figure 4(c). It is attributed to accelerated diffusion of carbon at higher temperatures [15]. Carbon starts diffusing at higher temperatures which lead to decrease in hardness and other mechanical properties as shown in Table 3. The microstructure of Batch 3 sample (shown in Figure 4 (d)) shows growth of carbide precipitates in the martensitic structure (coarsening) due to the longer time of tempering [15]. As a result of this, the material has lost its hardness as well as other mechanical properties with improved toughness.

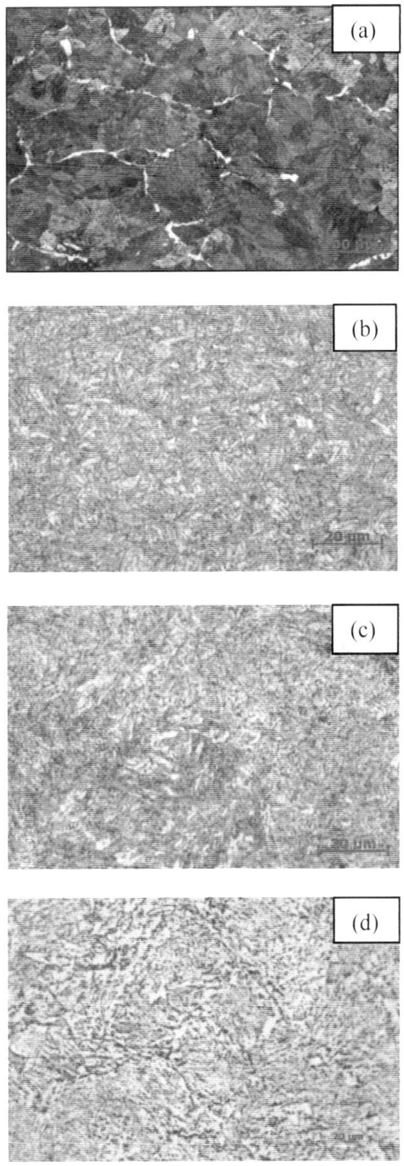

Figure 4: Microstructure of different material conditions: (a) Base material, (b)Tempering at 560°C for 1 hr., (c) Tempering at 680°C for 1 hr., (d) Tempering at 560°C for 8 hr

Fractography of tensile tested samples.

SEM fractographs of fractured surface of tensile tested samples for different material conditions are shown in Figure 5 (a-d). The base material shows dimples on fractured surface as shown in Figure 5 (a). The fractured surface of specimen, tempered at 560°C for 1 hr., shows riverline pattern and facets which confirms brittle mode of fracture. While, the fractured surface of specimen tempered at 680°C and specimen tempered at 560°C for 8 hr. show deep dimples which indicates a ductile mode of fracture. This observation corroborated the increase in ductility of the material due to higher tempering temperature.

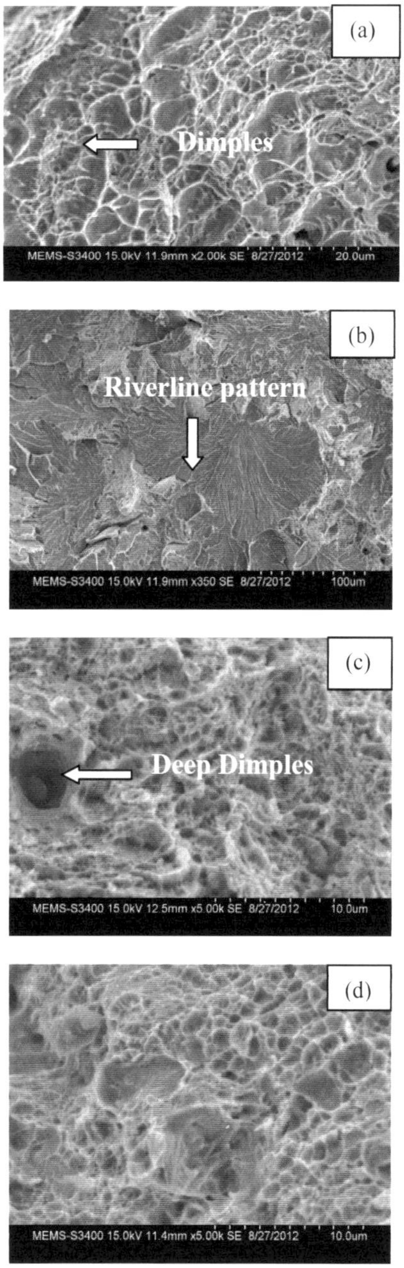

Figure 5: SEM Fractography of Tensile tested specimens: (a) Base material, (b) Tempered at 560°C for 1 hr., (c) Tempered at 680°C for 1 hr., (d) Tempered at 560°C for 8 hr.

Rotating bending test results

The results of RBF tests are presented in Table 3 and Figure 6. The results show that endurance strength increases on tempering at 560°C. This is attributed to the increase in tensile strength and hardness. However, when the tempering temperature was increased to 680°C, the endurance strength decreased. It is seen that the endurance strength of the material reduces by 2.5 to 3% when there is decrease in the hardness by 1 HRc and vice versa. The test data is plotted on the graph as shown in Fig 7 and it appears that, for a given material, endurance strength is directly proportional to hardness.

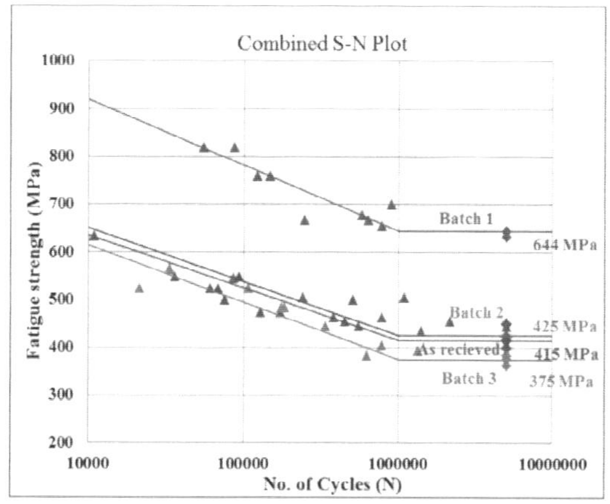

Figure 6: S-N curve for different material conditions

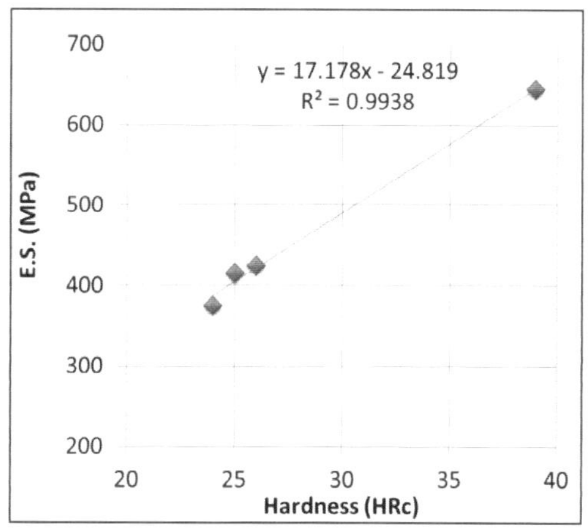

Figure 7: Relation between hardness and Fatigue limit.

Fractography of failed RBF samples

The fractured surface of failed RBF samples shows three typical characteristics of fatigue failure, i.e. crack initiation, crack propagation and final fracture shown in Figure 8 (a).The base material tested using RBF test shows striation marks at higher magnification separated by cliff wall at the grain boundaries shown in Figure 8 (b). Cleavage or brittle mode of fracture was observed at crack initiation, whereas final fracture shows the ductile mode of fracture shown in Figure 8 (c-d). Dimples of specimen tempered at 680°C are much wider and deeper than those of 560°C, revealing better toughness.

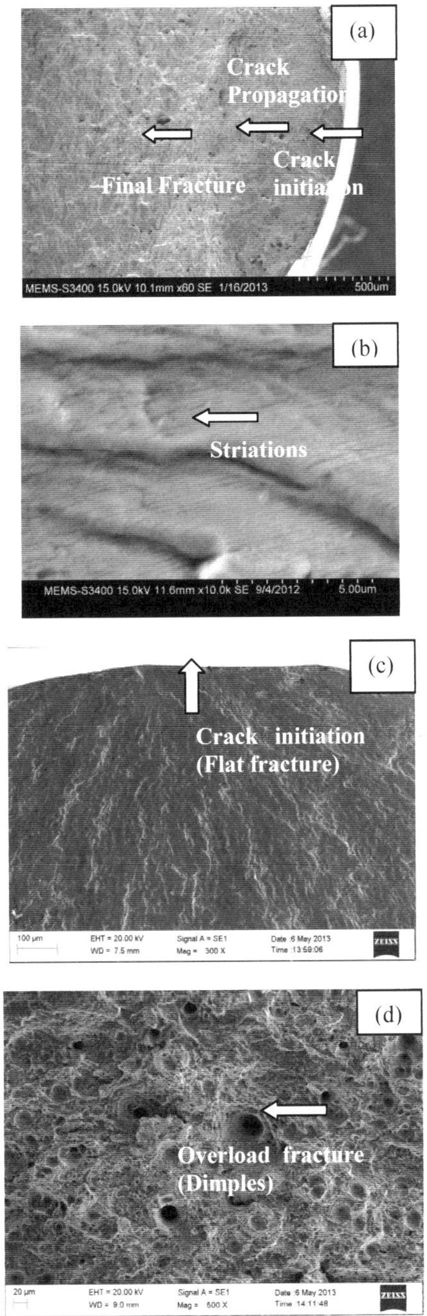

Figure 8: SEM Fractograph of rotating bending samples.

Fracture toughness test result

Linear elastic fracture mechanics test (K_{IC}) test

The tests were conducted under load controlled mode. The values of the plane strain fracture toughness (K_{IC}) were computed using below equation.

$$K_Q = \frac{PQ}{B\sqrt{w}} f\left(\frac{a}{w}\right)$$

and following criteria were checked for validity of K_{IC}

$$\frac{Pmax}{PQ} \leq 1.10$$

$$B, a \geq 2.5\left(\frac{KQ}{\sigma y}\right)^2$$

For both the tests all the above validity criteria's passed so, K_Q was considered as K_{IC}

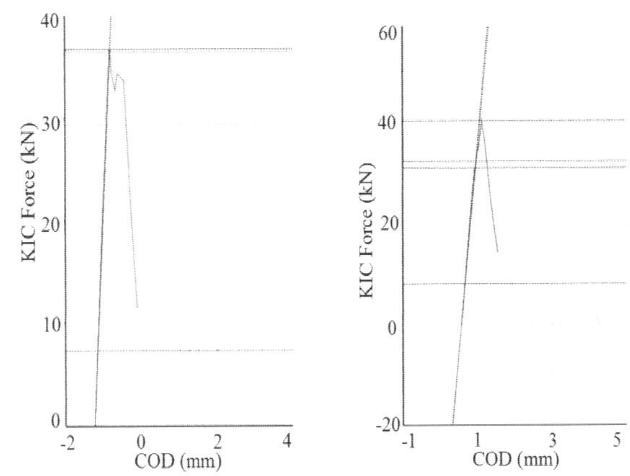

(a) (b)

Figure 9: Load Vs. COD: (a) Specimen K1 (b) Specimen K2

Table 4: K_{IC} Test Results for base material

Sample No.	K_{Ic} (MPa $m^{0.5}$)
K1	60.784
K2	61.064

Elastic plastic fracture mechanics test (J_{IC}) test.

Test was conducted under displacement controlled mode, using single specimen unloading compliance method, according to ASTM 1820 standard. From Fig. 10 load vs. COD, the J Vs. Δa curves for the two specimens J1 and J2 were plotted using the machine software which included construction line, exclusion line at 0.15 mm, offset line at 0.2 mm and exclusion line at 1.5 mm as shown in Fig. 11.

The initiation toughness (J_{IC}) was found at the point where the power law regression curve intercepts at 0.2 mm offset line. The J_{IC} values for two samples are listed in the Table 5. All the validity criteria's were passed according to ASTM 1820 standard. Using the relationship between J_{IC} and $K_{IC,}$ the values of K_{JC} were calculated from below equation.

$$J_{Ic} = \frac{K_{Ic}^{2}(1-\vartheta^2)}{E}$$

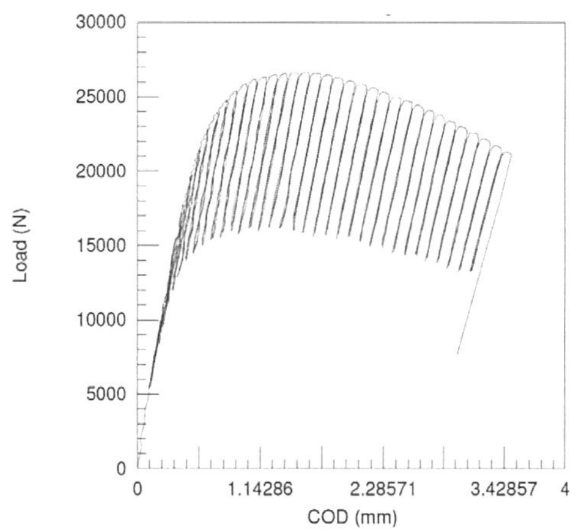

(a)

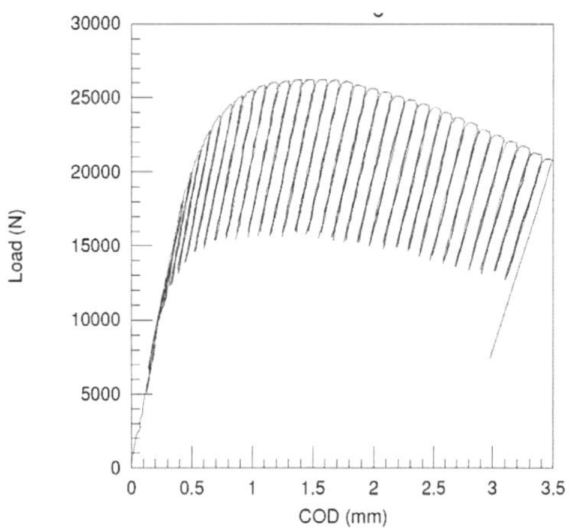

(b)

Figure 10: Load Vs COD : (a) Specimen J1 (b) Specimen J2

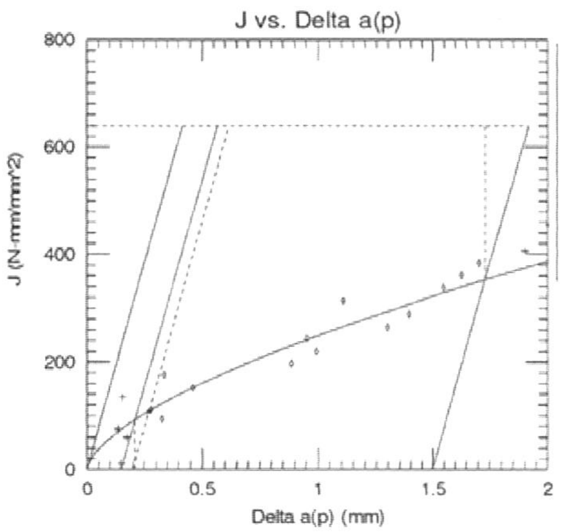

(a)

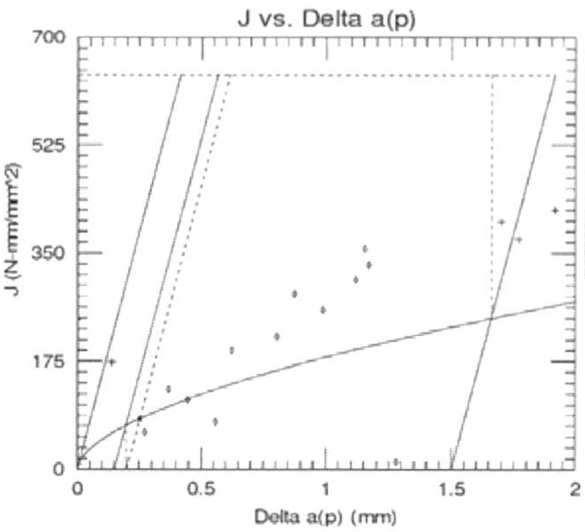

(b)

Figure 11: J Vs Delta Δa:(a) Specimen J1 (b) Specimen J2

Table 5: J_{IC} Test Results for Batch III material

Sample No.	J_{IC} KJ/m^2	K_{JC}^* MPa m$^{0.5}$
J1	108.604	158.30
J2	82.371	137.87
$^*K_{JC} = K_{IC}$ values estimated from J_{IC} values.		

Microstructure of steel has a strong influence on the fracture toughness (K_{IC}) of medium carbon steel. This has been clearly observed in the present investigation. The fracture toughness of heat treated sample has been found to be higher than as-received steel sample. This is attributed to change in microstructure from pearlite-ferrite to tempered martensite after heat treatment.

Presence of hard phases, distributed in microstructure, increases the fracture toughness of material. This happens due to deflection of crack path when it encounters the hard phase or particle [16]. The tempered martensite consist of harder martensitic structure with carbides (formed during tempering) distributed in the matrix [15]. The inherent randomness of martensitic structure coupled with the presence of carbides increases the crack path deflections and leads to tortuous path [17]. But, in case of pearlite-ferrite microstructure, the presence of softer ferrite phase makes crack propagation easier with less degree of crack path deviation. This phenomenon increases the fracture toughness of tempered martensitic structure as compared to the pearlite-ferrite structure.

Fractography of failed CT samples

The SEM fractographs were taken on three regions of cracked surface i.e. fatigue, overload and transition zone between the two. The SEM fractographs taken at region (A) with higher magnification, shows stepwise striation crack growth. As crack grows ahead of pre-cracked region, some void nucleation is observed at region (B). The region (C) shows dimple rupture which describes ductile fracture that occurs through the formation and coalescence of micro-voids along the fracture path as shown in Fig. 13.

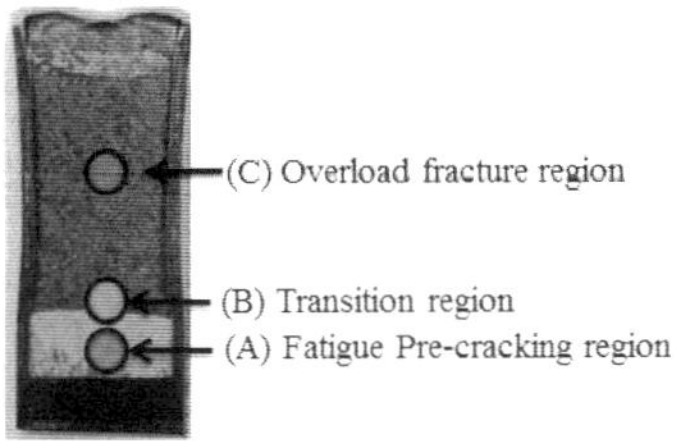

Figure12: Stereographic image of failed CT specimen.

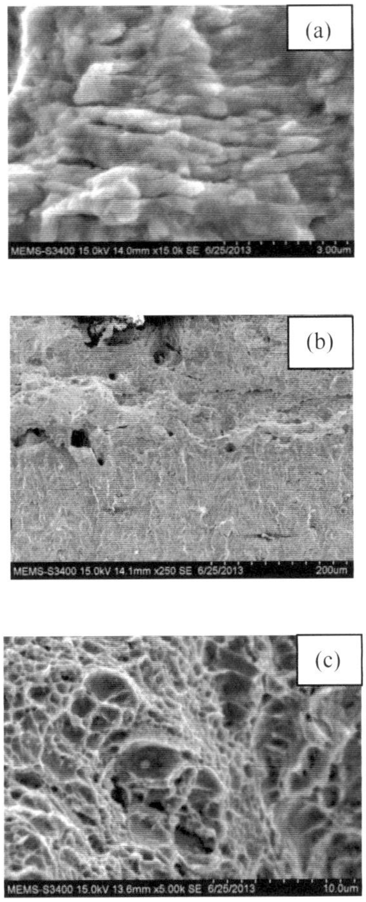

Figure13: Fractographs of CT specimen at different regions.

Conclusions

1. Mechanical properties are improved and Fatigue ratios (endurance strength to UTS) are increased by Quenching and tempering as compare to base material.
2. The fatigue strength was found to increase with decrease in tempering temperature and time.
3. The mechanical and fracture toughness properties of tempered martensite, is more favorable than those of ferritic and pearlitic microstructure even for the same hardness.
4. Endurance strength of the material reduces by 2.5 to 3% when the hardness is decreased by 1 HRc and vice versa.
5. Fracture surface of tensile specimen, dominated by dimples with different sizes, shows failure by microvoid coalescence
6. The fractography of fractured surface in bending fatigue shows typical fatigue striations represented by cliff wall at the grain boundary.
7. For AISI 4140 steel, the fracture toughness values significantly improve after heat treatment as compared to the as received material.
8. Fractured surface of failed CT sample shows stepwise striation crack growth. As crack grows ahead of pre-cracked region, some void nucleation is observed.

Acknowledgments

The authors would like to express sincere gratitude towards **Dr. RKP Singh** and **Shreyas Kirwai** for their invaluable guidance, support and constant encouragement throughout the present the research work.

The authors gratefully acknowledge the extended support provided to this work by KCTI (Kalyani Centre for Technology and Innovation) for providing financial funding, laboratory and library facilities. The authors also acknowledge the support provided by Bharat Forge Ltd, Pune and DSIR (Department of Scientific and Industrial Research), Govt. of India. Finally, the authors would like to express special thanks and gratitude to review committee and top management of Bharat Forge Ltd for granting the permission to publish/present the research work.

References

[1] Dieter, George E, "Mechanical Metallurgy," *SI Metric edition, McGraw- Hill Book*, 1988. pp. 391-431.

[2] Somer M et al**.**, **"**Effect of Heat Treatment on Fatigue Behaviour of (A193-51T-B7) Alloy steel," *Proceedings of the World Congress on Engineering*, 2007, Vol. II, pp. 2-6.

[3] W.Y. Chien et al., "Fatigue analysis of crankshaft sections under bending with consideration of residual stresses," *International journal of fatigue,* Vol. 27, 2005, pp. 1-19.

[4] Paul Spiteri et al., "Assessment of bending fatigue limit for crankshaft sections with inclusion of residual stresses," *International journal of fatigue*, Vol. 29, 2007, pp. 318-329.

[5] Xian-Kui Zhu et al., "Review of fracture toughness (G, K, J, CTOD, CTOA) testing and standardization," Engineering fracture mechanics, Vol.85, 2012, pp. 1-46.

[6] Tohru Arai et al., "Heat treating," *American society for metals handbook*, Vol. 4, 1991, pp. 160-306.

[7] Peter Andersen et al., "Fatigue and Fracture," *American society for metals handbook*, Vol. 19, 1996, pp. 963-996.

[8] A. Salemi et al., "A study on fracture properties of multiphase microstructures of a CrMo steel," *Material science and engineering,* Vol A 492, 2008 pp. 45-48

[9] I. Souki et al., "Influence of heat treatment on the fracture toughness and propagation in 5% Cr martensitic steel," *procedia Engineering*, Vol. 10, 2011, pp. 631-637.

[10] A.H. Meysami et al., "An investigation on the microstructure and mechanical properties of direct-quenched and tempered AISI 4140 steel", *Materials and design*, Vol. 31, 2010, pp. 1570-1575.

[11] Haftirman et al., "Effect of tempering temperature on fatigue strength of thyssen 6582," *Conference on applications and design of mechanical engineering,* October 2007, pp. 25-26.

[12] Ki-Hyoung Lee et al., "Evaluation of microstructural parameters controlling cleavage fracture toughness in Mn–Mo–Ni low alloy steels," *Materials Science & Engineering,* Vol. A 565, 2013, pp. 158-164.

[13] ASTM Standard E-399, "Standard Test Method for Linear-Elastic Plane-Strain Fracture Toughness KIc of Metallic Materials," 2010.

[14] ASTM Standard E-1820, "Standard Test Method for measurement of fracture toughness," 2011.

[15] A. Salemi et al., "The effect of tempering temperature on the mechanical properties and fracture morphology of a NiCrMoV steel," *Materials characterization*, Vol. 59, 2008, pp. 484-487.

[16] Akhmad A. et al., "Fatigue crack growth behavior in ferritic–pearlitic steels with networked and distributed pearlite structures," *International Journal of Fatigue*, Vol. 29, 2007, pp. 1140-1148.

[17] D. F. Laurito et al., "Microstructural effects on fatigue crack growth behavior of a microalloyed steel," *Procedia Engineering*, Vol. 2, 2010, pp. 1915-1925.

[18] Prashant Kumar, "Elements of Fracture Mechanics," *A.H. Wheeler & Co. Ltd, 1999.*

International Federation of Heat Treatment and Surface Engineering 2016
Proceedings of the 23rd IFHTSE Congress
April 18–21, 2016, Savannah, Georgia, USA

Manufacture, Examination and Application of Modern Composite Materials with a Nano-carbon Dispersion Phase

Tomasz Babul, Anna Olbrycht, Institute of Precision Mechanics, Warsaw, Poland
Corresponding author: E-mail: tomasz.babul@imp.edu.pl

Abstract

The paper presents the results of work on manufacturing and properties testing of composite materials with nano-carbon dispersion phase. The article describes a method for composite materials production and the results of thermal diffusivity as well as wear resistance measurements of these materials.

It was found, that the produced composite materials with nano-carbon dispersion phase are characterized by various advantageous properties such as increased hardness and wear resistance with respect to the matrix material and considerably higher thermal diffusivity.

The paper presents also the results of tests using composite materials with nano-carbon dispersion phase implementation for thermal spraying. There are also presented examples of coatings sprayed with detonation method on different substrates and the results of these coatings studies.

Introduction

There is a commonly accepted division of materials into four basic kinds: a) metals, b) ceramics, c) polymers, d) composites. Composites are the answer to an increasing demand for the materials which have got better construction and exploitation indexes and what is more, they make it possible to shape the desirable properties to such an extent which is unavailable in case of the three remaining types of materials. Composites are the materials which are composed of at least two components characterised by different properties. Properties of composites are never a sum or an average of their components. Most frequently one of the components is a matrix which guarantees its coherence, hardness, elasticity and compression resistance, while the other one, called a constructional component provides most of the remaining mechanical properties of a composite. By means of an appropriate and conscious matching of a matrix material with a dispersion phase it is possible to design unusually complex spatial structures which perform the most complicated functions. Nowadays composite materials are applied in many industrial fields and they are more and more frequently used in medicine [1, 2, 3].

Especially interesting, because of their exceptional properties and the possibilities of being used, are the composites having a metal matrix with a nano-carbon dispersion phase. In additional to the tested and widely used fullerenes and carbon nano-pipes, recently graphene has attracted a great interest of many world research centres because of its unique properties.

Graphene, which is built of a single layer of carbon atoms creating six-element connections, is unusually hard and resistant. It is more than twice as resistant as steel of the same thickness and in addition to this, it is unusually elastic. It can be stretched even by 20% without any obstacles. The tests, which were conducted in 2009, determined the elasticity limit within the range of 1 TPa and Young's modulus within the range of 0.5 TPa. It has also got unique properties as a thermal conductor. The measurements of thermal conductivity were from 4840 ± 440 to 5300 ± 480 W/mK, so above twice as much as for diamond (whose thermal conductivity is from 900 to 2320 W/mK). Graphene has also got small resistivity which is 10^{-8} Ω x m. A simple chemical treatment, such as an adsorption of hydrogen atoms, is enough in order to change it from an ideal conductor into an almost ideal non-conductor (so called: graphane). Furthermore, it is also possible to achieve all the indirect stages of current conductivity. In spite of the fact that it is almost totally transparent (it absorbs only 2.3% of light), even helium atoms do not go through its single layer. The speed of the electrons moving in graphene enables to apply this material in a lot of fields, such as: electronics, power engineering, transport, medicine [4, 5].

This article presents the examples of modern composite materials with a nano-carbon dispersion phase which are obtained by means of different methods.

Composite materials manufactured as a result of the consolidation of copper powders coated with graphene

Research methodology

The basis of this modern composite material is graphene coated on copper powders, so called the $3D^{IMP}$ Graphene which has been manufactured in the Institute of Precision Mechanics [6-9]. Forming of the graphene structures on the surface of the copper powder comes into being as a result of the thermal-chemical processes taking place in the atmosphere of hydrocarbons in a fluid deposit. The copper powder together with graphene was subjected to the tests by means of an electron scanning microscope (Fig.1) and a Raman spectrometer (Fig. 2). In the Raman spectra the 2D and G bands characteristic for graphene spectra are clearly visible, which confirms its presence in the tested powders (Fig. 2).

a

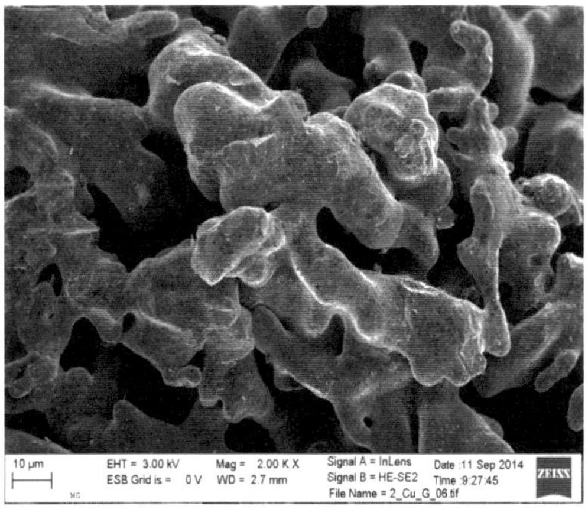

b

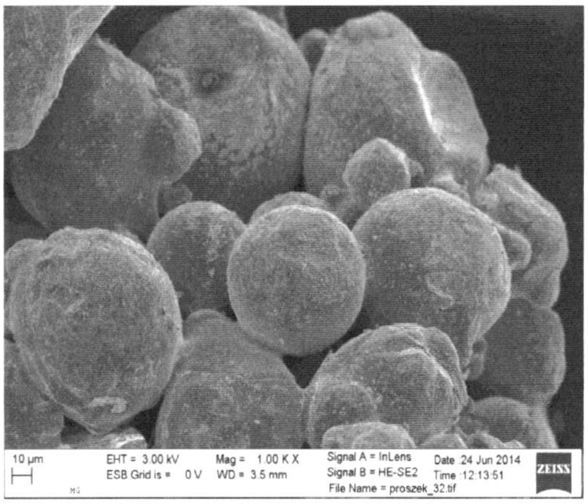

Figure 1: Copper powder particles after graphene processing (SEM) a. dendritic, b- spherical [6]

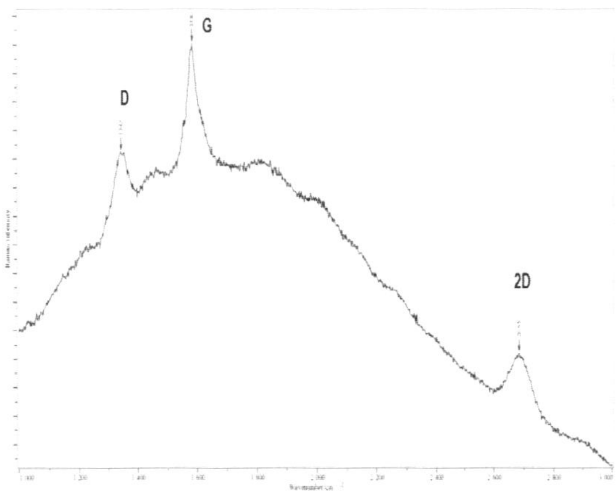

Figure 2: Raman spectra of graphene deposited on copper powder [6]

Due to the fact that there is no possibility of measuring the thermal diffusivity on a single copper grain coated with graphene, the powders were subjected to the cold isostatic pressing in the result of which the roller samples with a 30 mm radius were obtained [7]. For the comparison, the compacts of the copper powder without any graphene coating were made, accepting this variant as a reference one. Some different variants of the samples were made for the thermal diffusivity measurements, taking into consideration the kind of powder and their preparation technological processes (Table 1). The variant which is marked as "1" is different than the other ones; in this variant the samples from "0" variant were subjected to the process of graphene coating and then to another pressing in order to obtain the appropriate surface roughness. The measurements were conducted by means of the laser impulse analysis (LFA) with the application of the Netzach LFA 427 apparatus set, in the air atmosphere, at the ambient temperature (approximately 25^{0}C) and in the research direction in compliance with the pressing direction of the samples [7].

Table 1. Designation of samples for thermal diffusivity tests [6]

Sample Number	Kind of the Cu powder	Technological processes in a chronological order
0	dendritic	- pressing of the Cu powder (a basic variant)
1	dendritic	- pressing of the Cu powder - coating of the sample with graphene - repeated pressing of the sample
2	spherical	- pressing of the mixture of powders: 90% of the Cu powder + 10% of the Cu powder coated with graphene - coating of the sample with graphene after pressing
3	spherical	- pressing of the mixture of powders: 90% of the Cu powder + 10% of the Cu powder coated with graphene - coating of the sample with graphene after pressing
4	spherical	- pressing of the mixture of powders: 80% of the Cu powder + 20% of the Cu powder coated with graphene - coating of the sample with graphene after pressing
5	spherical	- pressing of the Cu powder coated with graphene - coating of the sample with graphene after pressing

The measurements of the abrasive wear were conducted on the samples manufactured from the pressed copper powders and the copper powders with graphene, at the stand called "Ball-tester" whose rule of activity and measurement methodology are presented in a publication [10]. In the conducted tests the diameter of the surface wear track made by a ball-tester ball was the measurement of the abrasive wear. Table 2 presents the marking and the variants of the preparation of the samples which were subjected to the abrasive wear.

Table 2. Designation of samples for abrasive wear tests [6]

Sample Number	Kind of the Cu powder	Technological processes in a chronological order
0	spherical	- pressing of the Cu powder sample (a basic variant)
1	spherical	- pressing of the Cu powder sample - coating of the sample with graphene
2	spherical	- pressing of the mixture of powders: 50% of the Cu powder + 50% the Cu - graphene powder - coating of the sample with graphene after pressing
3	dendritic	- pressing of the Cu powder sample (A density) - coating of the sample with graphene
4	dendritic	- pressing of the CU powder sample (B density) - coating of the sample with graphene

Research results of the thermal diffusivity and the abrasive wear resistance

The results of the thermal diffusivity are presented in Figure 3. Each result is an average value of three measurements.

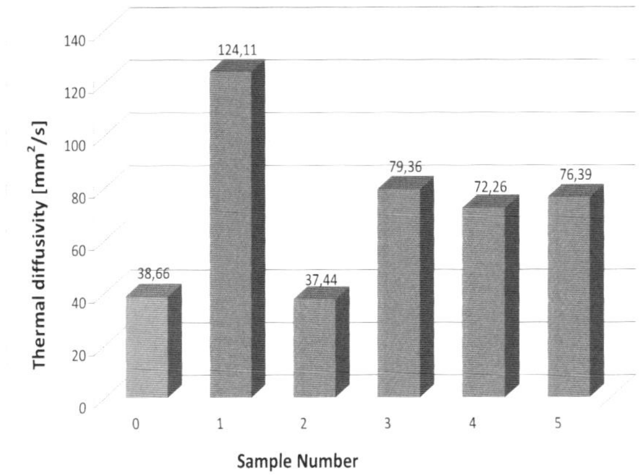

Figure 3. Measurement results of thermal diffusivity [6]

It arises from the measurement analysis that for the sample numbered 1, which was made of the pressed dendritic copper powder, coated with graphene and subjected to another pressing, the gained thermal diffusivity (124.11 mm^2/s) was more then three times as high as the one for the sample numbered 0 which was in 100% made of the same initial powder subjected to pressing (38.66 mm^2/s). For the samples numbered: 3, 4 and 5, which were made of the spherical powder and coated with graphene, another pressing enabled to gain about the two-time increase of the temperature conductivity. The lack of the thermal diffusivity increase was stated for the sample numbered 2 which was not subjected to coating with graphene after the initial pressing.

The results of the abrasive wear measurements were put together in Fig 4.

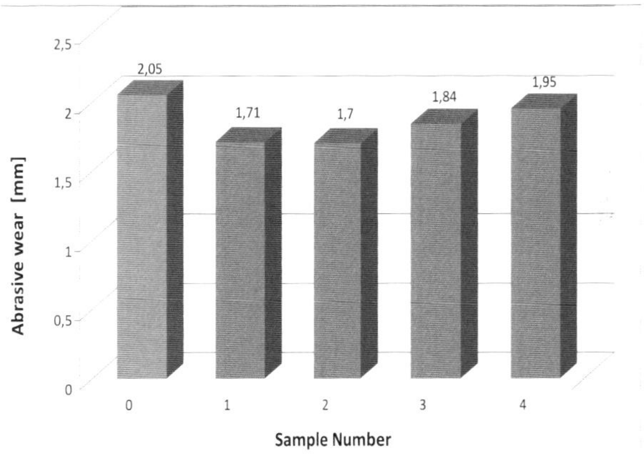

Fig.4. Measurement results of abrasive wear [6]

All the samples with the addition of graphene show the increased abrasive wear resistance while compared with the sample numbered 1 made in 100% of the copper powder. The samples made of the spherical powder showed larger abrasive wear resistance than the samples which were made of the dendritic powder – at the average by approximately 10%.

Composite coatings manufactured in the result of the detonation spraying of the copper powder coated with graphene

The copper powder coated with graphene – Graphene $3D^{IMP}$ was thermally sprayed by means of a detonation method in order to manufacture a composite coating. The detonation spraying is one of the methods of thermal spraying which enables to manufacture the coatings which are characterised by high hardness, low porosity, good adherence and increased wear resistance. The essence of the denotation spraying is the utilization of the explosive combustion energy of a gaseous mixture to heat the particles of the powdered coating material and to give them the specified high kinetic energy [11]. The following spraying parameters were used: oxygen pressure

– 0.06 MPa, propane-butane pressure – 0.001 MPa, nitrogen pressure – 0.008 MPa, detonation frequency – 4 Hz.

The pure copper powder is strongly oxidized during the process of spraying. A large number of oxides, which are included in the coating and are arranged in the bands between the copper grains, cause a significant decrease of the cohesion of the coating and make the connection with the substrate worse (Fig. 5). It is confirmed by the 90^0 angle bending test of the coating performed on the rod 10 mm in diameter. The Cu coating on the Al substrate becomes strongly cracked and on the S235JR steel substrate it becomes cracked and separated from the matrix (Fig. 6).

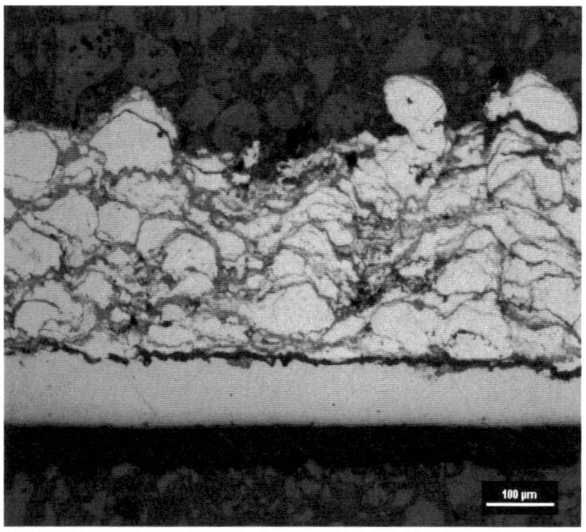

Figure 5: The coating sprayed with the copper powder by means of the detonation method

Figure 6: The Cu powder coating sprayed by means of the detonation method onto the S235JR steel substrate after the 90^0 angle bending test performed on the rod 10 mm in diameter

During the spraying of the copper powder with graphene, the carbon structures protect the particles of the copper powder against the excessive oxidization and thanks to this there are much fewer oxides in this coating than in the sprayed Cu powder coating (Fig. 7). The 90^0 angle bending test of the steel plate and the aluminium plate with the sprayed coating, which was performed on the mandrel 10 mm in diameter, proves that the cohesion of the coating, which was obtained from the copper powder with graphene, is good and that it is well connected with the substrate. This test showed good adherence and good deformability (there are no scratches and no cracks on the bending) of the sprayed coating (Fig. 8).

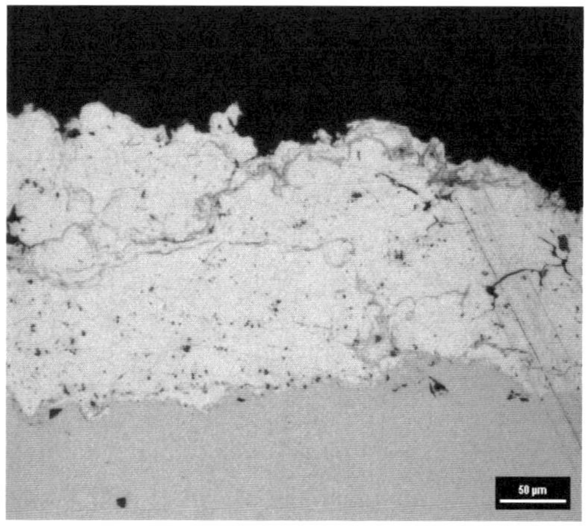

Figure 7: The coating sprayed with the Cu/G powder by means of the denotation method

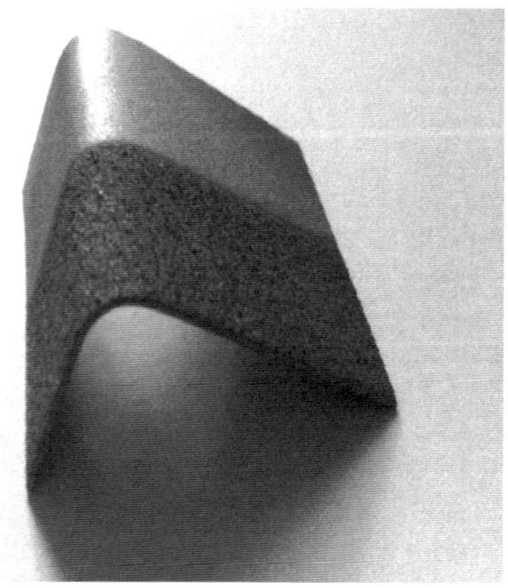

Figure 8: The Cu/G powder coating sprayed by means of the detonation method onto the S235JR steel substrate after the 90^0 angle bending test performed on the rod 10 mm in diameter

Figure 9 illustrates the SEM image of the surface of the Cu/G sprayed coating and figure 10 shows the diffraction pattern from the coat surface. The analysis of the chemical composition showed that there are: 99.92% of Cu, 5.60% C and 1.48% of O in the analysed area.

The fundamental advantage of the Cu-G composite coating compared with the Cu coating is the content and the cohesion of the structure which result from the much fewer number of oxides and from the good connection of the grains among one another. There is a good connection of the coating with the substrate. The average Vickers hardness number of the coating is 101.HV 0.1.

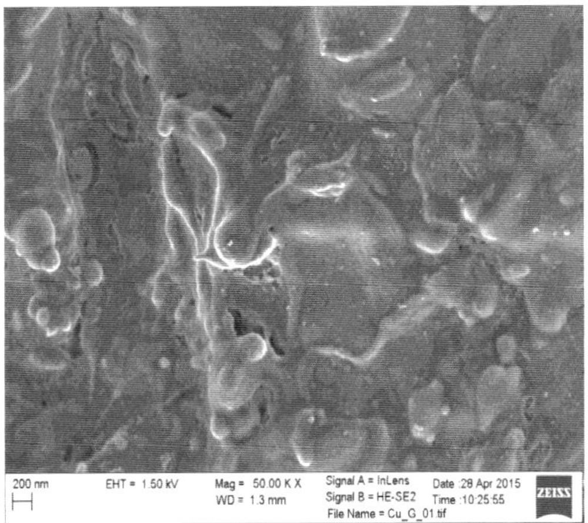

Figure 9: The SEM image of the surface of the Cu/G coating sprayed by means of the detonation method

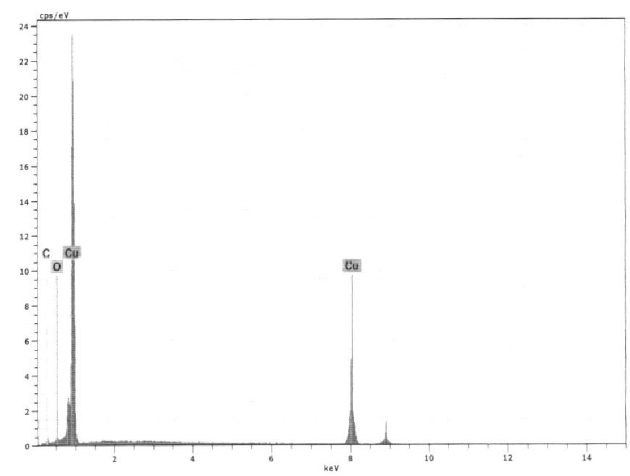

Figure 10: The diffraction pattern from the Cu/G coating sprayed by means of the detonation method

Composite coatings with the metal matrix and the graphene dispersion phase manufactured by means of the chemical and electro-chemical reduction methods

One of the methods of the manufacture of composite materials with the metal matrix and dispersion phases from different materials is the electro-crystallization, the basis of which are the oxidation and reduction reactions [12-15]. In this case the matrix is provided by nano-crystalline nickel deposited by means of the electro-chemical reduction method and the dispersion phase by the graphene flakes (Fig.11). In order to manufacture the composite layers, graphene in the form of the polydispersion powder was used. The EDS analysis results of the chemical composition of the Ni/graphene composite layer, which was manufactured in the bath containing 0.2 g/dm of graphene flakes, prove that graphene is built into the nickel layer. The Ni/graphene composite layer contains: 83.86% of Ni, 12.64% of C and 3.5% of O. The structure of the manufactured Ni/graphene layer is presented in Fig. 12. In case of the composite layer, the big sizes of the graphene flakes are the cause of obtaining the non-uniform thickness layers, because of which the surface of the layers is developed.

Figure 11: SEM image of graphene flakes [13]

Figure 12: SEM image of Ni/graphene layer surface [12]

Table 3 puts together the results of the micro-hardness and the R_a roughness parameter measurements of the surfaces of the manufactured nickel and Ni/graphene composite layers and of the substrate material (the DC01 steel).

Building of the graphene particles into the nickel matrix causes the increase of the R_a roughness parameter of the surfaces of the manufactured layers and increases the matrix material micro-hardness.

Table 3. Micro-hardness and surface roughness of the substrate and of the Ni, Ni/graphene layers [12]

Material	Hardness HV 0,025	Roughness parameter R_a [μm]
DC01 steel	106	0,13
Ni	520	0,20
Ni/ graphene	593	5,73

Conclusions

The properties of composite materials can be moulded by the appropriate choice of components, shape, kind and size of the dispersion phase particles and also the kind of their manufacturing process. The connection of graphene, which is characterised by unique properties and used as a dispersion phase in a composite material, with the metal, polymer or ceramic matrix creates the great possibilities of moulding the properties of these materials. Composite materials with the graphene dispersion phase are characterised by many advantageous properties. Their following qualities should be emphasised: increased hardness and abrasive wear resistance in comparison with the matrix material, significantly better thermal and electric conductivity in comparison with copper and increased corrosion resistance.

Modern composite materials with the graphene dispersion phase, because of a number of advantageous properties, create the possibilities which can be used in different fields of technology, such as for example: military technologies, power engineering, transport, electronics and others. Thanks to its very good current conductivity, graphene may replace silicon very soon and in power engineering it may increase the effectiveness of the conductivity of electricity, which will enable to decrease its loss. Thanks to the fact that graphene has got a high thermal conductivity coefficient, it can impart the properties of good heat abstraction to the corresponding composites. The appropriately selected composites used for building aeroplane construction plating can reduce its weight and as a consequence decrease the use of fuel.

References

[1] Barton J., Niemczyk A., Czaja K., Korach Ł., Sacher-Majewska B.: „Kompozyty, biokompozyty i nanokompozyty polimerowe. Otrzymywanie, skład, właściwości i kierunki zastosowań", Chemik, Vol. 68, (2014), No. 4, pp. 280-287

[2] Surowska B., „Materiały funkcjonalne i złożone w transporcie lotniczym", Eksploatacja i Niezawodność, No. 3 (2008) pp. 20-30

[3] Sobczak J., Wojciechowski S.:, „Współczesne tendencje praktycznego zastosowania kompozytów metalowych", Kompozyty Vol. 2 (2002) No. 3, pp. 24-37

[4] Jaworowska M., „Nowe materiały zapewnią rozwój elektroniki w zastosowaniach specjalistycznych. Elektronika", B2B. Portal informacyjno-biznesowy branży elektronicznej

[5] Hebda M., Łopata A.: Grafen – materiał przyszłości, Czasopismo Techniczne. Mechanika, R. 109, z. 8-M (2012), pp. 45-53

[6] Babul T., Obuchowicz Z., Kowalski S., Błaszczyk J., „Sposób wytwarzania struktur węglowych zawierających grafen na proszkach miedzi z wykorzystaniem obróbki cieplno-chemicznej", Zgłoszenie patentowe nr P.409141, 2014

[7] Obuchowicz Z., Baranowski M., Babul T., Sobczak N., „Kompozyty Cu-Grafen 3D^{IMP}", Inżynieria Powierzchni No. 2 (2015) pp. 10-14

[8] Babul T., Baranowski M., Sobczak N., Homa M., Leśniewski W., „Manufacturing Cu-matrix composites using Cu powder coated with Graphene 3DIMP produced by CarboTermoFluid® process. Journal of Materials Engineering and Performance

[9] Babul T., Trzaska M., Jeleńkowski J., Wojciechowski A., Potencjał Grafenu 3DIMP, Logistyka, 2015, No. 4, pp 2282-2291

[10] Betiuk M., Burzyński K., Jończyk S., Szczepański A., Niedźwiedzki K.: „Wirotest" i „Kulotester" – urządzenia kontrolno-pomiarowe do badań właściwości materiałów, Inżynieria Powierzchni No. 2/2008 pp. 50-56

[11] Babul T.: Podstawy procesu natryskiwania detonacyjnego powłok NiCrBSi i WC/Co. Monografia, Instytut Mechaniki Precyzyjnej, Warszawa, 2011

[12] Cieślak G., Mazurek A., Trzaska M.: Warstwy kompozytowe Ni/grafen wytwarzane metodą redukcji elektrochemicznej. Inżynieria Powierzchni No. 3,(2015) pp. 44-47.

[13] Trzaska M., Cieślak G.: Ni-P/graphene composite layers produced by chemical reduction method, Composites Theory and Practice, Vol.15 No.3 (2015) pp.137-140

[14] Szeptycka B., Gajewska-Midziałek A.: Nickel-graphene composite coatings. Composites Theory and Practice, No.2 (2015) pp.107-111

[15] Trzaska M., Cieślak G.: Powłoki z nowych nanokompozytowych materiałów nikiel/grafen o korzystnych właściwościach eksploatacyjnych. Book of Abstract, 8th International Warsaw Invention Show IWIS 2014, Warsaw, 14-16 October 2014, p.64.

International Federation of Heat Treatment and Surface Engineering 2016
Proceedings of the 23rd IFHTSE Congress
April 18–21, 2016, Savannah, Georgia, USA

Simulation of The Effect of Temperature and Time on The Tempering behavior of Carburized Steels

Yongjin Kim, Xiaoqing Cai, Lei Zhang, Haixuan Yu, Richard D. Sisson, Jr.[*]

Center of Heat Treating Excellence (CHTE), Material Science and Engineering
[1]Worcester Polytechnic Institute, Worcester, MA, USA

Abstract

The heat treating industry needs verified computer simulation tools to predict the carbon concentration profiles and microhardness profiles in carburized steels. Currently tools exist to predict the carbon concentration profiles for many carburization processes including endo-gas and vacuum. The models for the prediction of microhardness profiles as a function of tempering temperatures and times are being developed in this project using experimental results and Hollomon – Jaffe analysis. The experimental results show that the hardness increases with carbon concentration and decreases with an increase in tempering temperature and time. The results will be used to enhance CarbTool© to predict the microhardness profiles of carburized steel as a function of tempering temperature and time.

Introduction

Carburization is a surface hardening process, in which carbon is absorbed and diffused into the base steel. It is a chemical modification of a surface for the part by having a carbon rich atmosphere in a furnace at a high temperature [1]. The difference of the carbon potentials between the atmosphere and steel provides the driving force for absorption and diffusion.

As part of the process control and optimization study of industrial carburizing, Hollomon-Jaffe analysis is often used. Hollomon-Jaffe analysis helps to describe dependences of the properties of tempered steels on the temperature and time of the tempering treatment [2]. This analysis is an effective approach with lower cost and less time so that it can simplify processing of experimental data and reduce the number of experiments. The Hollomon-Jaffe equation is written in the form:

$$HJP = T(C + \log \tau)/1000 \qquad (1)$$

Where HJP is the Hollomon-Jaffe parameter, T is the temperature (Kelvin), τ is the time (seconds), and C is a constant dependent on the material being tempered. The constant C may vary depending on the carbon concentrations between 0.3 and 1.1% and is determined from experimental data.

User Interface of CarbTool©

CarbTool© was developed by CHTE to predict the carbon concentration profile as a function the process parameters. This tool has been experimentally verified [3]. In this paper, AISI 8620 steel was used to determine the effect of temperature and time on the tempering behavior of the carburized steel to enhance CarbTool© to predict the hardness profiles in tempered steel. The carburizing performance was experimentally studied in terms of the microhardness profiles and the carbon concentration profile against the case depth.

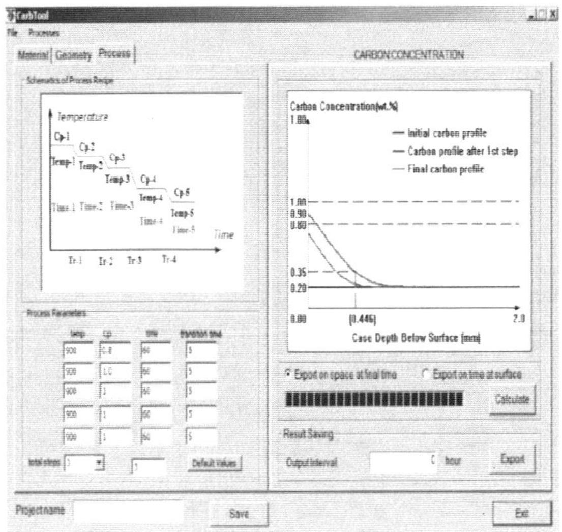

***Figure 1**: The interface of CarbTool©.*

Experimental procedure

In this work, AISI 8620 steel was investigated. The composition as measured by Optical Emission Spectroscopy (OES) is listed in Table 1.

***Table 1**: Alloy composition of AISI 8620 steel in wt. %*

Fe	C	Cr	Ni	Si	Mn	Mo	S	P
Rest	0.2	0.4	0.4	0.15	0.7	0.15	0.04	0.03

Initially, samples were carburized and quenched in oil at Bodycote in Worcester, MA. The carburized and quenched samples were tempered for five times (1hr, 2hrs, 4hrs, 9hrs and 16 hrs) and four temperatures (205°C, 315°C, 425°C ad 595°C). Table 2 shows an example of the sample numbers at one temperature and different times.

***Table 2**: The sample numbers of AISI 8620 at one temperature and different times.*

Temperature (°C)	Time (hour)	AISI 8620
205	1	8434
	2	8452
	4	8428
	9	8446
	16	8440

Then, these samples were placed onto OES to measure the carbon concentrations. The measurements were done to estimate the carbon concentration profile (carbon concentration vs. depth) on the surface of the sample. Carbon concentration measurements were done by grinding to decrease the depth and taken at four different locations on the surface of each layer. Tempered samples were prepared for microhardness testing by cutting, mounting, grinding and polishing. The microhardness measurement of the samples were conducted with a Vickers microhardness tester on a Clark, CM-400 AT using a 200gm load for a dwell time of 10 seconds. The first indentation on the edge of the sample was made and measured and then the following indentations were measured at a distance of 104 µm from the former indentation. Therefore, two lines of indentations were made in which there were twenty five indentations per each line. From obtained data, carbon concentrations and microhardness profiles were plotted and analyzed. Hollomon-Jaffe analysis was used to determine the relations among the hardness and the tempering temperatures and times as a function of carbon content.

Result and Discussion

Carbon Concentrations-Depth Profile Measurements

Optical Emission Spectroscopy analysis was performed to determine the carbon concentration profiles. From the experimental data, the carbon concentrations-depth measurements were determined and plotted. CarbTool© was then used to predict that carbon concentration profile. Figure 2 shows the carbon concentration profile from the experimental data compared to the prediction of CarbTool©. The experiment results correspond with the simulated results very well.

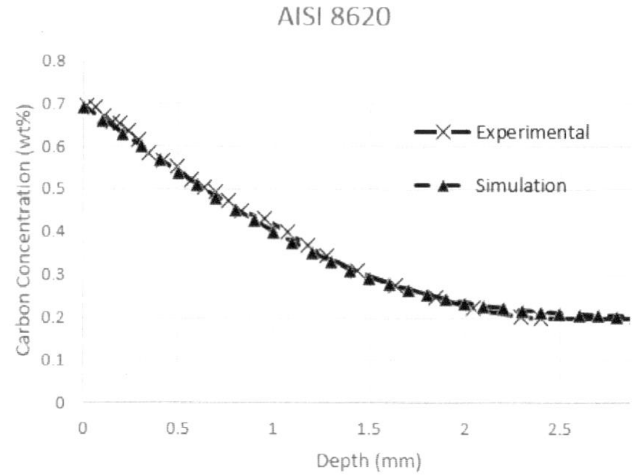

***Figure 2**: Carbon concentration (wt. %) vs. depth (mm) for AISI 8620, experimental and simulation carbon concentration profile.*

Microhardness-Depth Profile Measurements

Microhardness analysis was performed by two lines of indentations made on each sample. Figure 3 shows the microhardness profile of sample 8434 with respect to the depth. It was found that microhardness decreased with an increase in the depth. Regression analysis was performed on the two lines (Run 1 and Run 2) to determine the equations of best-fit polynomial trend. In order to study further investigation of microhardness analysis, two trends were investigates:

1) Varying tempering temperatures with a constant tempering time.

2) Varying tempering times with a constant tempering temperature.

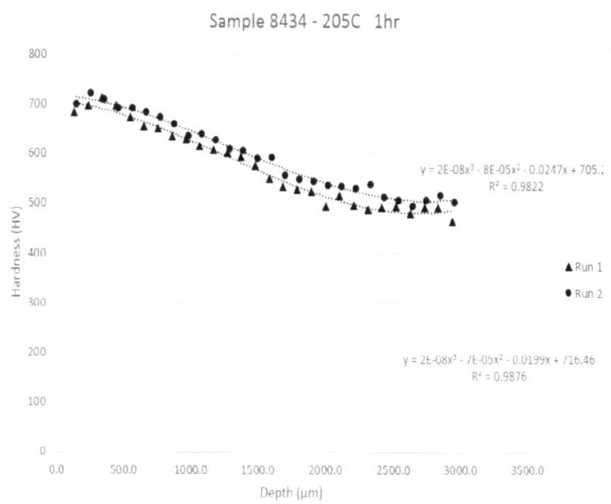

***Figure 3**: Microhardness (HV) vs. depth (µm) for sample 8434 tempered at 205°C 1hour.*

Figure 4 shows the compilation of all the microhardness of AISI 8620 as a function of the depth at a constant time of 1 hour with different tempering temperatures. It also includes the as-quenched microhardness. It was found that hardness decreased with increasing the depth for all the samples of AISI 8620. Hardness also decreased as the temperature increased from 205°C to 595°C. This indicated that tempered samples became relatively soften compared to sample without tempering.

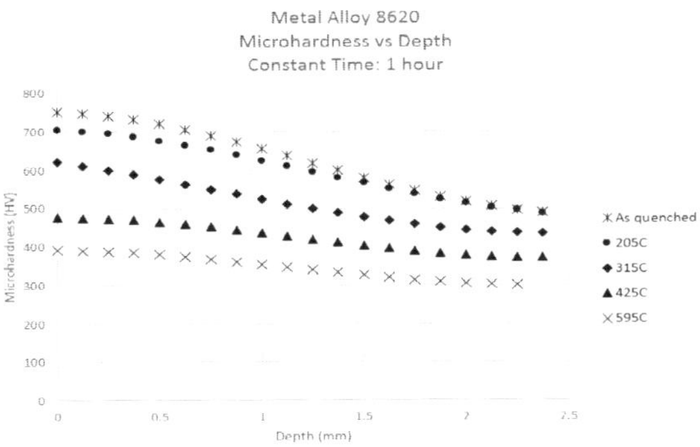

Figure 4: Microhardness (HV) vs. depth (mm) for AISI 8620 at a constant time with different tempering temperatures.

Figure 5 shows the microhardness profiles as a function of the depth for AISI 8620 at a constant temperature with different tempering times. This graph also shows hardness of tempered samples decreased as a function of the depth at a constant temperature of 205°C with different tempering times. In the case of the constant temperature, it can be seen that more and longer tempering time decreased sample hardness. As the time increased from 1 hour of tempering to 16 hours of that, sample hardness decreased. Moreover, Figure 4 shows a wide range of the hardness profiles, while Figure 5 showed a relatively smaller range of the hardness profiles. These results indicate that the effect of tempering temperature is more important than that of tempering time.

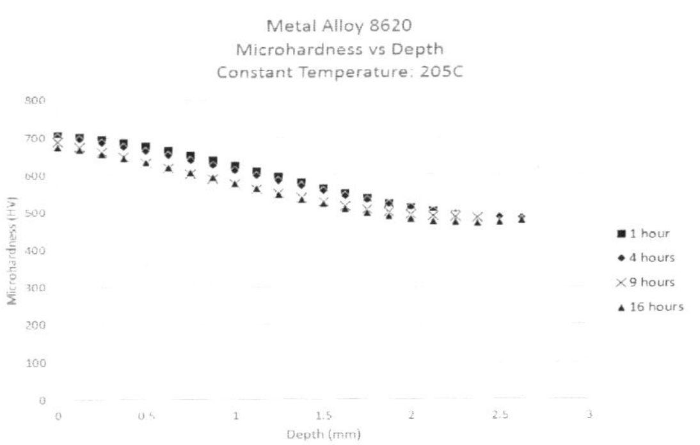

Figure 5: Microhardness (HV) vs. depth (mm) for AISI 8620 at a constant temperature with different tempering times.

Carbon Concentrations-Microhardness-Depth Profile Measurements

The carbon concentration and microhardness profiles of tempered sample were plotted with respect to the depth. The results obtained from testing of sample 8434 at 205°C for 1 hour were presented in Figure 6. It was found that hardness and carbon concentration of the tempered sample decreased with an increase in the depth. This was performed to show there was a correlation between the carbon concentration and the microhardness with respect to the depth. In other words, an arbitrary depth value can be used to estimate the microhardness to be plotted as a function of the carbon concentration. Regression analysis was also performed to determine the equations of best-fit polynomial trend. The microhardness with respect to the carbon concentration was plotted with the help of these equations. Figure 7 shows the microhardness of 8434 tempered sample to be plotted against carbon concentration.

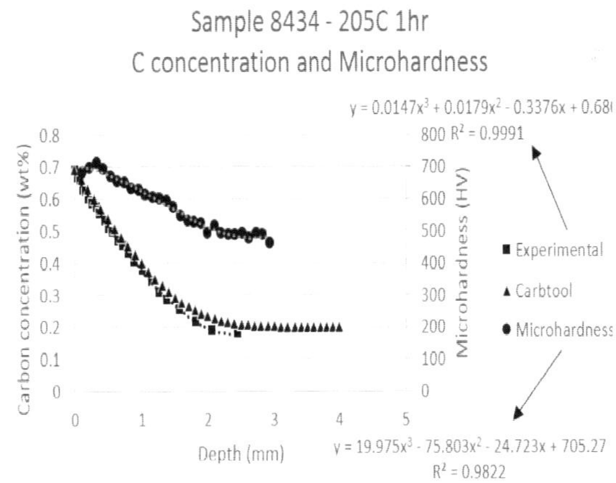

Figure 6: Carbon concentration (wt.%) and microhardness (HV) vs. depth (mm) for samples 8434 at 205°C 1hr

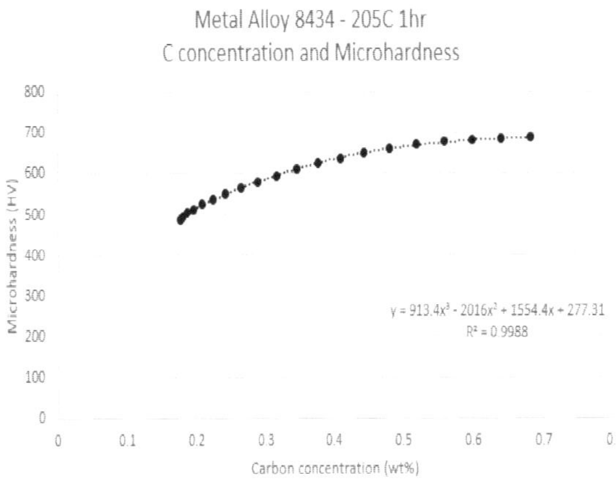

Figure 7: Microhardness (HV) vs. carbon concentration (wt.%) for sample 8434 at 205°C for 1hr.

On the basis of the results of microhardness-carbon concentration profiles, the compilation of all microhardness as a function of carbon concentration at a constant time with varying tempering temperatures was plotted. Figure 8 shows 8620 tempered samples hardness profiles against the carbon concentration at a constant time of 1 hour with different tempering temperatures. It was seen that hardness of the tempered samples increased with a decrease in tempering temperature. Moreover, hardness of that increased with an increase in the carbon concentration.

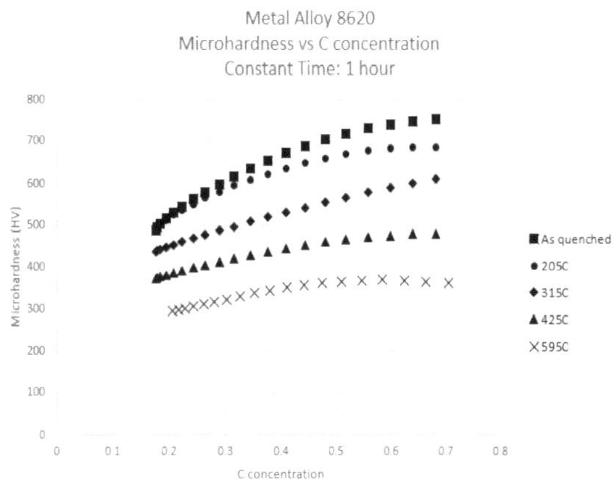

***Figure 8**: Microhardness (HV) vs. carbon concentration (wt. %) for AISI 8620 at a constant tempering time of 1 hour with varying tempering temperatures.*

Figure 9 shows 8620 tempered samples hardness profiles as a function of the carbon concentration at a constant temperature with different tempering times. At a constant tempering temperature of 205°C with different tempering times, the tempered sample hardness decreased with an increase in tempering time. Furthermore, it was found that hardness decreased as carbon concentration decreased. As seen in Figure 8 and Figure 9, it can be inferred that the decrease in hardness with either increasing tempering temperatures or times is associated with the decomposition of Martensite and the mobility of carbon atoms. In other words, as tempering time and temperature increase, Martensite decomposes into a combination of ferrite and cementite so that it can lead to a decrease in the volume of Martensite. As a result, this can result in a decrease in hardness.

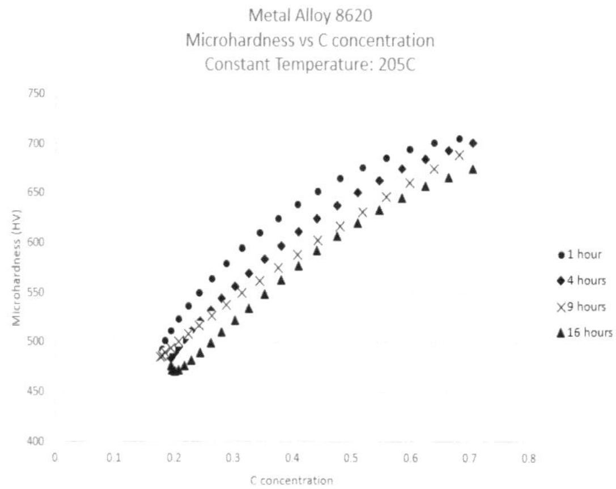

***Figure 9**: Microhardness (HV) vs. carbon concentration (wt. %) for AISI 8620 at a constant tempering temperature of 205°C with varying tempering times.*

Figure 10 shows a photograph of sample 8434 and the carbon concentration and microhardness of tempered sample were plotted as a function of the depth on the picture. These two profiles were plotted corresponding to the indentations that were made from the adjacent edges of the sample to subsequent indentations made at a distance of approximately 104 μm from the previous indentation.

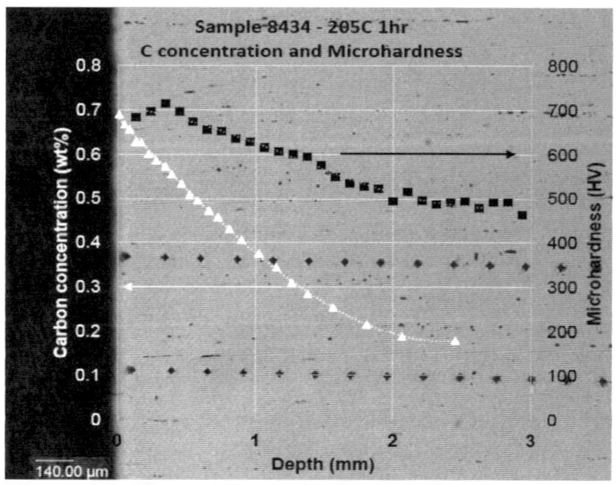

***Figure 10**: Photomicrograph of sample 8434 sample at a tempering temperature of 205°C with 1 hour.*

Hollomon-Jaffe Parameter and Constant C for CarbTool©

Microhardness was plotted against the $\log \tau$ in hours at different carbon concentrations (0.3, 0.5 and 0.7 wt.%). Figure 11 shows the hardness vs. the $\log \tau$ at 0.7 wt.% with different tempering temperatures. It was found that there were negative linear correlations.

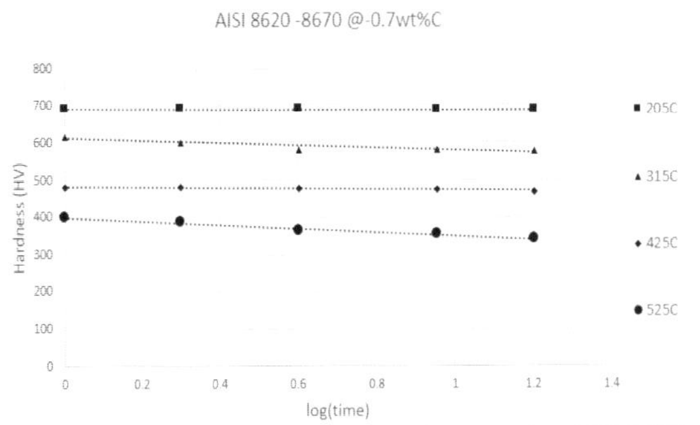

Figure 11: Microhardness (HV) vs. log time for AISI 8670.

With the microhardness measurements and the Hollomon-Jaffe temperature-time parameter, the constant C was determined for three selected carbon concentrations. The use of this value of C can be enhanced the existing CarbTool© for the process of tempering. The calculation of C can be determined with the help of equation (1). When the constant C has one constant value, then the same hardness will be determined: 1) at lower tempering temperature but higher tempering time or 2) at higher temperature but lower tempering time [4]. The constant C can also be estimated with the help of regression analysis.

G.A Thomas *et al.* have provided a detailed analysis of the origin and the determination of C in the Hollomon-Jaffe analysis [5]. Figure 12 shows a plot of microhardness versus the Hollomon-Jaffe parameter at 0.3% carbon. Figure 13 shows microhardness with respect to HJP for the three carbon concentrations.

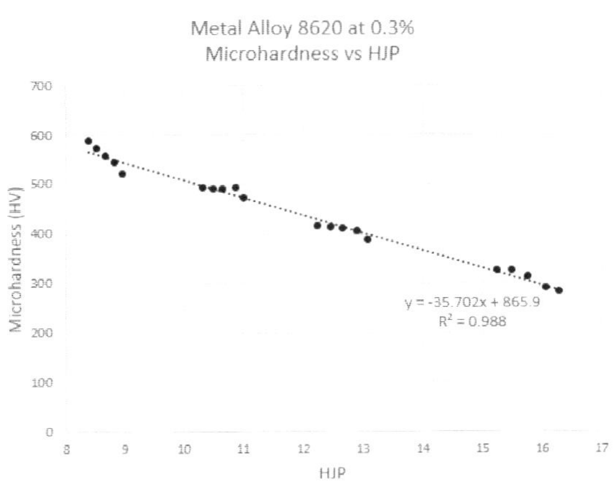

Figure 12: Microhardness (HV) vs. HJP for AISI 8620.

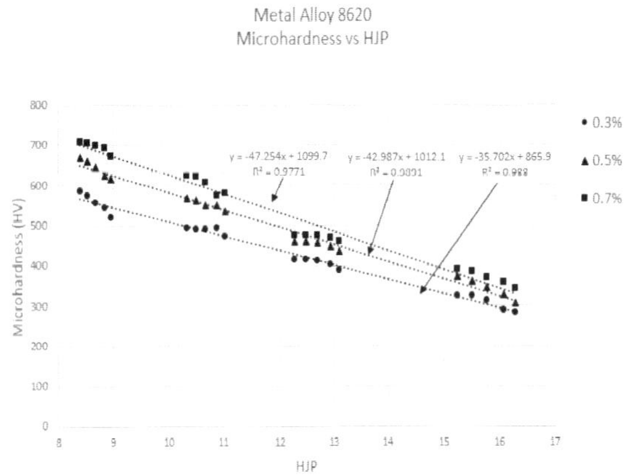

Figure 13: Microhardness (HV) vs. HJP for AISI 8620 for the three carbon concentrations.

To determine the value of C in the Hollomon parameter, arbitrary C constants were selected from 7 to 14 and the correlation coefficient (R^2) were calculated. Table 3 presents the results of the C constants corresponding to the R^2 for the three carbon compositions. As seen in Table 3, it was found that the R^2 was nearly constant for the values of C from 7 to 14 for each concentration. Therefore, the values of C for each carbon concentration did not change significantly the R^2 in the Hollomon-Jaffe parameter on the basis of all the experimental data. Grange and Baughman have also found that the value of c did not have a large affect the result [6].

Table 3: The results of the C constants for the three carbon concentrations.

0.30%		0.50%		0.70%	
C constant	R^2	C constant	R^2	C constant	R^2
7	0.9873	7	0.9883	7	0.9725
8	0.9882	8	0.9892	8	0.9743
9	0.9887	9	0.9897	9	0.9754
10	0.9888	10	0.9899	10	0.9762
11	0.9888	11	0.9898	11	0.9767
12	0.9886	12	0.9896	12	0.9769
13	0.9883	13	0.9894	13	0.9771
14	0.9880	14	0.9891	14	0.9771

Conclusions

This paper investigated the hardness profiles of AISI carburized 8620 as a function of tempering parameters. Hollomon-Jaffe analysis were used to calculate the constant C so that models for the prediction of microhardness profiles could be developed. The results show that microhardness of AISI 8620 decreased with an increase in the depth. As tempering temperature increased at a constant temperature, the hardness decreased. The longer is the tempering time at a constant temperature, the lower is the hardness. The analysis of the experimental data using the Hollomon-Jaffe parameter was done using different tempering temperature and time to determine the constant C. The constant C was estimated with the help of regression analysis. It was concluded that the R^2 was found to be a nearly constant as C was systematically varied. In the future, CarbTool© will be enhanced to include the calculation of the hardness profile.

Acknowledgements

We would like to thank CHTE for sponsoring the project. The support of CHTE at Worcester Polytechnic Institute and Body cote in Worcester, MA for carburizing and quenching the samples is gratefully acknowledged.

References

[1] Davis, Joseph R., ed. *Surface hardening of steels: understanding the basics*. ASM international, 2002.

[2] Hollomon, J. H., and L. D. Jaffe. "Time-temperature relations in tempering steel." *Trans. AIME* 162 (1945): 223-249.

[3] Ferguson, B. L., et al. "Intelligent Heat Treating: Simulation of Carburization Process." *Heat Treating 2011: Proceedings of the 26th Conference*. ASM International, 2011.

[4] Janjušević, Zoran, et al. "The investigation of applicability of the Hollomon-Jaffe equation on tempering the HSLA steel." *Chemical Industry and Chemical Engineering Quarterly* 15.3 (2009): 131-136.

[5] Thomas, G. A., et al. "Time–temperature equivalence in martensite tempering." *Proceedings of International Conference on Martensitic Transformations (ICOMAT-08)*. TMS, Warrendale, PA, USA. 2010.

[6] Grange, R. A., and R. W. Baughman. "Hardness of tempered martensite in carbon and low alloy steels." *Trans. Am. Soc. Met* 48 (1956): 165-197.

International Federation of Heat Treatment and Surface Engineering 2016
Proceedings of the 23rd IFHTSE Congress
April 18–21, 2016, Savannah, Georgia, USA

Fluidized-bed carburizing in chemically active powders of C22 steel

Aleksander Ciski, Tomasz Babul, Konrad Lankiewicz
Institute of Precision Mechanics, Warsaw, Poland
aleksander.ciski@imp.edu.pl, tel.: +48 22 56 02 942

Jana Nižňanská, Pavel Šuchmann
COMTES FHT a.s., Dobřany, Czech Republic
pavel.suchmann@comtesfht.cz, tel.: +420 377 197 305

Abstract

The paper presents the new method of fluidized-bed carburizing in chemically active micropowders. Basic principles of the thermo-chemical treatment method are described. Discussed in the article carburizing process combines the most important advantages of the pack and atmospheric fluidized-bed methods. The source of carbon in this case is based on a special mixture of active carbon, activator and chemically inert bed particles, which is also a heat accumulator ensuring the uniform temperature distribution in the treated part.
Results of the exemplary carburizing processes carried out at temperature of 930°C for 4 and 6.5 hours are presented. Samples of DIN C22 (AISI 1022M) unalloyed carbon steel were carburized, quenched in oil (after cooling to hardening temperature of 870°C) and tempered at 180°C for 2 hours. Two types of carburizing mixtures with different percentage of active carbon were used. Performed carburizing processes allowed forming of layers with surface hardness in the range of 730÷770 HV. Cross-section hardness measurements showed that, depending on the active carbon content, it is possible to obtain effective carburized layer thickness of about 0.8 to 1.1 mm. Resistance to wear of the carburized steel was characterized by 3 rollers-cone method. Carried out friction test demonstrated the superior tribological properties of the hardened layers. Metallographic observations proved the correct carburized layer structure consisting of lath martensite and retained austenite. It was stated that the obtained properties of carburized layers were comparable with properties of layers formed using conventional gas carburizing method.

Introduction

The carburizing of steel is used in the machine industry in order to obtain unique load-carrying capability connected with increased hardness, tribological and fatigue properties, impossible to be achieved by any other thermo-chemical treatment. Carburizing is commonly subjected to the parts such as gears, toothed shafts or pins. High resistance to wear properties of machine parts accompanied with their high fatigue strength, are especially important for parts subjected to heavy and varying loads.

The reasons for the increase of the properties of parts subjected to carburizing can be understood by observing the resulting steel microstructure. The subsurface diffusion zone is responsible for tribological properties (friction wear, resistance to galling) and fatigue strength of parts. Core of the material properties differ significantly from properties of the diffusion zone and determine the overall strength of the material.
Carburizing process can be carried out in several ways, including carburizing media like powders (pack method), salt baths, gaseous atmospheres (also in fluidized bed) or in so called vacuum/low pressure processes. Carburizing is the most widely used process of thermo-chemical treatment but still there are sought the new technological solutions for this kind of treatment, including, for example, low pressure/vacuum methods of carburizing described in [1, 2, 3].
The most popular process is carburizing in the gaseous atmospheres containing endothermic gas. In the gaseous processes the carbon source is carbon bearing gas (enrichment gas). Despite the widespread use of the gaseous atmosphere method, this technique is dangerous, both for the employees and natural environment, due to the produced exhaust gases and explosion risk of flammable atmosphere in the absence of appropriate precautions behavior. The method of carburizing in salt baths is used with a sodium or potassium cyanide solutions. These solutions represent a risk to human health and life because of their high toxicity to living organisms. Because of that, the salt baths carburizing technology is practically not used in the industry nowadays. Carburizing method using pack-method is devoid of above mentioned drawbacks, however this method is inconvenient to handle. Gaining in popularity vacuum/low pressure carburizing method requires very advanced equipment consisting of a vacuum chamber, pumping units, processes gas delivery systems, as well as complex process control systems.
Discussed in this article fluidized bed carburizing method in chemically active powder mixtures combines the most important advantages of pack method with fluidized bed atmospheric technique, eliminating the many disadvantages of these processes.
This technology does not use flammable gases and because of that thermo-chemical treatment is safe and economically efficient. The source of carbon in this case is based on a special mixture composed on the basis of active carbon, activator and chemically inert bed particles, which are also a

heat accumulator, which ensure uniform heat distribution of charge.

In the proposed method for the bed fluidization are used vibrations resulted from the forced mechanical vibrations which are subjected to furnace retort. In case of carburizing vibrations are used only during the stage of loading and removing of the charge from the soaking chamber. In this method retort with fluidized bed is equipped with electro-vibrator and the whole system is mounted elastically in the body of the furnace [4].

Because of vibrations, fluidized bed has similar characteristics to those obtained by introducing fluidization gas into it. Similar to gas, vibrations make that the fluidized bed moves in a furnace retort, heats up uniformly and is in direct contact with the entire surface of the workpiece, regardless of its shape [5]. The advantage of this technique, compared to the already used conventional methods in thermo-chemical treatments in the retort, continuous, vacuum and others furnaces, is that the workpiece can be removed and placed in the furnace at any time. This means that in one furnace can be carried out independently in parallel processing multiple parts treatments that require different processing time by inserting and removing them, depending on the technological requirements, without waiting for the completion of the processes carried out on other elements. The above technology, depending on the used fluidized bed mixture, allows also carrying out of nitrocarburizing [6] or nitriding [7] processes.

This article present the properties of DIN C22 (AISI 1022M) plain carbon steel subjected to carburizing by a new fluidized bed in chemically active powder methods.

Experimental

Samples of DIN C22 unalloyed carbon steel were carburized, quenched in oil (after cooling to hardening temperature of 870°C) and tempered at 180°C for 2 hours. Two types of carburizing mixtures with different percentage of active carbon were used. Mixture designated in the paper as a "low carbon content" was a mixture intended for use for parts requiring optimum combination of wear and fatigue resistance. Mixture named "high carbon content" had increased (doubled) active carbon contend in order to obtain high resistance to wear of steel.

As a reference mode a conventional gas carburizing process in the atmosphere of endogas enriched with earth gas was carried out (carbon potential Cp=0.8%). Heat and thermo-chemical modes and their processes parameters are given in table 1.

Table 1: Modes of heat and thermo-chemical treatment

Carburizing	Quench	Temper
Low carbon content – 930°C/4 h	870°C/oil	180°C
Low carbon content – 930°C/6.5 h	870°C/oil	180°C
High carbon content – 930°C/4 h	870°C/oil	180°C
Gas carburizing – 930°C/6.5 h	870°C/oil	180°C

Fluidized bed carburizing processes were carried out in a fluidized furnace FP700 with retort diameter ø110mm with attached electric vibrator type BM110/10, which was forcing mechanical vibrations of the circular type with amplitude approx. 4 mm (motor power 140 W and rotation speed 1500 rpm). Retort with vibration motor was placed in furnace axis and was based on a horizontal furnace plate with use of three brackets with coil springs, whose task was to suppress the vibration transmission to the body of the furnace (Fig. 1). Vibrations were used only at a stages of charge loading or removing from the retort.

Figure 1: View of the FP-700 furnace for fluidized bed thermo-chemical treatment in chemically active powders

Surface hardness measurements were made using Struers DuraScan-70 hardness tester under load of 5 kG.

Hardness profiles was were determined by laboratory hardness tester Struers DuraScan. Hardness profile was measured from the surface to the centre. Hardness HV0.1 was measured according to ČSN EN ISO 6507-1 (420374) Metallic materials – Vickers hardness test. Case Hardness Depth (CHD) was evaluated by samples after carburizing as a distance from surface to point where hardness is equal to 550 HV (EN ISO 2639).

The metallographic sections of samples of steel after carburizing underwent standard preparation procedure consisting of grinding followed by polishing. For microstructure observation the surfaces were etched by Nital solution. Metallographic observations were performed using Carl Zeiss light microscope Observer.Z1m.

The X-ray diffraction patterns measured at surfaces of carburized samples were recorded with aid of D8 ADVANCE X-Ray diffractometer with Co-Kα radiation (λ=0,179 [nm]).

Test of resistance to wear was carried out in a 3 rollers-cone system. Test was performed with a continuous lubrication of samples using Lux-10 oil, at a constant speed of counterspecimen rotating at a rate of 526 rpm and a constant unit pressure of 200 MPa. The depth of the wear, indicating the linear wear, was determined by measurements of

diameters of the ellipses formed at the surface of each of the worn out rollers. Total friction time test was 100 minutes, wear marks were measured after every 10 minutes. After each measurement the load was increased in order to achieve constant unit pressure at the worn out surface. Conical counterspecimen was made of 41Cr4 steel toughened to hardness 30 HRC and ground to a roughness Ra = 0.32 micrometers.
The impact strength measurements were performed with Charpy method using samples with a length of 55 mm, square cross section with a side of 10 mm, with U-shape notch depth 2 mm and the notch bottom fillet radius 1 mm. The test was performed on the WPM impact hammer with the pendulum initial energy of 50 J.

Results and Discussion

Microstructure
Metallographic observations of carburized layers formed in a process carried out in powders containing low carbon content showed the correct microstructure of the carburized layer, composed of lath martensite and retained austenite (Fig 2, 3). There were no presence of harmful grain-boundary carbide nets observed.
Precipitations in a form of large carbides with diameter of approx. 5 μm were observed as a result of carburizing process performed in high carbon content mixture (Fig. 4).

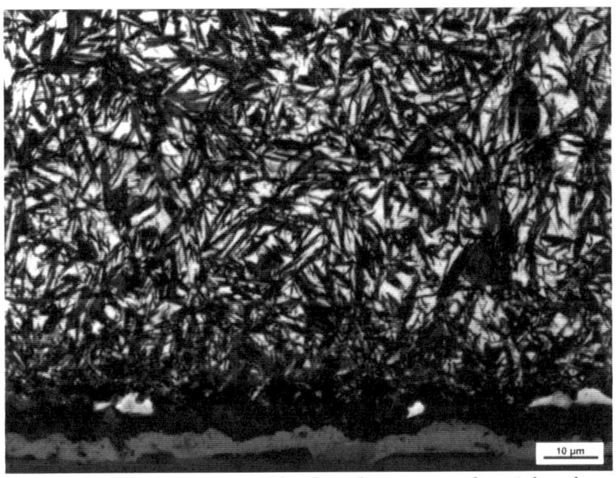

Figure 2: Microstructure of subsurface zone after 4 h – low carbon content carburizing process

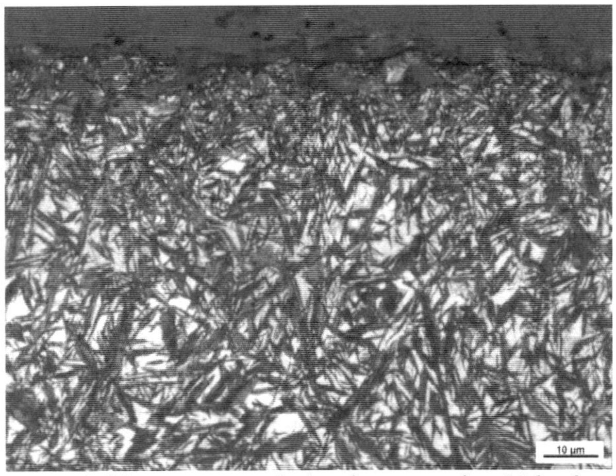

Figure 3: Microstructure of subsurface zone after 6.5 h – low carbon content carburizing process

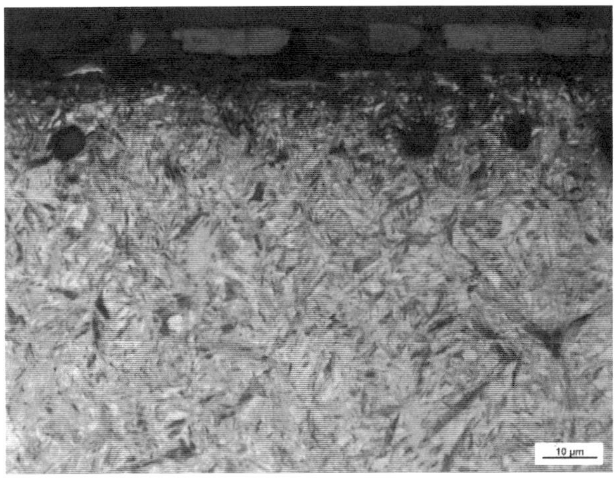

Figure 4: Microstructure of subsurface zone after 4 h – high carbon content carburizing process

The microscopic observations were supplemented with X-ray diffraction (XRD) analyses. While diffraction patterns obtained after carburizing in low carbon content mixtures revealed the diffraction peaks derived from martensite and austenite (Fig. 5), diffraction patterns after carburizing in high carbon content mixture revealed additional peaks coming from the Fe_3C cementite phase (Fig. 6). Large sub-surface precipitations of carbide phase may be unfavorable from the point of view of fatigue strength and cracking performance, so the additional tests of mixtures containing lowered content of active carbon should be performed.

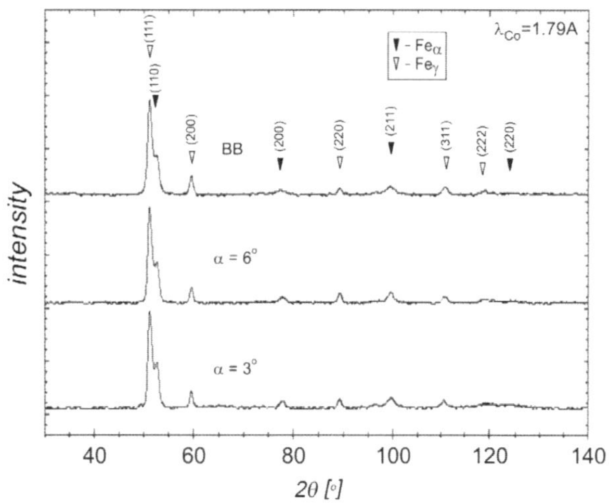

Figure 5: X-ray diffraction scans measured in Bragg–Brentano and SKP geometry for 6.5 h carburizing process in low carbon content mixture

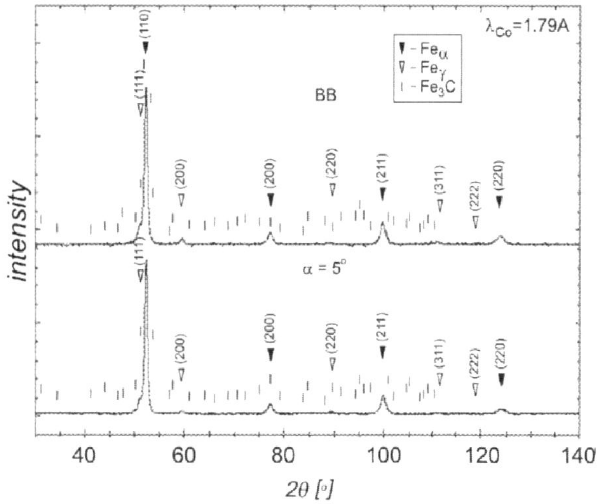

Figure 6: X-ray diffraction scans measured in Bragg–Brentano and SKP geometry for 4 h carburizing process in high carbon content mixture

Surface and cross section hardness

Performed carburizing processes allowed forming of layers with surface hardness in the range of 730÷770 HV5 (Fig. 7). Surprisingly, the highest hardness was reached by samples subjected to 4 hours carburizing in low carbon content mixture (766.4 HV5) while the lowest hardness was obtained after the 6.5 hours process in low carbon content mixture (729.6 HV5). Carburizing process in mixture with high carbon content allowed to obtain hardness of 749.6 HV5.

Lower hardness after longer carburizing process may be connected with lower saturating capacity of carburizing mixture which changes during the process time as a result of the depletion of active carbon and activator content.

Obtained surface hardness after carburizing in powder mixtures media is comparable with hardness obtained after conventional gaseous carburizing process (741.2 HV5).

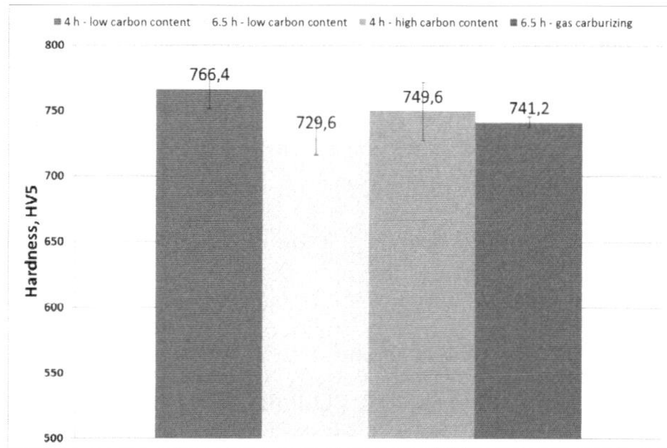

Figure 7: Surface hardness of samples subjected to different modes of carburizing

Cross-section hardness measurements showed that, depending on the active carbon content and duration of the process, it is possible to obtain effective carburized layer thickness of about 0.8 to 1.1 mm (Fig. 8).

Carburized layer with the highest thickness of approx. 1.1 mm was obtained after the process carried out for 6.5 hours in low carbon content mixture of powders. The lowest case depth (0.8 mm) was formed in a process carried out for 4 hours in high carbon content mixture. Despite of that, after this mode of treatment subsurface zone of steel had the highest hardness of about 1000 HV, while in case of other variants the hardness of subsurface zone was much lower (860÷875 HV). After 4 hours carburizing process in low carbon content mixture the effective case depth was about 0.9 mm and had the similar profile course as in the case of 6.5 hours process. Obtained case depths were slightly lower than the depths obtained after conventional carburizing.

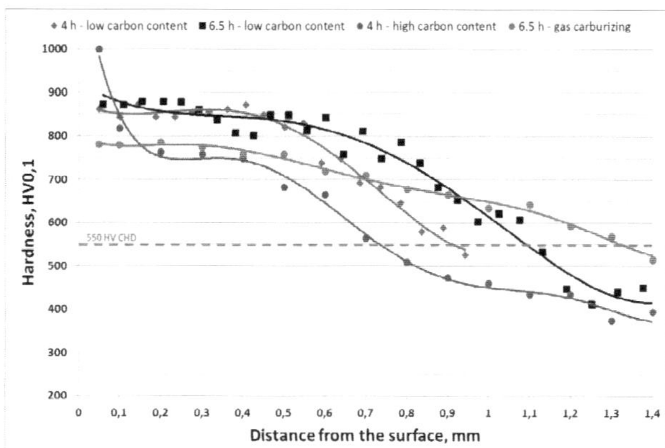

Figure 8: Distribution of hardness profile for processes of carburizing in mixtures containing different carbon content

Wear test

Three rollers-cone wear test allowed to select the most suitable from the point of view of wear resistance mode of thermo-chemical treatment. Courses of linear wear of samples subjected to conventional carburizing and chemically active fluidized bed carburizing modes are shown in figure 9.

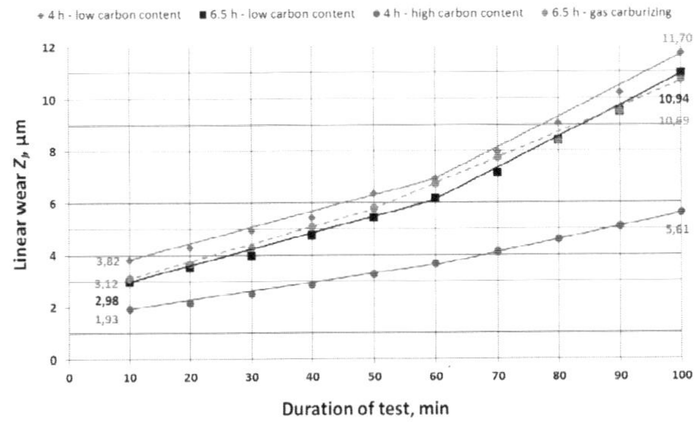

Figure 9: Wear courses of samples subjected to different modes of carburizing

All case carburized samples were characterized by the wear with a regular course, manifesting the occurrence of steady friction state in the entire course of wear, except running-in period.

Wear test demonstrated the superior tribological properties of the hardened layer formed in a process carried out in high active carbon content. For that mode of treatment linear wear after 100 minutes of test was 5.61 µm, while after the same duration of carburizing and lower carbon content the linear wear was above two times higher (11.70 µm). Conventional gas carburizing or carburizing in low carbon content powders mixture, carried out for 6.5 hours, had similar (insignificantly lower) wear rate as in case of 4 hours low carbon mixture carburizing.

The resulting low wear rate of samples subjected to thermo-chemical treatment in mixture of powder with a higher content of carbon is probably associated with achieved high carbon content in the subsurface zone after carburizing and formation of hard high-carbon martensite after quenching.

Impact strength

Impact strength test revealed much lower level of energy absorbed during fracture of samples subjected to 6.5 hours low carbon content carburizing process than other processes of carburizing in chemically active fluidized bed (Fig. 10). If we consider quite significant scattering of results data, the similar energy levels to break the sample were also absorbed in case of samples subjected to conventional carburizing process.

Samples subjected to 4 hours carburizing processes carried out in low and high carbon content powder mixtures had significantly higher values of impact strength, what may be connected with presence of thinner carburized layers.

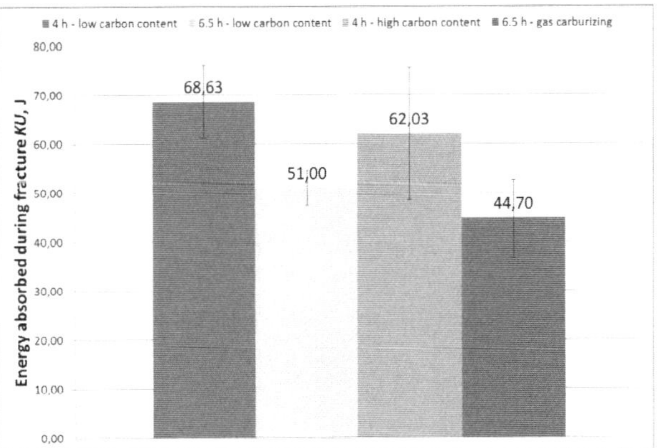

Figure 10: Impact strength of samples subjected to different modes of carburizing

Summary and Conclusions

It was stated that the obtained properties of carburized layers formed in processes carried out in chemically active fluidized bed were comparable with properties of layers formed using conventional gas carburizing method. All samples were characterized with proper surface hardness level, convenient hardness distribution profile and advantageous impact strength.

Based on the analysis of characteristics of layers formed during carburizing process in powder mixtures containing lower content of carbon, it can be said that the new method of thermo-chemical treatment allows obtaining a layers with correct structure, i.e. obtaining of diffusion zone with a suitable thickness and free of unfavorable carbide nets precipitations. Such process can be used as a replacement for the conventional carburizing with $C_p = 0.8$ %, carried out in gaseous media.

Despite of presence of precipitations of large carbides at subsurface zone of carburized layer, samples after carburizing in high carbon content powders mixture revealed very high resistance to wear. The wear resistance of samples measured under loading conditions of 200 MPa was about 2 times higher than in case of samples after conventional carburizing or other two modes of thermo-chemical treatment in chemically active fluidized bed. Further research should focus on the optimization of the carburizing powder mixture in order to avoid precipitation of undesirable carbide phase.

Discussed method seems to be an interesting alternative for bath, gas and low pressure/vacuum carburizing of parts requiring increased wear resistance properties. Considered processes can be used for carburizing of the parts of machines including gears, cams, rods, levers, axles, pivots, etc.

The obtained results confirm the possibility of implementation of the new type of carburizing in chemically active fluidized bed in the industrial practice, especially in small enterprises. The analyzed carburizing method offers a number of advantages and benefits for its use in the industrial practice, in comparison with other thermo-chemical treatment techniques. The most important advantage is an easy operation of the furnace and the lack of liquid or gaseous technological media, because for a proper treatment only the powder mixture with specified chemical composition is necessary. Other benefits are the low operating cost and environmental compatibility of the new fluidized bed technique.

Acknowledgments

This research was supported by Eureka Programme (Eureka E!7163 "TAMPFLU" Project).

References

[1] Stratton P. F., Bruce S., Cheetham V. "Low-pressure carburizing systems: A review of current technology", *BHM Berg- und Hüttenmännische Monatshefte*, Volume 151, Issue 11 (2006), pp 451-456.

[2] Kula P., Pietrasik R., Dybowski K. „Vacuum carburizing—process optimization", *Journal of Materials Processing Technology*, 164–165 (2005) 876–881

[3] http://www.amphtt.com/uploads/media/ULOTKA-email-PIRO-CARB-ENG.pdf

[4] N.G. Kucharieva, T. Babul, Z. Obuchowicz, A. Nakonieczny, Patent application no P-394532 12.04.2011 (in Polish).

[5] Babul T., Ciski A., Oleszczak P., "The New Fluidized Bed Thermo-Chemical Treatment in Chemically Active Powders – CFD Analysis of the Method", *Advanced Materials Research*, 902 (2014), 82-85.

[6] Ciski A., Babul T., Lankiewicz K., Obuchowicz Z., Šuchmann P., "New method of nitrocarburizing of X37CrMoV5-1 hot working tool steel in chemically active fluidized bed", *Journal of Achievements in Materials and Manufacturing Engineering*, 62/2 (2014) 81-87.

[7] Lankiewicz K., Ciski A., Babul T., Nižňanská J., "Fluidized bed nitriding in chemically active powders of HS 6-5-2 steel", *Proc European Conference on Heat Treatment 2015 & 22nd IFHTSE Congress Heat Treatment and Surface Engineering From tradition to innovation*, Venice, Italy, May 2015, electronic version.

International Federation of Heat Treatment and Surface Engineering 2016
Proceedings of the 23rd IFHTSE Congress
April 18–21, 2016, Savannah, Georgia, USA

An enhancement to the low pressure carburizing simulation

Lei Zhang, Richard D. Sisson Jr.
Center for Heat Treating Excellence, Worcester Polytechnic Institute,
100 Institute Road, Worcester, Massachusetts, 01609, U.S.A

Abstract

CarbTool© has been an effective tool for the simulation of both gas and low pressure carburizing (LPC) processes by industry. However, in some cases the LPC model prediction does not agree very well with the experimental results. The average carbon flux was initially used as the boundary condition which was calculated from experimental data. This boundary condition is modified to improve the LPC process simulation. An improvement using a carbon potential for LPC process is proposed. Both carbon potential and mass transfer coefficient will be calculated. Data will be used to optimize the modeling of the CarbTool ©. Experimental verification will also be discussed.

Introduction

CarbTool© is an effective tool developed by Center of Heat Treatment Excellence to predict the carbon concentration profile of steels [1-3]. The software can be used for the prediction of both gas and vacuum carburizing process. So far the materials in the data base includes 10XX, 48XX, 51XX, 86XX, 93XX, etc. In most cases, the simulation fits the experimental results pretty well. However, the prediction does not work well with 9310 steel for vacuum carburizing process. The vacuum carburizing process for 9310 steel was designed and compared with the experimental result. The experiment parameters are presented in Table 1. The Figure 1 shows the concentration profiles.

Table 1: The recipe for 9310 vacuum carburizing prcess

Specification: **Surface carbon: 0.70 ± 0.05 case depth: 0.889mm**					
Carbon flux (g/cm2/s)	1st Boost (min)	1st Diffuse (min)	2nd Boost (min)	2nd Diffuse (min)	Cooling (min)
1.63E-06	27	50	8.5	112	50

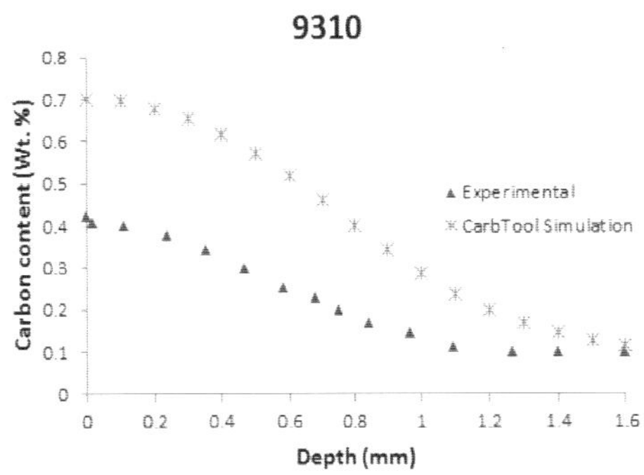

Figure 1: Carbon concentration profile for 9310 boost and diffusion vacuum carburizing process

In the simulation process, the carbon flux is calculated with the fitting of the experimental results. In Figure 1, with the input of carbon flux, 1.63E-06, the concentration profile is not closed to the experimental results. Carbon flux is the boundary condition with CarbTool ©. It is sensitive to the gas temperature, pressure and the surface condition. In our testing, the cyclohexane is used. The gas temperature and pressure remains constant. Besides, the surface of the steel are well cleaned.

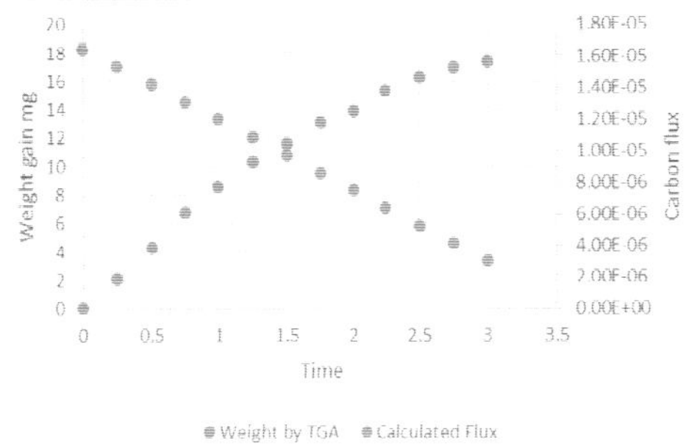

Figure 2: The weight gain during vacuum carburizing process

The weight gain has been studied by the thermalgravic analysis. It can provide us the data for the flux during the first 3 min boost step. The calculated carbon flux has presented that the flux is not constant. A better boundary condition is proposed. Piotr, Maciej have developed a boundary condition that use the multicycle boost-diffuse steps for vacuum carburizing process. In this method, constant carbon concentration was used and the boost step was extended because of the existing of the carbon deposit.

The surface carbon deposit layer was studied by carbon extracting method. Cementite and graphite is found as Figure 3 presents. Kouji and Hideaki has studied the formation of the cementite at the grain boundary [7].

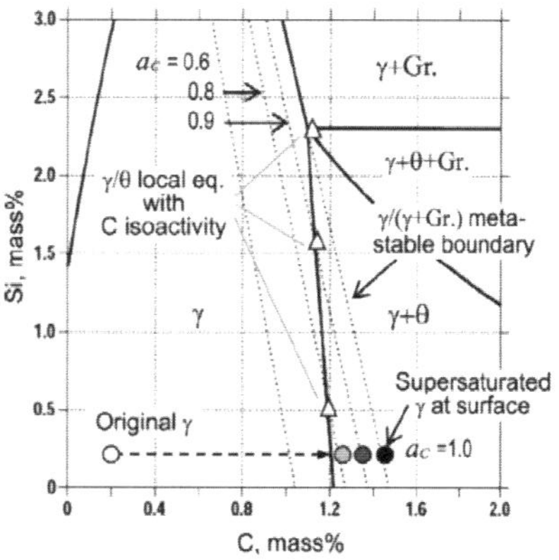

Figure 3: Dictra simulation shows the formation of the carbides

It may be that the carbides will form on the surface of the 9310 sample which will slow down the absorption of the carbon atoms towards the bulk. The Figure 4 present the solubility of 9310 at 927 °C which is about 1.0 wt%. Beyond the concentration, carbides starts to precipitate, so is the graphite.

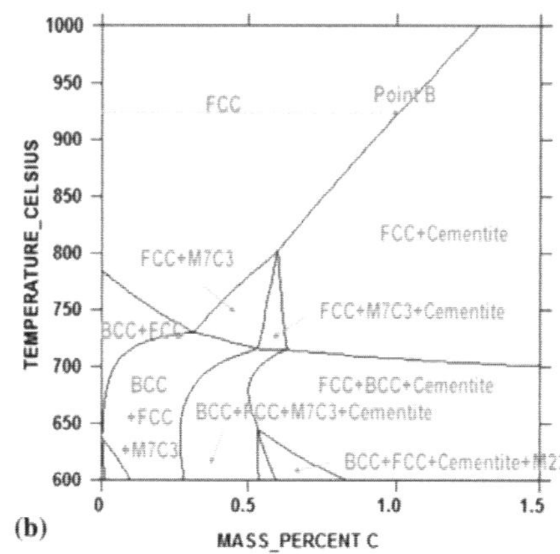

Figure 4: Isopleth for 93XX with carbon. Note: solubility limit at point B

Experiment

Samples preparation

To investigate the reaction during vacuum carburizing process of 9310. Two cycles are designed with boost steps only which last for 20min and 60 min. There is no carbides found with our previous experiments because the carbides can be dissolved during the diffusion step. The testing with boost step only will reveal the surface condition.

For next cycle, we would like to conduct three cycle with boost step only at 1700F using cyclohexane (C_6H_{12}). The boost time will be 20 min, 60min. Surface carbon and case depth calculated by CarbTool© are also presented.

Table 2: The suggested recipes for future vacuum process

Carbon flux	Diffusivity at start	Boost Time (min)	Surface carbon %C	Case depth mm
1.63E-06	1.670e-9	20	1.65	0.25
1.63E-06	1.670e-9	60	2.18	0.56

Figure 5 presents the carbon concentration profile for 9310 after vacuum carburizing process. For the process with 20min boost, surface carbon concentration is about 0.80 wt%. The surface carbon concentration of the other process is as high as 1.80 wt%, which leads to the carbides precipitation. The carbides can affect the absorption of carbon into the steel.

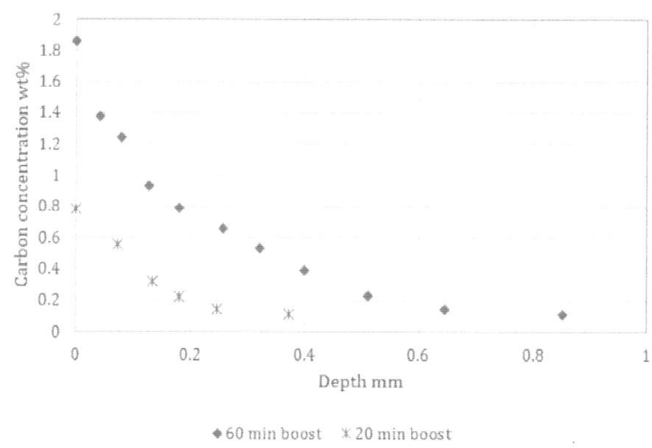

Figure 5: Carbon concentration profile for 20 and 60 minutes boost only processes at 927°C

Several experiments have been conducted to study the phase development at surface of the parts. The samples with only the boost stages for 20 and 60 minutes were analyzed for surface microstructural development and concentration profiles. The XRD is used for the surface phase identification. The samples were well cleaned before the XRD testing. Figure 6 and Figure 7 present the testing for both 20min and 60 min processes. At the surface area, a strong peak of graphite is found. There is also carbides found at the surface area. The 60 min boost step generate stronger carbides peaks than the 20 min boost step. There are more carbides found. It can also be found from the concentration profile as Figure 3 presents.

A thin layer is removed from the surface, which is around 15μm. The graphite peak disappeared. The carbides is hardly found for 20min boost process. There is still carbides left for the 60 min boost step. The graphite is attached to the surface of the samples. It could be a reason the slow diffusion of carbon atoms.

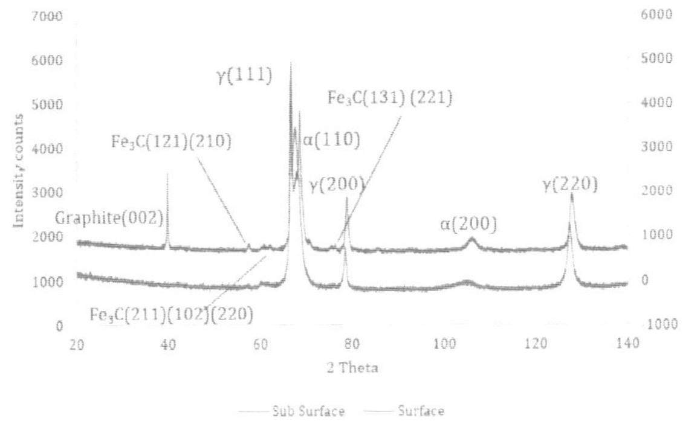

Figure 6: X-Ray Diffraction for 9310 of the 20 minutes boost only process, 15µm layer removed at bottom profile indicates

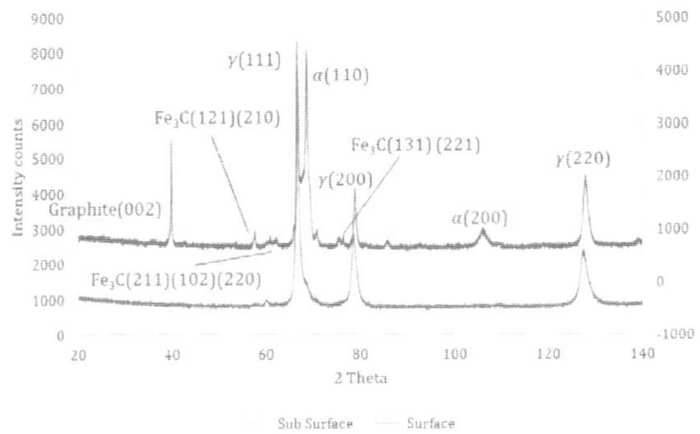

Figure 7: X-Ray Diffraction for 9310 of the 60 minutes boost only process, 15µm layer removed at bottom profile indicates

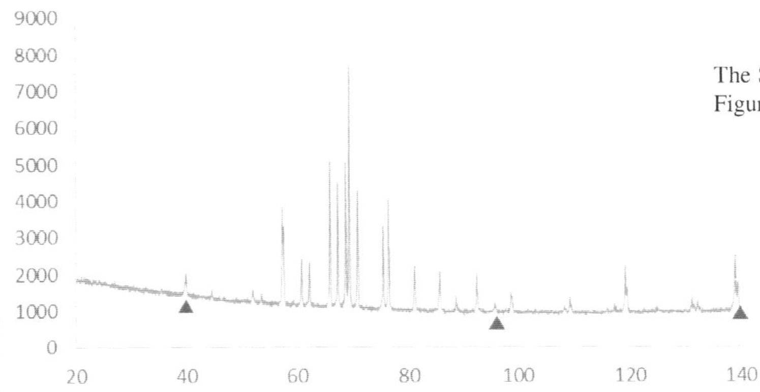

Figure 8: The extracted powder for carburized 9310 alloy with 60 Min boost

The SEM has be used for further study the surface of the carburized samples. Figure 9 presents the surface area of the sample with 20min boost step with SEM. EDS is applied on the white layer of the sample. The carbon centration is about 29.5wt% which indicates the existing of graphite. For the process with 60min boost step, as Figure 10 presents, the surface carbon concentration is as high as 52.7%.

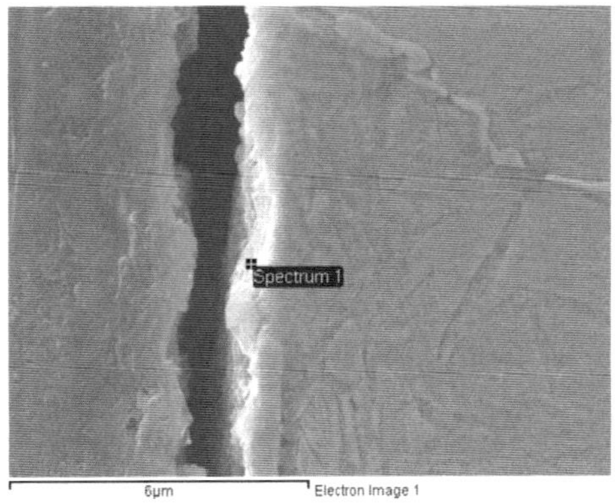

Figure 9: Scanning electric photomicrograph of cross section of the 20 minute boost at the surface. 20 wt% C austenite was measured by EDS

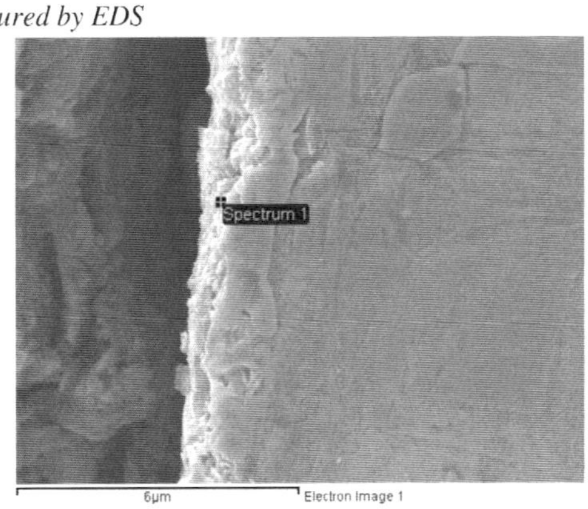

Figure 10: Scanning electric photomicrograph of cross section of the 60 minute boost at the surface. 50 wt% C austenite was measured by EDS

The SEM photomicrograph for original surface is presented in Figure 11. The graphite appearance can be seen on the surface.

Figure 11: Scanning electric photomicrograph of the surface after 60 minutes boost.

Figure 12 presents the carbides in the sub surface region. As explained previously, due to the high carbon concentration, these carbides form which can be dissolved during the diffusion process. The graphite and the carbides form a carbon reach layer that can provide enough carbon source to the vacuum carburizing process. On the other hand, the carbides will impede the diffusion process.

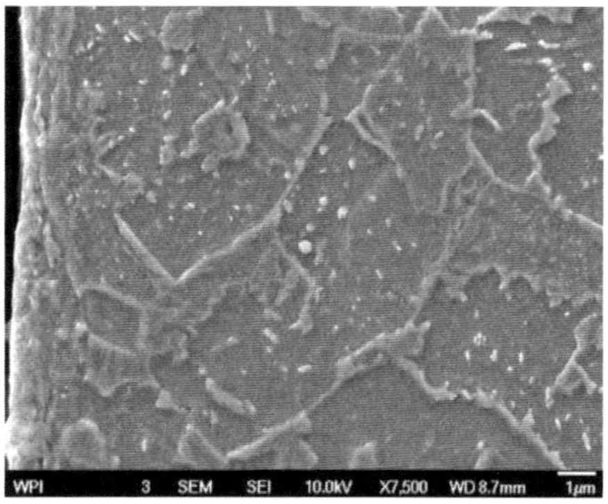

Figure 12: Scanning electric photomicrograph. Cross section of the 60 minute boost. Note: the carbides are found at the surface and at the austenite grain boundaries.

Kula proposed an assumption that during the boost step, it will form a thin carbon reach layer at the surface region. To simulate the process, a high carbon potential can be used [6]. The CarbTool© will be improved with the modification of the boundary condition for vacuum carburizing process.

Modeling development

Assumptions have been proposed for the modeling development. The precipitation and dissolving of the carbides do not influence the carbon diffusivity coefficient. The built-in diffusivity data is used. Carbon deposit layer will form during the vacuum carburizing process. A nominal carbon potential is proposed for the layer. Considered the mass transfer coefficient, the value is 1.6. The Armco foil testing has be done by Kula. The concentration of the foil reaches 1.3 wt% for 920°C[7].

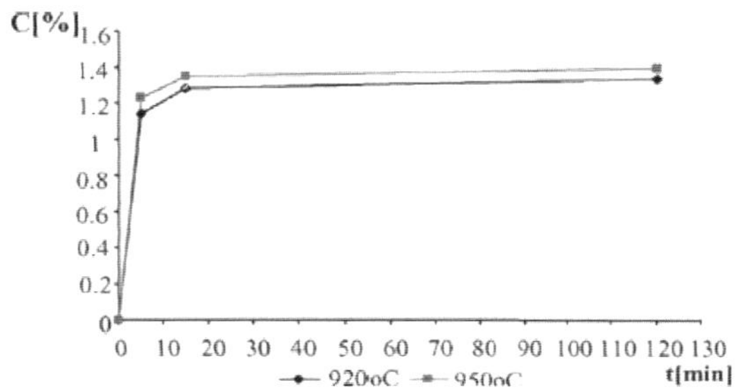

Figure 13. Carbon concentration in Armco foil testing in vacuum carburizing process

Calculation of the mass transfer coefficient
Assume the carbon deposit is providing carbon atom source for the vacuum carburizing process. The carbon activity is related to the pressure and the temperature of the atmosphere which is constant. When there is a balance at the surface during carburizing process:

$$\text{flux} = \beta\left(C_p - C_s\right) = -D\frac{\partial C}{\partial x}$$

The mass transfer coefficient can be calculated as:

$$\beta^t = \frac{(\Delta m/A)|_{t_0 \to t}}{t(C_p - C_s^t)}$$

With the weight gain data, the coefficient can be calculated as:
β =2.5E-05

With the input of the nominal carbon potential, both processes are simulated and presented in Figure 14 and Figure 15. They can meet the experimental well but the surface concentration. There is more carbon concentration at the surface because of the existing of the carbon deposit.

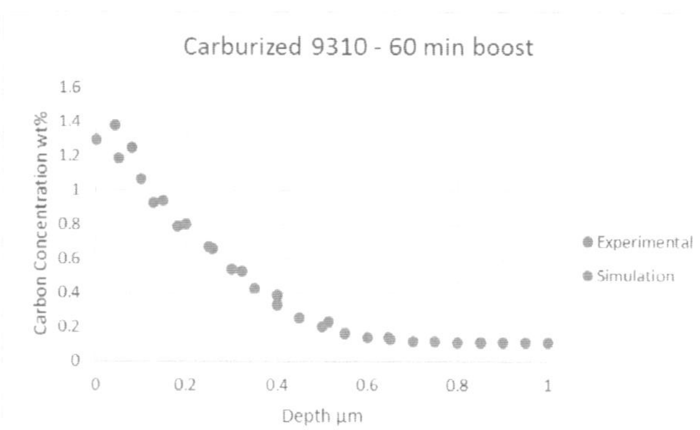

Figure 14: The carbon concentration simulation of carburized 9310 with 60 min boost

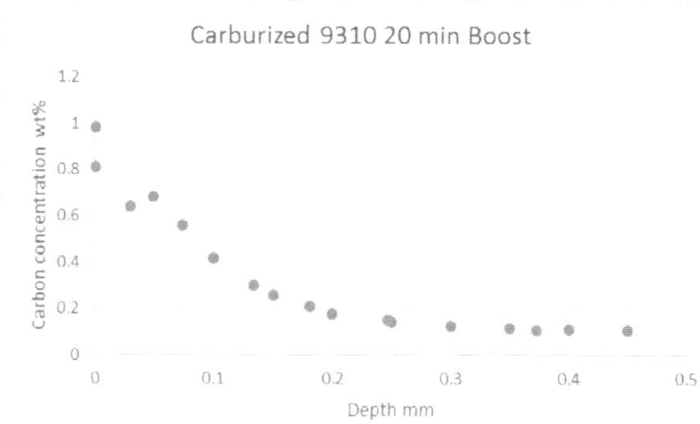

Figure 15: The carbon concentration simulation of carburized 9310 with 20 min boost

With the previous discussion, the carbon deposit layer will form soon during the boost step. The carbon deposit layer has a nominal carbon potential of 1.6. It takes a short time during diffuse step to dissolve the carbon deposit layer. With the calculated a value of mass transfer coefficient 2.5e-5. The carburized 9310 alloy in Figure 1 can have a better simulation as Figure 16 present.

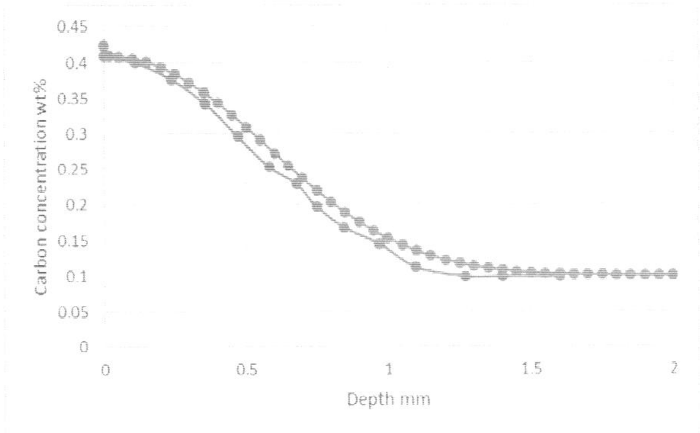

Figure 16: The optimized simulation for 9310 alloy

Conclusions

CarbTool © is an effective way to predict the carbon concentration. It is difficult to have good prediction for the 9310 steel. Experiments have been conducted to study the reason. The boost only vacuum carburizing process keeps the carbides and graphite on the surface without dissolve into the bulk as the diffusion step does.

The analysis from both XRD and SEM has proved that there exist carbides and graphite on the surface region which impede the further diffusion of carbon atoms.

The boundary condition for simulation vacuum carburizing process is optimized. Mass transfer coefficient is calculated and a nominal carbon potential is used for the simulation. With the novel model, the simulation result can meet the experimental result well.

References

[1] Downey, D. F. *et al.*, Ion Implantation Technology, Prentice-Hall (New York, 1993), pp. 65-67. [A book reference ...]

[2] Wei, Y., L. Zhang, and R. D. Sisson Jr. "Modeling of Carbon Concentration Profile Development During Both Atmosphere and Low Pressure Carburizing Processes." Journal of materials engineering and performance 22.7 (2013): 1886-1891.

[3] Karabelchtchikova, Olga, and Richard D. Sisson. "Carbon diffusion in steels: A numerical analysis based on direct integration of the flux." Journal of phase equilibria and diffusion 27.6 (2006): 598-604.

[4] Karabelchtchikova, O., Md Maniruzzaman, and R. D. Sisson. "Carburization Process Modeling and Sensitivity Analysis using Numerical Simulation."MATERIALS SCIENCE AND TECHNOLOGY-ASSOCIATION FOR IRON AND STEEL TECHNOLOGY- 2 (2006): 375.

[5] Goldstein, J. I., and A. E. Moren. "Diffusion modeling of the carburization process." Metallurgical Transactions A 9.11 (1978): 1515-1525.

[6] Zajusz, M., K. Tkacz-Śmiech, and M. Danielewski. "Modeling of vacuum pulse carburizing of steel." Surface and Coatings Technology 258 (2014): 646-651.

[7] Kula, Piotr, et al. "“Boost-diffusion” vacuum carburising–Process optimisation."Vacuum 99 (2014): 175-179.

International Federation of Heat Treatment and Surface Engineering 2016
Proceedings of the 23rd IFHTSE Congress
April 18–21, 2016, Savannah, Georgia, USA

Pre-Nit Technology For Accelerated Vacuum Carburizing

Michael Lister
SECO/WARWICK, Meadville, Pennsylvania

Abstract

The Pre-Nit carburization of carbon and alloy steels is a variation of vacuum or low pressure carburizing that involves a small addition of nitrogen during the heat up phase of parts prior to carburizing. This technology allows the parts to be heated to higher temperatures during carburization without the effects of grain growth in the case depth. The use of these higher temperatures, up to and above 1100°C allows for a significant reduction of carburizing time due to a greater diffusion coefficient of carbon in a face centered cubic structure.

Depending on the depth of the case, the Pre-Nit process can reduce the carburizing time of a part by nearly two-thirds when considering an effective case depth of 5.0 mm with no increase in plant floor space or additional equipment required to process parts.

This paper will explore the technological challenges, benefits, and examples for the appropriate applications of Pre-Nit technology as well as the commercial aspects for use of the technology. The paper will also detail the variations between endothermic carburizing, vacuum carburizing and Pre-Nit vacuum carburizing including the process, quenching, and final structure of parts processed in these diverse, yet similar, technologies.

Introduction

Case hardening is a popular method of improving the wear characteristics of a component while simultaneously providing the ability to tailor the inherent strength of the base material by means of improved hardness. Carburizing is one of the most common methods of improving the surface hardness and involves the introduction of carbon atoms into the surface of a part through a diffusion mechanism. The ending carbon concentration and the depth of the carburized layer (case depth) ultimately determine the product's final surface hardness. Examples of typical parts being carburized are: automotive transmissions gears, bearings, cam shafts, and fasteners.

Fig. 1 Example of parts for vacuum carburizing

The carburizing process has many different forms such as endothermic atmosphere, pack carburizing, and low pressure carburizing (also referred to as vacuum carburizing) among others. Each has its own advantages and disadvantages; however, the theories and principles behind diffusion, (most notably Fick's laws of diffusion) hold constant regardless of the method used. The diffusion process essentially limits the metallurgist and engineer to three variables – time, temperature, and carbon concentration.

Low pressure carburizing has gained popularity in recent years because of its advantages in today's Six Sigma, lean manufacturing, and environmentally friendly culture due to its quality, precision, repeatability, speed, and lack of carbon emissions. When combining this technology with various configurations of vessels, a wide range of materials and parts can be processed in a variety of different manners. Low pressure carburizing, unlike traditional batch atmosphere carburizing, can be configured to accept a pre-heated load in the main work chamber thus eliminating lengthy time periods while ramping up to a carburizing temperature. It also has the inherent advantage of all vacuum processing in that the oxygen free environment will not produce inter-granular oxidation. In addition, vacuum processing provides the ability to gas quench when the hardenability and specification of the material and part being carburized allow. Considering the environmental effects, there is no excess gas burn off as there would be in a traditional atmosphere furnace with a carbon potential – only the precise amount of hydrocarbons required for carburization are used.

Because low pressure carburizing is done in a vacuum furnace – it has significantly more flexibility when it comes to the quenching or hardening of the parts as there are more processing options than in conventional atmosphere carburizing. Variations of oil and gas quenching are available in vacuum furnaces such as directional quenching in gas or varying the pressure over oil to control distortion by means of controlling the vapor phase of the quench medium. [1]

Low Pressure Carburizing Process & Pre-Nit Carburizing

There are many methods available for Low Pressure Carburizing (LPC) that require hydrocarbons. This discussion focuses on carburizing with acetylene and ethylene, mixed with hydrogen. The hydrogen provides a tool to slow the reaction resulting in better control and increased cleanliness of the process. In most applications, the mixture of hydrocarbons is introduced directly into the heating chamber of a vacuum furnace where the gasses decompose on the surface of the material that is being carburized creating a carbon rich surface. This creates a concentration gradient and initiates the diffusion process of carbon into the material via the following reaction (1):

$$C_xH_y = XC + \frac{Y}{2}H_2 \quad (1)$$

In order to prevent the formation of soot within the confines of the vacuum furnace chamber, the introduction of gas is limited in both volume and duration based on the desired case depth of the parts, volume of the furnace, and the surface area available for carburizing. Unlike traditional atmospheric carburizing where a carbon potential is created, measured, and adjusted based on varying conditions; LPC gasses are precisely dosed to create the proper amount of free carbon for the available parts. The gas dosing rates are calculated prior to processing the parts while formulating the carburizing cycle recipe. The amount of gas is then measured during the injection to ensure proper mixing of gasses as well as ensure that the appropriate amount of free carbon is available for carburizing.

The interstitial diffusion of carbon in an iron based alloy is a function of temperature as well as the concentration of carbon. Given that the carbon concentration is fixed by the metallurgy of the part and the atmosphere of the furnace, the time to carburize is governed by the diffusion coefficient's correlation to temperature for a given case depth. Raising the temperature of the furnace has its own set of metallurgical challenges; namely, grain growth and intergranular oxidation in a traditional atmosphere furnace.

One way to control grain growth in the carburized case depth of the part is with a small addition of nitrogen during the thermal ramp of the recipe. As shown in Figure 2 in the yellow region, between approximately 400°C and 700°C, a precisely calculated and measured amount of ammonia can be introduced into the vacuum furnace. For a given case depth, material, and carburizing temperature the nascent nitrogen will diffuse into the material and stabilize the grains through forming small nitrides in the case. These nitrides restrict grain growth as well as act as nucleation sites for new austenitic grains. This ultimately allows the heat treater to carburize parts at a higher temperature; thereby, boosting the diffusion coefficient and reaching the targeted effective case depth (ECD) in a shorter period of time.

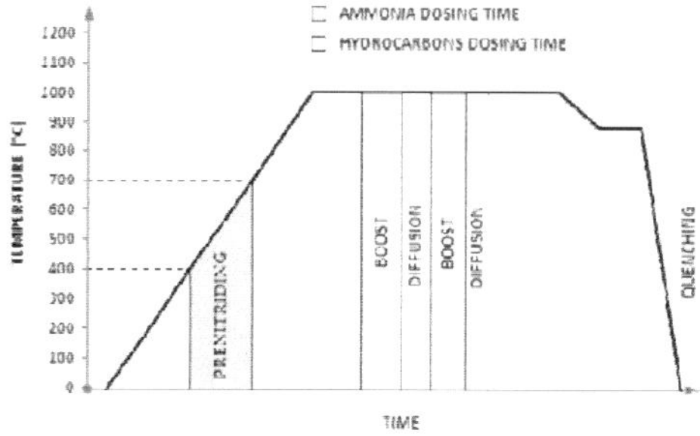

Fig. 2 Example of a Pre-Nit process showing the dosing of the furnace atmosphere versus time and temperature

Consider that for a targeted case depth of 0.61 mm, the carburizing time in a traditional endothermic atmosphere furnace would be approximately 180 minutes for 8620 material when processed at 925°C. By boosting the temperature to 1000°C, the carburizing time is only 46 minutes – over a 66% reduction in time spent carburizing. These gains are partially offset by an increased time to temperature and an increased soaking period to reduce the parts from their carburizing temperature to the hardening temperature, typically 840°C. When you factor in the total cycle time for this effective case depth, the total cycle time is slightly less than 75% of the endothermic atmospheric carburizing process at 925°C.

In figure 3 below, the effects on the grain structure can be seen – on an 18CrNiMo7-6 steel of carburizing at 1000°C with and without the PreNit technology. [2]

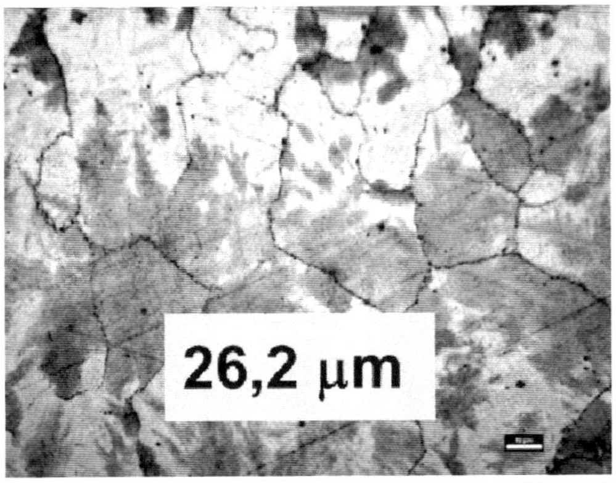

Fig. 3. The grain size in the carburized case following the LPC process for a 0.4 in case at a temperature of 1000°C of 18CrMNiMo7-6 steel without (top) and with PreNitLPC (bottom)

As shown in Table 1, with the increase in case depth, the total cycle time decreases due to the increase in the interstitial diffusion rate of carbon at higher temperatures. Equation A shows the dependence on the gradient of the carbon concentration (C_1 and C_2) at a given distance (dx) and diffusion coefficient D.

$$J = D\frac{C_1 - C_2}{dx}$$

Coefficient D correlates directly to the temperature as it is a function of exp(-k/T). When looking at carbon in an iron based matrix, Table 1 summarizes the reduction in diffusion time through the increase in temperature. With an increase from 925°C to 980°C, a reduction of the carburizing time by half is observed. Increasing the temperature another 55 degrees C, produces another reduction of greater than 50% or less than one quarter or the carburizing time at 925°C.

When coupled with the increased time required to achieve the required carburizing temperature and the increased time to return to an appropriate hardening temperature (commonly 840°C to avoid distortion and retained austenite in the carburized case), an overall decrease in cycle time, approximately 50% of the stated savings in carburizing time is still observed. That is to say that if the carburizing time is reduced 50%, the overall cycle time will be reduced by approximately 25%. If the carburizing time is reduced 75%, the overall cycle time will be reduced by about 40%. Since no special equipment is required, the PreNit technology allows for increased throughput with no additional footprint or equipment on the shop floor.

Table 1: The effect of temperature on carburizing time at various case depths.

Effective Case depth	LPC carburizing time			
[in]	925°C	950°C	980°C	1040°C
0.50 mm	83 min	57 min	39 min	19 min
1 mm	330 min	230 min	155 min	75 min
2 mm	1320min	910 min	620 min	290 min
% Of Baseline	100%	69%	47%	22%

5120 steel carburizing time for an effective case depth of 0.35%C, as a function of carburizing temperature.

Improved Quality

The quality of the low pressure carburized parts is superior to that of endothermic atmosphere carburized parts due to the lack of inter granular oxidation (IGO), the ability to carburize blind holes, and uniformity of case within a complex geometry.

Endothermic atmosphere gasses have small quantities of CO2 and H2O that can react with the boundaries of the austenite grains at the surface of the part creating intergranular oxidation (IGO). Based on the temperature of processing, quality of the atmosphere, and time at temperature, parts can experience IGO to significant depths which will adversely affect their mechanical properties. By carburizing in a chamber that has had the oxygen at ambient atmosphere removed there is not a sufficient amount of oxygen available to oxidize the grain boundaries though the purity of the carburizing and partial pressure gasses must be at a sufficiently high level in order to maintain the high quality of the atmosphere.

Another significant quality advantage is the ability to carburize intricate paths and blind holes. Because there is no atmosphere prior to the injection of gasses, the carburizing gasses evenly disperse (including blind holes and passages) without the need to displace atmosphere that would otherwise be difficult to displace in an atmosphere furnace. An excellent example of the blind hole would be an injection nozzle for diesel engines where the carburized case is achieved in a load with hundreds if not thousands of parts and has passages of less than 0.63mm diameter over a length of over 1 inch as seen in figure 4. [2]

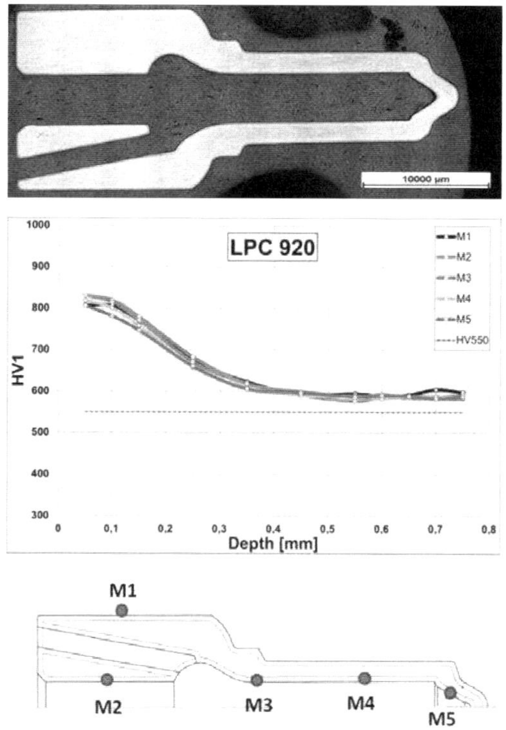

Fig. 4. A uniform carburized layer of a fuel injection nozzle with hardness profiles after LPC and HPGQ

Uniformity of the case is also a huge benefit with low pressure carburizing. In traditional atmospheric carburizing, parts are heated in a carbon rich atmosphere. During this critical ramp up phase a part such as a bevel gear would not reach the carburizing temperature evenly. Due to the carbon rich atmosphere and the uneven temperature of the component, the tooth of the gear will begin to accept carbon from the atmosphere before the root of the gear and will start to diffuse the carbon into the part at a faster rate due to the temperature. When you compound this effect with the surface to volume ratio of the tip of a gear tooth versus the root, the effect becomes amplified.

Compared to atmospheric carburizing, no carbon is available to carburize the part until the entire part has soaked to the appropriate carburizing temperature. Once the part is uniformly heated, the only variable to a non-uniform case would be the surface area to volume ratio over the geometry of the part.

Economic Impacts

Reducing the time of carburizing, in addition to the reduction in gasses used for carburizing in atmosphere, is paramount to the reduction in the cost of carburizing. The cost savings per pound are amplified as the case depth increases and time in the furnace decreases. Due to the regional nature of energy market, Table 2 should only be considered as an example of an integral quench furnace in an area with the average US utility costs as areas with low electric and high natural gas costs are likely to experience greater savings and areas with high electric costs and low natural gas costs are likely to experience reduced savings when compared to traditional endothermic atmospheres.

This table only addresses the utility costs of the process. If a business case were to be made regarding PreNit LPC processing of parts many overhead costs such as management, labor, and shop utilities would be spread out over a greater number of pounds or parts processed than in a conventional atmospheric furnace. When combined with the lower utility costs, deep case carburizing at elevated temperature makes a very compelling business case.

ECD [in]	**ENDO 925 °C** [$/lb]	**PreNitLPC** [$/lb]	
		980 °C	**1070 °C**
0.40mm	$0.40 (100%)	$0.42 (103%)	$045 (112%)
0.60mm	$0.46 (100%)	$0.43 (93%)	$0.45 (99%)
0.9mm	$0.55 (100%)	$0.50 (91%)	$0.50 (91%)
1.2mm	$0.69 (100%)	$0.60 (87%)	$0.55 (80%)
2.0mm	$1.25 (100%)	$0.96 (77%)	$0.73 (58%)
3.0mm	$2.15 (100%)	$1.57 (73%)	$1.12 (52%)
5.0mm	$5.35 (100%)	$3.69 (69%)	$2.19 (41%)

Table 2. Utility costs for varying effective case depths based on temperature

Summary

When considering a method for carburizing, low-pressure carburizing is capable of providing better quality parts at equivalent or better pricing compared to traditional endothermic atmospheric processing at case depths greater than 0.50 mm. By processing in a vacuum and at higher temperatures, low pressure carburizing is able to increase throughput within the same footprint as a traditional batch integral quench furnace. Also the carburizing technology leads to more quality and consistency by means of controlling the gas going in instead of controlling an ever changing carbon potential against a measurement.

By eliminating the oxygen from the environment around the parts, any concern of IGO is virtually eliminated and results in a higher quality of a finished part when post machining is not viable or feasible.

When coupled with an oil quench, low pressure carburizing becomes extremely versatile and offers more variables to control the distortion of parts being carburized.

Finally the absolute minimal use of combustible gasses makes low pressure carburizing a safer means of carburizing parts and a more environmentally friendly process as well.

References

[1] Herring, D. H., Sugiyama, M., and Uchigaito, M., Vacuum Furnace Oil Quenching – Influence of Oil Surface Pressure on Steel Hardness and Distortion, Industrial Heating Magazine, June 1986.

[2] M. Korecki, P. Kula, E. Wołowiec, M. Bazel, M. Sut: Low pressure carburizing and nitriding of fuel injection nozzles. Heat Processing 3 (2014) 59-62.

International Federation of Heat Treatment and Surface Engineering 2016
Proceedings of the 23rd IFHTSE Congress
April 18–21, 2016, Savannah, Georgia, USA

Influence of Ultrasonic Nanocrystal Surface Modification Treatment Temperature on Microstructure and Tribological Behavior of Bearing Steel

Auezhan Amanov, Jun-Hyong KIm and Young-Sik Pyun
Sun Moon University, Asan 31460, Rep. of Korea
aamanov@outlook.com

Abstract

In this study, a bearing steel AISI 52100 was subjected to ultrasonic nanocrystalline surface modification (UNSM) technique at ambient and high temperature of 500 ^{0}C. The objective of this study is to investigate the effectiveness of UNSM technique on the friction and wear behavior at ambient and high temperatures. The hardness with respect to depth from the top surface was measured using a microhardness. The findings from this preliminary study are expected to be implemented to the bearings and tools to increase the efficiency and performance of the components.

Introduction

The friction and wear behavior of any sliding elements significantly influences the performance, reliability and fatigue life of mechanical components and tools. Some previous investigations pointed out that some considerable efforts have been made to develop effective surface modification methods to increase the efficiency of mechanical components by controlling the strength and wear performance [1]. Relatively new surface modification techniques such as material removal, material modification and material adding are continuously being developed with the aim of saving energy and minimizing wear and damages. A number of alternative surface modification techniques were proposed and invented to improve the wear behavior and to extend the fatigue life of bulk materials. The ability of severe plastic deformation (SPD) surface modification techniques such as dynamic plastic deformation (DPD) [2], surface mechanical attrition treatment (SMAT) [3], shot peening (SP) [4], equal channel angular pressed (ECAP) [5] to improve the strength and wear behavior of mechanical components was invented several years ago. However, all the surface modification techniques are applied at an ambient temperature. In this study, a newly developed high-temperature ultrasonic nanocrystalline surface modification (HT-UNSM) technique and its preliminary results are presented. The main objective of this study is to investigate the hardness, and friction and wear behavior under dry conditions of bearing steel AISI 52100 treated at a high temperature of 500 ^{0}C. The newly developed HT-UNSM technique can serve as a materials strengthening device that can further increase the hardness of a material than the treated one at a room temperature.

Experimental Procedures

In this study, the specimens made of SAE 52100 steel with a diameter of 24.8 mm and a length of 8 mm were employed. The specimens were polished using a sand paper for microstructural characterization and friction and wear tests. The specimens subjected to UNSM technique at room and high temperatures under the parameters as listed in Table 1. More details of the UNSM technique can be found in our previous study [6].

Table 1: UNSM treatment parameters at room and high temperatures.

Frequency, kHz	20
Amplitude, μm	30
Horn speed, m/min	1
Load, N	20
Interval, μm	10
Ball diameter, mm	2.38

A commercially available ball-on-disk tribometer (CSM, Switzerland) was used to access the friction and wear behavior of the specimens under dry conditions at room temperature at a load of 10 N, with a stroke of 1 mm for 30 min of sliding at a reciprocating speed of 3 cm/s. A bearing steel SAE52100 ball with a diameter of 10.0 mm was used as a counter surface.

Results

The surface microstructure of the polished and UNSM-treated specimens is presented in Fig. 1. It can be seen from Fig. 1(a) that the microstructure of the polished specimen is not homogeneous with some defects and abrasive grooves on the surface. Clearly, regular UNSM treatment tracks with no abrasive grooves can be observed in Figs. 1(b and c). The average surface roughness of the polished, UNSM-treated at room and high temperatures specimens were found to be about 0.19, 0.12 and 0.17 μm. It is obvious that no difference in microstructure and surface roughness was found for the UNSM-treated at room and high temperatures specimens. However, a huge difference was found in hardness, which was increased at room temperature treatment, but decreased at high temperature treatment as shown in Fig. 2.

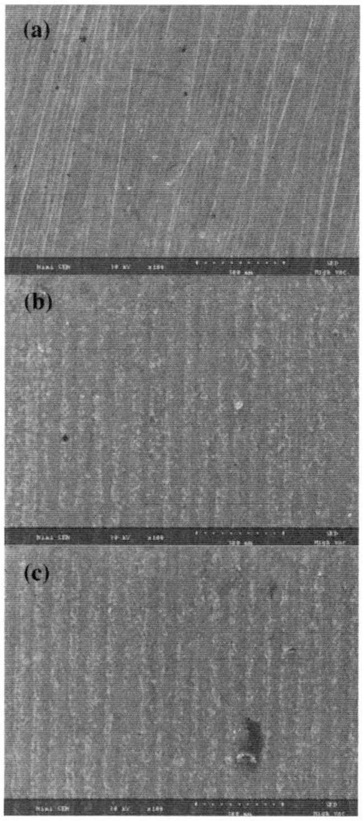

Figure 1: SEM images of the untreated (a), UNSM-treated at ambient (b) and high (c) temperatures.

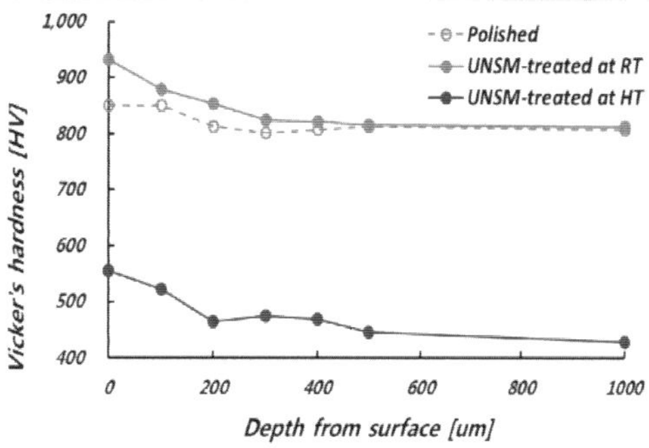

Figure 2: Comparison in Vicker's hardness of AISI 52100 with respect to depth from the top surface.

A comparison in friction coefficient and wear rate of the untreated, UNSM-treated specimens at room and high temperatures of the specimens are presented in Fig. 3. It can be seen that the friction coefficient and wear rate of the specimen improved by UNSM treatment at a room temperature and it further reduced by UNSM treatment at a high temperature. The reduction in friction coefficient and enhancement in wear resistance of UNSM-treated specimens may be mainly attributed to the increased hardness and surface roughness. It has been reported in our previous study that the increase in hardness by UNSM technique is responsible for the grain size refinement [6].

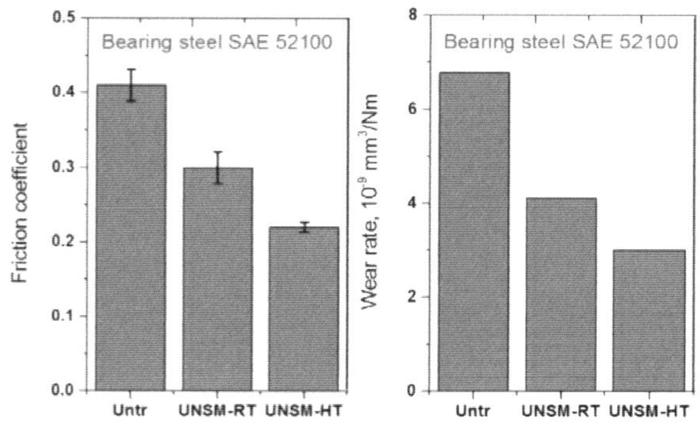

Figure 3: Comparison in friction coefficient and wear rate of the untreated, UNSM-treated specimens at room and high temperatures.

Conclusions

- Average surface roughness of the specimens reduced, while the surface hardness increased at ambient temperature and decreased at high temperature. Further temperature optimization needs to be made.
- The friction coefficient reduced by UNSM treatment and further reduced with increasing UNSM treatment temperature.
- In summary, a UNSM treatment temperature plays a critical role in increasing the hardness and improving the tribological properties of both steels. Further microstructural characterizations need to be performed in order to understand the mechanism of the specimens treated at a high temperature of 500 ^{0}C.

References

[1] Stanford, M.K., DellaCorte, C., "Friction and Wear Characteristics of Cu-4Al Foil Bearing Coating at 25 and 650 °C," NASA/TM–2004–212972 (2004).

[2] Yao, B., *et al.*, "Dry Sliding Tribological Properties and Subsurface Structure of Nanostructured Copper at Liquid Nitrogen Temperature," *Wear*, Vol. 301 (2013), pp. 608–614.

[3] Arifvianto, B., *et al.*, "Effects of Mechanical Attrition Treatment (SMAT) on a Rough Surface of AISI 316L Stainless Steel," *Appl. Surf. Sci.*, Vol. 258 (2012), pp. 4538–4543.

[4] Unal, O., Varol, R., "Surface Severe Plastic Deformation of AISI 304 via Conventional Shot Peening, Severe Shot Peening and Repeening," *Appl. Surf. Sci.*, Vol. 351, No. 1 (2015), pp. 289–295.

[5] Gallego, J., *et. al.*, "Microstructural Characterization of Ti-6Al-7Nb Alloy after Severe Plastic Deformation," *Mat. Res.*, Vol. 15, No. 5 (2012), pp. 786–791.

[6] Amanov, A., et al., "Micro-dimpled Surface by Ultrasonic Nanocrystal Surface Modification and Its Tribological Effects," *Wear*, Vols. 286–287 (2012), pp. 136–144.

International Federation of Heat Treatment and Surface Engineering 2016
Proceedings of the 23rd IFHTSE Congress
April 18–21, 2016, Savannah, Georgia, USA

The Erosion Behavior of Die-Steel Salt-Bath-Nitirocarburized with Lithium-Iron Compound Oxide Layer in Molten Al-Si-Cu Alloy

H. Ishizuka, Y. Hirai and Y. Watanabe
PARKER NETSUSHORI KOGYO CO., LTD, 3-13-10, Tamachi, Kawasaki-ku, Kawasaki-shi, Kanagawa-ken, Japan

Abstract

The erosion resistance of die steels in molten metal is a very important factor in determining the lifetime of a die-casting die. In this study, the effects of lithium-containing ferrous oxide (Li-Fe compound oxide; formed by cyanate- and lithium-ions-containing salt-bath nitrocarburizing) formed on the surface of a common die steel (JIS-SKD61) on the erosion resistance of the die steel in molten Al-Si-Cu alloy (JIS-ADC12) were investigated in comparison with a general ferrous oxide (largely Fe_3O_4) and nitrided layers. The erosion resistance was evaluated from weight loss of 16-mm-diameter cylindrical specimens after immersion (while rotating at 200 rpm) in liquid ADC12 metal at 953 K and 1023 K for up to 7.2 ks. The cross-sectional interface reaction layers between the specimens and the ADC12 alloy after the testing were analyzed by optical microscope, micro Vickers-hardness tester, FE-SEM with EDX and EBSD analyzers, and EPMA. The analysis results show that the Li-Fe compound-oxide layer improves the erosion resistance to a greater extent than the ferrous-oxide layer, and the erosion resistance is increased as the thickness of the compound-oxide layer increases, thereby inhibiting the reaction with the molten ADC12 metal. Also, the nitrogen-compound layer has limited effectiveness against erosion resistance because it can be decomposed in molten ADC12 metal at temperature over 923 K. Moreover, the nitrogen-diffusion layer is not thought to effectively improve the erosion resistance.

Introduction

The most-prominent example of steels being in direct contact with a high-temperature liquid metal is the aluminum die-casting process, which is generally used for manufacturing automotive parts made of aluminum alloy, such as engine blocks, automatic transmissions, and under-carriage parts. The use of aluminum die-casting has increased because of an increase in global automobile production and a growing need to replace steels by aluminum alloys with the aim of reducing the weight of cars.

Recently, aluminum die-casting dies have become increasingly used under severe conditions because the casting process has become a higher-cycle operation (using high-speed casting machines for improving productivity) in addition to practical use of high-vacuum die casting. Furthermore, demands for mold with longer lifetime are increasing with demands to decrease production cost of die casting.

To increase lifetime of die-casting dies, it is most important to improve the resistance to erosion of die steels in contact with molten aluminum alloy. Post-oxidation treatment with steam and/or air after nitriding or nitrocarburizing, by which a ferrous-oxide (composed largely of magnetite Fe_3O_4) layer with a thickness of less than 1 μm forms on the surface, is well-known to be effective for improving erosion resistance of steels [1]. However, post-oxidation is not enough to improve erosion resistance to be conducive to dramatically increase the lifetime of die steels because stably forming an oxidation layer with a thickness of more than 2 μm can hardly be expected.

It has been recently reported that the lifetime of a casting die can be significantly improved by developing new die materials, plasma nitriding, PVD (physical vapor deposition) coatings such as DLC (diamond-like carbon) film, and combining these processes [2]. However, these approaches might result in low productivity and increase production costs. Meanwhile, the erosion behavior of pure iron and steels (without surface modifications) in molten pure aluminum and aluminum alloy has been almost completely understood [3,4,5]. However, the erosion behavior of nitrided, nitrocarburized, and post-oxidized steels is not yet fully understood [6].

Cyanate-containing salt-bath nitrocarburizing can form a lithium (Li)-containing ferrous-oxide layer (hereinafter called "Li-Fe compound oxide layer") on the surface in concurrence with the nitrided layers (i.e., a compound layer composed of ε-$Fe_{2\text{-}3}N$ and γ'-Fe_4N phases and a nitrogen (N)-diffusion layer) by adding not only potassium (K^+) and sodium (Na^+) ions but also lithium ions (Li^+) to the molten salt and reducing the amount of cyanide ions (CN^-). Compared with a general ferrous-oxide layer, this Li-Fe compound oxide layer significantly improves resistances to corrosion and frictional wear [7].

In this study, aiming at increasing the lifetime of an aluminum die-casting die, the effects of a Li-Fe compound oxide layer on erosion behavior of salt-bath nitrocarburized die steel and partly nitrocarburized tool steel in molten aluminum-silicon (Al-Si)-based alloy were investigated in comparison to a general ferrous-oxide layer and nitrided layers.

Experimental procedure

Cylindrical specimens of die steel (JIS-SKD61), whose chemical composition is listed in Table 1, with diameters of 16 and 100 mm in length were prepared by machining after quenching followed by tempering at 873 K (to form hardness of around 47 HRC). Some of the specimens were then nitrocarburized at 853 to 873 K for 1.8 to 10.8 ks by using a lithium-added molten salt bath containing 15% CNO^- and less-than-0.1% CN^- (hereinafter, these specimens are referred to as "LCN").

Table 1: Chemical composition of die steel (JIS-SKD61) used (mass%).

C	Si	Mn	P	S	Cu	Ni	Cr	Mo	V
0.39	0.93	0.44	0.023	0.002	0.09	0.13	5.34	1.22	0.81

Other specimens were then nitrocarburized in a conventional salt bath at 853 K for 1.8 to 5.4 ks by using a molten salt with 35% CNO^- and 4% CN^- (hereinafter, these specimens are referred to as "CN"). After the conventional salt-bath nitrocarburizing, some specimens were post-oxidized by immersing in an oxidation bath at 673 K (hereinafter, these specimens are referred to as "CNO"). A specimen with a nitrogen-diffusion layer only (hereinafter, referred to as "CLL") was prepared by removing the oxide layer and compound layer (including a porous layer) by surface grinding after salt-bath nitrocarburizing at 803 K for 129.6 ks.

An erosion test (shown schematically in Fig. 1) was performed by immersing a 30-mm length of the specimens (rotated at 200 rpm) in molten Al-11.4mass%Si-1.9mass%Cu alloy, called JIS-ADC12 (whose chemical composition is listed in Table 2), at 953 K and 1023 K in an alumina crucible (86-mm diameter and 230-mm depth) for up to 7.2 ks. Weight loss of the specimens after erosion testing was measured (The ADC 12 alloy attached at the specimen surface was removed by dipping in sodium-hydroxide solution.). Erosion volumes were calculated from weight loss under the assumption that the density of SKD61 steel is $7.72 \times 10^{-3} g/mm^3$ [8]. The surface layers of the specimens and the cross section of the interfacial reaction layers with the molten ADC 12 alloy were examined by optical microscope, micro-Vickers-hardness tester, FE-SEM with EDX and EBSD analyzers, XRD, and EPMA.

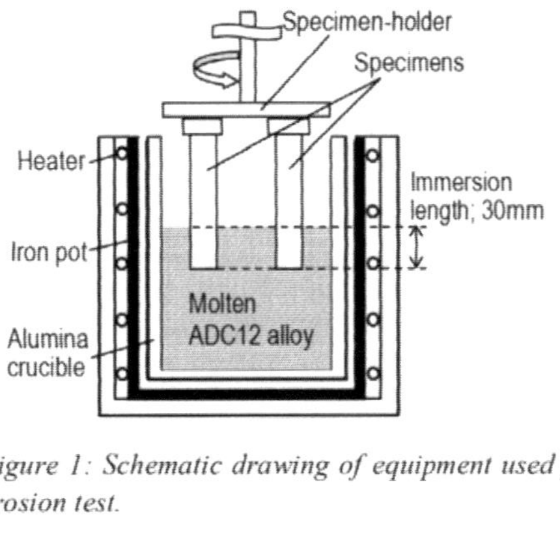

Figure 1: Schematic drawing of equipment used for erosion test.

Table 2: Chemical composition of Al-Si-Cu alloy (JIS-ADC12) used (mass%).

Si	Cu	Mg	Zn	Fe	Mn	Ni	Sn	Pb	Ti	Cr	Al
11.38	1.88	0.27	0.82	0.74	0.17	0.05	0.02	0.04	0.03	0.05	Bal.

Results

Surface microstructure

Cross-sectional surface microstructures of each specimen are shown in Fig. 2. Hardness profiles and nitrogen-concentration distributions are shown in Figs. 3 and 4, respectively. In the case of specimen LCN, shown in Fig. 2(a), a Li-Fe compound-oxide layer with a thickness of 3 to 4 μm is found clearly formed on the surface. The nitrided layers in specimen LCN include a 6-μm-thick compound layer and a nitrogen-diffusion layer with depth of up to 50 μm. Specimen CNO, shown in Fig. 2(b), has a ferrous-oxide layer (with thickness of approximately 0.6 μm) on the surface and a compound layer (with thickness of 4 to 6 μm) including a porous layer, and a nitrogen-diffusion layer with depth of up to 60 μm. As for specimen CN, shown in Fig. 2(c), the ferrous-oxide layer cannot be observed, and only a compound layer with thickness of 3 μm (including a porous layer) and a nitrogen-diffusion layer with thickness of 40 μm are formed. As shown in Fig. 2(d), specimen CLL only has a nitrogen-diffusion layer (with depth of approximately 80 μm) without both the oxide and compound layers on the surface.

Cross-sectional FE-SEM images of a Li-Fe compound-oxide layer formed on the surface of specimen LCN and a general ferrous-oxide layer on the surface of specimen CNO are shown in Figs. 5(a) and 5(b), respectively. Depth-concentration profiles of lithium, iron, nitrogen, and oxygen at depth up to 8 μm of specimen LCN, which were analyzed by GD-OES, are shown in Fig. 6. The Li-Fe compound-oxide layer formed on specimen LCN penetrates into the compound layer (which was previously thought to be a porous layer) and appears to be much thicker and uniformly denser than the ferrous-oxide layer formed on specimen CNO.

From the XRD results obtained from the surface of specimens LCN (7(a)) and CNO (7(b)) shown in Fig. 7, it is first revealed that the Li-Fe compound oxide consists almost entirely of α-$LiFeO_2$ with a NaCl-type crystal structure generated from FeO (called Wüstite) by replacing some of the iron atoms with lithium atoms. Because the replacement ratio is thought vary depending on depth from surface, the Li-Fe compound oxide should be described as "$Li_xFe_{1-x}O$" in this study. In contrast, the ferrous oxide in specimen CNO is recognized to be almost completely Fe_3O_4 (namely, "magnetite").

The result of EBSD analysis (shown in Fig. 8) also reveals that the Li-Fe compound oxide has remarkably uniform and fine grains (with size of less than 2 μm). This Li-Fe compound oxide forms a deep "sealing" porous-layer almost completely composed of ε-$Fe_{2\text{-}3}N$ phase.

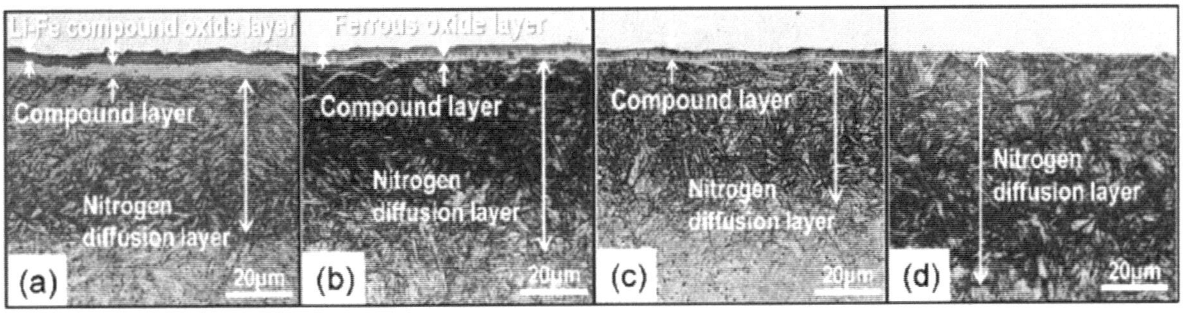

Figure 2: Optical micrographs showing cross-sectional surface microstructures of (a) LCN, (b) CNO, (c) CN, and (d) CLL.

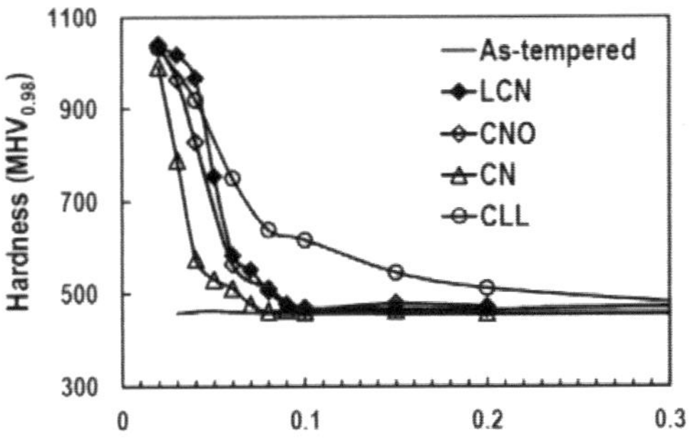

Figure 3: Hardness profile of several specimens.

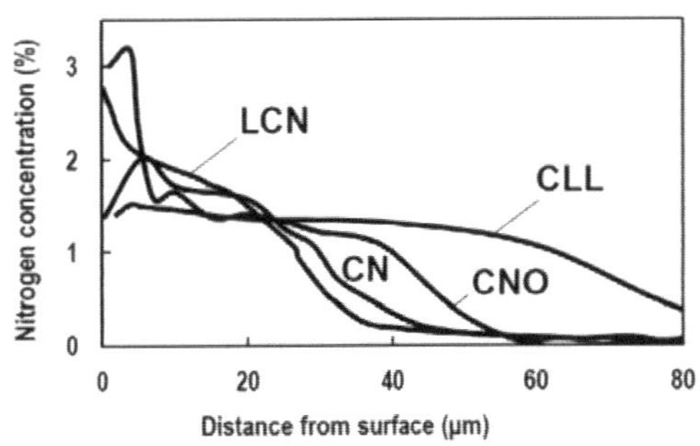

Figure 4: Nitrogen-concentration depth profiles of salt-bath nitrocarburized specimens analyzed by GD-OES.

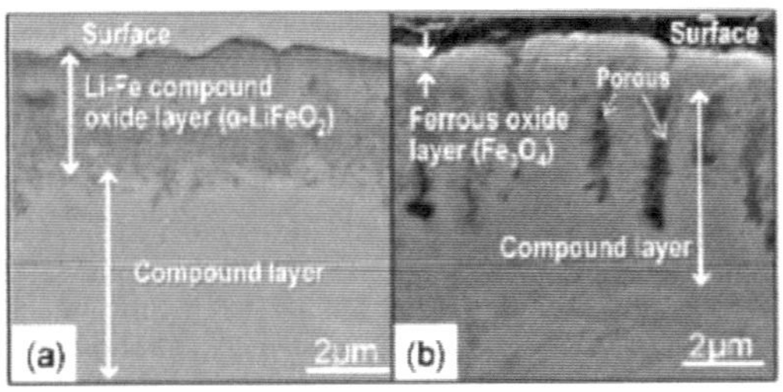

Figure 5: FE-SEM images of cross-sectional surface layers of specimens LCN and CNO.

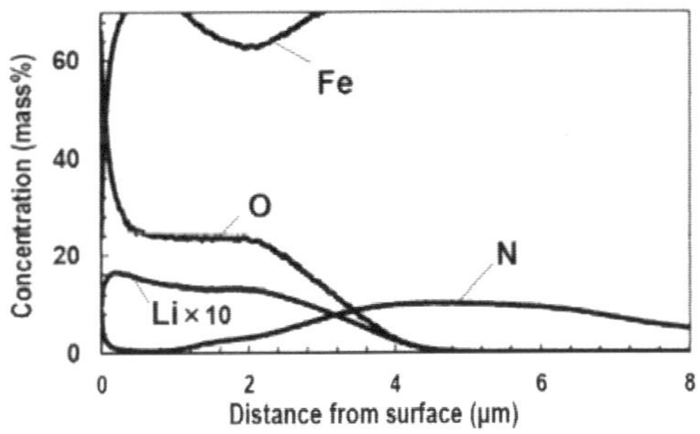

Figure 6: Concentration depth profiles of Li, Fe, O, and N in specimen LCN.

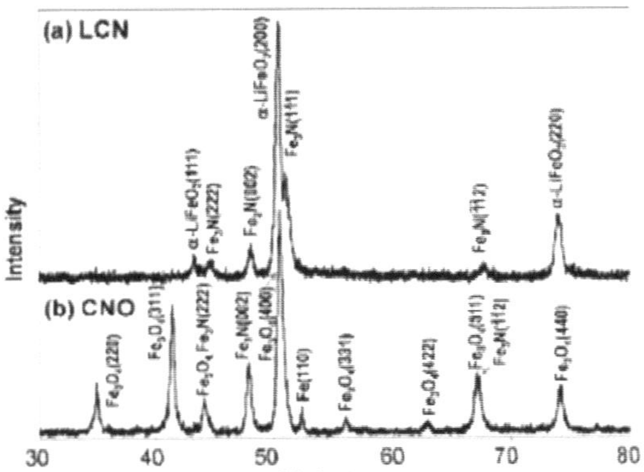

Figure 7: X-ray diffraction patterns of surface layers of (a) LCN and (b) CNO (Co Kα radiation).

Erosion behavior

The dissolved volume as a function of the immersion time of the specimens is shown in Fig. 9. The appearance of the specimens after erosion tests at 1023 K for 1.8 ks (as typical examples of erosion conditions) is shown in Fig. 10. At 1023 K (shown in Fig. 9(a)), the as-tempered specimens and specimen CLL started to dissolve rapidly soon after immersion and had a pronounced erosion after for 1.8 ks. Specimen CN dissolved relatively slowly, and specimen CNO (with a ferrous-oxide layer on its surface) dissolved partly as well. In contrast, specimen LCN (with a Li-Fe compound-oxide layer on the surface) did not dissolve for 1.8 ks and barely dissolved even after 7.2 ks. As shown in Fig. 9(b), the erosion of each specimen at 953 K showed the same tendency

Figure 8: Crystal orientation map by FE-SEM-based electron backscatter diffraction (EBSD) of the surface of Li-Fe compound oxide formed on specimen LCN.

as that at 1023 K, though the dissolved volumes of each specimen were much smaller than that at 1023 K. Specimen CNO also barely dissolved until 3.6 ks. Especially, specimen LCN almost did not dissolve, even after 7.2 ks, in the same manner as at 1023 K.

Under the assumption that the specimens dissolve uniformly

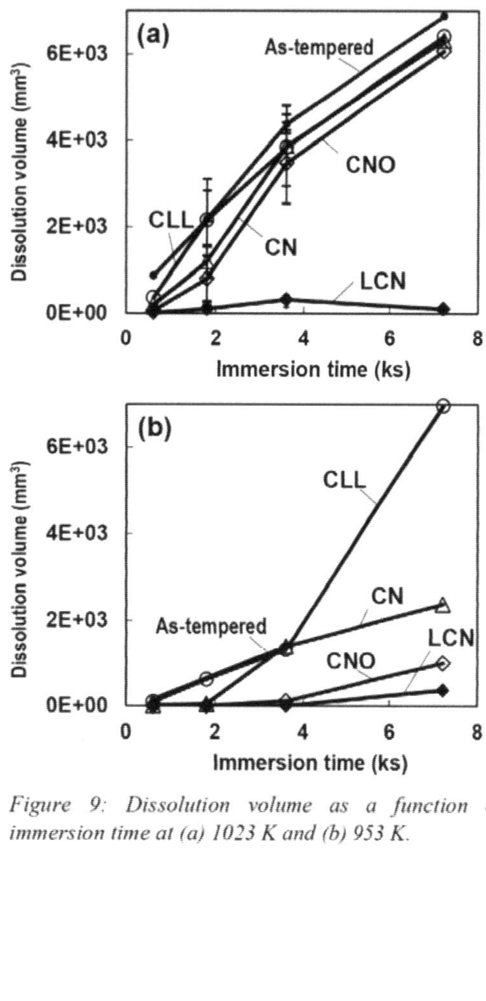

Figure 9: Dissolution volume as a function of immersion time at (a) 1023 K and (b) 953 K.

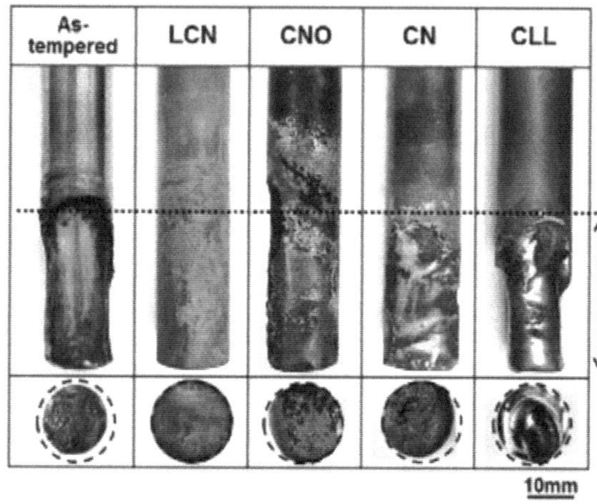

Figure 10: Appearance of specimens after erosion testing at 1023 K for 1.8 ks.

from the side-wall dissolution rate v (mm/s) at immersion time t (s) is defined as average rate of decrease in specimen radius r (mm) up to t (s). Dissolution volume V (mm^3) is expressed as

$$V = \pi r_0^2 l - \pi(r_0 - vt)^2 l \quad \text{------------------ (1)}$$

where v is derived from Eq. (1) as

$$v = r_0 t^{-1} - \sqrt{\frac{\pi r_0^2 l - V}{\pi l}} \cdot t^{-1} \quad \text{---------------(2)}$$

where r_0 is radius of the specimen, ℓ denotes immersion length (30 mm in this study), and V is given as the measured value. In Figure 11, v is plotted as a function of t. At 1023 K for 0.6 ks, the as-tempered specimen has the fastest dissolution rate of all the specimens. Subsequently, v increases in the order corresponding to specimens CLL and CN. Specimens LCN and CNO (with oxides on the surface) barely change in terms of dissolution rate. As immersion time increases, dissolution rate increases in the case of each specimen except LCN. Specimens CLL, CN, and CNO became similar in terms of dissolution rate to the as-tempered specimen after 1.8 ks, 3.6 ks, and 7.2 ks, respectively. In contrast to these behaviors, dissolution rate of specimen LCN stays very low, even after 7.2 ks. In case of 953 K, as early as 1.8 ks, specimen CLL shows a high dissolution rate. The dissolution rates of specimens CN, CNO, and CLL the specimen are very low up to 1.8 ks, but CN shows a similar dissolution rate to that of the as-tempered specimen after 3.6 ks.

Change in surface microstructures caused by erosion

FE-SEM images of cross-sectional microstructures of the interface between the specimen and ADC 12 metal after erosion testing at 1023 K for 1.8 ks are shown in Fig. 12. The formation of reaction layers (thought to be composed of intermetallic compound) was observed in the as-tempered specimen and CLL. Specimen CNO had a reaction layer composed partly of ferrous oxide remaining on its surface.

However, in the case of specimen LCN, the Li-Fe-compound-oxide layer remained completely, and the reaction layer did not form at all.

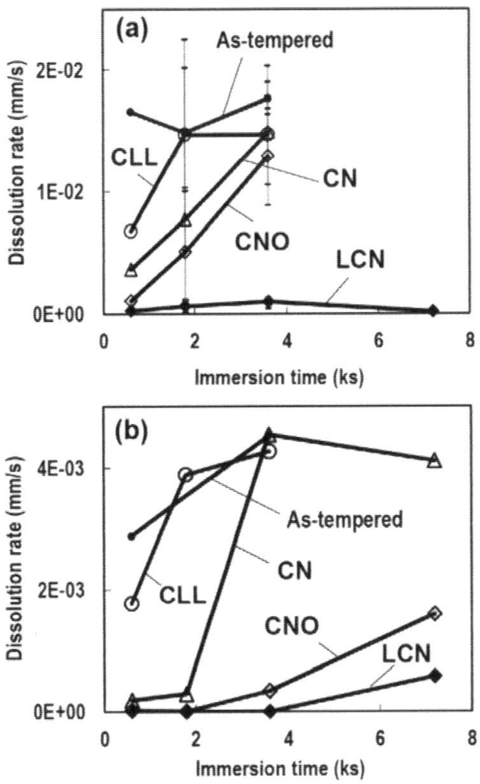

Figure 11: Dissolution rates as a function of immersion time at (a) 1023 K and (b) 953 K.

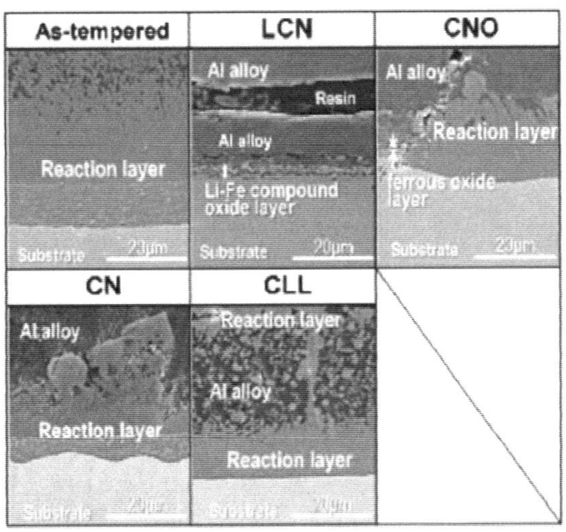

Figure 12: FE-SEM images showing cross-sectional interaction layers of the specimens and ADC12 metal after erosion testing at 1023 K for 1.8 ks.

An FE-SEM image of the as-tempered specimen (showing the cross-section of the reaction layer with ADC12 metal), the corresponding elemental mappings obtained by EPMA, and an illustrative drawing of the cross section, are shown in Figs. 13(a), 13(b), and 13(c), respectively. As indicated in Fig. 13(a), a micro Vickers-hardness test showed that layers A, B, C, and D have hardnesses of 100, 300 to 600, 700 to 900, and 900 to 1100 HV, respectively.

The hardness varies according to the composition of the intermetallic compound [9,10]. Accordingly, from the hardness values of each layer and the corresponding elemental mapping, it can be concluded that layer A is ADC12 metal, layers B and C are Al-Fe-Si intermetallic compounds, and layer D has a double-layer structure of Fe_2Al_5 and $FeAl_3$, as indicated in Fig. 13(c) [11]. These results also apply to the other specimens and are consistent with the results reported by Niinomi et al.

Discussion

Reaction between steel specimens and molten Al-Si-Cu alloy

Intermetallic compounds, such as Fe_2Al_5, $FeAl_3$, and α-(Fe,Al,Si), start to form at the specimen surface when iron concentration in the Al-Si-Cu liquid phase exceeds the solubility limit (approximately 3.5 mass% [12]) near the boundary between the specimen and molten Al-Si-Cu alloy, as shown in Fig. 13(a). It is considered that the erosion of the steel specimen increases with repeated formation and growth of intermetallic compounds at the interface with the molten Al-Si-Cu alloy, peeling of the intermetallic compounds, and dissolution of iron into the molten alloy.

The rate-determining process concerning erosion behavior was investigated by comparing the theoretical mass-transfer coefficient with the measured coefficient (calculated from the erosion-test results). Many studies on mass transfer suggest

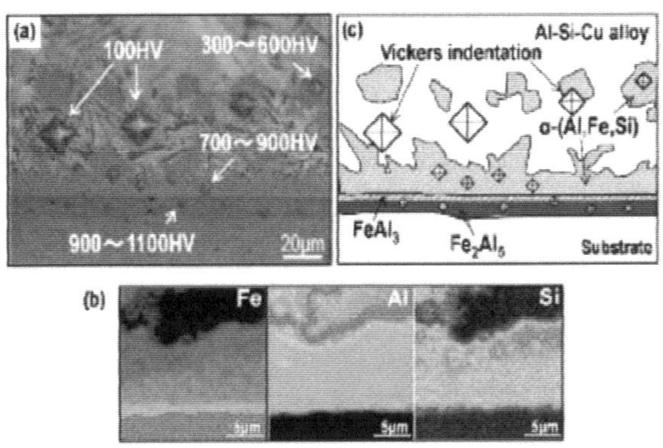

Figure 13: Cross-sectional interaction layer between as-tempered specimen and ADC12 metal: (a) optical micrograph, (b) elemental mappings obtained by EPMA, and (c) schematic diagram of (a).

that the erosion of steel in molten aluminum alloy generally includes the four following processes (with reference to Fig. 14) [10]:

(a) formation of intermetallic compounds,
(b) diffusion of iron in intermetallic compound layers,
(c) transfer of iron to solute-element diffusion-layer in molten aluminum alloy from intermetallic compound layers, and
(d) transfer of iron into the bulk of molten aluminum alloy through a solute-element diffusion layer in the molten metal.

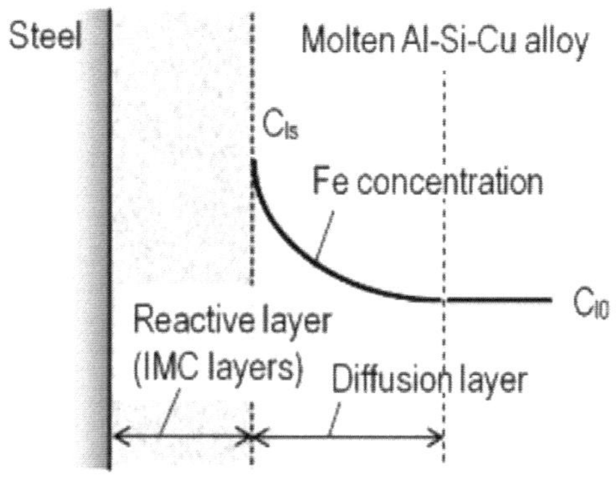

Figure 14: Schematic of boundary between the specimen and liquid Al-Si-Cu alloy, and distribution of iron concentration (C_{ls}; saturated solubility concentration of Fe, C_{l0} ; Fe concentration in liquid Al-Si-Cu alloy, IMC; Intermetallic compound).

In the case that dissolution of solid in fluid is rate controlled by transfer of solute elements in the fluid, mass-transfer coefficient J_D between the solid and liquid under forced convection can be expressed by the following dimensionless correlating equations [13]:

$$J_D = (k_c/u)(Sc)^b = c \cdot (Re)^a \text{ -------------- } (4)$$

$$u = 2\pi rR \text{ -------------------------------------- } (5)$$

$$Re = LU\rho/\mu \text{ ------------------------------------ } (6)$$

$$Sc = \mu/\rho D \text{ -------------------------------------- } (7)$$

where k_c (g/mm^2·s) denotes the theoretical mass-transfer coefficient, u (mm/s) is relative velocity, Sc is the Schmidt number, Re is the Reynolds number, and a, b, and c are constants. R (s^{-1}) is rotating speed of the specimen, μ (g/mm·s) is the viscosity coefficient, ρ (g/mm^3) is density of molten metal, L (mm) is diameter of the specimen, and D (mm^2/s) is the diffusion coefficient. Rearranging Equation (4) gives k_c as follows:

$$k_c = \frac{cu \cdot (Re)^a}{(Sc)^b} \text{ -------------------------------- } (8)$$

For the viscosity coefficient, diffusion coefficient, and density of molten metal, the values for pure aluminum were employed in this study [14,15,16]. And constants a=0.644, b=0.330, and C=0.112 (for measured mass transfer to molten pure iron from a rotating cylindrical specimen made of pure iron or AISI1045 steel) were employed [13,17,18]. Meanwhile, Re number used was 1.4×10^4. The effect of Re number on erosion morphology of the specimen was not considered [19]. In the case that dissolution of the specimen is regarded to be dominated by mass transfer in molten metal, the observed mass-transfer coefficient, k_{ob} is expressed as

$$k_{ob} = -\rho_s v / (\%x_{ls}\rho_{ls}/100 - \rho_1/100) \text{ ------- } (9)$$

where ρ_s (g/mm^3) is density of the steel specimen, v (mm/s) is dissolution rate, ρ_1 (g/mm^3) is density of molten metal, ρ_{ls} (g/mm^3) is density of molten metal saturated with solute atoms, and $\%x_{ls}$ (mass%) is concentration of saturated solution in molten metal. The rate-determining process of the erosion can be estimated by comparing the theoretical mass-transfer coefficient, k_c with the observed mass-transfer coefficient, k_{ob}. Specifically, if k_c is approximately equal to k_{ob} ($k_c \cong k_{ob}$), (d) can be considered as the controlling process, and if k_c is larger than k_{ob} ($k_c >> k_{ob}$), (a) to (c) can be expected to be the controlling processes [10].

In Figure 15, k_{ob} for each specimen is plotted in comparison with k_c at molten-Al-Si-Cu-alloy temperature of 1023 K for immersion times of 0.6 and 1.8 ks. From Equation (8), k_c was calculated to be 7.35×10^{-2}. When immersion time is 0.6 ks, k_{ob} of the as-tempered specimen, specimen CN, and specimen CLL are expected to be 1.2×10^{-1}, 2.6×10^{-2}, and 4.7×10^{-2}, respectively. In the case of these specimens, (d) is can thus be considered the rate-determining process. On the other hand, k_{ob} of specimens CNO and LCN are 7.4×10^{-3} and 1.8×10^{-3}, respectively, indicating $k_c >> k_{ob}$. This discrepancy is explained by the fact that the ferrous-oxide layer, especially the Li-Fe-compound-oxide layer on the surface, inhibits the reaction between the specimen and the molten Al-Si-Cu alloy and, thereby, inhibits the formation of the intermetallic compound on the specimen surface. In the case of specimen CNO, k_c can be considered almost commensurate with k_{ob} for 1.8 ks; that is, the rate-determining process is considered to shift to (d) after 1.8 ks.

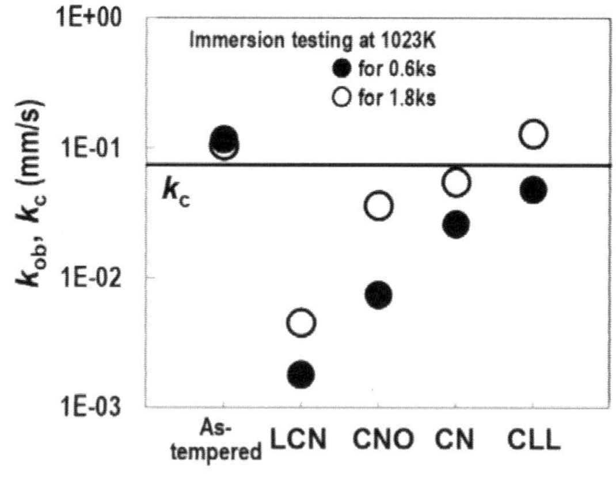

Figure 15: k_{ob} of each specimen compared with k_c at 1023 K for 0.6 and 1.8 ks.

Effect of surface-oxide layers on erosion behavior

The reason that the rate-determining process concerning erosion of specimens CNO and LCN (with an oxide-layer on their surface) is not (d) was investigated as follows. First, the dissolution rate after 3.6 ks increases to almost the same rate as that of specimen CN even though the dissolution rate is very small at 1.8 ks. This means that wear and/or spalling of the ferrous oxide-layer occurs because the layer is characterized by a porous-like and thinner morphology, so the reaction of specimen CNO with the molten Al-Si-Cu alloy starts. It can be concluded that the Li-Fe compound-oxide-layer of specimen LCN remains intact and inhibits the reaction with the molten Al-Si-Cu alloy effectively, even after the longer-time immersion, because the compound-oxide layer is much denser and thicker than the ferrous-oxide layer.

To understand the initial stage of erosion on specimen LCN, the change in the cross-sectional microstructure of the interface between the specimen and Al-Si-Cu alloy at 1023 K during immersion for 0.18, 0.6, 1.8, and 3.6 ks were examined more closely. The result shows that the intermetallic compounds are not formed because the reaction with the Al-Si-Cu alloy is moderated by the existence of the compound-oxide layer, but the intermetallic compounds are observed partially after 3.6 ks (see Fig. 16).

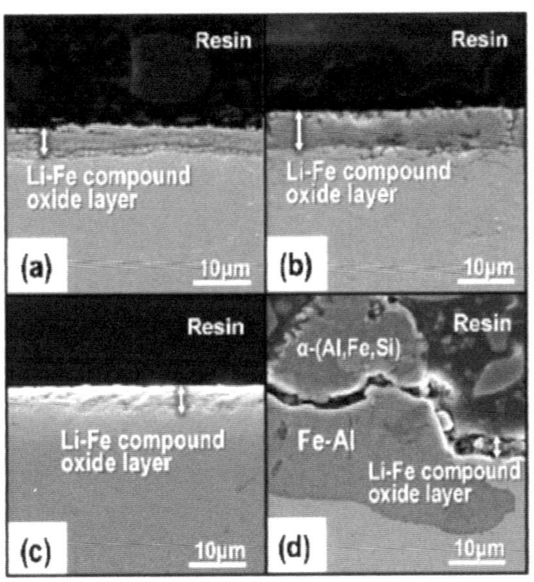

Figure 16: FE-SEM images showing cross-sectional interaction layers of LCN specimens: (a) immersion test for 0.18, (b) 0.6, (c) 1.8, and (d) 3.6 ks.

An FE-SEM image of the cross-section of the interface and corresponding GD-OES depth profiles of the specimen immersed for 1.8 ks are shown in Fig. 17. The appearance of a slight (but sufficient) aluminum concentration peak inside and just below the compound oxide layer indicates that a Fe-Al substitutional solid-solution phase is formed. The compound layer, composed of ε-$Fe_{2\text{-}3}N$ and γ'-Fe_4N phases, decomposed

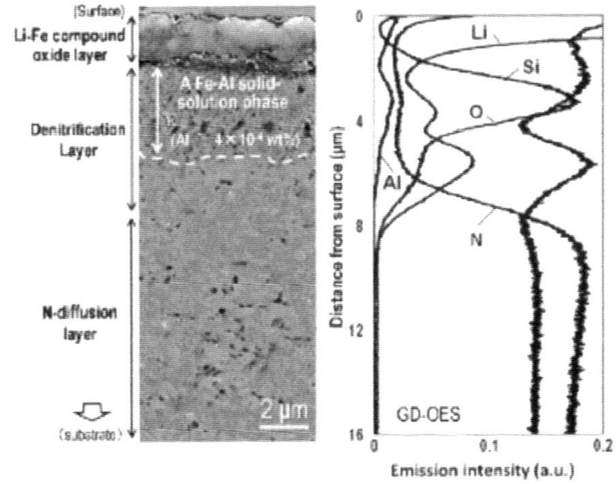

Figure 17: FE-SEM image showing cross-sectional microstructure and corresponding concentration depth profiles of specimen LCN analyzed by GD-OES after immersion test for 1.8 ks.

almost entirely at the melting temperature of Al-Si-Cu alloy and changed to a porous-like layer due to denitrification. The Fe-Al solid-solution phase is thought to be developed in and just under the porous-like layer by mutual diffusion of aluminum and iron atoms through the compound-oxide layer, which cause micro-cracks to develop in the compound-oxide layer because the phase volume of Fe-Al solid-solution increases (lattice constants are Al=4.0496 Å and Fe=2.8664 Å). Intermetallic compounds are then formed by the reaction of specimen LCN with the molten Al-Si-Cu alloy that penetrates through the micro-cracks. The erosion progresses with subsequent spalling of the compound-oxide layer. Therefore, the erosion resistance increases with increasing thickness of the Li-Fe compound-oxide layer, and the increased erosion resistance delays the formation of the Fe-Al solid-solution phase and intermetallic compounds (as shown in Fig. 18). However, a compound-oxide layer with thickness up to 9 μm might be appropriate to inhibit the erosion because the oxide layer easily causes spallation when it gets excessively thick.

The change in cross-sectional microstructures of the Li-Fe compound-oxide layer on the surface of specimen LCN during heating under a nitrogen-gas atmosphere at 1023 K is shown in Fig. 19. It is clear from the Fig. 19 that the compound layer disappears completely after 1.8 ks, but the Li-Fe compound-oxide layer is remains, even after 10.8 ks. This result means that the Li-Fe compound-oxide layer has a superior resistance to immersion at high temperatures.

Effect of nitrogen-compound layer and nitrogen-diffusion layer on erosion behavior

As shown in Fig. 20, increasing the thickness of the nitrogen-compound layer improves the erosion resistance of the specimen. This is because the nitrogen-compound layer inhibits the reaction between the steel and molten Al-Si-Cu alloy. However, in consideration that the nitrogen-compound

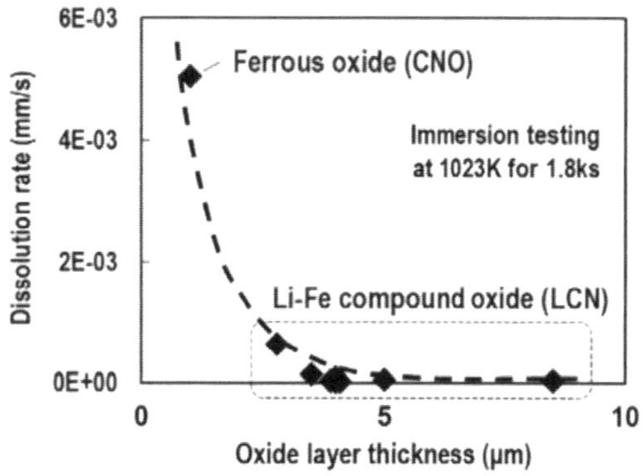

Figure 18: Effect of thickness of oxide layers on dissolution rate.

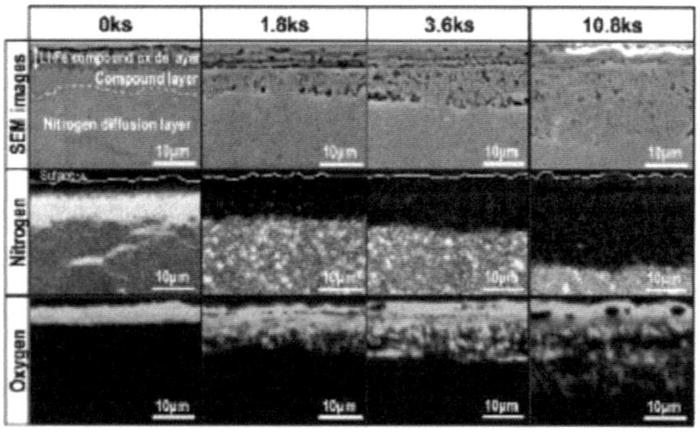

Figure 19: Changes in cross-sectional surface microstructures of specimen LCN, namely, Li-Fe compound-oxide layer, compound layer composed of ε/γ' phases, and nitrogen-diffusion layer at 1023 K for 1.8 to 10.8 ks under a nitrogen-gas atmosphere (analyzed by FE-SEM-EDS).

layer is easily decomposed by the molten metal at a high temperature (such as 923 K and over), the nitrogen-compound layer must have limited effectiveness against the erosion.

Specimen CLL (with only a nitrogen-diffusion layer on the surface) has better resistance to erosion than the as-tempered specimen in the early immersion stage; however, after 1.8-ks immersion, it has almost the same resistance as the as-tempered specimen. Hihara et al. [20] reported that the erosion is accelerated by discharge of nitrogen gas (originated from the solute nitrogen in the nitrogen-diffusion layer) at a molten-metal temperature. However, in consideration of the small amount of solute nitrogen in the nitrocarburized SKD61 steel used in this study (because diffusing nitrogen is consumed for precipitation of nitrides such as CrN, VN, and Si_3N_4), the erosion accelerative effect of the nitrogen-diffusion layer must be negligible. It can thus be considered that the nitrides (having excellent heat resistance) suppress the iron-diffusion flux into the molten Al-Si-Cu alloy and increase the viscosity of the molten Al-Si-Cu alloy [21], thereby improving erosion resistance at the early stage of immersion. Thus, the erosion process in the case of SKD61 steel (salt-bath nitrocarburized with a Li-Fe compound-oxide layer in the molten Al-Si-Cu alloy) can be summarized, in comparison with as-tempered specimens CN, CNO, and CLL, as illustrated in Fig. 21.

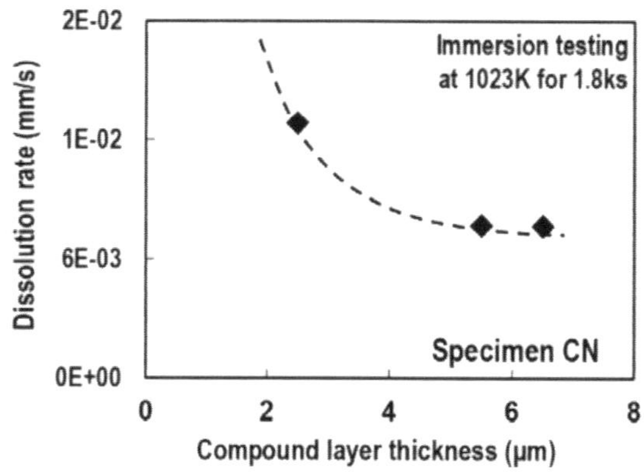

Figure 20: Effect of thickness of nitride-compound layer on dissolution rate.

Estimate of lifetime of die nitrocarburized by salt bath with lithium-iron compound-oxide layer

The effect of the Li-Fe compound oxide layer on the surface was focused on next; in particular, the lifetime of a die-casting die nitrocarburized by lithium-containing salt bath was estimated. First, it was supposed that a quenching-tempered die made of SKD61 steel is worn down by W=0.2 mm at the surface after 2×10^4 casting cycles (namely, lifetime of the die) under the following conditions: molten-aluminum-alloy temperature of 953 K, gate speed u of 40 m/s, and filling time T of 0.02 s. The relation between the relative velocity that is equivalent to the gate speed in the casting operation and dissolution rate can be obtained from the immersion-test at 953 K by using quenching-tempered (abbr. as-tempered) SKD61 steel. In this study, the appearance of erosion damage did not depend on the kind of surface treatment. Accordingly, under the assumption that the dissolution rate has the same positive correlation with relative velocity (that is, gate speed), for specimens LCN, CNO, and CN, as shown in Fig. 22, the dissolution rates v of specimens LCN, CNO, and CN can be derived respectively for u of 40 m/s.

Accordingly, the number of casting cycles (N) to lifetime of the dies nitrocarburized with or without Li-Fe-compound-oxide or ferrous-oxide layers on the surface can be estimated from the following equation:

$$N = \frac{W}{v \cdot T} \quad \text{--------------------- (9)}$$

where N (abbr. N_{CN}) of the die nitrocarburized without the oxide layer on the surface is calculated to be 2.8×10^6 casting cycles (increased by 14 times compared to that for as-tempered die), N_{CNO} of the die nitrocarburized with a ferrous-oxide layer is estimated to be 2.8×10^7 casting cycles (increased by 140 times compared to that for the as-tempered

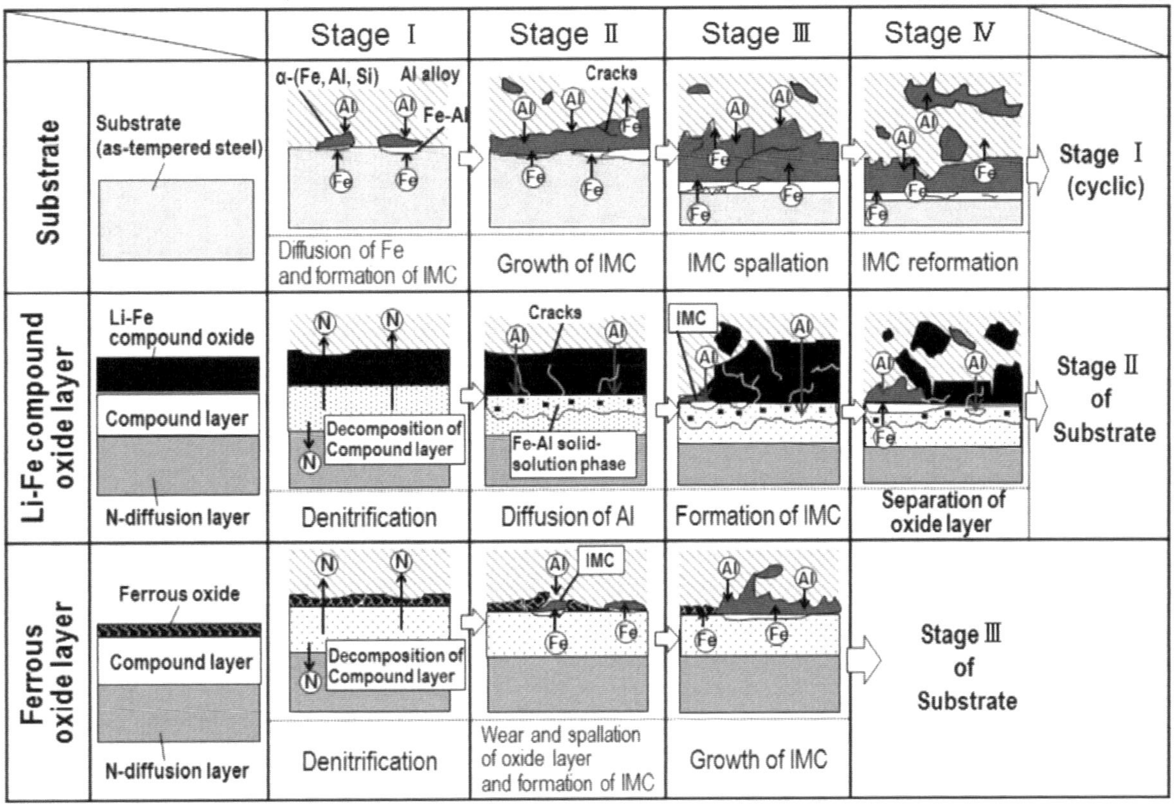

Figure 21: Schematic explanation of erosion processes for as-tempered steel (base steel as a substrate), Li-Fe compound-oxide layer, and a ferrous-oxide-layer (both formed by salt-bath nitrocarburizing of SKD61 steel).

die), and N_{LCN} of the die nitrocarburized with a Li-Fe compound-oxide layer is calculated to be 5×10^7 casting cycles.

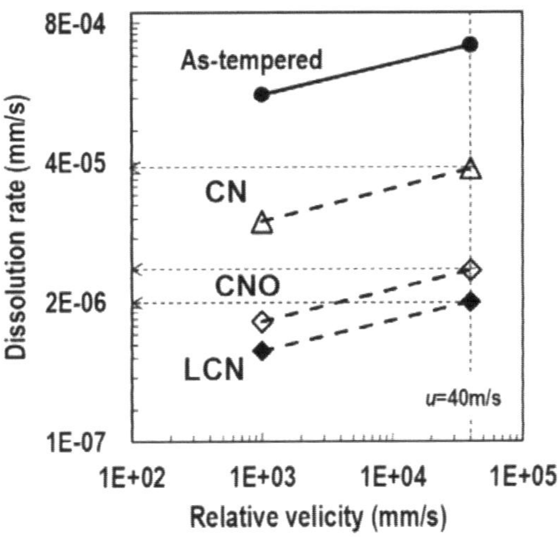

Figure 22: Positive correlations between dissolution rate and relative velocity for as-tempered specimen and specimens LCN, CNO, and LCN with various surface treatments (see text for details).

N_{CN} is therefore expected to be improved by as much as 250 times compared to that for the as-tempered die. However, under the practical casting conditions, the dies are subjected to more mechanical damage than that of the cylindrical-bar specimen used in this study. This severe casting condition might cause excessive spalling of the surface oxides and nitrogen-compound layers. These actual lifetime-improvements therefore estimated to be much smaller than the predicted values presented above.

Conclusions

With the aim of improving lifetime of an aluminum die-casting die, the erosion behavior of die-steel SKD61, nitrocarburized using a lithium-ion-containing salt bath, in molten Al-Cu-Si alloy (ADC12) was investigated. In particular, the effects of a Li-Fe compound-oxide layer in comparison with a ferrous-oxide layer on erosion resistance were investigated. The results of the investigation are summarized as the following three points.

1) The lithium-ion compound-oxide layer on the surface, which consists of an α-$LiFeO_2$ structure (described as $Li_xFe_{1-x}O$), improves the erosion resistance much more than the ferrous oxide (largely Fe_3O_4, formed generally by post-oxidation). This improved erosion resistance is explained by the fact that the lithium-iron compound-oxide layer is much

thicker (and denser) with a greater heat resistance than the ferrous-oxide layer, thereby inhibiting the interface reaction with molten Al-Si-Cu metal.

2) When the Li-Fe compound-oxide layer contacts the molten metal for a prolonged time, a Fe-Al solid-solution phase forms just under the oxide layer by diffusion of aluminum atoms through the oxide layer from the molten metal. This aluminum diffusion causes micro-cracking and spalling of the oxide layer, resulting in progress of erosion accompanied with formation of intermetallic compounds such as Fe_2Al_5, $FeAl_3$, and α-(Al, Fe, Si). Accordingly, the erosion resistance increases as the thickness of the Li-Fe compound-oxide layer increases.

3) A compound layer composed of ε-$Fe_{2\text{-}3}N$ and γ'-Fe_4N also has an inhibiting effect on the interface reaction with the molten metal; however, it has a limited effect against the erosion resistance because it can be decomposed at a molten-metal temperature over 923 K. It is concluded that the nitrogen-diffusion layer improves the erosion resistance a little only at the early stage of immersion in contact with the molten metal. Although this improvement of erosion resistance is thought to be caused by the presence of precipitated nitrides (such as CrN, VN, and Si_3N_4), it must be clarified by further investigation.

Acknowledgments

The authors express many thanks to our technical staff, namely, A. Nakazawa and K. Morita for making available the equipment for analyzing the test specimens. Special thanks are extended to our plant engineer T. Yamamura for his helpful discussions concerning salt-bath nitrocarburzing, Ootomo Seisakusho Company for providing a part of the ADC12 alloy ingot, and NIPPON STEEL & SUMIKIN TECHNOLOGY Co., Ltd. for their helpful discussions concerning the XRD results.

References

[1] S. H. CHANG et al., "Enhancement of Erosion Resistance on AISI H13 Tool Steel by Oxynitriding Treatment," *ISIJ International*, Vol. 49, No. 3 (2009), pp. 421-424.

[2] V. Joshia et al., "Dissolution and soldering behavior of nitride hot working steel with multilayer LAFAD PVD coatings," *Surface and Coatings Technology*, Vol 146-147 (2001), pp. 338-343.

[3] S. SHANKAR et al., "Die Soldering: Mechanism of the Interface Reaction between Molten Aluminum Alloy and Tool Steel," *Metallurgical and Materials Transactions B*, Vol. 33B (2002), pp. 465-476.

[4] M. S. Sidhu, "Liquid Aluminium corrosion Characteristics of Cast Iron and Steel," Canterbury University (2012), Ph. D. thesis.

[5] E. M. Khoonsari, "Reinforced Aluminum Structure Castings for Powertrain Automotive Applications," McGill University (2009), Ph. D. thesis.

[6] Ting-Cheng Chou et al., "The effect of nitriding and oxdation treatment on steel erosion resistance in liquid aluminum," *17th IFHTSE*, (2008), pp. 241-244.

[7] H. Ishizuka et al., "Effect of surface microstructures on corrosion resistance and wear behavior of salt-bath nitrocarburised low-carbon steels," *NETSUSYORI*, Vol. 55, No. 3 (2015), pp.196-204 (in Japanese).

[8] J. Ichikawa et al., "The Dissolution Behavior of SKD 61 in Molten B 390 Alloy under High Velocity Flow," *ELECTRIC FURNACE STEEL*, Vol. 60, No. 4 (1989), pp. 343-353 (in Japanese).

[9] K. Nishida et al., "Some Observations on the Inter-diffusion between Fe and Fe-Al Alloy," *Journal of the Japan Institute of Metals and Materials*, Vol. 35, No. 3 (1971), pp. 269-276 (in Japanese).

[10] M. Niinomi et al., "On the Alloy Layers Formed by the Reaction between Ferrous Alloys and Molten Aluminium," *Transactions of Japan Institute of Metals*, Vol. 23, No. 11 (1982), pp. 709-717.

[11] S. Kobayashi et al., "Formation of η-Fe_2Al_5 Grains Layer in Hot-dipped Aluminum Coated Iron," *17th IFHTSE* (2008), pp. 297-240.

[12] M. Hansen et al., *Constitution of Binary Alloys* (1958).

[13] M. Kosaka et al., "Mass-Transfer from Solid Metal Cylinder into Liquid Metal," *Tetsu-to-Hagane*, Vol. 52, No. 12 (1966), pp. 1748-1762 (in Japanese).

[14] T. P. Yao et al., "The Viscosity of Molten Tin, Lead, Zinc, Aluminium, and Some of Their Alloys," *J. Inst. Met.*, Vol. 81, No. 17 (1952).

[15] K. Uemura, "The Diffusion of Various Elements into Aluminium in Molten State," *Iron and Steel*, Vol. 25, No. 1 (1939), pp. 24-29 (in Japanese).

[16] J. F. Elliot et al., *Thermochemistry for Steelmaking*, Addison-Wesley, Vol. 1 (1960).

[17] R.N. Smith et al., "Turbulent transport to a Rotating Cylinder for Large Prandtl or Schmidt Numbers," *J. Heat Transfer*, Vol. 97, No. 4 (1975), pp. 594-597.

[18] YEONG-U Kim et al., "Mass transfer during dissolution of a solid into liquid in the iron-carbon system," *Metallurgical transaction*, Vol. 5 (1974), pp. 2527-2532.

[19] Kikai-Sekkei Binran (Handbook), (1973), pp1926.

[20] M. Hihara, 316th Sosei-Kakou Symposium (in Japanese).

[21] M. Ibuki et al.," Effect of Carbide Volume Fraction and Carbide Type on Dissolution of Steels in Molten Al Alloy," *ELECTRIC FURNACE STEEL*, Vol. 76, No. 4 (2005), pp. 241-248 (in Japanese).

International Federation of Heat Treatment and Surface Engineering 2016
Proceedings of the 23rd IFHTSE Congress
April 18–21, 2016, Savannah, Georgia, USA

DEVELOPMENT OF A LABORATORY INTENSIVE QUENCHING (IQ) SYSTEM

Luigi L. M. Albano[1], Fernando S. Misina[1], George E. Totten[2], Lauralice C.F. Canale[1]

1. University of São Paulo, São Carlos, SP, Brazil
lfcanale@sc.usp.br
2. Portland State University, Portland, OR
GETotten@gmail.com

Abstract

To perform intensive quenching (IQ), it is important that a sufficient mass of fluid is available to remove heat from the hot-metal interface so that the boundary temperature is at, or as close a possible, to the bath temperature. This condition requires a sufficiently high fluid mass flow rate and it is not equivalent to "high" or "very high" agitation rate. Adequate heat removal for the component being quenched is a necessary condition for the formation of optimally high surface compressive stresses, which is a fundamental characteristic of intensive quenching. Although a limited number of laboratory quenching systems that provide the necessary fluid flow/heat transfer conditions to provide optimal surface compressive stresses have been reported, most are relatively large requiring large quantities of quenchant. A brief overview of previously reported IQ systems will be provided here in addition to the details of the development of a smaller laboratory IQ system that has been shown to successfully produce very high surface compressive stresses will be provided here.

Keywords: Intensive quenching, residual stresses, agitation

Introduction

Metallurgists are generally taught that the potential for quench cracking increases with increasing cooling rate [1]. However, since the 1920's, there have been various, often little known, industrial heat treating processes that have been designated as intense, intensive, rapid, drastic, severe, or extreme quenching or shell-hardening methods [2,3]. The essence of these methods is to harden less hardenable steels using very fast cooling rates in order to impart high compressive stresses and improve fatigue properties of the quenched component [4]. Very high surface compressive stresses, typically in excess of those achievable by carburizing, induction heat treating and shot peening have been shown for IQ processed steels [4,5,6].

In 1964, Kobasko published the first of an extensive series of papers in which he used the term "Intensive Quenching" and showed experimental data which provided numerical evidence that showed that although it is true that increasing cooling rates result in increasing propensity for cracking as historically recognized, there does exist a critical cooling rate above which cracking propensity decreases [7]. A major contribution of Kobasko's lifetime of research is two-fold. The first has been to provide a rigorous numerical criterion that defines the quenching conditions that provide optimal surface compressive stresses [8,9] and the second was to develop quench system and process designs that achieve these effects for a broad range of steel and cast iron compositions. Subsequently, various industrial intensive quenching processes were developed and patented by Kobasko [10,11,12]. An invaluable reference for use in designing an intensive quenching system has been published by Kobasko, *et. al.* [13].

All of these reports have a common denominator and that is to provide very high and uniform fluid flow to maximize heat transfer from the steel surface at the appropriate time. In order to study the effects of intensive quenching, it would be desirable to utilize a relatively small quench system which would require relatively limited amounts of environmentally safe quenching fluids suitable for use in the laboratory. Recently, a project was conducted at the University of Sao Paulo to develop a small IQ quenching system suitable for laboratory use [14]. The design and construction of this system is the subject of this report. A brief overview of some other laboratory systems that have been reported for IQ process studies will also be provided.

Discussion

1. Review of Previously Reported IQ Systems

There have been numerous publications describing quench systems. Some historical references have noted that indicate that quench systems that provide very high ("extreme" or "violent") agitation rates with a Grossmann H-value of >6.0 are required for intensive quenching [15,16]. Another definition of intensive quenching that has been reported is the condition where the Kondratyev number (Kn) is: $0.8 \leq Kn \leq 1$ [17]. The value of Kn is calculated from:

Where: Ψ is the temperature field non-uniformity criterion which is defined as:

$$\psi = \frac{T_{sf} - T_m}{T_V - T_m} \quad (1)$$

Where: T_{sf} is the average surface temperature of the component being quenched, T_m is the temperature of the quenchant, and T_V is the average temperature over the volume of the component being quenched.

The value of Ψ may also be defined as a function of the generalized Biot criterion Bi_V:

$$K_n = \psi Bi_v = \frac{Bi_v}{\left(Bi_v^2 + 1.437 Bi_v + 1\right)^{1/2}} \quad (2)$$

And the Bi_V is defined as:

$$\frac{\alpha}{\lambda} L = \frac{\alpha}{\lambda} K \frac{S}{V} \quad (3)$$

Where: L is the size characteristic calculated from:

$$L = \frac{S}{V} K \quad (4)$$

In this equation, S is the surface area of the part being quenched, V is the volume of the part, K is the Kondratyev form coefficient (shape factor) which may be found in look-up tables such as those provided in References [8,13].

In a series of patents published from 1967 to 1971 [17-19], somewhat greater quantification was provided for the production of various machinery parts using a process designated as a "drastic" or "extreme" quench. In these processes, pressurized tanks (accumulators or "bladders") were used to deliver very high volumes, approximately 3000 gal/min, of quenchant to selected surface areas of the part being spray or press quenched. The purpose that was stated for performing such high-volume quenchant delivery processes to the hot austenitized steel surface was to achieve maximum hardness uniformity. One patent, USP 3,515,601, reported that there was a "critical cooling rate" which was not provided [19].

One laboratory IQ quench system specifically designed to provide a quench severity sufficient to produce optimal surface compressive stresses was illustrated by Liss, *et. al.* [4]. Unfortunately no design detain details were provided so it is not clear how such a system could be precisely reproduced.

The system that has been encountered in various literature references, especially more recently, is the system reported by Kobasko and co-workers and illustrated in Figure 1 [13, 20-22]. This system was reported to be based on a 800 gal. (3028 L) quench tank with a high–pressure 600 gal/min (2270 L/min) pump [20]. Parts to be quenched are heated in a salt bath or a box furnace and placed individually into the quench zone of the system shown. Kobasko has further reported that a flow velocity over the surface of a part being quenched should be 0.76 m/s (2.5 ft/s) for water or 1.07 m/s (3.5 ft/s) for oil. Although it is undeniable that this is an invaluable and successful system, the fluid volumes involved are rather large for routine quenching research in a university or industrial research laboratory using relatively small test sample size.

Figure 1 – Illustration of the Kobasko IQ "One-by-One" quench ststem

Another system that has been reported to produce essentially equivalent results to those published for the Kobasko system shown above was reported by Rath [23] and Lübben [24]. This system is illustrated schematically in Figure 2. This Rath-Lübben quench system utilizes a pneumatic transfer system for moving the hot metal "probe" to the quench zone. The desired flow rate is obtained within 2-3 seconds and flow rates may be varied from 7-18 L/s and the flow rate depends on the cross-section and alloy composition of the piece being quenched [23]. The flow rate must be sufficient to reduce the surface temperature of the steel below the Ms in order to maximize the temperature differential between the surface and the core in order to achieve conditions that will yield maximum surface compressive stresses [24].

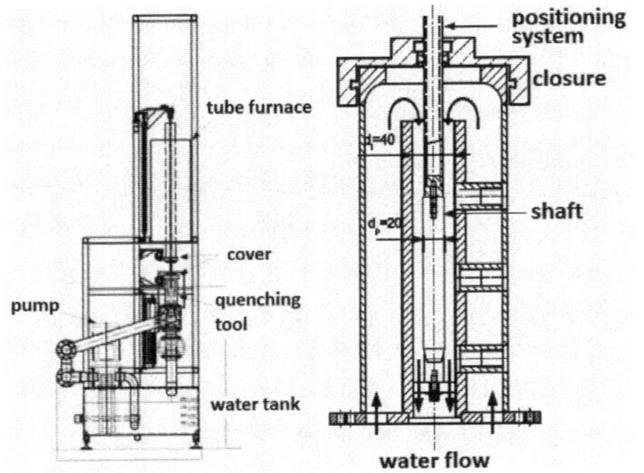

Figure 2 – Illustration of the Rath and Lübben IQ quench system.

A third laboratory IQ system that will be shown here was designed at the University of Sao Paulo [14] and is illustrated in Figure 3. The size of the quench tank used for this system is: 1.3 m (L) x 0.7 m (W) x 1 m (H) with a volume of 936 L (247 Gal) but the actual volume used for the IQ work was approximately 468 L (124 Gal) and the maximum flow rate through the quench zone is 1230 L/min (325.7 Gal/min). One of the primary differences in this system from those shown in Figure 1 and Figure 2 is that this system utilizes a very high flow impingement rate on the surface of the part being quenched whereas the previously discussed systems utilize a very high mass flow rate parallel to the cooling surface.

A disadvantage of these three IQ quenching systems is the high volume requirement and the footprint of the equipment in a laboratory that often has only limited space. So, at least for initial research, it would be highly desirable to design a much smaller system that will produce quench severities sufficient for the formation of maximum surface compressive stresses. The construction of such a unit is the subject of the remainder of this paper.

Figure 3 – High-Volume USP IQ Spray Quench System constructed at the University of São Paulo -- TOP Left shows the entire Large Volume IQ System and TOP Right Corner shows the spray quench zone and the picture underneath is an image of the spray head through which the quenchant is delivered into the quench zone when the flow is activated. BOTTOM: The steel is placed in a clamp in this assembly, the quench zone closed and then the spray quench is started. This figure shows the progression of cooling a 50 mm (2 inch) steel bar during the IQ spray quenching process.

1. Design of the USP Small Volume Laboratory IQ System

The small volume IQ system is illustrated in Figure 4. The tank was constructed from 2.0 mm (0.08 inch) AISI 1020 sheet stock and with dimensions of 40 cm x 50 cm x 30 cm in height and was assembled by welding (AWS 70). The pipes are ½ inch (12/7) PVC plastic. The total fluid capacity is 60 L (15 Gal) but typically the fluid volume used is 50 L (13 Gal). To provide corrosion protection, the system was electrostatically painted. Thus far, there have been no corrosion issues with the use of aqueous solutions of $NaNO_2$ as the quenchant which is used at ambient temperature (typically at 27°C).

Figure 4 – Relatively Small Laboratory IQ system designed and used at the University of Sao Paulo for IQ studies of steel and cast iron. The quenchant shown in the tank is a 9% aqueous solution of $NaNO_2$.

Fluid flow is provided by a DAB Model 30/16M centrifugal pump. The flow curve for this pump is shown in Figure 5. The pump is driven by a 0.5 Hp electric motor with a fluid flow rating of 0.6-2.16 m^3/h and a total head height (H) of 25 to 5 m (Hmax = 32.5 m; Hmin = 5 m). This means that the pump will deliver 0.6 m^3/h at a water head pressure of 25 meters at 0.5 Hp or 2.16 m^3/h at a water head pressure of 6 m at 0.5 Hp. For reference, the equation to convert water pressure in psi to water head pressure in meters is:

$$Water\ Head\ Pressure(m) = \frac{Water\ Pressure(psi)}{1.4203\ psi/m}$$

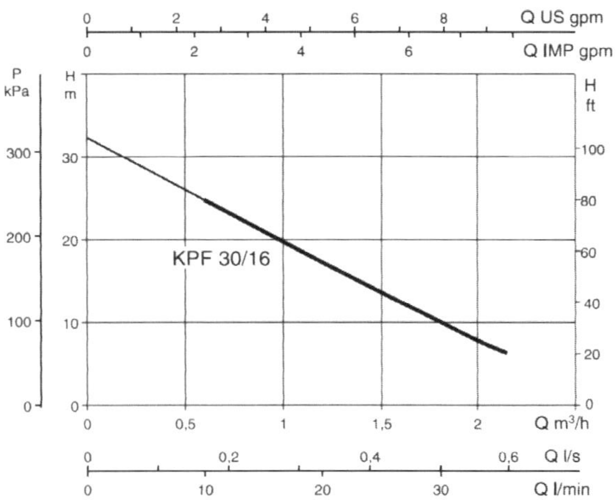

Figure 5 – The flow curve for a DAB Model 30/16M centrifugal pump.

Agitation is provided by a flow jet at each side of the tank as shown in Figure 6. The jets are placed at opposite sides of the tank in a staggered arrangement so that the two jets do not impinge each other directly.

Figure 6 – In the above illustration the quench tank is shown on the right and the fluid overflow tank is shown on the left. The fluid flow jets are also shown at the top and bottom of the quench tank. A drain is shown at the bottom of each tank.

The fluid flows into the tank through AMANCO PVC - poly(vinyl chloride - plastic plugs which act as flow jets, one on each side of the tank [27]. The inlet pipe diameter leading to the flow jets in ½ inch (12.7 mm). Each plug has a 1.5 mm dia. hole on each side of the plug (see Figure 7). The total volume flow rate through the horizontal spray nozzles was measured gravimetrically to be 135 g/s for the left nozzle and 145 g/s for the right nozzle.

Figure 7 – The fluid flow jets were AMANCO PVC plastic plugs with a 1.5 mm hole on each side of the plug as shown.

2. Validation of Intensive Quenching

The question at this point is if this Small Laboratory IQ system described here is able to provide sufficient agitation conditions to yield characteristically high surface compressive stresses. An initial answer to this question can be ascertained by comparing quenching results obtained by a study reported earlier where AISI 5160 was quenched under IQ conditions and residual stresses measured [26]. Those results were compared the results of the work conducted at USP on the same alloy. Although a different cross-section size and configuration was used for the earlier published work, Table 1 shows that the residual stresses obtained for AISI 5160 steel reported earlier are very close to those obtained in this work. Based on this limited comparison, it is concluded that the laboratory IQ system in conjunction with a manual time quenching process can be used to study the effect of intensive quenching on the properties attainable for other steel compositions.

Table 1- Comparison of Published Residual Stresses for AISI 5160 with Results Obtained Using the Small Volume Laboratory IQ System.

Steel	Component	Residual Stress (MPa)	Reference
5160 Oil Quench)[1]	Fracture Toughness Test Specimen	-50	This work
5160 (Oil Quench)	Steel Rod (27 mm dia x 127 mm)	-102	[26]
5160 (IQ)[2]	Fracture Toughness Test Specimen	-916	This work
5160 (IQ)	Steel Rod (27 mm dia x 127 mm)	-939	[26]

1. The oil quench for this work was performed by immersion of the austenitized test specimen into 30L of TP444 Houghton KB quench oil with mild manual agitation.
2. The IQ system used to obtain this data is the Small Laboratory IQ System described in this report.

Conclusions and future work

An overview of previously published quenching systems that have been reported for use to study IQ processes was provided here. In addition the successful construction and use of a SMALL VOLUME LABORATORY IQ System was also described.

Work is currently in progress to model the flow uniformity and heat transfer properties possible in the quench zone of this equipment. These studies will include potential modifications that will permit well-controlled flow variation as well as a study of quenchant compositions suitable **for** IQ processing. In addition, the properties of a range of steel and cast iron compositions that have been subjected to the IQ process is also in progress, particularly fatigue, fracture toughness and residual stress. The results of these studies will be reported when completed.

Acknowledgements

The authors acknowledge CAPES (Coordenação de Aperfeiçoamento Pessoal de Nível Superior) and Universidade de São Paulo (USP) for their financial support. The vital consultations with Roland J. Bishop Jr. of RJ Bishop & Associates, Inc. regarding assessment of the pumping and fluid flow systems.

References

1.G. Beck, "Internal Stresses Associated with Heat Treatment of Alloys", Mem. Etud. Sci. Rev. Metall., 1985, June, p. 269-282.

2. H. Kurz, "Process for Manufacture of Rails with Hardened Heads", USP 1,828,325, October 20,1931.

3. R.F. Kern, "Intense Quenching", Heat Treating, 1986, September, p. 19-23.

4. R.B. Liss, C.G. Massieon and A.S. McKloskey, "The Development of Heat Treat Stresses and Their Effect on Fatique Strength of Hardened Steels", SAE Technical Paper Series, Paper Number 650517, 1965. doi:10.4271/650517.

5. C.R. Hubbard, F. Tang, M.A. Arovov, N.I. Kobasko, J.A. Powell, B.L. Ferguson, Z. Li, "Effect of Intensive Quenching on Residual Stress", in *Proceedings of 23rd ASM Heat Treating Conference*, Eds. D. Herring, R. Hill, 2005, Pittsburgh, PA, USA, ASM International, Materials Park, OH, p. 338-342. ISBN: 0-87170-832-9.

6. N.I. Kobasko, M.A. Aronov, K. Ichitani, M. Hasegawa, K. Noguchi, "High Compressive Residual Stresses in Through Hardened Steel Parts as a Function of Biot Number", in *Recent Advances in Fluid Mechanics, Heat & Mass Transfer and Biology*, Rds. Manoj K. Jha, Myriam Lazard, Azami Zaharim, Kamaruzzaman Sopian, Havard University, Cambridge, MA, USA, January 25-27, 2012, WSEAS Press, p. 35-40. ISBN: 978-1-61804-065-7.

7. N.I. Kobasko, and N.I. Prokhorenko, "Cooling Rate Effect of Quenching on Crack Formation in 45 Steel", Metal Science and Heat Treatment, 1964, Vol. 6, No. 2, February, p. 104-105.

8. L.C.F. Canale, N.I. Kobasko and G.E. Totten, "Intensive Quenching: Part 1 – What Is It?", International Heat Treatment and Surface Engineering, 2007 , Vol. 1, No. 1, p. 30-33.

9. L.C.F. Canale, N.I. Kobasko and G.E. Totten, "Intensive Quenching: Part 2 – Formation of Optimal Surface Compressive Stresses", International Heat Treatment and Surface Engineering, 2007, Vol. 1, No. 2, p. 60-63.

10. N.I. Kobasko, *Steel Quenching in Liquid Media Under Pressure*, 1980, Naukova Dumka, Kiev, Ukraine.

11. A. Lisovoy, N.I. Kobasko, M.V. Kindrachuk, and A.A. Khalatov, "Stamp Quenching Method", USSR Patent 4,176,788/37-02, 01.06, 1987.

12. N.I. Kobasko, Quenching Apparatus and Method for Hardening Steel Parts", USP 6,364,974, April 2, 2002.

13. N.I. Kobasko, M.A. Aronov, J. Powell and G.E. Totten, *Intensive Quenching Systems: Engineering and Design*, 2010, ASTM International, West Conshohocken, PA. ISBN: 978-0-8031-7019-3

14. L.L.M. Albano, "Estudo Comparativo das Propriedades Mecâicas de aço AISI 5160 Submetidos á Têmpera Convencional e Têmpera Intensiva", MS Thesis, Universidad de São Paulo, Escola de Engenharia de São Carlos, SP, Brazil, 2013. (In Portuguese). Internet: http://www.teses.usp.br/teses/disponiveis/88/88131/tde-17042013-105945/pt-br.php

15. Mei Daming, "Intensive Quenching method for Preventing Quench Cracking", *Heat Treatment and Surface Coatings: New Processes and Application Experience Proceedings of the 7th International Congress of Heat Treatment and Technology of Surface Coating*, Moscow, Russia, December 11-14, 1990, Vol. 2, p. 62-71.

16. A. Morio, "Water Hardening", Kinzoku Zairyou, 1977,Vol. 17., No. 3, p. 45-53.

17. B. Paddock, "Quench Apparatus for Providing Pulsating and Sweeping Flow of Quenching Fluid", USP 3,517,676, June 30, 1970.

18. R.H. Hays, J.E. Sansom and K.D. Gladden, "Die-Quenched Crankshaft", USP 3,589,697, June 29, 1971.

19. J.E. Sansom, "Spray Closet Quench", USP 3,515,601, June 2, 1970.

20. M.A. Aronov, "Intensive Quenching Technology for Heat Treating and Forging Systems", Final Report to Department of Energy, DOE No. DE-FC36-031D 14463, Project Period April 2003-September 2005. Internet: http://www.osti.gov/scitech/servlets/purl/861669 (Accessed January 26, 2016)

21. M.A. Aronov, N.I. Kobasko, J.A. Powell, "Industrial-Scale Intensive Quenching Process for Tool Products Commercialization", Report Prepared for the Edison Material Technology Center, Report No. EMTEC CT-76-II, August 2005. Internet: https://www.forging.org/system/files/field_document/IntensiveQuenchingTechnologyforHeatTreatingandForgingIndustries.pdf (Accessed January 26, 2016)

22. IQ Technologies, Inc., Deformation Control Technology Inc., Air Flow Sciences Corporation, Benet Laboratories, NexTec Corporation, Edison Materials Technology Center, "Intensive Quenching Technology for Advanced Weapons Systems", Phase I Report Prepared for US Army, Report No. W15QKN-16-2-0105.

23. J. Rath, T. Lübben, F. Hoffmann, H-W. Zoch, "Generation of Compressive Residual Stresses by High-Speed Water Quenching", International Heat Treatment and Surface Engineering, 2010, Vol. 4, No. 4, p. 156-159.

24. T. Lübben, J. Rath, F. Krause, F. Hoffmann, U. Fritsching, H.W. Zoch, "Determination of Heat Transfer Coefficient During High-Speed Water Quenching", Int. J. Microstructure and Materials Properties, 2012, Vol. 7, Nos. 2/3, p. 106-124.

25. Further information regarding the design and composition of the PVS plugs from AMANCO in Brazil can be obtained at: http://www.amanco.com.br/produtos/predial/agua-fria/roscavel/plug-roscavel/ . (Last viewed on January 27, 2016.

26. C.R. Hubbard, F. Tang, M.A. Arovov, N.I. Kobasko, J.A. Powell, B.L. Ferguson, Z. Li, "Effect of Intensive Quenching on Residual Stress", in *Proceedings of 23rd ASM Heat Treating Conference*, Eds. D. Herring, R. Hill, 2005, Pittsburgh, PA, USA, ASM International, Materials Park, OH, p. 338-342. ISBN: 0-87170-832-9.

International Federation of Heat Treatment and Surface Engineering 2016
Proceedings of the 23rd IFHTSE Congress
April 18–21, 2016, Savannah, Georgia, USA

Effects of Quench and Partitioning Heat Treatment in mechanical properties and the microstructural characterization of AISI 300M

Ricardo Gratão Gregui
Instituto Federal de São Paulo - IFSP, Votuporanga, São Paulo, Brazil
gratao@yahoo.com.br

Lauralice de Campos Franceschini Canale
Universidade de São Paulo - USP, São Carlos, São Paulo, Brazil

Danilo Assad Ludewigs
Durferrit company – HEF Group, Diadema, São Paulo, Brazil

Abstract

The present work explored thermal processing routes of quenching followed by partitioning to obtain a multiphase microstructure combining brittle and ductile elements with excellent mechanical properties. According to each planned route, this work shows the effects in the mechanical behavior and multiphase microstructural analysis from the low alloyed high strength steel AISI 300M, aeronautical grade. The mechanical characterization was performed through traction tests, for determining strength and ductility values. It were also anaysed the toughness values through Charpy "V" impact strength testing, besides the hardness, measured in Rockwell "C" scale. For the microstructural characterization, the samples were prepared with several metallographic etchants and with the help of an optical microscope it were determined the microconstituents present, such as martensite and retained austenite. Through the thermal processing routes of quench and partitioning, it was possible to reach interesting mechanical properties and microstructural features combining high mechanical strength with increased toughness. The test results of the mechanical and microstructural behavior were discussed and confronted according to the proposed thermal processing situation. This work was supported by the academic institutions EESC – USP and IFSP Brazil, as well as by the company Durferrit from the HEF group.

Introduction

In recent decades, high strength low alloy steels that display the multi-phase microstructure attracted great interest from the industry, especially the automotive industry [1] for use in components, structural parts, for being increasingly resistant and lightweight, what provides lower fuel consumption [2] and increased efficiency of the vehicle.
The use of advanced high strength steels can generate reductions of at least 20% in weight [3], with better performance in crash tests due to high mechanical strength and superior toughness. The chemical composition of the steel and its different microconstituents are a major factor in the final mechanical properties required by project.
To meet this need this research explored heat treatment routes unconventional to the typical ones used by the industry, such as quenching and tempering. This is the promising treatment followed by quenching partition proposed recently [4] in Si-steel plates used in plastic deformation induced process, a phenomenon known as (TRIP). The concept of quenching and partitioning (Q & P) is well defined in the steel literature, but not fully clarified. With huge potential for exploitation and application [5], this new process has been proposed in order to improve the mechanical properties of the low alloyed high strength steel [6].
The motivation for this work was to evaluate the proposed new heat treatment, quenching and partition (Q & P), as well as its concepts, however, the austenitizing temperature used in the heat treatment was 50°C above the recommended standard, thereby promoting complete austenitizing of the material and the total dissolution of carbides and precipitates, what favored the presence of retained austenite after hardening process.
This work aims to achieve a high mechanical strength combined with high toughness of the AISI 300M steel through the (Q & P) process, what sets it apart from conventional quenching and tempering treatments, that require several treatment hours, promote high mechanical strength but in some cases also brittleness in the steel. (Q & P) also generates a multi-phase microstructure controlled by the presence of retained austenite and martensite that meets expectations in terms of mechanical strength and toughness.
After implementing the various routes of thermal processing and hardening partition (Q & P), a multi-phase microstructure that combines hard and ductile microconstituents was observed by optical microscopy, that showed significant fractions of retained austenite and martensite. In tests of mechanical tensile characterization and Charpy impact, specimens showed high mechanical strength combined in some cases with high impact resistance and considerable tenacity.

Experimental Procedure

The analyzed material, AISI 300M manufactured by Villares Metals, Sumaré-SP plant, obtained by vacuum induction melting and arc remelting, was awarded for this research in the hot rolled round bar format in 31,75 mm diameter and 0.5 m length in standard condition with hardness of 27 HRC with stripped and polished finish.

The steel chemical composition analysis is in accordance with the standard AMS 6419 / L, and as per the certificate issued by the quality control of Villares Metals the results are shown in Table 1.

Table 1 - Chemical Composition of the material (%) – 300M

C	Si	Mn	P	S	Co
0,42	1,69	0,86	0,005	0,001	0,01
Cr	**Mo**	**Ni**	**V**	**Cu**	**Al**
0,79	0,37	1,85	0,09	0,02	0,011

Heat treatment routes:

The specimens were subjected to heating up to 920ºC and kept at that temperature for 30 minutes under protection so that there was no decarburization (fig. 01).

Figure 01 - Heat treatment box with decarburization protection at the bottom: carburizing granulate type KG 6.2 (manufactured by Durferrit)

Then the speciments were submitted to annealing at subcritical temperature (240 °C), between the Ms and Mf temperatures. This salt bath temperature was adopted because it was the threshold that generated a better mechanical performance according to the literature [7]. In the next step (partition) two working temperatures were chosen, 300ºC and 400 °C (at the salt bath), varying the immersion time at 30, 60 and 120 seconds for each condition so as to promote the stabilization of retained austenite. Thereafter the samples were cooled in air to room temperature. Figure 02 shows the routes that have been explored.

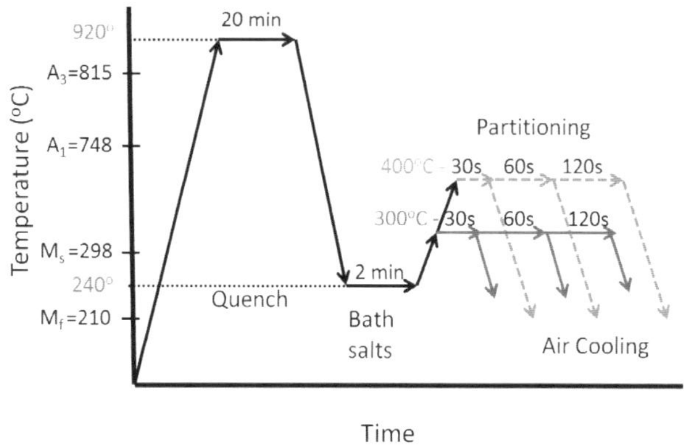

Figure 02 - Heat treatment routes

The heat treatments were performed in the research and development laboratory from the company Durferrit, located in Diadema - SP, who also provided the heat treatment salts, AS 140 (self-manufactured) and collaborated with this work. Two types of furnaces were used: the first a chamber furnace with inert atmosphere (Argon) injected into the heat treatment box, brand Brasimet type KOe 25/15/40, 6,51 kW and maximum temperature of 1100 ° C, and the second a salt bath pit furnace, also from Brasimet, type TOe 35/70, 31,4 kW and maximum temperature of 950 °C. Figure 03 illustrates the two types of furnaces.

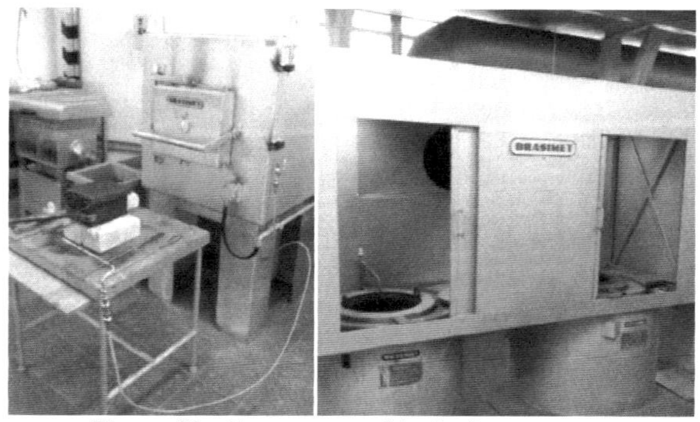

Figure 03 - Furnaces used in the heat treatment

Hardness test:

After the heat treatment, the specimens were hardness measured for each treatment condition on the Rockwell "C" scale (HRC) according to ASTM-E18-2014 standard, performed in a Leco brand durometer, RT-240 model. For all samples five Rockwell C mesurements were performed using a cone diamond indenter, with pre-load of 10 kgf and a total load of 150 kgf.

Impact test:

To determine the toughness values due to heat treatment Charpy impact tests with "V" notch were performed at room temperature, according to the recommendations of ASTM

E23-2006 in a Wolpert PW30 Instron machine model with a capacity of 300J.

Traction test:

To determine the strength and elongation limit values, tensile tests were performed for each treatment condition, at room temperature, according to the requirements of ASTM E8-M-2009. Samples were taken from the longitudinal direction in round billet.
The test was performed with a hydraulic servo machine type EMIC and speed of 1 mm / min, with a capacity of 10 tons. The mechanical characterization equipment used was available in the Department of Materials Engineering of EESC-USP.

Metallographic and phase analyzes:

To reveal the microstructures present in each stage of the test specimens was performed under the following etching solution: 75ml of distilled water, 55 ml of concentrated soap (detergent), 3.5 g of picric acid and 0.5 ml of inhibitor (Jar 3M). After chemical attacks the samples were analyzed under an optical microscope, Carl Zeiss AG brand, model Lab Axio A1, with an Axio cam ERc5s camera.
The phase fractions analysis were performed on the same machine using the Axion Vision 4.8 analysis image software.

Results and discussion

Hardness test:

Table 02 shows the results of testing the hardness in Rockwell scale "C" for each treatment condition.

Table 02: Results of hardness testing

	Partitioning Time		
Partitioning Temperature	**30s**	**60s**	**120s**
300° C	52 HRC	53 HRC	54 HRC
400° C	43 HRC	42 HRC	41 HRC

It can be noted that hardness of parts submitted to partition under 300°C increase with longer partition times. In this condition hardness approaches AISI 300M steel values in the hardened and tempered condition. When partitioning under 400°C there was a decrease in the hardness values for the longer partition times. Exactly the opposite occurred for partition done under 300 °C. It can also be verified that for the partition done at 400 °C the hardness values are much smaller than those obtained with 300 °C.

Impact test:

The results of the impact tests for each condition, energy absorbed are shown in Table 03.

Table 03: Results of Impact test (Joules)

	Partitioning Time		
Partitioning Temperature	**30s**	**60s**	**120s**
300° C	15	14	9
400° C	35	36	37

In the condition in which the partition temperature was 300 °C the values of absorbed impact energy were very low due to the extreme hardness, and displayed fragility. The values were decreased with increasing time of exposure. In partition with 120 seconds, the absorbed energy was 9 Joules.
However, with partition temperature of 400 °C the results were considered very good. Specimens showed excellent thoughness in that condition, and reached absorbed energy values of 37 Joules, when partition time was 120 seconds. It was noticed that the absorbed energy increased with increasing partition times at this temperature.

Tensile test:

The results of tensile tests of the specimens that underwent partition under 300 °C were considered weak, with little plastic deformation and elongation, although having achieved tensile strengths greater than 2000 MPa.
Due to the weakness presented by the specimens in this condition and the unavailability of suitable extensometer for the testing machine and due to the geometry of the specimens, it was not possible to determine all the mechanical properties that the tensile test could offer. Only tensile elongation and tensile strength parameters were mesured.
Under partition temperature of 400 ° C, the strength values were slightly lower compared to those generated under partition temperature of 300 °C, but showed greater elongation, which was a successful result compared to expectations, because it kept considerable mechanical strength allied with the ability to plastically deform. Table 04, 05 e 06 shows the values of the mechanical properties obtained in this test.

Table 04: Results of Tensile Tests for 30s partitioning

	Partitioning Time	
	30s	
PartitioningTemperature	**Tensile stress**	**True strain**
300° C	2028 MPa	3,19%
400° C	1649 MPa	18,95%

Table 05: Results of Tensile Tests for 60s partitioning

	Partitioning Time	
	60s	
Partitioning Temperature	**Tensile stress**	**True strain**
300° C	2138 MPa	3,58%
400° C	1620 MPa	18,53%

Table 06: Results of Tensile Tests for 120s partitioning

	Partitioning Time	
	120s	
Partitioning Temperature	**Tensile stress**	**True strain**
300º C	2214 MPa	6,29%
400º C	1622 MPa	19,35%

Figure 04 shows a graph obtained by 920ºC austenitizing temperature and 400°C partitioning. It was only possible to demonstrate this test condition because for the heat treatment done under 300 °C partition temperature the graph was not satisfactory.

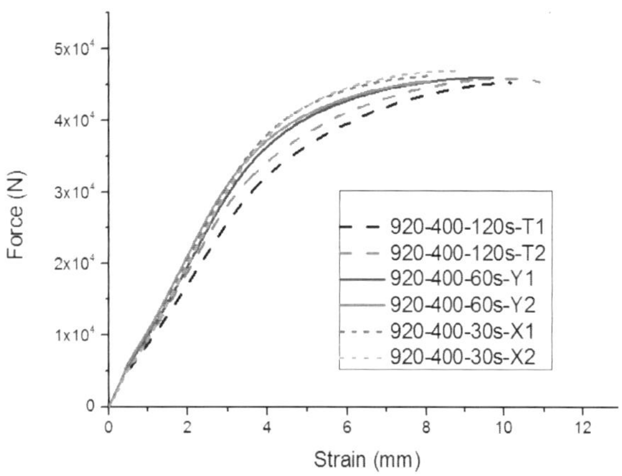

Figure 04 - Results of the tensile test in condition 400ºC partitioning

Metallographic and phase analyzes:

Figure 05 shows a metallographic image under 1000x magnification, from a specimen treated at 920ºC, 300ºC, 30 seconds, with virtually only presence of martensite.

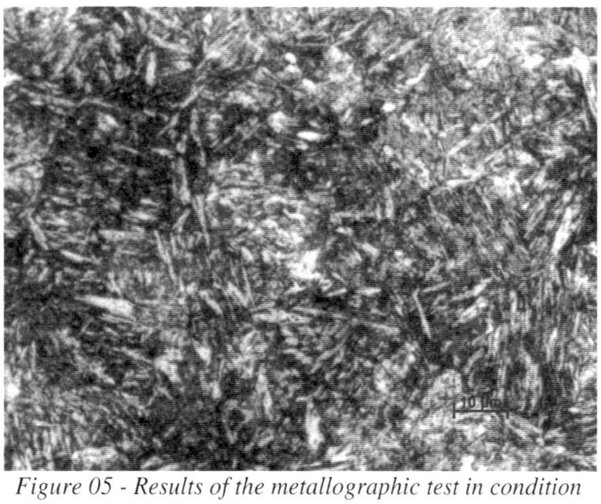

Figure 05 - Results of the metallographic test in condition 920ºC - 300ºC - 30s

Figure 06, shows analysis of retained austenite phase fraction performed by Axion Vision software. In this case the volume fraction was found to be 6.12%.

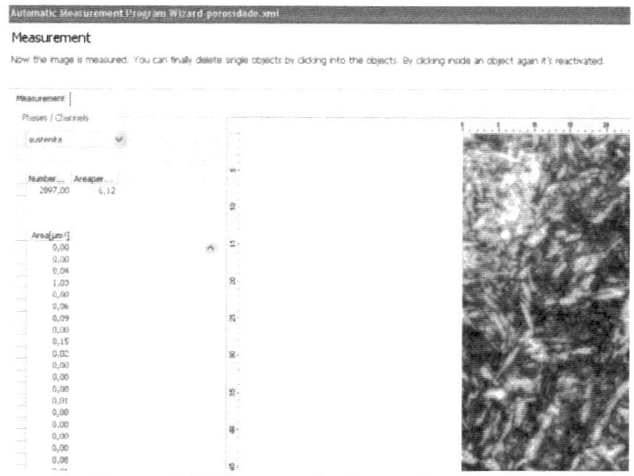

Figure 06-Phase analysis result, condition 920 °C - 300 °C -30s

Figure 07 shows a metallographic image under 1000x magnification, from a specimen treated at 920ºC, 300ºC, 60 seconds with virtually only presence of martensite.

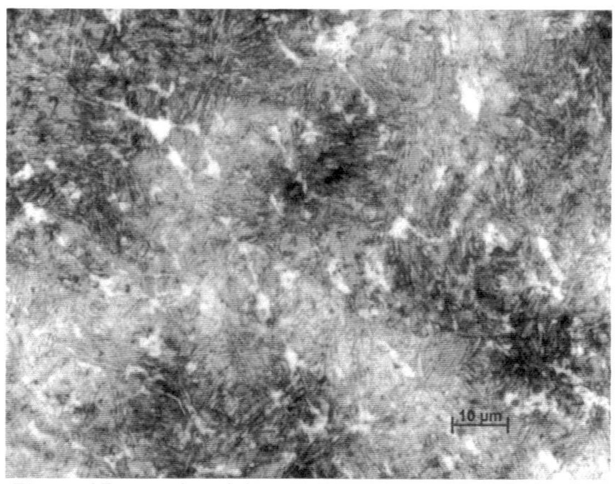

Figure 07- Results of the metallographic test in condition 920 º - 300ºC - 60s

Figure 08 shows analysis of fraction phase of the retained austenite performed by Axion Vision software. In this case the volume fraction was found to be 8.02%.

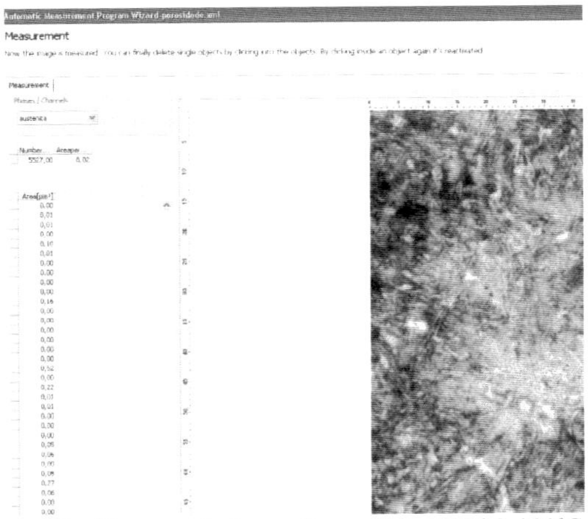

Figure 08 - Phase analysis result, condition 920°C-300°C-60s

Figure 09 shows shows a metallographic image under 1000x magnification, from a specimet treated at 920°C, 300°C, 120s condition.

Figure 09- Results of the metallographic test in condition 920°- 300°C - 120s

Figure 10 shows analysis fraction of retained austenite phase performed by AxionVision software. In this case the volume fraction was found to be 13.89%.

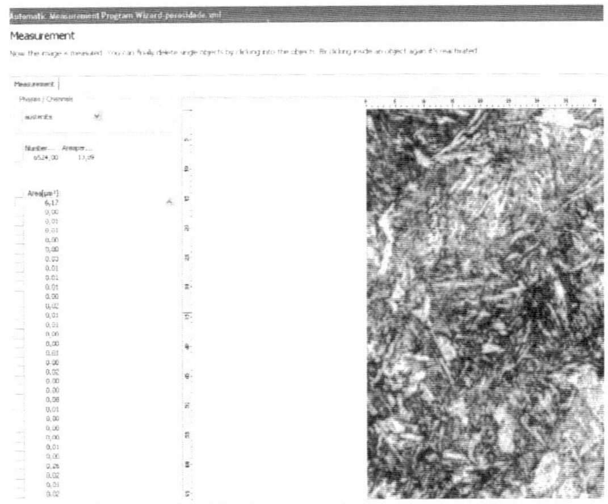

Figure 10 - Phase analysis result, condition 920°C - 300°C - 120s

Figure 11 shows a metallographic image under 1000x magnification, from a specimet treated at 920°C - 400°C - 30 seconds with significant fraction of retained austenite.

Figure 11 - Results of the metallographic test in condition 920°- 400°C -30s

Figure 12, shows analysis of fraction phase of retained austenite performed by AxionVision software. In this case the volume fraction was found to be 10,52%.

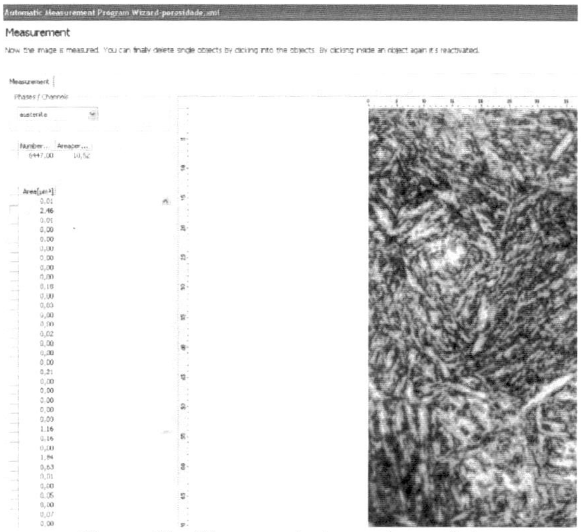

Figure 12 - Phase analysis result, condition 920°C - 400° C - 30s

Figure 13 shows a metallographic image under 1000x magnification, from a specimen treated at 920ºC - 400ºC-60 seconds.

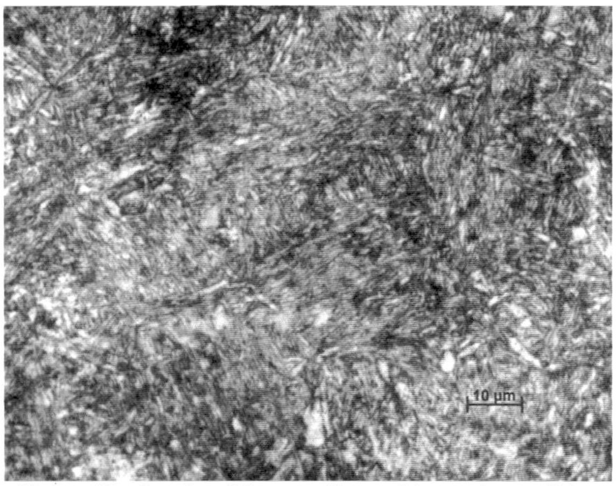

Figure 13 - Results of the metallographic test in condition 920º- 400ºC - 60s

Figure 14 shows analysis of fraction phase of the retained austenite performed by AxionVision software. In this case the volume fraction was found to be 17,75 %.

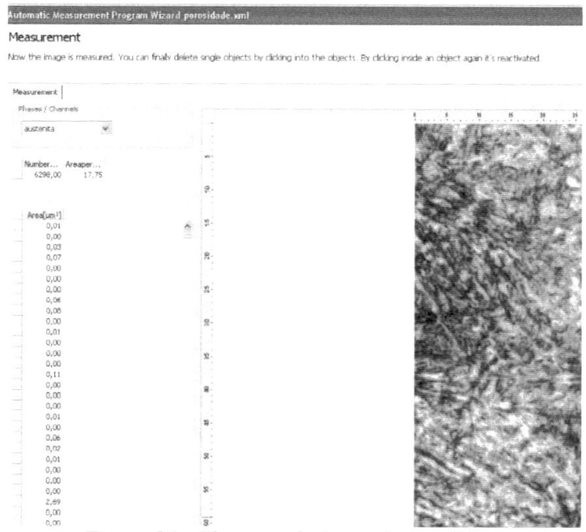

Figure 14 - Phase analysis result, condition 920°C - 400° C - 60s

Figure 15 shows a metallographic image under 1000x magnification, from a specimen treated at 920ºC - 400ºC - 120 seconds.

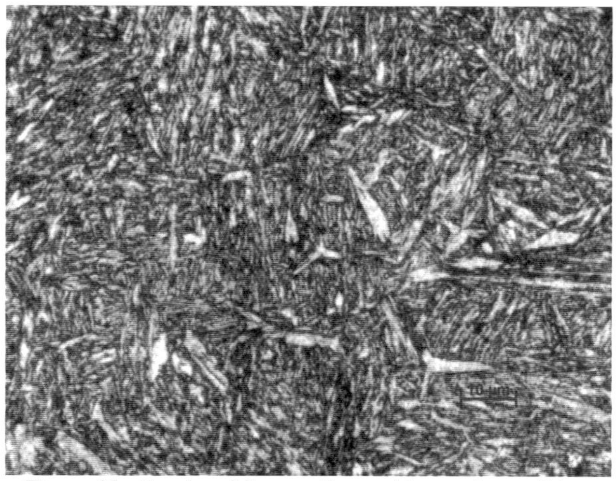

Figure 15 - Results of the metallographic test in condition 920º-400ºC -120s

Figure 16 shows analysis of fraction phase of the retained austenite performed by AxionVision software. In this case the volume fraction was found to be 13,28 %.

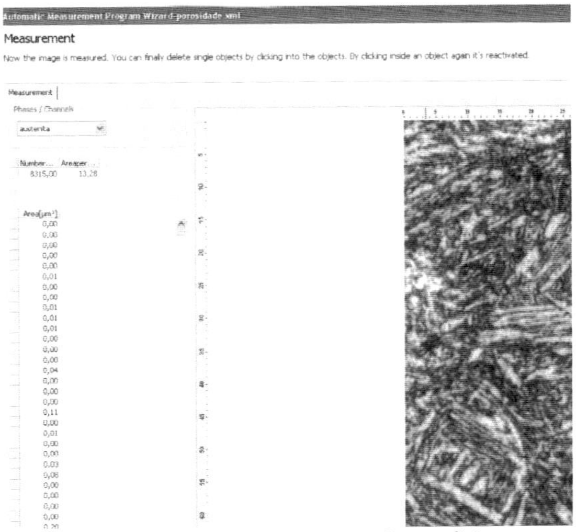

Figure 16 - Phase analysis result, condition 920°C - 400° C - 120s

Conclusion

The austenitizing temperature of 920ºC generated very brittle material when temperture partition was 300 ° C, what showed the prevalence of martensitic phase, and despite the high strength displayed it caused loss or reduction in ductility and toughness.

Tensile tests at 920ºC austenitizing temperature and partition temperature of 300 °C were not optimal, due to the small size of the specimen, the presence of machining risks likely warping of the specimen, combined with high fragility of the material, that splintered off during tests.

The heat treatments with austenitizing of 920ºC and 400ºC partition had excellent results with little loss of toughness strength when compared with the partition done under 300 °C.

The multiphase microstructures are complex, making it difficult to choose a compatible chemical attack, what caused many attempts of trial and error so that the phase could be revealed clearly.

The intrinsic details to the stabilization of retained austenite in the austenite interface / martensite by partition phenomenon could not be clarified at the moment and will be presented in a future work.

Acknowledgments

The Department of Materials Engineering School of Engineering of São Carlos - EESC-USP.

The company Durferrit, located in Diadema-SP from the HEF group for the collaboration and support.

The company Villares Metal located in Sumaré-SP .

The Federal Institute of Education – IFSP, Votuporanga-SP.

All the university staff, friends and colleagues who somehow contributed to this work.

References

[1] Abdalla, A.J., et al., Materials Sciense and Technology, 15 (1999) 1167.

[2] Schuberth, S. et al., "Next Generation Vehicle – Engineering guidelines for stainless steel in automotive applications", Proceedings of the 6th Stainless Steel Science and Market Conference, Helsinki (Finland), 10 – 13 June, 2008.

[3] Furusako, S.; Miyazaki, Y.; "Improvement of crashworthiness by application of high-strength steel", Nippon Steel Technical Report, n. 95, p. 35-38, 2007.

[4] Gallagher, M.F.; Speer, J.G.; Matlock, D.K.; Fonstein, N.M. "Microstructure development in TRIP-sheet steels containing Si, Al and P". In Proceedings of 44th Mechanical Working and Steel Processing Conference, AIST: Warrendale, PA, USA, 2002; 153-172

[5] Tomita, Y. "Development of fracture toughness of ultrahigh strength low alloy steels for aircraft and aerospace applications", Materials Science and Technology v. 7, pp. 481-489, Jun. 1991.

[6] Speer, J.G.; et al., "Carbon Partitioning into Austenite after Martensite Transformation". Acta Materialia, 51, p.2611-2622, 2003.

[7] Zhao, R; et al., "Effect of Q-P-T Process on the Microstructure and Mechanical Properties of 300M Steel" . Materials Science Forum. Vol. 749, p.287-293, 2013.

International Federation of Heat Treatment and Surface Engineering 2016
Proceedings of the 23rd IFHTSE Congress
April 18–21, 2016, Savannah, Georgia, USA

EFFECT OF QUENCHING AND TEMPERING IN MICROALLOYED STEELS CONTAINING BORON AND TITANIUM

André Itman Filho
Instituto Federal de Educação, Ciência e Tecnologia do Espírito Santo, Vitória-ES, Brazil. andrei@ifes.edu.br

Rosana Vilarim da Silva
Instituto Federal de Educação, Ciência e Tecnologia do Espírito Santo, Vitória-ES, Brazil.

Wagner Monteiro Souza
ArcelorMittal Tubarão, Serra-ES, Brazil.

João Batista Ribeiro Martins
ArcelorMittal Tubarão, Serra-ES, Brazil.

Luiz Carlos Casteletti
Universidade de São Paulo, São Carlos-SP, Brazil

George Edward Totten
Portland State University, Portland, Oregon, USA.

Abstract

The OCTG (Oil Country Tubular Goods) steels are used in the production of casing and tubing for drilling system in petroleum industry. These steels usually contain niobium, titanium and vanadium, which ensure good toughness, high tensile strength and resistance to hydrogen embrittlement. It has been known that titanium nitride can promote the refinement of austenite grains with a uniformly dispersion of precipitates in the steel matrix. Nowadays, several researches are being done with boron addition in the microalloyed steels. Boron increases hardenability and for the maximum effect, the concentration should be in the range 0.001 to 0.003 wt%. In the present study the hardness of one microalloyed steel was investigated after quenching and tempering in different temperatures. At first, the steel containing boron and titanium was manufactured by hot rolled until a thickness of 7,2 mm. Afterward, samples in longitudinal orientation with 15x150 mm were cut and heated at 850, 950 and 1050 °C for 30 minutes. Thereafter, all specimens were tempered in the range from 150 to 650 °C for 60 minutes. Both cooling were fulfilled with water at 23 °C. In all specimens, the averages of five measurements were obtained with a load of 100 N using a Vickers hardness tester. The results showed no significant deviations of the values of hardness, after quenching in three temperatures and tempering in the established range. It can be conclude that the titanium nitrides precipitate, refine the austenitic grain during hot rolling and keep characteristics of the martensitic transformation, even in different quenching temperatures. In addition, these nitrides favor the solubility of boron and carbon in the matrix, with hardness increasing of the steel. Though, applying small amounts of boron and titanium in the microalloyed steels, there was no significant variation of hardness after quenching in different temperatures.

Introduction

The discovery of oil in deeper regions of difficult access, such as the pre-salt, requires the development of materials with high hardness, mechanical properties and corrosion resistance. In the case of tubes used as support columns of the oil wells, the setting is designed according to the mechanical loading during operations. This casing string is formed by steel pipes OCTG connected to each other to support the loads during the oil extraction operations, preventing the collapse of the well and the groundwater contamination. These steels must show good toughness, high tensile strength, corrosion resistance and weldability as well wear resistance with the smallest thickness of the walls (Fig.1).

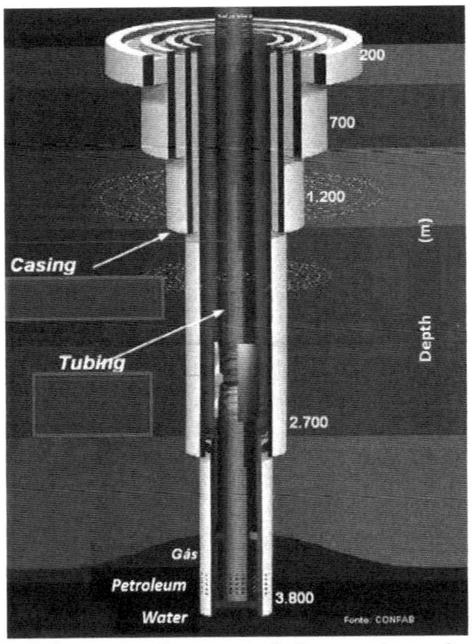

Figure 1 - Schematic diagram of the coating of an oil well column (CONFAB, 2005).

In the case of mechanical properties, thermal treatments are responsible for compliance with the standard. OCTG steels have martensitic structure with high hardness after quenching and tempering. Due to the high cost of chromium and molybdenum, boron can be used to replace these elements and increase the hardness. Nowadays, many researches are made with steels containing boron, because this element increases the hardenability and the yield strength, even at reduced levels in steel. The increase in hardness is due to boron's ability to segregate at grain boundaries, to inhibit the ferrite nucleation and to favor the martensite transformation. Simultaneously with the boron segregation, the phosphorus solubility increases in the ferrite and the steel hardness increases too [1].

Boron is an interstitial element and has poor solubility in the ferrite. Although with great potential for use in quenched and tempered steels, the boron content in the range of 10 to 30 ppm is hard to control, because it has a strong affinity with oxygen and nitrogen present in the steel. Only in the 70st it was possible to develop high-strength and low-cost steels containing boron [2]. Besides the economic advantage, boron facilitates mechanical forming and reduces cracks in the quenching of steels [3]. However, boron must be chemically free in the steel. In order to maintain the efficiency of boron in solution is necessary to add titanium to form titanium nitride. These precipitates remain in the matrix even at temperatures of 1250 °C and inhibit the growth of the austenitic grain [4].

Currently boron provides economically production of microalloyed steels with mechanical properties increasing of the steels, besides replaces partially nickel, chromium and molybdenum [5].

ArcelorMittal Tubarão manufactures hot-rolled boron steels for drilling system in petroleum industry.

Experimental Methods

At first, the steel containing boron and titanium was manufactured by hot rolled until a thickness of 7,2 mm. Afterward, samples in longitudinal orientation with 15x150 mm were cut and heated at 850, 950 and 1050 °C for 30 minutes. Thereafter, all specimens were tempered in the range from 150 to 650 °C for 60 minutes. Both cooling were fulfilled with water at 23 °C

After quenching and tempering, the cubic samples were prepared according to conventional metallographic methods. For etching, 2% Nital and Lepera were used, with an application time of approximately 30 seconds. The microstructures were characterized through backescattered electron images (BSE) and energy dispersive spectroscopy (EDS) on a Scanning Electron Microscope (SEM). In all specimens, the averages of five measurements were obtained with a load of 100 N using a Vicker hardness tester. The grain size after heat treatment was obtained according to the ASTM E112.

Results and Discussion

The chemical composition of the rolled microalloyed steel containing boron and nitrogen is shown in Table 1.

Table 1- Chemical composition of steel (% in weight).

C	Mn	P	S	B	N	Ti
0.23	1.26	0.013	0.0018	0.0013	0.005	0.019

It can be seen in Figure 2 that there was no variation in hardness values after the heat treatments. The same values are justified by the austenite grain size equal to ASTM 7 after quenching at different temperatures. It is expected that the preferential formation of titanium nitride was responsible for the maintenance of grain size during soaking. In the process, the precipitation of fine titanium nitride particles will start during hot rolling in the austenite phase. Most of the titanium nitride particles are smaller than 10 nm and are difficult to observe by conventional microscopy. However, it can be seen in Figure 10 one eutectic titanium nitride. The compositional mapping in Fig. 3 and the morphological feature reveal that the precipitate is titanium nitride. The industrial processing on a hot strip mill show that almost one-half of the total microalloyed addition remains in the form of eutectic or undissolved particles after soaking. In the control rolling of microalloyed steel, additional interphase precipitation may occur during the austenite to ferrite transformation, or during final cooling in the ferrite phase [6].

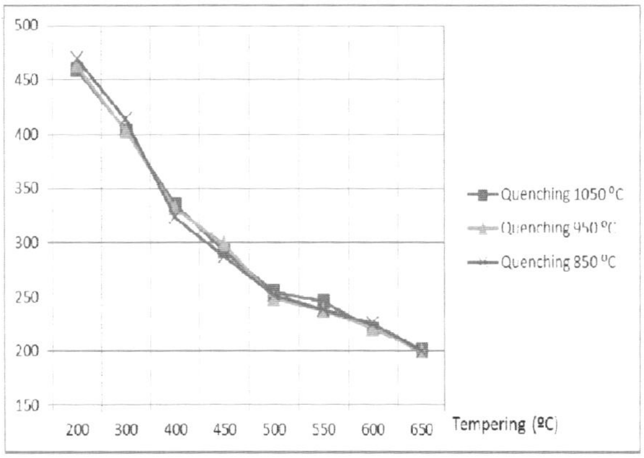

Figure 2- Hardness Vickers after quenching and tempering in different temperatures.

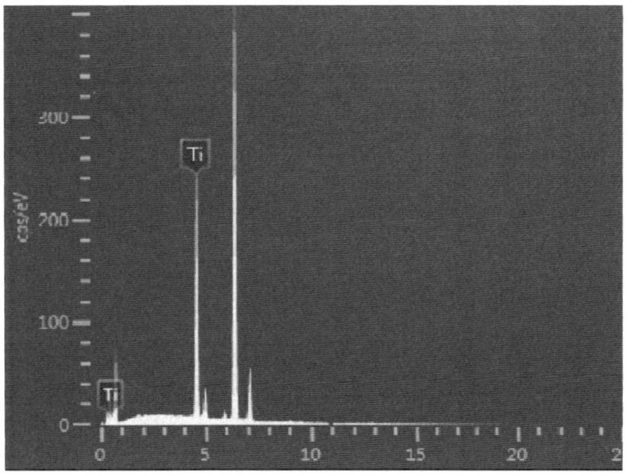

Figure 3- The compositional mapping of precipitated eutectic titanium nitride ((EDS).

Figures 4 to 15 present the microstructures of quenched and quenched plus tempering with cooling in water of microalloyed steel containing boron and nitrogen. These temperatures were chosen because they represent the highest, mean and lower hardness values.

After quenching at different temperatures the steel is fully martensitic. It can be noted that the martensitic laths have different thickness, due to the cooling rate. According to the tempering temperature the hardness reduces with similar values (Fig. 2). After tempering at 200 and 400 °C it can be noted little carbides in the matrix. It is possible to take place the coalescence of these carbides in grain boundary at 600 °C. (Fig. 7).

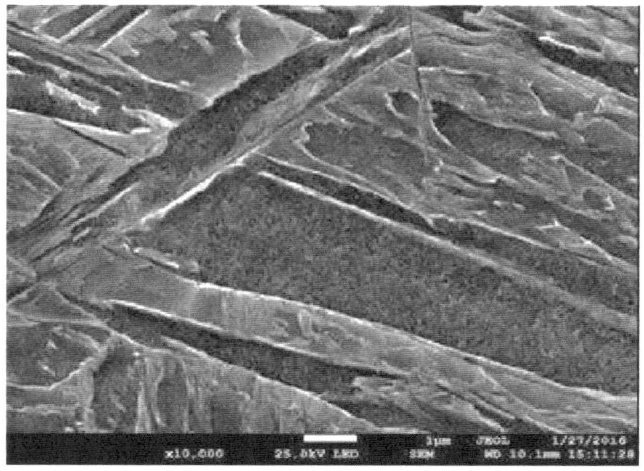

Figure 4- After quenching at 1050 °C the microstructure is fully martensitic. (Nital 2% - MEV).

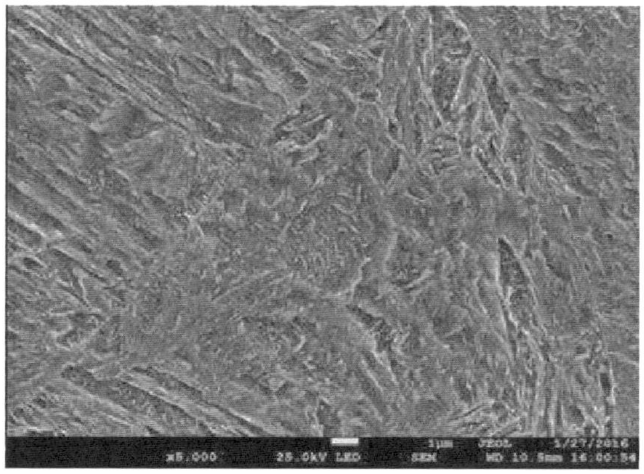

Figure 5- Quenching at 1050 °C plus tempering at 200 °C: microstructure is tempered martensitic. (Nital 2% - MEV).

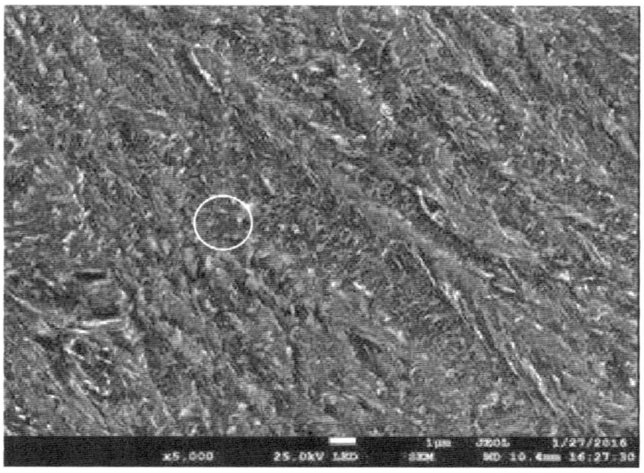

Figure 6- Quenching at 1050 °C plus tempering at 400 °C: tempered martensitic with little carbides. (Nital 2% - MEV).

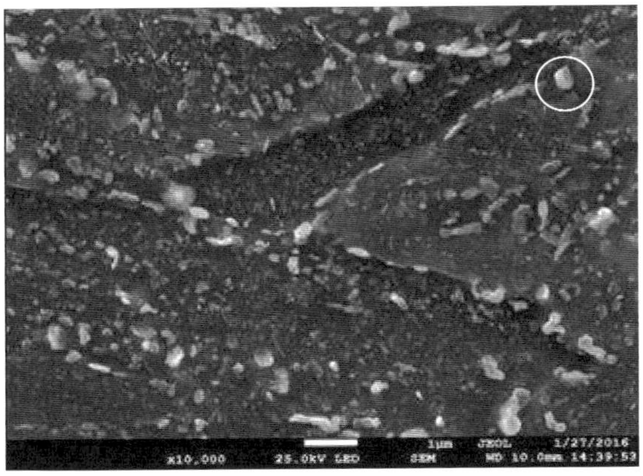

Figure 7- Quenching at 1050 °C plus tempering at 600 °C: carbides in grain boundary. (Nital 2% - MEV).

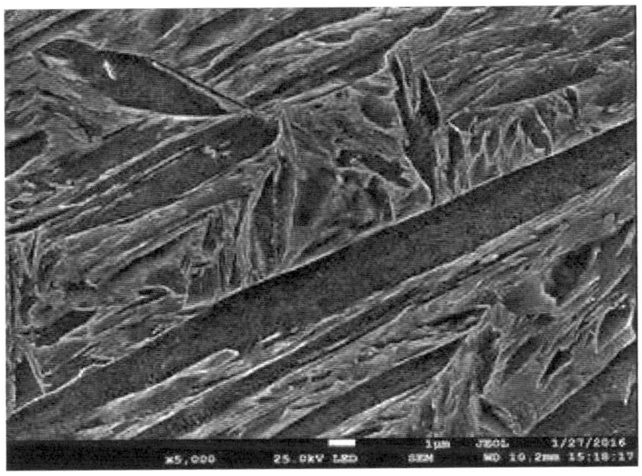

Figure 8- After quenching at 950 °C the microstructure is fully martensitic. (Nital 2% - MEV).

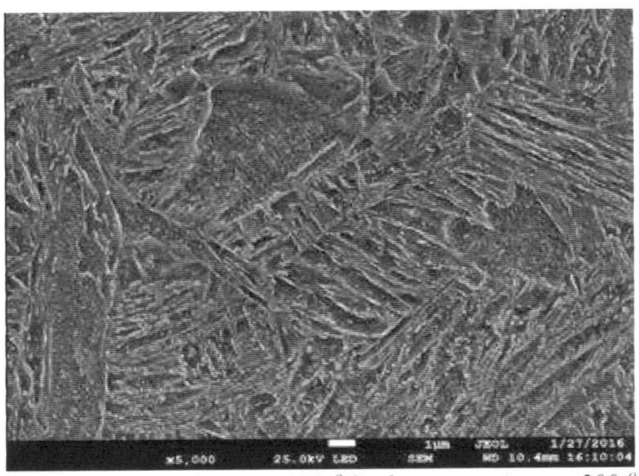

Figure 9- Quenching at 950 °C plus tempering at 200 °C: microstructure is tempered martensitic. (Nital 2% - MEV).

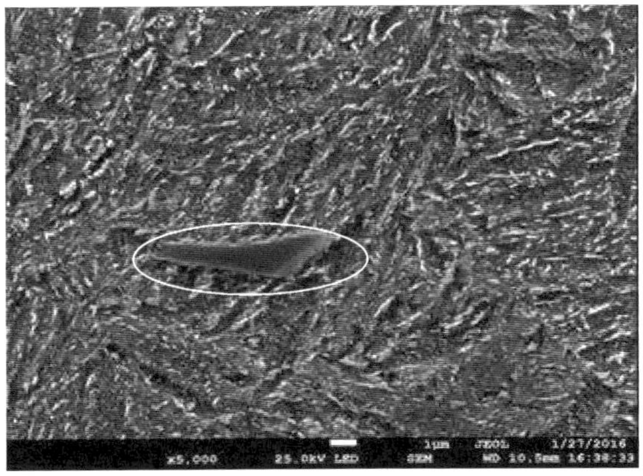

Figure 10- Quenching at 950 °C plus tempering at 400 °C: It can be noted an eutectic titanium nitride. (Nital 2% - MEV).

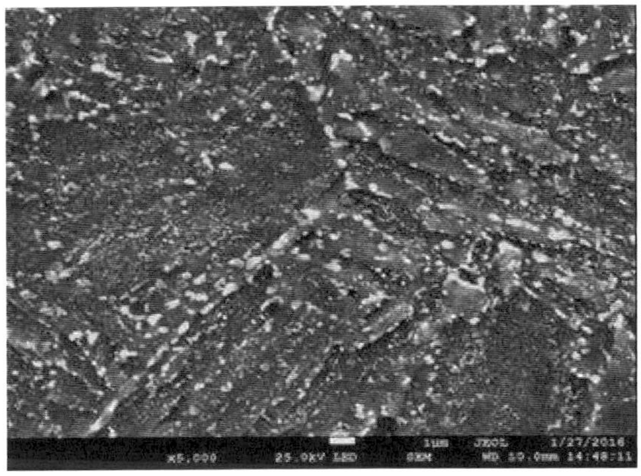

Figure 11- Quenching at 950 °C plus tempering at 600 °C: carbides in grain boundary. (Nital 2% - MEV).

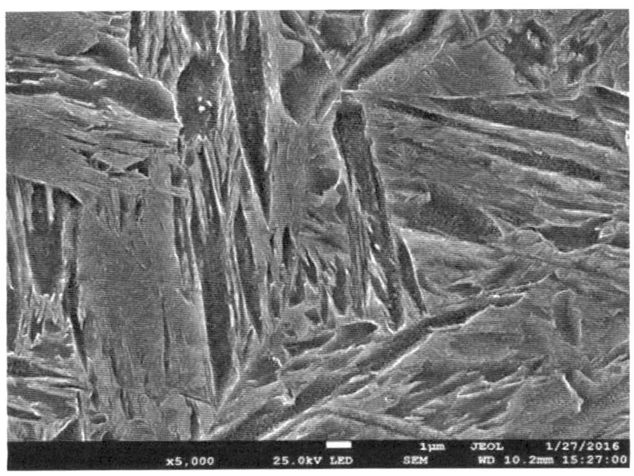

Figure 12- After quenching at 850 °C the microstructure is fully martensitic. (Nital 2% - MEV).

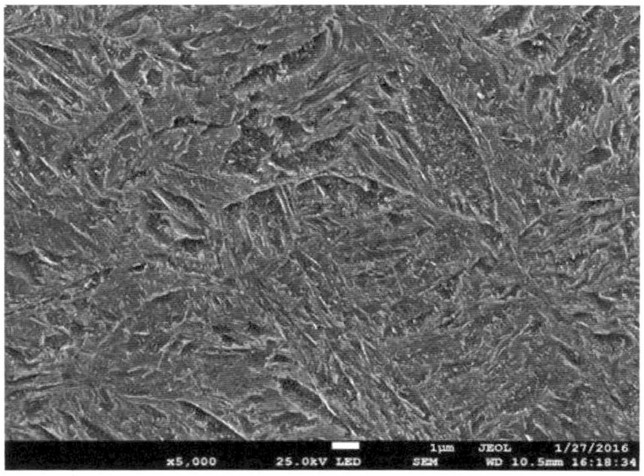

Figure 13- Quenching at 850 °C plus tempering at 200 °C: microstructure is tempered martensitic. (Nital 2% - MEV).

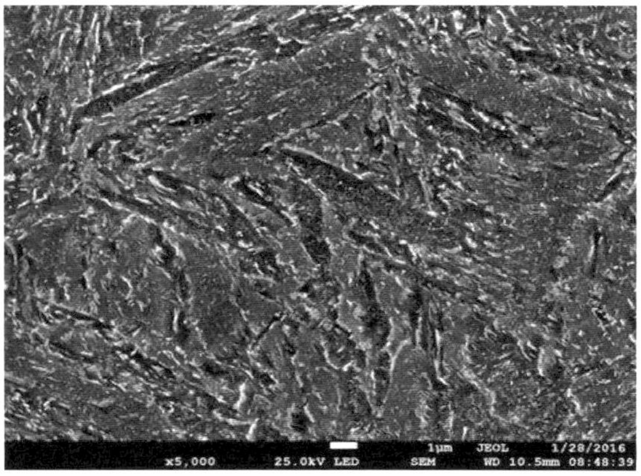

Figure 14- Quenching at 850 °C plus tempering at 400 °C: tempered martensitic with little carbides. (Nital 2% - MEV).

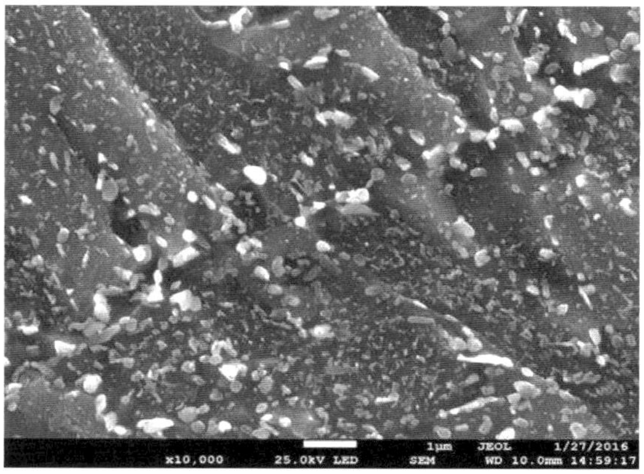

Figure 15- Quenching at 850 °C plus tempering at 600 °C: carbides in grain boundary.. (Nital 2% - MEV).

Conclusions

• Titanium nitride is mainly responsible for grain control after rolling;
• The grain size are similar after quenching in different temperatures;
• The hardness of steel is similar after different temperatures of quenching plus tempering;

Acknowledgments

The authors thank ArcelorMittal Tubarão for the samples and FINEP and FAPES for the financial support

References

[1] Ghali, S. N.; El-Faramawy, H. S. & Eissa, M. M. "Influence of Boron Additions on Mechanical Properties of Carbon Steel". *Scientif Research*, pp. Vol 11, (2012), pp. 995-999.

[2] Maitrepierre, P. "Hardenability concepts with applications to steel". *AIME*, (1978). pp. 421.

[3] Kapadia, B. "Hardenability Concepts with Applications to Steel". *AIME,* (1978), pp. 448.

[4] Panigrahi, B. K. "Processing of low carbon steel plate and hot strip". *Bull. Mater. Sci.* Vol. 24, nº 4, (2001), pp. 361–371.

[5] Zorin, A. I. & Podkorytov, A. Z. "Production of boron-bearing grades of steel with a regulated sulfur content at the Chelyabinsk metallurgical plant". *Metallurgist*, Vol. 49, (2005), n. 1-2.

[6] Itman, A; Cardoso, K. R. and Kestenbach, H. J. "Quantitative Study of Carbonitride Precipitation in Niobium and Titanium Microalloyed Hot Strip Steel". *Metallurgical Materials Science and Technology*, Volume 13, (1997), pp. 49-56.

International Federation of Heat Treatment and Surface Engineering 2016
Proceedings of the 23rd IFHTSE Congress
April 18–21, 2016, Savannah, Georgia, USA

Characterization of BN/TiBN/TiN Multilayer Coatings Prepared by Pulsed DC Plasma-Enhanced Chemical Vapor Deposition

Kazuki Kawata, Toru Kidachi
Oriental Engineering Co., Ltd., 2-8-49 Yoshinodai, Kawagoe-city, Saitama, Japan
E-mail: k-kawata@oriental-eg.co.jp, Tel: +81 49 225 5811

Abstract

BN/TiBN/TiN multilayer coatings were deposited onto steel specimens by pulsed dc plasma-enhanced chemical vapor deposition using mass-production-type equipment. Using glow discharge optical emission spectroscopy, we demonstrated that the coatings exhibited a three-layered structure composed of BN/TiBN/TiN coatings. The cross section of the upper BN coating exhibited a dense structure that differed from the columnar structures of the cross sections of the TiN and TiBN coatings. Analyses by X-ray diffraction and Raman spectroscopy demonstrated that the upper layer coating was hexagonal boron nitride. X-ray photoelectron spectroscopy analyses of the upper layer coating demonstrated that BN was present. Observations by transmission electron microscopy revealed that the upper BN coating was onion-structured and fibrous and that the coating was composed of particles with diameters of approximately 10-20 nm. The surface hardness of the BN/TiBN/TiN multilayer coatings was 4.5 GPa, which was lower than the hardnesses of the TiN and CrN coatings prepared by physical vapor deposition. The BN/TiBN/TiN multilayer coatings exhibited greater oxidation resistance and better tribological properties than the TiN and CrN coatings. Furthermore, compared to the other specimens investigated in this work, the BN/TiBN/TiN multilayer coatings exhibited superior soldering and erosion resistance in a molten aluminum die-casting alloy.

Introduction

Properties such as good erosion resistance, good soldering resistance, and good thermal fatigue resistance are required in aluminum die-casting dies. To achieve these performance requirements, various surface treatments have been applied to such dies. The physical vapor deposition (PVD) method has been used to deposit films onto die-casting pins and small insert dies because it is a low-temperature processing method. However, it is not suitable for depositing films onto main dies with complicated shapes involving slits or holes because the throwing power of the films tends to be poor. Furthermore, the PVD method tends to generate film defects such as droplets or pinholes and the thermal fatigue resistance of PVD-deposited films is low because no nitrogen diffusion layer is present. In contrast, pulsed dc plasma-enhanced chemical vapor deposition (PECVD) has been widely used to deposit films onto metal dies with complicated geometries, such as aluminum die-casting dies and aluminum extrusion dies, because of its good coating adhesion and good throwing power even at low coating temperatures [1]-[5]. Pulsed dc PECVD allows a duplex process (plasma nitriding (PN) and coating) to be performed in a single cycle using a single apparatus. Therefore, the process can be used to deposit coatings that satisfy requirements related to both erosion resistance and heat-checking resistance. In the present study, a duplex process (PN and BN/TiBN/TiN multilayer coatings) was used to treat high-speed tool steels (JIS-SKH51) and hot-work tool steels (JIS-SKD61) using mass-production-type pulsed dc PECVD equipment. We compared several properties of specimens prepared using the duplex process with those of nitrided specimens and specimens prepared by PVD.

Experimental

Plasma diffusion treatment and PECVD coating were performed using mass-production-type pulsed dc PECVD equipment. Figure 1 shows a schematic of the pulsed dc PECVD equipment used in the experiments. The equipment comprised a reaction chamber, an external heater, a vacuum system, a rotation system for a working table, a pulsed dc power supply, a gas supply system, and a computer-control system. The effective dimensions of the work pieces treated

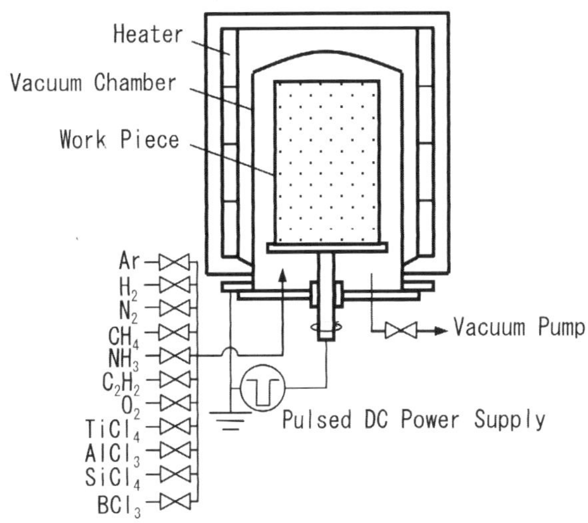

Figure 1: Schematic of the pulsed dc PECVD equipment.

using this equipment were a diameter of 540 mm and a height of 800 mm.

The duplex process was performed on steel substrates in two steps. PN with the formation of only a diffusion layer was followed by the deposition of BN/TiBN/TiN multilayer coatings. The diffusion-layer depth was approximately 50 μm. The duplex process was performed as a continuous process, where the hard coatings were applied one after another immediately after PN in a single facility.

The BN/TiBN/TiN multilayer coatings were deposited onto steel substrates at 550 °C using mixtures of $TiCl_4$, BCl_3, H_2, N_2, and Ar. In the duplex process, PN was performed to improve the heat-checking resistance of die-casting dies. The BN/TiBN/TiN multilayer coatings were deposited to improve the erosion resistance of the dies. In the multilayer coatings, the lower TiN coating was formed to improve the adhesion between the substrate and the middle TiBN coating. In the triple-layered coating, the middle TiBN coating was used to improve the adhesion between the lower TiN coating and the upper BN coating. Table 1 shows the processing conditions of PN and pulsed dc PECVD.

Table 1: PN and pulsed dc PECVD processing conditions.

	PN	**TiN**	**TiBN**	**BN**
Temperature (°C)	550	550	550	550
Duration (h)	1	2	2	2
Pressure (Pa)	100-270	100-270	100-270	100-270
Voltage (V)	450-600	500-700	500-700	500-700
Frequency (kHz)	20-50	20-50	20-50	20-50
Duty cycle (%)	10-50	10-50	10-50	10-50
H_2 (vol%)	1600	2000	2000	2000
N_2 (vol%)	400	500	500	500
Ar (vol%)	-	100	100	100
$TiCl_4$ (vol%)	-	60	10-60	-
BCl_3 (vol%)	-	-	10-60	60

The specimens used were prepared from Japanese Industrial Standards (JIS) SKH51 (high-speed tool steel) and SKD61 (hot-work tool steel). The chemical composition of the SKH51 was 0.89% C, 0.34% Si, 0.31% Mn, 4.16% Cr, 4.85% Mo, 5.88% W, and 1.89% V. The SKH51 specimens were quenched at 1150 °C and double tempered for 120 min at 550 °C in a vacuum furnace before the surface treatments were applied. The Rockwell hardness (HRC) of the untreated specimen was 62. The chemical composition of SKD61 was 0.40% C, 0.94% Si, 0.42% Mn, 5.11% Cr, 1.22% Mo, and 0.81% V. The SKD61 specimens were quenched at 1020 °C and double tempered for 120 min at 600 °C in a vacuum furnace before the surface treatments were applied. The Rockwell hardness (HRC) of the untreated specimen was 48. The SKH51 specimens for hardness measurements, structure observations, friction and wear examination and compositional analyses were 20 mm in diameter and 5 mm in thickness. The SKD61 specimens for oxidation resistance tests were 20 mm in diameter and 5 mm in thickness, whereas those for soldering tests and erosion tests were 10 mm in diameter and 100 mm in length. The properties of gas-nitriding specimens (compound layer thickness: 7 μm, diffusion layer depth: 0.15 mm), TiN coatings (thickness: 2.5 μm), and CrN coatings (thickness: 2.2 μm) prepared by PVD (cathodic vacuum arc deposition) were also evaluated for comparison with those of the BN/TiBN/TiN multilayer coatings. The multilayer coatings were analyzed using scanning ion microscopy (SIM), transmission electron microscopy (TEM), electron diffraction (ED), glow discharge optical emission spectroscopy (GDOS), X-ray photoelectron spectroscopy (XPS), Raman spectroscopy (RS), and X-ray diffraction (XRD). The hardness of these coatings was measured using a nanoindenter. The adhesion of these coatings was evaluated by the Rockwell C test. Oxidation resistance tests were conducted by exposing specimens to air for 60 min at high temperatures. The friction and wear properties were evaluated with a ball-on-disk-type tribometer (CSEM). The testing conditions were as follows: ball, Al (JIS-A1050);

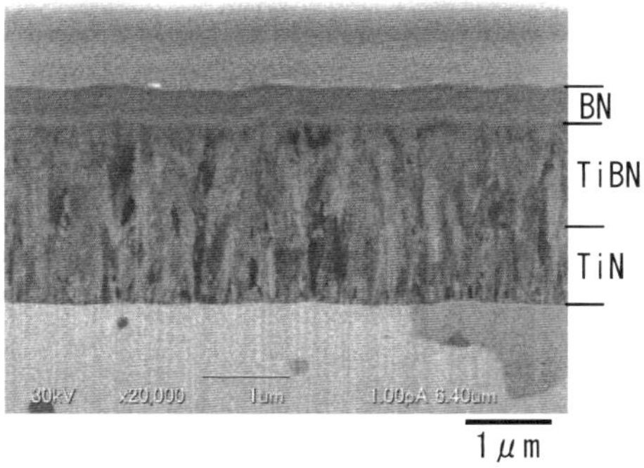

Figure 2: Cross-sectional SIM image of a BN/TiBN/TiN multilayer coating deposited by pulsed dc PECVD.

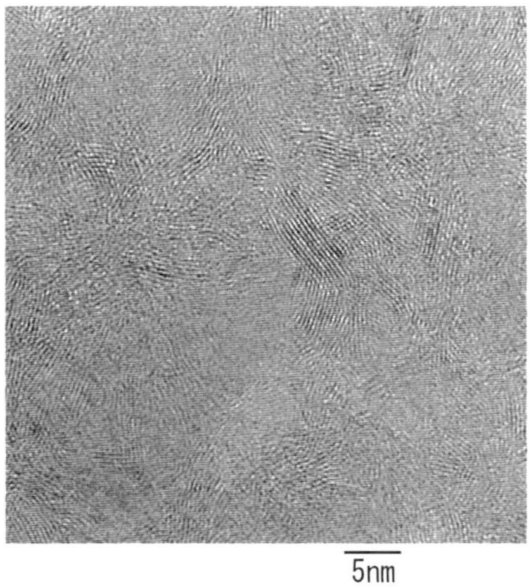

Figure 3: Cross-sectional TEM image of the upper BN coating in the BN/TiBN/TiN multilayer coatings deposited by pulsed dc PECVD.

diameter, 5.5 mm; load, 1 N; sliding speed, 50 mm/s; sliding distance, 100 m; and no lubricant was used. The erosion resistance of each specimen was investigated by immersing the specimens into a molten aluminum die-casting alloy (JIS-ADC12) at 700 °C. In the erosion resistance tests, the specimens were immersed in a molten aluminum alloy for 120 min and then cleaned in a NaOH solution until most of the soldered aluminum alloy was leached from their surfaces. The erosion loss for the specimens was measured on an electric balance. The procedure of immersing the specimens into the molten aluminum alloy was repeated twice; i.e., the immersion was performed intermittently and the specimens were immersed for a total of 360 min. Soldering resistance tests were performed after the specimens had been immersed in the molten aluminum alloy bath at 700 °C for 120 min and subsequently removed from the bath.

Results and Discussion

Morphology of Cross Sections

Figure 2 shows a cross-sectional SIM image of the BN/TiBN/TiN multilayer coating deposited by pulsed dc PECVD. The cross sections of the TiN and TiBN coatings prepared by pulsed dc PECVD exhibited columnar structures. The cross section of the upper BN coating in the BN/TiBN/TiN multilayer coating exhibited a dense structure. Figure 3 shows a cross-sectional TEM image of the upper BN coating in the BN/TiBN/TiN multilayer coatings prepared by pulsed dc PECVD. In this image, the upper BN coating was the onion structure and was fibrous; the coating was composed of particles with diameters of approximately 10-20 nm.

Compositional and Structural Analyses

Figure 4 shows an ED image of the upper BN coating. This image indicates that h-BN was formed in the upper coating. We considered the h-BN to be microcrystalline because of its weaker ring pattern. Figure 5 shows the GDOS concentration depth profiles for each element in the BN/TiBN/TiN multilayer

Figure 4: ED image of the upper BN coating.

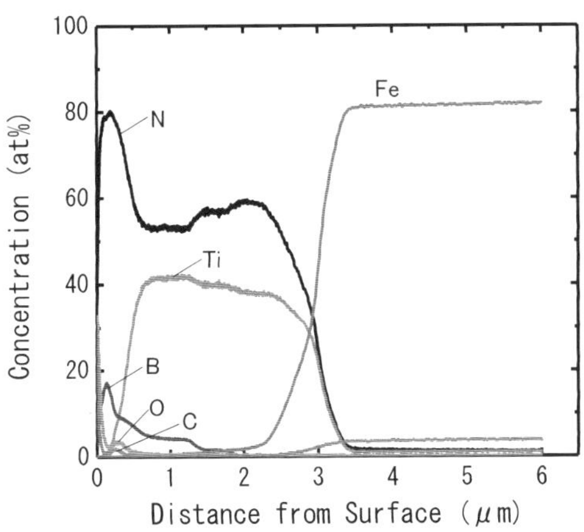

Figure 5: GDOS concentration depth profiles for each element in the BN/TiBN/TiN multilayer coatings.

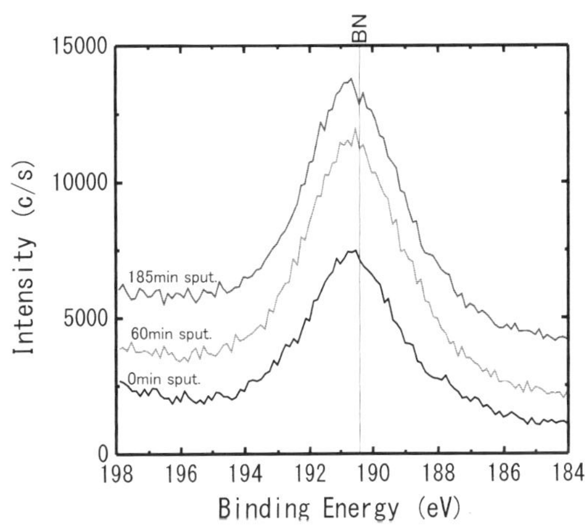

Figure 6: B1s XPS spectra of the upper BN coating produced by pulsed dc PECVD.

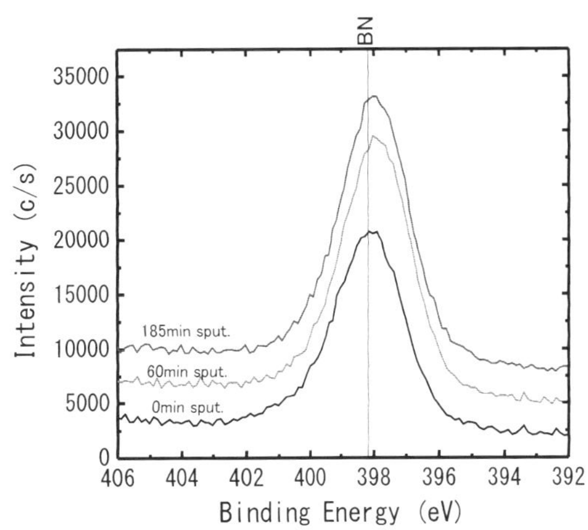

Figure 7: N1s XPS spectra of the upper BN coating produced by pulsed dc PECVD.

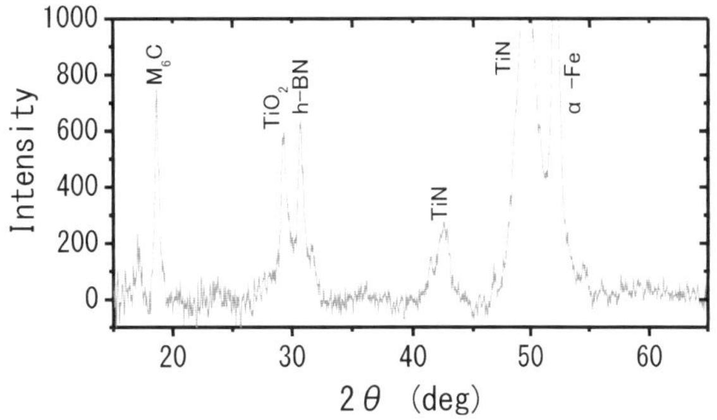

Figure 8: XRD pattern of the BN/TiBN/TiN multilayer coatings (Co-Kα, 40 kV, 135 mA).

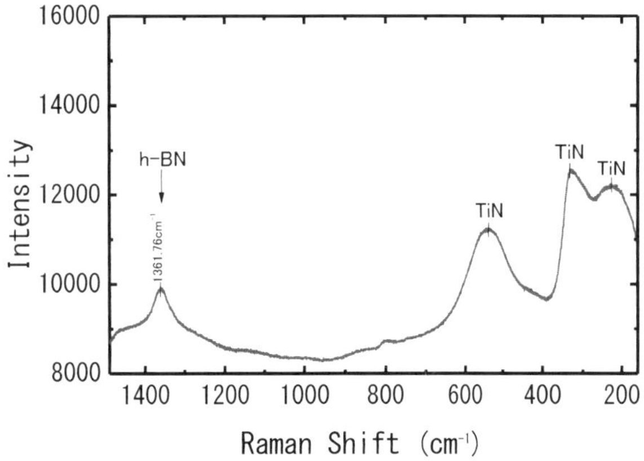

Figure 9: Raman spectroscopy analysis results for the BN/TiBN/TiN multilayer coatings.

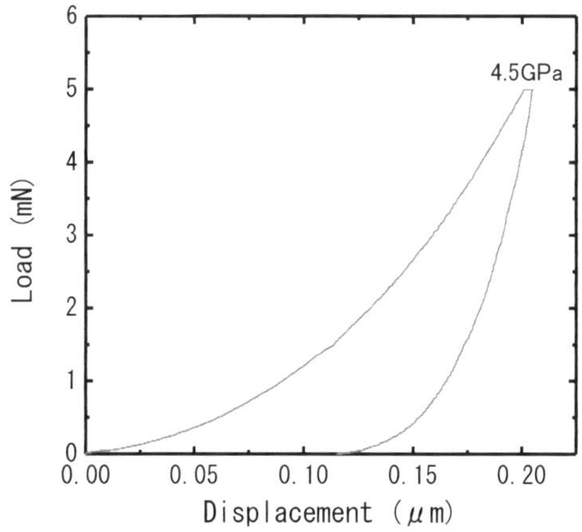

Figure 10: Load versus displacement curve for the BN/TiBN/TiN multilayer coatings, as obtained using the nanoindentation method.

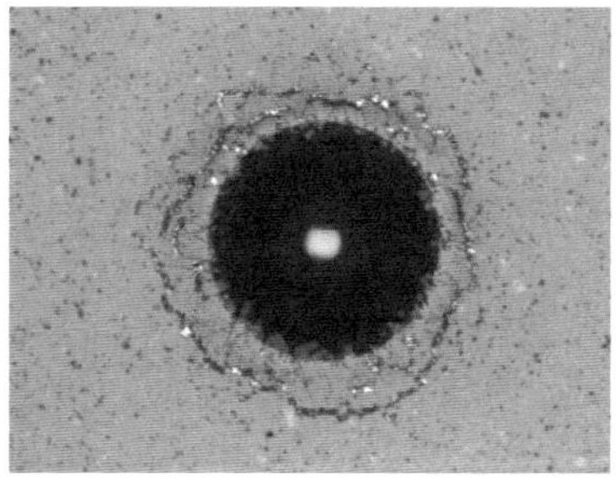

Figure 11: Result of Rockwell C adhesion test of the BN/TiBN/TiN multilayer coatings.

coating. On the basis of the concentration distributions of Ti, B and N, we confirmed that the coating was composed of three layers. Figure 6 and Figure 7 show the B1s XPS spectra and N1s XPS spectra of the upper BN coating produced by pulsed dc PECVD, respectively. Examination of the chemical bonding state of each element in the upper BN coating by XPS revealed that B was present as nitrides. Figure 8 shows the XRD pattern of the BN/TiBN/TiN multilayer coatings. The diffraction peak of the coating revealed that hexagonal boron nitride (h-BN) was formed. Figure 9 shows the Raman spectroscopy analysis results. The presence of the peak corresponding to h-BN at 1361.76 cm^{-1} was confirmed in the Raman spectra. The large width of this peak indicates that the structure exhibits low crystallinity.

Surface Hardness and Adhesion

Figure 10 shows the load versus displacement curve for the BN/TiBN/TiN multilayer coatings obtained using the nanoindentation method. The nanoindentation hardness was 4.5 GPa for the upper BN coating in the BN/TiBN/TiN multilayer coatings. Therefore, the surface hardness of the upper BN coating in the BN/TiBN/TiN multilayer coatings was lower than the hardnesses of the TiN and CrN coatings. Figure 11 shows the adhesion test results for the BN/TiBN/TiN multilayer coatings, as measured by the Rockwell C test. Slightly delamination of the coating in the indentation area was observed. We considered that the coatings exhibited relatively good adhesion, HF2, according to the VDI classification method [6].

Tribological Properties

Figure 12 shows the relationship between the friction coefficient and the sliding distance for the BN/TiBN/TiN multilayer coatings, the TiN coatings, and the CrN coatings. The friction coefficients in sliding distance of 100 m were approximately 0.45 for the BN/TiBN/TiN multilayer coatings, approximately 0.74 for the TiN coatings, and approximately 0.75 for the CrN coatings. Therefore, the friction coefficient of the BN/TiBN/TiN

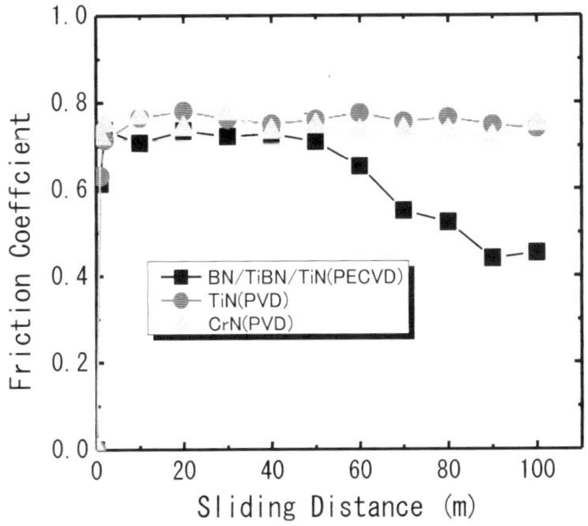

Figure 12: Relationship between the friction coefficient and the sliding distance for the BN/TiBN/TiN multilayer coatings (PECVD), the TiN coatings (PVD), and the CrN coatings (PVD) (ball, Al; load, 1 N; sliding speed, 50mm/s; unlubricated condition).

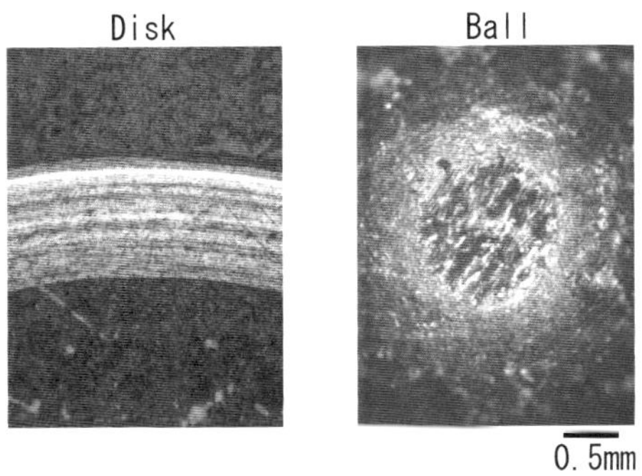

Figure 13: Disk-and-ball wear scars for the BN/TiBN/TiN multilayer coatings after the friction and wear tests.

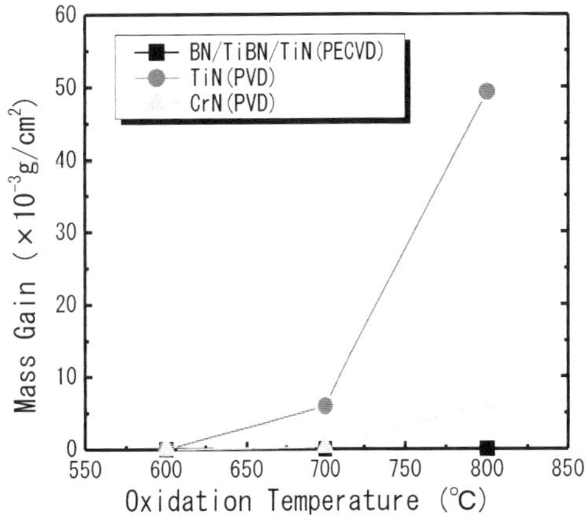

Figure 14: Relationship between the oxidation temperatures and the mass gain for the BN/TiBN/TiN multilayer coatings (PECVD), the TiN coatings (PVD), and the CrN coatings (PVD).

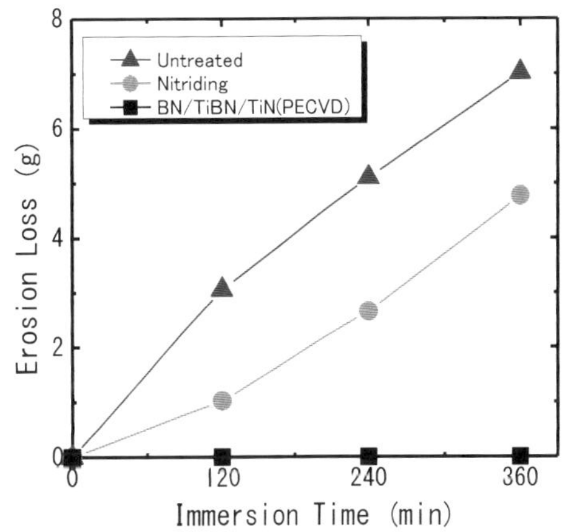

Figure 15: Relationship between the immersion time and the erosion loss for an untreated specimen, the nitrided specimen, and the BN/TiBN/TiN multilayer coatings.

multilayer coatings was the lowest among the investigated specimens. Figure 13 shows the disk-and-ball wear scars for the BN/TiBN/TiN multilayer coatings after the friction and wear tests. The BN/TiBN/TiN multilayer coatings did not exhibit oxidized adhesive wear. Moreover, the counterpart ball wear scar for the BN/TiBN/TiN multilayer coatings was relatively small compared to those for the other specimens. These results demonstrate the excellent tribological characteristics of the BN/TiBN/TiN multilayer coatings.

Oxidation Properties

Figure 14 shows the relationship between the oxidation temperatures and the mass gain for the BN/TiBN/TiN multilayer coatings, the TiN coatings, and the CrN coatings. The oxidation start temperature was 600 °C for the TiN coatings and 700 °C for the CrN coatings. Furthermore, the mass gain increased substantially with increasing oxidation temperature. In comparison, the mass gain for the BN/TiBN/TiN multilayer coatings was zero at 800 °C. Therefore, the BN/TiBN/TiN multilayer coatings exhibited the best oxidation resistance among the investigated specimens.

Erosion and Soldering Properties

Figure 15 shows the relationship between the immersion time and the erosion loss for an untreated specimen, a nitrided specimen, and the BN/TiBN/TiN multilayer coatings. The erosion of the untreated specimen and the nitrided specimen increased with increasing immersion time. However, the BN/TiBN/TiN coatings exhibited no erosion even after 360 min of immersion. The BN/TiBN/TiN multilayer coatings exhibited

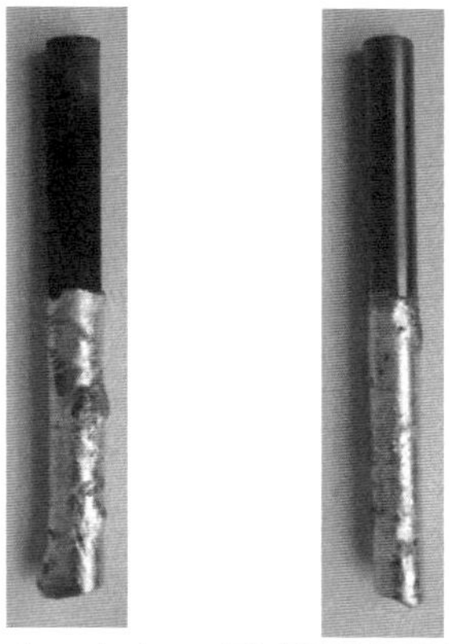

Figure 16: Appearance of the various specimens after the soldering resistance tests.

the best erosion resistance among the investigated specimens. Figure 16 shows the appearance of the various specimens after they were immersed in molten aluminum alloy for 120 min. The soldering resistance tests using molten aluminum alloy demonstrated that the BN/TiBN/TiN multilayer coatings exhibit the best soldering resistance among the investigated specimens.

Application

The duplex process (PN and BN/TiBN/TiN multilayer coatings prepared by pulsed dc PECVD) has already been applied to aluminum die-casting dies. In actual aluminum die-casting dies, the soldering resistance, erosion resistance, and thermal fatigue resistance of coatings prepared using the duplex process have been demonstrated to be better than those of coatings prepared using PVD (TiN, CrN, and TiAlN).

Conclusions

The following conclusions were drawn from these experiments:

1. The cross section of the upper BN coating in the BN/TiBN/TiN multilayer coating exhibited a dense structure. The upper BN coating was onion-structured, fibrous, and composed of particles approximately 10-20 nm in diameter.

2. XPS analyses of the surface of the upper BN coating demonstrated that B was present as nitrides.

3. Analyses by XRD and Raman spectroscopy demonstrated that the upper BN coating was h-BN.

4. The surface hardness of the upper BN coating in the BN/TiBN/TiN multilayer coatings was 4.5 GPa, which was lower than the hardnesses of the TiN and CrN coatings.

5. The BN/TiBN/TiN multilayer coatings exhibited relatively good film adhesion.

6. The BN/TiBN/TiN multilayer coatings exhibited excellent tribological properties.

7. The BN/TiBN/TiN multilayer coatings exhibited the best oxidation resistance, erosion resistance, and soldering resistance among the investigated specimens.

8. The duplex process (PN and BN/TiBN/TiN multilayer coatings prepared by pulsed dc PECVD) has already been applied to aluminum die-casting dies to improve soldering resistance, erosion resistance, and thermal fatigue resistance.

References

[1] D. Heim, F. Holler and C. Mitterer, "Hard coatings produced by PACVD applied to aluminium die casting" *Surf. Coat. Technol.*, Vol. 116/119 (1999), pp. 530-536.

[2] K. Kawata, Hiroyuki Sugimura and Osamu Takai, "Characterization of (Ti, Al)N films deposited by pulsed d.c. plasma-enhanced chemical vapor deposition," *Thin Solid Films*, Vol. 386 (2001), pp. 271-275.

[3] K. Kawata, H. Sugimura and O. Takai, "Characterization of multilayer films of Ti-Al-O-C-N system prepared by pulsed d.c. plasma-enhanced chemical vapor deposition," *Thin Solid Films*, Vol. 390 (2001), pp. 64-69.

[4] K. Kawata, Y. Sekiya and T. Kidachi, "Characterization of multilayer coatings of Ti-Al-Si-C-N-O system prepared by pulsed DC plasma-enhanced chemical vapor deposition," *Proc. 15th IFHTSE*, Vienna, September 2006, pp. 516-521.

[5] K. S. Klimek, A. Gebauer-Teichmann, P. Kaestner and K.-T. Rie, "Duplex-PACVD coating of surfaces for die casting tools," *Surf. Coat. Technol.*, Vol. 201 (2007), pp. 5628-5632.

[6] W. Heinke, A. Leyland, A. Matthews, G. Berg, C. Friedrich, E. Broszeit, "Evaluation of PVD nitride coatings, using impact, scratch and Rockwell-C adhesion tests," *Thin Solid Films*, Vol. 270 (1995), pp. 431-438.

International Federation of Heat Treatment and Surface Engineering 2016
Proceedings of the 23rd IFHTSE Congress
April 18–21, 2016, Savannah, Georgia, USA

Application of Nano-Structured Copper-Polymer Catalyst for Activation of Gas Nitriding of Steels

L.G. Petrova , I.M. Papisov , I.S. Belashova , V.A. Aleksandrov , G.Yu Ostaeva , A.I. Papisova , A.E. Perekrestov
Moscow Automobile and Road Construction State Technical University (MADI), Moscow, Russia
petrova_madi@mail.ru, +7 499 1550829

Abstract

In the paper forming of nanostructured catalytic systems of nitriding and of their precursors on the basis of copper nanoparticles is discussed. Recent investigation has discovered the catalytic effect of copper oxide film applied at the surface of low-carbon and medium-carbon steels before nitriding. It accelerates growth of a diffusion layer and mainly of a compound zone. It is known that the effect of catalytic systems is higher when nano-structured catalysts are used. It is the result of high reactivity of nano-particles having huge overall specific surface. An attempt was undertaken to increase the dispersion of particles of CuO-catalyst by synthesis of its precursor (metallic copper layer) at a steel surface with usage of polymers such as poly-N-vinilpirrolidone (PVP). Specimens of low-carbon steel were treated in $CuSO_4$ water solution with addition of PVP. After the treatment a coating of violet-brown color typical for copper sol was observed at a steel surface. Examination by AFM discovered nano-dimensional particles of composite structure formed as a result of reduction of copper from copper sulfate at the iron surface. It was shown by high-resolution TEM that composite consists of copper nano-particles (about 20 nm) surrounded by polymeric shell. According to the theory of pseudo-template synthesis the macromolecules of PVP act as protectors by shielding of metal surface, and prevent growth of copper particles. Catalytic activity of thermo-oxidized product (CuO) formed from such nanocomposite is higher than of oxide received from copper not modified by PVP. Nitriding of DC04 steel samples with nano-structured catalytic oxide film at 590°C during 120 minutes gives the compound zone in 1.5…1.85 times thicker comparing to the nitride layer formed under non-modified CuO-film. The regulating parameters of the catalytic nitriding process are: reacting time of deposition of nano-structured precursor, and molecular weight of used polymer influencing the morphology of nanocomposite.

Introduction

Low-temperature methods of thermo-chemical treatment (TCT) are the most required methods of surface strengthening; the share of items treated by TCT in the total quantity of surface strengthened machine parts increases significantly. Nitriding is supposed as one of the effective methods of TCT; it provides the highest set of working properties of treated items: surface hardness, wear resistance, heat and corrosion resistance, fatigue strength, and scoring resistance. Conformability of high-nitrogen ε-phase at a surface of a nitrided layer is the factor of its wear resistance; that is higher than wear resistance of quenched or carburized steels [1].

Nitriding occurs at low temperatures (as a rule, not higher than 600°C), so there are no phase transformations and no changes of geometry parameters of items. That is why mechanical grinding is not necessary after nitriding. Low temperatures of nitriding result in less energy consumption comparing to carburizing processes.

The main problem of nitriding consists in high duration of saturation process. The problem requires new solutions on process intensification; one of them consists in forming of catalytic films at a treated surface. Concentration of atomic nitrogen at the surface is a factor of acceleration of nitrided layer growth. The concentration of a saturating element in the atmosphere may be increased by activation of ammonia dissociation process.

It is known that copper oxide is one of the effective reductants of copper during ammonia dissociation; that is why it was suggested to use it as a catalytic film deposited on the surface of nitrided steel [2]. It is known that the influence of catalytic systems is higher when nano-structured catalysts are used. It is the result of high reactivity of nano-particles having huge overall specific surface. Investigations show that the process of nitriding may be accelerated upon deposition of copper oxide particles on a steel surface; the catalytic activity of copper oxide being increased with a decrease in the particle size [3].

To make copper be deposited on a steel surface as nanoparticles, we had to find an efficient method for restricting the size of copper particles during the formation. The usage of polymeric protectors during the synthesis of metal sols is a way of the implementation of this restriction. The macromolecules of an appropriate polymer can interact with nanoparticles of metal formed in a solution. This interaction results in a macromolecular screen being formed on the nanoparticle surface, and the particle growth ceases [4]. The reaction product is a sol of the nanocomposite comprising a complex of the nanoparticle with the polymer protector. These processes give small-size nanoparticles (1 to 10…20 nm) and, as a rule, narrow particle size distribution.

The purpose of this work is to study a possibility of forming of copper nano-particles at the iron (steel) surface by their reduction from sulfate solution with usage of protector properties of a polymer; and to compare catalytic properties of copper oxide films received on the basis of copper precursor synthesized with or without a polymer protector.

Thus, we intended to develop a new and a very simple way for the fabrication of nanocatalytic systems (and their precursors) based on metal nanoparticles. For this purpose, it was proposed for the first time to synthesize a metal—polymer nanocomposite containing nanoparticles of the target metal by reducing ions of this metal on the surface of another, more active metal in the presence of macromolecules controlling the nanoparticle size. We assumed that, in conformity with the theory of pseudo-matrix processes [5], the macromolecules should shield the nanoparticles of the target metal formed upon the reduction when the particles have reached a definite (critical) size and thus to terminate further growth.

This assumption is confirmed by investigations of processes of the reduction of copper ions by zinc in the presence of Poly-N-Vinylpyrrolidone (PVP) on a zinc surface [6]. The product consists of copper particles and the polymer; it has increased oxidation stability, specific surface layer morphology, and the ability to dispersion and formation of a sol upon the ultrasonic treatment of a reaction mixture. Sol particles consist of the polymer and metal nanoparticles; however, the majority of the reduced metal remains on the surface.

In this work, we have attempted to determine the effect of the polymer on the size of metal particles being formed on the surface of an active metal using a precipitated metal as a catalyst (or its precursor) of nitriding process. Iron was selected as a metal reductant of copper ions, because it is known that products of copper oxidation catalyze steel nitriding, with the catalytic activity of copper oxide increasing with a decrease in its particle size [3]. We assumed that the size of catalyst particles must depend on the size of precursor particles, i.e., copper metal nanoparticles.

Experimental Methods

Experimental study was made on low-carbon steel 08Yu containing 0.08%C *(GOST* (State Standard) 9045-93, Russia, corresponding to the European Standard of steel DC04/DC03 EN 10139).

Copper sulfate pentahydrate (reagent grade, Reakhim, Russia), PVP (Poly-N-Vinylpyrrolidone) (Acros Organics, United States), and twice-distilled water were used in the experiments. Steel plates with sizes of 10 x 8 x 2 mm were preliminarily degreased by washing with acetone. Samples to be examined were prepared by immersing the plates in solutions of copper sulfate or its mixture with PVP for necessary time at 20°C. The concentration of PVP was 0.02 base-moles per liter; the volume of the solutions was 50 mL. The control sample was immersed into a copper sulfate solution (0.01 moles per liter) without PVP.

The following varied parameters of the reaction process of synthesis of a precursor were foreseen:

- Deposition time in the copper sulfate solution $CuSO_4$ (5, 10, 15, 30 or 60 min) ;
- Molecular weight of the polymer PVP (10000, 60000, 360000);
- Concentration of copper ions in the solution (0.01 moles per liter or 0.02 moles per liter).

The surface structure of the samples after deposition of the copper coating was examined by an SMM-2000 multi-microscope (Proton, Russia). Scanning was performed by soft MSCT cantilevers for contact imaging mode (Veeco, USA) at a velocity of ~4 pm/s.

Electronic absorption spectra in the visible region were recorded with a Specord 210 Plus spectrophotometer (Analytik Jena AG, Germany) and processed with the use of the WinASPECT software. The measurements were performed in cells with a light path of 1 cm.

Transmission electron microscopy (TEM) examinations were performed using a JEM-1011 (Jeol, Japan) microscope equipped with a Gatan digital camera (United States) controlled by the Digital Micrograph software. Samples for TEM examinations were prepared by applying sol droplets onto formvar substrates and drying them at room temperature. Scanning electron microscopy (SEM) experiments were carried out using a JSM-6380 LA microscope equipped with an X-ray energy dispersive analyzer (Jeol, Japan). The samples were preliminarily washed with water, dried and coated with a gold layer by sputtering.

Solutions with steel plates immersed in them were subjected to ultrasonic treatment with a UKS-22 facility (designed at the Moscow Automobile and Road Construction State Technical University) in accordance with the scheme described in [7] at a frequency of 22 kHz and a power of 3-5 W/cm^2. The duration of treatment was 2 min.

Nitriding of steel samples covered by copper precursor of catalysts was implemented in 2 variants. Variant 1 included a stage of oxidizing of copper precursor in air atmosphere at temperature of 500°C during the time up to 10 min for achieving of specific black color of oxidized copper at a surface. Then the stage of nitriding in ammonia was carried out at temperatures between 570…590 °C during 2.5…3 hours. Variant 2 foresees soaking of samples at temperatures 580… 590 °C in flowing mixture of ammonia, carbon dioxide, and water vapour (the ratio of gas components was determined according to the method described in [8]) during the time necessary for copper oxidation, and subsequent nitriding in ammonia at the same temperature during the time of 2…3 hours.

The catalyst activity determines the thickness of modified layers formed during a certain period of time. These layers were studied by metallographic analysis, which comprised the manufacture of micro-cross-sections followed by etching and taking micrographs. To prepare the cross-sections, the samples were fastened in special holders with copper foil gaskets, then the sections were treated on an emery stone, ground on sandpapers with successively decreasing gritness, then polished on diamond emulsions deposited on a woven felt, and finished on a thin cloth with water. Polishing was performed on a grinding-and-polishing machine. With this procedure for manufacturing cross-sections, the area of chemical compounds is completely retained, layer spalling is prevented. Microscopic examination of the cross-sections was carried out using an AXIOVERT 25CA microscope (Carl Zeiss, Germany). Hardness equalizing in the cross-section served as the criterion for the layer thickness on the steel. Hardness distribution over the layer thickness and the effective layer thickness were determined for polished micro-cross-sections (AFFRI micro-hardness tester (Induno OLona Varese, Italia)) with a 1 N load according to GOST 2999-89.

Results and Discussion

Mechanism of catalytic activity of copper oxide during nitriding

In was shown in previous studies [2] that a Cu-film precursor of nitriding catalyst may be formed at iron surface by copper reduction from $CuSO_4$ according to the reaction:

$$Fe^0+Cu^{+2}\rightarrow Fe^{+2}+Cu^0. \quad (1)$$

The following oxidation of copper film by oxygen with receiving of CuO may be realized by two possible reactions: from pure Cu

$$2Cu + O_2\rightarrow 2CuO \quad (2)$$

or from Cu_2O if a surface film appears to be oxidized for oxide II

$$2Cu_2O + O_2 \rightarrow 4CuO. \quad (3)$$

The mechanism of forming of a nitriding layer under the CuO-film is connected with acceleration of dissociation of ammonia at nitriding temperatures; at the same time the reduction of pure copper from the oxide ie taking place:

$$2NH_3 + 3CuO \rightarrow 2N^- + 3Cu^{2+} + 3H_2O. \quad (4)$$

During nitriding ammonia dissociates with forming of nitrogen ions and hydrogen:

$$NH_3\rightarrow N^- +3/2H_2. \quad (5)$$

Hydrogen can reduce copper from the oxide at nitriding temperatures according to the reaction:

$$CuO+H_2=Cu+H_2O. \quad (6)$$

So at the surface pure copper is formed again, i.e. stages of reduction and oxidation alternate.

As a result of accelerated ammonia dissociation the concentration of adsorbed nitrogen ions at the surface dramatically increases. If a thickness of CuO- film is small enough, than nitrogen ions can penetrate through the film into the iron matrix. Due to diffusion movement of nitrogen ions into the metal different phases of nitrided layer are formed according to the Fe-N phase diagram:

$$N^-+Fe\rightarrow Fe_\alpha(N) \rightarrow \gamma'\text{-}Fe_4N \rightarrow \varepsilon\text{-}Fe_{2\text{-}3}N. \quad (7)$$

Early investigations have shown [2] that catalytic properties of CuO-film appear first of all in the increase of ε-nitride zone thickness. In definite frames this effect is stronger if the thickness of copper film is larger (supposing that the thickness of CuO layer corresponds to the thickness of Cu-precursor in case of total oxidation) (Fig. 1). The thickness of Cu film is determined by deposition time and by carbon content in steel (Fig. 2). The last effect caused by an ability of steel to assimilate absorbed nitrogen: the higher is carbon concentration in steel the lower is maximum nitrogen concentration at a surface (Table 1).

Table 1: Concentration of elements in near-surface layer of nitrided steels with previously formed CuO-catalyst

Concentration of an element, weight %	%C in steel		
	0,08	0,45	0,8
Fe	88.62	91.07	94.43
N	9.5	8.93	4.71
C	0	0	0.86
Cu	1.88	0	0

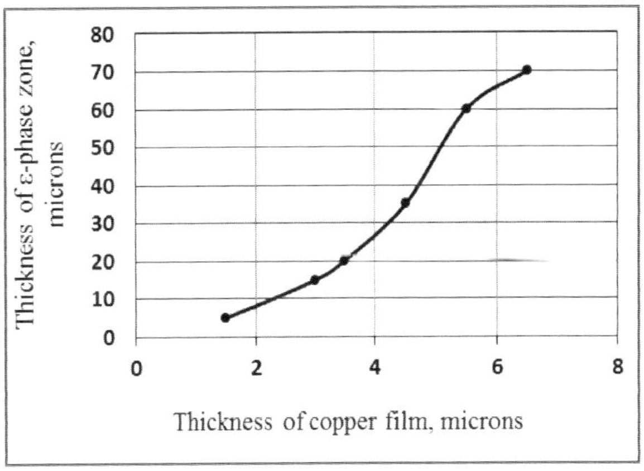

Figure 1: Dependence of thickness of ε-phase zone of nitrided layer in steel DC04 on the thickness of previously deposited copper film.

Mechanism of dispersion of copper particles by usage of polymer protectors

Further progress of studies brought to light a significant factor influencing the catalytic properties of a copper oxide film consisting in the dispersion of particles of copper precursor. Catalytic activity of nano-dispersed systems increases with the decrease of particles' size [9]. High effectivity of nano-structured catalytic systems usually is the reason for their huge specific surface. Besides the size of catalyst nano-particles can affect its selectivity and its ability to influence the rate of reactions [5]. That is why the highest catalytic activity of nitriding may be expected by usage of nano-dimensional particles of CuO – catalyst. This can be achieved by control of particles' size of initial copper precursor by its reduction at a steel surface with addition of a polymer.

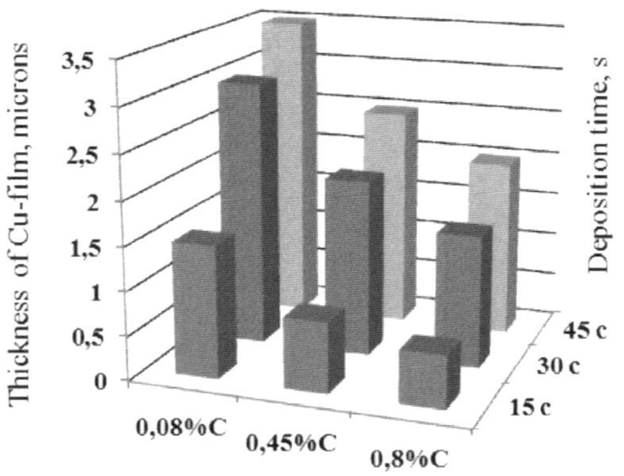

Figure 2: Dependence of copper film thickness on deposition time and on concentration of carbon in steels.

A mechanism of nano-structuring of copper particles by usage of polymer protectors is referred to a pseudo-template synthesis theory [10]. According to the theory a macromolecule of a polymer "screens" a nano-particle formed during synthesis of sols and stops its growth (Fig. 3a).

The theory of pseudo-template synthesis of sols is based on the idea that noncovalent interactions of macromolecules with nanoparticle surfaces have a cooperative nature. It predicts an exponential increase of the stability constant of a macromolecule-nanoparticle complex with an increase of nanoparticle surface area; provided that the macromolecule is sufficiently long to shield the entire surface of a nanoparticle. The development of the theory made it possible to efficiently control the size characteristics of nanoparticles directionally varying the synthesis conditions. It is reasonable to assume that the same principle of metal nanoparticle size limitation may be used not only for the production of free metal nanoparticles, but also upon their synthesis via the reduction of metal ions on the surface of a metal reductant (Fig. 3b).

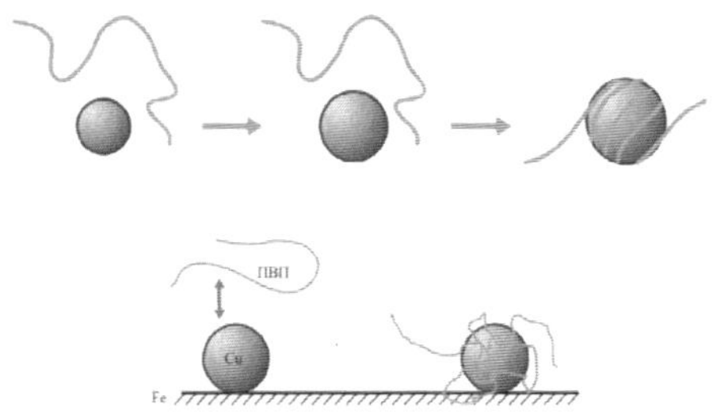

a)

b)

Figure 3: Schematic illustration of mechanisms of nano-structuring of copper particles by polymer protector: a) – shielding of a sol particle; b) – shielding of a particle reduced on metal surface.

Structure of copper precursor reduced with polymer protector

The experimental study has shown that reduction of copper ions with iron on the steel surface by exposing the samples in solutions of copper sulfate and its mixture with PVP yields coatings of different colors. Copper coatings received without PVP have specific red-brown color. Coatings received with PVP have violet-brown color. Note that a violet color is typical of copper sols in which metal nanoparticles are shielded by PVP macromolecules. Such sols result from the reduction of copper ions with boron hydride compounds in aqueous PVP solutions [3].

It was determined that the product resulting from the reduction of copper ions on a steel surface in the presence of PVP is a nanocomposite comprising metal nanoparticles and the polymer. Examinations in atomic-force microscope have shown that copper ion reduction in the presence of the polymer gave 50 nm and larger particles on the steel surface (Fig. 4). These were not just copper nanoparticles but nanocomposite particles comprising copper nanoparticles and the polymer shielding the nanoparticle surface. X-ray diffraction examination of treated surface shows the presence of reflections of Fe and Cu; it means that copper particles are isolated (i.e. separated from each other by areas of non-covered steel surface), and are not oxidized (Fig. 5).

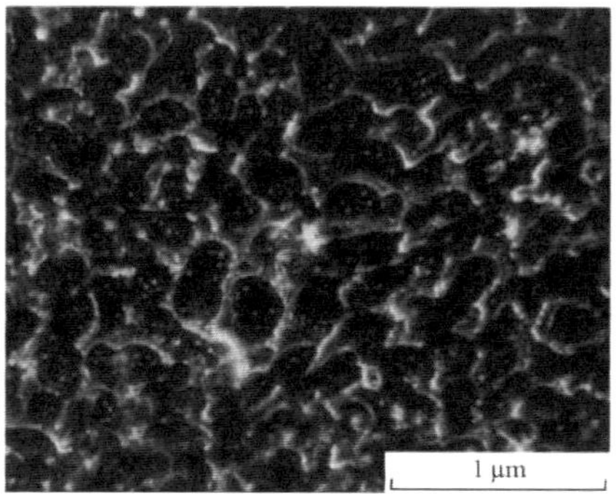

Figure 4: AFM-micrograph of copper-polymer nanocomposite formed on a steel surface in the presence of PVP.

For dispersion and examination of sols an ultrasonic treatment of reaction system was carried out directly during the reduction process. The results of studies of the sols formed upon ultrasonication of steel samples in solutions of copper sulfate and PVP mixtures directly during copper ion reduction prove the structures of nanocomposite. Study with high-resolution transmission electron microscopy shows that the particles of sols consist of nano-particles of metal (Cu) with the size of 10…20 microns (Fig. 6a) or of their agglomerates with the size of 100 microns or more (Fig. 6b) surrounding by the polymer shell.

It should be emphasized that the ultrasonic treatment of the system in which copper is reduced in the absence of the polymer does not yield sols. Moreover, the sols are not formed when steel plates preliminarily coated with the products of copper reduction in both the absence and presence of the polymer are immersed in water or a PVP solution and subjected to the ultrasonic treatment. It is possible that the formation of the sols, in these cases, is hindered by the mixed coagulation and crystallization structures formed in the products as a result of the sticking of copper particles.

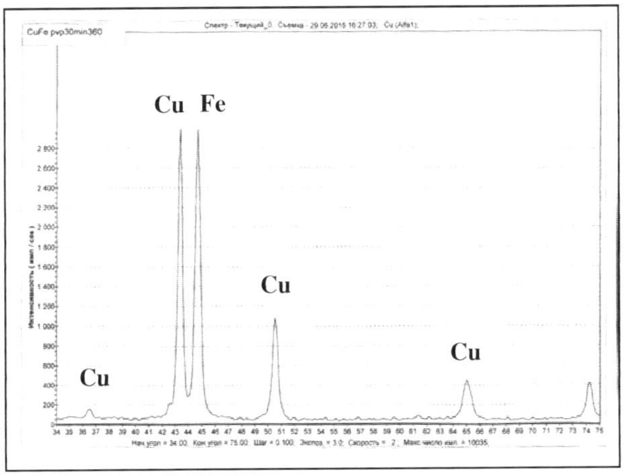

Figure 5: X-ray diffraction pattern of a steel surface after copper reduction with a presence of PVP.

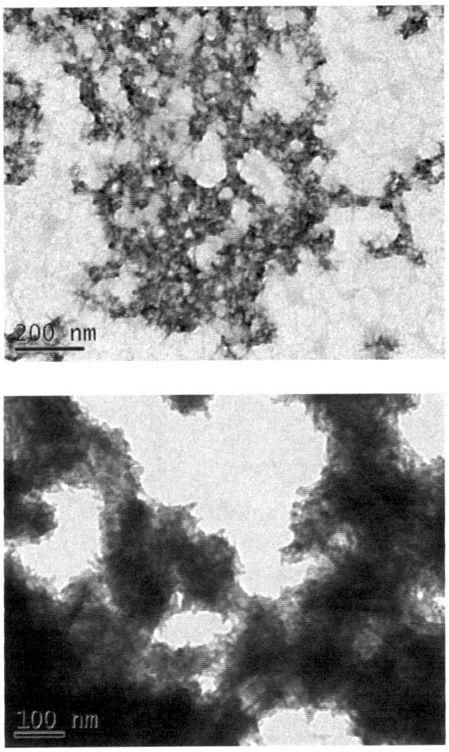

a)

b)

Figure 6: TEM images of sol particles resulting from ultrasonic treatment of a reaction system upon reduction of copper ions on a steel surface in the presence of PVP.

Examination in scanning electron microscope shows that agglomerates of copper particles at the steel surface are smaller and have more uniform dimensional distribution in the case of copper reduction with PVP (Fig. 7). Moreover particles (agglomerates) are finer if the molecular weight of PVP is higher (Fig. 7c).

a)

b)

c)

Figure 7: SEM images of products received by copper ions reduction at a steel surface: a) – without PVP; b) – with a presence of PVP with molecular weight 60000; c) – with a presence of PVP with molecular weight 360000.

Structure of nitrided layer in a steel surface with deposited catalytic nanocomposite coating

Nitriding of steel samples with CuO-film received by direct oxidation of copper precursor in the air atmosphere (Variant 1) confirmed it catalytic properties: the thickness of nitrided layer of ε-phase at the steel surface increases from 10 microns after classic nitriding to 19 microns after nitriding with receiving of copper film without PVP: and to 39 microns after nitriding with deposition of copper-polymer nanocomposite during reduction with PVP (Fig. 8).

Nitriding with oxidation of copper precursor in the mixture of ammonia, carbon dioxide and water vapour (Variant 2) allows to achieve more significant intensification of nitrided layer growth: total thickness of the compound zone increases to 39 microns after nitriding with catalytic Cuo-film without PVP. The thickness of the compound zone after nitriding with copper-polymer nanocomposite received with PVP depends on copper concentration in the initial solution of $CuSO_4$. In our experiment with the increase of Cu-concentration in the solution from 0.01 moles per liter to 0.02 moles per liter the thickness of the compound zone increases from 57 to 72 microns (Fig. 9). This effect can be explained by cyclic character of reduction-oxidation process.

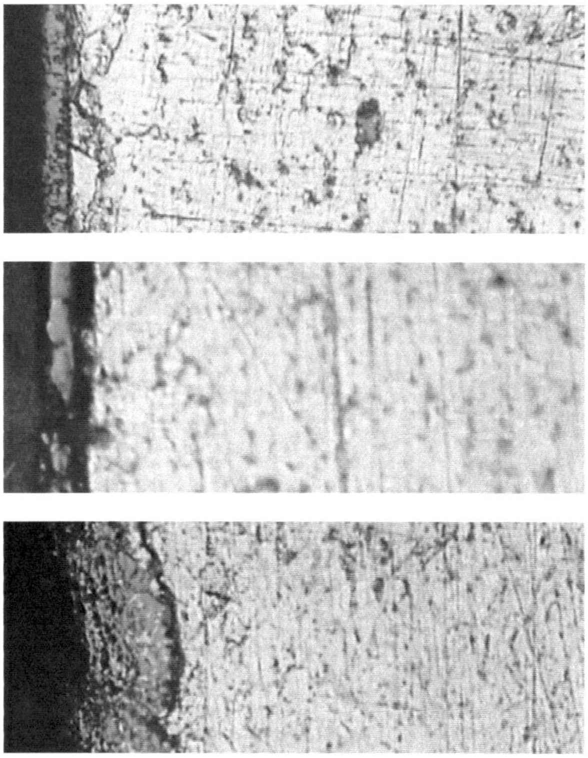

Figure 8: Micrographs of nitrided samples of steel DC04 after nitriding at 575°C during 150 min according to the Variant 1: a) – without Cu-precursor; b) with Cu- film received without addition of PVP; c) – with Cu-polymer nanocomposite received with a presence of PVP.

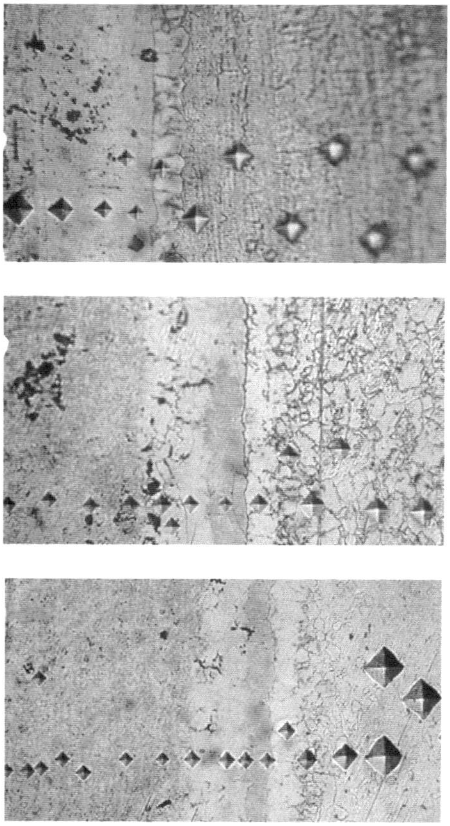

Figure 9: Micrographs of samples of steel DC04 nitrided with previous deposition of copper precursor according to the variant 2 at 590°C during 120 min: a) – with reduction of Cu-film without PVP; b) – with reduction at a presence of PVP and copper concentration in a solution 0.01 moles per liter; c) – with reduction at a presence of PVP and copper concentration in a solution 0.02 moles per liter.

Conclusions

Our investigations have shown that addition of PVP to the reaction system of copper reduction leads to significant changes in structure of a product, formed at a steel surface: during reduction of copper ions by iron with PVP in a solution nanocomposite is forming consisting of Cu nano-particles and polymer.

Catalytic activity of thermal oxidized product, received of the nanocomposite is higher than of copper oxide received without modification by polymer. Forming of CuO–catalyst by reduction of copper precursor from the solution with PVP gives the opportunity to accelerate additionally growth of nitrided layer during nitriding due to dispersion of particles of copper-polymer nanocomposite.

Regulating parameters of kinetics of catalytic nitriding process are: thickness of copper layer (i.e. time of deposition), molecular weight of the polymer, and copper concentration in the initial solution. Thickness of nitrided layer also depends on the ratio of treated surface areas covered by precursor and free of it. This may be controlled by variation of reduction conditions parameters.

Acknowledgments

This work was supported by the Ministry of Education and Science of the Russian Federation by financing the projects No. 16.301.2014/K and No. 11.1583.2014/K within the framework of a project part of the state assignment in the field of researches.

References

[1] Lakhtin, Yu. M. *et al.*, Theory and Technology of Nitriding, Metallurgia (Moscow, 1991), p. 320.

[2] Petrova, L. G. *et al.*, "Intensification of Nitriding Process of Structural Low-Carbon Steel by Formation of Catalytic Copper Oxide Film", *Strengthening Technologies and Coatings*, Vol. 123, No. 3 (2015), pp. 37-43.

[3] Petrova, L. G. *et al.*, "Forming of Composite Nano-Structured Coatings in Steel Parts by Methods of Thermo-Chemical Treatment", *Vestn. Kharkov. Nats. Avtomob.-Dor- ozhn. Univ.*, No. 51 (2010), pp. 7-13.

[4] Hirai, H. *et al.*, Polymeric Materials Encyclopedia, CRC Press (London, 1996), Vol. 2, pp. 1310-1312.

[5] Papisov, I. M. *et al.*, "On Recognition Phenomena in Polymer-Minute Particle Interactions and Pseudo-Matrix Processes", *Coll. Surf. A*, No. 151 (1999), pp. 399-408.

[6] Papisov, I. M., *et al.*, "Formation of a Metal–Polymer Nanocomposite During Contact of a Mixed Solution of Copper Sulfate and Poly(*N*-vinylpyrrolidone) with the Surface of Zinc", *Polym. Sci., Ser. B*, Vol. 57 (2015), p. 27.

[7] Prikhod'ko, V. M. *et al.*, "Effect of Ultrasonic Cavitation on Nitrided Steel Surface Layer Condition", *Met. Sci. Heat Treat.*, Vol. 57, No 5 (2015), pp. 300-303.

[8] Lakhtin, Yu. M. *et al.*, USSR Patent 831814

[9] Aleksandrov, V. A. *et al.*, "Synthesis of Copper–Polymer Nanocomposite on Steel Surface and Composite-Based Catalyst for Steel Nitriding", *Colloid Journal*, Vol. 77, No. 5 (2015), pp. 556-560.

[10] Litmanovich, O. E. *et al.*, "Hydrolysis of Poly (N-Vinyllactams) in their Nanocomposites with Metallic Copper", *Polym. Sci., Ser. A*, Vol. 42, No 9-10 (2000), pp. 242-243.

International Federation of Heat Treatment and Surface Engineering 2016
Proceedings of the 23rd IFHTSE Congress
April 18–21, 2016, Savannah, Georgia, USA

Effect of Deep Cryogenic Treatment on Fracture Toughness of 300M Steel Using Circumferentially Notched K_{IC} Test Specimens

Daniel Schuller
USP – Universidade de São Paulo, São Carlos,São Paulo State - Brazil
daniel.schuller@usp.br

Vojteh Leskovšek
IMT – Institute of Metals and Technology, Ljubljana – Slovenia.
Vojteh.Leskovsek@imt.si

Abstract

Deep cryogenic treatment (DCT) has been performed in 300M aeronautic steel quenched and tempered in conventional heat treatment parameters, in order to assess its effect on fracture toughness K_{IC} using circumferentially notched technique. DCT heat treatment was applied in one group of quenched and tempered specimens, before tempering, and fracture toughness was compared to other group, quenched and tempered, without DCT treatment. The results show a high influence of DCT treatment on fracture toughness of specimens and hardness too. It can be see a 16.37 MPa$\sqrt{m}$ loss on toughness and an increase around 2 HRC on hardness, for specimens that were submitted to DCT treatment in comparison with specimens without DCT.

Introduction

High Strength Steel

On aeronautic industry, the search for light, corrosion resistant and high strength materials is constant. Although, nowadays, many researchers have been developed searches in non-metallic materials for aeronautical applications, high strength parts that demands high reliability applications like landing gears, shafts, fittings, gears and others, are still made by steels. Some aeronautic steels with a large use in the aircrafts are the AISI 4340 and 300M steels. The 300M aeronautical steel is a development of the AISI 4340 with some changes in Silicon (1.6%), Molybdenum (0.4%) and Vanadium (0.5%) content. It combines high strength with high fracture toughness even in low temperatures, and then it is wide applied in aircraft industry [1, 2]. Vanadium alloy restrict the grain growth during austenitizing and, due to the silicon content that moves the TTT curve to the right, temperability and other effects are observed when it is compared with the AISI 4340. The tempering embrittlement temperature is higher (around 425°C for 300M and around 300°C for AISI4340) which allows large range of tempering temperature application for the best heat treatment combination [3]. The standard parameters for heat treatment of 300M, well applied in the industry, is 870°C for austenitizing for 1 hour and double tempering in 315°C.

Deep Cryogenic Treatment

Cryogenic treatment is a complementary treatment applied together other heat treatments, like quenching and tempering. It is usually applied after quenching exposing the parts at low temperature environment, in liquid nitrogen around -196°C, and is followed by tempering [4, 5]. This technique is largely used in materials with high carbon content, mainly tool steels, because its capacity to improve hardness and wear resistance of the material. During the deep cryogenic treatment a fine dispersed carbides precipitation is observed and it is due to martensite and austenite lattice contraction. The carbon atoms are forced to diffuse and make new carbide nucleus. This new nucleus increases the carbide percentage and the homogenous distribution [6]. Some studies point out that cryogenic treatment can improve wear resistance that are able to increase until 110% the tool life [7]. During this treatment, the amount of retained austenite decrease and, when it is applied before the tempering treatment, it also stabilizes carbides to precipitate during the tempering process. For this reason, it increases the hardness and wear resistance [6]. Deep Cryogenic Treatment (DCT) is not only applied on tool steels, Zhirafar et al. [8] analyzed this treatment effect on mechanical properties of low alloy AISI 4340. The retaied austenite content decrease from 5.7% to 4.2% to specimens that were only quenched and quenched and pass to cryogenic treatment respectively. At the same path, hardness increase from 54.5HRC to 55.3HRC and energy absorption decrease from 10.8J to 7.7J in Charpy test due to the increase of martesite content in the matrix. It is possible to analyze the energy absorption behavior for specimens with and without cryogenic test according tempering treatment in figure 1. It is possible to observe that increasing the tempering temperature, toughness first decrease around 11J at 300°C and then increase to higher values at 455°C. This behavior is due to the temper embrittlement for this material at this tempering temperature. The large differences between two treatments happened mainly at 200 an 455°C temperatures. Comparing both treatments, DCT have higher hardness which could have contributed to the lower toughness. At 455°C, DCT produces a slight increase in hardness (2.4%) at the expense of higher (14.3%) drop in toughness. Residual stress and carbides precipitation also increase this effect.

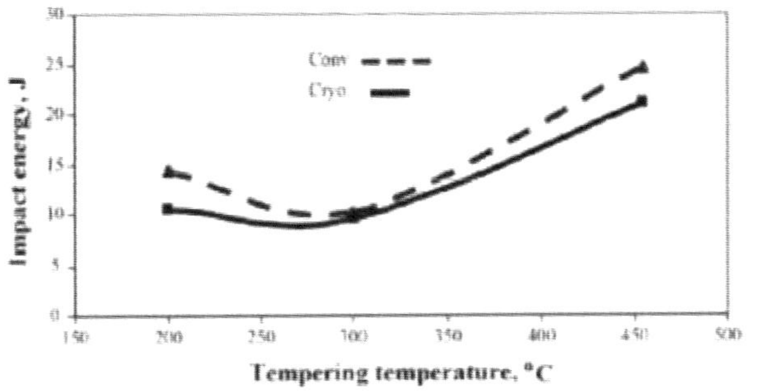

Figure 1: Impact energy absorption comparison for AISI 4340 steel with and without cryogenic treatment according tempering temperature for specimens quenched from 845°C [8].

Retained Austenite

In low alloy steel, some alloying elements like chrome (Cr), molybdenum (Mo) and manganese (Mn) get the final temperature of martensite transformation M_f lower than the environment temperature and, depending on the cooling medium it can be increased. This phenomenon causes a high amount of retained austenite in martensite matrix [9]. During some cryogenic processes, where parts are submitted to under zero temperatures, retained austenite can decrease due to transformation in martensite. During machining process or in work application of the part, this metastable phase can be transformed into martensite too, due to work hardening and then propitiate catastrophic failure [5]. A high amount of retained austenite in martensite microstructure can cause tensile and yield strength decrease, but in low alloy steels like 300M are not enough to increase the ductility to be possible to overcome tempering process [10]. The hardness decrease of the material, due to retained austenite in martensite matrix, cannot be considered that will straight improve fracture toughness of the material, because the secondary hardening effect during tempering. This effect can increase the secondary carbides precipitation and it is possible to find two specimens with the same hardness but different fracture toughness.

Fracture Toughness

Fracture toughness is a mechanical property, independent of geometry for the material, that describes the material ability to containing a crack to resist a fracture and is a very important material characteristic. ASTM E399 is the standard that describes the method to determine fracture toughness for materials. This standard method demands enough material to obtain plain strain fracture deformation. In some cases, it is pretty difficult to apply the standard method, mainly when it is about hard and brittle materials, with low ductility, in which it is hard to obtain the plain strain fracture or there is not enough material from the part to make the analysis. For these cases, many authors have been developed some alternative non-standardized methods to allow the fracture toughness measurement. Some examples of these methods are notched small punch tests [11], Vickers indentation method toughness measurement method [12] and three-point bending test for diamond [13]. One of the most promising methods is the use of circumferentially notched and fatigue pre-cracked tensile specimens. This technique, originally, was created to be used with high carbon and tool steels materials, because of the high difficulty to machine these steels after heat treatment, high difficulty to produce the exactly microstructure and heat treatment behavior using standardized toughness measurement method and also, in some cases, to be almost impossible to create a fatigue pre-crack in hardened test specimens. However, for circumferentially notched K_{IC} test specimens, fatigue pre-crack can be obtained in soft annealed state, that is, prior to the final heat treatment without producing any disturbing effect on the fracture toughness of the tool and high-speed steels [14]. Many studies have been developed using this technique for hard and weak materials [15, 16], but very good results with high precision were obtained in ductile materials too [17], when compared with the standardized test. The specimen used in this technique is shown in figure 2, where a radial load is applied on the specimen, still in an annealed condition, to produce the pre-crack. Afterwards, it was heat treated and then fractured under tensile load. The diameter of brittle fractured area (weak fracture) is measured, according to figure 3 and 4.

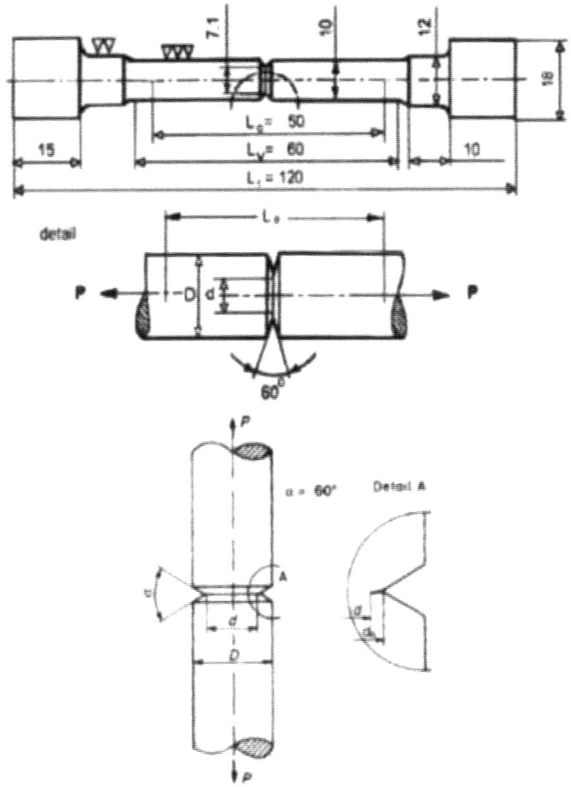

Figure 2: Circumferentially notched specimens with dimensions detail.

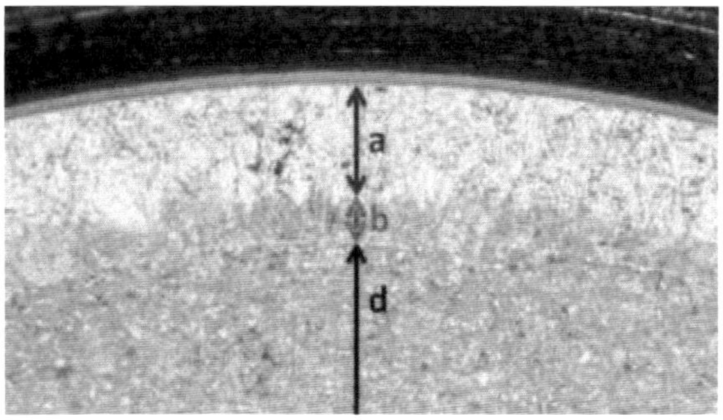

Figure 3: Pre-crack diameter "a"; ductile fracture diameter "b" (that is still a part of fatigue pre-crack), and brittle fracture diameter "d".

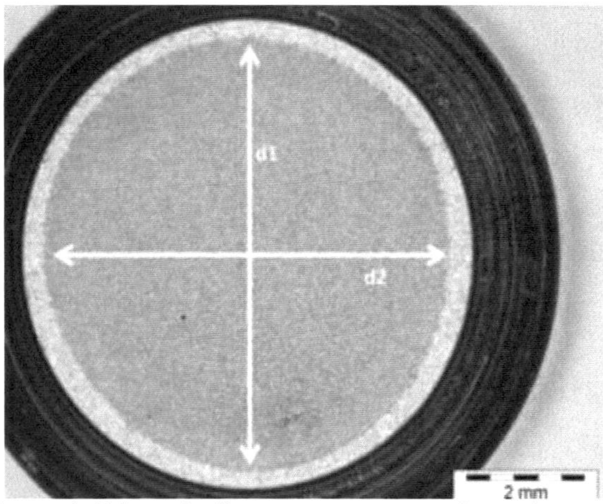

Figure 4: Schematic diameter measurement for each part of fractured specimens.

After all measurements, the fracture toughness is calculated using Equation I.

$$K_{IC} = \frac{F}{D^{3/2}}\left(-1.27 + 1.72\frac{D}{d}\right) \qquad \text{(I)}$$

Where F is the peak load at fracture, D is the specimen diameter and d is the measured diameter of the brittle fractured area. For this relationship to be valid, and plain fracture deformation to be reached, the specimen relations of formula II and III have to be reached.

$$D \geq 1.5\left(\frac{K_{IC}}{\sigma_y}\right)^2 \qquad \text{(II)}$$

Where σ_y is the yield strength of studied material. Other important correlation to be followed related with pre-crack formation is shown in equation III.

$$0.5 < \frac{d}{D} < 0.8 \qquad \text{(III)}$$

Equation III guarantee that a minimum pre-crack diameter was formed for plain-strain conditions to be reached and is enough to allow brittle fractured area diameters measurements, after tensile test. If all presented correlations are followed, then circumferentially notched and fatigue pre-cracked tensile specimens technique can be used.

Methods

All experimental procedure was made on IMT – Institute of Materials and Technology – Ljubljana, Slovenia. The raw material of 300M steel was donated by the company Villares Metals, located on Sumare, São Paulo State, Brazil. First, the specimens were machined, according to figure 2, from a cylindrical bar with diameter Ø 55mm, of 300M steel, and with initial hardness of 26 HRC. The chemical composition of the material, as received, is shown in table I.

Table I: Chemical composition for 300M steel as received.

Element	(%)	Element	(%)
C (%)	0.422	Ni (%)	1.61
Si (%)	1.71	Mo (%)	0.370
Mn (%)	0.870	Al (%)	0.017
P (%)	0.0036	Cu (%)	0.02
S (%)	0.004	Co (%)	0.018
Cr (%)	0.750	Ti (%)	0.021

Using a system installed on a turning machine and pre-determined radial load of 500 N, about 0.5 mm deep fatigue pre-crack was produced on all 26 specimens. Afterwards, heat treatments Q and V were applied according to table II, using a furnace *Ipsen* VTTC324-R, a vacuum furnace with high-pressure nitrogen gas system for quenching. It was used 13 specimens for each heat treatment.

Table II: Applied heat treatments and detail.

Heat Treatments	Detail
Q	Austenitizing at 870°C for 50 minutes, Quenching in nitrogen gas flow, double tempering at 300°C for 50 minutes.
V	Austenitizing at 870°C for 50 minutes, Quenching in nitrogen gas flow, Deep Cryogenic treatment at -196°C for 25 hours, double tempering at 300°C for 50 minutes.

After heat treatment application, specimens were submitted to tensile test until its fracture. It was performed using an *Instron* machine with a cross-head speed of 1.0 mm/min. The obtained curve until the fracture presents a linear elastic load-displacement relationship, confirming the plain-strain fracture dislocation behavior of the fracture. Rockwell C hardness was measured and also four measurements of fracture diameter was performed, according to figure 3 and 4. With all necessary parameters, and using equation I, fracture toughness was calculated.

Results

In Table III are listed the hardness HRC average values measured for the specimens of each heat treatment.

Table III: Hardness values for heat treatments Q and V.

Heat Treatment	Hardness (HRC)	Standard Deviation
Q	53.41	0.12
V	55.32	0.11

According to table III, the average hardness of Q and V heat treatments shows a significant difference of around 2 HRC. Converting the values, heat treatment Q presents a strength of 1880MPa and heat treatment V around 2000MPa, it is an increase of around 6,1% in strength. It is possible to see that the standard deviation for hardness measures are very low among the specimens, and this can be attributed to specimens shape, that allows a good and symmetric heat transfer along the parts and guarantee the same behavior and microstructure for all specimens. The rupture strength average values were also measured and the results are shown in table IV.

Table IV: Average load rupture for specimens of heat treatments Q and V.

Heat Treatment	Rupture strength average (MPa)	Standard Deviation
Q	1418	34.83
V	1060	49.60

The rupture strength, in this case, is slight related to fracture toughness of the material. Heat treatment Q results in lower strength than V, then it is more ductile and offers more resistance to crack propagation absorbing more energy on crack tip, for this reason, the average rupture strength is lower for heat treatment Q in comparison with heat treatment V. After the test, diameters of brittle fractured region were measured and the results are plotted in table V.

Table V: Brittle fracture diameter average (weak region)

Heat Treatment	Ø Diameter (mm)	Standard Deviation
Q	6.41	0.09
V	6.45	0.09

The low standard deviation for measured diameters also shows the advantage for this shape of specimen. The microstructure analysis doesn't present a high difference among the specimens, and then, high repeatability is reached with the use of good measurement equipment, microscope. Finally, with all data obtained, fracture toughness can be calculated using equation I and are shown in table VI.

Table VI: Fracture Toughness values average obtained for both heat treatments.

Heat Treatment	Fracture Toughness (MPa $\sqrt{m}$)	Standard Deviation
Q	64.67	1.62
V	48.30	2.23

According to the fracture toughness values, it is possible to see that the specimens of heat treatment V, submitted to deep cryogenic treatment, display values around 16,37 MPa $\sqrt{m}$ lower than the specimens of heat treatment Q, only quenched and tempered. It is a decrease of almost 34% in fracture toughness. The standard deviation is very low and guarantees a good accuracy for the results. This behavior is related mainly to the retained austenite reduction, when deep cryogenic treatment is applied, and then makes the specimens weaker due to increase in martensite amount. Other important characteristic that can influence the toughness behavior is the amount of precipitates in the matrix. It is showed by some authors [15, 18], that deep cryogenic treatment can facilitate carbide precipitation due to martensite saturation, where carbon atoms can migrate out of the lattice and form new carbides in the structure. Carbide amounts increasing in microstructure makes the distance among particles lower and then increase the crack propagation.

Fractography

Using a Scanning Electron Microscope (SEM), fractography analyses were performed on the specimens. Figure 5, *a* and *b*, shows the fracture zone with 50 times magnification. The pre-cracked region can be seen in figures 5 for both specimens, being the darker and relatively smooth-flat region. After this region, there is a ring around the whole specimen that presents the ductile fracture region created during the tensile test. It appears like a lighter and rough ring. Heat treatment V presents a smaller ductile region then specimens of heat treatment Q, what is observed on the toughness values too. After the border with the ductile diameter, it starts the brittle (fast and weak) fracture that grows until the center of the specimens. This region has a characteristic of very rough surface with light color due to the fast propagation

Figure 5: Fractography with 50 times magnification on SEM, for heat treatment Q and V (figure a and b respectively).

In figure 6, *a* and *b*, is shown the ductile region with 10000 times magnification.

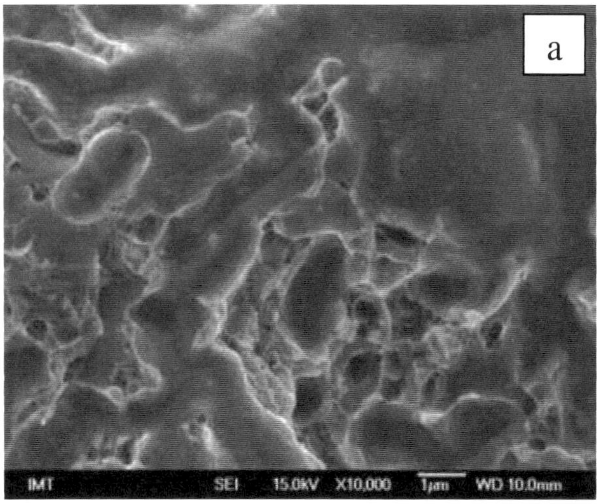

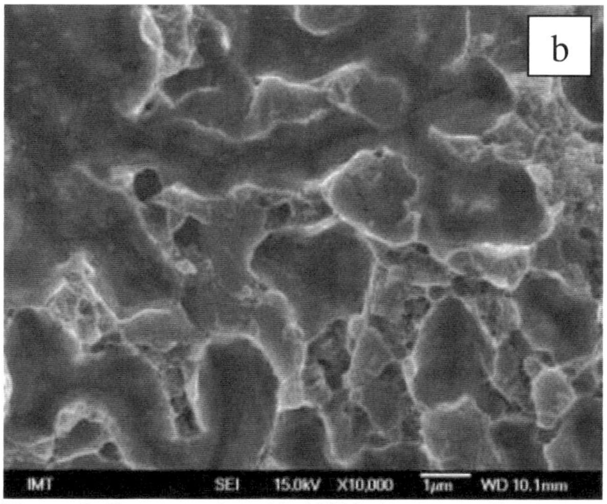

Figure 6: Fractography with 10000 times magnification on SEM, for heat treatment Q and V (figure a and b respectively).

Fractography of ductile region, showed in figure 6, indicates some dimple even with presence of some transgranular and flat fracture regions. The crack propagation in this region depends strongly on the material toughness. Figures also show some apparent light particles distributed around the flat surfaces with a higher amount observed for heat treatment V. These light particles can be related with non diluted particles of raw material and carbides precipitation after tempering and mainly due to cryogenic treatment. In figure 7, *a* and *b* it is possible to see the same region with a 50000 times magnification. In this region, inside the dimple regions, some apparent spherical particles can be seen for both heat treatments. These particles cannot be related to any known precipitates and, due to their dimension, it was not possible to analyze chemical composition. The size vary along the same part and there is no significant size or amount differences between specimens with heat treatment Q and V.

Figure 7: Fractography with 50000 times magnification on SEM, for heat treatment Q and V (figure a and b respectively).

Conclusions

- Circumferentially notched and fatigue pre-cracked tensile specimens have been used for fracture toughness measurement of 300M low alloy steel. The technique presented good results, when compared with other works using standardized method.
- Applying deep cryogenic treatment on 300M steel is possible to reach some tensile strength increase but with a high fracture toughness penalty.
- Carbide precipitation effect and mainly retained austenite transformation can be listed like the cause of this behavior.

Acknowledgments

Authors gratefully acknowledge IMT – Institute of Metals and Technology, from Ljubljana – Slovenia, for support on test execution and analysis.

References

[1] Zhang, S. et al., "The growth behavior of austenite grain in the heating process of 300M steel". Materials Science and Engineering, Vol. A528, (2011), pp 4967-4972.

[2] Abdalla, Antonio, et al. "Caracterização mecânica do aço 300M submetido a diferentes tratamentos térmicos".8° Congreso Iberoamericano de Ingenieria Mecanica, Cusco, Peru, October 2007, pp 130-137.

[3] Tomita, Y. and Okawa T., "Effect of modified heat treatment on mechanical properties of 300M steel", Vol. 11, (1993), pp. 245-251.

[4] Siva, R. et al., "Optimization of deep cryogenic treatment process for 100Cr6 bearing steel using the Gray-Taguchi method", Tribology Transactions, Vol. 55, (2012), pp. 854-862.

[5] Senthilkumar, D. et al., "Influence of shallow and deep cryogenic treatment on the residual state of stress of 4140 steel", Journal of Materials Processing Technology, Vol. 211, (2011), pp. 396-401.

[6] Li, S. et al., "Hardness and toughness investigation of deep cryogenic treatment cold work die steel", Cryogenics, Vol. 50, (2010), 89-92.

[7] Siva, R. et al., "Enhancing the wear resistance of 100Cr6 bearing steel using cryogenic treatment", Tribology Transactions, Vol. 55, (2012), pp. 387-393.

[8] Zhirafar, S. et al., "Effect of Cryogenic Treatment on the mechanical properties of steel and aluminum alloys", Journal of Materials Processing Technology, Vol. 186, (2007), pp. 298-303.

[9] Bakhsheshi-Rad, H. et al., "Effect of multi-step tempering on retained austenite and mechanical properties of low alloy steel", Journal of Iron and Steel Research International, Vol. 18, (2011), pp. 49-56.

[10] Youngblood, J., and Raghavan, M., "Correlation of microstructure with mechanical properties of 300M steel", Metallurgical Transactions A, Vol. 8A, (1977), pp. 1439-1448.

[11] Xu, Y.,and Guan, K., "Evaluation of fracture toughness by notched small punch test with Weibull stress method", Materials Design, Vol. 51, (2013), pp. 605-611.

[12] Nolan, D., et al., "Estimation of fracture toughness of nitride compound layers on tool steel by application of the Vickers indentation method", Surface & Coatings Technology, Vol. 201, (2006), pp. 186-188.

[13] Jiang, Z. et al., "Accurate measurement of fracture toughness of free standing diamond films by three-point bending tests with sharp pre-cracked specimens", Diamond and Related Materials, Vol. 9, (2000), pp. 1734-1738

[14] Leskovšek, V. and Podgornik, B., "Measurement and mechanical of fracture toughness of tool and high-speed steels", Encyclopedia of Iron, Steels, and Their Alloys, Vol. 99, (2015).

[15] Leskovšek, V. et al., "Relations between fracture toughness, hardness and microstructure of vacuum heat-treatment high-speed steel", Journal of Materials Processing Technology, Vol. 127, (2002), pp. 298-308.

[16] Leskovšek, V. et al., "The influence of austenitizing and tempering temperature on the hardness and fracture toughness of hot-worked H11 tool steel", Journal of Materials Processing Technology, Vol. 178, (2006), pp. 328-334.

[17] Baker, A. et al., "Relationships between microstructure and toughness in quenched and tempered ultrahigh-strength steels", Structure and Properties of Ultrahigh Strength Steels, American Soc. Testing Mater. Spec. tech, (1965), pp. 23-29.

[18] Carlson, M. et al., "The effect of austenitizing temperature upon the microstructure and mechanical properties of experimental Fe/Cr/C Steels", Vol. 10A, (1979), pp. 1273-1284.

International Federation of Heat Treatment and Surface Engineering 2016
Proceedings of the 23rd IFHTSE Congress
April 18–21, 2016, Savannah, Georgia, USA

Uphill Quenching of Aluminum Alloys: A Process Review

Wellington da S. Mattos - University of São Paulo
Materials Engineering Department, 1100, João Dagnone Avenue, São Carlos, SP, Zip Code 13563-120, Brazil
Aircrafts Maintenance Technology, Federal Institute of São Paulo, São Carlos, SP, Brazil
wels@sc.usp.br

George E. Totten – Portland State University
97201,Department of Mechanical and Materials Engineering, 1825, SW Broadway, Portland, OR, USA
getotten@gmail.com

D. Scott Mackenzie - Houghton International Inc.
P.O. Box 9300, Madison & Van Buren Avenues, Valley Forge, PA, USA, 19482
SMackenzie@houghtonintl.com

Lauralice de Campos F. Canale - University of São Paulo
Materials Engineering Department, 1100, João Dagnone Avenue, São Carlos, SP, Zip Code 13563-120, Brazil
lfcanale@sc.usp.br

Tom Croucher – Tom Croucher Metallurgical Service (Deceased)

Abstract

Uphill quenching is often not well understood and there are relatively few publications on the topic. Uphill quenching was originally developed by Alcoa approximately 50 years ago for aluminum alloys. It has also been referred to as "deep freezing" or "tri-cycle stress relieving". Uphill quenching has been reported to provide residual stress reduction which may exceed 80%..Therefore, uphill quenching is typically used to achieve dimensional stability in several critical types of aluminum parts. Uphill quenching is typically applied after quenching and before aging. The uphill quenching process consists of the immersion of the part into a cryogenic environment and after cooling is immediately followed by transferring to a hot-steam fixture to obtain a temperature gradient which maintains the mechanical properties gained with the heat treatment. When performed properly, uphill quenching results in low residual stresses and improved dimensional stability. This paper provided an overview of the uphill quenching process and its application in the heat treatment of critical aluminum alloy components such as large aerospace components.

Introduction

When aluminum is solution heat treated at elevated temperatures, typically in the range of 400-540°C, some alloying elements are redissolved to produce a solute-rich solid solution. The objective of the solutionizing process is to maximize the concentration of hardening elements which include copper, zinc, magnesium, and/or silicon in the solid solution. The concentration and rate of dissolution of these elements increases with increasing temperature, therefore, solutionizing temperatures are usually near the liquidus temperature of the alloy.

If an alloy is cooled from the solutionizing temperature, alloying elements are precipitated and diffuse from the solid solution to concentrate at the grain boundaries, small voids or undissolved particles, at dislocations and other imperfections in the aluminum lattice. To achieve optimal strength, toughness, and corrosion resistance, it is desirable to retard this diffusion process and maintain the elements in solid solution until the alloy is age hardened. This is accomplished by quenching from the solutionizing temperature. After quenching, aluminum is aged and during the ageing process, a fine dispersion of elements and compounds are precipitated that significantly increase the strength of the material.

Diffusion and precipitation kinetics are slower in some alloys than others, permitting lower cooling rates while still allowing high strengths and corrosion resistance to be obtained. Excessively slow cooling leads to excessive concentrations of alloying elements to develop in the grain boundaries which can aggravate intergranular corrosion. Therefore, cooling rates during quenching must be sufficiently fast to minimize precipitation. However, if the cooling rates are too fast, there may be an increased propensity for distortion. Therefore, one of the primary process design challenges is to select quenching conditions that optimize desirable as-quenched properties while minimizing others [1].

Age hardenable alloys such as 2024, 2219, 7075, 7050 and 6061 are often quenched in cold water [2]. However, cold-water quenching may produce unacceptable distortion due to the high thermal gradients produced. Alternatively, if distortion problems are encountered with cold water (10-32 °C) quenching, then ‘‘hot water’’ (60-71 °C) quenching is used

[2]. Although hot-water quenching has been used for many years to reduce quench distortion in aluminum alloys, excessive distortion still sometimes occurs. In such cases, an aqueous poly(alkylene glycol)-PAG copolymer solution may be used.

The use of PAG polymer quenchants to reduce quench distortion of aluminum alloys has been reported by Croucher [3,4] and others [1,5,22]. Furthermore, PAG quenchants have been reported to offer significantly greater residual stress reduction than hot water in a number of studies. [6]. Torgerson and Kropp conducted residual stress and other mechanical property measurements on specimens quenched in hot water and a PAG polymer quenchant at various concentrations using both AA7050-T736 forgings and plate stock. It was found that Type I quenchants, as defined by AMS 3025, provided minimum distortion while still meeting the design minimums for forgings up to 5 in. thick [7].

The thermal gradients caused by non-uniform quenching produce inelastic strains that lead to residual stresses that are compressive at the surface and tensile in the interior. This occurs because the surfaces of the parts cool faster than the interior because the thermal gradients which are produced by the non-uniform quenching causes different regions of the part to contract at different rates [8]. In the case of forgings, for example, these internal stresses may lead to: 1.) distortion when quenching thin sections; 2.) distortion during subsequent machining; and 3.) if sufficient material is removed to expose material in tension, reduced fatigue resistance and stress corrosion cracking may occur [9].

Although the ageing process which is performed after quenching will increase material strength, it exhibits no effect on the residual stresses in the part. Therefore, stress relief must be achieved after the solution heat treatment and quenching process by plastic deformation of the parts by further heat treatment or mechanical deformation. One relatively little-known heat treatment method which may be used for residual stress relief is the so-called Uphill Quenching process which was developed in the 1960 by Willey and co-workers [10,11]. This paper will provide an overview of uphill quenching process.

Discussion

Definition of the Process

Uphill quenching is also known as ''deep freezing'', "cold stabilization" or ''tricycle stress relieving'' . This process is conducted after the normal quenching process and prior to ageing by immersing quenched aluminum parts into dry ice or liquid nitrogen followed by immersion in boiling water or blasting with high pressure steam [12,13]. The uphill quench process, when performed optimally, is conducted in four steps [1]:

1. The quenched part is cooled to a ''sub-zero'' temperature, usually using liquid nitrogen.
2. The parts are held at the sub-zero temperature until they have achieved thermal equilibrium.
3. The parts are immediately transferred to the elevated temperature medium, preferably high-pressure steam.
4. The part is then aged in accordance with the alloy and desired temper.

The objective of uphill quenching is to offset residual stresses formed by rapid cooling by rapid heating thus producing a substantial decrease in residual stresses of the overall process. To be effective, the temperature gradients (ΔT) formed from the uphill quench must be greater than those formed by normal cryogenic cooling [14]. Furthermore, the subsequent reheating heating after cryogenic cooling must not be sufficient to affect the tensile properties [13].

When hot aluminum is cooled after solutionizing, the heat transfer does not occur uniformly over the surface of the workpiece which produces significant thermal stresses which then lead to distortion and, in some cases, cracking. If the hot aluminum is quenched cryogenically by immersion into liquid nitrogen or a dry-ice bath, a continuous film is formed around the workpiece providing an improvement in the uniformity of cooling but a significant thermal gradient is formed between the surface and the core of the metal. However, cryogenic cooling does not indicate a rapid reheating of the workpiece after quenching. Therefore, the terms *cryogenic cooling* and *uphill quenching* refer to very different processes and should not be confused with each other.

Uphill Quenching Conditions and Tensile/Yield Strength Results

Uphill quenching is most often used for precipitation hardenable alloys including aluminum-copper, aluminum-magnesium, and aluminum-zinc-magnesium alloys used for wrought or cast products. To evaluate the effect of varying uphill quenching conditions, test blocks (304.8 mm x 152.4 mm x 50.8 mm) were cut from hot-rolled plate of one of two alloys: 2014 with a composition of 4.4% Cu, 0.8% Si, 0.8% Mn, 0.4% Mg and the balance Al and 7075 which had a composition of 5.6% Zn, 2.5% Mg, 1.6% Cu, 0.3% Cr and the balance Al. The 2014 test blocks were solutionized for 3 hours at 504 ºC and the test blocks of 7075 were solutionized for 3 hours at 466 ºC. After solutionizing, all test blocks were quenched in cold water. One test block (a control block) of each alloy was then aged after precipitation hardening: 2014 was aged for 10 hours at 171 ºC immediately after quenching and the 7075 was aged for 36 hours at 121 ºC four (4) days after quenching. The remaining blocks were uphill quenched within 30 minutes to 1.5 hours after cold-water quenching.

The uphill quenching processes were conducted by immersing the blocks in either dry ice/trichloroethylene (-78 ºC) or by further cooling by transferring to liquid N_2 (-195 ºC) and holding at temperature for five (5) minutes after reaching equilibrium. After chilling the test blocks to either -78 °C or -195 °C, they were immersed into boiling water for 3-4 minutes or transferred to a steam cabinet where they were heated for 30 minutes with high-pressure steam (approximately 180 psig) jets positioned to assure uniform reheating [11,12,13]. In both reheating processes, boiling water or high-pressure steam, the final temperature was 99 °C. The reheating process is a critical part of the uphill quenching procedure since the reheating rate must be sufficiently rapid to cause plastic deformation and be sufficient to exceed the elastic limit within the metal [10].

After reheating, the test blocks were aged. AA2014 was aged for 10 hours at 171 ºC and AA7075 was aged for 36 hours at 121 ºC after which time the residual stresses and tensile properties were determined. The residual stresses obtained as a result of these treatments are summarized in Table 1 and the mechanical properties are summarized in Table 2 [10].

The residual stress range shown in Table 1 indicates the difference in the maximum compressive stress at the surface and the maximum tensile stress in the center of the test block. These data show that cooling in liquid nitrogen after cold-water quenching and reheating with high-pressure steam produced the greatest amount of reduction in residual stresses.

The mechanical properties which were obtained from these different uphill quenching conditions are shown in Table 2 [10]. These results show that even cooling to the lowest temperature and rapid reheating with high-pressure steam did not produce any significant effects on either tensile or yield strength. However, it was observed that Test Specimen 5, AA2014 which was cooled in liquid nitrogen and reheated with high-pressure steam, exhibited slightly lower elongation than that obtained with the other test specimens, however, this small difference was not expected to significantly affect the ductility [10].

Table 1.Residual Stress Results Obtained for Different Uphill Quenching Conditions.

Test Number	Alloy	Uphill Quenching Medium	Uphill Heating Medium	Temperature Range (ºC)	Residual Stress Range (MPa)	Reduction in Residual Stress (%)
1	2014	---	---	---	165	---
2	2014	Dry Ice	Boiling H_2O	-78 - 99	131	19
3	2014	Liquid N_2	Boiling H_2O	-195 - 99	131	19
4	2014	Dry Ice	Steam	-78 - 99	85.5	48
5	2014	Liquid N_2	Steam	-195 - 99	27.6	83
6	7075	---	---	---	234	---
7	7075	Dry Ice	Steam	-78 - 99	109	50
8	7075	Liquid N_2	Steam	-195 - 99	38.6	84

1. Test blocks (304.8 mm x 152.4 mm x 50.8 mm) were cut from hot-rolled plate. Alloy 2014 was solutionized for 3 hours at 504 ºC; Alloy 7075 was solutionized for 3 hours at 466 ºC. All test blocks were quenched in cold water. The control blocks were then aged after precipitation hardening: Alloy 2014 was aged for 10 hours at 171 ºC soon after quenching; Alloy 7075 was aged for 36 hours at 121 ºC four (4) days after quenching.
2. Composition of Alloy 2014 – 4.4% Cu, 0.8% Si, 0.8% Mn, 0.4% Mg and balance Al; Composition of Alloy 7075 – 5.6% Zn, 2.5% Mg, 1.6% Cu, 0.3% Cr and balance Al
3. Chilling was done by immersing the blocks in either dry ice/trichloroethylene (-77.8 ºC) or by further cooling by transferring to liquid N_2 (-195.5 ºC) and holding at temperature for five (5) minutes after reaching equilibrium.
4. After cooling in dry ice or liquid N2, block was individually transferred to a chamber where it was immediately heated by a blast of high-pressure steam (126 kg_f/m^2) from jets for 30 seconds which raised the temperature to (99 ºC). Other chilled blocks were immersed into boiling water for 3-4 minutes.

Table 2. Mechanical Properties of Uphill Quenched Aluminum Test Specimens.

Test Number[1]	Alloy Designation[2]	Tensile Stength (MPa)	Yield Strength (MPa)	Elongation (%)
1	2014 (blank)	479	429	7.5
2	2014	480	429	8.0
3	2014	480	430	7.0
4	2014	490	446	6.5
5	2014	481	441	4.0
6	7075 (blank)	581	514	10.0
7	7075	570	511	7.5
8	7075	585	518	8.5

1. Themal processing conditions for each test specimen are summarized in Table 1
2. Composition of Alloy 2014 – 4.4% Cu, 0.8% Si, 0.8% Mn, 0.4% Mg and balance Al; Composition of Alloy 7075 – 5.6% Zn, 2.5% Mg, 1.6% Cu, 0.3% Cr and balance Al

Effect of Uphill Quenching Process Variables

Four process variables will be discussed which were studied previously by Hill, et. al.: 1.) the temperature to which the part is cooled; 2.) the reheating method; 3.) yield strength and residual stress state of the aluminum alloy at the time the part is uphill quenched (which varies with the total time following the initial quenching process [11].

1.Effect of Cooling Temperature

The data in Table 1 shows that the greatest reduction in residual stress was obtained with the lowest cooling temperature, in this case, liquid nitrogen (-195 °C). Generally, residual stress reduction increases as the maximum temperature difference increases as shown in Figure 1 [11]. The maximum temperature differential must be sufficient to provide the necessary stress reduction to offset the residual stresses caused by the conventional thermal processing. To provide the maximum temperature differential for stress reduction, the reheating of the material must be initiated from as low a temperature as possible.

Willey reported that to obtain any significant advantage from uphill quenching, the maximum temperature difference should be at least 78 °C. In some cases, it may be advantageous to cool to even lower temperatures than that provided by liquid nitrogen, such as those achievable with liquid helium (-269 °C). However, when cooling in liquid gases such as nitrogen or helium, it may be advisable to first cool to an intermediate temperature of -70 °C to -130 °C in order to conserve gas.

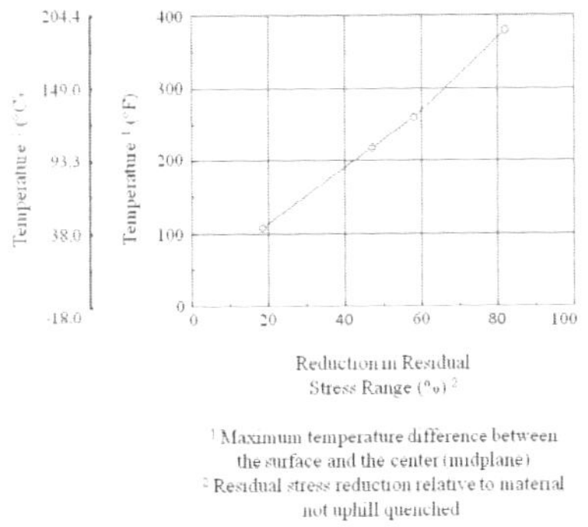

Fig. 1. Correlation of the residual stress reduction with the maximum temperature difference between the surface and the core of the test specimen during uphill quenching. These data relate to quenching of 2014 test specimens (50.8 mm x 152.4 mm x 308.4 mm) cooled to -195 °C within 1.5 hours after solutionizing and cold-water quenching. The test specimens were reheated using high-pressure steam and then aged to a T6 temper.

2.Reheating Methods

The first objective of uphill quench is to cool the material to as low a temperature as possible. In order to maximize the thermal gradient in the process, the second step is to reheat the material as fast as possible to a temperature just below the artificial aging temperature. This is illustrated in Table 3 where high-pressure steam provided a significant advantage over low-pressure steam and both modes of reheating provided considerably greater residual stress reduction relative to boiling water. It is important to note that the reheating must be as uniform as possible. Therefore, the steam is directed uniformly over all surfaces of the material being reheated.

Table 3. Effect of Reheating Method[1]

Reheating Method	Maximum Temperature Difference (°C)[2]	Residual Stress Reduction (%)[3]
High-velocity steam	371	82
Low-velocity steam	253	44
Boiling water (no agitation)	110	19

1. These data relate to quenching of 7075 test specimens (50.8 mm x 152.4 mm x 308.4 mm) cooled to -195 °C within 1.5 hours after solutionizing and cold-water quenching. The test specimens were quenched in liquid nitrogen reheated in the medium shown in the table and then aged to a T6 temper.
2. Between the surface and geometric center (mid-plane)
3. Relative to control test specimens not uphill quenched

Pellman, et. al. conducted uphill quenching uphill quenching processes using a perfluorocarbon in the vapor phase, preferably, perfluoromethyldecalin (bp = 142 °C) [15]. Advantages properties of these fluids for used with uphill quenching include: constant boiling point below the aging temperature of the alloy, high heat transfer rates, and sufficient thermal stability to provide a uniform reheating process. These perfluorocarbons provide a maximum thermal gradient significantly greater than that achievable with steam without increasing the temperature of the alloy above its aging temperature. For example, the temperature differential provided by perfluoromethyldecalin of an alloy cooled in liquid nitrogen is 257 °C versus 206 °C, the maximum thermal gradient reported in the literature for high-pressure steam [15]. A summary of residual stress reduction for 7075 using different cooling and reheating regimes is summarized in Table 4.

In an apparently similar work, Wang, et.al. described uphill quenching of 7050 aluminum forgings where the forgings were quenched in cold (8 °C) water after solutionizing and then immediately cooled to -196 °C in liquid nitrogen then reheated in a very hot (120 to ~175 °C) proprietary liquid bath and then tempered to a T74 condition. The results of this work also confirmed that the degree of stress relief was directly proportional to the magnitude of the maximum temperature difference as reported originally by Willey [10] and Hill , et. al. [11].

One common application for uphill quenching is to minimize the distortion of aluminum mirrors used in space applications. Ohl and co-workers studied the effect of various reheating media on residual stress and distortion produced during the heat treatment and water quenching and quenching with a Type 1 aqueous polymer quenchant [16,17] of a 6061 aluminum alloy [18,19]. The results of Ohl's study showed that uphill quenching of AA6061 that was quenched in either room-temperature water or in a Type I quenchant resulted in approximately 15% residual stress reduction when first cooled in liquid nitrogen then reheated in boiling water.

In addition to the known disadvantage of using boiling water relative to high pressure steam which was shown above, Croucher also showed that at least in some cases, uphill quenching could not be successfully performed on Type I polymer quenched material because of the lower residual stresses that are produced relative to cold water quenching [12]. Therefore, although uphill quenching may be effective for some parts, such as those quenched in cold water, the uphill quenching process may not be effective with parts having low as-quenched residual stresses such as aluminum alloy parts quenched in hot water or aqueous polymer quenchants [1,12].

3. Effect of Yield Strength and Stress State and Prior to Uphill Quenching

The objective of uphill quench is to counteract the unfavorable stresses in the part after quenching. The lower the surface compressive stresses, the greater the potential favorable effect of uphill quenching. Therefore, uphill quenching should be conducted after as soon after quenching as possible before natural aging increases the yield strength of the part. Uphill quenching of a tempered part, for example a 7075 in the T6 tempered condition, would not be effective for this reason.

Disadvantages

One of the primary disadvantages of uphill quenching is that it is a relatively expensive process that is often is considered too difficult, especially since it requires residual stresses to be measured on each part, to be a reliable stress-relieving process, especially when compared to mechanical deformation processes which can be performed at room temperature. However, processes like tensile deformation have been very successfully applied for relief of quenching stresses, it is typically limited to sheet and plate with a uniform cross-section in the stretching direction. While compression-tension loading may also be used, this method is limited by configuration, shape and size [9].

Table 4.Comparative Residual Stress Reduction of Uphill Quenched 7075 Using Different Cooling and Reheating Regimes[1]

Uphill Quenching Method		Temperature	Reduction in
Cooling Media	**Reheating Media**	**Differential (°C)**	**Residual Stress (%)**
Liquid Nitrogen[2]	Perfluoromethyldecalin	271	>90
Liquid Nitrogen[2]	Perfluoromethyldecalin	238	>90
Liquid Nitrogen[2]	Perfluoromethyldecalin	262	>90
Liquid Nitrogen[3]	High-velocity steam	206	82
Dry Ice[3]	High-velocity steam	122	48
Liquid Nitrogen[3]	Low-velocity steam	136	44
Liquid Nitrogen[3]	Boiling water	61	19
Dry Ice[3]	Boiling water	61	19

1. These data utilize 7075 test specimens (50.8 mm x 152.4 mm x 308.4 mm) solutionized for 3 hours at 466 °C. All test specimens were quenched in cold water after solutionizing, then uphill quenched using the media shown in Table 4 and then aged to a T6 temper.
2. Data reported by Pellman, et. al. [11].
3. Data reported previously by Croucher [12]

Conclusion

It has been shown that substantial relief of quenching stresses, in some cases, > 90%, can be achieved by a relatively little known process designated as uphill quenching. The degree of stress relief afforded by the uphill quenching process is directly proportional to increasing maximum temperature difference between the cryogenic cooling portion of the process and the maximum reheating temperature. Although this has traditionally been achieved by cooling in liquid nitrogen after cold water quenching and then subsequently reheating with high-pressure steam, variants of this process have successfully utilized immersion into fluorocarbon liquids at temperature higher than achievable with high-pressure steam thus providing even greater potential for residual stress reduction.

Acknowledgments

The authors acknowledge their appreciation to CAPES (Coordenação de Aperfeiçoamento de Pessoal de Nível Superior) and Universidade de São Paulo (USP) for the financial support of this work. Without their support, this work would not have been possible.

References

[1] Totten, G. E.; Webster, G. M.; Bates, C.E., Quenching, Marcel Dekker Inc., Vol.1, No. 20 (Nova York, 2003), pp. 971-1062. [A book reference *Handbook of Aluminum, Physical Metallurgy and Process*; Totten, G. E.; MacKenzie, D. S., Eds., 2003].

[2] Croucher, T., "Water Quenching Procedures for Aluminum Alloys", Heat Treating, 1982, September, pp. 18-19.

[3] Croucher, T., "Applying Synthetic Quenchants to High-Strength-Alloy Heat Treatment", Metals Engineering Quarterly, 1971, May, pp. 6-11.

[4] Croucher, T., "Synthetic Quenchants Eliminate Distortion", Metal Progress, 1973, November, pp. 52-55.

[5] Robinson, J.S.; Tanner, D.A. "Residual Stress Development and Relief in High Strength Aluminum Alloys Using Standard and Retrogression Thermal Treatments", Vol. 19, No. 4 (2003), pp. 512-518. [A reference to a presentation at Materials Science and Technology, 2003].

[6] Blackwood, R.R., Jarvis, L.M.,Totten, G.E.,Webster, G.M., Narumi, T., "Reducing Aluminum Distortion with Type I Quenchants", Metal Heat Treating, 1996, May/June, pp. 28-31.

[7] R.L. Torgerson and C.L. Kropp, "Improved Heat Treatment Processing of 7050 Aluminum Alloy Forgings Using Synthetic Quenchants", (1977), pp. 111-132. [A reference to a presentation at 22nd National SAMPE Symposium, 1977].

[8] Lados, D.; Apelian,; "The Effect of Residual Stress on the Fatigue Crack Growth Behavior of Al-Si-Mg Cast Alloys Mechanisms and Corrective Mathematical Models", Vol. 37A (2006), January, pp. 133-145.[A reference to Metallurgical and Materials Transactions].

[9] D.A. Tanner, J.S. Robinson, "Modelling Stress Reduction Techniques of Cold Compression and Stretching in Wrought Aluminum Alloy Products", Finite Elements in Analysis and Design, 2003, Vol. 39, pp. 369-386.

[10] Willey, A. L.; "Method of relieving residual stresses in light metal articles", Aluminium Company of America, Pittsburgh, USA (1958), pp.1-3. Patent: Willey, A. L.; Patent title: Method of relieving residual stresses in light metal articles, patent

number US 2949392, patented Aug. 16, 1960.
[11] Hill, H.N., Barker, R.S.; Willey, L.A., "The Thermo Mechanical Method for Relieving Residual Quenching Stresses in Aluminum Alloys", Transactions of the ASM, 1960, Vol. 52, pp. 657-674.
[12] Croucher, T., "Uphill quenching of Aluminum: Rebirth of a little-know process", Heat Treating, Norco, October, (1983), pp. 30-34.
[13] Croucher, T., "Minimizing Machining Distortion in Aluminum Alloys through Successful Application of Uphill Quenching — A Process Overview", Norco, Vol. 6, No. 7 (2009), pp. 1-14. [A reference to a journal article: Journal of ASTM International, Paper ID JAI101770].
[14] Il'yushko, E.G.; Bedarev,A.S., "Quenching Parts Made of Aluminum Alloys in Liquid Nitrogen", Metal Science and Heat Treatment, (1968), Vol. 10, No. 1, pp. 45-47.
[15] Pellman, M.A.; Kilhefner, P.T, III; Baxter, W.J.; Hahn, T.S.,"Vapor Phase Uphill Quenching of Metal Alloys using Fluorochemicals", Allentown, PA, USA, (1989), pp.1-10. Patent: Pellman et al.; Patent title: Vapor Phase Uphill Quenching of Metal Alloys using Fluorochemicals, patent number US 4968358, patented Nov. 6, 1990.
[16] Totten G.E., Bates C.E., Jarvis, L.M., "Type I Quenchants for Aluminum Heat Treating", Heat Treatment, (1991), December, pp. 16-19.
[17] Totten G.E., Webster, G.M., Bates C.E., "An Overview of Aluminum Quenching with Polymer Quenchants", Industrial Heating, (1998), November, pp. 61-64.
[18] Ohl, R., "Treatments to Produce Aluminum Mirrors for Cryogenic Uses", 2002, NASA Tech Briefs, GSC-14736-1. Paper Available from: NASA STI Help Desk, NASA Center for Aerospace Information, 7121 Standard Drive, Hanover, MD 21076-1320, E-Mail: help@sti.nasa.gov
[19] Ohl, R.G., *et al.,* "Comparison of Stress Relief Procedures for Cryogenic Aluminum Mirrors", SPIE (Society of Photographic Instrumentation Engineers), Proceedings Vol. 4822, pp. 51-71.
[20] Wang Qiu-cheng; Wang Li-tao; Peng Wei. "Thermal Stress Relief in 7050 Aluminum Forgings by Uphill Quenc*hing",* Materials Science Forum, 2005, Vol. 490-491, July, pp. 87-101.
[21] Wang Qiu-cheng; Wang Li-tao; Peng Wei., *Thermal Stress relief in 7050 Aluminum Forgings by Uphill Quenching*, No. 490-491 (2005), pp. 97-101. [A reference to a presentation at Materials Science Forum 2005].
[22] Simencio, E. C. A.; Canale, L. C. F.; Totten, G. E., *Uphil quenching of Aluminum: a process overview,* Vol. 5, No.1, 2011, pp. 26-30. [A reference to a presentation at a Conference: International Heat Treatment and Surface Engineering, 2011].

International Federation of Heat Treatment and Surface Engineering 2016
Proceedings of the 23rd IFHTSE Congress
April 18–21, 2016, Savannah, Georgia, USA

Wear Behavior of Layers Produced by Boriding and PVD + Boriding Treatments ON AISI H13 Tool Steel

Mr. Ricardo Gomes Pereira, Mr. Galtiere Corrêa Rêgo, Dr. Luiz C. Casteletti
Materials Engineering, EESCUSP (Escola de engenharia de São Carlos Universidade de São Paulo), São Carlos, Brazil

Dr. Amadeu N. Lombardi
Mechanical Engineering, Federal Technological University of Parana, Londrina, Brazil

Dr. George E. Totten
Department of Mechanical and Materials Engineering, Portland State University, Portland, OR

ABSTRACT

In this work, boride layers of high hardness and wear resistance were produced on AISI H13 tool steel through thermo-reactive treatment in borax bath with addition of 10% aluminum. Additionally, Physical Vapor Deposition (PVD) coating was performed over the boride layers for some samples. The samples were then characterized by optical microscopy (OM), scanning electron microscopy (SEM), Vickers hardness testing and free ball microabrasive wear tests. The boronizing was carried out at 900 °C with a bath residence time of 4 hours, followed by oil or air cooling. The boronizing treatment produced layers with a mix of FeB and Fe_2B borides and an average thickness of 70 μm. The layer hardness was in the range of 1500 to 1700 HV. The PVD treatment produced an AlTiCrN layer with a depth of 5 μm and with a hardness higher than 3000 HV. Samples with boriding-PVD duplex treatment showed the best wear performance, followed by the borided sample and the uncoated quenched and tempered AISI H13 steel.

Introduction

Abrasive wear corresponds to approximately 50% of the deterioration produced by all forms of wear, being one of the main industrial problems causing the replacement of components and equipment [1]. This type of wear is common in machine components that work in dusty environments, due to entrapment of particles between the surfaces, causing three bodies abrasive wear [2]. There is a constant necessity to recover or improve these components to withstand working conditions, increasing its service life and reducing costs. Moreover, the wear resistance is not an intrinsic property of all materials but the use of surface treatments techniques can substantially improve its tribological properties.

The increase of metals surface hardness by thermochemical treatment is associated with high wear resistance and is considered a good alternative to improve the material's tribological properties [3-5]. Various treatments have been used for that purpose, such as nitriding, CVD (Chemical Vapor Deposition), PVD (Physical Vapor Deposition) and boronizing, which can be successfully used for the production of wear resistant layers [5].

Boronizing or Boriding surface thermochemical treatment is done in high temperatures, generally from 850 ° C to 1000 ° C by the diffusion of boron which reacts with the substrate to form the boride layer. The thickness of the boride layer basically depends on the treatment temperature and exposure time [5].

Physical Vapor Deposition (PVD) is a coating forming method that uses atomistic deposition, wherein the material is vaporized from a solid or liquid source in the form of atoms or molecules and transported as a vapor by means of a vacuum or low pressure nitrogen gas (or plasma environment) to the substrate, where it condenses, forming a shallow hard layer (2-5 μm) [7, 8]. The restricted thickness of this layer limits its application due to the "egg shell" effect, when the substrate is not capable of properly support the layer which may spall off. The use of duplex treatments can overcome this problem by producing an intermediate thicker support layer [8], wherein the resulting duplex layer exhibit better surface properties, such as resistance to wear, corrosion and fatigue [4-8].

In this work simplex and duplex layers were produced on the surface of AISI H13 steel samples through the thermochemical treatment of boriding and boriding followed by PVD to verify the efficiency of the wear resistance and hardness of the layers.

Experimental Methods

The AISI H13 steel used had the composition 5.15Cr, 0.35C, 0.45Mo, 0.95V, 1.05Si, 0.38Mn, Fe bal. (in wt%). Test specimens of 20 x 20 x 7 mm were prepared. The boriding treatment was conducted in a molten borax bath ($Na_2B_4O_7$) with the addition of 10 wt % aluminum at 1000 °C. The treatment time was 4 hours, followed by quenching in air or oil and tempering at 540 °C for 5 hours. Using Physical Vapor Deposition (PVD), the borided duplex samples received the second coat at 500 °C with the complex nitride compound AlTiCrN. The samples were characterized by optical microscopy (OM), scanning electron microscopy (SEM), X-ray diffraction (XRD) and Vickers micro-hardness (0.98 N). The micro-abrasive wear testing was performed by a calotest type machine using an AISI 52100 steel ball with 25.4 mm in diameter, surface hardness of 63 HRC and a rotation speed of 800 rpm. An aqueous suspension of silicon carbide (SiC) with concentration of 0.5 g/ml was used as the abrasive. The test times were 5, 10, 15 and 20 minutes which were converted into sliding distances. The volume removed was obtained according to Equation 1 where R is the radius of the sphere and d the diameter of the cap when $d \lll R$ [10, 11].

$$V = \pi \, x \, d^4 / 64 \, x \, R; \; for \; d \lll R \tag{1}$$

Results and Discussion

Figure 1 shows the micrograph of the borided AISI H13 steel sample cross-section. The boride layer had an average thickness of 70 μm, displaying a "sawtooth" morphology followed by a sublayer with hardness lower than that of the substrate. The formation of the sub-layer is common in steels which have silicon content around 1 wt %. Its formation is attributed to the insolubility of silicon in the boride layer what pushes the Si toward the substrate, forming the low silicon sub-layer.

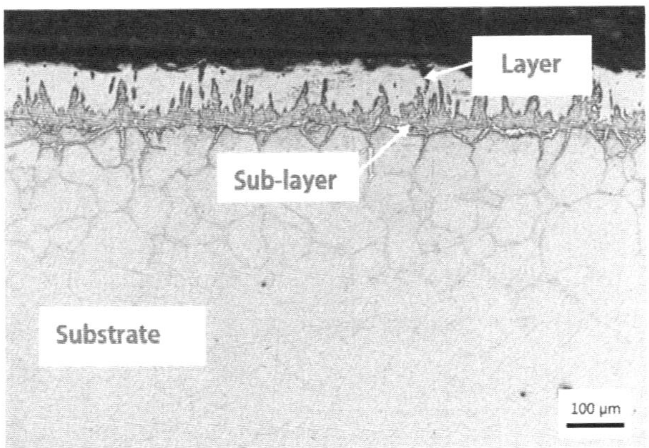

Figure 1: Optical micrograph steel AISI H13 boretado cold air.

Figure 2 shows a duplex AISI H13 steel sample cross-section borided and coated with PVD. This thin nitride layer (AlTiCrN) has a thickness of 5 μm and an average hardness 3300 HV.

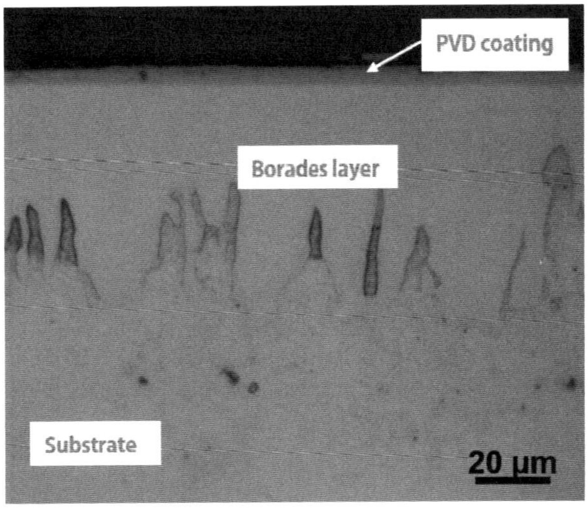

Figure 2: Optical Microscopy of the AISI H13 borided and PVD coated sample.

Figure 3 shows the Vickers hardness profiles of AISI H13 steel samples borided and air or oil cooled. It is observed that the hardness of the boride layers near the surface were around 1700 HV, with a progressive decrease of that value toward the substrate. Note that the hardness of the transition region is very close to the hardness of the substrate, indicating that there was the formation of a sublayer lower in hardness.

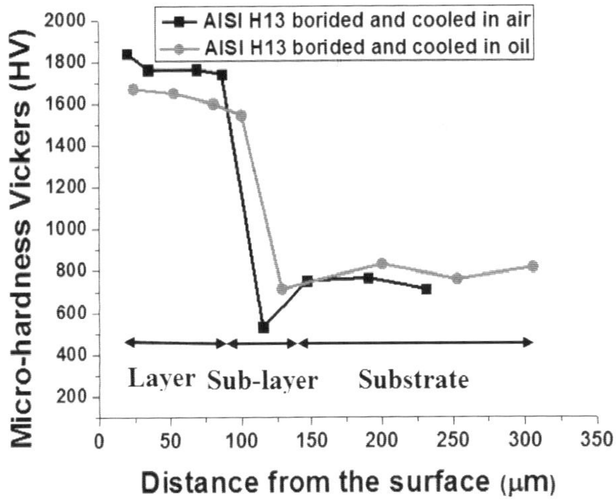

Figure 3: Vickers hardness profiles (100 gf load) of borided AISI H13 steel samples.

Figure 4 Shows optical micrographs of caps formed by the microabrasive wear test, exhibiting different wear mechanisms. It appears that all the produced layers had shallow wear caps, especially the sample that was borided and PVD coated, which had the shallowest cap presenting the least amount of wear and the very best wear performance of all the samples. Its superior performance can be attributed to the higher value of hardness and the well established low friction coefficient of the PVD layer. The morphology of the wear–caps presented three mechanisms of wear, parallel risks are the result of micro-cutting action of the abrasive particles which settle on the ball and are dragged through the contact surface.

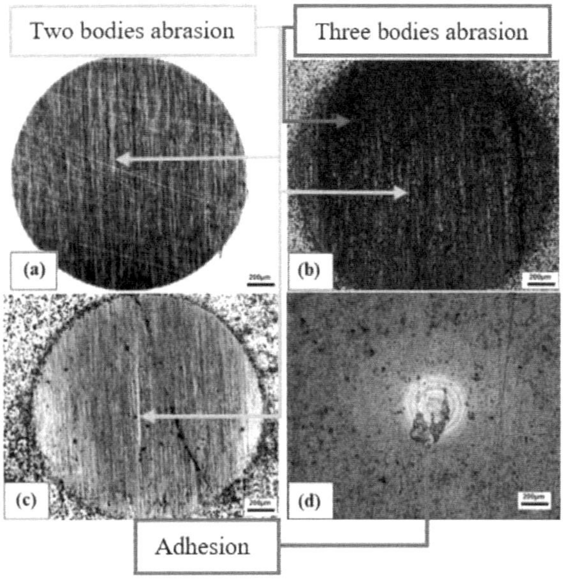

Figure 4: wear hubcaps for a sliding distance of 54 meters (a) AISI H13 hardened and tempered, (b) AISI H13 Boretado and cooled in oil, (c) AISI H13 Boretado and cooled in air, (d) AISI H13 Boretado + PVD.

Figure 5 shows the wear plots of all AISI H13 steel samples: hardened and tempered, borided and the duplex borided and PVD coated. Note that the boride and PVD coated sample showed a wear performance much superior than the other samples with the borided samples coming in second, performing better than the hardened and tempered sample.

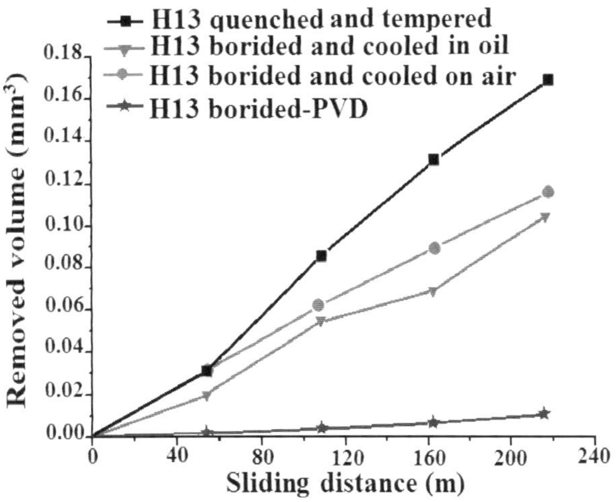

Figure 5: Micro-abrasive wear plots of all AISI H13 steel samples.

Conclusions

The boriding treatment provided the AISI H13 steel samples with a boride layer thickness of about 70 μm, hardness in the range of 1600 to 1800 HV, much higher than the hardness of the substrate which is around 700 HV.

The best wear performance, by far, was observed in the sample that was boride and coated with PVD.

Acknowledgments

The authors wish to thank the Conselho Nacional de Desenvolvimento Científico e Tecnológico (CNPq), Brazil, for the financial support under the process no. 305789/2012-0.

References

[1] Petrica, M., et al. "Study of abrasive wear phenomena in dry and slurry 3-body conditions." Tribology International, Vol. 64, 2013, pp. 196-203.

[2] Petrica, M., E. Badisch, and T. Peinsitt. "Abrasive wear mechanisms and their relation to rock properties." Wear, Vol. 308, n. 1, 2013, pp. 86-94.

[3] Zhang, Q. Y., et al. "Comparative research on dry sliding wear of hot-dip aluminized and uncoated AISI H13 steel." Wear, Vol. 344, 2015, pp. 22-31.

[4] Lee, Insup, and Kwang-Ho Jeong. "Plasma post oxidation of plasma nitrocarburized SKD 61 steel." Journal of Materials Science & Technology (2008).

[5] Garcia-Bustos, E., et al. "The wear resistance of boride layers measured by the four-ball test." Surface and Coatings Technology, Vol. 215, 2013, pp. 241-246.

[6] Kulka, M., et al. "Gradient boride layers formed by diffusion carburizing and laser boriding." Optics and Lasers in Engineering, Vol. 67, 2015, pp. 163-175.

[7] D.M.MATTOX, Handbook of Physical Vapor Deposition (PVD) Processing, 2010, 2ª Edition.

[8] Batista, J. C. A., C. Godoy, and A. Matthews. "Micro-scale abrasive wear testing of duplex and non-duplex (single-layered) PVD (Ti, Al) N, TiN and Cr–N coatings." Tribology International, Vol. 35.6, 2002, pp. 363-372.

[9] Jacobsen, S. D., et al. "Influence of current density on phase formation and tribological behavior of plasma nitrided AISI H13 steel." Surface and Coatings Technology, Vol. 286, 2016, pp. 129-139.

[10]Trezona, R.I., Allsopp, D.N. and Hutchings, I.M., "Transitions Between Two-Body and Three-Body Abrasive Wear: Influence of Test Conditions in the Microscale Abrasive Wear Test" Wear, Vol. 225-229, 1999, pp. 205-214.

[11]Allsopp, D.N. and Hutchings, I.M., "Micro-Scale Abrasion and Scrath Response of PVD Coatings at Elevated Temperatures, " Wear, Vol. 251, 2001, pp. 1308-1314.

Author Index

A

Ahrenberg, M. 94
Albano, L.M. 88, 523
Aleksandrov, V.A. 547
Amanov, A. 471, 511
Antoniou, A. 256
Arangi, S. 473
Arenas-García H. 457

B

Babul, T. 481, 493
Bader, D. 123
Barrallier, L. 46, 63
Beck, M. 94
Becquerelle, S. 25
Behera, R. 256
Belashova, I.S. 547
Bhadeshia, H.K.D.H. 1
Bottoli, F. 51
Bowen, P. 146
Brown, D.W. 313
Buchwalder, A. 55
Buscail, H. 139

C

Cai, X. 273, 404, 434, 487
Canale, L.C.F. 88, 132, 341, 523, 529, 560
Cardey, P.F. 139
Carroll, D. 256
Carron, D. 332
Casteletti, L.C. 536, 566
Castle, J. 313
Cerda-Rojas, R. 440, 464
Chavdar, B. 175, 248
Chobaut, N. 332
Christiansen, T.L. 51, 304
Chun, B.K. 313
Ciski, A. 493
Civera, C. 341
Clausen, B. 313
Colas, R. 348
Colás, R. 457
Combe, C. 139
Courleux, A. 25
Croucher, T. 560
Cryderman, R. 168
Czerwiec, T. 25

D

da Silva, R.V. 536
De Moor, E. 445
della Putta, P. 123
Deo, C. 256
Desmier, G. 217
Ding, P. 78
Doyon, G. 217
Drezet, J.-M. 332
Dulcy, J. 25

E

Ebner, R. 203
Epp, J. 410
Esmeralda, A.G. 457

F

Fabro, M. 123
Fallot, G. 63
Felde, I. 348
Ferguson, B.L. 78, 354, 398
Ferguson, L. 175, 248
Fernandez-Guzman, F. 440, 464
Filho, A.I. 536
Fillip, S. 196
Fraj, B.B. 451
Freborg, A.M. 78, 354, 398
Fürst, G. 425

G

Glavicic, M.G. 313
Godec, M. 38, 161
Goldstein, R. 175, 248
Goodwin, A. 196
Goulding, A.N. 362
Graham, G. 101
Green, J.D. 262
Gregui, R.G. 529
Guillot, B. 46
Guisbert, D.A. 223
Guisbert, D.E. 223
Guisbert, M.H. 223
Gujar, A. 184

H

Hattel, J.H. 51
Hawk, J.A. 279
Hebbes, M. 78
Hirai, Y. 513
Holden, T.M. 313

I

Ikhande, M. 473
Iljkić, D. 391
Ishizuka, H. 513
Issartel, C. 139

J

Jablonski, P.D. 279
Jacquot, N. 425
Jacquot, P. 235, 425
Jadhav, P. 473
Jégou, S. 46, 63
Jespersen, F.N. 51
Jung, A. 55

K

Karl, A. 235
Kattire, P. 190
Kawata, K. 541
Kessler, O. 383
Keßler, O. 94
Kidachi, T. 541
Kim, H.-D. 471
Kim, J.-H. 471, 511
Kim, Y. 487
Kim. S.-G. 19
Kimura, K. 289
Klose, N. 55
Knezevic, M. 313
Kogej, P. 38
Kong, J.-H. 19
Korecki, M. 71
Kozlov, N. 383
Kragl, U. 94

L

Lankiewicz, K. 493
Leduc-Lexama, L.A. ..440, 464
Leskovisek, V. 88
Leskovšek, V. 38, 161, 554
Li, H. 146
Li, W. 210
Li, Z. 78, 354, 398
Liang, J. 229
Licavoli, J.J. 279
Liscic, B. 132
Lister, M. 504
Lombardi, A.N. 566
Lu, Y. 229, 273, 404, 434
Lúcio, J.G. 88, 341
Ludewigs, D.A. 529

M

MacKenzie, D.S. 101, 110, 372, 560
Maher, J. 217
Mane, R. 473
Manthani, N. 184
Marcos, G. 25
Marsoner, S. 203
Martins, J.B.R. 536
Matijevic, B. 132
Mattos, W.S. 560
Mercado-Solís, R.D. ..440, 464
Minnick, R. 217
Misina, F.S. 88, 523
Mittemeijer, E.J. 11
Moore, A. 256

N

Nadolski, D. 325
Nalawadi, S.J. 123
Neise, C. 94
Neu, R.W. 362
Nižňanská, J. 493

O

Oh, J.Y. 313
Okumiya, M. 19, 418
Olbrycht, A. 481
Ostaeva, G.Y. 547
Otero, R.L.S. 341

P

Papisov, I.M. 547
Papisova, A.I. 547
Patil, V. 31
Pawar, V. 184
Pereira, R.G. 566
Perekrestov, A.E. 547
Pershing, M.A. 210
Petrova, L.G. 547
Pfaffenberger, J. 118
Pintér, G. 348
Podgornik, B. 161
Powar, A. 184
Prabhu, N. 190
Prasad, R.C. 473
Prevedel, P. 203
Puš, G. 161
Pyun, Y.-S 471, 511

Q

Quiñones-Salinas, D.I. 440, 464

R

Raab, G. 210
Rêgo, G.C. 566
Reich, M. 94
Ress, R. 313
Ressel, G. 203
Riffard, F. 139
Rivolta, B. 341
Rodríguez, A.F. 457
Rolland, R. 139
Rong, Y. 404
Rudnev, V. 196, 217, 242
Russell, C. 242

S

Saelzle, P. 332
Sanders, T.H. 362
Sawada, K. 289
Sawamiphakdi, K. 123
Schick, C. 94
Schneider, R.S.E. 418
Schuller, D. 554
Schulz, A. 325
Sherkar, J. 31
Shi, W. 348
Simmons, J. 313
Singh, R. 31, 184
Singh, R.K.P. 190
Sisneros, T. 313

Sisson, R.D. 153, 229, 273, 404, 434, 487, 499
Skiba, O. 25
Smoljan, B. 391
Somers, M.A.J. 51, 304
Souza, W.M. 536
Speer, J. 168
Spies, H.-J. 55
Srivatsa, S. 313
Stauder, B. 425
Steineder, K. 418
Steiner T. 11
Štic, L. 391
Šuchmann, P. 493
Supare, A. 31
Szénási, S. 348

T

Talamantes-Silva, J. 457
Thakare, S. 190
Toshima, Y. 418
Totten, G.E. 88, 132, 341, 523, 536, 560, 566
Tourki, Z. 451
Tsunekawa, Y. 19

V

Van Aken, D.C. 262
Vhavar, V. 190
Vieira, I. 445
Vieweg, A. 203
Villa, M. 304

W

Wägner, M. 235
Wang, A. 153
Watanabe, Y. 513
Watanuki, D. 146
Webb, T.O. 262
West, W. 196
Whitely, B. 168
Winther, G. 51
Wolowiec-Korecka, E. 71
Wu, W.-T. 313

X

Xu, Y. 229

Y

Yu, H. 273, 404, 434, 487

Z

Zapata-Hernández, O. 348
Zenker, R. 55
Zhang, L. 487, 499
Zoch, H.-W. 325